Copper-Catalyzed Amination of Aryl and Alkenyl Electrophiles

Copper-Catalyzed Amination of Aryl and Alkenyl Electrophiles

Kevin H. Shaughnessy, Engelbert Ciganek, and Rebecca B. DeVasher

WILEY

Published by John Wiley & Sons, Inc., Hoboken, New Jersey
Published simultaneously in Canada.

For general information on our other products and services or for technical support, please contact our Customer Care Department within the United States at (800) 762-2974, outside the United States at (317) 572-3993 or fax (317) 572-4002.

Wiley also publishes its books in a variety of electronic formats. Some content that appears in print may not be available in electronic formats. For more information about Wiley products, visit our web site at www.wiley.com.

Library of Congress Catalog Card Number: 42-20265
ISBN: 978-1-119-34598-5

Printed in the United States of America.

CONTENTS

FOREWORD

Chemical synthesis is an intellectually and technically challenging enterprise. Over the many decades of progress in this discipline, spectacular advances in methods have made once intimidating transformations now routine. However, as the frontier advances and the demands for ready access to greater molecular complexity increases, so does the sophistication of the chemical reactions needed to achieve these goals. With this greater sophistication (and the attendant expectation of enhanced generality, efficiency, and selectivity) comes the challenge of adapting these technologies to the specific applications needed by the practitioner. In its 75-year history, *Organic Reactions* has endeavored to meet this challenge by providing focused, scholarly, and comprehensive overviews of a given transformation.

The impact of organometallic catalysis in organic synthesis can hardly be overstated. The advent of newer and more efficient methods for the construction of carbon-carbon and carbon-heteroatom bonds has truly transformed the practice of making new compounds in academic and industrial settings. The ability to introduce new carbon-nitrogen bonds onto aromatic and heteroaromatic rings through the agency of palladium-catalyzed amination with various nitrogen-based nucleophiles revolutionized the synthesis of aromatic amines. Although the impact of this method cannot be overstated, the cost of palladium precatalysts and highly engineered ligands provided incentives to revisit the use of "earth-abundant" copper catalysts and simpler ligand systems.

The *Organic Reactions* series is fortunate to have published a comprehensive chapter on this important process that constituted Volume 85. This timely chapter was authored by one of the internationally recognized leaders in this field, Prof. Kevin Shaughnessy together with his student and coauthor Rebecca DeVasher with expert assistance from Engelbert Ciganek, a longtime member of the *Organic Reactions* family. Although many reviews and book chapters have been written on transition-metal catalyzed aminations, this massive chapter constitutes the definitive work in the field. Thus, in keeping with our educational mission, the Board of Editors of *Organic Reactions* has decided to publish this chapter as a separate, soft cover book to make the work available to a wider audience of chemists. In addition, to keep pace with the rapid development of this field, Prof. Shaughnessy has provided updated references that bring the literature coverage up to December 2015. These references are appended at the end of the original reference section and organized by the Tabular presentation of the different aromatic electrophiles.

The publication of this book represents the fifth, soft cover reproduction of single-volume *Organic Reactions* chapters. The success of the first four, soft cover books has convinced us that the availability of low-cost, high-quality publications that cover broadly useful transformations is addressing an unmet need in the organic synthesis community. Thus, we will continue to identify candidates for the compilation of such individual volumes as opportunities present themselves.

Scott E. Denmark
Urbana, Illinois

PREFACE

The metal-catalyzed amination of aryl and alkenyl electrophiles has developed into a widely used methodology for the synthesis of natural products, active pharmaceutical ingredients, agricultural chemicals, and materials for molecular electronics. Copper-catalyzed C–N coupling was first reported over a century ago and remained the state-of-the art for 90 years. Over the past 20 years, palladium-catalyzed C–N couplings largely supplanted copper-catalyzed reactions due to their increased generality and reliability. The development of more active ligand-supported copper catalysts has resulted in a resurgence of interest in the use of copper, however. Copper catalysts promote the coupling of a wide range of nitrogen nucleophiles, including amines, amides, and heteroaromatic nitrogen compounds with aryl and alkenyl halides. The reactivity profile of copper catalysts is complementary to that of palladium catalysts in many cases. Copper catalysts are highly effective with less nucleophilic nitrogen nucleophiles, such as amides and azoles, whereas palladium catalysts are more effective with more nucleophilic amine nucleophiles. Copper is an attractive alternative to palladium due to its significantly lower cost. In addition, high activity palladium catalysts require expensive and often air-sensitive ligands, whereas the modern copper systems use relatively stable and inexpensive diamine or amino acid ligands. Copper-catalyzed C–N coupling reactions are tolerant of a wide range of functional groups and have been applied to the synthesis of a variety of complex natural products. Significant work has also been done to understand the mechanism of these reactions. Current mechanistic understanding of these methodologies is covered in this monograph.

Optimal experimental conditions for the amination of aryl and alkenyl halides with all classes of nitrogen nucleophiles are presented. Specific experimental procedures from the literature are provided for the major classes of copper-catalyzed C–N coupling reactions. A tabular survey of all examples of Cu-catalyzed arylation and alkenylation of nitrogen nucleophiles is presented in 34 tables organized by nitrogen nucleophile and electrophilic coupling partner. Tables are organized by increasing carbon count of the nitrogen nucleophile then by carbon count of the organic halide.

The literature is covered through December 2015 and provides over 300 recent citations to supplement the 680 citations of the original hardbound chapter. These latest literature references have been collected in separate sections according to the sequence of the tables in the tabular survey section. In each of the sections, the individual citations have been arranged in alphabetic order of the author names.

Copper-Catalyzed Amination of Aryl and Alkenyl Electrophiles is intended to provide organic chemists with an accessible, but detailed, introduction to this important class of transformations.

COPPER-CATALYZED AMINATION OF ARYL AND ALKENYL ELECTROPHILES

KEVIN H. SHAUGHNESSY

Department of Chemistry, The University of Alabama, Box 870336, Tuscaloosa, Alabama 35487-0336, USA

ENGELBERT CIGANEK

121 Spring House Way, Kennett Square, Pennsylvania 19348, USA

REBECCA B. DEVASHER

Department of Chemistry, Rose-Hulman Institute of Technology, Terre Haute, Indiana 47803, USA

CONTENTS

kshaughn@ua.edu
Copper-Catalyzed Amination of Aryl and Alkenyl Electrophiles, by Keven H. Shaughnessy, Engelbert Ciganek, and Rebecca B. DeVasher
© 2017 Organic Reactions, Inc. Published 2017 by John Wiley & Sons, Inc.

ACKNOWLEDGMENTS

We thank Ms. Hannah Box for assistance in collecting references for this manuscript.

INTRODUCTION

Metal-catalyzed amination of aryl and alkenyl electrophiles has developed into a highly useful synthetic strategy for preparing *N*-aryl- and *N*-alkenyl- containing structures. Ullmann[1] first reported Cu-mediated *N*-arylation reactions of amines in 1903 (Scheme 1) followed closely by Goldberg's[2] report of amide *N*-arylations (Scheme 2). In these reactions, an aryl halide is condensed with an amine or amide in the presence of base and copper powder or a copper salt at high temperature. The need for stoichiometric amounts of copper and high reaction temperatures combined with the modest yields of product in most cases limits the application of these methods. Despite these limitations, Ullmann and Goldberg condensation reactions represented the state-of-the-art in metal-mediated C–N bond formation for nearly a century.[3] The development of efficient Pd-catalyzed C–N bond-forming reactions that could be carried out under mild conditions with less reactive substrates, such as aryl chlorides, resulted in palladium displacing copper as the preferred metal for aryl amination reactions in the late 1990s.[4–7]

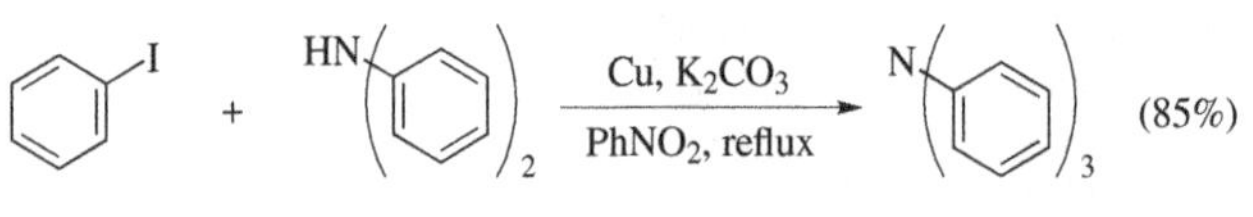

Scheme 1

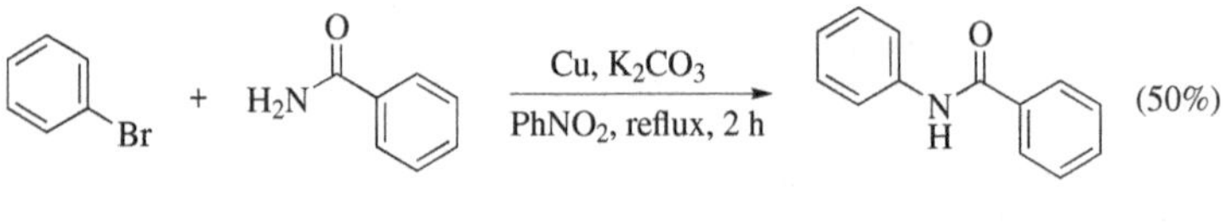

Scheme 2

Interest in Cu-mediated reactions was reinvigorated with the development of Cu(II)-mediated[8,9] and -catalyzed[10] oxidative couplings of carbon nucleophiles (e.g., arylboronic acids) and nitrogen nucleophiles (Scheme 3). Development of efficient, ligand-promoted, Cu-catalyzed coupling reactions of nitrogen nucleophiles and aryl or vinyl halide electrophiles further increased interest in the use of less expensive copper in place of palladium for C–N bond formation (Scheme 4).[11] The use of aryl halides is attractive due to their wider availability and lower cost compared to organoboron reagents. Copper-catalyzed N-arylation and alkenylation reactions are widely used in the synthesis of natural products, pharmaceuticals, electronics materials, and ligands.

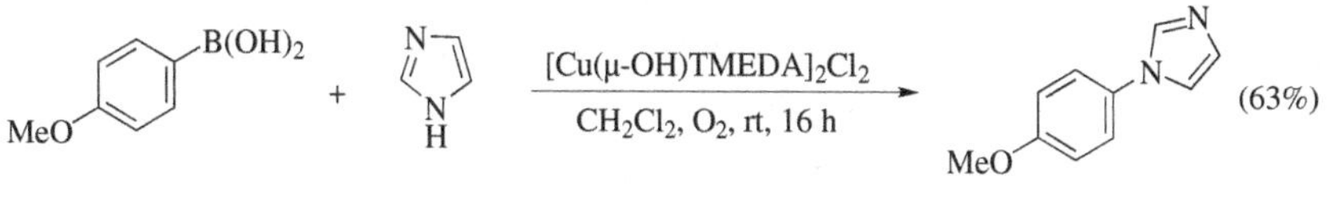

Scheme 3

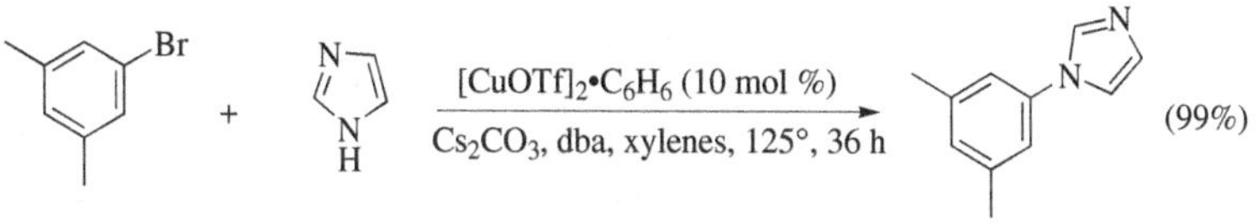

Scheme 4

Aromatic and heteroaromatic halides are the most common electrophiles in Cu-catalyzed C–N bond-forming reactions. The halide leaving group is typically an iodide or bromide, although chloride and even fluoride can be used in some cases. Alkenyl halides are also effective coupling partners in these reactions, with the iodides and bromides being most commonly used. Unlike Pd-catalyzed reactions, sulfonate leaving groups have not been successfully applied in Cu-catalyzed reactions. The nitrogen nucleophile can be nearly any class of compound with an N–H bond including aryl and alkyl amines, ammonia, azoles, amides, sulfonamides, carbamates, ureas, and guanidines. Copper-catalyzed systems are particularly useful with amide and azole substrates that prove challenging with Pd-catalyzed methods. In addition to amination reactions, Cu-catalyzed C–O, C–S, and C–C bond-forming reactions have been developed, but these reactions are outside the scope of this review.[12]

The catalysts used for the coupling of nitrogen nucleophiles and aryl or alkenyl halides can be broken down into two broad classes. Ligand-free catalyst systems based on the original Ullmann and Goldberg methods use copper powder or copper salts as the catalyst, and typically employ a polar aprotic solvent at high temperature (100–200°). Alternatively, copper salts in combination with ligands such as amines, ethers, or phosphines provide more active copper catalysts than ligand-free systems. Reactions using ligand-supported catalysts can often be carried out at lower temperature and with a wider range of substrates (i.e., aryl bromides and chlorides) than the ligand-free systems.

The goal of this chapter is to provide a critical and comprehensive review of the scope and mechanism of Cu-catalyzed amination of aryl and alkenyl electrophiles.

Copper-mediated or -catalyzed oxidative coupling reactions and couplings of preformed copper amides with aryl halides are not covered in this chapter, although they have previously been reviewed.[12] This review covers the literature through August 2010 with selected examples published through early 2011. Additional reactions reported mid-2011 through early 2014 are included only as references at the end of the bibliography, categorized according to the relevant table in the Tabular Survey. The patent literature is not covered in this chapter, but the relevant patent literature has been reviewed recently.[13] Several reviews have appeared in recent years that focus on the development of Cu-catalyzed amination reactions and their applications in complex target syntheses.[12,14−18]

MECHANISM

The mechanism of Cu-catalyzed C–N bond formation has proven more difficult to determine than that of the corresponding Pd-catalyzed cross-coupling reactions. The same general steps—nucleophile coordination to the metal center, activation of the C–X bond, and C–N bond formation—are involved in both the Pd- and Cu-catalyzed reactions, although the order of these steps is different. In the case of the Cu-catalyzed reactions, the order, exact mechanism of these steps, and identity of the active species continue to be debated.[19] The generally accepted sequence of steps begins with coordination of the deprotonated nucleophile to a Cu(I) center (Scheme 5). The Cu–amide complex then activates the organic halide to form the new C–N bond. The nature of the organic halide activation and C–N bond formation remains the least understood step in the catalytic cycle.

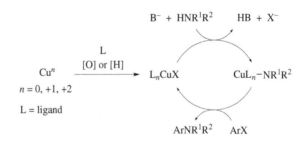

Scheme 5

Oxidation State of Catalytically Active Copper Species

The active form of copper is generally believed to be Cu(I) in these reactions. Copper precatalysts in 0, +1, and +2 oxidation states can be used to catalyze *N*-arylation reactions, however. When Cu(0) or Cu(II) sources are used, the active Cu(I) species is formed by in situ reduction or oxidation of the copper source. In kinetic studies of the Ullmann coupling of phenyl halides with diphenylamine, Cu(I) halides give reaction rates that are significantly higher than when Cu(II) salts are used.[20] When Cu(II) precursors are used, reduction to Cu(I) by alkoxide or amide reagents is observed.[20,21] Reduction of Cu(II) can be accelerated by ligands that stabilize the Cu(I) oxidation state, such as neocuproine.[22] Reduction of Cu(II) ions by amide or alkoxide ligands has been confirmed by electron paramagnetic resonance (EPR) spectroscopy of catalyst systems.[23,24] Amino acid complexes of Cu(I) ions are observed under catalytic conditions by electrospray–mass spectrometry (ESI–MS) analysis of reaction mixtures.[25] In the case of heterogeneous Cu(0) catalysts, the true catalyst may be formed by dissolving the Cu_2O coating found on copper metal catalysts.[20]

Nucleophile Coordination

In contrast to Pd-catalyzed coupling reactions where oxidative addition of the organic halide is the initial step, a strong body of evidence suggests that coordination of the nucleophile to copper occurs prior to activation of the organic halide. Support for this mechanism is provided by the ability of Cu–amido and –amidate complexes to react with aryl halides at catalytically competent rates. A *trans-N,N′*-dimethyl-1,2-diaminocyclohexane (DMCDA)-supported Cu–pyrrolidonate complex is kinetically competent in the arylation of pyrrolidone.[26] An NMR spectroscopic study of phosphine–copper complex **1** shows that this complex reacts with pyrazole in the presence of base to give a Cu–pyrazolate complex **2** (Scheme 6).[27] The pyrazolate complex reacts quantitatively with phenyl bromide to give 1-phenylpyrazole. Complex **1** is unreactive with phenyl bromide, however. Computational studies at the density functional theory (DFT) level suggest that a diamine Cu–amidate complex has a lower barrier to oxidative addition of the aryl halide than the analogous CuBr complex.[28]

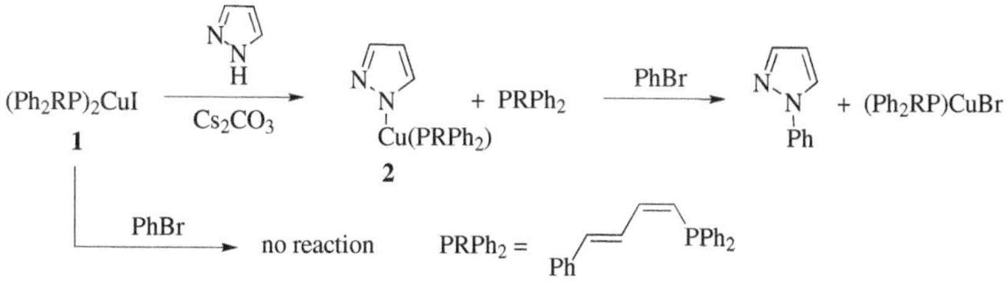

Scheme 6

A number of Cu(I)–amido and –amidate complexes have been prepared by metathesis of Cu(I) halides with nitrogen nucleophiles.[26,29–31] These complexes react stoichiometrically with aryl halides to give *N*-arylated products (Scheme 7), supporting the role of Cu(I) species in the catalytic cycle.[32] Copper–phthalimidate complexes react with aryl halides provided that the phthalimidate/Cu ratio is < 1.[29] With higher phthalimidate/Cu ratios, no reaction occurs. Part of the accelerating effect of chelating ligands in Cu-catalyzed coupling reactions appears to result from inhibition of the formation of catalytically inactive $[Cu(NR_2)_2]^-$ complexes (Scheme 8).[33,34]

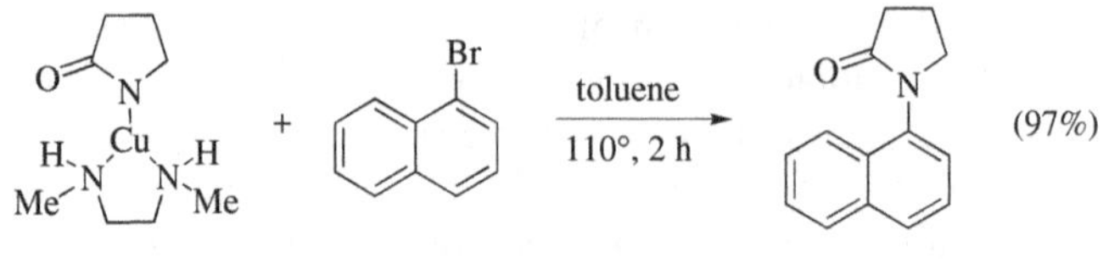

Scheme 7

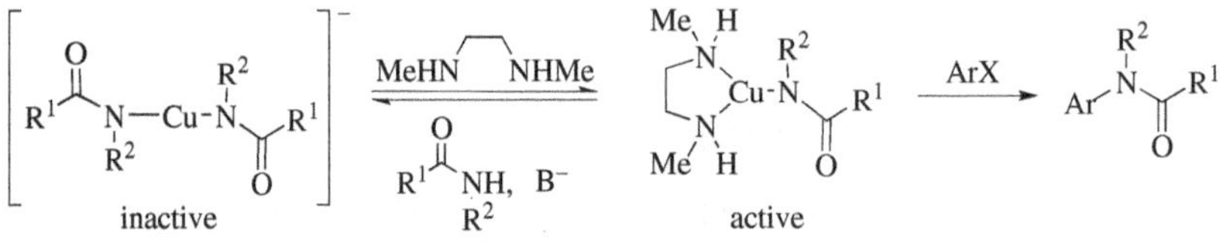

Scheme 8

Organic Halide Activation Step

The mechanism for the aryl halide activation and C–N bond formation steps remains a matter of debate. Four major mechanisms have been proposed for this process (Scheme 9): (1) oxidative addition/reductive elimination, (2) substitution promoted by single-electron transfer (SET) from Cu(I) to the aryl halide, (3) nucleophilic aromatic substitution via π-complexation of copper to the aryl halide, and (4) σ-bond metathesis via a four-centered transition state. Mechanistic studies on the original Ullmann systems provide support for several of these mechanisms.[19]

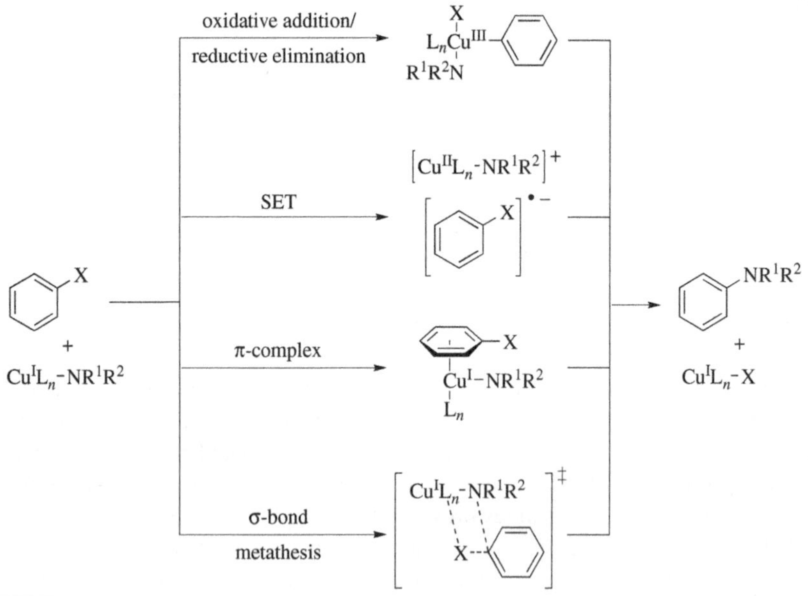

Scheme 9

Several authors have postulated SET mechanisms analogous to $S_{RN}1$ reactions. Key mechanistic support for the SET mechanism is the observation of EPR signals from Cu(II) and organic radicals in the Cu-catalyzed amination of 1-bromoanthraquinone.[35] Evidence that seems to rule out free-radical intermediates has also been reported. The photochemical $S_{RN}1$ reaction of phenylthiolate with 1-chloro-4-iodobenzene gives the disubstituted product exclusively, whereas the CuI-catalyzed reaction gives only the monosubstituted product (Scheme 10).[36] Substrates with radical clocks are not rearranged by the Cu-catalyzed reactions. Both of these results suggest that radical intermediates are not formed, or have very short lifetimes. Reactions carried out under N_2 with Cu(I) salts are faster than reactions run in air.[23] In addition, the Cu(I) reactions run under a N_2 atmosphere show very weak EPR signals. These results suggest that the observed Cu(II) species are not catalytically relevant, but are formed by air oxidation of the active Cu(I) catalyst.

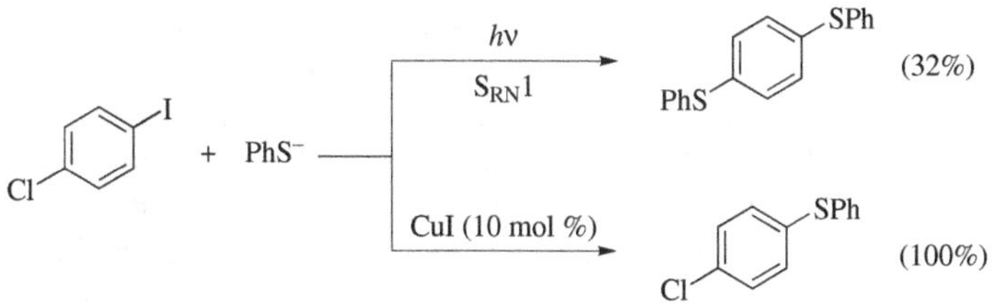

Scheme 10

Oxidative addition to give an aryl–Cu(III) intermediate followed by reductive elimination was first proposed on the basis of studies of the reaction of N,N-dimethyl o-halobenzamides with CuCl or CuCN.[37] Addition of benzoic acid results in hydrodehalogenation, which suggests an arylcopper intermediate is present that is protonated by the acid. A radical mechanism was discounted since no demethylation of the amide nitrogen occurs. The typical reactivity trend of ArI > ArBr > ArCl ≫ ArF is consistent with an oxidative addition mechanism and inconsistent with an S_NAr mechanism, which displays the opposite trend. In the amination of 1-haloanthraquinone, hydrodehalogenation occurs as a side reaction.[38] The halogen leaving group affects the rate of the reaction, but not the ratio of substitution to hydrodehalogenation. In addition, the hydrodehalogenation occurs by hydrogen transfer from the α-carbon of the amine. These results suggest a common $ArCu(III)(NR_2)$ intermediate that could form either the aryl amine product by reductive elimination or the hydrodehalogenation product by β-hydrogen elimination from the amine and then reductive elimination from the resulting aryl hydride complex.

The S_NAr mechanism via a copper π-complex was originally proposed in analogy to the known acceleration of S_NAr reactions in chromium–arene complexes as well as the observation of Cu–arene complexes.[23] This mechanism does not account for the aryl halide reactivity trend, however, which is opposite that of S_NAr reactions. In addition, this mechanism cannot explain the fact that o-halobenzoic acids undergo substitution at significantly higher rates than p-halobenzoic acids.[20] These substrates would be expected to react at similar rates in an S_NAr mechanism involving a copper π-complex. Substitution by a σ-bond metathesis mechanism has been proposed,[39] but there is little mechanistic evidence to support or refute this proposal due to the difficulty in differentiating the σ-bond metathesis mechanism from oxidative addition/reductive elimination.

Mechanistic Studies of Aryl Halide Activation by Ligand-Supported Copper Species

Mechanistic studies of ligand-promoted Cu-catalyzed C–N bond formation also largely support an oxidative addition/reductive elimination mechanism. Several pieces of evidence suggest that long-lived radicals or radical anions are not involved in C–N bond formation catalyzed by copper complexes of a tetradentate nitrogen ligand.[40] This reaction is not affected by radical traps or initiators. Reaction of 1,4-diiodobenzene with excess pyrazole forms 1-(4-iodophenyl)pyrazole as an observable intermediate, which then is converted to 1,4-dipyrazolylbenzene (Scheme 11). In contrast, a radical mechanism would predict formation of the disubstituted product without intermediate formation of 1-(4-iodophenyl)pyrazole. The 1-(4-iodophenyl)pyrazole radical anion intermediate formed after the initial substitution would rapidly lose iodide leading to the second substitution without significant buildup of 1-(4-iodophenyl)pyrazole. Vinyl halides undergo substitution with retention of configuration, whereas a radical mechanism would result in scrambling of the alkene geometry. A number of features support the oxidative addition/reductive elimination mechanism in this system. The reactivity trend of aryl halides (ArI > ArBr > ArCl) and modest electronic effect [electron-withdrawing group (EWG) > electron-donating group (EDG)] are consistent with expected trends for oxidative addition. The fact that steric bulk on either the aryl halide or azole reduces the reaction rate suggests that both the aryl halide and azole must coordinate to copper prior to the rate-limiting step.

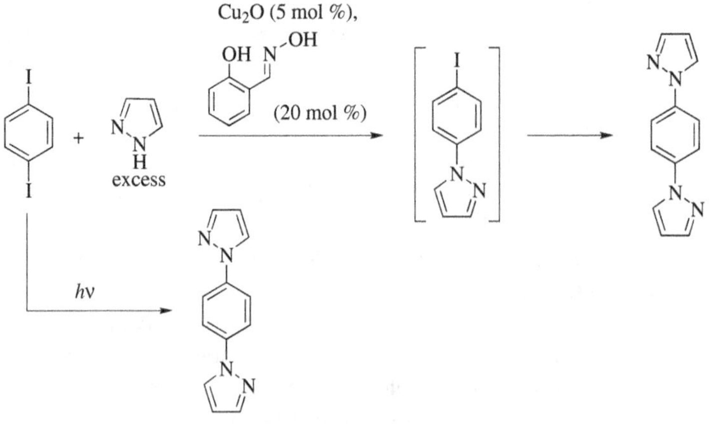

Scheme 11

The reactivities of 4-chlorobenzonitrile and 1-bromonaphthalene have been compared in the coupling of aryl halides and amides catalyzed by diamine copper complexes.[32] These substrates have similar reduction potentials and rates of halide dissociation from their radical anion forms. Therefore, these substrates would be expected to react at similar rates in an $S_{RN}1$-like mechanism. 1-Bromonaphthalene reacts with (DMEDA)Cu–pyrrolidonate to give N-1-naphthylpyrrolidone in 97% yield after 2 hours at 110° (Scheme 7). No reaction is seen when the copper complex is combined with 4-chlorobenzonitrile under identical conditions. In addition, coupling of o-allyloxyiodobenzene with pyrrolidone gives no cyclized product (Scheme 12). This result shows that any radical intermediate would have to be trapped significantly faster than the cyclization rate for the o-allyloxyphenyl radical (10^{12} s^{-1}). Evidence for reductive elimination from a Cu(III)(aryl)amido complex is provided by the rapid formation of N-arylamides by the reaction of preformed Cu(III)aryl complex **3** with amides (Scheme 13).[41,42]

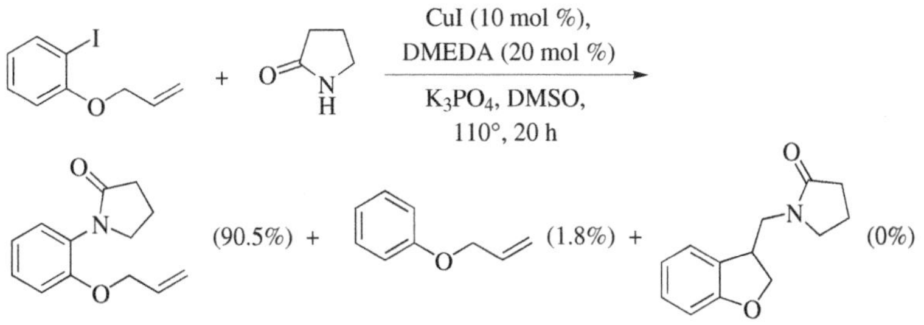

Scheme 12

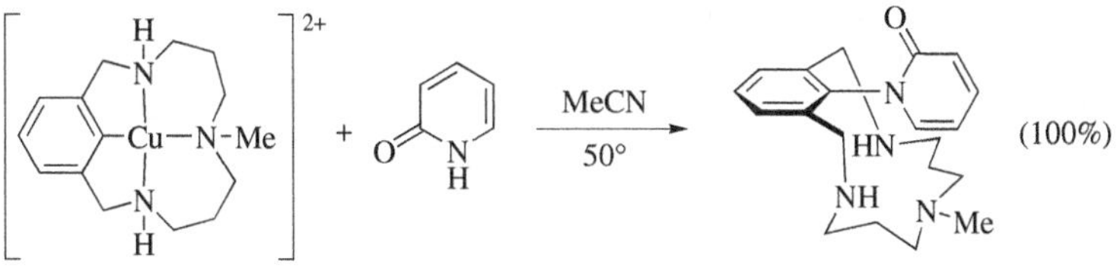

Scheme 13

Computational Studies of Copper-Catalyzed Amination Mechanisms

Computational studies show that the oxidative addition of aryl halides to ligand-supported copper amido or amidate complexes occurs with modest activation barriers that are consistent with experimentally determined activation barriers.[28,32,43–45] These studies only consider oxidative addition as a possible mechanism for the aryl halide activation step, so they cannot exclude the possibility of other mechanistic pathways. A computational study that evaluates oxidative addition/reductive elimination, σ-bond metathesis, SET, and atom-abstraction (IAT) mechanisms concludes that the SET pathway proceeding through a Cu(II) intermediate is the lowest-energy pathway for C–N bond formation catalyzed by copper phenanthroline and copper diketonate complexes.[46] The intermediates involved in the oxidative addition/reductive elimination and σ-bond metathesis mechanisms are calculated to be energetically inaccessible. The phenyl radical generated through SET from the Cu(I)-amido complex adds to the nitrogen to form the arylated Cu(I) amido complex (Scheme 14). The radical-trapping step in the SET mechanism is calculated to be strongly exothermic, suggesting that the caged Cu(II)-phenyl radical pair collapses at a very high rate. Rapid trapping of the caged radical would be consistent with experimental data that appears to exclude radical mechanisms.

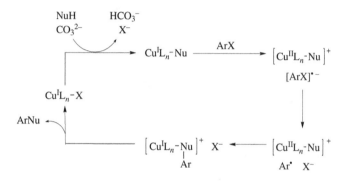

Scheme 14

Thus, despite significant effort, the exact nature of the aryl halide activation step remains an open question. It is likely that both oxidative addition and SET-type mechanisms are energetically accessible. The preferred pathway is likely dependent on the copper complex, ligands, substrates, and reaction conditions. In reactions where the SET mechanism is favored, rapid radical capture may make this mechanism experimentally indistinguishable from the oxidative addition mechanism.

Mechanistic Studies of Ligand Effects on Copper-Catalyzed Amination

Experimental studies on the effect of ligands in Cu-catalyzed C–N bond formation suggest that an important role is preventing the formation of catalytically inactive, anionic $[Cu(NR_2)_n]^{(n-1)}$-species. Kinetic studies in the diamine-promoted amide arylations show that the reaction is first order in diamine at low ligand concentration, but the dependence saturates at higher ligand concentration. This result is interpreted to indicate that at low ligand concentrations catalytically inactive anionic diamidate copper complexes dominate. The diamine ligand pushes the equilibrium to the catalytically active ligand–Cu–amidate complex (Scheme 8).[33] With phenanthroline as the ligand, $[Cu(phen)_2][Cu(NR_2)]$ complexes (NR_2 = pyrrolidonate) are formed rather than $Cu(phen)(NR_2)$.[32] Although the phenanthroline ligand is not coordinated to the Cu–amidate, its presence is critical for reaction with aryl halides. Tetrabutylammonium salts of $[Cu(NR_2)]^-$ are unreactive with aryl iodides. The $[Cu(phen)_2][Cu(NR_2)]$ complex is proposed to be in equilibrium with $Cu(phen)(NR_2)$, which is the true catalytically active species. Computational studies also show that anionic Cu–diamidate complexes have significantly higher barriers for oxidative addition than Cu–(diamine)(amidate) species.[28]

SCOPE AND LIMITATIONS

The Carbon Electrophile

Aromatic Halides and Sulfonates.   Aryl iodides and bromides are the most commonly employed substrates in Cu-catalyzed nitrogen arylations. The general reactivity trend for aryl halide substrates is ArI > ArBr > ArCl, whereas aryl fluoride and aryl sulfonate substrates are generally unreactive except in special cases. This order of reactivity is clearly demonstrated in a number of examples in which the more reactive halide can be selectively coupled in the presence of less reactive halide substituents. 1-Bromo-4-iodobenzene is coupled in high yield and selectivity at the iodide position with alkyl amine,[47] aniline,[48] imidazole,[40] and amide nucleophiles (Scheme 15).[49] Similarly, 1-bromo-4-chlorobenzene is selectively coupled at the bromine position in high yield with a variety of nitrogen nucleophiles.[47,49−51]

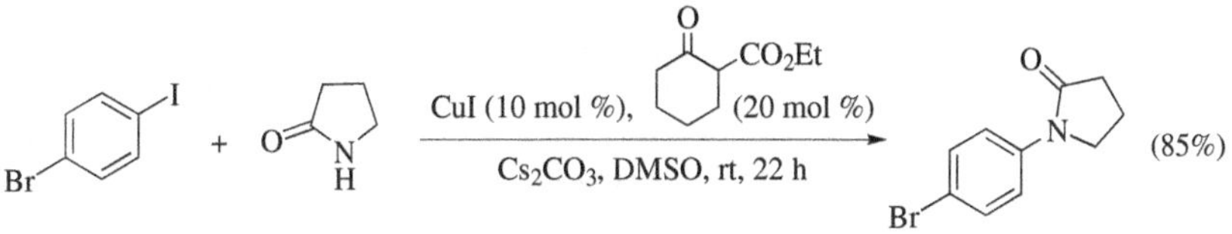

Scheme 15

Steric and electronic properties of the aryl halide substrate can affect reactivity, although there are relatively few careful studies of these effects. A kinetic study of the coupling of 4-substituted aryl iodides with n-hexylamine catalyzed by a CuI/diketonate catalyst at room temperature shows that the reaction time increases from 40 minutes for 4-iodobenzonitrile to 110 minutes for 4-iodoanisole.[52] In a competitive coupling of 4-substituted aryl iodides with 2-pyridone, ethyl 4-iodobenzoate (relative rate = 4.7) gives the highest rate followed by 4-iodotoluene (1.4) and 4-iodoanisole (1.0).[53] At the higher temperatures and longer reaction times typically used in methodology studies, these small differences in reactivity are generally not noticeable.

Substituents adjacent to the halide leaving group can inhibit the coupling reaction. The vast majority of examples reported are 4-substituted aryl halides. With the exception of the special case of 2-halobenzoic acid derivatives, few examples of Cu-catalyzed coupling of 2-substituted aryl halides have been reported. With aryl amine nucleophiles, 2-halotoluene substrates give good yields.[50,54−56] Coupling of 1-iodo-2,6-dimethylbenzene with the N6 of adenosine gives incomplete conversion

(65% yield) after 48 hours using stoichiometric amounts of CuI/DMEDA at 110° in DMSO (Scheme 16).[57] Unhindered aryl iodides give complete conversion after 1–2 hours under identical conditions. Couplings with N-heteroaromatic substrates appear to be more tolerant of steric hindrance in the aryl halide substrate. Numerous examples of 2-substituted aryl halides coupling with azoles are known. Coupling of iodomesitylene and imidazole using CuI (10 mol %) and DMEDA (40 mol %) at 170° in DMF gives 1-mesitylimidazole in 54% yield after 48 hours (Scheme 17).[58] Under the same conditions, 2,6-diisopropyliodobenzene gives only a 19% yield.

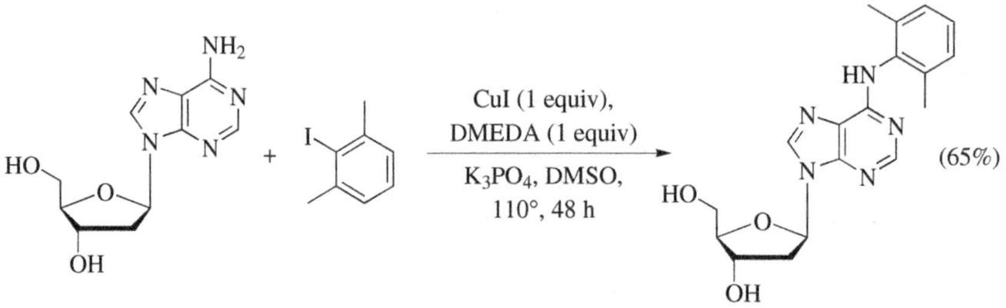

Scheme 16

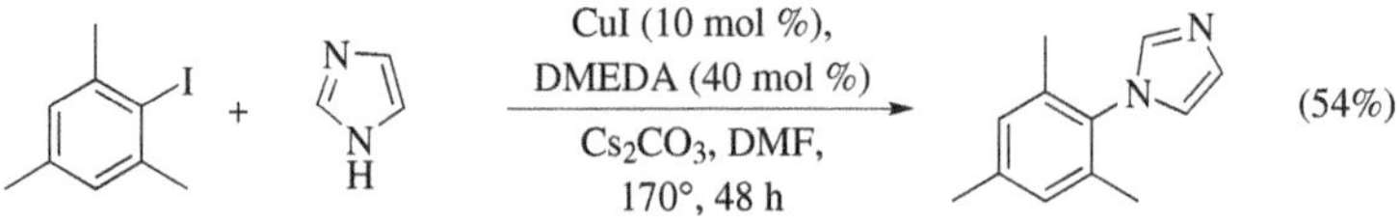

Scheme 17

Functional group tolerance is very high for these reactions. Nearly all classes of functional groups are tolerated by the copper catalyst systems. Electrophilic substituents such as esters, ketones, and aldehydes are unaffected. Free hydroxyl groups in carboxylic acids and alcohols do not need to be protected. Free amine or amide N–H bonds are generally tolerated provided they do not compete with the desired nitrogen nucleophile. As noted above, less reactive halogens than that in the desired reactive C–X bond are generally unaffected.

Aryl iodides are generally useful arylating agents for all classes of nitrogen nucleophiles and a wide range of copper catalyst systems. For ligand-free catalyst systems, reaction temperatures typically range from 80–150° for intermolecular couplings of unactivated aryl iodides. High yields are obtained at 40° for the coupling of aryl iodides with azoles using 20 mol % of CuI in DMF.[59] A coordinating solvent, such as DMF, DMSO, or NMP, is typically required for these ligand-free systems. Copper catalysts supported by proline,[60] acetylacetonate,[61] 2,2′-binaphthol,[62] or DMEDA[63] ligands promote coupling of electron-deficient and sterically unhindered aryl iodides at room temperature. Other ligand-promoted couplings of aryl iodides are generally carried out at 80–110°.

Although aryl bromides are less reactive than aryl iodides, they are generally useful electrophiles for Cu-catalyzed C–N bond-forming reactions with all classes of nitrogen nucleophiles. Ligand-free catalyst systems typically require temperatures between 100 and 140° when coordinating solvents are used. Ligand-supported copper catalysts are more general and the couplings can typically be carried out at lower temperatures (80–110°).

Aryl chlorides are much more challenging substrates for Cu-catalyzed C–N bond formation than aryl iodides or bromides. Although in some cases electron-deficient aryl chlorides can be coupled at relatively low temperatures (80–100°), temperatures between 110 and 150° are generally required. Substituent electronic properties have a significant effect on the ability of aryl chlorides to be coupled. Ligand-free copper catalyst systems are only effective for aryl chlorides that have strong electron-withdrawing substituents. In the few cases where non-activated chlorides are reported in ligand-free systems, the yields are typically much lower than for electron-deficient aryl chlorides.[64,65] For example, a Cu-exchanged fluorapatite catalyst (CuFAP) affords a 95% yield after six hours at 120° for the coupling of 4-chlorobenzonitrile and imidazole (Scheme 18).[66] In contrast, the coupling of 4-chloroanisole requires 36 hours and affords only a 52% yield under the same conditions. Ligand-free systems often give no conversion with unactivated aryl chlorides, such as chlorobenzene, under conditions where electron-deficient aryl chlorides are effectively coupled.[64,65]

R	Time (h)	Yield (%)
NC–	6	95
MeO	36	52

Scheme 18

Mixed results are obtained in the coupling of unactivated aryl chlorides and nitrogen nucleophiles with ligand-supported copper catalysts. Yields below 30% are obtained in the coupling of chlorobenzene with aniline derivatives using a CuI/pipecolinic acid catalyst system at 110°.[67] Chlorobenzene and 4-chlorotoluene give no conversion in coupling with indole or benzimidazole using a CuO/phenanthroline catalyst system at 140°.[68] A 95% yield is obtained in the arylation of 2-pyrrolidinone in neat 4-chlorotoluene using CuI/DMCDA (DMCDA = N,N'-dimethyl-*trans*-1,2-cyclohexanediamine) as the catalyst at 130°, but 4-chloroanisole gives only a 51% yield under identical conditions (Scheme 19).[69] Using the more electron-rich tributylphosphine ligand provides a more general catalyst for coupling of aryl chlorides and aniline derivatives at 135°.[56]

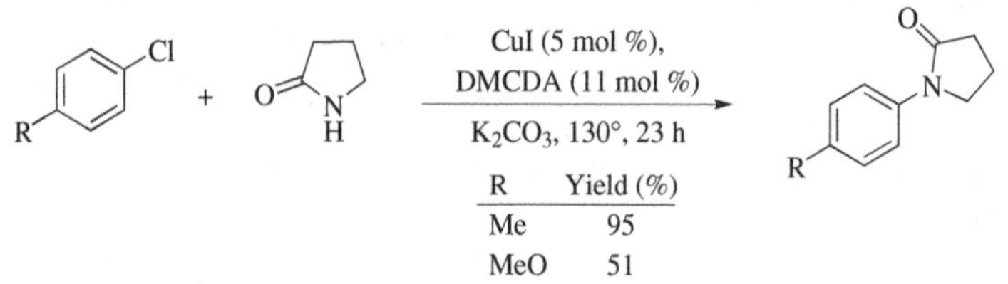

R	Yield (%)
Me	95
MeO	51

Scheme 19

Since Ullmann's original report,[1] 2-halobenzoic acids have been known to be particularly effective substrates for Cu-catalyzed C–N bond-forming reactions.[70–72] The proximity of the carboxylate group is critically important, since 4-halobenzoic acids do not give a similar acceleration. Therefore, the effect appears to be due to coordination rather than the electron-withdrawing property of the carboxylate group. The presence of a carboxylic acid group can reverse the normal reactivity order of carbon–halogen bonds. 5-Bromo-2-chlorobenzoic acid undergoes selective coupling with aniline at the chlorine position (Scheme 20).[71]

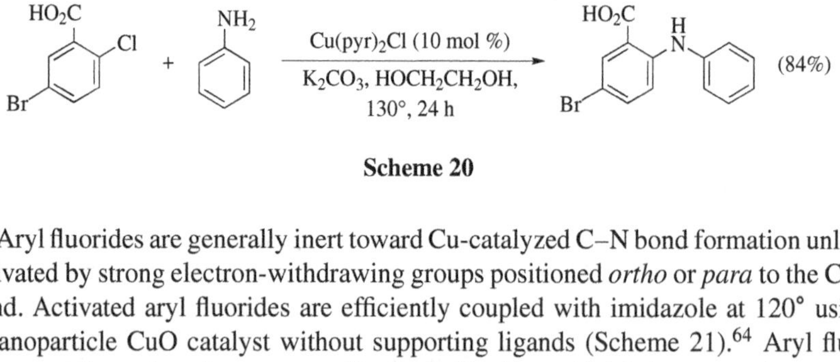

Scheme 20

Aryl fluorides are generally inert toward Cu-catalyzed C–N bond formation unless activated by strong electron-withdrawing groups positioned *ortho* or *para* to the C–F bond. Activated aryl fluorides are efficiently coupled with imidazole at 120° using a nanoparticle CuO catalyst without supporting ligands (Scheme 21).[64] Aryl fluorides react faster than the corresponding aryl chlorides, which is inverse to the normal reactivity trend. In addition, 2-chloro-1-fluoro-4-nitrobenzene reacts selectively at the fluorine position. No reaction occurs in the absence of the CuO catalyst precluding a Cu-free S_NAr mechanism. The reactivity trend suggests that the Cu-promoted C–F activation follows an S_NAr-like mechanism, however. Similar reactivity of activated aryl fluorides with azoles has been reported for sulfonated Cu–salen complexes[73] and a Cu-exchanged fluorapatite catalyst.[66] In both cases, activated aryl fluorides react faster than aryl chlorides.

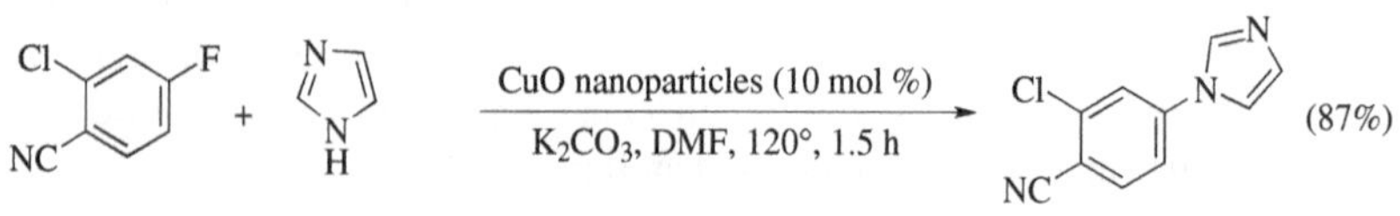

Scheme 21

Aryl sulfonates are generally unreactive in Cu-catalyzed C–N bond-forming reactions on the basis of a very small number of examples reported in the literature. No reaction is observed in the coupling of phenyl tosylate with benzamide or aniline using CuO oxide nanoparticles, although this system is also inactive for aryl bromides and chlorides.[74] Low yields of aniline derivatives are obtained in the coupling of electron-deficient aryl mesylates and sodium azide using a $CuSO_4$/proline catalyst in the presence of ascorbate as a reducing agent.[75]

Diaryliodonium ions can also serve as electrophiles in Cu-catalyzed C–N bond-forming reactions. The diaryliodonium salts are highly reactive towards nucleophilic substitution and oxidative addition. Coupling of diaryliodonium salts with alkyl or aryl amines occurs in good yield (65–83%) at room temperature using CuI without a supporting ligand, whereas azole nucleophiles require 50° (Scheme 22).[76] Amide nucleophiles give lower yields (30–70%).

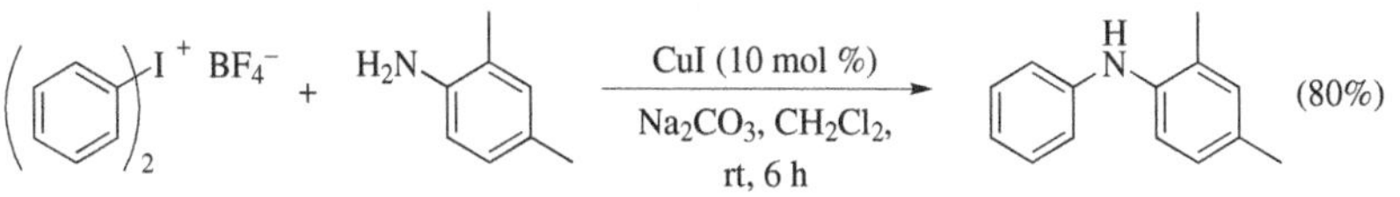

Scheme 22

Heteroaryl Halides.   Heteroaromatic halide substrates can present challenges in metal-catalyzed cross-coupling reactions due to their ability to act as ligands for the catalyst. Copper catalysts seem to be largely unaffected by potential coordination to heteroaromatic substrates containing N, O, S, or Se. The main effect of heteroatoms on the reactivity of heteroaryl halides appears to be electronic. Since pyridine rings are more electron-deficient than benzene, halopyridine derivatives tend to be more reactive than aryl halides. Each additional π-bonded nitrogen further increases the reactivity of the heteroaryl halide. In contrast, five-membered-ring heterocycles, such as pyrrole, furan, and thiophene, in which the heteroatom contributes a lone pair to the π-system, are more electron-rich than benzene. Therefore, these heteroaryl halides tend to be less reactive than simple aryl halides.

Halogenated pyridines, pyrimidines, and pyrazines are generally reactive substrates in Cu-catalyzed C–N bond-forming reactions. Because of the high reactivity of simple aryl iodides, there is usually not a noticeable difference in reactivity between aryl iodides and heteroaryl iodides. Some authors have noted improved reactivity of bromopyridines compared with non-activated aryl bromides.[77,78] The effect is strongest for 2- or 4-halopyridines. Coupling of 3-bromopyridine with imidazole requires 24 hours using a CuI/diketonate catalyst system at 110° compared with 12 hours for 2-bromopyridine.[79] 2,3-Dibromopyridine is substituted selectively at the 2-position by indole (Scheme 23).[77] Chloropyridine derivatives show a significantly increased activity compared to non-activated aryl chlorides. In cases where direct comparisons can be made, 2-chloropyrimidine gives improved yields and reacts faster than 2-chloropyridine, which is more reactive than chlorobenzene (Scheme 24).[79,80] Like other electron-deficient aryl chlorides, chloropyridine and chloropyrimidine derivatives can be coupled under conditions where non-activated aryl chlorides give no conversion.[68,81]

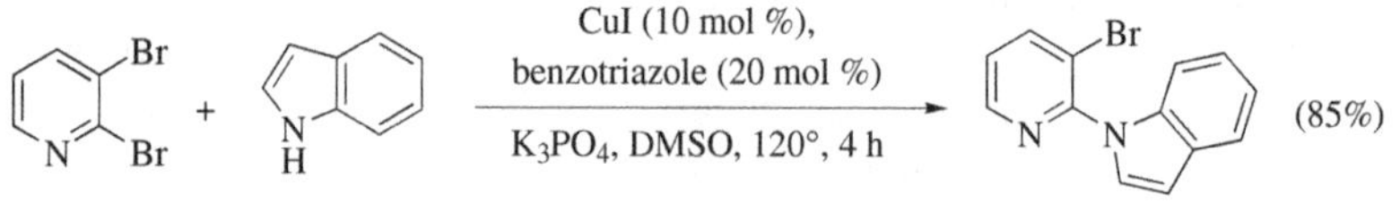

Scheme 23

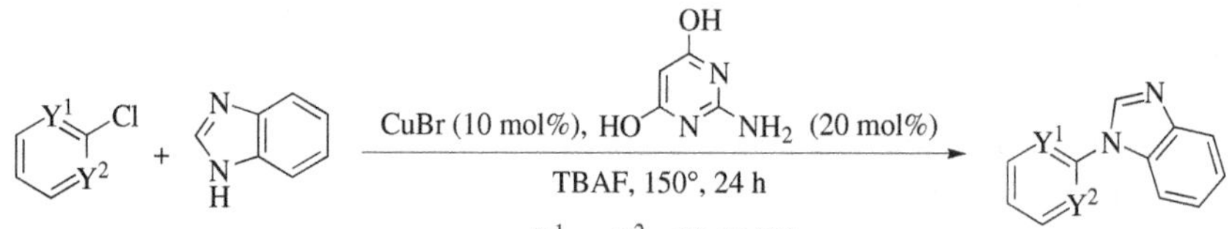

Y^1	Y^2	Yield (%)
CH	CH	40
N	CH	56
N	N	100

Scheme 24

Halogenated five-membered-ring heteroaromatic compounds have been used less extensively than pyridine derivatives due to their lower reactivity. The number of examples follows the trend in electron richness: halothiophenes/halothiazoles ≫ halofurans ≫ haloazoles. N-Methyl-4-iodopyrazole is coupled with pyrazole in good yield at 80° using a CuI/salicylaldehyde oxime catalyst system,[82] but no examples of bromo- or chloroazoles undergoing Cu-catalyzed coupling reactions are known. Only a few examples of the amination of 2- and 3-bromofurans are reported. Generally good yields are obtained in the coupling of 2- and 3-bromofurans with amides and carbamates using the CuI/DMEDA catalyst system at 110° (Scheme 25).[83] Coupling of 3-bromofuran with imidazole using Cu$_2$O/4,7-dimethoxyphenathroline as the catalyst affords a 66% yield of coupled product (Scheme 26).[84]

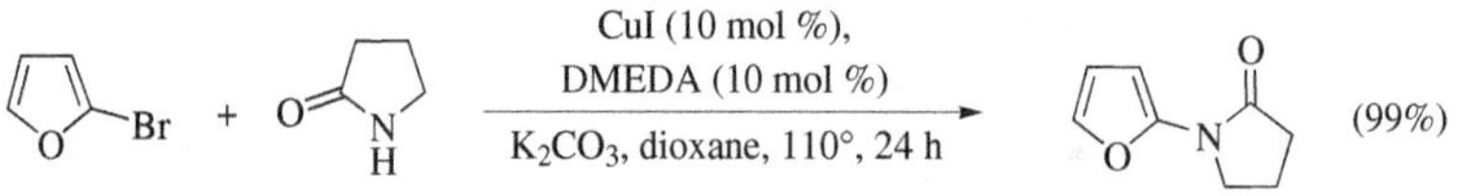

Scheme 25

Scheme 26

Couplings of iodo- and bromothiophenes and thiazoles have been reported with a variety of nitrogen nucleophiles. When a ligand-free Cu(OAc)$_2$ catalyst is used with DBU as a base, bromothiophenes require higher temperature (150° vs. 130°) and afford lower yields than bromopyridine derivatives.[85] 2-Bromothiophene provides a lower yield than 3-bromothiophene (Scheme 27). In another study, yields increase along the trend 2-bromothiophene < 2-bromo-3-thiazole < 4-bromotoluene < 2-bromopyridine for the coupling of the halide with imidazole using a CuI/aminopyrrole catalyst system.[86] 2-Iodoselenophene undergoes coupling with amide and azole nucleophiles in modest to excellent yields using CuI/EDA (EDA = ethylenediamine) as the catalyst (Scheme 28),[87] but no conversion occurs with aniline or morpholine. To our knowledge, no examples are on record for amination of chlorinated five-membered-ring heterocycles using copper catalysts. Given the low reactivity of copper catalysts with aryl chlorides, the electron-rich chloroazoles, chlorofurans, or chlorothiophenes are likely unreactive.

Ar	Yield (%)
2-thienyl	63
3-thienyl	80

Scheme 27

Scheme 28

Alkenyl Halides.   Alkenyl halides are also useful substrates in Cu-catalyzed C–N bond-forming reactions. Examples of these couplings are largely limited to amide and azole nucleophiles that form stable N-alkenylated products. Alkenyl iodides and bromides have been coupled with diarylamines to give stable enamine products, but primary amine substrates afford the imine tautomers as the final products.[88] The general reactivity trends for alkenyl halides mirror those of the aryl halides (I > Br > Cl). No significant limitations to substitution at the alkenyl center are noted based on the examples published to date. Good yields are obtained with mono-, di-, and tri-substituted halogenated alkenes. Electronic effects from substituents do not seem to inhibit activation of the vinyl halogen bond.

Alkenyl halides typically undergo Cu-catalyzed C–N bond formation with retention of alkene configuration. The strong preference for retention suggests a concerted C–X bond activation mechanism, rather than a process involving free-radical species. The coupling of (*E*)-β-halostyrenes with azole derivatives using a Cu_2O/β-keto ester catalyst system affords nearly exclusive formation of the (*E*)-substituted products (Scheme 29).[89] A few examples show a slight erosion of the alkene configurational purity [(*E*)/(*Z*) = 96:4], but no consistent trend accounts for the minor alkene isomerization. More hindered (*Z*)-haloalkenes also give (*Z*)-substituted products with complete retention of configuration in most cases (Scheme 30).[90] Axially chiral allenyl iodides undergo coupling with amides with complete retention of configuration as well. Coupling of (*M*)-1-iodo-1,2-butadiene [(*M*)/(*P*) = 85:15] with (*R*)-5-phenyloxazolidone using CuCN/DMEDA at 50° gives the coupled product as an 85:15 mixture of diastereomers (Scheme 31).[91]

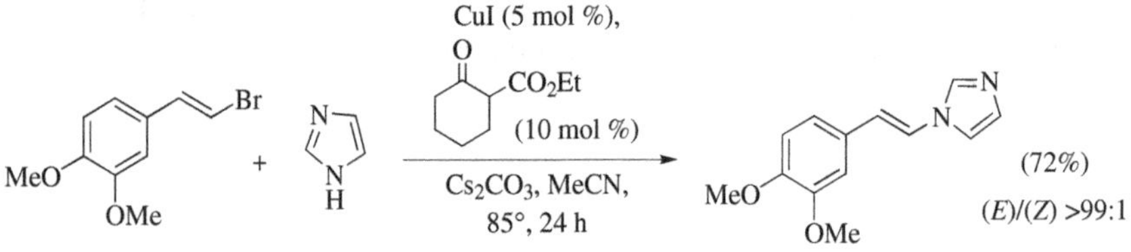

Scheme 29

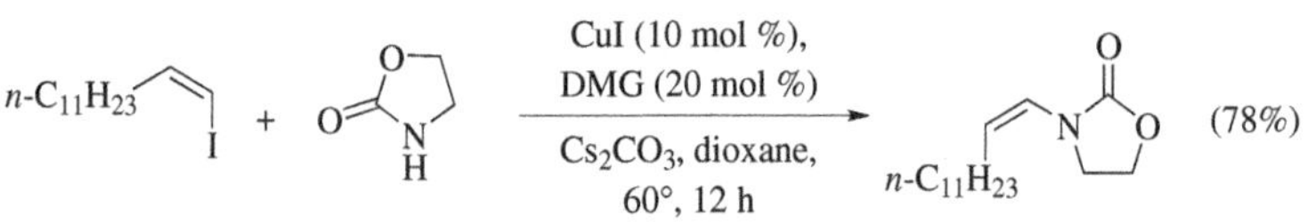

Scheme 30

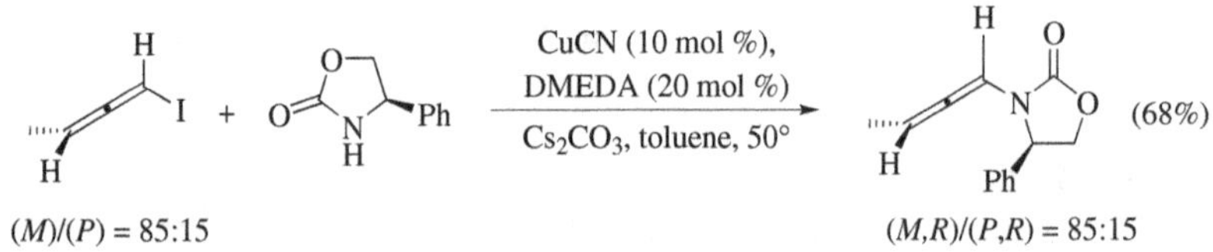

Scheme 31

Inversion of alkene configuration is observed in a few cases. In the coupling of conjugated amides with an (*E*)-β-iodoacryloyl amide, the thermodynamically more stable (*Z*)-product is obtained in low yield as the only coupling product (Scheme 32).[92] Coupling of (*Z*)-β-bromostyrene derivatives with amides using a CuI/DMG (DMG= *N,N*-dimethylglycine HCl) catalyst system in refluxing dioxane gives mixtures of (*Z*)- and (*E*)-alkenes in (*Z*)/(*E*) ratios ranging from 78:22 to 95:5 (Scheme 33).[93]

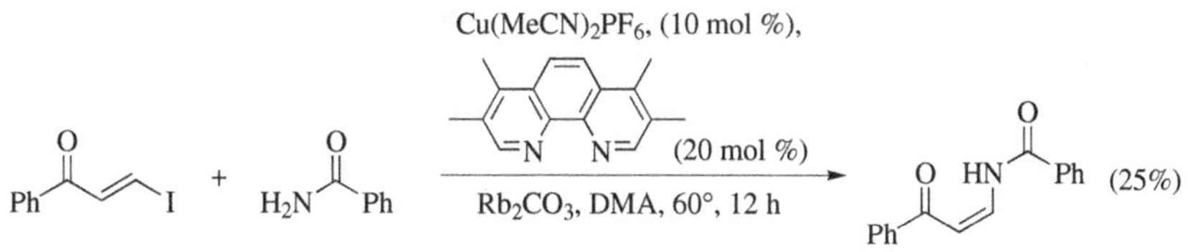

Scheme 32

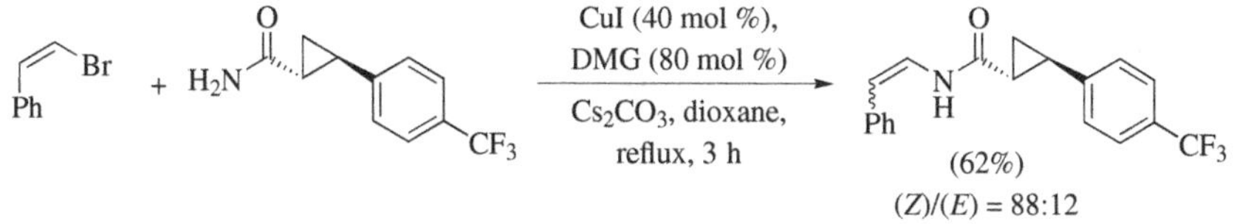

Scheme 33

Typical coupling temperatures for alkenyl iodides are between 50 and 100°, which is somewhat lower than the temperatures generally required for aryl iodides. Only a few examples of ligand-free Cu-catalyzed coupling of alkenyl iodides have been reported. Copper(II) oxide nanoparticles are an effective catalyst for coupling of (E)-β-iodostyrene and azole nucleophiles at 80°.[94] Coupling of vinyl iodides with amides catalyzed by copper thiophene-2-carboxylate (CuTC) without additional supporting ligands is used in the synthesis of several natural products.[95−97] More commonly, the catalyst derived from CuI and DMEDA is used for the coupling of alkenyl iodides with amides and azoles. Couplings of alkenyl iodides and amides are widely used in the synthesis of enamide natural products. Coupling of an alkenyl iodide with amides at room temperature with stoichiometric amounts of CuI/DMEDA is used in the synthesis of palmerolide A analogs.[98] Intramolecular coupling of an alkenyl iodide and amide catalyzed by CuI/DMEDA is used as the ring-closing step in the synthesis of abyssenine A.[99] These examples are discussed in more detail in the "Applications to Synthesis" section. The CuI/DMEDA system has also been used in the coupling of 1,4-diiodobutadienes and amides to give pyrroles (Scheme 34).[100]

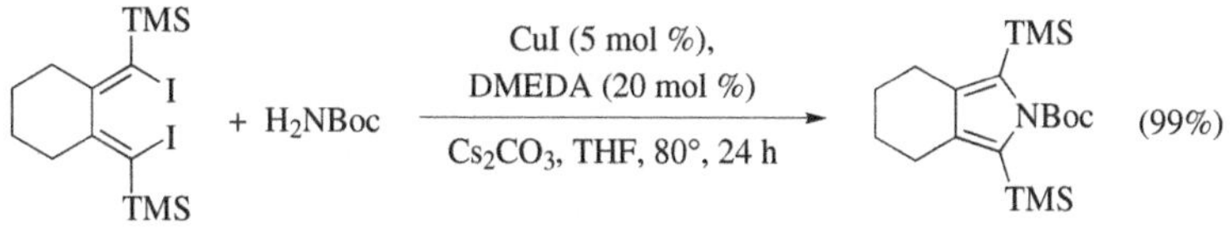

Scheme 34

Examples of coupling of alkenyl bromides are less common than those of iodides. Ligand-free couplings with CuO nanoparticles give good yields at 80°, but require approximately twice the reaction time compared to iodides.[94] Diamine-ligand-supported catalysts are most commonly used with alkenyl bromides. Typical reaction temperatures for these systems range from 100 to 120°. The scope of alkenyl halides is broad with (E) and (Z) 1-haloalkenes, 2-haloalkenes, and trisubstituted haloalkenes all giving good yields and complete retention of alkene configuration for couplings with amides and azoles (Scheme 35).[101,102] N,N-Dimethylglycine is also used as a ligand for coupling of alkenyl bromides and amides.[90,93]

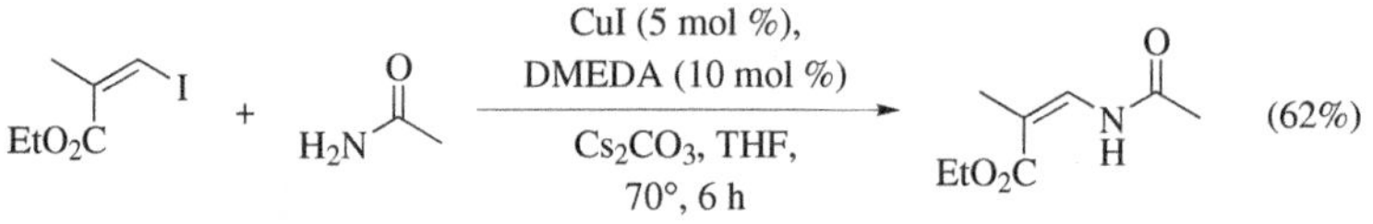

Scheme 35

The less reactive alkenyl chlorides have not been widely used in Cu-catalyzed C–N bond-forming reactions. Ligand-free catalyst systems give low yields in intermolecular couplings of alkenyl chlorides with azole nucleophiles in the few examples that have been reported.[94,103] Alkenyl chlorides show more utility in intramolecular couplings. Cyclization of 3-chloro-3-butenyl sulfonamide **4** provides a good yield of azetidine **5**, but the chloride substrate requires higher temperatures (100° vs. 68°) and higher catalyst loadings (20 vs. 10 mol % Cu) than the bromide substrate (Scheme 36).[104,105] Condensation of 2-bromo-β-chlorostyrene and O-tert-butyl carbamate provides a 56% yield of the Boc-protected indole in a cascade coupling reaction using Cu(OAc)₂/DMEDA as the catalyst (Scheme 37).[106] The yield is significantly lower than is obtained with the dibromo analog. Interestingly, switching the chloride and bromide positions (2-chloro-β-bromostyrene) leads to no conversion under identical conditions. No examples of vinyl fluoride or sulfonate substrates in Cu-catalyzed C–N couplings are known.

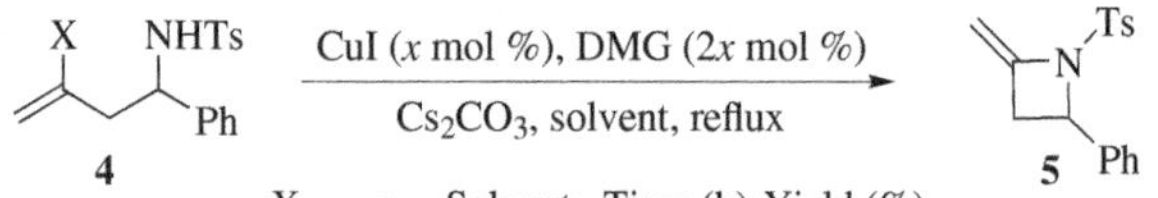

X	x	Solvent	Time (h)	Yield (%)
Br	10	THF	1	99
Cl	20	dioxane	2	99

Scheme 36

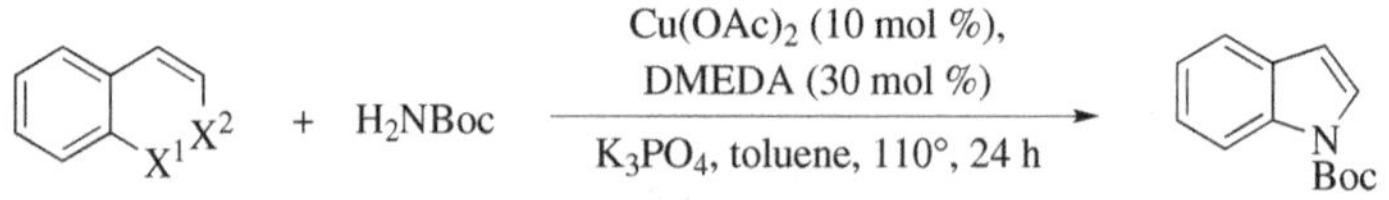

X¹	X²	Yield (%)
Br	Br	88
Br	Cl	56
Cl	Br	0

Scheme 37

The Nitrogen Nucleophile

A broad range of nitrogen nucleophiles can be arylated under copper catalysis. Similarly to Pd-catalyzed reactions, primary and secondary aryl or alkyl amines can be arylated. The Cu-catalyzed reactions provide more general reactivity than palladium catalysts with more acidic N–H functionalities including azoles, amides, ureas, carbamates, and sulfonamides. The copper systems are also more general for the arylation of ammonia than palladium-based systems. Specific features of the main classes of nitrogen nucleophiles are highlighted here.

The general order of reactivity for nitrogen nucleophiles using copper catalysts is that amides and azoles are more reactive than amine substrates. For example, coupling of 3-aminobenzamide with an aryl bromide using a CuI/DMEDA catalyst system gives amide arylation product **6** exclusively (Scheme 38).[107] The observed selectivity is opposite to the strong preference for arylated amine product **7** seen with palladium catalysts using the same substrates. Similarly, 5-aminoindole is arylated exclusively on the indole nitrogen using the copper catalyst (Scheme 39),[107] whereas the Pd-catalyzed coupling gives exclusive arylation on the amine. The increased reactivity of amides and azoles in Cu-catalyzed couplings may be due to the increased acidity of azoles and amides compared to amines and aniline derivatives, or it is possible that Cu–amidate and –azolate complexes undergo oxidative addition with aryl halides faster than Cu–anilido complexes. Highly acidic N–H bonds, such as that in tetrazole, are poor coupling partners, however, presumably due to the low nucleophilicity of the nitrogen.[40] In substrates that contain both alkyl amine and aniline nucleophiles, selective arylation of the alkyl amine is observed.[52,107] This selectivity may be due to the stronger coordinating ability of the alkyl amine nitrogen compared to that of the aryl amine.

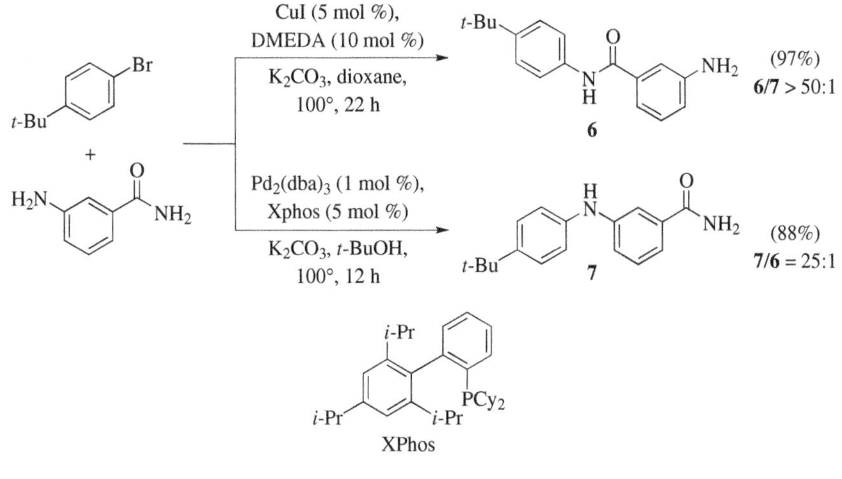

Scheme 38

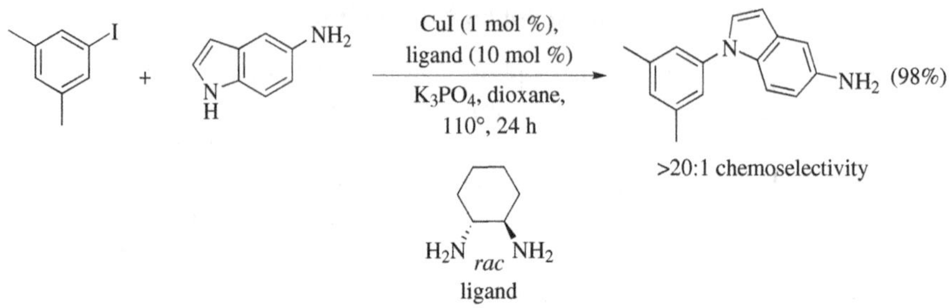

Scheme 39

Aromatic Amines.   Aniline derivatives show broad scope in Cu-catalyzed *N*-arylation reactions. Sterically demanding anilines, such as 2,6-dimethylaniline, require reaction times and give yields comparable to those of unhindered substrates in the few examples reported.[67,108] 2-Substituted anilines with large substituents, such as phenyl or *tert*-butyl, are also successfully coupled with aryl halides (Scheme 40).[109] Substituent electronic effects have little influence on the coupling efficiency. Electron-withdrawing substituents would be expected to decrease the nucleophilicity of the amine making it less reactive in the C–N bond-forming step. This effect may be counteracted by the increased acidity of the N–H bond, which makes formation of the Cu–amido complex occur more readily. Careful studies of the electronic effect on coupling have not been reported. In a few cases, nitroanilines give significantly lower yields than more nucleophilic amines.[56] Other examples of successful coupling of nitroanilines with aryl halides are known, however (Scheme 41).[67,108]

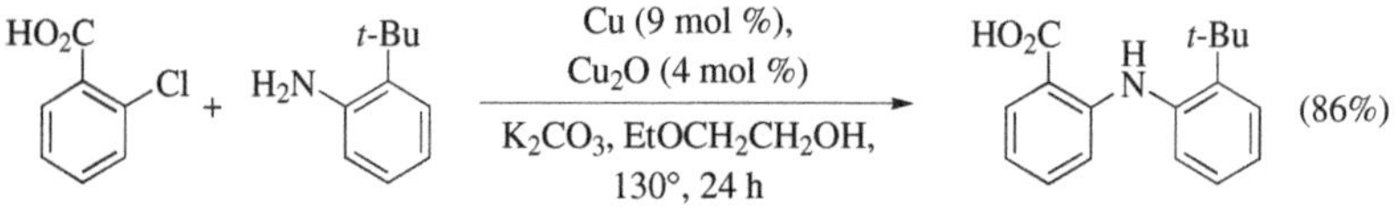

Scheme 40

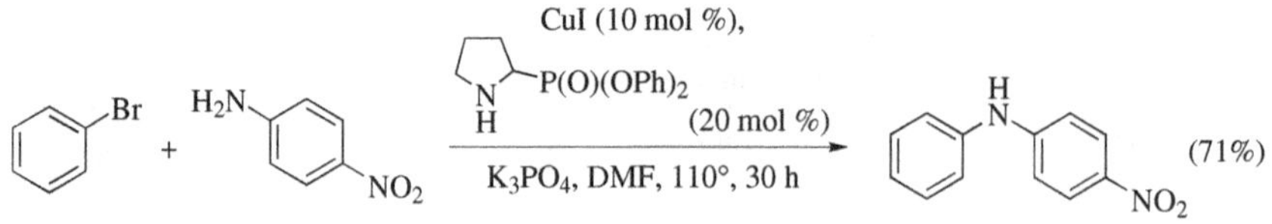

Scheme 41

In the arylation of primary anilines, both mono- and diarylation can occur to give either secondary or tertiary arylamines. To achieve monoarylation, the amine generally must be used in excess (2–4 equivalents) to avoid over-arylation. Under these conditions, high selectivity for the diaryl secondary amine is achieved. Use of a 2:1 ratio of aryl halide to aniline generally gives clean formation of the triarylamine product.

Ammonia, Hydrazine, and Hydroxylamine.   The direct synthesis of primary aryl amines from ammonia and aryl halides has been a long-sought synthetic objective. Palladium catalysts are generally ineffective for direct coupling with ammonia, although a few recent systems have shown promise.[110] In contrast, Cu-catalyzed arylations of ammonia have a long history.[111] Selectivity for the monoarylated product is achieved by using a large excess (typically ≥10 equivalents) of ammonia relative to the aryl halide. Anhydrous ammonia at high pressure (80–100 psi) can be used as the NH_3 source.[110,112] A more convenient source of ammonia, at least in academic settings, is aqueous ammonia. A number of copper systems catalyze the arylation of aqueous ammonia in good yields under mild conditions.[113,114] Ammonium chloride can also be used as the NH_3 source.[60]

No examples of N-arylation of hydrazine or hydroxylamine are found in the literature. The majority of examples of arylation of hydrazine derivatives involve arylation of hydrazides on the more acidic amide N–H bond, which is discussed below. Arylation of phenylhydrazine occurs on the unsubstituted nitrogen to afford N,N'-diarylhydrazine **8** with 2-bromotoluene, whereas less hindered aryl halides selectively arylate the more acidic phenyl-substituted nitrogen to give N,N-diaryl hydrazine **9** (Scheme 42).[48] A CuI/CDA (CDA = trans-1,2-diaminocyclohexane) catalyst gives selective arylation on the unsubstituted NH_2 of benzoylhydrazide or benzophenone hydrazone (Scheme 43).[115]

Ar	8/9	Yield (%)
2-MeC6H4	100:0	74
4-O2NC6H4	0:100	82

Scheme 42

Scheme 43

Alkyl Amines.   Primary and in some cases secondary alkyl amines are effective coupling partners in Cu-catalyzed C–N bond-forming reactions. Primary alkyl amines give high yields in couplings with aryl halides when the nitrogen is attached to primary or secondary carbons. Arylation of amino acids[116] and cyclohexylamine[117] has been successfully demonstrated with a number of systems. A smaller number of examples of arylation of amines attached to simple acyclic secondary carbons, such as isopropyl or *sec*-butyl amines are known.[118,119] Configurational purity at the amine α-carbon is retained (Scheme 44).[116,120] Examples of *N*-arylation of primary amines attached to tertiary carbons are rare. Arylation of 1-amino-1-hydroxymethylcyclopropane with 3,5-dichloro-1-iodobenzene gives a 53% yield of coupled product using CuI/ethylene glycol as the catalyst system (Scheme 45).[121]

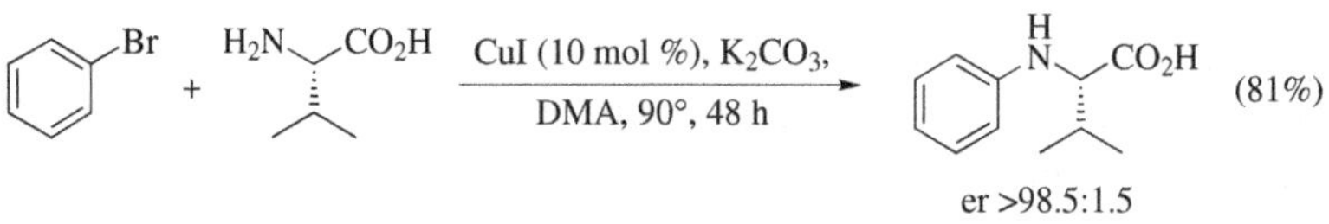

Scheme 44

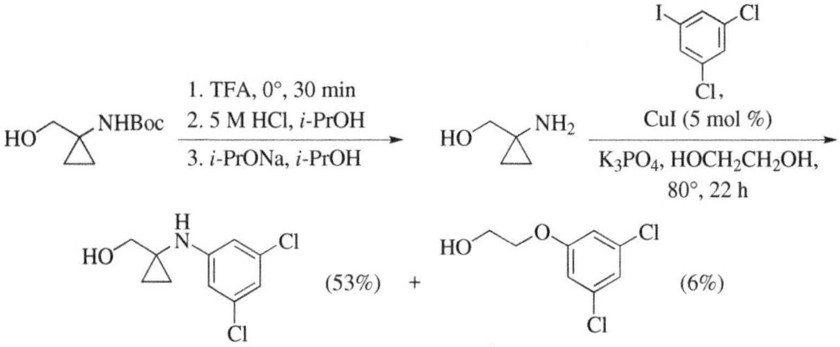

Scheme 45

Secondary alkyl amines often give significantly lower yields than primary amines or cyclic secondary amines.[47,122,123] Good yields for the coupling of dialkyl amines and activated aryl halides, such as 2-chloronitrobenzene, have been reported (Scheme 46).[117,122] In contrast to acyclic secondary amines, cyclic secondary amines, such as pyrrolidine, piperidine, and morpholine, generally give comparable yields to primary amines,[124,125] although some authors have noted lowered yields.[122,126] Even the sterically demanding 3,5-dimethylmorpholine can be arylated with phenyl iodide in 87% yield using a catalyst derived from CuBr and a phosphoramidite ligand (Scheme 47).[118]

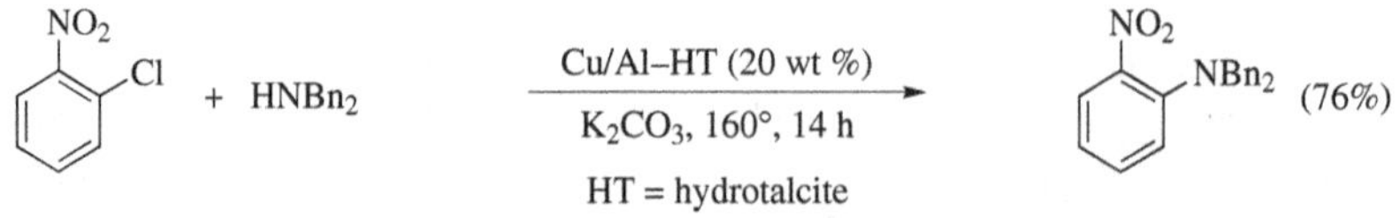

Scheme 46

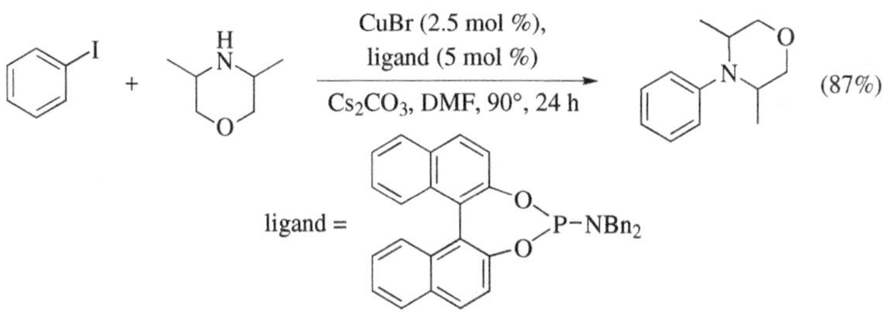

Scheme 47

Azole Nucleophiles.   Five-membered-ring *N*-heteroaromatic compounds (azoles) are highly effective substrates for Cu-catalyzed C–N bond-forming reactions. Copper catalysts tend to be highly selective for *N*-arylation, whereas palladium catalysts often give mixtures of *N*- and *C*-arylated products.[127] Nearly all classes of *N*-heterocyclic compounds with an N–H bond are reactive, including pyrroles, imidazoles, pyrazoles, triazoles, and their benzannulated analogs. The order of reactivity among azoles is pyrazole > imidazole > indole > pyrrole, > 1,2,4-triazole ≫5-phenyltetrazole for arylations catalyzed by Cu₂O/tetradentate Schiff-base ligand at 80° (Scheme 48).[40] This trend may represent a balance of the acidity of the N–H bond, which would promote Cu–azolate formation, and the nucleophilicity of the resulting azolate anion, which would favor C–N bond formation. On the

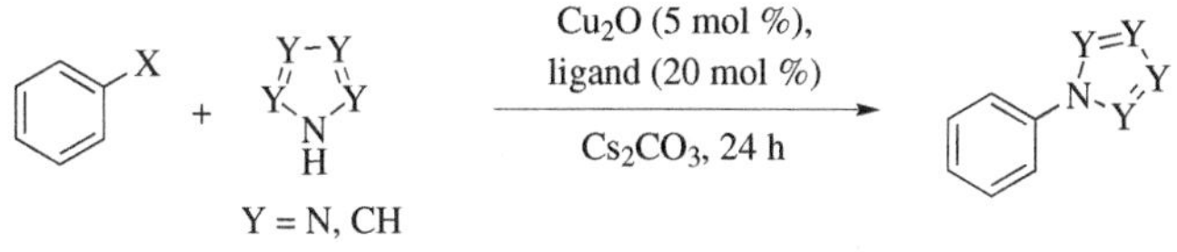

Azole	X	Solvent	Temp (°)	Yield (%)
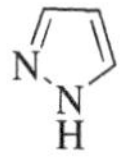 pyrazole	Br	MeCN	82	93
 imidazole	Br	MeCN	82	48
 pyrrole	I	MeCN	82	94
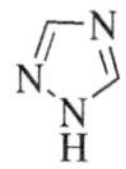 1,2,4-triazole	I	DMF	82	79
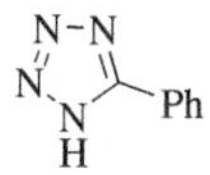 5-phenyltetrazole	I	DMF	110	0

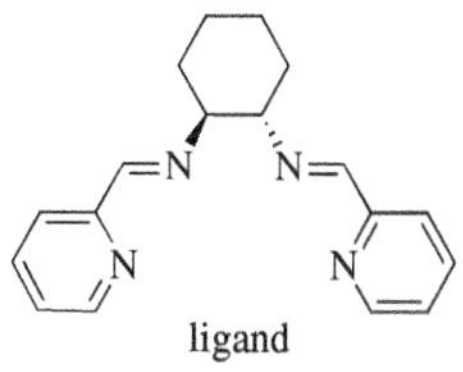

Scheme 48

basis of nucleophilicity, pyrrole and indole would be expected to be most reactive towards arylation, but their higher pK_a values apparently slow the formation of the Cu–azolate intermediate. Thus, the more acidic pyrazole and imidazole substrates are most reactive. The presence of additional nitrogen atoms in the ring results in a decrease in activity. Presumably the lower nucleophilicity of triazoles and tetrazoles has a larger effect on reactivity than the increased acidity of the N–H bond. No successful examples of tetrazole arylation have been reported.

Azole N-arylation reactions are highly tolerant of substitution on the azole ring, but do show similar steric and electronic effects to other nitrogen nucleophiles. Substitution next to the nucleophilic nitrogen can reduce the rate of coupling, thus requiring higher temperatures and/or catalyst loadings to achieve high yields. Coupling of 3,5-dimethylpyrazole and iodobenzene provides a 12% yield of coupled product at 80° using a Cu_2O/salicylaldehyde oxime catalyst system, whereas unhindered pyrazoles give high yields under the same conditions.[82] Increasing the reaction temperature to 110° and changing the ligand to a tetradentate bis(imine) increases the yield to 75% (Scheme 49), which is still lower than that observed with unhindered pyrazoles. Electron-withdrawing substituents such as ester, ketone, or trifluoromethyl groups are generally tolerated without a significant effect on the coupling yield.[128]

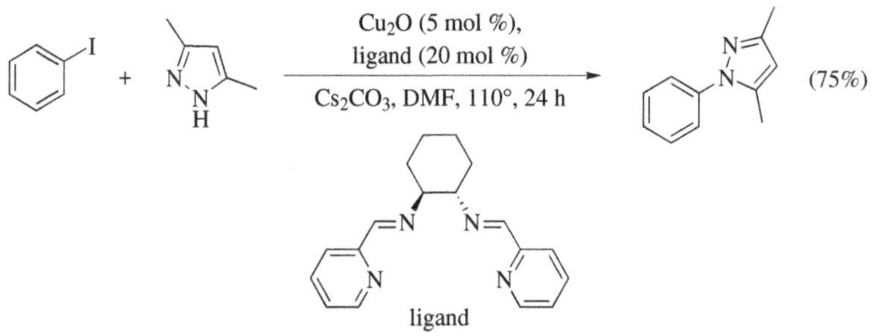

Scheme 49

Arylation of azoles is generally highly selective for N-arylation over C-arylation. For non-symmetrical azoles with multiple nitrogen centers, mixtures of products are often obtained, since each nitrogen atom can share the negative charge due to resonance delocalization. Unsymmetrical, 4-substituted imidazoles often give mixtures of products with N-arylation at the less hindered nitrogen being favored. The size of substituents on the imidazole as well as the steric demand of the aryl group affect the selectivity. The coupling of 4-methylimidazole and 1-bromo-4-methylthiobenzene gives a 4.4:1 ratio of N1 (product **10**) and N3 (product **11**) arylation using a catalyst derived from Cu_2O/4,7-dimethoxyphenanthroline (Scheme 50).[84] With 4-phenylimidazole as the substrate, a complete selectivity for product **10** is obtained. Similarly, coupling of 4-methylimidazole and 1-bromo-2-isopropylbenzene affords a 44:1 selectivity for the N1 product.

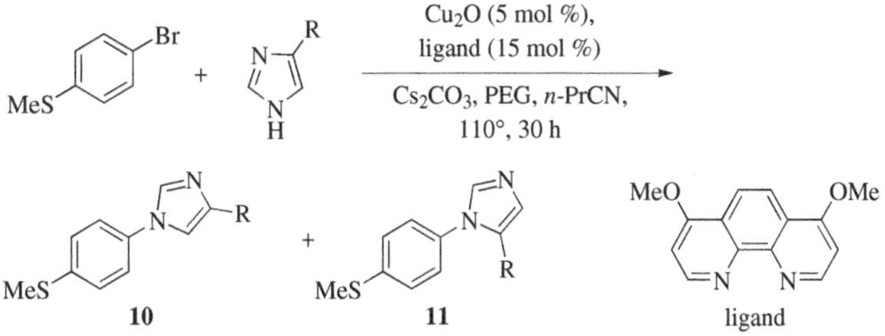

Scheme 50

R	Yield (%)	10/11
Me	97	4.4:1
Ph	96	99:1

Unsymmetrical pyrazoles can also give mixtures of products with the less-hindered nitrogen being the preferred site of arylation. Reaction of 3-methylpyrazole and iodobenzene gives a mixture of 3-methyl-1-phenylpyrazole and 3-methyl-2-phenylpyrazole in a 3.5:1 ratio using CuI/salicylaldehyde oxime as the catalyst system (Scheme 51).[82] Indazoles give mixtures of N1- and N2-arylated products. The selectivity is dependent upon the aryl halide. Aryl iodides give high selectivity (15–20:1) for the 1-arylindazole product using CuI/DMCDA as the catalyst system, whereas the corresponding aryl bromides give modest selectivities (<2:1) for the 1-arylindazole under the same conditions (Scheme 52).[128] Mixed results are obtained with 1,2,3-triazoles. A Cu_2O/phenanthroline-catalyzed coupling of iodobenzene and 1,2,3-triazole gives 1-phenyl-1,2,3-triazole as the sole product.[129] In contrast, the ligand-free Cu_2O/Fe(acac)$_3$ catalyst gives a nearly 1:1 mixture of 1-phenyl- and 2-phenyl-1,2,3-triazole (Scheme 53).[130] On the other hand, benzotriazole is arylated to give 1-arylbenzotriazoles with high selectivity in most reports,[128] although mixtures of products are obtained with a ligand-free catalyst system using TBAF as the base.[81] Complete selectivity for N1 is observed with 1,2,4-triazoles.[128]

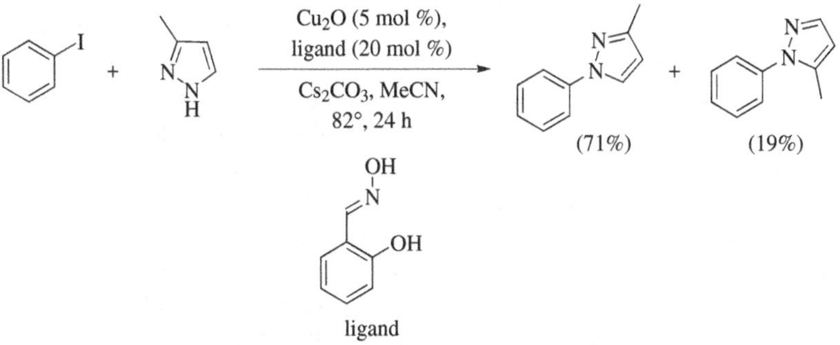

Scheme 51

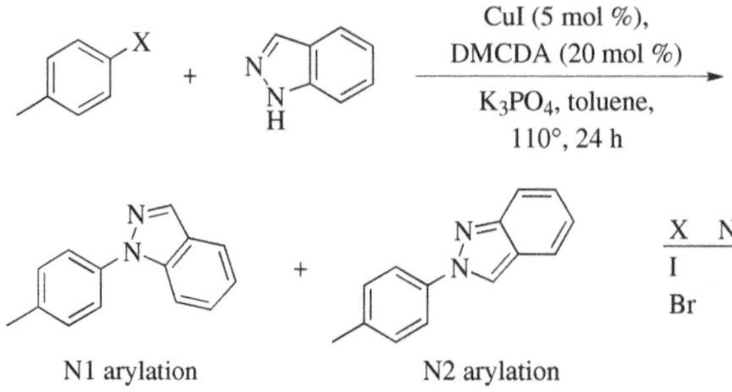

X	N1 + N2 Yield	N1/N2
I	85%	17:1
Br	75% conv	1.8:1

N1 arylation   N2 arylation

Scheme 52

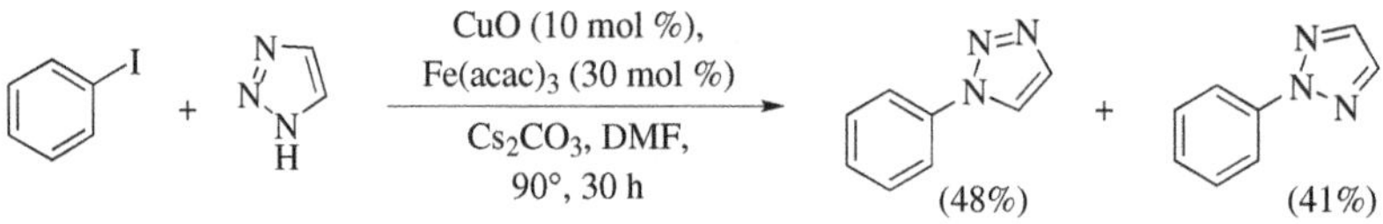

Scheme 53

Amides, Sulfonamides, and Related Compounds.   Amides and related compounds, including carbamates, ureas, guanidines, and sulfonamides, are generally useful substrates for Cu-catalyzed C–N bond-forming reactions. The more acidic N–H bond of amides and related compounds typically results in their being more reactive than aryl and alkyl amines. Like azoles, Cu-based catalysts have proven to be much more general for coupling of amides than palladium catalysts. Although both the nitrogen and oxygen can act as the nucleophile in amidate anions, the nitrogen is generally the exclusive site of arylation. However, pyridones give competitive O-arylation with sterically hindered aryl halides. Coupling of 2-pyridone with 2-iodotoluene gives a 4.8:1 ratio of O- to N-arylation using 8-hydroxyquinoline as the ligand, whereas 4,7-dimethoxyphenanthroline gives a 1.8:1 preference for N-arylation (Scheme 54).[131] Unhindered aryl halides give complete selectivity for N-arylation using 4,7-dimethoxyphenanthroline as the ligand.

Primary amides are the most widely reported class of amide substrates. Arylations of primary amides are largely unaffected by steric and electronic effects, although careful analysis of these effects under common conditions has not been reported. Electron-deficient amides, such as trifluoroacetamide, are arylated in good yields.[132] Hydrazides undergo N-arylation on the more acidic amide nitrogen (Scheme 55),[133,134] although this selectivity can be reversed with more hindered aryl halides (Scheme 56).[135] Arylation of acyclic secondary amides has been less widely reported than that of primary amides or lactams. Some authors report lower yields or reaction rates with secondary amides, such as acetanilide.[136–138] Optimized systems based on CuI/diamine (diamine = DMEDA or DMCDA) catalysts give comparable yields for primary and secondary amides, however.[69,115] Lactams typically show comparable reactivity to primary amides, with pyrrolidinone being a commonly

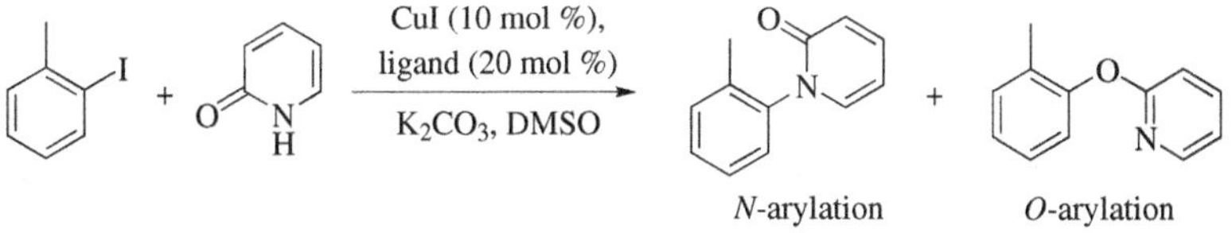

Ligand	Temp (°)	Time (h)	N-Aryl	O-Aryl	N-Aryl/O-Aryl
A	120	48	14	67	1.0:4.8
B	150	96	40	22	1.8:1.0

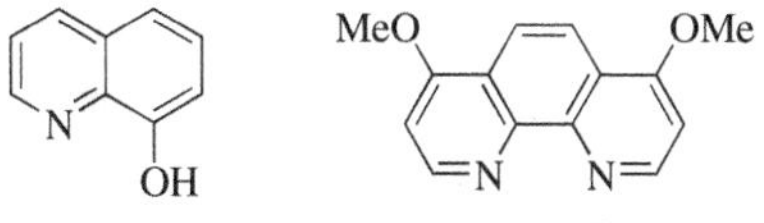

Scheme 54

used model substrate. Both 2- and 4-hydroxypyridines are also effective substrates that give *N*-arylpyridones upon *N*-arylation.[131] Amidines are used primarily in tandem processes leading to benzimidazoles and quinazoline derivatives (Scheme 57).[139−143]

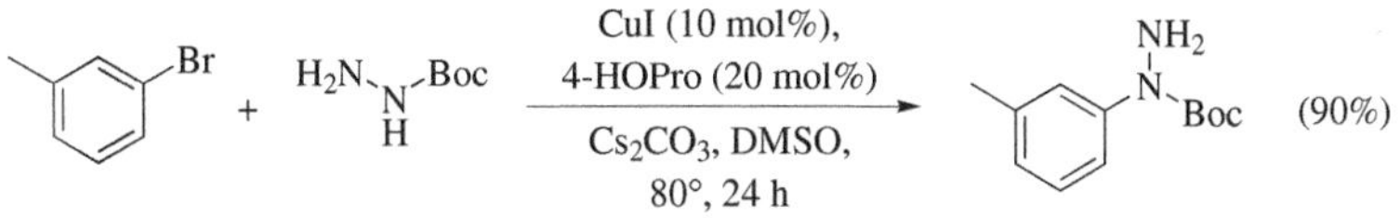

Scheme 55

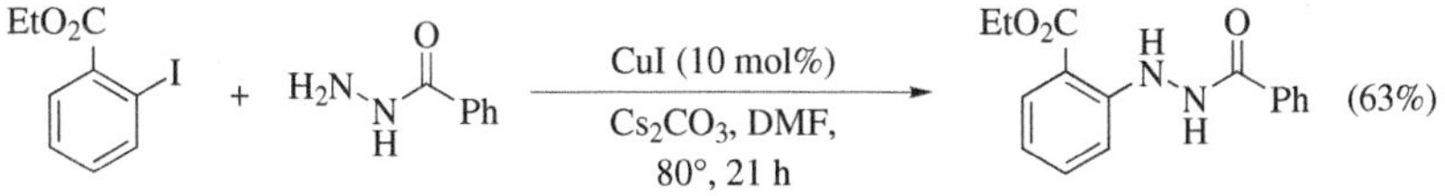

Scheme 56

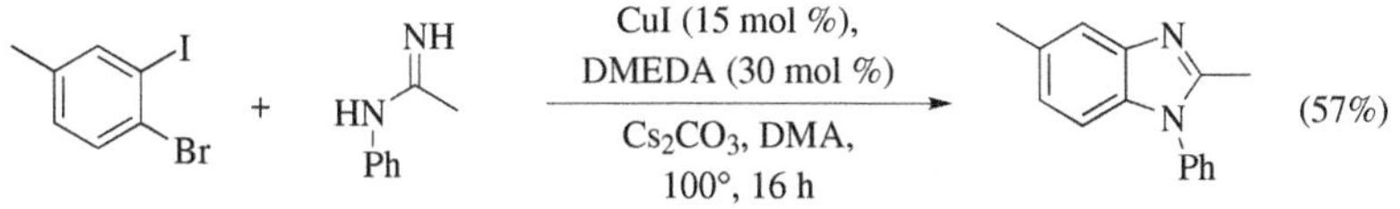

Scheme 57

N-Arylation of carbamates, ureas, and guanidines has been reported for a number of catalyst systems. Carbamates typically show comparable reactivity to structurally similar amides. Examples of acyclic carbamates are largely limited to Boc-protected ammonia[83] and hydrazine (Scheme 55),[133,134] and hydroxylamine ethers.[144] The majority of examples of carbamate arylations involve oxazolidinone derivatives. Oxazolidinone-based chiral auxiliaries can be N-arylated without degradation of enantiomeric purity (Scheme 58).[87,145] Notably in this example, the cyclic carbamate nitrogen is cleanly arylated in the presence of the secondary amide. Both acyclic and cyclic ureas are effective nitrogen nucleophiles in Cu-catalyzed C–N bond-forming reactions, although the reported yields are modest (40–70%).[146–148] Mono-N-substituted ureas undergo exclusive arylation on the unsubstituted nitrogen (Scheme 59).[146] For symmetric ureas, mono- or diarylation can occur. Monoarylation requires an excess of the urea substrate.[148] Diarylation of guanidinium nitrate to give N,N'-diarylguanidines occurs in modest to good yields (30–90%) using a CuI/salicylamide catalyst system.[149] Higher yields (75–90%) are reported for the monoarylation of an unsymmetrical guanidine derivative using CuI/DMEDA as the catalyst precursor (Scheme 60).[141]

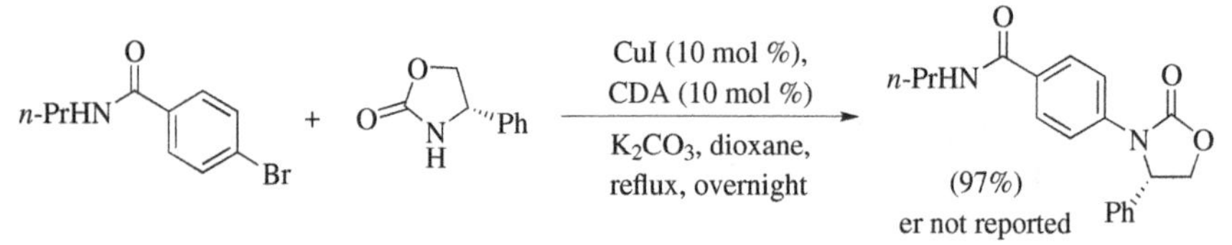

Scheme 58

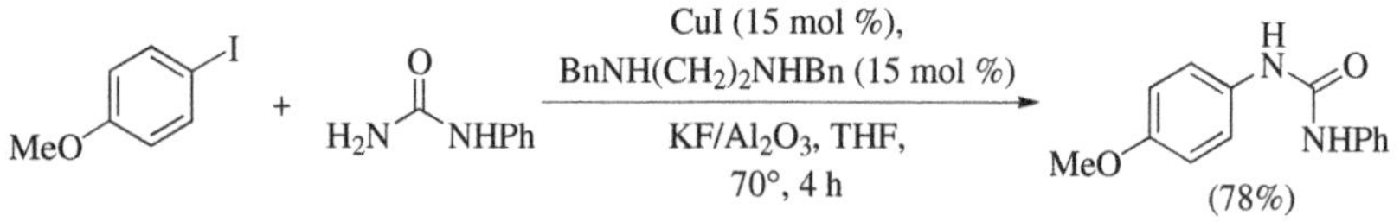

Scheme 59

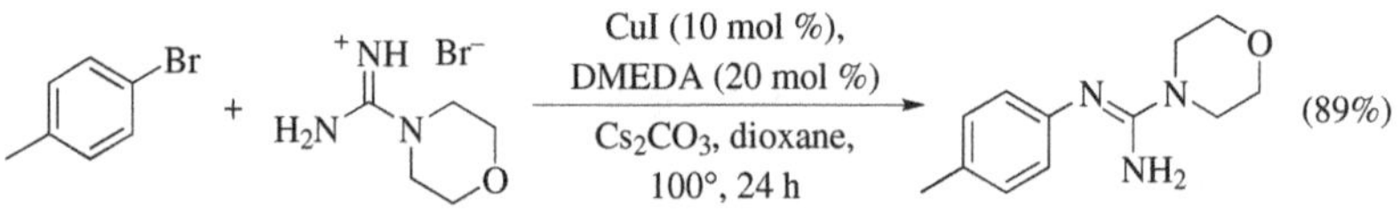

Scheme 60

Copper catalysts are effective for the N-arylation of sulfonamides. Primary and secondary sulfonamides are N-arylated with unhindered aryl halides in high yields.[150] Low yields are obtained with 2-substituted aryl halides and secondary sulfonamides, however. Sultams generally give modest to good yields (40–85%, Scheme 61).[151] In a direct comparison, a Pd/Xantphos (Xantphos = 4,5-bis(diphenylphosphino)-9,9-dimethyl-9H-xanthene) catalyst system provides higher yields than a Cu₂O/phen catalyst system for the N-arylation of 1,4-butanesultam.[152] A ligand-free system gives moderate yields for the coupling of aryl iodides and sulfonimidamides.[153]

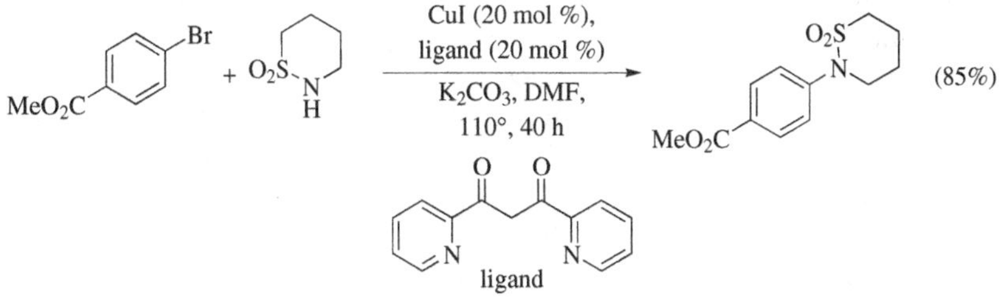

Scheme 61

Other Nitrogen Nucleophiles.   Other nitrogen nucleophiles that have been applied to Cu-catalyzed C–N bond-forming reactions include azides and sulfoximines. Aryl azides are useful precursors to a variety of compounds. Copper catalysts provide high yields for the coupling of both aryl and vinyl halides with sodium azide under mild conditions.[154,155] At higher temperatures, the resulting aryl azide can be decomposed to give aniline products (Scheme 62).[75,156,157] A stoichiometric amount of copper is required to promote the azide reduction. Thus, azide nucleophiles can be used as an alternative to ammonia for the synthesis of primary aryl amines. Both ligand-free and DMEDA-supported copper catalysts provide good yields in the N-arylation of sulfoximines (Scheme 63).[158,159]

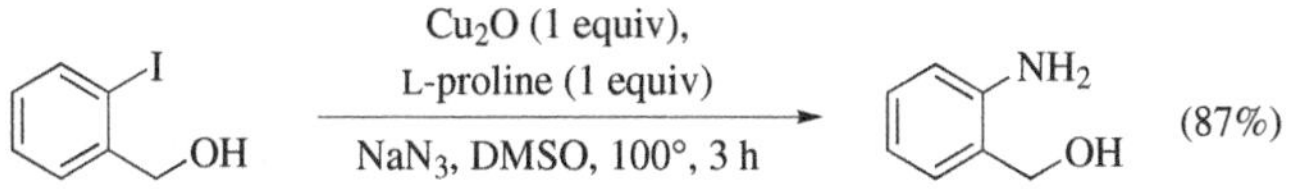

Scheme 62

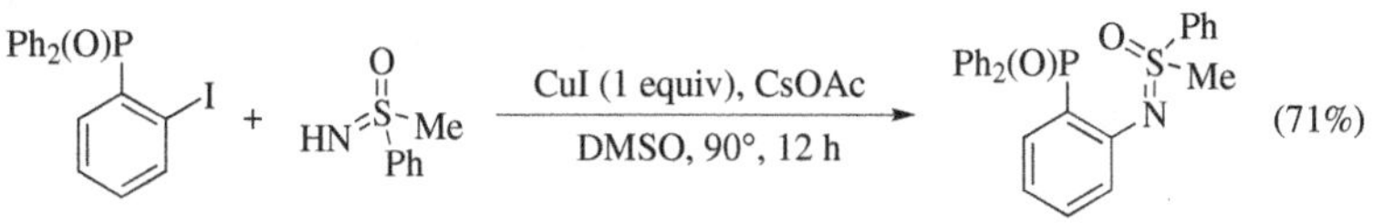

Scheme 63

Synthesis of Natural Products and Biologically Active Compounds

The ease and simplicity of Cu-catalyzed C–N bond formation along with the functional group tolerance of these reactions have made it a popular method in the synthesis of natural products.[16] Intramolecular and intermolecular amide arylation and vinylation reactions have been most widely used. Copper-catalyzed coupling of vinyl iodides and amides have been used in the synthesis of a large number of enamide-containing natural products. In this section, representative examples of the application of Cu-catalyzed C–N coupling reactions to the synthesis of natural products and pharmaceuticals are highlighted. Additional examples can be found in the "Tabular Survey" presented at the end of this chapter.

Amide arylation reactions have been used in the synthesis of a wide variety of alkaloid natural products. Intermolecular amide arylation is used in the early stages of the synthesis of the macrocyclic core of the cytotrienins **15** (Scheme 64).[160] Coupling of electron-rich aryl bromide **12** with functionalized amide **13** gives the aromatic segment **14**, which is further elaborated to provide **15**.

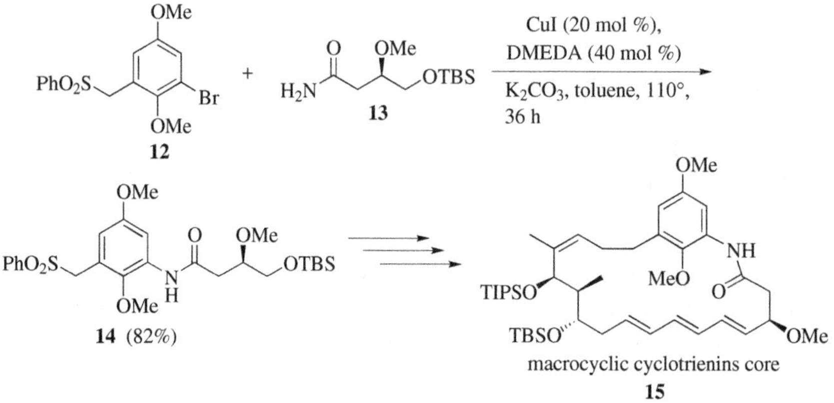

Scheme 64

A common application of Cu-catalyzed C–N bond formation is the synthesis of lactams through intramolecular amide arylation.[161] In the synthesis of chaetominine, the challenging intramolecular coupling of iodoindole **16** functionalized with a pendant hindered secondary amide affords a good yield of **17** (64%) using CuI/DMCDA as the catalyst in toluene at 110° with K_3PO_4 as base (Scheme 65).[162] No epimerization of the stereogenic centers is observed. Notably, there is no conversion when Pd/P(o-Tol)$_3$ is used as the catalyst.

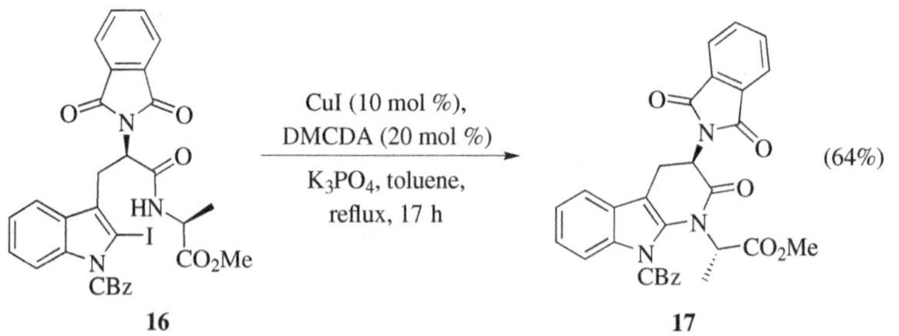

Scheme 65

Copper catalysts supported by diamine ligands are also effective at macrolactamization reactions. In the synthesis of reblastatin, intramolecular coupling of **18** gives 19-membered-ring lactam **19** in 83% yield using CuI/DMEDA (50 mol % Cu) at 100° for 36 hours at 0.015 M reaction concentration (Scheme 66).[163] This reaction is also used in the synthesis of the structurally similar macrolactam **21** in 81% yield from bromo amide **20** despite the presence of an isopropoxy group adjacent to the aryl bromide site (Scheme 67).[164] Deprotection of intermediate **21** provides geldanamycin.

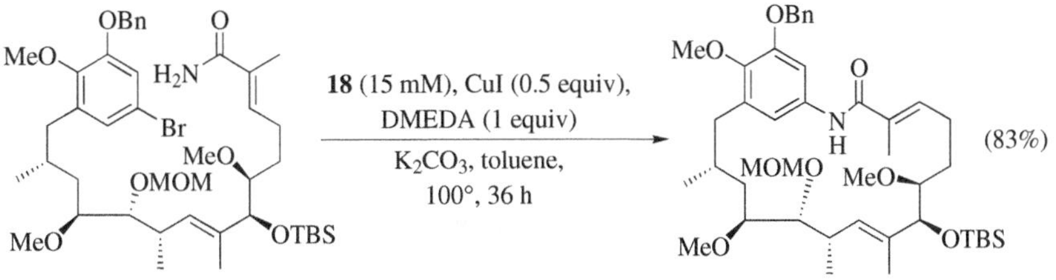

Scheme 66

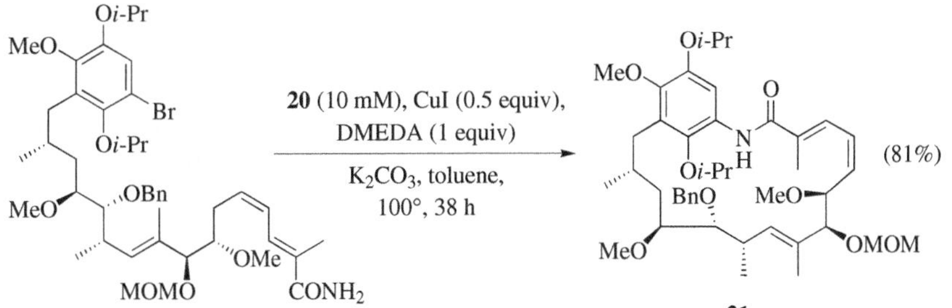

Scheme 67

Intermolecular coupling of vinyl iodides with amides is widely used in the synthesis of natural products with pendant enamide moieties. This approach was first used in the synthesis of lobatamide C (Scheme 68).[165] Coupling of enamide **23** with vinyl iodide **22** provides enamide **24**, which is then coupled with the lower portion of the macrocycle by sequential intermolecular esterification and Mitsunobu macrolactonization to provide lobatamide C.

Coupling of 3-methylbutenamide with late-stage intermediate macrocyclic vinyl iodide **25** is the final step of the synthesis of palmerolide A (Scheme 69).[166] This method is used in the synthesis of a variety of structural analogs of palmerolide.[98]

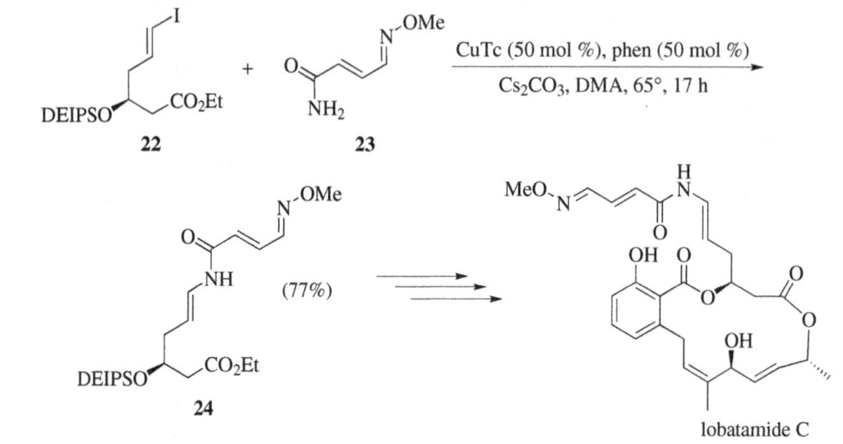

Scheme 68

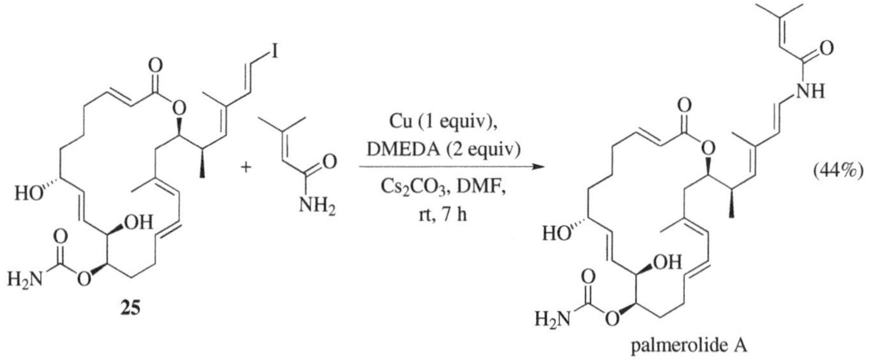

Scheme 69

The synthesis of apicularen A highlights the functional-group tolerance of the Cu-catalyzed methodology. Coupling of (2Z,4Z)-2,4-heptadienamide with (E)-vinyl iodide **26** containing free phenol and alcohol groups affords apicularen A in 40% yield, predominantly with retention of the vinyl iodide configuration [(E)/(Z) = 8:1, Scheme 70].[167]

Intramolecular coupling of vinyl iodides and amides has been used to prepare a family of cyclopeptide alkaloids, including paliurines E and F, ziziphines N and Q, abyssenine A, and mucronine E.[168] These compounds contain 13–15-membered-ring lactams. The paliurine core **28** is synthesized in 70% yield from acyclic peptide **27** using CuI/DMEDA (Scheme 71). A similar strategy is used to prepare the ziziphine, abyssenine, and mucronine cores. Alternative macrocyclization reactions for the synthesis of cyclic peptide **28** were explored (Scheme 72), including ring-closing metathesis of diene **29** with Grubbs' second-generation catalyst (49%), Wacker-type cyclization of styrene amide precursor **30** (21%), and acid-catalyzed intramolecular

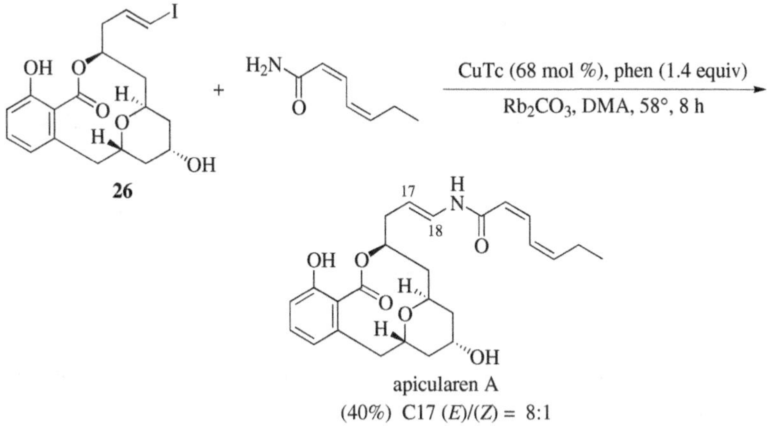

apicularen A
(40%) C17 (E)/(Z) = 8:1

Scheme 70

condensation of acetal-functionalized amide **31** (34%). The Cu-catalyzed cyclization proved superior to these alternative approaches under the conditions shown.

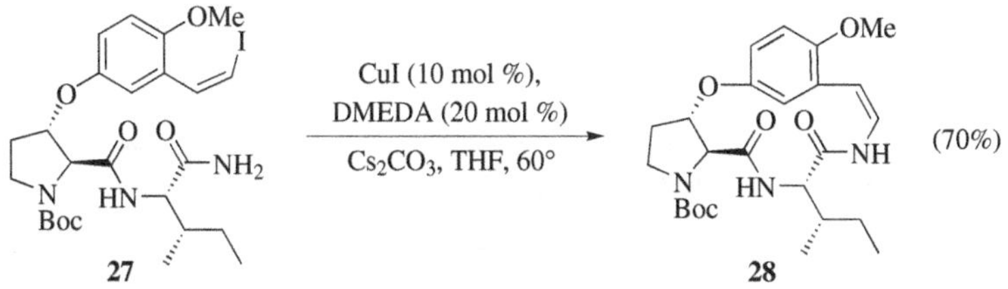

Scheme 71

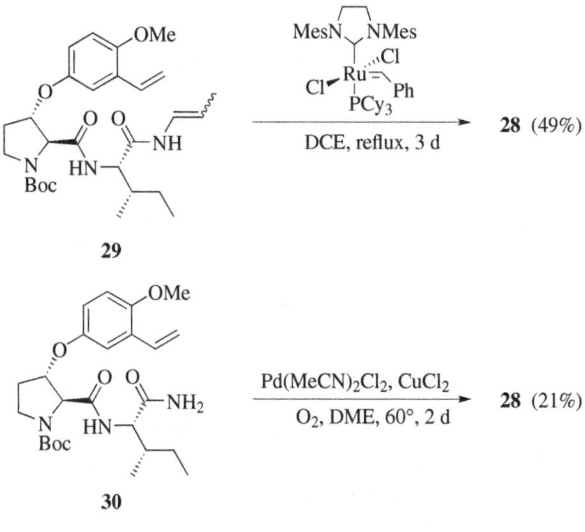

Scheme 72

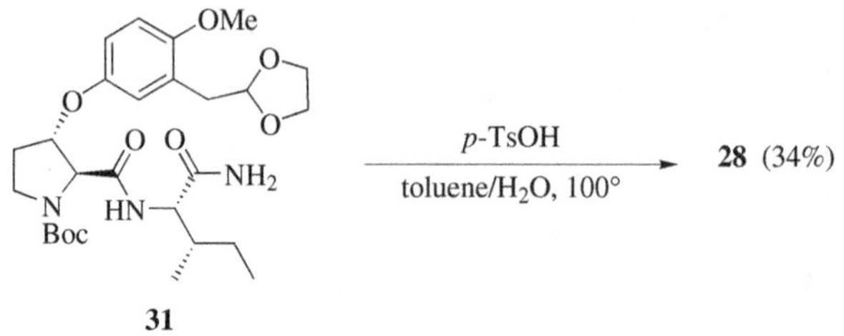

Scheme 72 (*Continued*)

Couplings with other nitrogen nucleophiles have received far less attention in the synthesis of natural products. Synthesis of the complex alkaloid psychotrimine involves two Cu-catalyzed C–N bond-forming reactions (Scheme 73).[169] The central ring system is formed by intramolecular cyclization of aryl bromide **32**. After further elaboration of pyrrolidinoindoline **33**, the final psychotrimine structure is formed by coupling of iodide **34** with indole **35** to afford protected psychotrimine **36**.

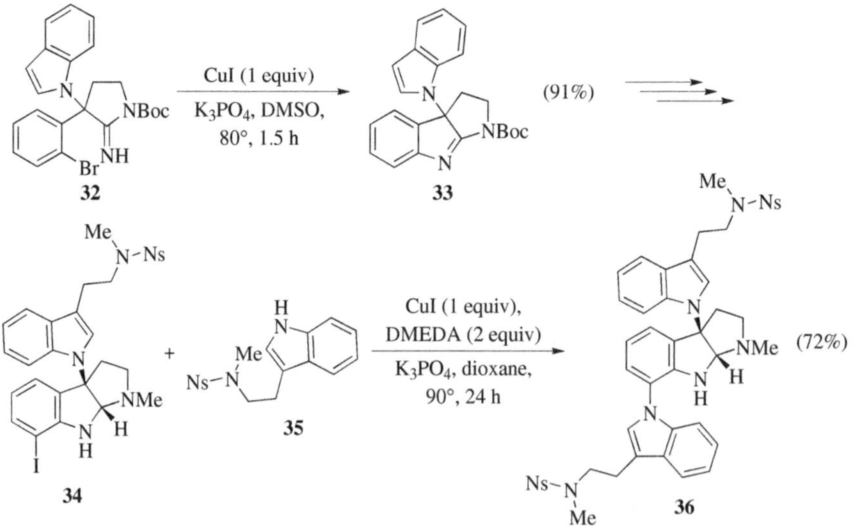

Scheme 73

Copper-catalyzed coupling of a 2-aminopyridine derivative and 2-bromo-4-nitrotoluene is used as the key step in a synthesis of the leukemia treatment imatinib (Scheme 74).[170] Arylation of ethyl (*S*)-3-amino-6-hydroxyhexanoate using CuI in aqueous DMF is used to prepare an early-stage intermediate in the synthesis of martinellic acid (Scheme 75).[171]

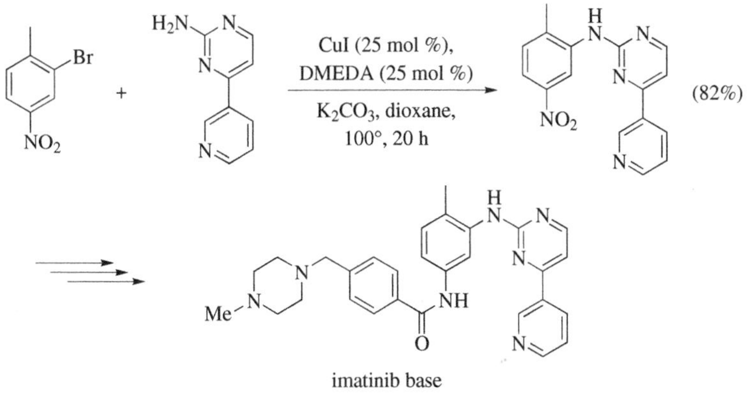

Scheme 74

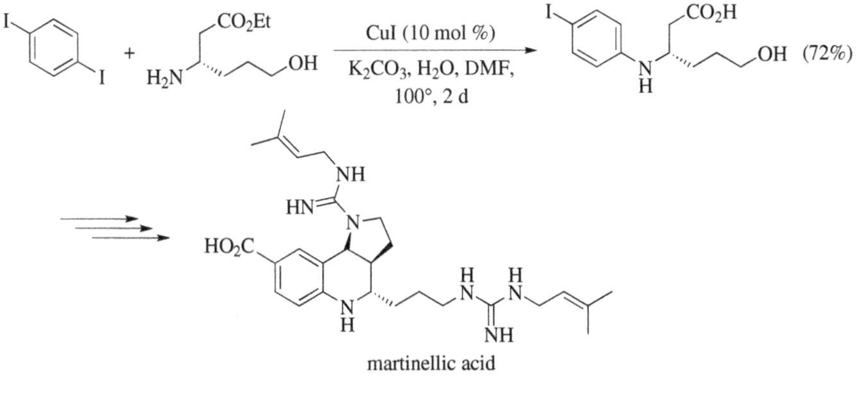

Scheme 75

Electronic Materials

Aromatic amines are important components in many electronic materials. Triaryl amines are effective hole transport materials with applications in areas such as xerography, organic light-emitting diodes (OLEDs), and photovoltaics. The Cu-catalyzed coupling of diiodofluorene oligomers **37** with carbazole **38** provides oligocarbazoles **39** for use in OLED devices (Scheme 76).[172] Copper-catalyzed C–N bond formation has also been applied to the synthesis of ligands for luminescent metal complexes.[173,174] The ability to tolerate nitrogen-containing heteroaromatic substrates allows for the facile synthesis of chelating pyridine ligands. Triarylamine-based redox gradient dendrimers that can serve as charge storage molecules can be synthesized by successive Cu-catalyzed arylation of aniline derivatives.[175] The final dendrimer is prepared by condensation of aryl iodide **40** and triamine **41** to give dendrimer **42** (Scheme 77).

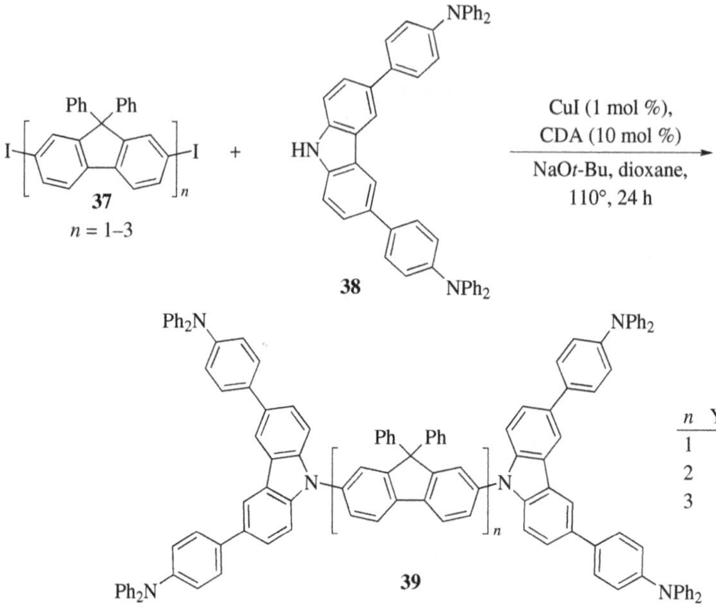

Scheme 76

Tandem Reactions for the Synthesis of Heterocyclic Compounds

Copper-catalyzed C–N bond formation coupled with other reactions, such as condensations, substitutions, or other metal-catalyzed bond formations, has been widely used in the synthesis of heteroaromatic compounds. A full collection of these useful transformations is available in Tables 27 and 35 of the "Tabular Survey." Specific examples are highlighted here.

The most common approach is to couple an aryl halide having a reactive group in the *ortho* position with a nitrogen nucleophile having a complementary reaction partner in close proximity to the nitrogen. Condensation of bifunctional amidine nucleophiles with 2-halobenzaldehydes provides quinazolines by a sequence involving imine formation and Cu-catalyzed C–N bond formation (Scheme 78),[140] whereas condensation with 2-halo benzoic acids provides quinazolinones through an amide formation.[142] Larger rings, such as benzodiazepines, can be prepared by the condensation of 2-halobenzyl amines and amino acids (Scheme 79).[176] The configurational homogeneity of the amino acid is retained in these reactions, allowing the natural pool of chiral amino acids to be used.

An interesting alternative approach to seven-membered (e.g., benzodiazepines) or larger rings involves the coupling of aryl halides having a pendant nitrogen nucleophile with β-lactams. Subsequent attack by the pendant amine on the lactam provides the ring-expanded product (Scheme 80).[177] Copper-catalyzed coupling of acetanilides with ammonia followed by intramolecular condensation provides

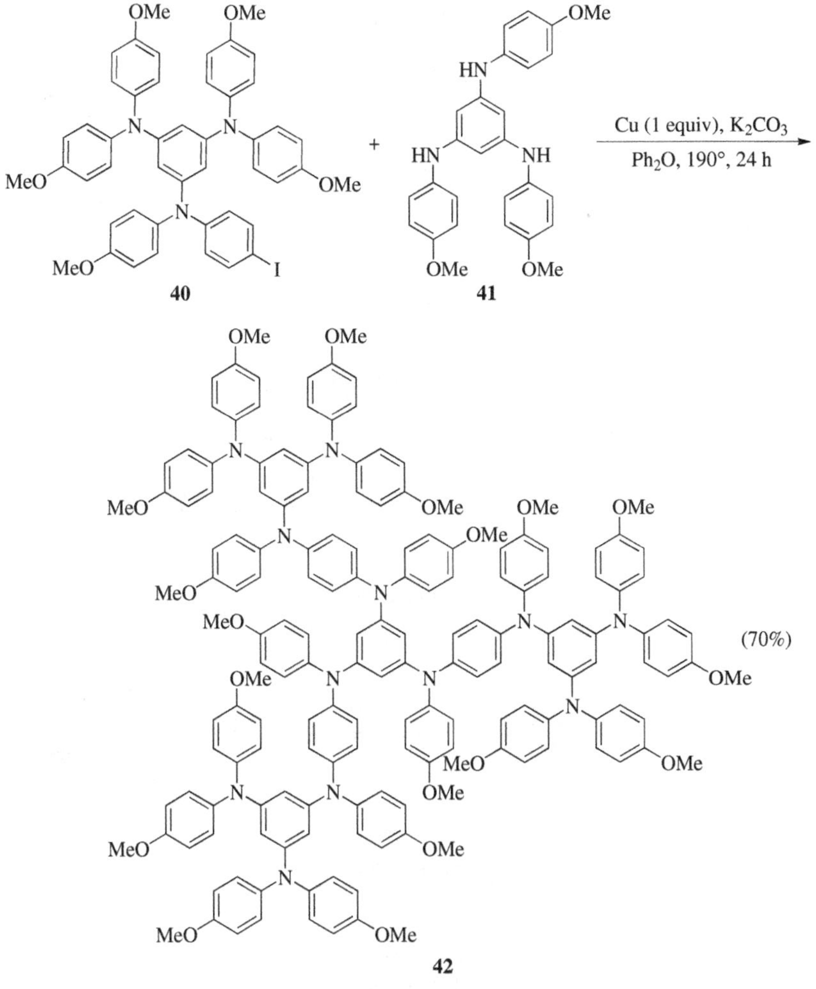

Scheme 77

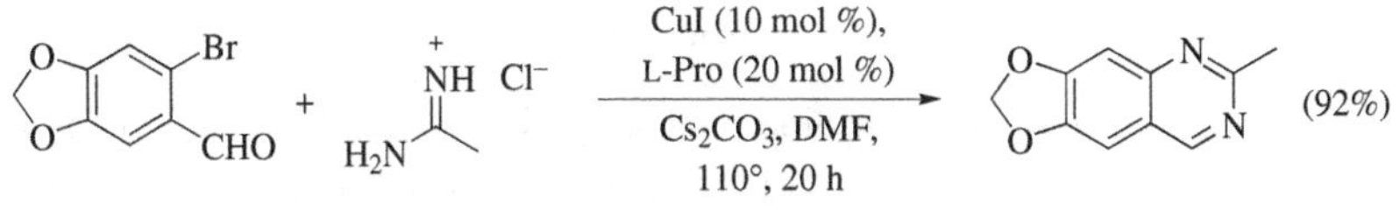

Scheme 78

benzimidazoles.[178] Condensation of α-halo acetamides with 2-halophenols is an effective way to prepare benzoxazin-3-(4*H*)-ones (Scheme 81).[179,180]

An alternative approach to tandem heterocycle synthesis is to generate the nitrogen nucleophile by initial nucleophilic attack on a nitrogen-containing electrophile,

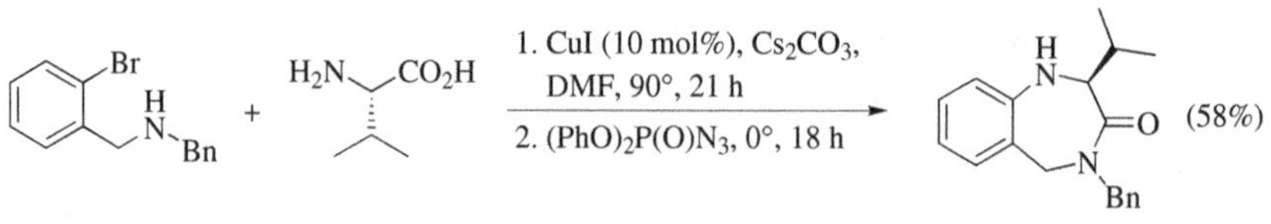

Scheme 79

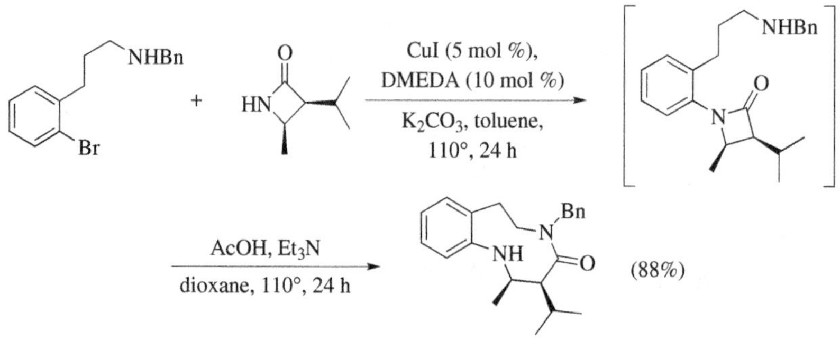

Scheme 80

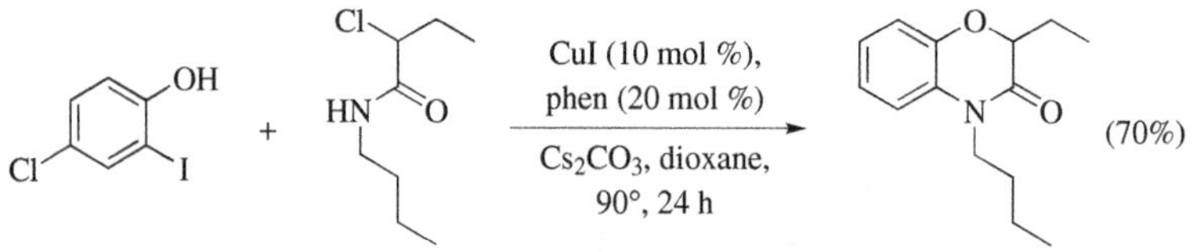

Scheme 81

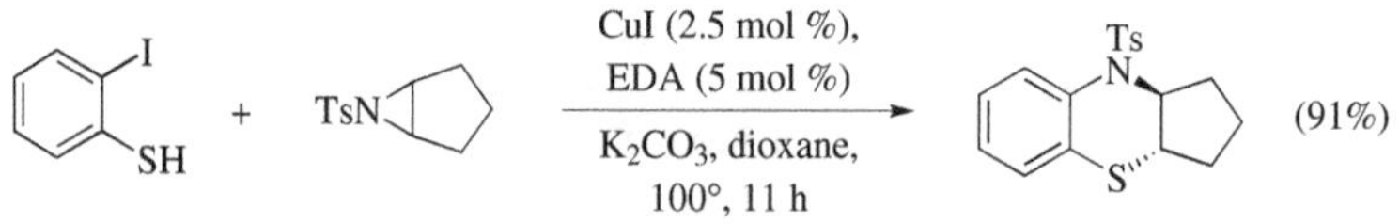

Scheme 82

such as a carbodiimide or aziridine. Nucleophilic attack by 2-halophenols or 2-halothiophenols on protected aziridines generates an amido ion that undergoes intramolecular coupling with the pendant aryl halide to provide tricyclic heterocycles (Scheme 82).[181,182] Tandem nucleophilic attack on a carbodiimide followed by Cu-catalyzed arylation of the newly generated amidate anion provides access to benzimidazole and benzoxazole rings (Scheme 83).[183] Alternatively, a 2-halophenol or 2-haloaniline derivative can react intermolecularly with a carbodiimide followed by Cu-catalyzed C–N bond formation to give benzoxazole and benzimidazole derivatives (Scheme 84).[184]

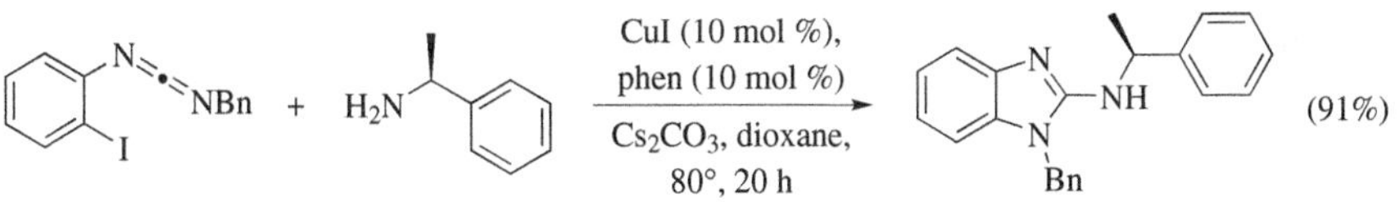

Scheme 83

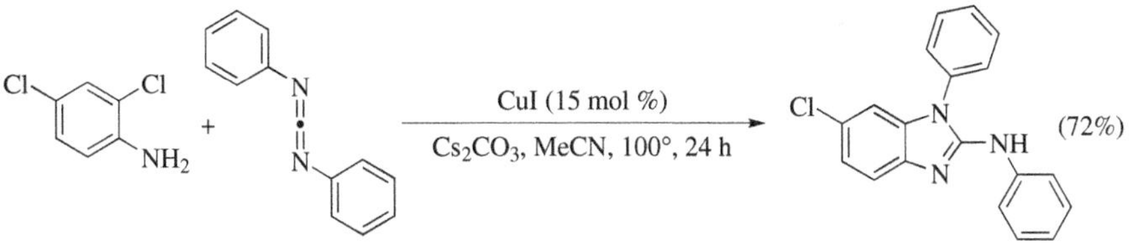

Scheme 84

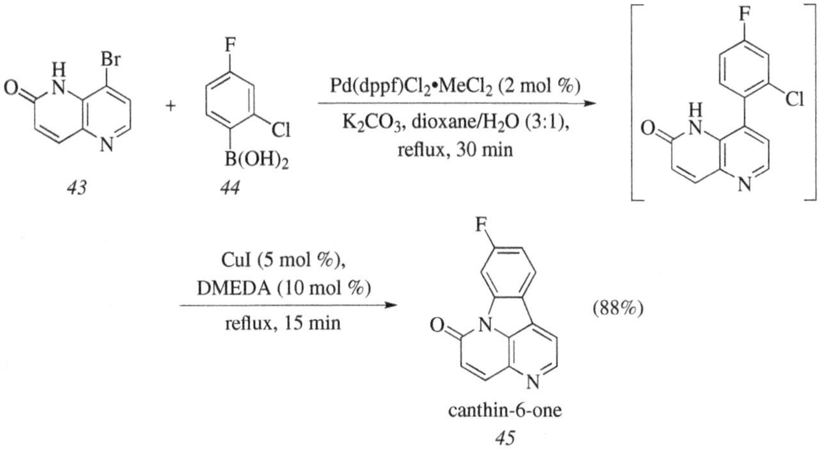

Scheme 85

The complementary reactivity of copper and palladium catalysts can be used to generate complex organic structures by tandem catalysis. Canthin-6-one (**45**) can be prepared in high yield by the condensation of 2-chlorophenylboronic acid **44** with naphthyridone **43** via tandem Pd-catalyzed Suzuki coupling and Cu-catalyzed ring closure through a lactam arylation (Scheme 85).[185] This procedure can be carried out in one pot. Through proper reaction sequencing, highly complex ring systems can be developed from relatively simple precursors. A tandem Cu- and Pd-catalyzed coupling of *ortho-gem*-dibromovinylphenyl isocyanate **46** with aniline provides access to pyrimido[1,6-*a*]indol-1(2*H*)-one (**49**) by a one-pot cascade process (Scheme 86).[186] The cascade reaction involves nucleophilic attack by aniline on isocyanate **46** followed by intramolecular coupling of the generated urea **47** with the vinyl bromide. The final ring is formed by Pd-catalyzed intramolecular coupling of the bromoindole moiety in **48** with the proximal aryl group. In this reaction three new bonds and two new rings are formed in 60–87% yields.

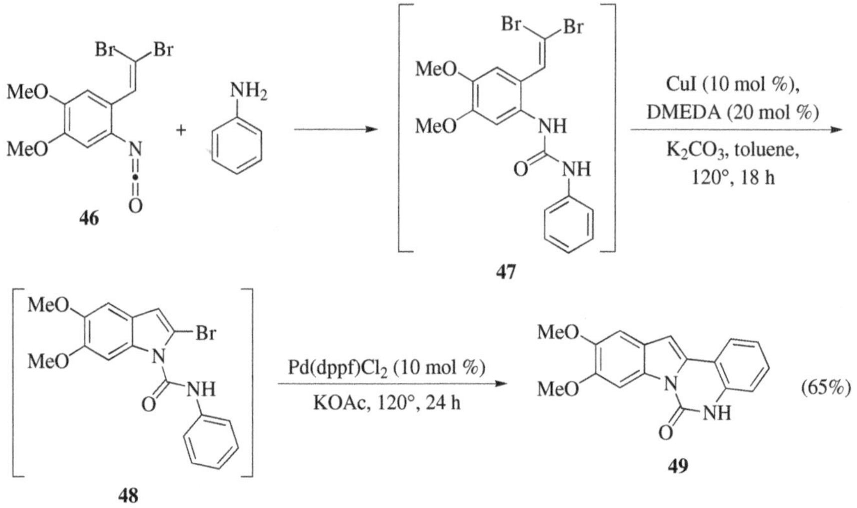

Scheme 86

SIDE REACTIONS

Relatively few side products have been identified in Cu-catalyzed C–N bond-forming reactions. Yields are generally high (>80%) for these reactions, but the identity of the remaining mass balance is usually not reported even in cases with lower yields. Hydrodehalogenation has been reported by a few authors. Deiodination is the only process observed in the reaction of a protected 5-iodouracil derivative with imidazole (Scheme 87), although coupling with heteroaryl and alkyl amines is achieved in good yields.[187] Dehydrobromination is observed in the coupling of 2-bromoanisole with piperidine using a CuCl/acac catalyst, whereas unhindered aryl bromides give the desired coupling product.[188] Homocoupled biaryl products are obtained in 26% yield in the coupling of 2-substituted aryl iodides with butylamine using stoichiometric amounts of CuI (Scheme 88).[189] In the coupling of an $(E)/(Z)$ mixture (90:10) of β-bromostyrene with pyrazole, the (E)-isomer gives complete conversion to (E)-β-pyrazolylstyrene, whereas the (Z)-isomer undergoes elimination of HBr to give phenylacetylene (Scheme 89).[190]

The lack of chemoselectivity in substrates that contain more than one potential nucleophile is the most common cause of side products in Cu-catalyzed C–N bond-forming reactions. Arylation of amine ligands has been noted in some cases. When CDA or EDA is used as a ligand, the N,N'-diarylated ligands are observed in 1–10% yield.[69,115,191] The arylated ligands appear to be ineffective at promoting the coupling reaction, which is why CDA and EDA are less effective than DMCDA or DMEDA in most cases. The more hindered secondary amine ligands are not prone to arylation. Similarly, when N-methylglycine is used as a ligand, approximately

5% of the *N*-arylated glycine is obtained.[192] This side reaction can be avoided by using *N,N*-dimethylglycine (DMG) or L-proline instead. Alcohol solvents can also be competitively arylated to give aryl ethers as side products (Scheme 90).[193–195] By proper choice of ligand, complete selectivity for *N*-arylation over *O*-arylation can be achieved (Scheme 91).[196] Using 2-isobutyrylcyclohexanone, selective *N*-arylation occurs to give product **50**, whereas using 3,4,7,8-tetramethylphenanthroline as the ligand gives the *O*-arylation product **51** with high selectivity.

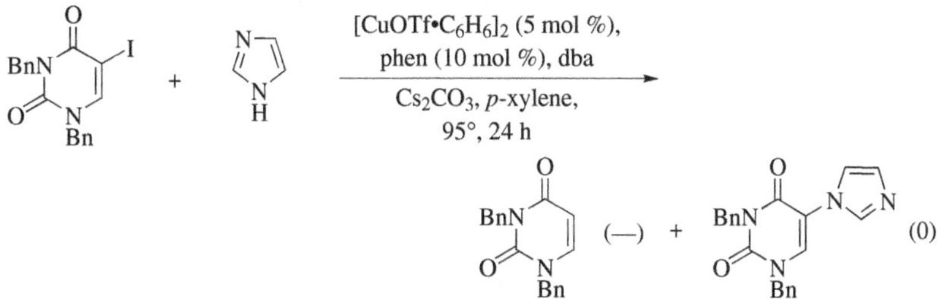

Scheme 87

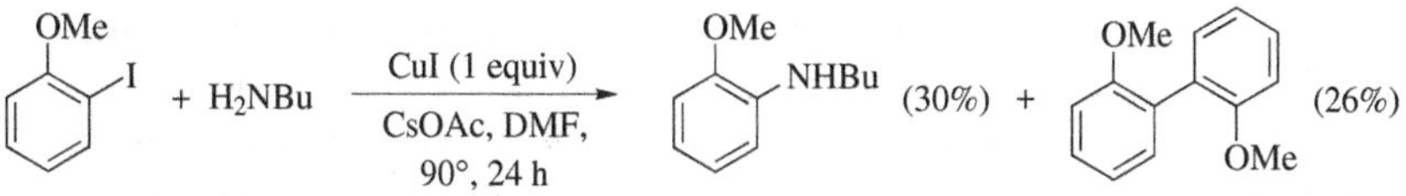

Scheme 88

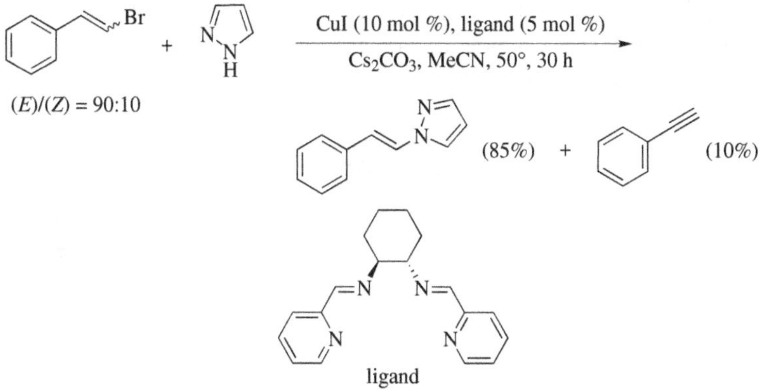

Scheme 89

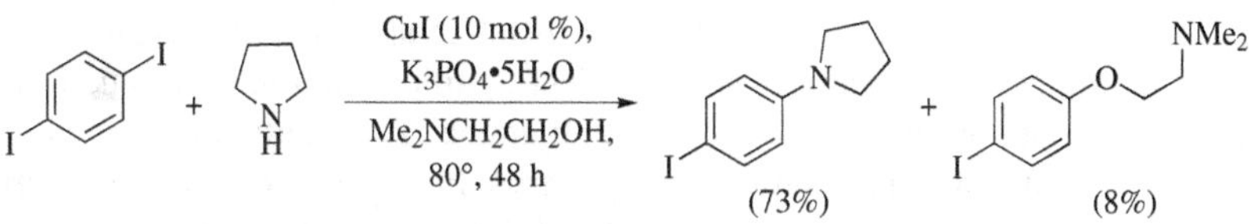

Scheme 90

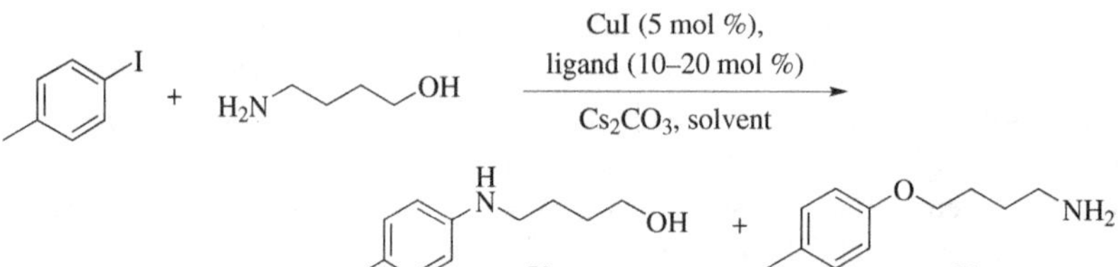

Ligand	Solvent	Temp	Yield (%)	**50/51**
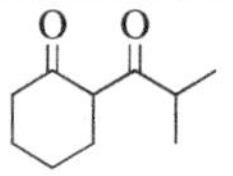	DMF	rt	99	>50:1
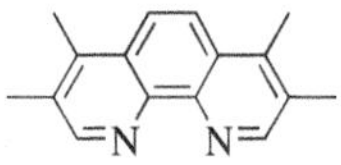	toluene	80°	91	1:18

Scheme 91

Overall, Cu-catalyzed reactions appear to be fairly selective compared to Pd-catalyzed C–N bond formation. Copper-catalyzed reactions are more selective for substrates with multiple halogens than palladium catalysts. In the intramolecular cyclization of **52**, iodotetrahydroquinoline **53** is formed in 81% yield with only a 10% yield of the deiodinated product **54** (Scheme 92).[197] Using a palladium catalyst, iodotetrahydroquinoline **53** is obtained in <3% yield with the remaining material being a mixture of deiodinated compounds and oligomers. In the coupling of aryl bromide **55** bearing a benzylic alcohol moiety, palladium catalysts afford mixtures of ketone **56** and other unidentified products (Scheme 93).[198] In contrast, CuI in combination with a salicylamide ligand provides the desired amine product **57** without oxidation of the benzylic alcohol moiety.

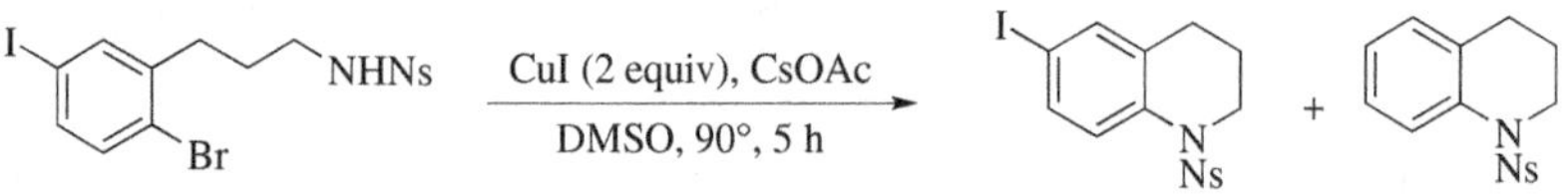

52 **53** (81%) **54** (10%)

Scheme 92

COMPARISON WITH OTHER METHODS

A number of classical routes are known for introduction of amine or amide nitrogens on aromatic rings, but these non-metal-catalyzed routes typically have significant limitations. A classic route is electrophilic nitration of aromatic rings followed by reduction of the nitro group to an amine. This method is limited to the normal electrophilic aromatic substitution patterns. In addition, only primary aniline derivatives can be made in this way. Nucleophilic addition of amines to electron-deficient aromatic rings via nucleophilic aromatic substitution provides a method to prepare a wider range of aryl amines (Scheme 94).[199] Ammonia and primary or secondary amines can be used in these reactions, but the aromatic ring must have one or more strongly electron-withdrawing groups positioned *ortho* and/or *para* to the leaving group. Nucleophilic substitution via the benzyne mechanism is possible with aryl halides lacking strong electron-withdrawing groups.[200] The benzyne mechanism requires strongly basic amido nucleophiles, which limits the functional-group tolerance of this methodology. In addition, for substituted aryl halides, mixtures of *ipso*- and *cine*-substitution products can be obtained (Scheme 95)[201] unless a substituent with an electronic directing effect is present. Naphthols can be converted into naphthylamines by the Bucherer reaction (Scheme 96).[202]

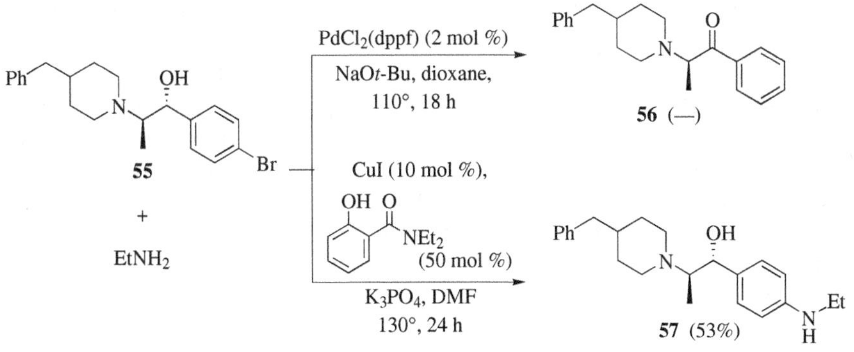

Scheme 93

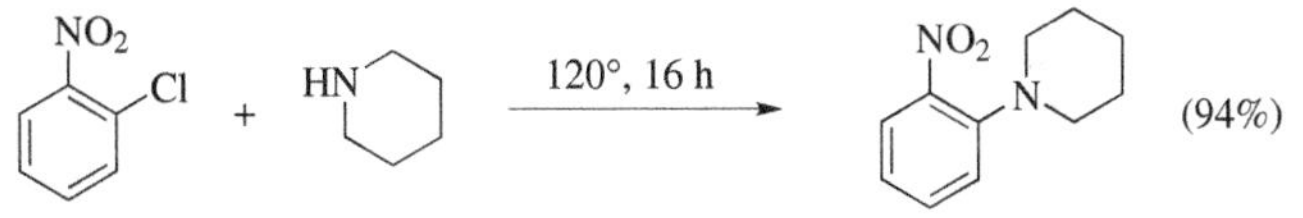

Scheme 94

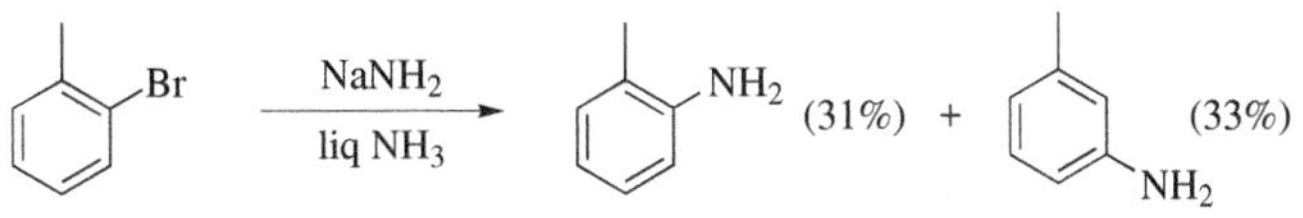

Scheme 95

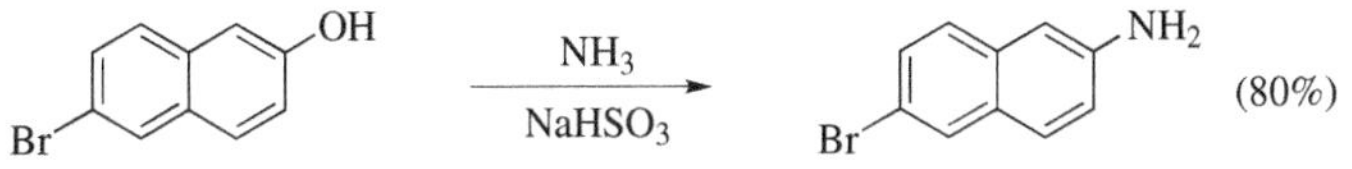

Scheme 96

Metal-catalyzed substitutions provide a more general approach to the synthesis of aryl and alkenyl amine derivatives. The classic Ullmann method represented the state-of-the-art for metal-mediated C–N bond formation until the 1990s, but the harsh conditions and modest yields limited its application. The development of Pd-catalyzed C–N bond formation largely replaced the use of traditional Ullmann-type couplings for arylation of amines. The mild conditions and improved scope of the Pd-catalyzed reactions, particularly with less reactive aryl bromide and chloride substrates, resulted in these methods becoming the dominant approach to amine N-arylation (Schemes 97 and 98).[6,7,203] Although the Pd-catalyzed method is highly useful, it does suffer from a number of drawbacks that has led to a resurgence of interest in the Cu-catalyzed methods. One obvious factor is cost. Palladium currently costs $20/g ($2,900/mole), whereas copper costs $0.01/g ($0.61/mole).[204] In addition, Pd-catalyst systems typically require expensive phosphines or N-heterocyclic carbene ligands, whereas Cu-catalyst systems rely on less expensive diamine or amino acid ligands. Even factoring in the use of higher catalyst loadings of copper (typically 5–20 mol %) compared to Pd (1–2 mol %), the cost difference is significant. Palladium catalysts tend to be more reactive toward aryl halides, so the reactions can generally be carried out at lower temperatures and lower catalyst loadings than are required for copper catalysts. Copper catalysts show limited reactivity with aryl chlorides and are ineffective with aryl sulfonate substrates, whereas optimized palladium systems can readily activate aryl chlorides and sulfonates.

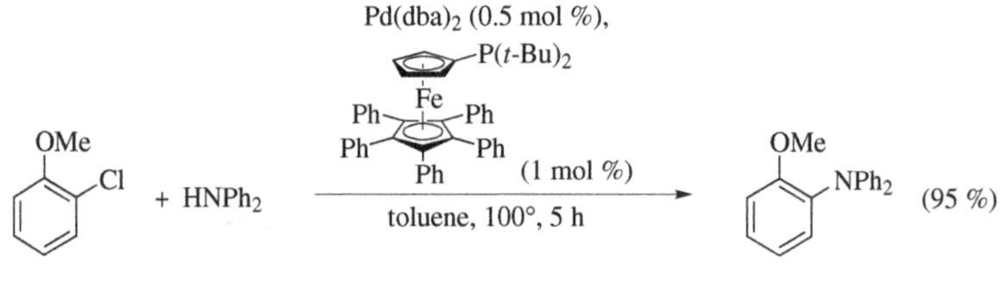

Scheme 97

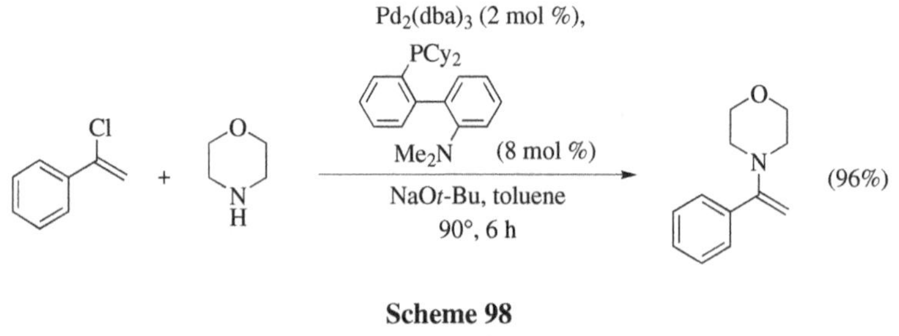

Scheme 98

Although there is overlap between the use of copper and palladium catalysts for nitrogen arylation, the two systems have complementary applications. Pd-catalyzed reactions are generally most useful for arylation of amines, but are less effective for the arylation of less nucleophilic nitrogen sources such as amides and azoles. In contrast, copper catalysts are highly effective for the coupling of amides and azoles. This complementary reactivity can be used to carry out highly selective functionalizations of substrates with two nitrogen nucleophiles. Using a CuI/DMEDA catalyst system, arylation of 2-, 3-, or 4-aminobenzamide with aryl bromides occurs selectively (>50:1) on the amide nitrogen (Scheme 38).[107] The copper catalyst also selectively arylates 5-aminoindole on the azolic nitrogen to afford **58** with >20:1 selectivity (Scheme 99). Arylation of the same substrate using a $Pd_2(dba)_3$/XPhos (XPhos = 2′,4′,6′-triisopropyl-2-dicylohexylphosphinobiphenyl) catalyst gives predominantly (>8:1) arylation on the aryl amine nucleophile (product **59**), rather than the azole nitrogen. Another example of this type of complementary reactivity between palladium and copper catalysts is seen in the arylation of oxindoles (Scheme 100).[205] Using Pd/XPhos, arylation occurs on the α-carbon to give arylated product **60**. With CuI/DMCDA, selective arylation of the amide nitrogen occurs to give N-arylated product **61**. DFT-level calculations show that the N-bound palladium intermediate is more stable than the C-bound intermediate by 4.8 kcal/mol, but that reductive elimination to form the C–C bond has a lower activation barrier than C–N bond formation by 2.4 kcal/mol, which is the selectivity-determining step. For the copper system, the Cu–amidate complex is more stable than the Cu–enolate complex by 14.1 kcal/mol. Thus, formation of the more stable amidate complex is the selectivity-determining step.

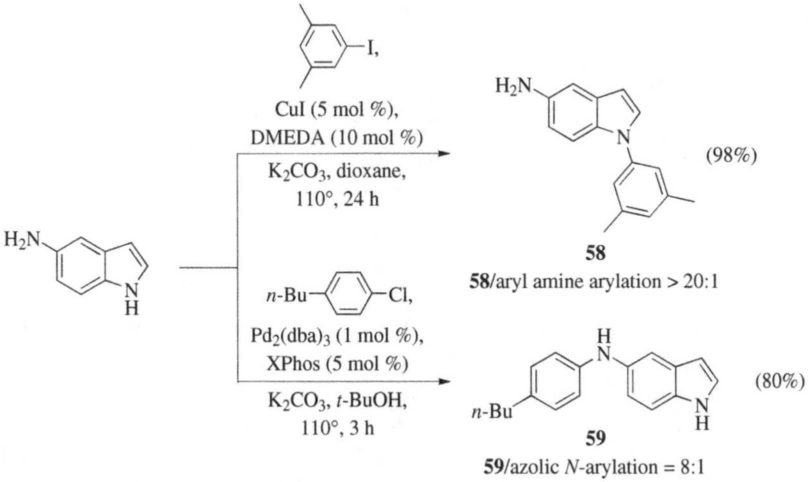

Scheme 99

Palladium and copper may also show different selectivities for aryl halide activation. 2-Halobenzoic acid derivatives show unusually high reactivity in Cu-catalyzed arylation reactions.[206] Coupling of 2,4-dibromobenzoic acid with amines using a copper catalyst leads to high selectivity for functionalization at the 2-position (Scheme 101). Use of a Pd/DiPPF (DiPPF = 1,1′-bis(diisopropylphosphino)ferrocene) catalyst gives selective arylation at the 4-position with >99:1 selectivity.

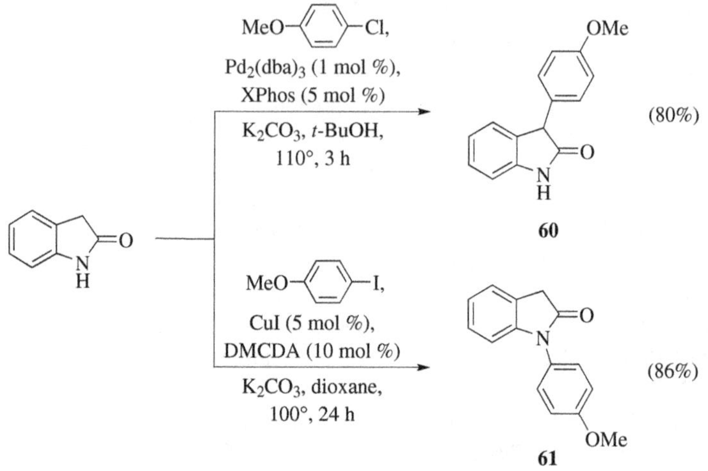

Scheme 100

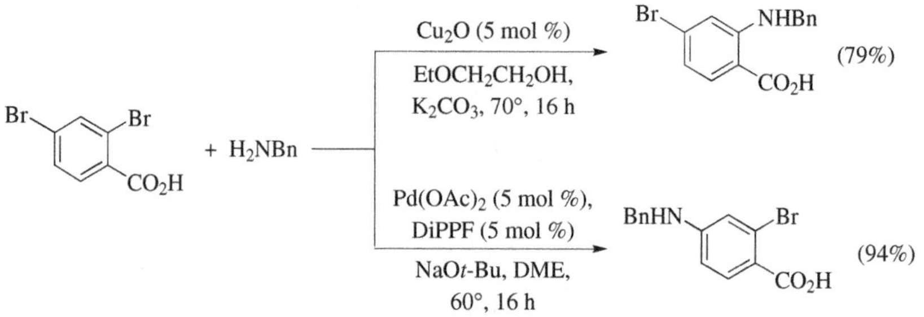

Scheme 101

The oxidative coupling of organoboronic acids with nitrogen nucleophiles catalyzed by copper under mild conditions revitalized interest in Cu-catalyzed cross-coupling.[12] The initially reported systems involve the reaction of arylboronic acids with a range of nitrogen nucleophiles using stoichiometric amounts of $Cu(OAc)_2$ to mediate the reaction in the presence of a base.[8,9] These systems allow the Cu-promoted C–N bond formation to occur at room temperature, which is a significant improvement on the harsh conditions required for traditional Ullmann couplings. However, the reaction requires an excess of the boronic acid coupling partner in addition to a stoichiometric amount of copper. Catalytic versions were soon developed that utilize stoichiometric amounts of secondary oxidants, such as oxygen or amine N-oxides (Scheme 102).[10,207] Although these systems allow for the use of a catalytic amount of copper, they still typically require a 2:1 ratio of boronic acid to the nitrogen nucleophile. Because arylboronic acids are typically derived from aryl halides, and are significantly more expensive, the need to use superstoichiometric amounts is undesirable. The development of efficient Cu-catalyzed couplings of aryl halides with nitrogen nucleophiles has largely supplanted the oxidative couplings with organometallic substrates.

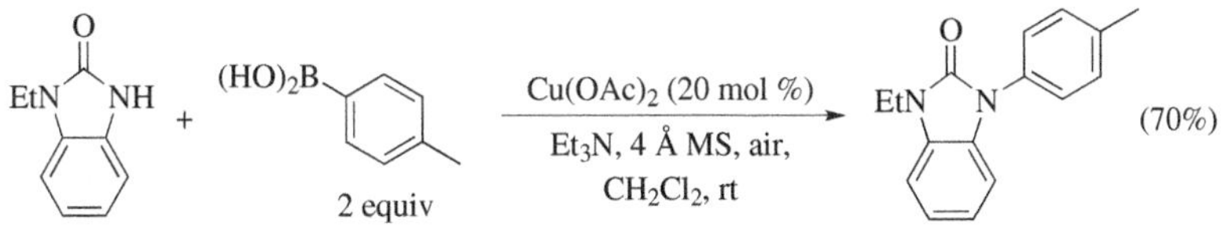

Scheme 102

A variety of other metals have been explored as potentially cheaper replacements for palladium catalysts for C–N bond formation. Nickel catalysts can also be effective in the arylation of amines. Ni(0) catalysts are generally more reactive with aryl chloride substrates than Pd(0) or Cu(I) catalysts.[208] In addition, nickel is significantly cheaper than palladium ($0.03/g, $1.54/mole).[204] Despite these advantages, nickel has received limited attention. In part, this may be attributable to the difficulty in reducing Ni(II) precatalysts or working with unstable Ni(0) precursors.[209,210] Iron-catalyzed C–N bond formation has received significant attention recently,[211–213] but the scope is limited to aryl iodides and yields are often modest. It appears that for some of these systems the active catalyst may be copper impurities in the iron salts used as precatalysts.[214] A catalyst derived from CoCl$_2$ and dppp (1,3-bis(diphenylphosphino)propane) provides modest yields in the coupling of amines with 2-chloropyridine derivatives.[215]

Although the majority of C–N bond-forming reactions involve reaction of a carbon electrophile with a nitrogen nucleophile, umpolung approaches in which the reaction polarity is reversed have also been developed. Diarylamines can be prepared by the reaction of two equivalents of an aryl Grignard reagent with nitroarenes followed by reductive workup with FeCl$_2$ and NaBH$_4$ (Scheme 103).[216] The first equivalent of the Grignard reagent reduces the nitro group to a nitroso group with formation of a phenoxide derivative.[217] The second equivalent then adds to the nitroso group. The need to use two equivalents of Grignard reagent can be eliminated by starting directly with the nitrosoarene. Arylazo tosylates are another class of nitrogen electrophiles that can be converted into diarylamines by reaction with Grignard reagents followed by reductive cleavage of the N–N bond (Scheme 104).[218]

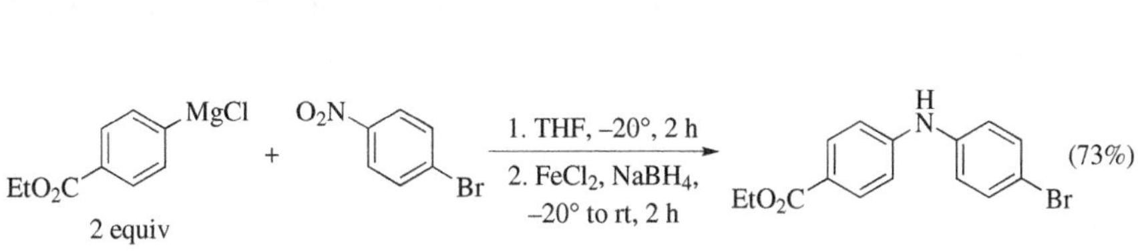

Scheme 103

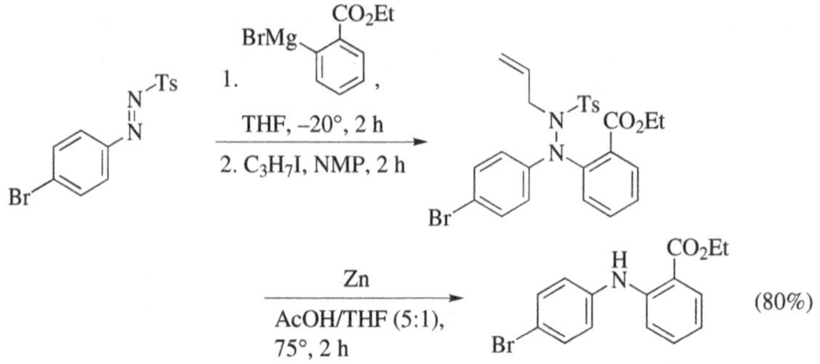

Scheme 104

EXPERIMENTAL CONDITIONS

Copper Precursors

Nearly any source of Cu(0), Cu(I), or Cu(II) can be used as a precatalyst for C–N bond formation. By far the most commonly used copper salts are the Cu(I) halides, with CuI being the most generally useful precatalyst. In a kinetic study of the coupling of 4-toluidine with two equivalents of 4-tolyl iodide, CuI, CuBr, and CuCl give nearly identical rates for the formation of tri(4-tolyl)amine.[219] Copper(I) oxide has also been used as a catalyst precursor with results comparable to Cu(I) halides. Copper thiophene-2-carboxylate (CuTC) has been used under both ligand-free conditions and with DMEDA for the coupling of vinyl halides and amides.[220] Copper(II) salts, such as CuO or CuSO$_4$, have also been successfully used under ligand-free conditions at high temperatures (150–200°). For ligand-promoted catalyst systems under milder conditions, the Cu(II) sources generally give lower yields than Cu(I) sources.[69,191,221] This lower activity is possibly due to slow reduction of the Cu(II) precatalyst to the Cu(I) active species. Copper(0) in the form of powder or bronze is commonly used in the traditional Ullmann conditions. The use of Cu(0) has largely been supplanted by the use of Cu(I) salts. Typical catalyst loadings range from 5–10 mol % of the copper source. Coupling of azoles with aryl iodides using CuCl$_2$ loadings as low as 10 ppm have been reported to give good yields.[222] Thus, it may be that the higher loadings commonly reported in the literature are not always required.

Commercially available CuO and Cu$_2$O nanoparticles have been shown to be effective catalysts under ligand-free conditions for the coupling of aryl and vinyl halides with azoles,[94,223] amines,[224] and amides.[225] The nanoparticle-based catalyst systems are generally limited to the coupling of aryl iodides, but a Cu$_2$O-coated Cu–nanoparticle system gives good yields with electron-deficient aryl chlorides.[223] Non-activated aryl chlorides, such as chlorobenzene, give no conversion with this system, however. An advantage of these nanoparticle-based systems is that they can be typically run in air. In addition, the insoluble nanoparticle catalyst can be recovered from the reaction mixture by centrifugation and reused.[224,225] A highly recyclable catalyst is formed by depositing CuO nanoparticles on acetylene black.[226]

The supported catalyst can be used for 10 cycles with no decrease in yield or conversion rate.

Copper catalysts supported by inorganic oxides have also been used successfully. Fluorapatite-supported copper catalysts are effective in the coupling of aryl iodides, bromides, chlorides, and even activated fluorides with azole nucleophiles.[66,227] An aluminum hydrotalcite-supported copper catalyst provides good activity for the coupling of aliphatic amines with non-activated aryl chlorides.[228] The heterogeneous catalyst can be recovered by centrifugation and reused for five reaction cycles with no degradation in product yield.

Ligands

The modern development of Cu-catalyzed C–N bond formation has relied on supporting ligands to provide more active and general catalysts that operate under milder conditions. Ligand-free systems are common, but generally require high temperatures, ultrasound activation, or highly reactive substrates, such as 2-halobenzoic acids. In addition, polar aprotic solvents are commonly used in "ligand-free" systems, presumably to help solubilize and stabilize the copper active species. The most commonly used supporting ligands are 1,2-diamines, such as EDA, CDA, DMEDA, or DMCDA, amino acids, such as proline or DMG, pyridine-based ligands, such as phenanthroline or 8-hydroquinoline, and oxygen-based ligands, such as diols and diketonates (Figure 1). All of the commonly used ligands, as well as ligand-free catalysts, have been successfully demonstrated with each class of nitrogen nucleophile. A 2:1 ligand/Cu ratio is most commonly used.

Careful studies comparing a wide range of ligand types for a given reaction class are rare. On the basis of comparisons of ligands and the data collected in the tables at the end of this chapter, certain trends can be deduced. Ligand-free catalysts are

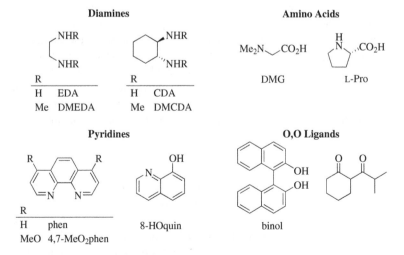

Figure 1.   Common ligands for Cu-catalyzed C–N bond formation.

most widely used for arylation of aniline derivatives, although the use of ligand-free catalysts for amine, azole, and amide nucleophiles is also well precedented. A study of the coupling of aniline and iodobenzene shows that pyrrole-2-carboxylate in combination with CuI gives the best yield of diphenylamine after 17 hours at 80°.[50] Proline, phenanthroline, DMEDA, and 2-isobutyrylcyclohexanone all give significantly lower yields. Arylations of alkyl amines benefit from acidic oxygen-based ligands. Amino acids and diketonate ligands are commonly used for these reactions, whereas nitrogen-based ligands (diamines and pyridines) are less commonly reported. A kinetic study of the coupling of 5-iodo-*m*-xylene and *n*-hexylamine shows that 2-isobutyrylcyclohexanone is the most active catalyst with CuI, whereas a TMHD (TMHD = 2,2,6,6-tetramethylheptan-2,4-dionate)-derived catalyst gives a high initial rate, but incomplete conversion.[52] *N,N*-Diethyl salicylamide is a low-activity ligand. Ethylene glycol is the most effective alcohol-based ligand in the coupling of iodobenzene and benzylamine.[194] A 2:1 glycol/CuI ratio affords the highest yield. Substituted vicinal diols, diols with longer backbones (e.g., 1,3-propanediol), and polyols all give much less active catalysts.

More electron-deficient pyridine and Schiff base ligands work well with azole substrates. 4,7-Dimethoxyphenanthroline is the most active ligand in a survey of a wide range of nitrogen-based ligands for the coupling of 4-bromo-*tert*-butylbenzene and 2-methylimidazole using Cu_2O (5 mol %) as the catalyst. Vicinal diamines based on ethylenediamine or 1,2-diaminocyclohexane appear to be particularly effective for the arylation and vinylation of amide nucleophiles. In an extensive survey of ethylenediamine and 1,2-diaminocyclohexane derivatives, DMEDA and DMCDA are the best ligands in combination with CuI for the coupling of 5-bromo-*m*-xylene with pyrrolidine or *N*-benzyl formamide.[69] No conversion occurs with TMEDA as the ligand, whereas EDA and CDA give much lower conversions than DMEDA or DMCDA for each reaction.

The mechanistic basis for the effectiveness of ligands with specific nucleophiles is not well understood. Computational studies comparing diamine and diketonate ligands at the DFT level suggest that the acidity of the nitrogen nucleophile determines which ligands work best.[28,43] More acidic substrates, such as amides, require charge neutral diamine ligands. Less acidic amine nucleophiles have lower reaction barriers with anionic ligands, such as amino acids or diketonates. Further computational study of the Cu(diamine)-promoted arylation of amides shows that secondary diamine ligands, such as DMEDA or DMCDA, provide copper complexes that have lower oxidative addition barriers than either primary or tertiary diamine ligands.[28]

Bases

Inorganic bases, such as carbonates and phosphates, are most commonly used for Cu-catalyzed C–N bond-formation reactions. Potassium and cesium carbonate are most widely used because of the higher solubility of the larger cations in organic solvents. Potassium phosphate is also used in many systems, but less commonly than carbonates. In couplings of amides, phosphate often provides higher rates than carbonate with aryl iodide substrates. With less reactive aryl bromides, carbonate bases often give better results than phosphate bases. This difference is thought to result from the need to match the rate of nucleophile deprotonation with that of aryl halide activation.[229] With aryl bromides, the more basic phosphate provides too high a concentration of the amidate anion, which deactivates the catalyst. For aryl iodides, the less basic carbonate provides too low a concentration of the Cu–amidate active species. Carbonate and phosphate bases are attractive because of their low cost and tolerance of a wide range of functional groups. Stronger bases, such as alkoxides and hydroxides, are sometimes used with less acidic nucleophiles, but their use increases the chance of undesired side reactions with base-sensitive functional groups.

Solvents

A wide range of solvents can be used in Cu-catalyzed C–N bond formation. The three most common classes of solvents are polar protic solvents, such as DMF, DMA, or DMSO; aromatics, such as toluene or xylene; and cyclic ethers, such as THF or dioxane. The choice of solvent depends in part on the catalyst system that is being used. Ligand-free catalyst systems most commonly are run with no solvent or a high-boiling dipolar aprotic solvent, such as DMF or DMSO. Because ligand-free systems often require high temperatures, high-boiling solvents are necessary. In addition, polar aprotic solvents can act as ligands to stabilize the copper catalyst species. Dipolar aprotic solvents are also the solvents of choice for amino acid- and diketonate-supported catalyst systems. With these anionic ligands, the polar solvent is likely needed to solubilize the ligand and the anionic copper intermediates. Catalysts supported by neutral diamine ligands give the best results with nonpolar solvents, such as toluene, dioxane, or THF.

Although the use of traditional organic solvents is most common, the use of non-traditional solvents has also been reported. Ionic liquids (IL) have been used as solvents for the arylation and vinylation of azoles using ligand-supported copper catalysts.[230–232] In the coupling of β-bromostyrene and imidazole, [bmin]BF$_4$ (bmin = 1-butyl-3-methylimidazolium) gives significantly higher yields than acetonitrile, DMF, or DMSO.[231] In contrast, little difference between [bmin]BF$_4$, DMF, and DMSO is seen in the coupling of benzimidazole with 2-bromothiophene.[230] The IL/Cu solution can be separated from the product and recycled up to four times without loss of activity.[230–232] Copper-catalyzed couplings can also be carried out in water. One approach is to use ligands with anionic substituents, such as sulfonates[233] or phenoxides.[78] Neutral, polar ligands with hydroxy, amino, or carbonyl substituents have also been used.[25,234–236] The advantage of using hydrophilic ligands is that the catalyst-containing aqueous phase can often be recovered and recycled.[233] Water-based systems are attractive for the arylation of ammonia, since aqueous ammonium hydroxide can be used as the ammonia source.[237,238]

Other Reaction Conditions

Microwave and ultrasound radiation have been used to accelerate Cu-catalyzed coupling reactions. Because of the high temperatures required in most ligand-free,

Cu-catalyzed couplings, the use of microwave irradiation can give complete conversion with short reaction times. Whereas traditional heating generally requires reaction times of 12–24 hours, microwave heating often allows reactions to be completed in 5–20 minutes.[25,85,239,240] Ultrasound can also be used to significantly accelerate Cu-catalyzed coupling reactions. Higher yields are achieved in the coupling of 2-halobenzoic acids and amines with copper powder using ultrasound irradiation at room temperature for 15 minutes than in reactions performed at reflux in DMF for 3 hours.[241,242]

EXPERIMENTAL PROCEDURES

N-(1-Naphthyl)anthranilic Acid (Ligand-Free Ullmann Arylation of an Aryl Amine).[70]

A mixture of 1-aminonaphthalene (1.3 g, 9.3 mmol), 2-bromobenzoic acid (1.8 g, 8.8 mmol), K_2CO_3 (1.2 g, 8.8 mmol), Cu powder (0.2–0.3 micron, 50 mg, 0.80 mmol), Cu_2O (<5 micron, 60 mg, 0.40 mmol) and 3 mL of 2-ethoxyethanol was heated at reflux at 130° for 24 h under nitrogen. The cooled reaction mixture was poured into 30 mL of water to which decolorizing charcoal was added. The mixture was filtered through Celite. The crude product was obtained by precipitation upon acidification of the filtrate with diluted HCl (the pH was adjusted to 5–6). The solid residue was dissolved in 100 mL of 5% aqueous Na_2CO_3. The solution was filtered through Celite and the final product was obtained by precipitation as described above. *N*-(1-Naphthyl)anthranilic acid was obtained as an off-white grey powder (2.2 g, 97%): ^{1}H NMR (300 MHz, $CDCl_3$) δ 9.60 (br s, 1H), 8.08 (m, 2H), 7.92–7.89 (m, 1H), 7.75 (d, $J = 7.6$ Hz, 1H), 7.51–7.46 (m, 4H), 7.29–7.23 (m, 1H), 6.85 (d, $J = 8.6$ Hz, 1H), 6.73 (dd, $J = 7.3$ Hz, 7.6 Hz, 1H); ^{13}C NMR (75 MHz, $CDCl_3$) δ 174.2, 151.2, 136.9, 136.0, 135.5, 133.1, 130.8, 129.1, 127.1, 126.5, 126.5, 123.5, 122.8, 117.5, 114.2, 110.5.

N-Phenylbenzamide (Ligand-Free Goldberg Arylation of an Amide).[74]

A mixture of benzamide (150 mg, 1.2 mmol), phenyl iodide (200 mg, 1.0 mmol), CuO nanoparticles (4 mg, 0.05 mmol), and KOH (84 mg, 1.5 mmol) in DMSO/t-BuOH (1:3, 1 mL) was stirred at 110°. The progress of the reaction was monitored by TLC using a mixture of EtOAc and hexane as the eluent. After completion, the reaction mixture was treated with EtOAc (10 mL) and water (3 mL). The organic layer was separated, and the aqueous layer was extracted with EtOAc (3 × 5 mL).

The combined organic solution was washed with brine (3 × 5 mL) and water (1 × 5 mL). The organic layer was dried over Na_2SO_4. The solvent was evaporated under reduced pressure to provide a residue, which was purified by passing through a short pad of silica gel using a mixture of EtOAc and hexane as eluent. The product was recovered as a colorless solid (146 mg, 74%): mp 163° (lit. mp 162–163°);[243] FTIR (KBr) 3310, 2995, 1659, 1604, 1533, 1420, 1315, 1233, 1185, 1090, 1023 cm^{-1}; ^{1}H NMR (400 MHz, CDCl$_3$) δ 7.84 (d, J = 7.2 Hz, 2H), 7.78 (s, 1H), 7.61 (d, J = 8.0 Hz, 2H), 7.52 (d, J = 7.6 Hz, 1H), 7.47 (t, J = 8.0 Hz, 2H), 7.36 (t, J = 8.0 Hz, 2H), 7.14 (t, J = 7.4 Hz, 1H); ^{13}C NMR (100 MHz, CDCl$_3$) δ 166.0, 138.1, 135.2, 132.0, 129.3, 129.0, 127.2, 124.8, 120.5. Anal. Calcd for $C_{13}H_{11}NO$: C, 79.16; H, 5.62; N, 7.10. Found: C, 79.23; H, 5.65; N, 7.06.

N-Phenyl-p-anisidine (CuI/L-Proline-Catalyzed Arylation of an Aniline Derivative).[47]

A mixture of phenyl iodide (610 mg, 30 mmol), p-anisidine (190 mg, 2.0 mmol), K$_2$CO$_3$ (550 mg, 4.0 mmol), CuI (38 mg, 0.20 mmol), and L-proline (40 mg, 0.40 mmol) in 3 mL of DMSO was heated at 90° for 36 h. The cooled mixture was partitioned between water and EtOAc. The organic layer was separated, and the aqueous layer was extracted with EtOAc. The combined organic layers were washed with brine, dried over Na$_2$SO$_4$, and concentrated under vacuum. The residual oil was loaded on a silica gel column and the product was eluted with EtOAc/petroleum ether (1:10 to 1:8) to afford N-phenyl-p-anisidine as a pale yellow solid (163 mg, 82%): mp 102–103°; ^{1}H NMR (400 MHz, CDCl$_3$) δ 7.21–7.19 (m, 2H), 7.09–7.05 (m, 2H), 6.91–6.77 (m, 5H), 5.49 (br s, 1H), 3.80 (s, 3H); EI–MS (m/z): 199 (M$^+$), 185, 184, 129, 128, 154, 77.

N-(3-Aminophenyl)-5-amino-1-pentanol (CuI/Diketonate-Catalyzed Arylation of an Alkyl Amine).[196]

An oven-dried, 10-mL Schlenk tube equipped with a Teflon valve was charged with a magnetic stir bar, CuI (9.5 mg, 0.050 mmol), and Cs$_2$CO$_3$ (650 mg, 2.0 mmol). The tube was evacuated and refilled with argon. Under a counterflow of argon, 3-iodoaniline (219 mg, 1.0 mmol) and 5-amino-1-pentanol (155 mg, 1.5 mmol) were added followed by DMF (0.5 mL) and 2-isobutyroylcyclohexanone (34 mg, 0.2 mmol). The tube was sealed and the mixture was stirred under argon at ambient temperature (22°) for 22 h. After the starting material was consumed, the reaction mixture was diluted with dichloromethane and filtered. The solvent was removed by rotary evaporation, with heating if necessary, to ensure the removal of most of the DMF. The residue was purified by column chromatography on silica gel (EtOAc/hexanes). The product-containing fractions were concentrated by rotary evaporation and dried under high vacuum for at least 1 h to remove residual solvent. N-(3-Aminophenyl)-5-amino-1-pentanol was recovered as a yellow oil (165 mg, 85%): ^{1}H NMR (CDCl$_3$) δ 6.95 (pseudo t, 1H), 6.07–6.04 (m, 2H), 5.94 (pseudo t, 1H), 3.60 (t, J = 7 Hz, 2H), 3.28 (br s, 4H), 3.06 (t, J = 6 Hz, 2H), 1.64–1.53 (m, 4H), 1.47–1.42 (m, 2H); ^{13}C NMR (CDCl$_3$) δ 149.8, 147.7, 130.2, 105.1, 104.3, 99.8, 62.8, 44.1, 32.6, 29.5, 23.6.

1-(3,5-Dichlorophenyl)-2-methyl-1*H*-imidazole (Cu/Phenanthroline-Catalyzed Arylation of an Azole).[84]   An oven-dried screw-capped test tube (4 × 100 mm) was charged with Cu$_2$O (7.2 mg, 0.050 mmol), 4,7-dimethoxyphenanthroline (36 mg, 0.15 mmol), 2-methylimidazole (100 mg, 1.2 mmol), poly(ethylene glycol) (PEG) (200 mg), Cs$_2$CO$_3$ (0.45 g, 1.4 mmol), and a magnetic stir bar, and the reaction vessel was fitted with a rubber septum. The vessel was evacuated and back-filled with argon, and this sequence was repeated a second time. 1,3-Dichloro-5-iodobenzene (273 mg, 1.00 mmol) and butyronitrile (0.5 mL) were then added successively. The reaction tube was sealed and the mixture was stirred in a preheated oil bath for 24 h at 110°. The reaction mixture was cooled to rt, diluted with CH$_2$Cl$_2$ (15 mL), and filtered through a plug of Celite, eluting with additional CH$_2$Cl$_2$ (20 mL). The filtrate was concentrated, and the resulting residue was purified by flash chromatography (hexane/EtOAc, 1:3) to provide the title compound as white needles (194 mg, 86%): mp 122–125°; IR (KBr disc) 1534, 1501, 1463, 1451, 1405, 1305, 1176, 1143, 1115, 1099, 985, 850, 781 cm^{-1}; ^{1}H NMR (300 MHz, CDCl$_3$) δ 7.44 (t, J = 1.8 Hz, 1H), 7.23 (d, J = 1.9 Hz, 2H), 7.05 (d, J = 1.2 Hz, 1H), 6.99 (d, J = 1.2 Hz, 1H), 2.40 (s, 3H); ^{13}C NMR (100 MHz, CDCl$_3$) δ 139.8, 135.8, 128.5, 128.4, 124.1, 14.0. Anal. Calcd for C$_{10}$H$_8$N$_2$Cl$_2$: C 52.89, H 3.55. Found: C 52.95, H 3.44.

N-(3-Hydroxymethylphenyl)-2-pyrrolidinone   (Cu/DMEDA-Catalyzed Arylation of an Amide).[69]   A 10-mL Schlenk tube was charged with CuI (9.6 mg, 0.050 mmol) and K$_3$PO$_4$ (430 mg, 2.03 mmol), evacuated, and backfilled with argon.

N,N'-Dimethylethylenediamine (11 μL, 0.10 mmol), 3-iodobenzyl alcohol (128 μL, 1.01 mmol), 2-pyrrolidinone (94 μL, 1.24 mmol), and toluene (1.0 mL) were added under argon. The Schlenk tube was sealed with a Teflon valve and the reaction mixture was stirred at 80° for 3 h. The resulting white suspension was allowed to reach rt and was filtered through a 0.5 × 1 cm pad of silica gel, eluting with 10 mL of Et$_2$O/MeOH (5:1). The filtrate was concentrated and the residue was purified by flash chromatography on silica gel (CH$_2$Cl$_2$/MeOH 25:1) to give the product as a white solid (180 mg, 93%): mp 120–121°; IR (neat) 3331, 1663 cm^{-1}; ^{1}H NMR (400 MHz, CDCl$_3$) δ 7.59 (m, 1H), 7.55–7.50 (m, 1H), 7.35 (t, *J* = 7.8 Hz, 1H), 7.17–7.12 (m, 1H), 4.68 (d, *J* = 5.8 Hz, 2H), 3.86 (t, *J* = 7.0 Hz, 2H), 2.65 (t, *J* = 5.8 Hz, 1H), 2.60 (t, *J* = 8.0 Hz, 2H), 2.16 (m, 2H); ^{13}C NMR (100 MHz, CDCl$_3$) δ 174.4, 141.8, 139.3, 128.9, 123.0, 119.0, 118.5, 64.9, 48.9, 32.7, 17.9. Anal. Calcd for C$_{11}$H$_{13}$NO$_2$: C, 69.09; H, 6.85. Found: C, 69.05; H, 6.81.

er 97.9:2.1

(2S)-2-(N-((1E)-6-(2-Furyl)hex-1-enyl)amino)-4-methylpentanamide (Cu/DMEDA-Catalyzed Vinylation of an Amide).[244] A solution of (*E*)-6-(2-furyl)-1-iodo-1-hexene (0.150 mg, 0.543 mmol) and DMEDA (6.0 μL, 54 μmol) in THF (1 mL) was prepared in a flame-dried 5-mL conical flask. Separately, an oven-dried 10-mL Schlenk tube fitted with a septum was charged with the L-leucine amide (141 mg, 1.09 mmol), CuI (5.2 mg, 27 μmol), and Cs$_2$CO$_3$ (265 mg, 0.815 mmol), and then was evacuated and back-filled with dry nitrogen three times. The vinyl iodide solution was transferred to the Schlenk tube, the septum rapidly exchanged for a glass stopper, and the vessel sealed with Teflon tape. The reaction tube was immersed in a preheated oil bath and maintained at 70° for 16 h. After cooling to rt, the reaction mixture was diluted with EtOAc (2 mL) and placed directly atop a previously prepared silica gel column. Elution with CH$_2$Cl$_2$/MeOH (9:1) afforded a colorless oil (139 mg, 87%, er 97.9:2.1): er determined by chiral-phase GC (Alltech Associates, Chirasil-Val column, 25 m x 0.25 mm); ^{1}H NMR (300 MHz, CDCl$_3$) δ 8.89 (br d, *J* = 9.7 Hz, 1H), 7.28 (dd, *J* = 1.9, 0.8 Hz, 1H), 6.71 (ddt, *J* = 14.0, 11.0, 1.4 Hz, 1H), 6.26 (dd, *J* = 3.1, 1.9 Hz, 1H), 5.96 (ddq, *J* = 3.1, 1.7, 0.9 Hz, 1H), 5.20 (dt, *J* = 14.3, 7.2 Hz, 1H), 3.41 (dd, *J* = 9.9, 3.6 Hz, 1H), 2.61 (t, *J* = 7.4 Hz, 2H), 2.06 (qd, *J* = 7.3, 1.4 Hz, 2H), 1.78–1.59 (m, 4H), 1.47–1.28 (m, 5H), 0.96 (d, *J* = 6.4 Hz, 3H), 0.93 (d, *J* = 6.2 Hz, 3H); ^{13}C NMR (151 MHz, CDCl$_3$) δ 172.6, 156.3, 140.7, 122.4, 113.2, 110.0, 104.7, 53.2, 43.9, 29.5, 29.4, 27.8, 27.5, 24.9, 23.4, 21.3; HRMS–ESI–MS (*m/z*): [M + H]$^+$ calcd for C$_{16}$H$_{26}$N$_2$O$_2$, 279.2067; found, 279.2065.

TABULAR SURVEY

The tables incorporate reactions reported in the literature through August 2010. Some key examples published through early 2011 are also included. Additional reactions reported mid-2011 through early 2014 are included only as references at the end of the bibliography, sorted according to the table in which they would belong. The literature was extensively searched using the CAS databases. All examples of Cu-catalyzed C–N bond formation involving the coupling of an aryl or vinyl halide or pseudohalide with a nitrogen nucleophile are included in the following tables. Yields reported are isolated yields, except when only NMR, GC, or HPLC yields are reported. An em-dash (—) indicates that the authors did not report a yield for this example, but provided sufficient evidence that the product was formed. Attempted reactions that did not produce the desired product are included with 0% yield reported. In cases where multiple sets of conditions were reported as part of an optimization study, only the conditions giving the highest yields are listed.

The list of tables can be found in the table of contents at the beginning of the chapter and is not repeated here. Tables 1A and 1B list reactions of all types of halogen electrophiles that lead to primary aryl and heteroaryl amines, respectively, in a single step. These two tables, as well as Table 25 (Preparation of Aryl and Heteroaryl Azides) and Table 33 (Preparation of Vinyl Azides) are organized by increasing carbon count of the halide electrophile, whereas in all other tables organization is by nitrogen nucleophile. Appropriate tables are sub-divided into *N*-arylations and *N*-heteroarylations. The latter contain electrophiles irrespective of whether the halide is attached to a heteroaryl group or an annulated aromatic ring. With the aim of grouping similar substrates together, protecting groups are excluded from the carbon count, as are simple groups on nitrogen, oxygen, and divalent sulfur, except when these groups are attached to the nucleophilic nitrogen. Tables 22A and 22B (*N*-Arylation and *N*-Heteroarylation of Ureas and Guanidines, respectively) also contain cyclic ureas such as 1*H*-imidazol-2(3*H*)-one and 1*H*-benzo[*d*]imidazol-2(3*H*)-one.

In addition to the standard abbreviations approved by the *Journal of Organic Chemistry*, the following abbreviations are used in the tables:

)))	ultrasound, sonication
L-4-HOPro	4-hydroxy L-proline
8-HOquin	8-hydroxyquinonline
AB	acetylene carbon black
Al–HT	aluminum hydrotalcite
An	4-methoxyphenyl (see also PMP)
BINAP	2,2′-bis(diphenylphosphino)-1,1′-binaphthyl
binol	1,1′-binaphthyl-2,2′-diol
BtH	benzotriazole
C_4mim	1-butyl-3-methylimidazolium
CDA	*trans*-1,2-diaminocyclohexane

dba	dibenzylideneacetone
DEIPS	diethylisopropylsilyl
DIPP	diisopropylphosphoramidite
DiPrPhDAB	N,N'-(2,6-diisopropylphenyl)-1,4-diazabutadiene
DMCDA	N,N'-dimethyl-*trans*-1,2-cyclohexanediamine
DMEDA	N,N'-dimethyl-1,2-diaminoethane
DMG	N,N-dimethylglycine HCl
DPP	diphenyl pyrrolidine-2-phosphonate
EDA	ethylenediamine
FAP	fluorapatite
IPr	1,3-di-(2,6-diisopropylphenyl)imidazol-2-ylidine
L-Pro	L-proline
4,7-(MeO)$_2$phen	4,7-dimethoxyphenanthroline
Mes*	2,4,6-tri-*tert*-butylphenyl
mes$_2$DAB	N,N'-dimesityl-1,4-diazabutadiene
MTBD	7-methyl-1,5,7-triazabicyclo-[4.4.0]dec-5-ene
Mtt	4-methylphenyldiphenylmethyl (4-methyltrityl)
MW	microwave heating
neocup	neocuproine
NMG	N-methylglycine
NP	nanoparticles
Np	naphthyl
PAnNF	poly(aniline) nanofiber
PEG	poly(ethylene glycol)
per-6-ABCD	per-6-amino-β-cyclodextrin
phen	1,10-phenanthroline
Phth	phthalimide
pip-2-CO$_2$H	piperidine-2-carboxylic acid
PMP	4-methoxyphenyl
PPAPM	pyrrolodine-2-phosphonic acid phenyl monoester
SEM	(2-trimethylsilylethoxy)methyl
TBAA	tetrabutylammonium adipate
TBPE	tetrabutylphosphonium acetate
TBPM	tetrabutylphosphonium malonate
TC	thiophene-2-carboxylic acid
TEAC	tetraethylammonium chloride
TMAH	tetramethylammonium hydroxide
THMD	2,2,6,6-tetramethylheptane-2,4-dionate
TM–BINAM	N,N,N',N'-tetramethyldiaminobinaphthalene
TMU	N,N,N',N'-tetramethylurea
xantphos	4,5-bis(diphenylphosphino)-9,9-dimethylxanthene

CHART 1. CATALYSTS USED IN TABLES

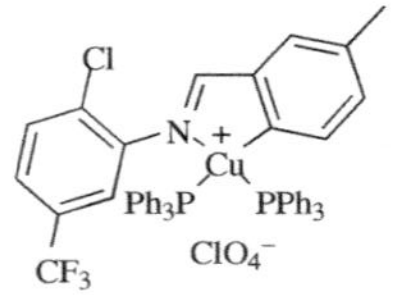

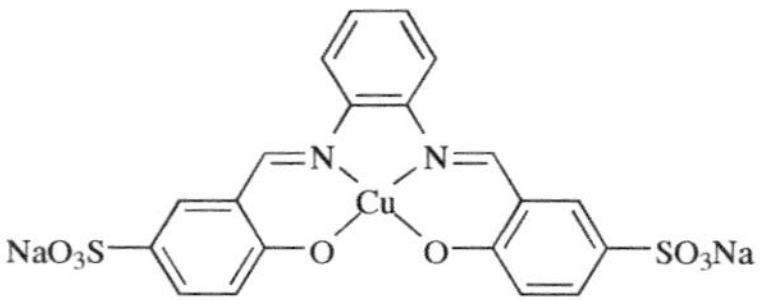

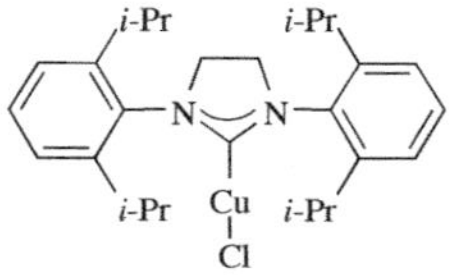

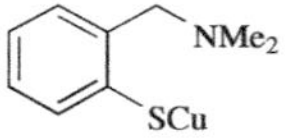

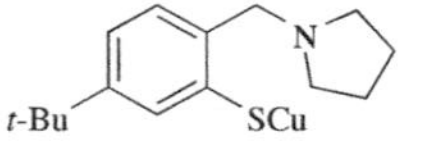

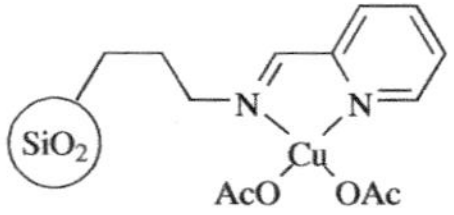

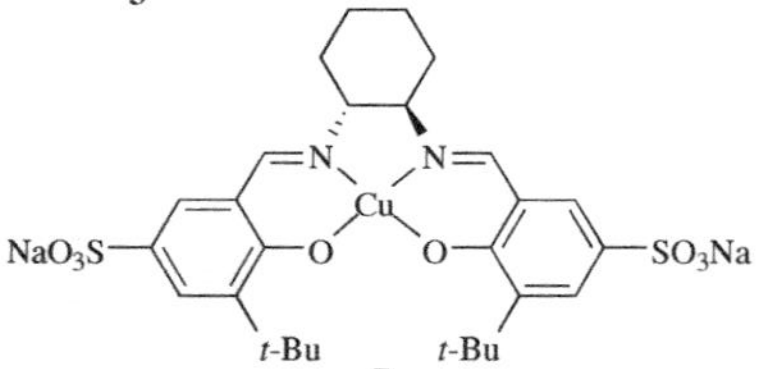

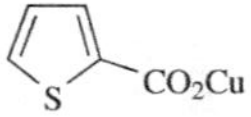

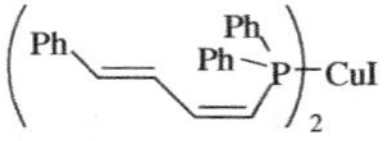

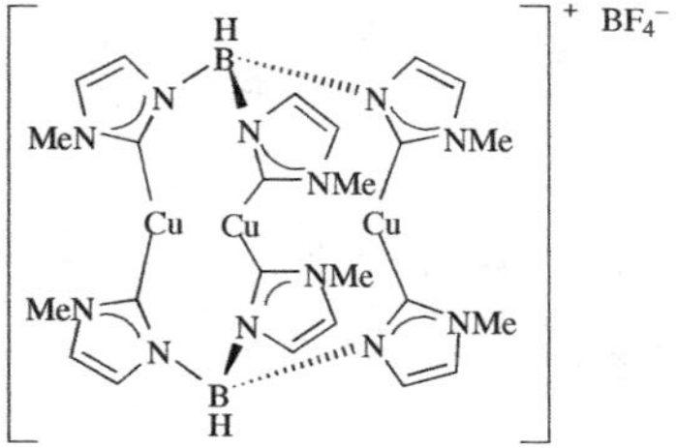

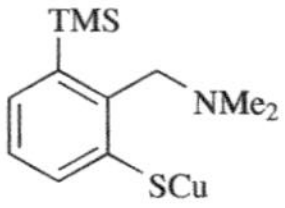

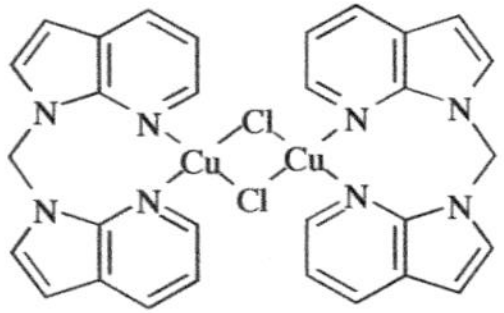

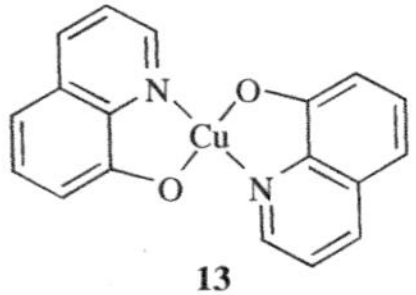

CHART 2. LIGANDS USED IN TABLES

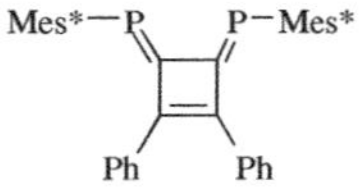

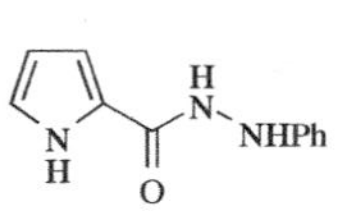

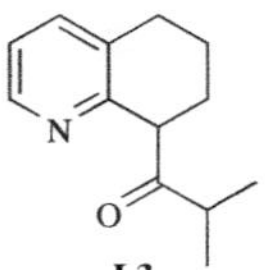

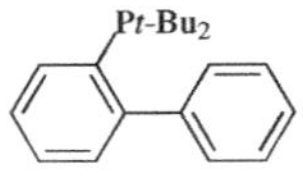

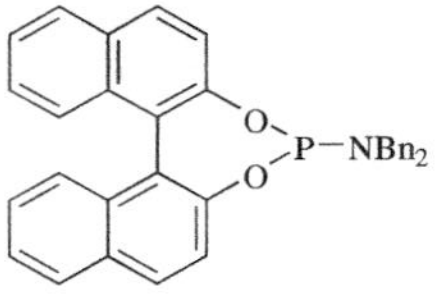

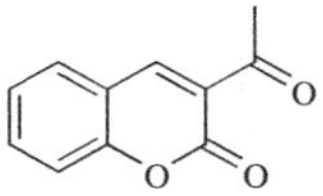

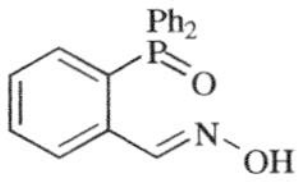

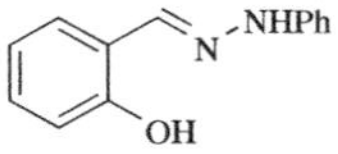

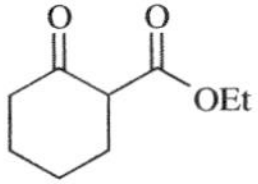

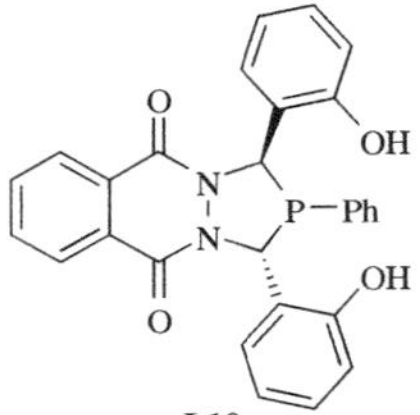

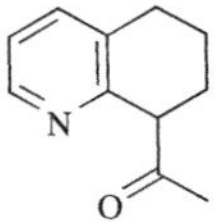

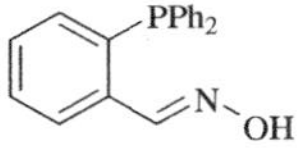

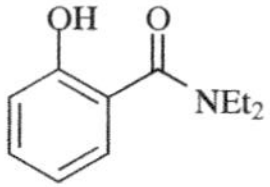

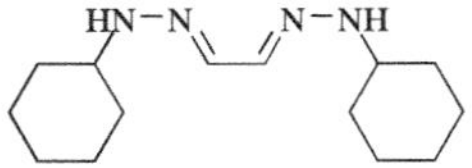

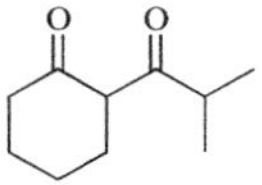

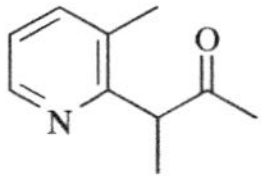

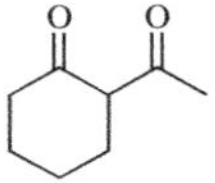

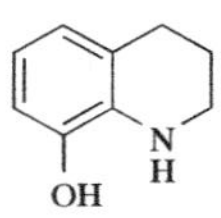

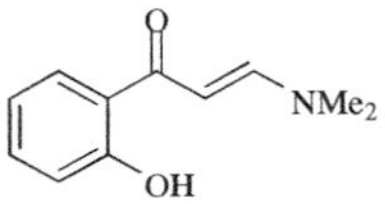

CHART 2. LIGANDS USED IN TABLES (*Continued*)

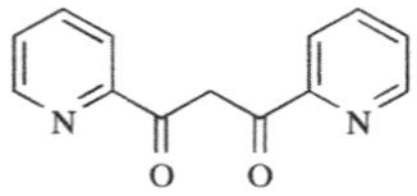

L22

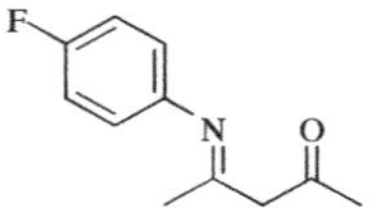

L23

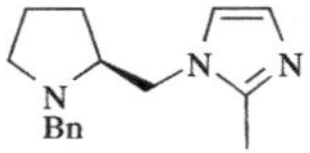

L24

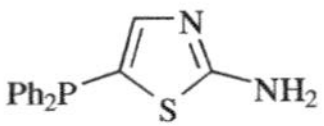

L25

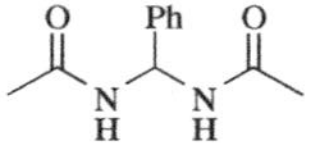

L26

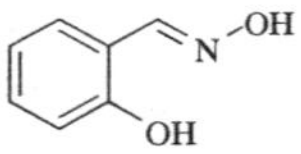

L27

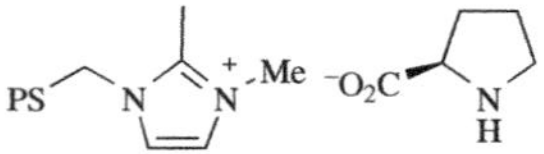

L28
PS = polystyrene

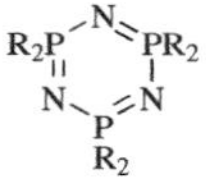

L29
R = dendron

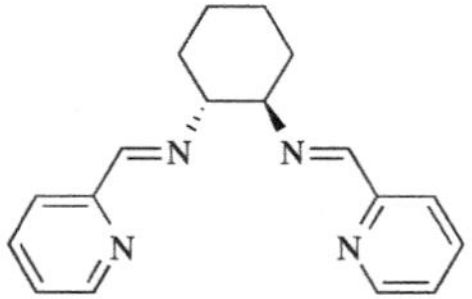

L30

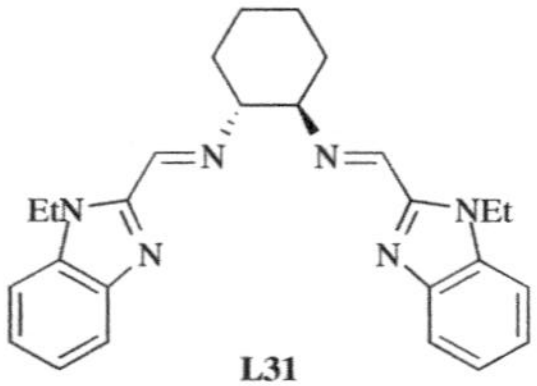

L31

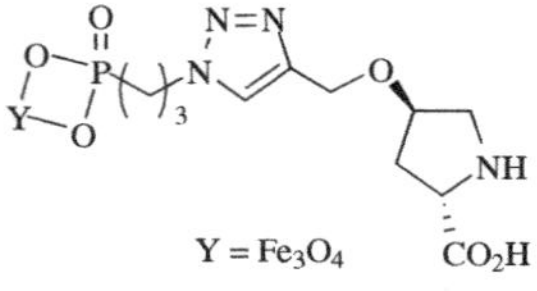

L32
Y = Fe$_3$O$_4$

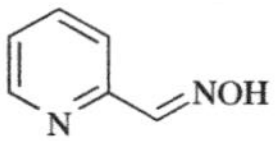

L33

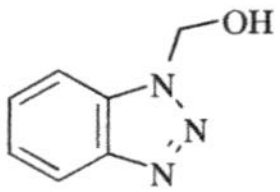

L34

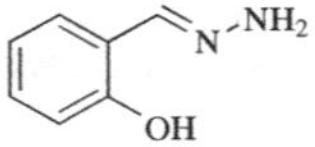

L35

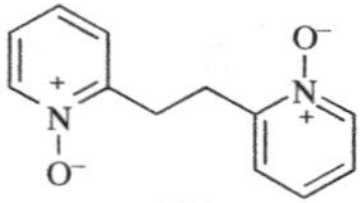

L36

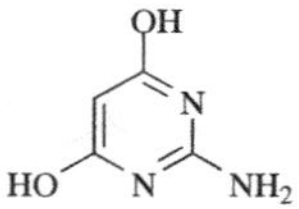

L37

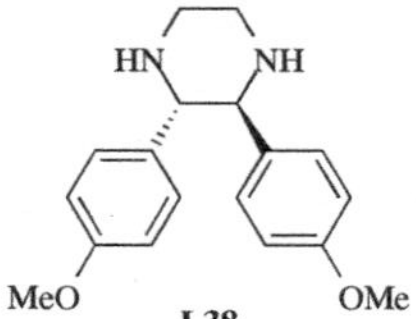

L38

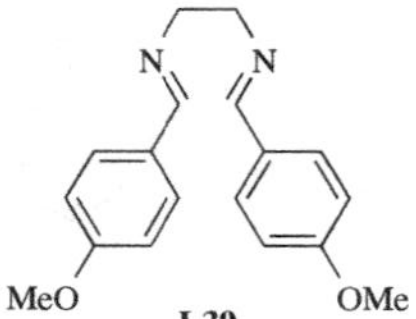

L39

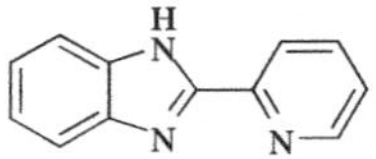

L43

L42

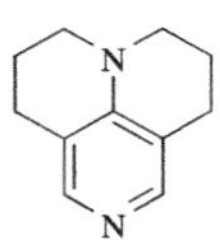

L41

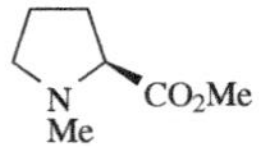

L40

L47

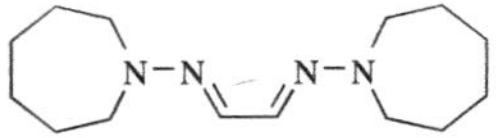

L46

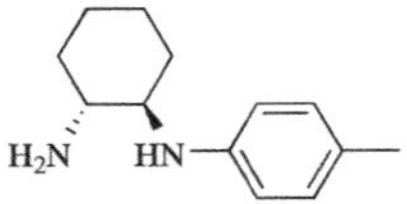

L45

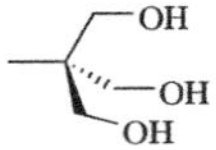

R = dendron

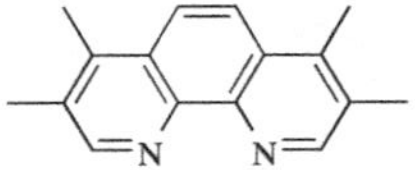

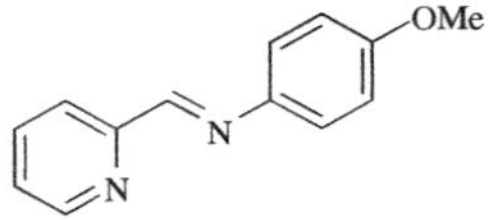

L44

Y=

Z=

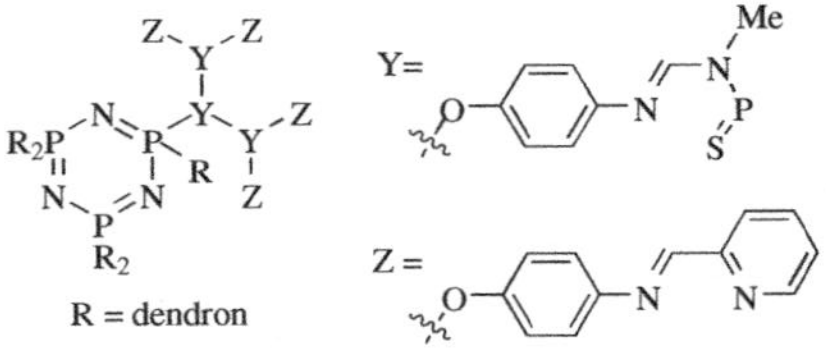

L48

TABLE 1A. PREPARATION OF PRIMARY ARYL AMINES

*Please refer to the charts preceding the tables for structures indicated by the **bold** numbers.*

C$_6$

Aryl Halide		Nitrogen Nucleophile	Conditions				Product(s) and Yield(s) (%)	Refs.
(C$_6$H$_5$X)		NH$_3$ (y)	Catalyst(s) (x amount)	Solvent(s)	Temp (°)	Time (h)	(C$_6$H$_5$NH$_2$)	

X	y	Catalyst(s)	x	Additive(s)		Solvent(s)	Temp (°)	Time (h)		Refs.
I	—	Cu, CuI	1 eq. 20 mol %	none		HOCH$_2$CH$_2$OH	50	10	(67)	237
Br	aq	CuBr	10 mol %	L3 (20 mol %), K$_3$PO$_4$		DMSO	110	24	(85)	245
I	aq	CuBr	10 mol %	L3 (20 mol %), K$_3$PO$_4$		DMSO	rt	24	(85)	245
I	aq	CuI	10 mol %	Fe$_2$O$_3$ (10 mol %), NaOH		EtOH	90	16	(96)	246
Cl	aq	Cu$_2$O	5 mol %	none		H$_2$O/NMP (1:1)	MW (150 W), 110	10	(93)	114
Br	aq	Cu$_2$O	5 mol %	none		H$_2$O/NMP (1:1)	80	15	(99)	114
Br	aq	Cu(acac)$_2$	10 mol %	acac (40 mol %), Cs$_2$CO$_3$		DMF	90	24	(78)	113
Br	aq	2	10 mol %	NaOH		H$_2$O	120	12	(85)	238
I	aq	2	5 mol %	NaOH		H$_2$O	120	12	(94)	238

iodobenzene	NH$_3$	Cu$_2$O (10 mol %), HOCH$_2$CH$_2$OH. 80°, 16 h	aniline NH$_2$ (64) + 2-phenoxyethanol (Ph-O-CH$_2$CH$_2$-OH) (10)	193

1-fluoro-2-iodobenzene	NH$_3$ (aq)	CuBr (5 mol %), L3 (10 mol %). K$_3$PO$_4$, DMSO, rt, 24 h	2-fluoroaniline NH$_2$ (93)	245

(2-fluoro-3-nitro-5-iodobenzene)	NaN$_3$	CuSO$_4$•5H$_2$O (20 mol %), L-Pro (20 mol %), Na ascorbate, DMSO/H$_2$O (9:1), 70°, 24 h	(NO$_2$/NH$_2$-substituted aryl iodide) (69)	75

66

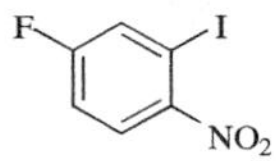

NaN$_3$

CuSO$_4$•5H$_2$O (20 mol %),
L-Pro (20 mol %), Na ascorbate,
DMSO/H$_2$O (9:1), 70°, 24 h

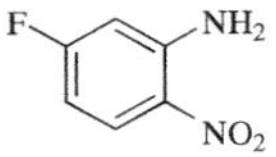

(67)

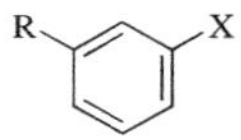

NH$_3$ (aq)

Catalyst (x mol %)

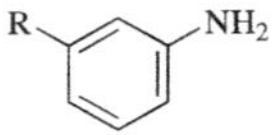

R	X	Catalyst	x	Additive(s)	Solvent(s)	Temp (°)	Time (h)		
Cl	Br	CuBr	10	**L3** (20 mol %), K$_3$PO$_4$	DMSO	110	24	(93)	245
Cl	I	CuBr	5	**L3** (20 mol %), K$_3$PO$_4$	DMSO	rt	24	(90)	245
Cl	Br	Cu$_2$O	5	none	H$_2$O/NMP (1:1)	80	15	(93)	114
Br	I	CuI	10	Fe$_2$O$_3$ (10 mol %), NaOH	EtOH	90	16	(89)	246

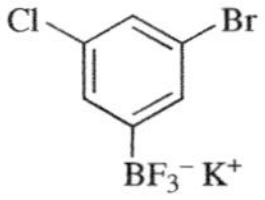

NaN$_3$

CuBr (10 mol %),
DMEDA (20 mol %), Cs$_2$CO$_3$,
DMSO, 90°, 12 h

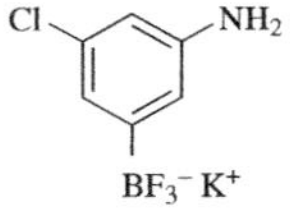

(91)

47

TABLE 1A. PREPARATION OF PRIMARY ARYL AMINES (Continued)

Aryl Halide		Nitrogen Nucleophile	Conditions			Product(s) and Yield(s) (%)			Refs.

*Please refer to the charts preceding the tables for structures indicated by the **bold** numbers.*

C_6

R–C6H4–X → R–C6H4–NH2 (para NH2)

R	X	N-Nucleophile	Catalyst(s)	x	Additive(s)	Solvent(s)	Temp (°)	Time (h)	Refs.
F	Br	NH$_3$ (aq)	**3**	10 mol %	NaOH	H$_2$O	120	12 (88)	238
Cl	Br	NH$_3$ (aq)	CuBr	10 mol %	**L3** (20 mol %), K$_3$PO$_4$	DMSO	110	24 (60)	245
Cl	I	NH$_3$ (aq)	CuBr	5 mol %	**L3** (10 mol %), K$_3$PO$_4$	DMSO	rt	24 (90)	245
Cl	Br	NH$_3$	CuI	10 mol %	DMG (20 mol %), TBPM	DMSO	rt	24 (54)	61
Cl	I	NH$_3$	CuI	10 mol %	DMG (20 mol %), TBPM	DMSO	rt	24 (86)	61
Cl	Br	NH$_3$ (aq)	**2**	10 mol %	NaOH	H$_2$O	120	12 (83)	238
Cl	I	NH$_3$ (aq)	**2**	5 mol %	NaOH	H$_2$O	120	12 (83)	238
Cl	Br	NH$_3$	**3**	5 mol %	K$_2$CO$_3$	MeOH/NMP	90	24 (72)	112
Cl	Br	NaN$_3$	CuI	1 eq	L-Pro (1 eq)	DMSO	100	16 (73)	156
Cl	Br	N$_3$TMS	CuF$_2$	2 eq	none	H$_2$N(CH$_2$)$_2$OH/DMA	95	24 (76)	157
Cl	I	MeC(NH$_2$)=NH$_2^+$ Cl$^-$	CuI	10 mol %	L-Pro (20 mol %), Cs$_2$CO$_3$	DMF	120	10 (72)	248
Br	I	NH$_4$Cl	CuI	20 mol %	L-Pro (40 mol %), K$_2$CO$_3$	H$_2$O/DMSO	rt	12 (80)	60
Br	I	NH$_3$	Cu, CuI	20 mol %, 1 eq	none	HOCH$_2$CH$_2$OH	50	10 (69)	237
Br	I	NH$_3$ (aq)	CuI	10 mol %	Fe$_2$O$_3$ (10 mol %), NaOH	EtOH	90	16 (82)	246
Br	Br	NH$_3$ (aq)	Cu(acac)$_2$	10 mol %	acac (40 mol %), Cs$_2$CO$_3$	DMF	90	36 (41)	113

H2N–C6H4–X → H2N–C6H4–NH2

Isomer	N-Nucleophile	X	Catalyst	x	Additive(s)	Solvent(s)	Temp (°)	Time (h)	Refs.
3	NH$_3$ (aq)	Br	Cu$_2$O	5	none	H$_2$O/NMP 1:1	80	24 (94)	114
4	NH$_3$ (aq)	I	CuI	10	Fe$_2$O$_3$ (10 mol %), NaOH	EtOH	90	16 (92)	246
4	MeC(NH$_2$)=NH$_2^+$ Cl$^-$	I	CuI	10	L-Pro (20 mol %), Cs$_2$CO$_3$	DMF	120	10 (94)	248

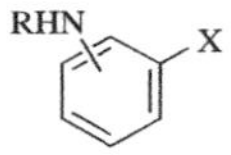

RHN–X   See table.   Catalyst (x mol %)   →   RHN–NH₂

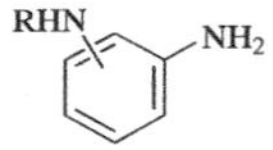

Isomer	R	X	N-Nucleophile	Catalyst	x	Additive	Solvent(s)	Temp (°)	Time (h)		
2	Ac	Br	NH₃	**3**	5	K_2CO_3	MeOH/NMP	90	24	(73)	112
2	Ac	Br	NaN₃	CuI	10	DMEDA (20 mol %), Cs_2CO_3	EtOH	95	36	(52)	249
2	Bz	Cl	NaN₃	CuI	10	DMEDA (20 mol %), Cs_2CO_3	EtOH	90	12	(68)	249
4	Ac	I	NH₃ (aq)	CuBr	5	**L3** (10 mol %), K_3PO_4	DMSO	rt	24	(90)	245

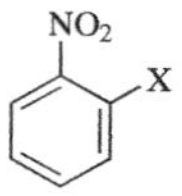

NO_2–X   See table.   Catalyst(s) (x amount) 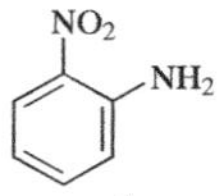   →   **I** or **II**

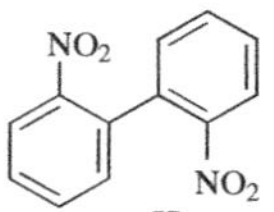

X	N-Nucleophile	Catalyst(s)	x	Additive	Solvent(s)	Temp (°)	Time (h)	I	II	
Br	NH₃ (aq)	CuI	20 mol %	L-4-HOPro (40 mol%), K_2CO_3	DMSO	70	24	(78)	(0)	133
I	NH₃ (aq)	Cu, CuOTf	1.5 mol %, 1 eq	none	Me₂CO/MeCN (20:1)	rt	—	(0)	(92)	111
I	NH₃ (aq)	**2**	5 mol %	NaOH	H_2O	120	12	(79)	(0)	238
Br	NaN₃	Cu	10 mol %	piperidine-2-carboxylic acid (30 mol %), ascorbic acid (20 mol %)	EtOH	100	3	(52)	(0)	250
Cl	NaN₃	Cu₂O	1 eq	L-Pro (1 eq)	DMSO	100	5	(75)	(0)	156
Cl	NaN₃	CuSO₄•5 H₂O	20 mol %	L-Pro (20 mol %), Na ascorbate	DMSO/H₂O (9:1)	70	24	(74)	(0)	75
I	NaN₃	CuSO₄•5 H₂O	20 mol %	L-Pro (20 mol %), Na ascorbate	DMSO/H₂O (9:1)	70	24	(76)	(0)	75
I	MeC(NH₂)=NH₂⁺ Cl⁻	CuI	10 mol %	L-Pro (20 mol %), Cs_2CO_3	DMF	110	10	(79)	(0)	251

TABLE 1A. PREPARATION OF PRIMARY ARYL AMINES (Continued)

Aryl Halide	Nitrogen Nucleophile	Conditions	Product(s) and Yield(s) (%)	Refs.

*Please refer to the charts preceding the tables for structures indicated by the **bold** numbers.*

C₆

O₂N–C₆H₄–X (meta)

Product: 3-nitroaniline (O₂N–C₆H₄–NH₂)

X	N-Nucleophile	Catalyst	x	Additive(s)	Solvent(s)	Temp (°)	Time (h)		Refs.
Br	NH₃ (aq)	CuI	20	L-4-HOPro (40 mol %), K₂CO₃	DMSO	50	24	(91)	133
I	NH₃ (aq)	CuI	20	L-Pro (40 mol %), K₂CO₃	DMSO	rt	12	(92)	60
I	NH₄Cl	CuI	20	L-Pro (40 mol %), K₂CO₃	DMSO	rt	12	(74)	60
Br	NH₃	**3**	5	K₂CO₃	MeOH/NMP	90	24	(93)	112
I	NaN₃	CuSO₄•5 H₂O	20	L-Pro (20 mol %), Na ascorbate	DMSO/H₂O (9:1)	70	24	(77)	75
I	MeC(NH₂)=NH₂⁺ Cl⁻	CuI	10	L-Pro (20 mol %), Cs₂CO₃	DMF	110	10	(92)	248
I	n-PrC(NH₂)=NH₂⁺ Cl⁻	CuI	10	L-Pro (20 mol %), Cs₂CO₃	DMF	110	10	(87)	248
I	PhC(NH₂)=NH₂⁺ Cl⁻	CuI	10	L-Pro (20 mol %), Cs₂CO₃	DMF	110	10	(65)	248

O₂N–C₆H₄–X (para)

Product: 4-nitroaniline (O₂N–C₆H₄–NH₂)

X	N-Nucleophile	Catalyst	x	Additive(s)	Solvent(s)	Temp (°)	Time (h)		Refs.
I	NH₃ (aq)	CuBr	5 mol %	**L3** (10 mol %), K₃PO₄	DMSO	rt	24	(92)	245
Br	NH₃ (aq)	CuI	20 mol %	L-4-HOPro (40 mol %), K₂CO₃	DMSO	50	24	(92)	133
I	NH₄Cl	CuI	20 mol %	L-Pro (40 mol %), K₂CO₃	H₂O/DMSO	rt	12	(94)	60
I	NH₃	CuI	20 mol %	DMG (20 mol %), TBPM	DMSO	rt	24	(86)	61
I	NH₃ (aq)	CuI	10 mol %	Fe₂O₃ (10 mol %), NaOH	EtOH	90	16	(61)	246
Cl	NH₃ (aq)	Cu₂O	5 mol %	none	H₂O/NMP (1:1)	MW (150 W), 110	15	(79)	114
Br	NH₃ (aq)	Cu₂O	5 mol %	none	H₂O/NMP (1:1)	80	15	(86)	114
I	NH₃ (aq)	Cu(acac)₂	10 mol %	acac (40 mol %), Cs₂CO₃	DMF	90	24	(63)	113
Br	NH₃ (aq)	**2**	5 mol %	NaOH	H₂O	120	12	(87)	238
I	NH₃ (aq)	**2**	5 mol %	NaOH	H₂O	120	12	(91)	238
Cl	NaN₃	Cu	2 eq	none	H₂N(CH₂)₂OH/DMA	95	24	(32)	157

X	N-Nucleophile	Catalyst		Additive(s)	Solvent	Temp (°)	Time (h)	(Yield)	Ref
Br	NaN_3	Cu	10 mol %	piperidine-2-carboxylic acid (30 mol %), ascorbic acid (20 mol %)	EtOH	100	3	(88)	250
Br	NaN_3	$CuSO_4 \cdot 5\ H_2O$	20 mol %	L-Pro (20 mol %), Na ascorbate	$DMSO/H_2O$ (9:1)	70	24	(76)	75
I	NaN_3	$CuSO_4 \cdot 5\ H_2O$	20 mol %	L-Pro (20 mol %), Na ascorbate	$DMSO/H_2O$ (9:1)	70	24	(86)	75
Br	N_3TMS	CuF_2	2 eq	none	$H_2N(CH_2)_2OH/DMA$	95	24	(95)	157
Br	$MeC(NH_2)=NH_2^+\ Cl^-$	CuI	10 mol %	L-Pro (20 mol %), Cs_2CO_3	DMF	120	10	(75)	251
I	$MeC(NH_2)=NH_2^+\ Cl^-$	CuI	10 mol %	L-Pro (20 mol %), Cs_2CO_3	DMF	110	10	(88)	248
I	$n\text{-}PrC(NH_2)=NH_2^+\ Cl^-$	CuI	10 mol %	L-Pro (20 mol %), Cs_2CO_3	DMF	110	10	(80)	248
I	$PhC(NH_2)=NH_2^+\ Cl^-$	CuI	10 mol %	L-Pro (20 mol %), Cs_2CO_3	DMF	110	10	(65)	248

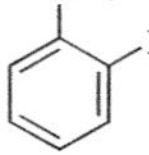

Catalyst (x mol %)   See table.

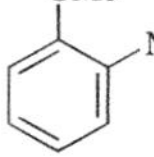

N-Nucleophile	Catalyst	x	Additive(s)	Solvent	Temp (°)	Time (h)	(Yield)	Ref
NH_3 (aq)	CuBr	10	**L3** (20 mol %), K_3PO_4	DMSO	110	24	(60)	245
NH_3 (aq)	CuI	20	L-4-HOPro (40 mol %), K_2CO_3	DMSO	70	24	(81)	133
NH_3 (aq)	**2**	5	NaOH	H_2O	120	12	(90)	238
NaN_3	Cu	10	piperidine-2-carboxylic acid (30 mol %), ascorbic acid (20 mol %)	EtOH	100	3	(73)	250

TABLE 1A. PREPARATION OF PRIMARY ARYL AMINES (*Continued*)

| Aryl Halide | | Nitrogen Nucleophile | | Conditions | | | Product(s) and Yield(s) (%) | | Refs. |

*Please refer to the charts preceding the tables for structures indicated by the **bold** numbers.*

C₆ — C6

MeO—C₆H₄—X   →   MeO—C₆H₄—NH₂

Catalyst (x amount)

N-Nucleophile	X	Catalyst	x	Additive(s)	Solvent(s)	Temp (°)	Time (h)	Refs.
NH_3 (aq)	Br	CuBr	10 mol %	**L3** (20 mol %), K_3PO_4	DMSO	110	24 (93)	245
NH_3 (aq)	Br	CuI	20 mol %	L-4-HOPro (40 mol %), K_2CO_3	DMSO	50	24 (83)	133
NH_3 (aq)	Br	Cu(acac)₂	10 mol %	acac (40 mol %), Cs_2CO_3	DMF	90	24 (85)	113
NH_3 (aq)	Br	**2**	10 mol %	NaOH	H_2O	120	12 (88)	238
N_3TMS	Br	Cu	2 eq	none	$H_2N(CH_2)_2OH$/DMA	95	24 (81)	157
$MeC(NH_2)=NH_2^+$ Cl⁻	I	CuI	10 mol %	L-Pro (20 mol %), Cs_2CO_3	DMF	120	10 (72)	248

MeO—C₆H₄—X   →   MeO—C₆H₄—NH₂

Catalyst(s) (x amount)

X	N-Nucleophile	Catalyst(s)	x	Additive(s)	Solvent(s)	Temp (°)	Time (h)	Refs.
Br	NH_3 (aq)	Cu, CuI	20 mol %, 1 eq	none	$HOCH_2CH_2OH$	50	10 (71)	237
Br	NH_3 (aq)	CuBr	10 mol %	**L3** (20 mol %), K_3PO_4	DMSO	110	24 (86)	245
I	NH_3 (aq)	CuBr	5 mol %	**L3** (10 mol %), K_3PO_4	DMSO	rt	24 (81)	245
Br	NH_3	CuI	10 mol %	DMG (20 mol %), TBPM	DMSO	rt	24 (35)	61
I	NH_3	CuI	10 mol %	DMG (20 mol %), TBPM	DMSO	rt	24 (78)	61
Br	NH_3 (aq)	CuI	20 mol %	L-Pro (40 mol %), K_2CO_3	DMSO	80	12 (44)	60
I	NH_3 (aq)	CuI	20 mol %	L-Pro (40 mol %), K_2CO_3	DMSO	rt	12 (77)	60
I	NH_4Cl	CuI	20 mol %	L-Pro (40 mol %), K_2CO_3	DMSO	rt	12 (32)	60
I	NH_3 (aq)	CuI	10 mol %	Fe_2O_3 (10 mol %), NaOH	EtOH	90	16 (81)	246
Cl	NH_3 (aq)	Cu_2O	5 mol %	none	H_2O/NMP (1:1)	MW (150 W), 110	20 (85)	114
Br	NH_3 (aq)	Cu_2O	5 mol %	none	H_2O/NMP (1:1)	80	20 (94)	114
I	NH_3 (aq)	Cu(acac)₂	10 mol %	acac (40 mol %), Cs_2CO_3	DMF	90	24 (80)	113
Br	NH_3	**3**	5 mol %	K_2CO_3	MeOH/NMP	90	24 (79)	112

Aryl–X	N-Nucleophile	Catalyst	x	Additive(s)	Solvent(s)	Temp (°)	Time (h)	(Yield)	Ref
Br	NH_3 (aq)	**2**	10 mol %	NaOH	H_2O	120	12	(85)	238
I	NH_3 (aq)	**2**	5 mol %	NaOH	H_2O	120	12	(90)	238
Br	NaN_3	Cu_2O	1 eq	L-Pro (1 eq)	DMSO	100	16	(46)	156
Br	N_3TMS	Cu	2 eq	none	$H_2N(CH_2)_2OH$/DMA	95	24	(48)	157
I	$MeC(NH_2)=NH_2^+\ Cl^-$	CuI	10 mol %	L-Pro (20 mol %), Cs_2CO_3	DMF	120	10	(78)	248

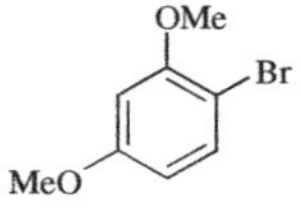

See table.

Catalyst (x amount)

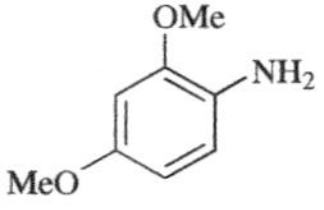

N-Nucleophile	Catalyst	x	Additive(s)	Solvent(s)	Temp (°)	Time (h)	(Yield)	Ref
NH_3	**3**	5 mol %	K_2CO_3	MeOH/NMP	90	24	(66)	112
NH_3 (aq)	CuI	20 mol %	L-Pro (40 mol %), K_2CO_3	DMSO	70	24	(82)	133
NaN_3	Cu_2O	1 eq	L-Pro (1 eq)	DMSO	100	6	(71)	156

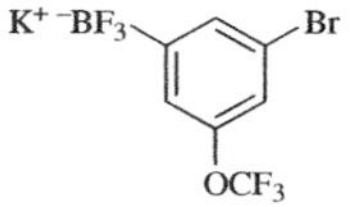

NaN_3 — CuBr (10 mol %), DMEDA (20 mol %), Cs_2CO_3, DMSO. 90°, 12 h — (83) — 247

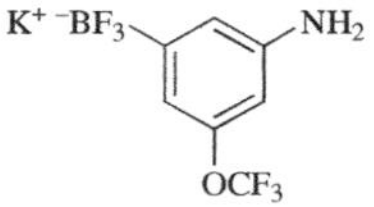

NaN_3 — Cu (10 mol %), piperidine-2-carboxylic acid (30 mol %), ascorbic acid (20 mol %), EtOH. 100°, 3 h — (64) — 250

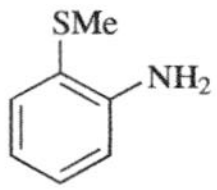

TABLE 1A. PREPARATION OF PRIMARY ARYL AMINES (Continued)

*Please refer to the charts preceding the tables for structures indicated by the **bold** numbers.*

C₇

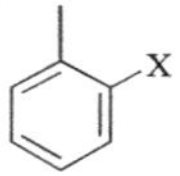

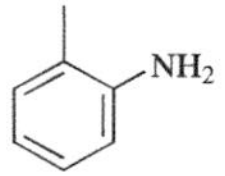

X	N-Nucleophile	Catalyst	x	Additive(s)	Solvent(s)	Temp (°)	Time (h)	Product(s) and Yield(s) (%)	Refs.
Br	NH₃ (aq)	CuBr	10 mol %	**L3** (20 mol %), K₃PO₄	DMSO	110	24	(52)	245
I	NH₃ (aq)	CuBr	5 mol %	**L3** (10 mol %), K₃PO₄	DMSO	80	24	(70)	245
Br	NH₃ (aq)	CuI	20 mol %	L-4-HOPro (40 mol %), K₂CO₃	DMSO	70	24	(55)	133
I	NH₃ (aq)	CuI	20 mol %	L-Pro (40 mol %), K₂CO₃	DMSO	80	12	(34)	60
I	NH₄Cl	CuI	20 mol %	L-Pro (40 mol %), K₂CO₃	DMSO	rt	12	(7)	60
Cl	NH₃ (aq)	Cu₂O	5 mol %	none	H₂O/NMP (1:1)	MW (150 W) 110	15	(83)	114
Br	NH₃ (aq)	Cu₂O	5 mol %	none	H₂O/NMP (1:1)	80	20	(99)	114
Br	NH₃ (aq)	**2**	5 mol %	NaOH	H₂O	120	12	(70)	238
I	NH₃ (aq)	**2**	5 mol %	NaOH	H₂O	120	12	(71)	238
Br	NH₃	**3**	5 mol %	K₂CO₃	MeOH/NMP	90	24	(80)	112
Br	NaN₃	CuI	1 eq	L-Pro (1 eq)	DMSO	100	18	(70)	156

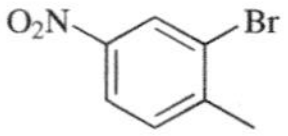

Aryl Halide	Nitrogen Nucleophile	Conditions	Product(s) and Yield(s) (%)	Refs.
(structure, Br, O₂N, CH₃)	N₃TMS	CuF₂ (2 eq), H₂NCH₂CH₂OH, DMA, 95°, 24 h	(75)	157
(structure, Br, MeO, CH₃)	NH₃ (aq)	CuI (20 mol %), K₂CO₃, DMSO, L-4-HOPro (40 mol %), 70°, 24 h	(78)	133

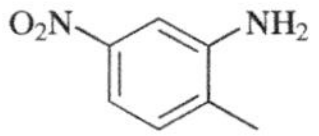

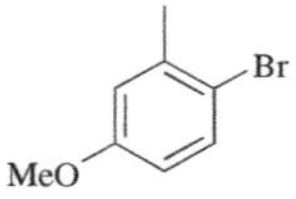

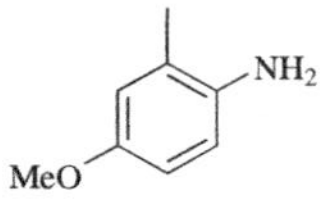

74

See table.

Catalyst(s) (x amount), 50°

R	N-Nucleophile	Catalyst(s)	x	Additives	Solvent	Temp (°)	Time (h)		ref
4-Br	NH$_3$ (aq)	CuI	20 mol %	L-4-HOPro (40 mol %), K$_2$CO$_3$	DMSO	50	24	(63)	133
6-NH$_2$	NH$_3$	Cu, CuI	20 mol %, 1 eq	none	HOCH$_2$CH$_2$OH	40	10	(37)	237
5-K$^+$-BF$_3$	NaN$_3$	CuBr	10 mol %	DMEDA (20 mol %), Cs$_2$CO$_3$	DMSO	90	8	(93)	247

See table.

Catalyst (x amount)

X	N-Nucleophile	Catalyst	x	Additive(s)	Solvent(s)	Temp (°)	Time (h)	ref
Br	NH$_3$ (aq)	CuBr	10 mol %	**L3** (20 mol %), K$_3$PO$_4$	DMSO	110	24 (86)	245
I	NH$_3$ (aq)	CuBr	10 mol %	**L3** (20 mol %), K$_3$PO$_4$	DMSO	rt	24 (83)	245
I	NH$_3$ (aq)	CuI	20 mol %	L-Pro (40 mol %), K$_2$CO$_3$	DMSO	rt	12 (80)	60
I	NH$_4$Cl	CuI	20 mol %	L-Pro (40 mol %), K$_2$CO$_3$	DMSO	rt	12 (56)	60
Br	NH$_3$ (aq)	CuI	20 mol %	L-4-HO-Pro (40 mol %), K$_2$CO$_3$	DMSO	50	24 (55)	133
I	NH$_3$ (aq)	CuI	10 mol %	Fe$_2$O$_3$ (10 mol %), NaOH	EtOH	90	16 (72)	246
Cl	NH$_3$ (aq)	Cu$_2$O	5 mol %	none	H$_2$O/NMP 1:1	MW (150 W), 110	12 (89)	114
Br	NH$_3$ (aq)	Cu$_2$O	5 mol %	none	H$_2$O/NMP 1:1	80	15 (98)	114
Br	NH$_3$ (aq)	Cu(acac)$_2$	10 mol %	acac (40 mol %), Cs$_2$CO$_3$	DMF	90	36 (79)	113
Br	NH$_3$ (aq)	**2**	10 mol %	NaOH	H$_2$O	120	12 (87)	238
I	NH$_3$ (aq)	**2**	5 mol %	NaOH	H$_2$O	120	12 (94)	238
Br	NaN$_3$	Cu$_2$O	1 eq	L-Pro (1 eq)	DMSO	100	16 (84)	156
I	MeC(NH$_2$)=NH$_2^+$ Cl$^-$	CuI	10 mol %	L-Pro (20 mol %), Cs$_2$CO$_3$	DMF	120	10 (72)	248

TABLE 1A. PREPARATION OF PRIMARY ARYL AMINES (Continued)

*Please refer to the charts preceding the tables for structures indicated by the **bold** numbers.*

C7

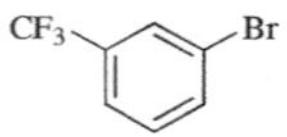

Aryl Halide	Nitrogen Nucleophile	Conditions	Product(s) and Yield(s) (%)	Refs.
(3-bromo-CF3-benzene)	NH3 (aq)	CuI (20 mol %), K2CO3, DMSO, L-4-HOPro (40 mol %), 50°, 24 h	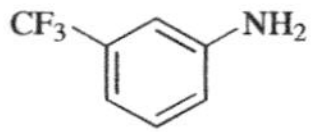 (90)	133
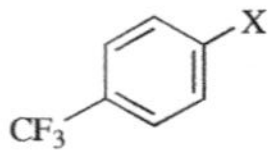 (X-CF3-benzene)	See table.	Catalyst (x amount)	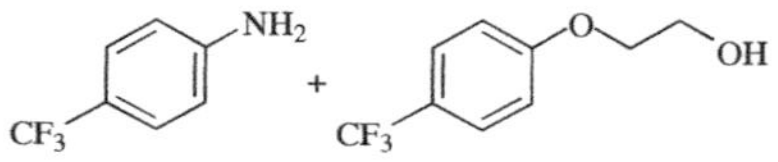 I + II	

X	N-Nucleophile	Catalyst	x	Additive(s)	Solvent(s)	Temp (°)	Time (h)	I	II	Refs.
Br	NH3	CuI	10 mol %	DMG (20 mol %), TBPM	DMSO	rt	24	(62)	(—)	61
Br	NH3 (aq)	CuI	20 mol %	L-4-HOPro (40 mol %), K2CO3	DMSO	50	24	(91)	(—)	133
I	NH4Cl	CuI	20 mol %	L-Pro (40 mol %), K2CO3	H2O/DMSO	rt	12	(80)	(—)	60
Br	NH3	Cu2O	10 mol %	none	HOCH2CH2OH	80	16	(54)	(18)	193
Br	NH3	**3**	5 mol %	K2CO3	MeOH/NMP	90	24	(95)	(—)	112
I	NaN3	CuSO4•5 H2O	20 mol %	L-Pro (20 mol %), Na ascorbate	DMSO/H2O (9:1)	70	24	(57)	(—)	75
Br	N3TMS	CuF2	2 eq	Et3N	DMA	95	24	(67)	(—)	157
I	N3TMS	CuF2	2 eq	Et3N	DMA	95	24	(60)	(—)	157

Aryl Halide	Nitrogen Nucleophile	Conditions	Product(s) and Yield(s) (%)	Refs.
(NO2, X-CF3-benzene)	NaN3	CuSO4•5H2O (20 mol %), L-Pro (20 mol %), Na ascorbate, DMSO/H2O (9:1), 70°, 24 h	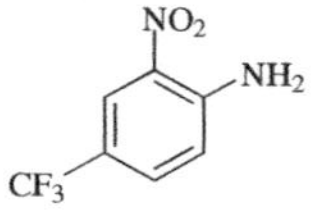 X: F (78), Cl (76), OMs (42)	75
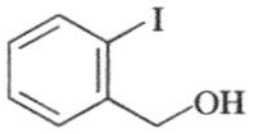 (2-iodobenzyl alcohol)	NaN3	Cu2O (1 eq), L-Pro (1 eq), DMSO, 100°, 16 h	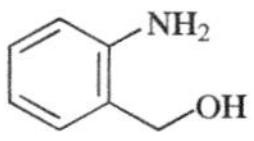 (87)	156

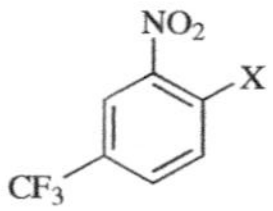

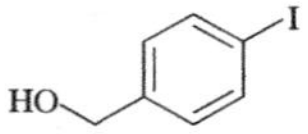

See table.   Catalyst (x mol %), DMSO, rt

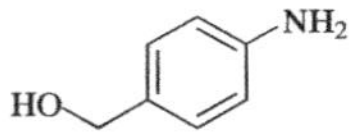

N-Nucleophile	Catalyst	x	Additives	Time (h)		
NH3 (aq)	CuBr	5	L3 (10 mol %), K3PO4	24	(79)	245
NH3 (aq)	CuI	20	L-Pro (40 mol %), K2CO3	12	(77)	60
NH4Cl	CuI	20	L-Pro (40 mol %), K2CO3	12	(57)	60

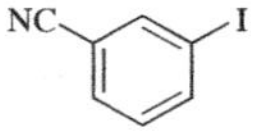

See table.

CuI (20 mol %),
L-Pro (40 mol %),
K2CO3, DMSO, rt, 12 h

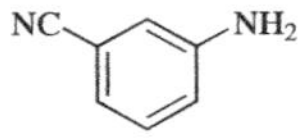

N-Nucleophile		
NH3 (aq)	(97)	60
NH4Cl/H2O	(89)	

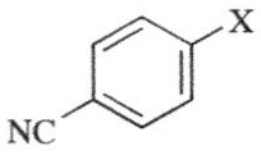

NH3 (y)   Catalyst(s) (x amount)

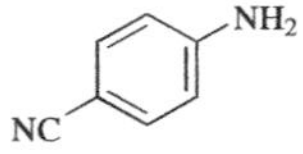

X	y	Catalyst(s)	x	Additive(s)	Solvent(s)	Temp (°)	Time (h)		
Br	—	Cu, CuI	20 mol %, 1 eq	none	HOCH2CH2OH	50	10	(59)	237
I	aq	CuBr	5 mol %	L3 (10 mol %), K3PO4	DMSO	rt	24	(93)	245
Br	aq	CuI	20 mol %	L-Pro (40 mol %), K2CO3	DMSO	80	12	(88)	60
Br	aq	CuI	20 mol %	L-4-HO-Pro (40 mol %), K2CO3	DMSO	50	24	(91)	133
Cl	aq	Cu2O	5 mol %	none	H2O/NMP 1:1	MW (150 W), 110	20	(86)	114
Br	aq	Cu2O	5 mol %	none	H2O/NMP 1:1	80	15	(81)	114
Br	aq	Cu(acac)2	10 mol %	acac (40 mol %), Cs2CO3	DMF	90	36	(92)	113
I	aq	Cu(acac)2	10 mol %	acac (40 mol %), Cs2CO3	DMF	90	24	(94)	113
Br	—	3	5 mol %	K2CO3	MeOH/NMP	90	24	(93)	112

TABLE 1A. PREPARATION OF PRIMARY ARYL AMINES (Continued)

Aryl Halide	Nitrogen Nucleophile	Conditions	Product(s) and Yield(s) (%)	Refs.

*Please refer to the charts preceding the tables for structures indicated by the **bold** numbers.*

C₇

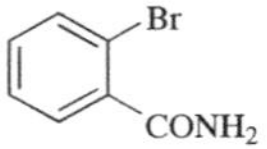

Aryl Halide	Nitrogen Nucleophile	Conditions	Product(s) and Yield(s) (%)	Refs.
(2-Br, CONH₂ benzene)	NH₃	Catalyst **3** (5 mol %), K₂CO₃, MeOH/NMP, 90°, 24 h	(2-NH₂, CONH₂ benzene) (93)	112
(Cl, CONH₂, R benzene)	NaN₃	CuI (10 mol %), DMEDA (20 mol %), Cs₂CO₃, EtOH, 95°	(R, NH₂, CONH₂ benzene)	249

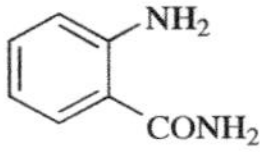

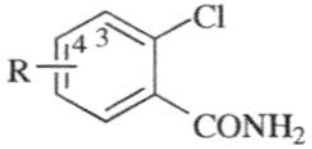

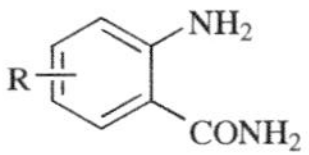

R	X	Time (h)	(%)
H	Cl	36	(70)
H	Br	36	(79)
5-Cl	Br	56	(66)
5-Br	Br	48	(64)
3-O₂N	Br	36	(71)

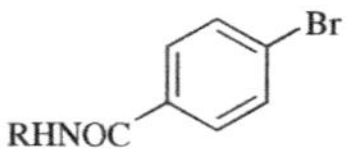

Aryl Halide	Nitrogen Nucleophile	Conditions	Product(s) and Yield(s) (%)	Refs.
(4-Br, RHNOC benzene)	NH₃	Catalyst(s) (x amount)	(4-NH₂, RHNOC benzene)	

R	Catalyst(s)	x	Additive	Solvent(s)	Temp (°)	Time (h)		Refs.
H	**3**	5 mol %	K₂CO₃	MeOH/NMP	90	24	(93)	112
n-Bu	Cu, CuI	20 mol %, 1 eq	none	HOCH₂CH₂OH	50	10	(75)	237

Aryl Halide	Nitrogen Nucleophile	Conditions	Product(s) and Yield(s) (%)	Refs.
(3-Br benzene with 1,3-dioxolane)	NH₃	Cu (20 mol %), CuI (1 eq), HOCH₂CH₂OH, 50°, 10 h	(3-NH₂ benzene with 1,3-dioxolane) (70)	237
(4-Br, OHC benzene)	NH₃ (aq)	Cu₂O (5 mol %), H₂O/NMP (1:1), 80°, 15 h	(4-NH₂, OHC benzene) (79)	114

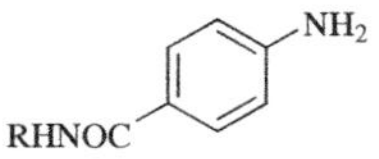

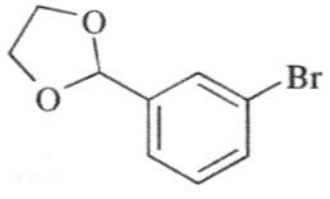

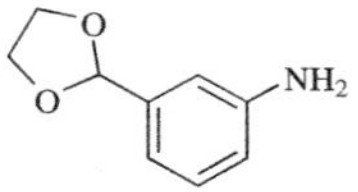

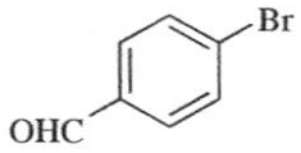

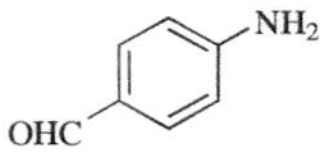

78

R	X	N-Nucleophile	Catalyst(s)	x	Additive(s)	Solvent(s)	Temp (°)	Time (h)		
			See table.							
H	Cl	NH$_3$ (aq)	Cu$_2$O	5 mol %	Cs$_2$CO$_3$	H$_2$O/NMP 1:1	80	24	(94)	114
H	Cl	NaN$_3$	CuI	10 mol %	DMEDA (20 mol %), Cs$_2$CO$_3$	EtOH	95	36	(90)	249
H	Br	NaN$_3$	CuI	10 mol %	Cs$_2$CO$_3$	EtOH	95	36	(91)	249
Me	Br	NH$_3$ (aq)	Cu$_2$O	5 mol %	none	H$_2$O/NMP 1:1	80	24	(92)	114
Me	I	NH$_3$ (aq)	Cu(acac)$_2$	10 mol %	acac (40 mol %), Cs$_2$CO$_3$	DMF	90	24	(23)	113
Me	I	NH$_3$ (aq)	Cu, CuOTf	1.5 mol %, 1 eq	none	Me$_2$CO/MeCN 20:1	rt	—	(80)	111
Me	Cl	NaN$_3$	Cu$_2$O	1 eq	L-Pro (1 eq)	DMSO	100	12	(40)	156
Me	I	NaN$_3$	Cu$_2$O	1 eq	L-Pro (1 eq)	DMSO	100	0.25	(64)	156
Et	Br	NaN$_3$	Cu	10 mol %	piperidine-2-carboxylic acid (30 mol %), ascorbic acid (20 mol %)	EtOH	100	3	(91)	250
Et	Br	TMSN$_3$	CuF$_2$	2 eq	none	H$_2$N(CH$_2$)$_2$OH/DMA	95	24	(80)	157

C$_{7-8}$

NaN$_3$, CuI (10 mol %), DMEDA (20 mol %), Cs$_2$CO$_3$, EtOH, 95°

R	X	Time (h)		
3-O$_2$N	Br	30	(95)	
4-O$_2$N	Cl	36	(91)	
5-O$_2$N	Cl	30	(79)	
4-Me	Br	48	(65)	249

C$_7$

Catalyst (x amount), 24 h

R	X	N-Nucleophile	Catalyst	x	Additive(s)	Solvent	Temp (°)		
			See table.						
Me	I	NH$_3$	CuI	10 mol %	DMG (20 mol %), TBPM	DMSO	rt	(93)	61
Me	Br	N$_3$Ts	CuF$_2$	2 eq	none	H$_2$N(CH$_2$)$_2$OH/DMA	95	(77)	157
Et	Br	N$_3$Ts	CuF$_2$	2 eq	none	H$_2$N(CH$_2$)$_2$OH/DMA	95	(80)	157

79

TABLE 1A. PREPARATION OF PRIMARY ARYL AMINES (Continued)

*Please refer to the charts preceding the tables for structures indicated by the **bold** numbers.*

C7

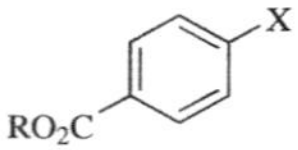

Aryl Halide		Nitrogen Nucleophile		Conditions					Product(s) and Yield(s) (%)		Refs.
RO2C–C6H4–X		See table.		Catalyst (x amount)					RO2C–C6H4–NH2		
R	X	N-Nucleophile	Catalyst	x	Additive(s)	Solvent(s)	Temp (°)	Time (h)			
Me	I	N3TMS	CuF2	2 eq	none	Et3N/DMA	95	24	(66)		157
Et	I	NH3 (aq)	CuBr	5 mol %	**L3** (10 mol %), K3PO4	DMSO	rt	24	(89)		245
Et	I	NH4Cl	CuI	20 mol %	L-Pro (40 mol %), K2CO3	DMSO	rt	12	(97)		60
Et	Cl	NH3 (aq)	Cu2O	5 mol %	none	H2O/NMP 1:1	MW (150 W), 110	10	(84)		114
Et	Br	NH3 (aq)	Cu2O	5 mol %	none	H2O/NMP 1:1	80	15	(83)		114
Et	I	NaN3	Cu	10 mol %	piperidine-2-carboxylic acid (30 mol %), ascorbic acid (20 mol %)	EtOH	100	3	(87)		250
Et	Br	NaN3	Cu2O	1 eq	L-Pro (1 eq)	DMSO	100	2	(93)		156
Et	Br	N3TMS	CuF2	2 eq	none	H2N(CH2)2OH/DMA	95	24	(99)		157

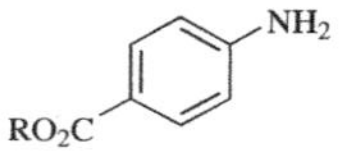

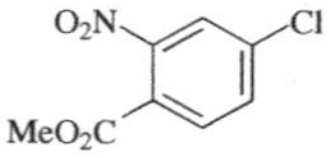

Additional entries:

O2N / MeO2C aryl chloride, NaN3, Cu2O (1 eq), L-Pro (1 eq), DMSO, 100°, 48 h → H2N...NH2, MeO2C product (20), 156

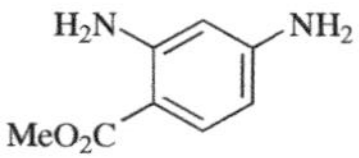

NO2 / Br / CONH2 aryl bromide, NaN3, CuI (10 mol %), Cs2CO3, EtOH, 95°, 36 h → NO2 / NH2 / CONH2 product (71), 249

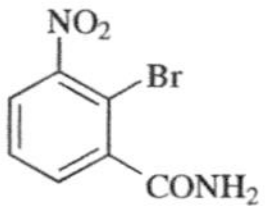

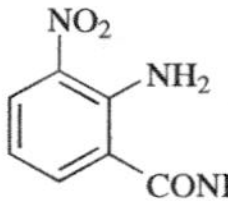

C$_8$

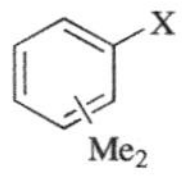

See table.

Catalyst(s) (x amount)

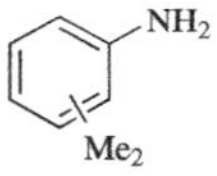

Isomer	X	N-Nucleophile	Catalyst(s)	x	Additive(s)	Solvent(s)	Temp (°)	Time (h)		
2,4	Br	NH$_3$	Cu, CuI	20 mol %, 1 eq	none	HOCH$_2$CH$_2$OH	50	10	(7)	237
2,6	Br	NH$_3$	3	5 mol %	K$_2$CO$_3$	MeOH/NMP	120	48	(7)	112
2,6	Br	NH$_3$ (aq)	2	5 mol %	NaOH	H$_2$O	120	12	(64)	238
3,5	I	NH$_3$ (aq)	CuBr	10 mol %	L3 (20 mol %), K$_3$PO$_4$	DMSO		24	(90)	245
3,5	I	NH$_3$ (aq)	CuI	20 mol %	L-Pro (40 mol %), K$_2$CO$_3$	DMSO	rt	12	(83)	60
3,5	I	NH$_4$Cl	CuI	20 mol %	L-Pro (40 mol %), K$_2$CO$_3$	DMSO	rt	12	(62)	60
3,5	Br	NH$_3$ (aq)	Cu$_2$O	5 mol %	none	H$_2$O/NMP 1:1	80	15	(99)	114

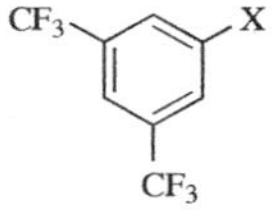

NaN$_3$

CuSO$_4$•5 H$_2$O (20 mol %),
L-Pro (20 mol %), Na ascorbate,
DMSO/H$_2$O (9:1), 70°, 24 h

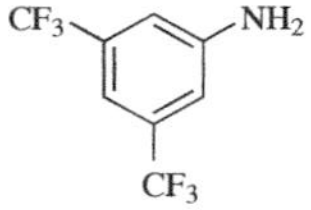

X	
I	(68)
MsO	(24)

75

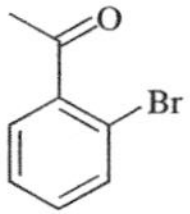

NH$_3$

Catalyst 3 (5 mol %), K$_2$CO$_3$,
MeOH/NMP, 90°, 24 h

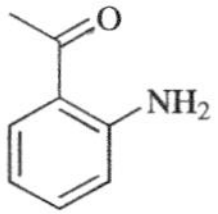

(92)

112

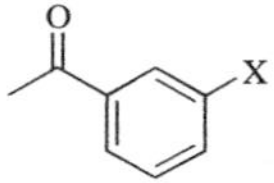

NH$_3$ (y)

Catalyst(s) (x amount)

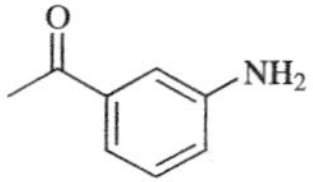

X	y	Catalyst(s)	x	Solvent(s)	Temp (°)	Time (h)		
Br	—	Cu, CuI	20 mol %, 1 eq	HOCH$_2$CH$_2$OH	50	10	(82)	237
Cl	aq	Cu$_2$O	5 mol %	H$_2$O/NMP (1:1)	80	20	(82)	114
Br	aq	Cu$_2$O	5 mol%	H$_2$O/NMP (1:1)	80	15	(92)	114

TABLE 1A. PREPARATION OF PRIMARY ARYL AMINES (Continued)

Aryl Halide	Nitrogen Nucleophile	Conditions	Product(s) and Yield(s) (%)	Refs.

*Please refer to the charts preceding the tables for structures indicated by the **bold** numbers.*

C₈

X	N-Nucleophile	Catalyst(s)	x	Additives	Solvent(s)	Temp (°)	Time (h)	I	II	Refs.
Br	NH₃	Cu, CuI	20 mol %, 1 eq	none	HOCH₂CH₂OH	50	10	(85)	(0)	237
Br	NH₃ (aq)	CuBr	10 mol %	L3 (10 mol %), K₃PO₄	DMSO	110	24	(96)	(—)	245
I	NH₃ (aq)	CuBr	5 mol %	L3 (10 mol %), K₃PO₄	DMSO	rt	24	(95)	(—)	245
Br	NH₃ (aq)	CuI	20 mol %	L-Pro (40 mol %), K₂CO₃	DMSO	80	12	(91)	(—)	60
I	NH₄Cl	CuI	20 mol %	L-Pro (40 mol %), K₂CO₃	DMSO	rt	12	(80)	(—)	60
I	NH₃ (aq)	CuI	10 mol %	Fe₂O₃ (10 mol %), NaOH	EtOH	90	16	(87)	(—)	246
Br	NH₃	Cu₂O	10 mol %	none	HOCH₂CH₂OH	80	16	(60)	(5)	193
Br	NH₃ (aq)	Cu(acac)	10 mol %	acac (40 mol %), Cs₂CO₃	DMF	90	36	(92)	(—)	113
Br	NH₃ (aq)	2	10 mol %	NaOH	H₂O	120	12	(95)	(—)	238
I	NH₃ (aq)	2	5 mol %	NaOH	H₂O	120	12	(95)	(—)	238
Br	NH₃	3	5 mol %	K₂CO₃	MeOH/NMP	90	24	(97)	(—)	112
Br	NaN₃	CuF₂	2 eq	none	Et₃N/DMA	95	24	(77)	(—)	157
Br	NaN₃	Cu₂O	1 eq	L-Pro (1 eq)	DMSO	100	7	(87)	(—)	156
I	N₃TMS	CuF₂	2 eq	none	Et₃N/DMA	95	24	(81)	(—)	157

C₉

X	N-Nucleophile	Catalyst	x	Additive(s)	Temp (°)	Time (h)		Refs.
I	NH₃ (aq)	CuBr	10 mol %	L3 (20 mol %), K₃PO₄	80	24	(60)	245
Br	NaN₃	Cu₂O	1 eq	DMEDA (1 eq)	100	10	(72)	156

Catalyst (x amount), DMSO

82

Br-C6H4-iPr + NH₃ (aq) →[Cu₂O (5 mol %), H₂O/NMP (1:1), 90°, 48 h] iPr-C6H4-NH₂ (82)   114

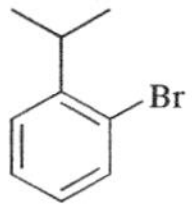

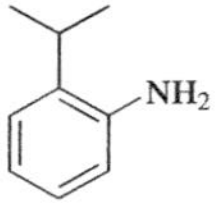

C₁₀₋₁₁

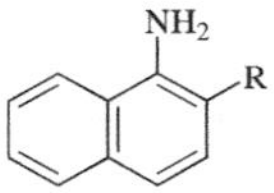

X	R	N-Nucleophile	Catalyst	x	Additive(s)	Solvent(s)	Temp (°)	Time (h)		
Br	H	NH₃ (aq)	CuBr	10 mol %	L3 (10 mol %), K₃PO₄	DMSO	110	24	(89)	245
Br	H	NH₃ (aq)	CuI	20 mol %	L-4-HOPro (40 mol %), K₂CO₃	DMSO	50	36	(76)	133
I	H	NH₃	CuI	10 mol %	DMG (20 mol %), TBPM	DMSO	rt	24	(74)	61
Br	H	NH₃ (aq)	Cu₂O	5 mol %	none	H₂O/NMP (1:1)	80	15	(90)	114
I	H	NH₃ (aq)	Cu₂O	5 mol %	none	H₂O/NMP (1:1)	80	15	(96)	114
Br	H	NH₃ (aq)	Cu(acac)	10 mol %	acac (40 mol %), Cs₂CO₃	DMF	90	24	(88)	113
I	H	NH₃ (aq)	Cu(acac)₂	10 mol %	acac (40 mol %), Cs₂CO₃	DMF	90	24	(90)	113
Br	H	NH₃ (aq)	2	5 mol %	NaOH	H₂O	120	12	(90)	238
Br	H	NaN₃	Cu₂O	1 eq	L-Pro (1 eq)	DMSO	100	12	(59)	156
I	H	MeC(NH₂)=NH₂⁺ Cl⁻	CuI	10 mol %	L-Pro (20 mol %), Cs₂CO₃	DMF	120	10	(67)	248
Br	OMe	NaN₃	Cu	10 mol %	piperidine-2-carboxylic acid (30 mol %), ascorbic acid (20 mol %)	EtOH	100	3	(69)	250
Br	Me	NaN₃	Cu₂O	1 eq	L-Pro (1 eq)	DMSO	100	14	(76)	156

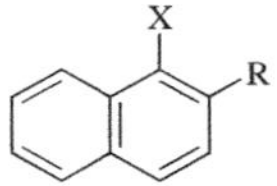

TABLE 1A. PREPARATION OF PRIMARY ARYL AMINES (Continued)

Please refer to the charts preceding the tables for structures indicated by the **bold** numbers.

C₁₀

Aryl Halide	Nitrogen Nucleophile		Conditions				Product(s) and Yield(s) (%)			Refs.

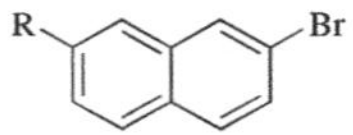

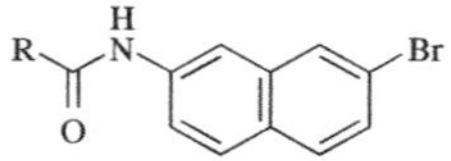

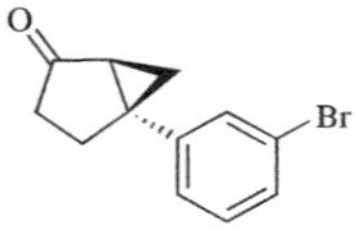

See table.

	N-Nucleophile	Catalyst	Catalyst (x amount)							
			x	Additive(s)	Solvent	Temp (°)	Time (h)			
H	NH₃	**3**	5 mol %	**L3** (5 mol %), K₂CO₃	MeOH/NMP	90	24	(75)		112
H	NaN₃	Cu₂O	1 eq	L-Pro (1 eq)	DMSO	100	12	(88)		156
K⁺⁻BF₃	NaN₃	CuBr	10 mol %	DMEDA (20 mol %), Cs₂CO₃	DMSO	90	12	(91)		247

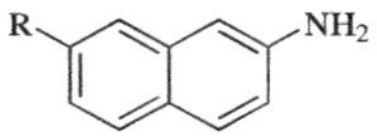

NH₃ (aq)

Cu₂O (30 mol %),
HOCH₂CH₂OH/co-solvent (1:1),
rt, 24 h

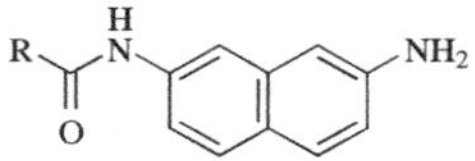

R	Co-solvent	
i-Pr	none	(65)
	BrCMe₂	DME	(0)
CH(Me)(CH₂)₃Me	DME	(87)
n-C₆H₁₃	DME	(53)
Ph	none	(23)
Bn	DME	(10)

252

C₁₂

BocNH₂

CuI (5 mol %), DMEDA (10 mol %),
K₂CO₃, toluene, 110°, 18 h

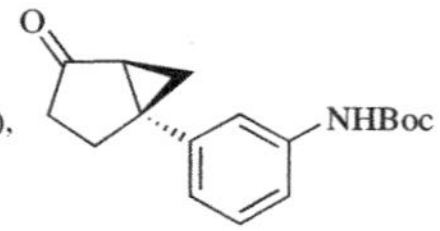

(45)

253

84

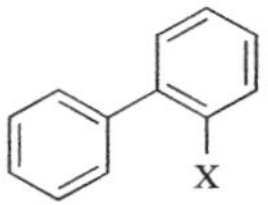

See table.

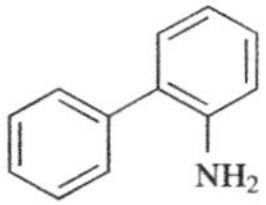

X	N-Nucleophile	Catalyst	x	Additive	Solvent(s)	Temp (°)	Time (h)		
Br	NH$_3$ (aq)	Cu$_2$O	5 mol %	none	H$_2$O/NMP (1:1)	90	48	(75)	114
Br	NH$_3$	3	5 mol %	K$_2$CO$_3$	MeOH/NMP	90	24	(81)	112
Cl	NaN$_3$	CuI	1 eq	DMEDA (1 eq)	DMSO	100	18	(73)	156

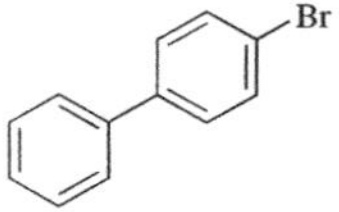

NH$_3$ (aq)

Catalyst (x mol %)

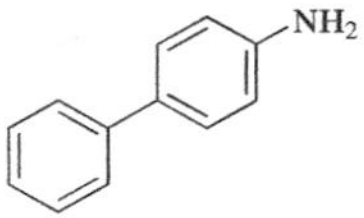

N-Nucleophile	Catalyst	x	Additive(s)	Solvent(s)	Temp (°)	Time (h)		
NH$_3$ (aq)	CuI	20 mol %	L-4-HOPro (40 mol %), K$_2$CO$_3$	DMSO	50	24	(91)	133
NH$_3$ (aq)	Cu(acac)	10 mol %	TMHD (40 mol %), Cs$_2$CO$_3$	DMF	90	36	(98)	113
NaN$_3$	Cu$_2$O	1 eq	L-Pro (1 eq)	DMSO	100	14	(82)	156
N$_3$TMS	Cu	2 eq	none	H$_2$N(CH$_2$)$_2$OH/DMA	95	24	(95)	157

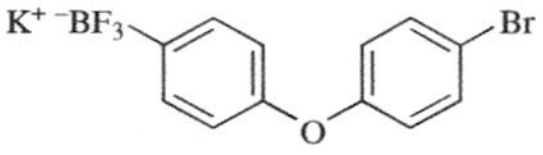

K$^+$ $^-$BF$_3$ — aryl

NaN$_3$

CuBr (10 mol %), DMEDA (20 mol %), Cs$_2$CO$_3$, DMSO, 90°, 12 h

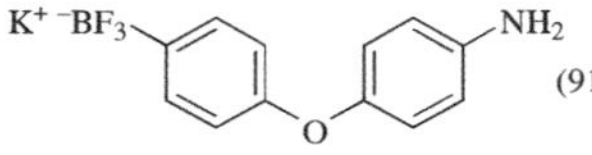

(91)   247

C$_{13}$

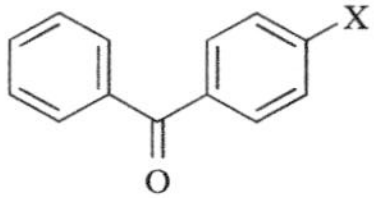

benzophenone with X

NH$_3$

CuI (10 mol %), DMG (20 mol %), TBPM, DMSO, rt, 24 h

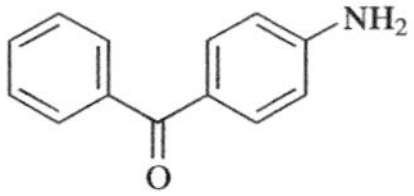

X	
Br	(71)
I	(86)

61

TABLE 1A. PREPARATION OF PRIMARY ARYL AMINES (*Continued*)

Aryl Halide	Nitrogen Nucleophile	Conditions	Product(s) and Yield(s) (%)	Refs.

*Please refer to the charts preceding the tables for structures indicated by the **bold** numbers.*

C₁₅

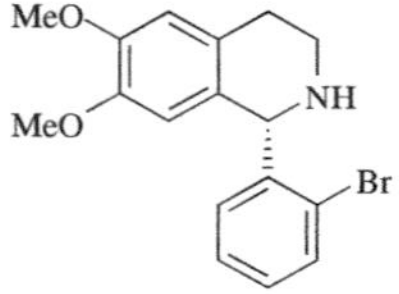

NH₃

Cu (40 mol %),
CuCl (25 mol %),
70°, 45 bar, 120 h

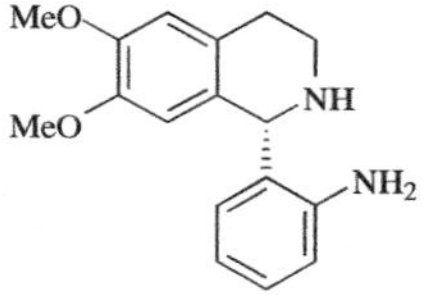

(82)

254

C₂₄

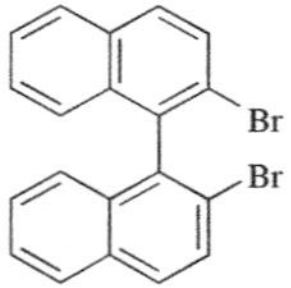

NaN₃

CuI (1 eq), DMEDA (1 eq),
DMSO, 100°, 48 h

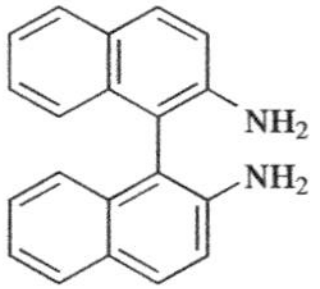

(48)

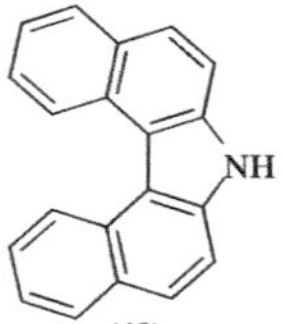

+

(48)

156

86

TABLE 1B. PREPARATION OF PRIMARY HETEROARYL AMINES

*Please refer to the charts preceding the tables for structures indicated by the **bold** numbers.*

Heteroaryl Halide	Nitrogen Nucleophile	Conditions	Product(s) and Yield(s) (%)	Refs.

C₃

(2-bromothiazole) | NH₃ | Cu₂O (10 mol %), HOCH₂CH₂OH, 80°, 16 h | (thiazol-2-amine) (94) | 193

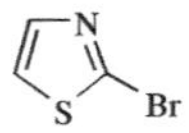

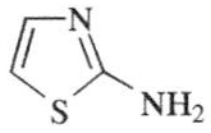

C₄

(3-iodofuran) | H₂NR | CuI (x mol %) | (NHR furan) | 83

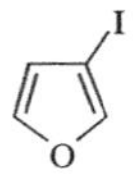

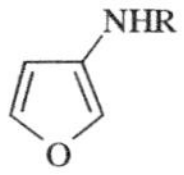

R	x	Additives	Solvent	Temp (°)	Time (h)
Boc	10	DMEDA (10 mol %), K₂CO₃	—	110	24 (97)
Cbz	5	phen (20 mol %), Cs₂CO₃	DMF	80	4 (27)

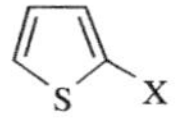

(2-halothiophene) | NH₃ (y) | Catalyst (x mol %) | (NH₂ thiophene) |

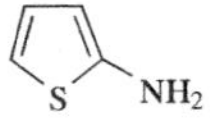

X	y	Catalyst	x	Additive(s)	Solvent(s)	Temp (°)	Time (h)	
I	aq	CuI	20	L-Pro (40 mol %), K₂CO₃	DMSO	80	12 (50)	60
Br	—	**3**	5	K₂CO₃	MeOH/NMP	120	48 (88)	112

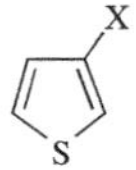

(3-halothiophene) | NH₂R | Catalyst (x mol %) | (NHR thiophene) |

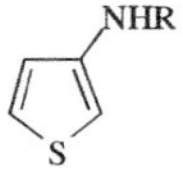

X	R	Catalyst	x	Additive(s)	Solvent(s)	Temp (°)	Time (h)	
I	H (aq)	CuBr	5	L3 (10 mol %), K₃PO₄	DMSO	rt	24 (79)	245
I	H (aq)	CuI	20	L-Pro (40 mol %), K₂CO₃	DMSO	80	12 (70)	60
Br	H	**3**	5	K₂CO₃	MeOH/NMP	120	48 (84)	112
I	Boc	CuI	10	DMEDA (10 mol %), K₂CO₃	—	110	24 (25)	83

Heteroaryl Halide	Nitrogen Nucleophile	Conditions	Product(s) and Yield(s) (%)	Refs.

*Please refer to the charts preceding the tables for structures indicated by the **bold** numbers.*

C4

| 2-bromopyrimidine | NH$_3$ (aq) | Cu$_2$O (5 mol %), DMEDA (10 mol %), K$_2$CO$_3$, HOCH$_2$CH$_2$OH, 60°, 16 h | 2-aminopyrimidine (85) | 255 |

C5

NH$_3$ (y) | Catalyst (x mol %)

Products: **I** (2-aminopyridine) + **II** (2-(2-hydroxyethoxy)pyridine)

X	y	Catalyst	x	Additive(s)	Solvent	Temp (°)	Time (h)	I	II	Refs.
I	aq	CuBr	5	**L3** (10 mol %), K$_3$PO$_4$	DMSO	rt	24	(89)	(—)	245
Cl	—	Cu$_2$O	10	none	HOCH$_2$CH$_2$OH	80	16	(0)	(0)	193
Br	—	Cu$_2$O	10	none	HOCH$_2$CH$_2$OH	80	16	(52)	(13)	193
Br	aq	Cu(acac)$_2$	10	acac (40 mol %), Cs$_2$CO$_3$	DMF	90	36	(82)	(—)	113
Br	aq	**2**	5	NaOH	H$_2$O	120	12	(92)	(—)	238

NH$_3$ (y) | Cu$_2$O (x mol %), HOCH$_2$CH$_2$OH

Products: **I** (R-2-aminopyridine) + **II**

R	X	y	x	Additive(s)	Temp (°)	Time (h)	I	II	Refs.
5-Br	Br	—	10	none	80	16	(56)	(6)	193
5-NO$_2$	Cl	—	10	none	80	16	(85)	(0)	193
5-NO$_2$	Br	—	10	none	80	16	(99)	(0)	193
6-Br	Br	aq	5	DMEDA (10 mol %), K$_2$CO$_3$	60	16	(68)	(0)	255
6-B(OH)$_2$	Br	—	2	none	100	24	(69)	(0)	256
6-MeO	Br	—	2	none	100	24	(82)	(0)	256
6-MeO	Br	—	10	none	80	16	(69)	(6)	193
6-MeO	Br	aq	5	DMEDA (10 mol %), K$_2$CO$_3$	80	16	(94)	(0)	255

X	N-Nucleophile	Catalyst(s)	x	Additive(s)	Solvent	Temp (°)	Time (h)	I	II	
Br	NH$_3$	Cu, CuI	20 mol %, 1 eq	none	HOCH$_2$CH$_2$OH	50	10	(65)	(0)	237
I	NH$_3$ (aq)	CuBr	5 mol %	**L3** (10 mol %), K$_3$PO$_4$	DMSO	rt	24	(79)	(—)	245
I	NH$_3$ (aq)	CuI	10 mol %	Fe$_2$O$_3$ (10 mol %), NaOH	EtOH	90	16	(94)	(—)	246
Br	NH$_3$ (aq)	Cu$_2$O	5 mol %	DMEDA (10 mol %), K$_2$CO$_3$	HOCH$_2$CH$_2$OH	80	16	(80)	(0)	255
Br	NH$_3$ (aq)	Cu$_2$O	10 mol %	none	HOCH$_2$CH$_2$OH	80	16	(81)	(4)	193
Br	NH$_3$ (aq)	Cu(acac)$_2$	10 mol %	acac (40 mol %), Cs$_2$CO$_3$	DMF	90	36	(84)	(—)	113
Br	NH$_3$ (aq)	**2**	5 mol %	NaOH	H$_2$O	120	12	(93)	(—)	238
I	NH$_3$ (aq)	**2**	5 mol %	NaOH	H$_2$O	120	12	(94)	(—)	238
I	NaN$_3$	**Cu**	10 mol %	piperidine-2-carboxylic acid (30 mol %), ascorbic acid (20 mol %)	EtOH	100	3	(76)	(—)	250

R	X	y	Catalyst(s)	x	Additive(s)	Solvent	Temp (°)	Time (h)	I	II	
Cl	Br	—	Cu, CuI	20 mol %, 1 eq	none	HOCH$_2$CH$_2$OH	50	10	(77)		237
Cl	I	aq	CuI	20 mol %	L-Pro (40 mol %), K$_2$CO$_3$	DMSO	80	12	(90)		60
MeO	Br	aq	CuBr	10 mol %	**L3** (20 mol %), K$_3$PO$_4$	DMSO	110	24	(86)		245

Cu$_2$O (x mol %), 80°

y	x	Solvent(s)	Time (h)	I	II	
aq	5	H$_2$O/NMP (1:1)	24	(74)	(—)	114
—	10	HOCH$_2$CH$_2$OH	16	(79)	(2)	193

89

TABLE 1B. PREPARATION OF PRIMARY HETEROARYL AMINES (Continued)

*Please refer to the charts preceding the tables for structures indicated by the **bold** numbers.*

Heteroaryl Halide	Nitrogen Nucleophile	Conditions	Product(s) and Yield(s) (%)	Refs.

C5

| | NaN3 | Cu (10 mol %), piperidine-2-carboxylic acid (30 mol %), ascorbic acid (20 mol %), EtOH, 100°, 3 h | (67) | 250 |

C6

| | NH3 (y) | Catalyst (x mol %) | I + II | |

R	X	y	Catalyst	x	Additive(s)	Solvent	Temp (°)	Time (h)	I	II	Refs.
3-Me	Br	aq	Cu2O	5	DMEDA (10 mol %), K2CO3	HOCH2CH2OH	60	16	(98)	(0)	255
5-CF3	Br	aq	Cu2O	5	DMEDA (10 mol %), K2CO3	HOCH2CH2OH	60	16	(86)	(0)	255
5-Me	Br	aq	CuBr	10	L3 (20 mol %), K3PO4	DMSO	110	24	(85)	(—)	245
6-Me	I	aq	CuBr	5	L3 (10 mol %), K3PO4	DMSO	rt	24	(81)	(—)	245
6-Me	Br	—	Cu2O	10	none	HOCH2CH2OH	80	16	(60)	(10)	193
6-Me	Br	—	Cu2O	2	none	HOCH2CH2OH	100	24	(62)	(0)	256
6-Me	Br	aq	Cu2O	5	DMEDA (10 mol %), K2CO3	HOCH2CH2OH	60	16	(88)	(0)	255

C7

| | NH3 (y) | Catalyst (x mol %) | | |

y	Catalyst	x	Additive(s)	Solvent(s)	Temp (°)	Time (h)		
aq	CuBr	10	L3 (20 mol %), K3PO4	DMSO	110	24	(81)	245
—	**3**	5	K2CO3	MeOH/NMP	120	48	(80)	112

C8-9

| | NH3 | Cu2O (2 mol %), HOCH2CH2OH, 100°, 24 h | R: i-Pr (66); t-Bu (75) | 256 |

C8

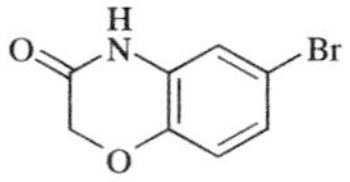

| | NaN3 | CuSO$_4$•5H$_2$O (20 mol %), L-Pro (20 mol %), Na ascorbate, DMSO/H$_2$O (9:1), 70°, 24 h | | | | (68) | 75 |

| | NaN3 | CuSO$_4$•5H$_2$O (20 mol %), L-Pro (20 mol %), Na ascorbate, DMSO/H$_2$O (9:1), 70°, 24 h | | | | (73) | 75 |

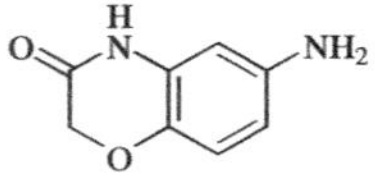

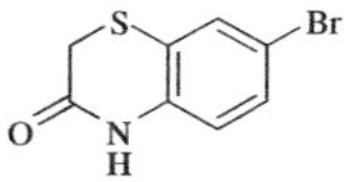

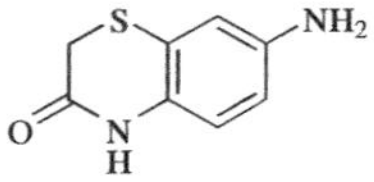

C9

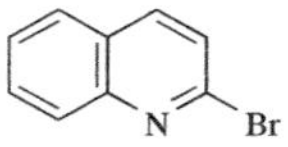

| | NH$_3$ | Cu$_2$O (10 mol %), HOCH$_2$CH$_2$OH, 80°, 16 h | | | | (76) + (8) | 112, 193 |

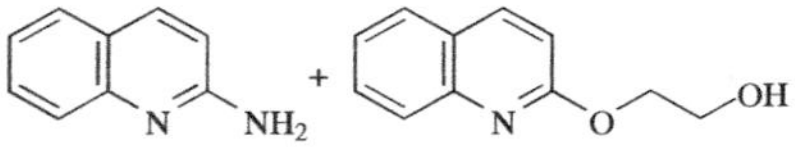

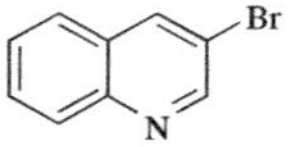

Catalyst (x mol %)

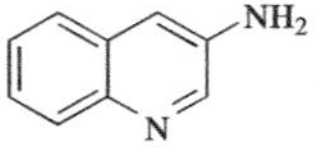

N-Nucleophile	Catalyst	x	Additive(s)	Solvent(s)	Temp (°)	Temp (°)		
NH$_3$	3	5	K$_2$CO$_3$	MeOH/NMP	120	120	(80)	112
NaN$_3$	Cu	10	piperidine-2-carboxylic acid (30 mol %), ascorbic acid (20 mol %)	EtOH	100	100	(82)	250

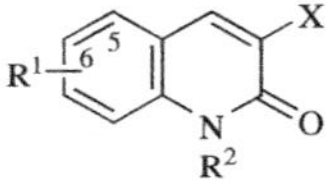

See table.

NaN$_3$, Cu (10 mol %), piperidine 2-carboxylic acid (30 mol %), ascorbic acid (20 mol %), EtOH, 100°, 3 h

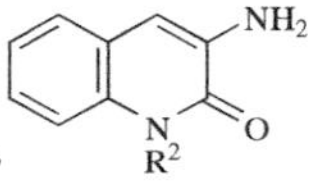

R^1	R^2	X	
H	H	Br	(50)
H	Me	Br	(98)
H	Et	I	(87)
H	EtO$_2$CCH$_2$	Br	(81)
H	PMB	Br	(90)
5-Br,	Me	Br	(65)
6-Br	Me	Br	(56)
6-MeO	Me	Br	(98)

250

TABLE 1B. PREPARATION OF PRIMARY HETEROARYL AMINES (*Continued*)

Heteroaryl Halide	Nitrogen Nucleophile	Conditions	Product(s) and Yield(s) (%)	Refs.
*Please refer to the charts preceding the tables for structures indicated by the **bold** numbers.*				
C_9				
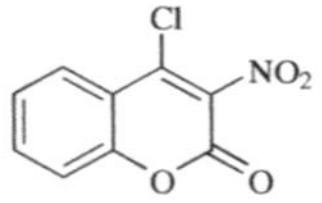	NaN_3	$CuSO_4$•$5H_2O$ (20 mol %), L-Pro (20 mol %), Na ascorbate, DMSO/H_2O (9:1), 70°, 24 h	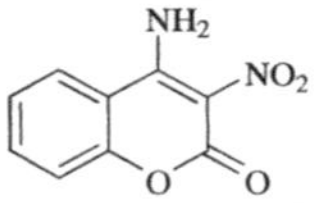 (76)	75
C_{9-10}				
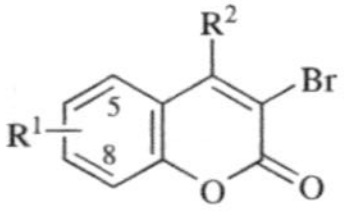	NaN_3	Cu (10 mol %), piperidine-2-carboxylic acid (30 mol %), ascorbic acid (20 mol %), EtOH, 100°, 3 h	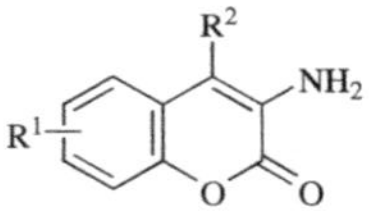 R^1 / R^2 H / H (73) H / Ph (96) 6-F / Ph (43) 6-Cl / H (42) 6-MeO / H (72) 8-MeO / H (63) 8-Me / H (70) 7-MeO, 8-Me / Me (52)	250
C_{10}				
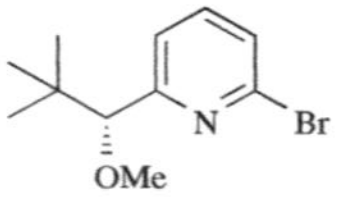	NH_3	Cu_2O (2 mol %), $HOCH_2CH_2OH$, 100°, 24 h	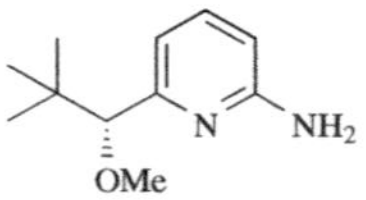 (68)	256
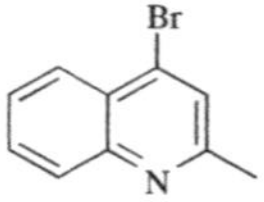	NH_3 (aq)	Cu_2O (5 mol %), H_2O/NMP (1:1), 90°, 15 h	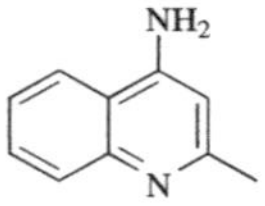 (81)	114
C_{11}				
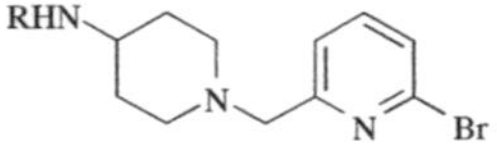	NH_3	Cu_2O (10 mol %), $HOCH_2CH_2OH$, 80°, 16 h	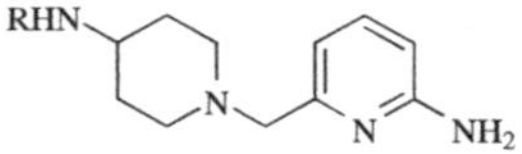 R Ac (91) Boc (85)	193

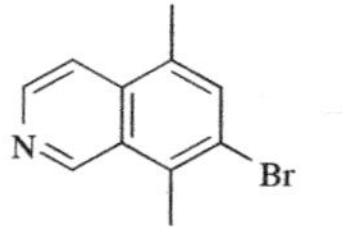

C₁₃		NH₃	(76)	Cu (40 mol %), CuCl (20 mol %), 195°, 24 h	257
		NaN₃	(64)	Cu (10 mol %), piperidine-2-carboxylic acid (30 mol %), ascorbic acid (20 mol %), EtOH. 100°, 3 h	250
C₁₅		NH₃	(68)	Cu₂O (2 mol %), HOCH₂CH₂OH. 100°, 24 h	256

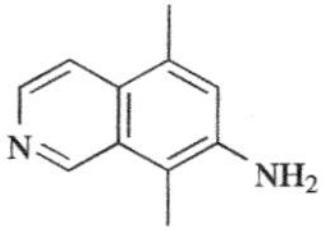

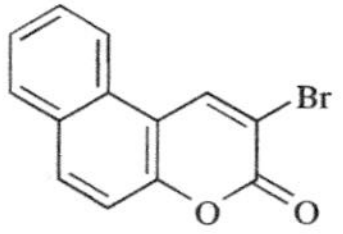

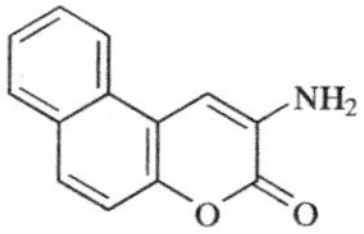

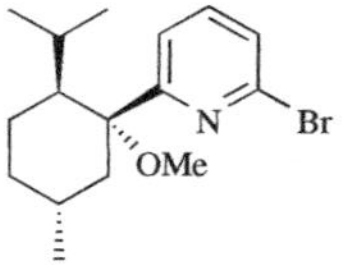

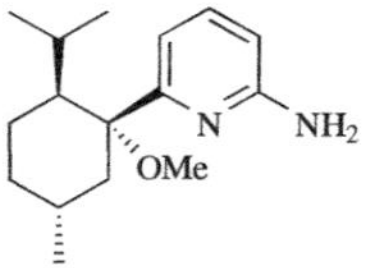

93

TABLE 2A. N-ARYLATION OF PRIMARY ALKYL AMINES

Nitrogen Nucleophile	Aryl Halide	Conditions	Product(s) and Yield(s) (%)	Refs.

*Please refer to the charts preceding the tables for structures indicated by the **bold** numbers.*

C₁

MeNH₂

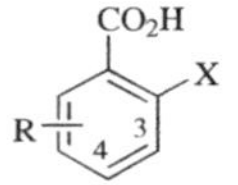

Aryl Halide: CO_2H, X, R (positions 3, 4)

Product: CO_2H, NHMe, R

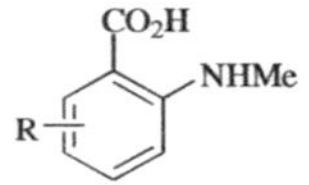

N-Nucleophile	R	X	Catalyst	x	Additive(s)	Solvent	Temp	Time		Refs.
free base	H	Cl	Cu	7.5	K₂CO₃	H₂O	))) (20 kHz), rt	20 min	(78)	241
free base	H	Cl	CuI	10	binol (20 mol %), K₃PO₄	DMF	rt	16 h	(95)	258
free base	H	I	CuI	10	binol (20 mol %), K₃PO₄	DMF	rt	16 h	(95)	258
HCl salt	H	Br	CuI	10	binol (20 mol %), K₃PO₄	DMF	rt	16 h	(96)	258
free base	5-Cl	Br	CuI	10	binol (20 mol %), K₃PO₄	DMF	rt	16 h	(84)	258
free base	5-Br	Br	CuI	10	binol (20 mol %), K₃PO₄	DMF	rt	16 h	(66)	258
free base	3-O₂N	Br	CuI	10	binol (20 mol %), K₃PO₄	DMF	rt	16 h	(68)	258
free base	4-O₂N	Br	CuI	10	binol (20 mol %), K₃PO₄	DMF	rt	16 h	(77)	258

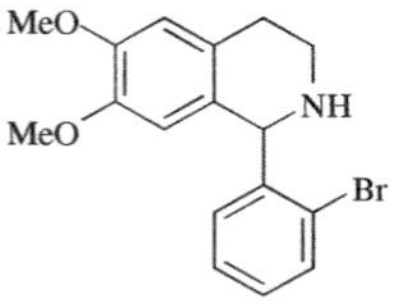

Aryl Halide: MeO, MeO, NH, Br

Conditions: Cu (50 mol %), CuCl (25 mol %), 70°, 120 h

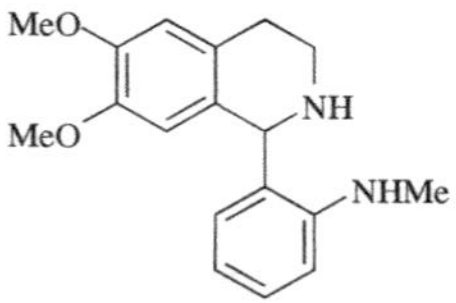

Product: MeO, MeO, NH, NHMe (80) — Refs. 254

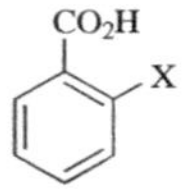

C₂

EtNH₂

Product: CO_2H, NHEt

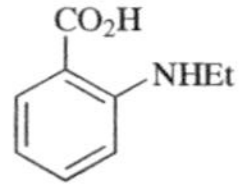

N-Nucleophile	X	Catalyst	x	Additive	Solvent	Temp	Time		Refs.
free base	Cl	Cu	7.5	K₂CO₃	H₂O	))) (20 kHz), rt	20 min	(74)	241
HCl salt	Br	CuI	10	binol (20 mol %), K₃PO₄	DMF	rt	16 h	(86)	258

94

H2N—OR^1 + R^2—X (aryl) → R^2-aryl—NH—OR^1

R^1	R^2	X	Catalyst	x	Additive(s)	Solvent(s)	Temp (°)	Time (h)		
H	H	I	CuBr	5	L10 (5 mol %), Cs$_2$CO$_3$	DMSO	80	30	(86)	259
H	H	I	CuI	10	L-Pro (20 mol %), TBAA	DMF	rt	24	(96)	61
H	H	I	CuI	10	K$_3$PO$_4$•H$_2$O	Me$_2$NCH$_2$CH$_2$OH/H$_2$O	90	—	(90)	260
H	4-Cl	I	CuI	10	DMG (20 mol %), TBPE	NMP	0	5	(88)	61
H	3-MeO	I	CuI	10	L-Pro (20 mol %), K$_2$CO$_3$	DMSO	80	29	(96)	47
H	4-MeO	I	CuBr	5	L10 (6 mol %), Cs$_2$CO$_3$	DMSO	80	30	(82)	259
H	2,4-(MeO)$_2$	I	CuI	10	L-Pro (20 mol %), K$_2$CO$_3$	DMSO	90	40	(93)	47
H	4-Me	I	CuI	5	Cs$_2$CO$_3$	DMF	rt	12	(92)	196
H	4-Me	I	CuBr	5	L10 (5 mol %), Cs$_2$CO$_3$	DMSO	55	24	(80)	259
H	4-Me	I	CuI	10	L11 (20 mol %), Cs$_2$CO$_3$	DMF	rt	8	(98)	261
H	4-CHO	I	CuI	10	L-Pro (20 mol %), K$_2$CO$_3$	DMSO	80	27	(91)	47
Me	H	I	Cu(acac)$_2$	10	Fe$_2$O$_3$ (20 mol %), Cs$_2$CO$_3$	DMSO/H$_2$O (1:1)	MW, 150	0.5	(90)	262
Me	4-Ac	Br	CuI	4	L13 (20 mol %), K$_3$PO$_4$	DMF	90	18	(81)	120

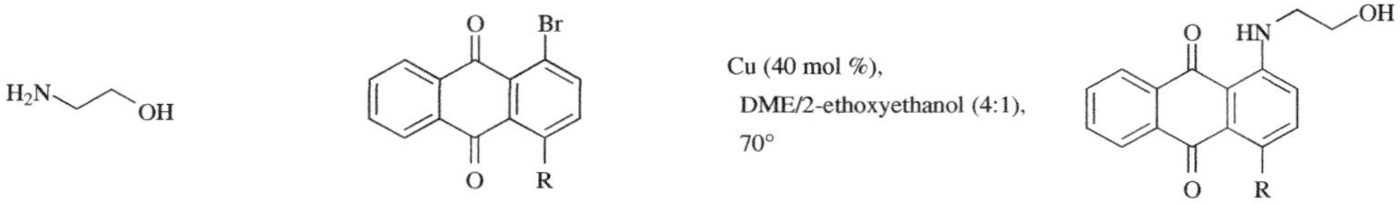

H2N—OH + (Br, R anthraquinone) → (HN—OH, R anthraquinone)

Cu (40 mol %),
DME/2-ethoxyethanol (4:1),
70°

R	
H	(43)
NH$_2$	(49)

263

TABLE 2A. N-ARYLATION OF PRIMARY ALKYL AMINES (Continued)

Nitrogen Nucleophile	Aryl Halide	Conditions	Product(s) and Yield(s) (%)	Refs.

*Please refer to the charts preceding the tables for structures indicated by the **bold** numbers.*

C_{2-12}

Nitrogen Nucleophile: $H_2N{\sim}R$

Aryl Halide: (4-Br-phenyl with OH and N-piperidine-Bn side chain)

Conditions: CuI (10 mol %), K_3PO_4, DMF, 24 h

Refs.: 198

R	Additive	Temp (°)		R	Additive	Temp (°)	
H_2N	none	90	(86)	$BocHN(CH_2)_4$	**L19** (20 mol %)	130	(0)
HO	**L19** (20 mol %)	90	(5)	$HO(CH_2)_4$	**L19** (20 mol %)	130	(51)
Me	**L19** (20 mol %)	90	(53)	$H_2N(CH_2)_6$	**L19** (20 mol %)	130	(58)
$H_2N(CH_2)_2$	**L19** (20 mol %)	90	(61)	$H_2N(CH_2)_8$	**L19** (20 mol %)	130	(57)
n-Pr	**L19** (20 mol %)	90	(50)	$H_2N(CH_2)_{10}$	**L19** (20 mol %)	130	(55)
$H_2N(CH_2)_4$	**L19** (20 mol %)	130	(62)				

C_2

Nitrogen Nucleophile: $H_2N{\sim}CO_2H$

Aryl Halide: R–(ring)–X

Conditions: Catalyst (x mol %)

Product: R–(ring)–NH–CH_2–CO_2H

R	X	Catalyst	x	Additive	Solvent(s)	Temp (°)	Time		Refs.
H	Br	CuCl	20	TBAH (aq)	MeCN	reflux	24 h	(41)	264
H	Br	CuI	10	$K_3PO_4{\cdot}H_2O$	$Me_2NCH_2CH_2OH/H_2O$	90	—	(62)	260
H	I	CuI	10	$K_3PO_4{\cdot}H_2O$	$Me_2NCH_2CH_2OH/H_2O$	90	—	(87)	260
H	Br	CuI	10	KI, K_2CO_3	H_2O	MW, 185	40 min	(66)	25
H	Br	CuI	10	K_2CO_3	DMF	MW, 140	20 min	(0)	265
$2\text{-}HO_2C$	Cl	Cu	7.5	K_2CO_3	H_2O	))) (20 kHz), rt	20 min	(87)	241
$3\text{-}HO_2C$	I	CuI	10	$K_3PO_4{\cdot}H_2O$	$Me_2NCH_2CH_2OH/H_2O$	90	—	(90)	260

C_3

Nitrogen Nucleophile: $H_2N{\sim}$ (propyl)

Aryl Halide: X–phenyl

Conditions: Cu_2O (5 mol %), NMP, 24 h

Product: phenyl–NH–propyl

Refs.: 266

X	Temp (°)	
Cl	100	(92)
Br	80	(80)
I	80	(92)

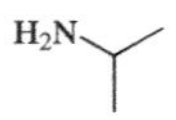

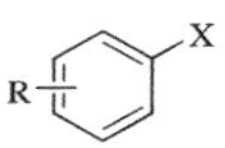

R—C₆H₄—X → Catalyst (x mol %) → R—C₆H₄—NH(propyl)

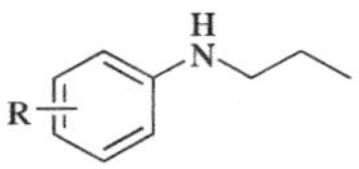

R	X	Catalyst	x	Additive(s)	Solvent	Temp (°)	Time		Ref
4-Br	Br	CuBr	5	L17 (20 mol %), TBAB, NaOH	H₂O	100	24 h	(81)	235
3-H₂N	Br	Cu₂O	5	none	NMP	110	24 h	(93)	266
3-O₂N	Cl	Cu₂O	5	none	NMP	100	24 h	(69)	266
3-O₂N	Br	Cu₂O	5	none	NMP	80	24 h	(88)	266
4-MeO	Br	Cu₂O	5	none	NMP	110	24 h	(81)	266
4-MeO	Br	CuO	25	L17 (50 mol %), TBAB, KOH	H₂O	MW (100 W), 130	5 min	(55)	267
4-MeS	Br	Cu₂O	5	none	NMP	110	24 h	(76)	266
3-Me	Cl	Cu₂O	5	NaOt-Bu	NMP	110	24 h	(87)	266
2-HO₂C	Cl	Cu	7.5	K₂CO₃	H₂O	))) (20 kHz), rt	20 min	(76)	241
2-HO₂C	Br	CuI	10	binol (20 mol %), K₃PO₄	DMF	rt	16 h	(68)	258
3-EtO₂C	Br	Cu₂O	5	none	NMP	80	24 h	(91)	266
1-Np	I	CuBr	5	L17 (20 mol %), TBAB, NaOH	H₂O	100	24 h	(86)	235
2-Np	Br	Cu₂O	5	none	NMP	80	24 h	(94)	266

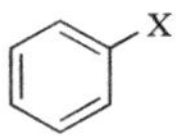

C₆H₅—X → See table. → C₆H₅—NH(isopropyl)

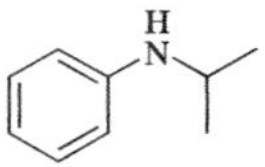

X	Catalyst	Additive(s)	Solvent	Temp (°)	Time		Ref
I	CuBr (2.5 mol %)	L5 (5 mol %), Cs₂CO₃	DMF	89	24 h	(96)	118, 119
Br	Cu(OAc)₂ (1 eq)	DBU	DMSO	MW, 130	10 min	(89)	85
I	Cu(OAc)₂ (1 eq)	DBU	DMSO	MW, 130	10 min	(90)	85

TABLE 2A. N-ARYLATION OF PRIMARY ALKYL AMINES (Continued)

Please refer to the charts preceding the tables for structures indicated by the **bold** numbers.

C₃

Nitrogen Nucleophile	Aryl Halide	Conditions	Product(s) and Yield(s) (%)	Refs.

H_2N–allyl

Aryl Halide: R-substituted aryl–X, CuI (x amount)

R	X	x	Additive(s)	Solvent	Temp (°)	Time (h)	Product/Yield	Refs.
H	I	10 mol %	L-Pro (20 mol %), K_2CO_3	DMSO	60	14	(80)	192
3-H_2N	I	10 mol %	L-Pro (20 mol %), K_2CO_3	DMSO	90	40	(84)	47
3-H_2N	I	5 mol %	**L16** (20 mol %), Cs_2CO_3	DMF	rt	1	(90)	52
4-H_2N	I	10 mol %	**L11** (20 mol %), Cs_2CO_3	DMF	rt	16	(80)	261
4-O_2N	I	1 eq	CsOAc	DMSO	90	24	(40)	189
2-MeO	I	10 mol %	L-Pro (20 mol %), K_2CO_3	DMSO	90	40	(98)	47
4-NC	I	10 mol %	L-Pro (20 mol %), K_2CO_3	DMSO	80	27	(96)	47
2-HO_2C	Cl	10 mol %	binol (20 mol %), K_3PO_4	DMF	rt	16	(85)	258
2-HO_2C	Br	10 mol %	binol (20 mol %), K_3PO_4	DMF	rt	16	(80)	258
2-HO_2C	I	10 mol %	binol (20 mol %), K_3PO_4	DMF	rt	16	(85)	258
2-HO_2C, 4-Cl	Br	10 mol %	binol (20 mol %), K_3PO_4	DMF	rt	16	(59)	258
2-HO_2C, 4-Br	Br	10 mol %	binol (20 mol %), K_3PO_4	DMF	rt	16	(59)	258
2-Br, 3-O_2N	Br	10 mol %	binol (20 mol %), K_3PO_4	DMF	rt	16	(86)	258
2-Br, 4-O_2N	Br	10 mol %	binol (20 mol %), K_3PO_4	DMF	rt	16	(65)	258

Aryl halide: R^1-substituted aryl with Br and $S(O_2)$–NHR^2 group

CuI (10 mol %), phen (20 mol %), Cs_2CO_3, DMSO, MW, 150°, 11 min

Product: R^1-substituted N-allyl aniline with $S(O_2)$–NHR^2

R^1	R^2	Yield	Refs.
H	Bn	(96)	268
F	$CH_2{=}CHCH_2$	(94)	

H_2N–propargyl

Aryl halide: Br aryl with $S(O_2)$–$NH(CH_2)_3Ph$ and CF_3

CuI (10 mol %), phen (20 mol %), Cs_2CO_3, DMSO, MW, 150°, 11 min

Product: N-propargyl aniline with $S(O_2)$–$NH(CH_2)_3Ph$ and CF_3 (69) — 268

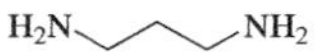

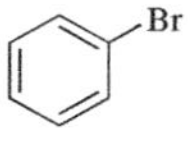

Bromobenzene + H₂N—(CH₂)₃—NH₂

$$\text{Cat. } \mathbf{8}\ (25\ \text{mol \%}),\ K_2CO_3,\ NMP,\ 160°,\ 16\ h$$

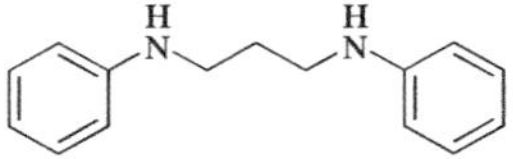

Product (PhNH—(CH₂)₃—NHPh) (43)   269

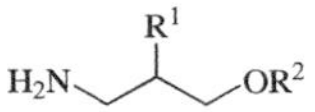

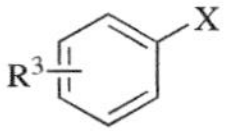

Catalyst (x mol %)

R^3-substituted aryl-X + amino alcohol → R^1/OR^2 aniline product

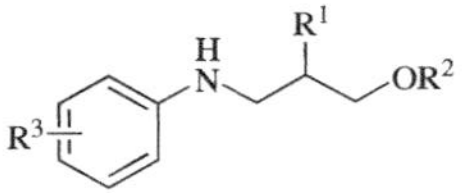

R^1	R^2	R^3	X	Catalyst	x	Additive(s)	Solvent(s)	Temp (°)	Time (h)		
H	H	4-Me	I	CuBr	5	L17 (20 mol %), TBAB. NaOH	H₂O	100	24	(86)	235
H	H	4-Me	I	CuI	5	L16 (20 mol %), Cs₂CO₃	DMF	rt	12	(96)	196
HO	H	H	I	CuI	10	K₃PO₄•H₂O	Me₂NCH₂CH₂OH.H₂O	90	—	(92)	260
H	H	H	Br	CuI	10	binol (20 mol %), K₃PO₄	DMF	rt	16	(71)	258
H	Me	H	I	Cu(acac)₂	10	Fe₂O₃ (20 mol %), Cs₂CO₃	DMSO/H₂O (1:1)	MW, 150	0.5	(93)	262

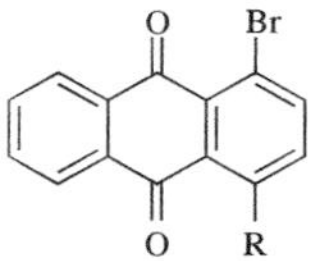

Anthraquinone (Br / R) + H₂N—(CH₂)₃—OH

Cu (40 mol %), DME/EtOCH₂CH₂OH (4:1), 70°

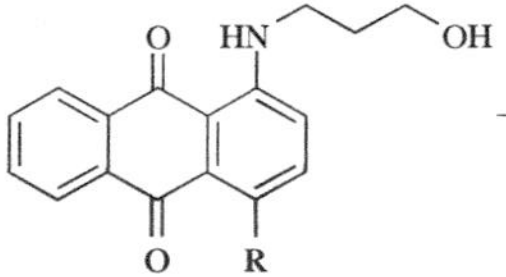

Anthraquinone product (HN / R)

R	
H	(31)
H₂N	(31)

263

Aryl-X + H₂N—(CH₂)₂—CO₂H → product

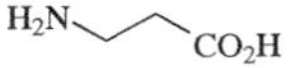

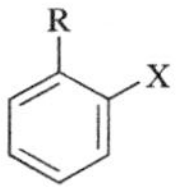

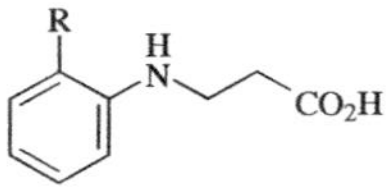

R	X	Additive(s)	Solvent(s)	Temp (°)	Time (h)		
H	I	K₃PO₄•H₂O	Me₂NCH₂CH₂OH/H₂O	90	—	(73)	260
HO₂C	Br	binol (20 mol %), K₃PO₄	DMF	rt	16	(71)	258

Iodobenzene + serine-type amino alcohol (OH / CO₂H)

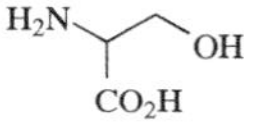

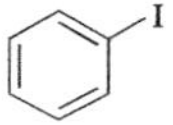

CuI (10 mol %), K₃PO₄•H₂O, Me₂NCH₂CH₂OH/H₂O, 90°

Product (OH / CO₂H aniline)   (49)   260

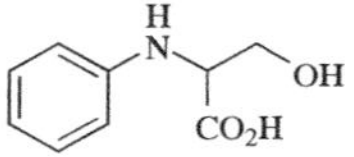

TABLE 2A. *N*-ARYLATION OF PRIMARY ALKYL AMINES (*Continued*)

Nitrogen Nucleophile	Aryl Halide	Conditions	Product(s) and Yield(s) (%)	Refs.

*Please refer to the charts preceding the tables for structures indicated by the **bold** numbers.*

C₃

H₂N–CH(CO₂H) (nitrogen nucleophile structure)

Aryl Halide: R–C₆H₄–X, Catalyst (x mol %)

Product: R–C₆H₄–NH–CH(CO₂H)

Config.	R	X	Catalyst	x	Additive(s)	Solvent(s)	Temp (°)	Time (h)	Yield	Refs.
(S)	H	I	CuBr	20	binol (20 mol %), K₃PO₄	DMF	40	10	(72)	62
(S)	H	Br	CuI	10	K₂CO₃	DMF/H₂O	90	24	(42)	270
(S)	H	Br	CuI	10	K₂CO₃	DMF/H₂O	90	48	(65)	116
(R,S)	H	Br	CuI	10	K₃PO₄•H₂O	Me₂NCH₂CH₂OH/H₂O	90	—	(88)	260
(R,S)	H	I	CuI	10	K₃PO₄•H₂O	Me₂NCH₂CH₂OH/H₂O	90	—	(85)	260
(S)	H	Cl	Cu₂O	5	NaO*t*-Bu	NMP	110	24	(76)	266
(S)	4-Cl	I	CuBr	5	**L17** (20 mol %), TBAB, NaOH	H₂O	100	24	(91)	235
(S)	4-Cl	I	CuI	10	**L8.** (20 mol %), K₃PO₄	DMF	30	3	(90)	271
(S)	2-O₂N	I	CuI	10	**L8.** (20 mol %), K₃PO₄	DMF	30	3	(68)	271

Nitrogen Nucleophile	Aryl Halide	Conditions	Product(s) and Yield(s) (%)	Refs.
H₂N–CH(CO₂H)	(4-methyl)C₆H₄–Br	1. CuI (10 mol %), K₂CO₃, DMF, 90°; 2. PhCOCH₂Br, Et₃N, EtOAc	N–CO₂CH₂COPh (36)	272

C₄

H₂N–CH₂CH₂CH₂CH₃ (nitrogen nucleophile), aryl iodide C₆H₅–I

Catalyst (x amount)

Product: Ph–NH–CH₂CH₂CH₂CH₃

Catalyst	x	Additives(s)	Solvent(s)	Temp (°)	Time (h)	Yield	Refs.
Cu	10 mol %	K₃PO₄•H₂O	Me₂NCH₂CH₂OH	70	14	(94)	195
CuBr	2.5 mol %	**L5** (5 mol %), Cs₂CO₃	DMF	89	24	(96)	118, 119
CuI	1 eq	CsOAc	DMSO	90	24	(96)	189
CuI	10 mol %	**L6** (20 mol %), Cs₂CO₃	DMF	rt	24	(13)	273
CuO nanoparticles	1.26 mol %	KOH	DMSO	110	13	(91)	224
Cu(acac)₂	10 mol %	Fe₂O₃ (20 mol %), Cs₂CO₃	DMSO/H₂O (1:1)	MW, 150	0.5	(76)	262

Catalyst (x amount)

R	X	Catalyst	x	Additive(s)	Solvent(s)	Temp (°)	Time		
2-F	I	CuI	10 mol %	CsOAc	DMSO	90	24 h	(37)	274
2-F	I	CuI	1 eq	CsOAc	DMF	90	24 h	(20)	189
4-Br	I	CuI	10 mol %	CsOAc	DMSO	90	24 h	(76)	274
4-Br	I	CuI	1 eq	CsOAc	DMF	rt	24 h	(89)^a	189
3-I	I	CuI	1 eq	CsOAc	DMSO	rt	24 h	(77)	189
3-I	I	CuI	10 mol %	CsOAc	DMSO	90	24 h	(50)	274
4-I	I	CuI	1 eq	CsOAc	DMSO	rt	24 h	(74)	189
4-H$_2$N	I	Cu	10 mol %	K$_3$PO$_4$•H$_2$O	Me$_2$NCH$_2$CH$_2$OH	85	11 h	(66)	195
2-O$_2$N	I	CuI	1 eq	CsOAc	DMF	90	24 h	(8)	189
2-O$_2$N	I	CuI	10 mol %	CsOAc	DMSO	90	24 h	(78)	274
3-O$_2$N	I	CuBr	2.5 mol %	L5 (5 mol %), Cs$_2$CO$_3$	DMF	89	24 h	(96)	118
3-O$_2$N	I	CuI	10 mol %	CsOAc	DMSO	90	24 h	(95)	274
4-O$_2$N	I	CuI	10 mol %	CsOAc	DMSO	90	24 h	(99)	274
4-O$_2$N	I	CuI	1 eq	CsOAc	DMF	90	24 h	(89)	189
2-MeO	I	CuI	10 mol %	CsOAc	DMSO	90	24 h	(49)	274
2-MeO	I	CuI	1 eq	CsOAc	DMF	90	24 h	(30)	189
3-MeO	I	Cu	10 mol %	K$_3$PO$_4$•H$_2$O	Me$_2$NCH$_2$CH$_2$OH	70	9 h	(89)	195
3-MeO	I	CuI	1 eq	CsOAc	DMSO	rt	24 h	(80)	189
3-MeO	I	CuI	10 mol %	CsOAc	DMSO	90	24 h	(89)	274

TABLE 2A. N-ARYLATION OF PRIMARY ALKYL AMINES (Continued)

Nitrogen Nucleophile	Aryl Halide	Conditions	Product(s) and Yield(s) (%)	Refs.

Please refer to the charts preceding the tables for structures indicated by the **bold** numbers.

Continued from previous page.

C₄

H₂N⌒⌒⌒ + R⬡—X → Catalyst (x amount) → R⬡—N(H)⌒⌒⌒

R	X	Catalyst	x	Additive(s)	Solvent(s)	Temp (°)	Time		Refs.
4-MeO	I	Cu	10 mol %	$K_3PO_4 \cdot H_2O$	$Me_2NCH_2CH_2OH$	70	9 h	(87)	195
4-MeO	I	CuI	1 eq	CsOAc	DMF	90	24 h	(85)	189
4-MeO	I	CuI	10 mol %	CsOAc	DMSO	90	24 h	(84)	274
4-MeO	Br	CuI	5 mol %	L2 (25 mol %), KOH, TBAB	H_2O	130	5 min	(80)	236
4-MeO	Br	CuO	5 mol %	L14 (50 mol %), cyclohexanone, KOH, TBAB	H_2O	MW (100 W), 130	5 min	(86)	51
4-MeO	Br	CuO	25 mol %	L15 (50 mol %), KOH, TBAB	H_2O	MW (100 W), 130	5 min	(50)	267
2,4-(MeO)₂	I	CuI	10 mol %	L-Pro (20 mol%), K_2CO_3	DMSO	90	40 h	(89)	47
2-Me	I	CuI	1 eq	CsOAc	DMSO	90	24 h	(20)	189
2-Me	I	CuI	10 mol %	CsOAc	DMSO	90	24 h	(21)	274
3-Me	I	CuI	1 eq	CsOAc	DMF	90	24 h	(77)	189
3-Me	I	CuI	10 mol %	CsOAc	DMSO	90	24 h	(99)	274
4-Me	I	Cu	10 mol %	$K_3PO_4 \cdot H_2O$	$Me_2NCH_2CH_2OH$	60	12 h	(93)	195
4-Me	I	CuI	1 eq	CsOAc	DMF	90	24 h	(76)	189
4-Me	I	CuI	10 mol %	CsOAc	DMSO	90	24 h	(86)	274
3-NC–	I	CuBr	2.5 mol %	L5 (5 mol %), Cs_2CO_3	DMF	89	24 h	(96)	118
2-HO₂C	Cl	Cu	7.5 mol %	K_2CO_3	H_2O	))) (20 kHz), rt	20 min	(81)	241
3-HO₂C	I	CuI	10 mol %	$K_3PO_4 \cdot H_2O$	$Me_2NCH_2CH_2OH$	90	—	(73)	260

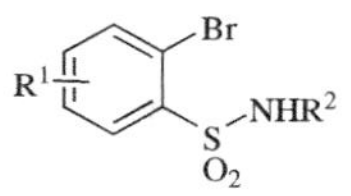

Reaction 1: CuI (10 mol %), phen (20 mol %), Cs$_2$CO$_3$, DMSO, MW, 150°, 11 min

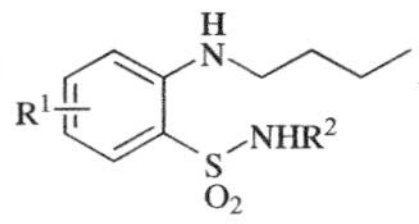

R^1	R^2	
3-F	n-Bu	(92)
4-CF$_3$	allyl	(95)

268

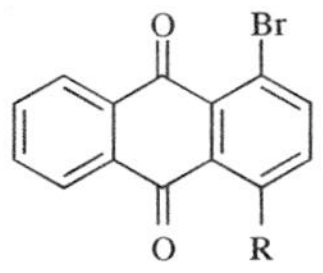

Reaction 2: Cu (40 mol %), DME/EtOCH$_2$CH$_2$OH (4:1), 70°

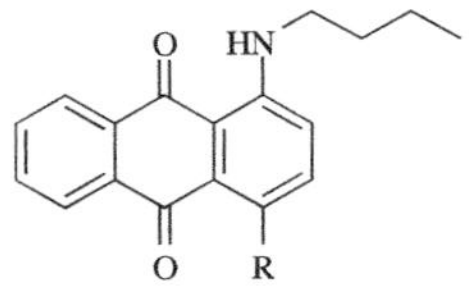

R	
H	(6)
NH$_2$	(9)

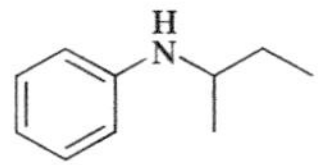

263

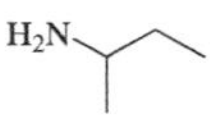

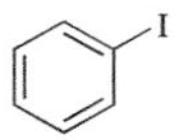

Table 1:

Catalyst	x	Additive(s)	Solvent	Temp (°)	Time (h)	
Cu	5	binol (10 mol %), Cs$_2$CO$_3$	DMSO	90	24	(77)	124
Cu	10	K$_3$PO$_4$•H$_2$O	Me$_2$NCH$_2$CH$_2$OH	80	26	(68)	195
CuBr	2.5	**L5** (5 mol %), Cs$_2$CO$_3$	DMF	90	24	(98)	118, 119
Cu$_2$O	5	**L7** (20 mol %), Cs$_2$CO$_3$	MeCN	80	18	(65)	125

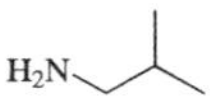

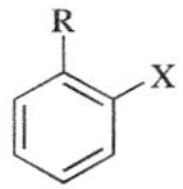

Table 2:

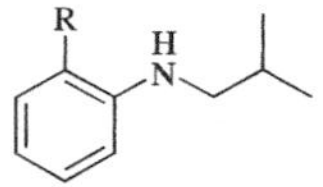

R	X	Catalyst	x	Additive(s)	Solvent	Temp (°)	Time	
H	I	Cu	5	binol (10 mol %), Cs$_2$CO$_3$	DMSO	90	24 h (77)	124
H	I	CuBr	2.5	**L5** (5 mol %), Cs$_2$CO$_3$	DMF	89	24 h (96)	118, 119
H	I	Cu$_2$O	5	**L7** (20 mol %), Cs$_2$CO$_3$	MeCN	80	18 h (80)	125
HO$_2$C	Cl	Cu	7.5	K$_2$CO$_3$	H$_2$O	))) (20 kHz), rt	20 min (80)	241

TABLE 2A. N-ARYLATION OF PRIMARY ALKYL AMINES (Continued)

Nitrogen Nucleophile	Aryl Halide	Conditions	Product(s) and Yield(s) (%)	Refs.

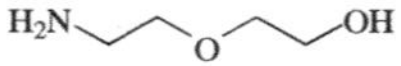

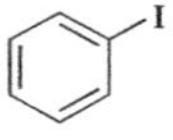

*Please refer to the charts preceding the tables for structures indicated by the **bold** numbers.*

C₄

First entry:

Nitrogen Nucleophile: H_2N-C(CH₃)₃ (tert-butylamine)

Aryl Halide: iodobenzene

Conditions: Cu (10 mol %), K₃PO₄•H₂O, Me₂NCH₂CH₂OH, 90°, 48 h

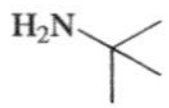

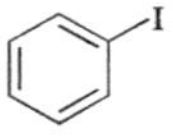

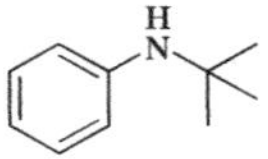

Product: PhNH-C(CH₃)₃ (3)

Refs.: 195

Second entry:

Nitrogen Nucleophile: H_2N~~~NHR¹

Aryl Halide: R²-C₆H₄-I

Conditions: CuI (x mol %)

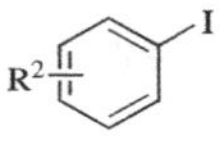

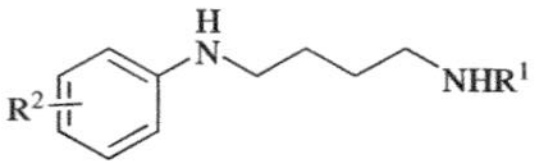

R¹	R²	x	Additive(s)	Solvent	Temp (°)	Time (h)	Product yield
Et	H	10	K₃PO₄•H₂O	Me₂NCH₂CH₂OH	90	—	(86)
Boc	3,5-Me₂	5	L16 (20 mol %), Cs₂CO₃	DMF	rt	4	(98)

Product: R²-C₆H₄-NH~~~NHR¹

Third entry:

Nitrogen Nucleophile: H_2N~O~OH

Aryl Halide: iodobenzene

Conditions: CuI (10 mol %), K₃PO₄•H₂O, Me₂NCH₂CH₂OH, H₂O, 90°

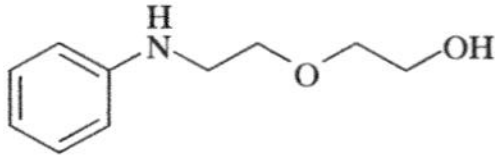

Product: Ph-N(H)~O~OH (77)

Refs.: 260

Fourth entry:

Nitrogen Nucleophile: H_2N~O~OMe

Aryl Halide: (aryl bromide with HO₂C, BocHN, OMe, OMe substituents)

Conditions: CuBr•Me₂S (20 mol %), L13 (40 mol %), K₃PO₄, DMF, 90°, 24 h

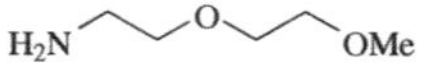

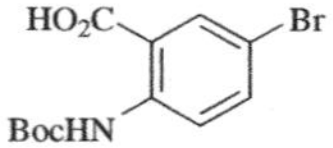

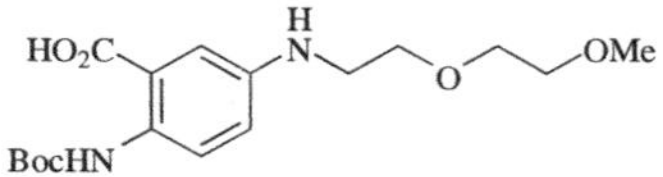

Product: (60)

Refs.: 122

Fifth entry:

Conditions: CuI (40 mol %), L16 (75 mol %), K₃PO₄, DMF, 90°, 24 h

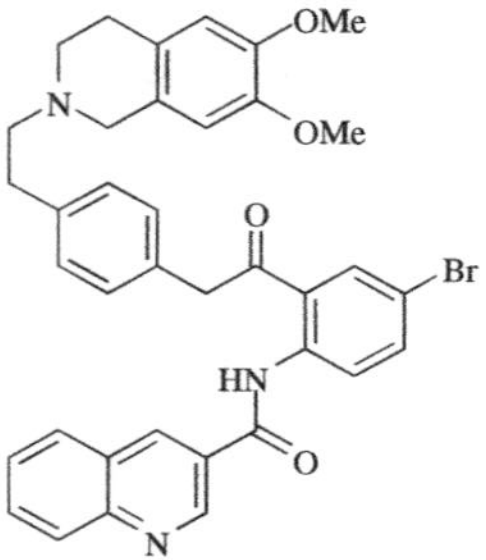

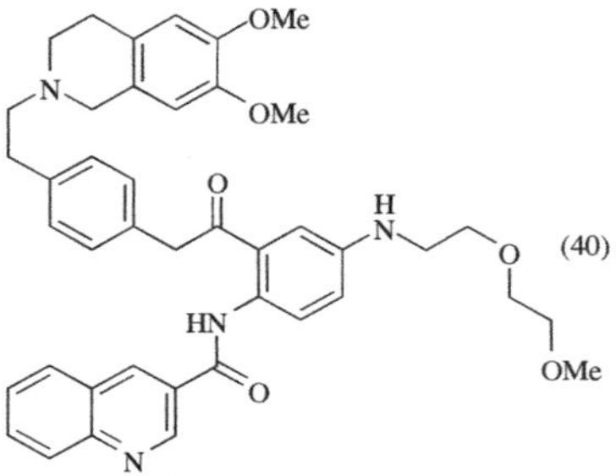

Product: (40)

Refs.: 122

104

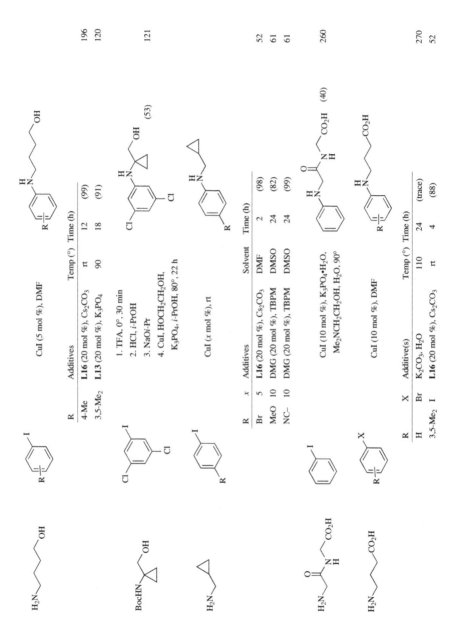

CuI (5 mol %), DMF

R	Additives	Temp (°)	Time (h)	
4-Me	**L16** (20 mol %), Cs_2CO_3	rt	12	(99)	196
3,5-Me_2	**L13** (20 mol %), K_3PO_4	90	18	(91)	120

1. TFA, 0°, 30 min
2. HCl, i-PrOH
3. NaOi-Pr
4. CuI, HOCH$_2$CH$_2$OH, K_3PO_4, i-PrOH, 80°, 22 h

(53)   121

CuI (x mol %), rt

R	x	Additives	Solvent	Time (h)	
Br	5	**L16** (20 mol %), Cs_2CO_3	DMF	2	(98)	52
MeO	10	DMG (20 mol %), TBPM	DMSO	24	(82)	61
NC–	10	DMG (20 mol %), TBPM	DMSO	24	(99)	61

CuI (10 mol %), K_3PO_4•H_2O, $Me_2NCH_2CH_2OH$, H_2O, 90°

CO_2H (40)   260

CuI (10 mol %), DMF

R	X	Additive(s)	Temp (°)	Time (h)	
H	Br	K_2CO_3, H_2O	110	24	(trace)	270
3,5-Me_2	I	**L16** (20 mol %), Cs_2CO_3	rt	4	(88)	52

105

Nitrogen Nucleophile	Aryl Halide	Conditions	Product(s) and Yield(s) (%)	Refs.

*Please refer to the charts preceding the tables for structures indicated by the **bold** numbers.*

C₄

Entry 1

Nitrogen Nucleophile: H₂N–CO₂H, SR¹

Aryl Halide: R²–C₆H₄–X, CuI (x mol %)

Product: (structure) R^2–C₆H₄–N(H)–CO₂H, SR¹

R¹	R²	X	x	Additive(s)	Solvent	Temp (°)	Time (h)		Refs.
Me	H	I	3	Cs₂CO₃	DMSO	90	48	(96)	275
Me	H	Br	10	K₂CO₃, H₂O	DMF	90	48	(76)	116
H	4-Cl	I	10	**L8** (20 mol %), K₃PO₄	DMF	30	3	(76)	271
H	2-O₂N	I	10	**L8** (20 mol %), K₃PO₄	DMF	30	3	(68)	271

Entry 2

Nitrogen Nucleophile: H₂N–CO₂H (ethyl)

Aryl Halide: iodobenzene (I)

Conditions: CuI (10 mol %), K₃PO₄·H₂O, Me₂NCH₂CH₂OH, H₂O, 90°

Product: (structure) Ph–N(H)–CO₂H (80); er 94.5:5.5

Refs: 260

Entry 3

Nitrogen Nucleophile: H₂N–CO₂H, CO₂H

Aryl Halide: R–C₆H₄–X

R	X	er
H	I	(37) 94.5:5.5
4-O₂N	Br	(32) —
2-MeO	Br	(51) —
3-MeO	Br	(86) —
4-MeO	Br	(35) —

Conditions: 1. CuCl (20 mol %), TBAH (aq). MeCN, reflux, 24 h; 2. AcCl, MeOH, reflux, 2 h

Product: (structure) R–C₆H₄–N(H)–CO₂Me, CO₂Me

R	X		er
2-*t*-BuN=CH	Br	(55)	95:5
2-*t*-BuN=CH	I	(32)	—
2-NC	Br	(31)	—
2-NC	I	(22)	—
2-(1,3-dioxolan-2-yl)	Br	(64) R = 2-CHO	91:9

Refs: 264

Entry 4

Nitrogen Nucleophile: H₂N–CO₂H, CO₂*t*-Bu

Aryl Halide: bromobenzene (Br)

Conditions: 1. CuCl (20 mol %), TBAH (aq). MeCN, reflux, 24 h; 2. AcCl, MeOH, reflux, 2 h

Product: (structure) Ph–N(H)–CO₂Me, CO₂*t*-Bu (27) er 97:3

Refs: 264

Entry 5

Nitrogen Nucleophile: H₂N–CO₂R (wavy)

Aryl Halide: X–C₆H₅

Conditions: CuI (10 mol %), K₂CO₃, H₂O, DMF, 100°, 48 h

Product: (structure) Ph–N(H)–CO₂H (wavy)

R	X	Config.	
H	Br	(*R,S*)	(64)
Et	Br	(*R,S*)	(62)
Et	I	(*R*)	(69)

Refs: 270

C_5

R	X	Catalyst(s)	x	Additive(s)	Solvent	Temp (°)	Time	
H	I	CuBr	20	binol (20 mol %), K_3PO_4	DMF	rt	4 h (65)	62
H	Cl	CuI–PAnNF	5	K_2CO_3	DMF	80	24 h (0)	276
H	I	CuI–PAnNF	5	K_2CO_3	DMF	rt	8 h (61)	276
H	Br	CuO	10	$FeCl_3$ (10 mol %), binol (20 mol %), Cs_2CO_3	DMF	100	12 h (76)	277
H	I	CuO	10	$FeCl_3$ (10 mol %), binol (20 mol %), Cs_2CO_3	DMF	80	12 h (84)	277
4-Cl	I	CuBr	20	binol (20 mol %), K_3PO_4	DMF	rt	5 h (85)	62
4-Cl	I	CuI	10	L8 (20 mol %), K_3PO_4	DMF	30	1 h (79)	271
4-Cl	I	CuI–PAnNF	5	K_2CO_3	DMF	rt	6 h (99)	276
4-Br	I	CuBr	20	binol (20 mol %), K_3PO_4	DMF	rt	4 h (80)	62
4-Br	I	CuI–PAnNF	5	K_2CO_3	DMF	rt	5 h (92)	276
4-Br	I	CuO	10	$FeCl_3$ (10 mol %), binol (20 mol %), Cs_2CO_3	DMF	80	12 h (87)	277
2-O_2N	I	CuI–PAnNF	5	K_2CO_3	DMF	rt	2 h (99)	276
3-O_2N	Br	CuO	10	$FeCl_3$ (10 mol %), binol (20 mol %), Cs_2CO_3	DMF	100	24 h (81)	277
4-O_2N	I	CuI–PAnNF	5	K_2CO_3	DMF	rt	4 h (85)	276
4-O_2N	Br	CuO	10	$FeCl_3$ (10 mol %), binol (20 mol %), Cs_2CO_3	DMF	90	12 h (86)	277
4-MeO	I	CuI–PAnNF	5	K_2CO_3	DMF	rt	6 h (99)	276
2-Me	I	CuI–PAnNF	5	K_2CO_3	DMF	rt	6 h (0)	276
4-Me	I	CuBr	20	binol (20 mol %), K_3PO_4	DMF	40	8 h (94)	62
4-Me	I	CuI	10	L8 (20 mol %), K_3PO_4	DMF	30	2 h (73)	271
4-Me	I	CuI–PAnNF	5	K_2CO_3	DMF	rt	9 h (56)	276
4-Me	I	CuO	10	$FeCl_3$ (10 mol %), binol (20 mol %), Cs_2CO_3	DMF	90	12 h (73)	277
2-NC–	I	CuI–PAnNF	5	K_2CO_3	DMF	rt	4 h (75)	276
4-NC–	I	CuI–PAnNF	5	K_2CO_3	DMF	rt	6 h (63)	276
2-HO_2C	Cl	Cu	7.5	K_2CO_3	H_2O	))) (20 kHz), rt	20 min (78)	241
2-HO_2C	Cl	Cu, Cu_2O	10.5	K_2CO_3	$EtOCH_2CH_2OEt$	130	24 h (85)	278
2-HO_2C	Br	Cu, Cu_2O	10.5	K_2CO_3	$EtOCH_2CH_2OEt$	130	24 h (97)	278
3-Ac	I	CuI	5	L16 (20 mol %), Cs_2CO_3	DMF	rt	6 h (92)	52

TABLE 2A. N-ARYLATION OF PRIMARY ALKYL AMINES (Continued)

Nitrogen Nucleophile	Aryl Halide	Conditions	Product(s) and Yield(s) (%)	Refs.

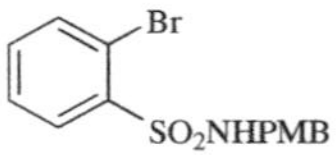

*Please refer to the charts preceding the tables for structures indicated by the **bold** numbers.*

C₅

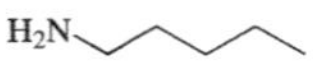

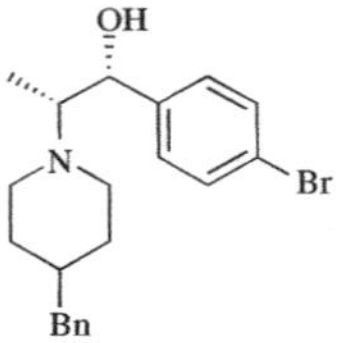

| | | CuI (40 mol %), **L19** (20 mol %), K₃PO₄, DMF, 90°, 24 h | 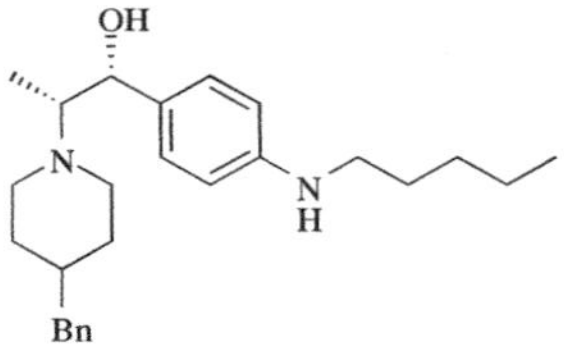 (61) | 198 |
| | | CuI–PAnNF (5 mol %), K₂CO₃, DMF, rt, 10 h | 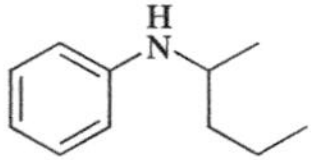 (71) | 276 |

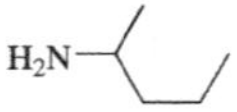

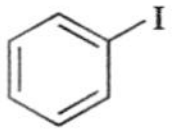

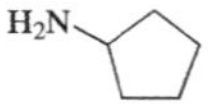

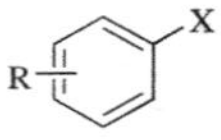

Catalyst (x amount)

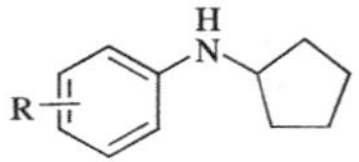

R	X	Catalyst	x	Additive(s)	Solvent	Temp (°)	Time (h)		
H	Cl	Cu/Al–HTB	2.5 mol %	K₂CO₃	DMF	100	16	(0)	228
H	I	CuBr	2.5 mol %	**L5** (5 mol %), Cs₂CO₃	DMF	89	24	(97)	118, 119
H	I	CuI–PAnNF	5 mol %	K₂CO₃	DMF	rt	5	(73)	276
H	I	Cu₂O	5 mol %	**L7** (20 mol %), Cs₂CO₃	MeCN	90	18	(88)	125
2-O₂N	Cl	Cu/Al–HTB	2.5 mol %	K₂CO₃	DMF	100	8	(94)	228
2-O₂N	I	Cu–PAnNF	5 mol %	K₂CO₃	DMF	rt	3	(62)	276
4-O₂N	Cl	Cu/Al–HTB	2.5 mol %	K₂CO₃	DMF	100	8	(90)	228
4-MeO	I	Cu₂O	5 mol %	**L7** (20 mol %), Cs₂CO₃	—	80	18	(64)	125
2-Me	Cl	Cu/Al–HTB	2.5 mol %	K₂CO₃	DMF	100	20	(0)	228
2-NC–	Cl	Cu/Al–HTB	20 wt %	K₂CO₃	—	100	6	(93)	117

| | CuI (10 mol %), phen (20 mol %), Cs₂CO₃, DMSO, MW, 150°, 11 min | 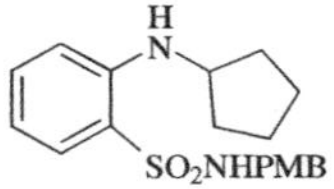 (90) | 268 |

Catalyst (x mol %)

H_2N–CO_2H (valine) + R–X (aryl halide) → product with CO_2H

Config.	R	X	Catalyst	x	Additives(s)	Solvent(s)	Temp (°)	Time		
(S)	H	I	CuBr	20	binol (20 mol %), K_3PO_4	DMF	40	10 h	(61)	62
(S)	H	Cl	CuI	10	K_2CO_3	H_2O/DMF	90	48 h	(3)	116
(S)	H	Br	CuI	10	K_2CO_3	H_2O/DMF	90	48 h	(81)	116
(S)	H	I	CuI	10	K_2CO_3	H_2O/DMF	90	48 h	(83)	116
(S)	H	Br	CuI	10	KI, K_2CO_3	H_2O	MW, 185	40 min	(58)	25
(S)	H	Br	CuI	10	K_2CO_3	DMF	MW, 140	15 min	(80)	265
(S)	H	I	CuI	10	L8 (20 mol %), K_3PO_4	DMF	30	3 h	(86)	271
(R)	H	Br	CuI	10	K_2CO_3	H_2O/DMF	90	48 h	(75)	116
(R,S)	H	Br	CuI	10	K_3PO_4•H_2O	$Me_2CH_2CH_2OH$/H_2O	90	—	(90)	260
(R,S)	H	I	CuI	10	K_3PO_4•H_2O	$Me_2CH_2CH_2OH$/H_2O	90	—	(82)	260
(S)	4-Cl	Br	CuI	10	K_2CO_3	H_2O/DMF	90	48 h	(83)	116
(S)	4-Cl	I	CuI	10	L8 (20 mol %), K_3PO_4	DMF	30	3 h	(88)	271
(S)	2-Br	I	CuI	10	L8 (20 mol %), K_3PO_4	DMF	30	3 h	(58)	271
(S)	2-O_2N	I	CuI	10	K_2CO_3	DMF	MW, 140	15 min	(90)	265
(S)	2-O_2N	I	CuI	10	L8 (20 mol %), K_3PO_4	DMF	30	3 h	(70)	271
(S)	2-HO	Br	CuI	10	K_2CO_3	H_2O/DMF	90	48 h	(75)	116
(S)	2-MeO	Br	CuI	10	K_2CO_3	H_2O/DMF	90	48 h	(75)	116
(S)	3-MeO	Br	CuI	10	KI, K_2CO_3	H_2O	MW, 185	40 min	(76)	25
(S)	3-MeO	Br	CuI	10	K_2CO_3	H_2O/DMF	90	48 h	(85)	116
(S)	4-Me	Br	CuI	10	K_2CO_3	DMF	90	—	(83)	272
(S)	4-Me	I	CuI	10	L8 (20 mol %), K_3PO_4	DMF	30	3 h	(84)	271
(S)	4-Me	Br	Cu(OAc)$_2$	10	K_2CO_3	DMF	MW, 140	25 min	(64)	265
(S)	2-$HOCH_2$	I	CuI	10	K_2CO_3	H_2O/DMF	90	48 h	(85)	116
(S)	2-$BnOCH_2$	I	CuI	10	K_2CO_3	H_2O/DMF	90	48 h	(47)	116

TABLE 2A. N-ARYLATION OF PRIMARY ALKYL AMINES (Continued)

Nitrogen Nucleophile	Aryl Halide	Conditions	Product(s) and Yield(s) (%)	Refs.

*Please refer to the charts preceding the tables for structures indicated by the **bold** numbers.*

Continued from previous page.

C₅

Nitrogen Nucleophile:

H₂N—CH(CO₂H)—CH(CH₃)₂

Aryl Halide:

R—C₆H₄—X

Product:

R—C₆H₄—NH—CH(CO₂H)—CH(CH₃)₂ (H on N, CO₂H shown)

Config.	R	X	Catalyst	x	Additive(s)	Catalyst (x mol %)	Solvent(s)	Temp (°)	Time		Refs.
(S)	2-HO₂C	Br	CuI	10	K₂CO₃		H₂O/DMF	90	48 h	(64)	116
(S)	2-HO₂C	I	CuI	10	K₂CO₃		H₂O/DMF	90	48 h	(64)	116
(S)	3-HO₂C	I	CuI	10	**L8** (20 mol %), K₃PO₄		DMF	30	3 h	(68)	271
(S)	3-HO₂C	Br	CuI	10	K₂CO₃		H₂O/DMF	90	48 h	(92)	116
(S)	4-HO₂C	Br	CuI	10	K₂CO₃		H₂O/DMF	90	48 h	(75)	116
(S)	2,5-Me₂	Br	CuI	10	K₂CO₃		H₂O/DMF	90	48 h	(74)	116
(S)	2,6-Me₂	Br	CuI	10	K₂CO₃		H₂O/DMF	90	48 h	(74)	116
(S)	3,5-Me₂	I	CuI	2.5	NaOH		DMSO/H₂O (2:1)	90	16.5 h	(88)	279
(S)	2-Ac	I	CuI	10	K₂CO₃		DMF	MW, 140	20 min	(78)	265
(S)	4-Ac	Br	CuI	10	K₂CO₃		DMF	MW, 140	20 min	(82)	265
(S)	4-Ac	I	CuI	10	K₂CO₃		DMF	MW, 140	18 min	(80)	265

Aryl Halide (bromobenzene):

C₆H₅—Br

Conditions:

1. CuCl (20 mol %), TBAH (aq),
   MeCN, reflux, 24 h
2. AcCl, MeOH, reflux, 2 h

Product:

C₆H₅—NH—CH(CO₂Me)—CH(CH₃)₂ (33) er 100:0 — 264

Nitrogen Nucleophile:

H₂N—CH[CH(CH₃)₂]—CH₂OH

Aryl Halide (3-iodoacetophenone):

Conditions:

CuI (2.5 mol %), NaOH,
DMSO/H₂O (2:1),
90°, 16.5 h

Product:

3-Ac-C₆H₄—NH—CH[CH(CH₃)₂]—CH₂OH (88) — 279

Nitrogen Nucleophile:

H₂N—CH(CO₂H)—(CH₂)₃—NH(4-ClC₆H₄)

Aryl Halide (4-chloroiodobenzene, Cl-C₆H₄-I):

Conditions:

CuI (10 mol %), **L8** (20 mol %),
K₃PO₄, DMF, 30°, 3 h

Product:

(4-ClC₆H₄)HN—(CH₂)₃—CH(CO₂H)—NH(4-ClC₆H₄) (56) — 271

110

H2N〜〜〜OH +

(R-C6H4-X)

→ R-C6H4-NH-(CH2)4-OH

Catalyst (x mol %)

R	X	Catalyst	x	Additive(s)	Solvent(s)	Temp (°)	Time (h)		
H	I	CuI	10	K3PO4•H2O	Me2N(CH2)2OH/H2O	90	—	(81)	260
4-Cl	I	CuBr	20	binol (20 mol %), K3PO4	DMF	rt	6	(62)	62
4-Cl	Br	CuI	10	L11 (20 mol %), Cs2CO3	DMF	65	17	(87)	261
3-Br	I	CuI	5	L16 (20 mol %), Cs2CO3	DMF	rt	12	(97)	196
3-H2N	I	CuI	5	L16 (20 mol %), Cs2CO3	DMF	rt	12	(84)	196
4-Me	I	CuI	5	L16 (20 mol %), Cs2CO3	DMF	rt	12	(97)	196
2-HO2C, 6-O2N	Br	CuBr	5	L17 (20 mol %), NaOH, TBAB	H2O	100	24	(81)	235

H2N-CH(CO2H)-CH2CH2CH2-CO2H +

(Ph-X)

CuI (10 mol %), 90°

→ Ph-NH-CH(CO2H)-CH2CH2-CO2H

Config.	X	Additive	Solvents	Time (h)	
(R,S)	I	K3PO4•H2O	Me2N(CH2)2OH/H2O	—	(32)
(S)	Br	K2CO3	H2O/DMF	48	(0)

C5-6

R1-furan-CH2-NH2 +

(R2-C6H4-X)

Catalyst (x amount)

→ R1-furan-CH2-NH-C6H4-R2

R1	R2	X	Catalyst	x	Additive(s)	Solvent	Temp (°)	Time (h)		
H	H	I	CuO nanoparticles	1.26 mol %	KOH	DMSO	110	17	(85)	224
H	O2N	I	CuI	1 eq	CsOAc	DMF	90	24	(40)	189
H	3,4-OCH2O	Br	CuI	5 mol %	L13 (20 mol %), K3PO4	DMF	90	18	(87)	120
Me	HO	I	CuI	5 mol %	L16 (20 mol %), Cs2CO3	DMF	rt	20	(79)	52

111

TABLE 2A. N-ARYLATION OF PRIMARY ALKYL AMINES (Continued)

*Please refer to the charts preceding the tables for structures indicated by the **bold** numbers.*

C₆

Nitrogen Nucleophile	Aryl Halide	Conditions					Product(s) and Yield(s) (%)	Refs.
$n\text{-}C_6H_{13}NH_2$								
	X	Catalyst (x amount)			Solvent	Temp (°)	Time	
X	Catalyst	x	Additive(s)					
Br	Cu	5 mol %	binol (10 mol %), Cs₂CO₃	DMSO	90	24 h	(69)	124
I	Cu	5 mol %	binol (10 mol %), Cs₂CO₃	DMSO	90	24 h	(99)	124
Cl	Cu (active)	5 eq	none	H₂O	MW (100 W)	6 min	(80)	280
Br	Cu (active)	5 eq	none	H₂O	MW (100 W)	6 min	(84)	280
I	Cu (active)	5 eq	none	H₂O	MW (100 W)	6 min	(87)	280
I	CuCl	10 mol %	**L9** (10 mol %), TBAH	DMSO	80	24 h	(87)	281
Br	CuCl	10 mol %	TMHD (25 mol %), Cs₂CO₃	NMP	120	10 h	(78)	188
I	CuCl	5 mol %	TBAH (aq)	—	80	24 h	(77)	282
I	CuBr	5 mol %	**L10** (6 mol %), Cs₂CO₃	DMSO	80	30 h	(91)	259
I	CuI	10 mol %	L-Pro (20 mol %), K₂CO₃	DMSO	60	14 h	(87)	192
I	CuI	10 mol %	L-Pro (20 mol %), TBAA	DMF	rt	24 h	(91)	61
Br	CuI	10 mol %	**L6** (20 mol %), Cs₂CO₃	DMF	80	24 h	(65)	273
I	CuI	10 mol %	**L6** (20 mol %), Cs₂CO₃	DMF	rt	24 h	(90)	273
I	CuI	10 mol %	DMG (20 mol %), TBPM	DMSO	rt	24 h	(94)	61
I	CuI	10 mol %	**L11** (20 mol %), Cs₂CO₃	DMF	rt	10 h	(97)	261
I	Cu₂O	5 mol %	**L12** (20 mol %), Cs₂CO₃	MeCN	60	18 h	(78)	125
I	Cu₂O	5 mol %	**L7** (20 mol %), Cs₂CO₃	MeCN	80	18 h	(88)	125

Product structure: $\overset{H}{N}\!-\!n\text{-}C_6H_{13}$ (N-hexylaniline)

R-[C6H4]-X → R-[C6H4]-NH-n-C6H13

Catalyst (x amount)

R	X	Catalyst	x	Additive(s)	Solvent	Temp (°)	Time		
4-Cl	Br	CuCl	5 mol %	TBAH (aq)	—	80	24 h	(62)	282
4-Cl	I	CuCl	5 mol %	TBAH (aq)	—	80	24 h	(72)	282
4-Cl	I	CuCl	10 mol %	L9 (10 mol %), TBAH	DMSO	80	24 h	(81)	281
4-Cl	I	CuI	10 mol %	L11 (20 mol %), Cs$_2$CO$_3$	DMF	rt	12 h	(91)	261
4-Cl	Br	CuI	10 mol %	L6 (20 mol %), Cs$_2$CO$_3$	DMF	80	24 h	(70)	273
4-Cl	I	CuI	10 mol %	L6 (20 mol %), Cs$_2$CO$_3$	DMF	rt	24 h	(91)	273
4-Br	I	CuI	10 mol %	L6 (20 mol %), Cs$_2$CO$_3$	DMF	rt	24 h	(92)	273
4-I	I	CuI	10 mol %	L10 (6 mol %), Cs$_2$CO$_3$	DMSO	80	30 h	(87)	259
4-I	I	CuI	10 mol %	L-Pro (20 mol %), K$_2$CO$_3$	DMSO	60	9 h	(86)	192
2-H$_2$N	I	CuI	10 mol %	L6 (20 mol %), Cs$_2$CO$_3$	DMF	rt	24 h	(80)	273
3-H$_2$N	Br	CuI	5 mol %	L13 (5 mol %), K$_3$PO$_4$	—	100	22 h	(90)	120
3-H$_2$N	Br	CuI	5 mol %	L13 (20 mol %), K$_3$PO$_4$	DMF	90	18 h	(80)	120
3-O$_2$N	I	CuBr	5 mol %	L10 (6 mol %), Cs$_2$CO$_3$	DMSO	80	30 h	(90)	259
3-O$_2$N	Br	CuI	5 mol %	L13 (20 mol %), K$_3$PO$_4$	DMF	90	18 h	(79)	120
4-O$_2$N	I	CuCl	10 mol %	L9 (10 mol %), TBAH	DMSO	80	24 h	(68)	281
2-MeO	Br	CuI	5 mol %	L13 (20 mol %), K$_3$PO$_4$	DMF	90	18 h	(88)	120
4-MeO	Br	Cu	5 mol %	binol (10 mol %), Cs$_2$CO$_3$	DMSO	90	40 h	(84)	124
4-MeO	I	Cu	5 mol %	binol (10 mol %), Cs$_2$CO$_3$	DMSO	90	24 h	(84)	124
4-MeO	I	CuBr	20 mol %	binol (20 mol %), K$_3$PO$_4$	DMF	50	6 h	(55)	62
4-MeO	I	CuBr	5 mol %	L10 (6 mol %), Cs$_2$CO$_3$	DMSO	80	30 h	(81)	259
4-MeO	I	CuI	10 mol %	L-Pro (20 mol %), K$_2$CO$_3$	DMSO	60	24 h	(79)	192
4-MeO	Br	CuI	10 mol %	L6 (20 mol %), Cs$_2$CO$_3$	DMF	80	24 h	(55)	273
4-MeO	I	CuI	10 mol %	L6 (20 mol %), Cs$_2$CO$_3$	DMF	rt	24 h	(68)	273
4-MeO	Br	CuI	10 mol %	DMCDA (20 mol %), K$_3$PO$_4$	[bmim]BF$_4$	110	12 h	(95)	232

TABLE 2A. N-ARYLATION OF PRIMARY ALKYL AMINES (Continued)

| Nitrogen Nucleophile | Aryl Halide | Conditions | | | | | | Product(s) and Yield(s) (%) | Refs. |

Please refer to the charts preceding the tables for structures indicated by the **bold** numbers.

Continued from previous page.

C_6

$n\text{-}C_6H_{13}NH_2$

Aryl Halide: R—C6H4—X   Product: R—C6H4—N(H)—$n\text{-}C_6H_{13}$

R	X	Catalyst	x	Additive(s)	Solvent	Temp (°)	Time		Refs.
4-MeO	I	CuI	10 mol %	L11 (20 mol %), Cs$_2$CO$_3$	DMF	rt	12 h	(97)	261
4-MeO	I	Cu$_2$O	5 mol %	L7 (20 mol %), Cs$_2$CO$_3$	toluene	80	18 h	(88)	125
4-Me	Br	Cu (active)	5 eq	none	H$_2$O	MW (100 W)	6 min	(78)	280
4-Me	I	CuCl	5 mol %	TBAH (aq)	—	80	24 h	(77)	282
4-Me	I	CuCl	10 mol %	L9 (10 mol %), TBAH	DMSO	80	24 h	(74)	281
4-Me	I	CuBr	5 mol %	L10 (6 mol %), Cs$_2$CO$_3$	DMSO	80	30 h	(87)	259
4-Me	I	CuI	10 mol %	L6 (20 mol %), Cs$_2$CO$_3$	DMF	rt	24 h	(85)	273
3-CF$_3$	I	Cu$_2$O	5 mol %	L7 (20 mol %), Cs$_2$CO$_3$	toluene	80	18 h	(76)	125
4-NC-	Br	CuI	5 mol %	L9 (20 mol %), K$_3$PO$_4$	DMF	90	18 h	(73)	120
4-NC-	I	CuI	10 mol %	L11 (20 mol %), Cs$_2$CO$_3$	DMF	rt	8 h	(95)	261
2-HOCH$_2$	Br	CuI	5 mol %	L13 (20 mol %), K$_3$PO$_4$	DMF	90	18 h	(88)	120
3-MeO$_2$C	I	CuI	10 mol %	L6 (20 mol %), Cs$_2$CO$_3$	DMF	rt	24 h	(80)	273
3-MeO$_2$C	I	CuCl	10 mol %	L9 (10 mol %), TBAH	DMSO	80	24 h	(59)	281
2-HO$_2$C, 4-Br	Br	Cu$_2$O	5 mol %	K$_3$PO$_4$	DMA	70	16 h	(83)	206
3,5-Me$_2$	I	CuI	10 mol %	L11 (20 mol %), Cs$_2$CO$_3$	DMF	rt	12 h	(97)	261
3,5-Me$_2$	Br	CuI	5 mol %	L13 (5 mol %), K$_3$PO$_4$	—	100	22 h	(90)	120
3,5-Me$_2$	Br	CuI	5 mol %	L13 (5 mol %), K$_3$PO$_4$	DMF	90	18 h	(91)	120
4-Ac	Br	CuI	10 mol %	L6 (20 mol %), Cs$_2$CO$_3$	DMF	80	24 h	(68)	273
4-Ac	Br	CuI	10 mol %	L11 (20 mol %), Cs$_2$CO$_3$	DMF	65	8 h	(95)	261

H2N–cyclohexane + Ph–X → Catalyst (x amount) → N-cyclohexylaniline

X	Catalyst	x	Additive(s)	Solvent(s)	Temp (°)	Time (h)	(yield)	ref
I	Cu/Fe–hydrotalcite	10 weight %	none	toluene	130	12	(81)	283
Br	CuBr	10 mol %	8-HOquin (20 mol %), Cs$_2$CO$_3$	DMSO	90	24	(82)	284
I	CuBr	2.5 mol %	L5 (5 mol %), Cs$_2$CO$_3$	DMF	89	24	(98)	118, 119
Br	CuI	10 mol %	piperidine-2-carboxylic acid (20 mol %), K$_2$CO$_3$	DMF	110	24	(86)	67
I	CuI	10 mol %	piperidine-2-carboxylic acid (20 mol %), K$_2$CO$_3$	DMF	110	20	(93)	67
I	CuI	10 mol %	DPP (20 mol %), K$_3$PO$_4$	DMF	90	24	(87)	108
I	CuI	10 mol %	L8 (20 mol %), K$_3$PO$_4$	DMF	30	10	(62)	271
I	CuI	10 mol %	L-Pro (20 mol %), K$_2$CO$_3$	DMSO	60	14	(80)	192
I	CuI	10 mol %	L6 (20 mol %), Cs$_2$CO$_3$	DMF	50	24	(64)	273
I	CuI	10 mol %	L-Pro (20 mol %), TBAA	DMF	rt	24	(94)	61
I	CuI–PAnNF	5 mol %	K$_2$CO$_3$	DMF	rt	5	(98)	276
I	CuO	10 mol %	FeCl$_3$ (10 mol %), binol (20 mol %), Cs$_2$CO$_3$	DMF	80	12	(89)	277
I	Cu$_2$O	5 mol %	NaOt-Bu	NMP	110	24	(69)	266
I	Cu(acac)$_2$	10 mol %	Fe$_2$O$_3$ (20 mol %), Cs$_2$CO$_3$	DMSO/H$_2$O (1:1)	MW, 150	0.5	(75)	262

115

TABLE 2A. N-ARYLATION OF PRIMARY ALKYL AMINES (Continued)

Nitrogen Nucleophile	Aryl Halide	Conditions	Product(s) and Yield(s) (%)	Refs.

*Please refer to the charts preceding the tables for structures indicated by the **bold** numbers.*

C₆ — cyclohexyl NH₂   ArX   cyclohexyl NHAr

Ar	X	Catalyst(s)	x	Additive(s)	Catalyst(s) (x amount)	Solvent(s)	Temp (°)	Time		Refs.
4-ClC₆H₄	I	CuBr	5 mol %		L17 (20 mol %), TBAB, NaOH	H₂O	100	24 h	(80)	235
4-ClC₆H₄	I	CuI	10 mol %		L11 (20 mol %), Cs₂CO₃	DMF	rt	19 h	(86)	261
4-ClC₆H₄	I	CuI	10 mol %		L8 (20 mol %), K₃PO₄	DMF	30	3 h	(70)	271
4-ClC₆H₄	I	CuO	10 mol %		binol (20 mol %), FeCl₃ (10 mol %), Cs₂CO₃	DMF	90	12 h	(90)	277
2-O₂NC₆H₄	Cl	Cu/Al–HTB	2.5 mol %		K₂CO₃	DMF	100	6 h	(99)	228
2-O₂NC₆H₄	Cl	Cu/Al–HTB	20 wt %		K₂CO₃	DMF	100	8 h	(91)	117
2-O₂NC₆H₄	I	CuI–PAnNF	5 mol %		K₂CO₃	—	rt	1 h	(99)	276
3-O₂NC₆H₄	Cl	CuI nanoparticles	1.25 mol %		Cs₂CO₃	DMF	110	3 h	(99)	285
3-O₂NC₆H₄	I	CuI–PAnNF	5 mol %		K₂CO₃	DMF	rt	6 h	(56)	276
3-O₂NC₆H₄	Br	CuO	10 mol %		binol (20 mol %), FeCl₃ (10 mol %), Cs₂CO₃	DMF	90	12 h	(88)	277
4-O₂NC₆H₄	Cl	Cu/Al–HTB	20 wt %		K₂CO₃	—	100	9 h	(71)	117
4-O₂NC₆H₄	Cl	Cu/Al–HTB	2.5 mol %		K₂CO₃	DMF	100	8 h	(92)	228
4-MeOC₆H₄	I	Cu/Fe-hydrotalcite	10 wt %		none	toluene	130	12 h	(80)	283
4-MeOC₆H₄	I	CuBr	5 mol %		L10 (6 mol %), Cs₂CO₃	DMSO	80	30 h	(70)	259
4-MeOC₆H₄	Br	CuI	10 mol %		piperidine 2-carboxylic acid (20 mol %), K₂CO₃	DMF	110	20 h	(81)	67
4-MeOC₆H₄	Br	CuI	5 mol %		L2 (25 mol %), TBAB, KOH, sealed tube	H₂O	130	5 min	(76)	236
4-MeOC₆H₄	Br	CuO	5 mol %		L14 (50 mol %), cyclohexanone, TBAB, KOH	H₂O	MW (100 W), 130	5 min	(86)	51
4-MeOC₆H₄	Br	CuO	25 mol %		L15 (50 mol %), TBAB, KOH	H₂O	MW (100 W), 130	5 min	(73)	267
4-MeC₆H₄	I	CuBr	5 mol %		L10 (6 mol %), Cs₂CO₃	DMSO	80	30 h	(75)	259
4-MeC₆H₄	I	CuI	10 mol %		L8 (20 mol %), K₃PO₄	DMF	30	10 h	(90)	271
2-NCC₆H₄	Cl	Cu/Al–HTB	20 wt %		K₂CO₃	—	100	9 h	(59)	117
3-NCC₆H₄	Cl	Cu/Al–HTB	2.5 mol %		K₂CO₃	DMF	100	14 h	(72)	228
4-HOCH₂C₆H₄	I	CuI	5 mol %		L16 (20 mol %), Cs₂CO₃	DMF	rt	7 h	(88)	52
2-CHO, 5-ClC₆H₃	Cl	Cu/Al–HTB	20 wt %		K₂CO₃	—	100	8 h	(56)	117
4-CHOC₆H₄	Cl	Cu/Al–HTB	20 wt %		K₂CO₃	—	100	8 h	(89)	117

Substrate	X	Catalyst	Amount	Base/Ligand	Solvent	Temp.	Time	(Yield) Ref.
2-HO$_2$CC$_6$H$_4$	Cl	Cu, Cu$_2$O	10. 5 mol %	K$_2$CO$_3$	EtOCH$_2$CH$_2$OEt	130	24 h	(93) 278
2-HO$_2$CC$_6$H$_4$	Br	Cu, Cu$_2$O	9, 4 mol %	K$_2$CO$_3$	EtOCH$_2$CH$_2$OH	130	24 h	(65) 70
2-HO$_2$CC$_6$H$_4$	Br	Cu, Cu$_2$O	10. 5 mol %	K$_2$CO$_3$	EtOCH$_2$CH$_2$OEt	130	24 h	(82) 278
2-HO$_2$CC$_6$H$_4$	Br	CuI	10 mol %	binol (20 mol %), K$_3$PO$_4$	DMF	rt	16 h	(85) 258
2-HO$_2$CC$_6$H$_4$	Br	Cu(acac)$_2$	10 mol %	binol (20 mol %), Fe$_2$O$_3$ (20 mol %), Cs$_2$CO$_3$	DMSO/H$_2$O (1:1)	MW, 130	0.5 h	(46) 262
2-HO$_2$C, 4-BrC$_6$H$_3$	Br	CuI	10 mol %	binol (20 mol %), K$_3$PO$_4$	DMF	rt	16 h	(40) 258
2-HO$_2$C, 5-BrC$_6$H$_3$	Br	Cu$_2$O	5 mol %	K$_3$PO$_4$	DMA	70	16 h	(78) 206
2-HO$_2$C, 6-O$_2$NC$_6$H$_3$	Br	CuI	10 mol %	binol (20 mol %), K$_3$PO$_4$	DMF	rt	16 h	(85) 258
4-HO$_2$CC$_6$H$_4$	Cl	Cu/Al–HTB	2.5 mol %	K$_2$CO$_3$	DMF	100	14 h	(72) 228
4-HO$_2$CC$_6$H$_4$	Cl	Cu/Al–HTB	20 weight %	K$_2$CO$_3$	—	100	8 h	(88) 117
4-HO$_2$CC$_6$H$_4$	I	CuI–PAnNF	5 mol %	K$_2$CO$_3$	DMF	rt	4 h	(75) 276
4-HO$_2$CC$_6$H$_4$	Cl	CuI nanoparticles	1.25 mol %	Cs$_2$CO$_3$	DMF	110	3 h	(98) 285
1-Np	Br	CuI	10 mol %	L11 (20 mol %), Cs$_2$CO$_3$	DMF	65	18 h	(83) 261
2-Np	Br	CuBr	5 mol %	L17 (20 mol %), NaOH, TBAB	H$_2$O	100	24 h	(81) 235

CuI (5 mol %), L16 (20 mol %), Cs$_2$CO$_3$, DMF; rt, 1 h

(98) 52

CuI (5 mol %), L16 (20 mol %), Cs$_2$CO$_3$, DMF; rt, 20 h

(82) 52

Nitrogen Nucleophile	Aryl Halide	Conditions	Product(s) and Yield(s) (%)	Refs.

*Please refer to the charts preceding the tables for structures indicated by the **bold** numbers.*

C₆

H₂N–CH₂CH₂–N(piperazine NH)	4-iodotoluene (I)	Cu (10 mol %), K₃PO₄•H₂O, Me₂NCH₂CH₂OH, 75°, 30 h	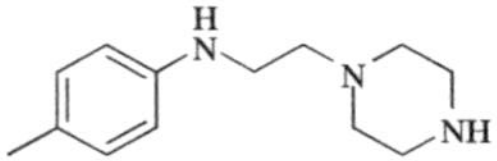 (84)	195
NH (piperidine)	4-*t*-Bu-C₆H₄Br	CuI (5 mol %), **L13** (20 mol %), K₃PO₄, DMF, 90°, 18 h	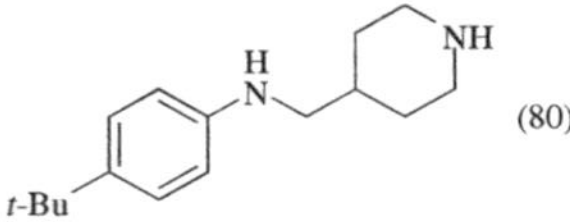 (80)	120
H₂N–(CH₂)₆–OH	R–C₆H₄I	CuI (*x* mol %), Cs₂CO₃, additive (20 mol %), DMF, rt, 12 h	see below	196, 261, 196

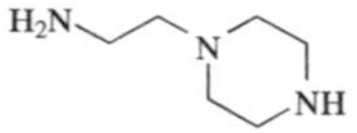

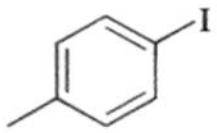

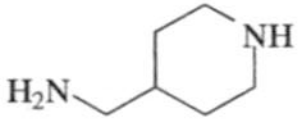

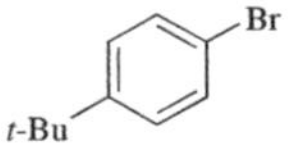

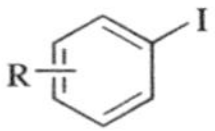

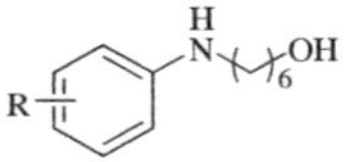

x R   Additive
5 3,4-F₂ **L16** (93)
10 4-Me **L11** (93)
5 4-Me **L16** (99)

| 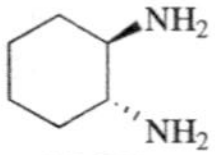 (1,2-diaminocyclohexane) NH₂/NH₂ | ArBr | CuBr (10 mol %), binol (20 mol %), K₃PO₄, DMF, 120°, 36 h | 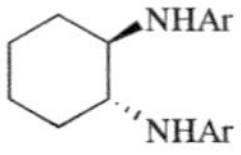 see below | 286 |

Ar	
Ph	(78)
4-O₂NC₆H₄	(49)
4-MeOC₆H₄	(71)
4-MeC₆H₄	(61)

| (1R,2R) NH₂/NH₂ | 4-Br-toluene (Br) | CuI (20 mol %), K₃PO₄, dioxane, 110°, 23 h | 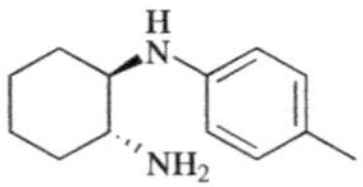 (66) | 115 |
| 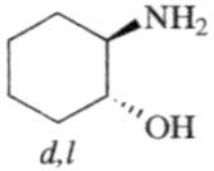 d,l (NH₂/OH) | 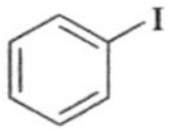 C₆H₅I | CuI (2.5 mol %), NaOH, DMSO/H₂O (2:1), 90°, 15 h | 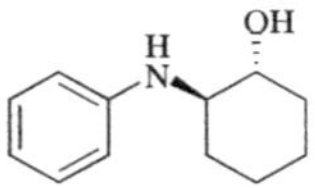 (87) | 279 |

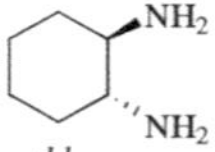

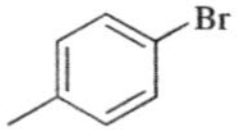

118

H2N—[cyclohexane]—OH (d,l)

CuI (5 mol %),
L16 (20 mol %),
Cs2CO3, DMF, rt, 12 h

[aryl-Cl]—NH—[cyclohexane]—OH (trans) (80)

H_2N–(CH2)5–CO2H

Ar I

CuI (10 mol %), K3PO4·H2O,
Me2N(CH2)2OH, H2O, 90°

$ArHN$–(CH2)5–CO2H

Ar	
H	(76)
3-MeOC6H4	(73)
4-MeOC6H4	(46)
2-HO2CC6H4	(76)
3-HO2CC6H4	(76)
4-HO2CC6H4	(88)

C6-7

H_2N–CH(R1)–CO2R2   + R^3—[C6H4]—Br   → R^3—[C6H4]—N(H)—CH(R1)—CO2H

Catalyst (x mol %)

R1	R2	N-Nucleophile	R3	X	Catalyst	x	Additive(s)	Solvent	Temp (°)	Time		
n-Bu	H	free base	H	I	CuI	10	**L8** (20 mol %), K3PO4	DMF	30	3 h	(80)	271
n-Bu	H	free base	2-O2N	I	CuI	10	**L8** (20 mol %), K3PO4	DMF	30	3 h	(73)	271
i-Bu	H	free base	H	Br	CuI	10	KI, K2CO3	H2O	MW, 185	40 min	(71)	25
i-Bu	H	free base	H	I	CuI	10	KI, K2CO3	H2O	MW, 185	40 min	(92)	25
i-Bu	H	HCl salt	H	Br	CuI	10	**L11** (20 mol %), Cs2CO3	DMF	65	12 h	(93)	261
i-Bu	H	HCl salt	H	Cl	Cu2O	5	NaOt-Bu	NMP	110	24 h	(76)	266
i-Bu	Me	HCl salt	H	Br	CuI	10	KI, K2CO3	H2O	MW, 185	40 min	(60)	25
i-Bu	t-Bu	HCl salt	H	Br	CuI	10	KI, K2CO3	H2O	MW, 185	40 min	(64)	25
i-Bu	H	free base	4-Cl	I	CuI	10	**L8** (20 mol %), K2CO3	DMF	30	3 h	(87)	271
i-Bu	H	free base	3-(1-imidazolylmethyl)	Br	CuI	10	KI, K2CO3	H2O	MW, 185	40 min	(67)	25
EtCH(MeCH2)	H	free base	4-Cl	I	CuI	10	**L8** (20 mol %), K3PO4	DMF	30	3 h	(88)	271

TABLE 2A. *N*-ARYLATION OF PRIMARY ALKYL AMINES (*Continued*)

Please refer to the charts preceding the tables for structures indicated by the **bold** numbers.

Nitrogen Nucleophile	Aryl Halide	Conditions	Product(s) and Yield(s) (%)	Refs.
C$_6$				
H$_2$N CO$_2$H (isobutyl)	4-Me-C$_6$H$_4$-Br	1. CuI (10 mol %), K$_2$CO$_3$, DMF, 90° 2. PhCOCH$_2$Br, Et$_3$N, EtOAc	CO$_2$CH$_2$COPh (53)	272
H$_2$N CO$_2$Et	I (C$_6$H$_5$I)	CuI (10 mol %), K$_2$CO$_3$, H$_2$O, DMF, 100°, 48 h	CO$_2$H (74)	270
H$_2$N CO$_2$Et (isopropyl)	R—⬡—X	CuI (10 mol %), K$_2$CO$_3$, H$_2$O, DMF, 100°, 48 h	CO$_2$H R / X H Br (62) H I (74) 2-Cl I (80) 4-Br Br (72) 4-I I (83) 4-H$_2$N I (0) 4-AcNH I (17) 4-O$_2$N I (87) 4-MeO Br (0) 4-MeO I (31) 4-Me I (75)	270
H$_2$N CO$_2$H / OH	1,4-diiodobenzene (I, I)	CuI (10 mol %), K$_2$CO$_3$, H$_2$O, DMF, 100°, 48 h	CO$_2$H / OH (72)	171
C$_7$				
n-C$_7$H$_15$NH$_2$	4-Me-C$_6$H$_4$-I	Cu (10 mol %), K$_3$PO$_4$•H$_2$O, Me$_2$NCH$_2$CH$_2$OH, 60°, 40 h	N *n*-C$_7$H$_{15}$ (88)	195

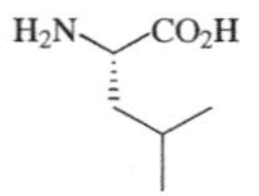

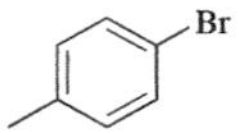

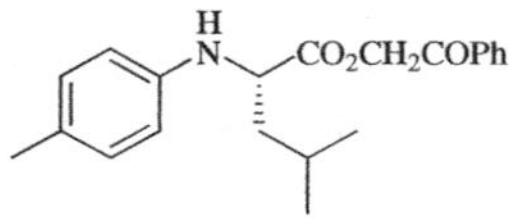

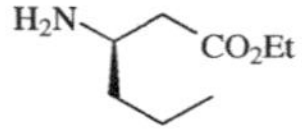

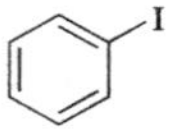

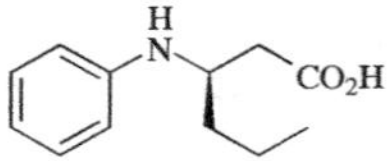

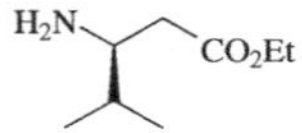

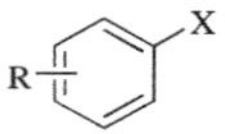

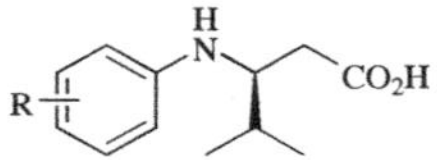

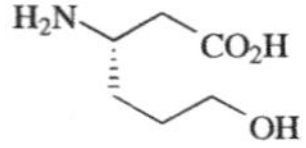

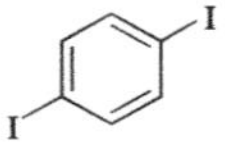

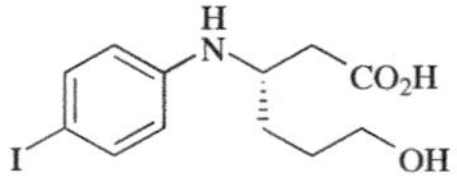

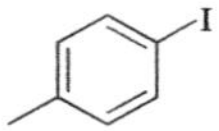

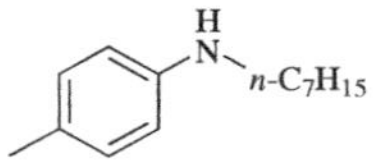

CuI (10 mol %), **L18** (20 mol %),
Cs$_2$CO$_3$, DMF, 65°, 17 h

(90) 261

Catalyst (x mol %)

X	Catalyst	x	Additive(s)	Solvent(s)	Temp (°)	Time		
Br	Cu	5	binol (10 mol %), Cs$_2$CO$_3$	DMSO	90	24 h	(95)	124
I	Cu/Fe–hydrotalcite	10 (wt)	none	toluene	130	12 h	(79)	283
Br	CuCl	10	TMHD (25 mol %), Cs$_2$CO$_3$	NMP	120	10 h	(86)	188
I	CuCl	10	**L9** (10 mol %), TBAH	DMSO	80	24 h	(91)	281
Br	CuBr	10	**L20** (20 mol %), Cs$_2$CO$_3$	DMSO	90	24 h	(92)	284
I	CuBr	2.5	**L5**, (5 mol %), Cs$_2$CO$_3$	DMF	89	24 h	(98)	118, 119
I	CuBr	5	**L10** (6 mol %), Cs$_2$CO$_3$	DMSO	55	24 h	(92)	259
Br	CuI	2.5	K$_2$CO$_3$	NMP	160	16 h	(50)	287
I	CuI	2.5	K$_2$CO$_3$	NMP	160	16 h	(9)	287
I	CuI	5	K$_3$PO$_4$	HOCH$_2$CH$_2$OH/i-PrOH	80	8 h	(91)	194
Br	CuI	10	K$_2$CO$_3$	H$_2$O/DMF	110	24 h	(<4)	270
I	CuI	10	DMG (20 mol %), TBPM	DMSO	rt	24 h	(92)	61
I	CuI	10	DMG (20 mol %), TBPE	NMP	0	5 h	(87)	61
I	CuI	10	L-Pro (20 mol %), TBAA	DMF	rt	24 h	(95)	61
Br	CuI	2	**L1** (2 mol %), KOt-Bu	—	100	12 h	(72)	288
I	CuI	10	**L6** (20 mol %), Cs$_2$CO$_3$	DMF	rt	24 h	(81)	273
I	CuI	10	**L8** (20 mol %), K$_3$PO$_4$	DMF	30	3 h	(82)	271
Cl	CuI	5	**L2** (25 mol %), KOH, TBAH	H$_2$O	MW (100 W), 130	5 min	(trace)	236
I	CuI	5	**L2** (25 mol %), KOH, TBAH	H$_2$O	MW (100 W), 130	5 min	(82)	236
I	CuI–PAnNF	5	K$_2$CO$_3$	DMF	50	10 h	(99)	276
I	CuO nanoparticles	1.26	KOH	DMSO	110	4 h	(90)	224
Cl	CuO	5	**L14** (50 mol %), cyclohexanone, KOH, TBAB	H$_2$O	MW (100 W), 130	5 min	(trace)	51
I	CuO	5	**L14** (50 mol %), cyclohexanone, KOH, TBAB	H$_2$O	MW (100 W), 130	5 min	(86)	51

TABLE 2A. N-ARYLATION OF PRIMARY ALKYL AMINES (Continued)

Nitrogen Nucleophile	Aryl Halide	Conditions	Product(s) and Yield(s) (%)	Refs.

*Please refer to the charts preceding the tables for structures indicated by the **bold** numbers.*

Continued from previous page.

C7

H_2N —CH2— (benzylamine)

X— (phenyl with X)

Catalyst (x mol %)

$\overset{H}{N}$ (N-benzylaniline)

X	Catalyst	x	Additive(s)	Solvent(s)	Temp (°)	Time		Refs.
Cl	CuO	25	L15 (50 mol %), KOH, TBAB	H2O	MW (100 W), 80	2 min	(<5)	267
Br	CuO	25	L15 (50 mol %), KOH, TBAB	H2O	MW (100 W), 130	5 min	(78)	267
I	CuO	25	L15 (50 mol %), KOH, TBAB	H2O	MW (100 W), 80	2 min	(78)	267
I	Cu2O	10	Cs2CO3	DMF	100	18 h	(0)	221
I	Cu2O	1	L7 (20 mol %), Cs2CO3	—	80	18 h	(95)	125
I	Cu(OAc)2•H2O	20	hippuric acid (20 mol %), Cs2CO3	DMF	140	20 h	(60)	289
I	Cu(OAc)2•H2O	15	(−)-sparteine, Cs2CO3	DMF	130	24 h	(75)	290
I	Cu(OAc)2•H2O	10	Fe2O3 (20 mol %), Cs2CO3	DMSO/H2O (1:1)	MW, 150	30 min	(77)	262
Br	cat. 4	25	K2CO3	NMP	160	16 h	(57)	269
Br	cat. 5	2.5	K2CO3	NMP	160	16 h	(79)	291
I	cat. 6	5	Cs2CO3	toluene	100	12 h	(90)	123

F— —I

CuI (10 mol %), DMSO

$\overset{H}{N}$—CH2—(phenyl), F substituted

Isomer	Additives	Temp (°)	Time (h)	
2	DMG (20 mol %), TBPM	rt	24	(70)
3	L-Pro (20 mol%), K2CO3	60	14	(80)
4	DMG (20 mol %), TBPM	rt	24	(81)

Cl— —Br

Cu (5 mol %), binol (10 mol %), Cs2CO3, DMSO, 90°, 24 h

Cl— $\overset{H}{N}$—CH2—(phenyl)   (93)   124

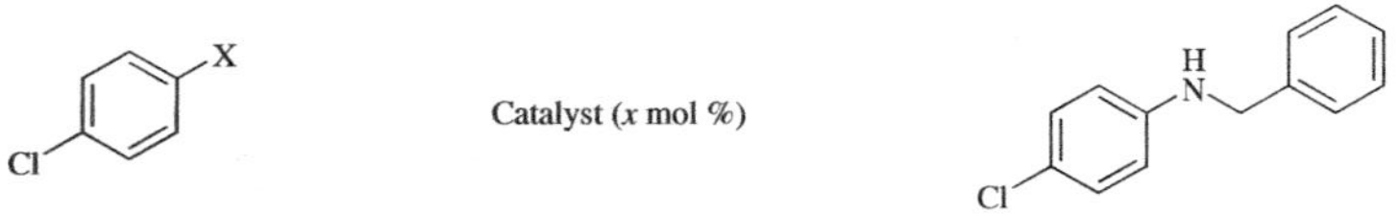

Catalyst (x mol %)

X	Catalyst	x	Additive(s)	Solvent(s)	Temp (°)	Time		
Br	CuCl	5	TBAH (aq)	—	80	24 h	(96)	282
I	CuCl	5	TBAH (aq)	—	80	24 h	(80)	282
I	CuCl	10	L9 (10 mol %), TBAH	DMSO	80	24 h	(78)	281
I	CuI	5	K$_3$PO$_4$	HOCH$_2$CH$_2$OH/i-PrOH	80	12 h	(84)	194
Br	CuI	10	L-Pro (20 mol %), K$_2$CO$_3$	DMSO	70	22 h	(76)	192
I	CuI	10	L8 (20 mol %), K$_3$PO$_4$	DMF	30	3 h	(81)	271
I	CuI	10	DMG (20 mol %), TBPE	NMP	0	5 h	(91)	61
I	CuI	10	DMG (20 mol %), TBPM	DMSO	rt	24 h	(98)	61
I	CuI–PAnNF	5	K$_2$CO$_3$	DMF	rt	3 h	(85)	276
Br	CuO	5	L14 (50 mol %), TBAB, cyclohexanone	H$_2$O	MW (100 W), 130	5 min	(86)	51
Br	CuO	25	L15 (50 mol %), TBAB	H$_2$O	MW (100 W), 130	5 min	(81)	267

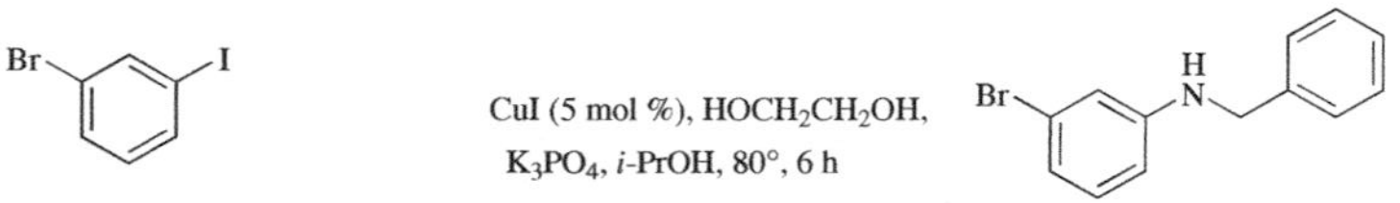

CuI (5 mol %), HOCH$_2$CH$_2$OH, K$_3$PO$_4$, i-PrOH, 80°, 6 h

(83) 194

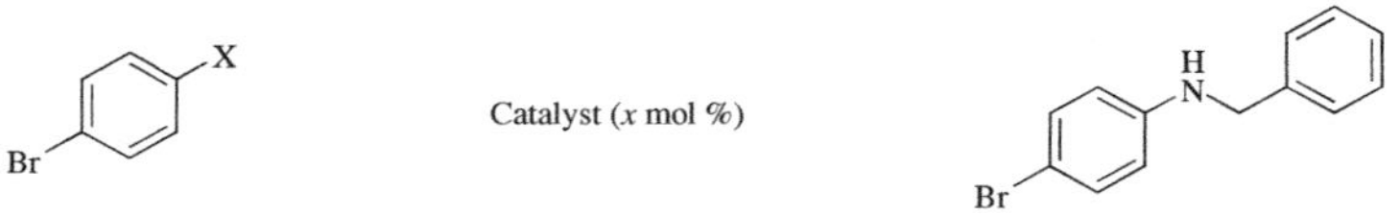

Catalyst (x mol %)

X	Catalyst	x	Additive(s)	Solvent	Temp (°)	Time (h)		
I	CuCl	10	L9 (10 mol %), TBAH	DMSO	80	24	(86)	281
I	CuCl	5	TBAH (aq)	—	80	24	(80)	282
Br	CuI	10	L-Pro (20 mol %), K$_2$CO$_3$	DMSO	70	22	(82)	192
I	CuI	10	L-Pro (20 mol %), K$_2$CO$_3$	DMSO	60	12	(81)	192, 47
I	CuI	10	L-Pro (20 mol %), TBAA	DMF	rt	24	(90)	61

TABLE 2A. N-ARYLATION OF PRIMARY ALKYL AMINES (Continued)

Nitrogen Nucleophile	Aryl Halide	Conditions	Product(s) and Yield(s) (%)	Refs.

*Please refer to the charts preceding the tables for structures indicated by the **bold** numbers.*

C_7

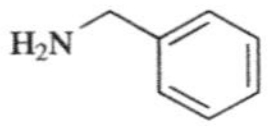

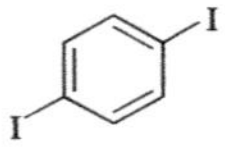

Catalyst (x mol %), DMSO

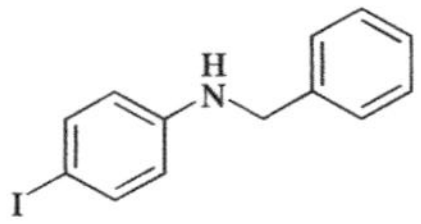

Catalyst	x	Additive(s)	Temp (°)	Time (h)	
CuBr	5	**L10** (6 mol %), Cs_2CO_3	55	24	(85)
CuI	10	NMG (20 mol %), K_2CO_3	40	13	(82)
CuI	10	L-Pro (20 mol %), K_2CO_3	60	14	(89)

259
47
192

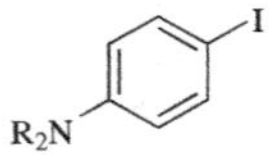

CuI (x mol %)

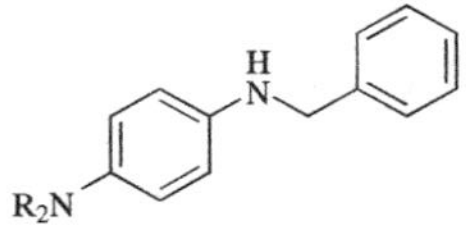

R	x	Additive(s)	Solvent(s)	Temp (°)	Time (h)	
H	5	K_3PO_4	$HOCH_2CH_2OH/i$-PrOH	90	38	(62)
Me	10	DMG (20 mol), TBPM	DMSO	rt	24	(85)

194
61

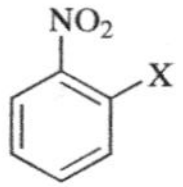

Catalyst (x amount), K_2CO_3

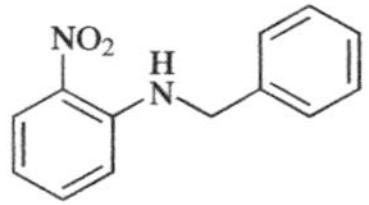

X	Catalyst	x	Solvent	Temp (°)	Time (h)	
Cl	Cu/Al–HTB	20 wt %	—	100	8	(80)
I	CuI–PAnNF	5 mol %	DMF	rt	4	(98)

117
276

124

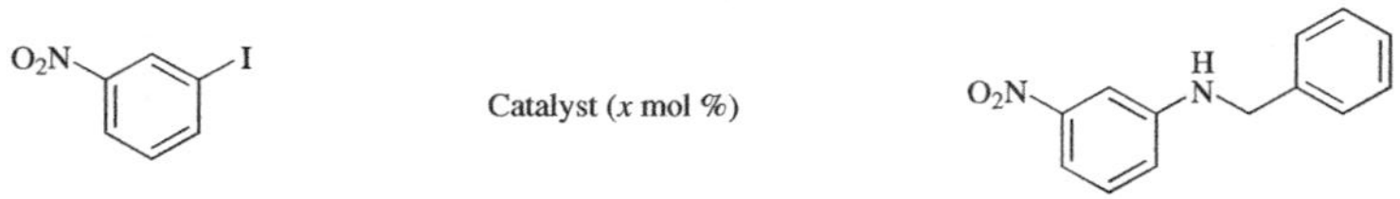

Catalyst (x mol %)

Catalyst	x	Additive(s)	Solvent(s)	Temp (°)	Time (h)	
CuBr	5	L10 (6 mol %), Cs$_2$CO$_3$	DMSO	55	24 (83)	259
CuI	10	NMG (20 mol %), K$_2$CO$_3$	DMSO	40	23 (80)	47, 192
CuI	5	K$_3$PO$_4$	HOCH$_2$CH$_2$OH/i-PrOH	80	12 (91)	194
CuI	10	L-Pro (20 mol %), TBAA	DMF	rt	24 (96)	61
Cu$_2$O	5	L7 (20 mol %), Cs$_2$CO$_3$	toluene	80	24 (87)	125

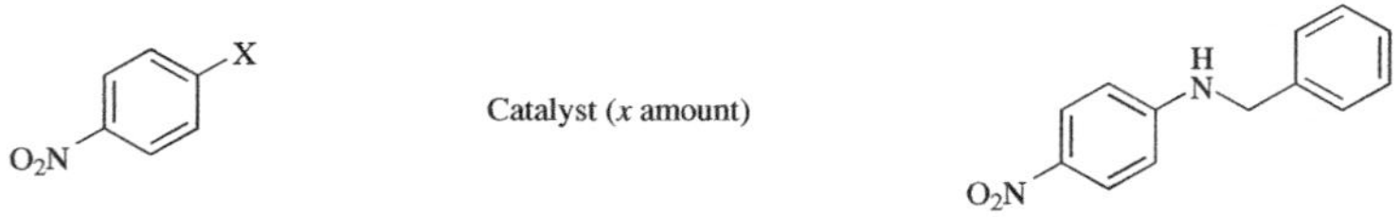

Catalyst (x amount)

X	Catalyst	x	Additive(s)	Solvent	Temp (°)	Time	
Cl	Cu/Al-HTB	20 wt %	K$_2$CO$_3$	—	100	16 h (77)	117
Br	CuO	5 mol %	L14 (50 mol %), cyclohexanone, KOH, TBAB	H$_2$O	MW (100 W), 130	5 min (63)	51
Br	CuO	25 mol %	L15 (50 mol %), KOH, TBAB	H$_2$O	MW (100 W), 130	5 min (50)	267

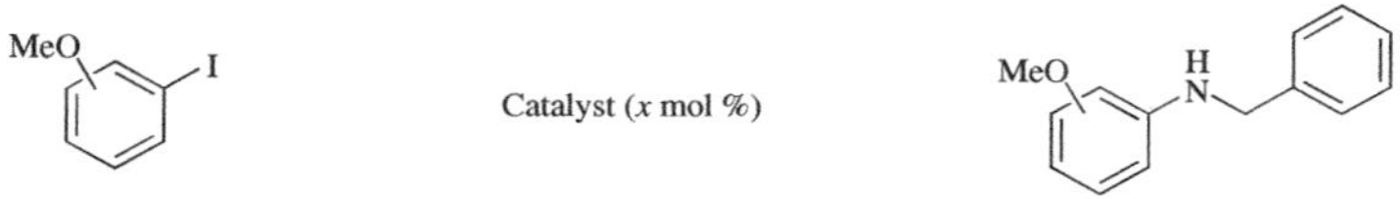

Catalyst (x mol %)

Isomer	X	Catalyst	x	Additive(s)	Solvent(s)	Temp (°)	Time(h)	
2	Br	CuCl	10	TMHD (25 mol %), Cs$_2$CO$_3$	NMP	120	10 (11)	188
2	I	CuI	5	K$_3$PO$_4$	HOCH$_2$CH$_2$OH, i-PrOH	100	72 (91)	194
3	I	CuI	10	DMG (20 mol %), TBPE	NMP	0	5 (79)	61
3	I	CuI	10	L-Pro (20 mol %), K$_2$CO$_3$	DMSO	80	30 (90)	47

TABLE 2A. N-ARYLATION OF PRIMARY ALKYL AMINES (Continued)

Nitrogen Nucleophile	Aryl Halide	Conditions	Product(s) and Yield(s) (%)	Refs.

*Please refer to the charts preceding the tables for structures indicated by the **bold** numbers.*

C₇

H_2N–CH₂–C₆H₅   MeO–C₆H₄–X   Catalyst (x amount)   H–N(benzyl)–C₆H₄–OMe

X	Catalyst	x	Additive(s)	Solvent(s)	Temp (°)	Time		Refs.
Br	Cu	5 mol %	binol (10 mol %), Cs₂CO₃	DMSO	90	24 h	(67)	124
I	Cu	5 mol %	binol (10 mol %), Cs₂CO₃	DMSO	90	24 h	(82)	124
I	Cu/Fe–hydrotalcite	10 wt %	none	toluene	130	12 h	(81)	283
I	CuCl	5 mol %	TBAH (aq)	—	80	24 h	(96)	282
Br	CuCl	10 mol %	TMHD (25 mol %), Cs₂CO₃	NMP	120	10 h	(81)	188
I	CuCl	10 mol %	L9 (10 mol %), TBAH	DMSO	80	24 h	(70)	281
I	CuBr	5 mol %	L17 (20 mol %), TBAH. NaOH	H₂O	100	24 h	(74)	235
I	CuBr	5 mol %	L10 (6 mol %), Cs₂CO₃	DMSO	55	24 h	(85)	259
Br	CuI	2.5 mol %	K₂CO₃	NMP	160	16 h	(25)	287
I	CuI	5 mol %	K₃PO₄	HOCH₂CH₂OH/i-PrOH	80	18 h	(89)	194
Br	CuI	10 mol %	L11 (20 mol %), Cs₂CO₃	DMF	65	14 h	(85)	261
I	CuI	10 mol %	L11 (20 mol %), Cs₂CO₃	DMF	rt	13 h	(91)	261
I	CuI	10 mol %	L-Pro (20 mol %), K₂CO₃	DMSO	60	16 h	(84)	192, 47
I	CuI	10 mol %	L-Pro (20 mol %), TBAA	DMF	rt	24 h	(83)	61
I	CuI–PAnNF	5 mol %	K₂CO₃	DMF	rt	11 h	(23)	276
Cl	CuO	5 mol %	L14 (50 mol %), cyclohexanone, TBAB, KOH	H₂O	MW (100 W), 130	5 min	(trace)	51
Br	CuO	5 mol %	L14 (50 mol %), cyclohexanone, TBAB, KOH	H₂O	MW (100 W), 130	5 min	(91)	51
I	CuO	5 mol %	L14 (50 mol %), cyclohexanone, TBAB, KOH	H₂O	MW (100 W), 130	5 min	(86)	51
Br	CuO	25 mol %	L15 (50 mol %), TBAB, KOH	H₂O	MW (100 W), 130	5 min	(82)	267
I	CuO	25 mol %	L15 (50 mol %), TBAB, KOH	H₂O	MW (100 W), 130	5 min	(82)	267
I	Cu₂O	1 mol %	L7 (20 mol %), Cs₂CO₃	MeCN	80	18 h	(90)	125
I	**6**	5 mol %	Cs₂CO₃	toluene	100	18 h	(72)	123

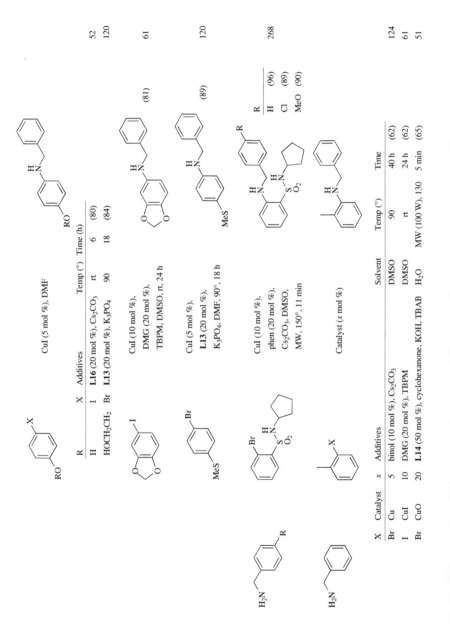

CuI (5 mol %), DMF

R	X	Additives	Temp (°)	Time (h)	
H	I	L16 (20 mol %), Cs₂CO₃	rt	6	(80)
HOCH₂CH₂	Br	L13 (20 mol %), K₃PO₄	90	18	(84)

52
120

CuI (10 mol %), DMG (20 mol %), TBPM, DMSO, rt, 24 h   (81)

61

CuI (5 mol %), L13 (20 mol %), K₃PO₄, DMF, 90°, 18 h   (89)

120

CuI (10 mol %), phen (20 mol %), Cs₂CO₃, DMSO, MW, 150°, 11 min

R	
H	(96)
Cl	(89)
MeO	(90)

268

Catalyst (x mol %)

X	Catalyst	x	Additives	Solvent	Temp (°)	Time	
Br	Cu	5	binol (10 mol %), Cs₂CO₃	DMSO	90	40 h	(62)
I	CuI	10	DMG (20 mol %), TBPM	DMSO	rt	24 h	(62)
Br	CuO	20	L14 (50 mol %), cyclohexanone. KOH, TBAB	H₂O	MW (100 W), 130	5 min	(65)

124
61
51

Nitrogen Nucleophile	Aryl Halide	Conditions	Product(s) and Yield(s) (%)	Refs.

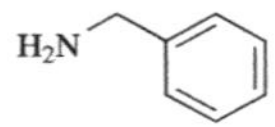

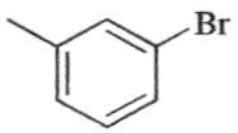

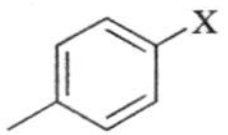

*Please refer to the charts preceding the tables for structures indicated by the **bold** numbers.*

C$_7$

Catalyst (x mol %), Cs$_2$CO$_3$

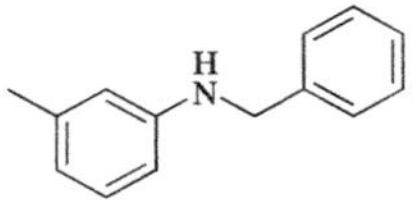

Catalyst	x	Additive	Solvent	Temp (°)	Time (h)		Refs.
Cu	5	binol (10 mol %)	DMSO	90	40	(93)	124
CuCl	10	TMHD (25 mol %)	NMP	120	10	(81)	188

Catalyst (x mol %)

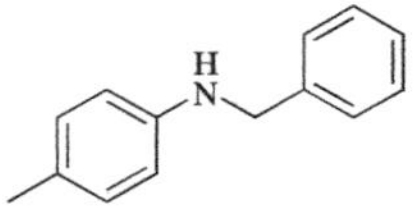

X	Catalyst	x	Additive(s)	Solvent	Temp (°)	Time		Refs.
Br	Cu	5	binol (10 mol %), Cs$_2$CO$_3$	DMSO	90	24 h	(92)	124
I	CuCl	10	L9 (10 mol %), TBAH	DMSO	80	24 h	(80)	281
I	CuBr	5	L10 (6 mol %), Cs$_2$CO$_3$	DMSO	55	24 h	(82)	259
Br	CuBr	10	8-HOQ (20 mol %), Cs$_2$CO$_3$	DMSO	90	24 h	(22)	284
Br	CuI	2.5	K$_2$CO$_3$	NMP	160	16 h	(35)	287
I	CuI	10	L8 (20 mol %), K$_3$PO$_4$	DMF	30	4 h	(76)	271
I	CuI	10	L-Pro (20 mol %), K$_2$CO$_3$	DMSO	60	16 h	(85)	192, 47
I	CuI–PAnNF	5	K$_2$CO$_3$	DMF	rt	14 h	(37)	276
Br	CuO	5	L14 (50 mol %), cyclohexanone, TBAB, KOH	H$_2$O	MW (100 W), 130	5 min	(85)	51
Br	CuO	25	L15 (50 mol %), TBAB, KOH	H$_2$O	MW (100 W), 130	5 min	(74)	267

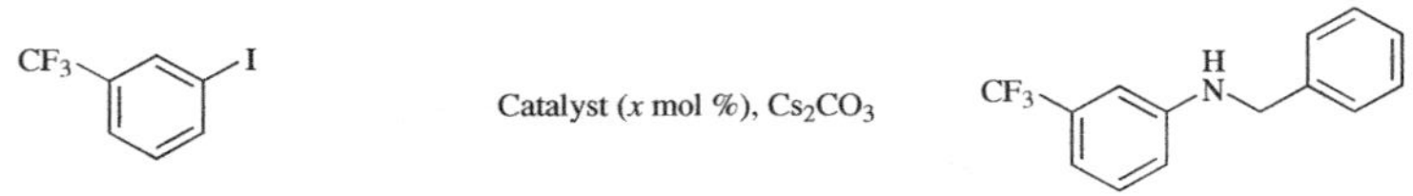

Catalyst (x mol %), Cs$_2$CO$_3$

Catalyst	x	Additives	Solvent	Temp (°)	Time (h)		
CuI	10	DMG (20 mol %), TBPE	NMP	0	5	(85)	61
Cu$_2$O	5	L7 (20 mol %), Cs$_2$CO$_3$	toluene	80	18	(91)	125

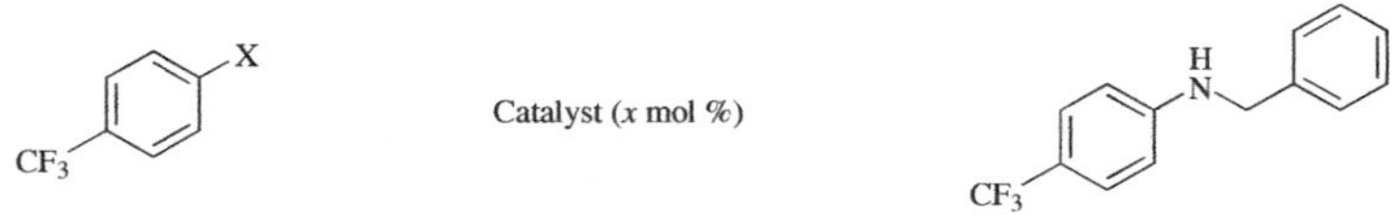

Catalyst (x mol %)

X	Catalyst	x	Additives	Solvent	Temp (°)	Time		
I	CuI	10	DMG (20 mol %), TBPM	DMSO	rt	24 h	(96)	61
Cl	CuO	5	L14 (50 mol %), cyclohexanone, TBAB, KOH	H$_2$O	MW (100 W), 130	5 min	(trace)	51
Br	CuO	5	L14 (50 mol %), cyclohexanone, TBAB, KOH	H$_2$O	MW (100 W), 130	5 min	(79)	51
Br	CuO	25	L15 (50 mol %), TBAB, KOH	H$_2$O	MW (100 W), 130	5 min	(57)	267

Catalyst (x mol %), Cs$_2$CO$_3$

R	X	Catalyst	x	Additive(s)	Solvent	Temp (°)	Time (h)		
H	I	6	5	none	toluene	100	12	(90)	123
MeO	Br	CuI	10	L11 (20 mol %), Cs$_2$CO$_3$	DMF	65	14	(85)	261

129

TABLE 2A. N-ARYLATION OF PRIMARY ALKYL AMINES (Continued)

*Please refer to the charts preceding the tables for structures indicated by the **bold** numbers.*

C₇

Nitrogen Nucleophile	Aryl Halide	Conditions	Product(s) and Yield(s) (%)	Refs.
H₂N–CH₂–(3,4,5-(OMe)₃C₆H₂)	R–C₆H₄–I	CuBr (5 mol %), **L10** (6 mol %), Cs₂CO₃, DMSO, 80°, 30 h	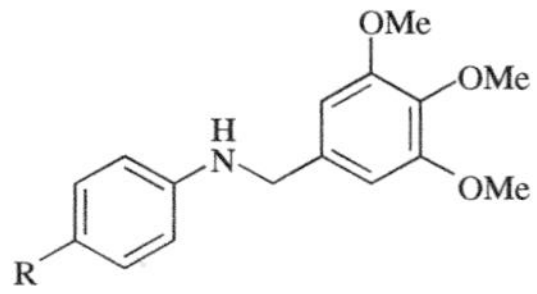 (3,4,5-(OMe)₃C₆H₂)CH₂NH–C₆H₄–R, R: H (95), MeO (80), Me (87)	259
H₂N–CH₂–C₆H₅	2-(HOCH₂)C₆H₄–I	CuI (5 mol %), HOCH₂CH₂OH, K₃PO₄, *i*-PrOH, 100°, 24 h	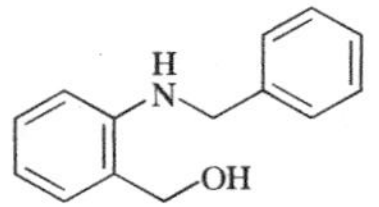 (95)	194

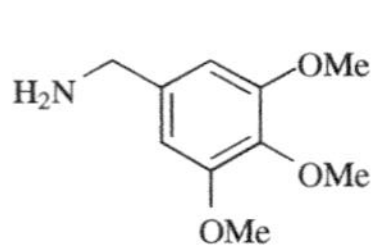

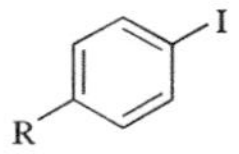

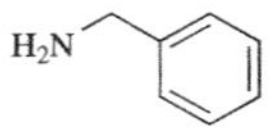

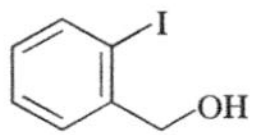

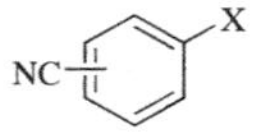

NC—C₆H₄—X   Catalyst (*x* amount)   →   NC—C₆H₄—NH—CH₂Ph

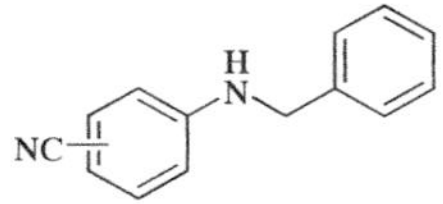

Isomer	X	Catalyst	*x*	Additive(s)	Solvent	Temp (°)	Time		Refs
2	Cl	CuI/Al–HTB	20 wt %	K₂CO₃	—	100	8	(52)	117
2	I	CuI–PAnNF	5 mol %	K₂CO₃	DMF	rt	3	(59)	276
3	I	CuBr	5 mol %	**L10** (6 mol %), Cs₂CO₃	DMSO	55	24	(87)	259
3	I	CuI	10 mol %	L-Pro (20 mol %), TBAA	DMF	rt	24	(99)	61
3	I	CuI	5 mol %	K₃PO₄	HOCH₂CH₂OH/*i*-PrOH	80	12	(80)	194
3	I	Cu₂O	5 mol %	**L7** (20 mol %), Cs₂CO₃	toluene	80	24	(85)	125
4	Cl	CuI/Al–HTB	20 wt %	K₂CO₃	—	100	16	(45)	117
4	Cl	CuCl	10 mol %	TMHD (25 mol %), Cs₂CO₃	NMP	120	10	(5)	188
4	Br	CuCl	10 mol %	TMHD (25 mol %), Cs₂CO₃	NMP	120	10	(76)	188
4	I	CuI	5 mol %	K₃PO₄	HOCH₂CH₂OH/*i*-PrOH	80	12	(79)	194
4	I	CuI–PAnNF	5 mol %	K₂CO₃	DMF	rt	5	(41)	276

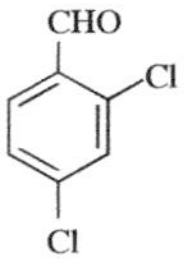

CHO H N (phenyl) (69) 117

Cu/Al–HTB (20 wt %),
K₂CO₃, 100°, 12 h

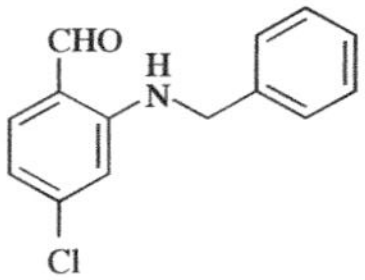

OMe N H OMe (92) 52

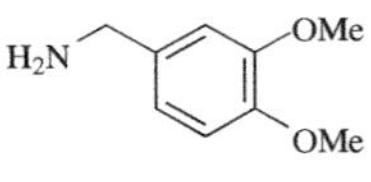

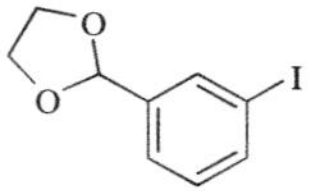

CuI (5 mol %),
L16 (20 mol %),
Cs₂CO₃, DMF, rt, 3 h

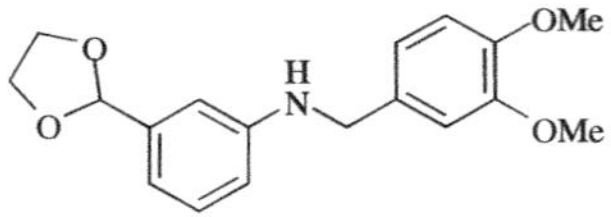

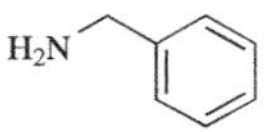

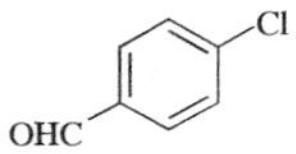

N H (phenyl) OHC (68) 117

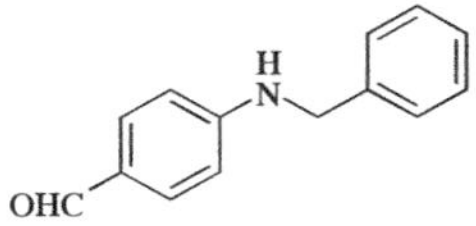

Cu/Al–HTB (20 wt %), K₂CO₃,
100°, 12 h

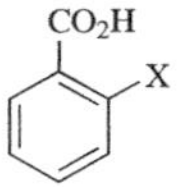

Catalyst(s) (x mol %)

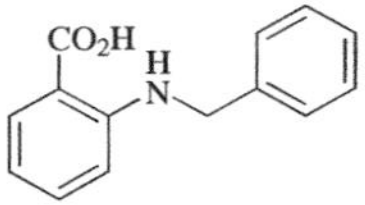

X	Catalyst(s)	x	Additive(s)	Solvent(s)	Temp(°)	Time		
Cl	Cu, Cu₂O	10, 5	K₂CO₃	EtOCH₂CH₂OEt	130	24 h	(99)	278
Br	Cu, Cu₂O	10, 5	K₂CO₃	EtOCH₂CH₂OEt	130	24 h	(95)	278
Br	Cu, Cu₂O	9, 4	K₂CO₃	EtOCH₂CH₂OEt	130	24 h	(82)	70
Br	Cu	5	binol (20 mol %), Cs₂CO₃	DMSO	90	24 h	(75)	124
Br	CuI	10	binol (20 mol %), K₃PO₄	DMF	rt	16 h	(80)	258
I	CuI	10	L-Pro (20 mol %), TBAA	DMF	rt	24 h	(89)	61
I	CuI	10	NMG (20 mol %), K₂CO₃	DMSO	40	12 h	(91)	192, 47
Cl	CuI	5	K₃PO₄	HOCH₂CH₂OH/i-PrOH	100	72 h	(48)	194
Br	CuI	5	K₃PO₄	HOCH₂CH₂OH/i-PrOH	100	48 h	(53)	194
I	CuI	5	K₃PO₄	HOCH₂CH₂OH/i-PrOH	80	18 h	(71)	194
Cl	CuO	25	**L15** (50 mol %), TBAB, KOH	H₂O	MW (100 W), 80	2 min	(55)	267

TABLE 2A. N-ARYLATION OF PRIMARY ALKYL AMINES (Continued)

Nitrogen Nucleophile	Aryl Halide	Conditions	Product(s) and Yield(s) (%)	Refs.

*Please refer to the charts preceding the tables for structures indicated by the **bold** numbers.*

C7

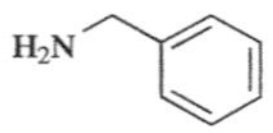

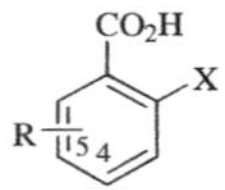

		Cu (10 mol %), Cu$_2$O (5 mol %), K$_2$CO$_3$, EtOCH$_2$CH$_2$OEt, 130°, 24 h	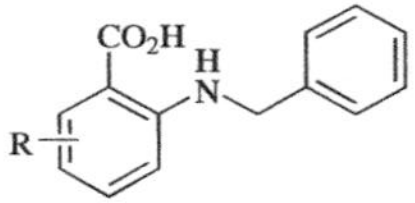 **R**: 4-F (92), 5-O$_2$N (81), 5-MeO (85), 5-HO$_2$C (91)	278
		Cu$_2$O (5 mol %), K$_3$PO$_4$, EtOCH$_2$CH$_2$OH, 70°, 16 h	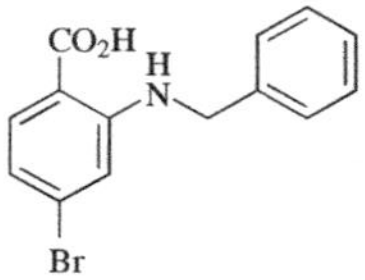 (94)	206
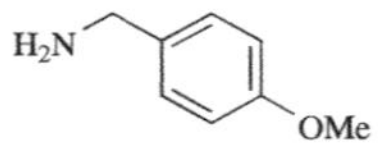	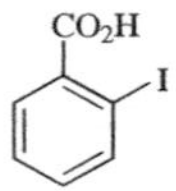	CuI (10 mol %), L11 (20 mol %), Cs$_2$CO$_3$, DMF, rt, 10 h	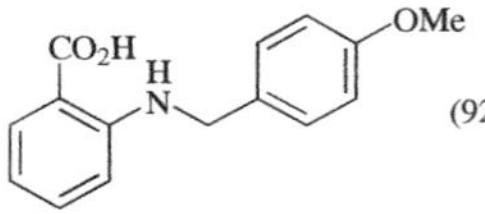 (92)	261
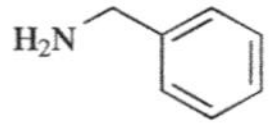	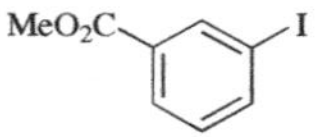	CuI (10 mol %), DMG (20 mol %), TBPE, NMP, 0°, 5 h	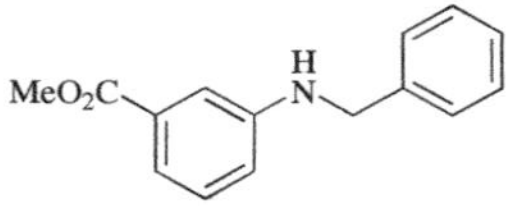 (94)	61

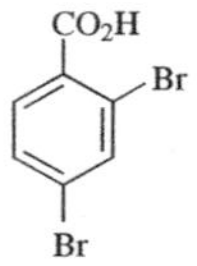

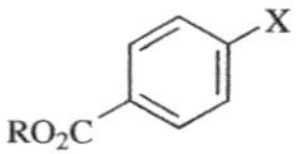

Catalyst (x amount)

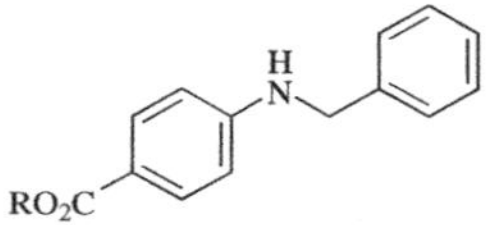

R	X	Catalyst	x	Additive	Solvent(s)	Temp (°)	Time (h)		
H	Cl	Cu/Al-HTB	20 wt %	K$_2$CO$_3$	—	100	16	(46)	117
Et	I	CuI	5 mol %	K$_3$PO$_4$	HOCH$_2$CH$_2$OH/i-PrOH	90	12	(58)	194

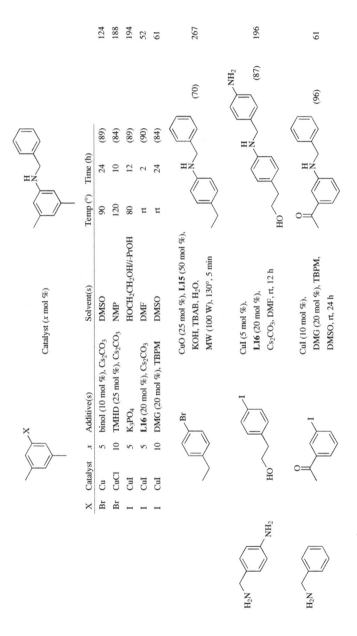

Catalyst (x mol %)

X	Catalyst	x	Additive(s)	Solvent(s)	Temp (°)	Time (h)		
Br	Cu	5	binol (10 mol %), Cs$_2$CO$_3$	DMSO	90	24	(89)	124
Br	CuCl	10	TMHD (25 mol %), Cs$_2$CO$_3$	NMP	120	10	(84)	188
I	CuI	5	K$_3$PO$_4$	HOCH$_2$CH$_2$OH/i-PrOH	80	12	(89)	194
I	CuI	5	L16 (20 mol %), Cs$_2$CO$_3$	DMF	rt	2	(90)	52
I	CuI	10	DMG (20 mol %), TBPM	DMSO	rt	24	(84)	61

CuO (25 mol %), L15 (50 mol %), KOH, TBAB, H$_2$O, MW (100 W), 130°, 5 min — (70) — 267

CuI (5 mol %), L16 (20 mol %), Cs$_2$CO$_3$, DMF, rt, 12 h — (87) — 196

CuI (10 mol %), DMG (20 mol %), TBPM, DMSO, rt, 24 h — (96) — 61

TABLE 2A. N-ARYLATION OF PRIMARY ALKYL AMINES (Continued)

*Please refer to the charts preceding the tables for structures indicated by the **bold** numbers.*

C7

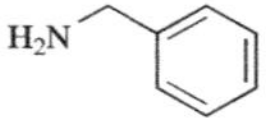

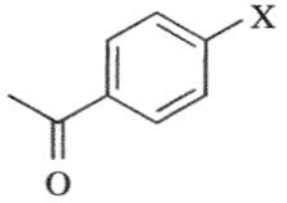

Nitrogen Nucleophile	Aryl Halide	Conditions		Product(s) and Yield(s) (%)		Refs.

H_2N — benzylamine structure

Aryl Halide: 4-bromo/iodo acetophenone (X-substituted)

Catalyst (x mol %)

Product: N-benzyl-4-acetylaniline structure

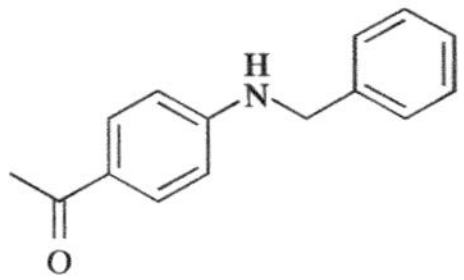

X	Catalyst	x	Additive(s)	Solvent(s)	Temp (°)	Time		Refs.
Br	Cu	5	binol (10 mol %), Cs$_2$CO$_3$	DMSO	90	40 h	(68)	124
I	CuCl	5	TBAH (aq)	—	80	24 h	(83)	282
I	CuCl	10	**L9** (10 mol %), TBAH	DMSO	80	24 h	(74)	281
I	CuI	5	K$_3$PO$_4$	HOCH$_2$CH$_2$OH/i-PrOH	80	5 h	(90)	194
I	CuI	10	L-Pro (20 mol %), K$_2$CO$_3$	DMSO	80	40 h	(88)	47
I	CuI	10	L-Pro (20 mol %), TBAA	DMF	rt	24 h	(90)	61
Br	CuO	5	**L14** (50 mol %), cyclohexanone, TBAB, KOH	H$_2$O	MW (100 W), 130	5 min	(84)	51
Br	CuO	25	**L15** (50 mol %), TBAB, KOH	H$_2$O	MW (100 W), 130	5 min	(50)	267

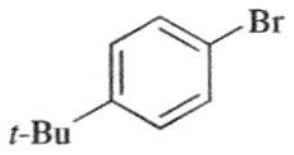

4-bromo-tert-butylbenzene (Br, t-Bu)

Cu (5 mol %), binol (10 mol %), Cs$_2$CO$_3$, DMSO, 90°, 24 h

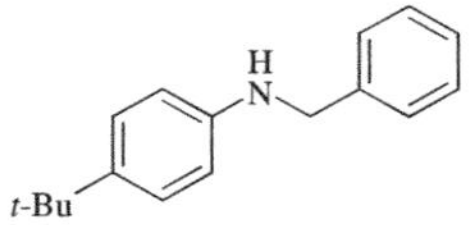

Product: N-benzyl-4-tert-butylaniline structure, (92), 124

4-bromophenyl oxazolidinone (Br, NHAc)

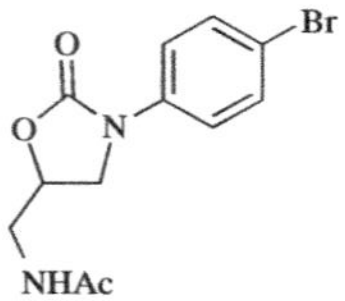

CuI (20 mol %), L-Pro (40 mol %), K$_2$CO$_3$, DMSO, 120°, 15 h

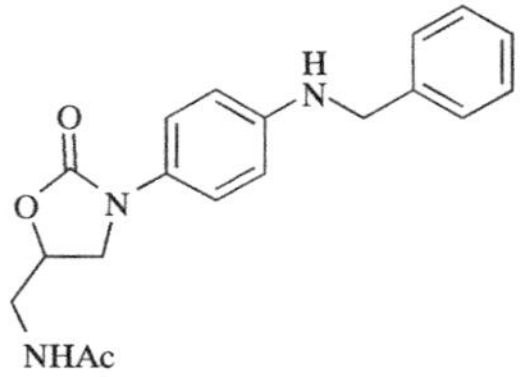

Product structure with N-benzyl, oxazolidinone, NHAc, (65), 292

134

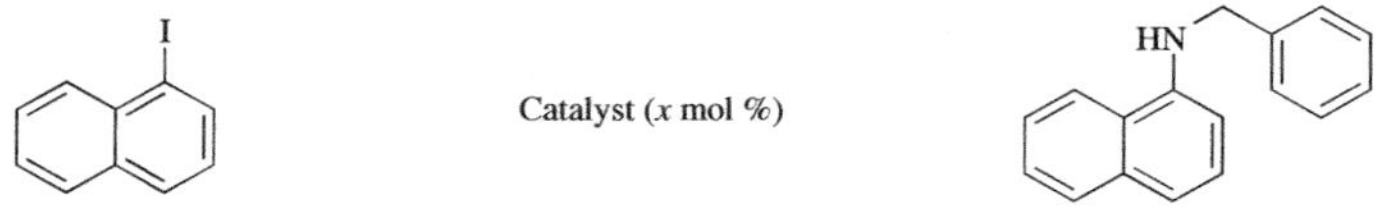

Catalyst (x mol %)

Catalyst	x	Additive(s)	Solvent(s)	Temp (°)	Time (h)		
CuCl	10	L9 (10 mol %), TBAH	DMSO	80	24	(68)	281
CuI	10	DMG (20 mol %), TBPM	DMSO	rt	24	(85)	61
CuI	5	K$_3$PO$_4$	HOCH$_2$CH$_2$OH/i-PrOH	90	12	(70)	194

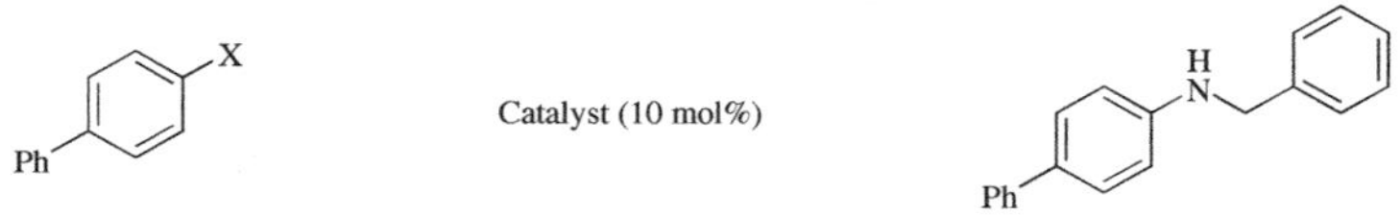

Catalyst (10 mol%)

X	Catalyst	Additives	Solvent	Temp (°)	Time (h)		
Br	CuCl	TMHD (25 mol %), Cs$_2$CO$_3$	NMP	120	10	(76)	188
Br	CuI	L-Pro (20 mol %), K$_2$CO$_3$	DMSO	60	12	(66)	192
I	CuI	L-Pro (20 mol %), K$_2$CO$_3$	DMSO	80	30	(81)	47
I	CuI	DMG (20 mol %), TBPM	DMSO	rt	24	(90)	61

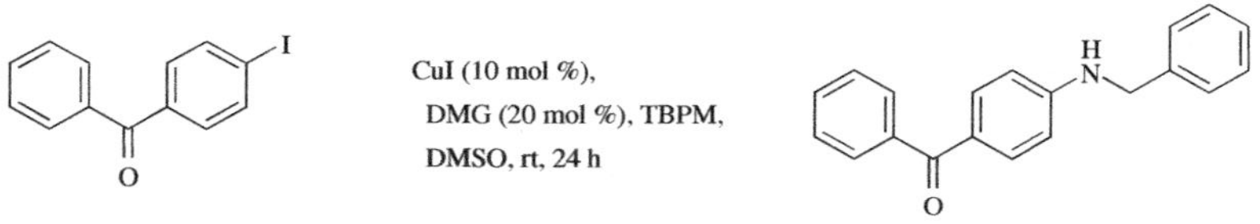

CuI (10 mol %),
DMG (20 mol %), TBPM,
DMSO, rt, 24 h

(97)   61

TABLE 2A. N-ARYLATION OF PRIMARY ALKYL AMINES (*Continued*)

*Please refer to the charts preceding the tables for structures indicated by the **bold** numbers.*

C$_8$

Nitrogen Nucleophile	Aryl Halide	Conditions	Product(s) and Yield(s) (%)	Refs.

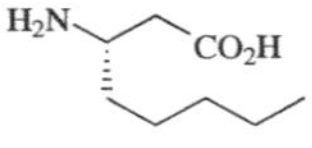

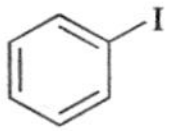

Nitrogen Nucleophile: n-C$_8$H$_{17}$NH$_2$

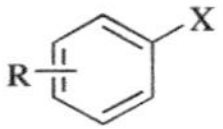

Aryl Halide: (R-substituted aryl–X)

Product: aryl–N(H)–n-C$_8$H$_{17}$

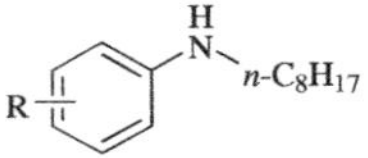

R	X	Catalyst	x	Additive(s)	Solvent	Temp (°)	Time	Yield (%)	Refs.
H	Cl	Cu/Al–HTB	2.5	K$_2$CO$_3$	DMF	100	24 h	(0)	228
2-O$_2$N	Cl	Cu/Al–HTB	2.5	K$_2$CO$_3$	DMF	100	6 h	(89)	228
4-O$_2$N	Cl	Cu/Al–HTB	2.5	K$_2$CO$_3$	DMF	100	8 h	(99)	228
2-BnNHSO$_2$	Br	CuI	10	phen (20 mol %), Cs$_2$CO$_3$	DMSO	MW, 150	11 min	(94)	268
2-Me	Cl	Cu/Al–HTB	2.5	K$_2$CO$_3$	DMF	100	32 h	(42)	228
4-Me	Cl	Cu/Al–HTB	2.5	K$_2$CO$_3$	DMF	100	28 h	(52)	228
2-NC–	Cl	Cu/Al–HTB	2.5	K$_2$CO$_3$	DMF	100	10 h	(75)	228
4-NC–	Cl	Cu/Al–HTB	2.5	K$_2$CO$_3$	DMF	100	12 h	(78)	228
2-CHO	Cl	Cu/Al–HTB	2.5	K$_2$CO$_3$	DMF	100	8 h	(95)	228
2-CHO, 4-Cl	Cl	Cu/Al–HTB	2.5	K$_2$CO$_3$	DMF	100	8 h	(95)	228
4-CHO	Cl	Cu/Al–HTB	2.5	K$_2$CO$_3$	DMF	100	10 h	(90)	228
2-HO$_2$C	Cl	Cu	7.5	K$_2$CO$_3$	H$_2$O	))) (20 kHz), rt	20 min	(88)	241
4-HO$_2$C	Cl	Cu/Al–HTB	2.5	K$_2$CO$_3$	DMF	100	8 h	(97)	228

Additional entries:

Nitrogen nucleophile: H$_2$N–(amino acid, CO$_2$H); Aryl halide: iodobenzene; Conditions: CuI (10 mol %), K$_2$CO$_3$, H$_2$O, DMF, 110°, 48 h; Product (anilino CO$_2$H): (73); Ref. 293

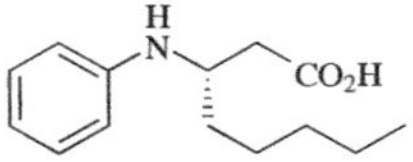

Nitrogen nucleophile: H$_2$N–(glycyl amino acid, CO$_2$H); Aryl halide: iodobenzene; Conditions: CuI (10 mol %), K$_3$PO$_4$•H$_2$O, Me$_2$NCH$_2$CH$_2$OH, H$_2$O, 90°; Product: (46); Ref. 260

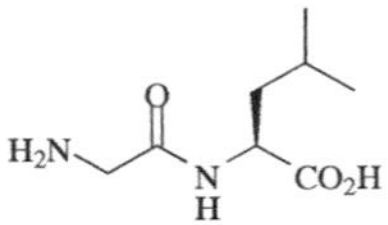

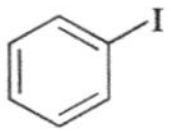

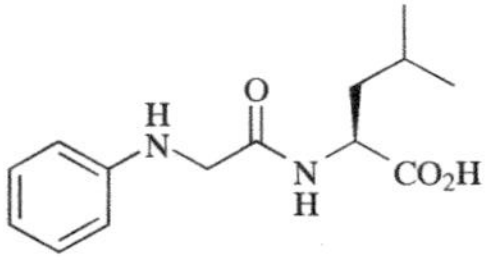

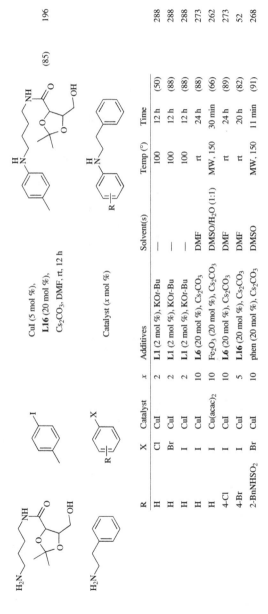

CuI (5 mol %),
L16 (20 mol %),
Cs$_2$CO$_3$, DMF, rt, 12 h   (85)   196

Catalyst (x mol %)

R	X	Catalyst	x	Additives	Solvent(s)	Temp (°)	Time		
H	Cl	CuI	2	**L1** (2 mol %), KOt-Bu	—	100	12 h	(50)	288
H	Br	CuI	2	**L1** (2 mol %), KOt-Bu	—	100	12 h	(88)	288
H	I	CuI	2	**L1** (2 mol %), KOt-Bu	—	100	12 h	(88)	288
H	I	CuI	10	**L6** (20 mol %), Cs$_2$CO$_3$	DMF	rt	24 h	(88)	273
H	I	Cu(acac)$_2$	10	Fe$_2$O$_3$ (20 mol %), Cs$_2$CO$_3$	DMSO/H$_2$O (1:1)	MW, 150	30 min	(66)	262
4-Cl	I	CuI	10	**L6** (20 mol %), Cs$_2$CO$_3$	DMF	rt	24 h	(89)	273
4-Br	I	CuI	5	**L16** (20 mol %), Cs$_2$CO$_3$	DMF	rt	20 h	(82)	52
2-BnNHSO$_2$	Br	CuI	10	phen (20 mol %), Cs$_2$CO$_3$	DMSO	MW, 150	11 min	(91)	268

TABLE 2A. N-ARYLATION OF PRIMARY ALKYL AMINES (Continued)

*Please refer to the charts preceding the tables for structures indicated by the **bold** numbers.*

C₈

Nitrogen Nucleophile	Aryl Halide	Conditions		Product(s) and Yield(s) (%)	Refs.

Nitrogen Nucleophile: H_2N–CH₂CH₂–C₆H₅ (2-phenylethylamine)

Aryl Halide:

Conditions header: Catalyst(s) (x mol %)

Product structure:

R	X	Catalyst(s)	x	Additive(s)	Solvent	Temp (°)	Time (h)		Refs.
H	Cl	Cu, Cu₂O	10, 5	K₂CO₃	EtOCH₂CH₂OEt	130	24	(91)	278
H	Br	Cu, Cu₂O	10, 4	K₂CO₃	EtOCH₂CH₂OEt	130	24	(91)	278
H	Br	Cu, Cu₂O	9, 4	K₂CO₃	EtOCH₂CH₂OEt	130	24	(82)	70
H	Br	CuI	10	binol (20 mol %), K₃PO₄	DMF	rt	16	(62)	258
4-F	Br	Cu, Cu₂O	10, 5	K₂CO₃	EtOCH₂CH₂OEt	130	24	(94)	278
4-Cl	Cl	Cu, Cu₂O	10, 5	K₂CO₃	EtOCH₂CH₂OEt	130	24	(82)	278
4-Cl	Br	Cu, Cu₂O	10, 5	K₂CO₃	EtOCH₂CH₂OEt	130	24	(94)	278
5-Cl	Cl	Cu, Cu₂O	10, 5	K₂CO₃	EtOCH₂CH₂OEt	130	24	(94)	278
5-Br	Cl	Cu, Cu₂O	10, 5	K₂CO₃	EtOCH₂CH₂OEt	130	24	(84)	278
5-Br	Br	Cu, Cu₂O	10, 5	K₂CO₃	EtOCH₂CH₂OEt	130	24	(88)	278
4-O₂N	Cl	Cu, Cu₂O	10, 5	K₂CO₃	EtOCH₂CH₂OEt	130	24	(86)	278
3,4-(MeO)₂	Cl	Cu, Cu₂O	10, 5	K₂CO₃	EtOCH₂CH₂OEt	130	24	(84)	278

Aryl Halide (bromobenzene with R):

Conditions: CuI (5 mol %), **L13** (x mol %), K₃PO₄, 100°, 22 h

Product (cyclohexenyl):

R	x		Refs.
4-Me	5	(79)	120
4-Me	20	(96)	
2,5-Me₂	20	(79)	

Nitrogen Nucleophile: H_2N–CH₂–CH(OH)–C₆H₅ (2-amino-1-phenylethanol)

Aryl Halide (iodide with R):

Conditions: CuI (2.5 mol %)

Product:

R	Additive	Solvents	Temp (°)	Time (h)		Refs.
2-H₂N	NaOH	DMSO/H₂O (2:1)	90	16.5	(86)	279
3-O₂N	K₃PO₄	HOCH₂CH₂OH/i-PrOH	75	16	(86)	

138

Catalyst (x mol %)

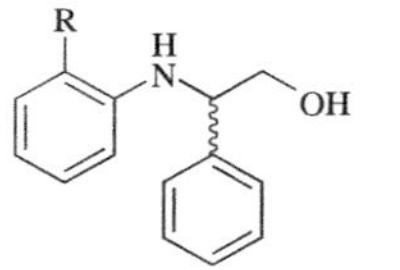

Config.	R	Catalyst	x	Additive(s)	Solvent	Temp (°)	Time (h)		
(R,S)	H	Cu₂O	1	L7 (20 mol %), Cs₂CO₃	MeCN	80	18	(91)	125
(R)	H	Cu(OAc)₂•H₂O	15	(–)-sparteine (30 mol %), Cs₂CO₃	DMF	130	24	(80)	290
(R)	Br	CuI	2.5	NaOH	i-PrOH	90	16	(84)	279

Catalyst (x mol %), 100°

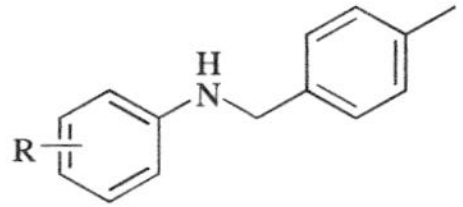

R	X	Catalyst	x	Additive(s)	Solvent	Time (h)		
H	Br	CuI	2	L1 (2 mol %), KOt-Bu	—	12	(77)	288
H	I	6	5	Cs₂CO₃	toluene	12	(82)	123
2-HO₂C, 4-O₂N	Br	CuBr	5	L17 (20 mol %), NaOH, TBAB	H₂O	24	(83)	235

Catalyst (x mol %)

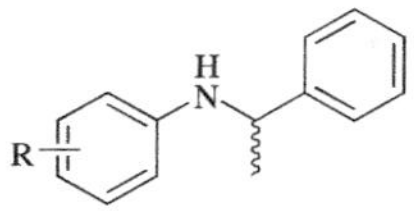

Config.	R	X	Catalyst	x	Additives	Solvent	Temp (°)	Time (h)		er	
(R)	H	I	CuI	10	L-Pro (20 mol %), K₂CO₃	DMSO	60	18	(65)	—	192
(R)	H	I	CuI	10	L-Pro (20 mol %), K₂CO₃	DMSO	60	12	(75)	—	47
(R)	H	I	Cu(OAc)₂•H₂O	20	hippuric acid (20 mol %), Cs₂CO₃	DMF	140	20	(48)	—	289
(R)	H	I	Cu(OAc)₂•H₂O	15	(–)-sparteine (30 mol %), Cs₂CO₃	DMF	130	24	(40)	—	290
(S)	H	I	CuI	10	L11 (20 mol %), Cs₂CO₃	DMF	rt	15	(85)	—	261
(R,S)	4-Cl	I	CuI	10	DMG (20 mol %), TBPE	NMP	0	5	(78)	—	61
(R,S)	3,5-Me₂	I	CuI	5	L16 (20 mol %), Cs₂CO₃	DMF	rt	10	(94)	—	52
(R)	3,5-Me₂	Br	CuI	5	L13 (20 mol %), K₃PO₄	DMF	90	18	(71)	99:1	120

Nitrogen Nucleophile	Aryl Halide	Conditions	Product(s) and Yield(s) (%)	Refs.

*Please refer to the charts preceding the tables for structures indicated by the **bold** numbers.*

C₉

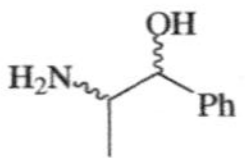	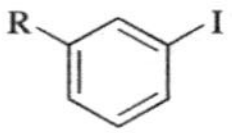	CuI (2.5 mol %), NaOH, *i*-PrOH, 90°, 16.5 h	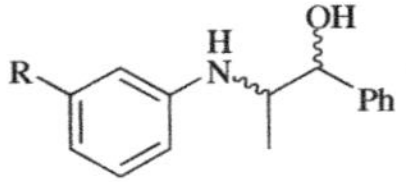 Config. R

(1*S*,2*R*)   MeO (84)   279
(1*R*,2*S*)   HOCH₂ (84)

R

n-C₆H₁₃ (76)   270
c-C₆H₁₁ (69)
4-MeOC₆H₄ (69)

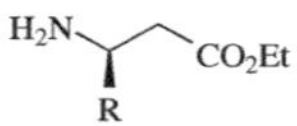

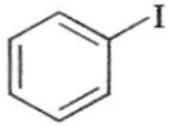

CuI (10 mol %), K₂CO₃, H₂O, DMF, 100°, 48 h

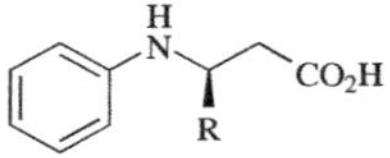

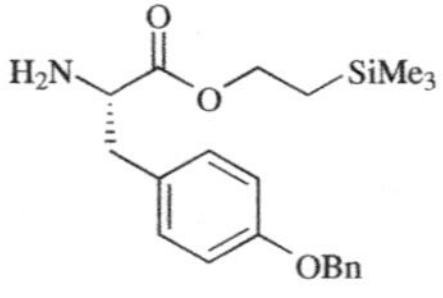

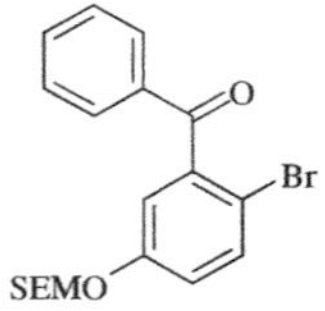

CuI (10 mol %), K₂CO₃, DMF, 90°, 72 h

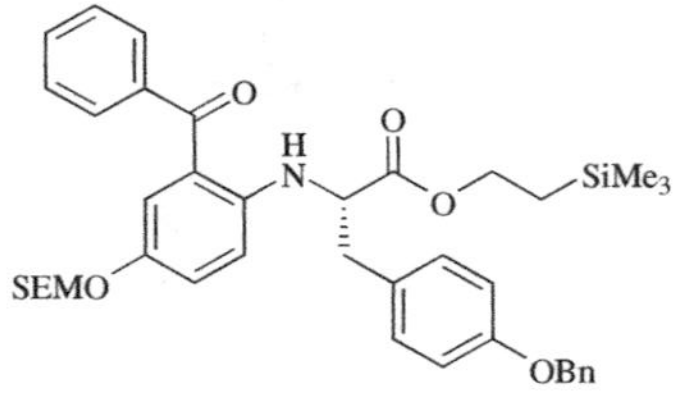

SiMe₃ (50)   294

OBn

Catalyst (x mol %)

R¹	N-Nucleophile	Config.	R²	X	Catalyst	x	Additive(s)	Solvent(s)	Temp (°)	Time		
H	free base	(R/S)	H	Br	CuI	10	K₃PO₄•H₂O	Me₂N(CH₂)₂OH/H₂O	90	—	(80)	260
H	free base	(R)	H	Br	CuI	10	KI, K₂CO₃	H₂O	MW, 185	40 min	(77)	25
H	free base	(S)	H	I	CuI	3	Cs₂CO₃	DMSO	90	48 h	(96)	275
H	free base	(S)	H	Br	CuI	10	K₂CO₃	H₂O/DMF	90	48 h	(92)	116
H	free base	(S)	H	Br	CuI	10	KI, K₂CO₃	H₂O	MW, 185	40 min	(78)	25
H	free base	(S)	H	Br	CuI	10	K₂CO₃	DMF	MW, 140	20 min	(65)	265
Et	HCl salt	(S)	H	Br	CuI	10	KI, K₂CO₃	H₂O	MW, 185	40 min	(96)	25
allyl	HCl salt	(S)	H	Br	CuI	10	KI, K₂CO₃	H₂O	MW, 185	40 min	(57)	25
allyl	tosylate	(S)	H	Br	CuI	10	KI, K₂CO₃	H₂O	MW, 185	40 min	(93)	25
t-Bu	HCl salt	(S)	H	Br	CuI	10	KI, K₂CO₃	H₂O	MW, 185	40 min	(66)	25
Bn	HCl salt	(S)	H	Br	CuI	10	KI, K₂CO₃	H₂O	MW, 185	40 min	(76)	25
H	free base	(S)	4-Cl	I	CuI	10	L8 (20 mol %), K₃PO₄	DMF	30	3 h	(86)	271
H	free base	(S)	4-Cl	I	La₀.₉Ce₀.₁Co₀.₆Cu₀.₄O₃	2.5	Cs₂CO₃	toluene	110	48 h	(22)	295
H	free base	(S)	3-MeO	Br	CuI	10	KI, K₂CO₃	H₂O	MW, 185	40 min	(79)	25
H	free base	(S)	2-Me	Br	CuI	10	KI, K₂CO₃	H₂O	MW, 185	40 min	(43)	25
H	free base	(S)	4-Me	I	CuI	10	L11 (20 mol %), Cs₂CO₃	DMF	rt	12 h	(85)	261
H	free base	(S)	4-Me	I	CuBr	5	L17 (20 mol %), TBAB, NaOH	H₂O	100	24 h	(89)	235
H	free base	(S)	3-NC-	Br	CuI	10	KI, K₂CO₃	H₂O	MW, 185	40 min	(53)	25
H	free base	(S)	3-HOCH₂	Br	CuI	10	KI, K₂CO₃	H₂O	MW, 185	40 min	(87)	25
H	free base	(S)	3-Ac	Br	CuI	10	KI, K₂CO₃	H₂O	MW, 185	40 min	(94)	25
H	free base	(S)	4-t-Bu	Br	CuI	10	KI, K₂CO₃	H₂O	MW, 185	40 min	(41)	25
H	free base	(S)	4-Bz	Br	CuI	10	KI, K₂CO₃	H₂O	MW, 185	40 min	(18)	25

141

TABLE 2A. *N*-ARYLATION OF PRIMARY ALKYL AMINES (*Continued*)

Nitrogen Nucleophile	Aryl Halide	Conditions	Product(s) and Yield(s) (%)	Refs.

*Please refer to the charts preceding the tables for structures indicated by the **bold** numbers.*

C₉

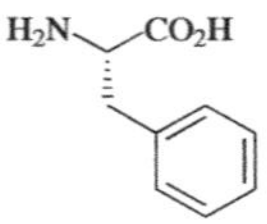

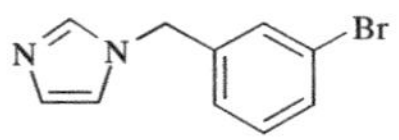

| | | CuI (10 mol %), K₂CO₃, KI, H₂O, MW, 185°, 40 min | | 25 |

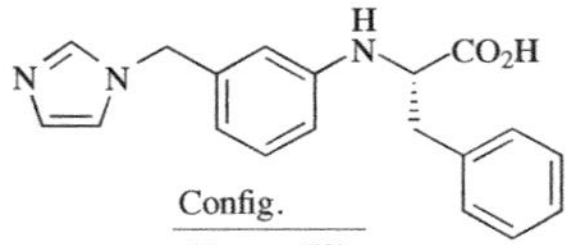

Config.
(R) (59)
(S) (50)

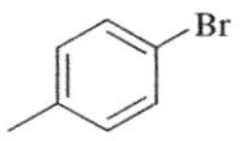

| | | 1. CuI (10 mol %), K₂CO₃, DMF; 90° 2. PhCOCH₂Br, Et₃N, EtOAc | 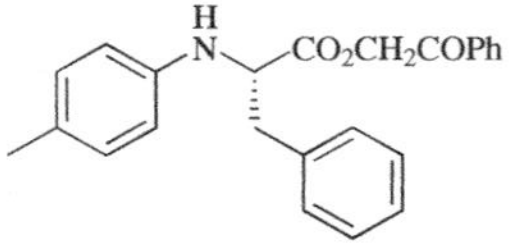 (59) | 272 |

| 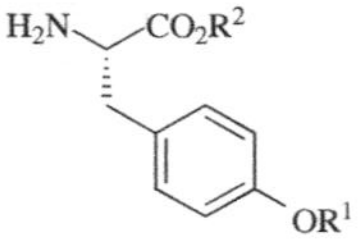 | 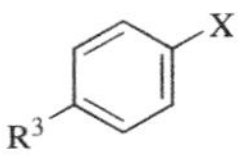 | CuI (10 mol %) | 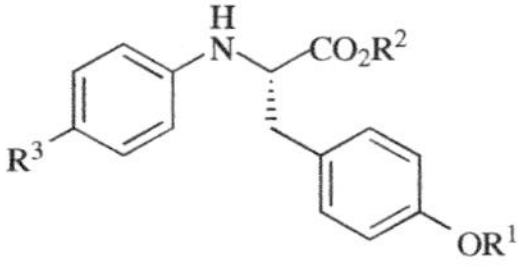 | |

R¹	R²	R³	X	Additives	Solvent(s)	Temp (°)	Time
H	H	H	I	**L8** (20 mol %), K₃PO₄	DMF	30	3 h (78)
H	H	Cl	I	**L8** (20 mol %), K₃PO₄	DMF	30	3 h (83)
H	Me	Cl	I	**L8** (20 mol %), K₃PO₄	DMF	30	3 h (71)
H	H	Me	I	**L8** (20 mol %), K₃PO₄	DMF	30	3 h (77)
allyl	H	*t*-Bu	Br	KI, K₂CO₃	H₂O/MeCN (4:1)	MW, 185	40 min (64)
allyl	H	Bz	Br	KI, K₂CO₃	H₂O/MeCN (4:1)	MW, 185	40 min (36)

142

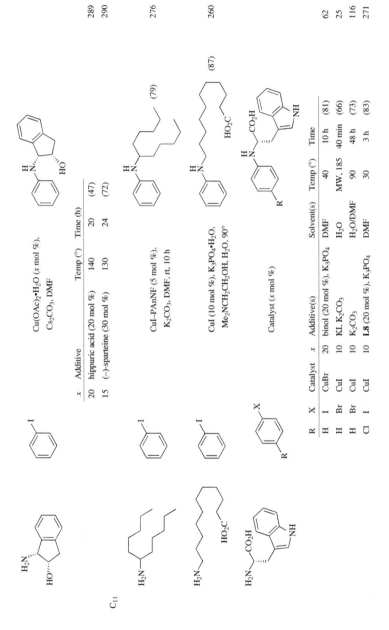

x	Additive	Temp (°)	Time (h)	
20	hippuric acid (20 mol %)	140	20	(47)	289
15	(–)-sparteine (30 mol %)	130	24	(72)	290

Cu(OAc)$_2$•H$_2$O (x mol %), Cs$_2$CO$_3$, DMF

CuI–PAnNF (5 mol %), K$_2$CO$_3$, DMF, rt, 10 h (79) 276

CuI (10 mol %), K$_3$PO$_4$•H$_2$O, Me$_2$NCH$_2$CH$_2$OH, H$_2$O, 90° (87) 260

Catalyst (x mol %)

R	X	Catalyst	x	Additive(s)	Solvent(s)	Temp (°)	Time		
H	I	CuBr	20	binol (20 mol %), K$_3$PO$_4$	DMF	40	10 h	(81)	62
H	Br	CuI	10	KI, K$_2$CO$_3$	H$_2$O	MW, 185	40 min	(66)	25
H	Br	CuI	10	K$_2$CO$_3$	H$_2$O/DMF	90	48 h	(73)	116
Cl	I	CuI	10	L8 (20 mol %), K$_3$PO$_4$	DMF	30	3 h	(83)	271

C$_{11}$

TABLE 2A. N-ARYLATION OF PRIMARY ALKYL AMINES (Continued)

Nitrogen Nucleophile	Aryl Halide	Conditions	Product(s) and Yield(s) (%)	Refs.

*Please refer to the charts preceding the tables for structures indicated by the **bold** numbers.*

C₁₂

n-C₁₂H₂₅NH₂

Aryl Halide: C₆H₅–X

Product: C₆H₅–NH–n-C₁₂H₂₅

X	Catalyst(s)	x	Additive(s)	Temp (°)	Time (h)		Refs.
I	CuBr	20	binol (20 mol %), K₃PO₄	rt	8	(83)	62
Br	CuI	10	DPP (20 mol %), K₃PO₄	90	24	(88)	108
Br	CuI	10	PPAPM (20 mol %), K₃PO₄	90	24	(87)	48
Cl	CuI–PAnNF	5	K₂CO₃	80	18	(7)	276
I	CuI–PAnNF	5	K₂CO₃	rt	4	(99)	276
I	CuO, FeCl₃	10, 10	binol (20 mol %), Cs₂CO₃	80	12	(76)	277

Aryl Halide: R–C₆H₄–X

Product: R–C₆H₄–NH–n-C₁₂H₂₅

R	X	Catalyst	x	Additive(s)	Solvent	Temp (°)	Time		Refs.
4-Cl	I	CuBr	20	binol (20 mol %), K₃PO₄	DMF	rt	6 h	(83)	62
4-Cl	I	CuBr	5	L17 (20 mol %), TBAB. NaOH	H₂O	100	24 h	(85)	235
4-Cl	I	CuI	10	L8 (20 mol %), K₃PO₄	DMF	30	1 h	(77)	271
4-Br	I	CuBr	20	binol (20 mol %), K₃PO₄	DMF	rt	10 h	(72)	62
2-O₂N	Cl	Cu/Al–HTB	2.5	K₂CO₃	DMF	100	6 h	(95)	228
2-O₂N	Cl	CuI–PAnNF	5	K₂CO₃	DMF	80	2 h	(95)	276
3-O₂N	I	CuBr	20	binol (20 mol %), K₃PO₄	DMF	rt	10 h	(72)	62
3-O₂N	I	CuI–PAnNF	5	K₂CO₃	DMF	rt	7 h	(76)	276
4-O₂N	Cl	Cu/Al–HTB	2.5	K₂CO₃	DMF	100	6 h	(99)	228
4-O₂N	Cl	CuI–PAnNF	5	K₂CO₃	DMF	80	5 h	(93)	276
4-O₂N	I	CuI–PAnNF	5	K₂CO₃	DMF	rt	4 h	(98)	276
4-MeO	Cl	CuI–PAnNF	5	K₂CO₃	DMF	80	10 h	(65)	276
4-MeO	I	CuI–PAnNF	5	K₂CO₃	DMF	rt	8 h	(68)	276
4-Me	I	CuI	10	L8 (20 mol %), K₃PO₄	DMF	30	5 h	(71)	271

	X	Catalyst	mol %	Base/additive	Solvent	Temp	Time	(Yield)	Ref
4-Me	I	CuI–PAnNF	5	K₂CO₃	DMF	rt	13 h	(68)	276
4-Me	I	CuO	10	binol (20 mol %), FeCl₃ (10 mol %), Cs₂CO₃	DMF	90	12 h	(72)	277
2-NC–	Cl	CuI–PAnNF	5	K₂CO₃	DMF	80	4 h	(78)	276
4-NC–	Cl	CuI–PAnNF	5	K₂CO₃	DMF	80	8 h	(64)	276
2-CHO, 5-Cl	Cl	CuI–PAnNF	5	K₂CO₃	DMF	80	4 h	(86)	276
3-CHO, 4-O₂N	Cl	CuI–PAnNF	5	K₂CO₃	DMF	80	1.5 h	(98)	276
4-CHO	Cl	CuI–PAnNF	5	K₂CO₃	DMF	80	4 h	(72)	276
2-HO₂C	Cl	Cu	7.5	K₂CO₃	H₂O	))) (20 kHz), rt	20 min	(88)	241
2-HO₂C	Br	CuBr	5	L17 (20 mol %), TBAB, NaOH	H₂O	100	24 h	(83)	235
2-HO₂C	Br	CuI	10	binol (20 mol %), K₃PO₄	DMF	rt	16 h	(64)	258
4-HO₂C	Cl	CuI–PAnNF	5	K₂CO₃	DMF	80	2 h	(85)	276

C₁₆

$n\text{-}C_{16}H_{33}NH_2$ + (Br–phenyl) → CuI (2 mol %), **L1** (2 mol %), KO*t*-Bu, 100°, 12 h →

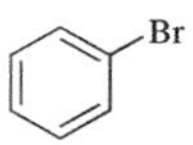

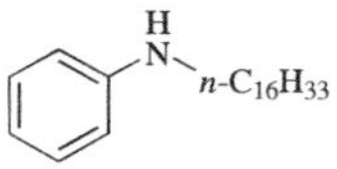

H–N–*n*-C₁₆H₃₃ (phenyl)   (65)   288

---

a 4-Iodo-*N*-(*n*-butyl)aniline was formed in 4% yield.

145

TABLE 2B. *N*-HETEROARYLATION OF PRIMARY ALKYL AMINES

*Please refer to the charts preceding the tables for structures indicated by the **bold** numbers.*

Nitrogen Nucleophile	Heteroaryl Halide	Conditions	Product(s) and Yield(s) (%)	Refs.

C₂

| H₂N⌢SMe | (I)pyridyl | CuI (10 mol %), **L11** (20 mol %), Cs₂CO₃, DMF; rt, 20 h | H-N(pyridyl)⌢SMe (95) | 261 |

C₃

| H₂N⌢(dioxolane) | NC—(Br)pyridyl | CuI (20 mol %), L-Pro (40 mol %), K₂CO₃, DMSO, 120°, 15 h | NC—pyridyl-NH—(dioxolane) (70) | 292 |

| H₂N⌢ (allyl) | | | R-NH⌢ (allyl) | |

R	X	x	Additives	Solvent	Temp (°)	Time (h)		
2-thienyl	I	5	**L11** (20 mol %), Cs₂CO₃	DMF	rt	17	(83)	52
3-pyridyl	Br	10	L-Pro (20 mol%), K₂CO₃	DMSO	90	40	(97)	47
quinolinyl	I	10	**L11** (20 mol %), Cs₂CO₃	DMF	rt	19	(87)	261

| (iodouracil with Bn, Bn) | (CuOTf)₂·C₆H₆ (5 mol %), phen (1 eq), dba (5 mol %), Cs₂CO₃, p-xylene, 95°, 24 h | | (thiazoline-NH-uracil) (0) | 187 |

C₄

| H₂N⌢⌢ (butyl) | | | R-NH⌢⌢ | |

R	X	Catalyst	x	Catalyst (x amount)	Additive(s)	Solvent	Temp (°)	Time (h)	
2-thienyl	I	Cu	10 mol %		K₃PO₄·H₂O	Me₂NCH₂CH₂OH	60	11	(62)	296, 195
3-thienyl	Br	Cu	10 mol %		K₃PO₄·H₂O	Me₂NCH₂CH₂OH	80	36	(85)	296
2-pyridyl	I	CuBr	2.5 mol %	**L5** (5 mol %), Cs₂CO₃		DMF	89	24	(95)	119
2-pyridyl	I	CuI	1 eq	CsOAc		DMSO	90	24	(80)	189
2-pyridyl	I	CuI	10 mol %	CsOAc		DMSO	90	24	(70)	274
3-pyridyl	I	CuBr	2.5 mol %	**L5** (5 mol %), Cs₂CO₃		DMF	89	24	(94)	119
3-pyridyl	I	CuI	10 mol %	CsOAc		DMSO	90	24	(96)	274

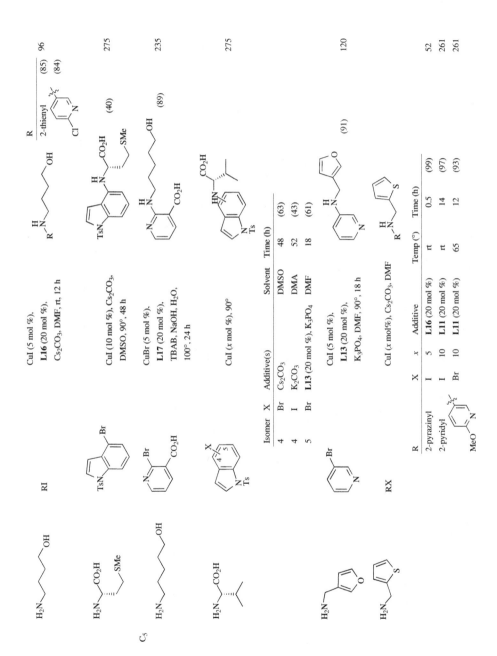

TABLE 2B. *N*-HETEROARYLATION OF PRIMARY ALKYL AMINES (*Continued*)

Nitrogen Nucleophile	Heteroaryl Halide	Conditions	Product(s) and Yield(s) (%)	Refs.

*Please refer to the charts preceding the tables for structures indicated by the **bold** numbers.*

C_6

$n\text{-}C_6H_{13}NH_2$

Heteroaryl Halide: (3-X-pyridine, X)

Catalyst (5 mol %)

X	Catalyst	Additive(s)	Solvent	Temp (°)	Time (h)		
Br	CuCl	TBAH (aq)	—	80	24	(70)	282
I	CuCl	TBAH (aq)	—	80	24	(75)	282
Br	CuI	**L13** (5 mol %), K_3PO_4	—	100	22	(90)	120
Br	CuI	**L13** (20 mol %), K_3PO_4	DMF	90	18	(92)	120

Product: $\text{NH}n\text{-}C_6H_{13}$ (3-substituted pyridine)

C_{6-7}

RNH_2

Heteroaryl Halide: (6-I-imidazo[1,2-a]pyridine-2-yl, 4-FC₆H₄)

Conditions: CuI (15 mol %), HOCH₂CH₂OH, K_3PO_4, *i*-PrOH, 85°, 48 h

Product: (6-NHR-imidazo[1,2-a]pyridine-2-yl, 4-FC₆H₄)

R	
c-C₆H₁₁	(72)
Bn	(85)

Refs. 126

Heteroaryl Halide: (8-I-6-methylimidazo[1,2-a]pyridine-2-yl, 4-FC₆H₄)

Conditions: CuI (15 mol %), HOCH₂CH₂OH, K_3PO_4, *i*-PrOH, 85°, 48 h

Product: (8-NHR-6-methylimidazo[1,2-a]pyridine-2-yl, 4-FC₆H₄)

R	Temp (°)	Time (h)	
n-C₆H₁₃	85	48	(69)
c-C₆H₁₁	112	20	(57)
Bn	85	20	(62)

Refs. 297

C_{6-8}

$H_2N\text{-}(CH_2)_3\text{-}R$

Heteroaryl Halide: (5-I-1,3-dibenzyluracil)

Conditions: (CuOTf)₂•C₆H₆ (50 mol %), phen (1 eq), dba, Cs₂CO₃, *p*-xylene, 95°, 24 h

Product: (5-NH-(CH₂)₃-R-1,3-dibenzyluracil)

R	
1-(1*H*-imidazolyl)	(76)
1-morpholino	(77)
1-piperidinyl	(78)

Refs. 187

148

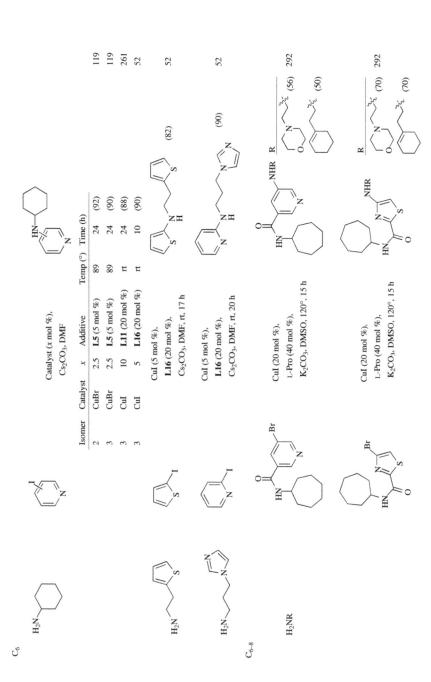

Isomer	Catalyst	x	Additive	Temp (°)	Time (h)		
2	CuBr	2.5	**L5** (5 mol %)	89	24	(92)	119
3	CuBr	2.5	**L5** (5 mol %)	89	24	(90)	119
3	CuI	10	**L11** (20 mol %)	rt	24	(88)	261
3	CuI	5	**L16** (20 mol %)	rt	10	(90)	52

149

TABLE 2B. *N*-HETEROARYLATION OF PRIMARY ALKYL AMINES (*Continued*)

Nitrogen Nucleophile	Heteroaryl Halide	Conditions	Product(s) and Yield(s) (%)	Refs.

*Please refer to the charts preceding the tables for structures indicated by the **bold** numbers.*

C₇

		Cu (10 mol %), K₃PO₄•H₂O, Me₂NCH₂CH₂OH, 80°, 36 h	(83)	296
$n\text{-}C_7H_{15}NH_2$				
		CuI (5 mol %), **L16** (20 mol %), Cs₂CO₃, DMF, rt, 20 h	(85)	52, 196
		CuI (10 mol %), **L11** (20 mol %), Cs₂CO₃, DMF, rt, 19 h	(87)	261
		Catalyst (10 mol %)		

X	Catalyst	Additive(s)	Solvent	Temp(°)	Time (h)		
Br	Cu	K₃PO₄•H₂O	Me₂NCH₂CH₂OH	80	36	(86)	296
I	CuI	DMG (20 mol %), TBPM	DMSO	rt	24	(42)	61

| | | CuI (5 mol %), **L13** (20 mol %), K₃PO₄, DMF, 90°, 18 h | (84) | 120 |

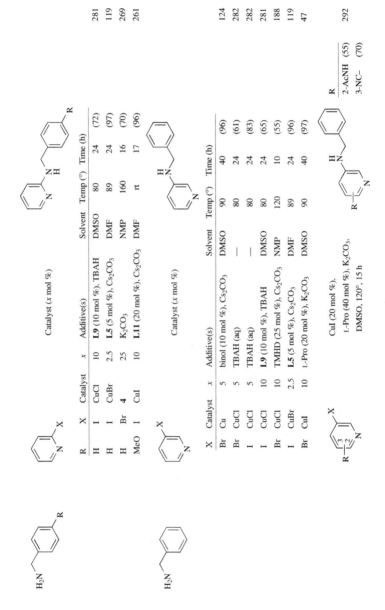

Catalyst (x mol %)

R	X	Catalyst	x	Additive(s)	Solvent	Temp (°)	Time (h)		
H	I	CuCl	10	L9 (10 mol %), TBAH	DMSO	80	24	(72)	281
H	I	CuBr	2.5	L5 (5 mol %), Cs$_2$CO$_3$	DMF	89	24	(97)	119
H	Br	4	25	K$_2$CO$_3$	NMP	160	16	(70)	269
MeO	I	CuI	10	L11 (20 mol %), Cs$_2$CO$_3$	DMF	rt	17	(96)	261

Catalyst (x mol %)

X	Catalyst	x	Additive(s)	Solvent	Temp (°)	Time (h)		
Br	Cu	5	binol (10 mol %), Cs$_2$CO$_3$	DMSO	90	40	(96)	124
Br	CuCl	5	TBAH (aq)	—	80	24	(61)	282
I	CuCl	5	TBAH (aq)	—	80	24	(83)	282
I	CuCl	10	L9 (10 mol %), TBAH	DMSO	80	24	(65)	281
Br	CuCl	10	TMHD (25 mol %), Cs$_2$CO$_3$	NMP	120	10	(55)	188
I	CuBr	2.5	L5 (5 mol %), Cs$_2$CO$_3$	DMF	89	24	(96)	119
Br	CuI	10	L-Pro (20 mol %), K$_2$CO$_3$	DMSO	90	40	(97)	47

CuI (20 mol %),
L-Pro (40 mol %), K$_2$CO$_3$,
DMSO, 120°, 15 h

R	
2-AcNH	(55)
3-NC—	(70)

292

Nitrogen Nucleophile	Heteroaryl Halide	Conditions	Product(s) and Yield(s) (%)	Refs.

*Please refer to the charts preceding the tables for structures indicated by the **bold** numbers.*

C7

(benzylamine)	(5-bromoindole, PhO2S)	CuI (20 mol %), L-Pro (40 mol %), K2CO3, DMSO, 120°, 15 h	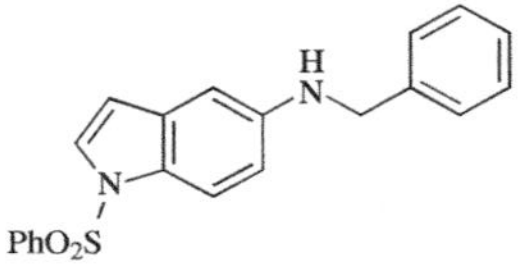 (65)	292
(morpholinopropylamine)	(3-bromothiophene)	CuI (10 mol %), L11 (20 mol %), Cs2CO3, DMF, rt, 14 h	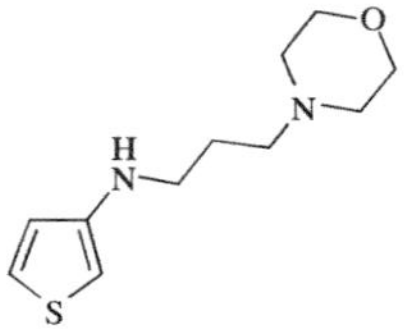 (90)	261

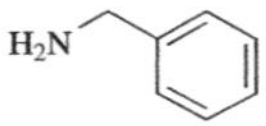

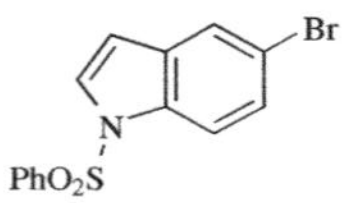

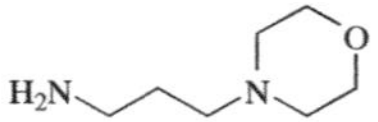

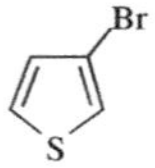

C8

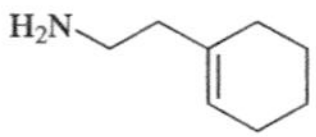

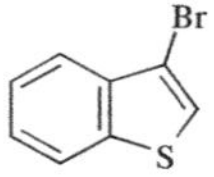

(cyclohexenylethylamine)	(3-bromobenzothiophene)	CuI (5 mol %), L13 (20 mol %), K3PO4, DMF, 90°, 18 h	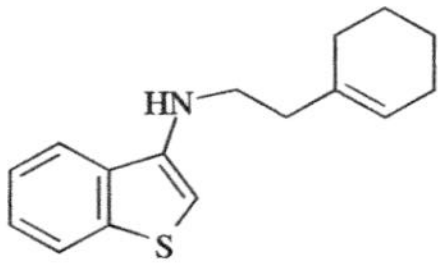 (85)	120
(4-methylbenzylamine)	(2-bromopyridine)	Cat. **4** (25 mol %), K2CO3, NMP, 160°, 16 h	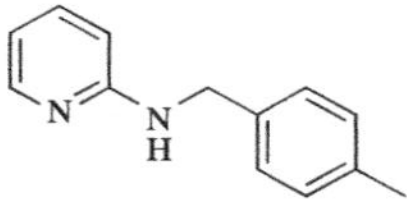 (90)	269

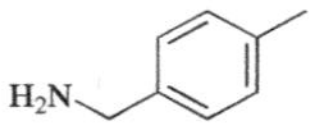

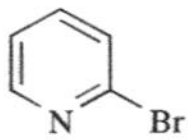

C9

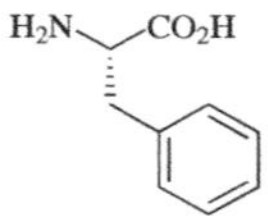 (H_2N–CH(CO$_2$H)–CH$_2$Ph)	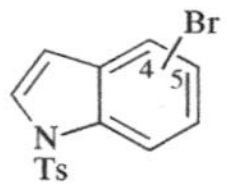 (bromoindole, Ts, 4/5)	CuI (x mol %), 90°	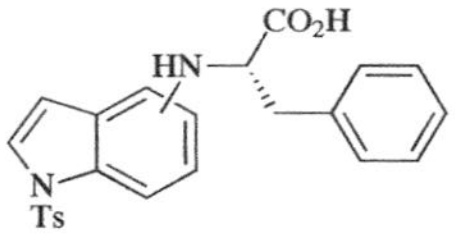	275

Isomer	x	Additive(s)	Solvent	Time (h)	
4	10	Cs$_2$CO$_3$	DMSO	48	(63)
5	5	L13 (20 mol %), K$_3$PO$_4$	DMF	18	(61)

TABLE 3A. N-ARYLATION OF ACYCLIC SECONDARY ALKYL AMINES

*Please refer to the charts preceding the tables for structures indicated by the **bold** numbers.*

Nitrogen Nucleophile	Aryl Halide	Conditions	Product(s) and Yield(s) (%)	Refs.

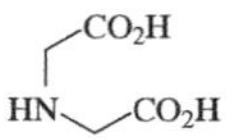

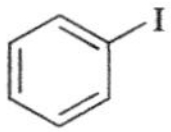

C₄

Et₂NH

Aryl Halide: 4-R-C₆H₄-X

Catalyst (x mol %)

Product: 4-R-C₆H₄-NEt₂

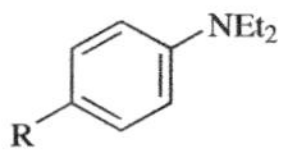

R	X	Catalyst	x	Additive(s)	Solvent	Temp (°)	Time (h)		Refs.
H	I	Cu	10	K₃PO₄•H₂O	Me₂NCH₂CH₂OH	85	60	(2)	195
H	Cl	Cu₂O	5	NaOt-Bu	NMP	110	24	(79)	266
H	Br	Cu₂O	5	NaOt-Bu	NMP	100	24	(84)	266
H	I	Cu₂O	5	NaOt-Bu	NMP	100	24	(71)	266
MeO	I	CuI	10	L-Pro (20 mol %), K₂CO₃	DMSO	90	40	(21)	47

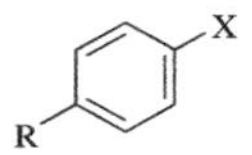

C₅

Nucleophile: HN(CH₂CO₂H)(CH₂CO₂H) — N,N-di(carboxymethyl)amine

Aryl Halide: iodobenzene

Conditions: CuI (10 mol %), K₃PO₄•H₂O, Me₂NCH₂CH₂OH, H₂O, 90°

Product: PhN(CH₂CO₂H)₂ (0)

Ref. 260

C₅

Nucleophile: MeHN–CH(CO₂H)–CH(CH₃)₂ (valine-derived)

Aryl Halide: 2-R-C₆H₄-Br

Conditions: CuI (10 mol %), K₂CO₃, DMA, 90°, 48 h

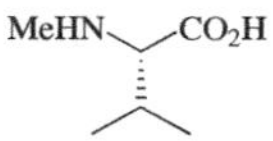

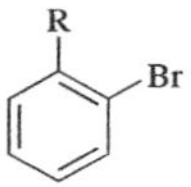

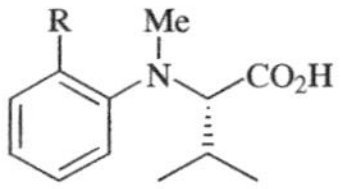

Product: 2-R-C₆H₄-N(Me)CH(CO₂H)CH(CH₃)₂

R	Yield
H	(46)
HOCH₂	(17)

Ref. 116

C₆

Nucleophile: (sugar-derived amine) Me-N with polyol chain (OH H, OH OH, H OH H H, CH₂OH)

Aryl Halide: bromobenzene

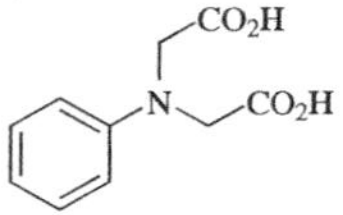

Conditions: CuI (10 mol %), K₃PO₄•H₂O, Me₂NCH₂CH₂OH, H₂O, 90°

Product: Ph-N(Me)- polyol chain (31)

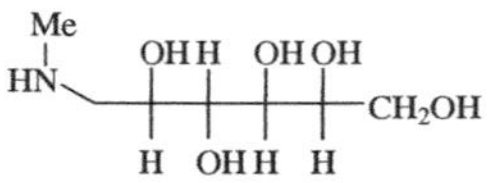

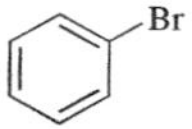

Ref. 260

C₈

Nucleophile: n-Bu₂NH

Aryl Halide: bromobenzene

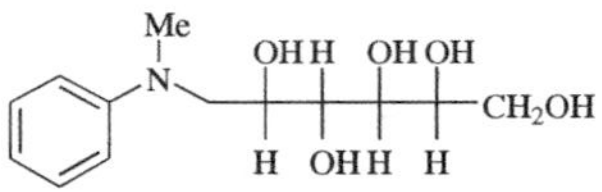

Conditions: CuCl (10 mol%), TMHD (25 mol %), Cs₂CO₃, NMP, 120°, 10 h

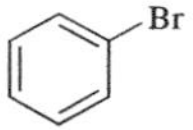

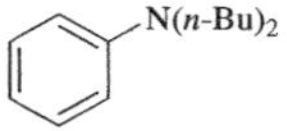

Product: PhN(n-Bu)₂ (9)

Ref. 188

TABLE 3A. *N*-ARYLATION OF ACYCLIC SECONDARY ALKYL AMINES (*Continued*)

Nitrogen Nucleophile	Aryl Halide	Conditions	Product(s) and Yield(s) (%)	Refs.

*Please refer to the charts preceding the tables for structures indicated by the **bold** numbers.*

C₈

n-Bu$_2$NH	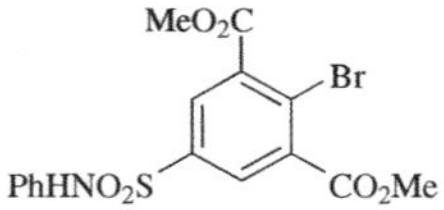 (MeO$_2$C, Br, CO$_2$Me, PhHNO$_2$S aryl halide)	Cu (15 mol %), K$_2$CO$_3$, dioxane, reflux, 24 h	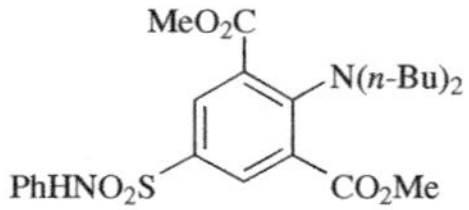 (MeO$_2$C, N(n-Bu)$_2$, CO$_2$Me, PhHNO$_2$S) (79)	298

C₁₂

(cyclohexyl)HN(cyclohexyl)	(Br, Br, CO$_2$H aryl halide)	Cu$_2$O (5 mol %), K$_3$PO$_4$, DMA, 70°, 16 h	(N,N-dicyclohexyl, Br, CO$_2$H) (0)	206

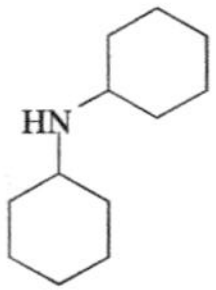

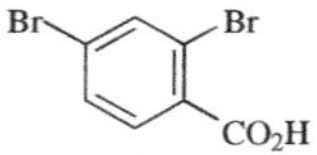

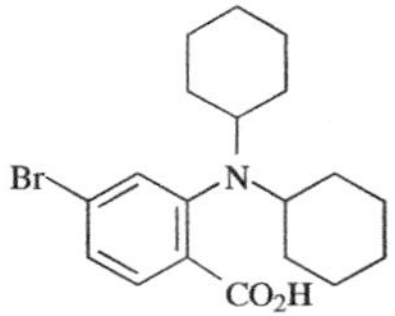

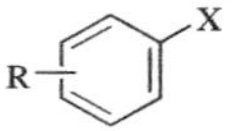

C₁₄

Bn$_2$NH   →   R–NBn$_2$ (from R–X)

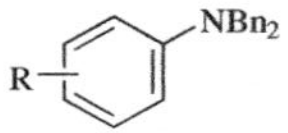

R	X	Catalyst	x	Additive(s)	Solvent	Temp (°)	Time (h)		Refs.
		Catalyst	x	Additive(s)					
H	I	**6**	5 mol %	Cs$_2$CO$_3$	xylene	135	16	(0)	123
2-O$_2$N	Cl	Cu/Al-HTB	1 eq	K$_2$CO$_3$	—	160	16	(76)	117
4-MeO	I	CuI	10 mol %	DMG (20 mol %), K$_2$CO$_3$	DMSO	90	40	(<10)	47

TABLE 3B. *N*-HETEROARYLATION OF ACYCLIC SECONDARY ALKYL AMINES

Nitrogen Nucleophile	Heteroaryl Halide	Conditions	Product(s) and Yield(s) (%)	Refs.
C₃ MeHN⌒OH	⟨S⟩-Br	Cu (x mol %), CuI (10 mol %), K₃PO₄•H₂O, 80°, 24 h	⟨S⟩-NMe⌒OH Isomer x / 2 — 10 (81) / 3 — 40 (90)	296
C₄ Et₂NH	⟨S⟩-Br	Cu (x mol %), CuI (10 mol %), K₃PO₄•H₂O, Me₂NCH₂CH₂OH	⟨S⟩-NEt₂ Isomer x Temp (°) Time (h) / 2 40 80 48 (39) / 3 10 85 60 (15)	296
EtHN⌒OH	⟨S⟩-Br	Cu (10 mol %), CuI (10 mol %), K₃PO₄•H₂O, 80°	⟨S⟩-NEt⌒OH Isomer Time (h) / 2 24 (81) / 3 45 (45)	296
C₆ *n*-Pr₂NH	⟨S⟩-Br	Cu (40 mol%), CuI (10 mol%), K₃PO₄•H₂O, Me₂NCH₂CH₂OH, 80°, 48 h	⟨S⟩-(n-Pr)₂ (35)	296
C₈ H N⌒OH (with phenyl)	⟨S⟩-Br	Cu (10 mol %), CuI (10 mol %), K₃PO₄•H₂O, 80°, 48 h	(15)	296

TABLE 4A. N-ARYLATION OF CYCLIC SECONDARY ALKYL AMINES

Nitrogen Nucleophile	Aryl Halide	Conditions	Product(s) and Yield(s) (%)	Refs.

*Please refer to the charts preceding the tables for structures indicated by the **bold** numbers.*

C₄

Pyrrolidine (N–H)

Catalyst (x mol %)

X	Catalyst	x	Additives(s)	Solvent	Temp (°)	Time (h)		Refs.
Br	Cu	10	K₃PO₄•H₂O	Me₃NCH₂CH₂OH	80	48	(61)	195
I	Cu	10	binol (10 mol %), Cs₂CO₃	DMSO	90	24	(96)	124
I	CuBr	2.5	**L5** (5 mol %), Cs₂CO₃	DMF	90	24	(90)	118
I	CuI	10	CsOAc	DMSO	90	24	(68)	274
Br	CuI	10	DMG (20 mol %), TBPM	DMSO	rt	24	(83)	61
I	CuI–PAnNF	5	K₂CO₃	DMF	rt	6	(67)	276
I	CuO	10	FeCl₃ (10 mol %), binol (20 mol %), Cs₂CO₃	DMF	90	12	(83)	277
I	CuO	1.26	KOH	DMSO	85	8	(93)	224

Catalyst (x mol %)

R	X	Catalyst	x	Additives(s)	Solvent	Temp (°)	Time (h)		Refs.
Cl	I	CuBr	20	BINAP (20 mol %), K₃PO₄	DMF	rt	5	(71)	62
Cl	Br	CuI	10	L-Pro (20 mol %), K₂CO₃	DMSO	90	27	(89)	47
Cl	I	CuI	10	**L8** (20 mol %), K₃PO₄	DMF	30	9	(68)	271
Cl	I	CuI–PAnNF	5	K₂CO₃	DMF	rt	6	(82)	276
Br	Br	Cu	10	K₃PO₄•H₂O	Me₃NCH₂CH₂OH	80	48	(28)	195
Br	I	CuBr	20	BINAP (20 mol %), K₃PO₄	DMF	rt	10	(67)	62
Br	I	CuI–PAnNF	5	K₂CO₃	DMF	rt	6	(75)	276
Br	I	CuO	10	FeCl₃ (10 mol %), binol (20 mol %), Cs₂CO₃	DMF	90	12	(85)	277

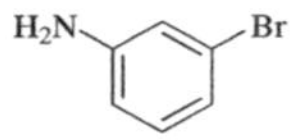

CuI (10 mol %),
L-Pro (20 mol %),
K₂CO₃, DMSO, 90°, 27 h

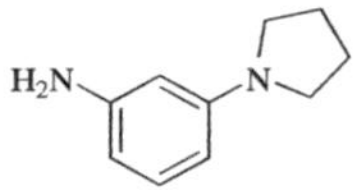

H₂N–C₆H₄–Br → H₂N–C₆H₄–N(pyrrolidine) (94)   47

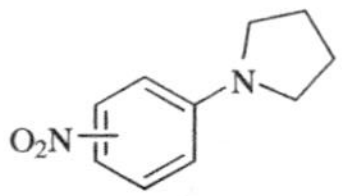

O₂N–C₆H₄–X → O₂N–C₆H₄–N(pyrrolidine)

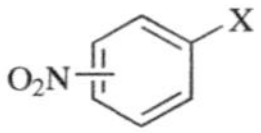

Catalyst (x amount %)

Isomer	X	Catalyst	x	Additive(s)	Solvent	Temp (°)	Time (h)		Ref
3	I	CuBr	2.5 mol %	L5 (5 mol %), Cs₂CO₃	DMF	90	24	(99)	119
3	I	Cu₂O	5 mol %	L7 (20 mol %), Cs₂CO₃	MeCN	90	18	(73)	125
4	Cl	CuI nanoparticles	1.25 mol %	K₂CO₃, air	DMF	110	5	(95)	285
4	Br	CuI	1 eq	CsOAc	DMF	90	24	(89)	189
4	Cl	CuI–PAnNF	5 mol %	K₂CO₃	DMF	80	9	(95)	276
4	I	CuI–PAnNF	5 mol %	K₂CO₃	DMF	rt	5	(71)	276

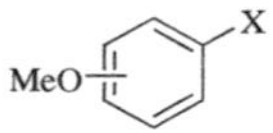

MeO–C₆H₄–X → MeO–C₆H₄–N(pyrrolidine)

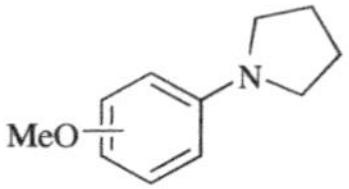

Catalyst (x mol %)

Isomer	X	Catalyst	x	Additives	Solvent	Temp (°)	Time		Ref
3	Br	CuI	10	L-Pro (20 mol %), K₂CO₃	DMSO	90	31 h	(99)	47
4	I	CuI	10	L-Pro (20 mol %), K₂CO₃	DMSO	65	27 h	(100)	47
4	I	CuI	5	L2 (25 mol %), TBAB, KOH	H₂O	MW (100 W), 130	5 min	(62)	236
4	Br	CuO	5	L14 (50 mol %), hexane-2,5-dione, KOH, TBAB	H₂O	MW (100 W), 120	5 min	(50)	51
4	I	Cu₂O	5	L7 (20 mol %), Cs₂CO₃	MeCN	90	18 h	(65)	125

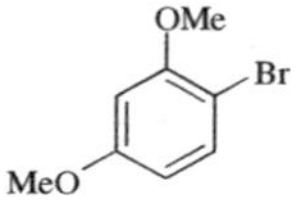

CuI (10 mol %),
L-Pro (20 mol %),
K₂CO₃, DMSO, 90°, 40 h

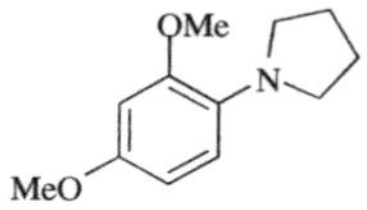

MeO–C₆H₃(OMe)–Br → MeO–C₆H₃(OMe)–N(pyrrolidine) (86)   47

TABLE 4A. N-ARYLATION OF CYCLIC SECONDARY ALKYL AMINES (*Continued*)

Nitrogen Nucleophile	Aryl Halide	Conditions	Product(s) and Yield(s) (%)	Refs.

*Please refer to the charts preceding the tables for structures indicated by the **bold** numbers.*

C4

Pyrrolidine (N–H) — Aryl Halide: phenyl with X, Me — Conditions: CuI (x mol %), DMF — Product: Me-substituted phenylpyrrolidine

Isomer	X	x	Additives	Temp (°)	Time		Refs.
3	I	5	L16 (20 mol %), Cs$_2$CO$_3$	rt	6 h	(90)	52
4	Br	10	L-Pro (20 mol %), K$_2$CO$_3$	MW (120 W)	15 min	(78)	299
4	I	10	L8 (20 mol %), K$_3$PO$_4$	30	10 h	(66)	271

Pyrrolidine (N–H) — Aryl Halide: R–X — Conditions: Catalyst (x mol %) — Product: R-substituted phenylpyrrolidine

R	X	Catalyst	x	Additives	Solvent	Temp (°)	Time		Refs.
3-CF$_3$	I	Cu$_2$O	5	L7 (20 mol %), Cs$_2$CO$_3$	MeCN	90	18 h	(72)	125
4-NC–	Br	CuI	10	L-Pro (20 mol %), K$_2$CO$_3$	DMSO	90	27 h	(89)	47
4-CHO	Br	CuI	10	L-Pro (20 mol %), K$_2$CO$_3$	DMF	MW (120 W)	15 min	(72)	299
4-Ac	Br	CuI	10	L-Pro (20 mol %), K$_2$CO$_3$	DMF	MW (120 W)	15 min	(88)	299
4-Ac	I	CuI	10	L-Pro (20 mol %), K$_2$CO$_3$	DMSO	65	20 h	(84)	47

Piperazine (HN–N–R^1) — Aryl Halide: phenyl with R^2 — Conditions: Catalyst (x mol %) — Product: N-aryl piperazine (NR^1, R^2)

R^1	R^2	X	Catalyst	x	Additives	Solvent	Temp (°)	Time (h)		Refs.
Me	H	Br	CuCl	10	TMHD (25 mol %), K$_2$CO$_3$	NMP	120	10	(17)	188, 195
Me	H	I	CuBr	2.5	L5 (5 mol %), Cs$_2$CO$_3$	DMF	90	24	(86)	118, 119
Et	H	I	CuBr	2.5	L5 (5 mol %), Cs$_2$CO$_3$	DMF	90	24	(90)	118, 119
Et	H	I	Cu$_2$O	5	L7 (20 mol %), Cs$_2$CO$_3$	MeCN	80	18	(72)	125
CO$_2$Et	MeO	I	Cu$_2$O	5	L7 (20 mol %), Cs$_2$CO$_3$	MeCN	90	24	(60)	125
COEt	H	I	CuBr	2.5	L5 (5 mol %), Cs$_2$CO$_3$	DMF	90	24	(83)	119

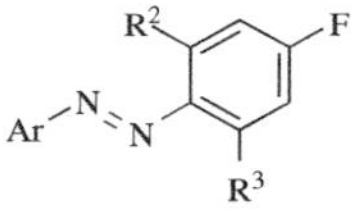

Ar–N=N– (aryl azo) substrate + piperazine → product with NR¹ piperazine

CuI (11 mol %), K₂CO₃, DMF, 100°, 24 h

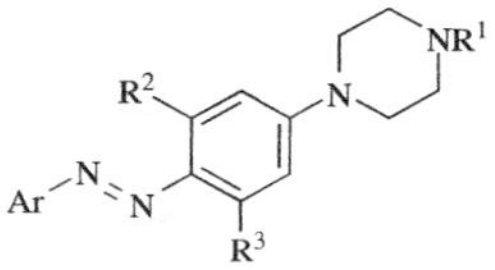

(—) 300

Ar = 4-O₂NC₆H₄

R¹	R²	R³
n-C₅H₁₁	F	H
n-C₅H₁₁	F	F
n-C₅H₁₁	Cl	H

R¹	R²	R³
n-C₆H₁₃	F	H
n-C₆H₁₃	F	F
n-C₆H₁₃	Cl	H

R¹	R²	R³
n-C₇H₁₅	F	H
n-C₇H₁₅	F	F
n-C₇H₁₅	Cl	H

R¹	R²	R³
n-C₈H₁₇	F	H
n-C₈H₁₇	F	F
n-C₈H₁₇	Cl	H

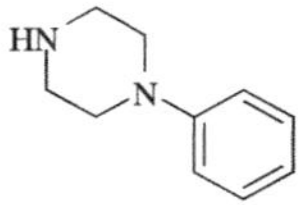

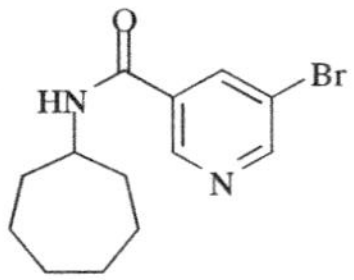

CuI (20 mol %), L-Pro (40 mol %), K₂CO₃, DMSO, 120°, 15 h

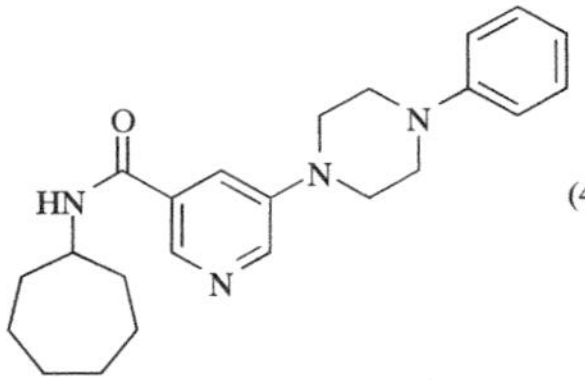

(42) 292

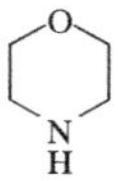

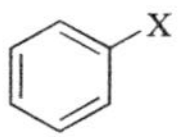

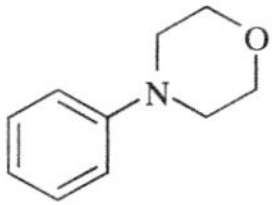

X	Catalyst	x	Additive(s)	Solvent(s)	Temp (°)	Time	
I	Cu	10 mol %	binol (10 mol %), Cs₂CO₃	DMSO	90	24 h	(84)
Cl	Cu/Al-HTB	2.5 mol %	K₂CO₃	DMF	100	8 h	(0)
I	Cu/Al-HTB	2.5 mol %	K₂CO₃	DMF	100	8 h	(90)
Br	CuCl	10 mol %	TMHD (25 mol %), Cs₂CO₃	NMP	120	10 h	(13)
I	CuBr	20 mol %	BINAP (20 mol %), K₃PO₄	DMF	rt	6 h	(73)
I	CuBr	2.5 mol %	L5 (5 mol %), Cs₂CO₃	DMF	90	24 h	(90)
I	CuI	10 mol %	CsOAc	DMSO	90	24 h	(83)

Catalyst (x amount)

124
228
228
188
62
118, 119
274

TABLE 4A. N-ARYLATION OF CYCLIC SECONDARY ALKYL AMINES (Continued)

Please refer to the charts preceding the tables for structures indicated by the **bold** numbers.

Nitrogen Nucleophile	Aryl Halide		Conditions					Product(s) and Yield(s) (%)		Refs.

C₄

Nitrogen Nucleophile: morpholine (N–H)

Aryl Halide: Ph–X

Catalyst (x amount)

Product: N-phenylmorpholine

X	Catalyst	x	Additive(s)	Solvent(s)	Temp (°)	Time		Refs.
I	CuI	10 mol %	L-Pro (20 mol %), K₂CO₃	DMSO	80	26 h	(61)	192
I	CuI	10 mol %	L-Pro (20 mol %), K₂CO₃	DMSO	90	40 h	(77)	47
Br	CuI	10 mol %	DMG (20 mol %), TBPM	DMSO	rt	24 h	(70)	61
I	CuI	10 mol %	**L9** (10 mol %), TBAH	DMSO	80	24 h	(77)	281
I	CuI	10 mol %	DPP (20 mol %), K₃PO₄	DMF	110	30 h	(78)	108
Cl	CuI	2 mol %	**L11** (2 mol %), KOt-Bu	—	100	12 h	(70)	281
Br	CuI	2 mol %	**L11** (2 mol %), KOt-Bu	—	100	12 h	(80)	281
I	CuI-PAnNF	5 mol %	K₂CO₃	DMF	rt	5 h	(83)	276
Br	CuO	10 mol %	FeCl₃ (10 mol %), binol (20 mol %), Cs₂CO₃	DMF	100	24 h	(82)	277
I	CuO	10 mol %	FeCl₃ (10 mol %), binol (20 mol %), Cs₂CO₃	DMF	90	12 h	(85)	277
I	CuO	1.26 mol %	KOH	DMSO	110	6.5 h	(89)	224
I	Cu₂O	5 mol %	**L12** (20 mol %), Cs₂CO₃	MeCN	80	18 h	(85)	125
Br	Cu(OAc)₂	1 eq	DBU	DMSO	MW, 130	10 min	(93)	85
I	Cu(OAc)₂	1 eq	DBU	DMSO	MW, 130	10 min	(92)	85
Cl	Cu(acac)₂	10 mol %	Fe₂O₃ (20 mol %), Cs₂CO₃	DMSO/H₂O	MW, 150	30 min	(0)	262
I	Cu(acac)₂	10 mol %	Fe₂O₃ (20 mol %), Cs₂CO₃	DMSO/H₂O	MW, 150	30 min	(85)	262

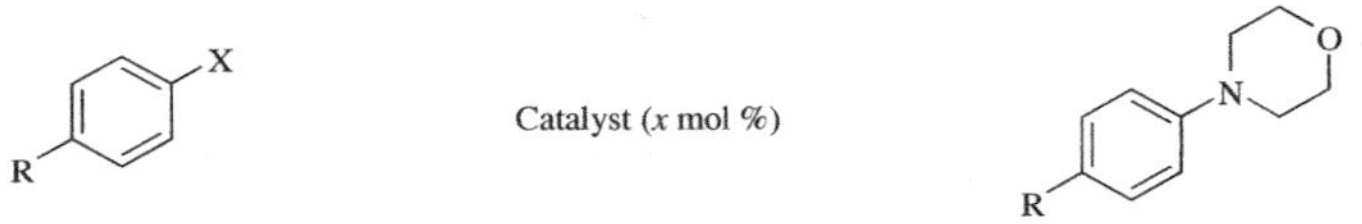

Catalyst (x mol %)

R	X	Catalyst	x	Additives(s)	Solvent(s)	Temp (°)	Time (h)	
Cl	I	Cu/Al–HTB	2.5	K$_2$CO$_3$	DMF	100	8	(90)	228
Cl	I	CuBr	20	BINAP (20 mol %), K$_3$PO$_4$	DMF	rt	6	(75)	62
Cl	I	CuI	10	L9 (10 mol %), TBAH	DMSO	80	24	(72)	281
Cl	Cl	CuI nanoparticles	1.25	K$_2$CO$_3$, air	DMF	110	2	(95)	285
Cl	I	CuO	10	FeCl$_3$ (10 mol %), binol (20 mol %), Cs$_2$CO$_3$	DMF	80	12	(90)	277
Cl	I	Cu(acac)$_2$	10	Fe$_2$O$_3$ (20 mol %), Cs$_2$CO$_3$	DMSO/H$_2$O	MW, 150	0.5	(11)	262
Br	I	Cu/Al–HTB	2.5	K$_2$CO$_3$	DMF	100	7	(99)	228
Br	I	CuI–PAnNF	5	K$_2$CO$_3$	DMF	rt	6	(85)	276
I	I	CuI	10	K$_3$PO$_4$•H$_2$O	Me$_3$NCH$_2$CH$_2$OH	55	36	(54)	195

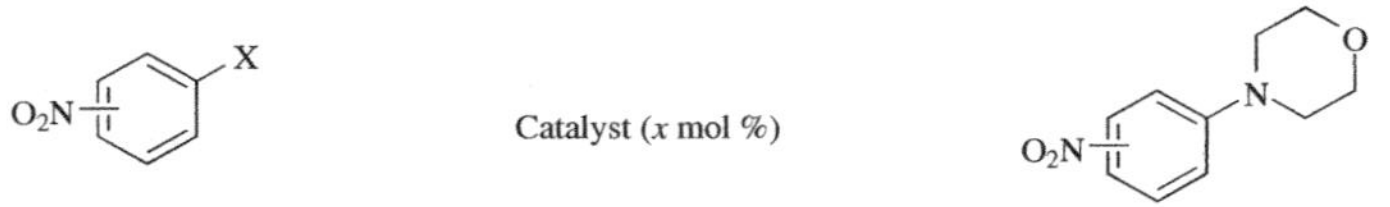

Catalyst (x mol %)

Isomer	X	Catalyst	x	Additive(s)	Solvent(s)	Temp (°)	Time (h)	
2	Cl	Cu/A–HTB	2.5	K$_2$CO$_3$	DMF	100	6	(99)	228
2	I	CuI–PAnNF	5	K$_2$CO$_3$	DMF	rt	3	(96)	276
3	I	CuBr	20	BINAP (20 mol %), K$_3$PO$_4$	DMF	rt	6	(73)	62
3	Br	CuO	10	binol (20 mol %), FeCl$_3$(10 mol %), Cs$_2$CO$_3$	DMF	100	12	(89)	277
4	Cl	Cu/Al–HTB	2.5	K$_2$CO$_3$	DMF	100	8	(92)	228
4	Cl	CuI–PAnNF	5	K$_2$CO$_3$	DMF	80	4	(98)	276
4	I	CuI–PAnNF	5	K$_2$CO$_3$	DMF	rt	4	(85)	276
4	I	Cu(acac)$_2$	10	Fe$_2$O$_3$ (20 mol %), Cs$_2$CO$_3$	DMSO/H$_2$O	MW, 150	0.5	(39)	262

TABLE 4A. N-ARYLATION OF CYCLIC SECONDARY ALKYL AMINES (*Continued*)

Nitrogen Nucleophile	Aryl Halide	Conditions	Product(s) and Yield(s) (%)	Refs.

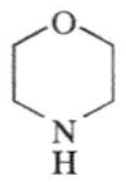

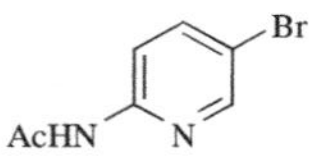

*Please refer to the charts preceding the tables for structures indicated by the **bold** numbers.*

C₄

CuI (20 mol %),
L-Pro (40 mol %),
K₂CO₃, DMSO, 120°, 15 h

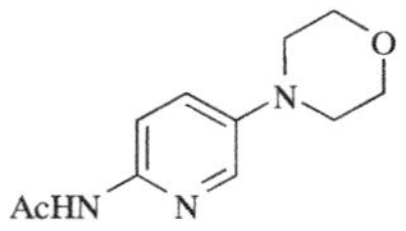

(58)   292

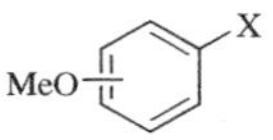

Catalyst (x mol %)

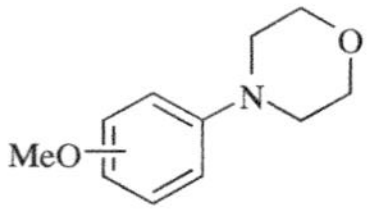

Isomer	X	Catalyst	x	Additive(s)	Solvent(s)	Temp (°)	Time		Refs.
2	I	Cu(acac)₂	10	Fe₂O₃ (20 mol %), Cs₂CO₃	DMSO/H₂O	MW, 150	30 min	(36)	262
3	I	Cu(acac)₂	10	Fe₂O₃ (20 mol %), Cs₂CO₃	DMSO/H₂O	MW, 150	30 min	(54)	262
4	I	CuBr	2.5	L5 (5 mol %), Cs₂CO₃	DMF	90	24 h	(80)	118, 119
4	I	CuI	5	L2 (25 mol %), TBAB, KOH	H₂O	MW (100 W), 130	5 min	(62)	236
4	I	CuI	10	piperidine-2-carboxylic acid (20 mol %), K₂CO₃	DMF	110	20 h	(80)	67
4	Br	CuI	5	EDA (10 mol %), K₂CO₃	PEG	80	4 h	(69)	301
4	I	CuI	10	PPAPM (20 mol %), K₃PO₄	DMF	90	30 h	(87)	48
4	I	CuI	10	DPP (20 mol %), K₃PO₄	DMF	110	30 h	(75)	108
4	Cl	CuI nanoparticles	1.25	K₂CO₃, air	DMF	110	7 h	(95)	285
4	I	CuO	5	L14 (50 mol %), hexane-2,5-dione, KOH, TBAB	H₂O	MW (100 W), 120	5 min	(60)	51
4	I	CuO	25	L15 (50 mol %), KOH, TBAB	H₂O	MW (100 W), 130	5 min	(44)	267
4	I	Cu₂O	5	L12 (20 mol %), Cs₂CO₃	MeCN	80	18 h	(85)	125
4	I	Cu(acac)₂	10	Fe₂O₃ (20 mol %), Cs₂CO₃	DMSO/H₂O	MW, 150	30 min	(80)	262

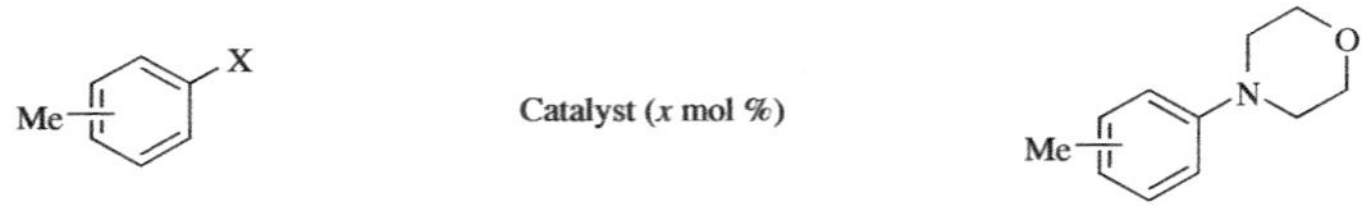

Catalyst (x mol %)

Isomer	X	Catalyst	x	Additive(s)	Solvent(s)	Temp (°)	Time		
2	I	Cu(acac)₂	10	Fe₂O₃ (20 mol %), Cs₂CO₃	DMSO/H₂O	MW, 150	30 min	(trace)	262
3	I	Cu(acac)₂	10	Fe₂O₃ (20 mol %), Cs₂CO₃	DMSO/H₂O	MW, 150	30 min	(45)	262
4	Cl	Cu/Al–HTB	2.5	K₂CO₃	DMF	100	28 h	(52)	228
4	I	CuBr	20	BINAP (20 mol %), K₃PO₄	DMF	40	6 h	(91)	62
4	Br	CuI	10	L-Pro (20 mol %), K₂CO₃	DMF	MW (120 W)	15 min	(49)	299
4	I	CuI-PAnNF	5	K₂CO₃	DMF	rt	4 h	(65)	276
4	I	CuO	10	FeCl₃ (10 mol %), binol (20 mol %), Cs₂CO₃	DMF	100	12 h	(75)	277
4	Cl	Cu(acac)₂	10	Fe₂O₃ (20 mol %), Cs₂CO₃	DMSO/H₂O	MW, 150	30 min	(0)	262
4	Br	Cu(acac)₂	10	Fe₂O₃ (20 mol %), Cs₂CO₃	DMSO/H₂O	MW, 150	30 min	(20)	262
4	I	Cu(acac)₂	10	Fe₂O₃ (20 mol %), Cs₂CO₃	DMSO/H₂O	MW, 150	30 min	(83)	262

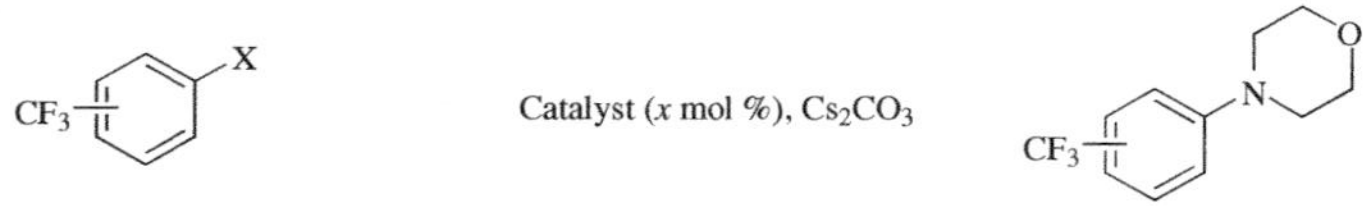

Catalyst (x mol %), Cs₂CO₃

Isomer	X	Catalyst	x	Additive	Solvent(s)	Temp (°)	Time (h)		
2	I	Cu(acac)₂	10	Fe₂O₃ (20 mol %)	DMSO/H₂O	MW, 150	0.5	(trace)	262
3	I	Cu₂O	5	L7 (20 mol %)	MeCN	90	18	(70)	125
3	I	Cu(acac)₂	10	Fe₂O₃ (20 mol %)	DMSO/H₂O	MW, 150	0.5	(62)	262
4	Br	Cu(acac)₂	10	Fe₂O₃ (20 mol %)	DMSO/H₂O	MW, 150	0.5	(11)	262
4	I	Cu(acac)₂	10	Fe₂O₃ (20 mol %)	DMSO/H₂O	MW, 150	0.5	(25)	262

TABLE 4A. N-ARYLATION OF CYCLIC SECONDARY ALKYL AMINES (*Continued*)

Nitrogen Nucleophile	Aryl Halide	Conditions	Product(s) and Yield(s) (%)	Refs.

*Please refer to the charts preceding the tables for structures indicated by the **bold** numbers.*

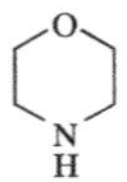

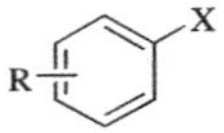

Catalyst (10 mol %)

C4

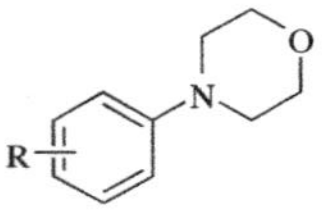

R	X	Catalyst	Additive(s)	Solvent	Temp (°)	Time		Refs.
4-NC—	Br	CuCl	TMHD (25 mol %), Cs$_2$CO$_3$	NMP	120	10 h	(32)	188
4-CHO	Br	CuI	L-Pro (20 mol %), K$_2$CO$_3$	DMF	MW (120 W)	15 min	(69)	299
3-MeO$_2$C	I	CuI	L9 (10 mol %), TBAH	DMSO	80	24 h	(67)	281
4-Ac	Br	CuI	L-Pro (20 mol %), K$_2$CO$_3$	DMF	MW (120 W)	15 min	(42)	299
4-Ph	Br	CuCl	TMHD (25 mol %), Cs$_2$CO$_3$	NMP	120	10 h	(23)	188

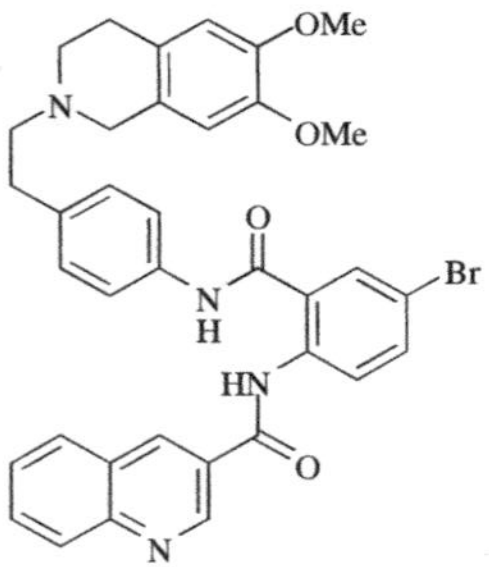

CuBr•DMS (20 mol %),
L-Pro (40 mol %),
K$_3$PO$_4$, DMSO, 90°

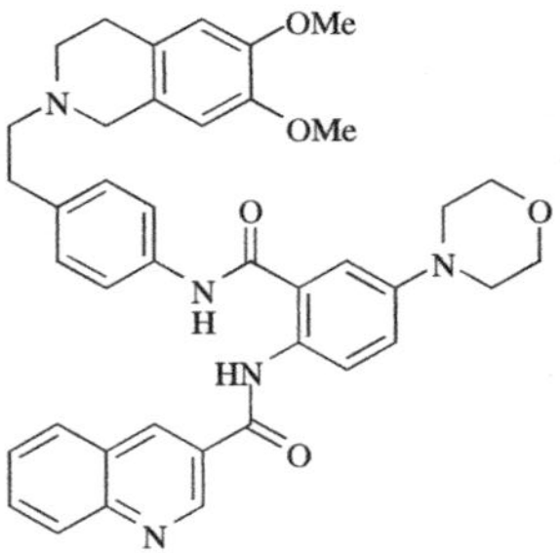

(75)

122

C5

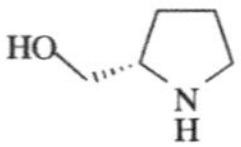

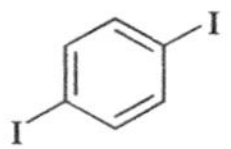

CuI (10 mol %), K$_3$PO$_4$•H$_2$O,
Me$_2$NCH$_2$CH$_2$OH, rt, 20 h;
then 40°, 28 h

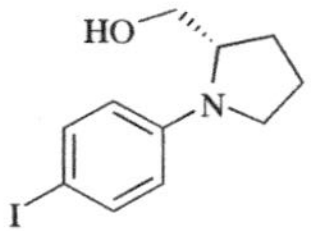

(49)

195

HO₂C pyrrolidine (proline) structure; CuI (x mol %)

R	X	x	Additive(s)	Solvent	Temp (°)	Time		
H	I	—	K₂CO₃	DMA	90	—	(—)	302
4-Cl	I	10	**L8** (20 mol %), K₃PO₄	DMF	30	12 h	(68)	271
2-O₂N	Br	10	K₂CO₃	DMF	MW, 140	15 min	(90)	265
4-O₂N	I	—	K₂CO₃	DMA	90	—	(—)	302
3-MeO	I	—	K₂CO₃	DMA	90	—	(—)	302
4-Me	Br	10	K₂CO₃	DMF	MW, 140	22 min	(65)	265
2-NC—	I	—	K₂CO₃	DMA	90	—	(—)	302
2-Ac	Br	10	K₂CO₃	DMF	MW, 140	20 min	(80)	265
4-Ac	Br	10	K₂CO₃	DMF	MW, 140	18 min	(80)	265
4-Ac	I	10	K₂CO₃	DMF	MW, 140	20 min	(83)	265

piperidine structure (N-phenylpiperidine product); Catalyst (x amount)

X	Catalyst	x	Additives(s)	Solvent	Temp (°)	Time		
Br	Cu	5 eq	none	H₂O	MW, 100	2 min	(91)	280
I	Cu	10 mol %	binol (10 mol %), Cs₂CO₃	DMSO	90	24 h	(88)	124
I	Cu	10 mol %	K₃PO₄•H₂O	Me₂NCH₂CH₂OH	70	9 h	(54)	195
Br	CuCl	10 mol %	TMHD (25 mol %), Cs₂CO₃	NMP	120	10 h	(39)	188
I	CuBr	2.5 mol %	**L5** (5 mol %), Cs₂CO₃	DMF	90	24 h	(87)	118, 119
I	CuI	10 mol %	L-Pro (20 mol %), K₂CO₃	DMSO	90	40 h	(83)	47
I	CuI	10 mol %	L-Pro (20 mol %), K₂CO₃	DMSO	80	26 h	(61)	192
I	CuI	10 mol %	**L6** (20 mol %), Cs₂CO₃	DMF	rt	24 h	(15)	273
I	CuI	10 mol %	**L9** (10 mol %), TBAH	DMSO	80	24 h	(82)	281
Cl	CuI	2 mol %	**L1** (2 mol %), KOt-Bu	—	100	12 h	(78)	281
Br	CuI	2 mol %	**L1** (2 mol %), KOt-Bu	—	100	12 h	(94)	281
I⁺Ph BF₄⁻	CuI	10 mol %	Na₂CO₃	CH₂Cl₂	rt	6 h	(65)	76
I	CuI–PAnNF	5 mol %	K₂CO₃	DMF	rt	6 h	(67)	276
I	CuO	1.26 mol %	KOH	DMF	110	7 h	(90)	224
I	CuFAP	—	K₂CO₃	DMF	110	12 h	(90)	227

HO₂C-C pyrrolidine N-H

piperidine N-H

TABLE 4A. N-ARYLATION OF CYCLIC SECONDARY ALKYL AMINES (*Continued*)

Nitrogen Nucleophile	Aryl Halide	Conditions	Product(s) and Yield(s) (%)	Refs.

*Please refer to the charts preceding the tables for structures indicated by the **bold** numbers.*

C5

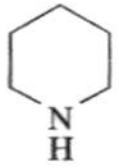

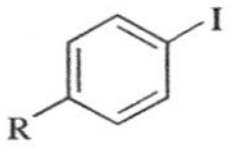

Nitrogen Nucleophile: piperidine (N–H)

Aryl Halide: R-substituted iodobenzene

Catalyst (x mol %)

Product: R-substituted phenylpiperidine

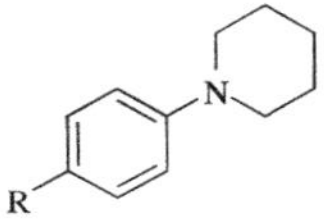

R	Catalyst	x	Additive(s)	Solvent	Temp (°)	Time (h)		Refs.
Cl	Cu/Al–HTB	2.5	K2CO3	DMF	100	9	(81)	228
Cl	CuBr	20	BINAP (20 mol %), K3PO4	DMF	rt	6	(67)	62
Cl	CuI	10	L8 (20 mol %), K3PO4	DMF	30	8	(64)	271
Br	Cu/Al–HTB	2.5	K2CO3	DMF	100	10	(88)	228
I	CuI	10	K3PO4•H2O	Me2NCH2CH2OH	45	20	(68)	195

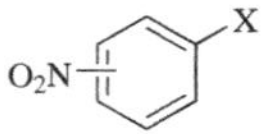

Aryl Halide: O2N–C6H4–X

Catalyst (x mol %)

Product: O2N–phenylpiperidine

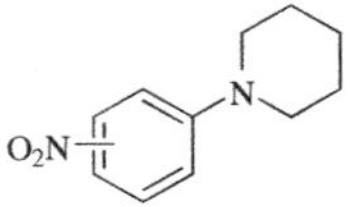

Isomer	X	Catalyst	x	Additive(s)	Solvent	Temp (°)	Time (h)		Refs.
3	I	Cu2O	5	L7 (20 mol %), Cs2CO3	MeCN	90	18	(68)	125
4	Cl	Cu/Al–HTB	2.5	K2CO3	DMF	100	8	(94)	228
4	I	CuI–PAnNF	5	K2CO3	DMF	80	7	(96)	276
4	F	CuFAP	—	K2CO3	DMF	120	1	(85)	66
4	Cl	CuFAP	—	K2CO3	DMF	120	5	(80)	66
4	F	cat. 7	1	K2CO3	DMF	110	1	(90)	73
4	Cl	cat. 7	1	K2CO3	DMF	110	5	(87)	73

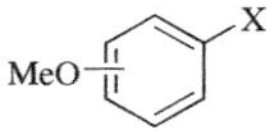

Catalyst (x mol %)

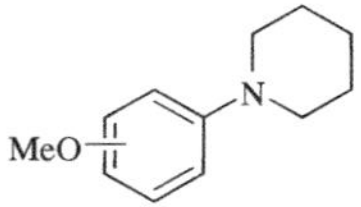

MeO—⟨X⟩ → MeO—⟨N-piperidine⟩

Isomer	X	Catalyst	x	Additive(s)	Solvent(s)	Temp (°)	Time		
2	Cl	Cu/Al-HTB	2.5	K₂CO₃	DMF	100	24 h	(48)	228
4	I	Cu	10	binol (10 mol %), Cs₂CO₃	DMSO	90	24 h	(75)	124
4	I	CuBr	2.5	L5 (5 mol %), Cs₂CO₃	DMF	90	24 h	(78)	118, 119
4	Br	CuI	10	DMEDA (20 mol %)	[C₄mim]BF₄	110	12 h	(87)	232
4	Br	CuI	5	L16 (20 mol %), K₃PO₄	DMF	100	16 h	(90)	52
4	I	CuO	25	L15 (50 mol %), KOH, TBAB	H₂O	MW (100 W), 130	5 min	(47)	267

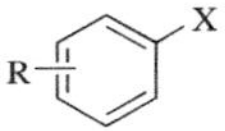

Catalyst (x amount)

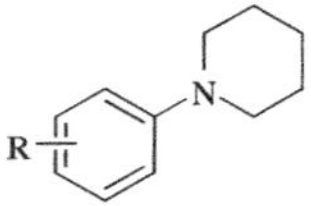

R—⟨X⟩ → R—⟨N-piperidine⟩

R	X	Catalyst	x	Additive(s)	Solvent	Temp (°)	Time		
4-Me	Br	Cu	5 eq	none	H₂O	MW (100 W)	2 min	(87)	280
4-NC–	Br	CuI	10 mol %	L-Pro (20 mol %), K₂CO₃	DMSO	90	40 h	(81)	47
3-MeO₂C	I	CuI	10 mol %	L9 (10 mol %), TBAH	DMSO	80	24 h	(70)	281

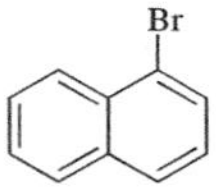

Br—⟨naphthalene⟩

CuI (5 mol %),
EDA (10 mol %),
PEG, K₂CO₃, 80°, 4 h

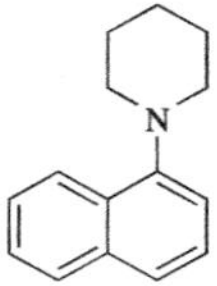

⟨1-(piperidin-1-yl)naphthalene⟩ (69)   301

TABLE 4A. *N*-ARYLATION OF CYCLIC SECONDARY ALKYL AMINES (*Continued*)

Please refer to the charts preceding the tables for structures indicated by the **bold** numbers.

Nitrogen Nucleophile	Aryl Halide	Conditions	Product(s) and Yield(s) (%)	Refs.
C₅				
	I⁺ BF₄⁻	CuI (10 mol %), Na₂CO₃, MeCl₂, rt, 6 h	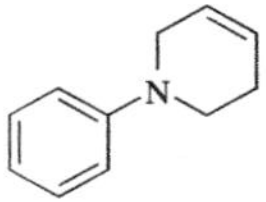 (66)	76
C₆				
	I	CuBr (2.5 mol %), **L5** (5 mol %), Cs₂CO₃, DMF, 90°, 24 h	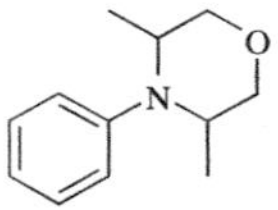 (87)	118, 119
C₇				
	I–R	CuI (10 mol %), K₃PO₄•H₂O, Me₂NCH₂CH₂OH, 60°, 40 h	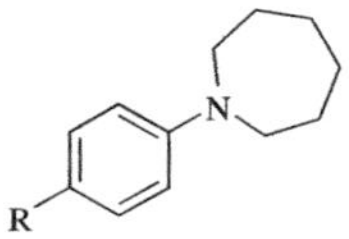 R Br (47) / I (50)	195
C₉				
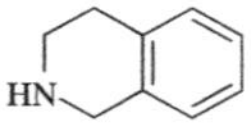	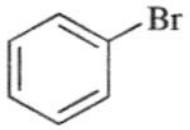 Br	CuI (2 mol %), **L1** (2 mol %), KO*t*-Bu, 100°, 12 h	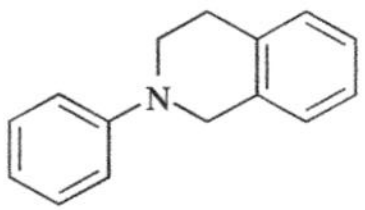 (94)	281

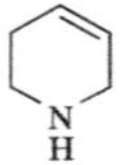

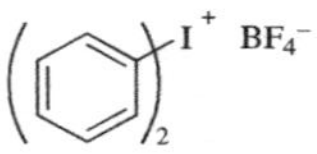

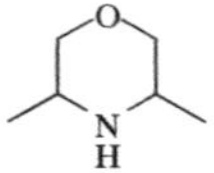

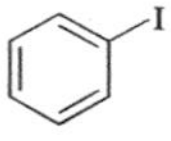

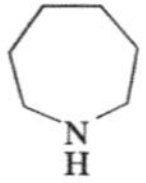

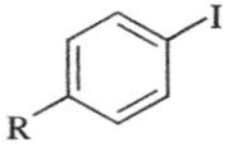

168

TABLE 4B. N-HETEROARYLATION OF CYCLIC SECONDARY ALKYL AMINES

*Please refer to the charts preceding the tables for structures indicated by the **bold** numbers.*

Nitrogen Nucleophile	Heteroaryl Halide	Conditions	Product(s) and Yield(s) (%)	Refs.

C₄

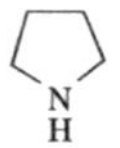

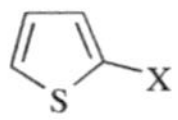

		Cu (10 mol %), CuI (10 mol %), K₃PO₄•H₂O, Me₂NCH₂CH₂OH	X Temp (°) Time (h)	
			Br 80 50 (81)	
			Br 80 49 (81)	296
			I 60 10 (91)	
			I 80 49 (91)	
		Cu (10 mol %), CuI (10 mol %), K₃PO₄•H₂O, Me₂NCH₂CH₂OH, 80°, 45 h	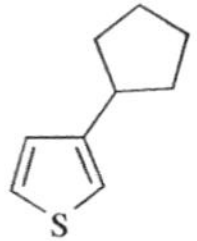 (83)	296
		CuBr (2.5 mol %), **L5** (5 mol %), Cs₂CO₃, DMF, 90°, 24 h	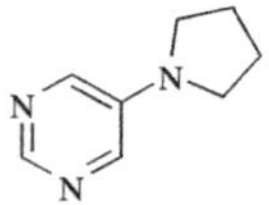 (83)	119
		CuI (10 mol %), L-Pro (20 mol %), K₂CO₃		
		Isomer Solvent Temp Time		
		2 DMF MW (120 W) 15 min (84)		299
		3 DMSO 90° 40 h (94)		47
		CuBr (2.5 mol %), **L5** (5 mol %), Cs₂CO₃, DMF, 90°, 24 h	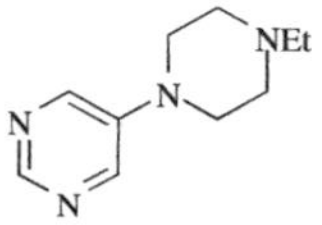 (80)	119
		CuBr (2.5 mol %), **L5** (5 mol %), Cs₂CO₃, DMF, 90°, 24 h	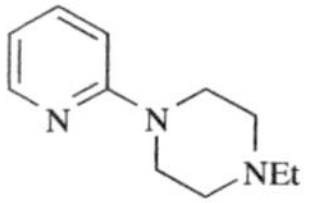 (85)	119

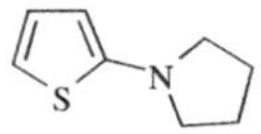

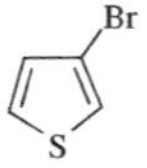

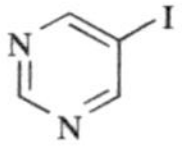

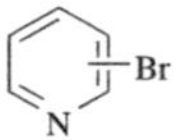

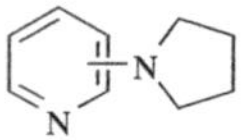

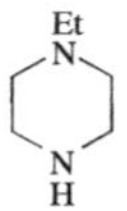

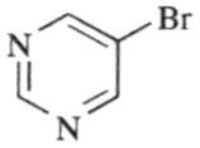

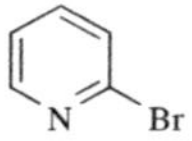

TABLE 4B. N-HETEROARYLATION OF CYCLIC SECONDARY ALKYL AMINES (Continued)

Nitrogen Nucleophile	Heteroaryl Halide	Conditions	Product(s) and Yield(s) (%)	Refs.

*Please refer to the charts preceding the tables for structures indicated by the **bold** numbers.*

C$_4$

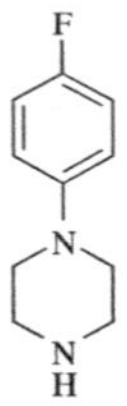

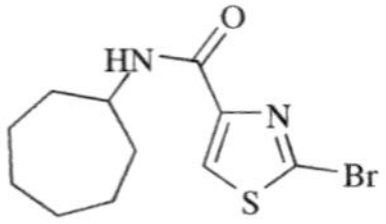

CuI (20 mol %), L-Pro (40 mol %), K$_2$CO$_3$, DMSO, 120°, 15 h

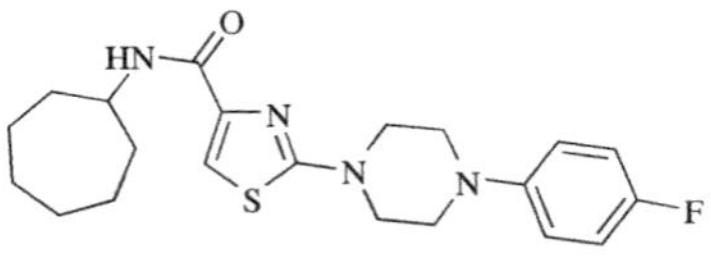

(85)

292

C$_{4-5}$

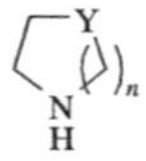

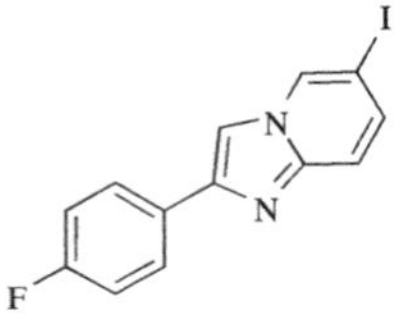

CuI (15 mol %), K$_3$PO$_4$, HOCH$_2$CH$_2$OH, i-PrOH, 85°

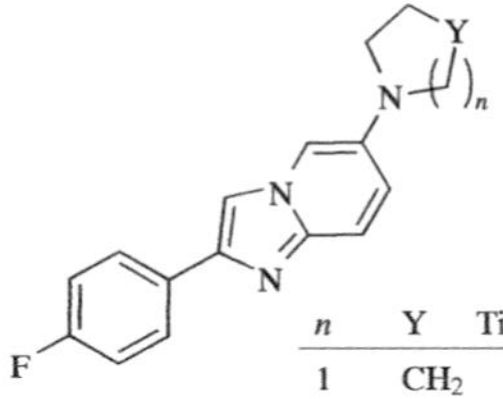

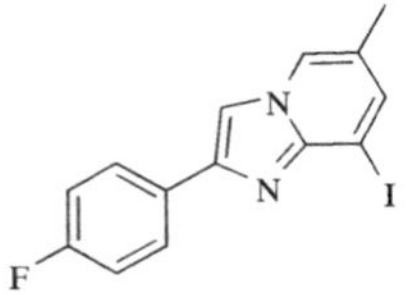

n	Y	Time (h)	
1	CH$_2$	20	(84)
2	O	48	(72)
2	CH$_2$	48	(70)

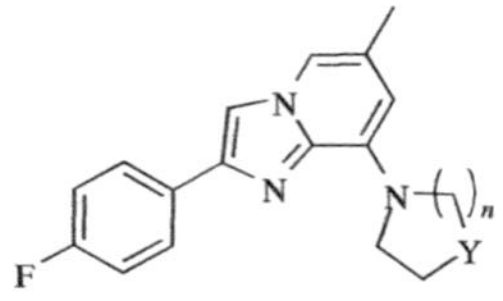

126

CuI (15 mol %), K$_3$PO$_4$, HOCH$_2$CH$_2$OH, i-PrOH

n	Y	Temp (°)	Time (h)	
1	CH$_2$	85	20	(60)
2	EtN	85	48	(69)
2	O	112	20	(55)
2	CH$_2$	85	48	(64)

297

C₄

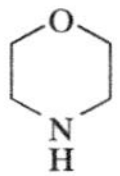

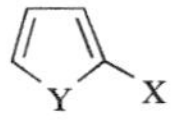

Morpholine (O[ ]N-H)

Catalyst(s) (10 mol %)

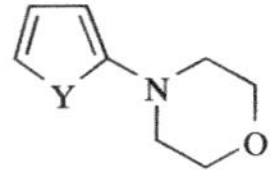

Y	X	Catalyst(s)	Additive(s)	Solvent	Temp (°)	Time (h)		Ref
S	Br	Cu, CuI	$K_3PO_4 \cdot H_2O$	$Me_2NCH_2CH_2OH$	80	50	(65)	296
Se	I	CuI	EDA (20 mol %), K_3PO_4	dioxane	reflux	24	(0)	87

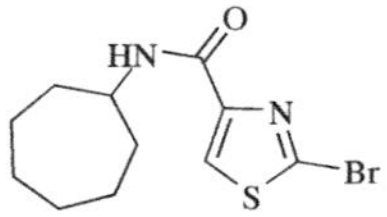

CuI (20 mol %), L-Pro (40 mol %), K_2CO_3, DMSO, 120°, 15 h — (65) — 292

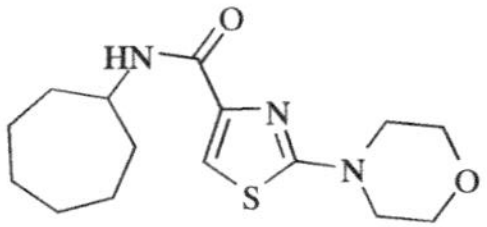

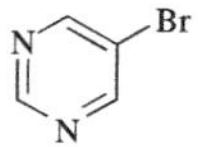

CuBr (2.5 mol %), L5 (5 mol %), Cs_2CO_3, DMF, 90°, 24 h — (86) — 119

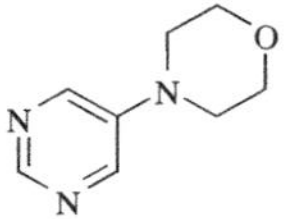

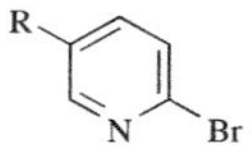

Catalyst (x mol %), 90°

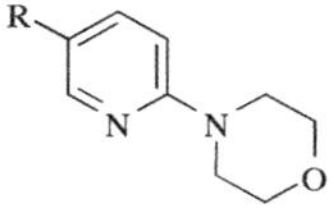

R	Catalyst	x	Additives	Solvent	Time (h)		Ref
H	CuBr	2.5	L5 (5 mol %), Cs_2CO_3	DMF	24	(82)	119
Br	CuI	10	L-Pro (20 mol%), K_2CO_3	DMSO	40	(81)	47

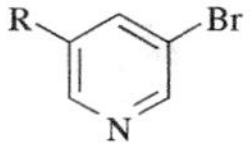

CuI (20 mol %), L-Pro (40 mol %), K_2CO_3, DMSO, 120°, 15 h — 292

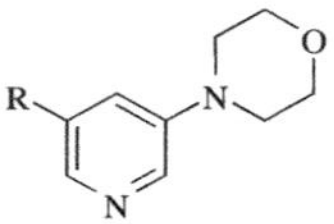

R	
H_2NCO	(52)
NC–	(65)

TABLE 4B. *N*-HETEROARYLATION OF CYCLIC SECONDARY ALKYL AMINES (*Continued*)

Nitrogen Nucleophile	Heteroaryl Halide	Conditions	Product(s) and Yield(s) (%)	Refs.

*Please refer to the charts preceding the tables for structures indicated by the **bold** numbers.*

C₄

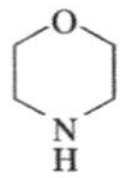

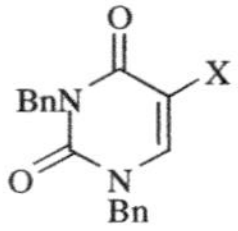

morpholine (O-containing ring with NH)

Catalyst, (*x* mol %)

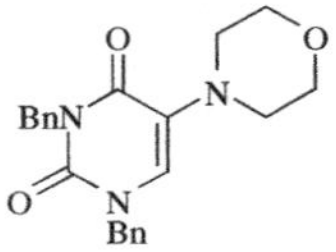

X	Catalyst	*x*	Additives
Br	CuI	10	L-Pro (20 mol %), K₂CO₃
I	(CuOTf)₂•C₆H₆	50	phen (1 eq), dba (50 mol %), Cs₂CO₃

Solvent	Temp (°)	Time (h)	
DMSO	90	40	(81)
p-xylene	95	24	(73)

Refs. 47, 187

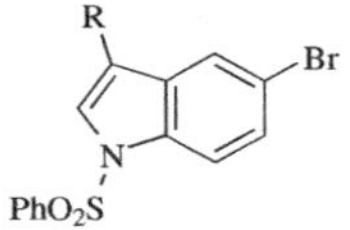

CuI (20 mol %),
L-Pro (40 mol %), K₂CO₃,
DMSO, 120°, 15 h

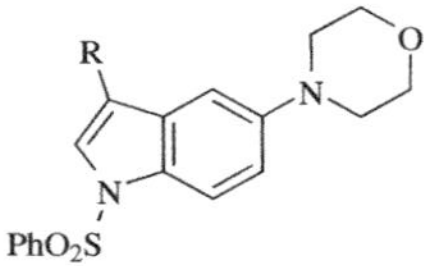

R	
H	(66)
CH₂CONHPh	(0)

Ref. 292

C₅

piperidine (N-containing ring with NH)

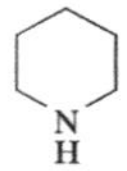

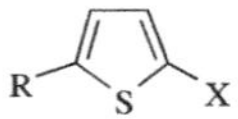

Cu (10 mol %), co-catalyst
(10 mol %), K₃PO₄•H₂O,
Me₂NCH₂CH₂OH

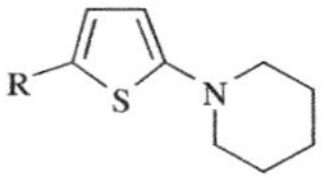

R	X	Co-Catalyst	Temp (°)	Time (h)	
H	Br	CuI	80	50	(73)
H	I	none	60	10	(91)
Me	Br	CuI	80	24	(71)
OHC	Br	CuI	65	48	(77)

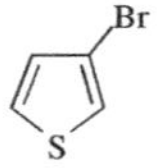

Ref. 296

Cu (10 mol %), CuI (10 mol %),
K₃PO₄•H₂O, Me₂NCH₂CH₂OH,
80°, 40 h

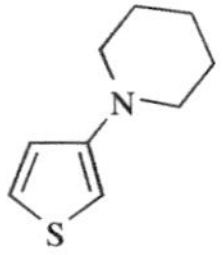

(73)

Ref. 296

172

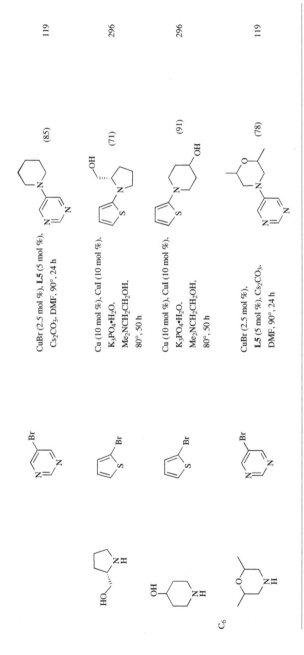

119

296

296

119

(85)

(71)

(91)

(78)

C₆

TABLE 5A. N-ARYLATION OF PRIMARY ARYL AMINES

Nitrogen Nucleophile	Aryl Halide	Conditions	Product(s) and Yield(s) (%)		Refs.

*Please refer to the charts preceding the tables for structures indicated by the **bold** numbers.*

C₆

H₂N–C₆H₅ + X–C₆H₅ → Catalyst (x amount) → Ph₂NH (**I**) + Ph₃N (**II**)

X	Catalyst	x	Additive(s)	Solvent	Temp (°)	Time	I	II	Refs.
Br[a]	Cu	3 eq	none	—	270	—	(21)	(4)	303
I[a]	Cu	3 eq	none	—	180	—	(31)	(46)	303
I[b]	Cu	3 eq	none	—	220	—	(0)	(86)	303
I	Cu (active)	5 eq	none	H₂O	MW (100 W)	6 min	(74)	(0)	280
Br	CuCl	10 mol %	2,5-pentanedione (25 mol %), K₂CO₃	NMP	130	18 h	(30)	(30)	188
I·Ph BF₄⁻	CuI	10 mol %	Na₂CO₃	CH₂Cl₂	rt	6 h	(75)	(0)	76
I	CuI	10 mol %	CsOAc	DMSO	90	24 h	(24)	(0)	274
Cl	CuI	2 mol %	KO*t*-Bu	toluene	135	14 h	(12)	(28)	304
Br	CuI	2 mol %	KO*t*-Bu	toluene	135	14 h	(12)	(45)	304
I	CuI	2 mol %	KO*t*-Bu	toluene	135	14 h	(7)	(70)	304
Cl	CuI	3.6 mol %	PBu₃ (7.2 mol %), KO*t*-Bu	toluene	135	10.5 h	(82)	(0)	41
I	CuI	3.5 mol %	bipy (3.5 mol %), KO*t*-Bu	toluene	115	3.5 h	(2)	(95)	304
I	CuI	3.6 mol %	bipy (3.6 mol %), KO*t*-Bu	toluene	115	3.5 h	(0)	(95)	305
Br	CuI	10 mol %	PPAPM (20 mol %), K₃PO₄	DMF	110	36 h	(77)	(0)	48
I	CuI	5 mol %	DiPrPhDAB (6 mol %), KO*t*-Bu	toluene	120	48 h	(97)	(0)	306
I	CuI	10 mol %	L-Pro (20 mol %), K₂CO₃	DMSO	90	27 h	(66)	(0)	192, 47
Br	CuI	10 mol %	DPP (20 mol %), K₃PO₄	DMF	110	36 h	(57)	(0)	108
I	CuI	10 mol %	DMEDA (10 mol %), K₂CO₃	dioxane	100	21 h	(84)	(0)	55
Cl	CuI	10 mol %	piperidine-2-carboxylic acid (20 mol %), K₂CO₃	DMF	110	36 h	(15)	(0)	67
Br	CuI	10 mol %	piperidine-2-carboxylic acid (20 mol %), K₂CO₃	DMF	110	20 h	(72)	(0)	67
I	CuI	10 mol %	piperidine-2-carboxylic acid (20 mol %), K₂CO₃	DMF	110	20 h	(79)	(0)	67
I	CuI	10 mol %	pyridine-2-carboxylic acid (20 mol %), K₃PO₄	DMSO	80	17 h	(68)	(0)	50

Cl	CuI	2 mol %	L1 (2 mol %), KOt-Bu	—	100	12 h	(80)	(0)	288
Br	CuI	2 mol %	L1 (2 mol %), KOt-Bu	—	100	12 h	(99)	(0)	288
Br	CuI	5 mol %	L2 (25 mol %), KOH, TBAB	H$_2$O	MW (100 W), 135	5 min	(79)	(0)	236
Cl	CuO nanoparticles (33 nm)	1.3 mol %	KOH, air	DMSO	110	18 h	(60)	(0)	224
Br	CuO nanoparticles (33 nm)	1.3 mol %	KOH, air	DMSO	110	10 h	(80)	(0)	224
I	CuO nanoparticles (33 nm)	1.3 mol %	KOH, air	DMSO	110	2 h	(95)	(0)	224
Cl	CuO nanoparticles	5 mol %	KOH	DMSO/t-BuOH (3:1)	110	18 h	(0)	(0)	74
Br	CuO nanoparticles	5 mol %	KOH	DMSO/t-BuOH (3:1)	110	18 h	(0)	(0)	74
I	CuO nanoparticles	5 mol %	KOH	DMSO/t-BuOH (3:1)	110	18 h	(88)	(0)	74
OTs	CuO nanoparticles	5 mol %	KOH	DMSO/t-BuOH (3:1)	110	18 h	(0)	(0)	74
Br	CuO	5 mol %	NaOt-Bu	NMP	100	—	(79)	(0)	266
I	CuO	5 mol %	NaOt-Bu	NMP	100	—	(74)	(0)	266
Br	CuO	10 mol %	FeCl$_3$ (10 mol %), rac-binol (20 mol %), Cs$_2$CO$_3$	DMF	110	12 h	(76)	(0)	277
I	CuO	10 mol %	FeCl$_3$ (10 mol %), rac-binol (20 mol %), Cs$_2$CO$_3$	DMF	100	24 h	(78)	(0)	277
Br	CuO	25 mol %	L15 (50 mol %), KOH, TBAB	H$_2$O	MW (100 W), 130	5 min	(49)	(0)	267
Br	CuO	25 mol %	L15 (50 mol %), KOH, TBAB	H$_2$O	reflux	10 h	(49)	(0)	267
I	CuO	25 mol %	L15 (50 mol %), KOH, TBAB	H$_2$O	MW (100 W), 130	5 min	(57)	(0)	267
I	CuO/AB	5 mol %	KOt-Bu	toluene	180	18 h	(26)	(0)	226
Br	CuO	5 mol %	L14 (50 mol %), hexane-2,5-dione, KOH, TBAB	H$_2$O	90	8 h	(70)	(0)	51
Br	CuO	5 mol %	L14 (50 mol %), hexane-2,5-dione, KOH, TBAB	H$_2$O	MW, 120	5 min	(75)	(0)	51
I	CuO	5 mol %	L14 (50 mol %), hexane-2,5-dione, KOH, TBAB	H$_2$O	MW, 120	5 min	(63)	(0)	51

TABLE 5A. N-ARYLATION OF PRIMARY ARYL AMINES (*Continued*)

Nitrogen Nucleophile	Aryl Halide	Conditions	Product(s) and Yield(s) (%)	Refs.

*Please refer to the charts preceding the tables for structures indicated by the **bold** numbers.*

Continued from previous page.

C₆

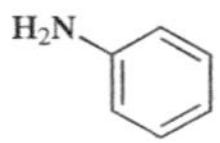

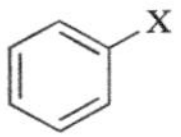

X	Catalyst	x	Additive(s)	Solvent	Temp (°)	Time	I	II	Refs.
I	Cu₂O	10 mol %	Cs₂CO₃	DMF	100	18 h	(0)	(0)	221
I	Cu(OAc)₂	4 mol %	phen (3 mol %), KOH	toluene	reflux	18 h	(0)	(—)	307
I	Cu(acac)₂	10 mol %	Fe₂O₃ (20 mol %), Cs₂CO₃	DMSO/H₂O (1:1)	MW, 150	5 min	(35)	(0)	262
I	Cu(TMHD)₂	20 mol %	KOt-Bu	toluene	120	40 h	(0)	(94)	308
Br	Cu/Fe-hydrotalcite	10 wt %	none	toluene	130	14 h	(79)	(0)	283
I	Cu/Fe-hydrotalcite	10 wt %	none	toluene	130	12 h	(81)	(0)	283
I	1	1.25 mol %	KOt-Bu	toluene	90	12 h	(0)	(65)	309

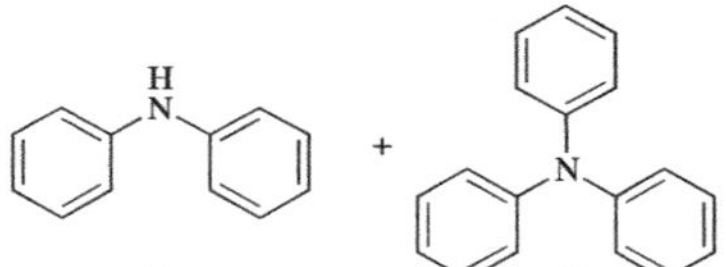

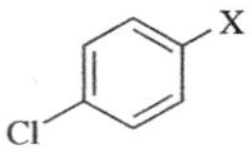

X	Catalyst	x	Additive(s)	Solvent	Temp (°)	Time		Refs.
Br	CuI	5 mol %	L2 (25 mol %), KOH, TBAB	H₂O	MW (100 W) 130	5 min	(66)	236
I	CuI	10 mol %	TBPM, 2,2'-biphenol (20 mol %)	dioxane	rt	12 h	(80)	61
Br	CuO	25 mol %	L15 (50 mol %), KOH, TBAB	H₂O	MW (100 W) 130	5 min	(81)	267
Br	CuO	5 mol %	KOH, TBAB, L14 (50 mol %), hexane-2,5-dione	H₂O	90	8 h	(59)	51
Br	CuO	5 mol %	KOH, TBAB, L14 (50 mol %), hexane-2,5-dione	H₂O	MW 120	5 min	(58)	51
Br	Cu/Fe-hydrotalcite	10 wt %	none	toluene	130	12 h	(76)	283

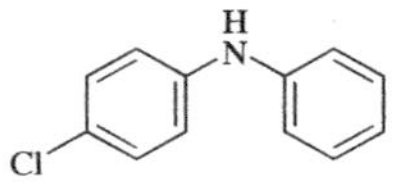

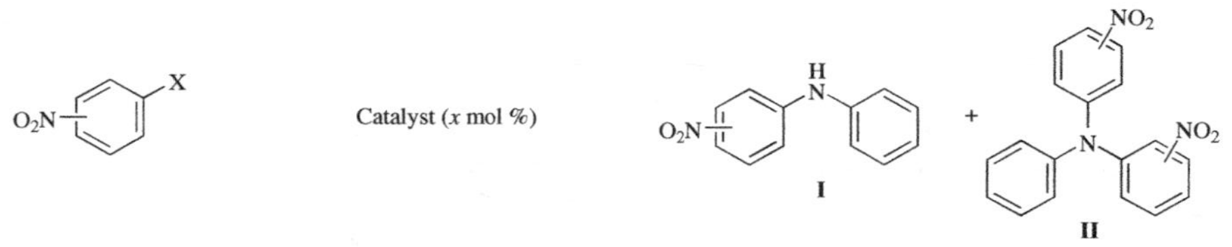

O_2N—C$_6$H$_4$—X + Catalyst (x mol %) → I + II

where I is a nitro-substituted diphenylamine and II is a dinitro-triphenylamine.

Isomer	X	Catalyst	x	Additives	Solvent(s)	Temp (°)	Time	I	II	
3	Br	CuI	10	DPP (20 mol %), K$_3$PO$_4$	DMF	110	30 h	(80)	(0)	108
3	Br	CuI	10	PPAPM (20 mol %), K$_3$PO$_4$	DMF	110	30 h	(89)	(0)	48
4	Cl	CuBr	2.5	L4 (2.5 mol %), K$_2$CO$_3$	—	190	4 h	(89)	(0)	310
4	Cl	CuI	3.6	PBu$_3$ (7.2 mol %), KOt-Bu	toluene	135	10.5 h	(0)	(41)	56
4	Br	CuI	5	L2 (25 mol %), KOH, TBAB	H$_2$O	MW (100 W) 130	5 min	(80)	(0)	236
4	Br	CuO	5	L14 (50 mol %), hexane-2,5-dione, KOH, TBAB	H$_2$O	MW 120	5 min	(56)	(0)	51
4	I	CuO nanoparticles	5	KOH, air	DMSO/t-BuOH (3:1)	110	4 h	(85)	(0)	74
4	I	CuO nanoparticles (33 nm)	1.3	PBu$_3$ (7.2 mol %), KOt-Bu	DMSO	110	1.2 h	(93)	(0)	224

TABLE 5A. N-ARYLATION OF PRIMARY ARYL AMINES (Continued)

Nitrogen Nucleophile	Aryl Halide	Conditions	Product(s) and Yield(s) (%)	Refs.

Please refer to the charts preceding the tables for structures indicated by the **bold** numbers.

C6

Isomer	X	Catalyst	x	Additive(s)	Solvents	Temp (°)	Time	I	II	Refs.
2	Br	CuI	2 mol %	L1 (2 mol %), KO-t-Bu	—	100	12 h	(99)	(0)	288
4	Cl	CuI	3.6 mol %	PBu3 (7.2 mol %), KO-t-Bu	toluene	135	10.5 h	(—)	(85)	56
4	Br	CuI	10 mol %	K2CO3, piperidine-2-carboxylic acid (20 mol %)	DMF	110	20 h	(67)	(0)	67
4	Br	CuI	2 mol %	L1 (2 mol %), KO-t-Bu	—	100	12 h	(99)	(0)	288
4	Br	CuI	3.6 mol %	PBu3 (7.2 mol %), KO-t-Bu	toluene	135	10.5 h	(0)	(85)	56
4	Br	CuI	10 mol %	1,2-CDA (20 mol %), K3PO4	[C4mim][BF4]	110	12 h	(83)	(0)	232
4	Br	CuI	5 mol %	L2 (25 mol %), KOH, TBAB	H2O	MW (100 W), 130	5 min	(78)	(0)	236
4-MeOC6H4I+		CuI	10 mol %	Na2CO3	CH2Cl2	rt	6 h	(38)	(0)	76
4	I	CuI	2 mol %	KO-t-Bu	toluene	135	14 h	(8)	(81)	304
4	I	CuI	10 mol %	TBPM, 2,2'-biphenol (20 mol %)	dioxane	rt	12 h	(65)	(0)	61
4	Br	CuO	25 mol %	L15 (50 mol %), KOH, TBAB	H2O	MW (100 W), 130	5 min	(52)	(0)	267
4	Br	CuO	25 mol %	L15 (50 mol %), KOH, TBAB	H2O	reflux	10 h	(52)	(0)	267
4	Br	CuO	5 mol %	L14 (50 mol %), hexane-2,5-dione, KOH, TBAB	H2O	MW, 120	5 min	(78)	(0)	51
4	I	CuO nanoparticles (33 nm)	1.3 mol %	KOH, air	DMSO	110	12 h	(22)	(0)	224
4	I	CuO nanoparticles	5 mol %	KOH	DMSO/t-BuOH (3:1)	110	24 h	(73)	(0)	74
4	Br	Cu/Fe-hydrotalcite	10 wt %	none	toluene	130	15 h	(80)	(0)	283
4	I	Cu/Fe-hydrotalcite	10 wt %	none	toluene	130	12 h	(82)	(0)	283

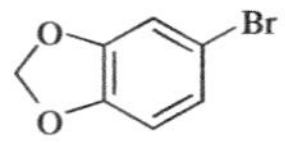

CuI (5 mol %),
L2 (25 mol %),
KOH, TBAB, H₂O,
MW (100 W), 130°, 5 min

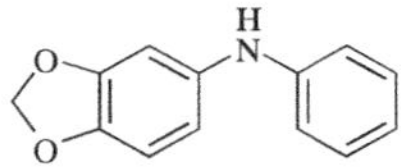

(80)

236

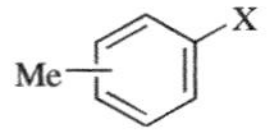

Catalyst (*x* amount)

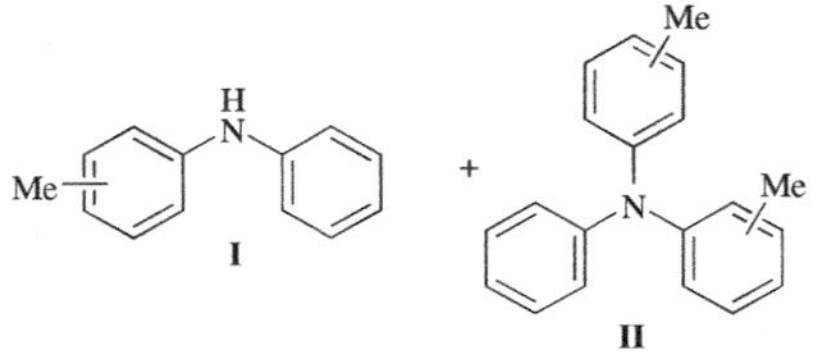

Isomer	X	Catalyst	x	Additive(s)	Solvent	Temp (°)	Time	I	II	
2	I	Cu	2 eq	K₂CO₃, 18-c-6	1,2-Cl₂C₆H₄	reflux	15 h	(0)	(95)	54
2	Cl	CuI	3.6 mol %	PBu₃ (7.2 mol %), KO*t*-Bu	toluene	135	10.5 h	(0)	(80)	56
4	Br	Cu (active)	5 mol %	none	H₂O	MW (100 W)	6 min	(71)	(0)	280
4	I	CuCl₂	4 mol %	phen (4 mol %), KOH	toluene	125	5 h	(0)	(73)	311
4	Br	CuI	7 mol %	L-Pro (14 mol %), K₂CO₃	DMSO	90	40 h	(51)	(0)	47
4	I	CuI	5 mol %	DiPrPhDAB (6 mol %), KO*t*-Bu	toluene	120	48 h	(92)	(0)	306
4	Cl	CuI	2 mol %	**L1** (2 mol %), KO*t*-Bu	—	100	12 h	(60)	(0)	288
4	Br	CuI	2 mol %	**L1** (2 mol %), KO*t*-Bu	—	100	12 h	(78)	(0)	288
4	Br	CuI	5 mol %	**L2** (25 mol %), KOH, TBAB	H₂O	MW (100 W), 130	5 min	(74)	(0)	236
4	I	CuO	10 mol %	FeCl₃ (10 mol%), *rac*-binol (20 mol %), Cs₂CO₃	DMF	90	12 h	(79)	(0)	277
4	Br	CuO	25 mol %	**L15** (50 mol %), KOH, TBAB	H₂O	reflux	10 h	(48)	(0)	51
4	Br	CuO	25 mol %	**L15** (50 mol %), KOH, TBAB	H₂O	MW (100 W), 130	5 min	(48)	(0)	51
4	Br	CuO	5 mol %	**L14** (50 mol %), hexane-2,5-dione, KOH, TBAB	H₂O	90	8 h	(70)	(0)	51
4	Br	CuO	5 mol %	**L14** (50 mol %), hexane-2,5-dione, KOH, TBAB	H₂O	MW, 120	5 min	(70)	(0)	51

TABLE 5A. N-ARYLATION OF PRIMARY ARYL AMINES (Continued)

Nitrogen Nucleophile	Aryl Halide	Conditions	Product(s) and Yield(s) (%)	Refs.

*Please refer to the charts preceding the tables for structures indicated by the **bold** numbers.*

C$_6$

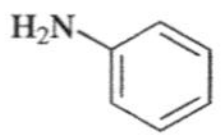

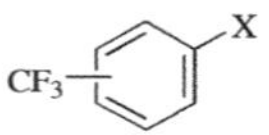

Isomer	X	Catalyst	x	Additive(s)	Solvent	Temp (°)	Time	I	II	
3	I	Cu(TMHD)$_2$	20	KOt-Bu	toluene	120	40 h	(0)	(82)	308
4	Br	CuI	5	L2 (25 mol %), KOH, TBAB	H$_2$O	MW (100 W) 130	5 min	(47)	(0)	236
4	Br	CuO	5	L14 (50 mol %), KOH, TBAB, hexane-2,5-dione	H$_2$O	MW 120	5 min	(45)	(0)	51

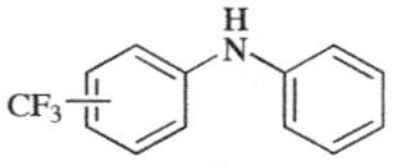

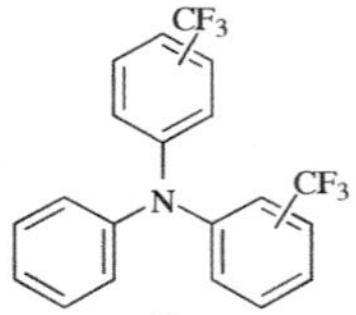

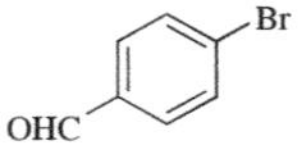

	Cu/Fe–hydrotalcite (10 wt %), toluene, 130°, 12 h	(79)	283
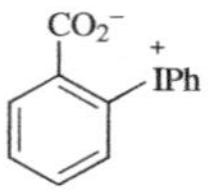	Cu(OAc)$_2$ (4 mol %), i-PrOH, reflux, 1 h	(68)	312
	Cu(OAc)$_2$ (4 mol %), i-PrOH, reflux, 48 h	(38)	312

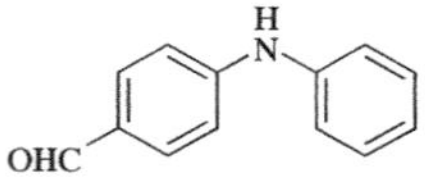

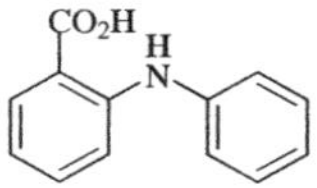

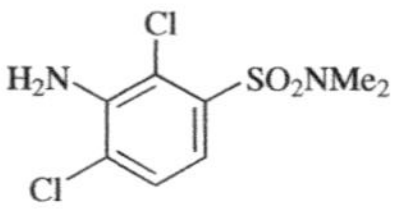

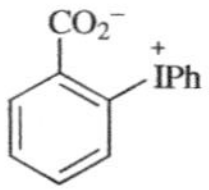

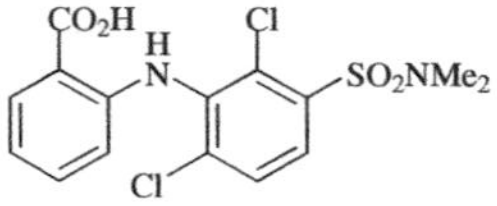

X	R	Catalyst(s)	x	Additive(s)	Solvent	Temp (°)	Time (h)		
Cl	H	Cu	8 mol %	pyridine (15 wt %), K_2CO_3	amyl alcohol	reflux	1h	(69)	313
Cl	H	Cu	1 eq	none	—	reflux	—	(83)	1
Cl	H	Cu powder	—	pyridine, K_2CO_3	H_2O	)))	20 min	(88)	314
Cl	H	Cu, CuI	2, 0.6 mol %	K_2CO_3	DMF	))) (20 kHz), reflux	20 min	(25)	315
Cl	H	Cu, CuI	4, 9 mol %	K_2CO_3	$EtO(CH_2)_2OH$	130	24 h	(83)	109
Br	H	Cu, CuI	9, 4 mol %	K_2CO_3	$EtO(CH_2)_2OH$	130	24 h	(86)	70
Cl	H	CuO	6 mol %	K_2O_3	—	reflux	—	(93)	316
Cl	H	$CuSO_4$	16 mol %	K_2CO_3	H_2O	MW, 100	5 min	(98)	240
Cl	H	CuOAc	10 mol %	NaOAc	H_2O	reflux	4 h	(91)	317
Br	H	CuOAc	10 mol %	NaOAc	H_2O	reflux	2.5 h	(94)	317
I	H	CuOAc	10 mol %	NaOAc	H_2O	reflux	0.5 h	(97)	317
Cl	H	$Cu(OAc)_2$	14 mol %	K_2CO_3	—	MW (700 W)	6 min	(87)	72
Br	H	$Cu(OAc)_2$	14 mol %	K_2CO_3	—	MW (700 W)	6 min	(93)	72
Cl	H	$Cu(pyr)_2Cl_2$	10 mol %	K_2CO_3	$HO(CH_2)_2OH$	130	24 h	(72)	71
Cl	H	$Cu(pyr)_2Cl_2$	11 mol %	K_2CO_3	$HO(CH_2)_2OH$	MW, 75	20 min	(70)	318
Br	H	$Cu(pyr)_2Cl_2$	10 mol %	K_2CO_3	$HO(CH_2)_2OH$	130	24 h	(69)	71
Br	H	$Cu(pyr)_2Cl_2$	11 mol %	K_2CO_3	$HO(CH_2)_2OH$	MW, 75	20 min	(71)	318
I	H	$Cu(pyr)_2Cl_2$	10 mol %	K_2CO_3	$HO(CH_2)_2OH$	130	24 h	(81)	71
I	H	$Cu(pyr)_2Cl_2$	11 mol %	K_2CO_3	$HO(CH_2)_2OH$	MW, 75	20 min	(97)	317

TABLE 5A. N-ARYLATION OF PRIMARY ARYL AMINES (Continued)

*Please refer to the charts preceding the tables for structures indicated by the **bold** numbers.*

Nitrogen Nucleophile	Aryl Halide	Conditions	Product(s) and Yield(s) (%)	Refs.
C6 H2N–Ph	CO2Me, I	Cu (1 mol %), K2CO3, 18-c-6, 1,2-Cl2C6H4, reflux, 48 h	CO2Me / MeO2C structure (81)	319

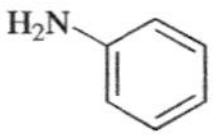

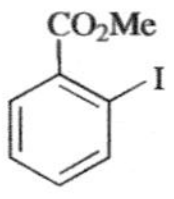

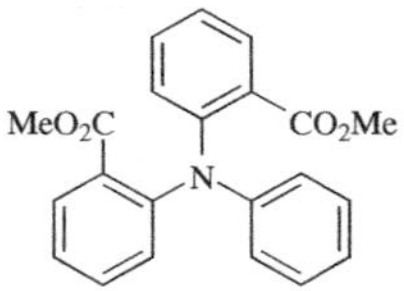

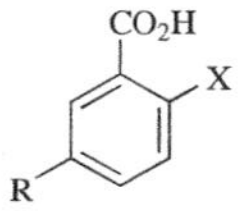

Aryl Halide: CO_2H with X and R substituents

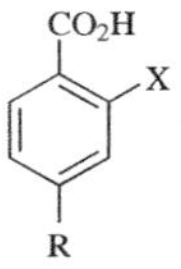

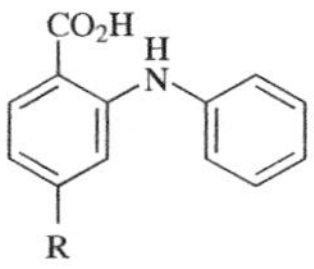

Conditions: Catalyst(s) (x mol %), K_2CO_3

Product: CO_2H H–N–Ph with R substituent

R	X	Catalyst(s)	x	Solvent	Temp (°)	Time		Refs.
F	Br	Cu, CuI	4, 9	EtOCH2CH2OH	130	24 h	(82)	70
Cl	Br	Cu, CuI	9, 4	EtOCH2CH2OH	130	24 h	(94)	70
Cl	Cl	CuSO4	16	H2O	MW, 100	10 min	(87)	240
Br	Cl	Cu(pyr)2Cl2	10	HOCH2CH2OH	130	24 h	(78)	71
Br	Br	Cu(pyr)2Cl2	10	HOCH2CH2OH	130	24 h	(83)	71
Br	I	Cu(pyr)2Cl2	10	HOCH2CH2OH	130	24 h	(88)	71

Aryl Halide: CO_2H with X and R substituents

Conditions: Catalyst(s) (x mol %), K_2CO_3

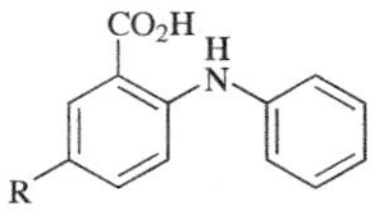

Product: CO_2H H–N–Ph with R substituent

R	X	Catalyst(s)	x	Solvent	Temp (°)	Time		Refs.
Cl	Cl	Cu, CuI	2, 0.6	DMF	))) (20 kHz), reflux	20 min	(33)	315
Br	Br	Cu, CuI	9, 4	EtOCH2CH2OH	130	24 h	(84)	70
Br	Cl	Cu(pyr)2Cl2	11	HOCH2CH2OH	MW, 75	20 min	(79)	318
Br	Br	Cu(pyr)2Cl2	11	HOCH2CH2OH	MW, 75	20 min	(91)	318
Br	I	Cu(pyr)2Cl2	11	HOCH2CH2OH	MW, 75	20 min	(96)	318

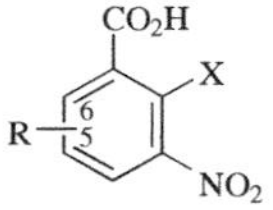

Cu (4 mol %), N-ethylmorpholine, 2,3-butanediol, 70°, 15 h

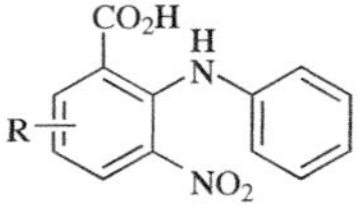

R	X	
H	Br	(78)
5-Cl	Cl	(59)
6-Cl	Cl	(52)
4-MeO	I	(50)
5-Me	I	(89)

320

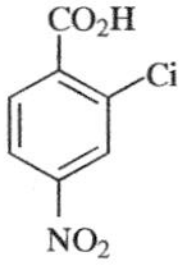

CuSO$_4$ (16 mol %), K$_2$CO$_3$, H$_2$O, MW, 100°, 5 min

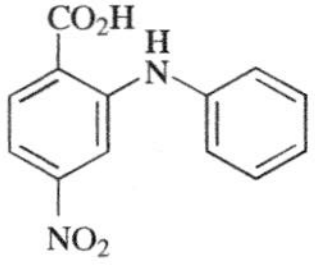

(95)

240

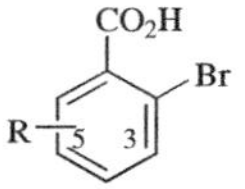

Cu (9 mol %), CuI (4 mol %), K$_2$CO$_3$, EtOCH$_2$CH$_2$OH, 130°, 24 h

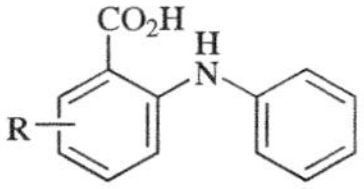

R		
3-Me	(58)	
5-HO$_2$C	(99)	

70

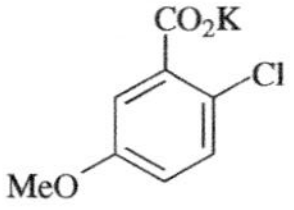

Cu (1 mol %), isoamyl alcohol, reflux, 18 h

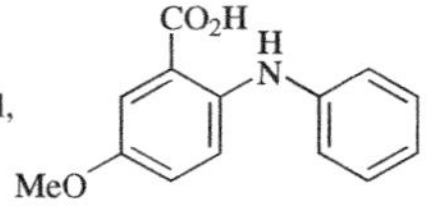

(—)

321

Nitrogen Nucleophile	Aryl Halide	Conditions	Product(s) and Yield(s) (%)	Refs.

*Please refer to the charts preceding the tables for structures indicated by the **bold** numbers.*

C$_{6-11}$

Row (Ref. 322)

Nitrogen Nucleophile: H$_2$N-C$_6$H$_4$-R

R	R	R
H	2,3,4-Cl$_3$	2,3-Br$_2$
2-Cl	2,3,5-Cl$_3$	3-Me$_2$N
3-Cl	2,4,5-Cl$_3$	3-O$_2$N
4-Cl	2,4,6-Cl$_3$	2-MeO, 5,6-Cl$_2$
2,3-Cl$_2$	3,4,5-Cl$_3$	3-MeO
2,4-Cl$_2$	2,3,4,5-Cl$_4$	3-EtO
2,5-Cl$_2$	2,3,4,6-Cl$_4$	3-MeS
3,4-Cl$_2$	2,3,4,5,6-Cl$_5$	3-Me$_2$NO$_2$S
3,5-Cl$_2$	3-Br	2-Me

Aryl Halide: (2-CO$_2$K, 3-Br benzene)

Conditions: Cu(OAc)$_2$ (6 mol %), DMF, 4-ethylmorpholine, 145°, 3 h

Product: (2-CO$_2$H, N-H diphenylamine)-R (—)

R	R	R	R
3-Me	2-Me, 3-O$_2$N	3-Et	3-Ac
4-Me	2-Me, 3-O$_2$N, 6-Cl	3-Et, 2,6-Cl$_2$	3,5-(CF$_3$)$_2$
2-Me, 3-Cl	2-Me, 3-Me$_2$NO$_2$S	2-H$_2$N, 3,6-Me$_2$	2,3-Me$_2$
2-Me, 6-Cl	3-CF$_3$	3-Me$_2$N, 2,6-Me$_2$	2,4-Me$_2$
3-Me, 2-Cl	3-CF$_3$, 2,6-Cl$_2$	3-O$_2$N, 2,6-Me$_2$	2,5-Me$_2$
3-Me, 2,6-Cl$_2$	3-CF$_3$, 2-Br	2-Me, 3-CF$_3$	2,6-Me$_2$
3-Me, 5,6-Cl$_2$	3-NC—	3-MeS, 2,6-Me$_2$	3,4-Me$_2$
3-Me, 2,4,6-Cl$_3$	3-NC—, 2,6-Cl$_2$	3-MeO$_3$S, 2,6-Me$_2$	2,5-Me$_2$, 3-Cl
2-Me, 4,5-Cl$_3$	3-NC—, 2-Br	4-Me$_2$NO$_2$S, 2,6-Me$_2$	2,6-Me$_2$, 2-Cl
		2-Me, 3-CN—	2,6-Me$_2$, 3-Br

R
3,6-Me$_2$, 2-Cl
2,4,5-Me$_3$
2,4,6-Me$_3$
3-*n*-Pr
3-*n*-Bu
3-NC—, 2,6-Me$_2$
2,3,5,6-Me$_4$
2,6-Me$_2$, 3-Et
2,6-Et$_2$, 3-Me$_2$NO$_2$S
2,6-Me$_2$, 3-*n*-Pr

Refs. 322

C$_6$

Row (Ref. 61)

Nitrogen Nucleophile: H$_2$N-C$_6$H$_5$

Aryl Halide: MeO$_2$C-C$_6$H$_4$-I (3-substituted)

Conditions: CuI (10 mol %), TBPM, 2,2'-biphenol (20 mol %), dioxane, rt, 12 h

Product: MeO$_2$C-C$_6$H$_4$-N(H)-C$_6$H$_5$ (85)

Refs. 61

Row (Ref. 267)

Aryl Halide: 4-ethyl-C$_6$H$_4$-Br

Conditions: CuO (25 mol %), **L15** (50 mol %), KOH, TBAB, H$_2$O, MW (100 W), 130°, 5 min

Product: C$_6$H$_5$-N(H)-C$_6$H$_4$-ethyl (49)

Refs. 267

Row (Ref. 307)

Aryl Halide: (3,4-dimethyl-C$_6$H$_3$-I)

Conditions: Cu(OAc)$_2$ (4 mol %), phen (3 mol %), KOH, toluene, reflux, 18 h

Product: C$_6$H$_5$-N(-Ar)$_2$ (59)

Refs. 307

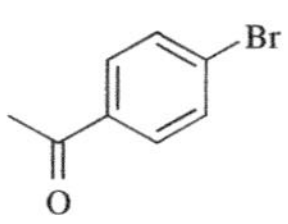

C_{6-9}

Catalyst (5 mol %),
H$_2$O, 5 min

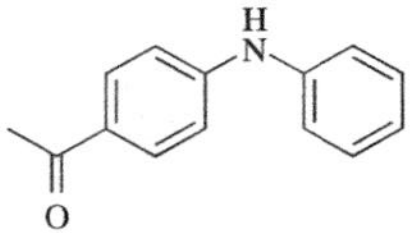

Catalyst	Additives	Temp (°)		
CuI	L2 (25 mol %), KOH, TBAB	MW (100 W), 130	(75)	236
CuO	L14 (50 mol %), hexane-2,5-dione, KOH, TBAB	MW, 120	(66)	51

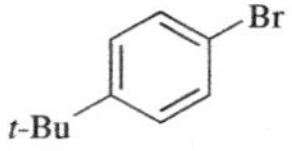

CuI (2 mol %), L1 (2 mol %),
KOt-Bu, 100°, 12 h

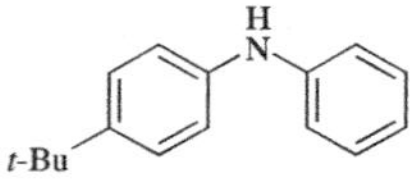

(90) 288

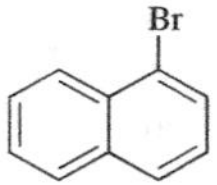

CuI (5 mol %),
EDA (5 mol %), K$_2$CO$_3$,
PEG-400, 80°, 24 h

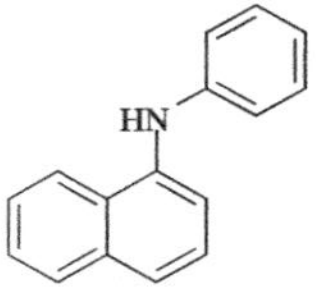

(58) 301

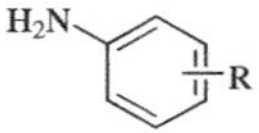

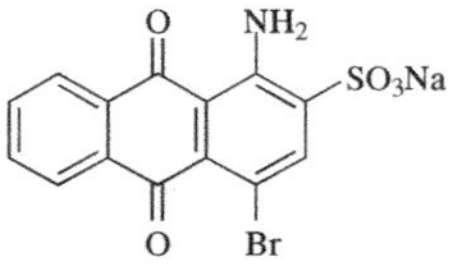

Cu (17 mol %), Na$_3$PO$_4$,
H$_2$O, pH 6-7, MW,
80–120°, 2–20 min

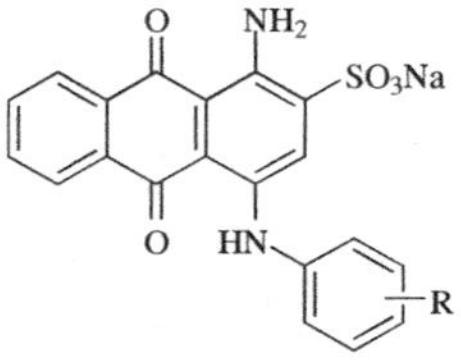

323

R		R		R	
H	(55)	4-MeO	(85)	2-HO$_2$C, 5-Cl	(64)
4-F	(76)	2-HO$_3$S	(30)	3-HO$_2$C	(76)
3-Cl	(56)	4-HO$_3$S	(40)	3-HO$_2$C, 5-H$_2$N	(70)
4-Cl	(87)	2-Me, 3-H$_2$N	(34)	3-HO$_2$C, 4-HO	(30)
3-Br	(41)	2-HO$_2$C	(83)	4-HO$_2$C	(70)
2-H$_2$N	(58)	2-HO$_2$C, 4-Cl	(43)	3,5-Me$_2$	(32)

R	
3-H$_2$N	(70)
4-H$_2$N	(90)
2-HO	(58)
4-HO	(72)
2-MeO	(79)
3-MeO	(67)

R	
4-HO$_2$CCH$_2$	(44)
2,4,6-Me$_3$	(77)

185

TABLE 5A. *N*-ARYLATION OF PRIMARY ARYL AMINES (*Continued*)

Nitrogen Nucleophile	Aryl Halide	Conditions	Product(s) and Yield(s) (%)	Refs.

*Please refer to the charts preceding the tables for structures indicated by the **bold** numbers.*

C$_{6-8}$

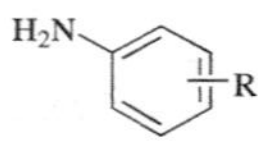

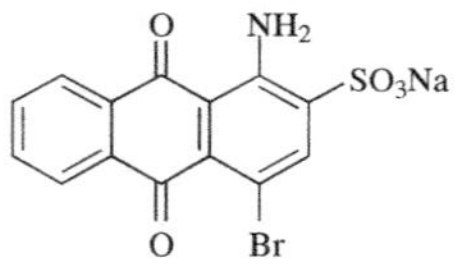

Nitrogen Nucleophile: H$_2$N—R

Aryl Halide: anthraquinone with NH$_2$, SO$_3$Na, Br

Conditions: Catalyst (x mol %), Na$_2$CO$_3$, H$_2$O

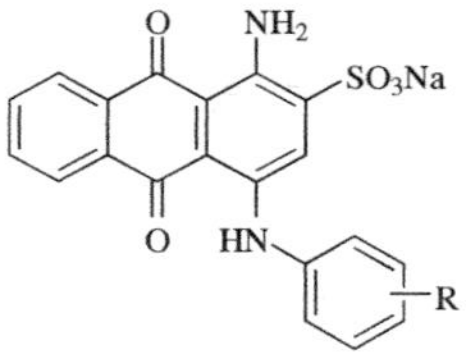

Product: anthraquinone with NH$_2$, SO$_3$Na, HN—R

R	Catalyst	x	Additive	Temp (°)	Time (h)	Yield	Refs.
3-HO$_2$S, 4-H$_2$N	CuCl	30	Na$_2$SO$_4$	80	8	(—)	324
3-HO$_2$S, 4-H$_2$N	Cu$_2$SO$_4$	50	Zn (50 mol %)	60	24	(75)	325
2-Me	CuCl	50	Na$_2$SO$_4$	80	8	(—)	324
3-Me	CuCl	50	Na$_2$SO$_4$	80	8	(—)	324
4-Me	CuCl	50	Na$_2$SO$_4$	80	8	(—)	324
2,6-Me$_2$	CuCl	50	Na$_2$SO$_4$	80	8	(—)	324

C$_6$

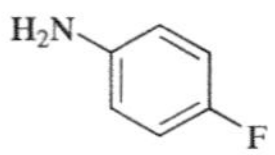

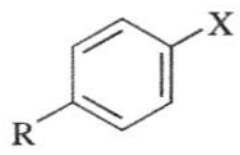

Nitrogen Nucleophile: H$_2$N—C$_6$H$_4$—F

Aryl Halide: X—C$_6$H$_4$—R

Conditions: Catalyst (x mol %)

Product: R—C$_6$H$_4$—N(H)—C$_6$H$_4$—F

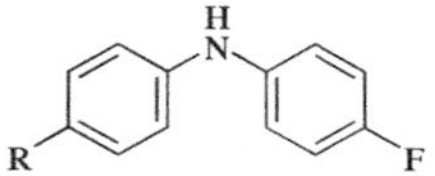

R	Catalyst	x	Additives	Solvent	Temp (°)	Time (h)	Yield	Refs.
F	Cu(OAc)$_2$	4	phen (3 mol %), KOH	toluene	reflux	18	(—)	307
MeS	CuI	10	pyridine 2-carboxylic acid (20 mol %), K$_3$PO$_4$	DMSO	100	24	(71)	50

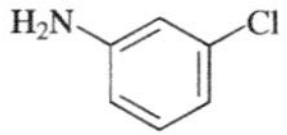

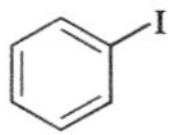

Nitrogen Nucleophile: H$_2$N—C$_6$H$_4$—Cl

Aryl Halide: iodobenzene

Conditions: Cu, K$_2$CO$_3$

Product: triphenylamine derivative with Cl, (50)

Refs. 326

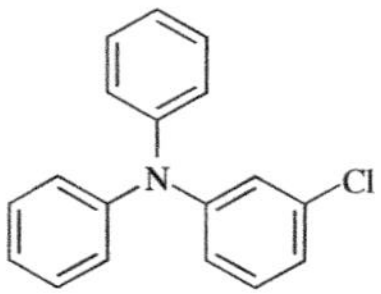

186

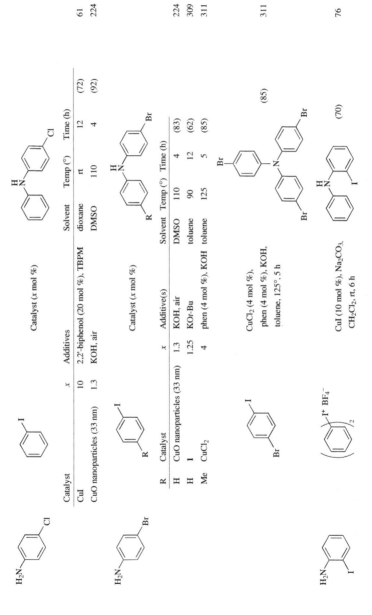

Catalyst	x	Additives	Solvent	Temp (°)	Time (h)	
CuI	10	2,2′-biphenol (20 mol %), TBPM	dioxane	rt	12	(72)	61
CuO nanoparticles (33 nm)	1.3	KOH, air	DMSO	110	4	(92)	224

R	Catalyst	x	Additive(s)	Solvent	Temp (°)	Time (h)	
H	CuO nanoparticles (33 nm)	1.3	KOH, air	DMSO	110	4	(83)	224
H	1	1.25	KOt-Bu	toluene	90	12	(62)	309
Me	CuCl$_2$	4	phen (4 mol %), KOH	toluene	125	5	(85)	311

CuCl$_2$ (4 mol %), phen (4 mol %), KOH, toluene, 125°, 5 h   (85)   311

CuI (10 mol %), Na$_2$CO$_3$, CH$_2$Cl$_2$, rt, 6 h   (70)   76

187

TABLE 5A. N-ARYLATION OF PRIMARY ARYL AMINES (*Continued*)

*Please refer to the charts preceding the tables for structures indicated by the **bold** numbers.*

C$_6$

Nitrogen Nucleophile	Aryl Halide	Conditions	Product(s) and Yield(s) (%)	Refs.
(structure: R-substituted aniline, H$_2$N)	(structure: 3,4-dimethyliodobenzene, I)	Catalyst (4 mol %), phen (x mol %), KOH, toluene	(structure: N-aryl product with R)	

R	Catalyst	x	Temp (°)	Time (h)		Refs.
2-F	Cu(OAc)$_4$	3	reflux	18	(—)	307
3-F	Cu(OAc)$_4$	3	reflux	18	(—)	307
4-F	Cu(OAc)$_4$	3	reflux	18	(—)	307
2,4-F$_2$	Cu(OAc)$_4$	3	reflux	18	(—)	307
4-Cl	Cu(OAc)$_4$	3	reflux	18	(—)	307
4-Br	CuCl$_2$	4	125	5	(83)	311
4-Br	Cu(OAc)$_4$	3	reflux	18	(—)	307

Nitrogen Nucleophile	Aryl Halide	Conditions	Product(s) and Yield(s) (%)	Refs.
(structure: H$_2$N-C$_6$H$_4$-Cl)	(structure: CO$_2$H-substituted aryl halide X)	Catalyst(s) (x mol %)	(structure: CO$_2$H diarylamine with Cl)	

X	Isomer	Catalyst(s)	x	Additive(s)	Solvent	Temp (°)	Time		Refs.
Cl	2	Cu, CuI	2, 0.6	K$_2$CO$_3$	DMF	(20 kHz), reflux	20 min	(75)	315
Cl	2	Cu(OAc)$_2$	14	K$_2$CO$_3$	—	MW (700 W)	15 min	(69)	72
Br	2	Cu(OAc)$_2$	14	K$_2$CO$_3$	—	MW (700 W)	14.5 min	(75)	72
Cl	3	Cu	8	pyridine (15 wt %), K$_2$CO$_3$	amyl alcohol	reflux	1 h	(56)	313
Cl	3	Cu	8	pyridine (15 wt %), K$_2$CO$_3$	H$_2$O	reflux	1 h	(60)	313
Br	3	Cu	9	K$_2$CO$_3$	EtO(CH$_2$)$_2$OH	130	24 h	(84)	70
Cl	3	CuI	9	K$_2$CO$_3$	EtO(CH$_2$)$_2$OH	130	24 h	(99)	109
Cl	3	Cu(OAc)$_2$	14	K$_2$CO$_3$	—	MW (700 W)	14 min	(73)	72
Br	3	Cu(OAc)$_2$	14	K$_2$CO$_3$	—	MW (700 W)	12.5 min	(81)	72

Structure (reactant):

CO$_2$H / X (substituted benzene)

H$_2$N–Br (4-bromoaniline)

Catalyst(s) (x mol %)

Product structure: 2-(arylamino)benzoic acid with CO$_2$H, NH, and Br substituents

X	Isomer	Catalyst(s)	x	Additive	Solvent	Temp (°)	Time	(yield)	ref
Cl	3	Cu(pyr)$_2$Cl$_2$	10	K$_2$CO$_3$	HO(CH$_2$)$_2$OH	130	24 h	(78)	71
Br	3	Cu(pyr)$_2$Cl$_2$	10	K$_2$CO$_3$	HO(CH$_2$)$_2$OH	130	24 h	(83)	71
I	3	Cu(pyr)$_2$Cl$_2$	10	K$_2$CO$_3$	HO(CH$_2$)$_2$OH	130	24 h	(88)	71
Cl	4	Cu, CuI	2, 0.6	K$_2$CO$_3$	DMF	)))(20 kHz), reflux	20 min	(33)	315
Cl	4	Cu	7.5	K$_2$CO$_3$	DMF	reflux	2 h	(82)	327
Cl	4	Cu(OAc)$_2$	14	K$_2$CO$_3$	—	MW (700 W)	13 min	(78)	72
Br	4	Cu(OAc)$_2$	14	K$_2$CO$_3$	—	MW (700 W)	10 min	(86)	72
Cl	4	Cu(pyr)$_2$Cl$_2$	11	K$_2$CO$_3$	HO(CH$_2$)$_2$OH	MW, 75	20 min	(80)	318
Br	4	Cu(pyr)$_2$Cl$_2$	11	K$_2$CO$_3$	HO(CH$_2$)$_2$OH	MW, 75	20 min	(94)	318
I	4	Cu(pyr)$_2$Cl$_2$	11	K$_2$CO$_3$	HO(CH$_2$)$_2$OH	MW, 75	20 min	(96)	318
Cl	2	Cu(OAc)$_2$	14	K$_2$CO$_3$	—	MW (900 W)	12 min	(65)	72
Br	2	Cu(OAc)$_2$	14	K$_2$CO$_3$	—	MW (900 W)	11 min	(69)	72
Br	3	Cu, CuI	9, 4	K$_2$CO$_3$	EtO(CH$_2$)$_2$OH	130	24 h	(81)	70
Cl	3	Cu(OAc)$_2$	14	K$_2$CO$_3$	—	MW (900 W)	11 min	(74)	72
Br	3	Cu(OAc)$_2$	14	K$_2$CO$_3$	—	MW (900 W)	11 min	(80)	72
Cl	3	Cu(pyr)$_2$Cl$_2$	10	K$_2$CO$_3$	HO(CH$_2$)$_2$OH	130	24 h	(76)	71
Br	3	Cu(pyr)$_2$Cl$_2$	10	K$_2$CO$_3$	HO(CH$_2$)$_2$OH	130	24 h	(80)	71
I	3	Cu(pyr)$_2$Cl$_2$	10	K$_2$CO$_3$	HO(CH$_2$)$_2$OH	130	24 h	(86)	71
Cl	4	Cu(pyr)$_2$Cl$_2$	11	K$_2$CO$_3$	HO(CH$_2$)$_2$OH	MW, 75	20 min	(83)	318
Br	4	Cu(pyr)$_2$Cl$_2$	11	K$_2$CO$_3$	HO(CH$_2$)$_2$OH	MW, 75	20 min	(91)	318
I	4	Cu(pyr)$_2$Cl$_2$	11	K$_2$CO$_3$	HO(CH$_2$)$_2$OH	MW, 75	20 min	(95)	318

TABLE 5A. N-ARYLATION OF PRIMARY ARYL AMINES (Continued)

Please refer to the charts preceding the tables for structures indicated by the **bold** numbers.

C₆

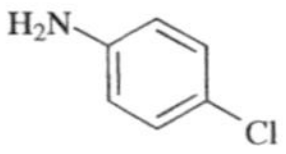

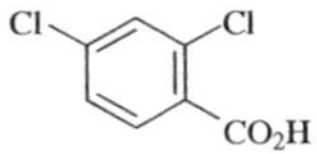

Nitrogen Nucleophile	Aryl Halide	Conditions	Product(s) and Yield(s) (%)	Refs.
		CuSO₄ (16 mol %), K₂CO₃, H₂O, MW, 100°, 7 min	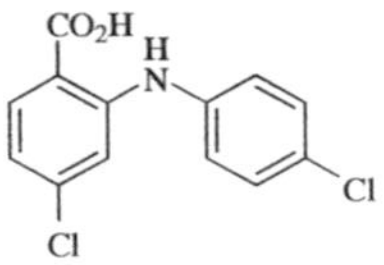 (95)	240

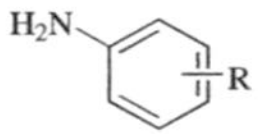	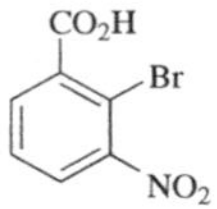	Cu (4 mol %), 4-ethylmorpholine, 2,3-butanediol, 70°, 15 h	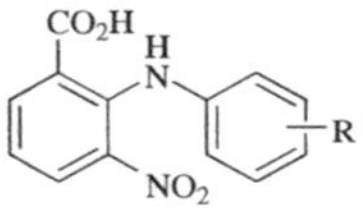
			 R R
			 2-F (67) 2-F, 5-Cl (31)
			 2,3-F₂ (57) 2,6-F₂ (58)
			 2,5-F₂ (62) 2-F, 6-Cl (12)
			   2-Cl (58)
			   3-Cl (40)
			   4-Cl (60)
				320

		CuI (2 eq), K₂CO₃, NMP, 100°, 24 h	 R¹ R² Cl H (61) F Me (41)	328
		Cu (10 mol %), K₂CO₃, xylene, 140°, 6 h	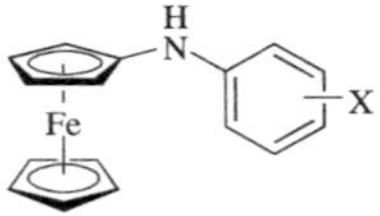 X 2-Cl (94) 4-Cl (0) 2,4-Cl₂ (0) 2-Br (46) 4-Br (0) 2,4-Br₂ (38)	329
		Cu powder (1 eq). K₂CO₃, Ph₂O, 190°, 36 h	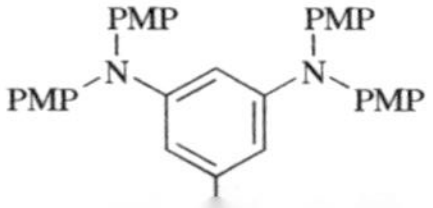 (52)	330

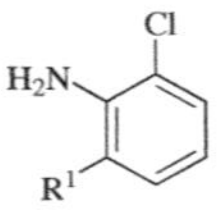

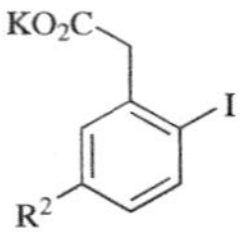

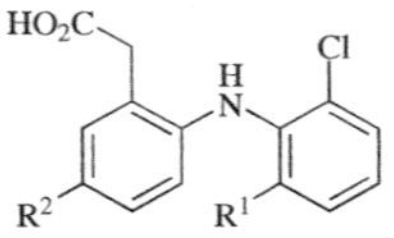

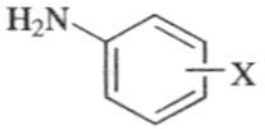

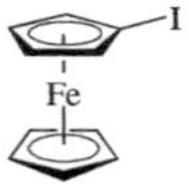

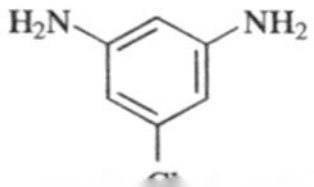

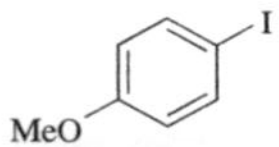

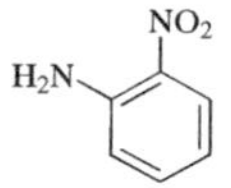

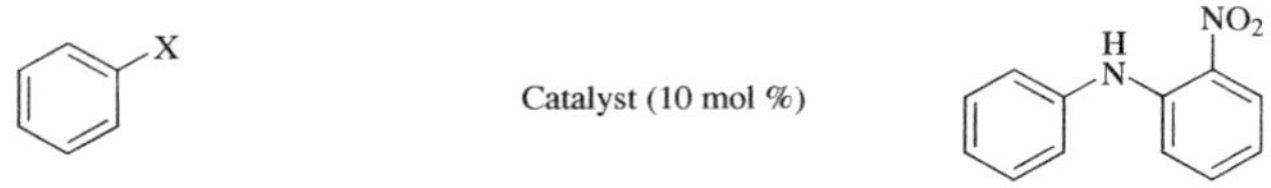

H2N–C6H4–NO2 (2-nitroaniline reagent)

X–C6H5 + Catalyst (10 mol %) → (PhNH–C6H4–NO2)

X	Catalyst	Additive(s)	Solvent	Temp (°)	Time (h)		
I	CuI	CsOAc	DMSO	90	24	(63)	274
Cl	CuI	PPAPM (20 mol %), K3PO4	DMF	120	48	(32)	48
Br	CuI	PPAPM (20 mol %), K3PO4	DMF	110	36	(90)	48
Cl	CuI	piperidine-2-carboxylic acid (20 mol %), K2CO3	DMF	110	36	(31)	67
Br	CuI	piperidine-2-carboxylic acid (20 mol %), K2CO3	DMF	110	20	(87)	67
Br	CuI	DPP (20 mol %), K3PO4	DMF	110	36	(75)	108
I	CuO	FeCl3 (10 mol %), rac-binol (20 mol %), Cs2CO3	DMF	90	12	(89)	277

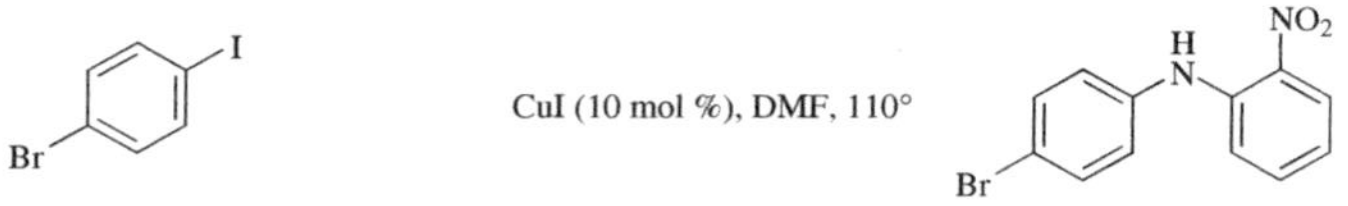

4-bromoiodobenzene $\xrightarrow{\text{CuI (10 mol \%), DMF, 110°}}$ (Br–C6H4–NH–C6H4–NO2)

Additives	Time (h)		
PPAPM (20 mol %), K3PO4	30	(98)	48
piperidine-2-carboxylic acid (20 mol %), K2CO3	24	(93)	67

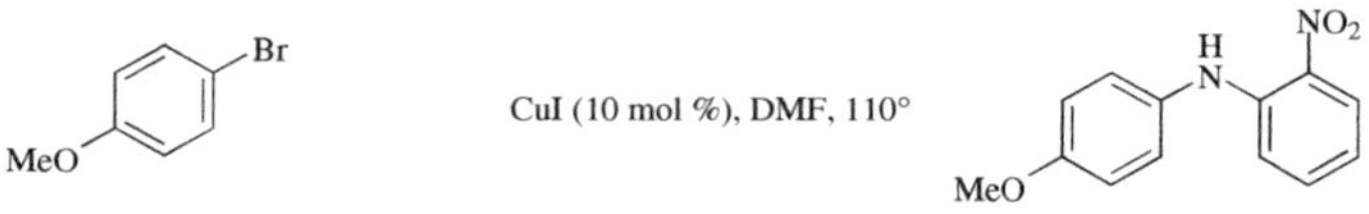

4-bromoanisole $\xrightarrow{\text{CuI (10 mol \%), DMF, 110°}}$ (MeO–C6H4–NH–C6H4–NO2)

Additives	Time (h)		
DPP (20 mol %), K3PO4	36	(70)	108
PPAPM (20 mol %), K3PO4	36	(82)	48
piperidine-2-carboxylic acid (20 mol %), K2CO3	20	(80)	67

191

TABLE 5A. N-ARYLATION OF PRIMARY ARYL AMINES (*Continued*)

Nitrogen Nucleophile	Aryl Halide	Conditions	Product(s) and Yield(s) (%)	Refs.

*Please refer to the charts preceding the tables for structures indicated by the **bold** numbers.*

C$_6$

Nitrogen Nucleophile: 2-aminobenzoic acid (CO_2H, H_2N-phenyl)

Aryl Halide: CO_2H-phenyl-X

Conditions: Catalyst (x mol %), K_2CO_3

Product: CO_2H/H-N-phenyl-NO_2

X	Catalyst	x	Additive	Solvent	Temp	Time (min)		
Cl	Cu powder	—	pyridine	H_2O	)))	20	(79)	314
Cl	Cu(OAc)$_2$	14	none	—	MW (700 W)	14	(70)	72
Br	Cu(OAc)$_2$	14	none	—	MW (700 W)	13	(78)	72

Nitrogen Nucleophile: 2-nitroaniline (NO_2, H_2N-phenyl)

Aryl Halide: bromobenzene (Br)

Conditions: Cu/Fe–hydrotalcite (10 wt %), toluene, 130°, 16 h

Product: NO_2/H-N-phenyl — (77) — Ref. 283

Nitrogen Nucleophile: 4-methoxy-2-nitroaniline (NO_2, OMe, H_2N-phenyl)

Aryl Halide: ferrocenyl iodide (Fe, I)

Conditions: Cu (10 mol %), K_2CO_3, xylene, 140°, 6 h — Ref. 329

Product: NO_2, OMe-N-ferrocenyl-R

R	
H	(85)
NO_2	(0)
Me	(89)

Nitrogen Nucleophile: 4-R-2-nitroaniline (NO_2, R, H_2N-phenyl)

Aryl Halide: X-phenyl

Conditions: Catalyst (10 mol %)

Product: NO_2/H-N-diphenyl

X	Catalyst	Additive(s)	Solvent	Temp (°)	Time (h)		
Cl	CuI	K_2CO_3, piperidine-2-carboxylic acid (20%)	DMF	110	36	(31)	67
Br	CuI	K_3PO_4, PPAPM (20 mol %)	DMF	110	36	(83)	48
Br	CuI	K_3PO_4, DPP (20 mol %)	DMF	110	36	(73)	108
I	CuI	CsOAc	DMSO	90	24	(12)	274
I	CuO	FeCl$_3$ (10 mol %), *rac*-binol (20 mol %), Cs$_2$CO$_3$	DMF	90	12	(81)	277

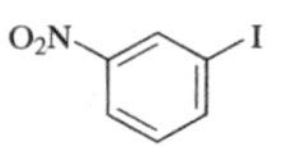

Cu powder, K$_2$CO$_3$,
1,2-Cl$_2$C$_6$H$_4$, reflux, 20 h

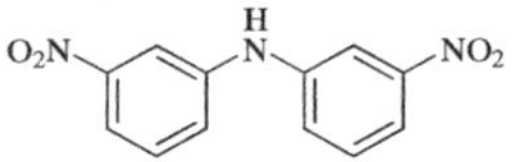

(43)   331

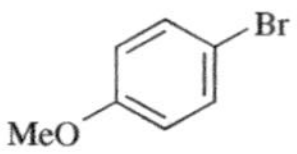

CuI (10 mol %),
piperidine-2-carboxylic acid
(20 mol %),
K$_2$CO$_3$, DMF, 110°, 20 h

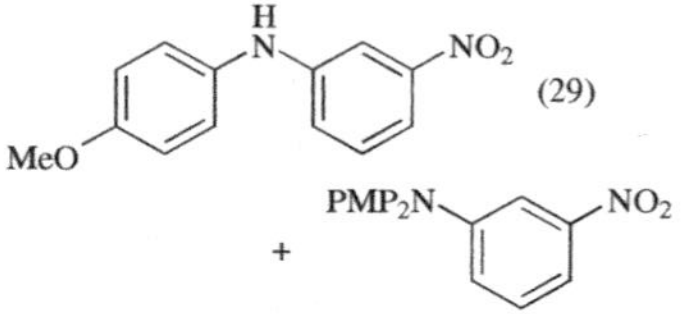

(29)

(23)

+

67

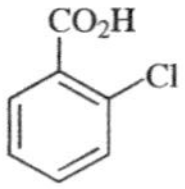

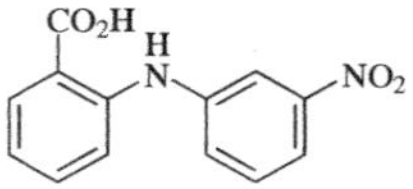

Additive(s)	Solvent	Time (h)		
KOH	H$_2$O	5	(49)	332
pyridine (15 wt %), K$_2$CO$_3$	H$_2$O	1	(62)	313
pyridine (15 wt %), K$_2$CO$_3$	amyl alcohol	1	(60)	313

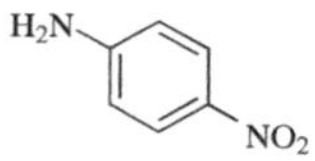

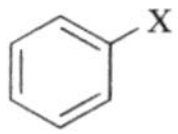

Cu (8 mol %), reflux

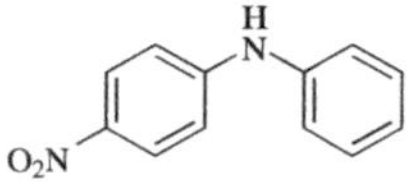

X	Catalyst	x	Additive(s)	Solvent	Temp (°)	Time (h)		
Cl	CuI	10	piperidine-2-carboxylic acid (20 mol %), K$_2$CO$_3$	DMF	110	36	(25)	67
Br	CuI	10	DPP (20 mol %), K$_3$PO$_4$	DMF	110	36	(71)	108
Br	CuI	10	PPAPM (20 mol %), K$_3$PO$_4$	DMF	110	36	(86)	48
I	CuI	10	CsOAc	DMSO	90	24	(64)	274
I	CuI	10	2,2'-biphenol (20 mol %), TBPM	dioxane	rt	12	(53)	61
Br	CuO	10	FeCl$_3$ (10 mol %), rac-binol (20 mol %), Cs$_2$CO$_3$	DMF	110	24	(82)	277
I	CuO	10	FeCl$_3$ (10 mol %), rac-binol (20 mol %), Cs$_2$CO$_3$	DMF	100	12	(85)	277
I	CuO nanoparticles (33 nm)	1.3	KOH, air	DMSO	110	1.5	(70)	224
I	Cu(acac)$_2$	10	Fe$_2$CO$_3$, (20 mol %), Cs$_2$CO$_3$	DMSO/H$_2$O (1:1)	MW, 150	0.5	(trace)	262

TABLE 5A. N-ARYLATION OF PRIMARY ARYL AMINES (Continued)

Nitrogen Nucleophile	Aryl Halide	Conditions	Product(s) and Yield(s) (%)	Refs.

*Please refer to the charts preceding the tables for structures indicated by the **bold** numbers.*

C₆

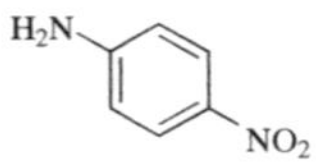	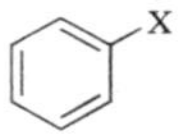	CuI (3.6 mol %), PBu₃ (7.2 mol %), KO*t*-Bu, toluene, 135°, 10.5 h	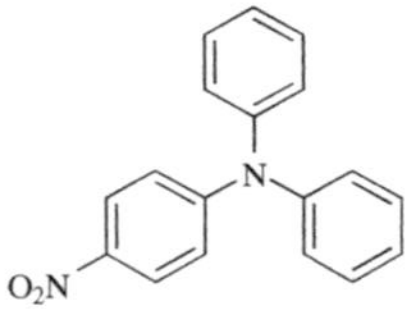 X: Cl (31), Br (35)	56
H₂N–⟨ ⟩–NO₂	⟨ ⟩–X			
	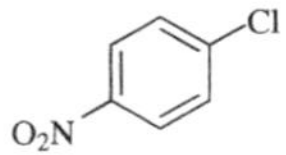 Cl–⟨ ⟩–NO₂	CuI (3.6 mol %), PBu₃ (7.2 mol %), KO*t*-Bu, toluene, 135°, 10.5 h	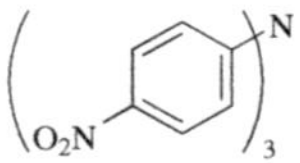 (29)	56

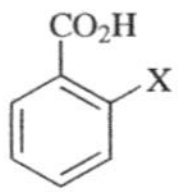

Anthranilic acid (CO₂H, X) — Catalyst(s) (x mol %), K₂CO₃

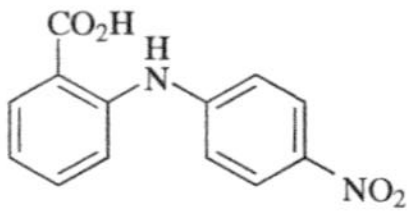

X	Catalyst(s)	x	Solvent	Temp (°)	Time	Yield	Refs.
Br	Cu, CuI	9, 4	EtO(CH₂)₂OH	130	24 h	(53)	70
Cl	CuI	9, 4	EtO(CH₂)₂OH	130	24 h	(87)	109
Cl	Cu(OAc)₂	14	—	MW (700 W)	11 min	(75)	72
Br	Cu(OAc)₂	14	—	MW (700 W)	10 min	(81)	72
Cl	Cu(pyr)₂Cl₂	10	HO(CH₂)₂OH	130	24 h	(56)	71
Br	Cu(pyr)₂Cl₂	10	HO(CH₂)₂OH	130	24 h	(58)	71
I	Cu(pyr)₂Cl₂	10	HO(CH₂)₂OH	130	24 h	(64)	71
Cl	Cu(pyr)₂Cl₂	11	HO(CH₂)₂OH	MW 75	20 min	(62)	318
Br	Cu(pyr)₂Cl₂	11	HO(CH₂)₂OH	MW 75	20 min	(82)	318
I	Cu(pyr)₂Cl₂	11	HO(CH₂)₂OH	MW 75	20 min	(81)	318

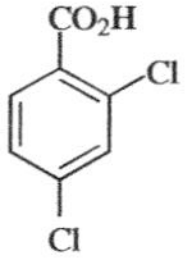

Catalyst (x mol %), K₂CO₃

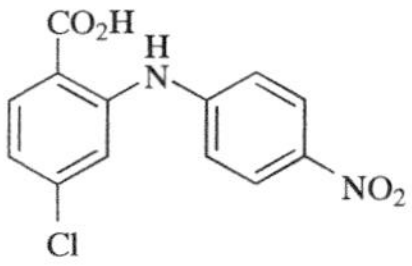

Catalyst	x	Solvent(s)	Temp	Time		
Cu powder	—	pyridine/H₂O	)))	20 min	(61)	314
Cu	7.5	DMF	reflux	2 h	(49)	327

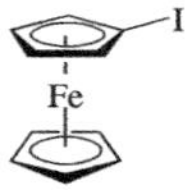

Cu (10 mol %), K_2CO_3, xylene, 140°, 6 h 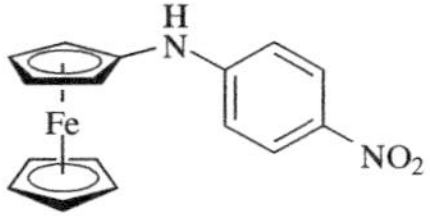   (0)   329

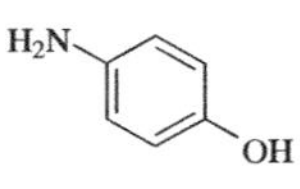

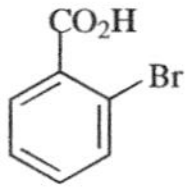

Cu (9 mol %), CuI (4 mol %), K_2CO_3, $EtOCH_2CH_2OH$, 130°, 24 h   (63)   70

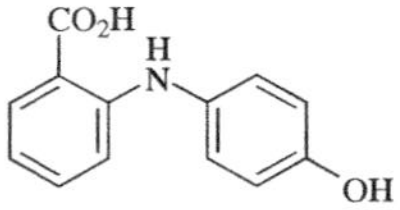

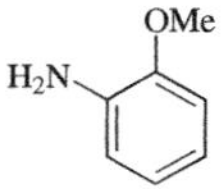

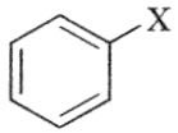

Catalyst (x amount)

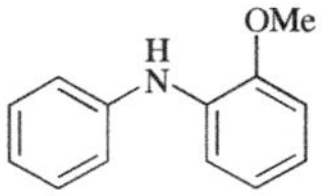

X	Catalyst	x	Additive(s)	Solvent	Temp (°)	Time (h)		
Br	CuI	10 mol %	piperidine-2-carboxylic acid (20%), K_2CO_3	DMF	110	36	(76)	67
Br	CuI	10 mol %	PPAPM (20 mol %), K_3PO_4	DMF	110	36	(60)	108
Br	CuI	10 mol %	DPP (20 mol %), K_3PO_4	DMF	110	36	(60)	108
Br	CuI	10 mol %	CsOAc	DMSO	90	24	(44)	274
Br	Cu/Fe-hydrotalcite	10 wt %	none	toluene	130	12	(75)	283
I	CuO	10 mol %	$FeCl_3$ (10 mol %), rac-binol (20 mol %), Cs_2CO_3	DMF	80	12	(80)	277
I	CuO nanoparticles (33 nm)	1.3 mol %	KOH, air	DMSO	110	1.7	(98)	224

195

Nitrogen Nucleophile	Aryl Halide	Conditions	Product(s) and Yield(s) (%)	Refs.

*Please refer to the charts preceding the tables for structures indicated by the **bold** numbers.*

C_6

Nitrogen Nucleophile	Aryl Halide	Conditions	Product(s) and Yield(s) (%)	Refs.
2-methoxyaniline (H₂N, OMe)	2-iodoanisole (I, OMe)	Cu powder, K_2CO_3, 1,2-Cl₂C₆H₄, reflux, 48 h	OMe–N< product (91)	331
	4-bromoanisole (Br, MeO)	Catalyst (x amount)	MeO–C₆H₄–N(H)–C₆H₄–OMe	

Catalyst	x	Additive(s)	Solvent	Temp (°)	Time (h)		Refs.
CuI	10 mol %	piperidine-2-carboxylic acid (20 mol %), K_2CO_3	DMF	110	36	(73)	67
Cu/Fe-hydrotalcite	10 wt %	none	toluene	130	12	(76)	283

Nitrogen Nucleophile	Aryl Halide	Conditions	Product(s) and Yield(s) (%)	Refs.
	4-iodotoluene	CuO (10 mol %), FeCl₃ (10 mol %), rac-binol (20 mol %), Cs₂CO₃, DMF, 90°, 12 h	product (82)	277

Nucleophile: 4-methoxyaniline (H₂N–C₆H₄–OMe)
Aryl Halide: 2-X-benzoic acid (CO₂H, X)
Conditions: Catalyst(s) (x mol %)
Product: CO₂H–C₆H₄–N(H)–C₆H₄–OMe

Isomer	X	Catalyst(s)	x	Additive(s)	Solvent	Temp (°)	Time		Refs.
2	Cl	Cu(OAc)₂	14	K_2CO_3	—	MW (700 W)	8 min	(83)	72
2	Br	Cu(OAc)₂	14	K_2CO_3	—	MW (700 W)	6 min	(91)	72
2	Cl	CuSO₄	16	K_2CO_3	H₂O	MW 100	7 min	(90)	240
3	Cl	Cu powder	—	K_2CO_3, pyridine	H₂O	))	20 min	(82)	314
3	Cl	Cu	7.5	K_2CO_3	DMF	reflux	2 h	(72)	327
3	Cl	Cu, CuI	2, 0.6	K_2CO_3	DMF	))) (20 kHz), reflux	20 min	(54)	315
4	Cl	Cu	8	KOH	H₂O	reflux	5 h	(89)	332
4	Cl	Cu	8	K_2CO_3, pyridine (15 wt %)	H₂O	reflux	1 h	(86)	332

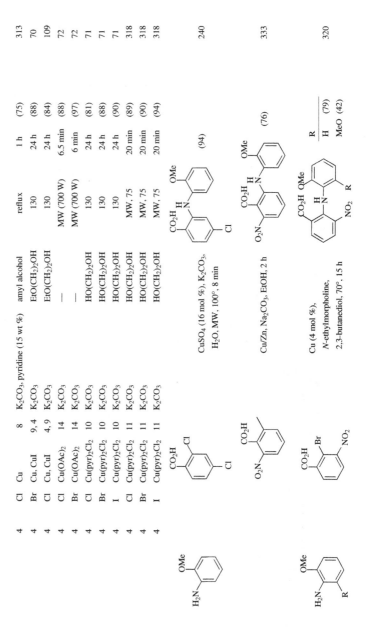

		Catalyst		Base	Solvent	Temp	Time	(Yield)	Ref
4	Cl	Cu	8	K$_2$CO$_3$, pyridine (15 wt %)	amyl alcohol	reflux	1 h	(75)	313
4	Br	Cu, CuI	9, 4	K$_2$CO$_3$	EtO(CH$_2$)$_2$OH	130	24 h	(88)	70
4	Cl	Cu, CuI	4, 9	K$_2$CO$_3$	EtO(CH$_2$)$_2$OH	130	24 h	(84)	109
4	Cl	Cu(OAc)$_2$	14	K$_2$CO$_3$	—	MW (700 W)	6.5 min	(88)	72
4	Br	Cu(OAc)$_2$	14	K$_2$CO$_3$	—	MW (700 W)	6 min	(97)	72
4	Cl	Cu(pyr)$_2$Cl$_2$	10	K$_2$CO$_3$	HO(CH$_2$)$_2$OH	130	24 h	(81)	71
4	Br	Cu(pyr)$_2$Cl$_2$	10	K$_2$CO$_3$	HO(CH$_2$)$_2$OH	130	24 h	(88)	71
4	I	Cu(pyr)$_2$Cl$_2$	10	K$_2$CO$_3$	HO(CH$_2$)$_2$OH	130	24 h	(90)	71
4	Cl	Cu(pyr)$_2$Cl$_2$	11	K$_2$CO$_3$	HO(CH$_2$)$_2$OH	MW, 75	20 min	(89)	318
4	Br	Cu(pyr)$_2$Cl$_2$	11	K$_2$CO$_3$	HO(CH$_2$)$_2$OH	MW, 75	20 min	(90)	318
4	I	Cu(pyr)$_2$Cl$_2$	11	K$_2$CO$_3$	HO(CH$_2$)$_2$OH	MW, 75	20 min	(94)	318

CuSO$_4$ (16 mol %), K$_2$CO$_3$, H$_2$O, MW, 100°, 8 min   (94)   240

Cu/Zn, Na$_2$CO$_3$, EtOH, 2 h   (76)   333

Cu (4 mol %), N-ethylmorpholine, 2,3-butanediol, 70°, 15 h   320

R	
H	(79)
MeO	(42)

TABLE 5A. N-ARYLATION OF PRIMARY ARYL AMINES (Continued)

Please refer to the charts preceding the tables for structures indicated by the **bold** numbers.

Nitrogen Nucleophile	Aryl Halide	Conditions	Product(s) and Yield(s) (%)	Refs.

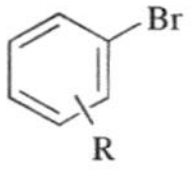

C₆

| 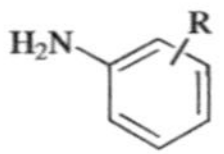 | 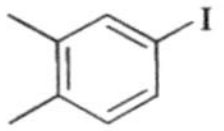 | Cu(OAc)₂ (4 mol %), phen (3 mol %), KOH, toluene, reflux, 18 h | 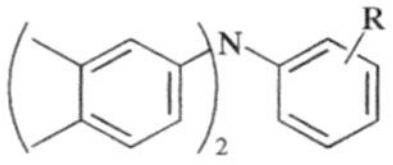 (—) | 307 |

R = 2-, 3-, or 4-MeO

| 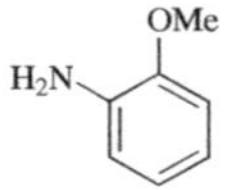 | 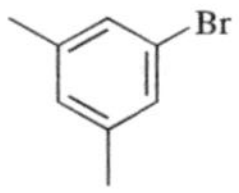 | CuI (10 mol %), pyridine-2-carboxylic acid (20 mol%), K₃PO₄, DMSO, 90°, 24 h | 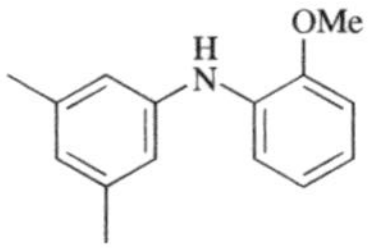 (70) | 50 |

| 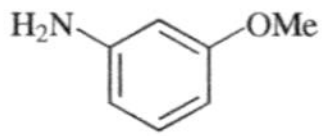 | 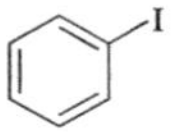 | Catalyst (x mol %) | 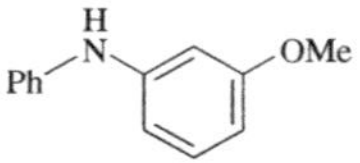 I 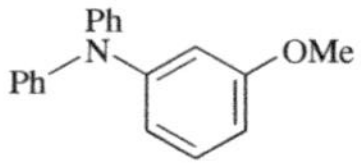 + II | 326 274 |

Catalyst	x	Additive	Solvent	Temp (°)	Time (h)	I	II
Cu	—	K₂CO₃	—	—		(0)	(60)
CuI	10	CsOAc	DMSO	90	24	(4)	(0)

| | | CuI (10 mol %), pyridine-2-carboxylic acid (20 mol %), K₃PO₄, DMSO, 100° | | 50 |

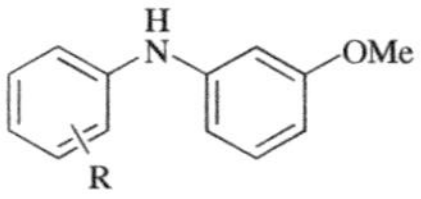

R	Time (h)	
2-Me	30	(67)
3-CF₃	24	(55)

| | | Cu (17 mol %), KI, K₂CO₃, DMF, 120°, 6 h | 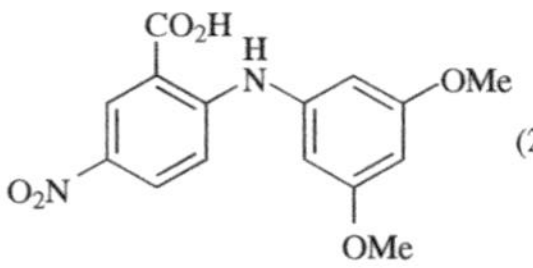 (25) | 334 |

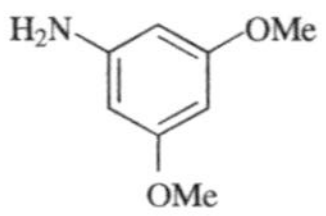

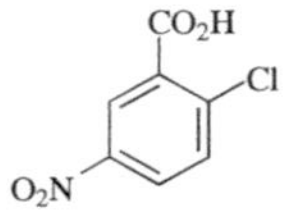

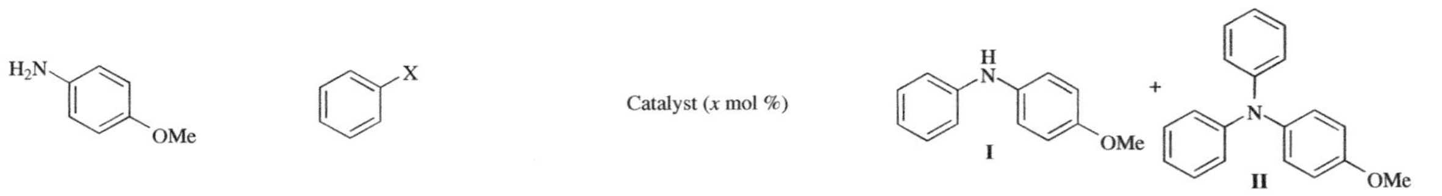

Catalyst (x mol %)

H_2N ‑OMe + Ph‑X →

I: Ph‑NH‑C6H4‑OMe
II: Ph2N‑C6H4‑OMe

X	Catalyst	x	Additive(s)	Solvent(s)	Temp (°)	Time (h)	I	II	
I	CuI	7	Cs₂CO₃	Si(OEt)₄	145	38	(85)	(0)	335
I	CuI	10	CsOAc	DMSO	90	24	(38)	(0)	274
I	CuI	2	KOt-Bu	toluene	135	14	(8)	(81)	304
Cl	CuI	3.6	PBu₃ (7.2 mol %), KOt-Bu	toluene	135	10.5	(0)	(85)	56
Br	CuI	3.6	PBu₃ (7.2 mol %), KOt-Bu	toluene	135	10.5	(0)	(86)	56
I	CuI	10	2,2'-biphenol (20 mol %), TBPM	dioxane	rt	12	(86)	(0)	61
Br	CuI	7	L-Pro (14 mol %), K₂CO₃	DMSO	90	40	(90)	(0)	47
I	CuI	10	L-Pro (20 mol %), K₂CO₃	DMSO	90	36	(82)	(0)	192, 336, 47
I	CuI	5	DiPrPhDAB (6 mol %), KOt-Bu	toluene	120	48	(76)	(0)	306
Br	CuI	2	L1 (2 mol %), KOt-Bu	—	100	12	(71)	(0)	288
I	CuO nanoparticles (33 nm)	1.3	KOH, air	DMSO	110	1.5	(94)	(0)	224
Br	CuO	10	FeCl₃ (10 mol %), rac-binol (20 mol %), Cs₂CO₃	DMF	110	24	(79)	(0)	277
I	CuO	10	FeCl₃ (10 mol %), rac-binol (20 mol %), Cs₂CO₃	DMF	90	12	(85)	(0)	277
I	Cu(acac)₂	10	Fe₂O₃ (20 mol %), Cs₂CO₃	DMSO/H₂O (1:1)	150	0.5	(51)	(0)	262
I	1	1.25	KOt-Bu	toluene	90	12	(69)	(0)	309

TABLE 5A. N-ARYLATION OF PRIMARY ARYL AMINES (*Continued*)

Nitrogen Nucleophile	Aryl Halide	Conditions	Product(s) and Yield(s) (%)	Refs.

*Please refer to the charts preceding the tables for structures indicated by the **bold** numbers.*

C$_6$

Nitrogen Nucleophile: H$_2$N–C$_6$H$_4$–OMe

Section 1

Aryl Halide: R–C$_6$H$_4$–X

Conditions: CuI (x mol %), DMSO

Product: R–C$_6$H$_4$–NH–C$_6$H$_4$–OMe

R	X	x	Additives	Temp (°)	Time (h)		Refs.
F	Br	10	pyridine-2-carboxylic acid (20 mol %), K$_3$PO$_4$	100	24	(76)	50
Cl	Br	7	L-Pro (14 mol %), K$_2$CO$_3$	90	40	(82)	47
Cl	Br	10	pyridine 2-carboxylic acid (20 mol %), K$_3$PO$_4$	100	24	(72)	50
Cl	I	5	pyridine 2-carboxylic acid (20 mol %), K$_3$PO$_4$	70	24	(82)	50

Section 2

Aryl Halide: O$_2$N–C$_6$H$_4$–Br

Conditions: CuO (10 mol %), FeCl$_3$ (10 mol %), rac-binol (20 mol %), Cs$_2$CO$_3$, DMF, 100°, 12 h

Product: O$_2$N–C$_6$H$_4$–NH–C$_6$H$_4$–OMe (86) — Refs. 277

Section 3

Aryl Halide: MeO–C$_6$H$_4$–X

Conditions: Catalyst (x mol %)

Products: I = PMP–NH–PMP + II = PMP–N(PMP)–PMP

X	Catalyst	x	Additive(s)	Solvent	Temp (°)	Time (h)	I	II	Refs.
Cl	CuI	3.6	bipy (3.6 mol %), KOt-Bu	toluene	115	3.5	(0)	(87)	305
Br	CuI	3.6	PBu$_3$ (7.2 mol %), KOt-Bu	toluene	135	10.5	(0)	(87)	56
I	CuI	2	KOt-Bu	toluene	135	14	(14)	(71)	304
I	CuI	7	L-Pro (14 mol %), K$_2$CO$_3$	DMSO	90	27	(81)	(0)	47
I	CuI	10	L-Pro (20 mol %), K$_2$CO$_3$	DMSO	90	27	(81)	(0)	192
I	CuI	20	Cs$_2$CO$_3$	Si(OEt)$_4$	145	32	(0)	(73)	335
I	Cu(OAc)$_2$	4	phen (3 mol %), KOH	toluene	reflux	18	(0)	(—)	307
I	Cu(TMHD)$_2$	20	KOt-Bu	toluene	120	40	(0)	(73)	308

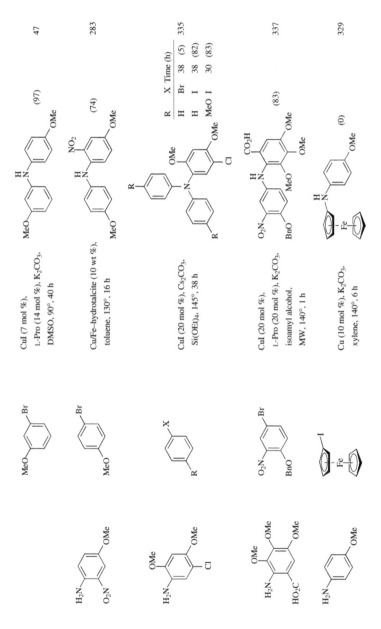

R	X	Time (h)
H	Br	38 (5)
H	I	38 (82)
MeO	I	30 (83)

CuI (7 mol %), L-Pro (14 mol %), K₂CO₃, DMSO, 90°, 40 h — (97) — 47

Cu/Fe–hydrotalcite (10 wt %), toluene, 130°, 16 h — (74) — 283

CuI (20 mol %), Cs₂CO₃, Si(OEt)₄, 145°, 38 h — 335

CuI (20 mol %), L-Pro (20 mol %), K₂CO₃, isoamyl alcohol, MW, 140°, 1 h — (83) — 337

Cu (10 mol %), K₂CO₃, xylene, 140°, 6 h — (0) — 329

TABLE 5A. *N*-ARYLATION OF PRIMARY ARYL AMINES (*Continued*)

Nitrogen Nucleophile	Aryl Halide	Conditions	Product(s) and Yield(s) (%)	Refs.

*Please refer to the charts preceding the tables for structures indicated by the **bold** numbers.*

C₆

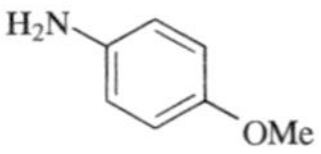

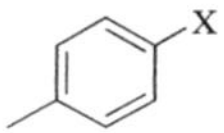

Nitrogen Nucleophile: H_2N–C₆H₄–OMe

Aryl Halide: X–C₆H₄–Me

Conditions: CuI (x mol %)

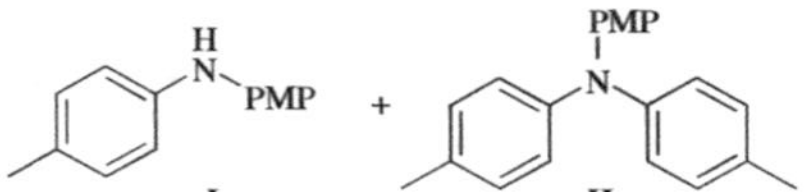

Products:

I: Me–C₆H₄–NH–PMP

II: (Me–C₆H₄)₂N–PMP

X	x	Additives	Solvent	Temp (°)	Time (h)	I	II	Refs.
Br	7	L-Pro (14 mol %), K₂CO₃	DMSO	90	40	(90)	(0)	47
I	5	DiPrPhDAB (6 mol %), KOt-Bu	toluene	120	48	(75)	(0)	306
I	5	**L41** (10 mol %), NaOt-Bu	toluene	110	40	(0)	(60)	338

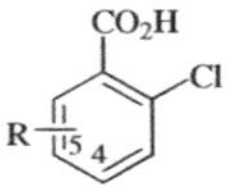

Aryl Halide: CO_2H, Cl substituted arene, R at positions 15, 4

Conditions: Catalyst(s) (x mol %), K₂CO₃

Product: R–arene with CO_2H and NH–C₆H₄–OMe

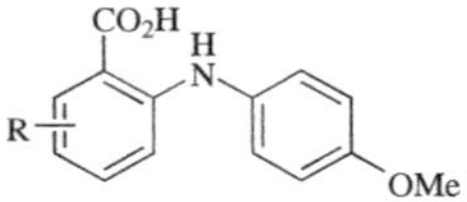

R	Catalyst(s)	x	Solvent(s)	Temp (°)	Time		Refs.
4-Cl	Cu powder	—	pyridine/H₂O	))	20 min	(90)	314
4-Cl	Cu, CuI	4, 9	EtOCH₂CH₂OH	130	24 h	(86)	109
5-Br	Cu, CuI	4, 9	EtOCH₂CH₂OH	130	24 h	(85)	109
4-O₂N	Cu, CuI	4, 9	EtOCH₂CH₂OH	130	24 h	(99)	109
4-O₂N	CuSO₄	16	H₂O	MW, 100	5 min	(93)	240

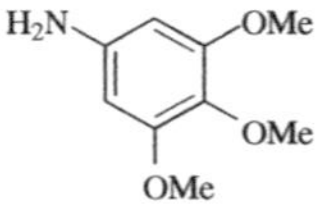

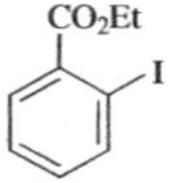

Nitrogen Nucleophile: H_2N–C₆H₂(OMe)₃ (2,3,4-OMe)

Aryl Halide: CO₂Et, I substituted arene

Conditions: CuI (20 mol %), Cs₂CO₃, Si(OEt)₄, 145°, 28 h

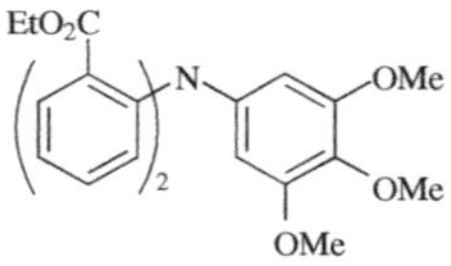

Product: (EtO₂C)₂(C₆H₄)₂N–C₆H₂(OMe)₃ (93), 335

C₇

H₂N-C₆H₄-Me + Ph-X →[Catalyst (x mol %)]→ **I** (PhNH-C₆H₄-Me) + **II** (Ph₂N-C₆H₄-Me)

Isomer	X	Catalyst	x	Additive(s)	Solvent	Temp (°)	Time (h)	I	II	
2	Cl	CuI	3.6	PBu₃ (7.2 mol %), KO*t*-Bu	toluene	135	10.5	(0)	(80)	56
2	I	CuI	2	KO*t*-Bu	toluene	135	14	(8)	(81)	304
2	I	CuO particles (33 nm)	1.3	KOH, air	DMSO	110	1.6	(96)	(0)	224
3	I	CuI	5	KO*t*-Bu, DiPrPhDAB (6 mol %)	toluene	120	48	(84)	(0)	306
4	I	CuI	5	KO*t*-Bu, DiPrPhDAB (6 mol %)	toluene	120	48	(82)	(0)	55
4	I	CuI	10	TBPM, 2,2′-biphenol (20 mol %)	dioxane	rt	12	(84)	(0)	61
4	I	CuO	10	FeCl₃ (10 mol %), *rac*-binol (20 mol %), Cs₂CO₃	DMF	80	12	(68)	(0)	277
4	I	Cu(PPh₃)₃Br	20	Cs₂CO₃	toluene	110	24	(88)	(0)	339

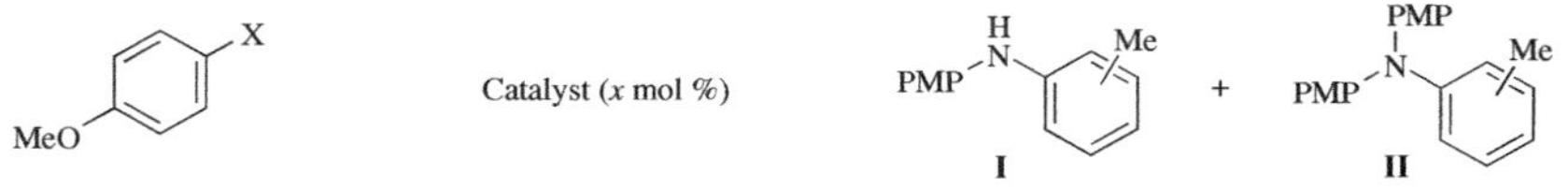

MeO-C₆H₄-X →[Catalyst (x mol %)]→ **I** (PMP-NH-C₆H₄-Me) + **II** (PMP-N(-PMP)-C₆H₄-Me)

Isomer	X	Catalyst	x	Additive(s)	Solvent	Temp (°)	Time (h)	I	II	
2	I	CuI	2	KO*t*-Bu	toluene	135	14	(14)	(71)	304
4	Br	CuI	5	L2 (25 mol %), TBAB, KOH	H₂O	MW (100 W), 130	5 min	(75)	(0)	236
4	Br	CuO	5	L14 (50 mol %), hexane-2,5-dione, KOH, TBAB	H₂O	MW, 120	5 min	(75)	(0)	51
4	Br	CuO	25	L15 (50 mol %), KOH, TBAB	H₂O	MW (100 W), 130	5 min	(55)	(0)	267

203

TABLE 5A. N-ARYLATION OF PRIMARY ARYL AMINES (*Continued*)

Nitrogen Nucleophile	Aryl Halide	Conditions	Product(s) and Yield(s) (%)	Refs.

Please refer to the charts preceding the tables for structures indicated by the bold numbers.

C7

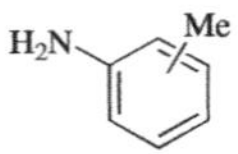

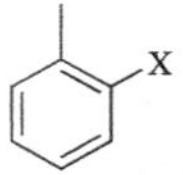

Catalyst (x mol %)

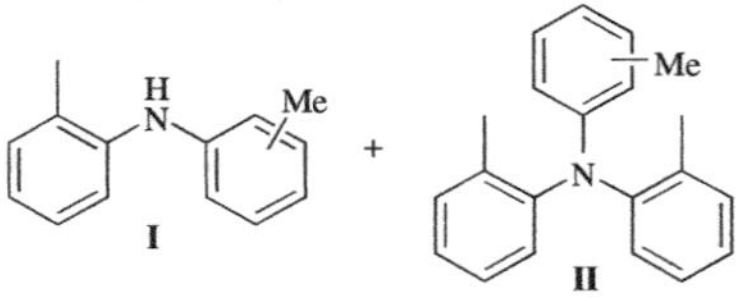

Isomer	X	Catalyst	x	Additives	Solvent	Temp (°)	Time (h)	I	II	
2	I	Cu	2	18-c-6, K2CO3	1,2-Cl2C6H4	reflux	15	(0)	(90)	54
3	Br	CuI	10	pyridine-2-carboxylic acid (20 mol %), K3PO4	DMSO	100	30	(74)	(0)	50
3	I	CuI	10	pyridine-2-carboxylic acid (20 mol %), K3PO4	DMSO	80	24	(71)	(0)	50

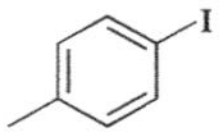

Catalyst (x mol %), toluene

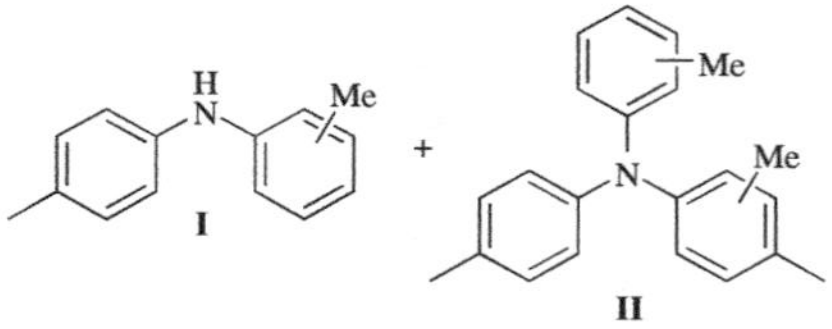

Isomer	Catalyst	x	Additive(s)	Temp (°)	Time (h)	I	II	
3	CuI	5	DiPrPhDAB (6 mol %), KOt-Bu	120	48	(73)	(0)	306
4	CuI	5	DiPrPhDAB (6 mol %), KOt-Bu	120	48	(82)	(0)	55, 170
4	Cu(OAc)2	4	phen (3 mol %), KOH	reflux	18	(0)	(—)	307
4	Cu(TMHD)2	20	KOt-Bu	120	40	(0)	(72)	308

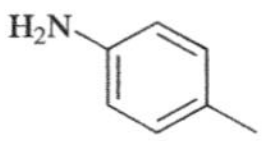

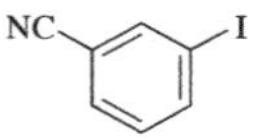

CuI (10 mol %),
pyridine-2-carboxylic acid
(20 mol %), K3PO4,
DMSO, 80°, 24 h

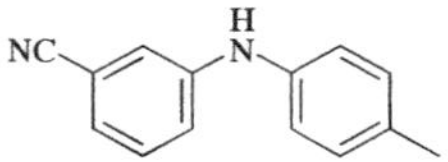

(73)   50

H₂N–⟨⟩–Me + ⟨CO₂H, X⟩ →(Catalyst(s) (x mol %))→ ⟨CO₂H, N(H)–⟨⟩–Me⟩

Isomer	X	Catalyst(s)	x	Additive(s)	Solvent	Temp (°)	Time	
2	Cl	Cu, CuI	2, 0.6	K_2CO_3	DMF	))) (20 kHz), reflux	20 min (70)	315
2	Cl	Cu, CuI	4, 9	Na_2CO_3	$EtO(CH_2)_2OH$	130	24 h (92)	109
2	Cl	Cu, Cu_2O	3, 1.3	K_2CO_3	$MeO(CH_2)_2OH$	reflux	2 h (55)	340
3	Cl	Cu	7.5	K_2CO_3	DMF	reflux	2 h (63)	327
3	Cl	Cu, CuI	4, 9	K_2CO_3	$EtO(CH_2)_2OH$	130	24 h (80)	109
3	Cl	$Cu(OAc)_2$	14	K_2CO_3	—	MW (700 W)	7.5 min (87)	72
3	Br	$Cu(OAc)_2$	14	K_2CO_3	—	MW (700 W)	6.6 min (95)	72
4	Cl	Cu	8	KOH	H_2O	reflux	5 h (98)	332
4	Cl	Cu	8	K_2CO_3, pyridine (15 wt %)	H_2O	reflux	1 h (75)	313
4	Cl	Cu	8	K_2CO_3, pyridine (15 wt %)	amyl alcohol	reflux	1 h (84)	313
4	I	Cu, CuI	25, 5	K_2CO_3	Bu_2O	170	120 h (70)	341
4	Cl	$Cu(OAc)_2$	14	K_2CO_3	—	MW (700 W)	6.6 min (85)	72
4	Br	$Cu(OAc)_2$	14	K_2CO_3	—	MW (700 W)	6.3 min (94)	72
4	Cl	$Cu(pyr)_2Cl_2$	10	K_2CO_3	$HO(CH_2)_2OH$	130	24 h (82)	71
4	Br	$Cu(pyr)_2Cl_2$	10	K_2CO_3	$HO(CH_2)_2OH$	130	24 h (90)	71
4	I	$Cu(pyr)_2Cl_2$	10	K_2CO_3	$HO(CH_2)_2OH$	130	24 h (89)	71
4	Cl	$Cu(pyr)_2Cl_2$	11	K_2CO_3	$HO(CH_2)_2OH$	MW, 75	20 min (80)	318
4	Br	$Cu(pyr)_2Cl_2$	11	K_2CO_3	$HO(CH_2)_2OH$	MW, 75	20 min (91)	318
4	I	$Cu(pyr)_2Cl_2$	11	K_2CO_3	$HO(CH_2)_2OH$	MW, 75	20 min (86)	318

TABLE 5A. *N*-ARYLATION OF PRIMARY ARYL AMINES (*Continued*)

Nitrogen Nucleophile	Aryl Halide	Conditions	Product(s) and Yield(s) (%)	Refs.

*Please refer to the charts preceding the tables for structures indicated by the **bold** numbers.*

C₇

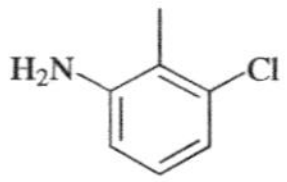

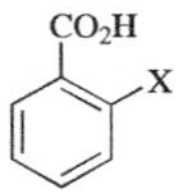

		Cu(OAc)₂ (14 mol %), K₂CO₃, MW (700 W)	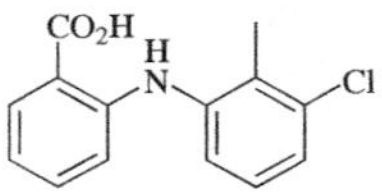 X — Time (min): Cl 14 (75); Br 12.5 (84)	72
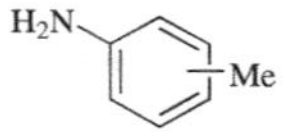	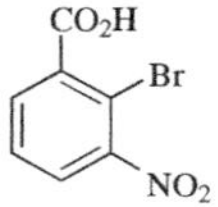	Cu (4 mol %), N-ethylmorpholine, 2,3-butanediol, 70°, 15 h	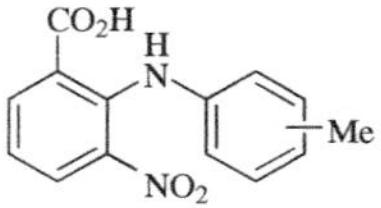 Isomer: 2 (67); 3 (58); 4 (63)	320
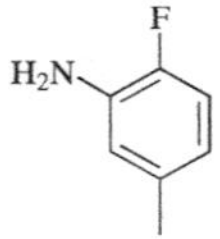	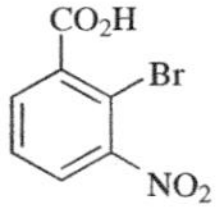	Cu (4 mol %), N-ethylmorpholine, 2,3-butanediol, 70°, 15 h	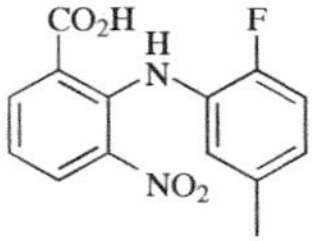 (64)	320
		Cu (7.5 mol %), K₂CO₃, DMF, reflux, 2 h	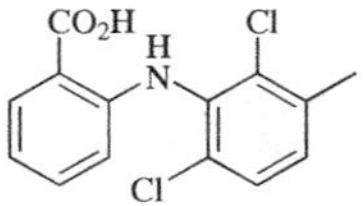 (18)	327
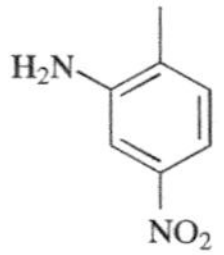	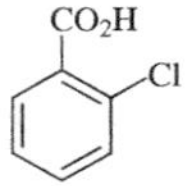	Cu (2 mol %), CuI (0.6 mol %), K₂CO₃, DMF,))) (20 kHz), reflux, 20 min	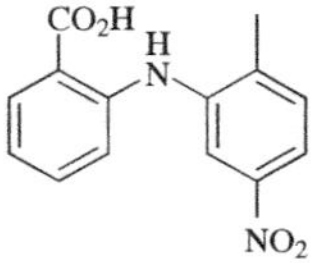 (14)	315
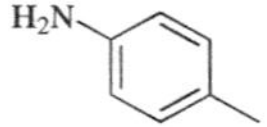	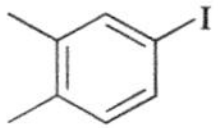	Cu(OAc)₂ (4 mol %), phen (3 mol %), KOH, toluene, reflux, 18 h	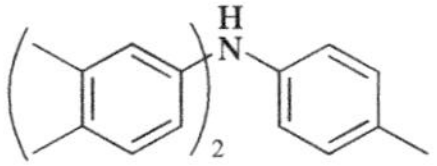 (—)	307

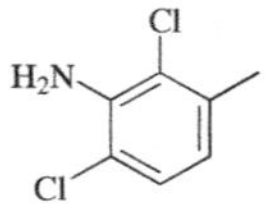

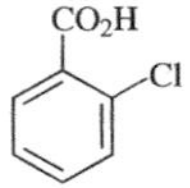

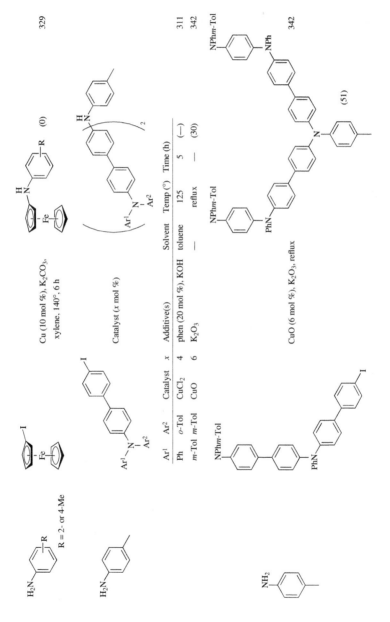

329

Cu (10 mol %), K$_2$CO$_3$,
xylene, 140°, 6 h

Catalyst (x mol %)

Ar^1	Ar^2	Catalyst	x	Additive(s)	Solvent	Temp (°)	Time (h)	
Ph	o-Tol	CuCl$_2$	4	phen (20 mol %), KOH	toluene	125	5	(—)
m-Tol	m-Tol	CuO	6	K$_2$O$_3$	—	reflux	—	(30)

311
342

CuO (6 mol %), K$_2$O$_3$, reflux

(51)

342

TABLE 5A. *N*-ARYLATION OF PRIMARY ARYL AMINES (*Continued*)

Nitrogen Nucleophile	Aryl Halide	Conditions	Product(s) and Yield(s) (%)	Refs.

*Please refer to the charts preceding the tables for structures indicated by the **bold** numbers.*

C_7

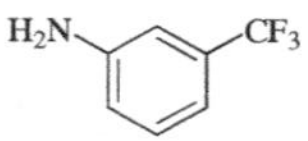

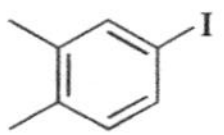

$Cu(OAc)_2$ (4 mol %), phen (3 mol %), KOH, toluene, reflux, 18 h

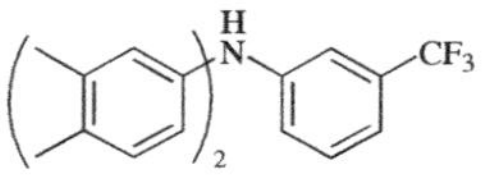

(—) 307

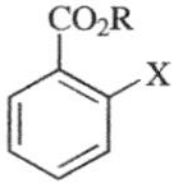

Catalyst (*x* mol %)

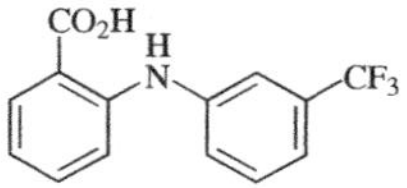

X	Catalyst(s)	*x*	Additive(s)	Solvent	Temp (°)	Time		Refs.
Cl	Cu, CuI	2, 0.6	K_2CO_3	DMF	))) (20 kHz), reflux	20 min	(24)	315
Cl	Cu	7.5	Na_2CO_3	DMF	reflux	2 h	(60)	327
I	Cu	25	K_2CO_3	H_2O	reflux	3 h	(61)	343
Cl	$Cu(OAc)_2$	14	K_2CO_3	—	MW (700 W)	14.5 min	(72)	72
Br	$Cu(OAc)_2$	14	K_2CO_3	—	MW (700 W)	13 min	(76)	72

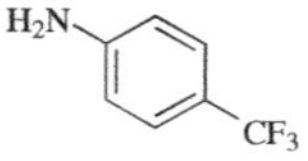

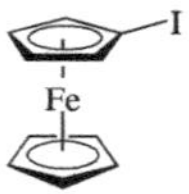

Cu (10 mol %), K_2CO_3, xylene, 140°, 6 h

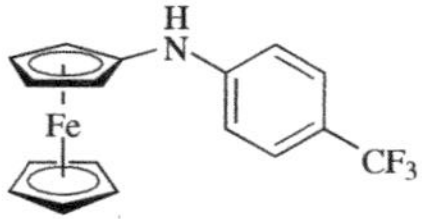

(0) 329

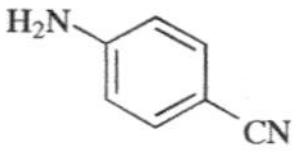

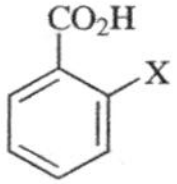

Catalyst(s) (*x* mol %), K_2CO_3

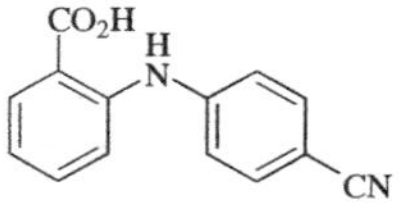

X	Catalyst(s)	*x*	Solvent	Temp (°)	Time		Refs.
Br	Cu, CuI	9, 4	$EtOCH_2CH_2OH$	130	24 h	(71)	70
Br	$Cu(pyr)_2Cl_2$	10	$HOCH_2CH_2OH$	130	24 h	(72)	71
Cl	$Cu(pyr)_2Cl_2$	11	$HOCH_2CH_2OH$	MW 75	20 min	(64)	318
Br	$Cu(pyr)_2Cl_2$	11	$HOCH_2CH_2OH$	MW 75	20 min	(75)	318
I	$Cu(pyr)_2Cl_2$	11	$HOCH_2CH_2OH$	MW 75	20 min	(89)	318

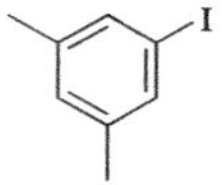

CuI (10 mol %),
pyridine-2-carboxylic acid
(20 mol %),
K₃PO₄, DMSO, 90°, 24 h

$$\text{CuI (10 mol \%), pyridine-2-carboxylic acid (20 mol \%), } K_3PO_4, \text{ DMSO, } 90°, 24 \text{ h}$$

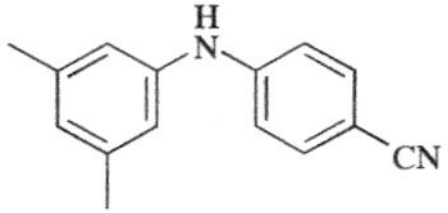

(52)   50

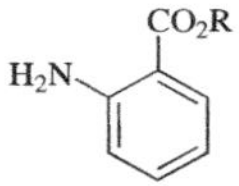

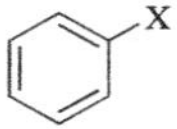

Catalyst (x mol %), 110°

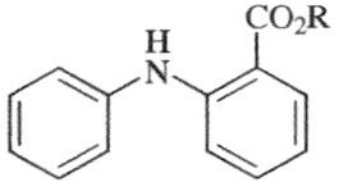

R	X	Catalyst	x	Additive(s)	Solvent	Time (h)	
H	Br	CuI	10	K₃PO₄, DPP (20 mol %)	DMF	30	(70)
Me	Br	CuI	10	K₃PO₄, PfAPM (20 mol%)	DMF	30	(84)
Me	I	Cu(PPh₃)₃Br	20	Cs₂CO₃	toluene	24	(70)

108
48
339

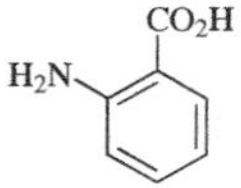

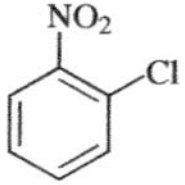

Cu (0.3 mol %),
K₂CO₃, 220°

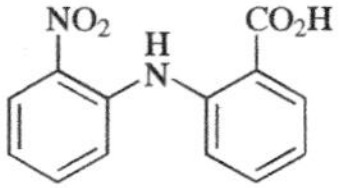

(90)   345

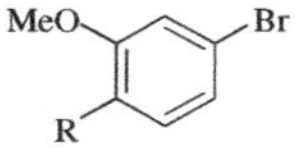

CuI (20 mol %),
L-Pro (20 mol %), K₂CO₃,
MW, 140°, 1 h

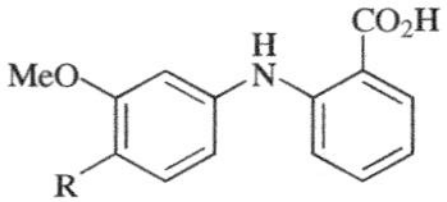

337

R	Solvent	
H	isoamyl alcohol	(78)
H	DMSO	(69)
BnO	isoamyl alcohol	(85)

209

TABLE 5A. N-ARYLATION OF PRIMARY ARYL AMINES (*Continued*)

*Please refer to the charts preceding the tables for structures indicated by the **bold** numbers.*

Nitrogen Nucleophile	Aryl Halide	Conditions	Product(s) and Yield(s) (%)	Refs.

C_7

Row 1:

Nitrogen Nucleophile: H_2N—(ring with CO_2H, positions **3 4**)—R

Aryl Halide: MeO, BnO substituted bromobenzene (Br)

Conditions: CuI (20 mol %), L-Pro (20 mol %), K_2CO_3, isoamyl alcohol, MW, 140°, 1 h

Product(s) and Yield(s): Diarylamine with CO_2H, MeO, BnO, R

R	
4,5-F_2	(38)
3-MeO	(70)

Refs.: 337

Row 2:

Nitrogen Nucleophile: H_2N—(ring with CO_2H, positions **6 5**)—R

Aryl Halide: O_2N, BnO substituted bromobenzene (Br)

Conditions: CuI (20 mol %), L-Pro (20 mol %), K_2CO_3, isoamyl alcohol, MW, 140°, 1 h

Product(s) and Yield(s): Diarylamine with CO_2H, O_2N, BnO, R

R	
H	(83)
4,5-F_2	(54)
6-MeO	(92)
4,5,6-(MeO)$_3$	(83)

Refs.: 337

Row 3:

Nitrogen Nucleophile: H_2N—(ring with CO_2H and NO_2)

Aryl Halide: R substituted bromobenzene (Br)

Conditions: CuI (20 mol %), L-Pro (20 mol %), K_2CO_3, MW, 1 h

R	Solvent(s)	Temp (°)	
H	DMF/H_2O	160	(82)
4-MeO	DMF/H_2O	160	(76)
3-BnO	isoamyl alcohol	140	(82)
4-BnO	DMF/H_2O	160	(69)

Product(s): Diarylamine with CO_2H, NO_2, R

Refs.: 337

Row 4:

Nitrogen Nucleophile: H_2N—(ring with CO_2Me)

Aryl Halide: Iodo-methylaniline (NH_2, I)

Conditions: Cu_2O (10 mol %), K_2CO_3, xylene, 145°, 8 h

Product(s) and Yield(s): Diarylamine with CO_2Me, NH_2 (70)

Refs.: 346

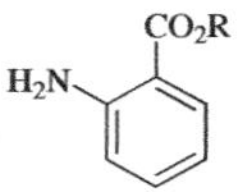

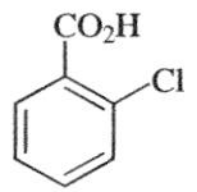

Catalyst(s) (x mol %)

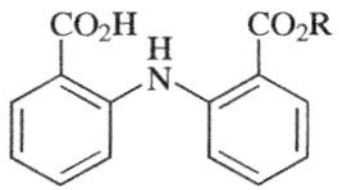

R	Catalyst(s)	x	Additive	Solvent(s)	Temp (°)	Time		Ref
H	Cu	7.5	K_2CO_3	DMF	reflux	2 h	(81)	327
H	Cu powder	—	K_2CO_3	pyridine/H_2O	)))	20 min	(81)	314
H	$CuSO_4$	16	K_2CO_3	H_2O	MW, 100	10 min	(82)	240
Me	Cu, CuI	8, 12	N-ethylmorpholine	2,3-butanediol	120	1 h	(87)	347

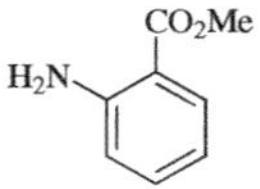

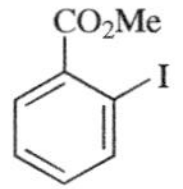

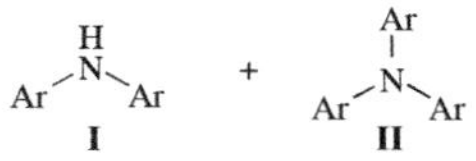

$Ar = 2\text{-}MeO_2CC_6H_4$

Catalyst (x mol %)

Catalyst	x	Additive(s)	Solvent	Temp (°)	Time (h)	I	II	Ref
Cu	1	K_2CO_3, 18-c-6	$1,2\text{-}Cl_2C_6H_4$	reflux	17	(0)	(55)	319
Cu	15	K_2CO_3	—	180	—	(70)	(0)	348
$Cu(PPh_3)_3Br$	20	Cs_2CO_3	toluene	110	24	(83)	(0)	339

TABLE 5A. N-ARYLATION OF PRIMARY ARYL AMINES (Continued)

Nitrogen Nucleophile	Aryl Halide	Conditions	Product(s) and Yield(s) (%)	Refs.

*Please refer to the charts preceding the tables for structures indicated by the **bold** numbers.*

C7

Nitrogen Nucleophile: H2N–C6H4–CO2Me

Aryl Halide: $R \overset{5,4}{-}$ C6H3(CO2H)(Cl)

Product: HO_2C-aryl-N(H)-aryl-CO_2Me with R

R	Catalyst(s)	x	Additive	Solvent	Temp (°)	Time	Yield	Refs.
4-Cl	Cu, CuI	8, 12	N-ethylmorpholine	2,3-butanediol	120	20 h	(83)	347
5-Cl	Cu, CuI	8, 12	N-ethylmorpholine	2,3-butanediol	120	20 h	(60)	347
4-O2N	Cu, CuI	8, 12	N-ethylmorpholine	2,3-butanediol	120	15 h	(64)	347
4-O2N	Cu	7.5	K2CO3	DMF	reflux	2 h	(80)	327
4-O2N	CuSO4	16	K2CO3	H2O	MW, 100	8 min	(85)	240
5-O2N	Cu, CuI	8, 12	N-ethylmorpholine	2,3-butanediol	120	2 h	(56)	347
4-MeO	Cu, CuI	8, 12	N-ethylmorpholine	2,3-butanediol	120	3 h	(67)	347
5-MeO	Cu, CuI	8, 12	N-ethylmorpholine	2,3-butanediol	120	2 h	(88)	347
6-MeO	Cu, CuI	8, 12	N-ethylmorpholine	2,3-butanediol	120	2 h	(51)	347
5-Me	Cu, CuI	8, 12	N-ethylmorpholine	2,3-butanediol	120	2 h	(70)	347
6-Me	Cu, CuI	8, 12	N-ethylmorpholine	2,3-butanediol	120	2 h	(69)	347
4-HO2C	Cu, CuI	8, 12	N-ethylmorpholine	2,3-butanediol	120	15 h	(61)	347

Nitrogen Nucleophile: CO2Me, H2N ($\overset{4,5}{-}NO_2$)

Aryl Halide: CO2H, I (2-iodobenzoic acid)

Conditions: Cu (8 mol %), CuI (12 mol %), N-ethylmorpholine, 2,3-butanediol, 120°, 3 h

Product: CO_2H-aryl-N(H)-aryl-CO_2Me, NO_2

Isomer	
4	(71)
5	(86)

Refs. 347

Nitrogen Nucleophile: CO2K, H2N

Aryl Halide: Br–C6H4–C(=O)CH3

Conditions: Cu (1 mol %), isoamyl alcohol, reflux, 21 h

Product: N(H)-aryl-CO_2H with acetyl — (56)

Refs. 321

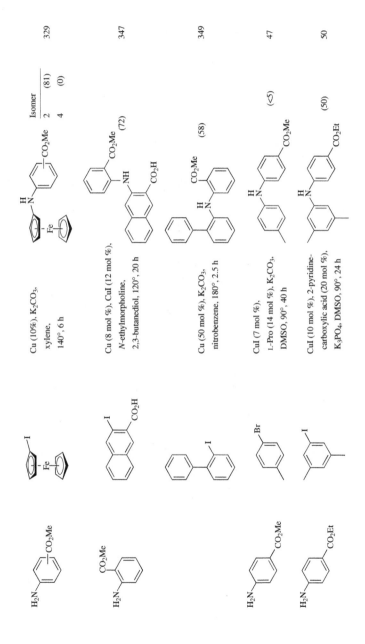

TABLE 5A. N-ARYLATION OF PRIMARY ARYL AMINES (Continued)

Nitrogen Nucleophile	Aryl Halide	Conditions	Product(s) and Yield(s) (%)	Refs.

*Please refer to the charts preceding the tables for structures indicated by the **bold** numbers.*

C₇

Nitrogen Nucleophile: 4-aminobenzoic acid (H_2N–C₆H₄–CO_2H)

Aryl Halide: 2-X-benzoic acid (X, CO_2H)

Product: diarylamine with CO_2H groups

X	Catalyst(s)	x	Additive(s)	Solvent	Temp (°)	Time	(%)	Refs.
Cl	Cu	7.5	K_2CO_3	DMF	reflux	2 h	(66)	327
Cl	Cu powder	—	K_2CO_3, pyridine	H_2O	)))	20 min	(65)	314
Cl	Cu, CuI	4.9	K_2CO_3	EtOCH₂CH₂OH	130	24 h	(98)	109
Br	Cu, CuI	9.4	K_2CO_3	EtOCH₂CH₂OH	130	24 h	(80)	70
Cl	Cu(pyr)₂Cl₂	10	K_2CO_3	HOCH₂CH₂OH	130	24 h	(59)	71
Br	Cu(pyr)₂Cl₂	10	K_2CO_3	HOCH₂CH₂OH	130	24 h	(61)	71
I	Cu(pyr)₂Cl₂	10	K_2CO_3	HOCH₂CH₂OH	130	24 h	(88)	71
Cl	Cu(pyr)₂Cl₂	11	K_2CO_3	HOCH₂CH₂OH	MW, 75	20 min	(61)	318
Br	Cu(pyr)₂Cl₂	11	K_2CO_3	HOCH₂CH₂OH	MW, 75	20 min	(72)	318
I	Cu(pyr)₂Cl₂	11	K_2CO_3	HOCH₂CH₂OH	MW, 75	20 min	(88)	318

C₇₋₈

Iodoferrocene nucleophile

Cu (10 mol %), K_2CO_3, xylene, 140°, 6 h

Product: ferrocenyl benzamide

R	
H	(96)
Me	(92)

Refs. 329

C₈

Nitrogen Nucleophile: 2,3-dimethylaniline (H_2N, 2,3-dimethylphenyl)

Aryl Halide: 2-X-benzoate (X, CO_2R)

Product: N-aryl anthranilic acid derivative (CO_2H H N, 2,3-dimethylphenyl)

Catalyst(s) (x mol %), K_2CO_3

X	Catalyst(s)	x	Solvent	Temp (°)	Time	(%)	Refs.
Cl	Cu, CuI	2, 0.6	DMF	))) (20 KHz) reflux	20 min	(44)	315
Cl	Cu	7.5	DMF	reflux	2 h	(70)	327
Cl	Cu(OAc)₂	14	—	MW (700 W)	7 min	(83)	72
Br	Cu(OAc)₂	14	—	MW (700 W)	6.6 min	(90)	72

214

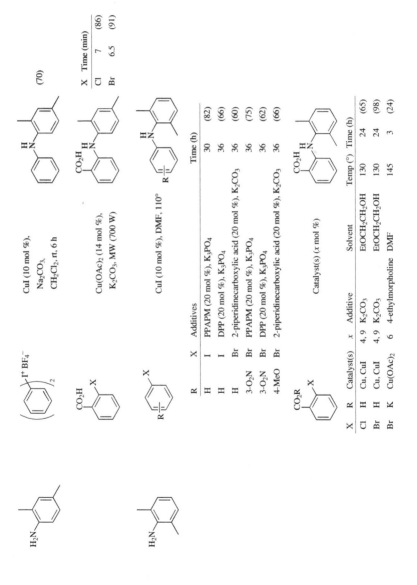

76

CuI (10 mol %),
Na$_2$CO$_3$,
CH$_2$Cl$_2$, rt, 6 h

(70)

72

Cu(OAc)$_2$ (14 mol %),
K$_2$CO$_3$, MW (700 W)

X	Time (min)	
Cl	7	(86)
Br	6.5	(91)

CuI (10 mol %), DMF, 110°

R	X	Additives	Time (h)	
H	I	PPAPM (20 mol %), K$_3$PO$_4$	30	(82)	48
H	I	DPP (20 mol %), K$_3$PO$_4$	36	(66)	108
H	Br	2-piperidinecarboxylic acid (20 mol %), K$_2$CO$_3$	36	(60)	67
3-O$_2$N	Br	PPAPM (20 mol %), K$_3$PO$_4$	36	(75)	48
3-O$_2$N	Br	DPP (20 mol %), K$_3$PO$_4$	36	(62)	108
4-MeO	Br	2-piperidinecarboxylic acid (20 mol %), K$_2$CO$_3$	36	(66)	67

Catalyst(s) (x mol %)

X	R	Catalyst(s)	x	Additive	Solvent	Temp (°)	Time (h)	
Cl	H	Cu, CuI	4, 9	K$_2$CO$_3$	EtOCH$_2$CH$_2$OH	130	24	(65)	109
Br	H	Cu, CuI	4, 9	K$_2$CO$_3$	EtOCH$_2$CH$_2$OH	130	24	(98)	70
Br	K	Cu(OAc)$_2$	6	4-ethylmorpholine	DMF	145	3	(24)	350

TABLE 5A. *N*-ARYLATION OF PRIMARY ARYL AMINES (*Continued*)

Nitrogen Nucleophile	Aryl Halide	Conditions	Product(s) and Yield(s) (%)	Refs.

*Please refer to the charts preceding the tables for structures indicated by the **bold** numbers.*

C_8

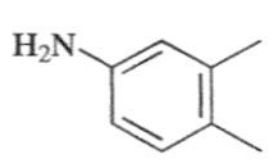

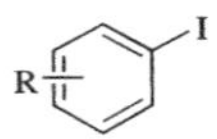

R = H, 4-F, 3-MeO, 4-MeO, 4-Me, 3-CF_3, 4-*t*-Bu

$Cu(OAc)_2$ (4 mol %), phen (3 mol %), KOH, toluene, reflux, 18 h

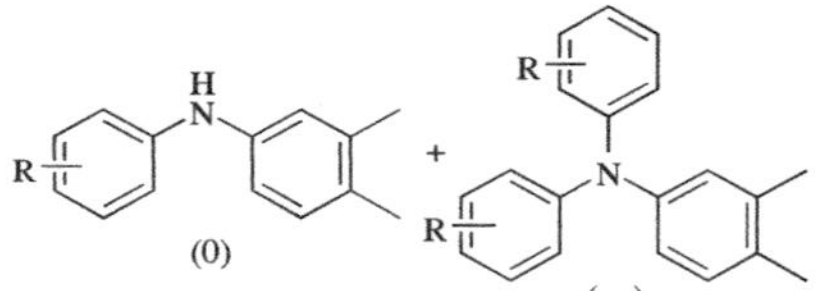

(0) + (—) — 307

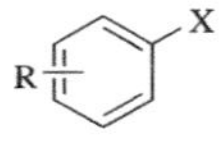

Catalyst (*x* mol %)

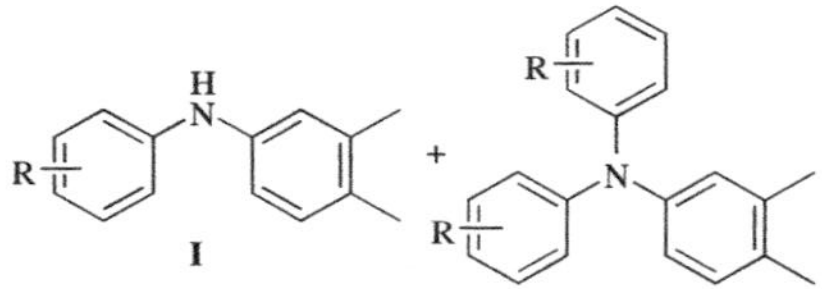

I + **II**

R	X	Catalyst	*x*	Additive(s)	Solvent	Temp (°)	Time	**I**	**II**	Refs.
2-HO_2C	Cl	$CuSO_4$	16	K_2CO_3	H_2O	MW, 100	7 min	(92)	(0)	240
3,4-Me_2	I	$CuCl_2$	4	phen (4 mol %), KOH	toluene	125	22 h	(0)	(70)	311

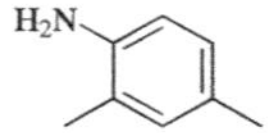

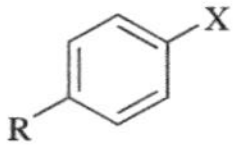

Catalyst (*x* mol %)

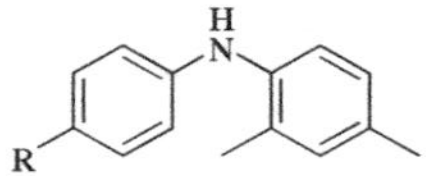

R	X	Catalyst	*x*	Additive(s)	Solvent	Temp (°)	Time (h)		Refs.
H	$I^+Ph\ BF_4^-$	CuI	10	Na_2CO_3	CH_2Cl_2	rt	6	(70)	76
H	I	CuO nanoparticles (33 nm)	1.3	KOH, air	DMSO	110	5	(92)	224
Me	$I^+C_6H_4$-4-Me BF_4^-	CuI	10	Na_2CO_3	CH_2Cl_2	rt	6	(83)	76

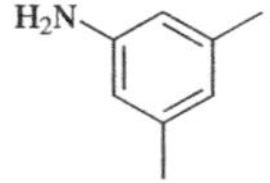

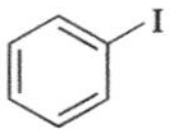

CuI (20 mol %), Cs_2CO_3, $Si(OEt)_4$, 145°, 24 h

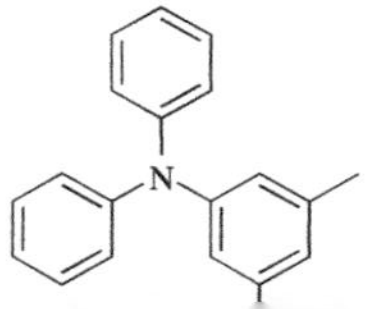

(85) — 335

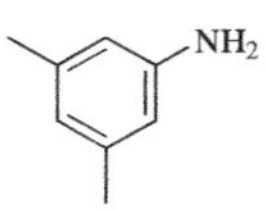

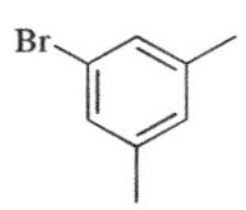

				50

CuI (10 mol %),
L16 (20 mol %), K$_3$PO$_4$,
DMF, 110°, 24 h

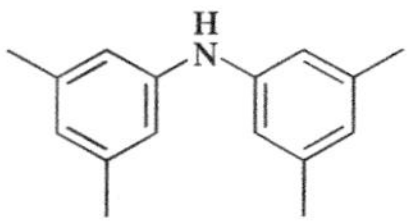

(75)

346

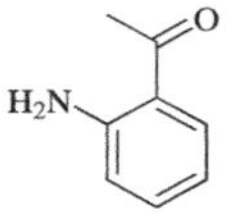

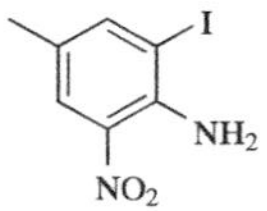

Cu$_2$O (10 mol %), K$_2$CO$_3$,
xylene, 145°, 12 h

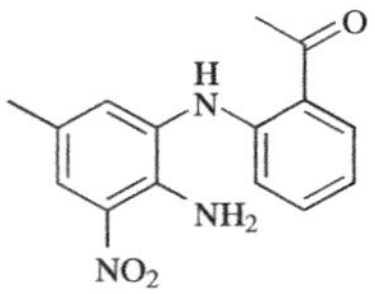

(85)

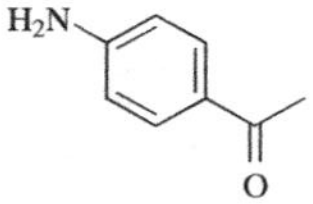

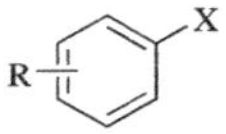

CuI (10 mol %), K$_3$PO$_4$

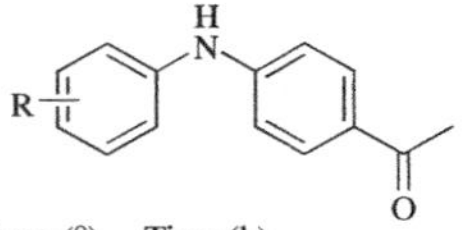

R	X	Additive	Solvent	Temp (°)	Time (h)		
H	Br	PPAPM (20 mol %)	DMF	110	36	(82)	48
H	Br	DPP (20 mol %)	DMF	110	36	(65)	108
4-Br	I	PPAPM (20 mol %)	DMF	110	36	(91)	48
4-Br	I	DPP (20 mol %)	DMF	110	36	(73)	108
3,5-Me$_2$	I	2-pyridinecarboxylic acid (20 mol %)	DMSO	90	24	(60)	50

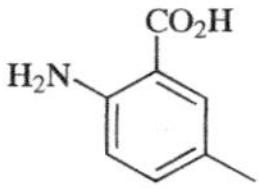

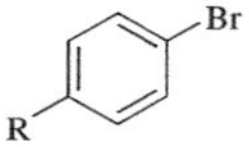

Catalyst(s) (x mol %)

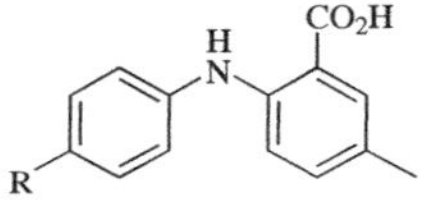

R	Catalyst(s)	x	Solvent	Temp (°)	Time (h)		
H	Cu	3	isoamyl alcohol	reflux	18	(83)	321
Me	Cu, CuI	1.5, 1.5	n-pentanol	150	5	(62)	351

TABLE 5A. N-ARYLATION OF PRIMARY ARYL AMINES (Continued)

Nitrogen Nucleophile	Aryl Halide	Conditions	Product(s) and Yield(s) (%)	Refs.

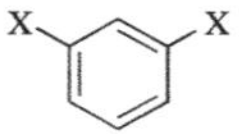

*Please refer to the charts preceding the tables for structures indicated by the **bold** numbers.*

C$_8$

| 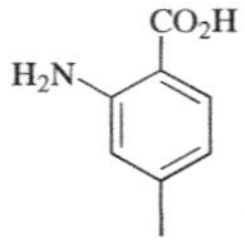 | 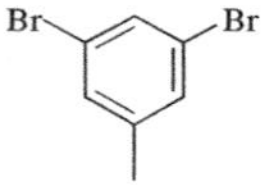 | Cu/CuI, K$_2$CO$_3$,
n-pentanol,
reflux | 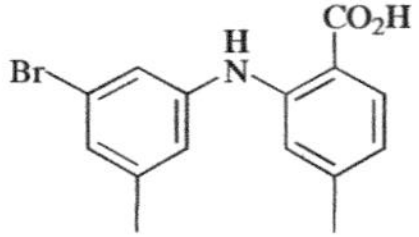 (25) | 352 |

| | | K$_2$CO$_3$,
n-pentanol, reflux | | |

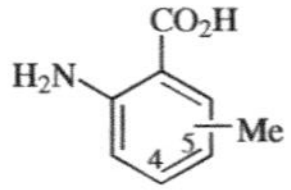

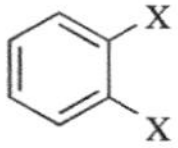

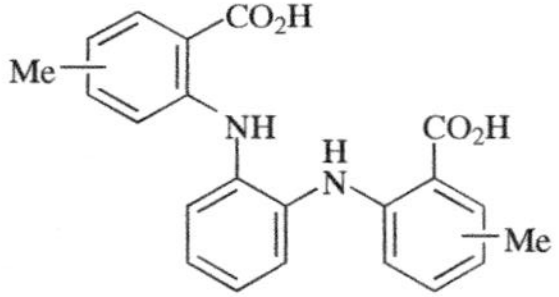

Isomer	X	Catalyst(s)	x	Time (h)	
4	Br	Cu, CuI	—	(10)	352
4	I	Cu, CuI	—	(30)	352
5	I	Cu	20	24	344
5	I	Cu, CuI	10, 8	24	344

| | | K$_2$CO$_3$, *n*-pentanol,
reflux | | |

Isomer	X	Catalyst(s)	x	Time (h)	
4	Br	Cu, CuI	—	(20)	352
4	I	Cu, CuI	—	(45)	352
5	I	Cu	20	24	(45)	344
5	I	Cu, CuI	10, 8	24	(65)	344

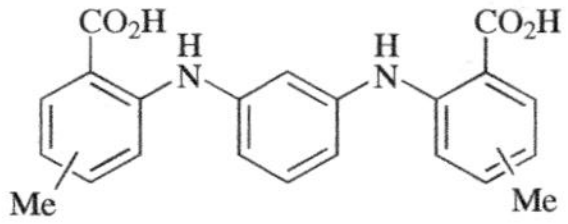

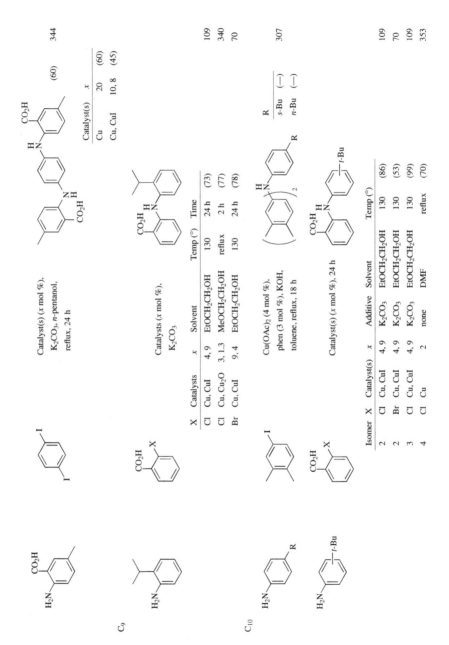

C9

Catalyst(s) (x mol %), K2CO3, n-pentanol, reflux, 24 h   (60)

Catalyst(s)	x
Cu	20 (60)
Cu, CuI	10, 8 (45)

344

Catalysts (x mol %), K2CO3

X	Catalysts	x	Solvent	Temp (°)	Time		
Cl	Cu, CuI	4, 9	EtOCH2CH2OH	130	24 h	(73)	109
Cl	Cu, Cu2O	3, 1.3	MeOCH2CH2OH	reflux	2 h	(77)	340
Br	Cu, CuI	9, 4	EtOCH2CH2OH	130	24 h	(78)	70

Cu(OAc)2 (4 mol %), phen (3 mol %), KOH, toluene, reflux, 18 h

307

C10

Catalyst(s) (x mol %), 24 h

Isomer	X	Catalyst(s)	x	Additive	Solvent	Temp (°)		
2	Cl	Cu, CuI	4, 9	K2CO3	EtOCH2CH2OH	130	(86)	109
2	Br	Cu, CuI	4, 9	K2CO3	EtOCH2CH2OH	130	(53)	70
3	Cl	Cu, CuI	4, 9	K2CO3	EtOCH2CH2OH	130	(99)	109
4	Cl	Cu	2	none	DMF	reflux	(70)	353

R	
s-Bu	(—)
n-Bu	(—)

TABLE 5A. N-ARYLATION OF PRIMARY ARYL AMINES (Continued)

Nitrogen Nucleophile	Aryl Halide	Conditions	Product(s) and Yield(s) (%)	Refs.

*Please refer to the charts preceding the tables for structures indicated by the **bold** numbers.*

C$_{10-11}$

Conditions: Cu(pyr)$_2$Cl$_2$ (11 mol %), K$_2$CO$_3$, HO(CH$_2$)$_2$OH, MW, 75°, 20 min

Product:

R	X	
H	Cl	(67)
H	Br	(79)
H	I	(90)
Me	Cl	(79)
Me	Br	(90)
Me	I	(94)

Ref. 318

C$_{10}$

Catalyst(s) (x mol %)

R	X	Catalyst(s)	x	Additive	Solvent	Temp (°)	Time		Refs.
H	Br	Cu, CuI	9, 4	K$_2$CO$_3$	EtOCH$_2$CH$_2$OH	130	24 h	(97)	70
H	Cl	Cu, CuI	4, 9	K$_2$CO$_3$	EtOCH$_2$CH$_2$OH	130	24 h	(96)	109
H	Cl	Cu(pyr)$_2$Cl$_2$	11	K$_2$CO$_3$	HOCH$_2$CH$_2$OH	MW, 75	20 min	(72)	318
H	Br	Cu(pyr)$_2$Cl$_2$	11	K$_2$CO$_3$	HOCH$_2$CH$_2$OH	MW, 75	20 min	(87)	318
H	I	Cu(pyr)$_2$Cl$_2$	11	K$_2$CO$_3$	HOCH$_2$CH$_2$OH	MW, 75	20 min	(98)	318
NO$_2$	Br	Cu	4	N-ethylmorpholine	2,3-butanediol	70	15 h	(35)	320

C$_{10}$

Catalyst(s) (x mol %)

R	X	Catalyst(s)	x	Additive	Solvent	Temp (°)	Time		Refs.
H	Br	Cu, CuI	9, 4	K$_2$CO$_3$	EtOCH$_2$CH$_2$OH	130	24 h	(90)	70
H	Cl	Cu(pyr)$_2$Cl$_2$	10	K$_2$CO$_3$	HOCH$_2$CH$_2$OH	130	24 h	(82)	71
H	Br	Cu(pyr)$_2$Cl$_2$	10	K$_2$CO$_3$	HOCH$_2$CH$_2$OH	130	24 h	(89)	71
H	I	Cu(pyr)$_2$Cl$_2$	10	K$_2$CO$_3$	HOCH$_2$CH$_2$OH	130	24 h	(91)	71
H	Cl	Cu(pyr)$_2$Cl$_2$	11	K$_2$CO$_3$	HOCH$_2$CH$_2$OH	MW, 75	20 min	(88)	318
H	Br	Cu(pyr)$_2$Cl$_2$	11	K$_2$CO$_3$	HOCH$_2$CH$_2$OH	MW, 75	20 min	(91)	318
H	I	Cu(pyr)$_2$Cl$_2$	11	K$_2$CO$_3$	HOCH$_2$CH$_2$OH	MW, 75	20 min	(91)	318
NO$_2$	Br	Cu	4	N-ethylmorpholine	2,3-butanediol	70	15 h	(62)	320

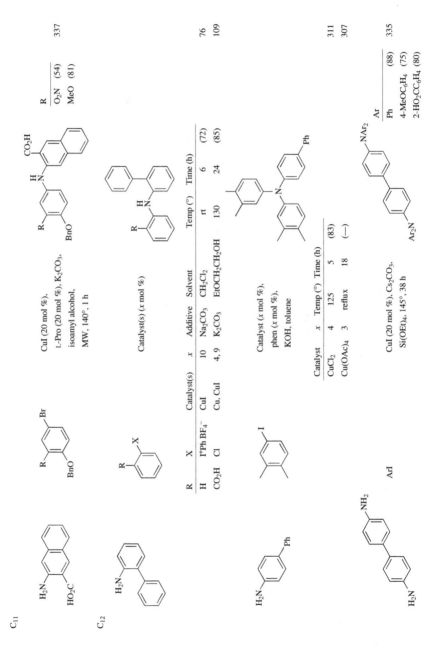

337

R
O$_2$N (54)
MeO (81)

CuI (20 mol %),
L-Pro (20 mol %), K$_2$CO$_3$,
isoamyl alcohol,
MW, 140°, 1 h

76
109

R	X	Catalyst(s)	x	Additive	Solvent	Temp (°)	Time (h)	
H	I⁺Ph BF$_4$⁻	CuI	10	Na$_2$CO$_3$	CH$_2$Cl$_2$	rt	6	(72)
CO$_2$H	Cl	Cu, CuI	4, 9	K$_2$CO$_3$	EtOCH$_2$CH$_2$OH	130	24	(85)

Catalyst(s) (x mol %)

Catalyst (x mol %),
phen (x mol %),
KOH, toluene

Catalyst	x	Temp (°)	Time (h)	
CuCl$_2$	4	125	5	(83)
Cu(OAc)$_4$	3	reflux	18	(—)

311
307

CuI (20 mol %), Cs$_2$CO$_3$,
Si(OEt)$_4$, 145°, 38 h

Ar
Ph (88)
4-MeOC$_6$H$_4$ (75)
2-HO$_2$CC$_6$H$_4$ (80)

335

C$_{11}$

C$_{12}$

Nitrogen Nucleophile	Aryl Halide	Conditions	Product(s) and Yield(s) (%)	Refs.

Please refer to the charts preceding the tables for structures indicated by the **bold** numbers.

C$_{12}$

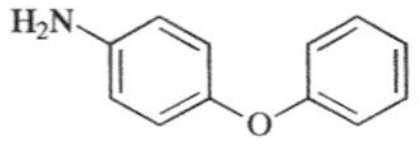

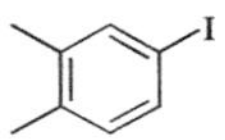

Cu(OAc)$_2$ (4 mol %), phen (3 mol %), KOH, toluene, reflux, 18 h

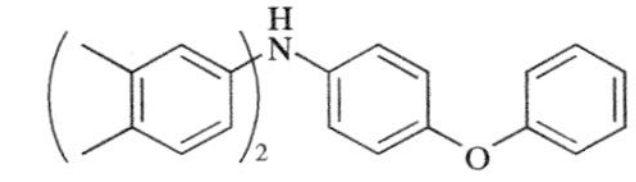

(—)

307

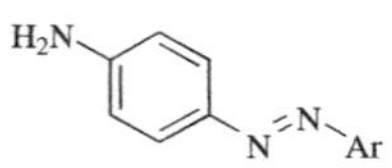

Ar = 4-O$_2$NC$_6$H$_4$

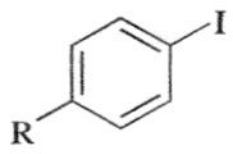

Cu, K$_2$CO$_3$, triglyme, 200–220°

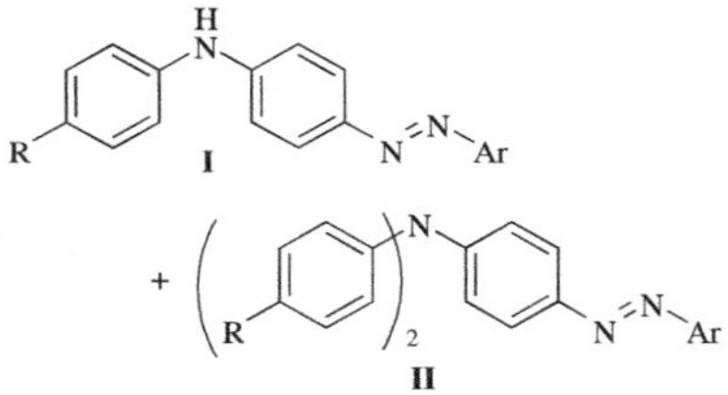

I

II

R	I	II
i-PrC(O)NH	(19)	(40)
MeO	(0)	(40)

354

C$_{13}$

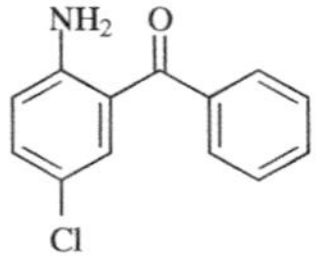

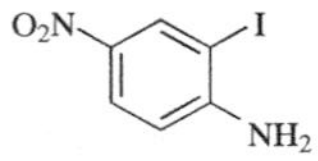

Cu$_2$O (10 mol %), K$_2$CO$_3$, xylene, 145°, 20 h

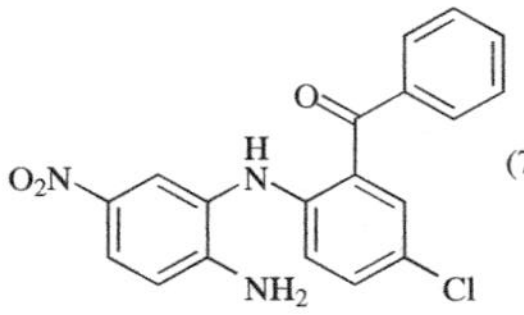

(75)

346

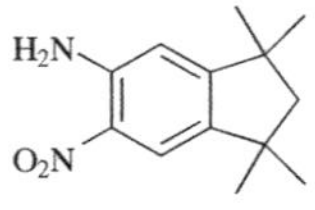

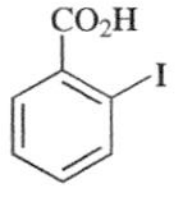

Cu (70 mol %), CaCO$_3$

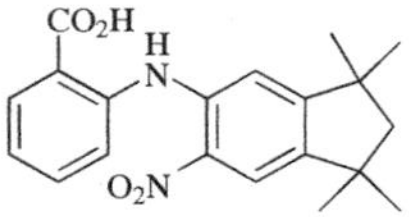

(24)

355

222

C$_{13-14}$

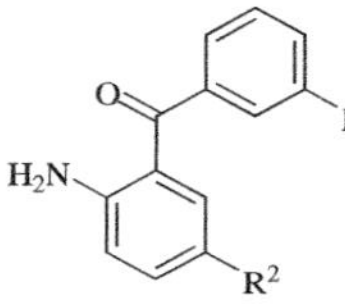

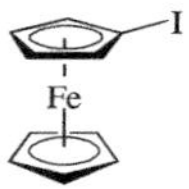

Cu (10 mol %), K$_2$CO$_3$,
xylene, 140°, 6 h

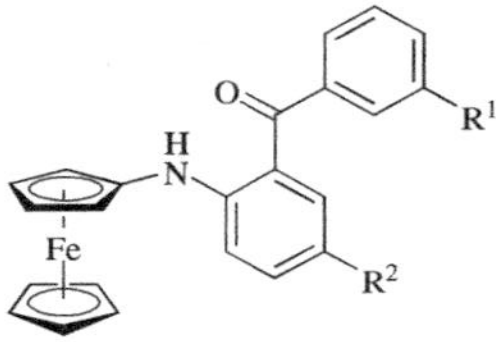

R^1	R^2		
H	Cl	(89)	329
H	Me	(84)	
Me	Cl	(86)	

C$_{14}$

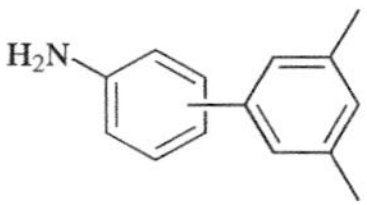

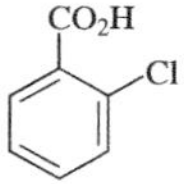

Catalysts (x mol %), K$_2$CO$_3$

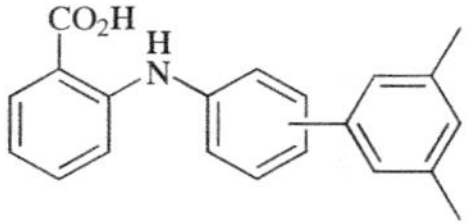

Isomer	Catalysts	x	Solvent	Temp (°)	Time (h)	
3	Cu, CuI	4, 9	EtOCH$_2$CH$_2$OH	130	24	(96)	109
4	Cu, Cu$_2$O	3, 1.3	MeOCH$_2$CH$_2$OH	reflux	2	(95)	356, 357

C$_{16}$

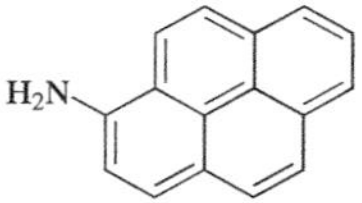

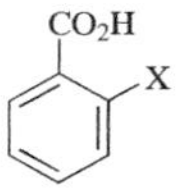

Catalyst(s) (x mol %), K$_2$CO$_3$

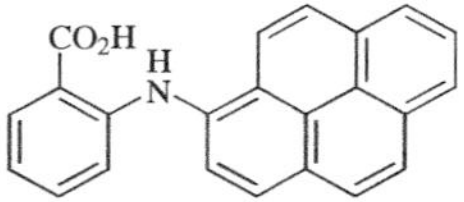

X	Catalyst(s)	x	Solvent	Temp (°)	Time	
Cl	Cu, CuI	4, 9	EtOCH$_2$CH$_2$OH	130	24 h	(73)	109
Br	Cu, CuI	9, 4	EtOCH$_2$CH$_2$OH	130	24 h	(55)	70
Cl	Cu(pyr)$_2$Cl$_2$	11	HOCH$_2$CH$_2$OH	MW, 75	20 min	(89)	318
Br	Cu(pyr)$_2$Cl$_2$	11	HOCH$_2$CH$_2$OH	MW, 75	20 min	(93)	318
I	Cu(pyr)$_2$Cl$_2$	11	HOCH$_2$CH$_2$OH	MW, 75	20 min	(96)	318

TABLE 5A. N-ARYLATION OF PRIMARY ARYL AMINES (Continued)

Nitrogen Nucleophile	Aryl Halide	Conditions	Product(s) and Yield(s) (%)	Refs.

Please refer to the charts preceding the tables for structures indicated by the **bold** numbers.

C₁₈

(H₂N–aryl)

MeO–C₆H₄–I

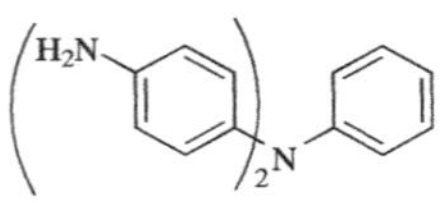

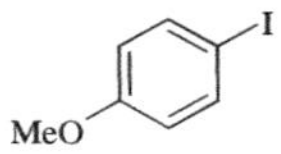

Cu (10 eq), 18-c-6 (3 eq), K₂CO₃, 1,2-Cl₂C₆H₄, reflux, 48 h

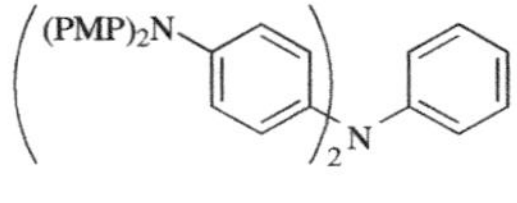

(PMP)₂N–C₆H₄–N()₂ (35)

358

C₁₉

MeO₂C, Ph, H, N

naphthyl–I

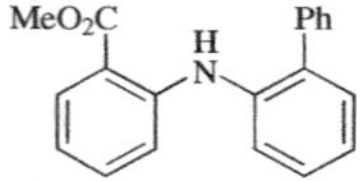

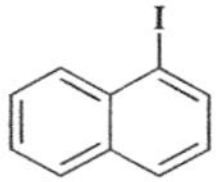

Cu (50 mol %), K₂CO₃, PhNO₂, 180°, 2.5 h

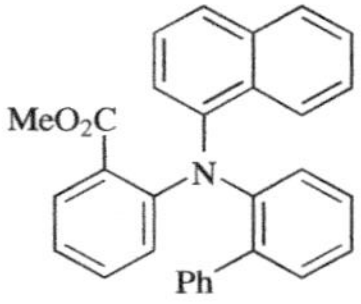

(18)

349

C₂₄

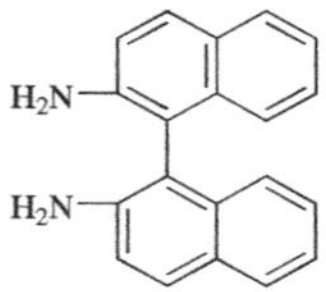

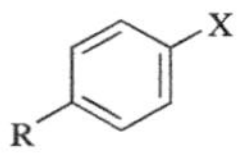

binaphthyl diamine

R–C₆H₄–X

Catalyst (x mol %)

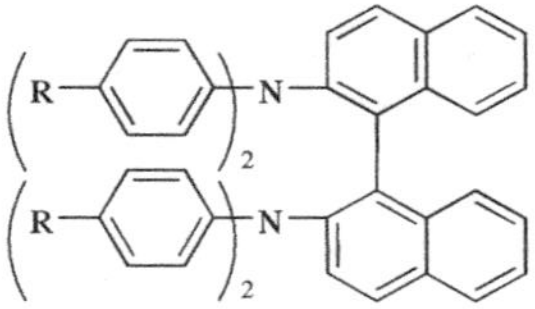

binaphthyl bis-aryl amine product

				er
Isomer	er			
(R)	99:1			
(R,S)	—			

R	X	Catalyst	x	Additive(s)	Solvent	Temp (°)	Time (h)		er	
H	Br	CuBr	10	rac-binol (20 mol %), K₃PO₄	DMF	120	36	(83)	56.5:45.4	286
MeO	I	CuI	20	Cs₂CO₃	Si(OEt)₄	145	120	(87)	—	335

224

C_40

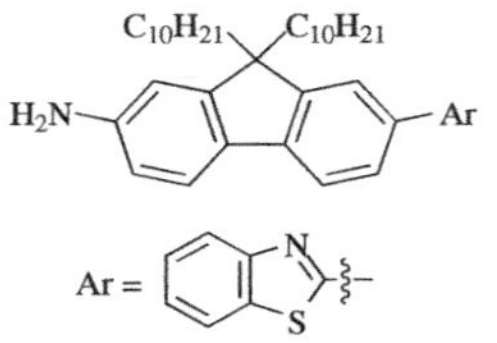

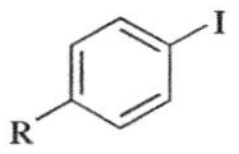

C_10H_21 C_10H_21

H_2N

$Ar = \begin{smallmatrix}N\\\\S\end{smallmatrix}$

I

R

R = H, MeO

Cu-bronze, 18-c-6,
K_2CO_3, 1,2-Cl_2C_6H_4,
180°, 43 h

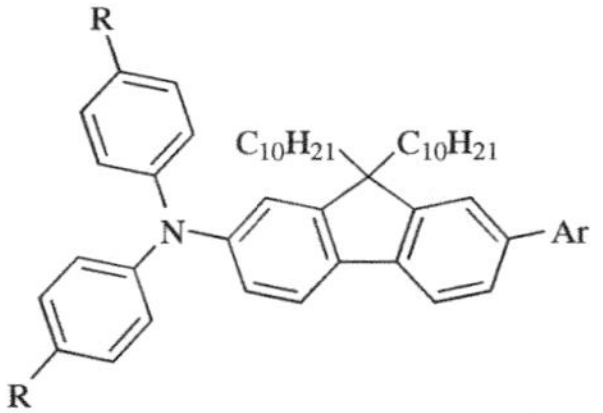

C_10H_21 C_10H_21

R

N

R

Ar

(82)   359

C_44

NH_2

H_2N

N N
Zn
N N

NH_2

H_2N

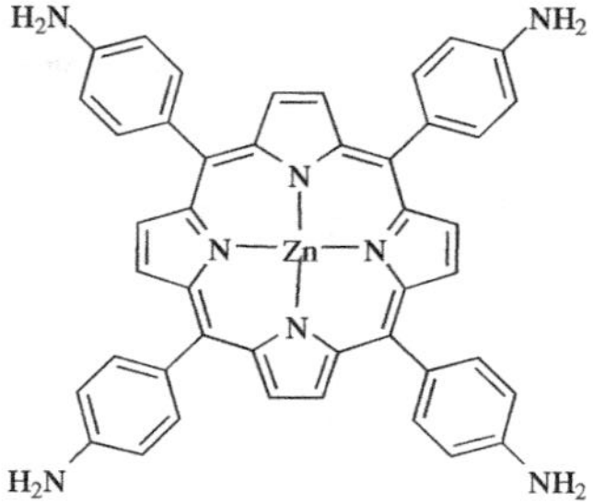

4-RC_6H_4I
R = MeO, (PMP)_2N

Cu (10 eq), 18-c-6 (3 eq),
K_2CO_3, 1,2-Cl_2C_6H_4,
reflux, 48 h

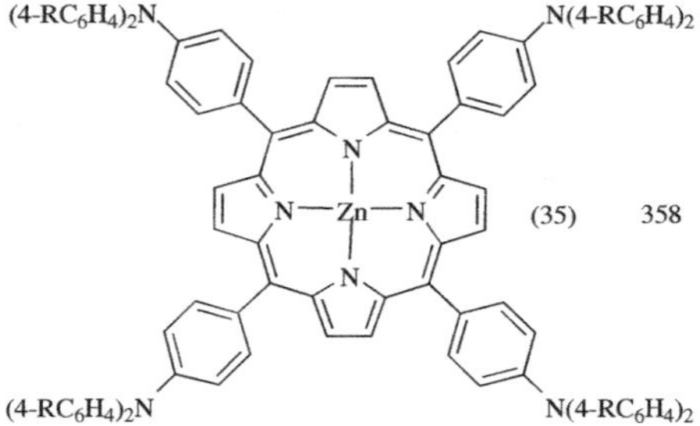

N(4-RC_6H_4)_2

(4-RC_6H_4)_2N

N N
Zn
N N

N(4-RC_6H_4)_2

(4-RC_6H_4)_2N

(35)   358

---

[a] The ratio of PhX to PhNH_2 was 2:1.

[b] The ratio of PhX to PhNH_2 was 29:1.

TABLE 5B. N-ARYLATION OF PRIMARY HETEROARYL AMINES

Please refer to the charts preceding the tables for structures indicated by the bold numbers.

Nitrogen Nucleophile	Aryl Halide	Conditions	Product(s) and Yield(s) (%)	Refs.
C₃				
		CuI (x amount), K₂CO₃, dioxane, 100°, 22 h	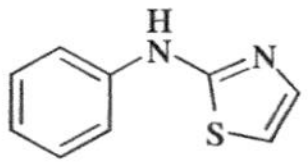	55
		X x Additive(s)		
		Br 50 mol % DMEDA (50 mol %), KI (62)		
		I 1 eq DMEDA (1 eq) (78)		
C₄				
		CuI (50 mol %), DMEDA (50 mol %), KI, K₂CO₃, dioxane, 100°	R Time (h) — H 22 (78); 2-MeO 24 (70)	55
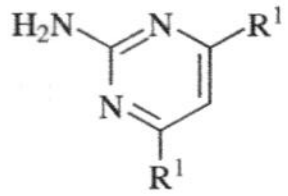	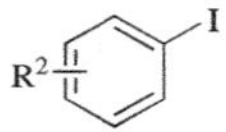	Catalyst (x mol %)	**I** or **II**	226 335
		R¹ R² Catalyst x Additive(s) Solvent Temp (°) Time (h) I II		
		H H Cu/AB 5 KOt-Bu toluene 180 18 (71) (0)		
		MeO 4-MeO CuI 20 Cs₂CO₃ Si(OEt)₄ 145 40 (0) (61)		
		CuI (35 mol %), DMEDA (35 mol %), K₂CO₃, KI, dioxane, 100°, 22 h	(74)	55

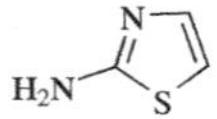

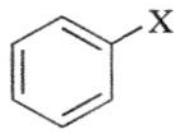

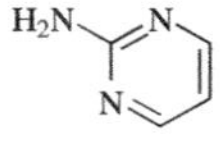

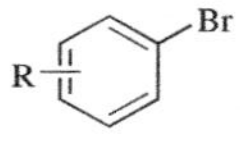

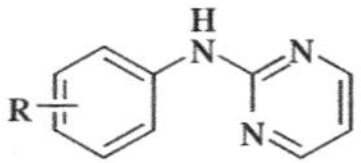

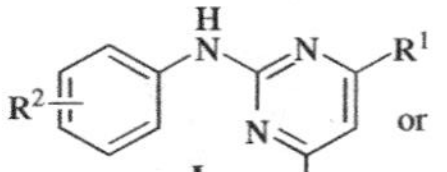

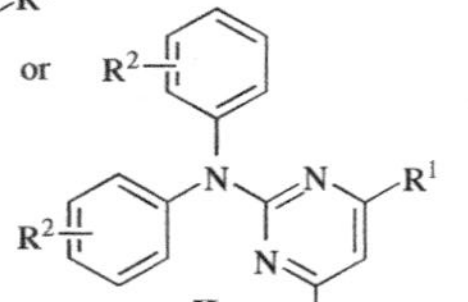

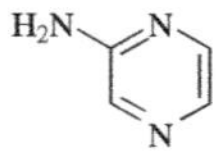

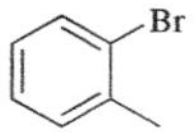

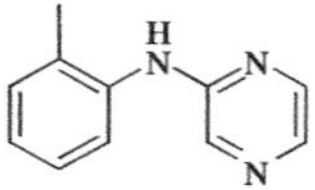

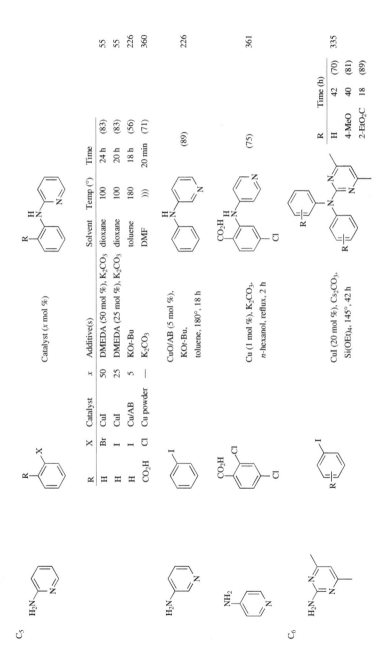

C₅ 2-aminopyridine (H₂N–pyridine)

R	X	Catalyst	x	Additive(s)	Solvent	Temp (°)	Time		
H	Br	CuI	50	DMEDA (50 mol %), K₂CO₃	dioxane	100	24 h	(83)	55
H	I	CuI	25	DMEDA (25 mol %), K₂CO₃	dioxane	100	20 h	(83)	55
H	I	Cu/AB	5	KO*t*-Bu	toluene	180	18 h	(56)	226
CO₂H	Cl	Cu powder	—	K₂CO₃	DMF	)))	20 min	(71)	360

3-aminopyridine

CuO/AB (5 mol %), KO*t*-Bu, toluene, 180°, 18 h — (89) — 226

4-aminopyridine

Cu (1 mol %), K₂CO₃, *n*-hexanol, reflux, 2 h — (75) — 361

C₆ 2-amino-4,6-dimethylpyrimidine

CuI (20 mol %), Cs₂CO₃, Si(OEt)₄, 145°, 42 h — 335

R	Time (h)	
H	42	(70)
4-MeO	40	(81)
2-EtO₂C	18	(89)

Nitrogen Nucleophile	Aryl Halide	Conditions	Product(s) and Yield(s) (%)	Refs.

Please refer to the charts preceding the tables for structures indicated by the bold numbers.

C₇

RNH₂ | (CO₂H, Br) | Cu (x mol %), K₂CO₃, 1-pentanol | (CO₂H, NHR) | |

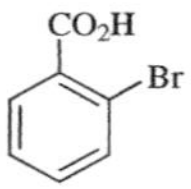

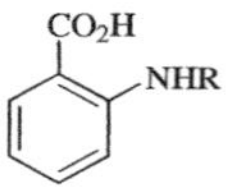

R	x	Additive	Time (h)		
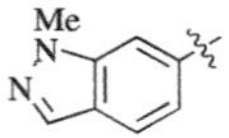 Me indazole	1.5	none	2.5	(68)	333
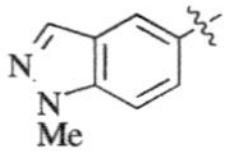 indazole (other)	1.5	none	2.5	(48)	333
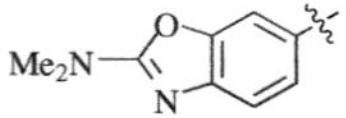 Me₂N-benzoxazole	13	none	3	(90)	362
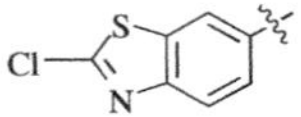 Cl-benzothiazole Me	1.5	KI	2	(76)	333

C₉

(H₂N-pyridine-pyrimidine) | (Br, NO₂) | CuI (50 mol %), DMEDA (50 mol %), K₂CO₃, KI, dioxane, 100°, 24 h | (82) | 55, 170 |

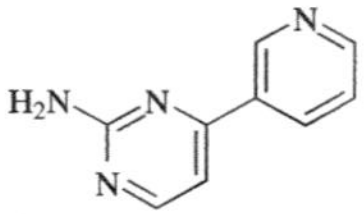

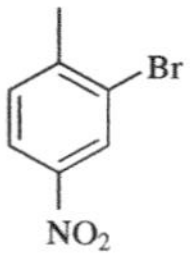

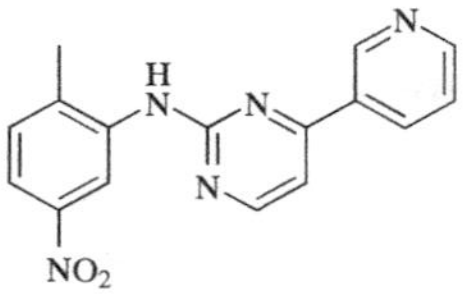

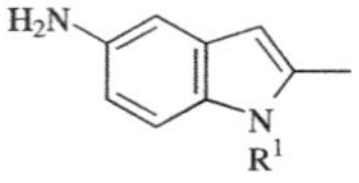

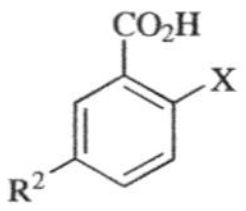

R¹	R²	X	Time (min)	
Me	H	Br	120	(82)
Me	H	I	30	(91)
H	MeO	Br	30	(81)
PhO₂S	Cl	I	30	(75)

(H₂N-indole, R¹) | (CO₂H, X, R²) | CuI (10 mol %), K₂CO₃, DMSO, 80° | (CO₂H, indole R¹, R²) | 363 |

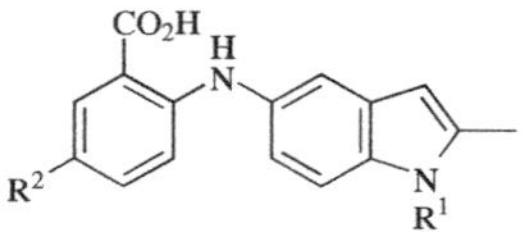

R¹	R²	X	Time (min)	
H	H	Br	120	(80)
H	H	I	30	(86)
PhO₂S	H	Cl	120	(65)
PhO₂S	H	Br	30	(68)
PhO₂S	H	I	15	(71)

R¹	R²	X	Time (min)	
PhO₂S	Br	I	15	(77)
Me	Cl	I	30	(87)
Me	Br	I	30	(79)
Me	O₂N	I	30	(84)

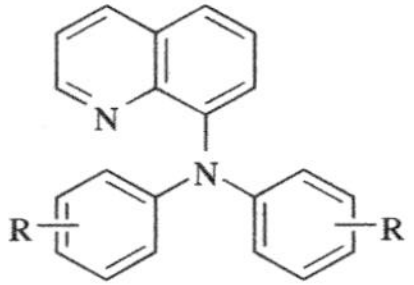

R	Time (h)	
H	38	(94)
4-MeO	30	(90)
2-EtO$_2$C	18	(90)

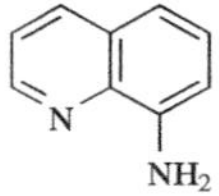

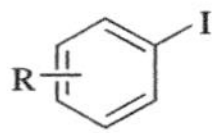

CuI (20 mol %), Cs$_2$CO$_3$,
Si(OEt)$_4$, 145°, 42 h

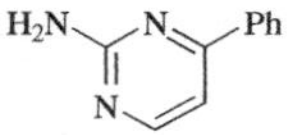

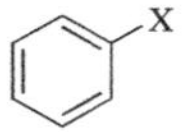

CuI (x amount),
DMEDA (x amount),
K$_2$CO$_3$, KI, dioxane, 100°

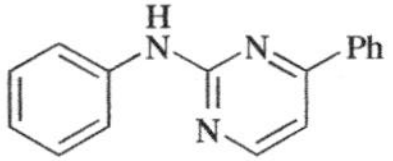

X	x	Time (h)	
Cl	1 eq	48	(<5)
Br	50 mol %	24	(77)

CuI (1 eq), DMEDA (1 eq),
K$_3$PO$_4$, DMSO, 110°

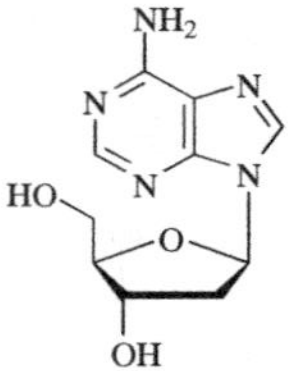

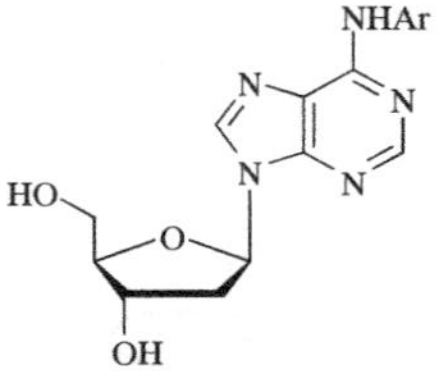

ArX

Ar	X	Additive	Time (h)	
4-ClC$_6$H$_4$	I	none	1.5	(70)
4-BrC$_6$H$_4$	I	none	1.5	(—)
4-O$_2$NC$_6$H$_4$	I	none	1.5	(73)
2-HOC$_6$H$_4$	I	none	2.5	(—)
4-MeOC$_6$H$_4$	Br	NaI	3.5	(84)
4-MeOC$_6$H$_4$	I	none	2.5	(84)
4-MeC$_6$H$_4$	Cl	NaI	24	(0)

Ar	X	Additive	Time (h)	
4-MeC$_6$H$_4$	I	none	2.5	(78)
4-MeC$_6$H$_4$	I	none	48	(65)
2,3-Me$_2$C$_6$H$_3$	I	none	2.5	(80)
2,6-Me$_2$C$_6$H$_3$	Br	none	6	(75)
3,4-Me$_2$C$_6$H$_3$	Br	none	6	(70)
9-phenanthryl	Br	NaI	3.5	(80)
1-pyrenyl	I	none	2.5	(88)

C$_{10}$

TABLE 5B. *N*-ARYLATION OF PRIMARY HETEROARYL AMINES (*Continued*)

Nitrogen Nucleophile	Aryl Halide	Conditions	Product(s) and Yield(s) (%)	Refs.

Please refer to the charts preceding the tables for structures indicated by the bold numbers.

C$_{10}$

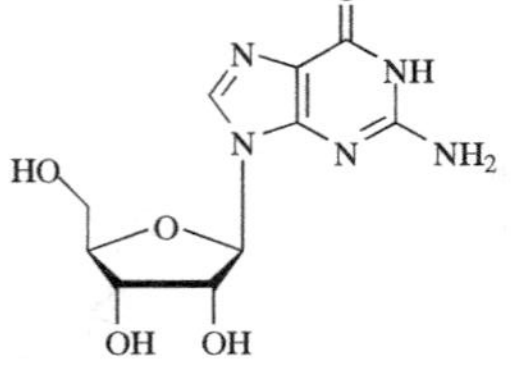

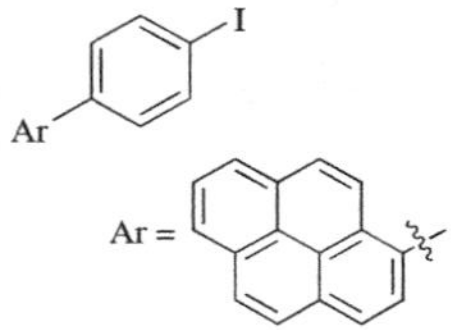

CuI (15 mol %), Cs$_2$CO$_3$, DMSO, 140°, 24 h

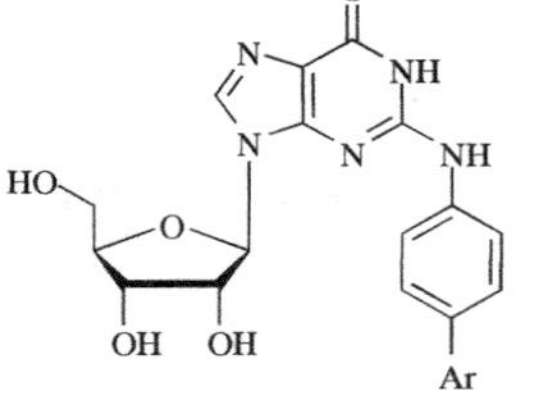

(59)

365

Ar =

C$_{12}$

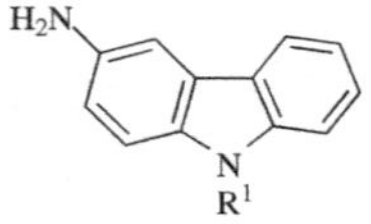

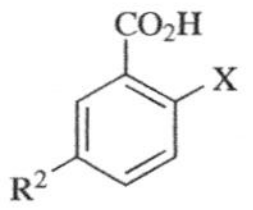

CuI (10 mol %), K$_2$CO$_3$, DMSO, 80°

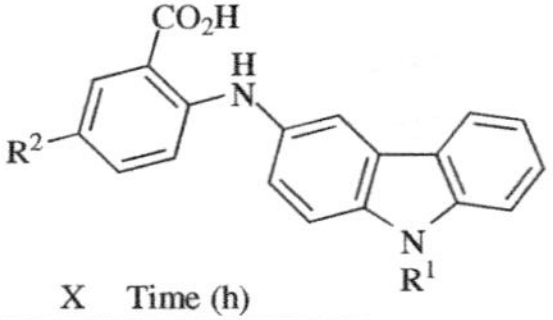

364

R^1	R^2	X	Time (h)	
H	H	I	1	(73)
Et	H	Br	3	(64)
Et	H	I	1	(73)

R^1	R^2	X	Time (h)	
Et	Cl	I	1	(72)
Et	Br	I	1	(70)
H	O$_2$N	I	0.5	(74)

TABLE 5C. N-HETEROARYLATION OF PRIMARY ARYL AND HETEROARYL AMINES

Nitrogen Nucleophile	Heteroaryl Halide	Conditions	Product(s) and Yield(s) (%)	Refs.
C₄		CuI (10 mol %), L-Pro (20 mol %), K₂CO₃, DMSO, 80°, 30 h	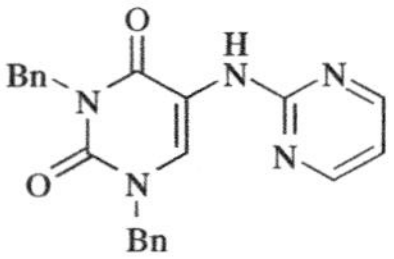 (77)	47
C₆		CuI (10 mol %), EDA (20 mol %), K₃PO₄, reflux, 24 h	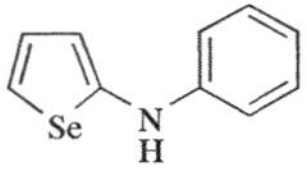 (0)	87
	Ar = 4-FC₆H₄	CuI (5 mol %), HO(CH₂)₂OH, K₃PO₄, i-PrOH, 85°, 20 h	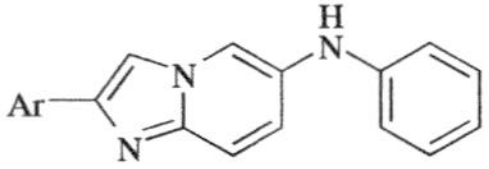 (30)	126
C₈		CuI (10 mol %), K₃PO₄, DMSO, 2-pyridine-carboxylic acid (20 mol %), 100°, 24 h	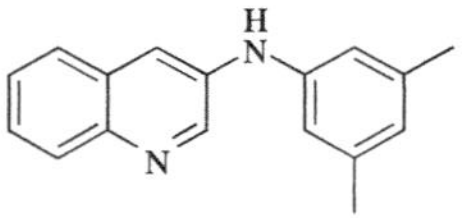 (51)	50
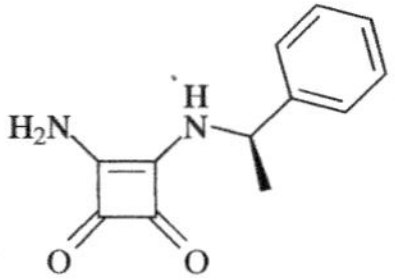 C₁₂	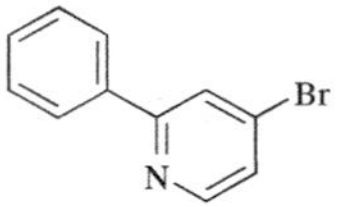	CuI (1 eq), DMEDA (1.5 eq). K₂HPO₃, DMF, 110°, 6.5 h	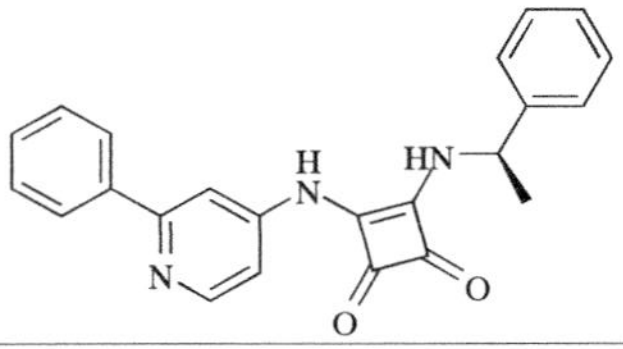 (—)	366

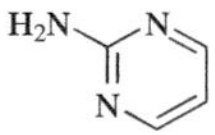

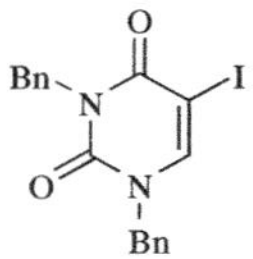

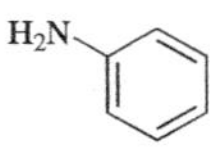

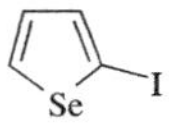

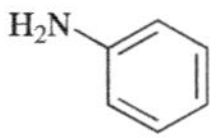

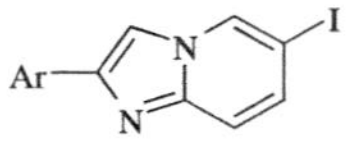

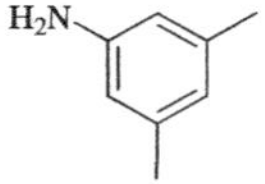

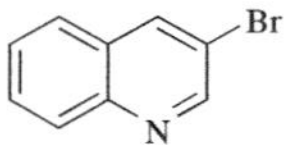

TABLE 6A. N-ARYLATION OF SECONDARY ARYL, ALKYL, AND DIARYL AMINES

Nitrogen Nucleophile	Aryl Halide	Conditions	Product(s) and Yield(s) (%)	Refs.

*Please refer to the charts preceding the tables for structures indicated by the **bold** numbers.*

C_7

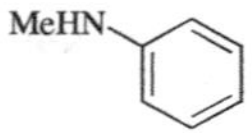

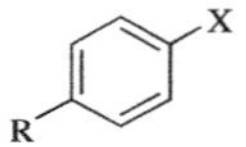

MeHN— (phenyl)

Aryl Halide: R—(phenyl)—X

Conditions: CuI (x mol %)

Product: R—(phenyl)—N(Me)(phenyl)

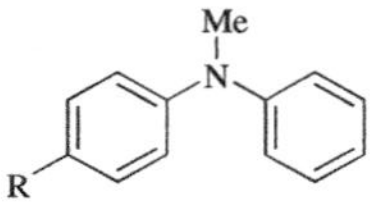

R	X	x	Additives	Solvent(s)	Temp (°)	Time (h)		
H	Cl	2	L1 (2 mol %), KO-t-Bu	—	100	12	(85)	288
H	Br	2	L1 (2 mol %), KO-t-Bu	—	100	12	(85)	288
Br	I	10	PPAPM (20 mol %), K₃PO₄	DMF	110	36	(60)	48
Br	I	10	DPP, K₃PO₄	DMF/H₂O (98:2)	110	36	(60)	108
MeO	Br	10	piperidine-2-carboxylic acid (20 mol %), K₂CO₃	DMF	110	36	(57)	67
Me	I	5	L41 (10 mol %), NaO-t-Bu	toluene	110	40	(71)	338

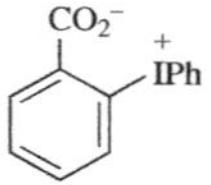

Aryl Halide:
CO_2^- + IPh (diaryliodonium structure)

Conditions: Cu(OAc)₂ (3 mol %), i-PrOH, reflux, 24 h

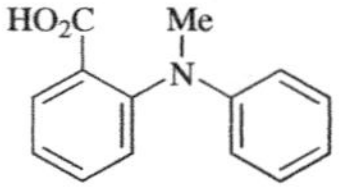

Product: HO₂C—(phenyl)—N(Me)(phenyl) (44)

Refs. 312

C_9

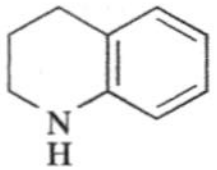

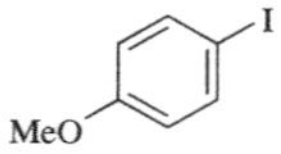

Nitrogen Nucleophile: (tetrahydroquinoline) N–H

Aryl Halide: I—(phenyl)—MeO

Conditions: Cu(THMD)₂ (20 mol %), KO-t-Bu, toluene, 120°, 12 h

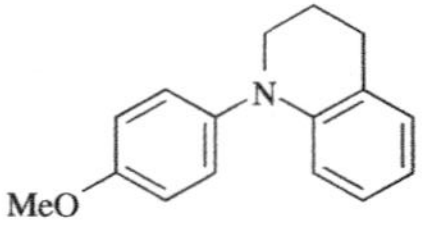

Product: (tetrahydroquinoline)—N—(phenyl)—OMe (83)

Refs. 308

Ph₂NH

X	Catalyst	x	Additive(s)	Solvent	Temp (°)	Time	
Br	Cu	4.7 eq	none	—	210	12 h	(86)	303
I	Cu	4.7 eq	none	—	210	12 h	(86)	303
I	Cu (active)	5 eq	none	none	MW (100 W)	6 min	(83)	280
I	Cu (active)	5 eq	none	H_2O	MW (100 W)	6 min	(85)	280
I	Cu	8 eq	KOH	n-$C_{16}H_{34}$	160	8 h	(88)	20
I	Cu	8 mol %	K_2CO_3	$PhNO_2$	reflux	24 h	(85)	367
I	CuI	5 mol %	mes₂DAB (6 mol %), KOt-Bu	toluene	120	20 h	(94)	306
Cl	Cu₂O	5 mol %	NaOt-Bu	NMP	110	24 h	(81)	266
Br	Cu₂O	5 mol %	NaOt-Bu	NMP	80	24 h	(83)	266
Cl	Cu(neocup)(PPh₃)Br	10 mol %	KOt-Bu	toluene	110	36 h	(49)	368
Br	Cu(neocup)(PPh₃)Br	10 mol %	KOt-Bu	toluene	110	36 h	(73)	368
I	Cu(neocup)(PPh₃)Br	10 mol %	KOt-Bu	toluene	110	6 h	(78)	368
I	Cu(THMD)₂	20 mol %	KOt-Bu	toluene	120	12 h	(95)	308
I	Cu(PPh₃)₃Br	20 mol %	Cs_2CO_3	toluene	120	24 h	(70)	339

X	R	Catalyst	x	Additive(s)	Solvent	Temp (°)	Time (h)	
2-Cl	H	Cu	8 eq	KOH	$C_{16}H_{34}$	160	27	(55)	20
2-I	H	Cu	8 eq	KOH	$C_{16}H_{34}$	160	27	(18)	20
3-Cl	H	Cu	2 eq	18-c-6, K_2CO_3	1,2-$Cl_2C_6H_4$	reflux	21	(87)	54
3-Cl	H	CuI	5 mol %	KOH	$C_{16}H_{34}$	160	24	(82)	20
4-Cl	H	Cu	2 eq	18-c-6, K_2CO_3	1,2-$Cl_2C_6H_4$	reflux	21	(78)	54
4-Br	H	CuI	1 mol %	CDA (10 mol %), NaOt-Bu	dioxane	reflux	24	(75)	172
4-Br	H	Cu(PPh₃)₃Br	20 mol %	Cs_2CO_3	toluene	120	24	(54)	339
4-I	H	Cu	8 eq	KOH	$C_{16}H_{34}$	160	16	(41)	20
4-I	H	Cu	1 eq	18-c-6, K_2CO_3	1,2-$Cl_2C_6H_4$	200	48	(35)	369
4-I	H	CuI	2.5 mol %	BuLi	xylenes	reflux	2	(48)	370

TABLE 6A. N-ARYLATION OF SECONDARY ARYL, ALKYL, AND DIARYL AMINES (*Continued*)

Nitrogen Nucleophile	Aryl Halide	Conditions	Product(s) and Yield(s) (%)	Refs.

*Please refer to the charts preceding the tables for structures indicated by the **bold** numbers.*

C12

Ph2NH

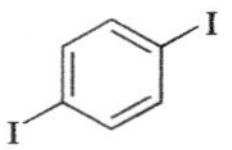

Catalyst (x amount)

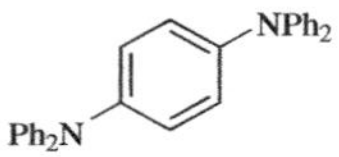

Catalyst	x	Additive	Solvent	Temp (°)	Time (h)		Refs.
Cu	8 eq	KOH	n-C16H34	160	24	(62)	20
CuI	2.5 mol %	BuLi	xylenes	reflux	2	(28)	370

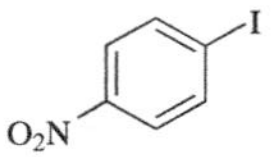

Product: Ph2N–C6H4–NPh2

Catalyst (20 mol %), KOt-Bu, toluene, 120°

Product: O2N–C6H4–NPh2

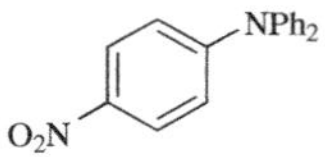

Catalyst	Time (h)		Refs.
Cu(THMD)2	12	(90)	308
Cu(PPh3)3Br	24	(62)	339

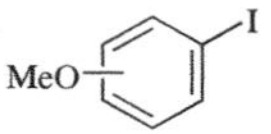

Catalyst (x amount)

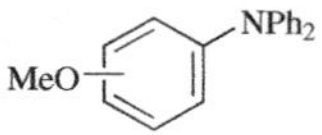

Product: MeO–C6H4–NPh2

Isomer	Catalyst	x	Additive(s)	Solvent	Temp (°)	Time (h)		Refs.
3	Cu	2 eq	18-c-6, K2CO3	1,2-Cl2C6H4	reflux	21	(78)	54
4	Cu	1 eq	K2CO3	triglyme	200	26.5	(84)	371
4	Cu	2 eq	18-c-6, K2CO3	1,2-Cl2C6H4	reflux	19	(90)	54
4	Cu	8 eq	KOH	C16H34	160	5	(79)	20
4	CuI	5 mol %	mes2DAB, KOt-Bu	toluene	120	20	(72)	306
4	CuI	5 mol %	9-AJ (10 mol %), NaOt-Bu	toluene	110	40	(96)	338
4	Cu(THMD)2	20 mol %	KOt-Bu	toluene	120	12	(90)	308

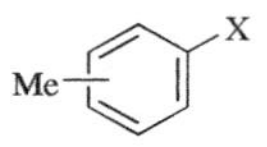

Me–⟨aryl⟩–X $\xrightarrow{\text{Catalyst } (x \text{ amount})}$ Me–⟨aryl⟩–NPh₂

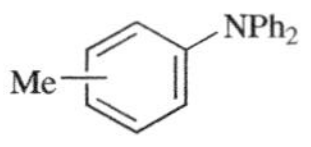

Isomer	X	Catalyst	x	Additive(s)	Solvent	Temp (°)	Time		
2	I	Cu	2 eq	18-c-6, K₂CO₃	1,2-Cl₂C₆H₄	reflux	18 h	(92)	54
2	I	Cu	8 eq	KOH	C₁₆H₃₄	160	27 h	(81)	20
2	I	CuI	5 mol %	L41 (10 mol %), NaOt-Bu	toluene	110	40 h	(85)	338
2	I	Cu(PPh₃)₃Br	20 mol %	Cs₂CO₃	toluene	120	24 h	(52)	339
2	I	Cu(neocup)(PPh₃)₃Br	10 mol %	KOt-Bu	toluene	110	36 h	(70)	368
3	I	Cu	2 eq	18-c-6, K₂CO₃	1,2-Cl₂C₆H₄	reflux	18 h	(93)	54
3	I	Cu	8 eq	KOH	n-C₁₆H₃₄	160	24 h	(82)	20
4	Br	Cu (active)	5 eq	none	H₂O	MW (100 W)	6 min	(81)	280
4	I	Cu	2 eq	18-c-6, K₂CO₃	1,2-Cl₂C₆H₄	reflux	16 h	(92)	54
4	I	CuI	5 mol %	L41 (10 mol %), NaOt-Bu	toluene	110	40 h	(96)	338
4	I	CuI	5 mol %	mes₂DAB (6 mol %), KOt-Bu	toluene	120	20 h	(79)	306
4	I	Cu(PPh₃)₃Br	20 mol %	Cs₂CO₃	toluene	120	24 h	(52)	339
4	I	Cu(THMD)₂	20 mol %	KOt-Bu	toluene	120	12 h	(90)	308
4	I	Cu(neocup)(PPh₃)₃Br	10 mol %	KOt-Bu	toluene	110	36 h	(70)	339

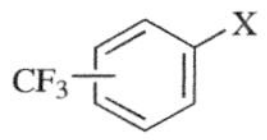

CF₃–⟨aryl⟩–X → CF₃–⟨aryl⟩–NPh₂

Cu (8 eq), KOH,
n-C₁₆H₃₄, 160°

Isomer	Time (h)	
3	8	(64)
4	24	(56)

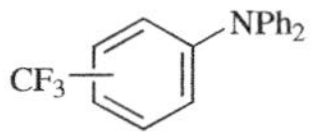

20

NC–⟨aryl⟩–I → NC–⟨aryl⟩–NPh₂

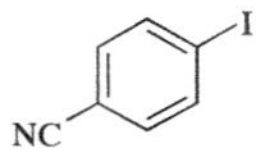

CuI (5 mol %),
L41 (10 mol %), NaOt-Bu,
toluene, 110°, 40 h   (96)

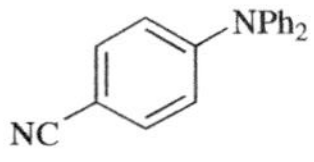

338

Nitrogen Nucleophile	Aryl Halide	Conditions	Product(s) and Yield(s) (%)	Refs.

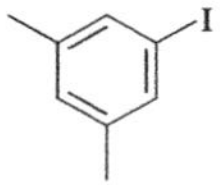

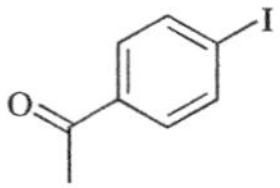

Please refer to the charts preceding the tables for structures indicated by the **bold** numbers.

C$_{12}$

Ph$_2$NH

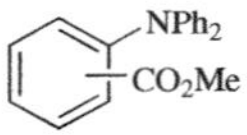

Isomer	Catalyst	x	Additive	Solvent	Temp (°)	Time (h)		
2	Cu	15	K$_2$CO$_3$	—	180	33	(67)	348
2	Cu(PPh$_3$)$_3$Br	20	Cs$_2$CO$_3$	toluene	120	24	(69)	339
4	Cu(PPh$_3$)$_3$Br	20	KOt-Bu	toluene	120	24	(62)	339

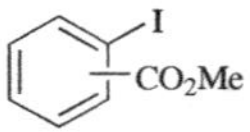

Catalyst (x eq)

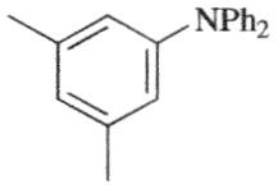

Catalyst	x	Additive(s)	Solvent	Temp (°)	Time (h)		
CuCl$_2$	4	phen (4 mol %), KOH	—	125	6	(85)	311
CuI	1	CDA (10 mol %)	dioxane	110	23	(77)	115

Cu(PPh$_3$)$_3$Br (20 mol %), Cs$_2$CO$_3$, toluene, 120°, 24 h

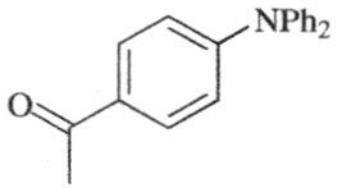

(68) — 339

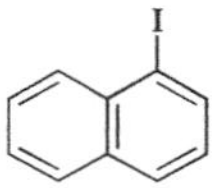

CuI (5 mol %), KOH, n-C$_{16}$H$_{34}$, 160°, 24 h

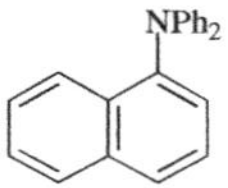

(59) — 20

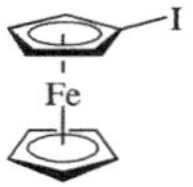

CuI (1 eq), K$_2$CO$_3$, DMSO, 90°, 18 h

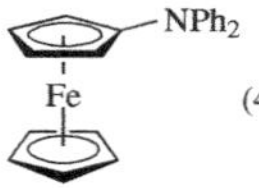

(43) — 372

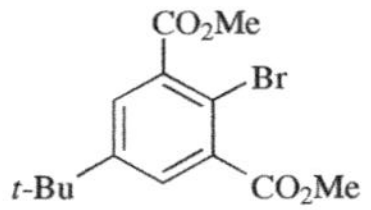

341

359

359

373

369
369
311
342

Cu (15 mol %), K₂CO₃, Ph₂O, 190°, 48 h

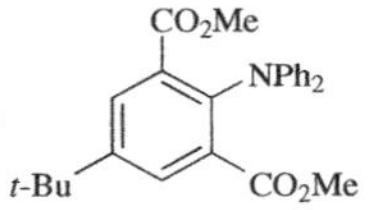

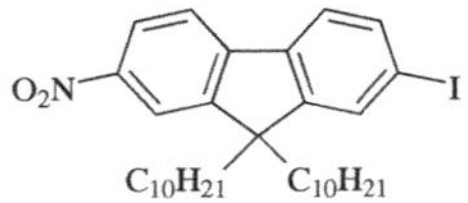

Cu-bronze, 18-c-6, K₂CO₃, 1,2-Cl₂C₆H₄, 180°, 43 h

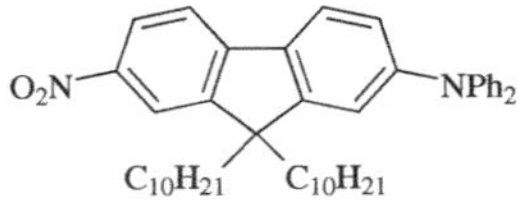

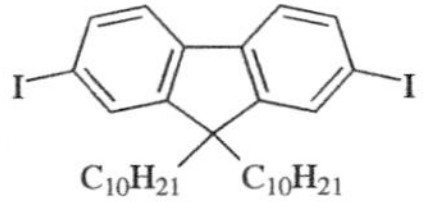

Cu-bronze, 18-c-6, K₂CO₃, 1,2-Cl₂C₆H₄, 180°, 43 h

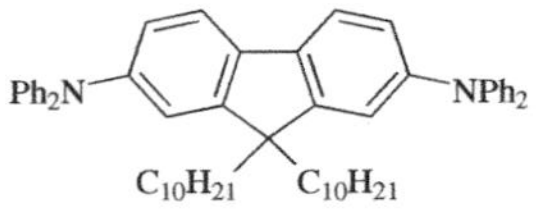

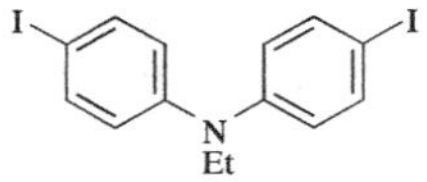

Cu (12 eq), 18-c-6, K₂CO₃, 1,2-Cl₂C₆H₄, reflux, 24 h

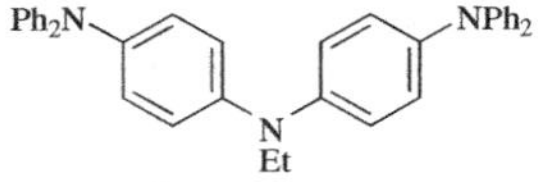

C_{12-14}

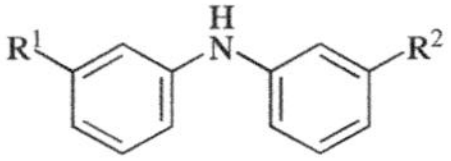

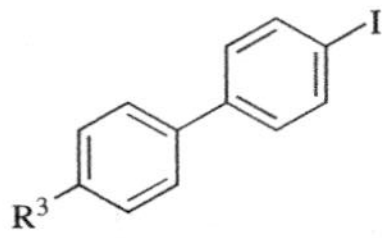

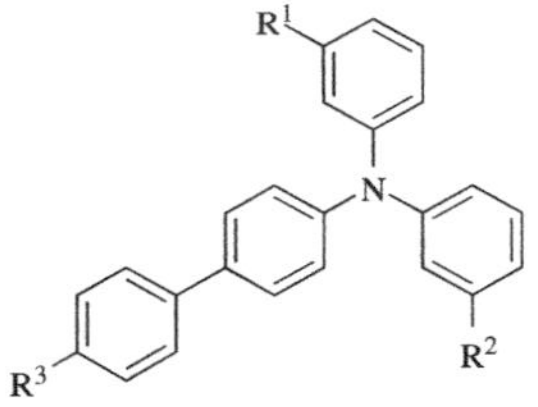

R^1	R^2	R^3	Catalyst	x	Additive(s)	Solvent	Temp (°)	Time (h)	
H	H	I	Cu	1 eq	18-c-6, K₂CO₃	1,2-Cl₂C₆H₄	200	48	(54)
H	Me	I	Cu	1 eq	18-c-6, K₂CO₃	1,2-Cl₂C₆H₄	200	48	(46)
H	Me	Br	CuCl₂	4 mol %	phen (4 mol %), KOH	—	125	5	(80)
Me	Me	I	Cu	—	18-c-6	xylene	160	40	(33)

Catalyst (x amount)

TABLE 6A. N-ARYLATION OF SECONDARY ARYL, ALKYL, AND DIARYL AMINES (Continued)

Nitrogen Nucleophile	Aryl Halide	Conditions	Product(s) and Yield(s) (%)	Refs.

*Please refer to the charts preceding the tables for structures indicated by the **bold** numbers.*

C_{12-14}

Ar^1Ar^2NH

(aryl halide with R^1, R^1, R^2, R^2 and two I)

Catalyst (x amount)

product with R^1, R^1, R^2, R^2, NAr^1Ar^2, Ar^1Ar^2N

Ar^1	Ar	R^1	R^2	Catalyst	x	Additive(s)	Solvent	Temp (°)	Time (h)		Refs.
Ph	Ph	H	H	CuI	5 mol %	**L41** (10 mol %), NaOt-Bu	toluene	110	40	(82)	338
Ph	Ph	H	H	CuI	5 mol %	mes₂DAB (6 mol %), KOt-Bu	toluene	120	20	(85)	306
Ph	Ph	H	H	Cu(THMD)₂	20 mol %	KOt-Bu	toluene	120	12	(81)	308
Ph	Ph	H	Me	Cu	1 eq	18-c-6, K₂CO₃	1,2-Cl₂C₆H₄	reflux	15	(35)	374
Ph	Ph	Me	Me	Cu	1 eq	18-c-6, K₂CO₃	1,2-Cl₂C₆H₄	reflux	15	(35)	374
Ph	2-MeOC₆H₄	H	H	CuCl₂	4 mol %	phen (4 mol %), KOH	—	130	5	(75)	311
Ph	2-MeC₆H₄	H	H	CuCl₂	4 mol %	phen (4 mol %), KOH	—	125	5	(85)	311
4-MeC₆H₄	4-MeC₆H₄	H	H	Cu	1 eq	18-c-6, K₂CO₃	1,2-Cl₂C₆H₄	reflux	15	(47)	374
4-MeC₆H₄	4-MeC₆H₄	H	H	CuI	5 mol %	mes₂DAB (6 mol %), KOt-Bu	toluene	120	20	(38)	306
4-MeC₆H₄	4-MeC₆H₄	H	Me	Cu	1 eq	18-c-6, K₂CO₃	1,2-Cl₂C₆H₄	reflux	15	(82)	374
4-MeC₆H₄	4-MeC₆H₄	Me	Me	Cu	1 eq	18-c-6, K₂CO₃	1,2-Cl₂C₆H₄	reflux	15	(35)	374

C_{12-13}

PhHN with R

aryl iodide; Ph–N(Ar)

Cu (1 eq), 18-c-6, K₂CO₃,
1,2-Cl₂C₆H₄, 200°, 48 h

R	Ar	
H	1-Np	(43)
Me	1-Np	(43)
H	2-Np	(38)
Me	2-Np	(44)

369

238

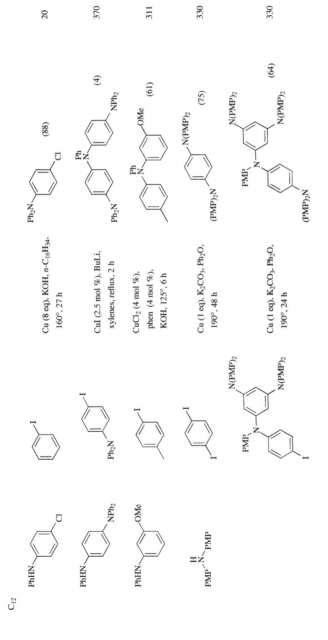

TABLE 6A. N-ARYLATION OF SECONDARY ARYL, ALKYL, AND DIARYL AMINES (Continued)

Nitrogen Nucleophile	Aryl Halide	Conditions	Product(s) and Yield(s) (%)	Refs.

*Please refer to the charts preceding the tables for structures indicated by the **bold** numbers.*

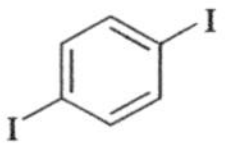

C₁₂

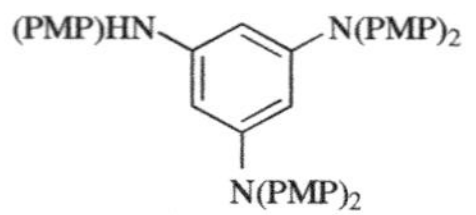

Cu (1 eq), K₂CO₃, Ph₂O, 190°, 48 h

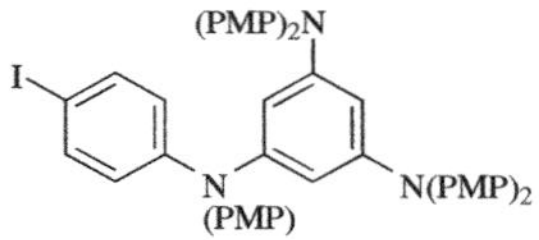

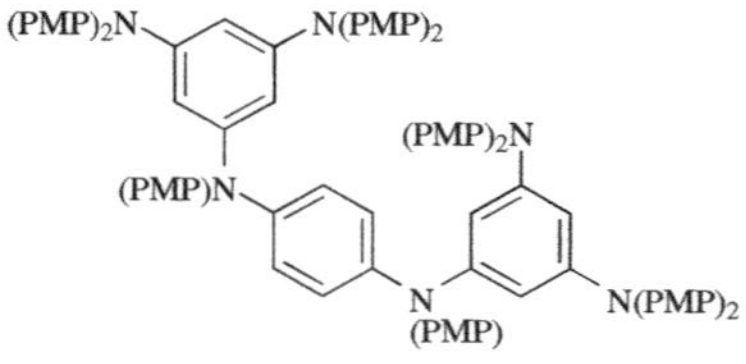

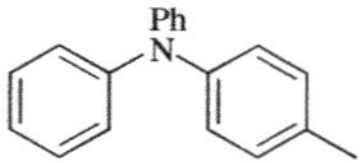

I/II	Time (h)	III	IV
1:4.5	24	(78)	(0)
2:1	48	(0)	(71)

330

C₁₃

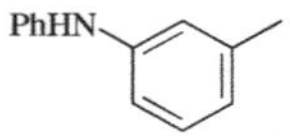

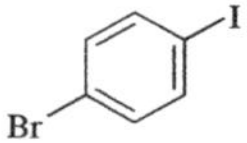

Cu (1 eq), 18-c-6, K₂CO₃, o-Cl₂C₆H₄, 200°, 48 h

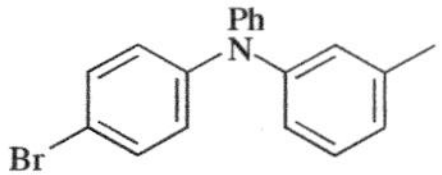

(30)

369

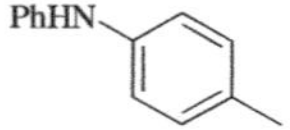

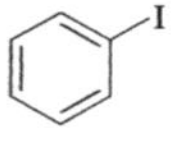

Catalyst (x amount)

Catalyst	x	Additive(s)	Solvent	Temp (°)	Time (h)		Refs.
Cu	—	K₂CO₃, 18-c-6	1,2-Cl₂C₆H₄	—	—	(29)	375
Cu	8 eq	KOH	n-C₁₆H₃₄	160	24	(88)	306
[Cu(phen)₂]Cl	8 mol %	K₂CO₃, 18-c-6	1,2-Cl₂C₆H₄	reflux	15	(75)	376

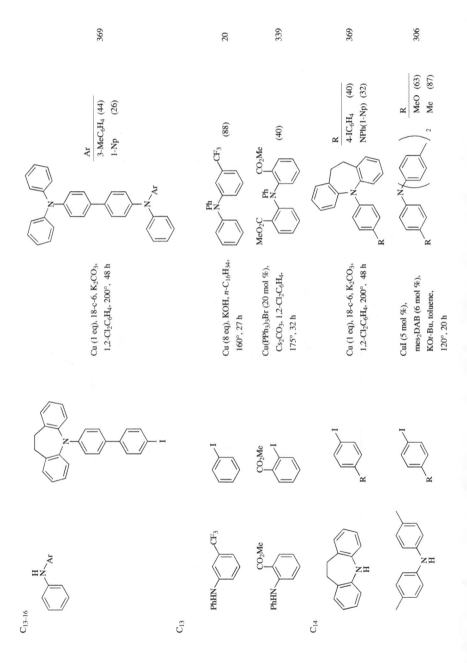

TABLE 6A. *N*-ARYLATION OF SECONDARY ARYL, ALKYL, AND DIARYL AMINES (*Continued*)

Nitrogen Nucleophile	Aryl Halide	Conditions	Product(s) and Yield(s) (%)	Refs.

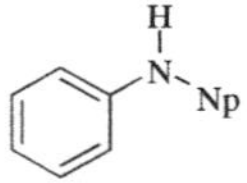

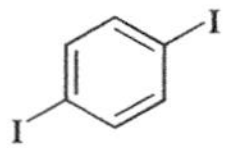

*Please refer to the charts preceding the tables for structures indicated by the **bold** numbers.*

C14

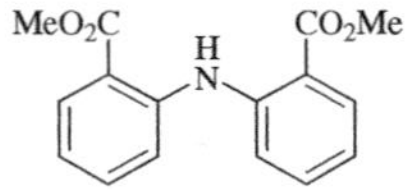

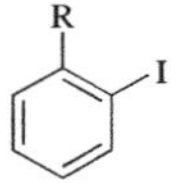

Cu (15 mol %), K2CO3, 180°, 33 h

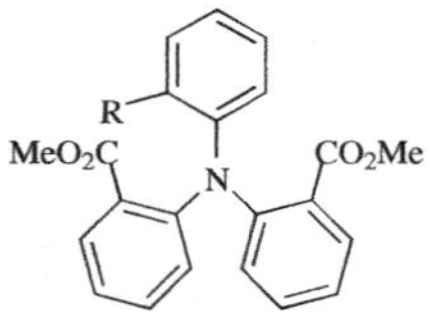

R	
H	(46)
CO2Me	(26)

348

C15

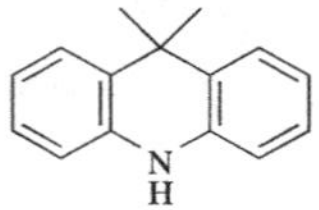

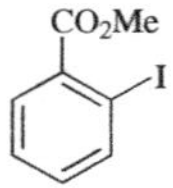

Cu (25 mol %), CaCO3

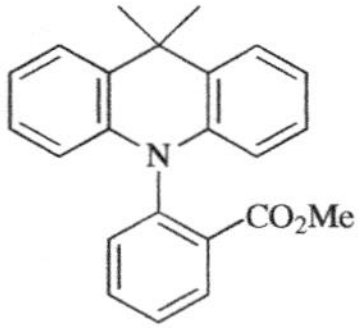

(52)

355

C16

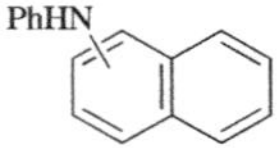

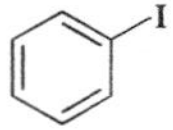

CuI (*x* mol %)

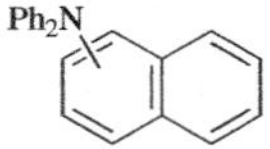

Isomer	*x*	Additive(s)	Solvent	Temp (°)	Time (h)	
1	4	NaH	*n*-C16H34	160	24	(23)
1	5	mes2DAB (6 mol %), KO*t*-Bu	toluene	120	20	(38)
2	5	mes2DAB (6 mol %), KO*t*-Bu	toluene	120	20	(88)

20
306
306

Cu (1 eq), 18-c-6, K2CO3, 1,2-Cl2C6H4, 200°, 48 h

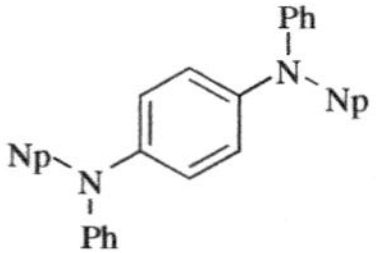

Isomer	
1-Np	(26)
2-Np	(52)

369

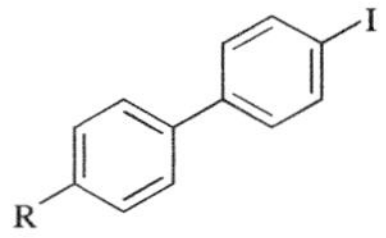

Catalyst (x eq), K$_2$CO$_3$

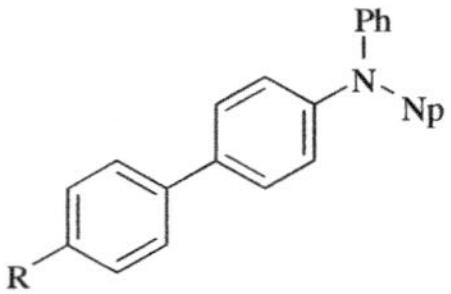

Isomer	R	Catalyst	x	Additive	Solvent	Temp (°)	Time (h)		
1-Np	I	Cu	1	18-c-6	1,2-Cl$_2$C$_6$H$_4$	200	48	(42)	369
1-Np	I	CuSO$_4$	—	none	—	—	—	(—)	377
2-Np	I	Cu	1	18-c-6	1,2-Cl$_2$C$_6$H$_4$	200	48	(50)	369
2-Np	NPh$_2$	Cu	1	18-c-6	1,2-Cl$_2$C$_6$H$_4$	200	48	(31)	369

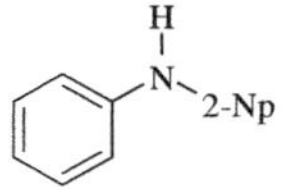

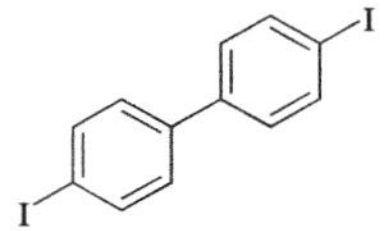

CuI (5 mol %), mes$_2$DAB (6 mol %), KOt-Bu, toluene, 120°, 20 h

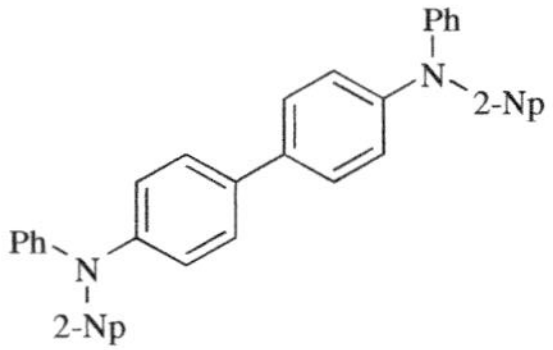

(87)

306

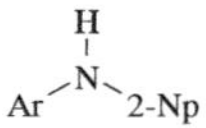

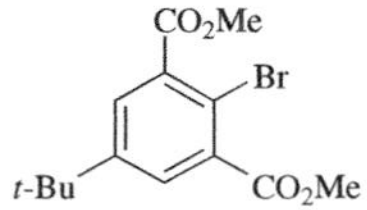

Cu (15 mol %), K$_2$CO$_3$, Ph$_2$O, 190°, 48 h

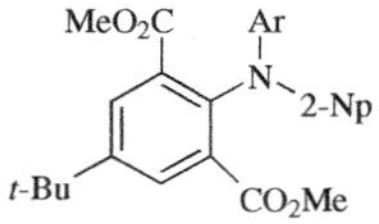

Ar	
Ph	(83)
2-Np	(83)

341

C$_{16-20}$

TABLE 6A. N-ARYLATION OF SECONDARY ARYL, ALKYL, AND DIARYL AMINES (Continued)

Nitrogen Nucleophile	Aryl Halide	Conditions	Product(s) and Yield(s) (%)	Refs.

*Please refer to the charts preceding the tables for structures indicated by the **bold** numbers.*

C_{18-20}

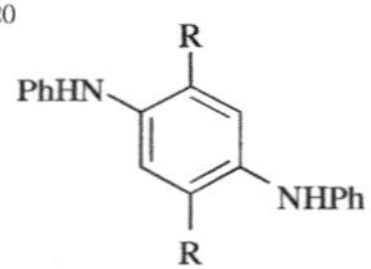

Nitrogen Nucleophile:

Aryl Halide: ArI

Conditions — Catalyst (x mol %)

Product:

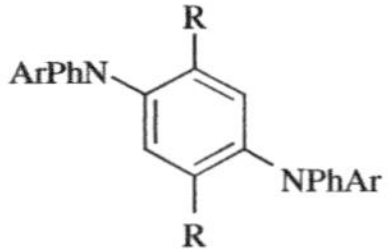

R	Ar	Catalyst	x	Additive	Solvent	Temp (°)	Time (h)		Refs.
H	Ph	CuI	2.5	BuLi	xylenes	reflux	2	(24)	370
H	4-$Ph_2NC_6H_4$	CuI	2.5	BuLi	xylenes	reflux	4	(4)	370
CO_2H	Ph	Cu	25	none	1-pentanol	195	3	(42)	378

C_{18}

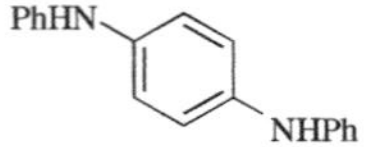

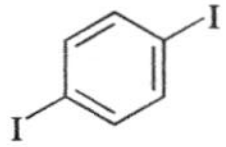

CuI (2.5 mol %), BuLi,
dihexyl ether, reflux, 6 h

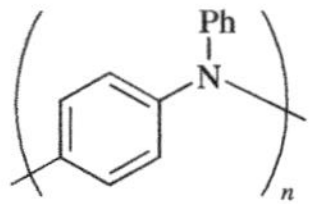

(48)

370

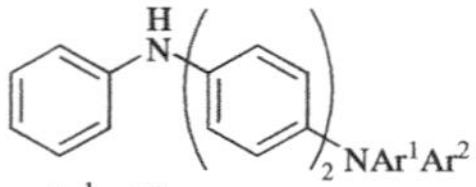

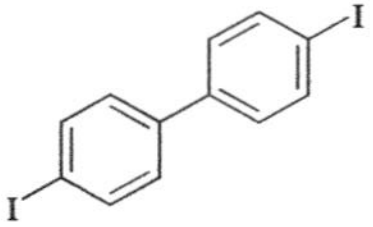

Cu, 18-c-6, xylene, 160°, 40 h

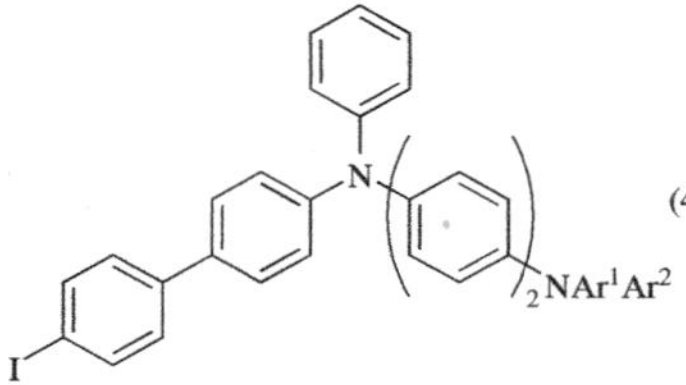

(47)

$Ar^1 = Ph$
$Ar^2 = 3\text{-}MeC_6H_4$

342

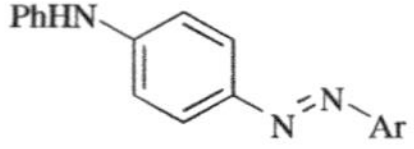

$Ar = 4\text{-}O_2NC_6H_4$

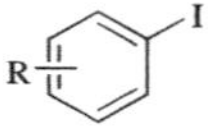

Cu, K_2CO_3, triglyme,
200–220°

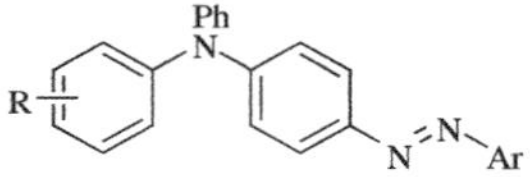

354

R		
H	(81)	
4-I	(74)	
4-H_2N	(61)	
4-i-PrCONH	(56)	
3-MeO	(55)	
4-MeO	(66)	
4-t-$BuPh_2SiO(CH_2)_2O$	(50)	

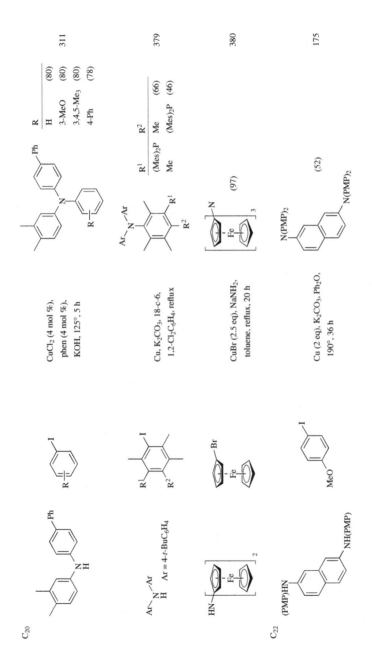

R	
H	(80)
3-MeO	(80)
3,4,5-Me₃	(80)
4-Ph	(78)

311

CuCl₂ (4 mol %),
phen (4 mol %),
KOH, 125°, 5 h

R¹	R²	
(Mes)₂P	Me	(66)
Me	(Mes)₂P	(46)

379

Cu, K₂CO₃, 18-c-6,
1,2-Cl₂-C₆H₄, reflux

(97)

380

CuBr (2.5 eq), NaNH₂,
toluene, reflux, 20 h

(52)

175

Cu (2 eq), K₂CO₃, Ph₂O,
190°, 36 h

C₂₀

Ar = 4-t-BuC₆H₄

C₂₂

TABLE 6A. *N*-ARYLATION OF SECONDARY ARYL, ALKYL, AND DIARYL AMINES (*Continued*)

Nitrogen Nucleophile	Aryl Halide	Conditions	Product(s) and Yield(s) (%)	Refs.
			*Please refer to the charts preceding the tables for structures indicated by the **bold** numbers.*	

C$_{22}$

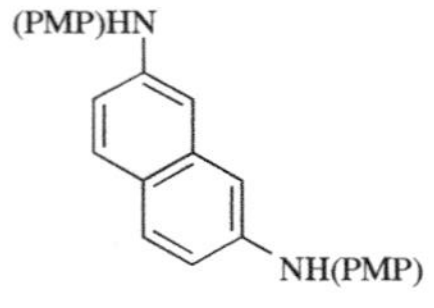

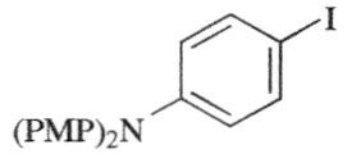

Cu (2 eq), K$_2$CO$_3$, Ph$_2$O, 190°, 36 h — (56) — 175

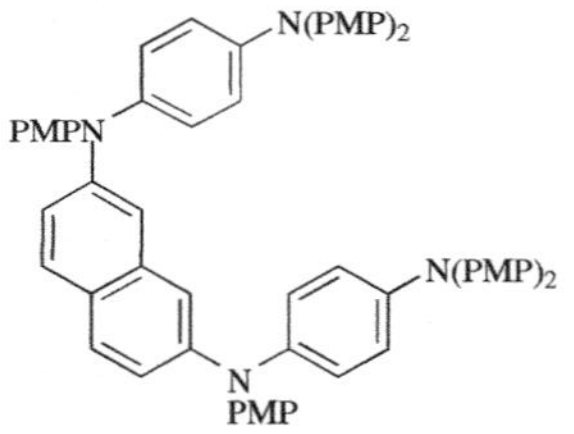

C$_{24}$

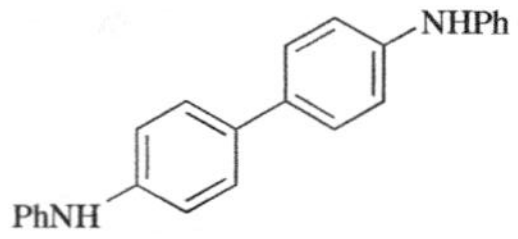

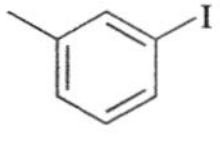

Cu, 18-c-6, xylene, 160°, 40 h — (16) — 342

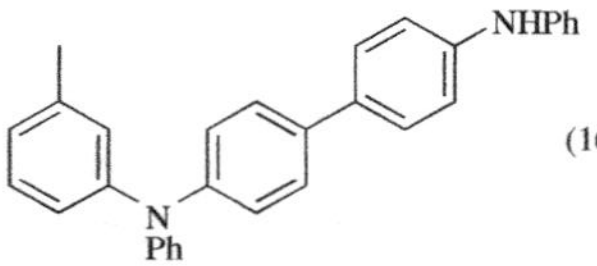

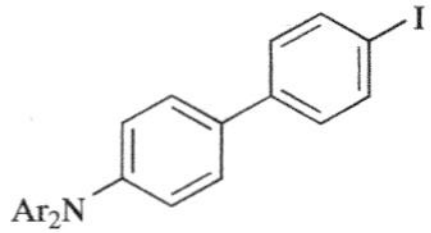

Cu, 18-c-6, xylene, 160°, 40 h — (75) — 342

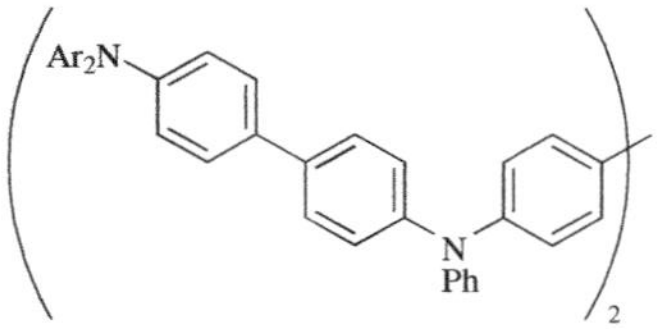

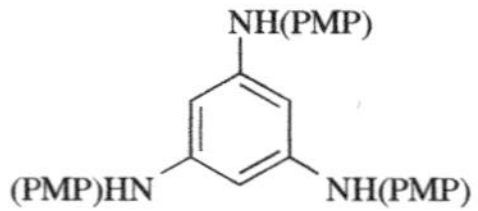

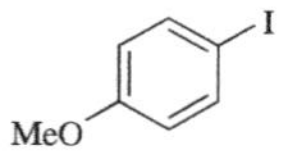

Cu (2 eq), K$_2$CO$_3$, Ph$_2$O, 190°, 36 h — (28) — 175

Ar = 3-MeC$_6$H$_4$

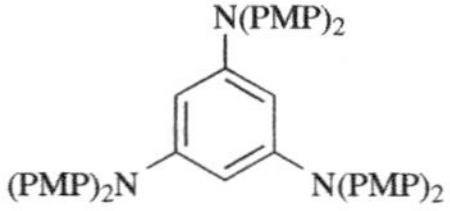

C$_{24}$

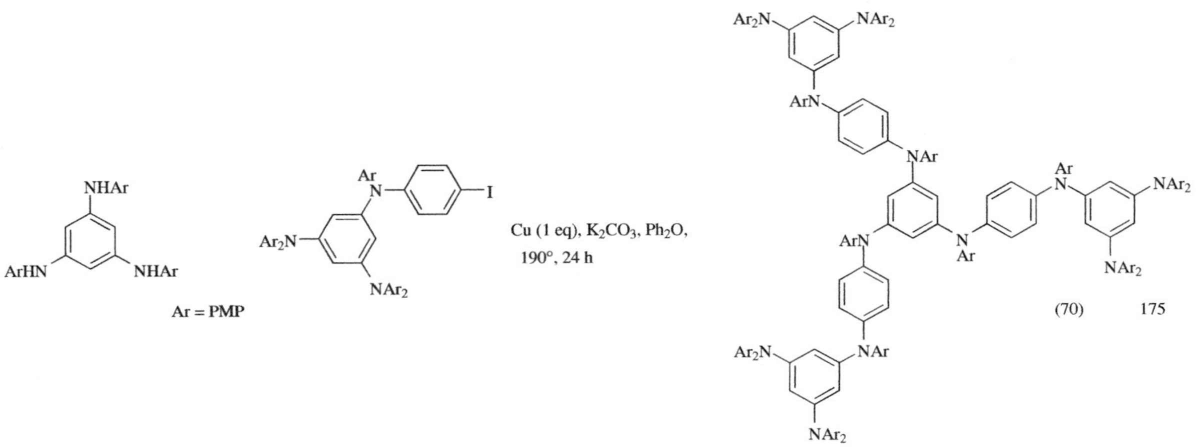

Cu (1 eq), K$_2$CO$_3$, Ph$_2$O, 190°, 24 h

Ar = PMP

(70)   175

Nitrogen Nucleophile	Aryl Halide	Conditions	Product(s) and Yield(s) (%)	Refs.
C$_{10}$				
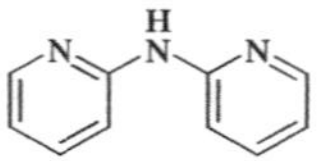	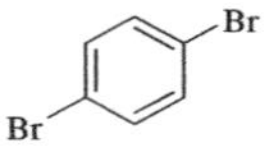 Br	CuI (10 mol %), K$_2$CO$_3$, DMF, piperidine-2-carboxylic acid (20 mol %), 110°, 36 h	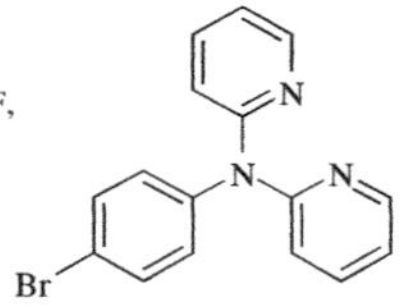 (57)	67
	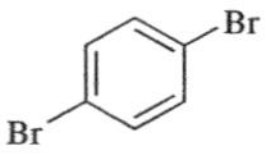 Br	CuSO$_4$ (2 mol %), KOH, 180°, 5 h	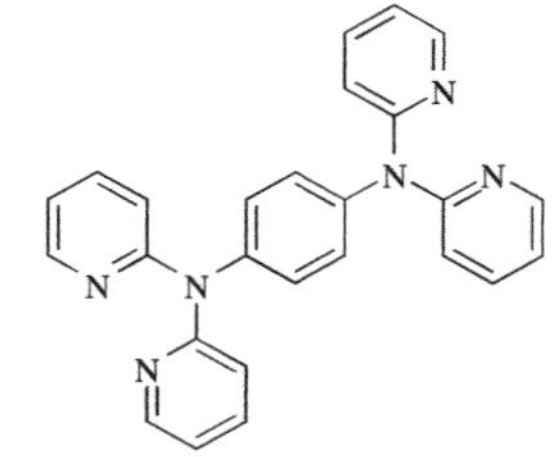 (29)	174
	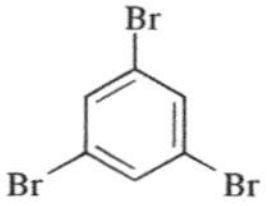 Br	CuSO$_4$ (5 mol %), K$_2$CO$_3$, 210°, 8 h	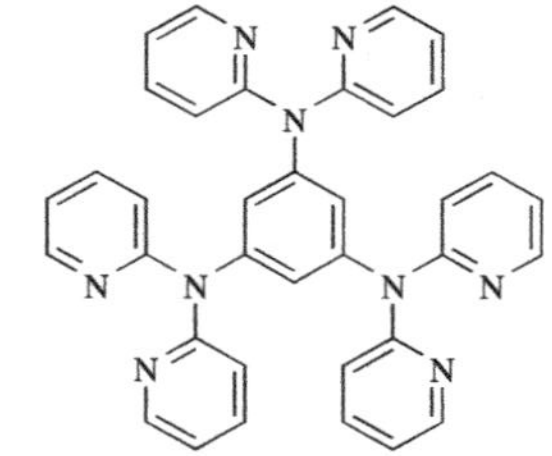 (45)	381

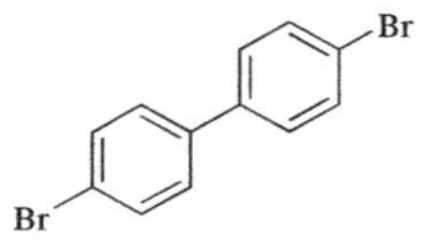

(33) 174

CuSO₄ (2 mol %), KOH, 180°, 5 h

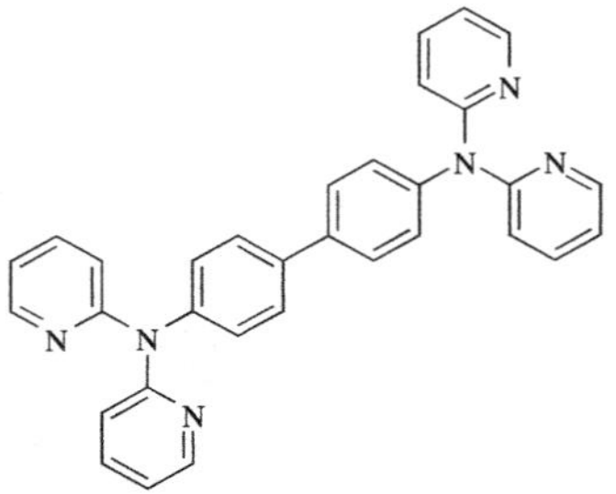

Y	
N	(60)	381
CH	(85)	382, 381

Ar = 2-pyridyl

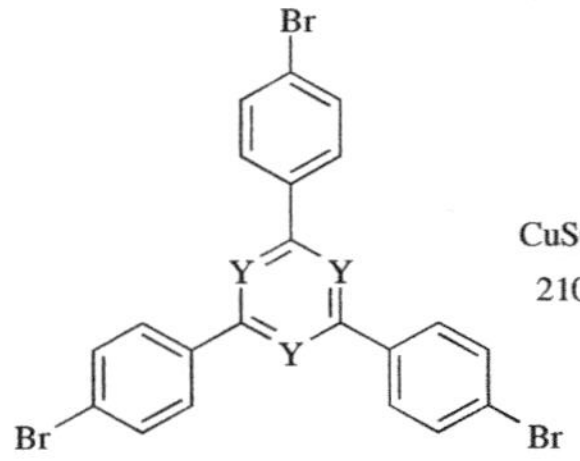

CuSO₄ (10 mol %), K₂CO₃, 210°, 8 h

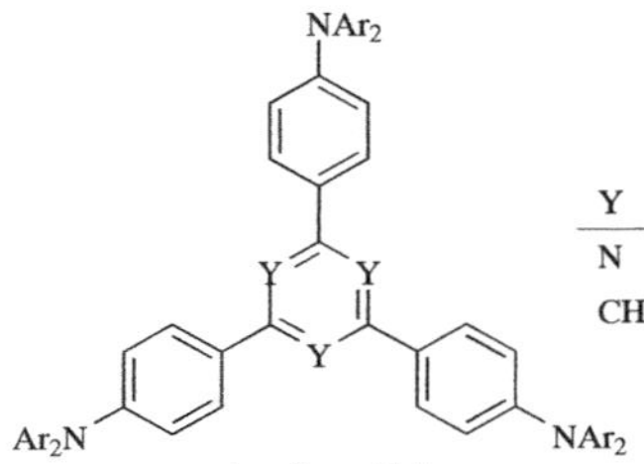

TABLE 6C. *N*-HETEROARYLATION OF SECONDARY ARYL AND HETEROARYL AMINES

	Nitrogen Nucleophile	Heteroaryl Halide	Conditions	Product(s) and Yield(s) (%)	Refs.
C_{10}			$CuSO_4$ (2 mol %), KOH, 180°, 6 h	(48)	383
			$CuSO_4$ (2 mol %), KOH, 180°, 6 h	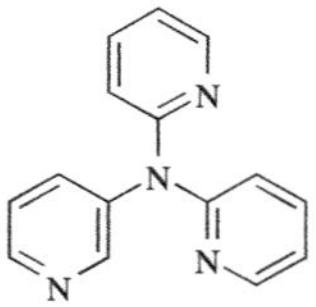 (48)	383
C_{12}	Ph_2NH		CuI (1 eq). NaH, DMPU	(40)	384
			See table.	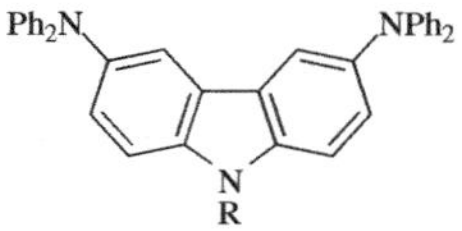	385

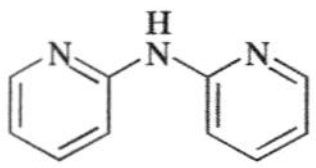

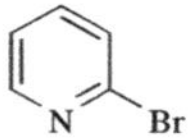

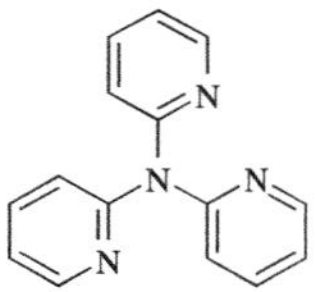

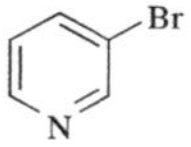

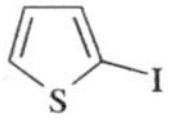

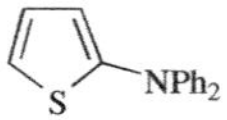

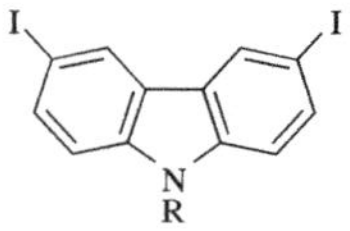

R	Catalyst	Additives	Solvent	Temp (°)	Time (h)	
Et	Cu (12 eq)	18-c-6, K_2CO_3	$1,2\text{-}Cl_2C_6H_4$	reflux	36	(74)	373
$4\text{-}MeC_6H_4$	CuI (10 mol %)	18-c-6 (1 mol %), K_2CO_3	DMPU	170	20	(70)	385

250

TABLE 7A. *N*-ARYLATION OF PYRROLES

Please refer to the charts preceding the tables for structures indicated by the **bold** numbers.

C₄

Nitrogen Nucleophile	Aryl Halide		Conditions		Product(s) and Yield(s) (%)			Refs.
			Catalyst (x amount)					
X	Catalyst	x	Additive(s)	Solvent	Temp (°)	Time (h)		
Br	Cu	20 mol %	Cs₂CO₃	*n*-PrCN	reflux	20	(81)	129
I	Cu	20 mol %	Cs₂CO₃	*n*-PrCN	reflux	7	(95)	129
I	CuCl	10 mol %	**L9** (10 mol %), TMAH	DMSO	80	24	(90)	281
I	CuCl	5 mol %	TBAH (40% aq)	H₂O	80	24	(86)	282
Br	CuBr	10 mol %	**L20** (20 mol %), Cs₂CO₃	DMSO	80	12	(77)	284
I	CuBr	20 mol %	binol (20 mol %), K₃PO₄	DMF	rt	8	(52)	62
Br	CuI	20 mol %	Cs₂CO₃	DMF	120	40	(66)	386
I	CuI	10 mol %	K₃PO₄	DMF	110	24	(88)	387
Br	CuI	20 mol %	L-Pro (20 mol %), K₃PO₄	DMF	140	48	(67)	388
I	CuI	5 mol %	L-Pro (20 mol %), K₃PO₄	dioxane	100	24	(95)	388
Br	CuI	10 mol %	L-His (20 mol %), K₂CO₃	DMSO	100	48	(56)	389
Br	CuI	10 mol %	phen (20 mol %), KF/Al₂O₃	toluene	reflux	9	(81)	390
I	CuI	10 mol %	phen (20 mol %), KF/Al₂O₃	toluene	reflux	8	(85)	390
I	CuI	5 mol %	TBAB (5 mol %), NaOH	toluene	reflux	22	(95)	391
Br	CuI	10 mol %	piperidine-2-carboxylic acid (20 mol %), K₂CO₃	DMF	110	24	(76)	67
I	CuI	5 mol %	oxazolidin-2-one (10 mol %), NaOMe	DMSO	80	10	(93)	392
Br	CuI	20 mol %	per-6-ABCD (10 mol %), K₂CO₃	DMSO	110	24	(92)	393
Br	CuI	10 mol %	**L21** (20 mol %), KI, Cs₂CO₃	MeCN	reflux	12	(45)	394
I	CuI	10 mol %	**L21** (20 mol %), Cs₂CO₃	MeCN	reflux	12	(92)	394
Br	CuI	10 mol %	**L22** (10 mol %), K₂CO₃	DMF	110	12	(71)	79
Br	CuI	5 mol %	**L24** (10 mol %), Cs₂CO₃	DMF	110	24	(72)	86
I	CuI	5 mol %	**L24** (10 mol %), Cs₂CO₃	DMF	90	24	(98)	86
Br	CuI	10 mol %	**L25** (20 mol %), Cs₂CO₃	MeCN	80	15	(80)	395
I	CuI	5 mol %	*N*-hydroxymaleimide (10 mol %), NaOMe	DMSO	90	12	(96)	396

TABLE 7A. N-ARYLATION OF PYRROLES (Continued)

*Please refer to the charts preceding the tables for structures indicated by the **bold** numbers.*
Continued from previous page.

C₄

Nitrogen Nucleophile	Aryl Halide	Conditions					Product(s) and Yield(s) (%)	Refs.
	X	Catalyst	x	Additive(s)	Solvent	Temp (°)	Time (h)	
I	CuI	10 mol %	**L26** (10 mol %), NaOMe	DMSO	120	12	(69)	397
I	CuI/PAnNF	5 mol %	K₂CO₃	DMF	rt	8	(80)	276
Br	CuO	10 mol %	FeCl₃ (10 mol %), rac-binol (20 mol %)	DMF	110	24	(75)	277
I	CuO	10 mol %	FeCl₃ (10 mol %), rac-binol (20 mol %)	DMF	90	12	(78)	277
I	CuO	10 mol %	Fe(acac)₃ (30 mol %), Cs₂CO₃	DMF	90	30	(90)	130
I	CuO	2.5 mol %	KOH	DMSO	110	24	(91)	224
I	CuO	1.26 mol %	KOH	DMSO	110	2.5	(93)	224
I	CuO	0.1 mol %	DMEDA (20 mol %), K₃PO₄•H₂O	toluene	135	24	(11)	222
Cl	CuO on acetylene black	5 mol %	KOt-Bu	toluene	180	18	(74)	226
Br	CuO on acetylene black	5 mol %	KOt-Bu	toluene	180	18	(80)	226
I	CuO on acetylene black	5 mol %	KOt-Bu	toluene	180	18	(96)	226
I	Cu₂O	10 mol %	Cs₂CO₃	DMF	100	18	(93)	221
I	Cu₂O	10 mol %	KOH	DMSO	120	24	(91)	398
I	Cu₂O	10 mol %	ninhydrin (20 mol %), KOH	DMSO	110	48	(91)	399
I	Cu₂O	5 mol %	**L27** (20 mol %)	MeCN	82	24	(94)	40
Br	Cu₂O	10 mol %	**L23** (20 mol %), Cs₂CO₃	DMF	110	24	(82)	400
I	Cu₂O	5 mol %	**L23** (10 mol %), KOt-Bu	DMF	80	24	(95)	400
Br	Cu(OAc)₂	1 eq	DBU	DMSO	MW, 130	10 min	(90)	85
I	Cu(OAc)₂	1 eq	DBU	DMSO	MW, 130	10 min	(91)	85
I	Cu(OAc)₂•H₂O	20 mol %	hippuric acid (20 mol %), Cs₂CO₃	DMF	140	30	(68)	289
I	Cu(OAc)₂•H₂O	15 mol %	(–)-sparteine (30 mol %), K₂CO₃	DMF	130	24	(78)	290
Br	Cu(TMHD)₂	20 mol %	KOt-Bu	DMF	120	24	(88)	308
I	Cu(TMHD)₂	20 mol %	KOt-Bu	DMF	120	24	(94)	308
Br	**8**	25 mol %	K₂CO₃	DMSO	135	35	(82)	65
I	**6**	5 mol %	Cs₂CO₃	toluene	100	10	(94)	123
Br	Cu fluorapatite	12.5 mol %	none	DMSO	110	10	(90)	227
I	Cu fluorapatite	12.5 mol %	none	DMSO	110	6	(92)	227

R	X	Catalyst	x	Additive(s)	Solvent	Temp (°)	Time (h)		
2-Br	I	Cu	20	Cs_2CO_3	n-PrCN	reflux	24	(83)	129
2-Br	I	CuI	10	benzotriazole (20 mol %), K_3PO_4	DMSO	100	30	(61)	77
2-I	I	CuI	10	benzotriazole (20 mol %), K_3PO_4	DMSO	100	30	(52)	77
3-Br	I	CuCl	5	L17 (10 mol %), NaOH, TBAB	H_2O	100	24	(75)	235
4-F	I	CuI	5	TBAB (5 mol %), NaOH	toluene	reflux	22	(77)	391
4-Cl	Br	Cu	20	Cs_2CO_3	n-PrCN	reflux	24	(77)	129
4-Cl	I	CuCl	5	TBAB (40% aq)	H_2O	80	24	(72)	282
4-Cl	I	CuCl	5	L9 (10 mol %), TMAH	DMSO	80	24	(83)	281
4-Cl	I	CuBr	20	binol (20 mol %), K_3PO_4	DMF	rt	8	(51)	62
4-Cl	Br	CuCl	5	TBAB (40% aq)	H_2O	80	24	(57)	282
4-Cl	I	CuI	5	TBAB (5 mol %), NaOH	toluene	reflux	22	(80)	391
4-Cl	I	CuI	5	N-hydroxyphthalimide (10 mol %), NaOMe	DMSO	90	12	(95)	396
4-Cl	Br	CuI	10	oxazolidin-2-one (20 mol %), NaOMe	DMSO	80	10	(81)	392
4-Cl	I	CuI	5	oxazolidin-2-one (20 mol %), NaOMe	DMSO	80	8	(86)	392
4-Cl	Br	CuI	10	L21 (20 mol %), KI, Cs_2CO_3	MeCN	reflux	12	(45)	394
4-Cl	I	CuI	10	L21 (20 mol %), Cs_2CO_3	MeCN	reflux	12	(92)	394
4-Cl	I	CuI	10	L8 (20 mol %), K_3PO_4	DMF	30	14	(67)	271
4-Cl	I	CuI	10	L26 (10 mol %), NaOMe	DMSO	130	12	(83)	397
4-Br	I	CuI	5	TBAB (5 mol %), NaOH	toluene	reflux	22	(90)	391
4-Br	I	CuI	10	L21 (20 mol %), KI, Cs_2CO_3	MeCN	reflux	12	(81)	394
4-Br	I	CuI	10	L26 (10 mol %), NaOMe	DMSO	130	12	(68)	397
4-Br	I	CuI	10	piperidine-2-carboxylic acid (20 mol %), K_2CO_3	DMF	110	24	(84)	67
4-Br	I	CuI	10	L-Pro (20 mol %), K_2CO_3	DMSO	90	36	(80)	47, 336
4-Br	I	CuI	5	L-Pro (10 mol %), K_2CO_3	DMSO	90	36	(80)	401
4-Br	I	CuO	10	$FeCl_3$ (10 mol %), rac-binol (20 mol %), $CsCO_3$	DMF	90	12	(89)	227

Catalyst (x mol %)

TABLE 7A. N-ARYLATION OF PYRROLES (Continued)

Nitrogen Nucleophile	Aryl Halide	Conditions	Product(s) and Yield(s) (%)	Refs.

Please refer to the charts preceding the tables for structures indicated by the **bold** numbers.

C4

Nitrogen Nucleophile: pyrrole (N–H)

Aryl Halide: R_2N–[ ]–X   Product: R_2N–[ ]–N(pyrrole)

Isomer	R	X	Catalyst	x	Additive(s)	Solvent	Temp (°)	Time (h)		Refs.
2	H	Br	CuI	5	DMCDA (20 mol %), K$_3$PO$_4$	toluene	110	24	(83)	128
3	H	I	CuI	5	oxazolidin-2-one (10 mol %), NaOMe	DMSO	80	10	(78)	392
3	H	I	CuI	5	N-hydroxysuccinimide (10 mol %), NaOMe	DMSO	90	12	(92)	396
4	H	I	Cu	20	Cs$_2$CO$_3$	n-PrCN	reflux	9	(85)	129
4	H	I	CuI	5	oxazolidin-2-one (10 mol %), NaOMe	DMSO	80	14	(92)	392
4	H	I	CuI	5	N-hydroxysuccinimide (10 mol %), NaOMe	DMSO	90	12	(98)	396
4	Me	Br	CuI	5	DMCDA (20 mol %), K$_3$PO$_4$	toluene	110	24	(94)	128

Aryl Halide: O_2N–[ ]–X   Product: O_2N–[ ]–N(pyrrole)

X	Catalyst	x	Additive(s)	Solvent	Temp (°)	Time (h)		Refs.
I	Cu	20	Cs$_2$CO$_3$	n-PrCN	reflux	7	(92)	129
I	CuCl	10	L9 (10 mol %), TMAH	DMSO	80	24	(74)	281
Br	CuCl	5	TBAH (40% aq)	H$_2$O	80	24	(43)	282
I	CuCl	5	TBAH (40% aq)	H$_2$O	80	24	(50)	282
Cl	CuI	5	N-hydroxymaleimide (10 mol %), NaOMe	DMSO	110	40	(45)	396
Cl	CuI	10	L26 (10 mol %), NaOMe	DMSO	110	12	(93)	397
Cl	CuI	10	oxazolidin-2-one (20 mol %), NaOMe	DMSO	120	24	(32)	392
Cl	CuO	8	K$_2$CO$_3$	pyridine	reflux	24	(1)	402
F	Cu fluorapatite	7	K$_2$CO$_3$	DMF	120	2	(85)	66
Cl	Cu fluorapatite	7	K$_2$CO$_3$	DMF	120	4	(80)	66
Br	Cu fluorapatite	12.5	none	DMSO	110	4	(95)	227
I	Cu fluorapatite	12.5	none	DMSO	110	3	(95)	227
Cl	7	1	K$_2$CO$_3$	DMF	110	5	(80)	73
I	7	1	K$_2$CO$_3$	DMF	110	1.5	(92)	73

Br, HO-phenyl-Br + CuO on acetylene black (5 mol %), KOt-Bu, toluene, 180°, 18 h → N-(4-hydroxyphenyl)pyrrole (HO) (70)

MeO, X-phenyl + Catalyst (x mol %) → N-(4-methoxyphenyl)pyrrole (MeO)

X	Catalyst	x	Additive(s)	Solvent	Temp (°)	Time (h)		
Br	Cu	20	Cs$_2$CO$_3$	n-PrCN	reflux	20	(79)	129
I	Cu	20	Cs$_2$CO$_3$	n-PrCN	reflux	20	(93)	129
I	CuI	5	TBAB (5 mol %), NaOH	toluene	reflux	22	(84)	391
Br	CuI	10	phen (20 mol %), KF/Al$_2$O$_3$	toluene	reflux	10	(80)	390
I	CuI	10	phen (20 mol %), KF/Al$_2$O$_3$	toluene	reflux	9	(85)	390
Br	CuI	10	piperidine-2-carboxylic acid (20 mol %), K$_2$CO$_3$	DMF	110	24	(71)	67
Br	CuI	10	DMG (20 mol %), Cs$_2$CO$_3$	DMSO	110	40	(46)	47
Br	CuI	20	L-Pro (20 mol %), K$_3$PO$_4$	DMF	140	48	(60)	388
I	CuI	5	L-Pro (20 mol %), K$_3$PO$_4$	dioxane	100	24	(98)	388
I	CuI	10	L-Pro (20 mol %), K$_2$CO$_3$	DMSO	90	36	(84)	47
I	CuI	5	L-Pro (20 mol %), K$_2$CO$_3$	DMSO	90	36	(84)	401
I	CuI	5	N-hydroxymaleimide (10 mol %), NaOMe	DMSO	90	12	(93)	396
Br	CuI	5	L34 (10 mol %), KOt-Bu	DMSO	110	—	(91)	403
Cl	CuI nanoparticles	1.25	K$_2$CO$_3$, air	DMF	110	5	(90)	285
I	CuI/PAnNF	5	K$_2$CO$_3$	DMF	rt	12	(55)	276
Br	CuO on acetylene black	5	KOt-Bu	toluene	180	18	(34)	226
I	8	25	K$_2$CO$_3$	DMSO	135	24	(82)	65
Br	Cu fluorapatite	12.5	none	DMSO	110	15	(85)	227
I	Cu fluorapatite	12.5	none	DMSO	110	14	(91)	227

TABLE 7A. N-ARYLATION OF PYRROLES (Continued)

Nitrogen Nucleophile	Aryl Halide	Conditions	Product(s) and Yield(s) (%)	Refs.

*Please refer to the charts preceding the tables for structures indicated by the **bold** numbers.*

C4

Aryl Halide: Me—C6H4—X
Product: Me—C6H4—N(pyrrole)

Isomer	X	Catalyst	x	Additive(s)	Solvent	Temp (°)	Time (h)		Refs.
2	Br	CuI	5	DMCDA (20 mol %), K$_3$PO$_4$	toluene	110	24	(87)	128
2	I	CuI	5	TBAB (5 mol %), NaOH	toluene	reflux	22	(62)	391
2	I	CuI	5	L-Pro (20 mol %), K$_3$PO$_4$	dioxane	110	24	(90)	388
3	I	CuI	5	TBAB (5 mol %), NaOH	toluene	reflux	22	(83)	391
4	Br	CuCl	5	TBAH (40% aq)	H$_2$O	80	24	(65)	282
4	I	CuCl	5	TBAH (40% aq)	H$_2$O	80	24	(63)	282
4	I	CuCl	10	L9 (10 mol %), TMAH	DMSO	80	24	(86)	281
4	I	CuI	5	TBAB (5 mol %), NaOH	toluene	reflux	22	(88)	391
4	I	CuI	10	phen (20 mol %), KF/Al$_2$O$_3$	toluene	reflux	8	(84)	390
4	Br	CuI	20	L-Pro (20 mol %), K$_3$PO$_4$	DMF	140	48	(57)	388
4	I	CuI	5	L-Pro (20 mol %), K$_3$PO$_4$	dioxane	100	24	(98)	388
4	I	CuI	5	L-Pro (20 mol %), K$_2$CO$_3$	DMSO	90	42	(82)	401
4	I	CuI	10	L8 (20 mol %), K$_3$PO$_4$	DMF	30	14	(64)	271
4	Br	CuI	10	L21 (20 mol %), KI, Cs$_2$CO$_3$	MeCN	reflux	12	(39)	394
4	I	CuI/PAnNF	5	K$_2$CO$_3$	DMF	rt	12	(63)	276
4	Cl	CuO on acetylene black	5	KOt-Bu	toluene	180	18	(40)	226
4	Br	Cu fluorapatite	12.5	none	DMSO	110	15	(88)	227
4	I	Cu fluorapatite	12.5	none	DMSO	110	14	(83)	227
4	Br	Cu(II)-NaY zeolite	10	K$_2$CO$_3$	DMF	120	36	(99)	404
4	I	Cu(II)-NaY zeolite	10	K$_2$CO$_3$	DMF	120	20	(99)	404

Catalyst (x mol %)

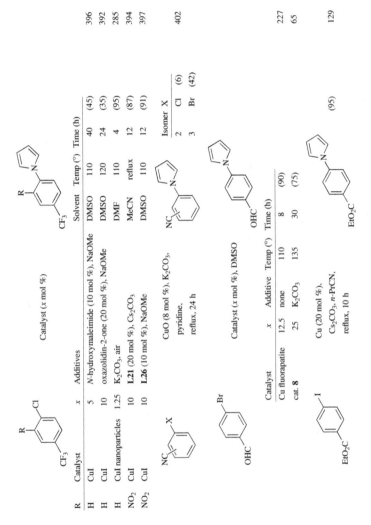

R	Catalyst	x	Additives	Solvent	Temp (°)	Time (h)	
H	CuI	5	N-hydroxymaleimide (10 mol %), NaOMe	DMSO	110	40	(45)	396
H	CuI	10	oxazolidin-2-one (20 mol %), NaOMe	DMSO	120	24	(35)	392
H	CuI nanoparticles	1.25	K_2CO_3, air	DMF	110	4	(95)	285
NO_2	CuI	10	L21 (20 mol %), Cs_2CO_3	MeCN	reflux	12	(87)	394
NO_2	CuI	10	L26 (10 mol %), NaOMe	DMSO	110	12	(91)	397

CuO (8 mol %), K_2CO_3, pyridine, reflux, 24 h

Isomer	X	
2	Cl	(6)
3	Br	(42)

402

Catalyst (x mol %), DMSO

Catalyst	x	Additive	Temp (°)	Time (h)	
Cu fluorapatite	12.5	none	110	8	(90)	227
cat. 8	25	K_2CO_3	135	30	(75)	65

Cu (20 mol %), Cs_2CO_3, n-PrCN, reflux, 10 h

(95)

129

TABLE 7A. N-ARYLATION OF PYRROLES (Continued)

Nitrogen Nucleophile	Aryl Halide	Conditions	Product(s) and Yield(s) (%)	Refs.

*Please refer to the charts preceding the tables for structures indicated by the **bold** numbers.*

C4

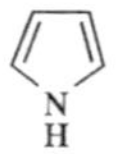

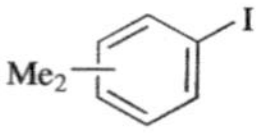

CuI (x mol %)

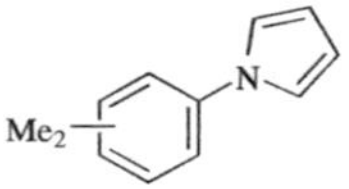

Isomer	x	Additives	Solvent	Temp (°)	Time (h)		
2,4	5	TBAB (5 mol %), NaOH	toluene	reflux	22	(63)	391
3,5	1	CDA (10 mol %), K₃PO₄	dioxane	110	24	(99)	115
3,5	5	L-Pro (10 mol%), K₂CO₃	DMSO	90	36	(85)	401
3,5	10	EDA (10 mol %), K₃PO₄	dioxane	110	24	(87)	405

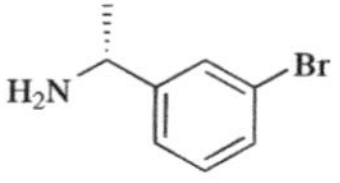

CuI (10 mol %),
K₂CO₃, NMP,
MW, 195°, 3 h

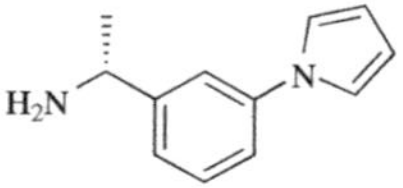

(66)   406

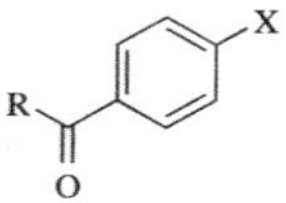

Catalyst (x mol %)

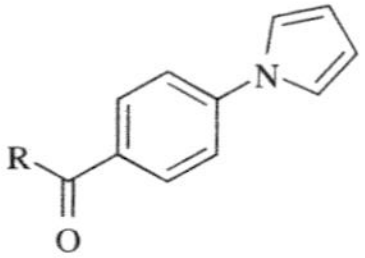

R	X	Catalyst	x	Additives	Solvent	Temp (°)	Time (h)		
Me	I	CuCl	10	**L9** (10 mol %), TMAH	DMSO	80	24	(70)	281
Me	Br	CuI	20	L-Pro (20 mol %), K₃PO₄	DMF	140	48	(68)	388
Me	I	CuI	5	L-Pro (20 mol %), K₃PO₄	dioxane	100	24	(99)	388
Et	Br	CuI	5	DMCDA (20 mol %), K₃PO₄	toluene	110	24	(92)	128

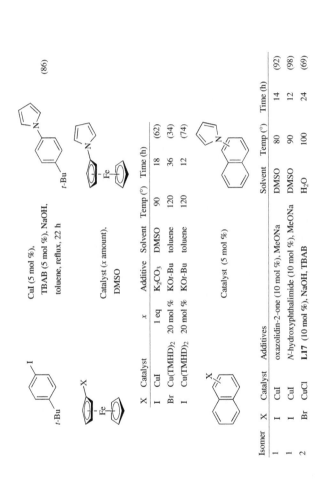

CuI (5 mol %),
TBAB (5 mol %), NaOH,
toluene, reflux, 22 h

(86)   391

Catalyst (x amount),
DMSO

X	Catalyst	x	Additive	Solvent	Temp (°)	Time (h)		
I	CuI	1 eq	K₂CO₃	DMSO	90	18	(62)	372
Br	Cu(TMHD)₂	20 mol %	KO*t*-Bu	toluene	120	36	(34)	407
I	Cu(TMHD)₂	20 mol %	KO*t*-Bu	toluene	120	12	(74)	407

Catalyst (5 mol %)

Isomer	X	Catalyst	Additives	Solvent	Temp (°)	Time (h)		
1	I	CuI	oxazolidin-2-one (10 mol %), MeONa	DMSO	80	14	(92)	392
1	I	CuI	N-hydroxyphthalimide (10 mol %), MeONa	DMSO	90	12	(98)	396
2	Br	CuCl	L17 (10 mol %), NaOH, TBAB	H₂O	100	24	(69)	235

TABLE 7A. N-ARYLATION OF PYRROLES (*Continued*)

*Please refer to the charts preceding the tables for structures indicated by the **bold** numbers.*

Nitrogen Nucleophile	Aryl Halide	Conditions	Product(s) and Yield(s) (%)	Refs.
C₄				
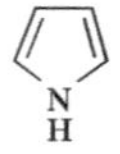 (pyrrole, N-H)	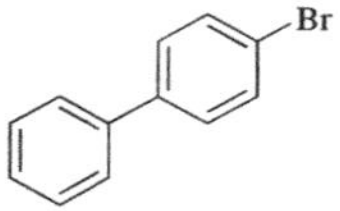 4-bromobiphenyl (Br)	CuI (5 mol %), NaOMe, DMSO, *N*-hydroxy-succinimide (10 mol %), 110°, 40 h	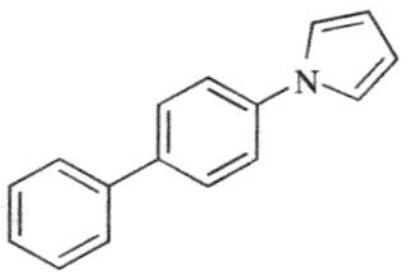 (82)	396
C₅				
(2-cyanopyrrole, N-H)	EtO₂C-(3-iodophenyl) (I)	CuI (5 mol %), DMCDA (20 mol %), K₃PO₄, toluene, 110°, 24 h	NC- / EtO₂C- product (85)	128
(2-formylpyrrole, CHO, N-H)	iodobenzene (I)	CuI (5 mol %), DMCDA (20 mol %), K₃PO₄, toluene, 110°, 24 h	OHC- product (88)	128
(2-carboethoxypyrrole, CO₂Et, N-H)	2-substituted iodobenzene (I, R)	CuI (5 mol %), DMEDA (10 mol %), K₂CO₃, toluene, 110°, 24 h	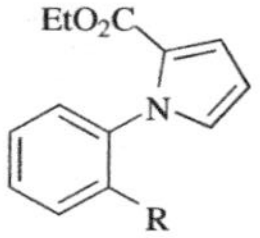 EtO₂C- product; R: MeO (96), Me (89)	128
C₆				
(2-ethylpyrrole, N-H)	MeO-(3-iodophenyl) (I)	CuI (5 mol %), DMCDA (20 mol %), K₃PO₄, toluene, 110°, 24 h	MeO- product (92)	128

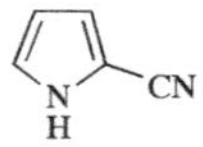

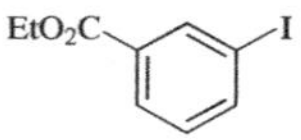

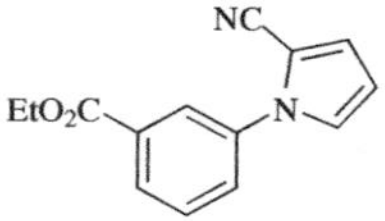

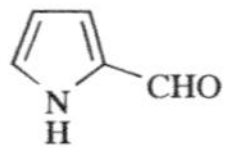

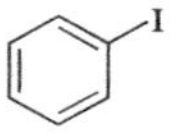

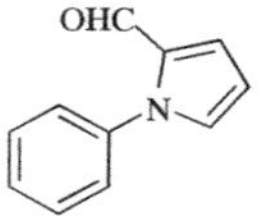

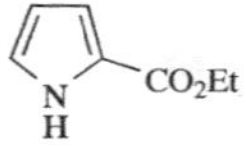

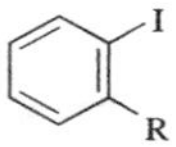

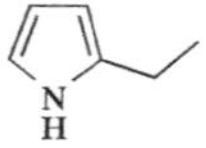

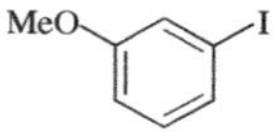

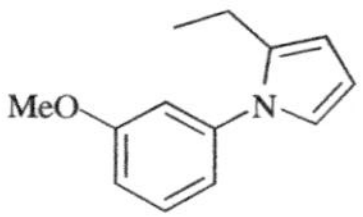

260

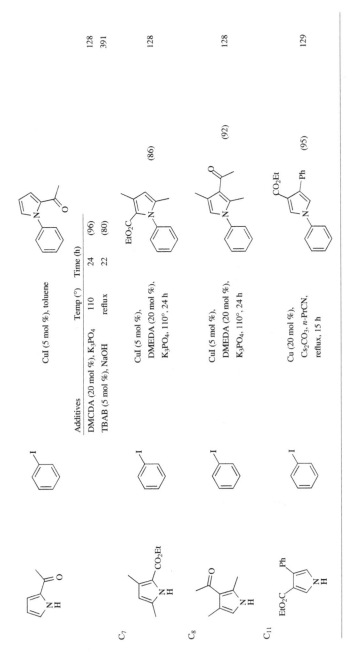

		Additives	Temp (°)	Time (h)		
C7		CuI (5 mol %), toluene				
		DMCDA (20 mol %), K3PO4	110	24	(96)	128
		TBAB (5 mol %), NaOH	reflux	22	(80)	391
		CuI (5 mol %), DMEDA (20 mol %), K3PO4, 110°, 24 h			(86)	128
C8		CuI (5 mol %), DMEDA (20 mol %), K3PO4, 110°, 24 h			(92)	128
C11		Cu (20 mol %), Cs2CO3, n-PrCN, reflux, 15 h			(95)	129

TABLE 7B. *N*-HETEROARYLATION OF PYRROLES

Nitrogen Nucleophile	Heteroaryl Halide	Conditions	Product(s) and Yield(s) (%)	Refs.

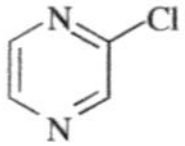

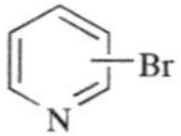

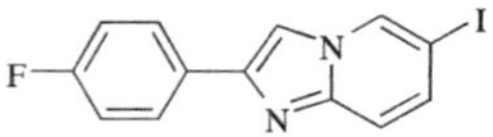

C4

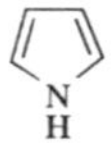

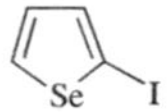

Nitrogen Nucleophile: pyrrole (N–H)

Entry 1

Heteroaryl Halide: 2-iodoselenophene

Conditions: CuI (10 mol %), EDA (20 mol %), K$_3$PO$_4$, dioxane, reflux, 24 h

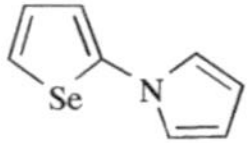

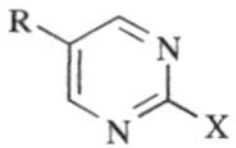

Product: (50) — 87

Entry 2

Heteroaryl Halide: R-pyrimidine–X

Conditions: Catalyst (x mol %)

Product: R-pyrimidine–pyrrole

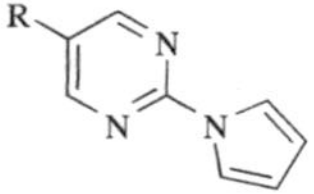

R	X	Catalyst	x	Additive(s)	Solvent	Temp (°)	Time (h)		Refs.
H	Br	Cu fluorapatite	12.5	none	DMSO	110	2	(98)	227
Br	I	CuI	5	phen (10 mol %), Cs$_2$CO$_3$	DMF	80	24	(84)	408

Entry 3

Heteroaryl Halide: 2-chloropyrazine

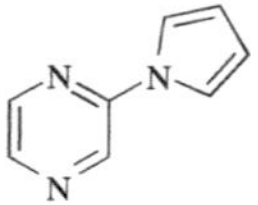

Conditions: CuI (x mol %), NaOMe, DMSO

Product: pyrazine–pyrrole

x	Additive	Temp (°)	Time (h)		Refs.
5	N-hydroxyphthalimide (10 mol %)	110	24	(90)	396
10	oxazolidin-2-one (20 mol %)	120	20	(78)	392

Entry 4

Heteroaryl Halide: bromopyridine

Conditions: Catalyst (x mol %), DMSO

Product: pyridine–pyrrole

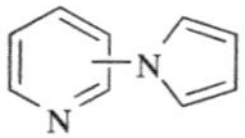

Isomer	Catalyst	x	Additive(s)	Temp (°)	Time (h)		Refs.
2	CuI	5	N-hydroxyphthalimide (10 mol %), NaOMe	110	24	(96)	396
2	CuI	10	oxazolidin-2-one (20 mol %), NaOMe	80	12	(88)	392
4	Cu fluorapatite	12.5	none	110	6	(92)	227

Entry 5

Heteroaryl Halide: 6-iodo-2-(4-fluorophenyl)imidazo[1,2-a]pyridine

Conditions: CuI (5 mol %), DMCDA (15 mol %), K$_3$PO$_4$, toluene, 112°, 24 h

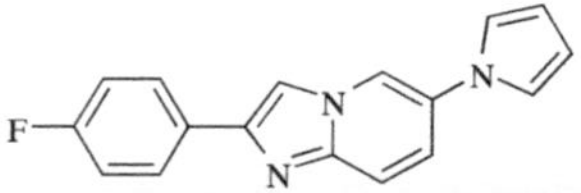

Product: (76) — 409

TABLE 8A. N-ARYLATION OF PYRAZOLES

Nitrogen Nucleophile	Aryl Halide	Conditions	Product(s) and Yield(s) (%)	Refs.

Please refer to the charts preceding the tables for structures indicated by the **bold** numbers.

C_3 — pyrazole (N-H)

Aryl Halide: phenyl-X

Product: 1-phenylpyrazole

X	Catalyst	x	Additive(s)	Solvent	Temp (°)	Time		Refs.
I	Cu	20 mol %	Cs₂CO₃	n-PrCN	reflux	8 h	(95)	129
I	CuBr	10 mol %	L10 (12 mol %), Cs₂CO₃	DMSO	80	24 h	(95)	259
Br	CuI	20 mol %	Cs₂CO₃	DMF	120	40 h	(84)	386
Br	CuI	20 mol %	phen (20 mol %), KF/Al₂O₃	xylene	140	15 h	(80)	410
I	CuI	10 mol %	phen (20 mol %), KF/Al₂O₃	xylene	140	13 h	(95)	411
Br	CuI	10 mol %	L-His (20 mol %), K₂CO₃	DMSO	100	48 h	(66)	389
I	CuI	10 mol %	DMG (20 mol %), TBPE	DMSO	rt	24 h	(81)	61
I	CuI	5 mol %	DMCDA (10 mol %), K₂CO₃	toluene	110	24 h	(93)	128
Br	CuI	20 mol %	per-6-ABCD (10 mol %), K₂CO₃	DMSO	110	24 h	(98)	393
Br	CuI	10 mol %	L25 (10 mol %), CTAB (20 mol %), Cs₂CO₃	H₂O	80	50 h	(67)	395
I	CuI	10 mol %	L25 (10 mol %), CTAB (20 mol %), Cs₂CO₃	H₂O	80	50 h	(67)	395
I	CuI	5 mol %	L2 (25 mol %), KOH, TBAB	H₂O	MW (100 W), 130	5 min	(63)	236
Br	CuI	10 mol %	L24 (10 mol %), Cs₂CO₃	DMF	110	24 h	(87)	86
Br	CuI	10 mol %	L28 (20 mol %), K₂CO₃	DMSO	120	60 h	(93)	412
Br	CuI	10 mol %	L29, Cs₂CO₃	MeCN	80	2 h	(80)	413
I	CuI	10 mol %	L29, Cs₂CO₃	MeCN	80	2 h	(95)	413
Br	CuO	10 mol %	Fe(acac)₃ (30 mol %), Cs₂CO₃	DMF	90	30 h	(94)	130
I	CuO	10 mol %	Fe(acac)₃ (30 mol %), Cs₂CO₃	DMF	90	30 h	(91)	130
I	CuO on acetylene black	5 mol %	KOt-Bu	toluene	180	18 h	(96)	226
Cl	Cu₂O	10 mol %	Cs₂CO₃	DMF	100	18 h	(0)	221
Br	Cu₂O	10 mol %	Cs₂CO₃	DMF	100	18 h	(93)	221
I	Cu₂O	10 mol %	Cs₂CO₃	DMF	100	18 h	(98)	221
I	Cu₂O	10 mol %	KOH	DMSO	120	24 h	(92)	398

TABLE 8A. N-ARYLATION OF PYRAZOLES (Continued)

Nitrogen Nucleophile	Aryl Halide	Conditions	Product(s) and Yield(s) (%)	Refs.

*Please refer to the charts preceding the tables for structures indicated by the **bold** numbers.*

Continued from previous page.

C3

Catalyst (x amount)

X	Catalyst	x	Additive(s)	Solvent	Temp (°)	Time		Refs
I	Cu2O	10 mol %	ninhydrin (20 mol %), KOH	DMSO	110	48 h	(92)	399
Br	Cu2O	5 mol %	L27 (20 mol %), Cs2CO3	MeCN	82	24 h	(96)	82
I	Cu2O	5 mol %	L27 (20 mol %), Cs2CO3	MeCN	82	24 h	(100)	82
I	Cu2O	5 mol %	L7 (20 mol %), Cs2CO3	MeCN	80	18 h	(94)	125
Br	Cu(OAc)2	1 eq	DBU	DMSO	MW, 130	10 min	(86)	85
I	Cu(OAc)2	1 eq	DBU	DMSO	MW, 130	10 min	(89)	85
Br	[Cu(μ-I)((−)-sparteine)]2	1 mol %	K2CO3	DMSO	115	7 h	(95)	414
Br	Cu fluorapatite	12.5 mol %	none	DMSO	110	10 h	(92)	227
I	6	5 mol %	Cs2CO3	toluene	100	12 h	(90)	123
Br	9	10 mol %	Cs2CO3	MeCN	reflux	4 h	(98)	27

Catalyst (x mol %)

X	R	Catalyst	x	Additives	Solvent	Temp (°)	Time (h)		Refs
Br	2,4-Cl2	Cu2O	5	L30 (20 mol %), Cs2CO3	DMF	110	48	(60)	82
I	4-Cl	CuI	10	DMG (20 mol %), TBPE	DMSO	rt	24	(98)	61
Br	4-Br	Cu2O	5	L27 (20 mol %), Cs2CO3	MeCN	82	30	(82)	82
I	4-Br	Cu2O	5	L27 (20 mol %), Cs2CO3	MeCN	82	30	(95)	82

CuI (5 mol %), L30 (5 mol %),
Cs2CO3, DMF, 100°, 24 h

Isomer	
1,3	(92)
1,4	(93)

415

CuI (20 mol %), K₂CO₃, PhNO₂,
reflux, 3 h

(90)

Catalyst (5 mol %)

Isomer	Catalyst	Additives	Solvent	Temp (°)	Time (h)		
2	CuI	DMCDA (10 mol %), K₂CO₃	toluene	110	24	(91)	416
4	Cu₂O	**L30** (20 mol %), Cs₂CO₃	MeCN	82	42	(91)	128
							82

Catalyst (x mol %)

Isomer	X	Catalyst	x	Additive(s)	Solvent	Temp (°)	Time (h)		
2	Cl	CuO	8	K₂CO₃	pyridine	reflux	24	(31)	402
3	I	CuBr	10	**L10** (12 mol %), Cs₂CO₃	DMSO	80	24	(95)	259
3	Cl	CuO	8	K₂CO₃	pyridine	reflux	24	(31)	402
3	Br	Cu₂O	10	Cs₂CO₃	DMF	110	18	(98)	221
3	Br	Cu₂O	5	**L27** (20 mol %), Cs₂CO₃	MeCN	82	24	(91)	82
4	Br	CuI	10	**L25** (20 mol %), Cs₂CO₃	MeCN	80	15	(92)	395
4	Br	CuI	10	**L32** (20 mol %), Cs₂CO₃	DMF	110	24	(97)	417
4	Cl	CuO	8	K₂CO₃	pyridine	reflux	24	(24)	402
4	Br	CuO	10	Fe(acac)₃ (30 mol %), Cs₂CO₃	DMF	90	30	(90)	130
4	I	CuO	10	Fe(acac)₃ (30 mol %), Cs₂CO₃	DMF	90	30	(90)	130
4	I	Cu₂O	5	**L30** (20 mol %), Cs₂CO₃	MeCN	82	24	(90)	82
4	F	**7**	1	K₂CO₃	DMF	110	1	(92)	73
4	Cl	**7**	1	K₂CO₃	DMF	110	3	(87)	73
4	F	Cu fluorapatite	7	K₂CO₃	DMF	120	1	(80)	66
4	Cl	Cu fluorapatite	7	K₂CO₃	DMF	120	12	(72)	66

TABLE 8A. N-ARYLATION OF PYRAZOLES (Continued)

| Nitrogen Nucleophile | Aryl Halide | Conditions | | Product(s) and Yield(s) (%) | Refs. |

*Please refer to the charts preceding the tables for structures indicated by the **bold** numbers.*

C₃

Nitrogen Nucleophile:

Aryl Halide:

Product:

Isomer	X	Catalyst	x	Additive(s)	Solvent	Temp (°)	Time	Product(s) and Yield(s) (%)	Refs.
2	I	CuI	10	phen (10 mol %), KF/Al₂O₃	—	140	17 h	(82)	411
2	I	Cu₂O	10	Cs₂CO₃	DMF	110	18 h	(99)	221
3	Br	CuI	10	L-Glu (20 mol %), K₃PO₄	—	MW (200 W), 150	10 h	(32)	418
3	Br	CuI	10	**L25** (20 mol %), Cs₂CO₃	MeCN	80	15 h	(93)	395
3	Br	CuO	10	Fe(acac)₃, Cs₂CO₃	DMF	120	24 h	(86)	130
4	Br	Cu/Al-hydrotalcite	2.5	K₂CO₃	DMF	100	20 h	(80)	228
4	I	CuF₂	50	phen (50 mol %), K₂CO₃	DMF	140	96 h	(98)	419
4	I	CuBr	10	**L10** (12 mol %), Cs₂CO₃	DMSO	80	24 h	(85)	259
4	I	CuBr	5	**L5** (10 mol %), Cs₂CO₃	DMF	90	24 h	(78)	119
4	I	CuI	20	K₃PO₄	DMF	40	40 h	(71)	59
4	Br	CuI	20	phen (20 mol %), KF/Al₂O₃	xylene	140	16 h	(88)	410
4	I	CuI	10	phen (20 mol %), KF/Al₂O₃	—	140	15 h	(94)	411
4	I	CuI	10	DMG (20 mol %), Bu₄POAc	DMSO	rt	24 h	(70)	61
4	Br	CuI	10	DMG (20 mol %), K₂CO₃	DMSO	110	45 h	(71)	47
4	Br	CuI	10	L-Lys (20 mol %), K₂CO₃	—	MW (200 W), 150	10 h	(83)	418
4	I	CuI	5	L-Pro (10 mol %), K₂CO₃	DMSO	90	65 h	(91)	401, 47
4	I	CuI	5	**L2** (25 mol %), KOH, TBAB	H₂O	MW (100 W), 130	5 min	(78)	236
4	Br	CuI	10	**L25** (20 mol %), Cs₂CO₃	MeCN	80	15 h	(82)	395
4	Cl	CuI nanoparticles	1.25	K₂CO₃, air	DMF	110	4 h	(90)	285
4	Br	CuO	10	Fe(acac)₃ (30 mol %), Cs₂CO₃	DMF	120	24 h	(80)	130
4	Br	Cu₂O	5	**L27** (20 mol %), Cs₂CO₃	MeCN	82	40 h	(93)	82
4	I	Cu₂O	5	**L7** (20 mol %), Cs₂CO₃	MeCN	80	18 h	(95)	125

Isomer	X	Catalyst	x	Additive(s)	Solvent	Temp (°)	Time (h)		ref
4	Br	[Cu(μ-I)((−)-sparteine)]₂	1	K₂CO₃	DMSO	115	12 h	(82)	414
4	I	[Cu(μ-I)((−)-sparteine)]₂	1	K₂CO₃	DMSO	115	7 h	(91)	414
4	Br	10	1	Cs₂CO₃	DMSO	100	24 h	(16)	420
4	I	10	1	Cs₂CO₃	DMSO	100	24 h	(50)	420
4	I	Cu(IPr)Cl	10	Cs₂CO₃	DMSO	100	24 h	(23)	421

Me—X $\xrightarrow{\text{Catalyst (}x\text{ mol \%)}}$ Me—(N–N-pyrazolyl)

Isomer	X	Catalyst	x	Additive(s)	Solvent	Temp (°)	Time (h)		ref
2	Br	CuI	5	DMCDA (10 mol %), K₂CO₃	—	110	24	(92)	128
2	Br	CuI	20	phen (20 mol %), KF/Al₂O₃	xylene	140	18	(81)	410
2	I	Cu₂O	10	Cs₂CO₃	DMF	110	18	(91)	221
2	Br	Cu₂O	5	L27 (20 mol %), Cs₂CO₃	DMF	110	48	(100)	82
2	I	Cu₂O	5	L30 (20 mol %), Cs₂CO₃	MeCN	82	70	(94)	82
3	Br	CuI	10	L-Glu (20 mol %), K₃PO₄	—	MW (200 W), 150	3	(27)	418
4	Br	CuI	10	L-Glu (20 mol %), K₃PO₄	—	MW (200 W), 150	2	(37)	418
4	Br	CuI	10	L25 (20 mol %), Cs₂CO₃	MeCN	80	15	(80)	395
4	Br	CuI	10	L32 (20 mol %), Cs₂CO₃	DMF	110	24	(98)	417
4	Br	CuO	10	Fe(acac)₃ (30 mol %), Cs₂CO₃	DMF	120	24	(57)	130
4	Br	Cu₂O	5	L27 (20 mol %), Cs₂CO₃	MeCN	82	36	(95)	82
4	I	Cu(IPr)Cl	10	Cs₂CO₃	DMSO	100	24	(54)	421
4	Br	Cu(II)–NaY zeolite	10	K₂CO₃	DMF	120	36	(99)	404
4	I	Cu(II)–NaY zeolite	10	K₂CO₃	DMF	120	22	(99)	404
4	I	10	1	Cs₂CO₃	DMSO	100	24	(21)	420
4	Br	9	10	Cs₂CO₃	MeCN	reflux	4	(97)	27
3,5	I	CuI	1	CDA (10 mol %), K₂CO₃	dioxane	110	24	(89)	115

TABLE 8A. N-ARYLATION OF PYRAZOLES (Continued)

Nitrogen Nucleophile	Aryl Halide	Conditions	Product(s) and Yield(s) (%)	Refs.

*Please refer to the charts preceding the tables for structures indicated by the **bold** numbers.*

C₃ — pyrazole (N–N–H)

Aryl Halide: CF₃-substituted aryl–X

Catalyst (x mol %)

Product: 1-aryl(CF₃)pyrazole

Isomer	X	Catalyst	x	Additive(s)	Solvent	Temp (°)	Time (h)		Refs.
3	I	Cu₂O	5	L7 (20 mol %), Cs₂CO₃	MeCN	80	18	(81)	125
4	Br	CuO	10	Fe(acac)₃ (30 mol %), Cs₂CO₃	DMF	140	24	(40)	130
4	Br	Cu₂O	5	L27 (20 mol %), Cs₂CO₃	MeCN	82	36	(96)	82
4	Cl	**9**	10	Cs₂CO₃	DMF	140	24	(70)	27

Aryl Halide: NC-substituted aryl–X

Catalyst(s) (x mol %)

Product: 1-aryl(CN)pyrazole

Isomer	X	Catalyst(s)	x	Additive(s)	Solvent	Temp (°)	Time (h)		Refs.
3	I	CuBr	10	L10 (12 mol %), Cs₂CO₃	DMSO	80	24	(89)	259
4	Br	Cu, CuI	10, 5	binol (20 mol %), Cs₂CO₃	DMSO	110	36	(89)	124
4	Br	CuI	10	L-Pro (20 mol %), K₂CO₃	DMSO	75	45	(96)	47
4	Br	CuI	10	L25 (20 mol %), Cs₂CO₃	MeCN	80	15	(94)	395
4	Br	CuO	8	K₂CO₃	pyridine	reflux	24	(47)	402
4	Br	CuO	10	Fe(acac)₃ (30 mol %), Cs₂CO₃	DMF	90	30	(98)	130
4	I	Cu₂O	10	Fe(acac)₃ (30 mol %), Cs₂CO₃	DMF	90	30	(98)	130
4	Cl	Cu₂O	10	Cs₂CO₃	DMF	110	18	(98)	221
4	Br	Cu₂O	5	L30 (20 mol %), Cs₂CO₃	MeCN	82	24	(91)	82
4	Br	**9**	10	Cs₂CO₃	MeCN	reflux	4	(98)	27

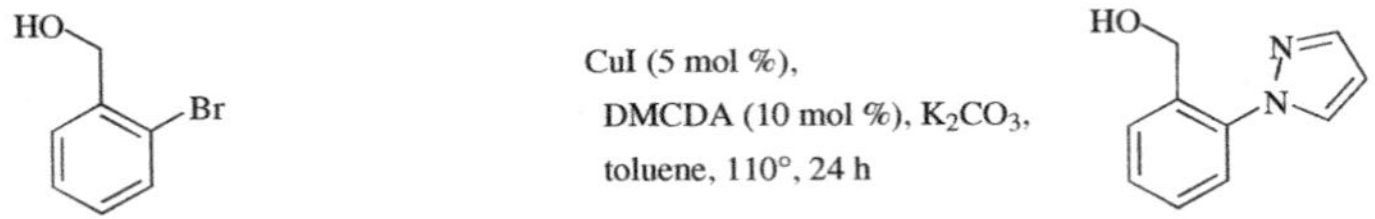

CuI (5 mol %),
DMCDA (10 mol %), K$_2$CO$_3$,
toluene, 110°, 24 h   (96)   128

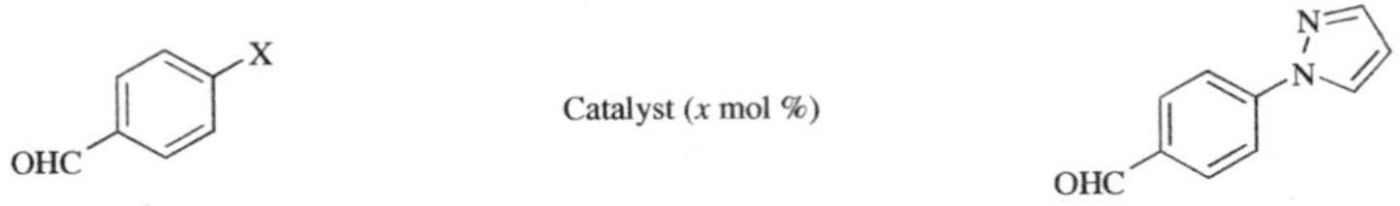

Catalyst (x mol %)

X	Catalyst	x	Additive(s)	Solvent	Temp (°)	Time (h)		
Br	CuI	5	L31 (5 mol %), Cs$_2$CO$_3$	DMF	100	24	(72)	415
Cl	[Cu(μ-I)((–)-sparteine)]$_2$	1	K$_2$CO$_3$	DMSO	125	19	(65)	414
Br	[Cu(μ-I)((–)-sparteine)]$_2$	1	K$_2$CO$_3$	DMSO	115	8	(77)	414

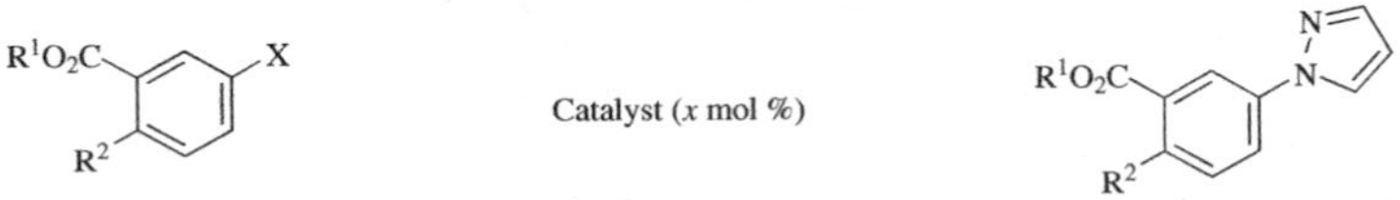

Catalyst (x mol %)

R^1	R^2	X	Catalyst	x	Additives	Solvent	Temp (°)	Time (h)		
Me	H	I	CuI	10	Me$_2$Gly (20 mol %), Bu$_4$POAc	DMSO	rt	24	(93)	61
Et	H	Br	Cu$_2$O	5	L30 (20 mol %), Cs$_2$CO$_3$, 3 Å MS	MeCN	82	48	(50)	82
Me	Cl	I	CuI	5	DMCDA (10 mol %), K$_2$CO$_3$	toluene	110	24	(89)	128

TABLE 8A. N-ARYLATION OF PYRAZOLES (Continued)

Nitrogen Nucleophile	Aryl Halide	Conditions	Product(s) and Yield(s) (%)	Refs.

*Please refer to the charts preceding the tables for structures indicated by the **bold** numbers.*

C₃

Nitrogen Nucleophile: pyrazole (N–N–H)

Aryl Halide: Ac–C₆H₄–X

Catalyst (x mol %)

Product: Ac–C₆H₄–pyrazole

Isomer	X	Catalyst	x	Additive(s)	Solvent	Temp (°)	Time		Refs.
3	Br	CuI	10	L-Lys (20 mol %), K₃PO₄	—	MW (200 W), 150	10 h	(44)	418
3	Br	CuO	8	K₂CO₃	pyridine	reflux	48 h	(62)	402
4	I	Cu/Al–hydrotalcite	2.5	K₂CO₃	DMF	100	8 h	(80)	228
4	Br	CuI	10	L-Glu (20 mol %), K₃PO₄	—	MW (200 W), 150	5 h	(59)	418
4	I	CuI	5	L2 (25 mol %), KOH, TBAB	H₂O	MW (100 W), 130	5 min	(54)	236
4	Br	CuI	10	L25 (20 mol %), Cs₂CO₃	MeCN	80	15 h	(95)	395
4	Br	CuI	5	L31 (5 mol %), Cs₂CO₃	DMF	100	24 h	(91)	415
4	Br	CuI	10	L32 (20 mol %), Cs₂CO₃	DMF	110	24 h	(98)	417
4	Br	CuO	8	K₂CO₃	pyridine	reflux	48 h	(75)	402
4	Br	CuO	10	Fe(acac)₃ (30 mol %), Cs₂CO₃	DMF	90	30 h	(81)	130
4	Br	Cu₂O	5	L27 (20 mol %), Cs₂CO₃	MeCN	82	24 h	(91)	82
4	Cl	Cu(IPr)Cl	10	Cs₂CO₃	DMSO	100	24 h	(50)	421
4	Br	Cu(IPr)Cl	10	Cs₂CO₃	DMSO	100	24 h	(60)	421
4	I	Cu(IPr)Cl	10	Cs₂CO₃	DMSO	100	24 h	(90)	421
4	Cl	**10**	1	Cs₂CO₃	DMSO	100	24 h	(44)	420
4	Br	**10**	1	Cs₂CO₃	DMSO	100	24 h	(93)	420
4	I	**10**	1	Cs₂CO₃	DMSO	100	24 h	(93)	420

Br–C₆H₄–C(O)Et

CuI (5 mol %),
DMCDA (10 mol %), K₂CO₃,
toluene, 110°, 24 h

Et(O)C–C₆H₄–pyrazole (98) 128

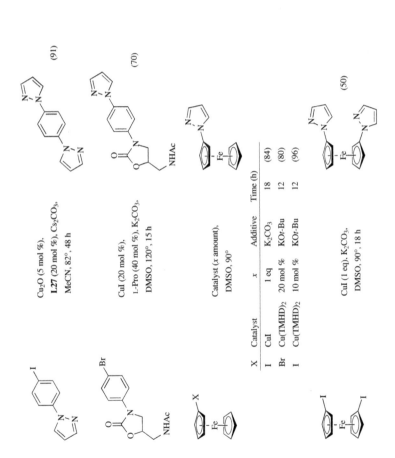

X	Catalyst	x	Additive	Time (h)	
I	CuI	1 eq	K₂CO₃	18	(84)
Br	Cu(TMHD)₂	20 mol %	KOt-Bu	12	(80)
I	Cu(TMHD)₂	10 mol %	KOt-Bu	12	(96)

TABLE 8A. *N*-ARYLATION OF PYRAZOLES (*Continued*)

Nitrogen Nucleophile	Aryl Halide	Conditions	Product(s) and Yield(s) (%)	Refs.

*Please refer to the charts preceding the tables for structures indicated by the **bold** numbers.*

C₃

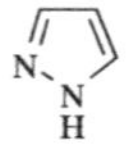

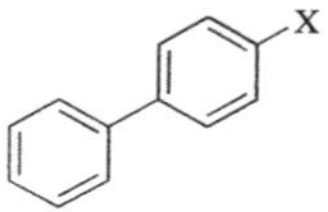

Catalyst (10 mol %), Cs₂CO₃

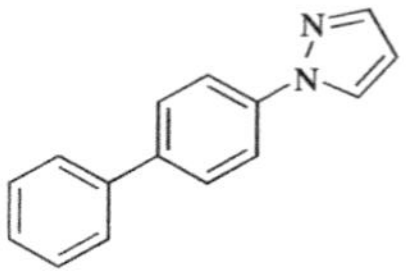

X	Catalyst	Additive	Solvent	Temp (°)	Time (h)		
Br	CuI	**L25** (20 mol %)	MeCN	80	15	(98)	395
Br	CuO	Fe(acac)₃ (30 mol %)	DMF	120	24	(93)	130
I	CuO	Fe(acac)₃ (30 mol %)	DMF	120	24	(93)	130

C₃₋₅

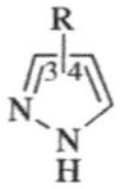

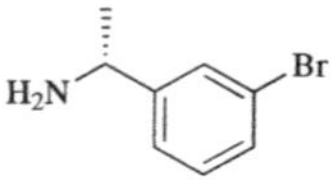

CuI (10 mol %), K₂CO₃,
NMP, MW, 195°

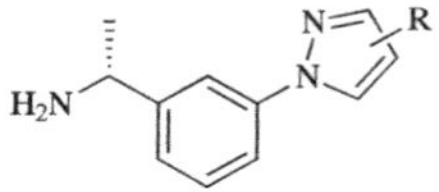

R	Time (h)	
H	2	(71)
3-Me	17	(32) (3-Me) + (32) (1-Me)
4-Me	1	(90)
3,5-Me₂	22	(49)

406

272

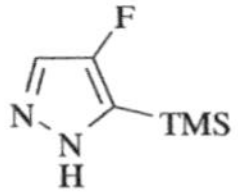

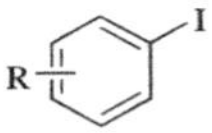

CuI (10 mol %),
L33 (20 mol %),
Cs$_2$CO$_3$, MeCN

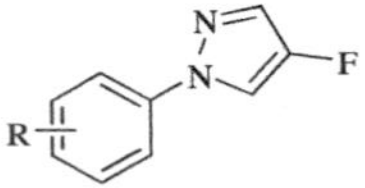

R	Temp (°)	Time (h)	
H	60	24	(76)
4-O$_2$N	60	24	(98)
3-MeO	60	24	(67)
4-NC	60	24	(98)
4-CF$_3$	60	24	(86)

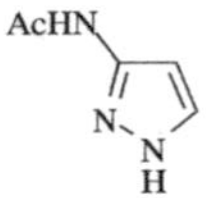

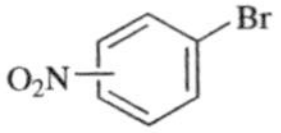

Cu$_2$O (12 mol %), K$_2$CO$_3$,
pyridine, reflux, 24 h

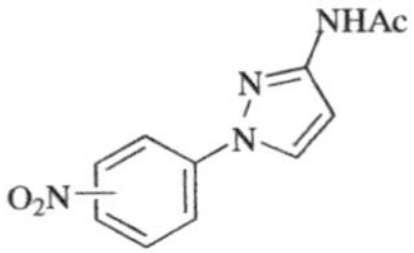

R	Temp (°)	Time (h)	
2-MeO$_2$C	60	24	(89)
4-EtO$_2$C	60	24	(76)
2,4-Me$_2$	70	144	(trace)
2-Et	70	144	(69)
4-Ac	60	24	(82)

Isomer

2	(45)
3	(39)
4	(40)

I + **II**

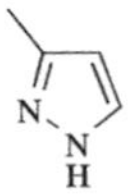

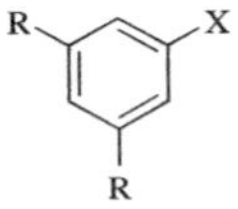

Catalyst (x mol %)

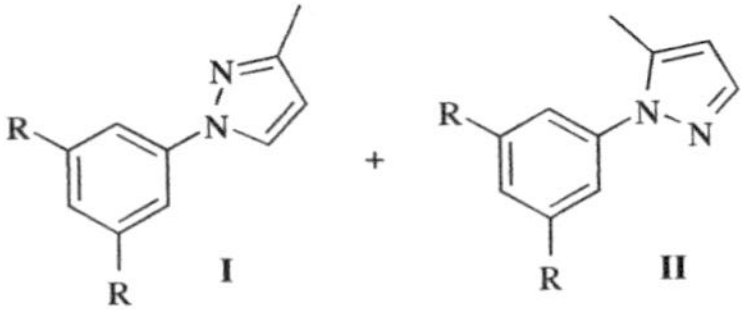

R	X	Catalyst	x	Additive(s)	Solvent	Temp (°)	Time (h)	I	II	
H	I	CuI	20	Cs$_2$CO$_3$	DMF	120	40	(64)	(21)	386
H	Br	Cu$_2$O	5	**L27** (20 mol %), Cs$_2$CO$_3$	MeCN	82	96	(52)	(16)	82
H	I	Cu$_2$O	5	**L27** (20 mol %), Cs$_2$CO$_3$	MeCN	82	96	(71)	(19)	82
Me	I	CuI	1	CDA (10 mol %), K$_2$CO$_3$	dioxane	110	24	(84)	(0)	115

C$_4$

TABLE 8A. *N*-ARYLATION OF PYRAZOLES (*Continued*)

Nitrogen Nucleophile	Aryl Halide	Conditions	Product(s) and Yield(s) (%)	Refs.

*Please refer to the charts preceding the tables for structures indicated by the **bold** numbers.*

C₄

Nucleophile: 4-Me pyrazole (N–N–H)
Aryl Halide: *para*-substituted benzene (R, X)
Product: 1-aryl-4-methylpyrazole

Isomer	R	X	Catalyst	x	Additives	Solvent	Temp (°)	Time (h)	Yield	Refs.
4	H	I	CuI	5	DMCDA (20 mol %), Cs₂CO₃	DMF	110	24	(74)	128
4	H	I	CuI	10	phen (10 mol %), KF/Al₂O₃	—	140	13	(95)	411
4	H	Br	Cu₂O	5	L27 (20 mol %), Cs₂CO₃	MeCN	82	96	(100)	82
4	H	I	Cu₂O	5	L27 (20 mol %), Cs₂CO₃	MeCN	82	24	(98)	82
5	HO	Br	CuI	5	DMCDA (10 mol %), K₂CO₃	toluene	110	24	(78)	128

Nucleophile: 4-CF₃ pyrazole (N–N–H)
Aryl Halide: halobenzene (X)
Conditions: Cu₂O (5 mol %), L27 (20 mol %), Cs₂CO₃, MeCN, 82°, 96 h
Product: 3-CF₃-1-phenylpyrazole

X	Yield
Br	(81)
I	(81)

Refs. 82

Nucleophile: 4-(CH₂NMe₂) pyrazole (N–N–H)
Aryl Halide: 2-fluoro-4-amino iodobenzene
Conditions: CuI, 8-HOquin, K₂CO₃, DMSO, 130°
Product: N-aryl pyrazole with NMe₂, F, H₂N

Isomer	Yield
4	(—)
5	(—)

Refs. 424

Nucleophile: 4-CO₂Et pyrazole (N–N–H)
Aryl Halide: R-substituted halobenzene (X)
Conditions: CuI (5 mol %), DMCDA (20 mol %), K₃PO₄, toluene, 110°, 24 h
Product: N-aryl-4-(ethoxycarbonyl)pyrazole

R	X	Yield
2-Me	Br	(77)
4-EtO₂C	I	(79)

Refs. 128

274

C₅

Catalyst (x amount)

Pyrazole (R¹ at 4-position, 3,5-dimethyl, N–H) + aryl halide (R²-C₆H₄-X) → 1-aryl pyrazole product (R¹, R², substituted on N-phenyl)

R¹	R²	X	Catalyst	x	Additive(s)	Solvent	Temp (°)	Time (h)		
H	H	I	CuI	20 mol %	Cs₂CO₃	DMF	120	40	(63)	386
H	H	I	CuI	5 mol %	DMCDA (20 mol %), Cs₂CO₃	DMF	110	24	(98)	128
H	H	I	CuI	10 mol %	L22 (10 mol %), K₂CO₃	DMF	90	12	(71)	79
H	H	I	Cu₂O	5 mol %	L30 (20 mol %), Cs₂CO₃	MeCN	82	54	(94)	82
H	4-MeO	I	CuF₂	25 mol %	phen (25 mol %), K₂CO₃	DMF	140	96	(81)	419
Br	4-MeO	I	CuF₂	50 mol %	phen (50 mol %), K₂CO₃	DMF	140	96	(83)	419
H	4-Me	I⁺4-MeC₆H₄ BF₄⁻	Cu(acac)₂	1 eq	K₂CO₃	toluene	50	6	(80)	76
H	2-CH₂NH₂	I	CuI	5 mol %	DMCDA (10 mol %), K₂CO₃	—	110	24	(71)	128

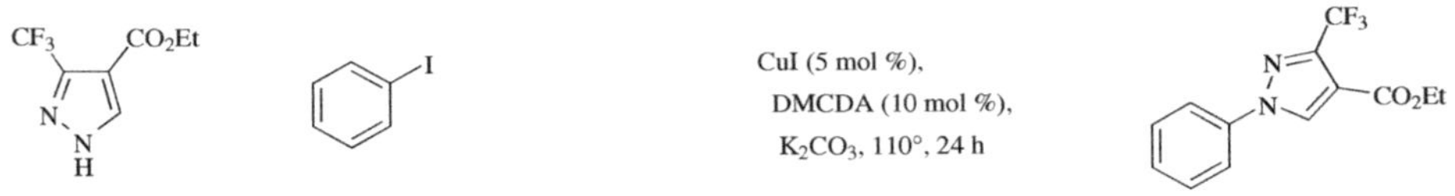

CF_3-, CO_2Et-pyrazole (N–H) + iodobenzene, CuI (5 mol %), DMCDA (10 mol %), K₂CO₃, 110°, 24 h → 1-phenyl-3-CF₃-4-CO₂Et-pyrazole (77)   128

C₉

Ph-, NH₂-pyrazole (N–H) + iodobenzene, CuI (5 mol %), DMCDA (10 mol %), K₂CO₃, 110°, 24 h → 3-Ph-5-NH₂-1-phenylpyrazole (68)   128

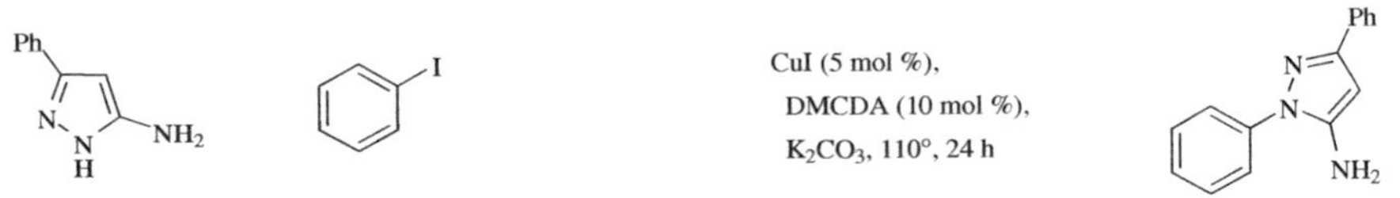

TABLE 8B. N-HETEROARYLATION OF PYRAZOLES

Nitrogen Nucleophile	Heteroaryl Halide	Conditions	Product(s) and Yield(s) (%)	Refs.

*Please refer to the charts preceding the tables for structures indicated by the **bold** numbers.*

C₃

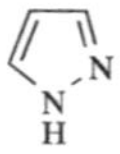

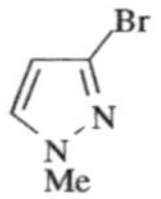

First entry: pyrazole (N–H) / 3-bromo-N-methylpyrazole

Cu₂O (5 mol %),
L27 (20 mol %), Cs₂CO₃,
MeCN, 82°, 54 h

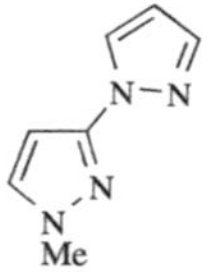

Product (92) — Refs. 82

Second entry: thiophene bromide (2-bromothiophene)

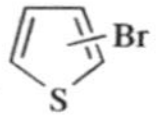

Cu₂O (x mol %), Cs₂CO₃

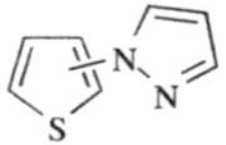

Isomer	x	Additive	Solvent	Temp (°)	Time (h)		Refs.
2	5	**L27** (20 mol %)	MeCN	82	24	(91)	82
2	10	none	DMF	110	18	(95)	221
3	5	**L27** (20 mol %)	MeCN	82	54	(92)	82

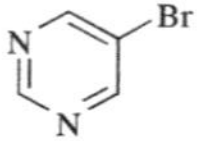

Third entry: 5-bromopyrimidine

Cu (10 mol %), CuI (5 mol %),
binol (20 mol %), Cs₂CO₃,
DMSO, 110°, 40 h

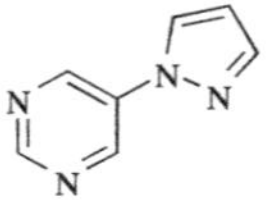

Product (82) — Refs. 124

Fourth entry: 2-halopyridine

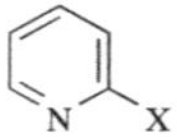

Catalyst (x mol %)

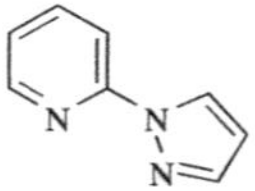

X	Catalyst	x	Additive(s)	Solvent	Temp (°)	Time (h)		Refs.
Cl	Cu	5	binol (20 mol %), Cs₂CO₃	DMSO	110	36	(89)	124
Br	CuI	10	L-Pro (20 mol %), K₃CO₃	DMSO	60	45	(94)	47
Br	CuI	10	**L25** (20 mol %), Cs₂CO₃	MeCN	80	15	(94)	395
Br	CuI	5	**L31** (5 mol %), Cs₂CO₃	DMF	100	24	(85)	415
Br	CuI	10	**L32** (20 mol %), Cs₂CO₃	DMF	110	24	(95)	417
Br	CuO	8	K₂CO₃	pyridine	reflux	16	(42)	402
Br	Cu₂O	5	**L27** (20 mol %), Cs₂CO₃	MeCN	82	24	(93)	82

276

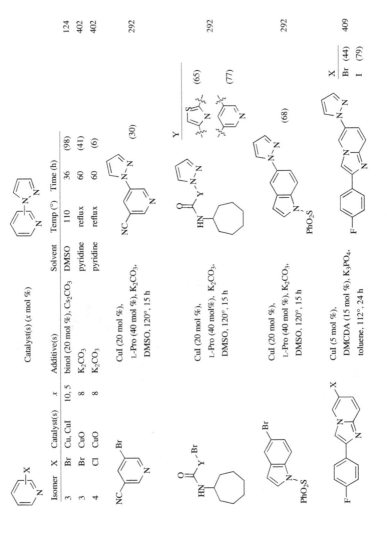

Catalyst(s) (x mol %)

Isomer	X	Catalyst(s)	x	Additive(s)	Solvent	Temp (°)	Time (h)		
3	Br	Cu, CuI	10, 5	binol (20 mol %), Cs$_2$CO$_3$	DMSO	110	36	(98)	124
3	Br	CuO	8	K$_2$CO$_3$	pyridine	reflux	60	(41)	402
4	Cl	CuO	8	K$_2$CO$_3$	pyridine	reflux	60	(6)	402

CuI (20 mol %), L-Pro (40 mol %), K$_2$CO$_3$, DMSO, 120°, 15 h (30) 292

CuI (20 mol%), L-Pro (40 mol%), K$_2$CO$_3$, DMSO, 120°, 15 h

Y	
	(65)
	(77)

292

CuI (20 mol %), L-Pro (40 mol %), K$_2$CO$_3$, DMSO, 120°, 15 h (68) 292

CuI (5 mol %), DMCDA (15 mol %), K$_3$PO$_4$, toluene, 112°, 24 h

X	
Br	(44)
I	(79)

409

TABLE 8B. N-HETEROARYLATION OF PYRAZOLES (Continued)

Nitrogen Nucleophile	Heteroaryl Halide	Conditions	Product(s) and Yield(s) (%)	Refs.

*Please refer to the charts preceding the tables for structures indicated by the **bold** numbers.*

C$_{3-16}$

Nitrogen Nucleophile:

R^2 OR^1 pyrazole with R^3, N–N–H

Heteroaryl Halide: 2-bromopyridine

Conditions: Cu$_2$O (0.5 mol %), **L30** (1 mol %), Cs$_2$CO$_3$, 4 Å MS, MeCN, MW, 180°

Products: **I** (R^1, R^2, R^3 pyridylpyrazole with OR^1) + **II**

R^1	R^2	R^3	Isomer	Time (h)	I	II
Et	H	H	2	6	(70)	(0)
Et	H	Me	2	2	(55)	(26)
i-Pr	H	Me	2	2	(53)	(25)
Et	H	Ph	2	6	(36)	(41)
Et	H	Bn	2	2	(41)	(25)
Et	F	Me	2	6	(67)	(9)

R^1	R^2	R^3	Isomer	Time (h)	I	II
Et	EtO$_2$C	H	2	2	(72)	(0)
Et	EtO$_2$CCH$_2$	H	2	6	(7)	(0)
Et	Ph	Me	2	6	(79)	(5)
Et	Bn	Ph	2	6	(66)	(7)
i-Pr	H	Me	3	6	(30)	(2)
i-Pr	H	Me	4	6	(17)	(10)

425

C$_3$

Nitrogen Nucleophile: F, pyrazole with TMS, N–N–H

Heteroaryl Halide: 2-iodothiophene

Conditions: CuI (10 mol %), **L33** (20 mol %), Cs$_2$CO$_3$, MeCN, 70°, 144 h

Product: 4-F pyrazole N-linked to thiophene (46)

422

C$_4$

Nitrogen Nucleophile: EtO$_2$C pyrazole, N–N–H

Heteroaryl Halide: 3-bromopyridine

Conditions: CuI (5 mol %), DMCDA (10 mol %), K$_2$CO$_3$, toluene, 110°, 24 h

Product: CO$_2$Et pyrazole N-linked to pyridine (84)

128

C$_5$

Nitrogen Nucleophile: R, dimethylpyrazole, N–N–H

Heteroaryl Halide: 2-iodothiophene

Conditions: CuF$_2$ (x mol %), phen (x mol %), K$_2$CO$_3$, DMF, 140°, 96 h

Product: R, dimethylpyrazole N-linked to thiophene

R	x	
H	50	(65)
Br	25	(37)

419

C$_6$

	Conditions	Ref.
(36)	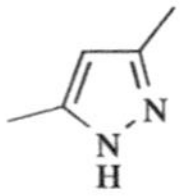 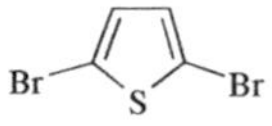 CuF$_2$ (50 mol %), phen (50 mol %), K$_2$CO$_3$, DMF, 140°, 96 h	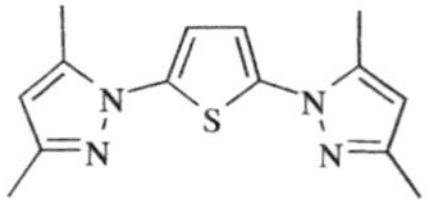 419
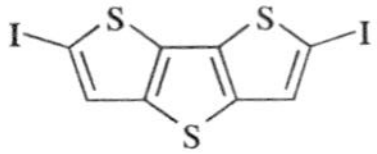 (20)	CuF$_2$ (50 mol %), phen (50 mol %), K$_2$CO$_3$, DMF, 140°, 96 h	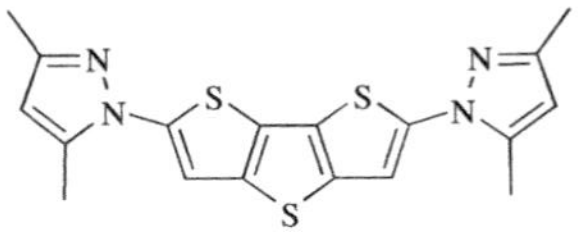 419
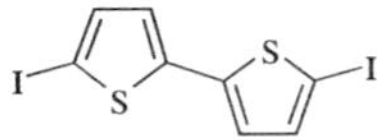 (87)	CuF$_2$ (50 mol %), phen (50 mol %), K$_2$CO$_3$, DMF, 140°, 96 h	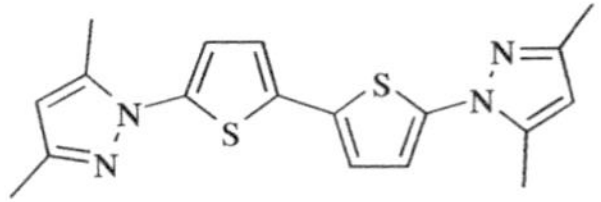 419
R: H (67), Br (71)	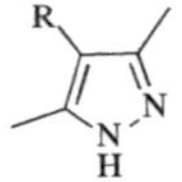 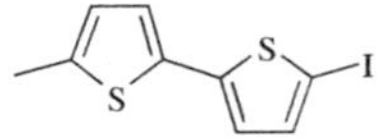 CuF$_2$ (25 mol %), phen (25 mol %), K$_2$CO$_3$, DMF, 140°, 96 h	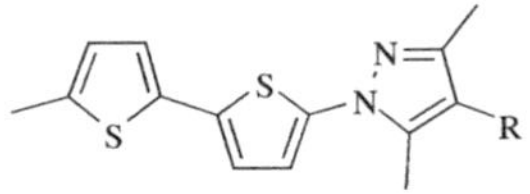 419
(75)	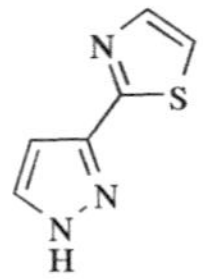 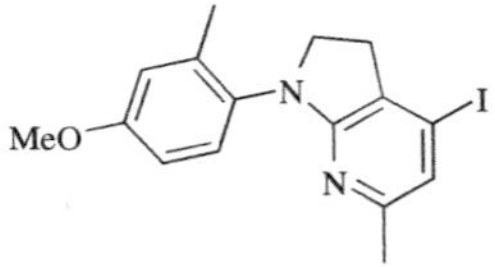 CuI (10 mol %), DMG (30 mol %), DBU, DMSO, 110°, 7 h	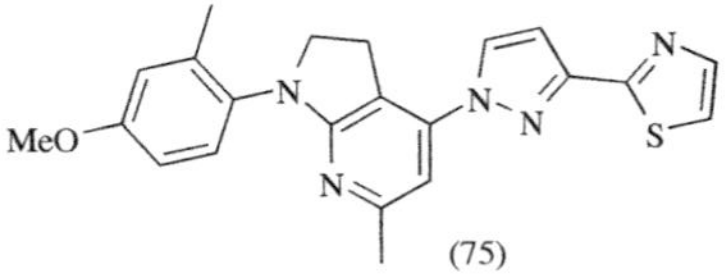 426
(90)	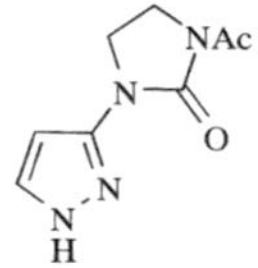 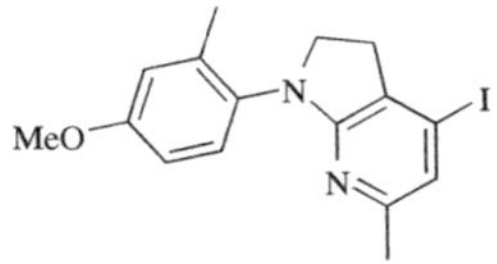 CuI (10 mol %), DMG (30 mol %), DBU, DMSO, 110°, 7 h	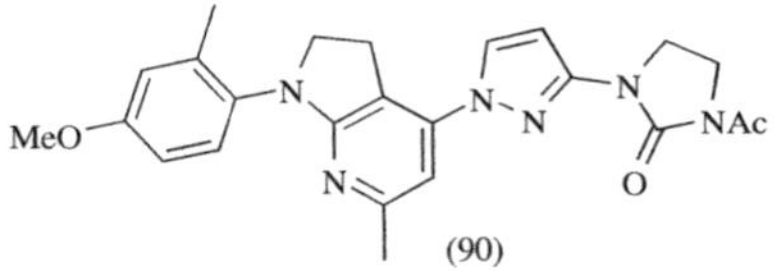 426

TABLE 9A. *N*-ARYLATION OF IMIDAZOLES

*Please refer to the charts preceding the tables for structures indicated by the **bold** numbers.*

Nitrogen Nucleophile	Aryl Halide	Conditions	Product(s) and Yield(s) (%)	Refs.

C₃

X	Catalyst	x	Additive(s)	Solvent	Temp (°)	Time		Refs.
I	Cu	20 mol %	Cs₂CO₃	n-PrCN	reflux	9 h	(95)	129
Cl	Cu/Al-hydrotalcite	2.5 mol %	K₂CO₃	DMF	100	18 h	(80)	228
Br	Cu/Al-hydrotalcite	2.5 mol %	K₂CO₃	DMF	100	18 h	(92)	228
I	Cu/Al-hydrotalcite	2.5 mol %	K₂CO₃	DMF	100	18 h	(92)	228
Cl	Cu on cellulose	2 mol %	K₂CO₃	DMSO	130	48 h	(20)	427
Br	Cu on cellulose	1 mol %	K₂CO₃	DMSO	130	24 h	(70)	427
I	Cu on cellulose	1 mol %	K₂CO₃	DMSO	130	12 h	(95)	427
Br	CuBr	2.5 mol %	K₂CO₃	NMP	160	16 h	(86)	287
I	CuBr	2.5 mol %	K₂CO₃	DMSO	160	16 h	(86)	287
Br	CuBr	10 mol %	L9 (20 mol %), Cs₂CO₃	DMSO	75	22 h	(87)	49
I	CuBr	10 mol %	L9 (20 mol %), Cs₂CO₃	DMSO	45	25 h	(96)	49
Br	CuBr	10 mol %	L20 (20 mol %), Cs₂CO₃	DMSO	80	12 h	(92)	284
I	CuBr	10 mol %	L10 (12 mol %), Cs₂CO₃	DMSO	80	24 h	(98)	259
I	CuI	20 mol %	K₃PO₄	DMF	40	40 h	(93)	59
I	CuI	10 mol %	K₃PO₄	DMF	110	24 h	(91)	387
Br	CuI	20 mol %	Cs₂CO₃	DMF	120	40 h	(91)	386
I	CuI	5 mol %	TBAB, KOH	—	110	6 h	(85)	428
I	CuI	5 mol %	TBAB, NaOH	toluene	reflux	22 h	(84)	391
Br	CuI	20 mol %	phen (20 mol %), KF/Al₂O₃	xylene	140	15 h	(72)	410
I	CuI	10 mol %	phen (10 mol %), KF/Al₂O₃	—	140	13 h	(92)	411
Br	CuI	20 mol %	per-6-ABCD, K₂CO₃	DMSO	110	24 h	(98)	393
I	CuI	10 mol %	PPAPM (20 mol %), K₃PO₄	DMF	100	24 h	(86)	48
I	CuI	5 mol %	N-hydroxysuccinimide (10 mol %), NaOMe	DMSO	90	12 h	(98)	396
I	CuI	5 mol %	oxazolidin-2-one (10 mmol %), NaOMe	DMSO	80	10 h	(91)	392

Halide	Catalyst	Loading	Ligand/Additives	Solvent	Temp	Time	(Yield)	Ref
I	CuI	10 mol %	DMG (20 mol %), TBPE	DMSO	rt	24 h	(90)	61
I	CuI	5 mol %	L-Pro (20 mol %), K$_3$PO$_4$	dioxane	100	24 h	(90)	388
Br	CuI	10 mol %	L-His (20 mol %), K$_2$CO$_3$	DMSO	100	36 h	(82)	389
I	CuI	10 mol %	L-His (20 mol %), K$_2$CO$_3$	DMSO	100	18 h	(91)	389
Br	CuI	20 mol %	D-glucosamine (40 mol %), Cs$_2$CO$_3$	DMSO	110	24 h	(84)	429
I	CuI	10 mol %	L6 (20 mol %), Cs$_2$CO$_3$	DMF	rt	24 h	(93)	273
Br	CuI	10 mol %	L22 (10 mol %), K$_2$CO$_3$	DMF	110	24 h	(90)	79
Br	CuI	5 mol %	L24 (10 mol %), Cs$_2$CO$_3$	DMF	110	24 h	(96)	86
Br	CuI	10 mol %	L21 (20 mol %), KI, Cs$_2$CO$_3$	MeCN	reflux	12 h	(43)	394
I	CuI	10 mol %	L21 (20 mol %), Cs$_2$CO$_3$	MeCN	reflux	12 h	(88)	394
I	CuI	10 mol %	L26 (10 mol %), NaOMe	DMSO	120	12 h	(85)	397
I	CuI	5 mol %	L31 (5 mol %), Cs$_2$CO$_3$	DMF	80	24 h	(82)	415
Cl	CuI nanoparticles	1.25 mol %	K$_2$CO$_3$, air	DMF	110	16 h	(56)	285
I	CuI nanoparticles	1.25 mol %	K$_2$CO$_3$, air	DMF	110	4 h	(95)	285
I	CuI/PAnNF	5 mol %	K$_2$CO$_3$	DMF	rt	6 h	(99)	276
I	CuO on acetylene black	10 mol %	Fe(acac)$_3$ (30 mol %), Cs$_2$CO$_3$	DMF	90	30 h	(90)	130
I	CuO nanoparticles	5 mol %	KOt-Bu	toluene	180	18 h	(100)	226
Cl	CuO nanoparticles	10 mol %	K$_2$CO$_3$	DMF	120	24 h	(0)	64
Br	CuO	5 mol %	L14 (50 mol %), TBAB, 2,5-hexanedione, K$_3$PO$_4$	H$_2$O	MW (100 W), 140	5 min	(67)	51
I	CuO	5 mol %	L14 (50 mol %), TBAB, 2,5-hexanedione, K$_3$PO$_4$	H$_2$O	MW (100 W), 140	5 min	(88)	51
Br	Cu$_2$O	20 mol %	KOH	DMSO	130	24 h	(85)	398
Cl	Cu$_2$O	20 mol %	ninhydrin (30 mol %), KOH	DMSO	130	48 h	(49)	399
Br	Cu$_2$O	10 mol %	ninhydrin (20 mol %), KOH	DMSO	130	24 h	(91)	399
Cl	Cu$_2$O	10 mol %	phen (20 mol %), TBAF	—	145	24 h	(48)	68
Br	Cu$_2$O	10 mol %	phen (20 mol %), TBAF	—	145	24 h	(98)	68
I	Cu$_2$O	5 mol %	phen (10 mol %), TBAF	—	115	24 h	(98)	68
I	Cu$_2$O	2.5 mol %	(MeO)$_2$phen (7.5 %), PEG, Cs$_2$CO$_3$	n-PrCN	110	24–48 h	(95)	84
I	Cu$_2$O	2.5 mol %	(MeO)$_2$phen (7.5 %), PEG, Cs$_2$CO$_3$	NMP	110	3 h	(92)	430
I	Cu$_2$O	5 mol %	(MeO)$_2$phen (7.5 %), PEG, Cs$_2$CO$_3$	DMF	110	3 h	(95)	430, 84

Nitrogen Nucleophile	Aryl Halide	Conditions	Product(s) and Yield(s) (%)	Refs.

Please refer to the charts preceding the tables for structures indicated by the **bold** numbers.

C3

Catalyst (x mol %), DMSO

R	X	Catalyst	x	Additive(s)	Temp (°)	Time (h)		Refs.
H	Br	CuI	20	D-glucosamine (40 mol %), Cs$_2$CO$_3$	110	22	(90)	429
H	Cl	Cu$_2$O/Cu NP	5	Cs$_2$CO$_3$	150	18	(97)	223
F	Br	CuBr	10	**L20** (20 mol %), Cs$_2$CO$_3$	80	12	(94)	284

Catalyst (x mol %)

X	Catalyst	x	Additive(s)	Solvent	Temp (°)	Time		Refs.
Br	Cu/Al-hydrotalcite	2.5	K$_2$CO$_3$	DMF	100	8 h	(98)	228
I	Cu/Al-hydrotalcite	2.5	K$_2$CO$_3$	DMF	100	6 h	(99)	228
I	CuCl	5	**L17** (10 mol %), NaOH, TBAB	H$_2$O	100	24 h	(92)	235
Br	CuBr	10	**L9** (20 mol %), Cs$_2$CO$_3$	DMSO	75	24 h	(93)	49
I	CuBr	10	**L9** (20 mol %), Cs$_2$CO$_3$	DMSO	45	23 h	(91)	49
Cl	CuI nanoparticles	1.25	K$_2$CO$_3$, air	DMF	110	3 h	(95)	285
I	CuI	5	DMEDA (10 mol %), CsF	THF	60	24 h	(82)	63
Br	CuI	10	oxazolidin-2-one (20 mmol %), NaOMe	DMSO	120	24 h	(78)	392
Br	CuI	20	per-6-ABCD, K$_2$CO$_3$	DMSO	110	24 h	(99)	393
Br	CuI	30	L-Pro (60 mol %)	[C$_4$mim][BF$_4$]	110	24 h	(80)	230
I	CuI	5	8-HOquin (10 mol %), CsF	DMSO	100	1 h	(87)	63
I	CuI	10	**L26** (10 mol %), NaOMe	DMSO	130	12 h	(87)	397
Br	CuI	10	**L22** (10 mol %), K$_2$CO$_3$	DMF	110	24 h	(87)	79
Br	Cu$_2$O	10	**L23** (20 mol %), KOt-Bu	DMF	130	24 h	(95)	400
Cl	Cu$_2$O/Cu NP	5	Cs$_2$CO$_3$	DMSO	150	18 h	(69)	223

Reaction scheme (middle of page):

1-(4-bromophenyl)-1H-imidazole → **Catalyst (x mol %)** → **I** + **II**

(**I** = 1-(4-bromophenyl)imidazole; **II** = 1,4-di(1H-imidazol-1-yl)benzene)

X	Catalyst	x	Additive(s)	Solvent	Temp (°)	Time	I	II	Ref
I	CuI nanoparticles	1.25	K_2CO_3, air	DMF	110	1 h	(99)	(0)	285
I	CuI	20	K_3PO_4	DMF	40	40 h	(87)	(0)	59
I	CuI	10	PPAPM (20 mol %), K_3PO_4	DMF	100	24 h	(89)	(0)	48
I	CuI	5	BtH (10 mol %), KOt-Bu	DMSO	110	8 h	(96)	(0)	431
I	CuI	5	L-Pro (20 mol %), K_3PO_4	dioxane	100	24 h	(80)	(0)	388
Br	CuI	5	**L24** (10 mol %), Cs_2CO_3	DMF	110	24 h	(87)	(0)	86
I	CuI	1	**L21** (20 mol %), Cs_2CO_3	MeCN	reflux	12 h	(83)	(0)	394
Br	CuI	10	**L26** (10 mol %), NaOMe	DMSO	130	12 h	(43)	(0)	397
I	CuI	10	**L26** (10 mol %), NaOMe	DMSO	130	12 h	(76)	(0)	397
I	CuI/PAnNF	5	K_2CO_3	DMF	rt	8 h	(75)	(0)	276
Br	Cu_2O	2.5	(MeO)$_2$phen (7.5 mol %), PEG, Cs_2CO_3	NMP	110	24 h	(0)	(97)	430
Br	Cu_2O	5	(MeO)$_2$phen (15 mol %), PEG, Cs_2CO_3	NMP	110	48 h	(0)	(97)	84
Br	Cu_2O	2.5	(MeO)$_2$phen (7.5 mol %), PEG, Cs_2CO_3	n-PrCN	110	24 h	(82)[a]	(0)	430
I	Cu_2O	5	(MeO)$_2$phen (7.5 mol %), PEG, Cs_2CO_3	n-PrCN	110	24 h	(84)	(0)	430, 84
Br	Cu_2O	10	(MeO)$_2$phen (15 mol %), PEG, Cs_2CO_3	n-PrCN	110	24 h	(0)	(97)	430, 84
I	Cu_2O	5	**L35** (20 mol %)	MeCN	50	36 h	(89)	(0)	40
I	$Cu(OAc)_2$	100	DBU	DMSO	MW, 130	10 min	(89)	(0)	85
I	$CuSO_4$	10	**L36** (20 mol %), Cs_2CO_3	H_2O	120	24 h	(90)	(0)	234
I	**2**	2	NaOH, TBAB	H_2O	100	12 h	(86)	(0)	233

Additional rows:

X	Catalyst	x	Additive(s)	Solvent	Temp (°)	Time	I	II	Ref
Br	$Cu(OAc)_2$	100	DBU	DMSO	MW, 130	10 min	(85)	(85)	85
I	$Cu(OAc)_2$	100	DBU	DMSO	MW, 130	10 min	(92)	(92)	85
Br	$Cu(OAc)_2$•H_2O	20	hippuric acid (20 mol %), Cs_2CO_3	DMF	140	30 h	(53)	(53)	289
I	$Cu(OAc)_2$•H_2O	20	hippuric acid (20 mol %), Cs_2CO_3	DMF	140	20 h	(99)	(99)	289
I	Cu(II)–NaY zeolite	10	K_2CO_3	DMF	120	40 h	(99)	(99)	404
I	**6**	5	Cs_2CO_3	toluene	100	10 h	(90)	(90)	123

TABLE 9A. N-ARYLATION OF IMIDAZOLES (Continued)

Nitrogen Nucleophile	Aryl Halide	Conditions	Product(s) and Yield(s) (%)	Refs.

*Please refer to the charts preceding the tables for structures indicated by the **bold** numbers.*

C₃

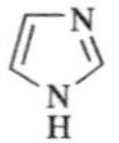

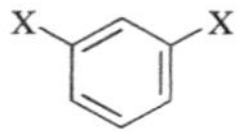

Catalyst (x mol %)

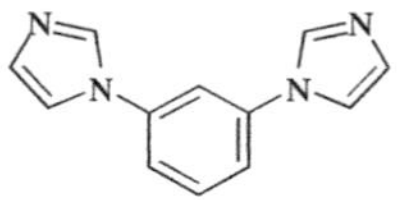

X	Catalyst	x	Additive(s)	Solvent	Temp (°)	Time (h)		Refs.
Br	CuO	25	K₂CO₃	DMSO	150	48	(77)	432
Br	**11**	2.5	K₂CO₃	NMP	180	16	(98)	291
I	CuI	10	CDA (40 mol %), Cs₂CO₃	dioxane	95	24	(95)	58
I	CuI	5	**L31** (5 mol %), Cs₂CO₃	DMF	100	24	(94)	415

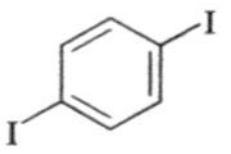

I + **II**

Catalyst (x mol %)

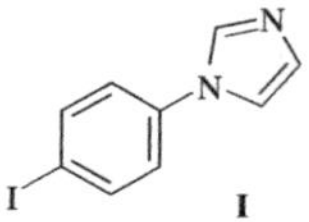

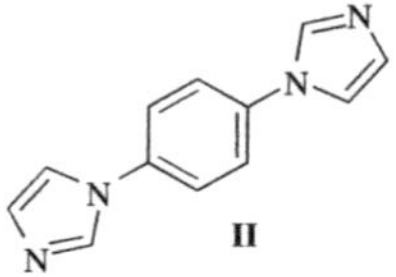

Catalyst	x	Additive(s)	Solvent	Temp (°)	Time (h)	I	II	Refs.
CuBr	10	**L10** (12 mol %), Cs₂CO₃	DMSO	80	24	(85)	(0)	259
CuI	5	**L31** (5 mol %), Cs₂CO₃	DMF	100	24	(0)	(97)	415
Cu₂O	5	**L35** (20 mol %)	MeCN	82	48	(0)	(89)	40
6	5	Cs₂CO₃	toluene	100	16	(90)	(0)	123

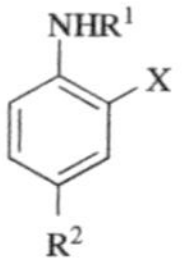

Catalyst (x mol %)

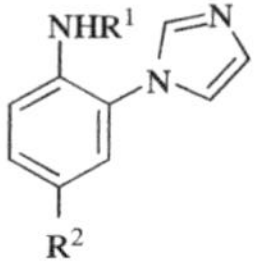

R¹	R²	X	Catalyst	x	Additives	Solvent	Temp (°)	Time (h)		Refs.
Ac	H	I	Cu₂O	5	phen (10 mol %), TBAF	—	115	48	(80)	68
CF₃CO	Me	Br	CuI	20	D-glucosamine (40 mol %), Cs₂CO₃	DMSO	60	20	(90)	429

H₂N–(C₆H₄)–X + imidazole → Catalyst (x mol %) → H₂N–(C₆H₄)–(imidazol-1-yl)

X	Catalyst	x	Additive(s)	Solvent	Temp (°)	Time (h)		
Br	CuI	20	Cs₂CO₃	DMF	120	40	(80)	386
I	CuI	20	K₃PO₄	DMF	40	40	(66)	59
Br	CuI	20	per-6-ABCD (10 mol %), K₂CO₃	DMSO	110	36	(90)	393
Br	CuI	10	L28 (20 mol %), K₂CO₃	DMSO	120	60	(99)	412
Br	CuI	5	L24 (10 mol %), Cs₂CO₃	DMF	110	24	(85)	386
I	Cu₂O	10	KOH	DMSO	125	24	(85)	398
Br	Cu₂O	10	(MeO)₂phen (15 mol %), Cs₂CO₃, PEG	n-PrCN	110	24	(92)	430, 84

R₂N–(C₆H₄)–X + imidazole → Catalyst (x mol %) → R₂N–(C₆H₄)–(imidazol-1-yl)

R	X	Catalyst	x	Additive(s)	Solvent	Temp (°)	Time (h)		
H	Br	CuBr	10	L20 (20 mol %), Cs₂CO₃	DMSO	80	12	(78)	284
H	I	CuBr	5	L5 (10 mol %), Cs₂CO₃	DMF	90	24	(69)	119
H	I	CuI	20	K₃PO₄	DMF	40	40	(73)	59
H	Br	CuI	20	Cs₂CO₃	DMF	120	40	(79)	386
H	I	CuI	5	N-hydroxysuccinimide (10 mol %), NaOMe	DMSO	90	12	(98)	396
H	Br	CuI	10	L28 (20 mol %), K₂CO₃	DMSO	120	60	(86)	412
H	Br	CuI	10	L22 (10 mol %), K₂CO₃	DMF	110	24	(83)	79
H	Br	CuI	10	L24 (10 mol %), Cs₂CO₃	DMF	110	24	(82)	86
H	Br	Cu₂O	10	L23 (20 mol %), KOt-Bu	DMF	130	24	(85)	400
H	Br	CuSO₄	10	L36 (20 mol %), Cs₂CO₃	H₂O	120	24	(46)	234
H	Br	cat. 2	2	NaOH, TBAB	H₂O	100	24	(64)	233
Me	Br	Cu₂O	10	L23 (20 mol %), KOt-Bu	DMF	130	24	(98)	400

TABLE 9A. N-ARYLATION OF IMIDAZOLES (Continued)

Nitrogen Nucleophile	Aryl Halide	Conditions	Product(s) and Yield(s) (%)	Refs.

Please refer to the charts preceding the tables for structures indicated by the **bold** numbers.

C_3

Imidazole (N–N–H)

Aryl Halide: AcHN–C₆H₄–X

Conditions: CuI (x mol %),

Product: AcHN–phenyl–imidazole

X	x	Additive(s)	Solvent	Temp (°)	Time (h)		Refs.
Br	20	D-glucosamine (40 mol %), Cs₂CO₃	DMSO	110	24	(72)	429
Br	10	**L22** (10 mol %), K₂CO₃	DMF	110	24	(82)	79
I	20	K₃PO₄	DMF	40	40	(98)	59

Aryl Halide: NO₂–C₆H₄–X (ortho)

Conditions: Catalyst (x mol %)

Product: nitro-phenyl-imidazole

X	Catalyst	x	Additive(s)	Solvent	Temp (°)	Time (h)		Refs.
Cl	Cu/Al–hydrotalcite	2.5	K₂CO₃	DMF	100	6	(92)	228
I	Cu/Al–hydrotalcite	2.5	K₂CO₃	DMF	100	6	(99)	228
Cl	CuI nanoparticles	1.25	K₂CO₃, air	DMF	110	2	(99)	285
I	CuI/PAnNF	5	K₂CO₃	DMF	rt	4	(99)	276
F	CuO nanoparticles	10	K₂CO₃	DMF	120	2	(89)	64
Cl	CuO nanoparticles	10	K₂CO₃	DMF	120	4	(86)	64
Cl	CuO	8	K₂CO₃	pyridine	reflux	24	(64)	402
Cl	Cu fluorapatite	7	K₂CO₃	DMF	120	3	(88)	66
Cl	[Cu(μ-I)((−)-sparteine)]₂	1	K₂CO₃	DMSO	125	12	(92)	414
Cl	**8**	25	K₂CO₃	DMSO	135	24	(90)	65
Br	**8**	25	K₂CO₃	DMSO	135	18	(90)	65
F	**7**	1	K₂CO₃	DMF	110	1.5	(93)	73
Cl	**7**	1	K₂CO₃	DMF	110	3	(91)	73

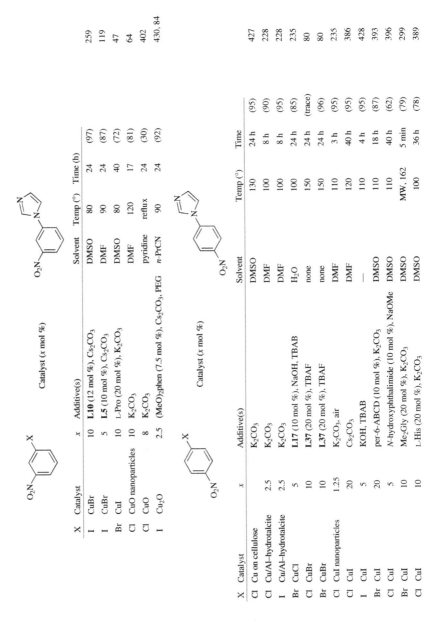

Catalyst (x mol %)

X	Catalyst	x	Additive(s)	Solvent	Temp (°)	Time (h)		
I	CuBr	10	L10 (12 mol %), Cs₂CO₃	DMSO	80	24	(97)	259
I	CuBr	5	L5 (10 mol %), Cs₂CO₃	DMF	90	24	(87)	119
Br	CuI	10	L-Pro (20 mol %), K₂CO₃	DMSO	80	40	(72)	47
Cl	CuO nanoparticles	10	K₂CO₃	DMF	120	17	(81)	64
Cl	CuO	8	K₂CO₃	pyridine	reflux	24	(30)	402
I	Cu₂O	2.5	(MeO)₂phen (7.5 mol %), Cs₂CO₃, PEG	n-PrCN	90	24	(92)	430, 84

Catalyst (x mol %)

X	Catalyst	x	Additive(s)	Solvent	Temp (°)	Time		
Cl	Cu on cellulose		K₂CO₃	DMSO	130	24 h	(95)	427
Cl	Cu/Al–hydrotalcite	2.5	K₂CO₃	DMF	100	8 h	(90)	228
I	Cu/Al–hydrotalcite	2.5	K₂CO₃	DMF	100	8 h	(95)	228
Br	CuCl	5	L17 (10 mol %), NaOH, TBAB	H₂O	100	24 h	(85)	235
Cl	CuBr	10	L37 (20 mol %), TBAF	none	150	24 h	(trace)	80
Br	CuBr	10	L37 (20 mol %), TBAF	none	150	24 h	(96)	80
Cl	CuI nanoparticles	1.25	K₂CO₃, air	DMF	110	3 h	(95)	235
Cl	CuI	20	Cs₂CO₃	DMF	120	40 h	(95)	386
I	CuI	5	KOH, TBAB	—	110	4 h	(95)	428
Br	CuI	20	per-6-ABCD (10 mol %), K₂CO₃	DMSO	110	18 h	(87)	393
Cl	CuI	5	N-hydroxyphthalimide (10 mol %), NaOMe	DMSO	110	40 h	(62)	396
Br	CuI	10	Me₂Gly (20 mol %), K₂CO₃	DMSO	MW, 162	5 min	(79)	299
Cl	CuI	10	L-His (20 mol %), K₂CO₃	DMSO	100	36 h	(78)	389

TABLE 9A. *N*-ARYLATION OF IMIDAZOLES (*Continued*)

*Please refer to the charts preceding the tables for structures indicated by the **bold** numbers.*

Continued from previous page.

C₃

Nitrogen Nucleophile	Aryl Halide	Conditions	Product(s) and Yield(s) (%)	Refs.

X	Catalyst	x	Additive(s)	Solvent(s)	Temp (°)	Time		Refs.
Br	CuI	10	L-His (20 mol %), K₂CO₃	DMSO	100	48 h	(92)	389
I	CuI	10	L-His (20 mol %), K₂CO₃	DMSO	100	30 h	(91)	389
Br	CuI	10	L-Glu (20 mol %), K₃PO₄	—	MW (200 W), 130	7 h	(51)	418
Br	CuI	30	L-Pro (60 mol %)	[C₄mim][BF₄]	110	24 h	(62)	230
Cl	CuI	10	L26 (10 mol %), NaOMe	DMSO	110	12 h	(91)	397
Cl	CuI	5	L24 (10 mol %), Cs₂CO₃	DMF	110	24 h	(98)	86
Cl	CuI	10	L28 (20 mol %), K₂CO₃	DMSO	120	30 h	(74)	412
Br	CuI	10	L22 (10 mol %), K₂CO₃	DMF	110	24 h	(97)	79
Br	CuI	10	L32 (20 mol %), Cs₂CO₃	DMF	110	24 h	(96)	417
Cl	Cu/PAnNF	5	K₂CO₃	DMF	80	6 h	(93)	276
F	CuO nanoparticles	10	K₂CO₃	DMF	120	30 min	(91)	64
Cl	CuO nanoparticles	10	K₂CO₃	DMF	120	1.5 h	(88)	64
I	CuO nanoparticles	5	KOH	DMSO/t-BuOH (1:3)	110	11 h	(89)	74
Cl	CuO	8	K₂CO₃	pyridine	reflux	24 h	(54)	402
Br	Cu₂O	10	L23 (20 mol %), Cs₂CO₃	DMF	130	24 h	(70)	400
I	Cu(OAc)₂	100	DBU	DMSO	MW, 130	10 min	(89)	85
Br	Cu(OAc)₂•H₂O	20	hippuric acid (20 mol %), Cs₂CO₃	DMF	140	30 h	(70)	289
I	Cu(OAc)₂•H₂O	20	hippuric acid (20 mol %), Cs₂CO₃	DMF	140	30 h	(99)	289
Cl	Cu(OAc)₂•H₂O	15	(−)-sparteine (30 mol %), K₂CO₃	DMF	130	36 h	(48)	290
Br	Cu(OAc)₂•H₂O	15	(−)-sparteine (30 mol %), K₂CO₃	DMF	130	36 h	(83)	290
Br	CuSO₄	10	L36 (20 mol %), Cs₂CO₃	H₂O	120	24 h	(95)	234
Cl	Cu(II)-NaY zeolite	10	K₂CO₃	DMF	120	20 h	(92)	404
Cl	[Cu(μ-I)((−)-sparteine)]₂	1	K₂CO₃	DMSO	125	12 h	(91)	414

Table (continued):

Isomer	Catalyst	x	Base/Additive	Solvent	Temp (°)	Time	(Yield %)	Ref
F	Cu fluorapatite	7	K$_2$CO$_3$	DMF	120	30 min	(85)	66
Cl	Cu fluorapatite	7	K$_2$CO$_3$	DMF	120	1 h	(90)	66
Br	Cu fluorapatite	12.5	none	DMSO	110	4 h	(95)	227
Br 3		2	NaOH, TBAB	H$_2$O	100	12 h	(85)	233
I 3		2	NaOH, TBAB	H$_2$O	100	12 h	(93)	233
Br 6		5	Cs$_2$CO$_3$	toluene	100	12 h	(90)	123
I 6		5	Cs$_2$CO$_3$	toluene	100	3 h	(90)	123
Cl 8		25	K$_2$CO$_3$	DMSO	135	24 h	(95)	65
Br 8		25	K$_2$CO$_3$	DMSO	135	18 h	(96)	65
F 7		1	K$_2$CO$_3$	DMF	110	1 h	(95)	73
Cl 7		1	K$_2$CO$_3$	DMF	110	1.5 h	(94)	73

Catalyst (x mol %), K$_2$CO$_3$, DMF

Catalyst	x	Temp (°)	Time (h)		Ref
Cu fluorapatite	7	120	1	(85)	66
7	1	110	3	(91)	73

Catalyst (x mol %)

Isomer	X	Catalyst	x	Additive(s)	Solvent	Temp (°)	Time (h)		Ref
2	Br	CuI	5	L24 (10 mol %), Cs$_2$CO$_3$	DMF	110	48	(61)	86
2	Br	CuI	10	L28 (20 mol %), K$_2$CO$_3$	DMSO	120	60	(74)	412
4	Br	CuI	20	Cs$_2$CO$_3$	DMF	120	40	(85)	386
4	I	CuI	20	K$_3$PO$_4$	DMF	40	40	(63)	59
4	Br	CuI	10	L28 (20 mol %), K$_2$CO$_3$	DMSO	120	60	(97)	412
4	Br	CuI	5	L24 (10 mol %), Cs$_2$CO$_3$	DMF	110	24	(70)	86
4	Br	Cu$_2$O	5	(MeO$_2$)$_2$phen (15 mol %), Cs$_2$CO$_3$, PEG	n-PrCN	110	24	(84)	430, 84
4	Br	Cu$_2$O	10	L23 (20 mol %), KOt-Bu	DMF	130	24	(95)	400
4	Br	CuSO$_4$	10	L36 (20 mol %), Cs$_2$CO$_3$	H$_2$O	120	24	(38)	234
4	Br	cat. 2	2	NaOH, TBAB	H$_2$O	100	24	(58)	233

TABLE 9A. N-ARYLATION OF IMIDAZOLES (Continued)

Nitrogen Nucleophile	Aryl Halide	Conditions	Product(s) and Yield(s) (%)	Refs.

*Please refer to the charts preceding the tables for structures indicated by the **bold** numbers.*

C3

R	X	Catalyst	x	Additive(s)	Solvent	Temp (°)	Time		Refs.
H	I	Cu on cellulose	1	K2CO3	DMSO	130	12 h	(40)	427
H	I	CuI	20	K3PO4	DMF	40	40 h	(78)	59
H	Br	CuI	20	Cs2CO3	DMF	120	40 h	(57)	386
H	Br	CuI	10	L-Lys (20 mol %), K3PO4	—	MW (200 W), 130	6.5 h	(95)	418
H	I	CuI	5	BtH (10 mol %), KOt-Bu	DMSO	110	8 h	(93)	431
H	Br	CuI	5	L24 (10 mol %), Cs2CO3	DMF	110	48 h	(78)	386
H	Br	CuI	10	L28 (20 mol %), K2CO3	DMSO	120	6 h	(90)	412
H	I	Cu2O	10	ninhydrin (20 mol %), KOH	DMSO	110	24 h	(90)	399
H	Br	Cu2O	10	L23 (20 mol %), KOt-Bu	DMF	130	24 h	(72)	400
H	I	Cu(OAc)2	100	DBU	DMSO	MW, 130	10 min	(92)	85
H	Br	[Cu(μ-I)((−)-sparteine)]2	1	K2CO3	DMSO	115	24 h	(72)	414
H	I	[Cu(μ-I)((−)-sparteine)]2	1	K2CO3	DMSO	115	12 h	(92)	414
Na	I	CuI	10	none	DMF	150	4 h	(47)	433

X	Catalyst	x	Additive(s)	Solvent	Temp (°)	Time		Refs.
I	CuI	10	DMG (20 mol %), TBPE	DMSO	rt	24 h	(93)	61
Br	CuI	10	L-Lys (20 mol %), K3PO4	—	MW (200 W), 130	5 h	(99)	418
Br	CuI	20	per-6-ABCD (10 mol %), K2CO3	DMSO	110	18 h	(90)	393
Br	Cu2O	10	ninhydrin (20 mol %), KOH	DMSO	130	24 h	(90)	399
Br	Cu2O	10	L23 (20 mol %), KOt-Bu	DMF	130	24 h	(91)	400
I	Cu(OAc)2	100	DBU	DMSO	MW, 130	10 min	(90)	85

MeO–C6H4–X + imidazole → MeO–C6H4–(N-imidazole)   Catalyst (x mol %)

X	Catalyst	x	Additive(s)	Solvent(s)	Temp (°)	Time	
I	Cu on cellulose	1	K2CO3	DMSO	130	12 h (89)	427
Cl	Cu/Al–hydrotalcite	2.5	K2CO3	DMF	100	8 h (59)	228
I	Cu/Al–hydrotalcite	2.5	K2CO3	DMF	100	8 h (90)	228
I	CuF2	50	phen (50 mol %), K2CO3	DMF	140	96 h (80)	419
Br	CuBr	2.5	K2CO3	NMP	160	16 h (91)	287
Br	CuBr	10	L20 (20 mol %), Cs2CO3	DMSO	80	12 h (85)	284
I	CuBr	5	L5 (10 mol %), Cs2CO3	DMF	90	24 h (78)	119
Cl	CuI nanoparticles	1.25	K2CO3, air	DMF	110	5 h (95)	285
Br	CuI	20	Cs2CO3	DMF	120	40 h (89)	386
I	CuI	20	K3PO4	DMF	40	40 h (95)	59
I	CuI	5	KOH, TBAB	—	110	6 h (93)	428
I	CuI	5	BtH (10 mol %), KOt-Bu	DMSO	110	8 h (95)	431
I	CuI	10	PPAPM (20 mol %), K3PO4	DMF	100	24 h (85)	48
Br	CuI	5	N-hydroxysuccinimide (10 mol %), NaOMe	DMSO	110	40 h (88)	396
Br	CuI	10	oxazolidin-2-one (20 mol %), NaOMe	DMSO	120	12 h (83)	392
I	CuI	10	CDA(20 mol %), K3PO4	[C4mim][BF4]	110	12 h (75)	232
Br	CuI	20	phen (20 mol %), KF/Al2O3	xylene	140	16 h (89)	410
I	CuI	10	phen (10 mol %), KF/Al2O3	—	140	15 h (94)	411
Br	CuI	10	DMG (20 mol %), K2CO3	DMSO	110	45 h (95)	47
Br	CuI	10	DMG (20 mol %), K2CO3	DMSO	MW, 171	0.25 h (61)	299
Br	CuI	20	d-glucosamine (40 mol %), Cs2CO3	DMSO	110	24 h (73)	429
Br	CuI	10	L-Lys (20 mol %), K3PO4	—	MW (200 W), 130	5 h (99)	418
Br	CuI	10	L-His (20 mol %), K2CO3	DMSO	100	48 h (70)	389
I	CuI	10	L-His (20 mol %), K2CO3	DMSO	100	30 h (86)	389
Br	CuI	20	L-Pro (20 mol %), K3PO4	DMF	140	48 h (75)	388

TABLE 9A. N-ARYLATION OF IMIDAZOLES (Continued)

Nitrogen Nucleophile	Aryl Halide	Conditions	Product(s) and Yield(s) (%)	Refs.

*Please refer to the charts preceding the tables for structures indicated by the **bold** numbers.*

Continued from previous page.

C₃

X	Catalyst	x	Additive(s)	Solvent(s)	Temp (°)	Time		Refs.
I	CuI	5	L-Pro (20 mol %), K₃PO₄	dioxane	100	24 h	(40)	388
I	CuI	5	L-Pro (10 mol %), K₂CO₃	DMSO	90	36 h	(91)	401, 47
Br	CuI	10	L22 (10 mol %), K₂CO₃	DMF	110	36 h	(85)	79
Br	CuI	5	L24 (10 mol %), Cs₂CO₃	DMF	110	24 h	(88)	86
Br	CuI	10	L28 (20 mol %), K₂CO₃	DMSO	120	60 h	(97)	412
Br	CuI	10	L32 (20 mol %), Cs₂CO₃	DMF	110	24 h	(30)	417
I	CuI	10	L6 (20 mol %), Cs₂CO₃	DMF	rt	24 h	(81)	273
I	CuI/PAnNF	5	K₂CO₃	DMF	rt	12 h	(65)	276
Br	CuO	5	L14 (50 mol %), TBAB, 2.5-hexanedione, K₃PO₄	H₂O	MW (100 W), 140	5 min	(89)	51
I	CuO	5	L14 (50 mol %), TBAB, 2.5-hexanedione, K₃PO₄	H₂O	MW (100 W), 140	5 min	(77)	51
I	CuO nanoparticles	5	KOH	t-BuOH/DMSO (3:1)	110	11 h	(64)	74
I	Cu₂O	10	KOH	DMSO	125	24 h	(90)	398
Br	Cu₂O	10	phen (20 mol %), TBAF	—	145	24 h	(89)	68
I	Cu₂O	5	phen (10 mol %), TBAF	—	115	48 h	(90)	68
Br	Cu₂O	10	ninhydrin (20 mol %), KOH	DMSO	110	24 h	(89)	399
Br	Cu₂O	10	L23 (20 mol %), KOt-Bu	DMF	130	24 h	(92)	400
I	Cu₂O	5	L27 (20 mol %)	MeCN	82	72 h	(97)	40
Cl	Cu₂O/Cu NP		Cs₂CO₃	DMSO	150	18 h	(0)	223
Br	Cu(OAc)₂	100	DBU	DMSO	MW, 130	10 min	(87)	85
I	Cu(OAc)₂	100	DBU	DMSO	MW, 130	10 min	(92)	85
I	Cu(OTf)₂•C₆H₆	10	phen (1 eq), dba, Cs₂CO₃	xylenes	110	48 h	(96)	11
Br	CuSO₄	10	L36 (20 mol %), Cs₂CO₃	H₂O	120	24 h	(32)	234
I	Cu(IPr)Cl	10	Cs₂CO₃	DMSO	100	24 h	(67)	421
Cl	Cu fluorapatite	7	K₂CO₃	DMF	120	36 h	(52)	66
Br	[Cu(μ-I)((−)-sparteine]₂	1	K₂CO₃	DMSO	115	12 h	(85)	414
I	[Cu(μ-I)((−)-sparteine]₂	1	K₂CO₃	DMSO	115	7 h	(95)	414

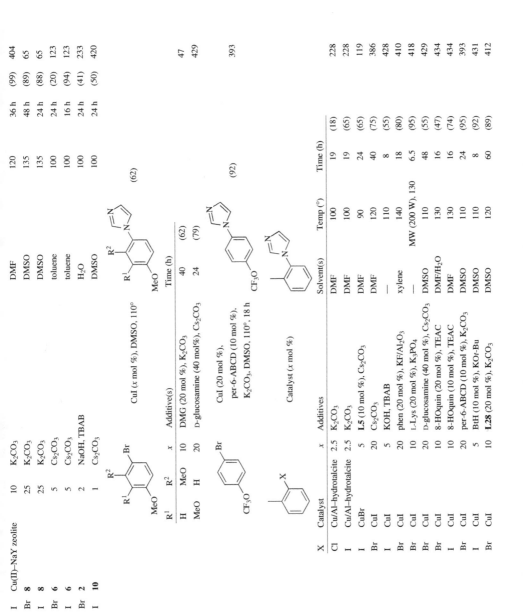

I	Cu(II)–NaY zeolite	10	K₂CO₃	DMF	120	36 h	(99)	404
Br	**8**	25	K₂CO₃	DMSO	135	48 h	(89)	65
I	**8**	25	K₂CO₃	DMSO	135	24 h	(88)	65
Br	**6**	5	Cs₂CO₃	toluene	100	24 h	(20)	123
I	**6**	5	Cs₂CO₃	toluene	100	16 h	(94)	123
Br	**2**	2	NaOH, TBAB	H₂O	100	24 h	(41)	233
I	**10**	1	Cs₂CO₃	DMSO	100	24 h	(50)	420

CuI (x mol %), DMSO, 110°

R^1	R^2	x	Additive(s)	Time (h)		
H	MeO	10	DMG (20 mol %), K₂CO₃	40	(62)	47
MeO	H	20	D-glucosamine (40 mol%), Cs₂CO₃	24	(79)	429

CuI (20 mol %),
per-6-ABCD (10 mol %),
K₂CO₃, DMSO, 110°, 18 h | (92) | 393

Catalyst (x mol %)

X	Catalyst	x	Additives	Solvent(s)	Temp (°)	Time (h)		
Cl	Cu/Al-hydrotalcite	2.5	K₂CO₃	DMF	100	19	(18)	228
I	Cu/Al-hydrotalcite	2.5	K₂CO₃	DMF	100	19	(65)	228
I	CuBr	5	**L5** (10 mol %), Cs₂CO₃	DMF	90	24	(65)	119
Br	CuI	20	Cs₂CO₃	DMF	120	40	(75)	386
I	CuI	5	KOH, TBAB	—	110	8	(55)	428
Br	CuI	20	phen (20 mol %), KF/Al₂O₃	xylene	140	18	(80)	410
Br	CuI	10	L-Lys (20 mol %), K₃PO₄	—	MW (200 W), 130	6.5	(95)	418
Br	CuI	20	D-glucosamine (40 mol %), Cs₂CO₃	DMSO	110	48	(55)	429
Br	CuI	10	8-HOquin (20 mol %), TEAC	DMF/H₂O	130	16	(47)	434
I	CuI	10	8-HOquin (10 mol %), TEAC	DMF	130	16	(74)	434
Br	CuI	20	per-6-ABCD (10 mol %), K₂CO₃	DMSO	110	24	(95)	393
I	CuI	5	BtH (10 mol %), KOt-Bu	DMSO	110	8	(92)	431
Br	CuI	10	**L28** (20 mol %), K₂CO₃	DMSO	120	60	(89)	412

TABLE 9A. N-ARYLATION OF IMIDAZOLES (Continued)

Nitrogen Nucleophile	Aryl Halide	Conditions	Product(s) and Yield(s) (%)	Refs.

*Please refer to the charts preceding the tables for structures indicated by the **bold** numbers.*

Continued from previous page.

C3

X	Catalyst	x	Additive(s)	Solvent(s)	Temp (°)	Time (h)		Refs.
Br	CuI	5	**L24** (10 mol %), Cs$_2$CO$_3$	DMF	110	24	(78)	86
Br	CuI	10	**L22** (10 mol %), K$_2$CO$_3$	DMF	110	24	(81)	79
I	Cu$_2$O	10	KOH	DMSO	125	24	(88)	398
Br	Cu$_2$O	10	phen (20 mol %), TBAF	—	145	24	(84)	68
I	Cu$_2$O	5	phen (10 mol %), TBAF	—	115	48	(81)	68
Br	Cu$_2$O	10	(MeO)$_2$phen (15 mol %), Cs$_2$CO$_3$	NMP	110	24	(85)	430, 84
I	Cu$_2$O	10	ninhydrin (20 mol %), KOH	DMSO	110	24	(89)	399
I	Cu$_2$O	5	**L35** (20 mol %)	MeCN	82	36	(92)	40
Br	Cu$_2$O	10	**L23** (20 mol %), KO*t*-Bu	DMF	130	24	(92)	400
I	Cu(TMHD)$_2$	20	KO*t*-Bu	DMF	120	24	(74)	308
Br	cat. **8**	25	K$_2$CO$_3$	DMSO	135	48	(60)	65
I	cat. **8**	25	K$_2$CO$_3$	DMSO	135	38	(75)	435
I	cat. **6**	5	Cs$_2$CO$_3$	toluene	100	24	(79)	123

Catalyst (x mol %)

X	Catalyst	x	Additive(s)	Solvent(s)	Temp (°)	Time (h)		Refs.
I	CuBr	5	**L5** (10 mol %), Cs$_2$CO$_3$	DMF	90	24	(75)	119
I	CuI	5	BtH (10 mol %), KO*t*-Bu	DMSO	110	8	(98)	431
Br	CuI	30	L-Pro (60 mol %)	[C$_4$mim][BF$_4$]	110	24	(75)	230
Cl	CuI	10	8-HOquin (20 mol %), TEAC	DMF/H$_2$O	130	60	(40)	434
I	Cu$_2$O	10	KOH	DMSO	125	24	(90)	398

Catalyst (x mol %)

X	Catalyst	x	Additive(s)	Solvent	Temp (°)	Time (h)		
Cl	Cu/Al–hydrotalcite	2.5	K$_2$CO$_3$	DMF	100	12	(27)	228
I	Cu/Al–hydrotalcite	2.5	K$_2$CO$_3$	DMF	100	12	(89)	228
I	CuCl	5	L17 (10 mol %), NaOH, TBAB	H$_2$O	100	24	(88)	235
Br	CuBr	2.5	K$_2$CO$_3$	NMP	160	16	(89)	287
I	CuBr	10	L10 (12 mol %), Cs$_2$CO$_3$	DMSO	80	24	(85)	259
Br	CuBr	10	L9 (20 mol %), Cs$_2$CO$_3$	DMSO	75	28	(90)	49
Br	CuBr	10	L20 (20 mol %), Cs$_2$CO$_3$	DMSO	80	12	(80)	284
Cl	CuI nanoparticles	1.25	K$_2$CO$_3$, air	DMF	110	14	(36)	285
Br	CuI	20	Cs$_2$CO$_3$	DMF	120	40	(80)	386
I	CuI	20	K$_3$PO$_4$	DMF	40	40	(80)	59
I	CuI	5	KOH, TBAB	–	110	8	(70)	428
I	CuI	5	BtH (10 mol %), KOt-Bu	DMSO	110	8	(96)	431
Br	CuI	5	N-hydroxyphthalimide (10 mol %), NaOMe	DMSO	110	40	(73)	396
Br	CuI	20	L-Pro (20 mol %), K$_3$PO$_4$	DMF	140	48	(89)	388
I	CuI	5	L-Pro (20 mol %), K$_3$PO$_4$	dioxane	100	24	(88)	388
Br	CuI	30	L-Pro (60 mol %)	[C$_4$mim][BF$_4$]	110	24	(82)	230
Br	CuI	10	L-His (20 mol %), K$_2$CO$_3$	DMSO	100	48	(71)	389
I	CuI	10	L-His (20 mol %), K$_2$CO$_3$	DMSO	110	30	(89)	389
Br	CuI	20	L-glucosamine (40 mol %), Cs$_2$CO$_3$	DMSO	110	24	(80)	429
Br	CuI	10	L-Lys (20 mol %), K$_3$PO$_4$	–	MW (200 W), 130	2.5	(49)	418
Br	CuI	20	per-6-ABCD (10 mol %), K$_2$CO$_3$	DMSO	110	24	(96)	393
I	CuI	10	L26 (10 mol %), NaOMe	DMSO	130	12	(48)	397
I	CuI	10	L6 (20 mol %), Cs$_2$CO$_3$	DMF	rt	24	(90)	273
Br	CuI	10	L21 (20 mol %), KI, Cs$_2$CO$_3$	MeCN	reflux	12	(43)	394
Br	CuI	10	L22 (10 mol %), K$_2$CO$_3$	DMF	110	24	(84)	79

TABLE 9A. N-ARYLATION OF IMIDAZOLES (Continued)

Nitrogen Nucleophile	Aryl Halide	Conditions	Product(s) and Yield(s) (%)	Refs.

Please refer to the charts preceding the tables for structures indicated by the **bold** numbers.

Continued from previous page.

C₃

Catalyst (x mol %)

X	Catalyst	x	Additive(s)	Solvent	Temp (°)	Time (h)		Refs.
Br	CuI	5	**L24** (10 mol %), Cs₂CO₃	DMF	110	24 h	(89)	86
Br	CuI	10	**L28** (20 mol %), K₂CO₃	DMSO	120	60 h	(98)	412
I	CuI/PAnNF	5	K₂CO₃	DMF	rt	14 h	(55)	276
Br	CuO	5	**L14** (50 mol %), TBAB, 2.5-hexanedione. K₃PO₄	H₂O	MW (100 W), 140	5 min	(76)	51
Br	Cu₂O	20	KOH	DMSO	130	24 h	(88)	398
I	Cu₂O	10	KOH	DMSO	125	24 h	(91)	398
I	Cu₂O	5	phen (10 mol %), TBAF	—	115	24 h	(94)	68
Cl	Cu₂O	20	ninhydrin (30 mol %), KOH	DMSO	130	48 h	(46)	399
I	Cu₂O	10	ninhydrin (20 mol %), KOH	DMSO	110	24 h	(91)	399
Br	Cu₂O	10	**L23** (20 mol %), KOt-Bu	DMF	130	24 h	(99)	400
I	Cu(OAc)₂	100	DBU	DMSO	MW, 130	10 min	(91)	85
Br	CuSO₄	10	**L36** (20 mol %), Cs₂CO₃	H₂O	120	48 h	(47)	234
I	CuSO₄	10	**L36** (20 mol %), Cs₂CO₃	H₂O	120	24 h	(78)	234
Br	Cu(II)–NaY zeolite	10	K₂CO₃	DMF	120	36 h	(90)	404
I	Cu(II)–NaY zeolite	10	K₂CO₃	DMF	120	24 h	(99)	404
Br	[Cu(μ-I)((−)-sparteine)]₂	1	K₂CO₃	DMSO	115	12 h	(88)	414
Cl	Cu fluorapatite	7	K₂CO₃	DMF	120	36 h	(85)	66
Br	**8**	25	K₂CO₃	DMSO	135	24 h	(91)	65
I	**6**	5	Cs₂CO₃	toluene	100	8 h	(90)	123
Br	**2**	2	NaOH, TBAB	H₂O	100	24 h	(47)	233
I	**2**	2	NaOH, TBAB	H₂O	100	12 h	(78)	233
I	**10**	3	Cs₂CO₃	DMSO	100	24 h	(93)	421
I	**10**	1	Cs₂CO₃	DMSO	100	24 h	(20)	420

CuI (20 mol %),
per-6-ABCD (10 mol%),
K$_2$CO$_3$, DMSO,
110°, 24 h

(81)

Isomer	X	Catalyst	x	Additive(s)	Solvent	Temp (°)	Time		
2	Cl	Cu/Al-hydrotalcite	2.5	K$_2$CO$_3$	DMF	100	7 h	(97)	228
2	I	Cu(OAc)$_2$	100	DBU	DMSO	MW, 130	10 min	(trace)	85
3	I	CuBr	5	L5 (10 mol %), Cs$_2$CO$_3$	DMF	90	24 h	(85)	119
3	I	CuI	10	phen (10 mol %), KF/Al$_2$O$_3$	—	140	11 h	(98)	411
3	Br	CuI	20	per-6-ABCD (10 mol %), K$_2$CO$_3$	DMSO	110	18 h	(93)	393
3	I	Cu$_2$O	10	KOH	DMSO	110	16 h	(92)	398
3	I	Cu$_2$O	10	ninhydrin (20 mol %), KOH	DMSO	90	16 h	(92)	399
3	I	Cu(OAc)$_2$	100	DBU	DMSO	MW, 130	10 min	(94)	85
3	I	Cu(OTf)$_2$•C$_6$H$_6$	10	phen (1 eq), dba, Cs$_2$CO$_3$	xylenes	110	24 h	(94)	11
4	Cl	CuI nanoparticles	1.25	K$_2$CO$_3$, air	DMF	110	5 h	(95)	285
4	Cl	CuI	5	N-hydroxymaleimide (10 mol %), NaOMe	DMSO	110	24 h	(48)	396
4	Br	CuI	10	oxazolidin-2-one (20 mol %), NaOMe	DMSO	120	24 h	(45)	392
4	Br	CuI	20	D-glucosamine (40 mol %), Cs$_2$CO$_3$	DMSO	110	24 h	(72)	429
4	Cl	Cu$_2$O	10	KOH	DMSO	130	24 h	(88)	398
4	Br	Cu$_2$O	10	KOH	DMSO	110	24 h	(90)	398
4	Cl	Cu$_2$O	10	ninhydrin (20 mol %), KOH	DMSO	130	24 h	(90)	399
4	Br	Cu$_2$O	10	ninhydrin (20 mol %), KOH	DMSO	110	24 h	(91)	399
4	Br	Cu$_2$O	5	L27 (20 mol %)	MeCN	82	72 h	(94)	40
4	I	Cu$_2$O	5	L35 (20 mol %)	MeCN	50	24 h	(100)	40
4	Cl	Cu$_2$O/Cu NP	5	Cs$_2$CO$_3$	DMSO	150	18 h	(90)	223
4	Br	Cu(OAc)$_2$	100	DBU	DMSO	MW, 130	10 min	(87)	85
4	I	Cu(OAc)$_2$	100	DBU	DMSO	MW, 130	10 min	(92)	85
4	Br	CuSO$_4$	10	L36 (20 mol %), Cs$_2$CO$_3$	H$_2$O	120	24 h	(88)	234
4	Cl	Cu(II)-NaY zeolite	10	K$_2$CO$_3$	DMF	120	30 h	(97)	404
4	Cl	[Cu(μ-I)((−)-sparteine)]$_2$	1	K$_2$CO$_3$	DMSO	125	15 h	(89)	414
4	Cl	8	25	K$_2$CO$_3$	DMSO	135	30 h	(85)	65
4	Br	2	2	NaOH, TBAB	H$_2$O	100	12 h	(88)	233

TABLE 9A. N-ARYLATION OF IMIDAZOLES (*Continued*)

Nitrogen Nucleophile	Aryl Halide	Conditions	Product(s) and Yield(s) (%)	Refs.

*Please refer to the charts preceding the tables for structures indicated by the **bold** numbers.*

C₃

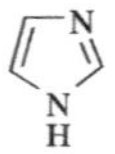

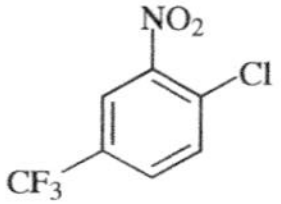

CuI (10 mol %), 12 h

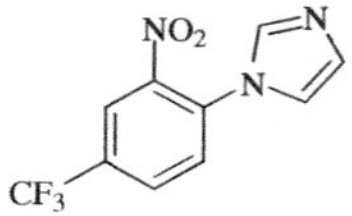

Additives	Solvent	Temp (°)		
L21 (20 mol %), Cs₂CO₃	MeCN	reflux	(79)	394
L21 (20 mol %), Cs₂CO₃	MeCN	reflux	(43)	397
L26 (10 mol %), NaOMe	DMSO	130	(90)	397

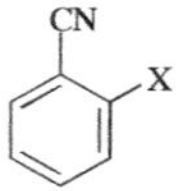

Catalyst (*x* mol %)

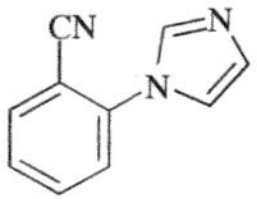

X	Catalyst	*x*	Additive(s)	Solvent	Temp (°)	Time (h)		Refs.
Cl	Cu/Al–hydrotalcite	2.5	K₂CO₃	DMF	100	12	(99)	228
Cl	CuI nanoparticles	1.25	K₂CO₃, air	DMF	110	2	(87)	285
Cl	CuI	5	N-hydroxysuccinimide (10 mol %), NaOMe	DMSO	110	24	(52)	396
F	CuO nanoparticles	10	K₂CO₃	DMF	120	1.5	(91)	64
Cl	CuO nanoparticles	10	K₂CO₃	DMF	120	12	(91)	64
Cl	CuO	8	K₂CO₃	pyridine	reflux	50	(43)	402
F	Cu fluorapatite	7	K₂CO₃	DMF	120	1	(82)	66
Cl	[Cu(μ-I)(−)-sparteine)]₂	1	K₂CO₃	DMSO	125	14	(86)	414
F	**7**	1	K₂CO₃	DMF	110	1.5	(92)	73
Cl	**7**	1	K₂CO₃	DMF	110	5	(88)	73
Cl	**8**	25	K₂CO₃	DMSO	135	24	(84)	65

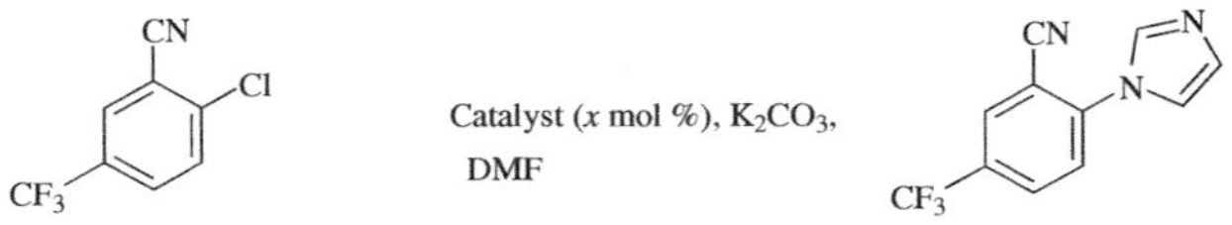

Reaction 1

2-Chloro-5-(trifluoromethyl)benzonitrile → 2-(imidazol-1-yl)-5-(trifluoromethyl)benzonitrile

Catalyst (x mol %), K₂CO₃, DMF

Catalyst	x	Temp (°)	Time (h)		
Cu fluorapatite	7	120	3	(92)	66
cat. 7	1	110	2	(95)	73

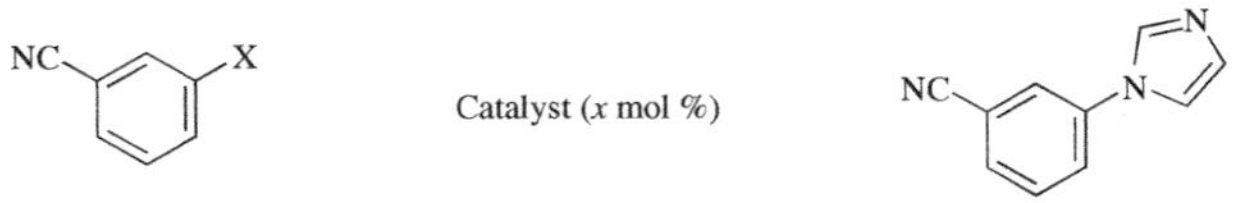

Reaction 2

3-halobenzonitrile → 3-(imidazol-1-yl)benzonitrile

Catalyst (x mol %)

X	Catalyst	x	Additive(s)	Solvent	Temp (°)	Time (h)		
I	CuBr	10	**L10** (12 mol %), Cs₂CO₃	DMSO	80	24	(93)	259
I	CuI	5	8-HOquin (10 mol %), CsF	DMSO	80	4	(85)	63
Br	CuI	20	per-6-ABCD (10 mol %), K₂CO₃	DMSO	110	18	(99)	393
I	Cu₂O	5	(MeO)₂phen (7.5 mol %), Cs₂CO₃, PEG	MeCN	80	24	(95)	430, 84
Br	CuO	8	K₂CO₃	pyridine	reflux	50	(13)	402

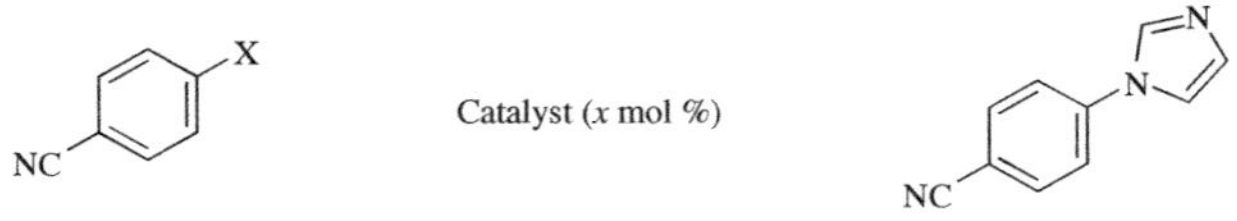

Reaction 3

4-halobenzonitrile → 4-(imidazol-1-yl)benzonitrile

Catalyst (x mol %)

X	Catalyst	x	Additive(s)	Solvent	Temp (°)	Time (h)		
Cl	Cu/Al–hydrotalcite	2.5	K₂CO₃	DMF	100	8	(95)	228
Br	CuBr	10	**L20** (20 mol %), Cs₂CO₃	DMSO	80	12	(94)	284
Cl	CuI nanoparticles	1.25	K₂CO₃, air	DMF	110	4	(65)	285
Cl	CuI	20	Cs₂CO₃	DMF	120	40	(90)	386
Br	CuI	20	Cs₂CO₃	DMF	120	40	(98)	386
I	CuI	10	L-Pro (20 mol %), K₂CO₃	DMSO	90	36	(93)	336
I	CuI	5	L-Pro (10 mol %), K₂CO₃	DMSO	80	34	(93)	401

TABLE 9A. *N*-ARYLATION OF IMIDAZOLES (*Continued*)

Please refer to the charts preceding the tables for structures indicated by the **bold** numbers.

C₃

Nitrogen Nucleophile: imidazole (N–H)

Aryl Halide: NC–C₆H₄–X

Product(s): NC–C₆H₄–N(imidazolyl)

X	Catalyst	x	Additive(s)	Solvent	Temp (°)	Time	Product(s) and Yield(s) (%)	Refs.
Br	CuI	10	L-Pro (20 mol %), K₂CO₃	DMSO	80	48 h	(97)	47
I	CuI	10	L-Pro (20 mol %), K₂CO₃	DMSO	80	34 h	(93)	47
Br	CuI	10	L-His (20 mol %), K₂CO₃	DMSO	100	36 h	(43)	389
Cl	CuI	5	**L24** (10 mol %), Cs₂CO₃	DMF	110	24 h	(92)	86
Br	CuI	5	**L24** (10 mol %), Cs₂CO₃	DMF	110	24 h	(95)	86
Cl	CuI	10	**L28** (20 mol %), K₂CO₃	DMSO	120	30 h	(97)	412
Br	CuI	10	**L28** (20 mol %), K₂CO₃	DMSO	90	60 h	(95)	412
Br	CuI	10	**L32** (20 mol %), Cs₂CO₃	DMF	110	24 h	(68)	417
F	CuO nanoparticles	10	K₂CO₃	DMF	120	12 h	(83)	64
Cl	CuO nanoparticles	10	K₂CO₃	DMF	120	12 h	(84)	64
Br	CuO	8	KOH, 4 Å MS	pyridine	reflux	50 h	(73)	402
I	Cu₂O	10		DMSO	100	24 h	(88)	398
Br	Cu(OAc)₂	100	DBU	DMSO	MW, 130	10 min	(92)	85
I	Cu(OAc)₂	100	DBU	DMSO	MW, 130	10 min	(94)	85
Br	CuSO₄	10	**L36** (20 mol %), Cs₂CO₃	H₂O	120	24 h	(82)	234
Cl	Cu fluorapatite	7	K₂CO₃	DMF	120	6 h	(95)	66
Cl	Cu(II)-NaY zeolite	10	K₂CO₃	DMF	120	24 h	(99)	404
Cl	[Cu(μ-I)((−)-sparteine)]₂	1	K₂CO₃	DMSO	125	14 h	(89)	414
Br	[Cu(μ-I)((−)-sparteine)]₂	1	K₂CO₃	DMSO	115	7 h	(95)	414
Br	**2**	2	NaOH, TBAB	H₂O	100	12 h	(82)	233
F	**7**	1	K₂CO₃	DMF	110	1 h	(96)	73
Cl	**7**	1	K₂CO₃	DMF	110	6 h	(90)	73

CuO nanoparticles (10 mol %), K₂CO₃; product (fluoro-substituted *N*-aryl imidazole) (85), 64

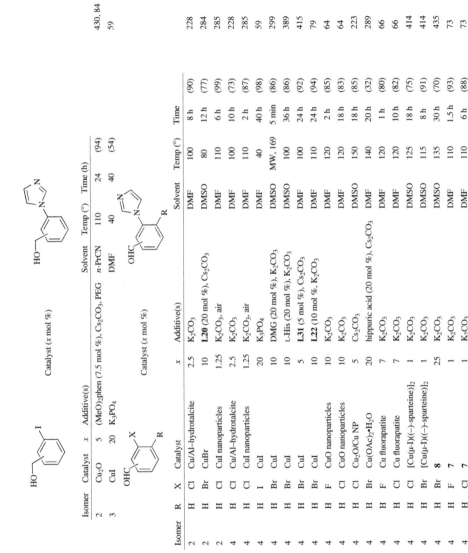

Isomer	Catalyst	x	Additive(s)	Solvent	Temp (°)	Time (h)		
2	Cu_2O	5	$(MeO)_2$phen (7.5 mol %), Cs_2CO_3, PEG	n-PrCN	110	24	(94)	430, 84
3	CuI	20	K_3PO_4	DMF	40	40	(54)	59

Isomer	R	X	Catalyst	x	Additive(s)	Solvent	Temp (°)	Time		
2	H	Cl	Cu/Al-hydrotalcite	2.5	K_2CO_3	DMF	100	8 h	(90)	228
2	H	Br	CuBr	10	L20 (20 mol %), Cs_2CO_3	DMSO	80	12 h	(77)	284
2	H	Cl	CuI nanoparticles	1.25	K_2CO_3, air	DMF	110	6 h	(99)	285
4	H	Cl	Cu/Al-hydrotalcite	2.5	K_2CO_3	DMF	100	10 h	(73)	228
4	H	Cl	CuI nanoparticles	1.25	K_2CO_3, air	DMF	110	2 h	(87)	285
4	H	I	CuI	20	K_3PO_4	DMF	40	40 h	(98)	59
4	H	Br	CuI	10	DMG (20 mol %), K_2CO_3	DMSO	MW, 169	5 min	(86)	299
4	H	Br	CuI	10	L-His (20 mol %), K_2CO_3	DMSO	100	36 h	(86)	389
4	H	Br	CuI	5	L31 (5 mol %), Cs_2CO_3	DMF	100	24 h	(92)	415
4	H	Br	CuI	10	L22 (10 mol %, K_2CO_3	DMF	110	24 h	(94)	79
4	H	F	CuO nanoparticles	10	K_2CO_3	DMF	120	2 h	(85)	64
4	H	Cl	CuO nanoparticles	10	K_2CO_3	DMF	120	18 h	(83)	64
4	H	Cl	Cu_2O/Cu NP	5	Cs_2CO_3	DMSO	150	18 h	(85)	223
4	H	Br	$Cu(OAc)_2 \cdot H_2O$	20	hippuric acid (20 mol %), Cs_2CO_3	DMF	140	20 h	(32)	289
4	H	F	Cu fluorapatite	7	K_2CO_3	DMF	120	1 h	(80)	66
4	H	Cl	Cu fluorapatite	7	K_2CO_3	DMF	120	10 h	(82)	66
4	H	Cl	[Cu(μ-I)((−)-sparteine)]$_2$	1	K_2CO_3	DMSO	125	18 h	(75)	414
4	H	Br	[Cu(μ-I)((−)-sparteine)]$_2$	1	K_2CO_3	DMSO	115	8 h	(91)	414
4	H	Br	8	25	K_2CO_3	DMSO	135	30 h	(70)	435
4	H	F	7	1	K_2CO_3	DMF	110	1.5 h	(93)	73
4	H	Cl	7	1	K_2CO_3	DMF	110	6 h	(88)	73
4	Cl	F	CuO nanoparticles	10	K_2CO_3	DMF	120	2 h	(94)	64

TABLE 9A. N-ARYLATION OF IMIDAZOLES (Continued)

Nitrogen Nucleophile	Aryl Halide	Conditions	Product(s) and Yield(s) (%)	Refs.

*Please refer to the charts preceding the tables for structures indicated by the **bold** numbers.*

C₃

(imidazole, N–H)

			Isomer	
(3-Br phenyl dioxolane)	1. NaH, DMF 2. Cu (10 mol %), 150°, 4 h		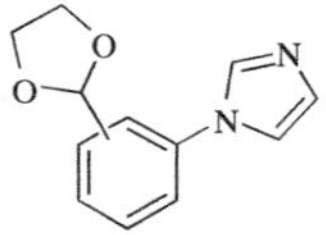 2 (87) 3 (84) 4 (65)	436
(4-Br phenyl dioxolane)	CuI (10 mol %), DMG (20 mol %), K₂CO₃, DMSO, MW, 173°, 15 min	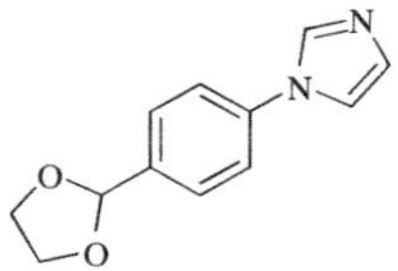 (69)		299
(CO₂H, 2-Br)	CuCl (5 mol %), **L17** (10 mol %), TBAB, NaOH, H₂O, 100°, 24 h	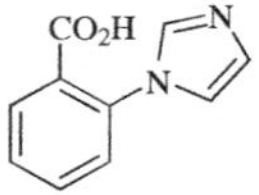 (97)		235

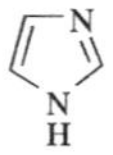

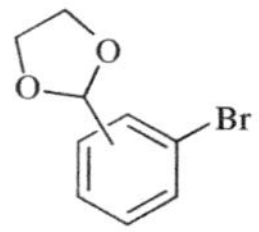

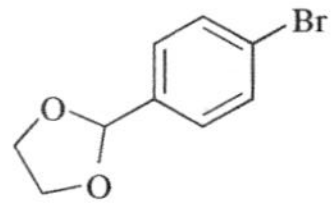

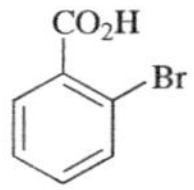

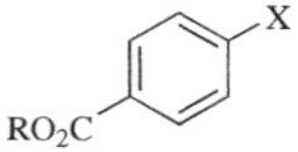

(RO₂C–C₆H₄–X)   Catalyst (x mol %)

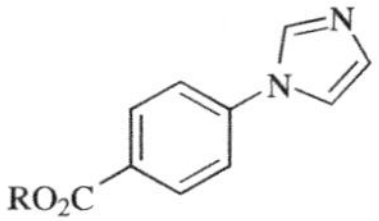

R	X	Catalyst	x	Additive(s)	Solvent	Temp (°)	Time		Refs.
Me	I	Cu(OAc)₂	100	DBU	DMSO	MW, 130	10 min	(89)	85
Et	Br	CuI	20	Cs₂CO₃	DMF	120	40 h	(69)	386
Et	I	CuI	20	K₃PO₄	DMF	40	40 h	(91)	59
Et	Br	CuI	5	**L24** (10 mol %), Cs₂CO₃	DMF	110	24 h	(61)	86
Et	Br	CuI	5	8-HOquin (10 mol %), CsF	DMSO	100	7 h	(65)	63
Et	I	Cu₂O	10	KOH, 4 Å MS	DMSO	100	24 h	(85)	398
Et	I	Cu₂O	2.5	(MeO)₂phen (7.5 mol %), Cs₂CO₃, 3 Å MS, PEG	MeCN	80	24–48 h	(87)	430, 84

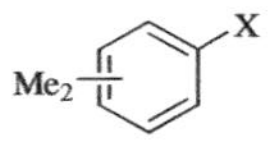

The substrate is Me_2-substituted aryl halide (X) reacting to give the Me_2-substituted N-aryl imidazole.

Catalyst (x mol %)

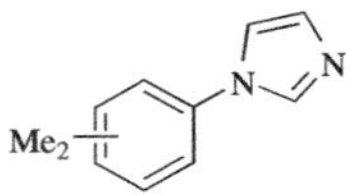

Isomer	X	Catalyst	x	Additive(s)	Solvent(s)	Temp (°)	Time (h)		
2,4	I	CuI	5	KOH, TBAB	–	110	7	(65)	428
2,5	Br	Cu(OTf)$_2$•C$_6$H$_6$	10	phen (1 eq), dba, Cs$_2$CO$_3$	xylenes	125	48	(79)	11
3,4	Br	CuI	30	L-Pro (60 mol %)	[C$_4$mim][BF$_4$]	110	24	(70)	230
3,5	I	CuCl	5	L17 (10 mol %), NaOH, TBAB	H$_2$O	100	24	(87)	235
3,5	Br	CuI	10	8-HOquin (10 mol %), Cs$_2$CO$_3$	DMF/H$_2$O	130	16	(80)	434
3,5	I	CuI	5	phen (10 mol %), Cs$_2$CO$_3$	dioxane	110	24	(91)	128, 115
3,5	Br	Cu$_2$O	20	KOH	DMSO	130	24	(90)	398
3,5	Br	Cu$_2$O	10	phen (20 mol %), TBAF	–	145	24	(86)	68
3,5	Br	Cu(OTf)$_2$•C$_6$H$_6$	10	phen (1 eq), dba, Cs$_2$CO$_3$	xylenes	125	36	(99)	11
3,5	I	Cu(OTf)$_2$•C$_6$H$_6$	10	phen (1 eq), dba, Cs$_2$CO$_3$	xylenes	110	30	(96)	11

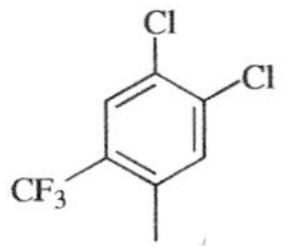

CuI (10 mol %), L26 (10 mol %), NaOMe, DMSO, 130°, 12 h → (50)   397

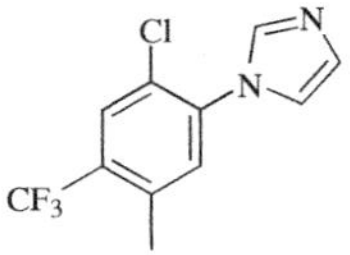

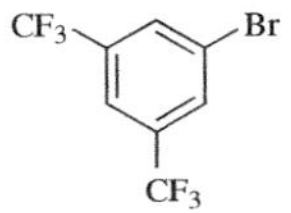

CuI (x mol %), Cs$_2$CO$_3$ → 429, 394

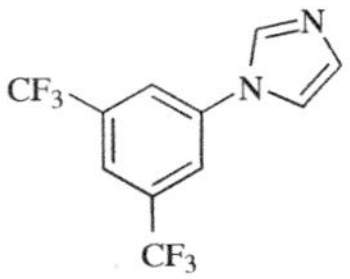

x	Additive	Solvent	Temp (°)	Time (h)		
20	D-glucosamine (40 mol %)	DMSO	110	20	(91)	429
10	L21 (20 mol %)	MeCN	reflux	12	(39)	394

TABLE 9A. N-ARYLATION OF IMIDAZOLES (Continued)

Nitrogen Nucleophile	Aryl Halide	Conditions	Product(s) and Yield(s) (%)	Refs.

Please refer to the charts preceding the tables for structures indicated by the **bold** numbers.

C₃

Nitrogen Nucleophile: imidazole (N, N-H)

Aryl Halide: Br-substituted, NC

Conditions: CuI (20 mol%), per-6-ABCD (10 mol %), K₂CO₃, DMSO, 110°, 18 h

Product(s) and Yield(s): (85)

Refs.: 393

Catalyst (x mol %)

Isomer X	Catalyst	x	Additive(s)	Solvent	Temp (°)	Time		Refs.
2 Br	CuO	8	K₂CO₃	pyridine	reflux	7 h	(33)	402
3 Br	CuI	10	L-Lys (20 mol%), K₃PO₄	—	MW (200 W), 130	5 h	(73)	418
3 Br	CuO	8	K₂CO₃	pyridine	reflux	48 h	(68)	402
4 Cl	Cu on cellulose	1	K₂CO₃	DMSO	130	24 h	(50)	427
4 Br	Cu on cellulose	1	K₂CO₃	DMSO	130	24 h	(70)	427
4 I	Cu/Al-hydrotalcite	2.5	K₂CO₃	DMF	100	18 h	(95)	228
4 Br	CuBr	10	L20 (20 mol %), Cs₂CO₃	DMSO	80	12 h	(95)	284
4 Cl	CuI nanoparticles	1.25	K₂CO₃, air	DMF	110	5 h	(99)	285
4 Br	CuI	20	Cs₂CO₃	DMF	120	40 h	(92)	386
4 Br	CuI	20	per-6-ABCD (10 mol %), K₂CO₃	DMSO	110	18 h	(94)	393
4 Br	CuI	10	DACA (20 mol %), K₃PO₄	[C₄mim][BF₄]	110	12 h	(100)	232
4 Br	CuI	10	Me₂Gly (20 mol %), K₂CO₃	DMSO	MW, 169	5 min	(88)	299
4 I	CuI	10	Me₂Gly (20 mol %), Bu₄POAc	DMSO	rt	24 h	(96)	61
4 Br	CuI	10	L-Glu (20 mol %), K₃PO₄	—	MW (200 W), 130	6.5 h	(69)	418
4 Br	CuI	20	L-Pro (20 mol %), K₃PO₄	DMF	140	48 h	(90)	388
4 I	CuI	5	L-Pro (20 mol %), K₃PO₄	dioxane	100	24 h	(97)	388
4 Br	CuI	10	L-Pro (20 mol %), K₂CO₃	DMSO	95	40 h	(93)	47
4 Br	CuI	10	L-His (20 mol %), K₂CO₃	DMSO	100	36 h	(83)	389
4 I	CuI	10	L6 (20 mol %), Cs₂CO₃	DMF	rt	24 h	(95)	273
4 Br	CuI	5	L24 (10 mol %), Cs₂CO₃	DMF	110	24 h	(99)	86
4 Br	CuI	10	L28 (20 mol %), K₂CO₃	DMSO	90	60 h	(92)	412
4 Br	CuI	10	L22 (10 mol %), K₂CO₃	DMF	110	24 h	(98)	79
4 Br	CuI	10	L32 (20 mol %), Cs₂CO₃	DMF	110	24 h	(98)	417
4 Br	CuI	5	L31 (5 mol %), Cs₂CO₃	DMF	100	24 h	(84)	415

304

4	F	CuO nanoparticles	10	K_2CO_3	DMF	120	2 h	(83)	64
4	Cl	CuO nanoparticles	10	K_2CO_3	DMF	120	12 h	(89)	64
4	Br	CuO	8	K_2CO_3	pyridine	reflux	48 h	(82)	402
4	Br	CuO	5	L14 (50 mol %), TBAB, 2,5-hexanedione, K_3PO_4	H_2O	MW (100 W), 140	5 min	(77)	51
4	I	CuO	5	L14 (50 mol %), TBAB, 2,5-hexanedione, K_3PO_4	H_2O	MW (100 W), 140	5 min	(84)	51
4	Cl	Cu_2O/Cu NP	5	Cs_2CO_3	DMSO	150	18 h	(91)	223
4	Br	Cu_2O	10	phen (20 mol %), TBAF	—	145	24 h	(100)	68
4	Br	Cu_2O	10	(MeO)$_2$phen (15 mol %), PEG, Cs_2CO_3	n-PrCN	110	24 h	(90)	430, 84
4	Br	Cu_2O	10	L23 (20 mol %), Cs_2CO_3	DMF	130	24 h	(73)	400
4	Br	Cu(OAc)$_2$•H_2O	20	hippuric acid (20 mol %), Cs_2CO_3	DMF	140	20 h	(49)	289
4	Br	$CuSO_4$	10	L36 (20 mol %), Cs_2CO_3	H_2O	120	24 h	(69)	234
4	Cl	Cu(IPr)Cl	10	Cs_2CO_3	DMSO	100	24 h	(56)	421
4	Br	Cu(IPr)Cl	10	Cs_2CO_3	DMSO	100	24 h	(86)	421
4	I	Cu(IPr)Cl	10	Cs_2CO_3	DMSO	100	24 h	(86)	421
4	Cl	[Cu(μ-I)(−)-sparteine)]$_2$	1	K_2CO_3	DMSO	125	17 h	(79)	414
4	Br	[Cu(μ-I)(−)-sparteine)]$_2$	1	K_2CO_3	DMSO	115	7 h	(97)	414
4	Cl	Cu(II)–NaY zeolite	10	K_2CO_3	DMF	120	48 h	(95)	404
4	Br	Cu(II)–NaY zeolite	10	K_2CO_3	DMF	120	20 h	(99)	404
4	Br	2	2	NaOH, TBAB	H_2O	100	12 h	(79)	233
4	Br	6	5	Cs_2CO_3	toluene	100	18 h	(90)	123
4	I	6	5	Cs_2CO_3	toluene	100	4 h	(92)	123
4	Cl	8	25	K_2CO_3	DMSO	135	40 h	(81)	65
4	Br	8	25	K_2CO_3	DMSO	135	30 h	(85)	65
4	Cl	10	1	Cs_2CO_3	DMSO	100	24 h	(50)	420
4	Br	10	1	Cs_2CO_3	DMSO	100	24 h	(70)	420
4	I	10	1	Cs_2CO_3	DMSO	100	24 h	(99)	420

TABLE 9A. N-ARYLATION OF IMIDAZOLES (*Continued*)

Nitrogen Nucleophile	Aryl Halide	Conditions	Product(s) and Yield(s) (%)	Refs.

*Please refer to the charts preceding the tables for structures indicated by the **bold** numbers.*

C$_{3-4}$

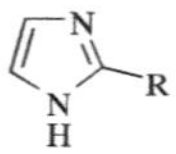

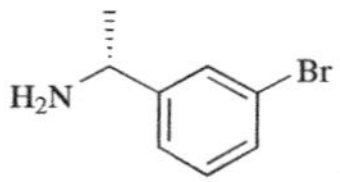

CuI (10 mol %), K$_2$CO$_3$, NMP, MW, 195°, 2 h

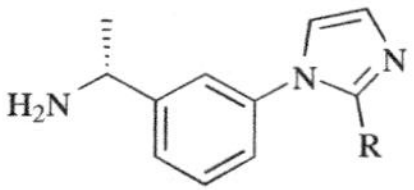

R	Time (h)	
H	2	(76)
Me	3	(68)

406

C$_3$

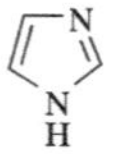

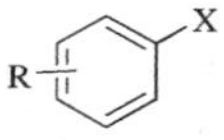

Catalyst (x mol %)

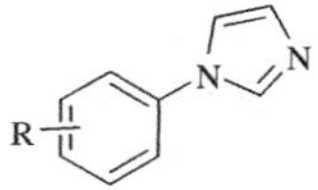

R	X	Catalyst	x	Additive(s)	Solvent	Temp (°)	Time (h)		Refs.
4-Et	Br	CuI	10	L28 (20 mol %), K$_2$CO$_3$	DMSO	90	60	(91)	412
4-HC≡C	I	CuI	20	K$_3$PO$_4$	DMF	40	40	(82)	59
4-n-Pr	Br	CuBr	10	L20 (20 mol %), Cs$_2$CO$_3$	DMSO	80	12	(84)	284
2-i-Pr	I	Cu$_2$O	5	(MeO)$_2$phen (7.5 mol %), Cs$_2$CO$_3$, PEG	n-PrCN	110	24	(94)	430, 84
3-i-Pr	I	Cu$_2$O	5	(MeO)$_2$phen (7.5 mol %), Cs$_2$CO$_3$, PEG	n-PrCN	110	24	(94)	430
4-t-Bu	Br	Cu$_2$O	5	(MeO)$_2$phen (15 mol %), Cs$_2$CO$_3$, PEG	n-PrCN	110	48	(93)	84
4-t-Bu	I	(CuOTf)$_2$•C$_6$H$_6$	10	phen (1 eq). dba, Cs$_2$CO$_3$	xylenes	110	24	(97)	11

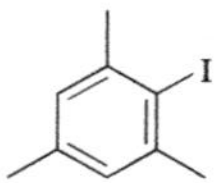

Catalyst (x mol %), Cs$_2$CO$_3$

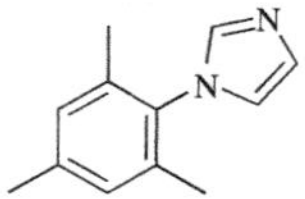

Catalyst	x	Additive(s)	Solvent	Temp (°)	Time (h)		Refs.
CuI	10	DMEDA (40 mol %)	DMF	170	48	(54)	58
CuI	10	EDA (40 mol %)	DMF	170	48	(50)	58
Cu$_2$O	2.5	(MeO)$_2$phen (7.5 mol %), PEG	DMSO	150	24	(44)	84

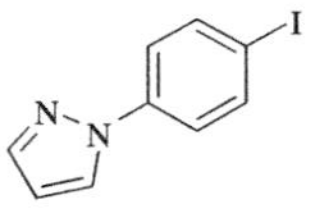

Cu₂O (5 mol%),
L35 (20 mol %),
MeCN, 82°, 24 h

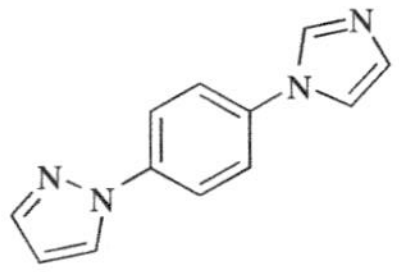

(100)

40

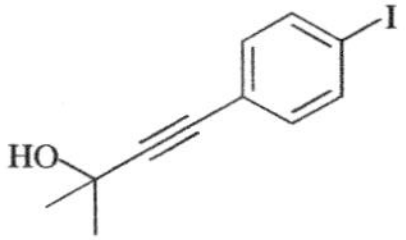

CuI (20 mol %), K₃PO₄,
DMF, 40°, 40 h

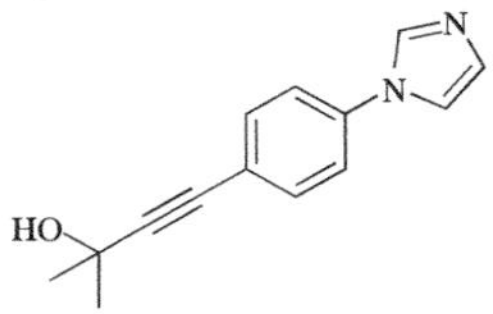

(93)

59

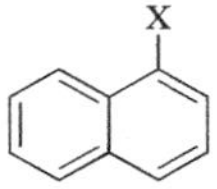

Catalyst (x mol %)

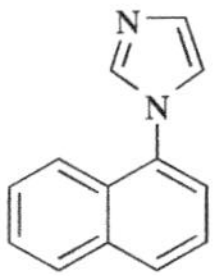

X	Catalyst	x	Additive(s)	Solvent	Temp (°)	Time (h)	
Br	CuI	20	Cs₂CO₃	DMF	120	40 (56)	386
Br	CuI	10	CDA (40 mol %), Cs₂CO₃	dioxane	95	24 (57)	58
Br	CuI	20	per-6-ABCD (10 mol %), K₂CO₃	DMSO	110	24 (86)	393
I	CuI	10	phen (10 mol %), KF/Al₂O₃	—	140	15 (74)	411
Br	CuI	20	D-glucosamine (40 mol %), Cs₂CO₃	DMSO	110	24 (68)	429
Br	CuI	10	L22 (10 mol %), K₂CO₃	DMF	110	48 (77)	79
Br	CuI	5	L24 (10 mol %), Cs₂CO₃	DMF	110	24 (66)	86
Br	CuI	10	L28 (20 mol %), K₂CO₃	DMSO	120	60 (62)	412
I	Cu₂O	10	KOH	DMSO	110	24 (90)	398
I	Cu₂O	10	ninhydrin (20 mol %), KOH	DMSO	100	24 (90)	399
I	Cu₂O	10	(MeO)₂phen (15 mol %), Cs₂CO₃	NMP	110	24 (93)	430, 84
Br	Cu₂O	10	L20 (20 mol %), Cs₂CO₃	DMSO	80	12 (68)	284
Br	Cu₂O	10	L23 (20 mol %), KOt-OBu	DMF	130	24 (92)	400

TABLE 9A. N-ARYLATION OF IMIDAZOLES (Continued)

Nitrogen Nucleophile	Aryl Halide	Conditions	Product(s) and Yield(s) (%)	Refs.

*Please refer to the charts preceding the tables for structures indicated by the **bold** numbers.*

C₃

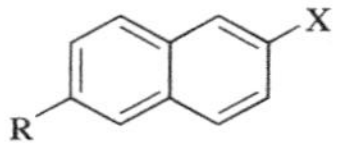

Nitrogen Nucleophile: imidazole (N–H)

Aryl Halide: naphthalene (R, X)

Conditions: CuI (x mol %)

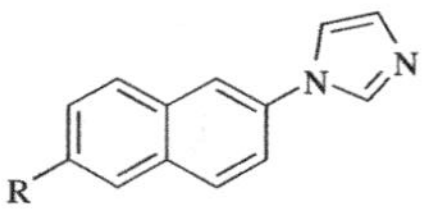

R	X	x	Additive(s)	Solvent	Temp (°)	Time (h)		Refs.
H	Br	10	EDA (40 mol %), Cs₂CO₃	DMF	170	48	(80)	58
H	Br	10	DMEDA (40 mol %), Cs₂CO₃	DMF	170	48	(80)	437
H	Br	20	per-6-ABCD (10 mol %), K₂CO₃	DMSO	110	24	(90)	393
H	Br	20	D-glucosamine (40 mol %), Cs₂CO₃	DMSO	110	22	(86)	429
H	I	10	L26 (10 mol %), NaOMe	DMSO	130	24	(48)	397
MeO	Br	30	L-Pro (60 mol %)	[C₄mim][BF₄]	110	24	(85)	230

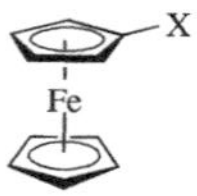

Aryl Halide: ferrocene (X)

Conditions: Catalyst (x amount), DMSO, 90°

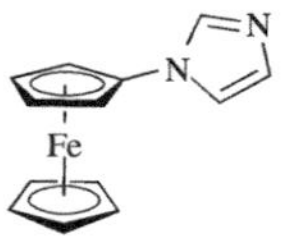

X	Catalyst	x	Additive	Time (h)		Refs.
I	CuI	1 eq	K₂CO₃	18	(31)	372
Br	Cu(TMHD)₂	20 mol %	KOt-Bu	12	(22)	407
I	Cu(TMHD)₂	10 mol %	KOt-Bu	12	(46)	407

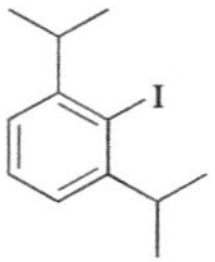

Aryl Halide: 2,6-diisopropyliodobenzene

Conditions: CuI (10 mol %), EDA (40 mol %), Cs₂CO₃, DMF, 170°, 48 h

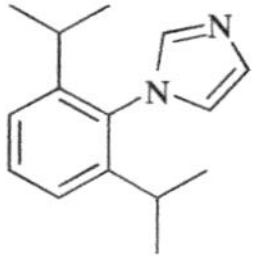

Product: (19) — Refs. 58

Aryl Halide: 2-phenyliodobenzene

Conditions: Cu₂O (10 mol %), (MeO)₂phen (15 mol %), Cs₂CO₃, PEG,

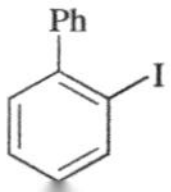

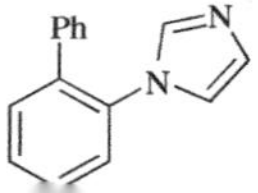

Product: (94) — Refs. 430, 84

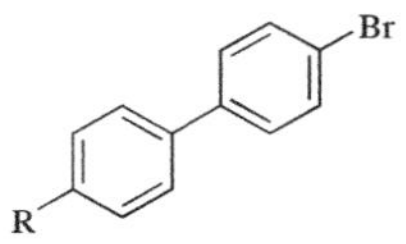

CuI (10 mol %), DMSO

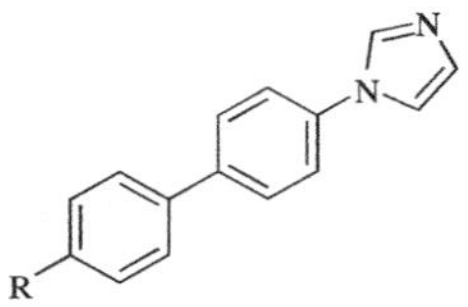

R	Additives	Temp (°)	Time (h)		
H	DMG (20 mol %), K$_2$CO$_3$	110	40	(81)	47
Br	L26 (10 mol %), NaOMe	130	12	(40)	397

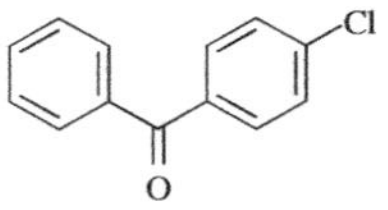

Catalyst (x mol %), K$_2$CO$_3$, DMF

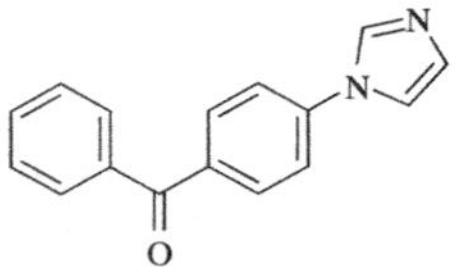

Catalyst	x	Temp (°)	Time (h)		
CuO nanoparticles	10	120	11	(87)	64
Cu fluorapatite	7	120	6	(88)	66
cat. 7	1	110	5	(88)	73

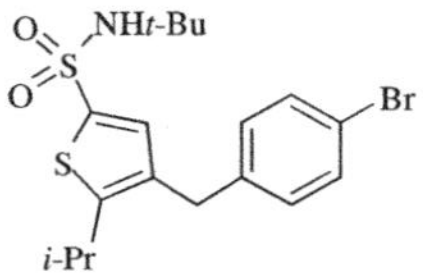

CuI (1 mol %), DMCDA (4 mol %), Cs$_2$CO$_3$, DMF, MW. 180°, 40 min

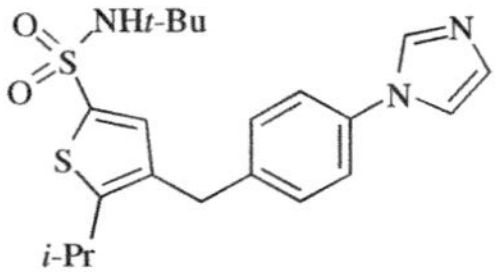

(27)   438

TABLE 9A. *N*-ARYLATION OF IMIDAZOLES (*Continued*)

Nitrogen Nucleophile	Aryl Halide	Conditions	Product(s) and Yield(s) (%)	Refs.

*Please refer to the charts preceding the tables for structures indicated by the **bold** numbers.*

C4

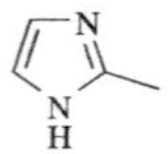

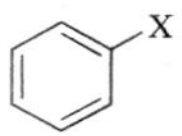

Catalyst (*x* amount)

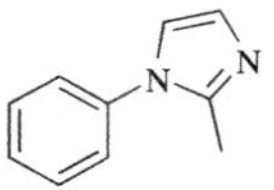

X	Catalyst	*x*	Additive(s)	Solvent	Temp (°)	Time		Refs.
Br	CuBr	10 mol %	**L20** (20 mol %), Cs$_2$CO$_3$	DMSO	90	24 h	(73)	284
I	CuI	5 mol %	TBAB, KOH	—	110	7 h	(78)	428
Br	CuI	10 mol %	**L22** (10 mol %), K$_2$CO$_3$	DMF	110	24 h	(63)	79
Br	CuI	20 mol %	per-6-ABCD (10 mol %), K$_2$CO$_3$	DMSO	110	24 h	(86)	393
Br	CuI	10 mol %	**L28** (20 mol %), K$_2$CO$_3$	DMSO	120	60 h	(68)	412
Br	Cu(OAc)$_2$	1 eq	DBU	DMSO	MW, 130	10 min	(53)	85
I	Cu(OAc)$_2$	1 eq	DBU	DMSO	MW, 130	10 min	(62)	85

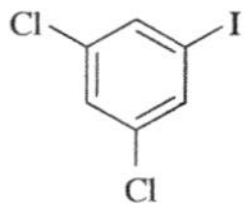

Catalyst (5 mol %), 24 h

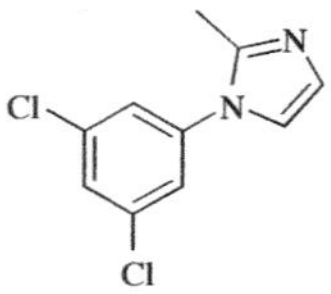

Catalyst	Additives	Solvent	Temp (°)		Refs.
CuI	8-HOquin (10 mol %), CsF	DMSO	100	(78)	63
Cu$_2$O	(MeO)$_2$phen (7.5 mol %), Cs$_2$CO$_3$, **PEG**	*n*-PrCN	110	(84)	430, 84

C4–8

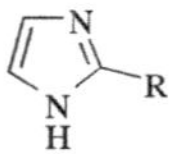

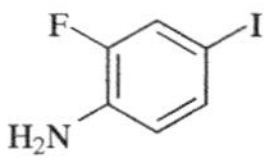

CuI (20 mol %), K$_2$CO$_3$,
DMSO, 120°, 12 h

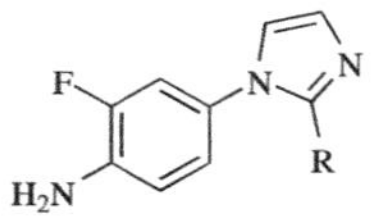

439

R	
Me	(53)
Et	(55)
i-Pr	(50)
CH$_2$NMe$_2$	(67)
CH$_2$NMeCbz	(65)
 N	(55)

C₄

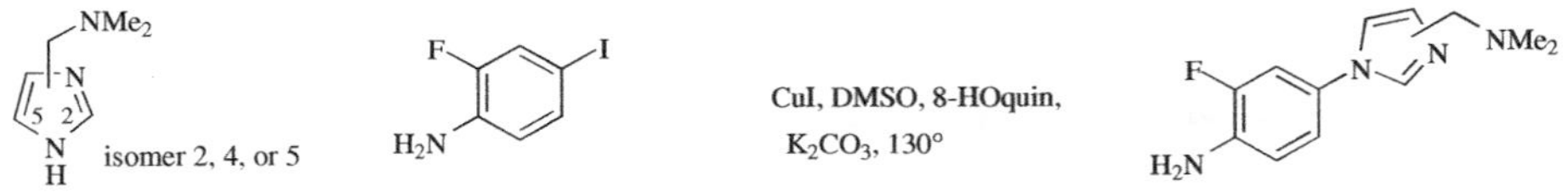

isomer 2, 4, or 5

CuI, DMSO, 8-HOquin,
K₂CO₃, 130°

(—)   424

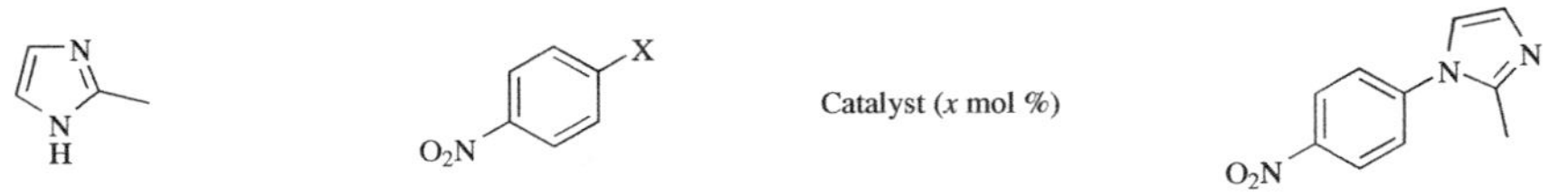

Catalyst (x mol %)

X	Catalyst	x	Additive(s)	Solvent(s)	Temp (°)	Time (h)	
Cl	CuI	10	L26 (10 mol %), NaOMe	DMSO	130	12 (43)	397
I	CuO nanoparticles	5	KOH	t-BuOH/DMSO (3:1)	110	11 (83)	74
Br	CuSO₄	10	L36 (20 mol %), Cs₂CO₃	H₂O	120	24 (85)	234
Br	cat. 2	2	NaOH, TBAB	H₂O	100	12 (82)	233

Catalyst (x mol %)

R	X	Catalyst	x	Additive(s)	Solvent(s)	Temp (°)	Time	
H	I	CuO nanoparticles	10	KOH	t-BuOH/DMSO (3:1)	110	11 h (86)	74
H	Br	CuO	5	L14 (50 mol %), TBAB, 2,5-hexanedione, K₃PO₄	H₂O	MW (100 W), 140	5 min (70)	51
H	I	Cu(OTf)·C₆H₆	10	phen (1 eq), dba, Cs₂CO₃	xylenes	125	48 h (62)	11
Br	I	Cu₂O	5	(MeO)₂phen (7.5 mol %), Cs₂CO₃	MeCN	110	24 h (76)	84

TABLE 9A. *N*-ARYLATION OF IMIDAZOLES (*Continued*)

Nitrogen Nucleophile	Aryl Halide	Conditions	Product(s) and Yield(s) (%)	Refs.

*Please refer to the charts preceding the tables for structures indicated by the **bold** numbers.*

C₄

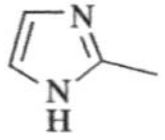

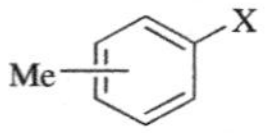

Catalyst (*x* mol %)

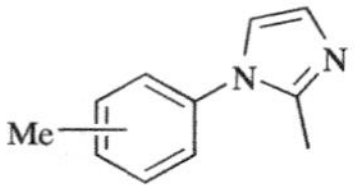

Isomer	X	Catalyst	*x*	Additives	Solvent(s)	Temp (°)	Time (h)		Refs.
2	I	Cu₂O	2.5	(MeO)₂phen (7.5 mol %), Cs₂CO₃, PEG	NMP	150	24	(86)	84
3,5	Br	CuI	10	8-HOquin (10 mol %), TEAC	DMF/H₂O	130	16	(64)	434
2,4,6	I	Cu₂O	2.5	(MeO)₂phen (7.5 mol %), Cs₂CO₃, PEG	DMSO	150	24	(44)	84

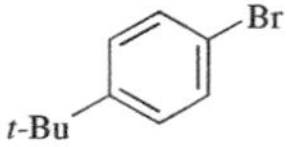

Cu₂O (5 mol %),
(MeO)₂phen (15 mol %),
Cs₂CO₃, PEG,
n-PrCN, 110°, 24 h

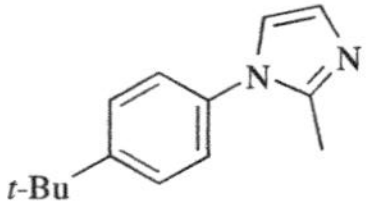

(95)   430, 84

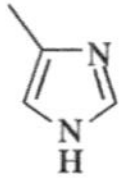

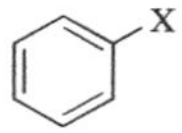

Catalyst (*x* amount)

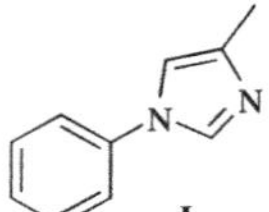

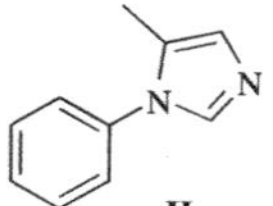

X	Catalyst	*x*	Additive(s)	Solvent	Temp (°)	Time	**I**	**II**	Refs.
Br	CuI	20 mol %	Cs₂CO₃	DMF	120	40 h	(50)	(12)	386
Br	CuI	20 mol %	phen (20 mol %), KF/Al₂O₃	xylene	140	15 h	(91)	(0)	410
I	CuI	10 mol %	**L21** (20 mol %), Cs₂CO₃	MeCN	reflux	12 h	(63)	(10)	394
Br	Cu(OAc)₂	1 eq	DBU	DMSO	MW, 130	10 min	(91)	(0)	85
I	Cu(OAc)₂	1 eq	DBU	DMSO	MW, 130	10 min	(90)	(0)	85
I	Cu(OTf)₂•C₆H₆	10 mol %	phen (1 eq), dba, Cs₂CO₃	xylenes	110	36 h	(77)	(17)	11

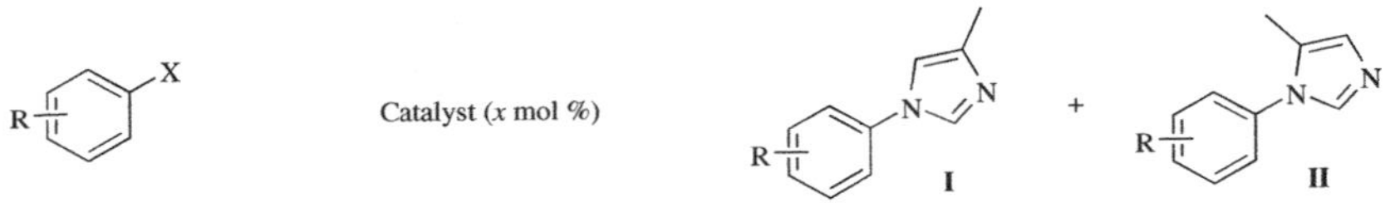

Catalyst (x mol %)

R	X	Catalyst	x	Additive(s)	Solvent(s)	Temp (°)	Time (h)	I	II	
4-F	Br	Cu₂O	5	(MeO)₂phen (15 mol %), Cs₂CO₃, PEG	n-PrCN	110	30	(78)	(15)	84
2-Cl	I	Cu₂O	2.5	(MeO)₂phen (7.5 mol %), Cs₂CO₃, PEG	n-PrCN	110	24	(96)	(0)	84
2-Cl	I	Cu₂O	2.5	(MeO)₂phen (7.5 mol %), Cs₂CO₃, PEG	NMP	110	48	(90)	(6)	84
4-MeO	I	CuI	20	K₃PO₄	DMF	40	40	(62)	(4)	59
4-MeS	Br	Cu₂O	5	(MeO)₂phen (15 mol %), Cs₂CO₃, PEG	n-PrCN	110	24	(97)	(0)	84
2-Me	I	Cu₂O	2.5	(MeO)₂phen (7.5 mol %), Cs₂CO₃, PEG	NMP	110	48	(86)	(5)	84
2-i-Pr	I	Cu₂O	2.5	(MeO)₂phen (7.5 mol %), Cs₂CO₃, PEG	NMP	110	48	(82)	(0)	84

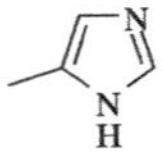

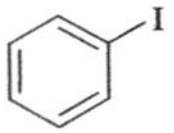

CuO (2.5 mol %), KOH, DMSO, 110°, 24 h 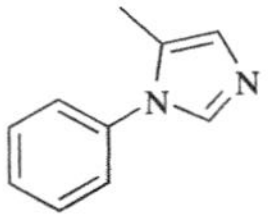 → (96) 224

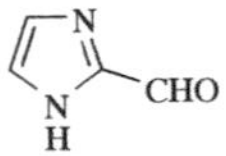

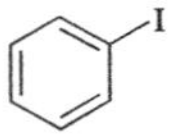

CuI (5 mo l%), DMCDA (20 mol %), K₃PO₄, toluene, 110°, 24 h → (88) 128

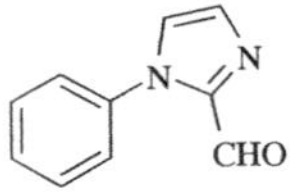

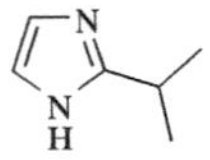

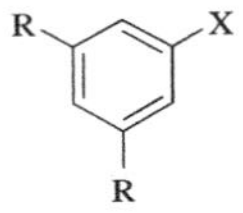

Catalyst (x amount)

R	X	Catalyst	x	Additive	Solvent(s)	Temp (°)	Time		
H	Br	Cu(OAc)₂	1 eq	DBU	DMSO	MW, 130	10 min	(34)	85
H	I	Cu(OAc)₂	1 eq	DBU	DMSO	MW, 130	10 min	(42)	85
Me	Br	CuI	10 mol %	8-HOquin (10 mol %), Cs₂CO₃	DMF/H₂O	130	16 h	(40)	434

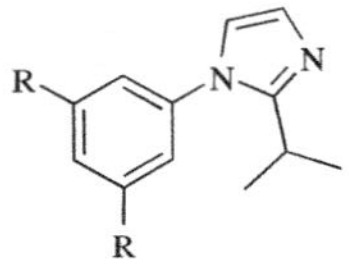

C₆

313

TABLE 9A. N-ARYLATION OF IMIDAZOLES (Continued)

Nitrogen Nucleophile	Aryl Halide	Conditions	Product(s) and Yield(s) (%)	Refs.

*Please refer to the charts preceding the tables for structures indicated by the **bold** numbers.*

C₆

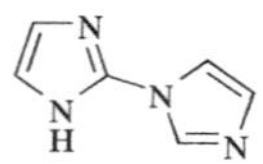

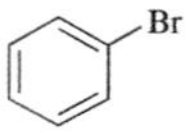

Conditions: CuI (x mol %), Cs₂CO₃, DMF			
x	Additive	Temp (°)	Time (h)
20	none	120	40 (53)
5	**L24** (10 mol %)	110	24 (62)

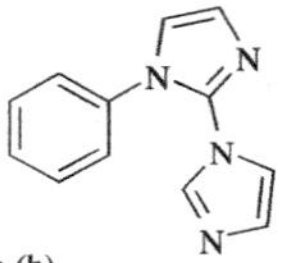

386

Product:

Conditions: (CuOTf)₂•C₆H₆ (5 mol %), phen (5 mol %), dba (5 mol %), Cs₂CO₃, DMF/xylene (1:5), 115°, 48 h

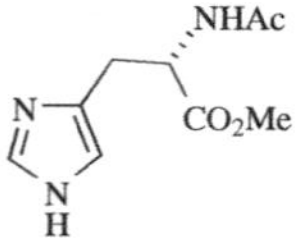

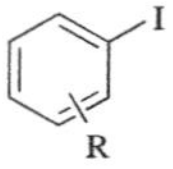

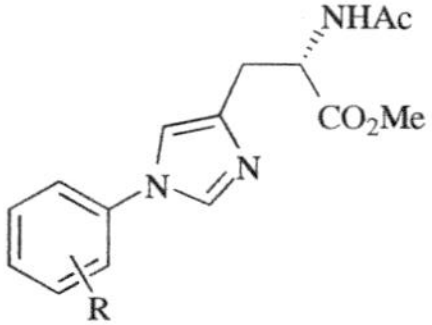

R	
H	(78)
4-Cl	(10–15)
3-Br	(10–15)
4-Br	(10–15)
4-MeO	(10–15)
3-Me	(10–15)
4-Me	(10–15)

440

C₉

Product:

Conditions: Catalyst (x mol %)

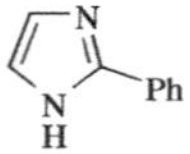

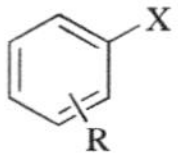

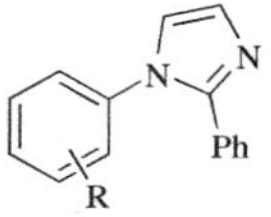

R	X	Catalyst	x	Additives	Solvent(s)	Temp (°)	Time (h)		Refs.
H	Br	CuI	10	8-HOquin (10 mol %), TEAC	DMF/H₂O	130	36	(trace)	434
H	Br	Cu₂O	20	(MeO)₂phen (30 mol %), Cs₂CO₃, PEG	n-PrCN	120	24	(92)	430, 84
4-F	I	Cu₂O	5	(MeO)₂phen (7.5 mol %), Cs₂CO₃, PEG	n-PrCN	110	24	(85)	430, 84, 441
4-O₂N	Cl	CuBr	10	**L37** (20 mol %), TBAF	none	150	24	(56)	80
4-O₂N	Br	Cu₂O	10	phen (20 mol %), TBAF	—	145	24	(63)	68

314

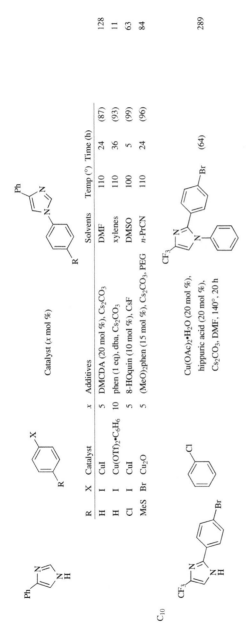

		Catalyst (x mol %)						
R	X	Catalyst	x	Additives	Solvents	Temp (°)	Time (h)	
H	I	CuI	5	DMCDA (20 mol %), Cs_2CO_3	DMF	110	24	(87)	128
H	I	$Cu(OTf)_2 \cdot C_6H_6$	10	phen (1 eq), dba, Cs_2CO_3	xylenes	110	36	(93)	11
Cl	I	CuI	5	8-HOquin (10 mol %), CsF	DMSO	100	5	(99)	63
MeS	Br	Cu_2O	5	$(MeO)_2$phen (15 mol %), Cs_2CO_3, PEG	n-PrCN	110	24	(96)	84

C_{10}

$Cu(OAc)_2 \cdot H_2O$ (20 mol %), hippuric acid (20 mol %), Cs_2CO_3, DMF, 140°, 20 h

(64)

289

---

[a] The product was a 6:1 mixture of 4-bromo- and 4-iodoimidazoylbenzene.

TABLE 9B. N-HETEROARYLATION OF IMIDAZOLES

Nitrogen Nucleophile	Heteroaryl Halide	Conditions	Product(s) and Yield(s) (%)	Refs.

*Please refer to the charts preceding the tables for structures indicated by the **bold** numbers.*

C₃

(imidazole, N–H)

Heteroaryl Halide: 2-bromothiazole

Conditions: CuI (x mol %)

x	Additive(s)	Solvent	Temp (°)	Time (h)
20	Cs₂CO₃	DMF	120	40
5	L24 (10 mol %), Cs₂CO₃	DMF	110	24
10	L28 (20 mol %), K₂CO₃	DMSO	120	60

Product (thiazol-2-yl imidazole): (83), (85), (86)

Refs: 386, 86, 412

Heteroaryl Halide: 3-bromofuran

Conditions: Cu₂O (5 mol %), Cs₂CO₃, PEG, (MeO)₂phen (15 mol %), DMSO, 110°, 24 h

Product (furan-3-yl imidazole): (60)

Ref: 84

Heteroaryl Halide: 2-halothiophene (X)

Conditions: Catalyst (x mol %)

X	Catalyst	x	Additive(s)	Solvent	Temp (°)	Time		Refs.
I	CuF₂	50	phen (50 mol %), K₂CO₃	DMF	140	96 h	(59)	419
Br	CuI	20	Cs₂CO₃	DMF	120	40 h	(72)	386
Br	CuI	30	L-Pro (60 mol %)	[C₄mim][BF₄]	110	24 h	(74)	230
Br	CuI	10	oxazolidin-2-one (20 mol %), NaOMe	DMSO	120	12 h	(79)	392
Br	CuI	20	D-glucosamine (40 mol %), Cs₂CO₃	DMSO	110	22 h	(82)	429
Br	CuI	10	L22 (10 mol %), K₂CO₃	DMF	110	48 h	(76)	79
Br	CuI	5	L24 (10 mol %), Cs₂CO₃	DMF	110	24 h	(75)	86
Br	CuI	10	L28 (20 mol %), K₂CO₃	DMSO	120	60 h	(60)	412
Br	CuI	10	L32 (20 mol %), Cs₂CO₃	DMF	110	24 h	(98)	417
I	Cu₂O	2.5	(MeO)₂phen (7.5 %), PEG, Cs₂CO₃	DMSO	110	24 h	(83)	84
Br	Cu(acac)₂	100	DBU	DMSO	MW, 150	10 min	(63)	85

Product (thiophen-2-yl imidazole)

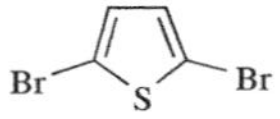

2,5-dibromothiophene	CuI (0.9 eq), L-Pro (1.8 eq), [C$_4$mim][BF$_4$], 115°, 50 h	2,5-di(imidazol-1-yl)thiophene	(77)				230
2-bromo-5-methylthiophene	CuI (30 mol%), L-Pro (60 mol %), [C$_4$mim][BF$_4$], 110°, 48 h	product	(68)				230

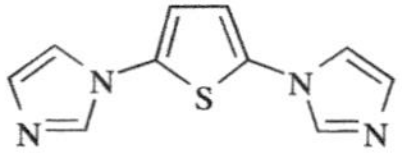

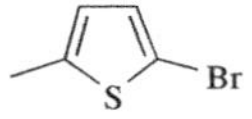

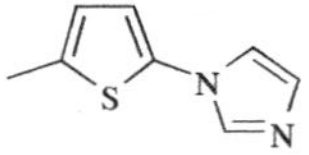

3-bromothiophene → Catalyst (x mol %)

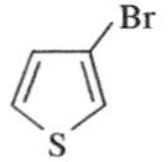

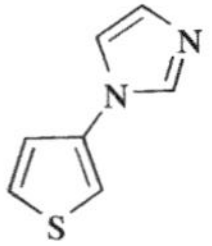

Catalyst	x	Additive(s)	Solvent	Temp (°)	Time		
CuI	20	D-glucosamine (40 mol %), Cs$_2$CO$_3$	DMSO	110	20 h	(83)	429
CuI	20	per-6-ABCD (10 mol %), K$_2$CO$_3$	DMSO	110	24 h	(97)	393
CuI	10	L22 (10 mol %), K$_2$CO$_3$	DMF	110	12 h	(68)	79
Cu(OAc)$_2$	100	DBU	DMSO	MW, 150	10 min	(80)	85

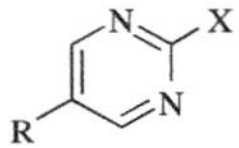

CuI (x mol %), 24 h

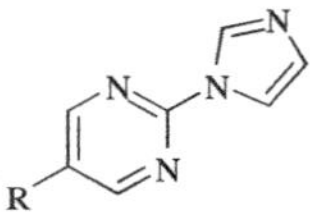

R	X	x	Additives	Solvent	Temp (°)		
H	Br	20	per-6-ABCD (10 mol %), K$_2$CO$_3$	DMSO	110	(98)	393
Br	I	5	phen (10 mol %), Cs$_2$CO$_3$	DMF	80	(65)	408

TABLE 9B. N-HETEROARYLATION OF IMIDAZOLES (*Continued*)

Nitrogen Nucleophile	Heteroaryl Halide	Conditions	Product(s) and Yield(s) (%)	Refs.

*Please refer to the charts preceding the tables for structures indicated by the **bold** numbers.*

C₃

Nitrogen Nucleophile:

imidazole (N–H)

Heteroaryl Halide:

2-X-pyrimidine

Catalyst (x mol %)

Product(s) and Yield(s):

2-(imidazol-1-yl)pyrimidine

X	Catalyst	x	Additive(s)	Solvent	Temp (°)	Time		Refs.
Cl	CuBr	10	**L37** (20 mol %), TBAF	none	150	24 h	(90)	80
Cl	CuI	20	Cs₂CO₃	DMF	120	40 h	(97)	386
Br	CuI	20	D-glucosamine (40 mol %), Cs₂CO₃	DMSO	110	18 h	(95)	429
Cl	CuI	10	**L22** (10 mol %), K₂CO₃	DMF	110	24 h	(87)	79
Cl	CuI	5	**L24** (10 mol %), Cs₂CO₃	DMF	110	24 h	(97)	86
Cl	CuI	10	**L28** (20 mol %), K₂CO₃	DMSO	120	60 h	(94)	412
Cl	CuO nanoparticles	10	K₂CO₃	DMF	120	1 h	(94)	64
Cl	Cu₂O	10	phen (20 mol %), TBAF	—	145	24 h	(72)	68
Br	Cu₂O	10	phen (20 mol %), TBAF	—	145	24 h	(86)	68
Br	Cu(OAc)₂	100	DBU	DMSO	MW, 130	10 min	(90)	85
Cl	Cu fluorapatite	7	K₂CO₃	DMF	120	1 h	(90)	66
Cl	**2**	1	K₂CO₃	DMF	110	1 h	(87)	73

Nitrogen Nucleophile:

imidazole (N–H)

Heteroaryl Halide:

3-chloropyridine

Conditions: CuI (x mol %), NaOMe, DMSO

Product(s) and Yield(s): 3-(imidazol-1-yl)pyridine

x	Additive	Temp (°)	Time (h)		Refs.
10	oxazolidin-2-one (20 mol%)	120	16	(67)	392
5	N-hydroxyphthalimide (10 mol%)	110	24	(94)	396
10	**L26** (10 mol %)	130	12	(86)	397

Catalyst (x mol %)

X	Catalyst	x	Additive(s)	Solvent	Temp (°)	Time (h)	
Cl	Cu$_2$O-coated Cu nanoparticles	5	Cs$_2$CO$_3$	DMSO	150	18 (88)	223
Br	CuBr	5	L5 (10 mol %), Cs$_2$CO$_3$	DMF	90	24 (76)	119
Br	CuBr	10	L37 (20 mol %), TBAF	none	150	24 (96)	80
Br	CuI	20	Cs$_2$CO$_3$	DMF	120	40 (98)	386
Br	CuI	5	BtH (10 mol %), KOt-Bu	DMSO	110	8 (98)	431
Br	CuI	10	L-Pro (20 mol %), K$_2$CO$_3$	DMSO	60	45 (93)	47
Br	CuI	30	L-Pro (60 mol %)	[C$_4$mim][BF$_4$]	110	18 (74)	230
Br	CuI	20	D-glucosamine (40 mol %), Cs$_2$CO$_3$	DMSO	110	24 (78)	429
Br	CuI	5	N-hydroxysuccinimide (10 mol %), NaOMe	DMSO	110	24 (98)	396
Br	CuI	20	phen (20 mol %), KF/Al$_2$O$_3$	xylene	140	15 (90)	410
Br	CuI	10	oxazolidin-2-one (20 mol %), NaOMe	DMSO	80	15 (85)	392
Br	CuI	20	per-6-ABCD (10 mol %), K$_2$CO$_3$	DMSO	110	24 (92)	393
Br	CuI	10	DACA (20 mol %), K$_3$PO$_4$	[C$_4$mim][BF$_4$]	110	12 (100)	232
Br	CuI	10	L21 (20 mol %), Cs$_2$CO$_3$	MeCN	reflux	12 (79)	394
Cl	CuI	10	L22 (10 mol %), K$_2$CO$_3$	DMF	110	24 (65)	79
Br	CuI	10	L22 (10 mol %), K$_2$CO$_3$	DMF	110	12 (99)	79
Br	CuI	5	L24 (10 mol %), Cs$_2$CO$_3$	DMSO	110	24 (98)	86
Br	CuI	10	L26 (10 mol %), NaOMe	DMSO	130	12 (92)	397
Br	CuI	10	L28 (20 mol %), K$_2$CO$_3$	DMSO	120	60 (99)	412
Br	CuI	5	L31 (5 mol %), Cs$_2$CO$_3$	DMF	100	24 (94)	415
I	CuI	5	L31 (5 mol %), Cs$_2$CO$_3$	DMF	90	24 (86)	415
Br	CuI	10	L32 (20 mol %), Cs$_2$CO$_3$	DMF	110	24 (68)	417
Br	CuO	8	K$_2$CO$_3$	pyridine	reflux	48 (37)	402
Cl	CuO nanoparticles	10	K$_2$CO$_3$	DMF	120	16 (87)	64

Nitrogen Nucleophile	Heteroaryl Halide	Conditions	Product(s) and Yield(s) (%)	Refs.

Please refer to the charts preceding the tables for structures indicated by the bold numbers.

Continued from previous page.

C₃

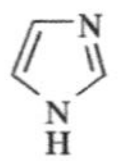

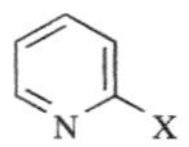

Catalyst (x mol %)

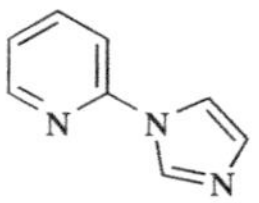

X	Catalyst	x	Additive(s)	Solvent	Temp (°)	Time (h)		Refs.
Cl	Cu₂O	20	KOH	DMSO	130	24	(89)	398
Br	Cu₂O	10	KOH	DMSO	110	24	(91)	398
Br	Cu₂O	10	phen (20 mol %), TBAF	—	145	24	(53)	68
Cl	Cu₂O	10	ninhydrin (20 mol %), KOH	DMSO	130	24	(91)	399
Br	Cu₂O	10	ninhydrin (20 mol %), KOH	DMSO	110	24	(92)	399
Br	CuSO₄	10	**L36** (20 mol %), Cs₂CO₃	H₂O	120	24	(74)	234
Br	Cu(OAc)₂	100	DBU	DMSO	MW, 130	10 min	(89)	85
Cl	Cu fluorapatite	7	K₂CO₃	DMF	120	7	(92)	66
Br	**2**	2	NaOH, TBAB	H₂O	100	12	(81)	233
Cl	**7**	1	K₂CO₃	DMF	110	2	(95)	73

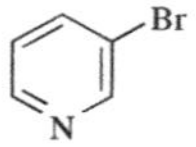

Catalyst (x mol %)

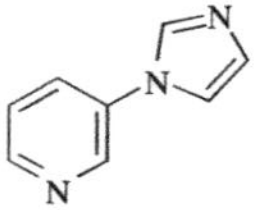

Catalyst	x	Additive(s)	Solvent	Temp (°)	Time (h)		Refs.
CuBr	5	**L5** (10 mol %), Cs₂CO₃	DMF	90	24	(74)	119
CuI	10	**L22** (10 mol %), K₂CO₃	DMF	110	24	(92)	79
CuI	20	D-glucosamine (40 mol %), Cs₂CO₃	DMSO	110	24	(78)	429
CuO	8	K₂CO₃	pyridine	reflux	16	(51)	402

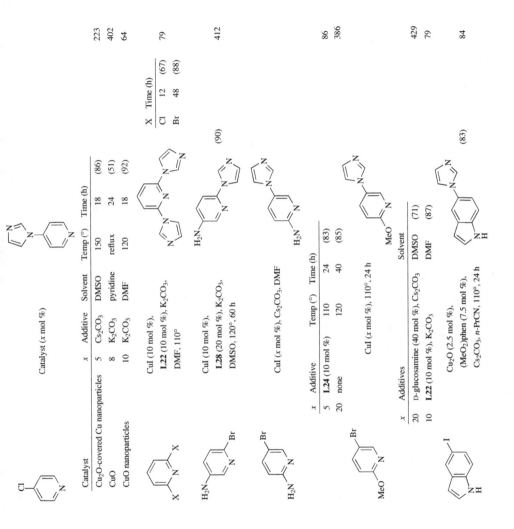

Catalyst (x mol %)

Catalyst	x	Additive	Solvent	Temp (°)	Time (h)		
Cu₂O-covered Cu nanoparticles	5	Cs₂CO₃	DMSO	150	18	(86)	223
CuO	8	K₂CO₃	pyridine	reflux	24	(51)	402
CuO nanoparticles	10	K₂CO₃	DMF	120	18	(92)	64

CuI (10 mol %),
L22 (10 mol %), K₂CO₃,
DMF, 110°

(67)   79

X	Time (h)	
Cl	12	(67)
Br	48	(88)

CuI (10 mol %),
L28 (20 mol %), K₂CO₃,
DMSO, 120°, 60 h

(90)   412

CuI (x mol %), Cs₂CO₃, DMF

x	Additive	Temp (°)	Time (h)	
5	L24 (10 mol %)	110	24	(83)
20	none	120	40	(85)

86
386

CuI (x mol %), 110°, 24 h

x	Additives	Solvent	
20	D-glucosamine (40 mol %), Cs₂CO₃	DMSO	(71)
10	L22 (10 mol %), K₂CO₃	DMF	(87)

429
79

Cu₂O (2.5 mol %),
(MeO₂)phen (7.5 mol %),
Cs₂CO₃, n-PrCN, 110°, 24 h

(83)   84

321

TABLE 9B. *N*-HETEROARYLATION OF IMIDAZOLES (*Continued*)

Nitrogen Nucleophile	Heteroaryl Halide	Conditions	Product(s) and Yield(s) (%)	Refs.

*Please refer to the charts preceding the tables for structures indicated by the **bold** numbers.*

C₃

		Cu₂O (5 mol %), (MeO)₂phen (15 mol %), Cs₂CO₃, PEG, DMSO, 110°, 24 h	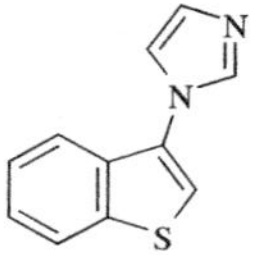 (79)	84
		CuI (30 mol %), L-Pro (60 mol %), [C₄mim][BF₄], 110°, 48 h	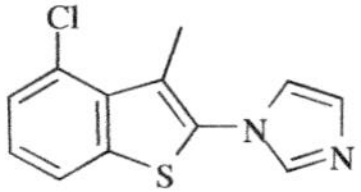 (19)	230
		Cu₂O (5 mol %), (MeO)₂phen (15 mol %), Cs₂CO₃, PEG, DMSO, 110°, 24 h	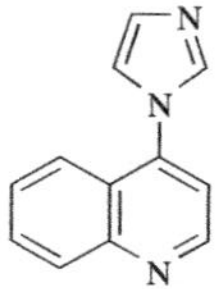 (85)	84
		CuI (5 mol %), DMCDA (15 mol %), K₃PO₄, toluene, 112°, 24 h	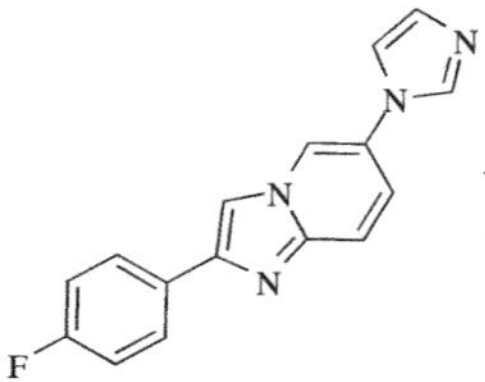 X — Br (18), I (55)	409
	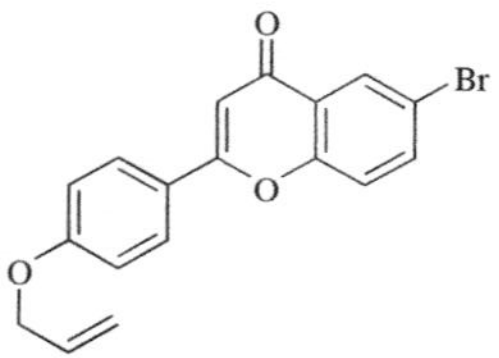	CuI (20 mol %), L-Lys (40 mol %), K₃PO₄, MW (200 W), 140°, 10 h	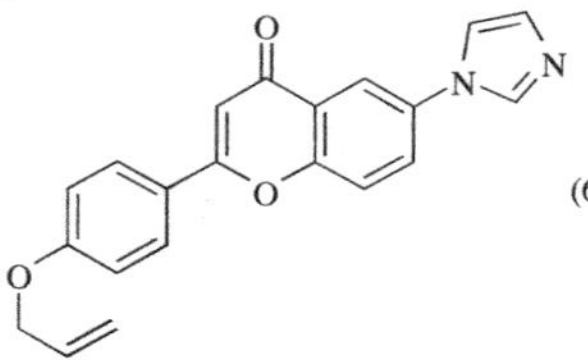 (64)	418

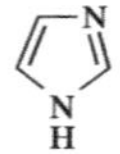

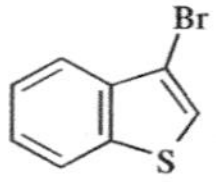

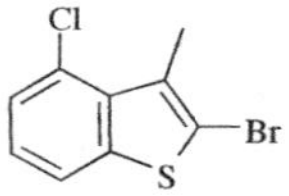

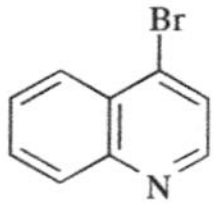

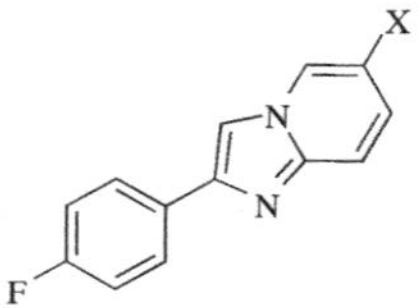

C₄

| 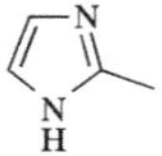 | 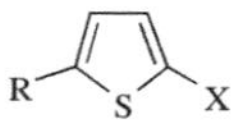 | Catalyst (x mol %), 110° | 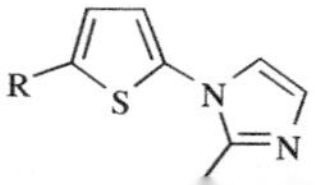 | |

322

R	X	Catalyst	x	Additive(s)	Solvent	Time (h)		
H	Br	CuI	30	L-Pro (60 mol %)	[C$_4$mim][BF$_4$]	48	(62)	230
H	I	Cu$_2$O	2.5	(MeO)$_2$phen (7.5 mol %), PEG, Cs$_2$CO$_3$	DMSO	24	(70)	84
Me	Br	CuI	30	L-Pro (60 mol %)	[C$_4$mim][BF$_4$]	48	(46)	230

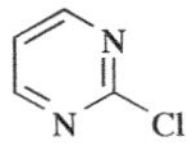

Catalyst (10 mol %), TBAF, 24 h

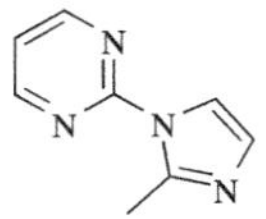

Catalyst	Additive	Solvent	Temp (°)		
CuBr	L37 (20 mol %)	none	150	(100)	80
Cu$_2$O	phen (20 mol %)	—	145	(98)	68

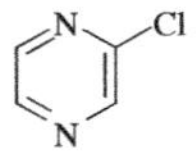

CuI (10 mol %),
L26 (10 mol %), NaOMe,
DMSO, 130°, 12 h 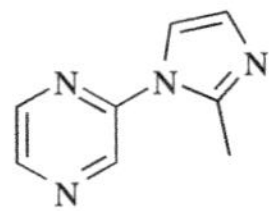   (85)   397

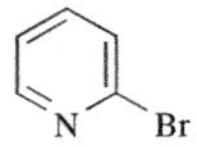

Catalyst (x mol %)

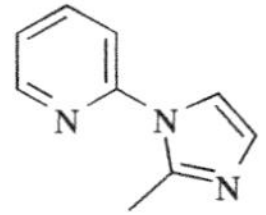

Catalyst	x	Additive(s)	Solvent	Temp (°)	Time (h)		
CuI	30	L-Pro (60 mol %)	[C$_4$mim][BF$_4$]	110	48	(71)	230
CuI	10	oxazolidin-2-one (10 mol %), NaOMe	DMSO	80	12	(81)	392
CuI	10	L26 (10 mol %), NaOMe	DMSO	130	12	(82)	297
CuSO$_4$	10	L36 (20 mol %), Cs$_2$CO$_3$	H$_2$O	120	24	(72)	234
2	2	NaOH, TBAB	H$_2$O	100	12	(80)	233

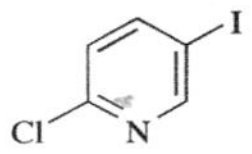

Cu$_2$O (2.5 mol %),
(MeO)$_2$phen (7.5 mol %),
Cs$_2$CO$_3$, PEG,
DMSO, 110°, 24 h   (76)   84

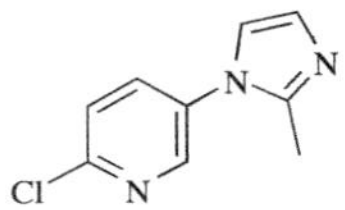

TABLE 9B. *N*-HETEROARYLATION OF IMIDAZOLES (*Continued*)

Nitrogen Nucleophile	Heteroaryl Halide	Conditions	Product(s) and Yield(s) (%)	Refs.

Please refer to the charts preceding the tables for structures indicated by the **bold** numbers.

C4–5

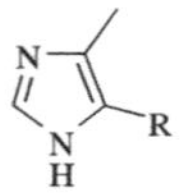

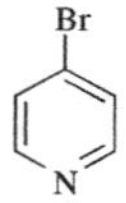

CuO (10 mol %), K2CO3, reflux, 20 h

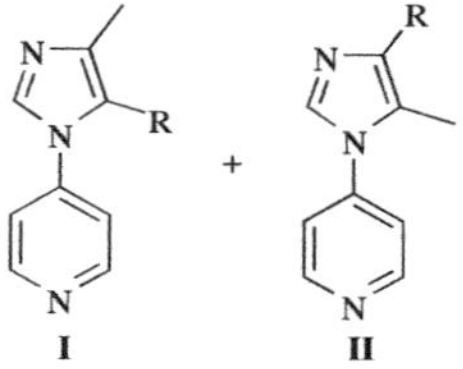

I + II

R	I	II
H	(38)	(7)
Me	(63)	(0)

442

C5–6

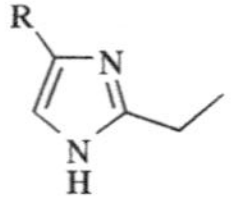

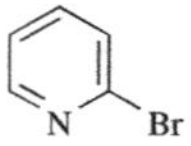

Catalyst (x mol %)

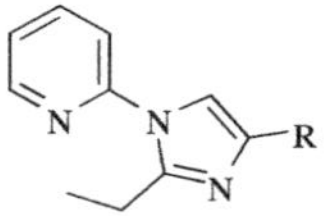

R	Catalyst	x	Additive(s)	Solvent	Temp (°)	Time (h)			Refs.
H	CuI	5	N-hydroxyphthalimide (10 mol %), NaOMe	DMSO	110	24	(92)		396
H	CuSO4	10	**L36** (20 mol %), Cs2CO3	H2O	120	24	(64)		234
Me	**2**	2	NaOH, TBAB	H2O	100	12	(73)		233

C9

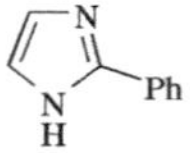

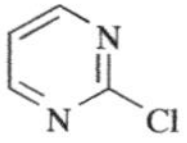

Catalyst (10 mol %), TBAF, 24 h

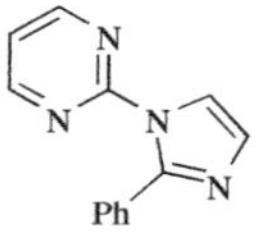

Catalyst	Additive	Solvent	Temp (°)		Refs.
CuBr	**L37** (20 mol %)	none	150	(trace)	80
Cu2O	phen (20 mol %)	—	145	(trace)	68

324

TABLE 10A. N-ARYLATION OF TRIAZOLES

Nitrogen Nucleophile	Aryl Halide	Conditions	Product(s) and Yield(s) (%)	Refs.

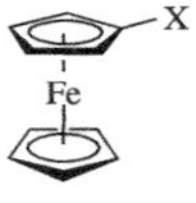

*Please refer to the charts preceding the tables for structures indicated by the **bold** numbers.*

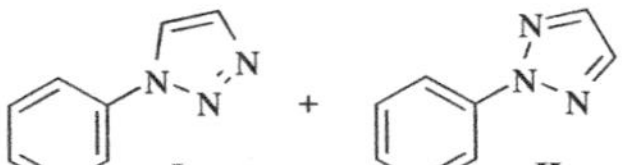

1,2,3-Triazoles

C₂

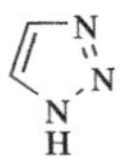

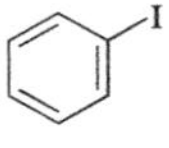

Catalyst	x	Additive(s)	Solvent	Temp (°)	Time (h)	I	II
Cu	20	Cs₂CO₃	n-PrCN	reflux	48	(45)	(—)
CuO	10	Fe(acac)₃ (30 mol %), Cs₂CO₃	DMF	90	30	(48)	(41)
Cu₂O	5	phen (10 mol %), TBAF	—	115	48	(85)	(—)

							129
							130
							68

Cu(TMHD)₂ (10 mol %), KOt-Bu, DMSO, 90°, 12 h

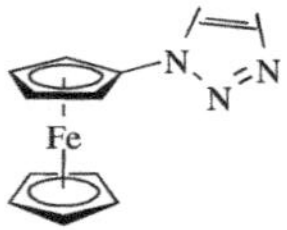

X	
Br	(14)
I	(28)

407

C₈

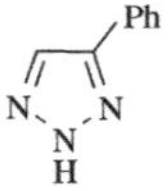

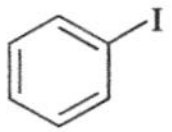

CuCl (10 mol %), L-Pro (20 mol %), K₂CO₃, DMSO, MW, 160°

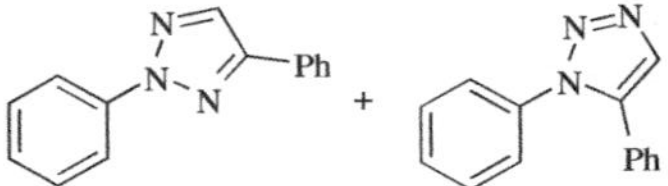

(68)

+

(11)

443

C₁₄

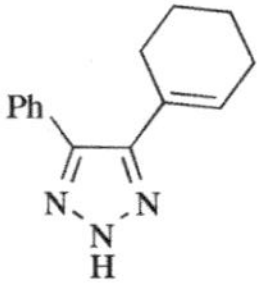

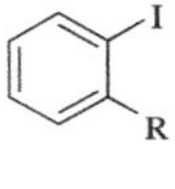

CuCl (10 mol %), L-Pro (20 mol %), K₂CO₃, DMSO, MW, 160°

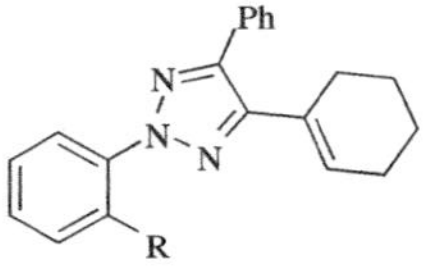

R	
H	(85)
Me	(70)

443

C₁₄₋₁₅

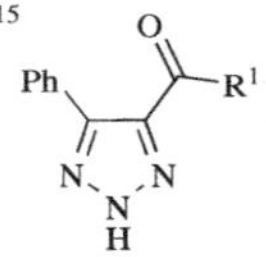

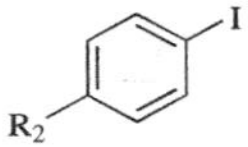

CuCl (10 mol %), L-Pro (20 mol %), K₂CO₃, DMSO, MW, 160°

R¹	R²	
HO(CH₂)₅	H	(60)
Ph	H	(85)
Ph	O₂N	(88)

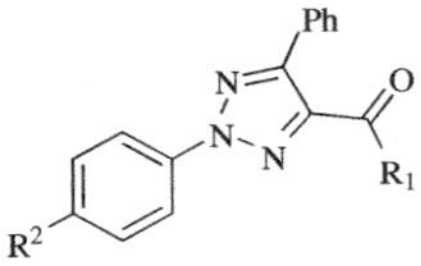

443

TABLE 10A. *N*-ARYLATION OF TRIAZOLES (*Continued*)

Nitrogen Nucleophile	Aryl Halide	Conditions	Product(s) and Yield(s) (%)	Refs.

*Please refer to the charts preceding the tables for structures indicated by the **bold** numbers.*

C₁₆

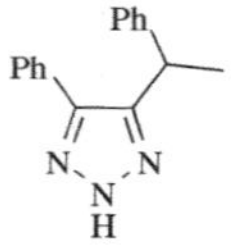

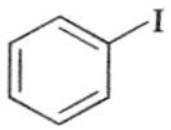

| | | CuCl (10 mol %), L-Pro (20 mol %), K₂CO₃, DMSO, MW, 160° | 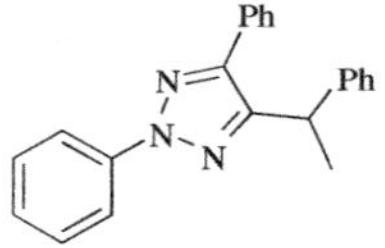 (85) | 443 |

| 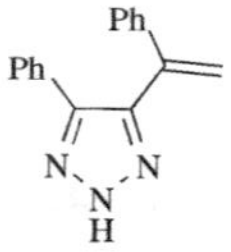 | ArI | CuCl (10 mol %), L-Pro (20 mol %), K₂CO₃, **DMSO, MW, 160°** | Ar: Ph (87); 4-MeOC₆H₄ (85); 4-MeC₆H₄ (86); 1-Np (70) | 443 |

| | | CuCl (10 mol %), L-Pro (20 mol %), K₂CO₃, DMSO, MW, 160° | 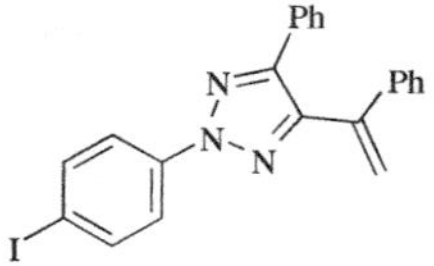 (67) 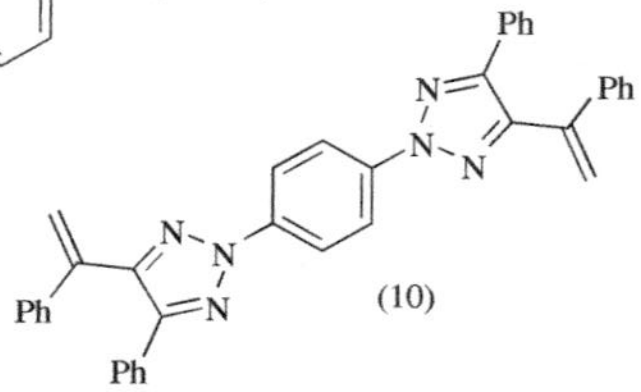 + (10) | 443 |

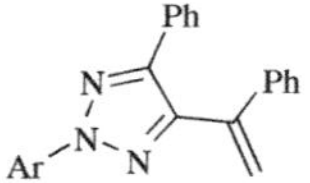

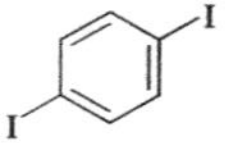

C₁₈

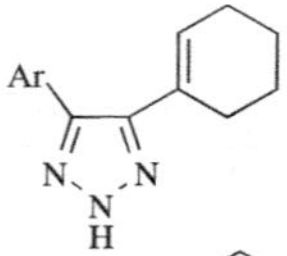

| | | CuCl (10 mol %), L-Pro (20 mol %), K₂CO₃, **DMSO, MW, 160°** | (40) + (45) | 443 |

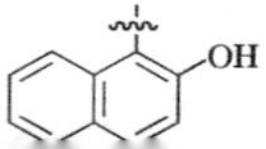

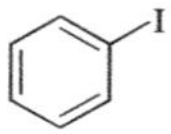

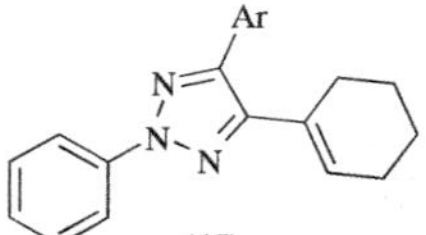

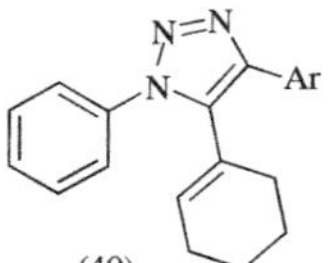

Ar =

C₂₀

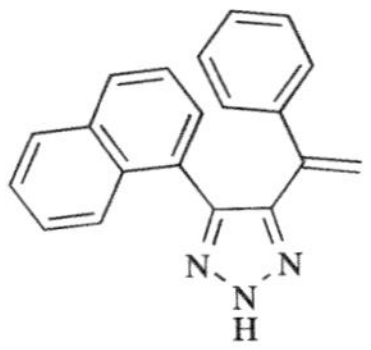

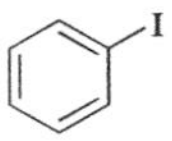

443

CuCl (10 mol %),
L-Pro (20 mol %), K₂CO₃,
DMSO, MW, 160°

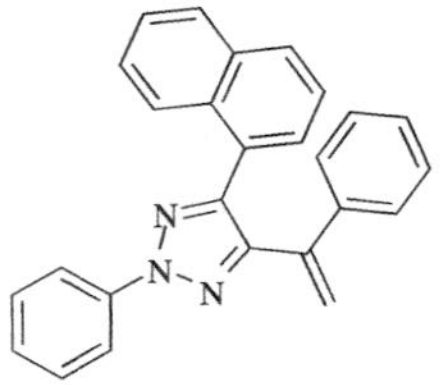

(87)

1,2,4-Triazoles

C₂

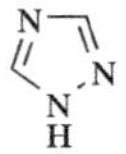

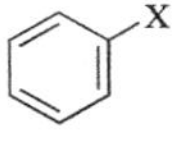

Catalyst (x mol %)

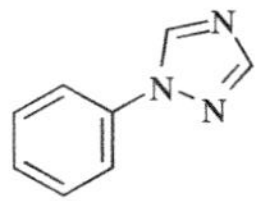

X	Catalyst	x	Additive(s)	Solvent	Temp (°)	Time (h)	
I	Cu	20	Cs₂CO₃	n-PrCN	reflux	48 (54)	129
Br	CuI	20	Cs₂CO₃	DMF	120	40 (85)	386
I	CuI	5	DMCDA (10 mol %), K₃PO₄	DMF	110	24 (89)	128
Br	CuI	20	per-6-ABCD (10 mol %), K₂CO₃	DMSO	110	24 (98)	393
I	CuI	10	**L28** (20 mol %), K₂CO₃	DMSO	120	60 (93)	412
I	CuI	5	**L24** (10 mol %), Cs₂CO₃	DMF	90	24 (96)	86
I	CuO	10	Fe(acac)₃ (30 mol %), Cs₂CO₃	DMF	90	30 (83)	130
I	Cu₂O	10	Cs₂CO₃	DMF	100	18 (76)	221
Br	Cu₂O	5	phen (10 mol %), TBAF	—	110	24 (55)	68
I	Cu₂O	5	phen (10 mol %), TBAF	—	110	24 (90)	68
I	Cu₂O	5	**L30** (20 mol %)	DMF	82	48 (91)	40
I	Cu(TMHD)₂	20	KOt-Bu	DMF	120	24 (51)	308

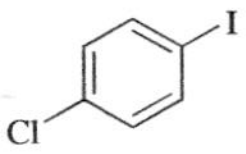

CuI (5 mol %),
8-HOquin (10 mol %), CsF,
DMSO, 100°, 7 h

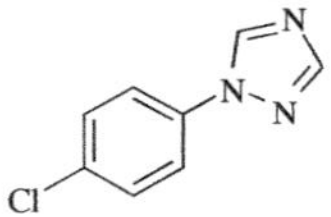

(85)

63

Nitrogen Nucleophile	Aryl Halide	Conditions	Product(s) and Yield(s) (%)	Refs.

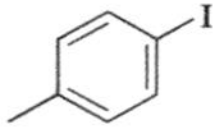

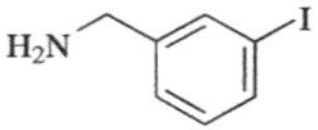

*Please refer to the charts preceding the tables for structures indicated by the **bold** numbers.*

C₂

Aryl halide: O₂N–C₆H₄–Cl

Conditions: CuO (8 mol %), K₂CO₃, pyridine, reflux, 24 h

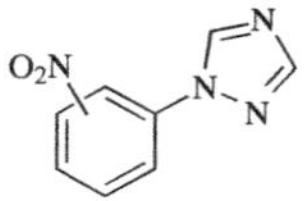

Product: O₂N-phenyl-triazole

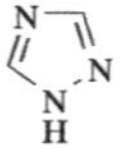

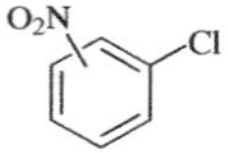

Isomer	
2	(68)
3	(6)
4	(13)

Refs. 402

Aryl halide: MeO–C₆H₄–I

Conditions: Catalyst (x mol %)

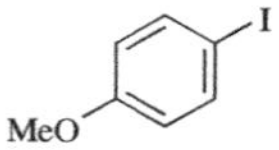

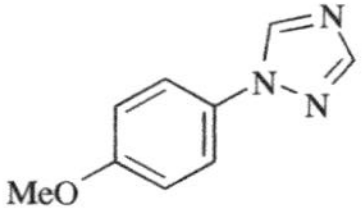

Product: MeO-phenyl-triazole

Catalyst	x	Additive(s)	Solvent	Temp (°)	Time (h)		Refs.
CuI	20	K₃PO₄	DMF	50	40	(84)	59
CuI	5	DMCDA (10 mol %), K₃PO₄	DMF	110	24	(82)	128
Cu₂O	5	phen (10 mol %), TBAF	—	145	48	(85)	68
Cu(TMHD)₂	20	KO*t*-Bu	DMF	120	24	(51)	308
10	1	Cs₂CO₃	DMSO	100	24	(84)	420

Aryl halide: 4-iodotoluene

Conditions: Catalyst (x mol %)

Product: tolyl-triazole

Catalyst	x	Additive(s)	Solvent	Temp (°)	Time (h)		Refs.
CuI	5	phen (10 mol %), TBAF	—	145	48	(61)	68
10	1	Cs₂CO₃	DMSO	100	24	(22)	420

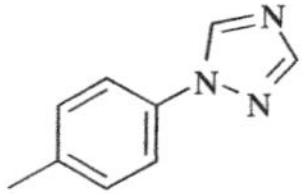

Nitrogen Nucleophile: H₂N–CH₂–C₆H₄–I

Conditions: CuI (5 mol %), DMEDA (10 mol %), K₃PO₄, DMF, 110°, 24 h

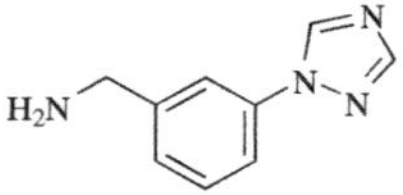

Product: (79)

Refs. 128

Aryl halide: NC–C₆H₄–Cl

Conditions: CuO (8 mol %), K₂CO₃, pyridine, reflux

Product: NC-phenyl-triazole

Isomer	Time (h)	
2	165	(10)
3	64	(8)
4	50	(12)

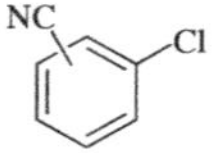

Refs. 402

Nitrogen Nucleophile (C₂): triazole

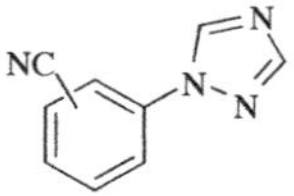

$$
\begin{array}{c}
\text{N} \\
\text{N—N} \\
\text{H}
\end{array}
$$

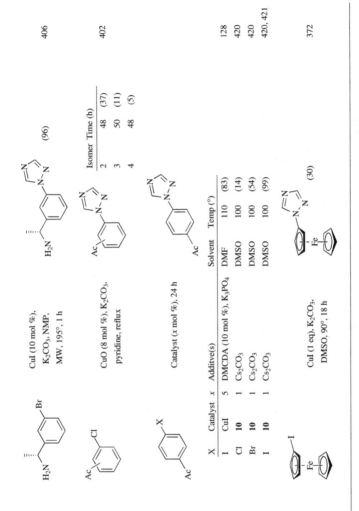

CuI (10 mol %),
K$_2$CO$_3$, NMP,
MW, 195°, 1 h

(96)

406

CuO (8 mol %), K$_2$CO$_3$,
pyridine, reflux

Isomer	Time (h)	
2	48	(37)
3	50	(11)
4	48	(5)

402

Catalyst (x mol %), 24 h

X	Catalyst	x	Additive(s)	Solvent	Temp (°)	
I	CuI	5	DMCDA (10 mol %), K$_3$PO$_4$	DMF	110	(83)	128
Cl	10		Cs$_2$CO$_3$	DMSO	100	(14)	420
Br	10		Cs$_2$CO$_3$	DMSO	100	(54)	420
I	10		Cs$_2$CO$_3$	DMSO	100	(99)	420, 421

CuI (1 eq), K$_2$CO$_3$,
DMSO, 90°, 18 h

(30)

372

TABLE 10B. *N*-HETEROARYLATION OF TRIAZOLES

Nitrogen Nucleophile	Heteroaryl Halide	Conditions	Product(s) and Yield(s) (%)	Refs.
1,2,3-Triazoles				
C₂		CuI (5 mol %), DMCDA (15 mol %), K₃PO₄, DMF, 112°, 24 h	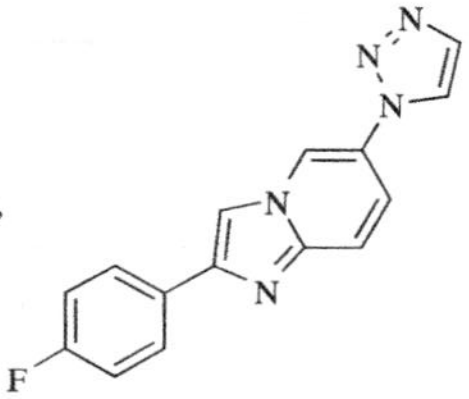 (26) 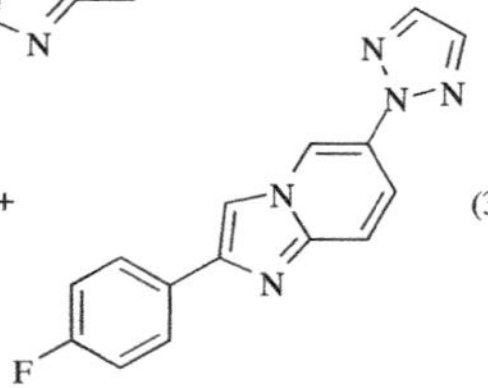 + (33)	409
C₁₅		CuCl (10 mol %), L-Pro (20 mol %), K₂CO₃, DMSO, MW, 160°	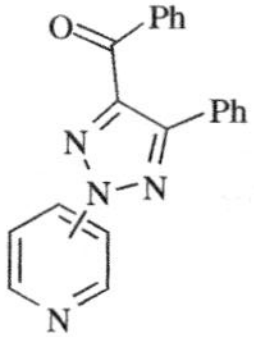 Isomer 2 (88) 4 (80)	443
C₁₆		CuCl (10 mol %), L-Pro (20 mol %), K₂CO₃, DMSO, MW, 160°	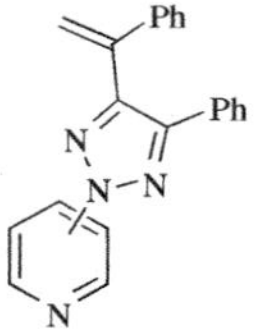 Isomer 2 (85) 3 (70) 4 (70)	443

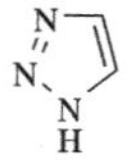

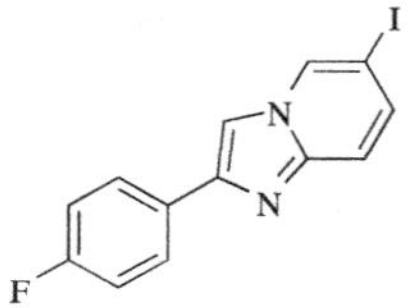

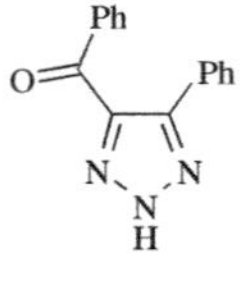

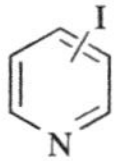

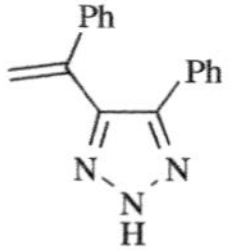

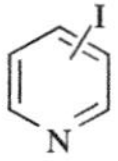

1,2,4-Triazoles

C_2

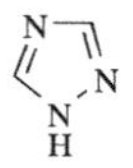

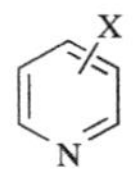

Isomer	X	Time (h)	I	II	
2	Br	16	(29)	(3)	402
3	Br	17	(11)	(0.4)	
4[a]	Cl	22	(6)	(2)	

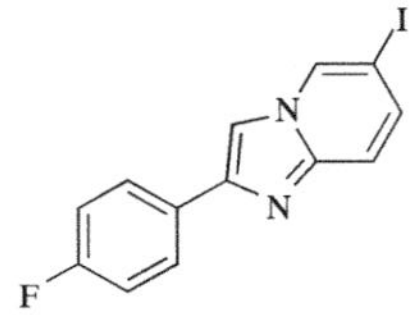

CuO (8 mol %), K₂CO₃, pyridine, reflux

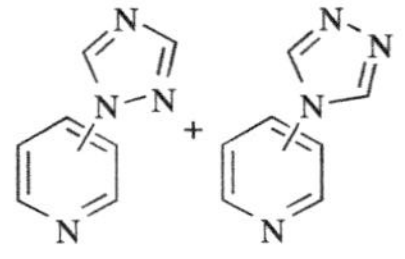

CuI (5 mol %), DMCDA (15 mol %), K₃PO₄, DMF, 112°, 24 h

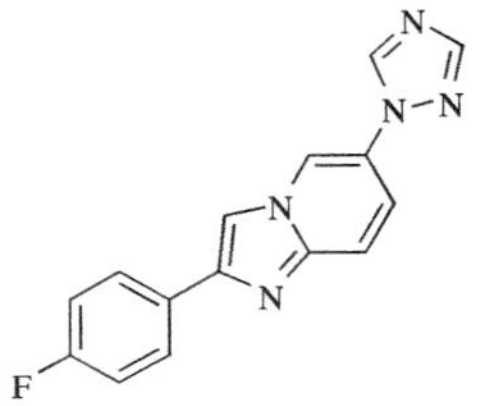

(45) 409

[a] 4-Chloropyridine hydrochloride was used as the halide.

TABLE 11A. *N*-ARYLATION OF INDOLES

*Please refer to the charts preceding the tables for structures indicated by the **bold** numbers.*

C₈

Nitrogen Nucleophile	Aryl Halide		Conditions			Product(s) and Yield(s)			Refs.
			Catalyst(s) (x amount)						
	X								
X	Catalyst(s)	x	Additive(s)	Solvent	Temp (°)	Time (h)	Yield(s) (%)		
I	Cu	20 mol %	Cs₂CO₃	n-PrCN	reflux	10	(95)		129
I	Cu on cellulose	1 mol %	K₂CO₃	DMSO	130	24	(60)		427
Br	Cu/Fe-hydrotalcite	10 wt %	none	toluene	130	12	(85)		283
I	Cu/CuI	1.3 eq/7 mol %	K₂CO₃	—	210	8	(87)		444
I	CuBr	10 mol %	**L10** (12 mol %), Cs₂CO₃	DMSO	80	24	(71)		259
Br	CuBr	10 mol %	**L20** (20 mol %), Cs₂CO₃	DMSO	90	24	(84)		284
I⁺Ph BF₄⁻	CuI	10 mol %	K₂CO₃	DMF	150	7	(80)		445
I	CuI	5 mol %	K₃PO₄	DMF	110	24	(95)		387
I	CuI	5 mol %	NaOH, TBAB (5 mol %)	toluene	reflux	22	(80)		391
I	CuI	5 mol %	EDA (10 mol %), K₂CO₃	PEG-400	80	12	(86)		301
I	CuI	10 mol %	PPAPM (20 mol %), K₃PO₄	DMF	100	24	(96)		48
I	CuI	10 mol %	TM-BINAM (10 mol %), K₂CO₃	MeCN	reflux	26	(98)		446
I	CuI	0.2 mol %	DMCDA (1 mol %), K₃PO₄	toluene	110	48	(98)		191
Br	CuI	20 mol %	per-6-ABCD (10 mol %), K₂CO₃	DMSO	110	36	(93)		393
Br	CuI	10 mol %	phen (20 mol %), KF/Al₂O₃	toluene	reflux	8	(86)		390
I	CuI	10 mol %	phen (20 mol %), KF/Al₂O₃	toluene	reflux	7	(90)		390
I	CuI	5 mol %	*N*-hydroxysuccinimide (10 mol %), NaOMe	DMSO	90	12	(97)		396
Br	CuI	10 mol %	L-Pro (20 mol %), Cs₂CO₃	DMSO	90	40	(33)		47
I	CuI	10 mol %	L-Pro (20 mol %), K₂CO₃	DMSO	90	40	(85)		47
I	CuI	5 mol %	L-Pro (10 mol %), K₂CO₃	DMSO	90	40	(85)		401
I	CuI	20 mol %	L-Pro (20 mol %), K₃CO₄	DMF	140	48	(92)		388
Cl	CuI	10 mol %	piperidine-2-carboxylic acid (20 mol %), K₂CO₃	DMF	110	24	(36)		67
Br	CuI	10 mol %	piperidine-2-carboxylic acid (20 mol %), K₂CO₃	DMF	110	24	(91)		67
I	CuI	10 mol %	piperidine-2-carboxylic acid (20 mol %), K₂CO₃	DMF	110	24	(95)		67

Halide	Cu source	mol %	Ligand / Base	Solvent	Temp	Time	(Yield)	Ref
I	CuI	5 mol %	oxazolidin-2-one (10 mol %), KI, Cs$_2$CO$_3$	DMSO	80	13	(90)	392
Br	CuI	10 mol %	L21 (10 mol %), Cs$_2$CO$_3$	MeCN	reflux	12	(42)	394
I	CuI	10 mol %	L21 (10 mol %), Cs$_2$CO$_3$	MeCN	reflux	12	(78)	394
Br	CuI	10 mol %	L22 (10 mol %), K$_2$CO$_3$	DMF	110	48	(85)	79
I	CuI	5 mol %	L24 (10 mol %), Cs$_2$CO$_3$	DMF	110	24	(98)	86
Br	CuI	10 mol %	L25 (20 mol %), Cs$_2$CO$_3$	MeCN	80	15	(78)	395
I	CuI	10 mol %	L26 (10 mol %), NaOMe	DMSO	90	12	(65)	397
Br	CuI	10 mol %	L28 (20 mol %), K$_2$CO$_3$	DMSO	120	60	(68)	412
I	CuI	10 mol %	L28 (20 mol %), K$_2$CO$_3$	DMSO	120	60	(99)	412
I	CuI	10 mol %	L38 (10 mol %), K$_3$PO$_4$	toluene	reflux	24	(96)	447
I	CuI	10 mol %	L39 (10 mol %), K$_3$PO$_4$	toluene	reflux	24	(95)	447
Br	CuO	6 mol %	K$_2$CO$_3$	DMF	reflux	24	(50)	448
I	CuO	1.3 mol %	KOH	DMSO	110	3.5	(94)	224
I	CuO	10 mol %	Fe(acac)$_3$ (30 mol %), Cs$_2$CO$_3$	DMF	90	30	(94)	130
Br	CuO	10 mol %	rac-binol (20 mol %), FeCl$_3$ (10 mol %), Cs$_2$CO$_3$	DMF	90	12	(89)	277
I	CuO	10 mol %	rac-binol (20 mol %), FeCl$_3$ (10 mol %), Cs$_2$CO$_3$	DMF	90	12	(89)	277
I	Cu$_2$O	10 mol %	KOH	DMSO	120	24	(90)	398
I	Cu$_2$O	10 mol %	Cs$_2$CO$_3$	DMF	100	18	(95)	221
I	Cu$_2$O	10 mol %	ninhydrin (20 mol %), KOH	DMSO	110	24	(90)	399
Cl	Cu$_2$O	10 mol %	phen (20 mol %), TBAF	—	145	48	(trace)	68
Br	Cu$_2$O	10 mol %	phen (20 mol %), TBAF	—	145	24	(91)	68
I	Cu$_2$O	5 mol %	L7 (20 mol %), Cs$_2$CO$_3$	MeCN	80	18	(64)	125
Br	Cu$_2$O	10 mol %	L23 (20 mol %), KOt-Bu	DMF	130	24	(40)	431
I	Cu$_2$O	5 mol %	L23 (10 mol %), KOt-Bu	DMF	110	24	(95)	431
I	Cu$_2$O	5 mol %	L30 (20 mol %)	MeCN	82	24	(92)	40

TABLE 11A. *N*-ARYLATION OF INDOLES (*Continued*)

Nitrogen Nucleophile	Aryl Halide	Conditions	Product(s) and Yield(s) (%)	Refs.

*Please refer to the charts preceding the tables for structures indicated by the **bold** numbers.*
Continued from previous page.

C$_8$

Catalyst(s) (x amount)

X	Catalyst	x	Additive(s)	Solvent	Temp (°)	Time (h)		Refs.
I	Cu(OAc)$_2$	1.1 eq	none	DMA	160	48	(76)	127
Br	Cu(OAc)$_2$·H$_2$O	15 mol %	(−)-sparteine (30 mol %), K$_2$CO$_3$	DMF	130	30	(63)	290
I	Cu(OAc)$_2$·H$_2$O	15 mol %	(−)-sparteine (30 mol %), K$_2$CO$_3$	DMF	130	24	(90)	290
I	Cu(OAc)$_2$·H$_2$O	20 mol %	hippuric acid (20 mol %), Cs$_2$CO$_3$	DMF	140	30	(91)	289
I	CuSO$_4$	10 mol %	L36 (20 mol %), Cs$_2$CO$_3$	H$_2$O	120	24	(90)	234
I	Cu(TMHD)$_2$	20 mol %	KO*t*-Bu	DMF	120	24	(92)	308
Br	Cu fluorapatite	12.5 mol %	none	DMSO	110	15	(88)	227
Br	2	2 mol %	NaOH, TBAB	H$_2$O	100	24	(94)	233
I	6	5 mol %	Cs$_2$CO$_3$	toluene	100	14	(88)	123

CuI (x mol %), K$_3$PO$_4$

R	x	Additive	Solvent	Temp (°)	Time (h)	
Br	5	DMCDA (20 mol %)	toluene	110	24	(72)	191
Br	10	BtH (20 mol %)	DMSO	120	30	(65)	77
I	10	BtH (20 mol %)	DMSO	120	30	(65)	77

Catalyst (x mol %)

R	Catalyst	x	Additives	Solvent	Temp (°)	Time (h)	
F	CuI	5	L-Pro (10 mol %), K$_2$CO$_3$	DMSO	80	34	(88)	401
F	CuI	10	L-Pro (20 mol %), K$_2$CO$_3$	DMSO	80	34	(88)	47
Cl	CuCl	5	L17 (10 mol %), NaOH, TBAB	H$_2$O	100	24	(93)	235
Cl	CuI	5	L40 (5 mol %), K$_3$PO$_4$	DMF	110	24	(89)	449

Catalyst(s) (x amount)

X	Catalyst(s)	x	Additive(s)	Solvent	Temp (°)	Time (h)		
I	Cu/CuI	1.3 eq/7 mol %	K$_2$CO$_3$	—	210	8	(82)	444
I$^+$C$_6$H$_4$Cl-4 BF$_4$$^-$	CuI	10 mol %	K$_2$CO$_3$	DMF	80	3	(92)	445
I	CuI	5 mol %	DMEDA (10 mol %), CsF	THF	60	24	(86)	63
Br	CuI	10 mol %	oxazolidin-2-one (20 mol %), NaOMe	DMSO	80	18	(85)	392
I	CuI	5 mol %	N-hydroxyphthalimide (10 mol %), NaOMe	DMSO	90	12	(99)	396
I	CuI	5 mol %	N-hydroxymaleimide (10 mol %), NaOMe	DMSO	110	24	(48)	396
Br	CuI	10 mol %	L21 (20 mol %), Cs$_2$CO$_3$	MeCN	reflux	12	(52)	394
I	CuI	10 mol %	L21 (20 mol %), Cs$_2$CO$_3$	MeCN	reflux	12	(89)	394
I	CuI	5 mol %	L40 (5 mol %), K$_3$PO$_4$	DMF	110	24	(86)	449
I	Cu(OAc)$_2$•H$_2$O	20 mol %	hippuric acid (20 mol %), Cs$_2$CO$_3$	DMF	140	30	(43)	289

CuI (x mol %)

X	x	Additive(s)	Solvent	Temp (°)	Time (h)		
I$^+$Ph BF$_4$$^-$	10	K$_2$CO$_3$	DMF	80	3	(81)	445
I	5	K$_3$PO$_4$	DMF	110	24	(76)	387
Br	10	PPAPM (20 mol %), K$_3$PO$_4$	DMF	100	36	(87)	48
I	10	PPAPM (20 mol %), K$_3$PO$_4$	DMF	100	24	(96)	48
I	5	oxazolidin-2-one (10 mol %), NaOMe	DMSO	80	16	(83)	392
I	5	L-Pro (20 mol %), K$_3$PO$_4$	dioxane	100	24	(60)	388
I	5	N-hydroxysuccinimide (10 mol %), NaOMe	DMSO	110	24	(86)	396
I	10	piperidine-2-carboxylic acid (20 mol %), K$_2$CO$_3$	DMF	110	24	(94)	67
I	10	L21 (20 mol %), KI, Cs$_2$CO$_3$	MeCN	reflux	12	(91)	394

Please refer to the charts preceding the tables for structures indicated by the **bold** numbers.

C8

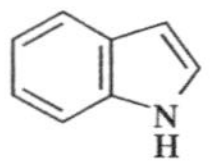

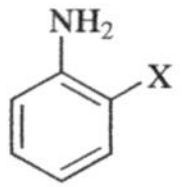

Nitrogen Nucleophile	Aryl Halide	Conditions	Product(s) and Yield(s) (%)	Refs.

Nitrogen Nucleophile: indole (with N–H)

Aryl Halide: 2-(NH2)-C6H4-X   **Conditions:** CuI (5 mol %)

X	Additives	Solvent	Temp (°)	Time (h)		
Br	EDA (10 mol %), K2CO3	PEG-400	80	24	(79)	301
I	EDA (10 mol %), K2CO3	PEG-400	80	18	(84)	301
I	DMEDA (20 mol %), K3PO4	toluene	110	24	(88)	191

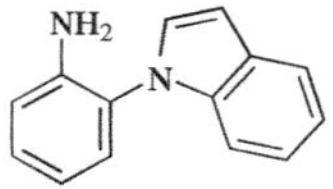

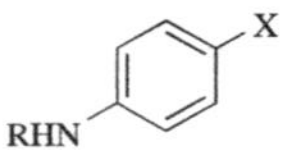

Product: N-(2-aminophenyl)indole

Aryl Halide: 4-(RHN)-C6H4-X   **Conditions:** CuI (x mol %)

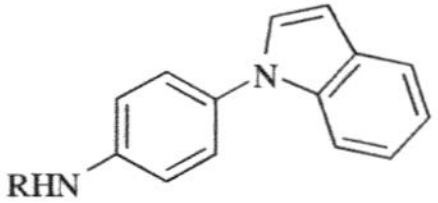

R	X	x	Additive(s)	Solvent	Temp (°)	Time (h)		
Ac	I+C6H4NHAc·4 BF4^-	10	K2CO3	DMF	80	3	(72)	445
Ac	I	5	DMCDA (20 mol %), K3PO4	toluene	110	24	(72)	191
Boc	I	5	DMCDA (20 mol %), K3PO4	toluene	110	24	(75)	191

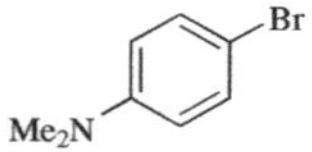

Product: N-(4-RHN-phenyl)indole

Aryl Halide: 4-(Me2N)-C6H4-Br   **Conditions:** CuI (5 mol %), DMCDA (20 mol %), K3PO4, toluene, 110°, 24 h

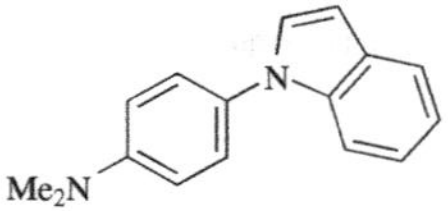

Product: N-(4-Me2N-phenyl)indole   (93)   191

Aryl Halide: 2-(NO2)-C6H4-X   **Conditions:** Catalyst (x mol %)

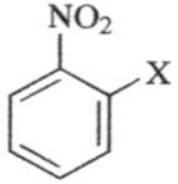

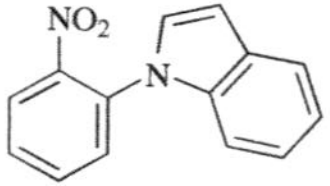

X	Catalyst	x	Additive(s)	Solvent	Temp (°)	Time (h)		
I	CuI	5	L-Pro (20 mol %), K3PO4	dioxane	100	24	(<5)	388
Cl	CuO	6	K2CO3	DMF	reflux	24	(48)	448
Cl	CuO	8	K2CO3	pyridine	reflux	18	(50)	402

Product: N-(2-nitrophenyl)indole

Catalyst (x mol %)

O$_2$N— (aryl)—X → O$_2$N— (aryl)—N(aryl with NO$_2$)

X	Catalyst	x	Additive(s)	Solvent	Temp (°)	Time (h)		
I·C$_6$H$_4$NO$_2$·3 BF$_4^-$	CuI	10	K$_2$CO$_3$	DMF	80	3	(76)	445
I	CuI	5	L-Pro (10 mol %), K$_2$CO$_3$	DMSO	75	42	(93)	401
I	CuI	10	L-Pro (20 mol %), K$_2$CO$_3$	DMSO	75	42	(93)	47
I	CuI	10	TM-BINAM (10 mol %), K$_2$CO$_3$	MeCN	reflux	20	(95)	446
Br	CuO	6	K$_2$CO$_3$	DMF	reflux	24	(38)	448
Cl	CuO	8	K$_2$CO$_3$	pyridine	reflux	48	(80)	402
I	CuO	10	rac-binol (20 mol %), FeCl$_3$ (10 mol %), Cs$_2$CO$_3$	DMF	100	12	(89)	277

Catalyst(s) [x amount(s)]

O$_2$N— (aryl)—X → O$_2$N— (indol-1-yl aryl)

X	Catalyst(s)	x	Additive(s)	Solvent(s)	Temp (°)	Time (h)		
Br	Cu/CuI	1.3 eq/7 mol %	K$_2$CO$_3$	—	210	8	(60)	444
I	CuBr$_2$	10 mol %	TBAF	—	145	4	(64)	81
Br	CuI	10 mol %	phen (20 mol %), KF/Al$_2$O$_3$	toluene	reflux	9	(89)	390
I	CuI	10 mol %	phen (20 mol %), KF/Al$_2$O$_3$	toluene	reflux	8	(94)	390
I	CuI	1 mol %	CDA (10 mol %), K$_3$PO$_4$	dioxane	110	24	(99)	115
Cl	CuI	5 mol %	N-hydroxyphthalimide (10 mol %), NaOMe	DMSO	110	40	(62)	396
Cl	CuI	10 mol %	L26 (10 mol %), NaOMe	DMSO	90	12	(57)	397
I	CuO nanoparticles	5 mol %	KOH	t-BuOH/DMSO (3:1)	110	11	(45)	74
Cl	CuO nanoparticles	10 mol %	K$_2$CO$_3$	DMF	120	9	(89)	64
Cl	CuO	8 mol %	K$_2$CO$_3$	pyridine	reflux	19	(15)	402
Br	CuO	6 mol %	K$_2$CO$_3$	DMF	reflux	24	(77)	448
I	Cu$_2$O	10 mol %	Cs$_2$CO$_3$	PEG-3400	MW, 150	1	(47)	450
I	CuOAc	1.1 eq	none	DMA	160	48	(66)	127
Cl	Cu(OAc)$_2$•H$_2$O	20 mol %	hippuric acid (20 mol %), Cs$_2$CO$_3$	DMF	140	20	(55)	289
I	Cu(OAc)$_2$•H$_2$O	20 mol %	hippuric acid (20 mol %), Cs$_2$CO$_3$	DMF	140	30	(43)	289
Cl	Cu(OAc)$_2$•H$_2$O	15 mol %	(−)-sparteine (30 mol %), K$_2$CO$_3$	DMF	130	36	(34)	290
Br	Cu(OAc)$_2$•H$_2$O	1 mol %	(−)-sparteine (30 mol %), K$_2$CO$_3$	DMF	130	36	(71)	290
I	Cu(TMHD)$_2$	20 mol %	KOt-Bu	DMF	120	24	(87)	308
F	7	1 mol %	K$_2$CO$_3$	DMF	110	5	(85)	73
Cl	7	1 mol %	K$_2$CO$_3$	DMF	110	1	(82)	73

TABLE 11A. N-ARYLATION OF INDOLES (*Continued*)

Nitrogen Nucleophile	Aryl Halide	Conditions	Product(s) and Yield(s) (%)	Refs.

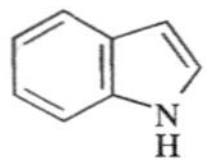

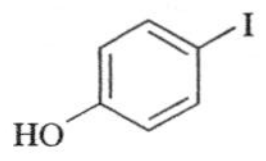

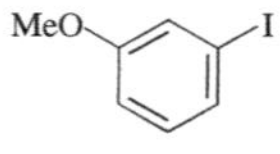

*Please refer to the charts preceding the tables for structures indicated by the **bold** numbers.*

C₈

(indole, N–H)

Aryl Halide: 4-iodophenol (HO–C₆H₄–I)

Conditions: CuOAc (1.1 eq), DMA, 160°, 48 h

Product: (49)

Refs.: 127

Second entry:

Aryl Halide: 2-OMe-aryl-X

Conditions: Catalyst (x mol %)

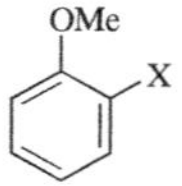

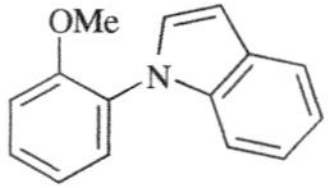

X	Catalyst	x	Additive(s)	Solvent	Temp (°)	Time (h)		Refs.
Br	CuI	10	phen (20 mol %), KF/Al₂O₃	toluene	reflux	12	(73)	390
I	CuI	10	phen (20 mol %), KF/Al₂O₃	toluene	reflux	11	(78)	390
I	CuI	1	CDA (10 mol %), K₃PO₄	dioxane	110	24	(100)	115
I	CuI	10	BtH (20 mol %), K₃PO₄	DMSO	120	30	(68)	77
I	CuI	10	L-Pro (20 mol %), K₂CO₃	DMSO	90	48	(79)	47
I	CuI	5	L-Pro (10 mol %), K₂CO₃	DMSO	90	48	(79)	401
I	Cu₂O	10	Cs₂CO₃	PEG-3400	MW, 150	1	(83)	450

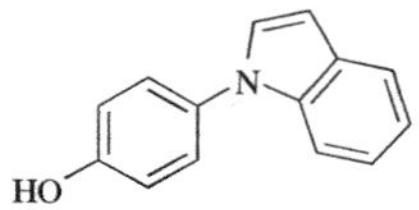

Product (for 2-OMe entries):

Aryl Halide: 3-MeO-iodobenzene (MeO–C₆H₄–I)

Conditions: CuI (10 mol %), K₂CO₃, TM–BINAM (10 mol %), MeCN, reflux, 40 h

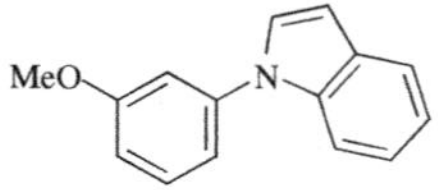

Product: (90)

Refs.: 446

Aryl Halide: 5-bromo-1,3-benzodioxole (Br–benzodioxole)

Conditions: CuI (5 mol %), EDA (10 mol %), K₂CO₃, PEG-400, 80°, 24 h

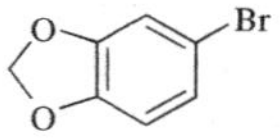

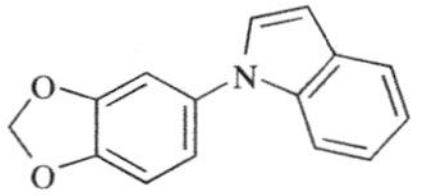

Product: (60)

Refs.: 301

Aryl Halide: 3,4,5-trimethoxyphenyl iodide (MeO, MeO, OMe substituted aryl–I)

Conditions: CuOAc (1.1 eq), DMA, 160°, 48 h

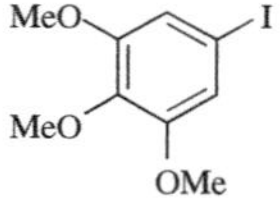

Product: (36)

Refs.: 127

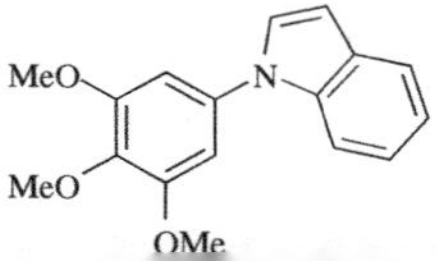

338

Catalyst(s) [x amount(s)]

X	Catalyst(s)	x	Additive(s)	Solvent(s)	Temp (°)	Time		
I	Cu/CuI	1.3 eq/7 mol %	K₂CO₃	—	210	8 h	(79)	444
I	CuF₂	25 mol %	phen (25 mol %), K₂CO₃	DMF	140	96 h	(72)	419
I	CuBr₂	10 mol %	TBAF	—	145	24 h	(47)	81
I⁺C₆H₄OMe-4 BF₄⁻	CuI	10 mol %	K₂CO₃	DMF	80	3 h	(80)	445
I	CuI	10 mol %	K₃PO₄	DMF	110	24 h	(81)	387
Br	CuI	10 mol %	phen (20 mol %), KF/Al₂O₃	toluene	reflux	9 h	(84)	390
I	CuI	10 mol %	phen (20 mol %), KF/Al₂O₃	toluene	reflux	8 h	(87)	390
Br	CuI	20 mol %	L-Pro (20 mol %), K₃PO₄	DMF	140	48 h	(85)	388
I	CuI	5 mol %	L-Pro (20 mol %), K₃PO₄	dioxane	100	24 h	(99)	388
I	CuI	10 mol %	BtH (20 mol %), K₃PO₄	DMSO	120	30 h	(87)	77
Br	CuI	5 mol %	EDA (10 mol %), K₂CO₃	PEG-400	80	24 h	(79)	301
I	CuI	10 mol %	TM–BINAM (10 mol %), K₂CO₃	MeCN	reflux	36 h	(76)	446
Br	CuI	5 mol %	L34 (5 mol %), KOt-Bu	DMSO	110	—	(94)	403
Cl	CuI	10 mol %	L39 (10 mol %), K₃PO₄	DMF	reflux	24 h	(trace)	447
I	CuI	5 mol %	L40 (5 mol %), K₃PO₄	DMF	110	24 h	(83)	449
Br	CuO	5 mol %	L14 (50 mol %), TBAB, 2,5-hexanedione. K₃PO₄	H₂O	MW (100 W), 140	5 min	(52)	51
I	Cu₂O	10 mol %	Cs₂CO₃	PEG-3400	MW, 150	1 h	(76)	450
Br	Cu₂O	10 mol %	phen (20 mol %), TBAF	—	145	24 h	(74)	68
Br	Cu–Fe hydrotalcite	10 wt %	none	toluene	130	12 h	(87)	283
I	Cu(TMHD)₂	20 mol %	KOt-Bu	DMF	120	24 h	(86)	308
I	6	5 mol %	Cs₂CO₃	toluene	100	24 h	(82)	123

TABLE 11A. N-ARYLATION OF INDOLES (Continued)

Nitrogen Nucleophile	Aryl Halide	Conditions	Product(s) and Yield(s) (%)	Refs.

*Please refer to the charts preceding the tables for structures indicated by the **bold** numbers.*

C₈

Nitrogen Nucleophile: indole (N–H)

Aryl Halide: 2-methylphenyl halide (X)

Conditions: Catalyst (x mol %)

X	Catalyst	x	Additive(s)	Solvent	Temp (°)	Time (h)	Yield	Refs.
I	Cu	200	18-c-6 (10 mol %), K₂CO₃	1,2-Cl₂C₆H₄	reflux	36	(87)	444
Br	CuI	10	phen (20 mol %), KF/Al₂O₃	toluene	reflux	10	(76)	390
I	CuI	10	phen (20 mol %), KF/Al₂O₃	toluene	reflux	9	(81)	390
Br	CuI	5	DMCDA (20 mol %), K₃PO₄	toluene	110	24	(91)	191
I	CuI	5	**L40** (5 mol %), K₃PO₄	DMF	110	24	(73)	449

Aryl Halide: 3-methylphenyl iodide (I)

Conditions: Cu (1.3 eq), CuI (7 mol %), K₂CO₃, 210°, 8 h

Product: N-(3-methylphenyl)indole (84) — Refs. 444

Aryl Halide: 4-methylphenyl halide (X)

Conditions: Catalyst(s) [x amount(s)]

Product: N-(4-methylphenyl)indole

X	Catalyst(s)	x	Additive(s)	Solvent(s)	Temp (°)	Time (h)	Yield	Refs.
I	Cu/CuI	1.3 eq/7 mol %	K₂CO₃	—	210	8	(84)	444
I⁺C₆H₄Me-4 BF₄⁻	CuI	10 mol %	K₂CO₃	DMF	80	3	(87)	445
Br	CuI	1 mol %	DMCDA (5 mol %), K₃PO₄	toluene	110	24	(96)	191
I	CuI	10 mol %	TM-BINAM (10 mol %), K₂CO₃	MeCN	reflux	36	(76)	446
Br	CuI	20 mol %	L-Pro (20 mol %), K₃PO₄	DMF	140	48	(90)	388
I	CuI	5 mol %	L-Pro (20 mol %), K₃PO₄	dioxane	100	24	(80)	388
I	CuI	5 mol %	NaOH, TBAB (5 mol %)	toluene	reflux	22	(64)	391
Br	CuI	10 mol %	oxazolidin-2-one (20 mol %), NaOMe	DMSO	120	24	(52)	392
I	CuI	10 mol %	phen (20 mol %), KF/Al₂O₃	toluene	reflux	7	(89)	390

X	Catalyst	amount	Additive(s)	Solvent	Temp (°)	Time (h)	(%)	Ref
Br	CuI	10 mol %	L21 (20 mol %), KI, Cs$_2$CO$_3$	MeCN	reflux	12	(51)	394
Br	CuI	10 mol %	L38 (10 mol %), K$_3$PO$_4$	DMF	reflux	24	(95)	447
Br	CuI	10 mol %	L39 (10 mol %), K$_3$PO$_4$	DMF	reflux	24	(95)	447
I	CuI	5 mol %	L40 (5 mol %), K$_3$PO$_4$	DMF	110	24	(83)	449
I	CuI	5 mol %	L41 (10 mol %), NaOt-Bu	toluene	reflux	40	(99)	451
I	Cu$_2$O	10 mol %	Cs$_2$CO$_3$	PEG-3400	MW, 150	1	(57)	450
Br	Cu(II)–NaY zeolite	10 mol %	Cs$_2$CO$_3$	DMF	120	20	(99)	404
I	Cu(II)–NaY zeolite	10 mol %	Cs$_2$CO$_3$	DMF	120	22	(99)	404

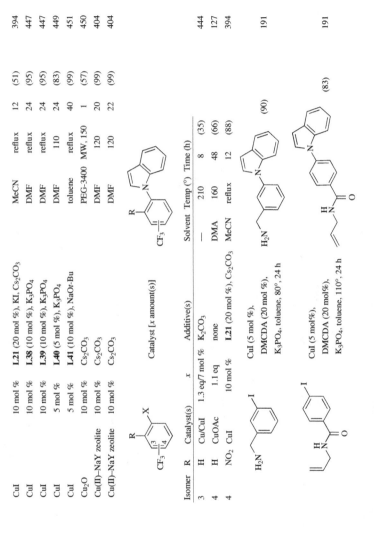

Catalyst [x amount(s)]

Isomer	R	Catalyst(s)	x	Additive(s)	Solvent	Temp (°)	Time (h)	(%)	Ref
3	H	Cu/CuI	1.3 eq/7 mol %	K$_2$CO$_3$	—	210	8	(35)	444
4	H	CuOAc	1.1 eq	none	DMA	160	48	(66)	127
4	NO$_2$	CuI	10 mol %	L21 (20 mol %), Cs$_2$CO$_3$	MeCN	reflux	12	(88)	394

CuI (5 mol %), DMCDA (20 mol %), K$_3$PO$_4$, toluene, 80°, 24 h (90) 191

CuI (5 mol%), DMCDA (20 mol%), K$_3$PO$_4$, toluene, 110°, 24 h (83) 191

TABLE 11A. N-ARYLATION OF INDOLES (Continued)

Nitrogen Nucleophile	Aryl Halide	Conditions	Product(s) and Yield(s) (%)	Refs.

*Please refer to the charts preceding the tables for structures indicated by the **bold** numbers.*

C8

Nitrogen Nucleophile: indole (N–H)

Aryl Halide: NC–C6H4–X ; Conditions: Catalyst (x mol %)

Isomer	X	Catalyst	x	Additive	Solvent	Temp (°)	Time (h)	Product(s) and Yield(s) (%)	Refs.
2	Cl	CuO nanoparticles	10	K_2CO_3	DMF	120	18	(83)	64
4	Cl	CuO nanoparticles	10	K_2CO_3	DMF	120	18	(76)	64
4	Br	CuO	8	K_2CO_3	pyridine	reflux	39	(77)	402
4	I	Cu_2O	10	Cs_2CO_3	PEG-3400	MW, 150	1	(80)	450

(Product: NC-phenyl-indole)

Aryl Halide	Conditions	Product(s) and Yield(s) (%)	Refs.
3-bromophenyl-1,3-dioxolane	CuI (5 mol %), DMCDA (20 mol %), K_3PO_4, toluene, 80°, 24 h	(78)	191
4-iodo-EtO_2C-benzene	CuI (5 mol %), DMCDA (20 mol %), K_3PO_4, toluene, 110°, 24 h	(90)	191

Aryl Halide: 3,5-dimethylphenyl–X ; Conditions: Catalyst (x mol %)

X	Catalyst	x	Additives	Solvent	Temp (°)	Time (h)	Product(s) and Yield(s) (%)	Refs.
I	CuI	10	TM–BINAM (10 mol %), K_2CO_3	MeCN	reflux	36	(60)	446
I	CuI	10	EDA (10 mol %), Cs_2CO_3	dioxane	110	24	(96)	405
I	CuI	1	CDA (10 mol %), K_3PO_4	dioxane	110	24	(99)	115
Br	$[CuOTf]_2 \cdot C_6H_6$	5	4,7-Cl_2-phen (10 mol %), Cs_2CO_3	NMP	125	41	(94)	452

Aryl Halide	Conditions	Product(s) and Yield(s) (%)	Refs.
3-bromophenyl-CH(H_2N)CH3	CuI (10 mol %), K_2CO_3, NMP, MW, 195°, 22 h	(49)	406

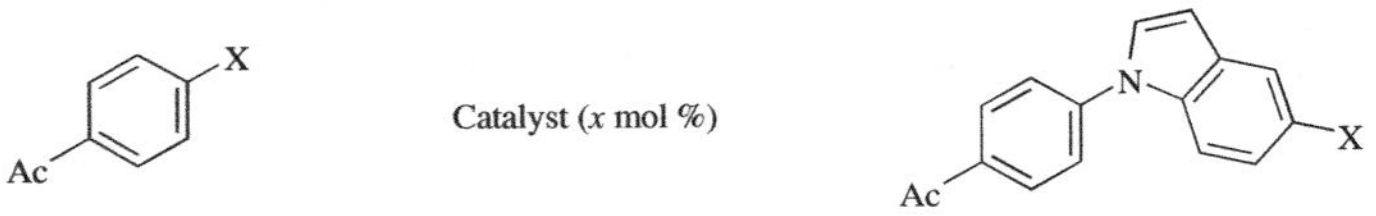

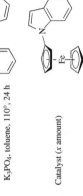

Catalyst (x mol %)

X	Catalyst	x	Additive(s)	Solvent	Temp (°)	Time (h)		
I	CuI	10	TM–BINAM (10 mol %), K_2CO_3	MeCN	reflux	28	(94)	446
Br	CuI	20	L-Pro (20 mol %), K_3PO_4	DMF	140	48	(91)	388
I	CuI	5	L-Pro (20 mol %), K_3PO_4	dioxane	100	24	(99)	388
I	CuI	10	L-Pro (20 mol %), K_2CO_3	DMSO	75	42	(90)	47
I	CuO	5	L-Pro (10 mol %), K_2CO_3	DMSO	80	22	(90)	401
Br	CuO	8	K_2CO_3	pyridine	reflux	45	(93)	402
Br	CuI	10	**L32** (20 mol %), Cs_2CO_3	DMF	110	24	(85)	417

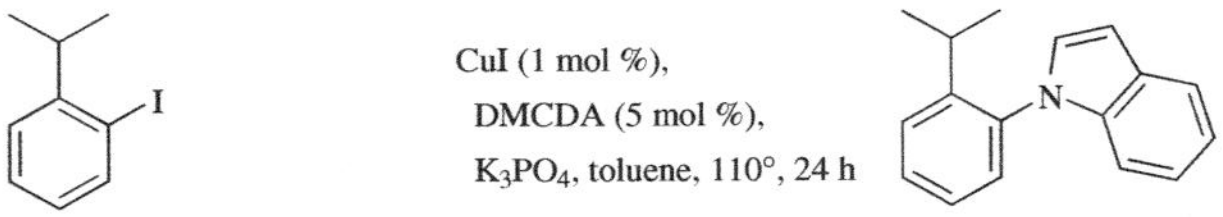

CuI (1 mol %),
DMCDA (5 mol %),
K_3PO_4, toluene, 110°, 24 h

(86)   191

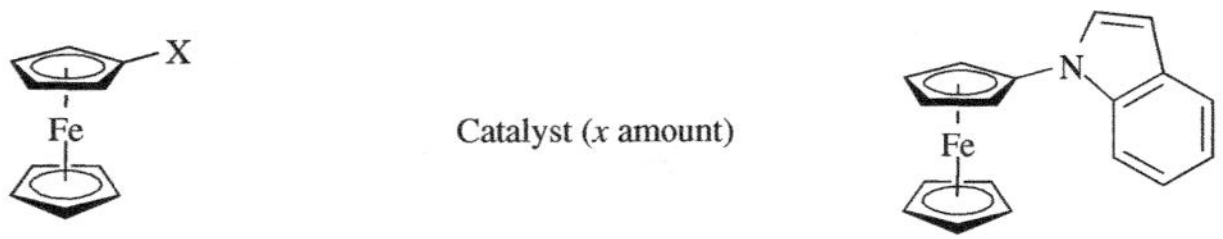

Catalyst (x amount)

X	Catalyst	x	Additive	Solvent	Temp (°)	Time (h)		
I	CuI	1 eq	K_2CO_3	DMSO	90	18	(73)	372
Br	Cu(TMHD)$_2$	20 mol %	KOt-Bu	toluene	120	36	(61)	407
I	Cu(TMHD)	10 mol %	KOt-Bu	DMSO	90	12	(95)	407

TABLE 11A. N-ARYLATION OF INDOLES (*Continued*)

Nitrogen Nucleophile	Aryl Halide	Conditions	Product(s) and Yield(s) (%)	Refs.

*Please refer to the charts preceding the tables for structures indicated by the **bold** numbers.*

C8

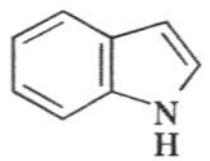

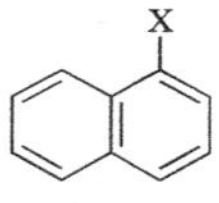

Catalyst(s) [x amount(s)]

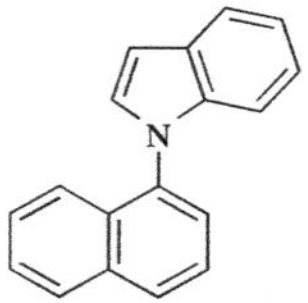

X	Catalyst(s)	x	Additive(s)	Solvent(s)	Temp (°)	Time (h)	Product/Yield	Refs.
I	Cu/CuI	1.3 eq/7 mol %	K_2CO_3	—	210	8	(84)	444
I	CuI	5 mol %	L-Pro (20 mol %), K_3PO_4	dioxane	100	24	(85)	388
Br	CuI	20 mol %	L-Pro (20 mol %), K_3PO_4	DMF	140	48	(80)	388
Br	CuI	5 mol %	EDA (10 mol %), K_2CO_3	PEG-400	80	24	(92)	301
Br	CuI	10 mol %	**L38** (10 mol %), K_3PO_4	DMF	reflux	24	(82)	447
Br	CuI	10 mol %	**L39** (10 mol %), K_3PO_4	DMF	reflux	24	(82)	447

(product: 1-(1-naphthyl)indole)

Aryl Halide	Conditions	Product/Yield	Refs.
4-bromobiphenyl	CuI (10 mol %), oxazolidin-2-one (20 mol %), MeONa, DMSO, 120°, 24 h	(45)	392

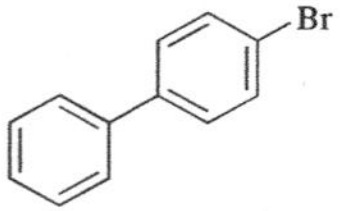

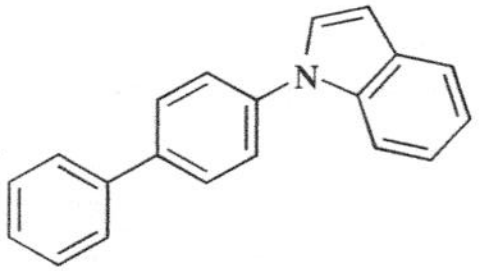

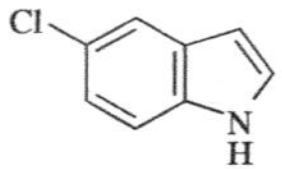

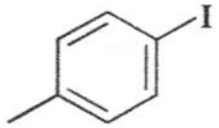

Nitrogen nucleophile: 5-chloroindole

Aryl Halide	Conditions	Product/Yield	Refs.
4-iodotoluene	CuI (5 mol %), **L40** (5 mol %), K_3PO_4, DMF, 110°, 24 h	(82)	449

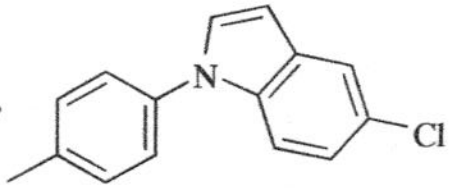

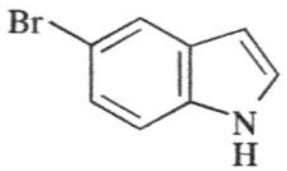

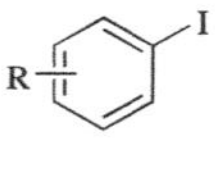

Nitrogen nucleophile: 5-bromoindole

Aryl Halide	Conditions	Refs.
R-substituted iodobenzene	CuI (10 mol%), K_2CO_3, TM–BINAM (10 mol %), MeCN, reflux	446

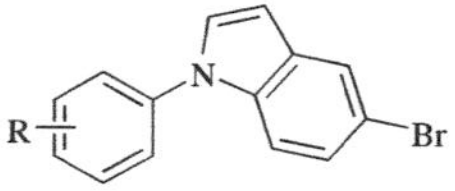

R	Time (h)	
H	36	(34)
3-MeO	72	(65)
4-MeCO	28	(93)

344

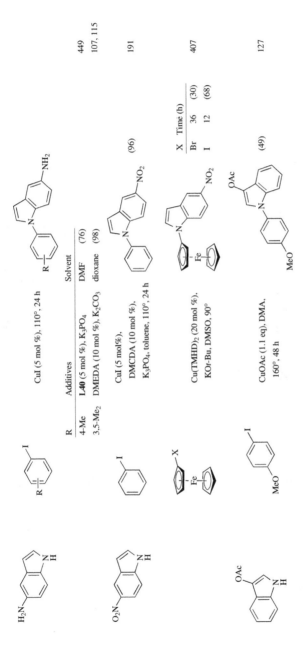

CuI (5 mol %), 110°, 24 h

R	Additives	Solvent	
4-Me	L40 (5 mol %), K₃PO₄	DMF	(76)
3,5-Me₂	DMEDA (10 mol %), K₂CO₃	dioxane	(98)

449, 107, 115

CuI (5 mol%), DMCDA (10 mol %), K₃PO₄, toluene, 110°, 24 h

(96)

191

Cu(TMHD)₂ (20 mol %), KOt-Bu, DMSO, 90°

X	Time (h)	
Br	36	(30)
I	12	(68)

407

CuOAc (1.1 eq), DMA, 160°, 48 h

(49)

127

TABLE 11A. N-ARYLATION OF INDOLES (Continued)

Nitrogen Nucleophile	Aryl Halide	Conditions	Product(s) and Yield(s) (%)	Refs.

Please refer to the charts preceding the tables for structures indicated by the bold numbers.

C8

Nitrogen Nucleophile: 5-MeO-indole (MeO-substituted indole, N–H)

Aryl Halide: R—C6H4—X

Product: 5-OMe indole, N-aryl (R-substituted)

R	X	Catalyst	x	Additives	Solvent	Temp (°)	Time (h)	Yield (%)	Refs.
H	I	CuI	10	TM–BINAM (10 mol %), K2CO3	MeCN	reflux	36	(66)	446
2-Br	I	CuI	10	BtH (20 mol %), K3PO4	DMSO	120	30	(70)	77
2-I	I	CuI	10	BtH (20 mol %), K3PO4	DMSO	120	30	(68)	77
3-MeO	I	CuI	10	TM–BINAM (10 mol %), K2CO3	MeCN	reflux	72	(97)	446
4-Me	I	CuI	5	L40 (5 mol %), K3PO4	DMF	110	24	(83)	449
3,5-Me2	Br	(Cu(OTf)2•C6H6	5	4,7-Cl2phen (10 mol %), Cs2CO3	NMP	125	89	(100)	452
3,5-Me2	I	CuI	10	CDA (10 mol %), K3PO4	dioxane	110	24	(100)	115

Nitrogen Nucleophile: MeO-indole (N–H)

Aryl Halide: R—C6H4—X

Conditions: CuCl (5 mol %), L17 (10 mol %), TBAB, NaOH, H2O, 100°, 24 h

Product: MeO-indole, N-aryl (R-substituted)

R	X	Yield	Refs.
Br	Br	(79)	235
O2N	I	(96)	

C9

Nitrogen Nucleophile: 2-methylindole (N–H)

Aryl Halide: R—C6H4—X

Product: 2-methyl indole, N-aryl (R-substituted)

Catalyst: CuI (x mol %)

R	X	x	Additive(s)	Solvent	Temp (°)	Time (h)	Yield (%)	Refs.
H	Br	40	8-HOquin (10 mol %), TEAC	H2O	130	36	(45)	434
H	I	20	Cs2CO3	DMF	120	40	(65)	386
4-MeO	I	10	BtH (20 mol %), K3PO4	DMSO	100	30	(87)	77
4-MeO	I	10	L-Pro (20 mol %), K2CO3	DMSO	90	40	(75)	47
4-MeO	I	5	L-Pro (10 mol %), K2CO3	DMSO	90	40	(75)	401
3-Me	I	10	BtH (20 mol %), K3PO4	DMSO	80	8	(90)	77
4-Me	I	10	BtH (20 mol %), K3PO4	DMSO	100	30	(89)	77
4-Me	I	5	L40 (5 mol %), K3PO4	DMF	110	24	(23)	449

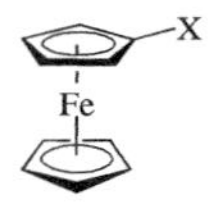

Cu(TMHD)₂ (20 mol %),
KOt-Bu, toluene, 120°

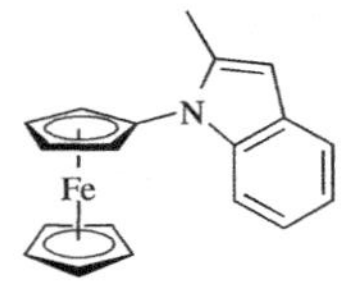

X	Time (h)	
Br	36	(52)
I	12	(88)

407

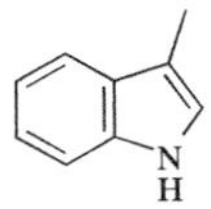

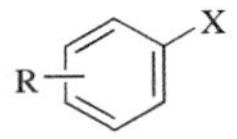

CuI (x mol %), K₃PO₄

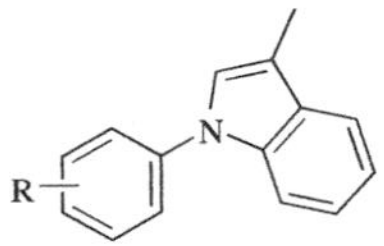

R	X	x	Additive	Solvent	Temp (°)	Time (h)		
H	I	10	BtH (20 mol %)	DMSO	100	30	(85)	77
2-Br	I	10	BtH (20 mol %)	DMSO	120	30	(72)	77
2-I	I	10	BtH (20 mol %)	DMSO	120	30	(70)	77
2-O₂N	I	10	BtH (20 mol %)	DMSO	80	8	(98)	77
4-MeO	I	10	BtH (20 mol %)	DMSO	100	30	(90)	77
4-Me	Br	10	BtH (20 mol %)	DMSO	120	30	(55)	77
4-Me	I	10	BtH (20 mol %)	DMSO	120	30	(91)	77
4-Me	Br	1	DMCDA (5 mol %)	toluene	110	24	(85)	191
4-Me	I	5	**L40** (5 mol %)	DMF	110	24	(83)	449
2-NC–	I	10	BtH (20 mol %)	DMSO	80	8	(98)	77

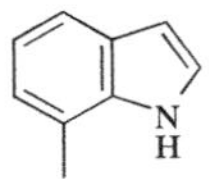

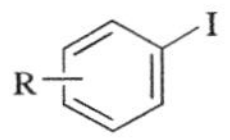

CuI (x mol %), K₃PO₄,
110°

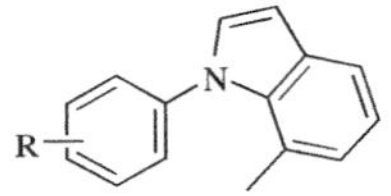

R	x	Additive	Solvent	Time (h)		
H	5	DMCDA (20 mol %)	toluene	24	(92)	191
4-Me	5	**L40** (5 mol %)	DMF	24	(92)	449
3,5-Me₂	1	CDA (10 mol %)	dioxane	48	(62)	115

TABLE 11A. N-ARYLATION OF INDOLES (Continued)

Nitrogen Nucleophile	Aryl Halide	Conditions	Product(s) and Yield(s) (%)	Refs.

*Please refer to the charts preceding the tables for structures indicated by the **bold** numbers.*

C9

				Refs.
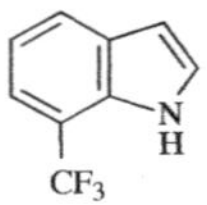 7-CF_3 indole	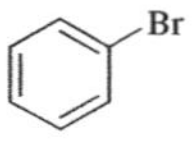 PhBr	Cu_2Br_2 (14 mol %), K_2CO_3, pyridine, 160°, 18 h	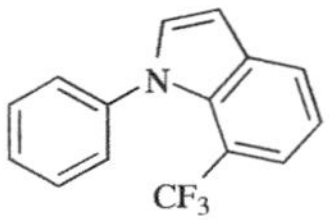 (3)	453
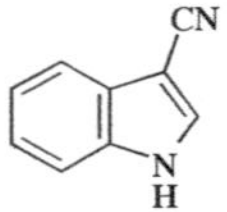 3-CN indole	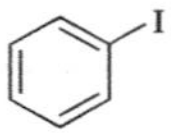 PhI	CuI (5 mol %), DMCDA (20 mol %), K_3PO_4, toluene, 110°, 24 h	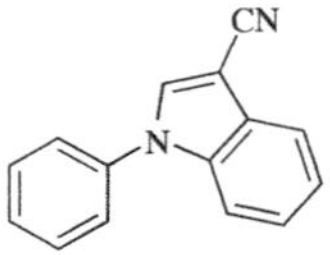 (92)	191
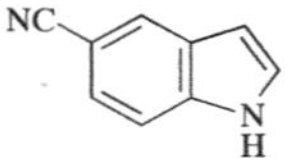 5-CN indole	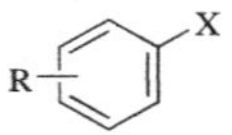 R-C_6H_4-X	Catalyst (5 mol %)	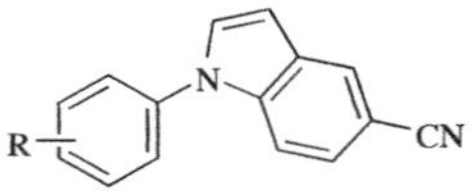 (see below)	

R	X	Catalyst	Additives	Solvent	Temp (°)	Time (h)		
4-Me	I	CuI	**L40** (5 mol %), K_3PO_4	DMF	110	24	(88)	449
3,5-Me_2	Br	$(CuOTf)_2·C_6H_6$	4,7-Cl_2phen (10 mol %), Cs_2CO_3	NMP	125	89	(100)	452

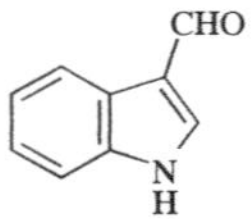

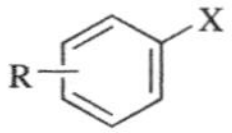

3-CHO indole; R-C_6H_4-X; Catalyst (x amount)

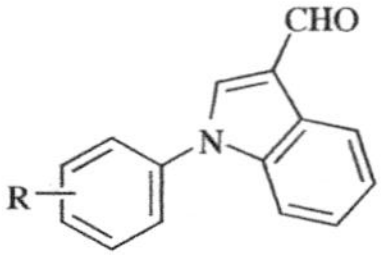

R	X	Catalyst	x	Additive	Solvent	Temp (°)	Time (h)		
H	Br	CuO	6 mol %	K_2CO_3	DMF	reflux	24	(60)	448
2-O_2N	Cl	CuO	6 mol %	K_2CO_3	DMF	reflux	24	(56)	448
3-O_2N	Br	CuO	6 mol %	K_2CO_3	DMF	reflux	24	(76)	448
4-O_2N	Br	CuO	6 mol %	K_2CO_3	DMF	reflux	24	(88)	448
4-O_2N	I	CuOAc	1.1 eq	none	DMA	160	48	(55)	127
4-MeO	I	CuOAc	1.1 eq	none	DMA	160	48	(54)	127

348

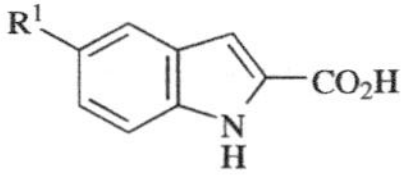

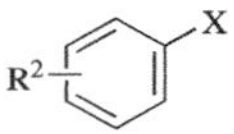

Table 1

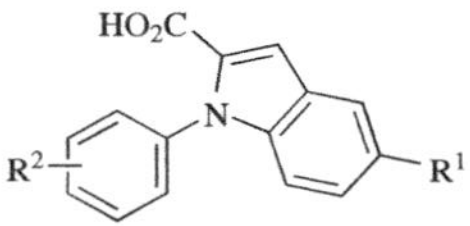

R^1	R^2	X	Catalyst	x	Additive(s)	Solvent	Temp (°)	Time (h)		Ref
H	H	Br	CuO	6	K_2CO_3	DMF	reflux	24	(67)	448
MeO	H	Br	CuO	30	KOH	DMF	reflux	6	(93)	454
H	3-O_2N	Br	CuO	6	K_2CO_3	DMF	reflux	24	(50)	448
H	4-O_2N	Br	CuO	6	K_2CO_3	DMF	reflux	24	(45)	448
BnO	4-MeO	I	CuF_2	50	phen (50 mol %), K_2CO_3	DMF	140	96	(60)	419
H	3-CF_3	Br	Cu_2Br_2	14	K_2CO_3	pyridine	150	9	(68)	453
H	4-CF_3	Br	Cu_2Br_2	14	K_2CO_3	pyridine	150	9	(95)	453

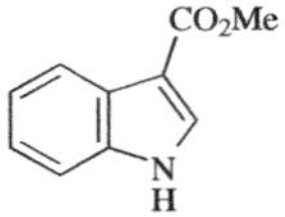

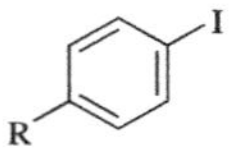

Table 2

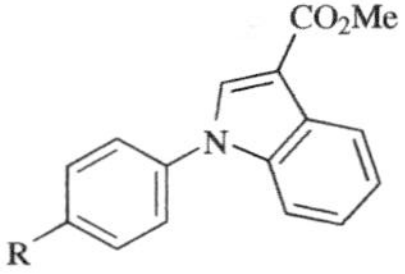

R	Catalyst	x	Additive(s)	Solvent	Temp (°)	Time (h)		Ref
H	CuI	5 mol %	DMCDA (10 mol %), K_3PO_4	toluene	110	24	(96)	191
MeO	CuOAc	1.1 eq	none	DMA	160	48	(60)	127
Me	CuI	5 mol %	**L40** (5 mol %), K_3PO_4	DMF	110	24	(97)	449

Table 3

C_{9-10}

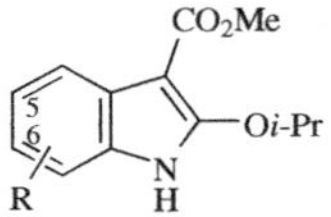

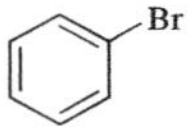

CuBr (10 mol %), K_2CO_3, KOH, 140°, 21 h

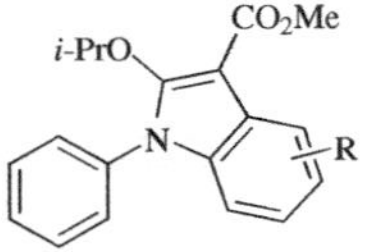

R		Ref
5,6-Cl_2	(93)	454
6-Br	(—)	
6-MeO	(—)	
6-Me	(—)	

TABLE 11A. *N*-ARYLATION OF INDOLES (*Continued*)

Nitrogen Nucleophile	Aryl Halide	Conditions	Product(s) and Yield(s) (%)	Refs.

*Please refer to the charts preceding the tables for structures indicated by the **bold** numbers.*

C₉

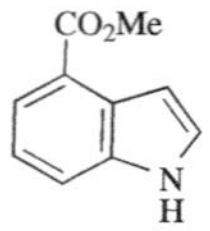

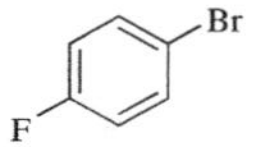

CuO, K₂CO₃, reflux

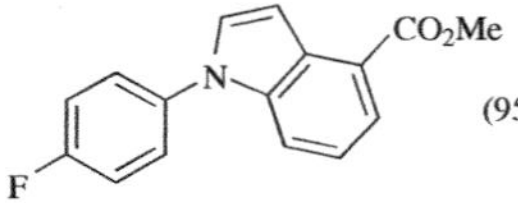

(95)

435

C₁₀

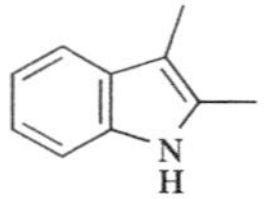

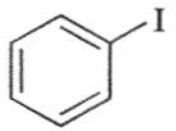

CuI (5 mol %),
DMCDA (10 mol %),
K₃PO₄, toluene, 110°, 24 h

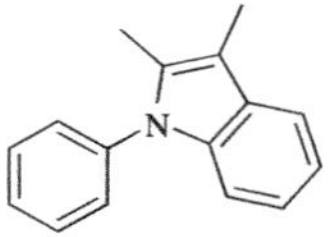

(98)

191

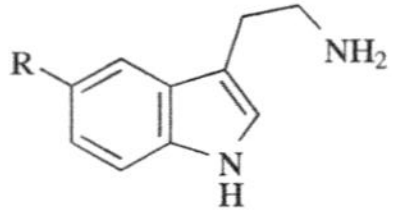

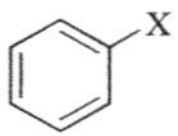

CuI (5 mol %),
DMCDA (20 mol %),
K₃PO₄, toluene, 110°, 24 h

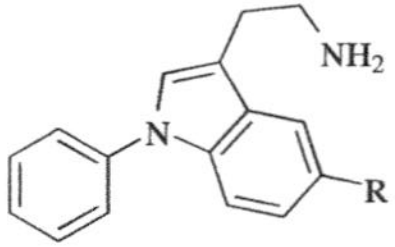

X	R
I	H (90)
Br	MeO (76)

191

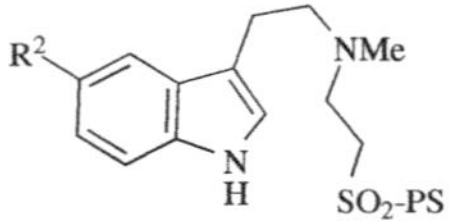

PS = polystyrene

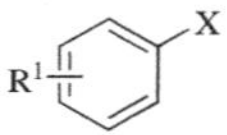

1. CuI (1 eq), CDA (1 eq),
   KO-*t*-Bu, dioxane, 80°, 24 h
2. MeI, DMF
3. *i*-Pr₂NEt, CH₂Cl₂

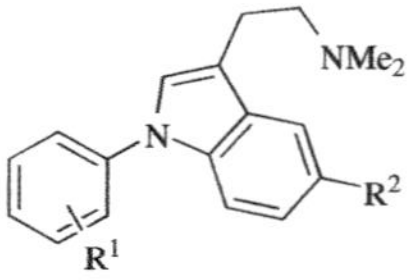

R¹	R²	
H	H	(10–20)
2-F	H	(10–20)
3,4-Cl₂	H	(10–20)
3-HO	H	(10–20)
4-MeO	MeO	(74)
4-PhO	H	(10–20)
4-Me	H	(10–20)
3-Ph	H	(10–20)
4-Ph	H	(10–20)

455

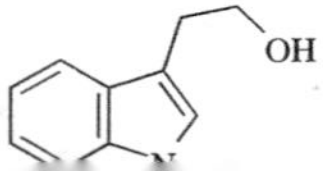

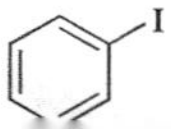

CuI (5 mol %), DMCDA
(20 mol %), K₃PO₄,

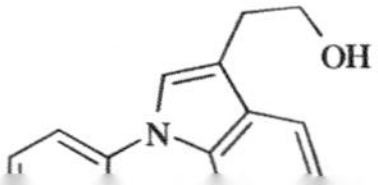

(90)

191

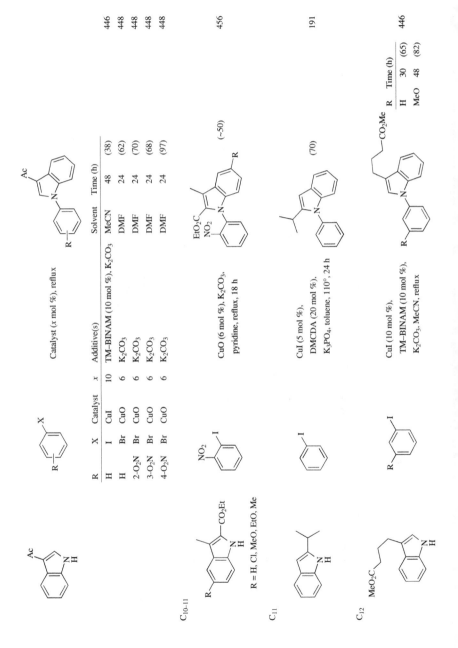

R	X	Catalyst	x	Additive(s)	Solvent	Time (h)		
H	I	CuI	10	TM–BINAM (10 mol %), K$_2$CO$_3$	MeCN	48	(38)	446
H	Br	CuO	6	K$_2$CO$_3$	DMF	24	(62)	448
2-O$_2$N	Br	CuO	6	K$_2$CO$_3$	DMF	24	(70)	448
3-O$_2$N	Br	CuO	6	K$_2$CO$_3$	DMF	24	(68)	448
4-O$_2$N	Br	CuO	6	K$_2$CO$_3$	DMF	24	(97)	448

CuO (6 mol %), K$_2$CO$_3$, pyridine, reflux, 18 h (~50) 456

CuI (5 mol %), DMCDA (20 mol %), K$_3$PO$_4$, toluene, 110°, 24 h (70) 191

CuI (10 mol %), TM–BINAM (10 mol %), K$_2$CO$_3$, MeCN, reflux

R	Time (h)	
H	30	(65)
MeO	48	(82)

446

C$_{10–11}$

R = H, Cl, MeO, EtO, Me

C$_{11}$

C$_{12}$

TABLE 11A. N-ARYLATION OF INDOLES (Continued)

Nitrogen Nucleophile	Aryl Halide	Conditions	Product(s) and Yield(s) (%)	Refs.

*Please refer to the charts preceding the tables for structures indicated by the **bold** numbers.*

C14

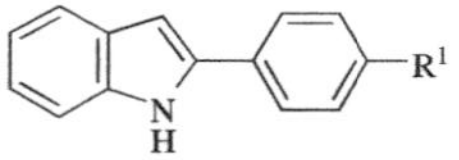

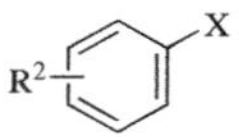

Catalyst (x mol %), reflux

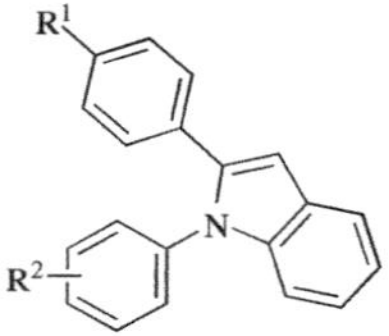

R1	R2	X	Catalyst	x	Additive(s)	Solvent	Temp (°)	Time (h)		Refs.
H	H	I	CuI	5	DMCDA (10 mol %), K$_3$PO$_4$	toluene	110	24	(67)	191
H	2-O$_2$N	Cl	CuO	6	K$_2$CO$_3$	DMF	reflux	24	(52)	448
H	3-O$_2$N	Br	CuO	6	K$_2$CO$_3$	DMF	reflux	24	(34)	448
H	4-O$_2$N	Br	CuO	6	K$_2$CO$_3$	DMF	reflux	24	(48)	448
H	4-Me	I	CuI	5	**L40** (5 mol %), K$_3$PO$_4$	DMF	110	24	(38)	449
H	3,5-Me$_2$	I	CuI	1	CDA (10 mol %), K$_3$PO$_4$	dioxane	110	48	(72)	115
F	3,5-Me$_2$	I	CuI	1	CDA (10 mol %), K$_3$PO$_4$	dioxane	110	48	(65)	115

TABLE 11B. *N*-HETEROARYLATION OF INDOLES

Nitrogen Nucleophile	Heteroaryl Halide	Conditions	Product(s) and Yield(s) (%)	Refs.

*Please refer to the charts preceding the tables for structures indicated by the **bold** numbers.*

C$_8$

(indole nucleophile)

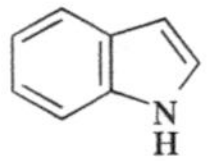	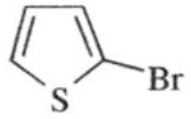 (2-bromothiophene)	Cu$_2$O (10 mol %), phen (20 mol %), TBAF, 145°, 24 h	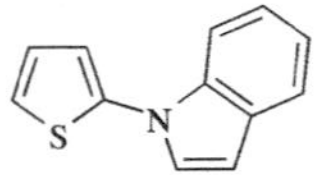 (74)	68
	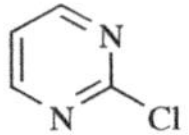 (2-chloropyrimidine)	CuO nanoparticles (10 mol %), K$_2$CO$_3$, DMF, 120°, 1.5 h	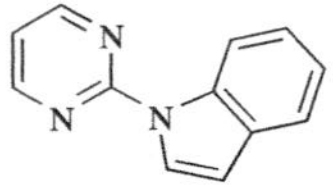 (94)	64
	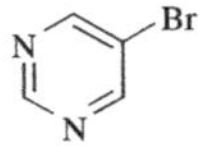 (5-bromopyrimidine)	[Cu(OTf)]$_2$•C$_6$H$_6$ (5 mol %), 4,7-Cl$_2$phen (10 mol %), Cs$_2$CO$_3$, NMP, 125°, 17 h	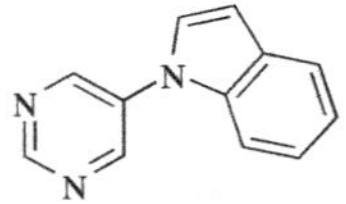 (100)	452
	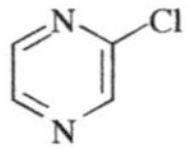 (2-chloropyrazine)	Catalyst (10 mol %)	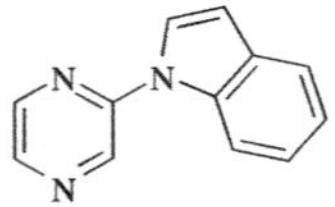	

Catalyst	Additive(s)	Solvent	Temp (°)	Time (h)	
CuBr$_2$	TBAF	—	145	2.5	(58)	81
CuI	L26 (10 mol %), NaOMe	DMSO	130	12	(70)	397

| | 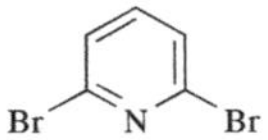 (2,6-dibromopyridine) | CuI (10 mol %), L22 (10 mol %), K$_2$CO$_3$, DMF, 110°, 24 h | 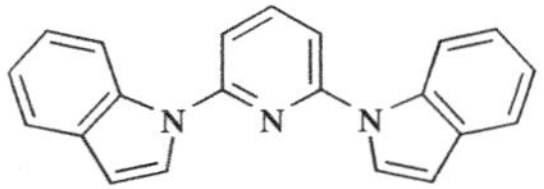 (76) | 79 |

353

TABLE 11B. N-HETEROARYLATION OF INDOLES (*Continued*)

Nitrogen Nucleophile	Heteroaryl Halide	Conditions	Product(s) and Yield(s) (%)	Refs.

*Please refer to the charts preceding the tables for structures indicated by the **bold** numbers.*

C₈ — indole

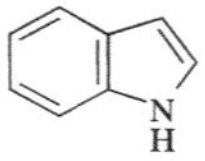

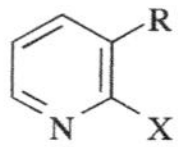

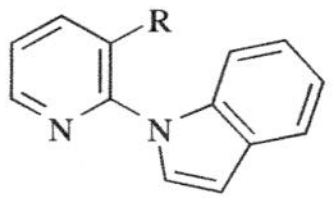

R	X	Catalyst	x	Additive(s)	Solvent	Temp (°)	Time (h)		Refs.
H	I	CuI	10	BtH (20 mol %), K₃PO₄	DMSO	100	30	(92)	77
H	Br	CuI	5	N-hydroxysuccinimide (10 mol %), NaOMe	DMSO	110	24	(68)	396
H	Br	CuI	10	phen (20 mol %), KF/Al₂O₃	toluene	reflux	7	(91)	390
H	Br	CuI	10	L21 (20 mol %), Cs₂CO₃	MeCN	reflux	12	(92)	394
H	Cl	CuO nanoparticles	10	K₂CO₃	DMF	120	23	(77)	64
H	Br	CuO	8	K₂CO₃	pyridine	reflux	17	(79)	402
H	Br	Cu₂O	10	phen (20 mol %), TBAF	—	145	24	(86)	68
Br	Br	CuI	10	BtH (20 mol %), K₃PO₄	DMSO	120	30	(85)	77

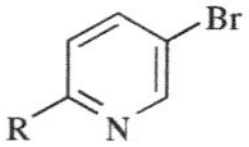

Catalyst (x mol %), 24 h

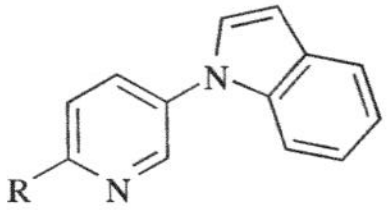

R	Catalyst	x	Additive(s)	Solvent	Temp (°)		Refs.
H	CuI	5	DMCDA (20 mol %), K₃PO₄	toluene	80	(93)	191
H	CuO	8	K₂CO₃	pyridine	reflux	(75)	402
H₂N	CuI	5	DMCDA (20 mol %), K₃PO₄	toluene	110	(78)	191

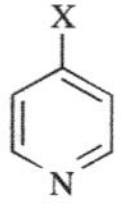

Catalyst (x mol %)

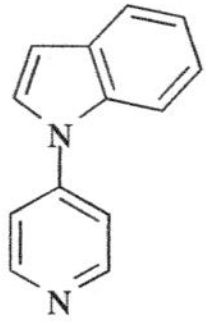

X	Catalyst	x	Additive(s)	Solvent	Temp (°)	Time (h)		Refs.
Cl	CuO	8	K₂CO₃	pyridine	reflux	22	(3)	402

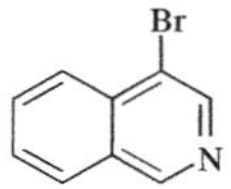

C9

3-methylindole (structure)

CuI (5 mol %),
DMCDA (20 mol %),
K_3PO_4, toluene, 110°, 24 h

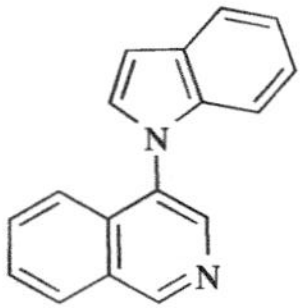

4-bromoisoquinoline (structure, Br)

(72)

191

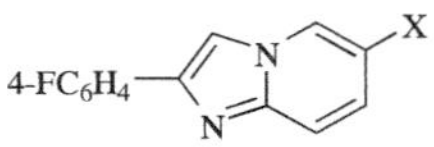

CuI (5 mol %),
DMCDA (15 mol %),
K_3PO_4, toluene, 112°, 24 h

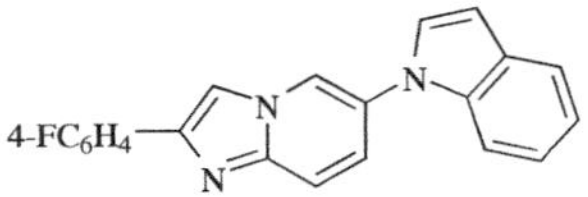

4-FC$_6$H$_4$ imidazopyridine (X) (structure)

4-FC$_6$H$_4$ (structure)

X	
Br	(72)
I	(74)

409

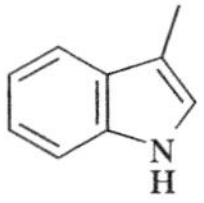

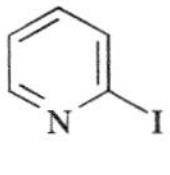

CuI (10 mol %),
BtH (20 mol %),
K_3PO_4, DMSO, 100°, 30 h

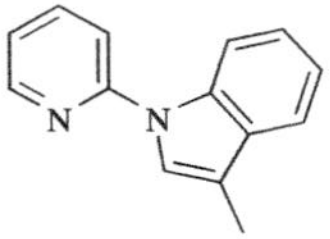

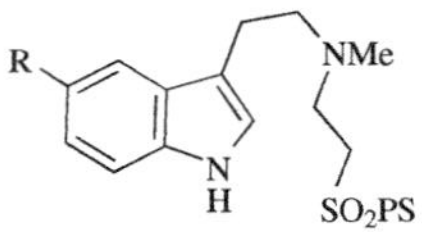

2-iodopyridine (structure, I)

(93)

77

C10

R—indole—NMe (structure)
SO$_2$PS

PS = polystyrene

1. CuI (1 eq), CDA (1 eq),
KO-t-Bu, dioxane, 80°, 24 h
2. MeI, DMF
3. i-Pr$_2$NEt, CH$_2$Cl$_2$

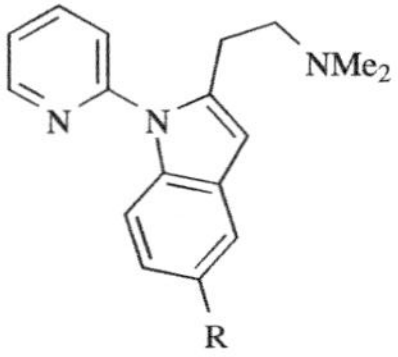

ArBr

R—indole—CH$_2$CH$_2$NMe$_2$ (structure)
R

R	Ar
H	2-thiazolyl
H	2-furyl
H	3-pyrimidyl
H	2-pyridyl
BnO	2-pyridyl
H	3-quinolyl

R	Ar
H	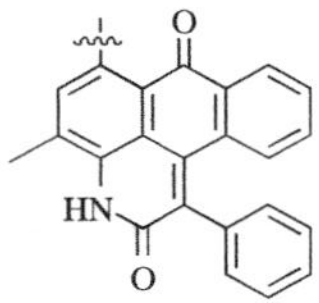

455

TABLE 11B. *N*-HETEROARYLATION OF INDOLES (*Continued*)

Nitrogen Nucleophile	Heteroaryl Halide	Conditions	Product(s) and Yield(s) (%)	Refs.

*Please refer to the charts preceding the tables for structures indicated by the **bold** numbers.*

C$_{10}$

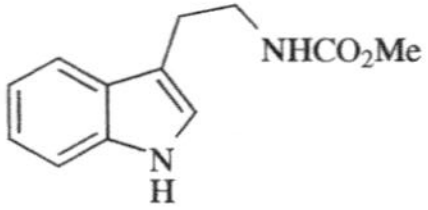

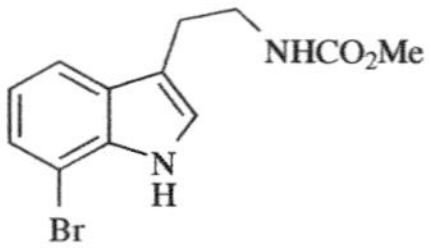

CuI (30 mol %),
DMCDA (60 mol %),
K$_2$CO$_3$, dioxane, 101°, 20 h

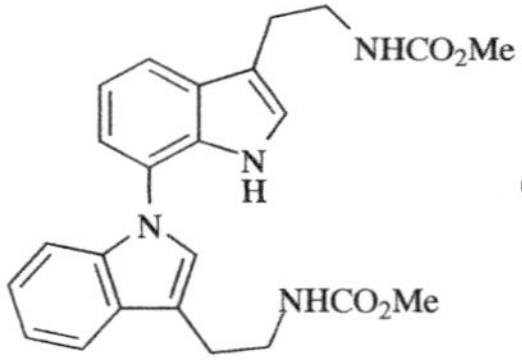

(45)

457

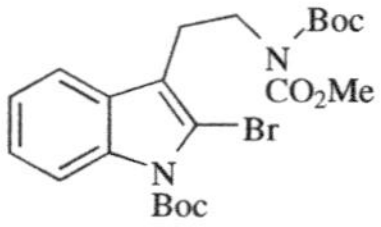

1. CuI (1 eq),
DMCDA (20 mol %),
K$_3$PO$_4$, dioxane, 101°, 24 h
2. TFA/MeCl$_2$ (1:4), rt, 1 h

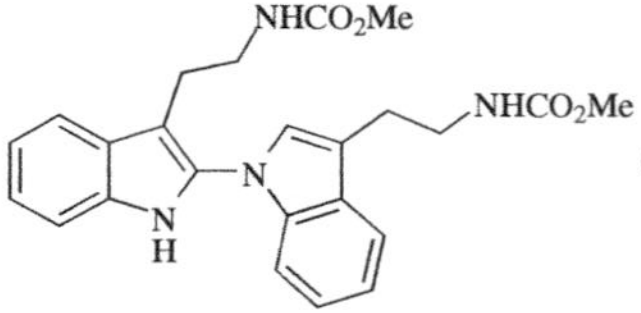

(22)

457

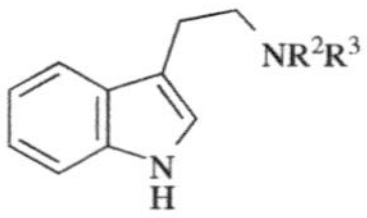

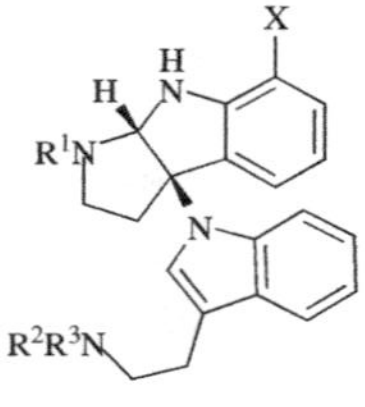

CuI (x amount), dioxane

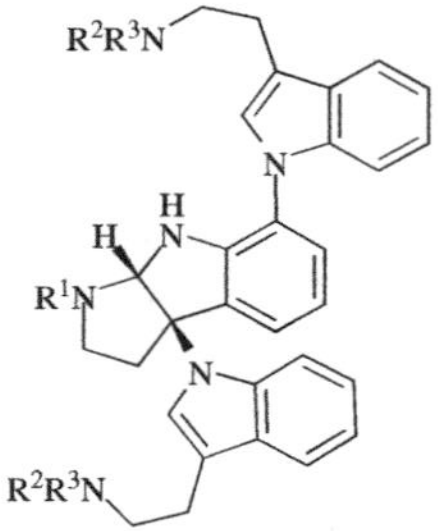

R^1	R^2	R^3	X	x	Additives	Temp (°)	Time (h)	
Me	Me	2-O$_2$NC$_6$H$_4$SO$_2$	I	1 eq	DMEDA (2 eq), K$_3$PO$_4$	90	24	(72)
MeO$_2$C	MeO$_2$C	H	Br	30 mol %	DMCDA (60 mol %), K$_2$CO$_3$	101	9	(89)

169
457

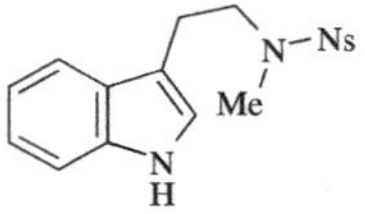

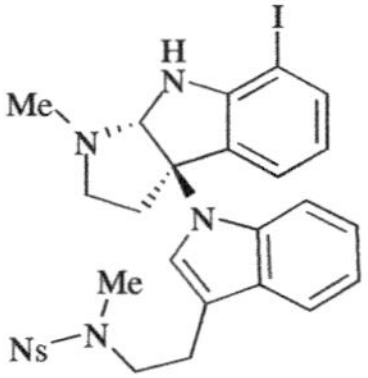

169

(72)

CuI (1 eq), DMEDA (2 eq),
K$_3$PO$_4$, dioxane, 90°

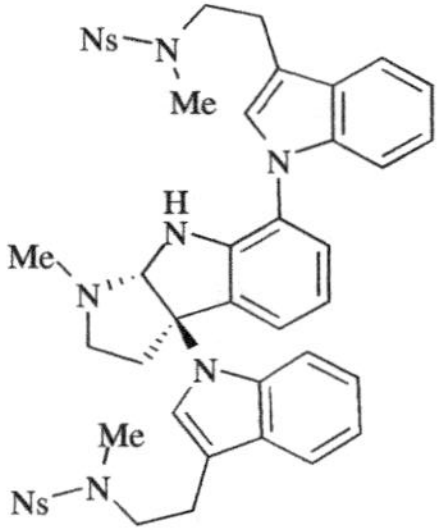

TABLE 12A. N-ARYLATION OF INDAZOLES

Nitrogen Nucleophile	Aryl Halide	Conditions	Product(s) and Yield(s) (%)	Refs.
C7				
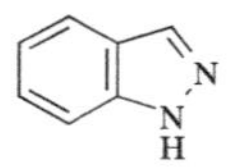 (indazole, N–H)	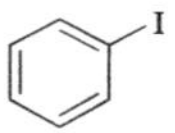 iodobenzene	CuI (5 mol %), TBAB (5 mol %), NaOH, toluene, reflux, 22 h	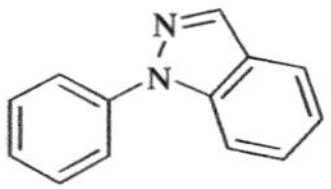 (N-phenylindazole) (50)	391
(indazole, N–H)	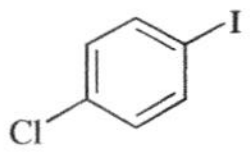 4-chloroiodobenzene	CuI (5 mol %), DMEDA (10 mol %), CsF, THF, 60°, 24 h	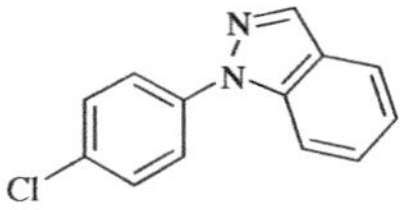 (N-(4-chlorophenyl)indazole) (88)	63
(indazole, N–H)	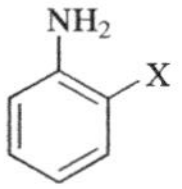 2-amino(NH2) aryl halide (X)	CuI (7 mol %), DMCDA (20 mol %), K3PO4, toluene, 110°, 24 h	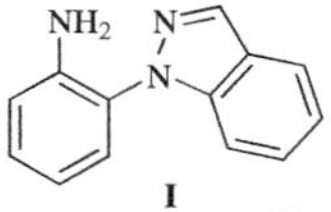 I 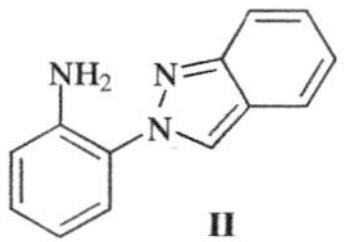 (1-substituted) + II (2-substituted) X / I / II Br (41) (29) I (80) (0)	128
(indazole, N–H)	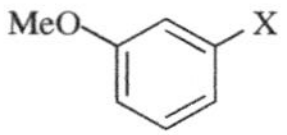 MeO-substituted aryl halide (X)	CuI (7 mol %), DMCDA (20 mol %), K3PO4, toluene, 110°, 24 h	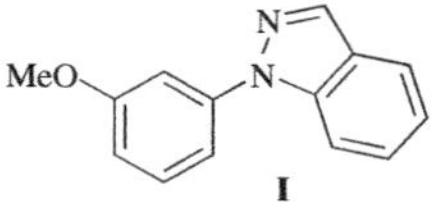 I 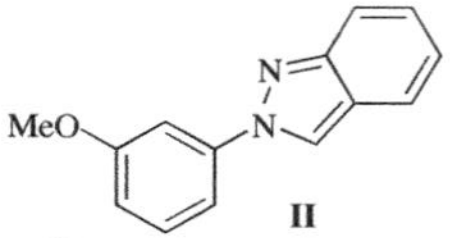 + II + MeO X / I / II Br (49) (25) I (92) (0)	128
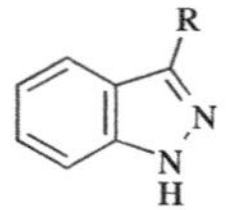 R-substituted indazole (N–H)	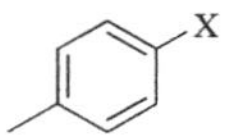 4-methyl aryl halide (X)	CuI (7 mol %), DMCDA (20 mol %), K3PO4, toluene, 110°, 24 h	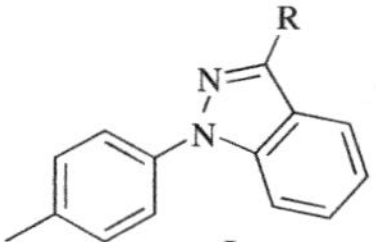 I 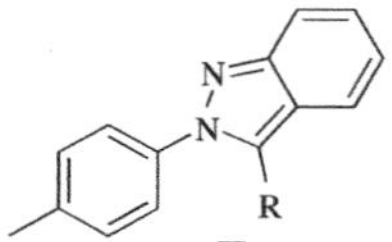 + II X / R / I / II Br H (48) (27) I H (85) (0) Br Cl (86) (0)	128

C$_8$

(91)		115
(72)		406
		458
		458

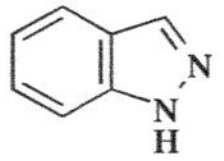

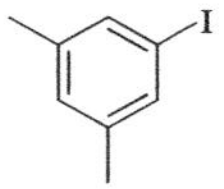

CuI (1 mol %), CDA (10 mol %), K$_3$PO$_4$, dioxane, 110°, 24 h

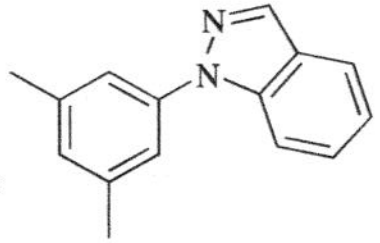

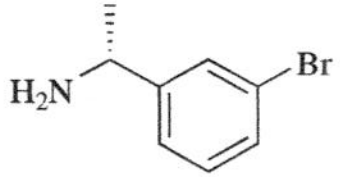

CuI (10 mol %), K$_2$CO$_3$, NMP, MW, 195°, 2 h

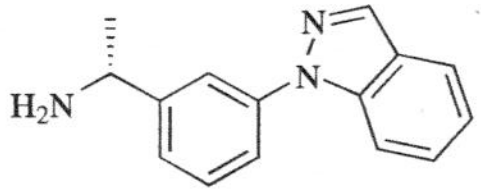

CuI, DMCDA, K$_3$PO$_4$, dioxane, 110°, 12 h

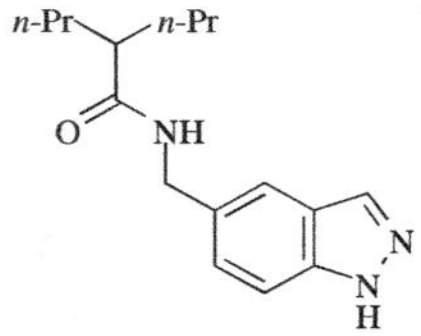

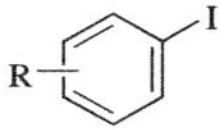

R = H, 2-F, 3-F, 4-F, 4-Cl, 2-MeO, 3-MeO, 4-MeO, 2-Me, 3-Me, 4-Me, 2,3-Me$_2$, 2,6-Me$_2$, 4-CO$_2$t-Bu, 4-Ac

CuI, DMCDA, K$_3$PO$_4$, dioxane, 110°, 12 h

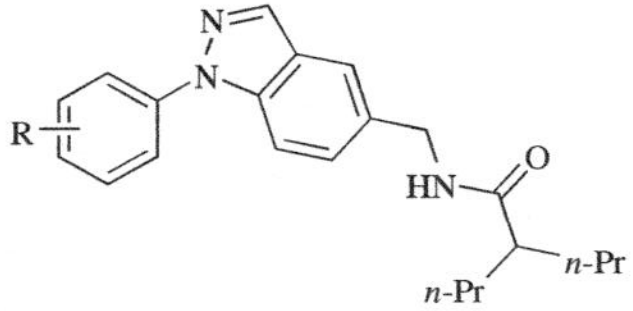

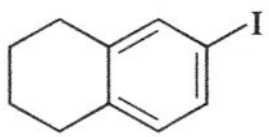

359

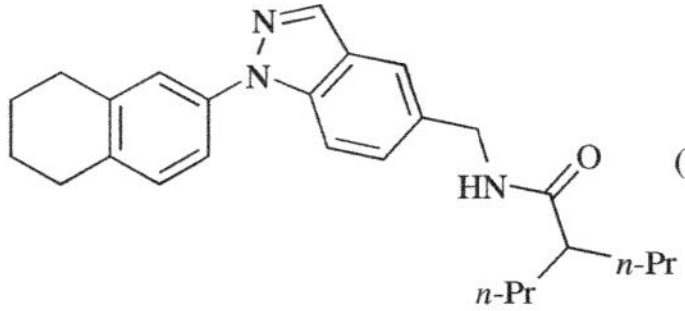

Nitrogen Nucleophile	Aryl Halide	Conditions	Product(s) and Yield(s) (%)	Refs.
C₁₃		CuI (7 mol %), DMCDA (20 mol %), K₃PO₄, toluene, 110°, 18 h	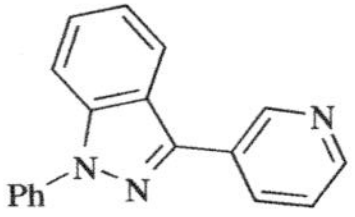 (94)	459
C₁₃₋₁₄ 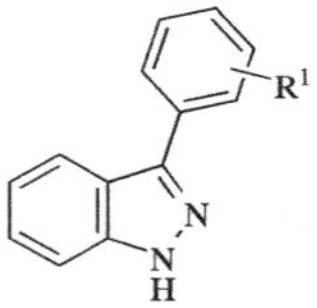	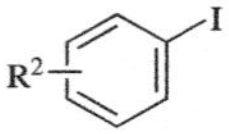	CuI (7 mol %), DMCDA (20 mol %), K₃PO₄, toluene, 110°, 18 h	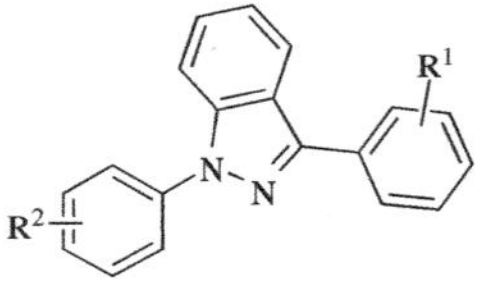	459

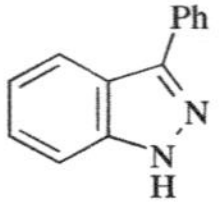

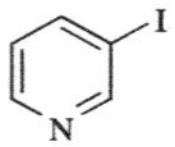

R^1	R^2	
H	H	(86)
2,4-F₂	H	(98)
3-Cl	H	(93)
3-Me	H	(80)
H	2-F	(28)
H	3-F	(65)
H	4-F	(75)
H	2-H₂N	(73)
H	3-H₂N	(86)
H	4-H₂N	(75)
H	3-O₂N	(97)
H	2-MeO	(30)
H	3-MeO	(62)
H	4-MeO	(84)
H	4-Me	(68)

TABLE 12B. *N*-Heteroarylation of Indazoles

Nitrogen Nucleophile	Heteroaryl Halide	Conditions	Product(s) and Yield(s) (%)	Refs.
C7				
		CuI (5 mol %), DMCDA (15 mol %), K3PO4, toluene, 112°, 24 h	I + II 	409
			X I II	
			Br (36) (16)	
			I (83) (3)	
C8				
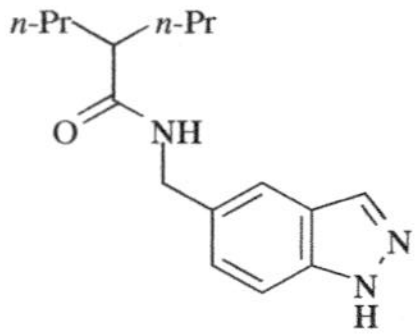	ArI Ar = 2-pyrimidinyl, 2-pyrazinyl, 2-pyridinyl, 4-pyridinyl, 3-(5-MeO-pyridinyl), 3-(2-MeO-pyridinyl)	CuI, DMCDA, K3PO4, dioxane, 110°, 12 h	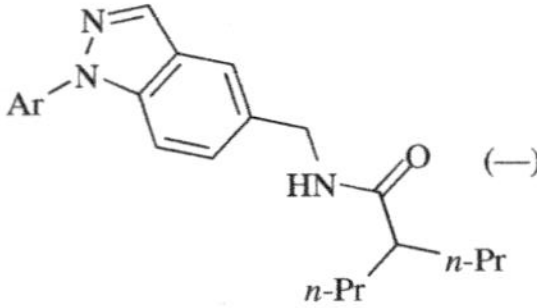 (—)	458
C13				
		CuI (7 mol %), DMCDA (20 mol %), K3PO4, toluene, 110°, 18 h	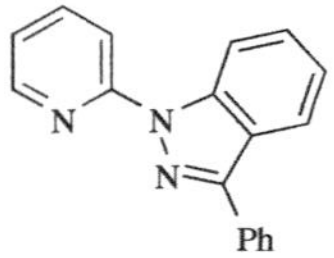 (98)	459

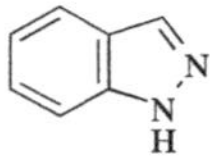

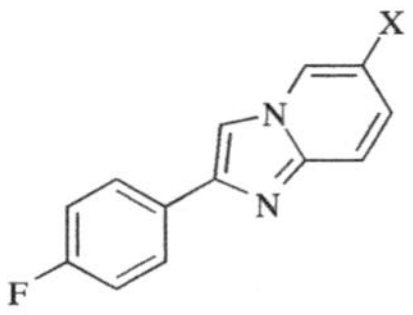

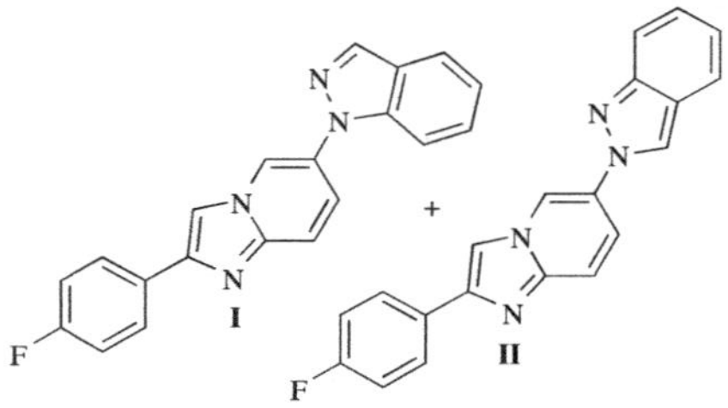

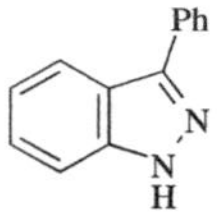

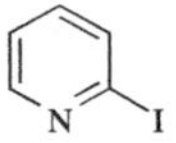

TABLE 13A. N-ARYLATION OF BENZIMIDAZOLES

Nitrogen Nucleophile	Aryl Halide	Conditions						Product(s) and Yield(s) (%)	Refs.

*Please refer to the charts preceding the tables for structures indicated by the **bold** numbers.*

C_7

Nitrogen Nucleophile: benzimidazole (N–H)

Aryl Halide: C$_6$H$_5$–X

Product: 1-phenyl-1H-benzimidazole

X	Catalyst	x	Additive(s)	Solvent	Temp (°)	Time (h)	Yield (%)	Refs.
I	Cu on cellulose	1	K$_2$CO$_3$	DMSO	130	24	(40)	427
I	CuBr	5	**L5** (10 mol %), Cs$_2$CO$_3$	DMF	90	24	(82)	119
Br	CuBr	10	**L9** (20 mol %), Cs$_2$CO$_3$	DMSO	75	28	(72)	49
I	CuBr	10	**L9** (20 mol %), Cs$_2$CO$_3$	DMSO	60	25	(85)	49
I	CuBr	10	**L10** (12 mol %), Cs$_2$CO$_3$	DMSO	80	24	(78)	259
Br	CuBr	10	**L20** (20 mol %), Cs$_2$CO$_3$	DMSO	90	24	(87)	284
Cl	CuBr	10	**L37** (20 mol %), TBAF	none	150	24	(40)	80
Br	CuBr	10	**L37** (20 mol %), TBAF	none	150	24	(86)	80
I$^+$Ph BF$_4^-$	CuI	10	K$_2$CO$_3$	DMF	80	3	(80)	460
I	CuI	20	Cs$_2$CO$_3$	DMF	120	40	(91)	386
I	CuI	5	K$_3$PO$_4$	DMF	110	24	(72)	387
I	CuI	5	KOH, TBAB	—	110	6	(72)	428
Br	CuI	20	phen (20 mol %), KF/Al$_2$O$_3$	xylene	140	15	(90)	410
I	CuI	10	phen (20 mol %), KF/Al$_2$O$_3$	—	140	13	(94)	410
I	CuI	10	CDA (20 mol %), K$_3$PO$_4$	[C$_4$bmim][BF$_4$]	110	12	(96)	232
Br	CuI	10	L-His (20 mol %), K$_2$CO$_3$	DMSO	100	36	(78)	389
I	CuI	5	oxazolidin-2-one (10 mol %), NaOMe	DMSO	80	15	(87)	392
Br	CuI	10	8-HOquin (20 mol %), TEAC	DMF	130	24	(85)	434
Br	CuI	20	per-6-ABCD (10 mol %), K$_2$CO$_3$	DMSO	110	24	(99)	393
Br	CuI	10	**L21** (20 mol %), KI, Cs$_2$CO$_3$	MeCN	reflux	12	(29)	394
I	CuI	10	**L21** (20 mol %), Cs$_2$CO$_3$	MeCN	reflux	12	(73)	394
Br	CuI	10	**L22** (10 mol %), K$_2$CO$_3$	DMF	110	24	(73)	79
Br	CuI	5	**L24** (10 mol %), Cs$_2$CO$_3$	DMF	110	24	(98)	86
I	CuI	10	**L26** (10 mol %), NaOMe	DMSO	120	12	(52)	397

Br	CuI	10	L28 (20 mol %), K$_2$CO$_3$	DMSO	120	60 (69)	412
I	CuI	10	L28 (20 mol %), K$_2$CO$_3$	DMSO	120	60 (99)	412
I	CuO	2.5	KOH	DMSO	110	29 (98)	224
I	Cu$_2$O	10	Cs$_2$CO$_3$	DMF	100	18 (86)	221
I	Cu$_2$O	10	KOH	DMSO	120	24 (88)	398
Cl	Cu$_2$O	10	phen (20 mol %), TBAF	—	145	24 (82)	68
I	Cu$_2$O	10	ninhydrin (20 mol %), KOH	DMSO	110	24 (88)	399
I	Cu$_2$O	5	L7 (20 mol %), Cs$_2$CO$_3$	MeCN	80	18 (60)	125
Br	Cu$_2$O	10	L23 (20 mol %), KOt-Bu	DMF	110	24 (90)	400
I$^+$Ph BF$_4^-$	Cu(acac)$_2$	100	K$_2$CO$_3$	toluene	50	6 (80)	76
I	Cu(OAc)$_2$•H$_2$O	15	(−)-sparteine (30 mol %), K$_2$CO$_3$	DMF	130	24 (85)	290
Br	Cu(TMHD)$_2$	20	KOt-Bu	DMF	120	24 (70)	308
I	Cu(TMHD)$_2$	20	KOt-Bu	DMF	120	24 (90)	308
Br	8	25	K$_2$CO$_3$	DMSO	135	35 (85)	65

TABLE 13A. N-ARYLATION OF BENZIMIDAZOLES (Continued)

Nitrogen Nucleophile	Aryl Halide	Conditions	Product(s) and Yield(s) (%)	Refs.

*Please refer to the charts preceding the tables for structures indicated by the **bold** numbers.*

C7

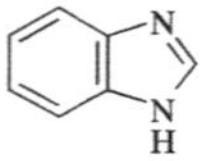

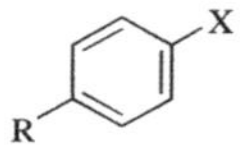

Catalyst (x mol %)

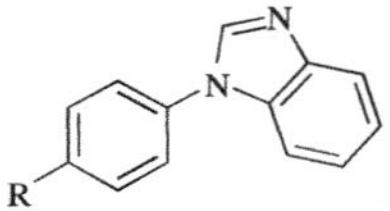

R	X	Catalyst	x	Additive(s)	Solvent(s)	Temp (°)	Time (h)		Refs.
F	I	CuCl	5	**L17** (10 mol %), NaOH, TBAB	H₂O	100	24	(83)	235
F	Br	Cu₂O	5	(MeO)₂phen (15 mol %), Cs₂CO₃	n-PrCN/PEG	110	30	(81)	84
Cl	I⁺C₆H₄-4-Cl BF₄⁻	CuI	10	K₂CO₃	DMF	80	3	(68)	460
Cl	I	CuI	5	8-HOquin (10 mol %), CsF	DMSO	100	24	(88)	63
Cl	Br	CuI	10	**L21** (20 mol %), KI, Cs₂CO₃	MeCN	reflux	12	(33)	394
Cl	I	CuI	10	**L21** (20 mol %), Cs₂CO₃	MeCN	reflux	12	(63)	394
Br	I⁺C₆H₄-4-Br BF₄⁻	CuI	10	K₂CO₃	DMF	80	3	(68)	460
Br	I	CuI	5	K₃PO₄	DMF	110	24	(76)	387
Br	I	CuI	10	**L21** (20 mol %), KI, Cs₂CO₃	MeCN	reflux	12	(61)	394

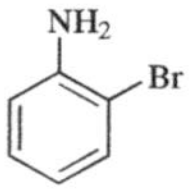

Catalyst (10 mol %), 110°, 24 h

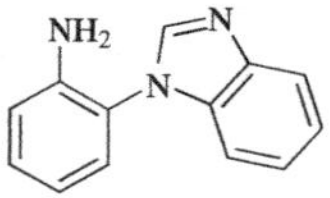

Catalyst	Additives	Solvents		Refs.
CuI	8-HOquin (20 mol %), TEAC	DMF/H₂O	(67)	434
Cu₂O	(MeO)₂phen (20 mol %), Cs₂CO₃	DMSO/PEG	(71)	84

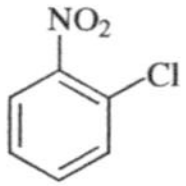

Conditions	Yield	Refs.
CuO (8 mol %), K₂CO₃, pyridine, reflux, 7 h	(45)	402

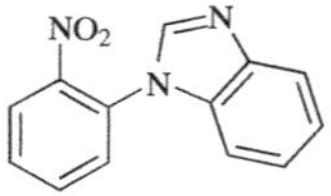

Catalyst (x mol %)

X	Catalyst	x	Additive(s)	Solvent	Temp (°)	Time (h)	Ref
Br	CuCl	5	L17 (10 mol %), KI, NaOH, TBAB	H$_2$O	100	24 (66)	235
I	CuBr	5	L5 (10 mol %), Cs$_2$CO$_3$	DMF	90	24 (90)	119
I$^+$C$_6$H$_4$NO$_2$·3 BF$_4^-$	CuI	10	K$_2$CO$_3$	DMF	80	3 (54)	460
Cl	CuO	8	K$_2$CO$_3$	pyridine	reflux	46 (8)	402

Catalyst (x mol %)

X	Catalyst	x	Additive(s)	Solvent	Temp (°)	Time (h)	Ref
Cl	Cu on cellulose	1	K$_2$CO$_3$	DMSO	130	24 (85)	427
Cl	CuBr	10	L37 (20 mol %), TBAF	none	150	24 (74)	80
Br	CuBr	10	L37 (20 mol %), TBAF	none	150	24 (95)	80
Cl	CuBr$_2$	10	TBAF	—	145	3 (81)	81
I	CuBr$_2$	10	TBAF	—	145	4.5 (86)	81
I	CuI	100	CsOAc	DMF	90	24 (73)	189
Cl	CuI	10	L26 (10 mol %), NaOMe	DMSO	130	12 (48)	397
I	CuO nanoparticles	5	KOH	t-BuOH/DMSO (3:1)	110	10 (74)	74
Cl	CuO	8	K$_2$CO$_3$	pyridine	reflux	3.5 (28)	402
Cl	Cu$_2$O	10	phen (20 mol %), TBAF	—	145	24 (98)	68
Br	CuSO$_4$	10	L36 (20 mol %), Cs$_2$CO$_3$	H$_2$O	120	24 (87)	234
I	Cu(TMHD)$_2$	20	KOt-Bu	DMF	120	24 (74)	308
Cl	[Cu(µ-I)(−)-sparteine]$_2$	1	K$_2$CO$_3$	DMSO	110	12 (83)	414
Br	[Cu(µ-I)(−)-sparteine]$_2$	1	K$_2$CO$_3$	DMSO	115	6 (95)	414
F	Cu fluorapatite	7	K$_2$CO$_3$	DMF	120	2 (85)	66
Cl	Cu fluorapatite	7	K$_2$CO$_3$	DMF	120	4 (85)	66
Cl	8	25	K$_2$CO$_3$	DMSO	135	24 (89)	65

TABLE 13A. N-ARYLATION OF BENZIMIDAZOLES (Continued)

Nitrogen Nucleophile	Aryl Halide	Conditions	Product(s) and Yield(s) (%)	Refs.

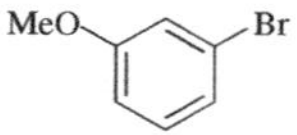

*Please refer to the charts preceding the tables for structures indicated by the **bold** numbers.*

C7

| 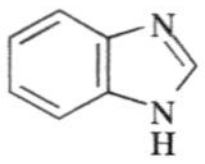 | 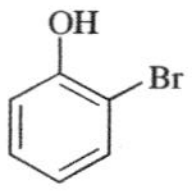 | CuI (10 mol %), 8-HOquin (20 mol %), TEAC, DMF, H2O, 120°, 70 h | 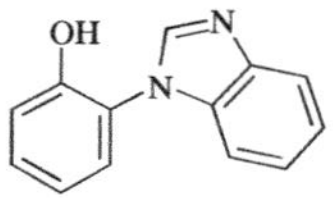 (85) | 434 |

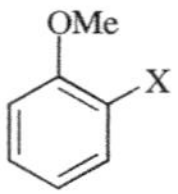

Aryl Halide: OMe, X-substituted — Conditions: CuI (x mol %),

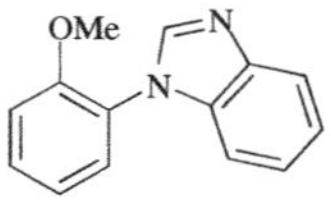

X	x	Additive(s)	Solvent(s)	Temp (°)	Time (h)	Refs.
I	20	K3PO4	DMF	60	40 (51)	59
Br	20	phen (20 mol %), KF/Al2O3	xylene	140	18 (71)	410
I	10	phen (10 mol %), KF/Al2O3	—	140	17 (76)	411
Br	10	8-HOquin (20 mol %), TEAC	DMF/H2O	110	64 (64)	434

Aryl Halide: MeO—(3-bromobenzene) — Conditions: Catalyst (x mol %)

Product: N-(3-MeO-phenyl)benzimidazole

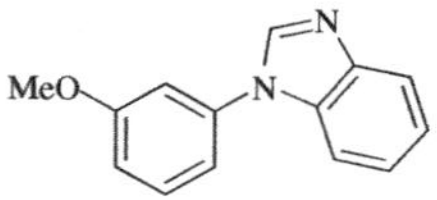

Catalyst	x	Additive(s)	Solvent(s)	Temp (°)	Time (h)	Refs.
CuI	20	phen (20 mol %), KF/Al2O3	xylene	140	16 (91)	410
CuI	10	8-HOquin (20 mol %), TEAC	DMF/H2O	130	16 (87)	434
Cu2O	10	(MeO)2phen (20 mol %), MTBD	DMSO	130	24 (82)	84

366

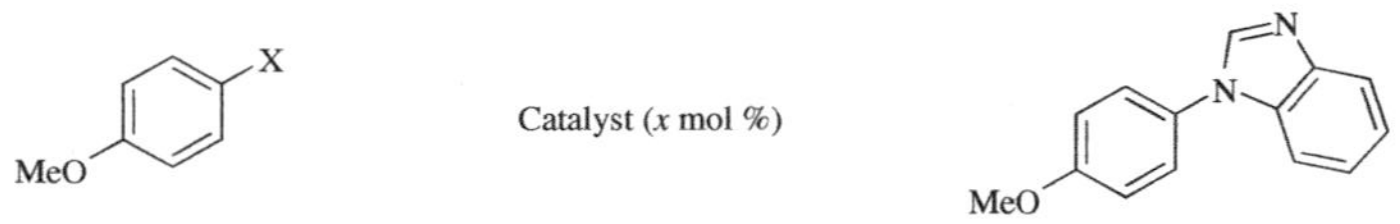

Catalyst (x mol %)

X	Catalyst(s)	x	Additive(s)	Solvent(s)	Temp (°)	Time (h)	
I	CuF₂	50	phen (50 mol %), K₂CO₃	DMF	140	96	(16)	419
I	CuBr	5	L5 (10 mol %), Cs₂CO₃	DMF	90	24 (80)	119
Br	CuBr	10	L37 (20 mol %), TBAF	none	150	24 (81)	80
Cl	CuBr₂	10	TBAF	—	145	15 (0)	81
Br	CuBr₂	10	TBAF	—	145	24 (8)	273
I	CuBr₂	10	TBAF	—	145	24 (18)	81
Cl	CuI nanoparticles	1.25	K₂CO₃, air	DMF	110	9 (78)	285
I⁺C₆H₄OMe-4 BF₄⁻	CuI	10	K₂CO₃	DMF	80	3 (71)	460
Br	CuI	20	phen (20 mol %), KF/Al₂O	xylene	140	17 (91)	410
I	CuI	10	phen (10 mol %), KF/Al₂O	—	140	15 (86)	411
Br	CuI	10	8-HOquin (20 mol %), TEAC	DMF/H₂O	130	16 (88)	434
	CuI	5	oxazolidin-2-one (10 mol %), NaOMe	DMSO	80	12 (92)	392
I	CuI	5	BtH (10 mol %), KO-t-Bu	DMSO	110	8 (94)	431
I	CuI	10	CDA (20 mol %), K₃PO₄	[C₄bmim][BF₄]	110	12 (97)	232
I	CuO nanoparticles	10	KOH	t-BuOH/DMSO (3:1)	110	11 (10)	74
Br	Cu₂O	10	phen (20 mol %), TBAF	—	145	24 (66)	68
I	Cu₂O	5	phen (10 mol %), TBAF	—	115	48 (85)	68
Br	[Cu(μ-I){(−)-sparteine}]₂	1	K₂CO₃	DMSO	115	13 (88)	414
I	[Cu(μ-I){(−)-sparteine}]₂	1	K₂CO₃	DMSO	115	7 (95)	414
I	Cu/Al–hydrotalcite	2.5	K₂CO₃	DMF	100	18 (85)	228
Br	8	25	K₂CO₃	DMSO	135	48 (85)	435

TABLE 13A. N-ARYLATION OF BENZIMIDAZOLES (Continued)

Nitrogen Nucleophile	Aryl Halide	Conditions	Product(s) and Yield(s) (%)	Refs.

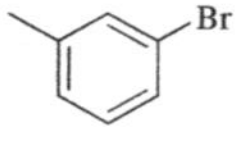

*Please refer to the charts preceding the tables for structures indicated by the **bold** numbers.*

C7

| 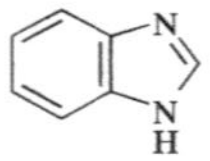 | 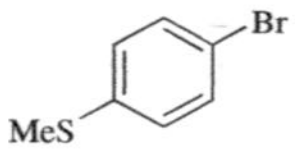 | CuI (10 mol %), 8-HOquin (20 mol %), TEAC, DMF, H2O, 130°, 16 h | 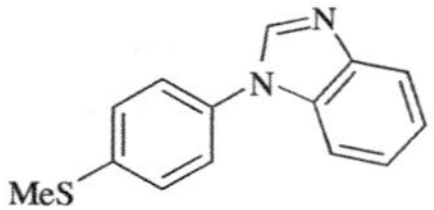 (83) | 434 |

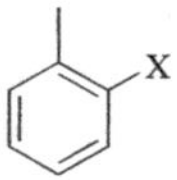

Catalyst (5 mol %)

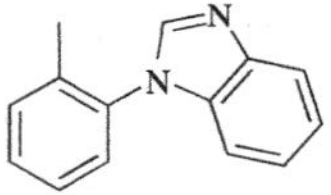

X	Catalyst	Additive	Solvent(s)	Temp (°)	Time (h)		
I	CuBr	**L5** (10 mol %), Cs2CO3	DMF	90	24	(67)	119
I	CuI	BtH (10 mol %), KOt-Bu	DMSO	110	8	(95)	431
Br	Cu2O	(MeO)2phen (15 mol %), Cs2CO3	DMSO/PEG	110	24	(76)	84

| | | CuBr (5 mol %), **L5** (10 mol %), Cs2CO3, DMF, 90°, 24 h | 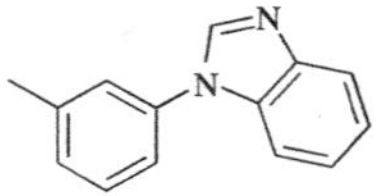 (76) | 119 |

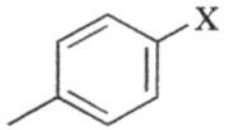

Catalyst (x mol %)

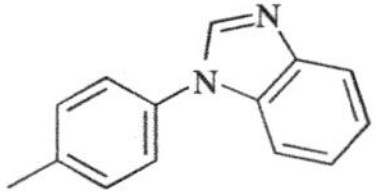

X	Catalyst	x	Additive(s)	Solvent	Temp (°)	Time (h)		Refs.
I⁺C6H4Me-4 BF4⁻	CuI	10	K2CO3	DMF	80	3	(76)	460
I	CuI	5	BtH (10 mol %), KOt-Bu	DMSO	110	8	(95)	431
Br	CuI	10	**L21** (20 mol %), Cs2CO3	MeCN	reflux	12	(34)	394
Cl	Cu2O	10	phen (20 mol %), TBAF	—	145	24	(trace)	68
Br	Cu2O	10	phen (20 mol %), TBAF	—	145	24	(89)	68
Br	CuSO4	10	**L36** (20 mol %), Cs2CO3	H2O	120	24	(71)	234
I⁺C6H4Me-4 BF4⁻	Cu(acac)2	100	K2CO3	toluene	50	6	(94)	76
Br	Cu(II)–NaY zeolite	10	K2CO3	DMF	120	48	(99)	404
I	Cu(II)–NaY zeolite	10	K2CO3	DMF	120	36	(99)	404
Br	**2**	2	NaOH, TBAB	H2O	100	12	(81)	233

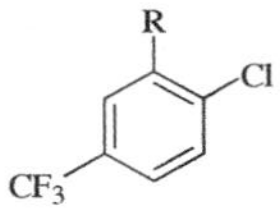

CuI (10 mol %)

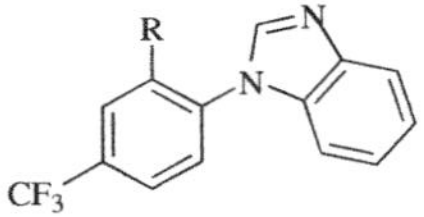

R	Additives	Solvent(s)	Temp (°)	Time (h)		Ref
H	8-HOquin (20 mol %), TEAC	DMF/H$_2$O	130	60	(50)	434
NO$_2$	L21 (20 mol %), Cs$_2$CO$_3$	MeCN	reflux	12	(90)	394

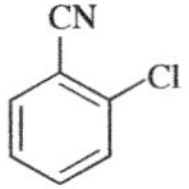

[Cu(μ-I)((−)-sparteine)]$_2$
(1 mol %), K$_2$CO$_3$,
DMSO, 125°, 16 h

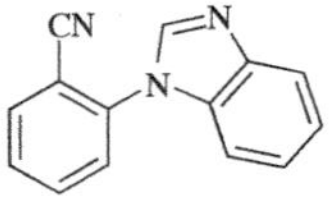

(82)   414

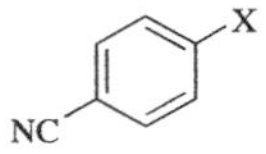

Catalyst (x mol %)

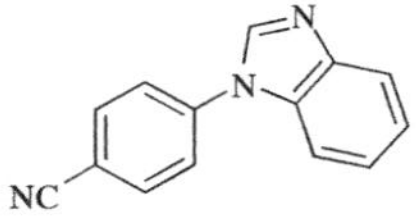

X	Catalyst	x	Additive(s)	Solvent(s)	Temp (°)	Time (h)		Ref
Br	CuI	10	L-Pro (20 mol %), K$_2$CO$_3$	DMSO	95	42	(99)	47
Br	CuI	10	CDA (20 mol %), K$_3$PO$_4$	[C$_4$mim][BF$_4$]	110	12	(70)	232
Br	CuI	10	8-HOquin (20 mol %), TEAC	DMF/H$_2$O	110	24	(87)	434
Br	CuO	8	K$_2$CO$_3$	pyridine	reflux	101	(42)	402
Br	CuSO$_4$	10	L36 (20 mol %), Cs$_2$CO$_3$	H$_2$O	120	24	(82)	234
Cl	[Cu(μ-I)((−)-sparteine)]$_2$	1	K$_2$CO$_3$	DMSO	125	14	(86)	414
Br	2	2	NaOH, TBAB	H$_2$O	100	12	(92)	233
Cl	8	25	K$_2$CO$_3$	DMSO	135	24	(90)	435

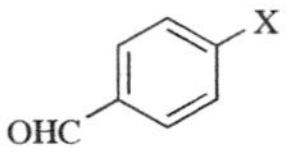

[Cu(μ-I)((−)-sparteine)]$_2$
(1 mol %), K$_2$CO$_3$,
DMSO, 115°, 8 h

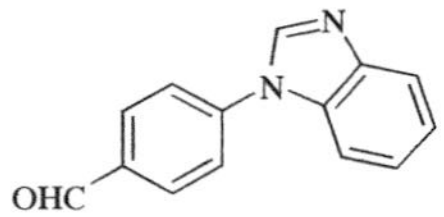

X	Time (h)		Ref
Cl	19	(77)	414
Br	8	(90)	

TABLE 13A. N-ARYLATION OF BENZIMIDAZOLES (*Continued*)

Nitrogen Nucleophile	Aryl Halide	Conditions	Product(s) and Yield(s) (%)	Refs.

*Please refer to the charts preceding the tables for structures indicated by the **bold** numbers.*

C$_7$

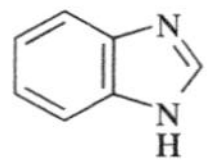

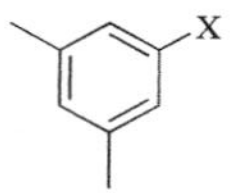

Nitrogen Nucleophile:

Aryl Halide:

Conditions: Catalyst (x mol %)

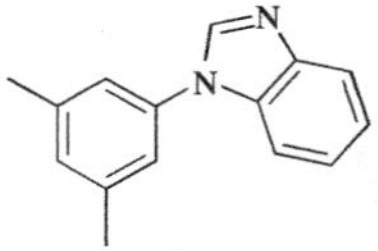

X	Catalyst	x	Additives	Solvent(s)	Temp (°)	Time (h)	Product(s) and Yield(s) (%)	Refs.
I	CuI	10	phen (20 mol %), Cs$_2$CO$_3$	dioxane	110	24	(91)	115
I	CuI	10	phen (20 mol %), Cs$_2$CO$_3$	DMF	110	24	(80)	128
Br	CuI	10	8-HOquin (10 mol %), Cs$_2$CO$_3$	DMF/H$_2$O	120	70	(90)	434
Br	Cu(OAc)•H$_2$O	5	4,7-Cl$_2$phen (10 mol %), KO*t*Bu	NMP	125	22	(99)	452
I	(CuOTf)$_2$•C$_6$H$_6$	10	phen (1 eq). dba, Cs$_2$CO$_3$	xylenes	125	48	(91)	11

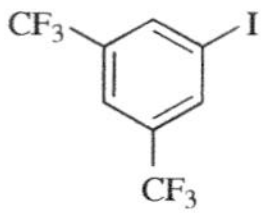

CuI (10 mol %), DMEDA (20 mol %), TEAC, DMF, 110°, 10 h — (76) — 434

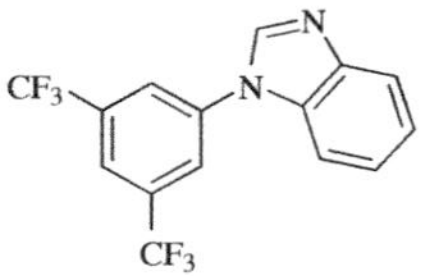

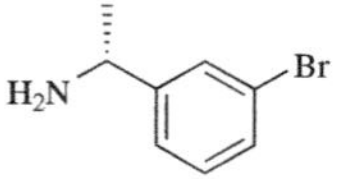

CuI (10 mol %), K$_2$CO$_3$, NMP, MW, 195°, 2 h — (88) — 406

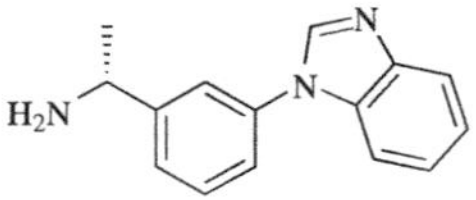

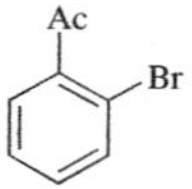

CuO (8 mol %), K$_2$CO$_3$, pyridine, reflux, 3 h — (46) — 402

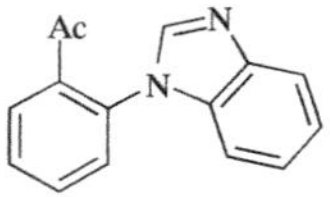

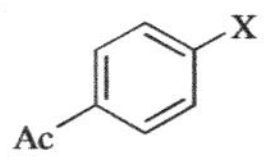

Catalyst (x mol %)

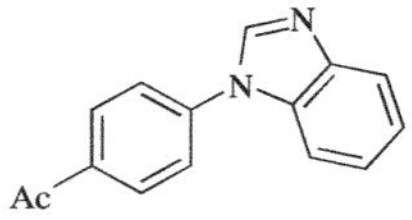

X	Catalyst	x	Additive(s)	Solvent	Temp (°)	Time (h)		
Br	CuCl	5	L17 (10 mol %), NaOH, TBAF	H$_2$O	100	24	(62)	235
Br	CuBr	10	L37 (20 mol %), TBAF	none	150	24	(92)	80
Br	CuI	10	DMG (20 mol %), K$_2$CO$_3$	DMSO	110	40	(75)	47
Br	CuI	10	L32 (20 mol %), Cs$_2$CO$_3$	DMF	110	24	(80)	417
Br	CuO	8	K$_2$CO$_3$	pyridine	reflux	80	(48)	402
I	Cu/Al–hydrotalcite	2.5	K$_2$CO$_3$	DMF	100	15	(88)	228
Br	[Cu(μ-I)((−)-sparteine)]$_2$	1	K$_2$CO$_3$	DMSO	115	8	(92)	414
Br	8	25	K$_2$CO$_3$	DMSO	135	30	(82)	65

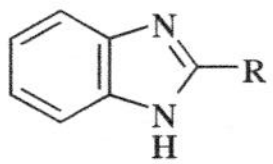

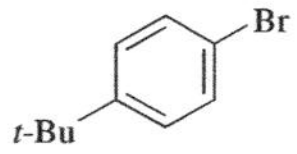

Cu$_2$O (5 mol %), MeO$_2$phen (15 mol %), Cs$_2$CO$_3$, PEG, DMSO, 110°, 24 h

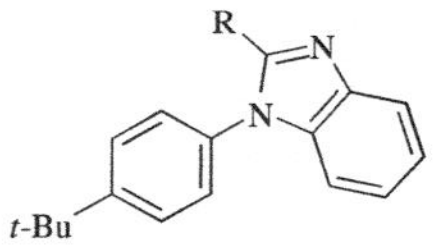

R	
H	(98)
Me	(82)

84

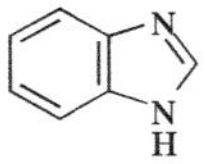

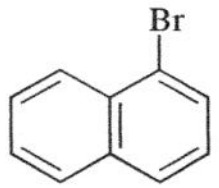

CuI (20 mol %), phen (20 mol %), KF/Al$_2$O$_3$, xylene, 140 °C, 17 h

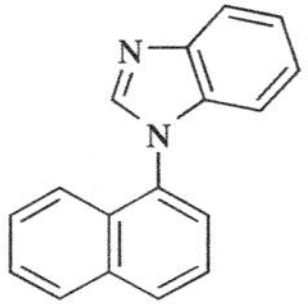

(71)

410

C$_{7-8}$

C$_7$

TABLE 13A. *N*-ARYLATION OF BENZIMIDAZOLES (*Continued*)

Nitrogen Nucleophile	Aryl Halide	Conditions	Product(s) and Yield(s) (%)	Refs.

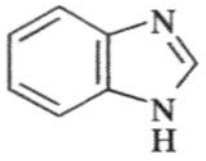

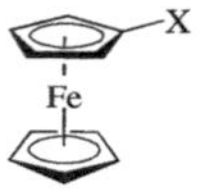

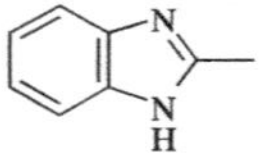

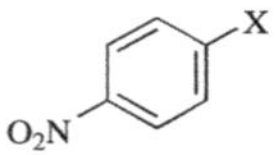

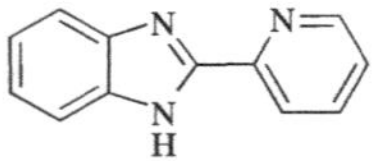

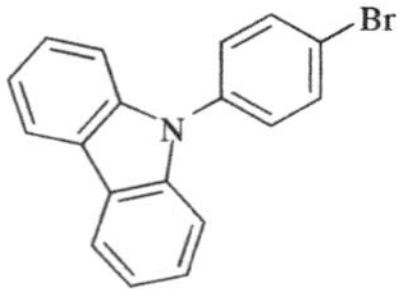

*Please refer to the charts preceding the tables for structures indicated by the **bold** numbers.*

C₇ refers to the chart structures.

Cu(TMHD)₂ (20 mol %),
KO*t*-Bu, DMSO, 90°, 12 h

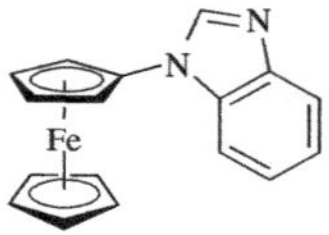

X	
Br	(18)
I	(35)

407

C₈

Catalyst (10 mol %)

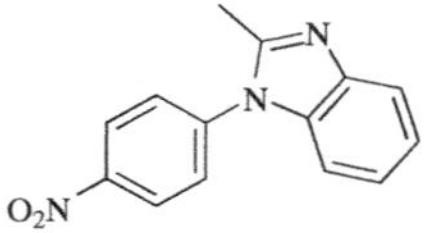

X	Catalyst	Additive(s)	Solvent	Temp (°)	Time (h)	
Cl	CuI	**L26** (10 mol %), NaOMe	DMSO	110	12	(50)
Cl	Cu₂O	phen (20 mol %), TBAF	—	145	24	(trace)
Br	Cu₂O	phen (20 mol %), TBAF	—	145	24	(63)

397
68
68

C₁₂

CuI (10 mol %),
18-c-6 (3 mol %), K₂CO₃,
DMPU, 230°, 13 h

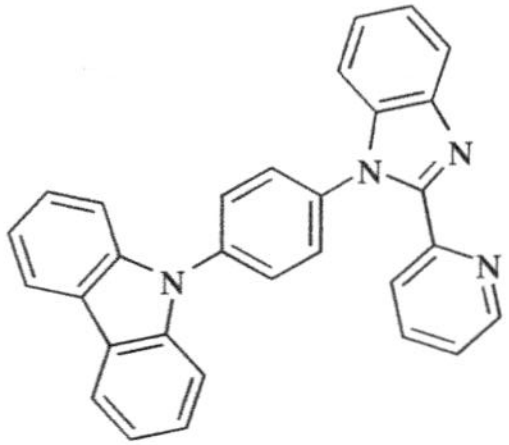

(67)

461

Nitrogen Nucleophile	Heteroaryl Halide	Conditions	Product(s) and Yield(s) (%)	Refs.

*Please refer to the charts preceding the tables for structures indicated by the **bold** numbers.*

C₇

Nitrogen nucleophile: benzimidazole

Heteroaryl Halide	Conditions	Product(s) and Yield(s) (%)	Refs.
2-bromothiazole	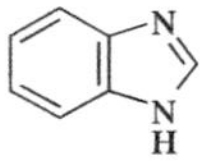 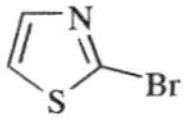 CuBr (10 mol %), L37 (20 mol %), TBAF, 150°, 24 h	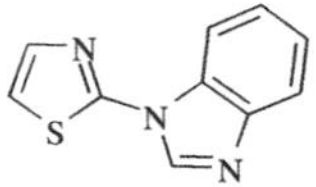 (55)	80

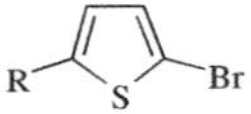

5-R-2-bromothiophene; Catalyst (x mol %)

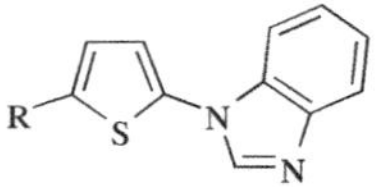

R	Catalyst	x	Additive(s)	Solvent	Temp (°)	Time (h)	Yield	Refs.
H	CuBr	10	L37 (20 mol %), TBAF	none	150	24	(62)	80
H	CuI	30	L-Pro (60 mol%)	[C₄mim][BF₄]	110	24	(76)	230
H	CuI	10	L32 (20 mol %), Cs₂CO₃	DMF	110	24	(80)	417
Me	CuI	30	L-Pro (60 mol%)	[C₄mim][BF₄]	110	36	(73)	230

Heteroaryl Halide	Conditions	Product(s) and Yield(s) (%)	Refs.
3-bromothiophene	CuI (30 mol %), L-Pro (60 mol %), [C₄mim][BF₄], 110°, 36 h	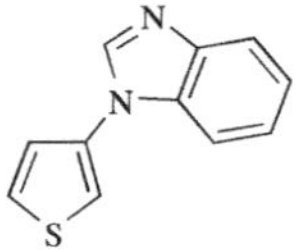 (75)	230

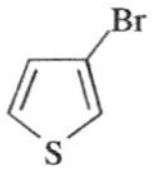

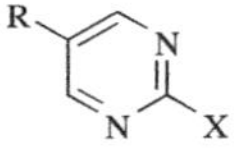

5-R-2-X-pyrimidine; Catalyst (x mol %), 24 h

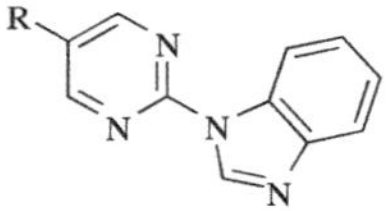

R	X	Catalyst	x	Additives	Solvent	Temp (°)	Yield	Refs.
H	Cl	CuBr	10	L37 (20 mol %), TBAF	none	150	(100)	80
H	Br	CuBr	10	L37 (20 mol %), TBAF	none	150	(100)	80
H	I	CuI	5	phen (10 mol %), Cs₂CO₃	DMF	110	(50)	408
H	Br	CuI	5	L30 (10 mol %), Cs₂CO₃	MeCN	reflux	(78)	408
H	Cl	Cu₂O	10	phen (10 mol %), TBAF	—	145	(77)	68
Br	I	CuI	5	phen (10 mol %), Cs₂CO₃	DMF	80	(57)	408

TABLE 13B. *N*-HETEROARYLATION OF BENZIMIDAZOLES (*Continued*)

Nitrogen Nucleophile	Heteroaryl Halide	Conditions	Product(s) and Yield(s) (%)	Refs.

*Please refer to the charts preceding the tables for structures indicated by the **bold** numbers.*

C7

			Conditions	Product	Refs.
	5-bromopyrimidine		**CuBr** (5 mol %), **L5** (10 mol %), Cs$_2$CO$_3$, DMF, 90°, 24 h	(73)	119

Catalyst (x mol %)

X	Catalyst	x	Additive(s)	Solvent	Temp (°)	Time (h)		
Cl	CuBr$_2$	10	TBAF	—	145	1	(76)	81
I	CuI	5	phen (10 mol %), Cs$_2$CO$_3$	DMF	110	24	(50)	408
Cl	CuI	10	**L26** (10 mol %), NaOMe	DMSO	130	12	(88)	397

Catalyst (x mol %),

R	X	Catalyst	x	Additive(s)	Solvent	Temp (°)	Time (h)		
H	Cl	Cu	5	binol (20 mol %), Cs$_2$CO$_3$	DMSO	110	36	(88)	124
H	Br	CuCl	5	**L17** (10 mol %), NaOH, TBAB	H$_2$O	100	24	(87)	235
H	Br	CuBr	5	**L5** (10 mol %), Cs$_2$CO$_3$	DMF	90	24	(78)	119
H	Cl	CuBr	10	**L37** (20 mol %), TBAF	none	150	24	(56)	80
H	Br	CuBr	10	**L37** (20 mol %), TBAF	none	150	24	(99)	80
H	Br	CuBr$_2$	10	TBAF	—	145	4	(90)	81
H	Br	CuI	30	L-Pro (60 mol %)	[C$_4$mim][BF$_4$]	100	10	(87)	230
H	Br	CuI	5	BtH (10 mol %), KO*t*-Bu	DMSO	110	8	(98)	431
H	Br	CuI	10	oxazolidin-2-one (20 mol %), NaOMe	DMSO	80	12	(94)	392
H	Br	CuI	10	CDA (20 mol %), K$_3$PO$_4$	[C$_4$mim][BF$_4$]	110	12	(78)	232
H	Br	CuI	10	**L26** (10 mol %), NaOMe	DMSO	110	12	(90)	397

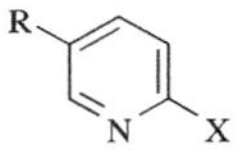

Catalyst (x mol %),

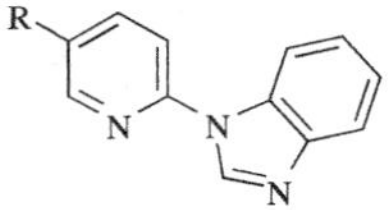

R	X	Catalyst	x	Additive(s)	Solvent	Temp (°)	Time (h)		
H	Br	CuO	8	K$_2$CO$_3$	pyridine	reflux	24	(46)	402
H	Cl	Cu$_2$O	10	phen (20 mol %), TBAF	—	145	24	(95)	68
H	Br	CuSO$_4$	10	L36 (20 mol %), Cs$_2$CO$_3$	H$_2$O	120	24	(78)	234
H	Br	2	2	NaOH, TBAB	H$_2$O	100	12	(58)	233
Br	I	CuI	5	phen (10 mol %), Cs$_2$CO$_3$	DMF	80	24	(90)	408

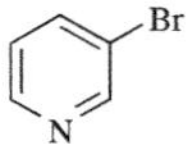

Catalyst (x mol %), 24 h

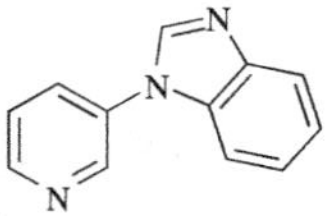

Catalyst	x	Additive(s)	Solvent	Temp (°)		
CuBr	5	L5 (10 mol %), Cs$_2$CO$_3$	DMF	90	(77)	119
CuO	8	K$_2$CO$_3$	pyridine	reflux	(33)	402

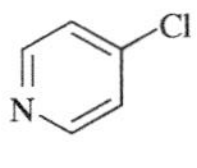

CuO (8 mol %), K$_2$CO$_3$, pyridine, reflux, 19.5 h

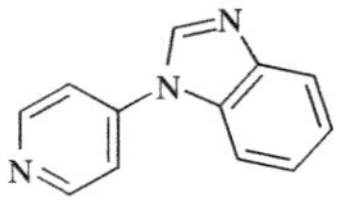

(21) 402

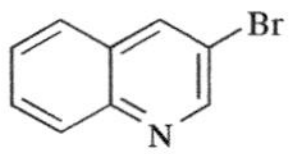

CuBr (10 mol %), L37 (20 mol %), TBAF, neat, 150°, 24 h

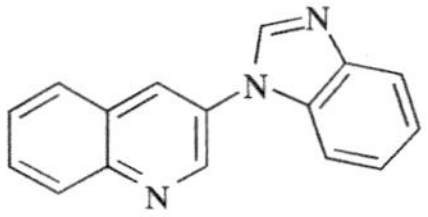

(63) 80

TABLE 13B. *N*-HETEROARYLATION OF BENZIMIDAZOLES (*Continued*)

Nitrogen Nucleophile	Heteroaryl Halide	Conditions	Product(s) and Yield(s) (%)	Refs.

*Please refer to the charts preceding the tables for structures indicated by the **bold** numbers.*

C₇

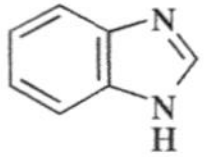

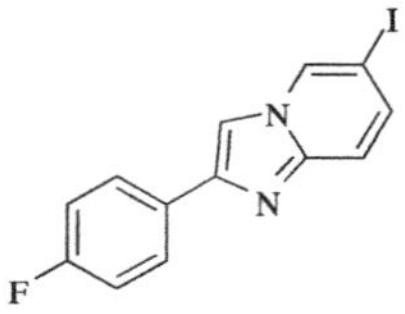

CuI (5 mol %), DMCDA (15 mol %), K₃PO₄, toluene, 112°, 24 h — (9) — 409

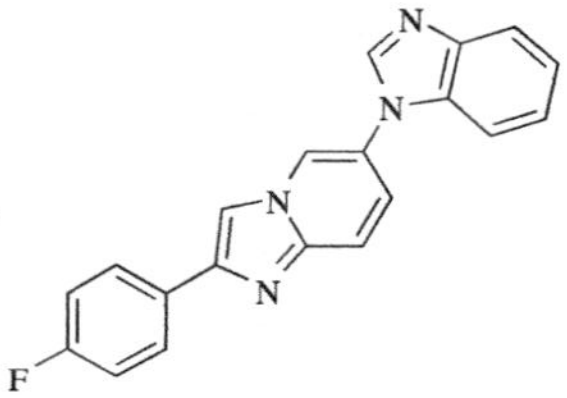

C₈

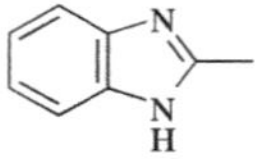

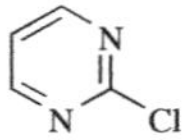

Catalyst (10 mol %), 24 h

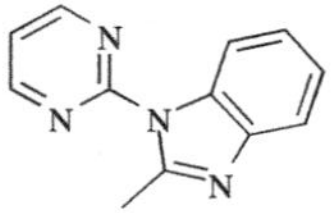

Catalyst	Additives	Solvent	Temp (°)	
CuBr	L37 (20 mol %), TBAF	none	150	(78)	80
Cu₂O	phen (20 mol %), TBAF	—	145	(66)	68

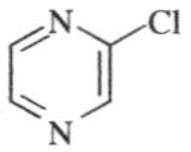

CuI (10 mol %), L26 (10 mol %), NaOMe, DMSO, 130°, 12 h — (66) — 397

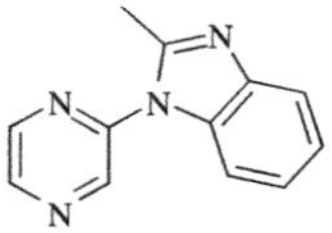

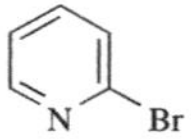

Catalyst (10 mol %),

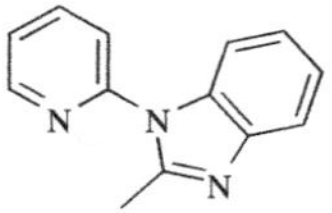

Catalyst	Additive(s)	Solvent	Temp (°)	Time (h)	
CuBr	L37 (20 mol %), TBAF	none	150	24	(95)	80
CuI	L26 (10 mol %), NaOMe	DMSO	130	12	(88)	397

TABLE 14A. N-ARYLATION OF CARBAZOLES

*Please refer to the charts preceding the tables for structures indicated by the **bold** numbers.*

Nitrogen Nucleophile	Aryl Halide	Conditions	Product(s) and Yield(s) (%)	Refs.

C_{12}

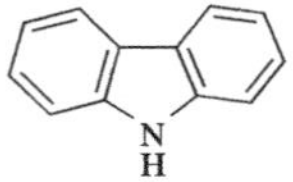

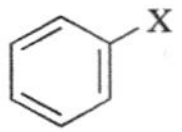

Catalyst (x mol %)

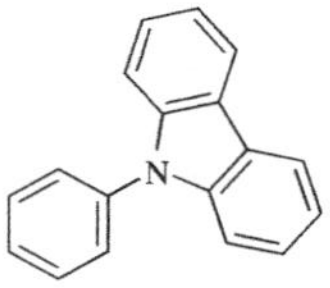

X	Catalyst	x	Additive(s)	Solvent	Temp (°)	Time (h)		
I	Cu	20	Cs$_2$CO$_3$	n-PrCN	reflux	20	(81)	129
I	Cu	20	K$_2$CO$_3$	PhNO$_2$	reflux	24	(99)	348
Br	CuI	10	phen (20 mol %), KF/Al$_2$O$_3$	toluene	reflux	7	(90)	390
I	CuI	10	phen (20 mol %), KF/Al$_2$O$_3$	toluene	reflux	7	(90)	390
I	CuO	0.1	DMEDA (20 mol %), K$_3$PO$_4$•H$_2$O	toluene	135	24	(25)	222

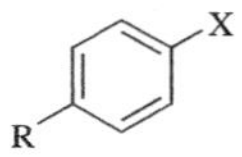

CuI (x mol %)

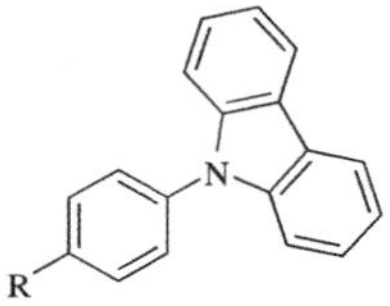

R	X	x	Additives	Solvent	Temp (°)	Time (h)		
Cl	I	5	DMEDA (10 mol %), CsF	THF	60	25	(88)	63
Br	Br	10	18-c-6 (3 mol %), K$_2$CO$_3$	DMPU	170	13	(34)	461

C_{12-36}

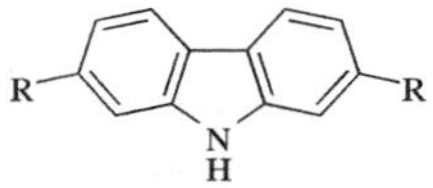

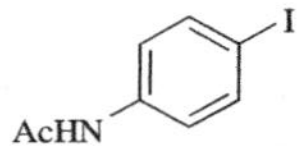

Cu$_2$O (1.5 eq), DMA, 160°, 24 h

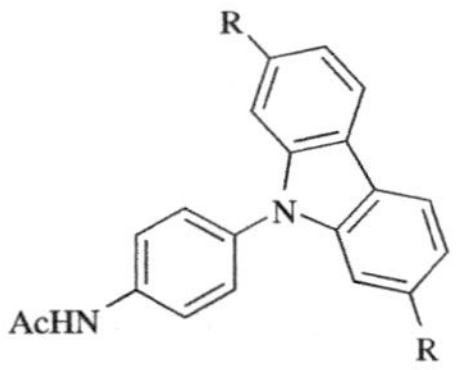

R		
H	(55)	462
1-carbazolyl	(99)	

TABLE 14A. N-ARYLATION OF CARBAZOLES (Continued)

Nitrogen Nucleophile	Aryl Halide	Conditions	Product(s) and Yield(s) (%)	Refs.

*Please refer to the charts preceding the tables for structures indicated by the **bold** numbers.*

C$_{12}$

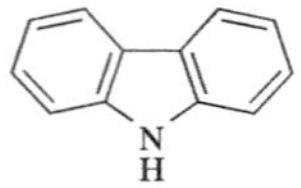

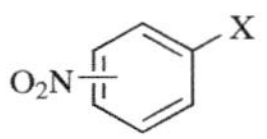

Catalyst (x mol %)

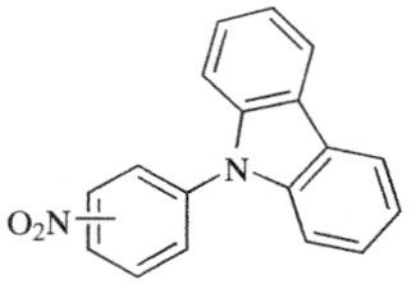

Isomer	X	Catalyst	x	Additive(s)	Solvent	Temp (°)	Time (h)		
2	Cl	CuO	8	K$_2$CO$_3$	pyridine	reflux	16	(90)	402
3	Cl	CuO	8	K$_2$CO$_3$	pyridine	reflux	16	(90)	402
4	Br	CuCl	5	L17 (10 mol %), NaOH, TBAB	H$_2$O	100	24	(85)	235
4	Cl	CuO	8	K$_2$CO$_3$	pyridine	reflux	18	(0.2)	402

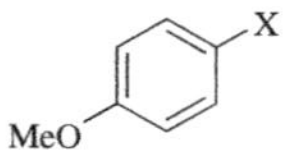

Catalyst (x amount)

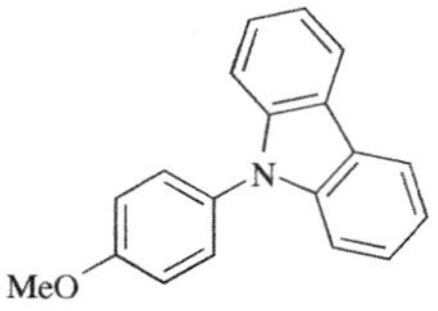

X	Catalyst	x	Additive(s)	Solvent	Temp (°)	Time (h)		
I	CuI	10 mol %	L-Pro (10 mol %), K$_2$CO$_3$	DMSO	90	40	(75)	47
I	CuI	5 mol %	L-Pro (10 mol %), K$_2$CO$_3$	DMSO	90	40	(93)	401
Br	CuI	10 mol %	phen (20 mol %), KF/Al$_2$O$_3$	toluene	reflux	10	(83)	390
I	CuI	10 mol %	phen (20 mol %), KF/Al$_2$O$_3$	toluene	reflux	8	(89)	390
I	Cu$_2$O	2 eq	none	DMA	160	24	(82)	463

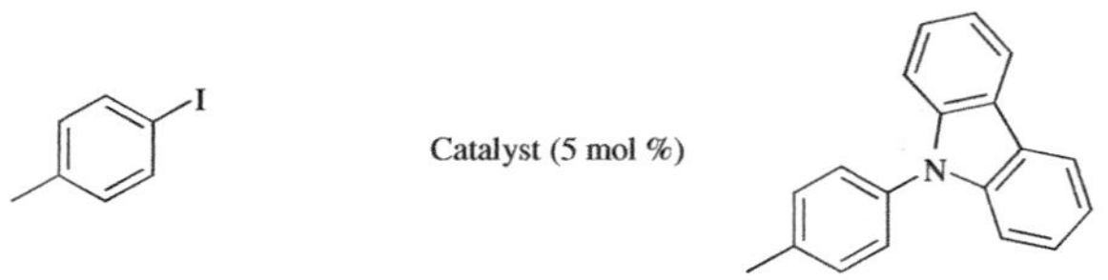

Catalyst (5 mol %)

Catalyst	Additives	Solvent	Temp (°)	Time (h)		
CuCl	L17 (10 mol %), NaOH, TBAB	H₂O	100	24	(89)	235
CuI	L41 (10 mol %), NaO*t*-Bu	toluene	reflux	40	(86)	338

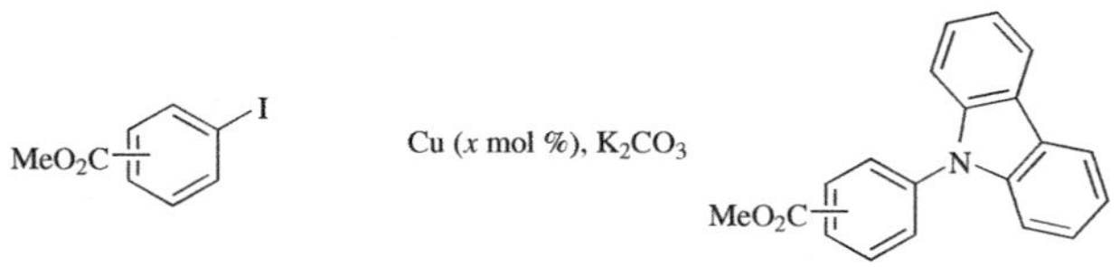

Cu (x mol %), K₂CO₃

Isomer	x	Solvent	Temp (°)	Time (h)		
2	50	PhNO₂	180	40	(59)	349
2	20	PhNO₂	reflux	24	(60)	348
4	3	xylene	200	1	(60)	464

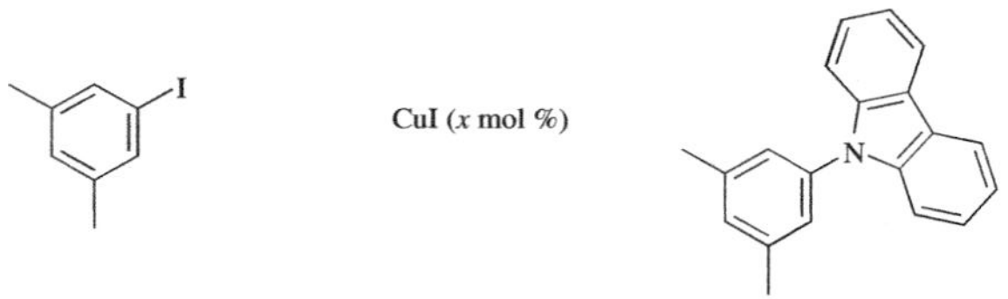

CuI (x mol %)

x	Additives	Solvent	Temp (°)	Time (h)		
1	CDA (10 mol %), K₃PO₄	dioxane	110	24	(90)	115
5	L-Pro (10 mol%), K₂CO₃	DMSO	90	40	(90)	401

TABLE 14A. N-ARYLATION OF CARBAZOLES (Continued)

Nitrogen Nucleophile	Aryl Halide	Conditions	Product(s) and Yield(s) (%)	Refs.

*Please refer to the charts preceding the tables for structures indicated by the **bold** numbers.*

C$_{12-108}$

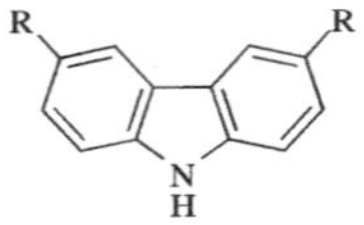

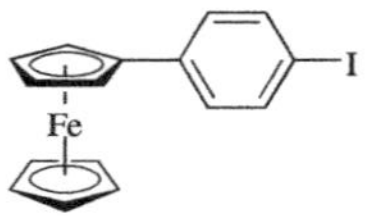

| | | CuI (10 mol %), CDA (15 mol %), K$_3$PO$_4$, dioxane, 110°, 24 h | | 465 |

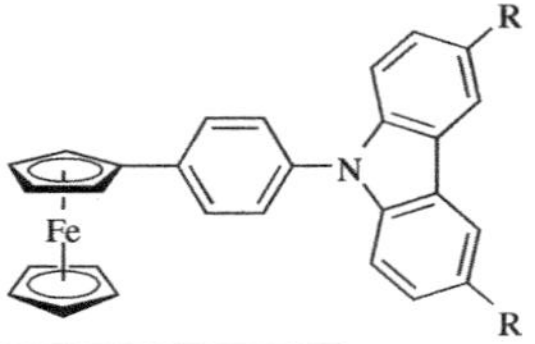

R	
H	(77)
1-carbazolyl	(47)
3,6-bis(1-carbazolyl)-1-carbazolyl	(54)
3,6-bis[3,6-bis(1-carbazolyl)-1-carbazolyl]-1-carbazolyl	(54)

C$_{12}$

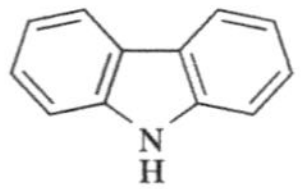

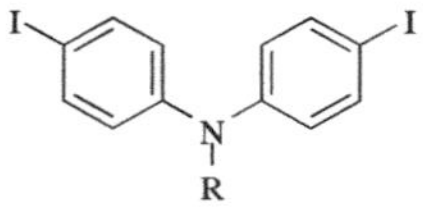

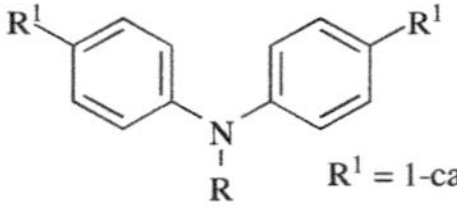

R^1 = 1-carbazolyl

Cu (x eq), K$_2$CO$_3$, 24 h

R	x	Additive	Solvent	Temp (°)		Refs
n-Bu	2	18-c-6 (10 mol %)	1,2-Cl$_2$C$_6$H$_4$	reflux	(69)	373
4-MeOC$_6$H$_4$	1	none	triglyme	200	(74)	371

C$_{12-108}$

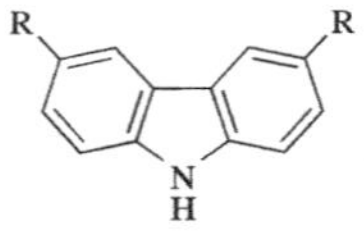

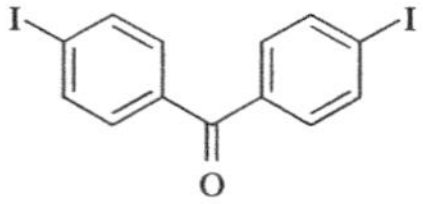

Catalyst (x amount)

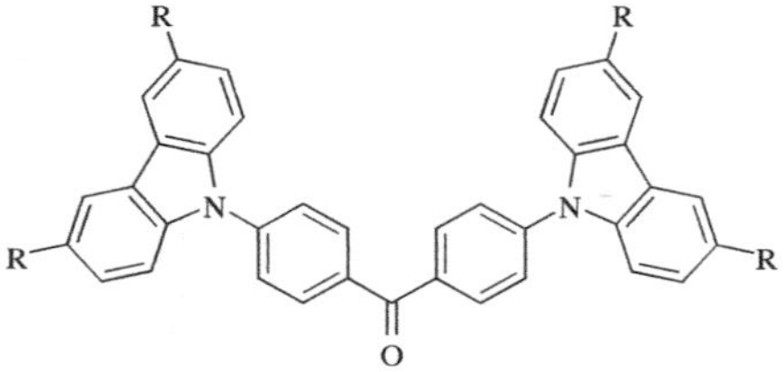

R	Catalyst	x	Additives	Solvent	Temp (°)	Time (h)		Refs
H	Cu$_2$O	3 eq	none	DMA	165	48	(83)	466
H	CuI	5 mol %	CDA (10 mol %), K$_3$PO$_4$	dioxane	110	24	(80)	465
1-carbazolyl	Cu$_2$O	3 eq	none	DMA	165	48	(46)	466
1-carbazolyl	CuI	5 mol %	CDA (10 mol %), K$_3$PO$_4$	dioxane	110	24	(77)	465
3,6-bis(1-carbazolyl)-1-carbazolyl	CuI	5 mol %	CDA (10 mol %), K$_3$PO$_4$	dioxane	110	24	(85)	465
2,6-bis[3,6-bis(1-carbazolyl)-1-carbazolyl]-1-carbazolyl	CuI	5 mol %	CDA (10 mol %), K$_3$PO$_4$	dioxane	110	24	(32)	465

C_{12}

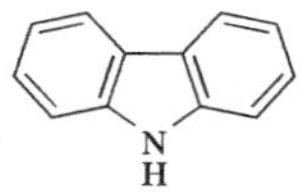

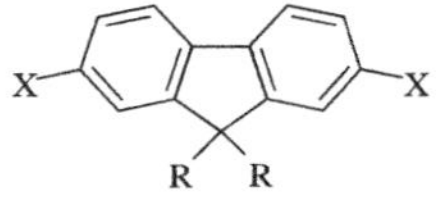

Catalyst (x mol %), 24 h

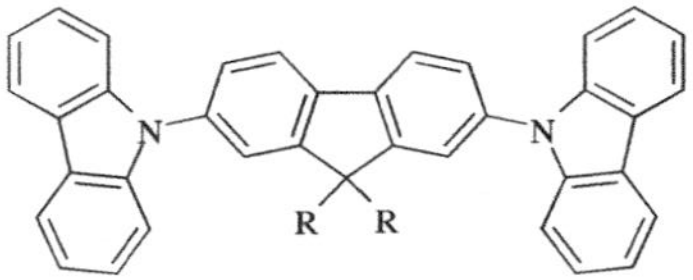

R	X	Catalyst	x	Additive(s)	Solvent	Temp (°)		
Et	I	CuI	20	D-glucosamine (40 mol %), Cs_2CO_3	DMSO	110	(48)	467
n-C_6H_{13}	Br	Cu	45	K_2CO_3	$PhNO_2$	reflux	(98)	468
n-C_8H_{17}	Br	Cu	44	K_2CO_3	$PhNO_2$	reflux	(93)	469

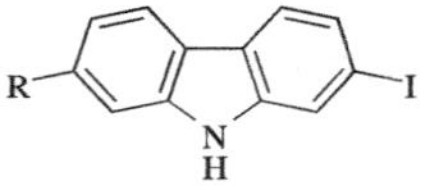

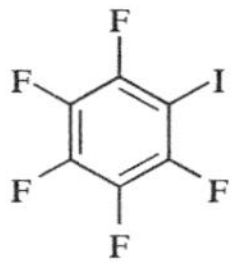

Cu, 18-c-6, K_2CO_3, reflux

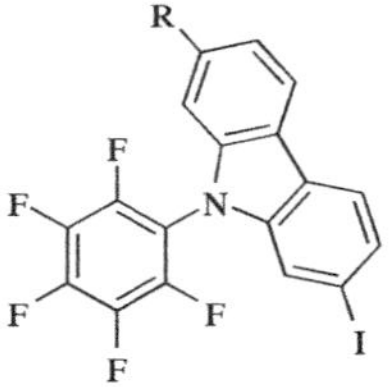

R		
H	(78)	470
I	(65)	471

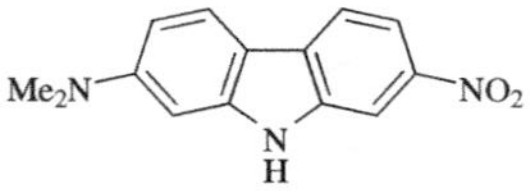

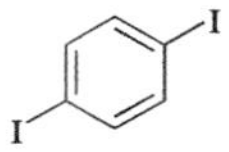

Cu (10 mol %), Na_2SO_4, K_2CO_3, $PhNO_2$, 180°, 23 h

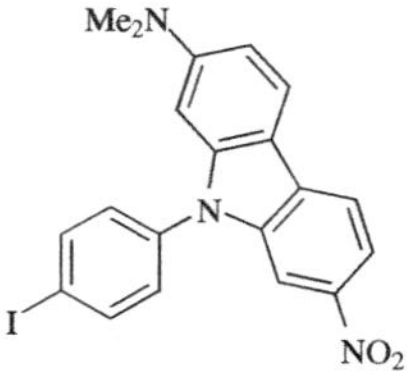

(50) 472

TABLE 14A. *N*-ARYLATION OF CARBAZOLES (*Continued*)

*Please refer to the charts preceding the tables for structures indicated by the **bold** numbers.*

Nitrogen Nucleophile	Aryl Halide	Conditions	Product(s) and Yield(s) (%)	Refs.
C₁₂				
		Cu-bronze (14 mol %), K₂CO₃, DMA, 160°, 12 h	R: O₂N (98); MeO (98)	473
		Cu-bronze (14 mol %), K₂CO₃, DMA, 160°, 12 h	R: Me (98); n-C₁₈H₃₇ (98)	473
R¹ = H or C₉H₁₉; R² = C₉H₁₉		Cu-bronze (14 mol %), K₂CO₃, DMA, 160°, 12 h	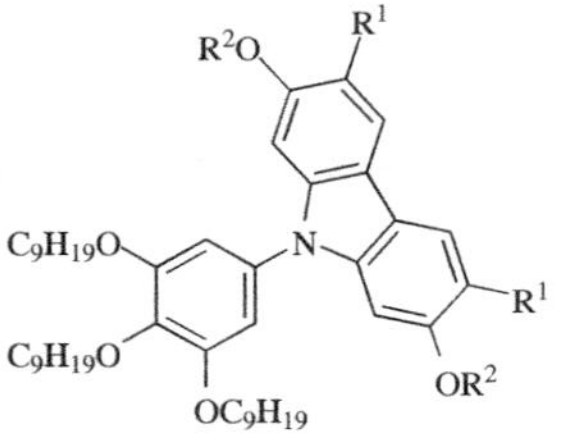 (94)	473
R = C₉H₁₉		Cu-bronze (14 mol %), K₂CO₃, DMA, 160°, 12 h	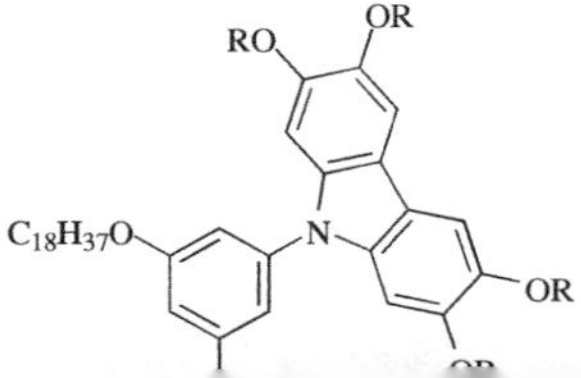 (85)	473

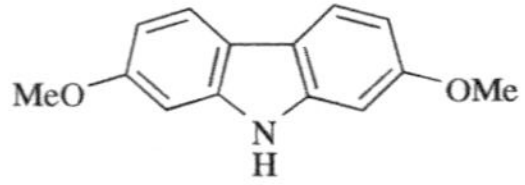

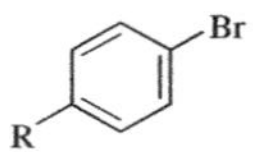

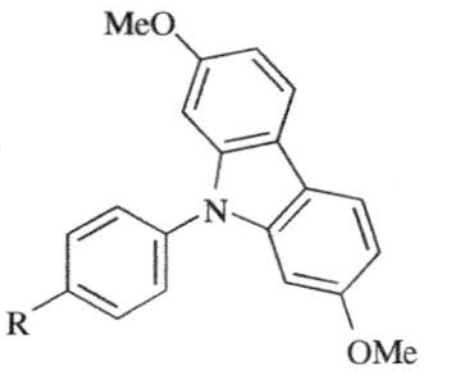

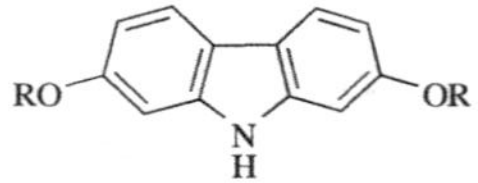

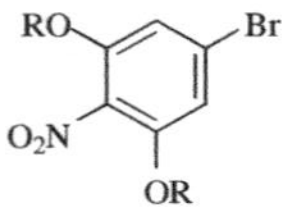

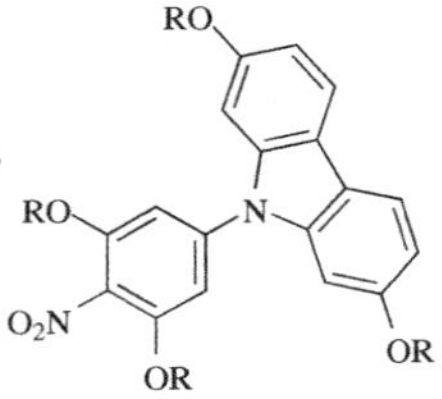

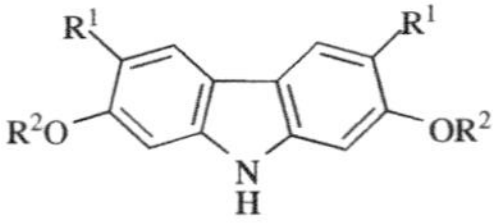

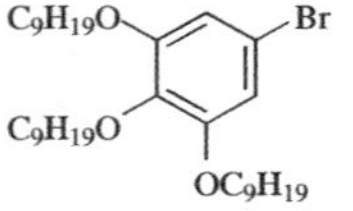

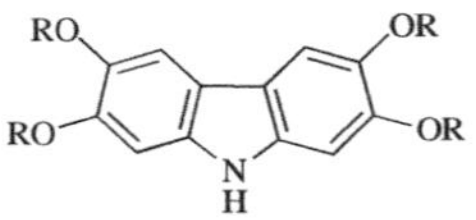

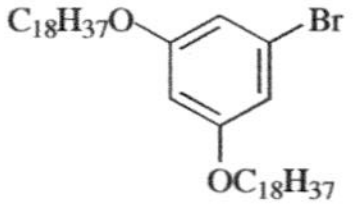

382

C_{13}

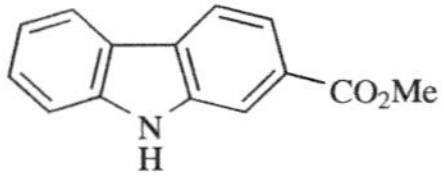

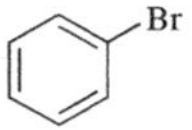

Cu-bronze (14 mol %), K_2CO_3, DMA, 160°, 12 h

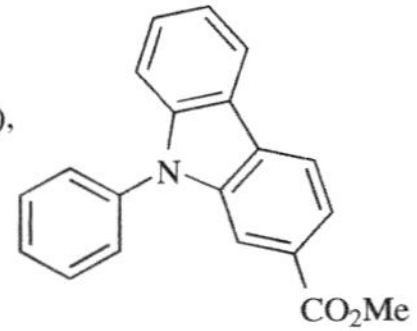

(72) 473

C_{20}

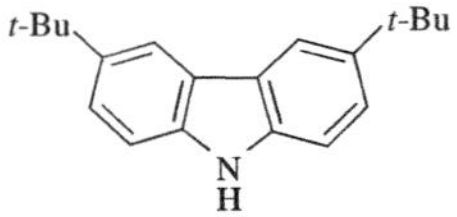

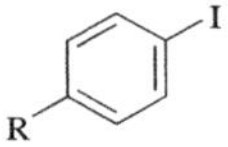

Catalyst (*x* eq), 24 h

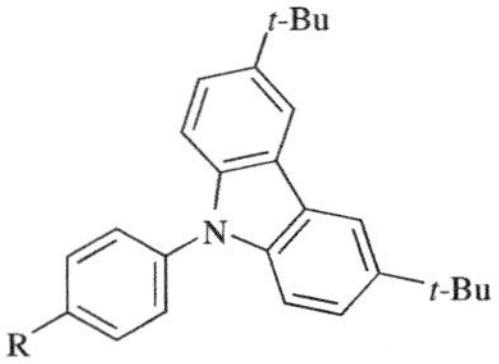

R	Catalyst	*x*	Additive	Solvent	Temp (°)		
H	Cu	1	K_2CO_3	$PhNO_2$	170	(92)	474
O_2N	Cu	1	K_2CO_3	$PhNO_2$	170	(92)	444
MeO	Cu_2O	2	none	DMA	160	(83)	463
CHO	Cu	1	K_2CO_3	$PhNO_2$	170	(80)	444

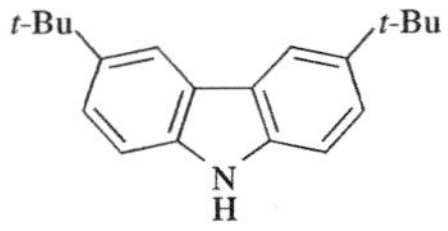

I

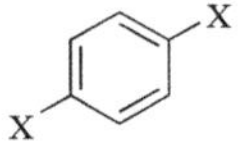

II

Cu (1 eq), K_2CO_3, $PhNO_2$, 170°, 24 h

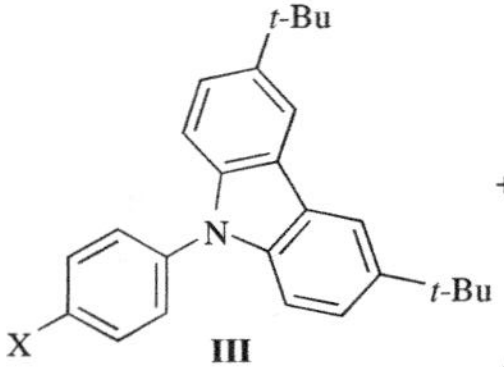

III

+

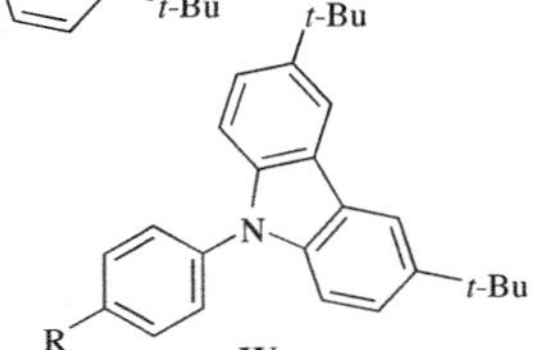

IV

R =

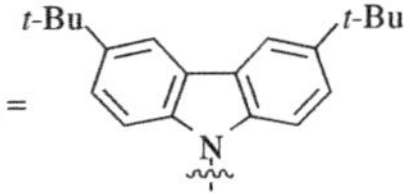

474

X	I/II	III	IV
Br	excess of **II**	(55)	(—)
I	excess of **II**	(80)	(—)
I	2:1	(—)	(55)

TABLE 14A. N-ARYLATION OF CARBAZOLES (*Continued*)

*Please refer to the charts preceding the tables for structures indicated by the **bold** numbers.*

Nitrogen Nucleophile	Aryl Halide	Conditions	Product(s) and Yield(s) (%)	Refs.

C$_{20}$

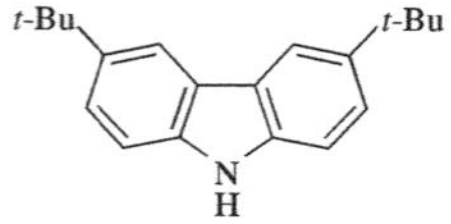

I

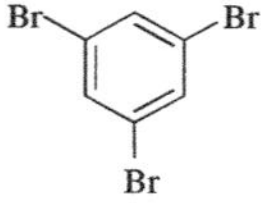

II

Cu (1 eq), K$_2$CO$_3$,
PhNO$_2$,
170°, 24 h

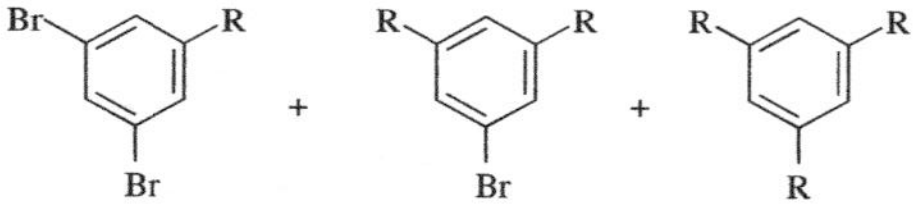

III

+

IV

+

V

R =

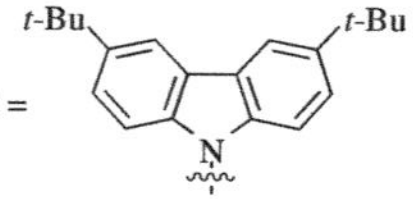

I/II	III	IV	V
1:2	(50)	(—)	(—)
2:1	(—)	(27)	(—)
slight excess of **II**	(—)	(—)	(0)

474

C$_{20-52}$

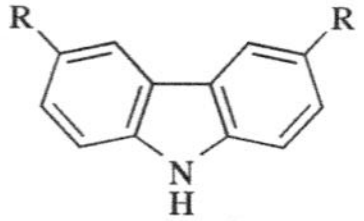

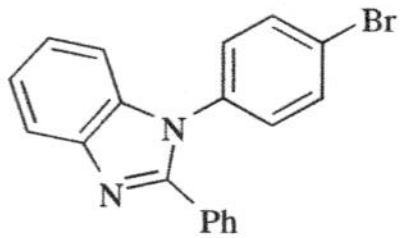

CuI (5 mol %),
18-c-6 (5 mol %),
K$_2$CO$_3$, DMPU,
190°, 24 h

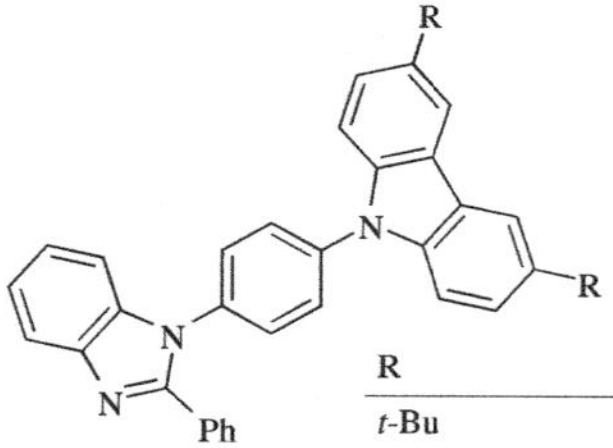

R	
t-Bu	(86)
3,6-(*t*-Bu)$_2$-1-carbazolyl	(72)

475

C$_{24}$

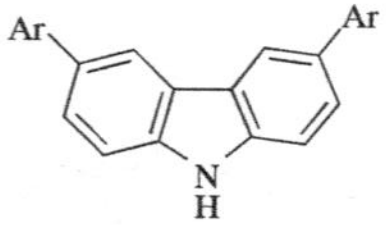

Ar = 4-Ph$_2$NC$_6$H$_4$

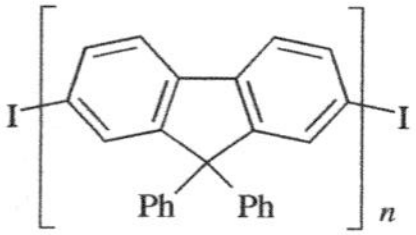

CuI (1 mol %), CDA
(10 mol %), NaO-*t*-Bu,
dioxane, 110°, 24 h

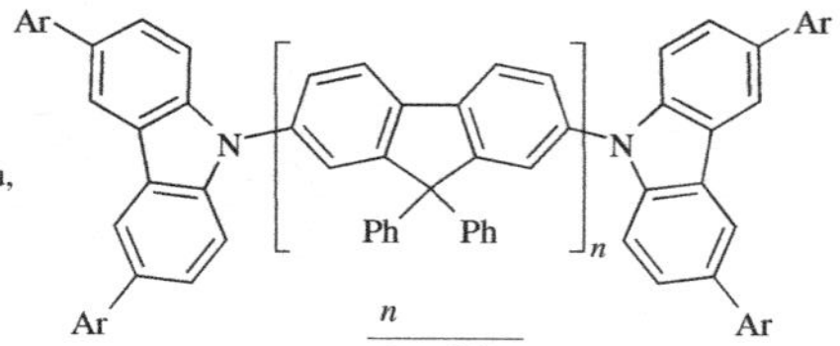

n	
1	(45)
2	(51)
3	(55)

172

384

C_{36–180}

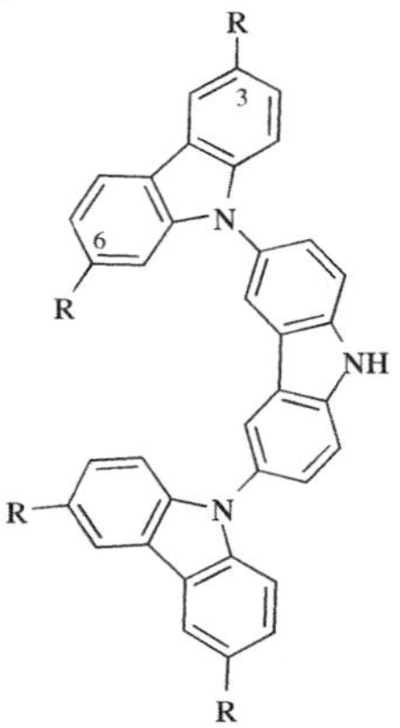

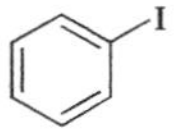

C_{36-180}

CuI (5 mol %),
CDA (10 mol %),
K₃PO₄, dioxane.
110°, 24 h

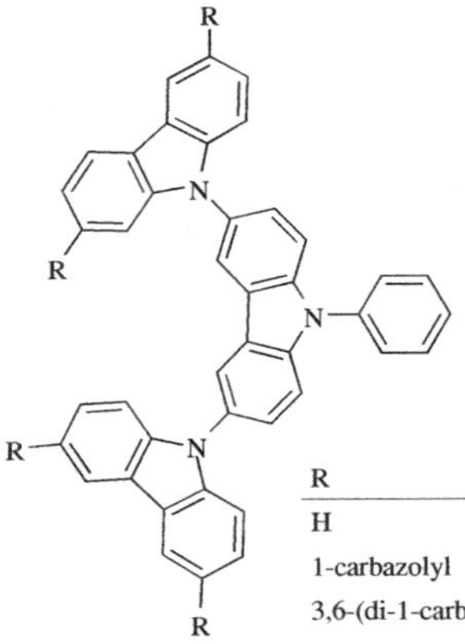

R	
H	(40)
1-carbazolyl	(32)
3,6-(di-1-carbazolyl)-1-carbazolyl	(47)

C_{52}

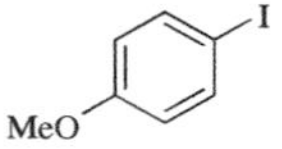

Cu₂O (2 eq), DMA,
160°, 24 h

(76)   463

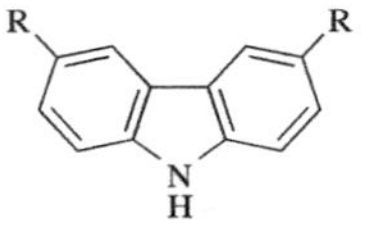

R = 3,6-di(t-Bu)₂-1-carbazolyl

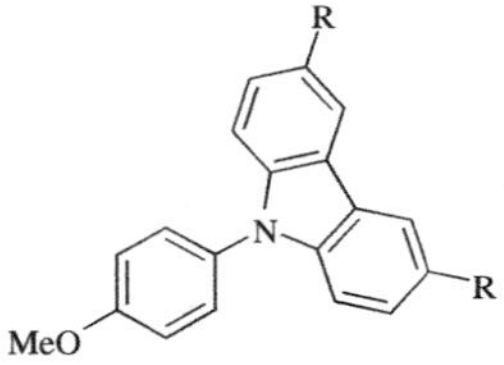

TABLE 14B. N-HETEROARYLATION OF CARBAZOLES

Nitrogen Nucleophile	Heteroaryl Halide	Conditions	Product(s) and Yield(s) (%)	Refs.
C$_{12}$				
		CuO (8 mol %), K$_2$CO$_3$, pyridine, reflux, 26 h	(59)	402
	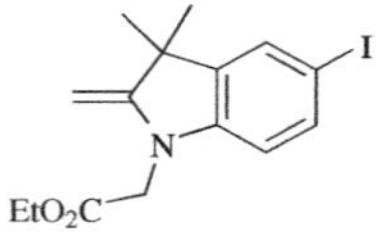	CuI (6 mol %), DMCDA (12 mol %), Cs$_2$CO$_3$, dioxane, 100°, 17 h	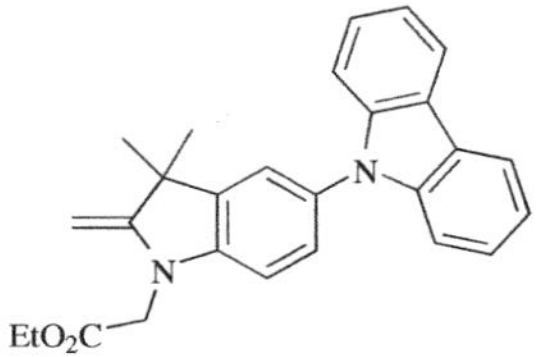 (53)	476

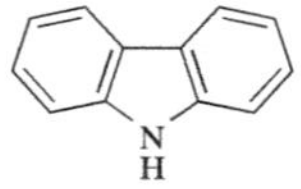

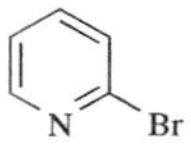

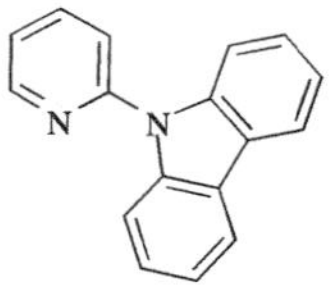

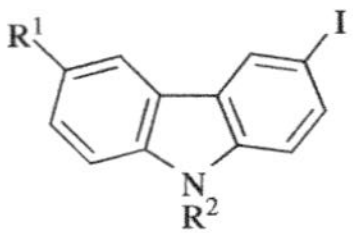

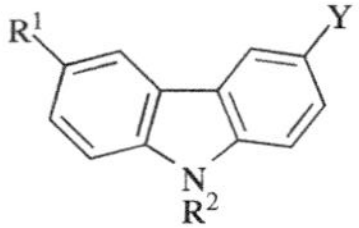

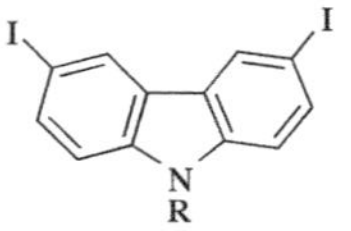

Y = 1-carbazolyl

| | | Catalyst (x eq) | | |

R^1	R^2	Catalyst	x	Additives	Solvent	Temp (°)	Time (h)		
H	n-C$_6$H$_{13}$	Cu	2	18-6-c (10 mol %), K$_2$CO$_3$	1,2-Cl$_2$C$_6$H$_4$	reflux	36	(63)	477
H	Ts	Cu$_2$O	—	none	DMA	170	24	(76)	478
CHO	n-Bu	Cu	2.5	none	DMA	190	24	(63)	479

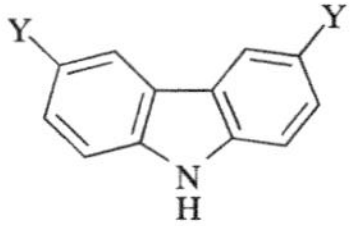

Y = 1-carbazolyl

R	Conditions		
Ac	1. Cu$_2$O (1.5 eq), DMA, 160°, 24 h	(55)	462
	2. KOH, H$_2$O, DMSO, THF		
Boc	1. CuI (10 mol %), CDA (15 mol %), K$_3$PO$_4$, dioxane, 110°, 24 h	(58)	465

See table.

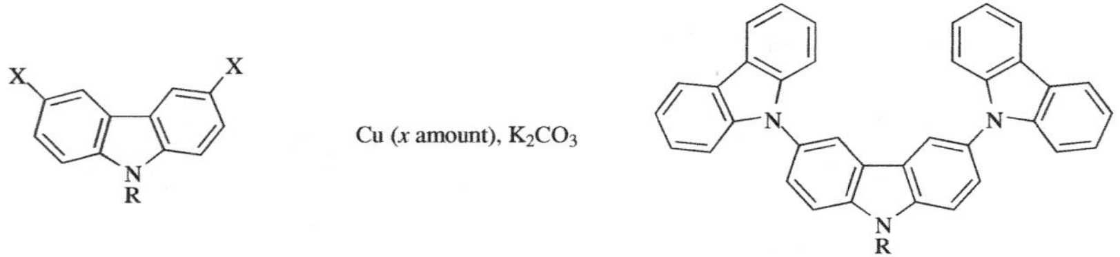

Cu (x amount), K$_2$CO$_3$

R	X	x	Additive	Solvent	Temp (°)	Time (h)		
n-Bu	I	2 eq	18-c-6 (10 mol %)	1.2-Cl$_2$C$_6$H$_4$	reflux	24	(69)	373
Bn	I	—	18-c-6	DMF	170	24	(67)	480
n-C$_8$H$_{17}$	Br	44 mol %	none	PhNO$_2$	reflux	24	(87)	469
n-C$_8$H$_{17}$	I	2 eq	18-c-6 (10 mol %)	1.2-Cl$_2$C$_6$H$_4$	reflux	36	(77)	477
n-C$_{10}$H$_{21}$	I	2 eq	18-c-6 (10 mol %)	1.2-Cl$_2$C$_6$H$_4$	reflux	36	(63)	477

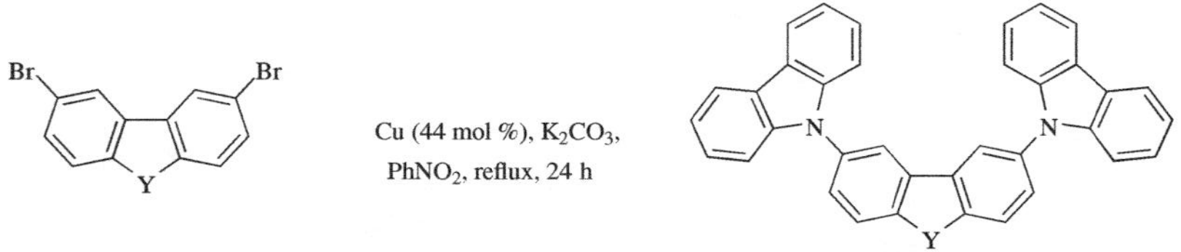

Cu (44 mol %), K$_2$CO$_3$, PhNO$_2$, reflux, 24 h

Y		
O	(70)	469
S	(68)	

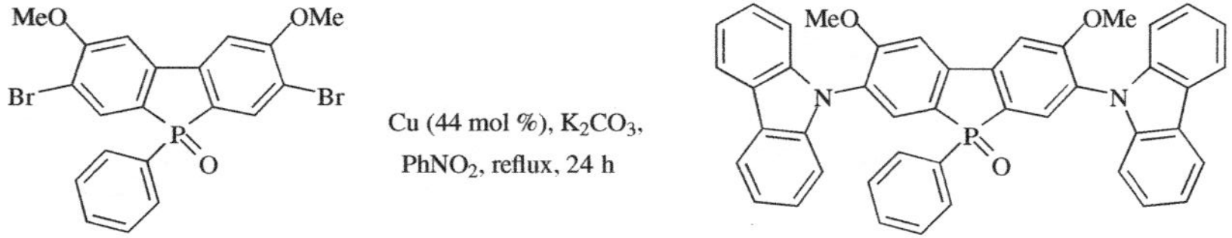

Cu (44 mol %), K$_2$CO$_3$, PhNO$_2$, reflux, 24 h

(55) 469

Nitrogen Nucleophile	Heteroaryl Halide	Conditions	Product(s) and Yield(s) (%)	Refs.
C$_{12}$				
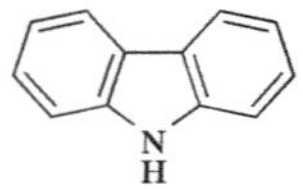	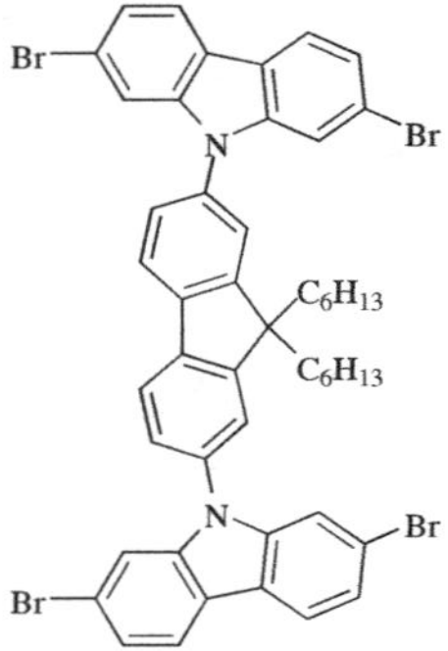	Cu (45 mol %), K$_2$CO$_3$, PhNO$_2$, reflux, 24 h	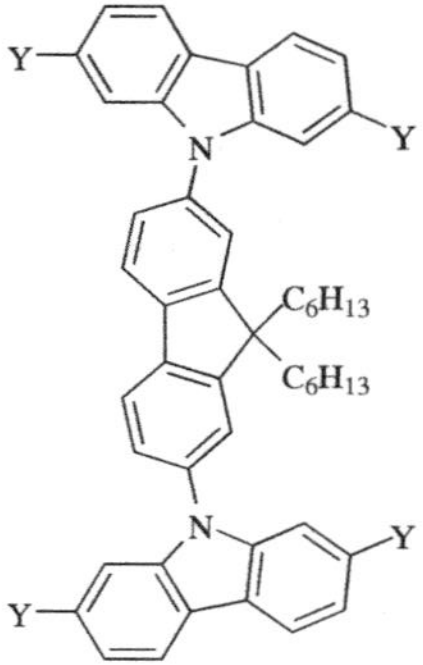 (55) Y = 1-carbazolyl	468
C$_{20}$		Cu (1 eq), K$_2$CO$_3$, PhNO$_2$, 170°, 24 h	(52)	474
	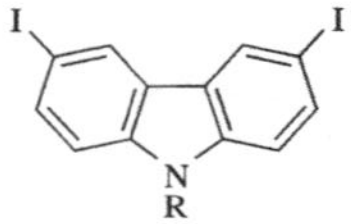	Cu (*x* eq), K$_2$CO$_3$, PhNO$_2$, 170°, 24 h	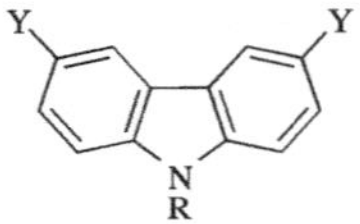 Y = 3,6-(*t*-Bu)$_2$-1-carbazolyl R *x* Bn 1 (82) Ts 1 (66) Ts 2.3 (66)	474 474 481
	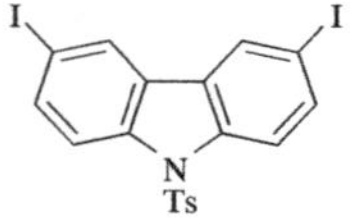	1. Cu$_2$O (2 eq), DMA, 180°, 24 h 2. KOH, H$_2$O, DMSO, THF reflux, 4 h	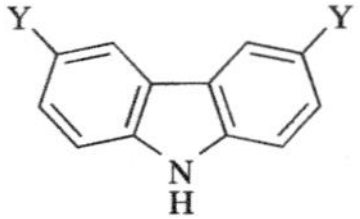 (83) Y = 3,6-(*t*-Bu)$_2$-1-carbazolyl	463

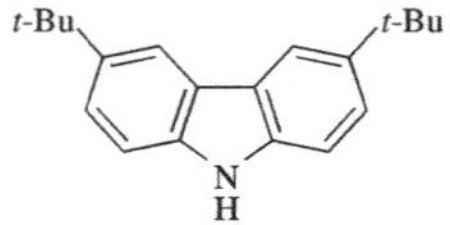

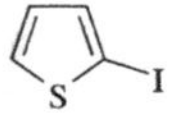

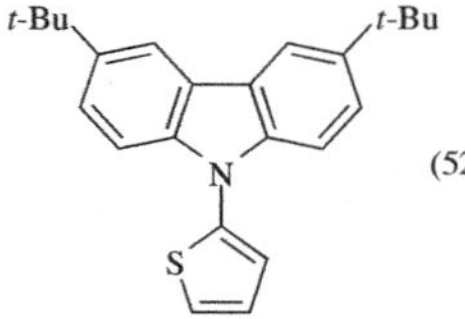

388

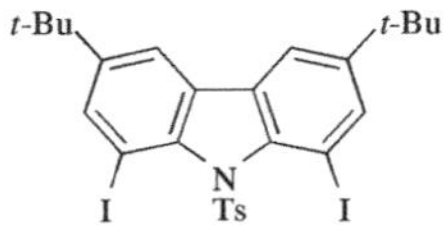

474

474

482

478

(65)

t-Bu

t-Bu

t-Bu

Y = 3,6-(t-Bu)$_2$-1-carbazolyl

(74)

R

R

R

n	
1	(58)
2	(49)
3	(82)
4	(49)

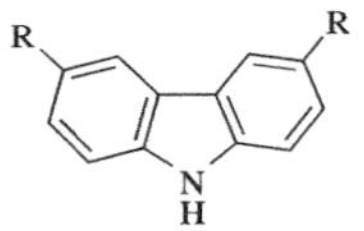

R

R

R

R

n + 1

n	
1	(38)
3	(35)
5	(22)

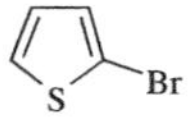

Cu (1 eq), K$_2$CO$_3$, PhNO$_2$, 170°, 24 h

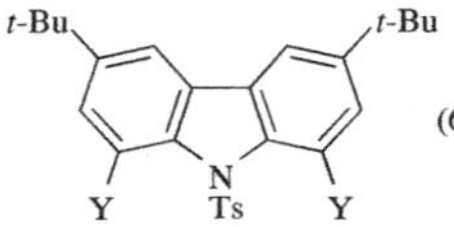

CuI (25 mol %), CDA (2.5 eq), NaOt-Bu, dioxane, 110°, 40 h

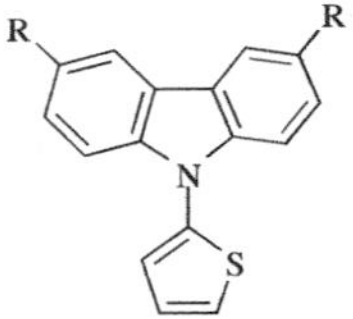

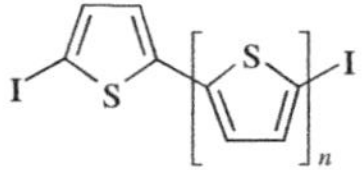

CuI (25 mol %), CDA (2.5 eq), NaOt-Bu, dioxane, 110°, 40 h

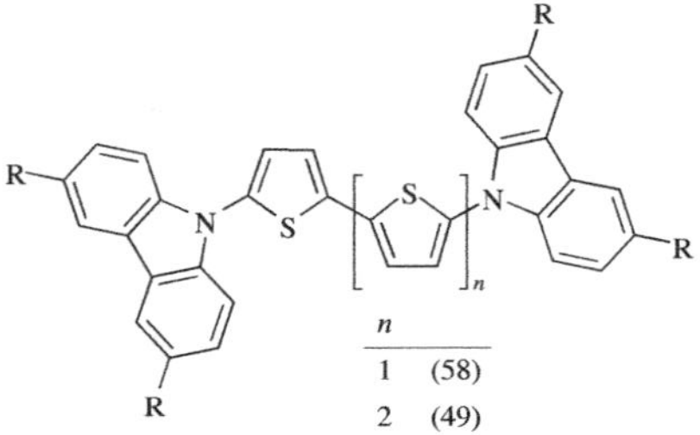

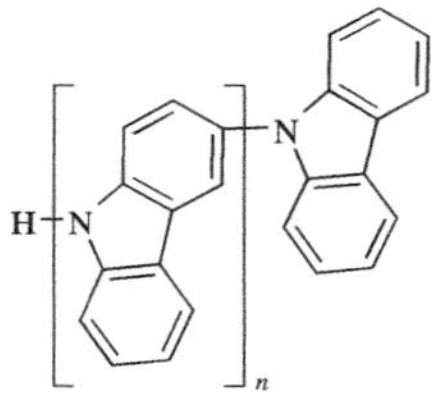

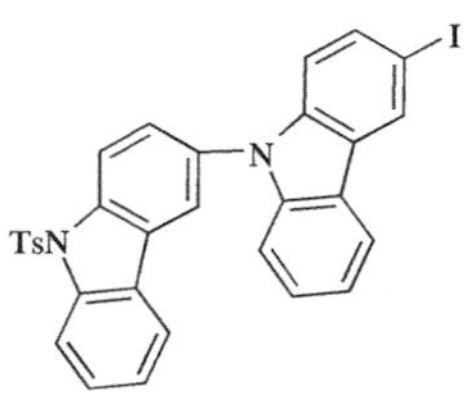

1. Cu$_2$O, DMA, 170°, 24 h
2. KOH, H$_2$O, DMSO, THF, reflux

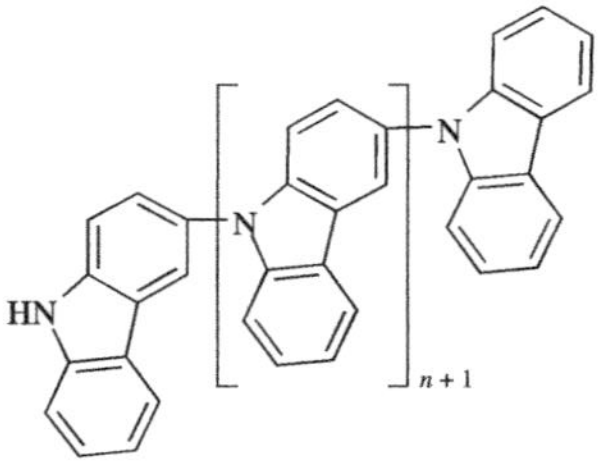

t-Bu

t-Bu

n

n

R

R

R = 4-Ph$_2$NC$_6$H$_4$

C$_{24}$

C$_{24-72}$

TABLE 14B. N-HETEROARYLATION OF CARBAZOLES (Continued)

Nitrogen Nucleophile	Heteroaryl Halide	Conditions	Product(s) and Yield(s) (%)	Refs.
C$_{36}$				
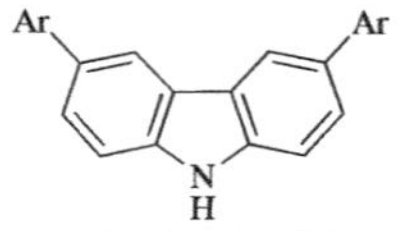 Ar = 1-carbazolyl	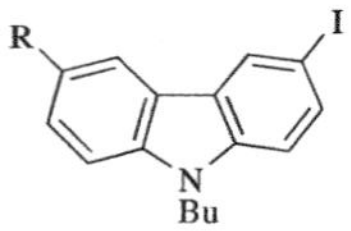	Cu$_2$O (2.5 eq), DMAc, 190°, 24 h	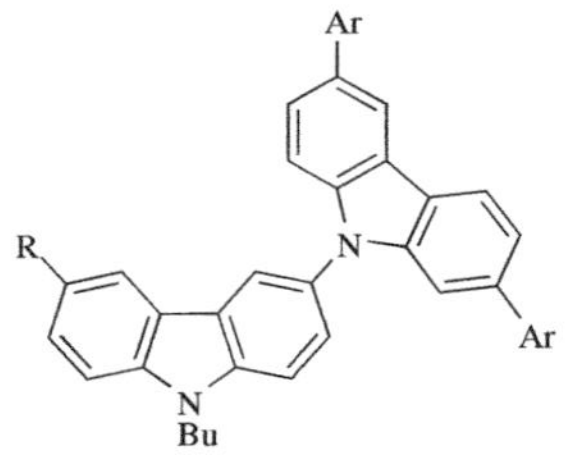 R: H (68), CHO (51)	479
C$_{36-52}$		Cu$_2$O (x eq), DMA, 24 h		
			Z: 4-AcNHC$_6$H$_4$, x = 1.5, Temp 160, (37)	462
			Z: 4-MeOC$_6$H$_4$, x = 2, Temp 160, (64)	463
C$_{36-180}$		1. CuI (10 mol %), CDA (15 mol %), K$_3$PO$_4$, dioxane, 110°, 24 h 2. TFA, H$_2$O, anisole, toluene	R: 1-carbazolyl (75); 3,6-bis(1-carbazolyl)-1-carbazolyl (75); 3,6-bis[3,6-bis(1-carbazolyl)-1-carbazolyl]-1-carbazolyl (47)	466

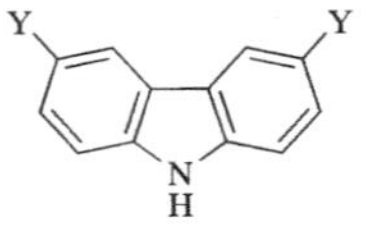

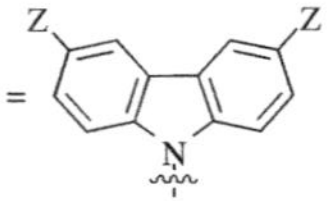

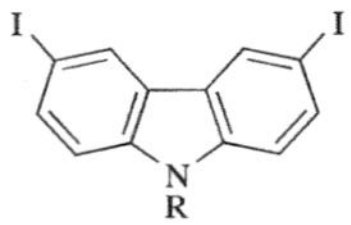

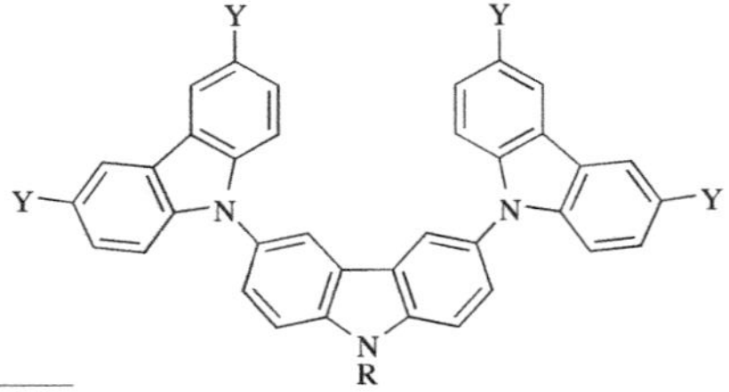

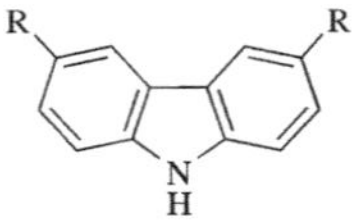

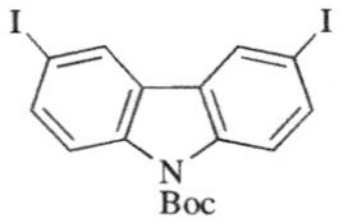

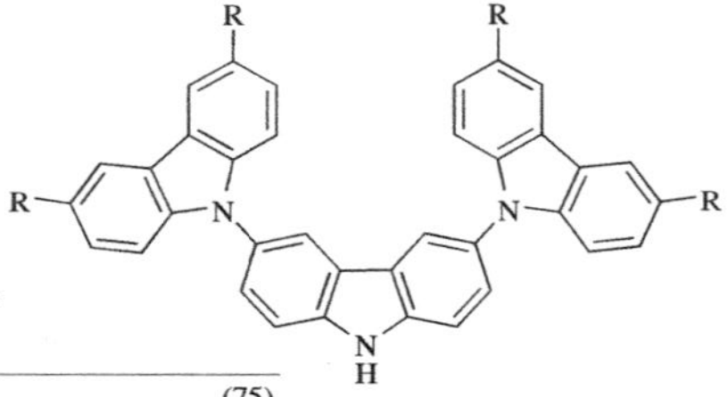

390

TABLE 15A. N-ARYLATION OF MISCELLANEOUS HETEROAROMATIC NITROGEN NUCLEOPHILES

Nitrogen Nucleophile	Aryl Halide	Conditions	Product(s) and Yield(s) (%)	Refs.

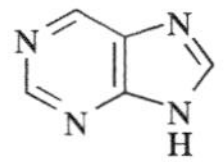

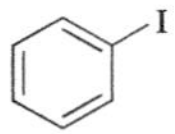

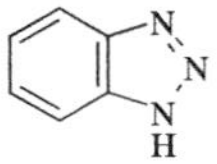

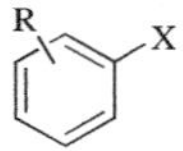

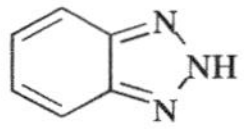

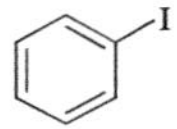

*Please refer to the charts preceding the tables for structures indicated by the **bold** numbers.*

9H-Purines

C₅

| | | CuI (5 mol %), DMCDA (20 mol %), Cs₂CO₃, DMF, 70°, 24 h | 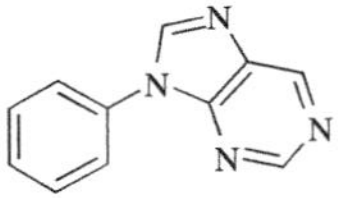 (66) | 128 |

1H-Benzo[d][1,2,3]triazoles

C₆

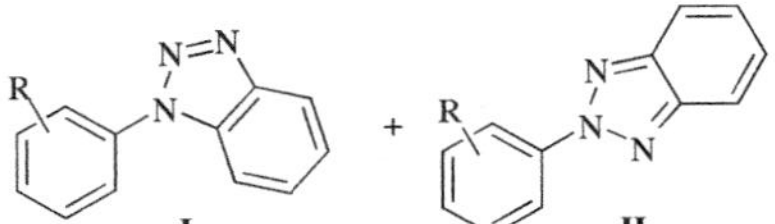

I + II

R	X	Catalyst	x	Additive(s)	Solvent	Temp (°)	Time (h)	I	II	Refs.
H	Br	CuI	10	**L28** (20 mol %), K₂CO₃	DMSO	120	60	(93)	(—)	412
H	I	CuI	5	DMCDA (10 mol %), K₃PO₄	DMF	110	24	(89)	(—)	128
H	I	Cu₂O	5	phen (10 mol %), TBAF	—	145	48	(19)	(—)	68
H	I⁺Ph BF₄⁻	Cu(acac)₂	100	K₂CO₃	toluene	50	6	(80)	(—)	76
4-O₂N	I	CuBr₂	10	TBAF	—	145	24	(35)	(13)	81
4-O₂N	Cl	CuBr₂	10	TBAF	—	145	15	(38)	(30)	81
4-O₂N	Cl	CuI	10	**L26** (10 mol %), NaOMe	DMSO	130	12	(65)	(—)	397
4-Me	I⁺C₆H₄Me-4 BF₄⁻	Cu(acac)₂	100	K₂CO₃	toluene	50	6	(94)	(—)	76
2-O₂N, 4-CF₃	Cl	CuI	10	**L21** (20 mol %), Cs₂CO₃	MeCN	reflux	12	(95)	(—)	397
2-O₂N, 4-CF₃	Cl	CuI	10	**L21** (20 mol %), Cs₂CO₃	MeCN	reflux	12	(89)	(—)	394
4-Ac	I	CuI	5	DMCDA (20 mol %), K₃PO₄	DMF	110	24	(67)	(—)	128

2H-Benzo[d][1,2,3]triazoles

C₆

| | | CuCl (10 mol %), L-Pro (20 mol %), K₂CO₃, DMSO, MW, 160° | (60) + Ph–N (8) | 443 |

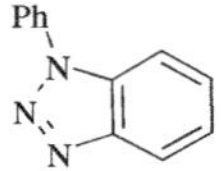

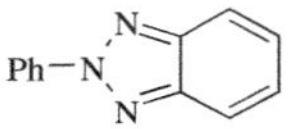

TABLE 15A. *N*-ARYLATION OF MISCELLANEOUS HETEROAROMATIC NITROGEN NUCLEOPHILES (*Continued*)

Nitrogen Nucleophile	Aryl Halide	Conditions	Product(s) and Yield(s) (%)	Refs.

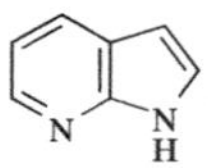

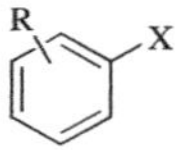

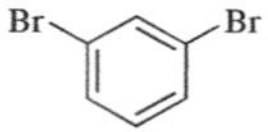

Please refer to the charts preceding the tables for structures indicated by the bold numbers.

2*H*-Tetrazoles

C₇

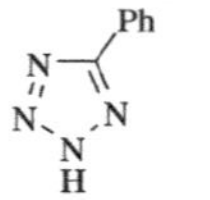	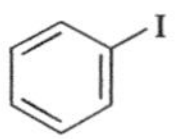	Cu₂O (5 mol %), L30 (20 mol %), DMF, 110°, 24 h	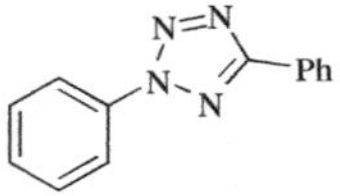 (0)	40

1*H*-Pyrrolo[3.2-*b*]pyridines

C₇

Catalyst (x mol %)

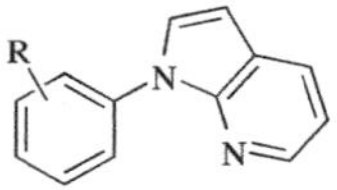

R	X	Catalyst	x	Additive(s)	Solvent	Temp (°)	Time (h)		Refs.
H	I	CuO	0.001	DMEDA (20 mol %), K₃PO₄•H₂O	toluene	135	24	(86)	425
4-F	I	CuI	10	LiCl, K₂CO₃	DMF	120	24	(75)	483
3,5-F₂	I	CuI	10	LiCl, K₂CO₃	DMF	120	24	(78)	483
4-Cl	I	CuI	5	DMEDA (10 mol %), CsF	THF	rt	24	(88)	63
4-Br	I	CuI	10	LiCl, K₂CO₃	DMF	120	24	(60)	483
4-Br	Br	CuSO₄	5	K₂CO₃	—	220	6	(43)	76, 484
2-EtO₂C	I	CuI	10	LiCl, K₂CO₃	DMF	120	36	(92)	483
3,5-Me₂	I	CuI	1	CDA (10 mol %), K₃PO₄	dioxane	110	24	(99)	115

		CuSO₄ (2 mol %), K₂CO₃, 210°, 6 h	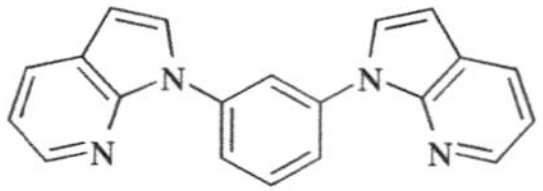 (85)	437

392

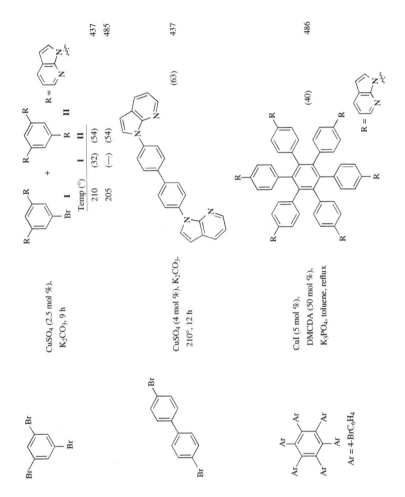

Temp (°)	I	II
210	(32)	(54)
205	(—)	(54)

(63)

(40)

CuSO₄ (2.5 mol %),
K₂CO₃, 9 h

$CuSO_4$ (2.5 mol %),
K_2CO_3, 9 h

CuSO₄ (4 mol %), K₂CO₃,
210°, 12 h

CuI (5 mol %),
DMCDA (50 mol %),
K₃PO₄, toluene, reflux

Ar = 4-BrC₆H₄

437
485

437

486

TABLE 15A. N-ARYLATION OF MISCELLANEOUS HETEROAROMATIC NITROGEN NUCLEOPHILES (*Continued*)

Nitrogen Nucleophile	Aryl Halide	Conditions	Product(s) and Yield(s) (%)	Refs.

*Please refer to the charts preceding the tables for structures indicated by the **bold** numbers.*

1*H*-Pyrrolo[3,2-c]pyridines

C₇

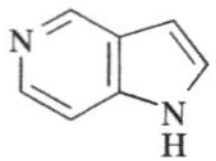

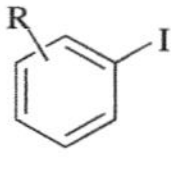

CuI (10 mol %),
LiCl, K₂CO₃,
DMF, 120°, 24 h

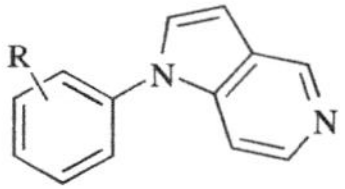

R	
H	(70)
4-Cl	(85)
4-MeO	(65)
2-EtO₂C	(88)

483

4,5,6,7-Tetrahydro-4,7-methano-2-indazoles

C₁₁

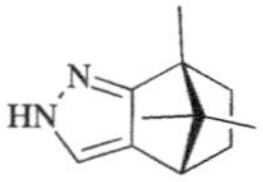

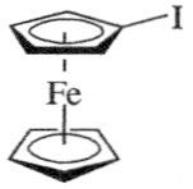

CuI (1 eq), K₂CO₃,
DMSO, 90°, 18 h

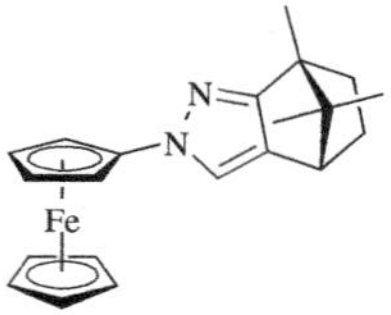

(70)

372

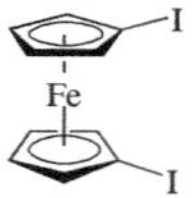

CuI (1 eq), K₂CO₃,
DMSO, 90°, 18 h

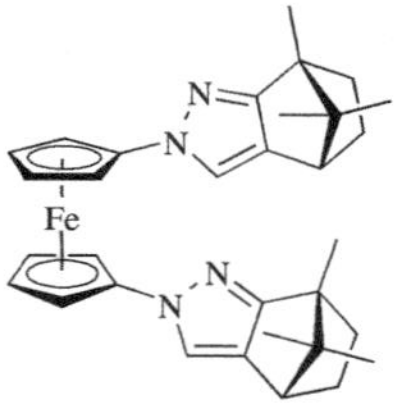

(31)

372

2,3,4,5-Tetrahydro-1H-pyrido[4,3-b]indoles

C$_{11-12}$

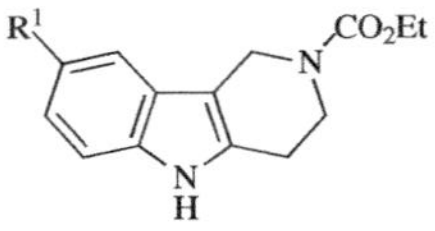

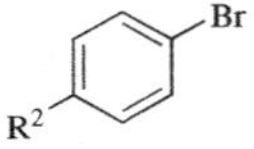

CuBr (10 mol %), Na$_2$CO$_3$, DMF, 200°, 8 h

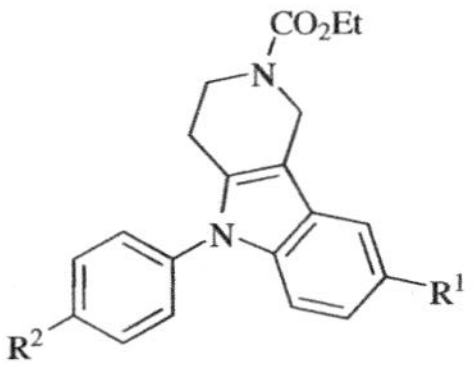

R^1	R^2	
H	H	(40)
F	H	(61)
Cl	H	(65)
Br	H	(46)
MeO	H	(47)
Me	H	(62)
F	F	(78)
Cl	F	(69)
MeO	F	(47)

487

1H-Perimidines

C$_{11}$

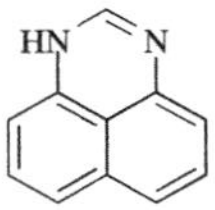

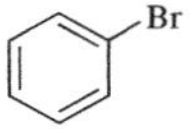

CuI (10 mol %), L22 (10 mol %), K$_2$CO$_3$, DMF, 110°, 24 h

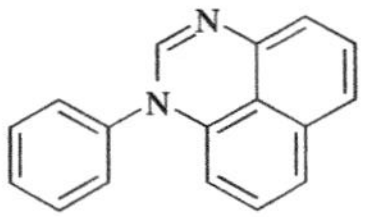

(53)

79

2,3,4,9-Tetrahydro-1H-carbazoles

C$_{12}$

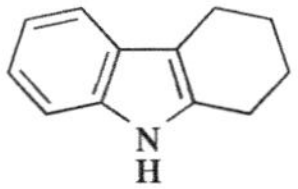

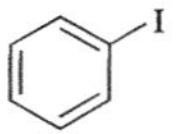

Catalyst (x mol %), reflux

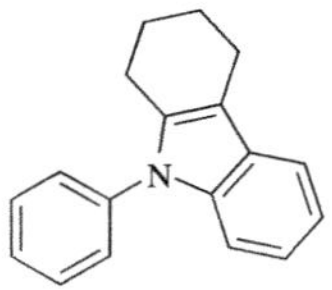

Catalyst	x	Additive(s)	Solvent	Time (h)	
Cu	20	Cs$_2$CO$_3$	n-PrCN	10	(95)
CuI	10	TM-BINAM (10 mol %), K$_2$CO$_3$	MeCN	36	(72)

129
446

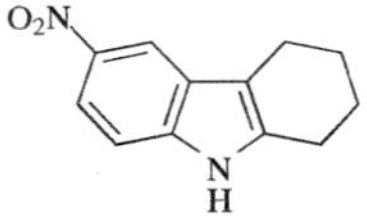

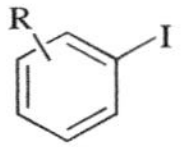

R = 2-EtO$_2$C, 3-EtO$_2$C, 4-EtO$_2$C

CuI, DMEDA, K$_2$CO$_3$, toluene, reflux

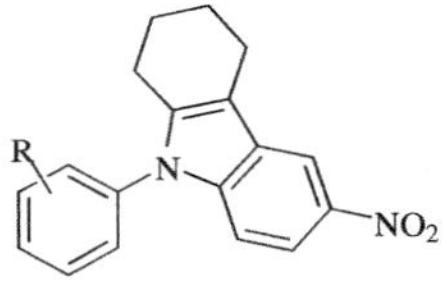

(—)

488

395

TABLE 15B. *N*-HETEROARYLATION OF MISCELLANEOUS HETEROAROMATIC NITROGEN NUCLEOPHILES

Nitrogen Nucleophile	Heteroaryl Halide	Conditions	Product(s) and Yield(s) (%)	Refs.

*Please refer to the charts preceding the tables for structures indicated by the **bold** numbers.*

1*H*-Benzo[*d*][1,2,3]triazoles

C₆

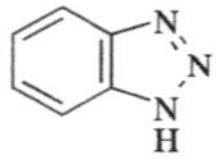

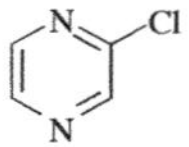

Catalyst (10 mol %)

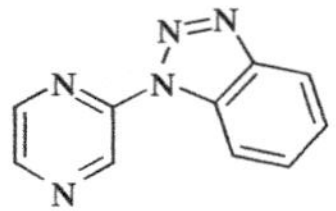

Catalyst	Additive(s)	Solvent	Temp (°)	Time (h)		
CuBr₂	TBAF	—	145	2.5	(76)	81
CuI	**L26** (10 mol %), NaOMe	DMSO	130	12	(94)	397

1*H*-Pyrrolo[3,2-*b*]pyridines

C₇

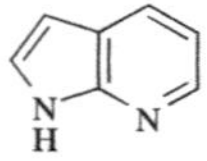

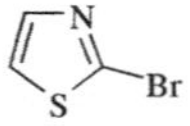

CuI (10 mol %), LiCl, K₂CO₃, DMF, 120°, 48 h

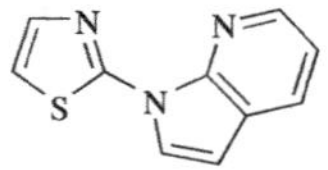

(40)

483

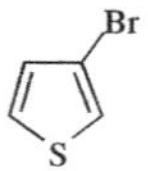

CuI (10 mol %), LiCl, K₂CO₃, DMF, 120°, 48 h

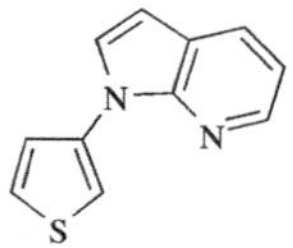

(40)

483

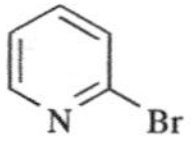

CuI (*x* mol %)

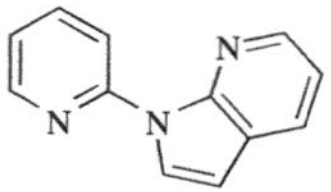

x	Additives	Solvent	Temp (°)	Time (h)		
10	LiCl, K₂CO₃	DMF	120	36	(85)	483
20	D-glucosamine (40 mol %), Cs₂CO₃	DMSO	110	22	(64)	467

396

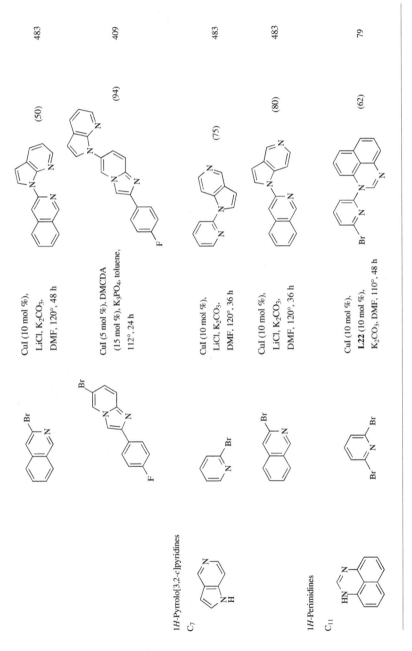

483

409

483

483

79

(50)

(94)

(75)

(80)

(62)

CuI (10 mol %), LiCl, K₂CO₃, DMF, 120°, 48 h

CuI (5 mol %), DMCDA (15 mol %), K₃PO₄, toluene, 112°, 24 h

CuI (10 mol %), LiCl, K₂CO₃, DMF, 120°, 36 h

CuI (10 mol %), LiCl, K₂CO₃, DMF, 120°, 36 h

CuI (10 mol %), **L22** (10 mol %), K₂CO₃, DMF, 110°, 48 h

1*H*-Pyrrolo[3,2-*c*]pyridines

C₇

1*H*-Perimidines

C₁₁

TABLE 16A. N-ARYLATION OF PRIMARY AMIDES

*Please refer to the charts preceding the tables for structures indicated by the **bold** numbers.*

Nitrogen Nucleophile	Aryl Halide	Conditions	Product(s) and Yield(s) (%)	Refs.

C$_{1-8}$

Nitrogen Nucleophile: H_2N–C(=O)–R

Aryl Halide: (octahydrophenanthrene with OMe and I, MeO$_2$C)

Conditions: CuI (1 mol %), DMEDA (2 mol %), K$_2$CO$_3$, dioxane, 100°, 24 h

Product I (amide-substituted octahydrophenanthrene, OMe, MeO$_2$C):

R	
H	(82)
Me	(84)
Et	(81)
n-Pr	(84)
Ph	(80)
Bn	(92)

Refs. 489

C$_2$

Nitrogen Nucleophile: NH$_2$Ac

Aryl Halide: C$_6$H$_5$–X

Conditions: Catalyst (x amount)

Products: I (PhNHAc) + II (Ph–N(Ac)–Ph)

X	Catalyst	x	Additive(s)	Solvent	Temp (°)	Time (h)	I	II	Refs.
Br	Cu	3 eq	none	—	210	12	(79)	(0)	303
I	Cu	3 eq	none	—	180	12	(64)	(0)	303
I	CuBr	10 mol %	**L9** (20 mol %), Cs$_2$CO$_3$	DMSO	75	28	(84)	(0)	49
I	CuI	10 mol %	K$_3$PO$_4$	DMF	110	24	(65)	(0)	387
I	CuI	5 mol %	KOH, TBAB	—	110	6	(87)	(0)	428
I	CuI	10 mol %	phen (10 mol %), **KF/Al$_2$O$_3$**	toluene	110	5	(97)	(0)	490
I	CuI	5 mol %	Gly (20 mol %), K$_3$PO$_4$	dioxane	100	24	(95)	(0)	136
I	CuI	15 mol %	L-Pro (15 mol %), KF/Al$_2$O	toluene	110	5	(80)	(0)	491
I	CuI	10 mol %	**PPAPM** (20 mol %), K$_3$PO$_4$	—	110	30	(89)	(0)	48
I	CuI	7.5 mol %	**DMEDA** (7.5 mol %), KF/Al$_2$O$_3$	toluene	110	2	(95)	(0)	137
I	CuI	10 mol %	oxazolidin-2-one (20 mol %), Na$_2$CO$_3$	DMSO	120	24	(88)	(0)	492
Cl	CuI	10 mol %	piperidine-2-carboxylic acid (20 mol %), K$_2$CO$_3$	DMF	110	30	(24)	(0)	67
Br	CuI	10 mol %	piperidine-2-carboxylic acid (20 mol %), K$_2$CO$_3$	DMF	110	36	(78)	(0)	67
I	CuI	10 mol %	piperidine-2-carboxylic acid (20 mol %), K$_2$CO$_3$	DMF	110	36	(86)	(0)	67
I	CuI	5 mol %	**L43** (10 mol %), K$_3$PO$_4$	DMSO	110	5	(50)	(11)	138
I	Cu$_2$O	5 mol %	**L42** (10 mol %), K$_3$PO$_4$	DMSO	110	5	(76)	(0)	493
I	Cu$_2$O nanoparticles	5 mol %	KOH	PEG 4000	120	3	(73)	(0)	225
I	Cu$_2$O	5 mol %	Cs$_2$CO$_3$	NMP	80	18	(88)	(0)	266
Br	Cu$_2$O	5 mol %	NaOt-Bu	NMP	110	18	(72)	(0)	266
I	Cu$_2$O	5 mol %	**L30** (20 mol %), Cs$_2$CO$_3$, 3 Å MS	DMF	82	75	(81)	(7)	40

R	X	x	Additive(s)	Solvent	Temp (°)	Time (h)		
2-Cl	I	10	oxazolidin-2-one (20 mol %), K$_2$CO$_3$	DMSO	120	24	(80)	492
4-Br	I	10	PPAPM (20 mol %), K$_3$PO$_4$	—	110	20	(94)	48
4-Br	I	5	Gly (20 mol %), K$_3$PO$_4$	dioxane	100	24	(91)	136
2-NO$_2$	I	5	Gly (20 mol %), K$_3$PO$_4$	dioxane	100	24	(88)	136
2-MeO	I	1	CDA (10 mol %), K$_3$PO$_4$	dioxane	90	23	(94)	115
2-MeO	I	7.5	DMEDA (7.5 mol %), KF/Al$_2$O$_3$	toluene	110	4	(90)	137
2-MeO	I	15	L-Pro (15 mol %), KF/Al$_2$O$_3$	toluene	110	5.5	(70)	137
4-MeO	Br	20	DMG (20 mol %), K$_3$PO$_4$	DMF	reflux	48	(85)	494
4-MeO	I	7.5	DMEDA (7.5 mol %), KF/Al$_2$O$_3$	toluene	110	2	(90)	137
4-MeO	I	0.2	CDA (5 mol %), K$_3$PO$_4$	dioxane	110	23	(99)	115
4-MeO	I	5	Gly (20 mol %), K$_3$PO$_4$	dioxane	100	24	(90)	136
4-MeO	I	15	L-Pro (15 mol %), KF/Al$_2$O$_3$	toluene	110	5.5	(90)	491
4-MeO	I	10	phen (10 mol %), KF/Al$_2$O$_3$	toluene	110	9.5	(90)	490
4-MeO	I	10	oxazolidin-2-one (20 mol %), K$_2$CO$_3$	DMSO	120	24	(80)	492
2-Me	I	15	L-Pro (15 mol %), KF/Al$_2$O$_3$	toluene	110	6	(0)	491
2-Me	I	1	DMEDA (10 mol %), K$_3$PO$_4$	DMF	80	23	(95)	69
2-Me	I	7.5	DMEDA (7.5 mol %), KF/Al$_2$O$_3$	toluene	110	4	(90)	137
4-Me	I	5	Gly (20 mol %), K$_3$PO$_4$	dioxane	100	24	(92)	136
4-Me	I	7.5	DMEDA (7.5 mol %), KF/Al$_2$O$_3$	toluene	110	2	(90)	137
4-Me	I	15	L-Pro (15 mol %), KF/Al$_2$O$_3$	toluene	110	5	(90)	491
4-Ac	Br	20	DMG (20 mol %)	DMF	reflux	48	(86)	494
4-Ac	I	5	Gly (20 mol%)	dioxane	100	24	(62)	136

1. CuI (15 mol %),
   DMG (20 mol %), K$_3$PO$_4$,
   DMSO, 80°, 24 h
2. ROH, 120°, 48 h

R	
Ph	(65)
4-ClC$_6$H$_4$	(60)
2-MeOC$_6$H$_4$	(65)
4-MeOC$_6$H$_4$	(70)
2-Np	(66)

495

TABLE 16A. N-ARYLATION OF PRIMARY AMIDES (Continued)

Nitrogen Nucleophile	Aryl Halide	Conditions	Product(s) and Yield(s) (%)	Refs.

Please refer to the charts preceding the tables for structures indicated by the **bold** numbers.

C₂

NH₂Ac

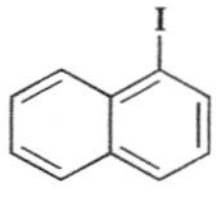

(structure: naphthalene with I)

CuI (10 mol %),
oxazolidin-2-one
(20 mol %), K₂CO₃,
DMSO, 120°, 24 h

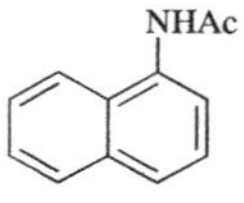

(structure: naphthalene with NHAc) (76)

492

(structure: ferrocene with I)

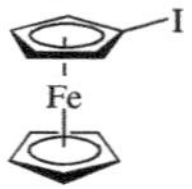

CuI (1 eq), K₂CO₃, DMSO,
90°, 18 h

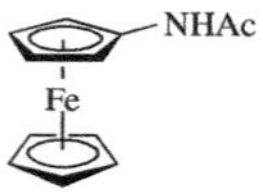

(structure: ferrocene with NHAc) (70)

372

(structure, H₂N–C(=O)–CH₂F)

(structure: 4-chloro-iodobenzene)

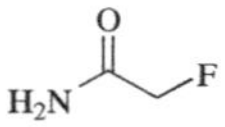

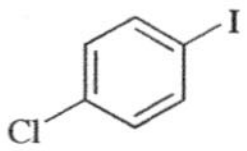

CuI (5 mol %),
DMEDA (10 mol %),
CsF; THF, rt, 24 h

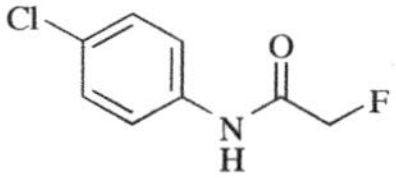

(structure: Cl-phenyl-NH-C(=O)-CH₂F) (98)

63

(structure, H₂N–C(=O)–CF₃)

(structure: R-phenyl-X)

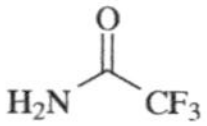

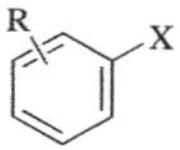

CuI (5 mol %),
DMEDA (10 mol %),
K₂CO₃, 4 Å MS,
DMF, 36 h

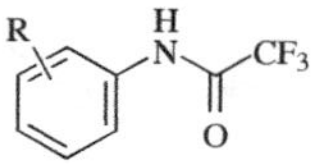

(structure: R-phenyl-NH-C(=O)-CF₃)

R	X	Temp (°)	
H	Br	75	(96)
H	I	45	(95)
4-Br	I	45	(94)
2-H₂N	I	45	(6)
4-Me₂N	Br	75	(99)
4-O₂N	I	45	(94)
2-MeO	I	45	(26)
4-MeO	Br	75	(93)
4-MeO	I	45	(93)
2-Me	I	45	(15)
4-Me	I	45	(94)
3-NC–	Br	75	(83)
4-NC–	Br	75	(84)
4-EtO₂C	I	45	(83)
4-Ac	Br	75	(81)
4-Ac	I	45	(90)

132

400

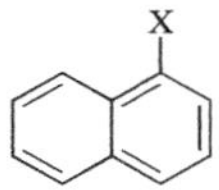

C₃

Entry 1

ArX (1-X-naphthalene)

CuI (10 mol %), DMEDA (10 mol %), K₃PO₄, 4 Å MS, DMF, 36 h

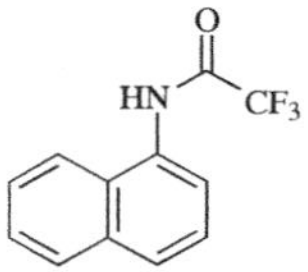

Product: N-(naphthalen-1-yl)-2,2,2-trifluoroacetamide (HN–C(O)CF₃ on naphthalene)

X	Temp (°)	
Br	75	(61)
I	45	(21)

132

Entry 2

ArX

1. CuI (5 mol %), K₃PO₄, DMEDA (10 mol %), 4 Å MS, DMF, temp, 36 h
2. MeOH/H₂O, rt, 12 h

ArNH₂

Ar	X	Temp (°)	
Ph	Br	75	(93)
Ph	I	45	(91)
4-BrC₆H₄	I	45	(80)
4-O₂NC₆H₄	Br	75	(93)
4-MeOC₆H₄	Br	75	(87)
4-MeC₆H₄	Br	75	(91)
4-MeC₆H₄	I	45	(95)
3-NCC₆H₄	Br	75	(95)
4-NCC₆H₄	Br	75	(99)
4-EtO₂CC₆H₄	I	45	(99)
4-AcC₆H₄	I	45	(98)
4-PhC₆H₄	Br	75	(93)
1-Np	Br	75	(50)

132

Entry 3

(2-chlorophenyl pyrroline–quinoline)

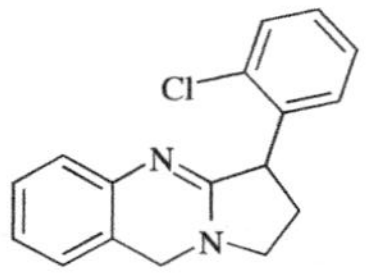

1. CuI (25 mol %), DCDA (50 mol %), K₂CO₃, dioxane, 140°, 21 h
2. HCl, H₂O

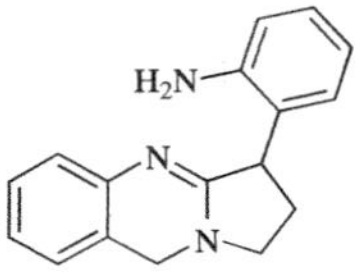

H₂N-substituted product (37)

496

Entry 4

(1-iodo-4-chlorobenzene)

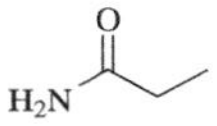

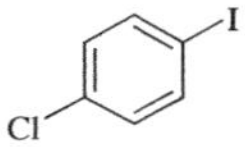

CuI (5 mol %), DMEDA (10 mol %), CsF, THF, rt, 24 h

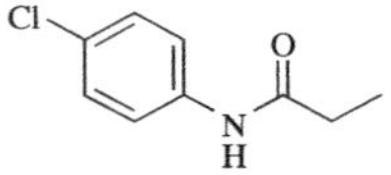

propanamide (H₂N–C(O)Et)

N-(4-chlorophenyl)propanamide (94)

63

TABLE 16A. N-ARYLATION OF PRIMARY AMIDES (Continued)

Nitrogen Nucleophile	Aryl Halide	Conditions	Product(s) and Yield(s) (%)	Refs.
*Please refer to the charts preceding the tables for structures indicated by the **bold** numbers.*				
C₃				
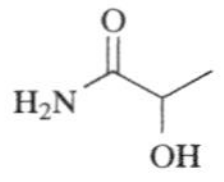 (H₂N–CO–CH(OH)CH₃)	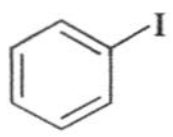 (Ph–I)	CuI (1 mol %), DMEDA (10 mol %), K₃PO₄, DMF, 60°, 23 h	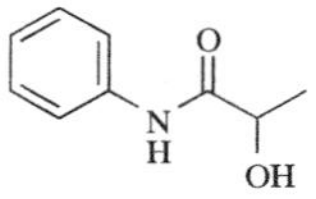 (90)	69
C₄				
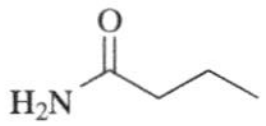 (H₂N–CO–CH₂CH₂CH₃)	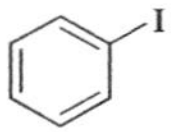 (Ph–I)	CuO (0.1 mol %), DMEDA (20 mol %), K₂CO₃, toluene, 135°, 24 h	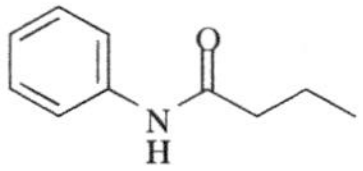 (56)	222
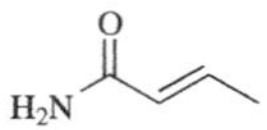 (H₂N–CO–CH=CH–CH₃)	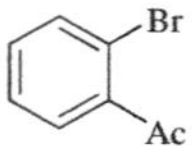 (2-Ac-C₆H₄–Br)	CuI (10 mol %), DMCDA (20 mol %), K₂CO₃, 5 Å MS, 110°, 24 h	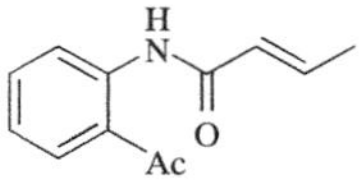 (68)	497
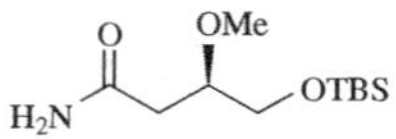 (H₂N–CO–CH₂–CH(OMe)–CH₂OTBS)	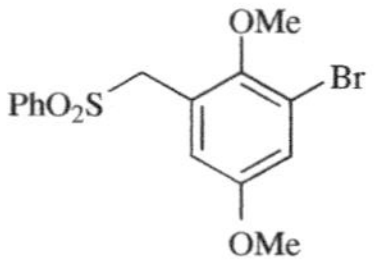 (2,5-(OMe)₂-3-(CH₂SO₂Ph)-C₆H₂–Br)	CuI (20 mol %), DMEDA (40 mol %), K₂CO₃, toluene, 110°	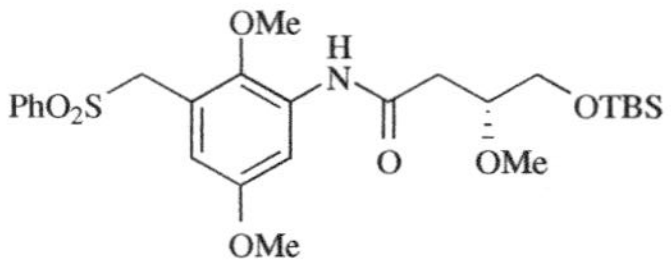 (82)	160
C₅				
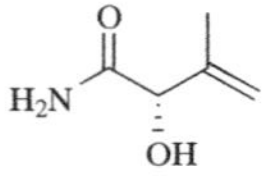 (H₂N–CO–CH(OH)–C(=CH₂)CH₃)	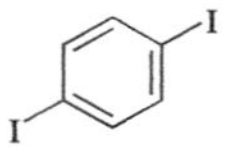 (1,4-diiodobenzene)	CuI (1 eq), DMEDA (2 eq), K₃PO₄, dioxane, 50°, 22 h	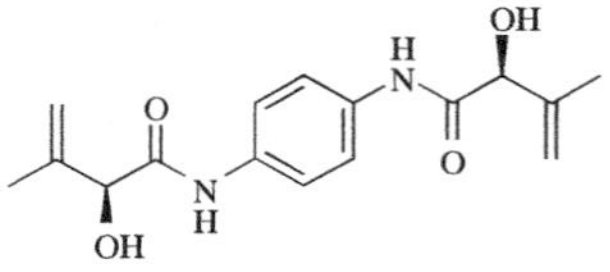 (63)	498
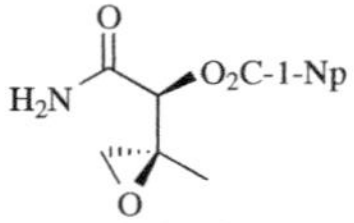 (H₂N–CO–CH(O₂C-1-Np)–C(CH₃)(O epoxide))	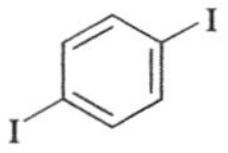 (1,4-diiodobenzene)	CuI (1 eq), DMEDA (2 eq), K₃PO₄, dioxane, 50°, 22 h	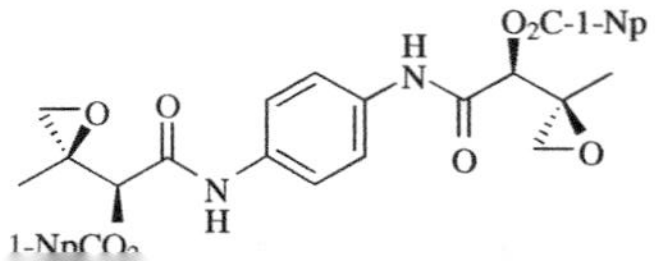 (10)	498

402

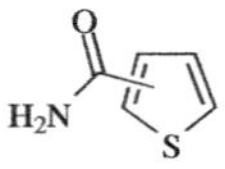

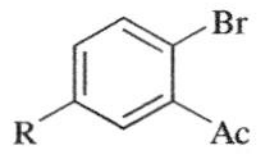

C$_6$

CuI (10 mol %),
DMEDA (20 mol %),
K$_2$CO$_3$, 110°, 24 h

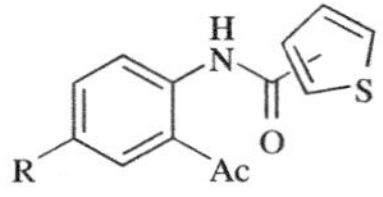

Isomer	R	Additive	
2	H	none	(81)	497
2	MeO	5 Å MS	(72)	428
3	H	none	(81)	225

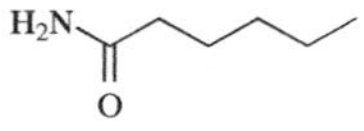

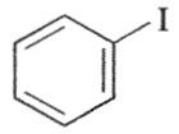

Catalyst (5 mol %), KOH

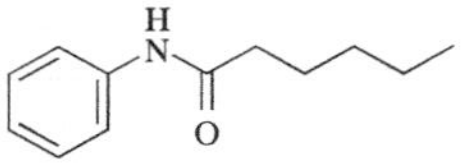

Catalyst	Additive(s)	Solvent	Temp (°)	Time (h)		
CuI	TBAB	—	110	4	(75)	428
Cu$_2$O nanoparticles	none	PEG 4000	120	3	(75)	225

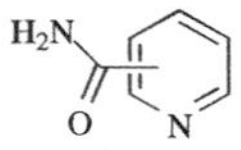

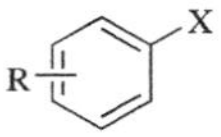

Catalyst (10 mol %)

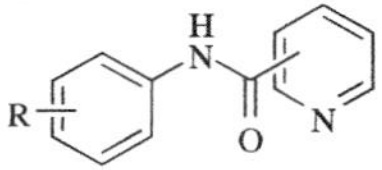

Isomer	R	X	Catalyst	Additives	Solvent	Temp (°)	Time (h)		
2	4-MeO	I	CuI	EDA (10 mol %), K$_3$PO$_4$	dioxane	110	24	(59)	405
2	3,5-Me$_2$	I	CuI	EDA (10 mol %), K$_3$PO$_4$	dioxane	110	24	(50)	405
2	2-Ac, 5-F	Br	CuI	DMEDA (20 mol %), K$_2$CO$_3$, 5 Å MS	—	110	24	(77)	497
3	H	I	CuBr	L9 (20 mol %), Cs$_2$CO$_3$	DMSO	60	24	(83)	49
3	2-Ac	Br	CuI	DMEDA (20 mol %), K$_2$CO$_3$, 5 Å MS	—	110	24	(86)	497
4	2-Ac	Br	CuI	DMEDA (20 mol %), K$_2$CO$_3$	—	110	24	(80)	497
4	2-Ac	Br	CuI	DMEDA (20 mol %), K$_3$PO$_4$, 5 Å MS	—	110	42	(71)	497

C$_7$

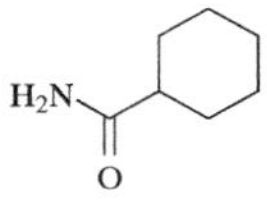

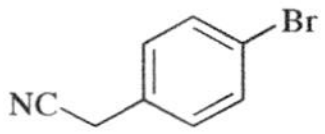

CuI (5 mol %),
DMEDA (10 mol %),
K$_2$CO$_3$, toluene, 110°, 24 h

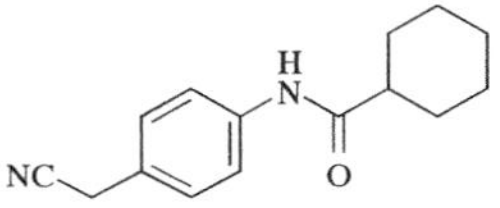

(83)   69

403

TABLE 16A. N-ARYLATION OF PRIMARY AMIDES (Continued)

*Please refer to the charts preceding the tables for structures indicated by the **bold** numbers.*

Nitrogen Nucleophile: H₂N–C(=O)–Ph

Aryl Halide: Ph–X

Product(s): PhNH–C(=O)–Ph

C₇

X	Catalyst	x	Additive(s)	Solvent(s)	Temp (°)	Time (h)	Yield(s) (%)	Refs.
Br	Cu	—	K₂CO₃	—	reflux	2	(50)	2
Br	Cu	300	none	—	210	12	(79)	303
I	Cu	300	none	—	180	12	(64)	303
I⁺Ph BF₄⁻	CuI	5	K₂CO₃	toluene	50	6	(42)	76
I	CuI	5	Gly (20 mol %), K₃PO₄	dioxane	100	24	(97)	136
I	CuI	5	KOH, TBAB	—	110	5	(75)	428
Br	CuI	20	DMG (20 mol %), K₃PO₄	DMF	reflux	48	(84)	494
I	CuI	5	EDA (5 mol %), K₂CO₃	PEG-400	80	8	(75)	301
I	CuI	7.5	DMEDA (7.5 mol %), KF/Al₂O₃	toluene	110	1.5	(95)	137
I	CuI	10	oxazolidin-2-one (20 mol %), NaOMe	DMSO	120	24	(70)	492
I	CuI	10	phen (10 mol %), KF/Al₂O₃	toluene	110	1.5	(99)	490
Br	CuI	15	L-Pro (15 mol %), KF/Al₂O₃	toluene	110	8	(25)	491
I	CuI	15	L-Pro (15 mol %), KF/Al₂O₃	toluene	110	4	(80)	491
I	CuI	5	L43 (10 mol %), K₃PO₄	DMSO	110	5	(71)	138
Br	CuI	10	L21 (20 mol %), KI, Cs₂CO₃	MeCN	82	12	(32)	394
I	CuI	10	L21 (20 mol %), Cs₂CO₃	MeCN	82	12	(79)	394
I	CuI	5	L42 (10 mol %), K₃PO₄	DMSO	110	5	(76)	493
Cl	Cu₂O nanoparticles	5	KOH	t-BuOH/DMSO (3:1)	110	24	(0)	74
Br	Cu₂O nanoparticles	5	KOH	t-BuOH/DMSO (3:1)	110	24	(0)	74
I	Cu₂O nanoparticles	5	KOH	t-BuOH/DMSO (3:1)	110	24	(73)	74
TsO	Cu₂O nanoparticles	5	KOH	t-BuOH/DMSO (3:1)	110	24	(0)	74
I	Cu₂O nanoparticles	5	KOH	PEG-4000	120	10	(73)	225
Br	Cu₂O	5	NaOt-Bu	NMP	105	32	(97)	266
I	Cu₂O	5	Cs₂CO₃	NMP	90	24	(97)	266
I	Cu₂O	5	L30 (20 mol %), Cs₂CO₃, 3 Å MS	DMF	82	48	(91)	40

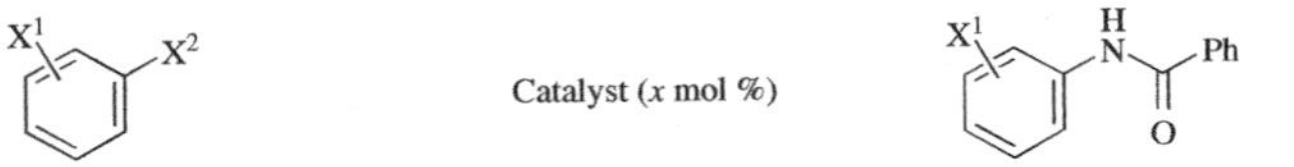

Catalyst (x mol %)

X^1 / X^2 benzene → X^1-C₆H₄-NH-C(O)-Ph

X^1	X^2	Catalyst	x	Additive(s)	Solvent	Temp (°)	Time (h)	
4-F	Br	CuI	10	L21 (20 mol %), Cs₂CO₃	MeCN	82	12 (46)	394
4-F	I	CuI	10	L21 (20 mol %), Cs₂CO₃	MeCN	82	12 (81)	394
2-Cl	I	CuI	10	oxazolidin-2-one (20 mol %), K₂CO₃	DMSO	120	24 (86)	492
4-Cl	I	CuI	7.5	DMEDA (7.5 mol %), KF/Al₂O₃	toluene	110	4.5 (80)	491
4-Br	I	CuI	5	KOH, TBAB	—	110	5 (75)	428
4-Br	I	CuI	15	L-Pro (15 mol %), KF/Al₂O₃	toluene	110	5.5 (75)	491
4-Br	I	Cu₂O nanoparticles	5	KOH	PEG-4000	120	12 (76)	225

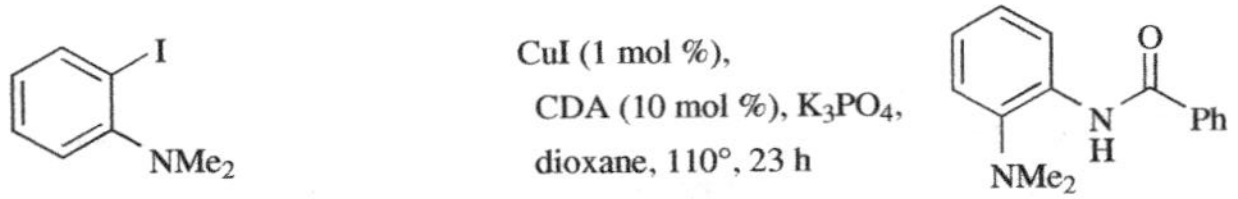

2-iodo-N,N-dimethylaniline

CuI (1 mol %),
CDA (10 mol %), K₃PO₄,
dioxane, 110°, 23 h

→ 2-(NMe₂)-C₆H₄-NH-C(O)-Ph   (96)   115, 69

Catalyst (x mol %)

O₂N-C₆H₄-X → O₂N-C₆H₄-NH-C(O)-Ph

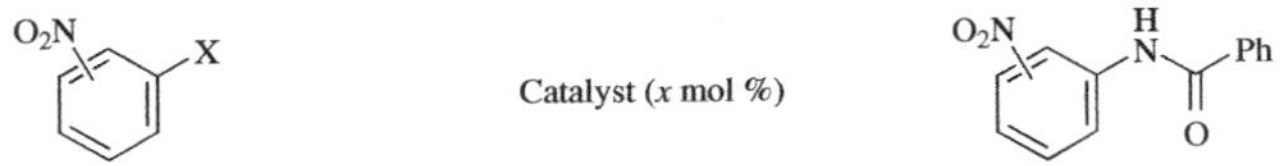

Isomer	x	Additives	Solvent	Temp (°)	Time (h)	
2	5	DMEDA (10 mol %), K₃PO₄	toluene	80	7 (88)	69
2	1	CDA (10 mol %), K₃PO₄	dioxane	110	23 (69)	115
4	7.5	DMEDA (7.5 mol %), KF/Al₂O₃	toluene	110	2 (95)	137
4	15	L-Pro (15 mol %), KF/Al₂O₃	toluene	110	6 (75)	491

TABLE 16A. N-ARYLATION OF PRIMARY AMIDES (Continued)

*Please refer to the charts preceding the tables for structures indicated by the **bold** numbers.*

C₇

Nitrogen Nucleophile: H_2N–C(O)–Ph

Aryl Halide: MeO-phenyl-X

Product: MeO-phenyl-NH-C(O)-Ph

Isomer	X	Catalyst	x	Additive(s)	Solvent	Temp (°)	Time (h)	Product(s) and Yield(s) (%)	Refs.
2	I	CuI	5	KOH, TBAB	—	110	8	(72)	428
2	I	CuI	7.5	DMEDA (7.5 mol %), KF/Al₂O₃	toluene	110	4	(90)	137
2	I	CuI	1	DMCDA (10 mol %), K₂CO₃	dioxane	110	15	(84)	115
2	I	CuI	15	L-Pro (15 mol %), KF/Al₂O₃	toluene	110	5	(80)	491
2	I	Cu₂O nanoparticles	5	KOH	PEG-4000	120	17	(70)	225
2	Br	Cu₂O	5	NaO-t-Bu	NMP	105	24	(84)	266
3	I	Cu₂O	5	Cs₂CO₃	NMP	90	24	(92)	266
4	I	CuI	5	KOH, TBAB	—	110	6	(82)	428
4	Br	CuI	5	EDA (5 mol %), K₂CO₃	PEG-400	80	8	(70)	301
4	I	CuI	10	EDA (10 mol %), K₃PO₄	dioxane	110	24	(60)	405
4	I	CuI	7.5	DMEDA (7.5 mol %), KF/Al₂O₃	toluene	110	1.5	(95)	137
4	I	CuI	10	oxazolidin-2-one (20 mol %), NaOMe	DMSO	120	24	(85)	492
4	I	CuI	10	DMCDA (20 mol %), K₃PO₄	[C₄mim][BF₄]	110	12	(83)	232
4	Br	CuI	20	DMG (20 mol %), K₃PO₄	DMF	reflux	48	(73)	494
4	Br	CuI	15	L-Pro (15 mol %), KF/Al₂O₃	toluene	110	5.5	(trace)	491
4	I	CuI	15	L-Pro (15 mol %), KF/Al₂O₃	toluene	110	4	(85)	491
4	I	CuI	5	Gly (20 mol %), K₃PO₄	dioxane	100	24	(71)	136
4	I	CuI	10	phen (10 mol %), KF/Al₂O₃	toluene	110	3.5	(99)	490
4	I	Cu₂O nanoparticles	5	KOH	PEG-4000	120	15	(75)	225
4	I	**10**	3	Cs₂CO₃	DMSO	100	24	(<1)	420

Aryl Halide: 3,5-(MeO)(OMe)-phenyl-X

Conditions: CuI (10 mol %), **L21** (20 mol %), Cs₂CO₃, MeCN, 82°, 12 h

Product: 3,5-dimethoxy-N-phenyl benzamide

X	Additive	
Br	KI	(42)
I	none	(78)

Refs.: 394

First reaction (2-methyl / 4-methyl aniline → N-aryl benzamide, Me-substituted)

Catalyst (x mol %)

Isomer	X	Catalyst	x	Additive(s)	Solvent	Temp (°)	Time		
2	I	CuI	7.5	DMEDA (7.5 mol %), KF/Al₂O₃	toluene	110	4 h	(90)	137
2	I	CuI	15	L-Pro (15 mol %), KF/Al₂O₃	toluene	110	5 h	(80)	491
4	Br	Cu (active)	500	none	—	MW, 100 W	10 min	(71)	280
4	I	CuI	5	KOH, TBAB	—	110	5 h	(75)	428
4	Cl	CuI	5	DMEDA (11 mol %), K₂CO₃	—	110	23 h	(93)	115
4	I	CuI	7.5	DMEDA (7.5 mol %), KF/Al₂O₃	toluene	110	2 h	(90)	137
4	I	CuI	15	L-Pro (15 mol %), KF/Al₂O₃	toluene	110	3.5 h	(90)	491
4	Cl	CuI	5	DMCDA (11 mol %), K₂CO₃	—	110	23 h	(93)	69
4	Br	CuI	10	L21 (20 mol %), KI, Cs₂CO₃	MeCN	82	12 h	(41)	394
4	I	Cu₂O nanoparticles	5	KOH	PEG-4000	120	12 h	(75)	225

Second reaction (CF₃-substituted anilide)

Cu₂Br₂ (1 mol %), K₂CO₃

Isomer	Temp (°)	Time (h)		
2	160	55	(35)	453
3	180	26	(75)	

(4-iodobenzoate allyl ester → N-aryl amide)

CuI (1 mol %), CDA (10 mol %), K₃PO₄, dioxane, 110°, 23 h   (86)   115, 69

Third reaction (Me₂-substituted aniline → N-aryl benzamide)

Catalyst (x mol %)

Isomer	X	Catalyst	x	Additive(s)	Solvent(s)	Temp (°)	Time (h)		
2,3	Br	CuI	10	L21 (20 mol %), Cs₂CO₃	MeCN	82	12	(36)	394
2,3	I	CuI	10	L21 (20 mol %), Cs₂CO₃	MeCN	82	12	(71)	394
2,4	I	CuI	5	KOH, TBAB	—	110	6	(91)	428
2,4	I	Cu₂O nanoparticles	5	KOH	PEG-4000	120	16	(72)	225
3,5	I	CuI	5	DMEDA (10 mol %), Cs₂CO₃	H₂O. THF	rt	7	(99)	69
3,5	I	CuI	1	DMCDA (10 mol %), K₂CO₃	dioxane	110	15	(90)	115
3,5	I	CuI	5	L44 (10 mol %), Cs₂CO₃	dioxane	rt	46	(96)	115
3,5	I	Cu₂O	5	Cs₂CO₃	NMP	90	24	(98)	266

TABLE 16A. N-ARYLATION OF PRIMARY AMIDES (Continued)

Nitrogen Nucleophile	Aryl Halide	Conditions	Product(s) and Yield(s) (%)	Refs.

*Please refer to the charts preceding the tables for structures indicated by the **bold** numbers.*

C₇

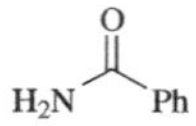

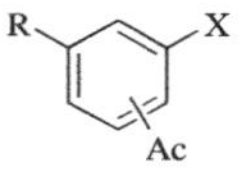

Catalyst (x mol %)

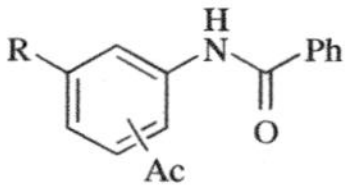

Isomer	R	X	Catalyst	x	Additive(s)	Solvent	Temp (°)	Time (h)		Refs.
2	H	Br	CuI	10	DMEDA (20 mol %), K₂CO₃, 5 Å MS	—	90	24	(86)	497
2	F	Br	CuI	10	DMEDA (20 mol %), K₂CO₃, 5 Å MS	—	110	24	(78)	497
4	H	I	CuI	15	L-Pro (15 mol %), KF/Al₂O₃	toluene	110	5.5	(75)	491
4	H	Cl	**10**	3	Cs₂CO₃	DMSO	100	24	(10)	420
4	H	Br	**10**	3	Cs₂CO₃	DMSO	100	24	(31)	420
4	H	I	**10**	3	Cs₂CO₃	DMSO	100	24	(73)	420

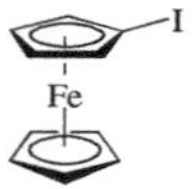

CuI (1 eq), K₂CO₃, DMSO. 90°, 18 h 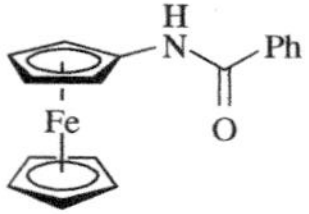 → (57) — 372

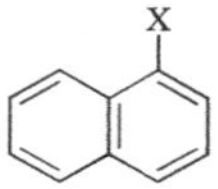

Catalyst (x mol %)

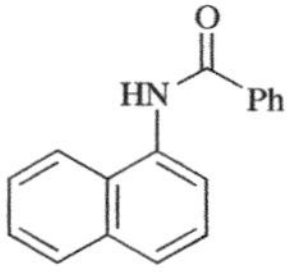

X	Catalyst	x	Additive(s)	Solvent	Temp (°)	Time (h)		Refs.
Br	CuI	5	EDA (5 mol %), K₂CO₃	PEG-400	80	8	(73)	301
I	CuI	10	oxazolidin-2-one (20 mol %), NaOMe	DMSO	120	24	(79)	492
I	CuI	7.5	DMEDA (7.5 mol %), KF/Al₂O₃	toluene	110	6	(60)	491
I	Cu₂O nanoprticles	5	KOH	PEG-4000	120	12	(80)	225

394

C$_{7-9}$

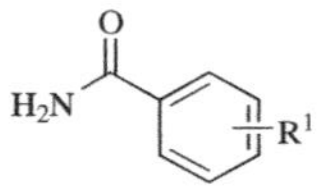

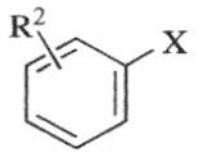

Conditions: CuI (10 mol %), **L21** (20 mol %), Cs$_2$CO$_3$, MeCN, 82°, 12 h

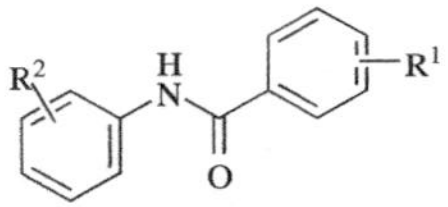

R^1	R^2	X	
4-F	4-Br	I	(81)
4-F	3,5-(CF$_3$)$_2$	Br	(42)
4-F	4-Me	Br	(40)
4-Cl	4-F	Br	(81)
4-Cl	4-F	I	(81)
4-Cl	3,5-(MeO)$_2$	Br	(37)
4-Cl	3,5-(MeO)$_2$	I	(83)

R^1	R^2	X	
3,5-(MeO)$_2$	4-Br	I	(67)
3,5-(MeO)$_2$	4-Me	Br	(45)
2,3-Me$_2$	4-Br	Br	(36)
2,3-Me$_2$	4-Me	I	(71)

C$_7$

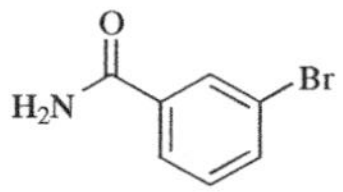

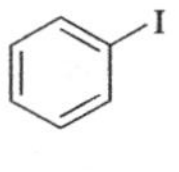

Conditions: CuI (5 mol %), TBAB, KOH, 110°, 6 h

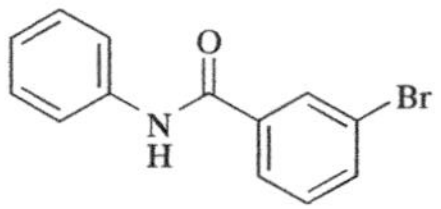

(69)

428

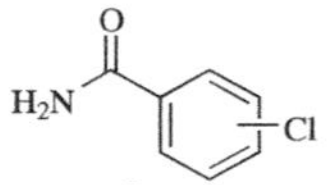

I

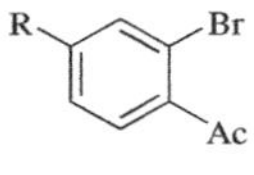

Conditions: CuI (10 mol %), DMEDA (20 mol %), K$_2$CO$_3$, 5 Å MS, 110°, 24 h

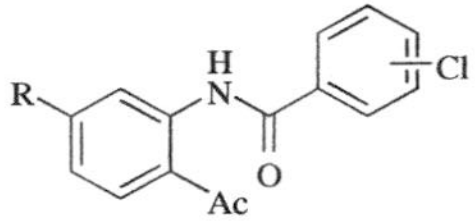

Isomer	R	
2	F	(67)
3	H	(83)
3	MeO	(72)

497

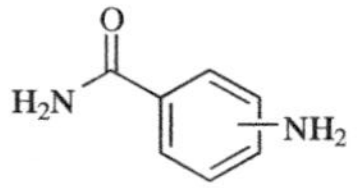

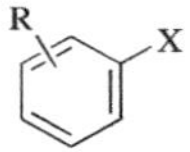

Conditions: CuI (x mol %), dioxane, 110°

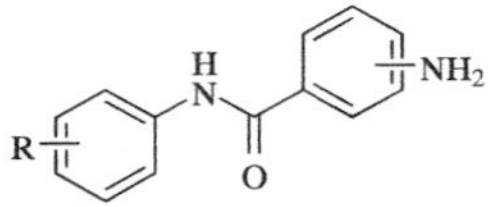

Isomer	R	X	x	Additives	Time (h)		
2	H	Br	5	DMEDA (10 mol %), K$_2$CO$_3$	24	(97)	107
3	4-t-Bu	Br	5	DMEDA (10 mol %), K$_2$CO$_3$	24	(97)	107
4	3,5-Me$_2$	Br	5	DMEDA (10 mol %), K$_2$CO$_3$	24	(97)	107
4	3,5-Me$_2$	I	1	CDA (10 mol %), K$_3$PO$_4$	23	(81)	115, 69

TABLE 16A. N-ARYLATION OF PRIMARY AMIDES (Continued)

Nitrogen Nucleophile	Aryl Halide	Conditions	Product(s) and Yield(s) (%)	Refs.

*Please refer to the charts preceding the tables for structures indicated by the **bold** numbers.*

C₇

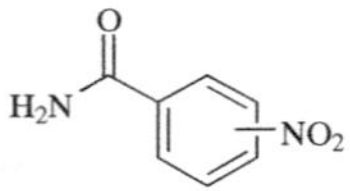

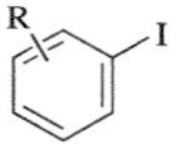

CuI (x mol %), 110°

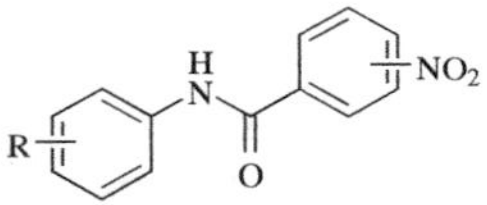

Isomer	R	x	Additives	Solvent	Time (h)		
2	4-MeO	10	EDA (10 mol %), K₃PO₄	dioxane	24	(61)	405
2	3,5-Me₂	10	EDA (10 mol %), K₃PO₄	—	24	(68)	405
3	H	5	TBAB, KOH	—	7	(63)	428
4	H	5	TBAB, KOH	—	6	(58)	428
4	4-MeO	10	EDA (10 mol %), K₃PO₄	dioxane	24	(67)	405

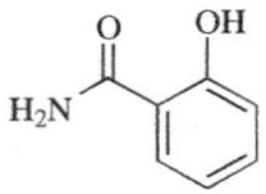

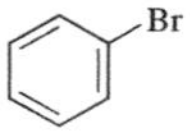

Cu, K₂CO₃, reflux, 2 h 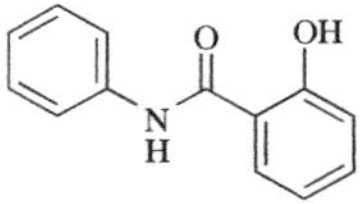   (56)   2

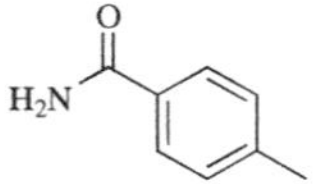

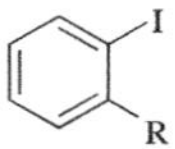

CuI (x mol %)

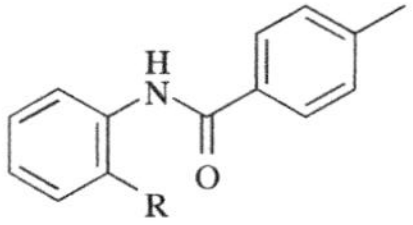

R	x	Additives	Solvent	Temp (°)	Time (h)	
H	5	TBAB, KOH	—	110	5	(76)
Cl	10	oxazolidinone (20 mol %), K₂CO₃	DMSO	120	24	(78)

428
492

C₈

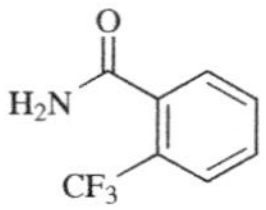

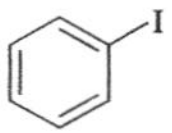

Cu₂Br₂ (1 mol %), K₂CO₃, 180°, 26 h 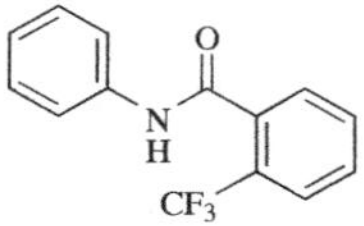   (79)   453

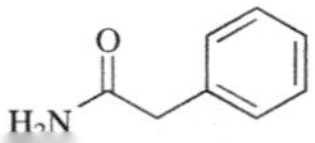

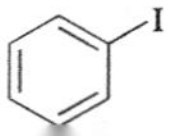

CuI (10 mol %), NaOMe, DMSO, oxazolidinone   (69)   492

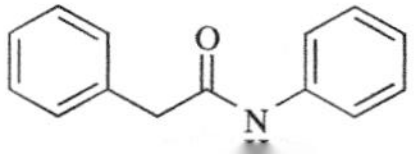

C$_9$

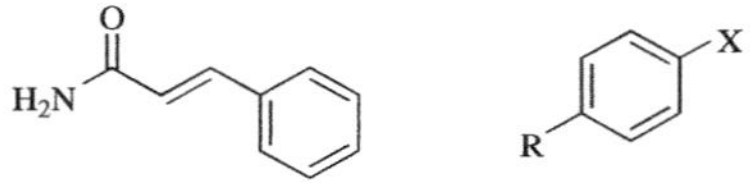

CuI (5 mol %), 110°

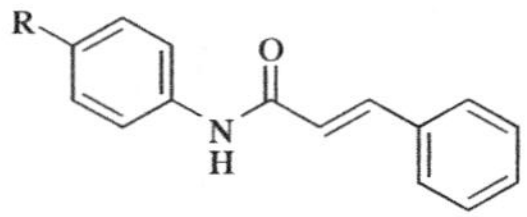

R	X	Additives	Solvent	Time (h)		
H	I	KOH, TBAB	—	5	(52)	428
Me$_2$N	Br	DMCDA (10 mol %), K$_2$CO$_3$	toluene	24	(99)	69

C$_{19}$

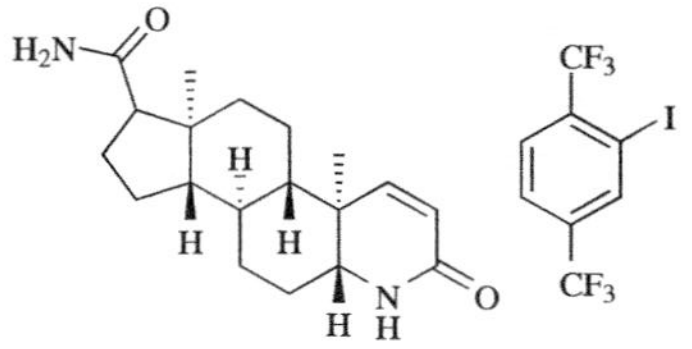

Cu (3 eq), K$_2$CO$_3$,
o-xylene, 140°, 52 h

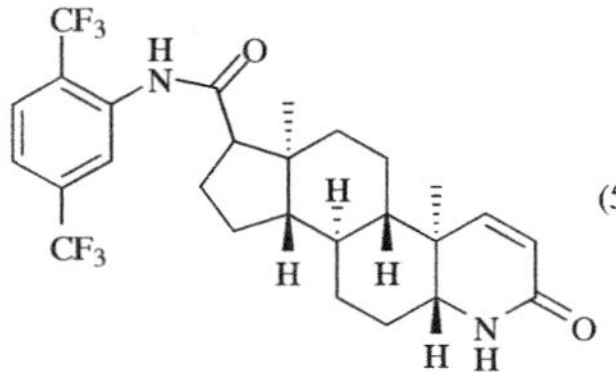

(53)   499

TABLE 16B. N-HETEROARYLATION OF PRIMARY AMIDES

Nitrogen Nucleophile	Heteroaryl Halide	Conditions	Product(s) and Yield(s) (%)	Refs.
C₂ NH₂Ac		CuI (10 mol %), EDA (20 mol %), K₃PO₄, dioxane, reflux, 24 h	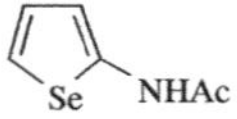 (48)	87
C₂₋₇		CuI (10 mol %), DMEDA (10 mol %), K₂CO₃, toluene, 110°, 20 h	R ‾‾‾‾‾ Me (85) Ph (99)	500
C₂		1. CuI (25 mol %), DMCDA (50 mol %), K₃CO₃, dioxane, 140°, 21 h 2. HCl, H₂O	(37)	496
C₄		CuI (10 mol %), DMEDA (10 mol %), K₂CO₃, dioxane, 110°, 24 h	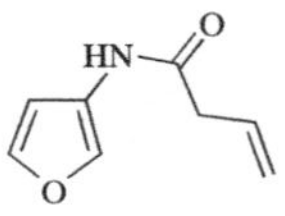 (12) 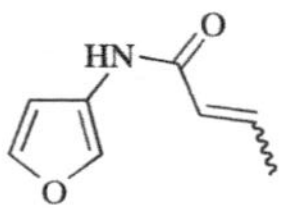 + (12)	83
C₅		CuI (10 mol %), DMEDA (10 mol %), K₃PO₄, dioxane, 110°, 24 h	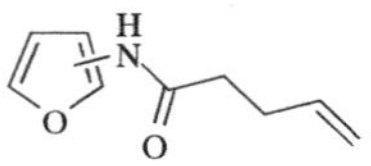 Isomer ‾‾‾‾‾ 2 (43) 3 (82)	501

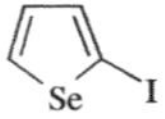

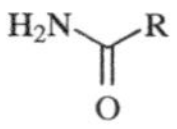

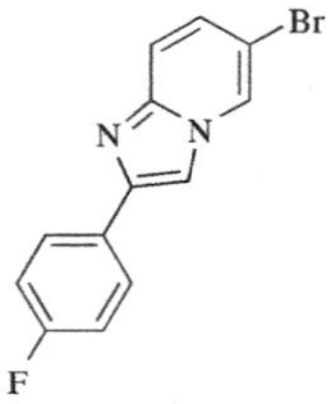

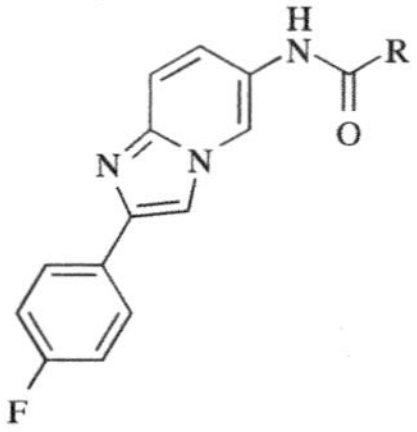

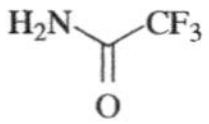

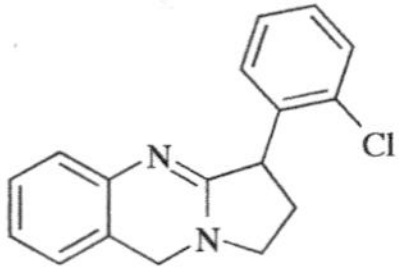

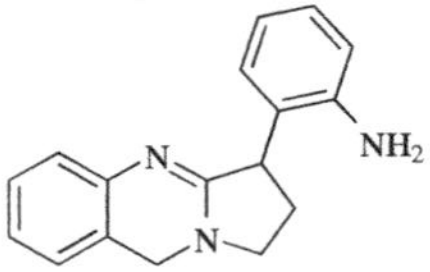

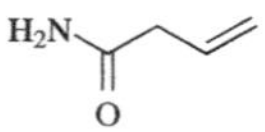

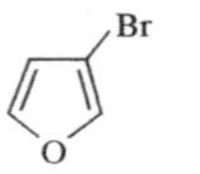

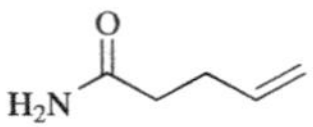

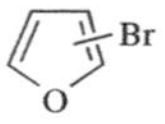

412

C₇

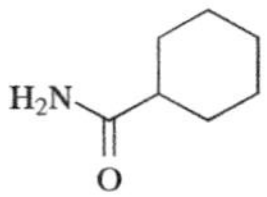

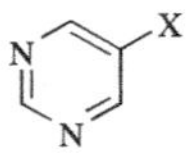

H_2N–C(=O)–Ph (cyclohexanecarboxamide structure)

CuI (x mol %),
DMEDA (y mol %)

Product: cyclohexanecarboxamide of pyrimidine (N-pyrimidinyl cyclohexanecarboxamide)

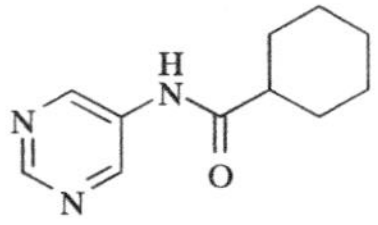

X	x	y	Additive	Solvent	Temp (°)	Time (h)		Ref
Br	10	20	K₂CO₃	—	110	23	(87)	441
Br	10	10	K₃PO₄	dioxane	110	24	(86)	69
I	2	10	K₃PO₄	diglyme	120	24	(99)	115

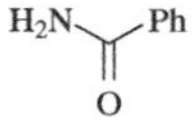

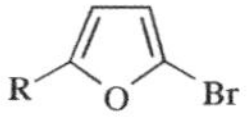

H_2N–C(=O)–Ph (benzamide)

CuI (10 mol %),
DMEDA (10 mol %),
dioxane, 110°, 24 h

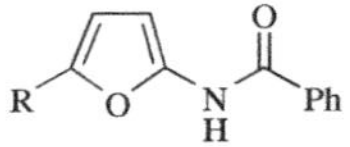

Product: N-(furan-2-yl)benzamide

R	Additive		Ref
CHO	K₂CO₃	(71)	83
CHO	K₃PO₄	(98)	501
MeO₂C	K₃PO₄	(67)	501

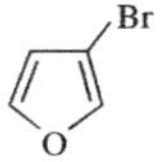

CuI (10 mol %),
DMEDA (10 mol %), K₂CO₃,
dioxane, 110°, 24 h

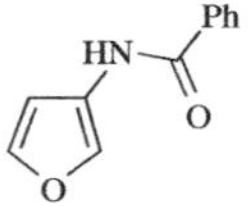

Product: N-(furan-3-yl)benzamide (98)   83, 501

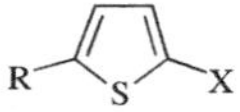

CuI (10 mol %),
additive (10 mol %), K₃PO₄,
dioxane, 110°, 24 h

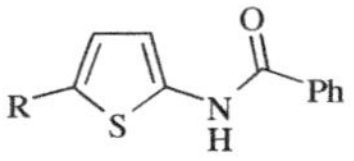

Product: N-(thiophen-2-yl)benzamide

R	X	Additive		Ref
H	I	EDA	(42)	405
Me	Br	DMEDA	(66)	83

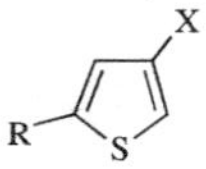

CuI (10 mol %),
additive (10 mol %), K₃PO₄,
dioxane, 110°, 24 h

Product: N-(thiophen-3-yl)benzamide

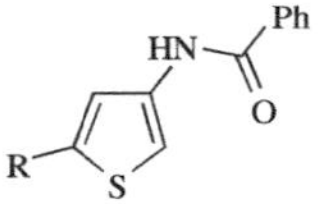

R	X	Additive		Ref
H	Br	CDA	(97)	69
H	I	DMCDA	(99)	115
CHO	Br	DMEDA	(39)	83

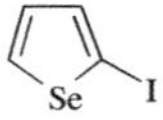

CuI (10 mol %),
EDA (20 mol %), K₃PO₄,
dioxane, reflux, 24 h

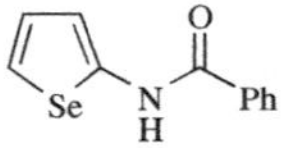

Product: N-(selenophen-2-yl)benzamide (77)   87

TABLE 16B. *N*-HETEROARYLATION OF PRIMARY AMIDES (*Continued*)

Nitrogen Nucleophile	Heteroaryl Halide	Conditions	Product(s) and Yield(s) (%)	Refs.
C$_7$				
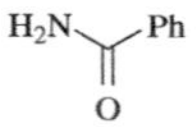	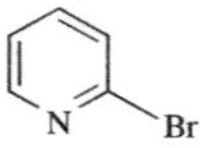	CuI (10 mol %), oxazolidinone (20 mol %), NaOMe. DMSO, 120°, 24 h	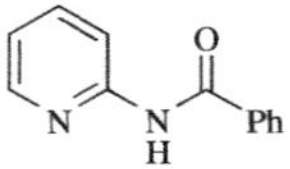 (80)	492
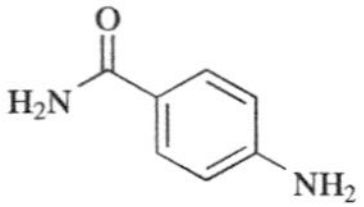	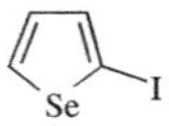	CuI (10 mol %), DMEDA (10 mol %), K$_3$PO$_4$, dioxane, 110°, 24 h	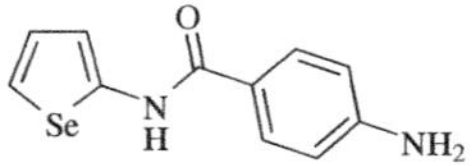 (45)	87
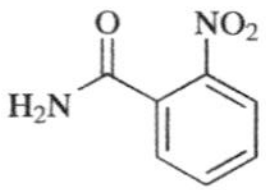	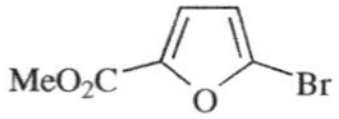	CuI (10 mol %), DMEDA (10 mol %), K$_3$PO$_4$, dioxane, 110°, 24 h	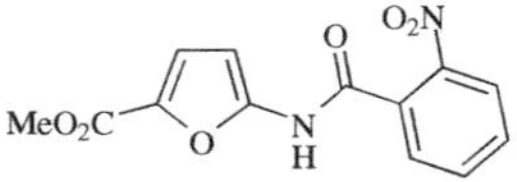 (30)	83
	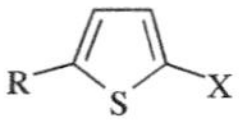	CuI (10 mol %), additive (10 mol %), K$_3$PO$_4$, dioxane, 110°, 24 h	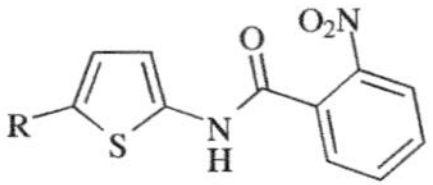 R X Additive H I EDA (52) Me Br DMEDA (33)	405 83
		CuI (10 mol %), DMEDA (10 mol %), K$_3$PO$_4$, dioxane, 110°, 24 h	(86)	83, 501
C$_8$				
	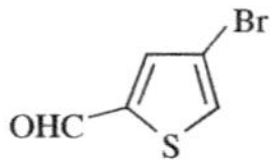	CuI (10 mol %), DMEDA (10 mol %), K$_3$PO$_4$, dioxane, 110°, 24 h	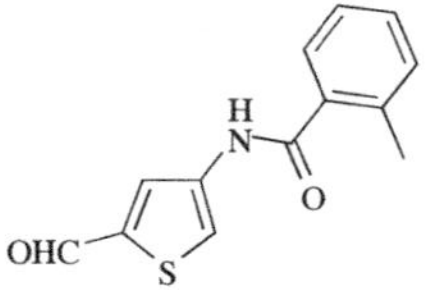 (43)	83

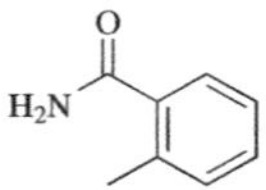

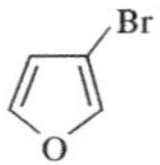

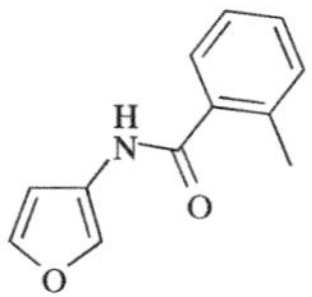

414

TABLE 17A. N-ARYLATION OF ACYCLIC SECONDARY AMIDES

Nitrogen Nucleophile	Aryl Halide	Conditions	Product(s) and Yield(s) (%)	Refs.

Please refer to the charts preceding the tables for structures indicated by the bold numbers.

C₂

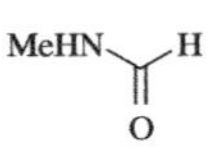

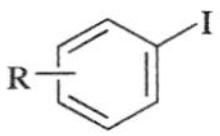

$MeHN\text{-}CHO$; Aryl Halide: $R\text{-}C_6H_4\text{-}I$; Conditions: CuI (x mol %) ; Product: $R\text{-}C_6H_4\text{-}N(Me)\text{-}CHO$

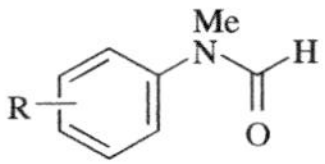

R	x	Additives	Solvent	Temp (°)	Time (h)		Refs.
H	5	DMEDA (20 mol %), K₃PO₄	toluene	110	23	(87)	69
4-Cl	5	DMEDA (10 mol %), CsF	THF	rt	24	(98)	63
2-MeO	1	CDA (10 mol %), K₃PO₄	dioxane	110	23	(95)	115, 69
4-MeO	0.2	CDA (10 mol %), K₃PO₄	dioxane	110	23	(99)	69

C₃

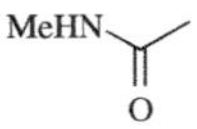

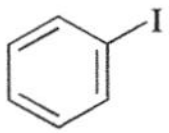

$MeHN\text{-}C(O)Me$; Phenyl iodide ; Conditions: Catalyst (5 mol %) ; Product: $Ph\text{-}N(Me)\text{-}C(O)Me$

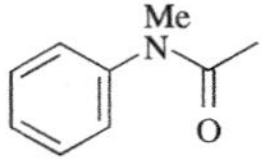

Catalyst	Additive(s)	Solvent	Temp (°)	Time (h)		Refs.
CuI	L42 (10 mol %), K₃PO₄	DMSO	110	5	(79)	493
CuI	L43 (10 mol %), K₃PO₄	DMSO	110	5	(25)	138
Cu₂O	NaOt-Bu	NMP	100	18	(81)	266

C₄

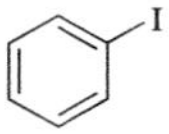

$HO\text{-}NHAc$; Phenyl iodide ; CuI (5 mol %), Gly (20 mol %), K₃PO₄, dioxane, 100°, 24 h ; Product: Ac–N(CH₂CH₂OH)–Ph (84) ; Refs. 136

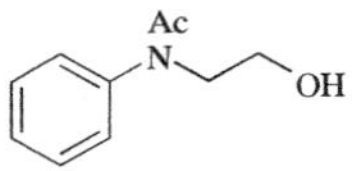

C₆₋₂₈

$R^1HN\text{-}...\text{-}NHR^2$ (squaramide) ; ArBr ; CuI (25 mol %), L-Pro (25 mol %), K₂CO₃, DMF, 120°, 18 h ; Products I + II ; Refs. 502

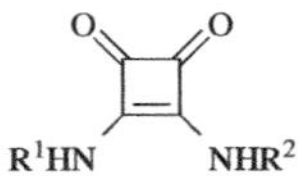

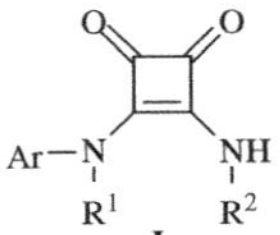

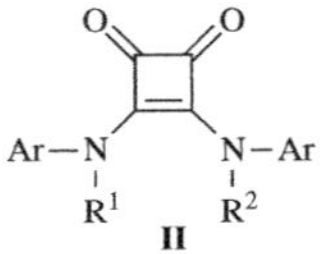

R¹	R²	Ar	I	II
Me	Me	Ph	(12)	(65)
Me	Me	4-ClC₆H₄	(0)	(44)
Me	Me	4-Me₂NC₆H₄	(0)	(55)
Me	Me	4-O₂NC₆H₄	(0)	(0)

TABLE 17A. *N*-ARYLATION OF ACYCLIC SECONDARY AMIDES (*Continued*)

Nitrogen Nucleophile	Aryl Halide	Conditions	Product(s) and Yield(s) (%)	Refs.

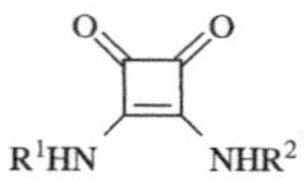

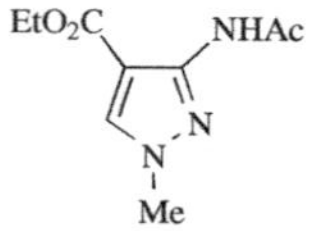

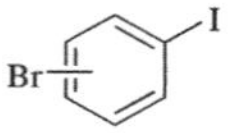

*Please refer to the charts preceding the tables for structures indicated by the **bold** numbers.*

Continued from previous page.

C_{6-28}

R^{1}HN ⟶ NHR^2 (squarate structure) | ArBr | CuI (25 mol %), L-Pro (25 mol %), K$_2$CO$_3$, DMF, 120°, 18 h | (structure **I**) Ar—N(R^1)— with NH—R^2 + (structure **II**) Ar—N(R^1)— with N(R^2)—Ar | 502

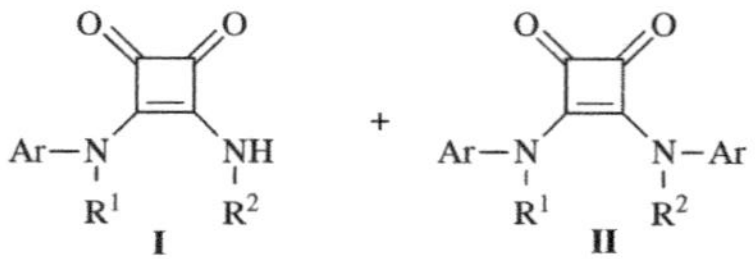

R^1	R^2	Ar	I	II
Me	Me	2-MeC$_6$H$_4$	(0)	(20)
Me	Me	4-MeC$_6$H$_4$	(0)	(70)
Me	Me	3-MeO$_2$CC$_6$H$_4$	(0)	(10)
Me	Me	4-MeOCC$_6$H$_4$	(0)	(10)
Me	Me	4-EtO$_2$CCH$_2$C$_6$H$_4$	(0)	(45)
Me	Me	1-Np	(10)	(35)
Me	Me	9-anthracenyl	(12)	(0)
Me	Me	1-phenanthrenyl	(30)	(14)
Me	Me	1-pyrenyl	(55)	(0)
t-Bu	t-Bu	Ph	(96)	(0)
c-C$_6$H$_{11}$	c-C$_6$H$_{11}$	Ph	(85)	(0)
n-C$_{12}$H$_{25}$	n-C$_{12}$H$_{25}$	4-EtO$_2$CCH$_2$C$_6$H$_4$	(35)	(18)

C_6

EtO$_2$C — (pyrazole) NHAc, N–N–Me | 4-Br-iodobenzene | CuI (10 mol %), DMCDA (20 mol %), K$_3$PO$_4$, dioxane, 110°, 22 h | (product with Ac–N, CO$_2$Et pyrazole, Br-phenyl, N–N–Me)

Isomer	
2	(0)
3	(67)
4	(70)

503

EtO$_2$C — (pyrazole) NHAc, N–N–Me + (2-pyrrolidinone, N–H, O=) | 4-Br-iodobenzene | CuI (20 mol %), DMCDA (40 mol %), K$_3$PO$_4$, dioxane, 110°, 22 h | (product with CO$_2$Et pyrazole, Ac–N, phenyl with pyrrolidinone N, N–N–Me) (53) | 504

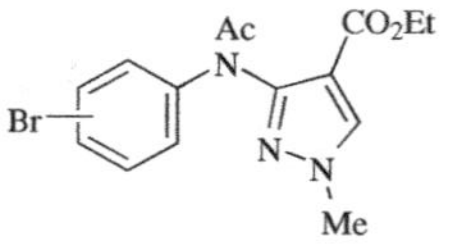

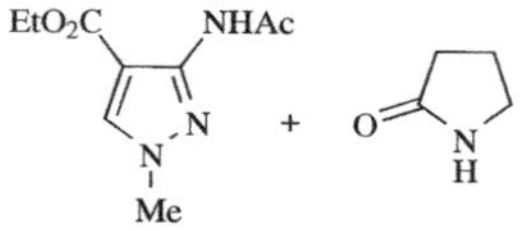

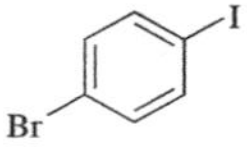

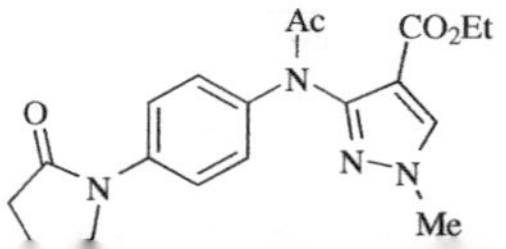

416

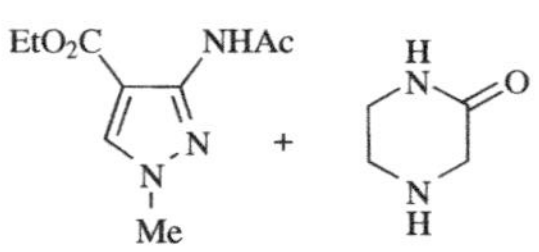

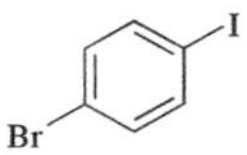

C_7

CuI (20 mol %),
DMCDA (40 mol %),
K_3PO_4, dioxane, 110°, 22 h

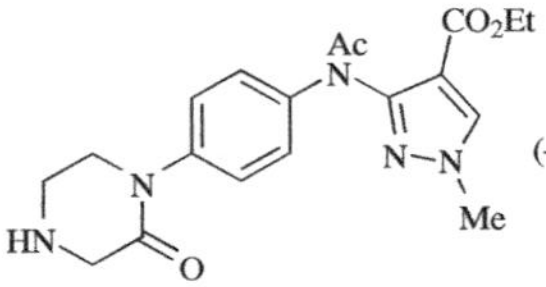

(—)   504

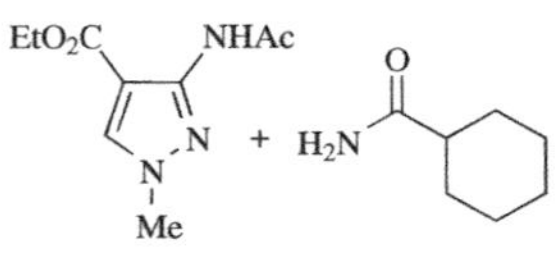

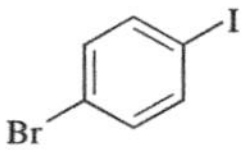

CuI (20 mol %),
DMCDA (40 mol %),
K_3PO_4, dioxane,
110°, 22 h

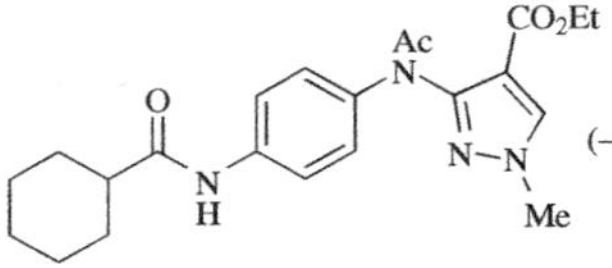

(—)   504

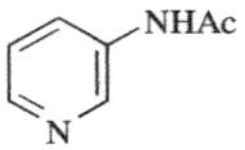

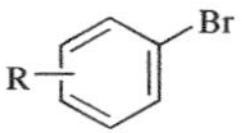

CuCl (10 mol %), K_2CO_3,
N(CH_2CH_2OCH_2–
CH_2OMe)_3,
xylene, reflux, 22 h

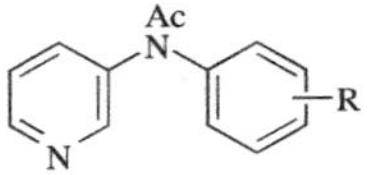

505

R	X	
H	Br	(74)
2-F	Br	(44)
3-F	Br	(68)
4-F	Br	(74)
2,3,4,5,6-F_5	Br	(0)
2-Cl	Cl	(82)
3-Cl	Cl	(56)
4-Cl	Br	(73)
2-CF_3	Br	(22)
3-CF_3	Br	(69)
4-CF_3	Br	(73)
4-CF_3O	Br	(53)

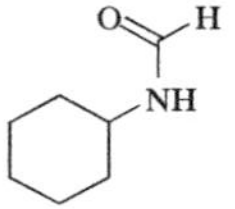

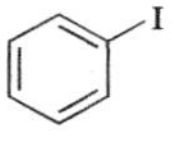

CuI (5 mol %),
CDA (20 mol %), K_3PO_4,
toluene, 110°, 25 h

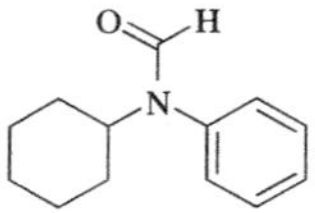

(63)   69

417

TABLE 17A. N-ARYLATION OF ACYCLIC SECONDARY AMIDES (Continued)

Nitrogen Nucleophile	Aryl Halide	Conditions	Product(s) and Yield(s) (%)	Refs.

*Please refer to the charts preceding the tables for structures indicated by the **bold** numbers.*

C$_7$

Nitrogen Nucleophile: Ph–HN–CHO

Aryl Halide: R—(ring)—I

Conditions: CuI (5 mol %)

R	X	Additives	Solvent	Temp (°)	Time (h)	Product(s) and Yield(s) (%)	Refs.
H	I	CDA (20 mol %), K$_3$PO$_4$	toluene	80	4	(93)	69
4-Cl	I	DMEDA (10 mol %), CsF	THF	rt	24	(98)	63
3-Ac	Br	DMCDA (10 mol %), K$_2$CO$_3$	toluene	110	24	(91)	69

Product: R—(ring)—N(Ph)—CHO

C$_8$

Nitrogen Nucleophile: H–N(Ac)–Ph

Aryl Halide: X—(ring)

Conditions: Catalyst (x mol %)

X	Catalyst	x	Additive(s)	Solvent	Temp (°)	Time	Product(s) and Yield(s) (%)	Refs.
Br	Cu bronze	100	none	—	180	—	(36)	506
I	CuI	10	K$_3$PO$_4$	DMF	110	24 h	(90)	387
Br	CuI	2.5	K$_2$CO$_3$	NMP	MW (250 W)	40 min	(76)	239
Br	CuI	20	DMG (20 mol %), K$_3$PO$_4$	DMF	reflux	48 h	(88)	494
I	CuI	15	L-Pro (15 mol %), KF/Al$_2$O$_3$	toluene	110	6 h	(30)	137
I	CuI	7.5	DMEDA (7.5 mol %), KF/Al$_2$O$_3$	toluene	110	3 h	(50)	137
I	CuI	10	phen (10 mol %), KF/Al$_2$O$_3$	toluene	110	6 h	(95)	490
I	CuI	5	Gly (20 mol %), K$_3$PO$_4$	dioxane	100	24 h	(84)	136
I	CuI	5	L42 (10 mol %), K$_3$PO$_4$	DMSO	110	5 h	(75)	493
I	CuI	5	L43 (10 mol %), K$_3$PO$_4$	DMSO	110	5 h	(30)	138
I	CuO	0.1	DMEDA (20 mol %), K$_2$CO$_3$	toluene	135	24 h	(38)	222

Product: Ac–N(Ph)$_2$

C$_{8–10}$

Nitrogen Nucleophile: H–N(Ac)–(ring)–R

Aryl Halide: Br—(ring)

Conditions: 1. CuI, K$_2$CO$_3$, PhNO$_2$, reflux, 15 h; 2. HCl, EtOH, reflux, 3 h

Product: R—(ring)—NH—Ph

R	Yield (%)
H	(60)
3-O$_2$N	(80)
4-Me	(33)
2,4-Me$_2$	(80)

Refs.: 507

C$_8$

NHAc (benzene structure)

R—(benzene)—X → Catalyst (x amount) → R—(benzene)—N(Ac)(phenyl)

R	X	Catalyst	x	Additive(s)	Solvent	Temp (°)	Time		
2-F	Br	Cu bronze	1 eq	none	—	180	—	(42)	506
4-F	Br	Cu bronze	1 eq	none	—	180	—	(35)	506
4-Br	I	CuI	5 mol %	Gly (20 mol %), K$_3$PO$_4$	dioxane	100	24 h	(80)	136
4-H$_2$N	I	CuI	1 mol %	CDA (1 mol %), K$_3$PO$_4$	dioxane	110	23 h	(81)	115, 69
2-O$_2$N	I	CuI	5 mol %	Gly (20 mol %), K$_3$PO$_4$	dioxane	100	24 h	(98)	136
4-O$_2$N	Br	Cu bronze	1 eq	none	—	180	—	(35)	506
2-MeO	Br	CuI	2.5 mol %	K$_2$CO$_3$	NMP	MW (250 W)	40 min	(56)	239
2-MeO	I	CuI	7.5 mol %	DMEDA (7.5 %), KF/Al$_2$O$_3$	toluene	110	8 h	(40)	137
2-MeO	I	CuI	15 mol %	L-Pro (15 mol %), KF/Al$_2$O$_3$	toluene	110	8 h	(trace)	491
2,6-(MeO)$_2$	Br	Cu	10 mol %	I$_2$, K$_2$CO$_3$	PhNO$_2$	reflux	24 h	(3)	508
4-MeO	Br	Cu bronze	1 eq	none	—	180	—	(19)	506
4-MeO	I	CuI	5 mol %	Gly (20 mol %), K$_3$PO$_4$	dioxane	100	24 h	(88)	136
4-MeO	I	CuI	15 mol %	L-Pro (15 mol %), KF/Al$_2$O$_3$	toluene	110	6 h	(trace)	137
4-MeO	I	CuI	7.5 mol %	DMEDA (7.5 mol %), KF/Al$_2$O$_3$	toluene	110	3 h	(40)	137
4-MeO	Br	CuI	20 mol %	DMG (20 mol %), K$_3$PO$_4$	DMF	reflux	48	(81)	494
4-MeO	I	CuI	10 mol %	phen (10 mol %), KF/Al$_2$O$_3$	toluene	110	8 h	(95)	490
4-MeO	I	CuI	10 mol %	DMCDA (20 mol %), K$_3$PO$_4$	[C$_4$min][BF$_4$]	110	12 h	(87)	232
3-MeS	Br	Cu	5 mol %	K$_2$CO$_3$	—	220	2 h	(64)	509
3-CF$_3$S	Br	Cu	5 mol %	K$_2$CO$_3$	—	220	2 h	(72)	509
3-MeO$_2$S	Br	Cu	5 mol %	K$_2$CO$_3$	—	220	2 h	(30)	509
2-Me	Br	Cu bronze	1 eq	none	—	180	—	(60)	506
2-Me	I	CuI	15 mol %	L-Pro (15 mol %), KF/Al$_2$O$_3$	toluene	110	9 h	(trace)	491
2-Me	I	CuI	15 mol %	L-Pro (15 mol %), KF/Al$_2$O$_3$	toluene	110	5 h	(20)	137
4-Me	I	CuI	5 mol %	Gly (20 mol %), K$_3$PO$_4$	dioxane	100	24 h	(82)	136
4-Me	I	CuI	15 mol %	L-Pro (15 mol %), KF/Al$_2$O$_3$	toluene	110	7 h	(30)	491
4-Me	I	CuI	7.5 mol %	DMEDA (7.5 mol %), KF/Al$_2$O	toluene	110	2 h	(40)	137
4-NC–	Br	Cu bronze	1 eq	none	—	180	—	(35)	506
4-CHO	Br	Cu bronze	1 eq	none	—	180	—	(38)	506
2,5-Me$_2$	Br	Cu bronze	1 eq	none	—	180	—	(27)	506

Please refer to the charts preceding the tables for structures indicated by the **bold** numbers.

Continued from previous page.

C₈ → C_8

Nitrogen Nucleophile	Aryl Halide				Conditions				Product(s) and Yield(s) (%)		Refs.

Nitrogen Nucleophile: phenyl–NHAc

Aryl Halide: R–X

R	X	Catalyst	x	Additives	Solvent	Temp (°)	Time				
3,5-Me₂	I	CuI	1 mol %	CDA (10 mol %), K₃PO₄	dioxane	110	24 h	(96)			115, 69
4-Ac	I	CuI	5 mol %	Gly (20 mol %), K₃PO₄	dioxane	100	24 h	(62)			136

Product: Ac–N(Ph)–C₆H₄–R

Aryl Halide: 4-bromo iodobenzene (I and Br substituted benzene)

Conditions:
1. CuI (15 mol %), DMG (20 mol %), K₃PO₄, DMSO, 80°, 24 h
2. ROH, 120°, 48 h

Product: Ac–N(Ph)–C₆H₄–OR

R	
Ph	(77)
4-ClC₆H₄	(71)
2-MeOC₆H₄	(80)
4-MeOC₆H₄	(81)
2-Np	(70)

Refs. 495

Aryl Halide: Ac–N(Ph)–C₆H₃(Br)–NAc

Conditions: CuI, K₂CO₃, 215°, 4 d

Product: Ac–N(Ph)–C₆H₄–NAc(Ph) (45)

Refs. 510

Aryl Halide: 4-bromo benzene with R (Br, R substituted)

Conditions: Cu bronze (1 eq), 180°

Product: Ac–N(C₆H₄F)–C₆H₄–R

R	
H	(13)
2-Me	(19)
2,5-Me₂	(21)

Refs. 506

Nitrogen Nucleophile: F–C₆H₄–NHAc

Aryl Halide: 4-bromo benzene (Br, R substituted)

Conditions:
1. Cu (30 mol %), CuI (10 mol %), KI, I₂, K₂CO₃, PhNO₂, reflux, 17 h
2. KOH, EtOH, reflux, 3 h

Product: F–C₆H₄–NH–C₆H₄–R

Isomer	R	
2	4-MeO	(70)
2	4-Me	(65)
4	2-Cl	(72)
4	2-MeO	(80)
4	4-MeO	(63)
4	4-Me	(75)

Refs. 511

C$_{8-11}$

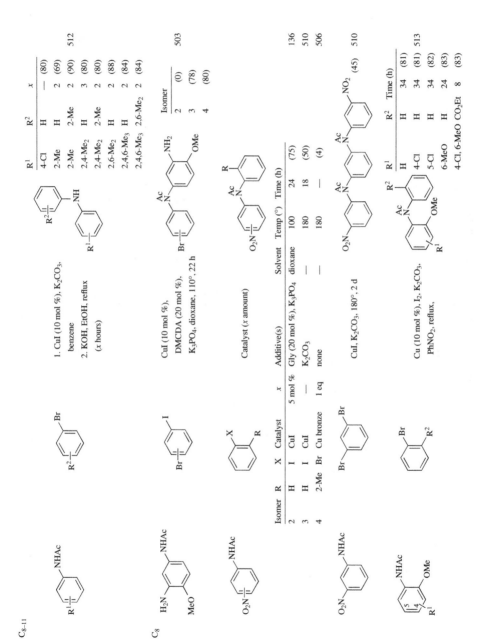

(top scheme)

1. CuI (10 mol %), K$_2$CO$_3$, benzene
2. KOH, EtOH, reflux (x hours)

R^1	R^2	x	
4-Cl	H	—	(80)
2-Me	H	2	(69)
2-Me	2-Me	2	(90)	512
2,4-Me$_2$	H	3	(80)
2,4-Me$_2$	2-Me	2	(80)
2,6-Me$_2$	H	2	(88)
2,4,6-Me$_3$	H	2	(84)
2,4,6-Me$_3$	2,6-Me$_2$	2	(84)

C$_8$

CuI (10 mol %), DMCDA (20 mol %), K$_3$PO$_4$, dioxane, 110°, 22 h

Isomer	
2	(0)
3	(78)	503
4	(80)

Catalyst (x amount)

Isomer	R	X	Catalyst	x	Additive(s)	Solvent	Temp (°)	Time (h)	
2	H	I	CuI	5 mol %	Gly (20 mol %), K$_3$PO$_4$	dioxane	100	24 (75)	136
3	H	I	CuI	—	K$_2$CO$_3$	—	180	18 (50)	510
4	2-Me	Br	Cu bronze	1 eq	none	—	180	— (4)	506

CuI, K$_2$CO$_3$, 180°, 2 d   (45)   510

Cu (10 mol %), I$_2$, K$_2$CO$_3$, PhNO$_2$, reflux,

R^1	R^2	Time (h)	
H	H	34	(81)
4-Cl	H	34	(81)	513
5-Cl	H	34	(82)
6-MeO	H	24	(83)
4-Cl, 6-MeO	CO$_2$Et	8	(83)

Nitrogen Nucleophile	Aryl Halide	Conditions	Product(s) and Yield(s) (%)	Refs.

*Please refer to the charts preceding the tables for structures indicated by the **bold** numbers.*

C₈

Nitrogen Nucleophile: MeO–C₆H₄–NHAc
Aryl Halide: R–C₆H₄–X
Conditions: Catalyst (x amount)
Product: MeO–C₆H₄–N(Ac)–C₆H₄–R

R	X	Catalyst	x	Additive(s)	Solvent	Temp (°)	Time (h)	Yield	Refs.
H	I	CuI	5 mol %	TBAB, KOH	—	110	4	(79)	428
4-O₂N	Br	Cu bronze	1 eq	none	—	180	—	(40)	506
2-Me	Br	Cu bronze	1 eq	none	—	180	—	(58)	506

Nitrogen Nucleophile: O=CH–NH–CH₂Ph
Conditions: CuI (x mol %), 110°
Product: O=CH–N(CH₂Ph)–C₆H₄–R

R	X	x	Additives	Solvent	Time (h)	Yield	Refs.
4-MeS	Br	5	DMEDA (10 mol %), K₂CO₃	toluene	24	(95)	69
3,5-Me₂	I	1	CDA (10 mol %), K₃PO₄	dioxane	23	(93)	115, 69

C₉

Nitrogen Nucleophile: 4-Me–C₆H₄–NHAc
Aryl Halide: R–C₆H₄–X
Conditions: CuI (x mol %), K₃PO₄
Product: 4-Me–C₆H₄–N(Ac)–C₆H₄–R

R	X	x	Additive	Solvent	Temp (°)	Time (h)	Yield	Refs.
H	Br	20	DMG (20 mol %)	DMF	reflux	48	(85)	494
H	I	5	Gly (20 mol %)	dioxane	100	24	(74)	136
4-MeO	Br	20	DMG (20 mol %)	DMF	reflux	48	(83)	494
2-Me	Br	20	DMG (20 mol %)	DMF	reflux	48	(72)	494
4-Ac	Br	20	DMG (20 mol %)	DMF	reflux	48	(87)	494

Nitrogen Nucleophile: MeS–CH₂–C₆H₃(Cl)–NHAc
Aryl Halide: PhBr
Conditions: Cu (10 mol %), I₂, K₂CO₃, PhNO₂, reflux, 24 h
Product: (structure with SMe, N(Ac), Cl) (85) — 508

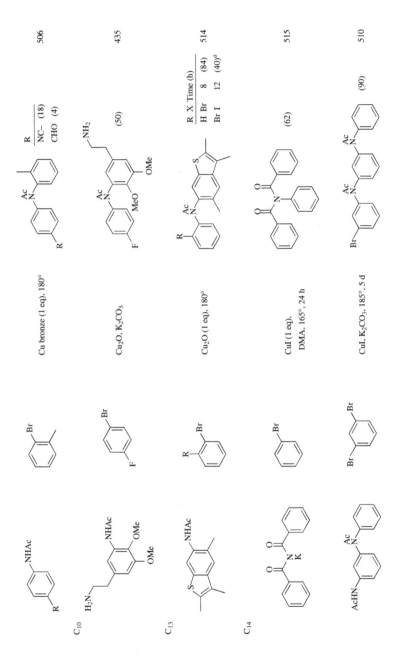

TABLE 17A. *N*-ARYLATION OF ACYCLIC SECONDARY AMIDES (*Continued*)

Nitrogen Nucleophile	Aryl Halide	Conditions	Product(s) and Yield(s) (%)	Refs.

Please refer to the charts preceding the tables for structures indicated by the **bold** numbers.

C$_{21}$

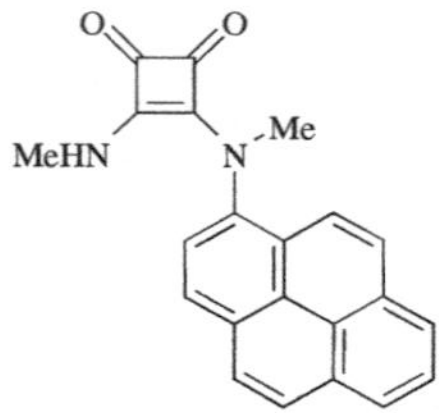

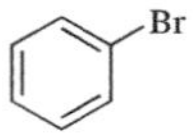

CuI (25 mol %),
L-Pro (25 mol %), K$_2$CO$_3$,
DMF, 120°, 18 h

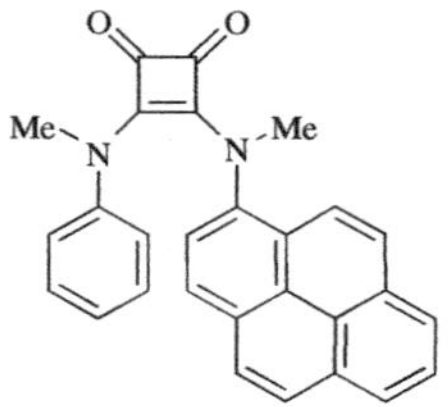

(15)

502

C$_{22}$

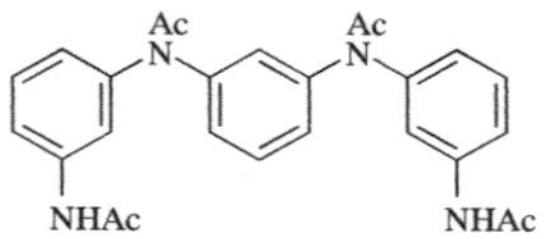

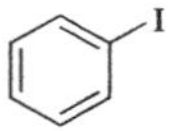

CuI, K$_2$CO$_3$, 190°, 4 d

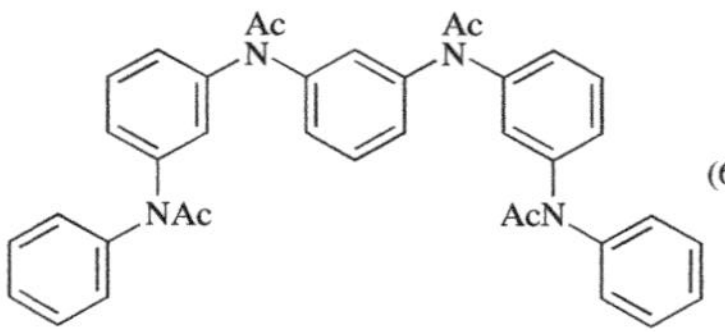

(60)

510

^{a} A mixture of R = Br and R = I was formed.

TABLE 17B. *N*-HETEROARYLATION OF ACYCLIC SECONDARY AMIDES

Nitrogen Nucleophile	Heteroaryl Halide	Conditions	Product(s) and Yield(s) (%)	Refs.
C₂				
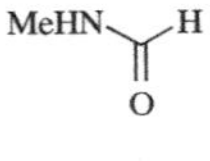	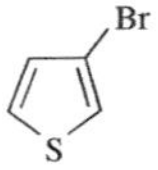	CuI (5 mol %), DMCDA (10 mol %), K₃PO₄, dioxane, 110°, 24 h	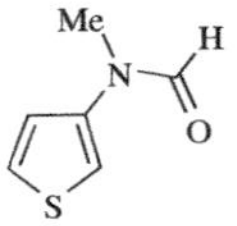 (99)	115
C₈				
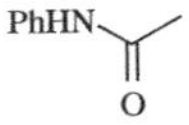	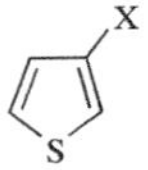	CuI (10 mol %), additive (x mol %), K₃PO₄, dioxane, 110°, 24 h	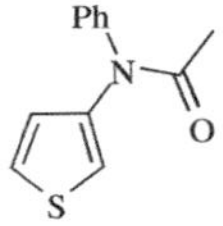 X, Additive, x: Br CDA 20 (81); I DMCDA 10 (99)	69, 115
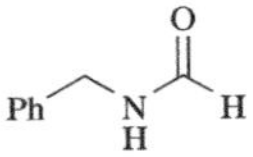	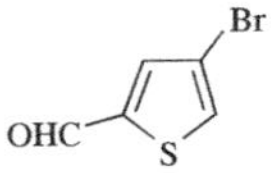	CuI (10 mol %), DMEDA (10 mol %), K₂CO₃, dioxane, 110°, 24 h	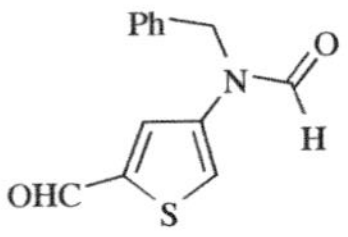 (43)	83
	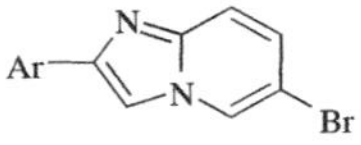 Ar = 4-FC₆H₄	CuI (10 mol %), DMCDA (10 mol %), K₃PO₄, toluene, 110°, 20 h	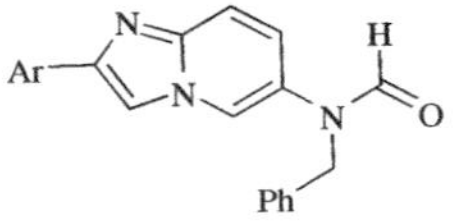 (71)	500
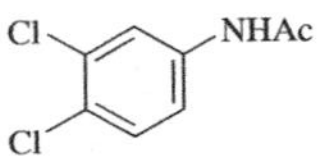	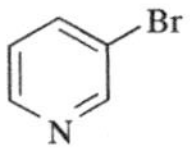	CuCl (10 mol %), K₂CO₃, xylene, N(CH₂CH₂OCH₂CH₂OMe)₃, reflux, 27 h	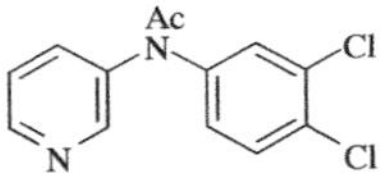 (61)	505
C₁₃				
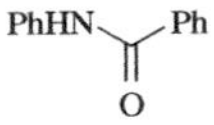	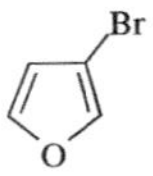	CuI (10 mol %), DMEDA (10 mol %), K₂CO₃, dioxane, 110°, 24 h	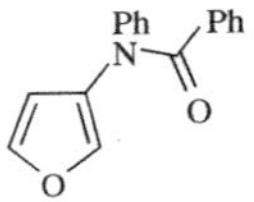 (11)	83

TABLE 17B. *N*-HETEROARYLATION OF ACYCLIC SECONDARY AMIDES (*Continued*)

Nitrogen Nucleophile	Heteroaryl Halide	Conditions	Product(s) and Yield(s) (%)	Refs.
C₁₃				
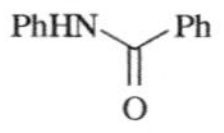 PhHN—C(Ph)=O	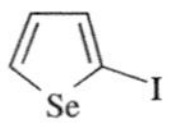 Se-I (selenophene)	CuI (10 mol %), EDA (20 mol %), K$_3$PO$_4$, dioxane, reflux, 24 h	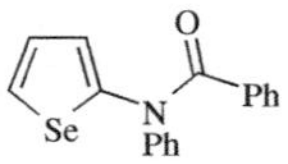 (selenophene)–N(Ph)C(=O)Ph (70)	87

Nitrogen Nucleophile	Aryl Halide	Conditions	Product(s) and Yield(s) (%)	Refs.

*Please refer to the charts preceding the tables for structures indicated by the **bold** numbers.*

C₃

R	X	Catalyst	x	Additive(s)	Solvent	Temp (°)	Time (h)		Refs.
H	I⁺Ph BF₄⁻	Cu(acac)₂	5	K₂CO₃	toluene	50	6	(70)	76
4-MeO	Br	CuI	1	DMCDA (10 mol %), K₂CO₃	dioxane	110	24	(90)	69
4-MeO	I	CuI	1	DMCDA (10 mol %), K₂CO₃	dioxane	110	15	(99)	115
3,5-Me₂	I	CuI	1	CDA (10 mol %), K₃PO₄	dioxane	110	23	(99)	115, 69

R	X	Catalyst	x	Additive(s)	Solvent(s)	Temp (°)	Time (h)		Refs.
H	Br	Cu/SiO₂	10	KOAc	—	reflux	4	(73)	516
H	Br	CuI	3	CDA (10 mol %), K₂CO₃	dioxane	110	15	(85)	517
H	I	CuI	5	KOH, TBAB	—	110	4	(85)	428
H	I	CuO nanoparticles	5	KOH	t-BuOH/DMSO (3:1)	110	24	(90)	74
3-F	I	CuI	10	DMEDA (10 mol %), K₂CO₃	MeCN	80	21	(99)	518
4-I	Br	CuI	10	DMCDA (20 mol %), K₃PO₄	dioxane	110	22	(95)	503
4-O₂N	Br	CuI	3	CDA (10 mol %), K₂CO₃	dioxane	110	15	(50)	517
4-O₂N	I	CuO nanoparticles	5	K₂CO₃	t-BuOH/DMSO (3:1)	110	24	(95)	74
2-MeO	I	CuI	10	DMEDA (10 mol %), K₂CO₃	MeCN	80	21	(99)	518
2-MeO	Br	CuI	3	CDA (10 mol %), K₂CO₃	dioxane	110	15	(83)	517
4-MeO	I	CuI	10	DMEDA (10 mol %), K₂CO₃	MeCN	80	21	(99)	518
4-MeO	Br	CuI	5	CDA (10 mol %), K₂CO₃	dioxane	110	15	(82)	517
4-MeO	I	CuO nanoparticles	10	K₂CO₃	t-BuOH/DMSO (3:1)	110	24	(54)	74
4-MeS	Br	CuI	3	CDA (10 mol %), K₂CO₃	dioxane	110	15	(86)	517
2-Me	I	CuI	10	DMEDA (10 mol %), K₂CO₃	MeCN	80	21	(99)	518
3-Me	Br	CuI	3	CDA (10 mol %), K₂CO₃	dioxane	110	15	(80)	517
4-NC–	Br	CuI	3	CDA (10 mol %), K₂CO₃	dioxane	110	15	(81)	517
4-CHO	Br	CuI	3	CDA (10 mol %), K₂CO₃	dioxane	110	15	(30)	517
4-MeO₂C	I	CuI	10	DMEDA (10 mol %), K₂CO₃	MeCN	80	21	(92)	518
3-Ac	I	CuI	10	DMEDA (10 mol %), K₂CO₃	MeCN	110	21	(93)	518
4-Ac	Br	CuI	3	CDA (10 mol %), K₂CO₃	dioxane	110	15	(98)	517

TABLE 18A. N-ARYLATION OF LACTAMS, OXAZOLIDINONES, AND CYCLIC IMIDES (Continued)

Nitrogen Nucleophile	Aryl Halide	Conditions						Product(s) and Yield(s) (%)	Refs.

Please refer to the charts preceding the tables for structures indicated by the bold numbers.

C4

Nitrogen Nucleophile: pyrrolidin-2-one (N–H lactam)

Aryl Halide: C6H5–X

Conditions — Catalyst (x mol %)

X	Catalyst	x	Additive(s)	Solvent(s)	Temp (°)	Time (h)	Product/Yield	Refs.
Br	Cu/SiO2	14	KOAc	—	reflux	—	(91)	516
I	CuBr	10	L9 (20 mol %), Cs2CO3	DMSO	rt	22	(96)	49
I	CuI	1	K2CO3	DMF	150	6	(89)	519
I+Ph BF4−	CuI	5	K2CO3	toluene	50	6	(65)	76
I	CuI	5	K3PO4	DMF	110	24	(98)	387
Br	CuI	20	DMG (20 mol %), K3PO4	DMF	reflux	48	(90)	494
I	CuI	5	Gly (20 mol %), K3PO4	dioxane	100	24	(98)	136
I	CuI	10	oxazolidin-2-one (20 mol %), NaOMe	DMSO	120	24	(85)	492
I	CuI	10	L21 ((20 mol %), Cs2CO3	MeCN	82	12	(86)	394
I	CuI	10	L45 (10 mol %), Cs2CO3	dioxane/DMF (9:1)	110	24	(98)	520
Cl	CuI	5	L42 (10 mol %), K3PO4	DMSO	110	5	(3)	493
Br	CuI	5	L42 (10 mol %), K3PO4	DMSO	110	5	(11)	493
I	CuI	5	L42 (10 mol %), K3PO4	DMSO	110	5	(90)	493
I	CuO	1	Fe(acac)3 (30 mol %), Cs2CO3	DMF	90	30	(81)	130
I	Cu2O	5	L30 (20 mol %), Cs2CO3; 3 Å MS	DMF	82	40	(92)	40
I	12	5	K3PO4	dioxane	110	24	(99)	521

Nitrogen Nucleophile	Aryl Halide	Conditions	Product(s) and Yield(s) (%)	Refs.
pyrrolidin-2-one + H2N–C(O)–OMe	4-iodo-1-bromobenzene	CuI (20 mol %), DMCDA (40 mol %), K3PO4, dioxane, 110°, 22 h	(60)	504
pyrrolidin-2-one + piperazin-2-one (N–H)	4-iodo-1-bromobenzene	CuI (20 mol %), DMCDA (40 mol %), K3PO4, dioxane, 110°, 22 h	(70)	504

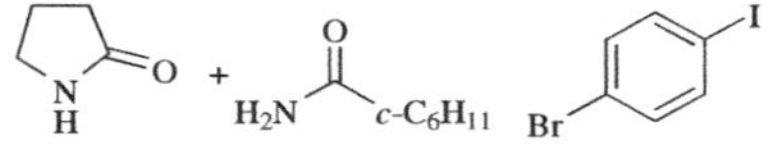

Reaction scheme:

pyrrolidinone + H_2N–C(O)–c-C_6H_{11} (with p-iodo bromobenzene)

CuI (20 mol %), DMCDA (40 mol %), K_3PO_4, dioxane, 110°, 22 h

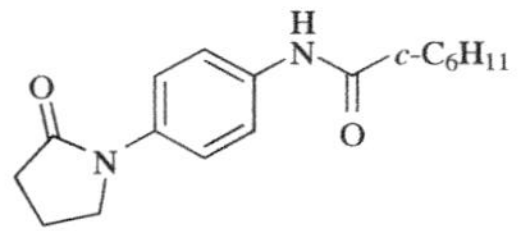

→ product (64)   504

CuI (x mol %), CDA (10 mol %), K_2CO_3, dioxane

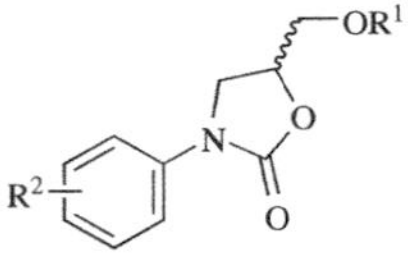

R^1	Config.	R^2	x	Temp (°)	Time (h)		
H	(R,S)	H	5	110	15	(20)	517
THP	(R,S)	H	5	110	15	(85)	517
THP	(R,S)	3-F, 4-(1-morpholino)	5	110	15	(85)	517
CPh_3	(S)	4-Ac	10	reflux	24	(95)	145

C_{4-5}

Catalyst (x mol %)

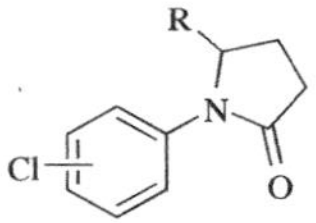

Isomer	R	X	Catalyst	x	Additive(s)	Solvent	Temp (°)	Time (h)		
2	H	Cl	Cu/SiO₂	14	KOAc	—	reflux	—	(37)	516
3	H	Br	Cu/SiO₂	14	KOAc	—	reflux	—	(44)	516
4	H	I	CuBr	10	**L9** (20 mol %), Cs_2CO_3	DMSO	rt	23	(90)	49
4	H	I	CuI	5	DMEDA (10 mol %), CsF	EtOAc	rt	24	(99)	63
4	H	Br	CuI	10	**L21** (20 mol %), Cs_2CO_3	MeCN	82	12	(61)	394
4	H	I	CuI	10	**L21** (20 mol %), Cs_2CO_3	MeCN	82	12	(92)	394
4	H	I	**12**	5	K_3PO_4	dioxane	110	24	(98)	521
4	Me	I	CuI	5	DMEDA (10 mol %), CsF	THF	rt	24	(83)	63

TABLE 18A. *N*-ARYLATION OF LACTAMS, OXAZOLIDINONES, AND CYCLIC IMIDES (*Continued*)

Nitrogen Nucleophile	Aryl Halide	Conditions	Product(s) and Yield(s) (%)	Refs.

*Please refer to the charts preceding the tables for structures indicated by the **bold** numbers.*

C4

(pyrrolidinone, N-H)

(1-bromo-4-iodobenzene)

Catalyst (x mol %)

Catalyst	x	Additive(s)	Solvent	Temp (°)	Time (h)		
CuBr	10	L9 (20 mol %), Cs$_2$CO$_3$	DMSO	rt	22	(85)	49
CuI	5	DMEDA (10 mol %), CsF	EtOAc	rt	24	(99)	63
CuI	5	Gly (20 mol %), K$_3$PO$_4$	dioxane	100	24	(93)	136
CuI	5	L42 (10 mol %), K$_3$PO$_4$	DMSO	110	5	(89)	493
12	5	K$_3$PO$_4$	dioxane	110	24	(70)	521

(N-(4-bromophenyl)pyrrolidinone)

(4-iodoaniline, H$_2$N)

Catalyst (5 mol %), 24 h

Catalyst	Additive(s)	Solvent	Temp (°)		
CuI	DMEDA (10 mol %), CsF	THF	rt	(99)	63
12	K$_3$PO$_4$	dioxane	110	(72)	521

(N-(4-aminophenyl)pyrrolidinone, H$_2$N)

(O$_2$N, X)

Catalyst (x mol %)

Isomer	X	Catalyst	x	Additive(s)	Solvent(s)	Temp (°)	Time (h)		
2	Cl	Cu/SiO$_2$	14	KOAc	—	reflux	—	(45)	516
2	I	CuI	5	Gly (20 mol %), K$_3$PO$_4$	dioxane	100	24	(95)	136
3	Cl	Cu/SiO$_2$	14	KOAc	—	reflux	—	(41)	516

(N-(nitrophenyl)pyrrolidinone, O$_2$N)

(MeO, X)

Catalyst (x mol %)

Isomer	X	Catalyst	x	Additive(s)	Solvent(s)	Temp (°)	Time (h)		
2	I	CuI	10	L45 (10 mol %), Cs$_2$CO$_3$	dioxane/DMF (9:1)	110	24	(91)	520
2	I	cat. **12**	5	K$_3$PO$_4$	dioxane	110	24	(93)	521
4	Br	CuBr	10	L5 (20 mol %), Cs$_2$CO$_3$	DMSO	75	30	(76)	49

(N-(methoxyphenyl)pyrrolidinone, MeO)

Catalyst (x mol %)

Isomer	X	Catalyst	x	Additive(s)	Solvent	Temp (°)	Time (h)	
4	I	CuBr	10	L5 (20 mol %), Cs$_2$CO$_3$	DMSO	75	25 (85)	49
4	I	CuI	10	EDA (10 mol %), K$_3$PO$_4$	dioxane	110	24 (85)	405
4	Cl	CuI	5	DMEDA (11 mol %), K$_2$CO$_3$	—	130	23 (51)	115
4	Cl	CuI	5	DMCDA (11 mol %), K$_2$CO$_3$	—	130	23 (51)	69
4	Br	CuI	10	oxazolidin-2-one (10 mol %), NaOMe	DMSO	120	24 (74)	492
4	I	CuI	20	DMG (20 mol %), K$_3$PO$_4$	DMF	reflux	48 (89)	494
4	I	CuI	5	Gly (20 mol %), K$_3$PO$_4$	dioxane	100	24 (95)	136
4	I	CuI	5	L42 (10 mol %), K$_3$PO$_4$	DMSO	110	5 (88)	493
4	I	CuI	5	L43 (10 mol %), K$_3$PO$_4$	DMSO	110	5 (82)	138
4	I	CuI	10	L45 (10 mol %), Cs$_2$CO$_3$	dioxane/DMF (9:1)	110	24 (96)	520
4	I	cat. 12	5	K$_3$PO$_4$	dioxane	110	24 (89)	521

Isomer	X	Catalyst	x	Additive(s)	Solvent	Temp (°)	Time (h)	
2	Br	CuI	20	DMG (20 mol %), K$_3$PO$_4$	DMF	reflux	48 (78)	494
2	I	CuI	5	L42 (10 mol %), K$_3$PO$_4$	DMSO	110	5 (86)	493
2	I	CuI	5	L43 (10 mol %), K$_3$PO$_4$	DMSO	110	5 (86)	138
4	Br	CuBr	10	L9 (20 mol %), Cs$_2$CO$_3$	DMSO	75	26 (81)	49
4	I	CuBr	10	L9 (20 mol %), Cs$_2$CO$_3$	DMSO	60	24 (86)	49
4	I$^+$C$_6$H$_4$Me-4 BF$_4^-$	CuI	5	K$_2$CO$_3$	toluene	50	6 (68)	76
4	Cl	CuI	5	DMEDA (11 mol %), K$_2$CO$_3$	—	130	23 (95)	115
4	Cl	CuI	5	DMCDA (11 mol %), K$_2$CO$_3$	—	130	23 (95)	69
4	I	CuI	5	Gly (20 mol %), K$_3$PO$_4$	dioxane	100	24 (95)	136
4	Br	CuI	10	oxazolidin-2-one (20 mol %), NaOMe	DMSO	120	24 (76)	492
4	Br	CuI	10	L21 (20 mol %), KI, Cs$_2$CO$_3$	MeCN	82	12 (58)	394
4	I	CuI	5	L42 (10 mol %), K$_3$PO$_4$	DMSO	110	5 (93)	493
4	I	CuI	5	L43 (10 mol %), K$_3$PO$_4$	DMSO	110	5 (91)	138

431

TABLE 18A. N-ARYLATION OF LACTAMS, OXAZOLIDINONES, AND CYCLIC IMIDES (Continued)

Nitrogen Nucleophile	Aryl Halide	Conditions	Product(s) and Yield(s) (%)	Refs.

*Please refer to the charts preceding the tables for structures indicated by the **bold** numbers.*

C_4

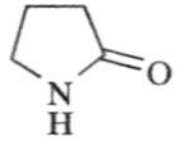

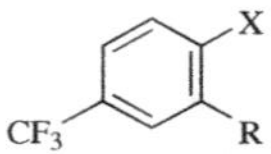

Nitrogen nucleophile: 2-pyrrolidinone (N–H lactam)

Aryl Halide: aryl ring bearing CF_3, R, and X substituents.

Catalyst (x mol %)

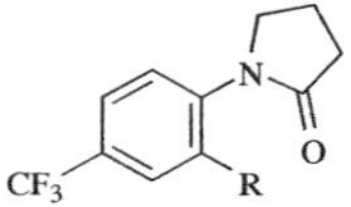

R	X	Catalyst	x	Additive(s)	Solvent	Temp (°)	Time (h)		Refs.
H	I	CuI	5	L42 (10 mol %), K$_3$PO$_4$	DMSO	110	5	(89)	493
H	I	CuI	5	L43 (10 mol %), K$_3$PO$_4$	DMSO	110	5	(81)	138
H	I	12	5	K$_3$PO$_4$	dioxane	110	24	(99)	521
O$_2$N	Cl	CuI	10	L21 (20 mol %), Cs$_2$CO$_3$	MeCN	82	12	(22)	394

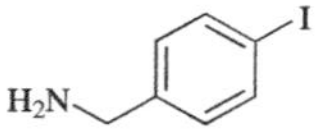

Aryl Halide (H$_2$N–CH$_2$ benzyl iodide):
CuI (5 mol %), DMEDA (10 mol %), CsF, THF, rt, 24 h — Product (93), 63

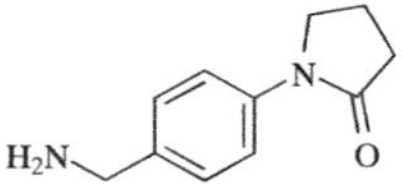

Aryl Halide (HO–CH$_2$ benzyl iodide):
CuI (5 mol %), DMEDA (10 mol %)

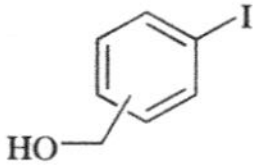

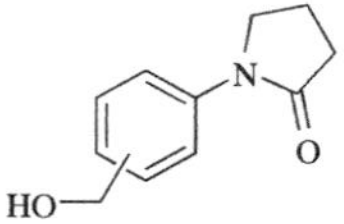

Isomer	Additive	Solvent	Temp (°)	Time (h)		Refs.
3	K$_3$PO$_4$	toluene	80	23	(97)	69
4	CsF	THF	rt	24	(95)	63

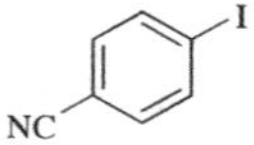

Aryl Halide (NC aryl iodide):
CuI (5 mol %), DMEDA (10 mol %), CsF, THF, rt, 24 h — Product (89), 63

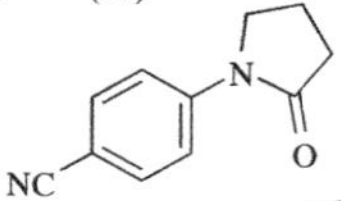

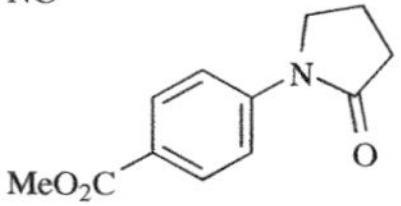

Aryl Halide (MeO$_2$C aryl):
CuI (5 mol %)

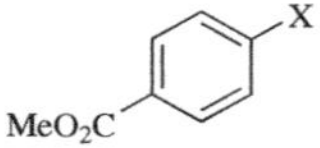

X	Additives	Solvent	Temp (°)	Time (h)		Refs.
Cl	DMEDA (11 mol %), K$_2$CO$_3$	—	130	23	(62)	115

432

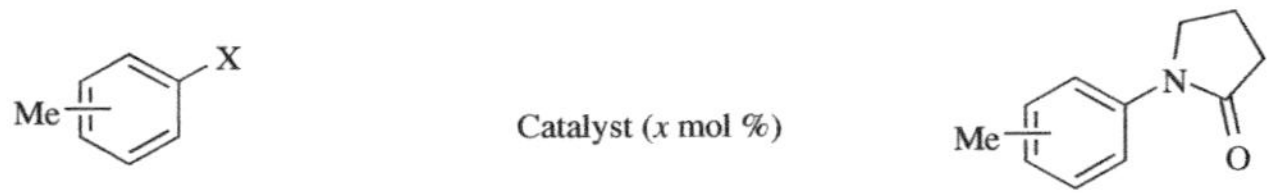

Catalyst (x mol %)

Isomer	X	Catalyst	x	Additive(s)	Solvent	Temp (°)	Time (h)	
3,5	I	CuI	10	EDA (10 mol %), Cs₂CO₃	dioxane	110	24 (68)	405
3,5	I	CuI	1	CDA (10 mol %), K₃PO₄	dioxane	110	23 (99)	115
3,5	I	**12**	5	K₃PO₄	dioxane	110	24 (85)	521
2,4,6	Br	Cu/SiO₂	14	KOAc	—	reflux	— (48)	516

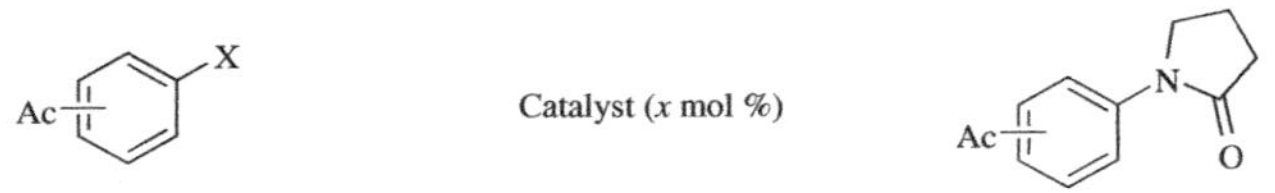

Catalyst (x mol %)

Isomer	X	Catalyst	x	Additive(s)	Solvent(s)	Temp (°)	Time (h)	
2	Br	CuI	5	DMEDA (10 mol %), K₂CO₃, 5 Å MS	—	110	24 (89)	497
4	I	CuI	5	Gly (20 mol %), K₃PO₄	dioxane	100	24 (98)	136
4	Br	CuI	20	DMG (20 mol %), K₃PO₄	DMF	reflux	48 (90)	494
4	I	CuI	5	**L42** (10 mol %), K₃PO₄	DMSO	110	5 (90)	493
4	I	CuI	5	**L43** (10 mol %), K₃PO₄	DMSO	110	5 (62)	138
4	I	CuI	10	**L45** (10 mol %), Cs₂CO₃	dioxane/DMF (9:1)	110	34 (86)	520
4	Br	**10**	3	Cs₂CO₃	DMSO	100	24 (<1)	420
4	I	**10**	3	Cs₂CO₃	DMSO	100	24 (100)	420

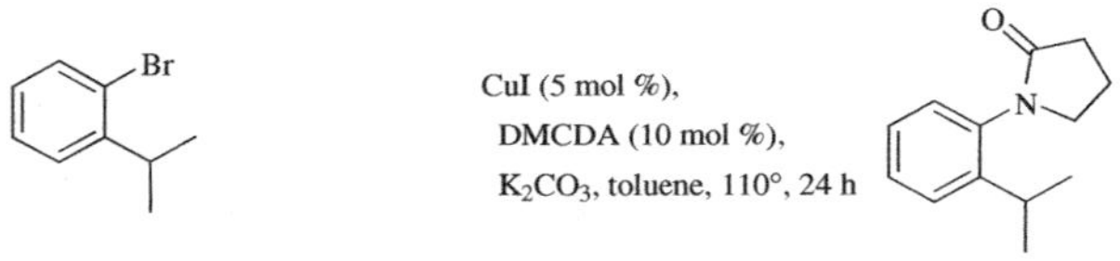

CuI (5 mol %),
DMCDA (10 mol %),
K₂CO₃, toluene, 110°, 24 h

(94)

69

TABLE 18A. *N*-ARYLATION OF LACTAMS, OXAZOLIDINONES, AND CYCLIC IMIDES (*Continued*)

Nitrogen Nucleophile	Aryl Halide	Conditions	Product(s) and Yield(s) (%)	Refs.

*Please refer to the charts preceding the tables for structures indicated by the **bold** numbers.*

C₄

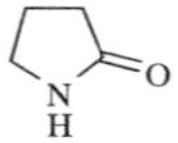

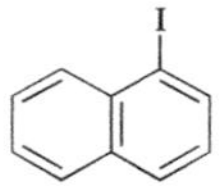

First entry:

CuI (5 mol %), additive (10 mol %), K₃PO₄, DMSO, 110°, 5 h

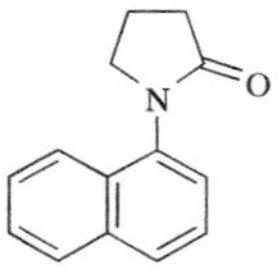

Additive	
L42	(93)
L43	(92)

Refs. 493, 138

Second entry:

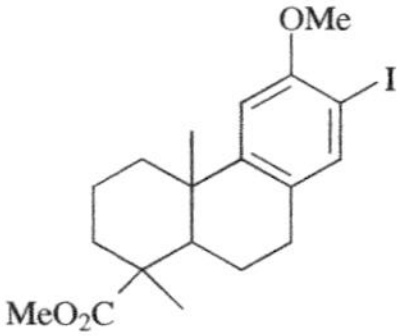

CuI (1 mol %), DMEDA (2 mol %), K₂CO₃, dioxane, 100°, 24 h

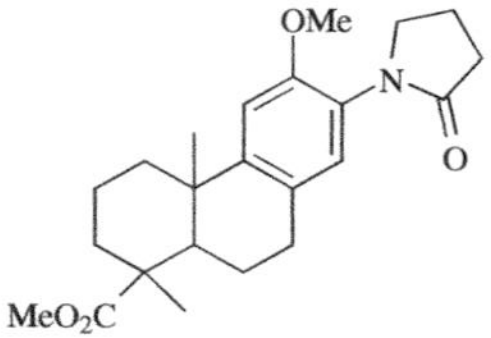

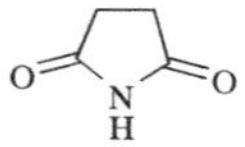

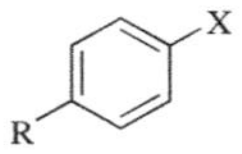

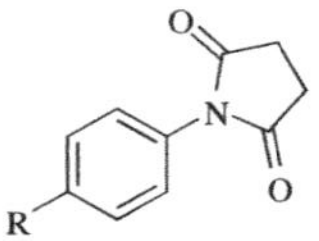

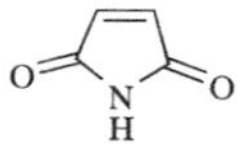

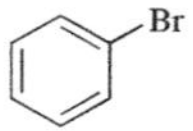

(81) — Ref. 489

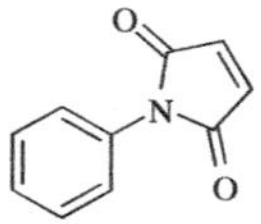

Third entry — Catalyst (*x* eq):

R	X	Catalyst	*x*	Solvent	Temp (°)	Time		Refs.
H	Cl	Cu	3	—	210	12 h	(100)	303
H	Br	Cu	3	—	210	12 h	(100)	303
H	I	Cu	3	—	150	12 h	(97)	303
H	I	Cu (active)	5	—	MW (100 W)	10 min	(87)	280
H	Br	CuI	1	DMA	165	24 h	(62)	515
Me	Br	Cu (active)	5	—	MW (100 W)	6 min	(83)	280

Fourth entry:

CuI (1 eq), DMA, 165°, 24 h

(62) — Ref. 515

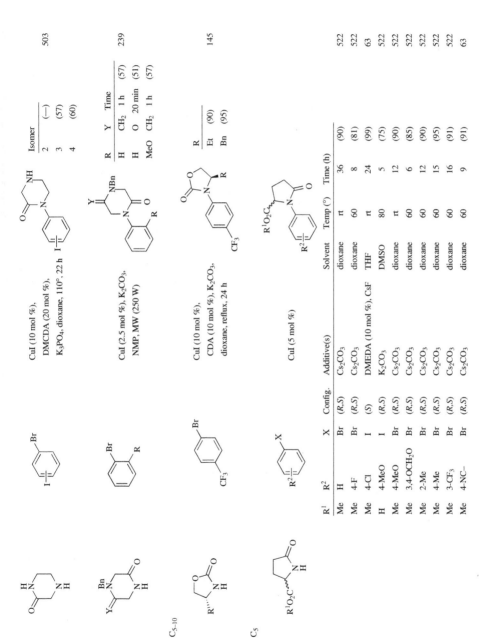

CuI (10 mol %), DMCDA (20 mol %), K₃PO₄, dioxane, 110°, 22 h — 503

Isomer		
2	(—)	
3	(57)	
4	(60)	

CuI (2.5 mol %), K₂CO₃, NMP, MW (250 W) — 239

R	Y	Time	
H	CH₂	1 h	(57)
H	O	20 min	(51)
MeO	CH₂	1 h	(57)

CuI (10 mol %), CDA (10 mol %), K₂CO₃, dioxane, reflux, 24 h — 145

R	
Et	(90)
Bn	(95)

CuI (5 mol %)

R^1	R^2	X	Config.	Additive(s)	Solvent	Temp (°)	Time (h)	
Me	H	Br	(R,S)	Cs₂CO₃	dioxane	rt	36	(90)	522
Me	4-F	Br	(R,S)	Cs₂CO₃	dioxane	60	8	(81)	522
Me	4-Cl	I	(S)	DMEDA (10 mol %), CsF	THF	rt	24	(99)	63
H	4-MeO	I	(R,S)	K₂CO₃	DMSO	80	5	(75)	522
Me	4-MeO	Br	(R,S)	Cs₂CO₃	dioxane	rt	12	(90)	522
Me	3,4-OCH₂O	Br	(R,S)	Cs₂CO₃	dioxane	60	6	(85)	522
Me	2-Me	Br	(R,S)	Cs₂CO₃	dioxane	60	12	(90)	522
Me	4-Me	Br	(R,S)	Cs₂CO₃	dioxane	60	15	(95)	522
Me	3-CF₃	Br	(R,S)	Cs₂CO₃	dioxane	60	16	(91)	522
Me	4-NC–	Br	(R,S)	Cs₂CO₃	dioxane	60	9	(91)	63

C₅₋₁₀

C₅

Nitrogen Nucleophile	Aryl Halide	Conditions	Product(s) and Yield(s) (%)	Refs.

*Please refer to the charts preceding the tables for structures indicated by the **bold** numbers.*

C₅

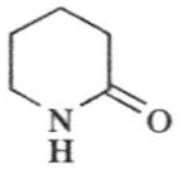

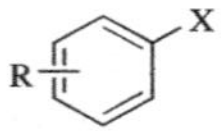

CuI (x mol %)

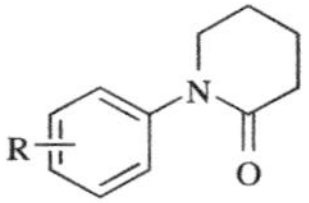

R	X	x	Additive(s)	Solvent	Temp (°)	Time (h)		Refs.
H	I	10	phen (10 mol %), KF/Al₂O₃	toluene	110	8	(99)	490
H	I	5	**L42** (10 mol %), K₃PO₄	DMSO	110	5	(81)	493
H	I	5	**L43** (10 mol %), K₃PO₄	DMSO	110	5	(74)	138
4-Br	I	5	K₃PO₄	DMF	110	24	(95)	387
2-Me	Br	5	DMEDA (20 mol %), K₂CO₃	toluene	110	24	(49)	69
4-Me	I	5	K₃PO₄	DMF	110	24	(93)	387
4-H₂NCH₂	I	5	DMEDA (10 mol %), K₂CO₃	toluene	100	18	(95)	69
4-EtO₂C	Br	5	DMEDA (10 mol %), K₂CO₃	toluene	110	24	(94)	69

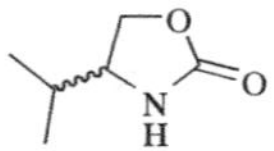

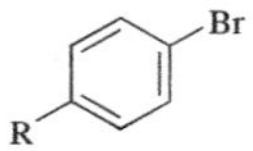

CuI (x mol %),
CDA (10 mol %),
K₂CO₃, dioxane

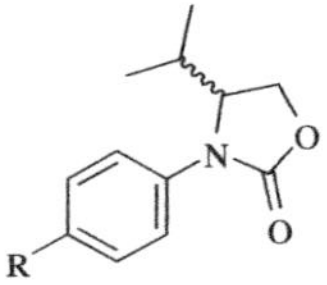

Config.	R	x	Temp (°)	Time		Refs.
(R,S)	MeO	3	110	15	(50)	517
(R,S)	NC–	3	110	15	(99)	517
(R)	Ac	10	reflux	24	(80)	145

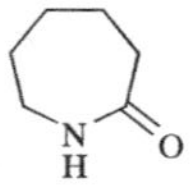

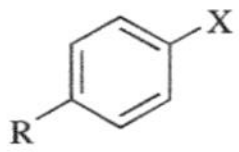

C₆

CuI (x mol %)

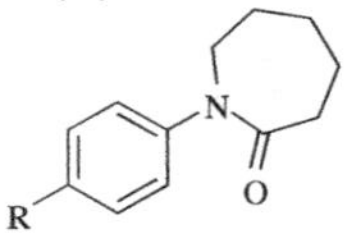

R	X	x	Additive(s)	Solvent	Temp (°)	Time (h)		Refs.
H	Br	10	K₃PO₄	DMF	110	24	(trace)	387
H	I	5	K₃PO₄	DMF	110	24	(98)	387
H	Cl	5	Gly (20 mol %), K₃PO₄	dioxane	100	24	(26)	136
H	Br	5	Gly (20 mol %), K₃PO₄	dioxane	100	24	(62)	136

									Refs.
H	I	5	Gly (20 mol %), K$_3$PO$_4$	dioxane	100	24	(97)		136
H	I	10	PPAPM (20 mol %), K$_3$PO$_4$	—	110	30	(89)		48
H	Br	20	DMG (20 mol %), K$_3$PO$_4$	DMF	reflux	48	(87)		494
H	Br	10	piperidine-2-carboxylic acid (20 mol %), K$_2$CO$_3$	DMF	110	36	(70)		67
Br	I	5	K$_3$PO$_4$	toluene	110	24	(94)		387
MeO	Br	20	DMG (20 mol %), K$_3$PO$_4$	DMF	reflux	48	(88)		494
MeO	I	10	DMCDA (20 mol %), K$_3$PO$_4$	[C$_4$min][BF$_4$]	110	12	(78)		232
MeO	I	5	Gly (20 mol %), K$_3$PO$_4$	dioxane	100	24	(92)		136
Ac	I	10	K$_3$PO$_4$	toluene	110	24	(95)		387
Ac	I	5	Gly (20 mol %), K$_3$PO$_4$	dioxane	110	24	(98)		136

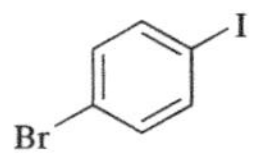

1. CuI (15 mol %),
   DMG (20 mol %), K$_3$PO$_4$,
   DMSO, 80°, 24 h
2. ArOH, 120°, 48 h

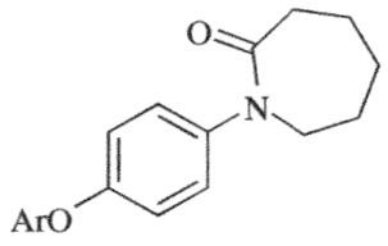

Ar	
Ph	(82)
4-ClC$_6$H$_4$	(75)
2-MeOC$_6$H$_4$	(82)
4-MeOC$_6$H$_4$	(87)
4-HO$_2$CC$_6$H$_4$	(0)
4-AcC$_6$H$_4$	(45)
2-Np	(85)

495

TABLE 18A. N-ARYLATION OF LACTAMS, OXAZOLIDINONES, AND CYCLIC IMIDES (Continued)

Nitrogen Nucleophile	Aryl Halide	Conditions	Product(s) and Yield(s) (%)	Refs.

*Please refer to the charts preceding the tables for structures indicated by the **bold** numbers.*

C_8

Conditions: CuI (x mol %), K₂CO₃

R	X	x	Additive(s)	Solvent	Temp (°)	Time (h)	Yield	Refs.
H	Br	10	DMEDA (10 mol %)	MeCN	80	21	(43)	518
4-F	I	1	DMCDA (4 mol %)	dioxane	100	24	(94)	205
4-Cl	Br	10	DMCDA (20 mol %)	dioxane	100	24	(77)	205
3,4-Cl₂	Br	10	DMEDA (10 mol %)	MeCN	80	21	(50)	518
4-Br	I	5	DMCDA (10 mol %)	dioxane	40	24	(61)	205
4-Me₂N	Br	10	DMCDA (20 mol %)	dioxane	100	24	(73)	205
4-HO	I	10	DMEDA (10 mol %)	MeCN	80	21	(73)	518
4-MeO	I	5	DMCDA (10 mol %)	dioxane	100	24	(86)	205
3,4-OCH₂O	Br	10	DMEDA (10 mol %)	MeCN	80	21	(50)	518
2-F, 4-Me	Br	10	DMEDA (10 mol %)	MeCN	80	21	(50)	518
4-CF₃	Br	10	DMEDA (10 mol %)	MeCN	80	21	(50)	518
4-CF₃	I	1	DMCDA (4 mol %)	dioxane	80	24	(69)	518
4-NC–	I	5	DMCDA (10 mol %), 4 Å MS	dioxane	80	24	(72)	205
3-MeO₂C	Br	10	DMEDA (10 mol %)	MeCN	80	21	(40)	518
3-MeO₂C	I	10	DMEDA (10 mol %)	MeCN	80	21	(74)	518
3,4-Me₂	I	10	DMEDA (10 mol %)	MeCN	80	21	(86)	518
3,5-Me₂	Br	10	DMCDA (20 mol %)	dioxane	100	24	(71)	205
3,5-Me₂	I	10	DMEDA (10 mol %)	MeCN	80	21	(86)	518

C_8–14

Conditions: CuI (x mol %), K₂CO₃

R¹	R²	X	x	Additive(s)	Solvent	Temp (°)	Time (h)	Yield	Refs.
5-F	3-MeO₂C	I	10	DMEDA (10 mol %)	MeCN	80	21	(61)	518
4-Cl	3-EtO₂C	I	5	DMCDA (10 mol %), 4 Å MS	dioxane	80	24	(87)	205
5-Cl	4-n-Pr	Br	10	DMEDA (10 mol %)	MeCN	80	21	(50)	518

438

				Catalyst (x amount)					
6-Cl	3-MeO₂C	I	10	DMEDA (10 mol %)	MeCN	80	21	(82)	518
6-Cl	3,5-Me₂	I	10	DMEDA (10 mol %)	MeCN	80	21	(69)	518
6-Br	3,4-Me₂	I	10	DMEDA (10 mol %)	MeCN	80	21	(72)	518
6-Me	3-O₂N	I	5	DMCDA (10 mol %)	dioxane	80	24	(87)	205
6-Me	3-MeO	Br	10	DMCDA (20 mol %)	dioxane	100	24	(62)	205
6-Me	2,4-Me₂	Br	10	DMEDA (10 mol %)	MeCN	80	24	(50)	518
6-CF₃	3-MeO₂C	I	10	DMEDA (10 mol %)	MeCN	80	21	(65)	518
6-CF₃	3,4-Me₂	I	10	DMEDA (10 mol %)	MeCN	80	21	(77)	518
6-MeO₂C	3,4-Me₂	I	10	DMEDA (10 mol %)	MeCN	80	21	(66)	518
6-Ph	3,4-Me₂	Br	10	DMEDA (10 mol %)	MeCN	80	21	(50)	518

C_8

R	X	Catalyst	x	Additive(s)	Solvent	Temp (°)	Time	
H	I	Cu (active)	5 eq	none	—	MW (100 W)	8 min (81)	280
H	Br	Cu	3 eq	none	—	240	12 h (100)	303
H	Br	CuI	1 eq	none	DMA	165	24 h (62)	515
H	I	CuI	10 mol %	K₂CO₃	DMF	150	6 h (32)	519
H	Br	CuI	20 mol %	DMG (20 mol %), K₃PO₄	DMF	reflux	48 h (60)	494
H	I	CuI	5 mol %	Gly (20 mol %), K₃PO₄	dioxane	100	24 h (53)	136
H	Br	Cu₂O	60 mol %	none	collidine	—	— (80)	523
H	I	Cu₂O	60 mol %	none	collidine	—	— (92)	523
F	Br	Cu	3 eq	none	—	240	12 h (98)	303
MeO	Br	Cu	3 eq	none	—	240	12 h (86)	303
Me	Br	Cu	3 eq	none	—	240	12 h (100)	303

TABLE 18A. *N*-ARYLATION OF LACTAMS, OXAZOLIDINONES, AND CYCLIC IMIDES (*Continued*)

Nitrogen Nucleophile	Aryl Halide	Conditions	Product(s) and Yield(s) (%)	Refs.

*Please refer to the charts preceding the tables for structures indicated by the **bold** numbers.*

C8

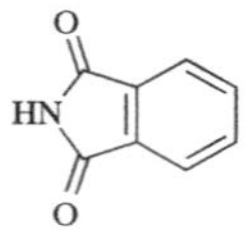

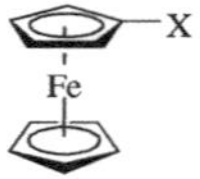

Cu2O (60 mol %)

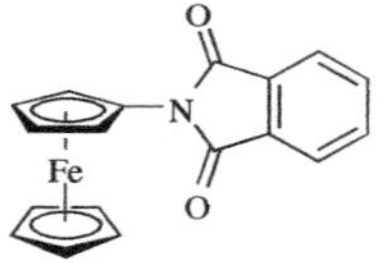

X	Solvent	
Br	4-picoline	(71)
I	pyridine	(73)

523

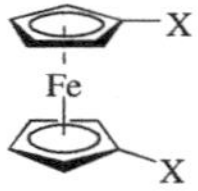

Cu2O (60 mol %),
4-picoline,
reflux, 24 h

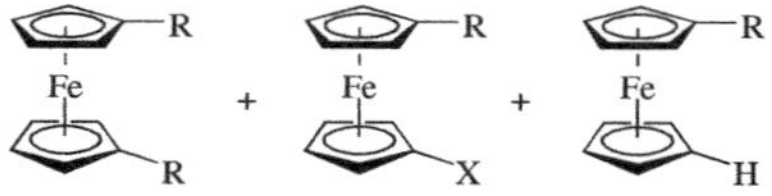

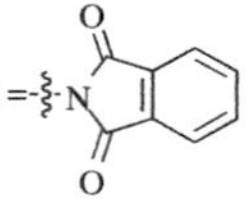

X	I	II	III
Br	(47)	(0)	(8)
I	(35)	(11)	(21)

R = ξ–N

523

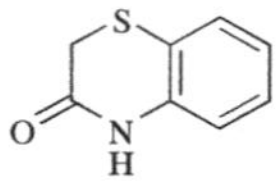

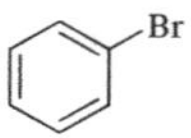

Cu/SiO2 (14 mol %),
KOAc, reflux

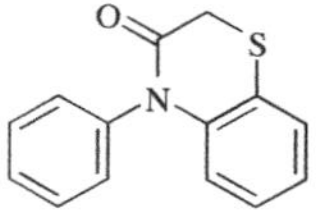

(72)

516

C9

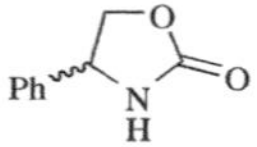

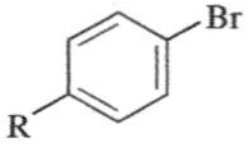

CuI (x mol %),
CDA (10 mol %),
K2CO3, dioxane

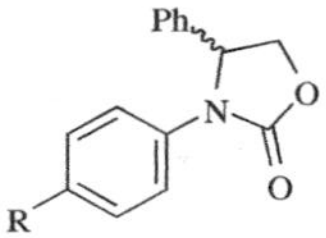

Config.	R	x	Temp (°)	Time (h)	
(R)	H	10	reflux	24	(95)
(R,S)	MeO	3	110	15	(98)
(R,S)	Me	10	reflux	24	(85)
(R,S)	NC–	3	110	15	(98)
(R)	NC–	10	reflux	24	(82)

145
517
145
517
145

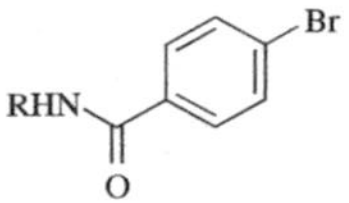

CuI (x mol %), K_2CO_3

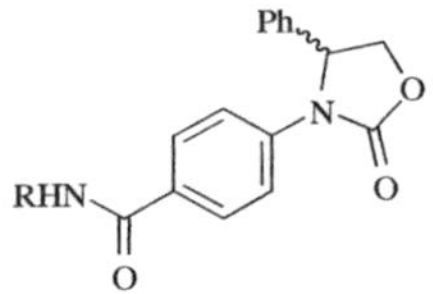

Config.	R	x	Additive	Solvent	Temp (°)	Time (h)		
(*R*)	*n*-Pr	10	CDA (10 mol %)	dioxane	reflux	24	(97)	145
(*R*)	*i*-Pr	—	DMEDA	—	—	—	(97)	524
(*S*)	*i*-Pr	10	CDA (10 mol %)	dioxane	reflux	24	(95)	145

C_{9-10}

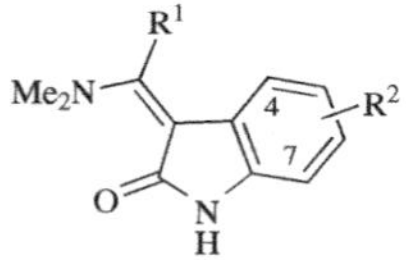

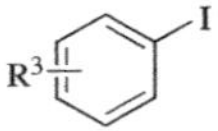

1. NaH
2. CuBr (1 eq), DMF, 100°, 24 h

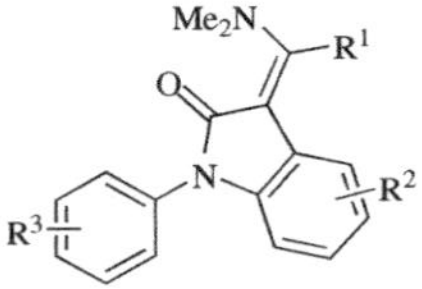

525

R^1	R^2	R^3	
H	H	2-Cl	(12)
H	4-Cl	3-Cl	(5)
H	5-Cl	3-Cl	(24)
H	6-Cl	3-Cl	(2)
H	7-Cl	3-Cl	(31)
H	5-MeO	3-Cl	(9)
H	H	4-Cl	(10)
H	H	2,5-Cl_2	(19)
H	H	3,4-Cl_2	(28)
H	H	3,5-Cl_2	(72)
H	H	3-Br	(24)
H	H	4-Me_2N	(32)
H	H	3-O_2N	(51)
H	H	3-HO	(83)
H	H	3-MeO	(83)
H	H	4-MeO	(42)
H	H	3-F, 4-MeO	(38)

R^1	R^2	R^3	
H	H	3-Cl, 4-MeO	(38)
H	H	3-Cl, 6-MeO	(27)
H	H	3,4-$(MeO)_2$	(6)
H	H	3-EtO	(17)
H	H	3-Cl, 4-MeO, 5-Me	(33)
H	H	4-MeS	(32)
H	H	2-Me	(10)
H	H	3-Me	(38)
Me	H	3-Me	(8)
H	H	4-Me	(23)
H	H	3-Me, 4-MeO	(18)
H	H	3-CF_3	(21)
H	H	3-NC–	(22)
H	H	3-Me_2NCO	(14)
H	H	3-CHO	(18)
H	H	3-Et	(39)
H	H	3-Ph	(39)

TABLE 18A. *N*-ARYLATION OF LACTAMS, OXAZOLIDINONES, AND CYCLIC IMIDES (*Continued*)

Nitrogen Nucleophile	Aryl Halide	Conditions	Product(s) and Yield(s) (%)	Refs.

*Please refer to the charts preceding the tables for structures indicated by the **bold** numbers.*

C$_9$

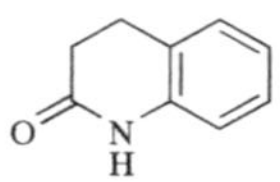

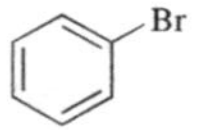

CuI (x mol %), K$_2$CO$_3$

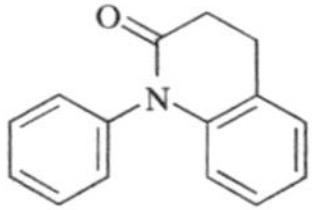

x	Additive	Solvent	Temp (°)	Time (h)	
2.5	none	NMP	reflux	—	(77)
20	DMCDA (20 mol %)	dioxane	125	—	(80)

239
526

C$_{10}$

CuI, CDA, K$_3$PO$_4$, DMF, 110°, 12 h

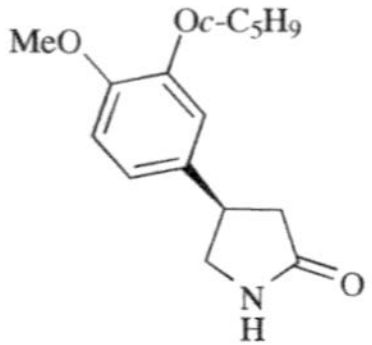

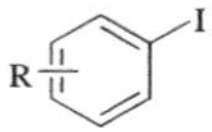

R = H;

2-H$_2$N, 3-H$_2$N, 4-H$_2$N, 5-H$_2$N, 3,5-(H$_2$N)$_2$;

3-F, 3-O$_2$N;

3-MeSO, 3-MeSO$_2$, 4-MeSO, 4-MeSO$_2$;

2-Me, 3-Me, 4-Me;

2-CF$_3$, 3-CF$_3$, 4-CF$_3$;

2-NC–, 3-NC–, 4-NC–

2 R groups = 3-H$_2$N and 2-Me; 3-H$_2$N and 4-Me

(—)

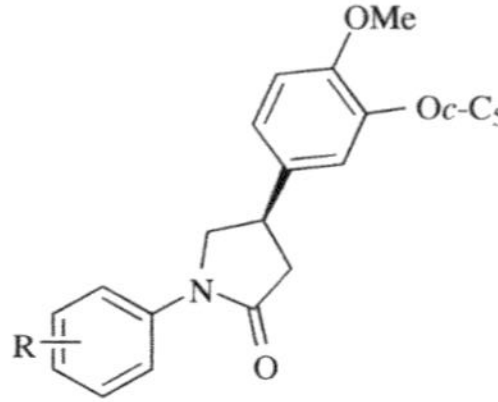

527

CuI, CDA, K$_3$PO$_4$, DMF, 110°, 12 h

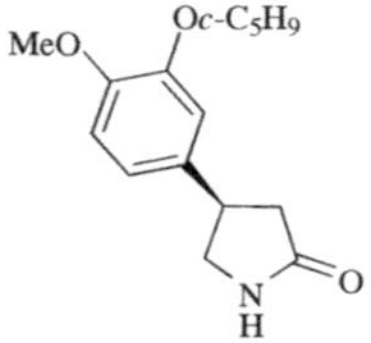

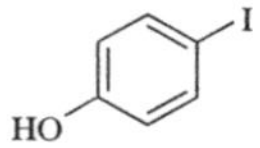

(92)

527

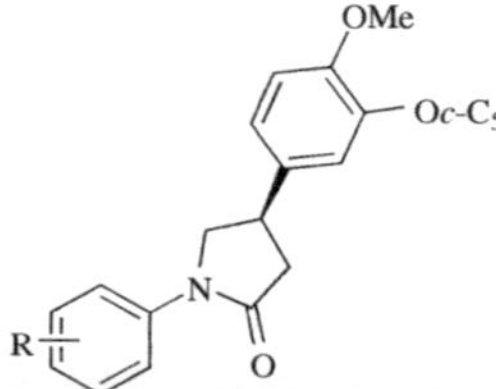

442

C_{12}

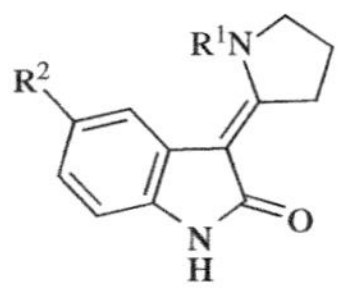

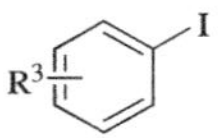

1. NaH
2. CuBr (1 eq), DMF, 100°, 24 h

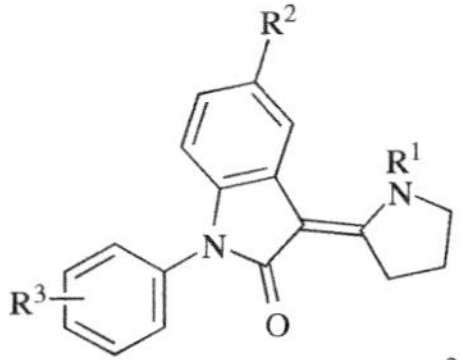

R^1	R^2	R^3	
Me	Cl	3-Cl	(43)
Me	H	4-Cl	(48)
Me	H	3-MeO	(20)
Me	H	3-NC–	(16)
Me	H	3-Me$_2$NCO	(17)
4-ClC$_6$H$_4$CH$_2$	H	3-Cl	(29)

525

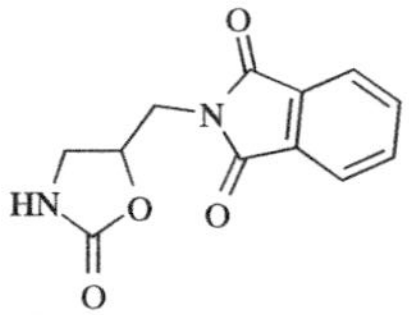

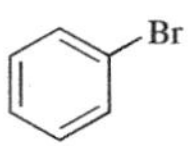

C_{12-13}

CuI (5 mol %),
CDA (10 mol %), K$_2$CO$_3$,
dioxane, 110°, 15 h

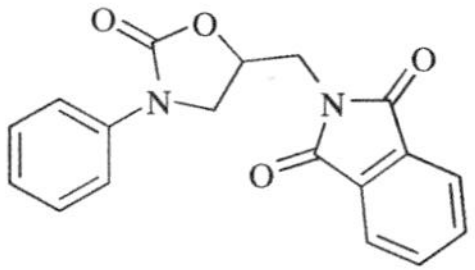

(65)

517

1. CuI (20 mol %),
CDA (20 mol %),
K$_2$CO$_3$, dioxane, 125°
2. TFA, CH$_2$Cl$_2$, rt, 2 h

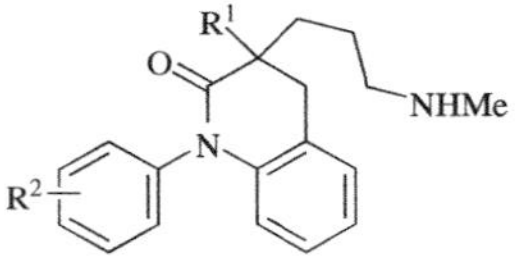

(>80)

526

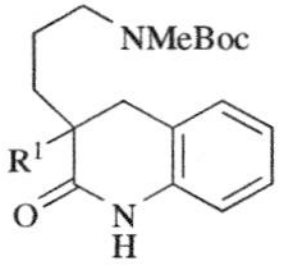

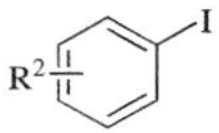

R^1	R^2	R^1	R^2
H	H	H	3,4-Cl$_2$
Me	H	Me	3,5-Cl$_2$
H	3-F	Me	3-Me
H	4-F	H	4-Me
Me	3,4-F$_2$	Me	4-Me
Me	3,5-F$_2$	H	4-CF$_3$
H	3-Cl	H	4-Et
H	4-Cl	Me	4-i-Pr

TABLE 18A. *N*-ARYLATION OF LACTAMS, OXAZOLIDINONES, AND CYCLIC IMIDES (*Continued*)

*Please refer to the charts preceding the tables for structures indicated by the **bold** numbers.*

Nitrogen Nucleophile	Aryl Halide	Conditions	Product(s) and Yield(s) (%)	Refs.

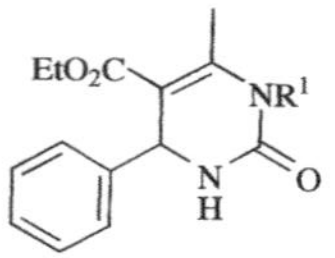

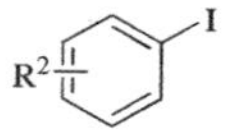

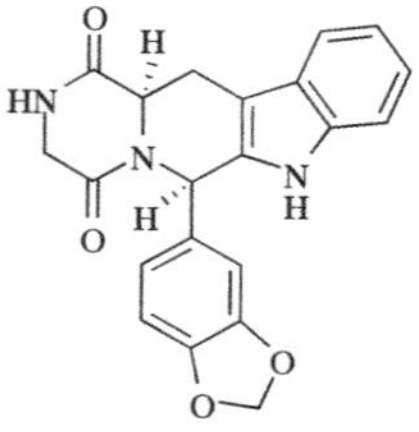

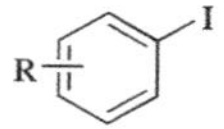

C$_{12}$

CuI (20 mol %), Cs$_2$CO$_3$,
DMF, 180°, 40 min

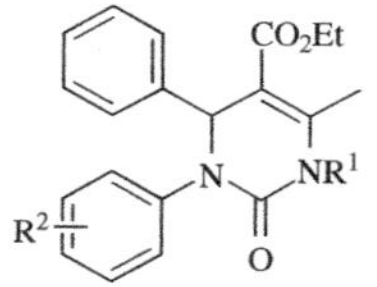

R^1	R^2	
H	H	(36)
H	4-EtO$_2$C	(13)
Me	H	(69)
Me	4-Cl	(70)
Me	4-O$_2$N	(49)
Me	3-MeO	(63)
Me	4-Me	(34)
Me	2-MeO$_2$C	(83)
Me	4-EtO$_2$C	(80)
Me	4-MeCO	(67)

528

C$_{20}$

CuI (10 mol %),
CDA (10 mol %), K$_3$PO$_4$,
dioxane, 14°, 14 d

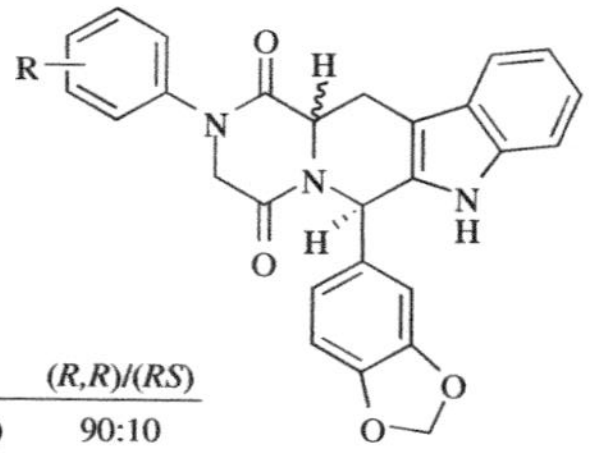

R		(*R,R*)/(*RS*)
H	(80)	—^{a}
3-Cl	(81)	80:20
4-Cl	(43)	66:34
4-O$_2$N	(60)	17:83

R		(*R,R*)/(*RS*)
4-MeO	(52)	90:10
3-Me	(52)	—^{a}
4-CF$_3$	(50)	60:40
4-*n*-Bu	(63)	76:24

529

^{a} Only the (*R,R*)-isomer was isolated.

TABLE 18B. N-HETEROARYLATION OF LACTAMS, OXAZOLIDINONES, AND CYCLIC IMIDES

*Please refer to the charts preceding the tables for structures indicated by the **bold** numbers.*

Nitrogen Nucleophile	Heteroaryl Halide	Conditions	Product(s) and Yield(s) (%)	Refs.

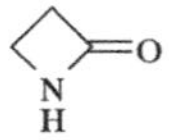

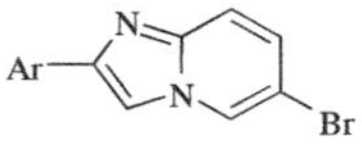

C₃

| | | CuI (10 mol %), DMEDA (10 mol %), K₂CO₃, toluene, 110°, 20 h | 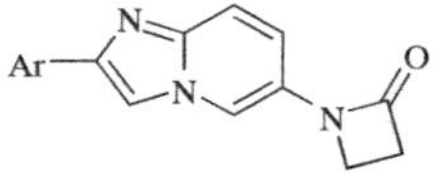 (80) | 500 |
| | | CuI (10 mol %), DMEDA (10 mol %), K₂CO₃, MeCN, 80°, 21 h | 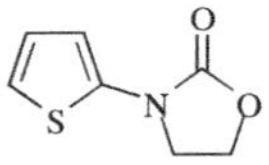 (99) | 518 |

| | 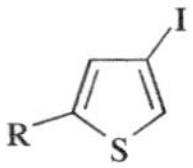 | CuI (x mol %) | 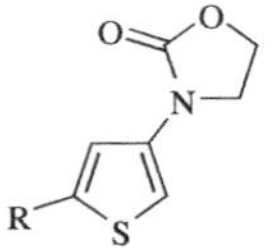 | |

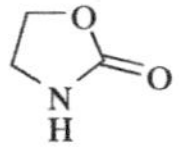

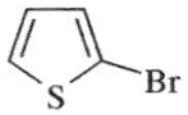

R	x	Additives	Solvent	Temp (°)	Time (h)		
H	1	DMEDA (1 mol %), K₂CO₃	dioxane	110	24	(99)	501
H	10	DMEDA (10 mol %), K₂CO₃	—	110	24	(99)	83
CHO	1	phen (10 mol %), Cs₂CO₃	DMF	80	21	(85)	501
CHO	10	DMEDA (10 mol %), K₂CO₃	—	110	24	(85)	83

C₃₋₁₀

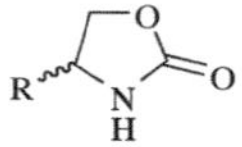

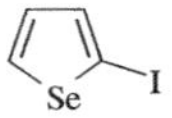

| | | CuI (10 mol %), EDA (20 mol %), K₃PO₄, dioxane, reflux, 24 h | | |

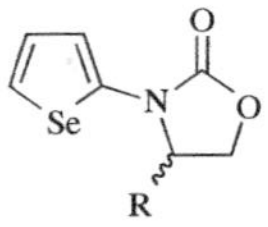

R	Config.	
H	—	(90)
i-Pr	(S)	(43)
(S)-s-Bu	(R)	(65)
Bn	(S)	(95)

87

C₄

| 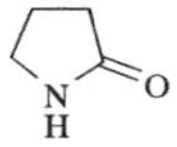 | 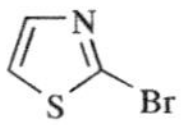 | CuI (10 mol %), DMEDA (10 mol %), dioxane, 110°, 24 h | 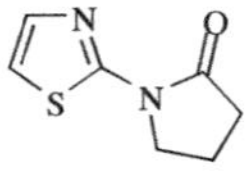 | |

	Additive	
	K₂CO₃	(58)
	K₃PO₄	(58)

83
501

TABLE 18B. *N*-HETEROARYLATION OF LACTAMS, OXAZOLIDINONES, AND CYCLIC IMIDES (*Continued*)

Nitrogen Nucleophile	Heteroaryl Halide	Conditions	Product(s) and Yield(s) (%)	Refs.

*Please refer to the charts preceding the tables for structures indicated by the **bold** numbers.*

C₄

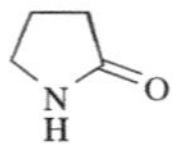

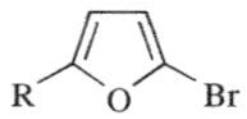

	Conditions	Product	Refs.
	CuI (10 mol %), DMEDA (10 mol %), dioxane, 110°, 24 h	 R / Additive H / K₂CO₃ (99) H / K₃PO₄ (82) MeO₂C / K₂CO₃ (77) MeO₂C / K₃PO₄ (77)	83 501 83 501
	CuI (10 mol %), DMEDA (10 mol %), dioxane, 110°, 24 h	 Additive K₂CO₃ (80) K₃PO₄ (80)	83 501

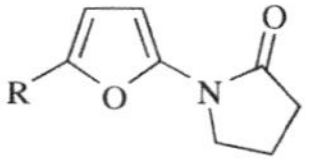

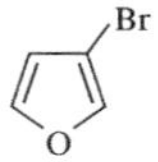

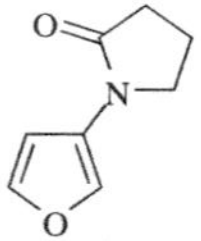

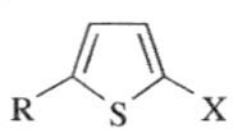

CuI (*x* mol %), dioxane, 110°

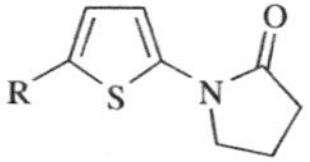

R	X	*x*	Additives	Time (h)	
H	Br	10	CDA (10 mol %), K₃PO₄	24	(96)
H	I	1	CDA (10 mol %), K₃PO₄	23	(97)
H	Br	10	DMEDA (10 mol %), K₂CO₃	24	(99)
H	I	10	EDA (10 mol %), K₃PO₄	24	(95)
H	I	1	DMCDA (10 mol %), K₃PO₄	24	(99)
Me	Br	10	DMEDA (10 mol %), K₂CO₃	24	(90)

Product:

Refs.: 69, 69, 83, 405, 115, 83, 501

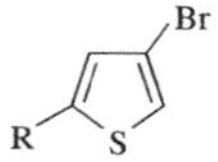

| | CuI (10 mol %), DMEDA (10 mol %), K₂CO₃, dioxane, 110°, 24 h |

R / Additive
H / K₂CO₃ (99)
CHO / K₃PO₄ (48) | 83
501 |
| | CuI (10 mol %), EDA (20 mol %), K₃PO₄, | (77) | 87 |

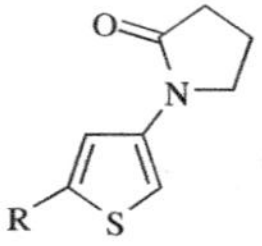

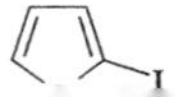

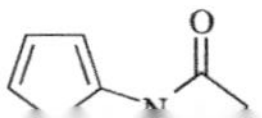

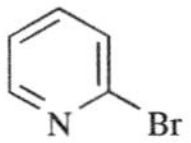

C₅

Me labeled pyrrolidin-2-one (positions 3, 5)

MeO₂C-pyrrolidin-2-one

MeO₂C-pyrrolidin-2-one

Catalyst (x mol %)

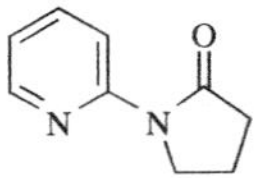

2-bromopyridine

Catalyst	x	Additive(s)	Solvent	Temp (°)	Time (h)		
Cu/SiO₂	14	KOAc	—	reflux		(51)	516
CuBr	10	L9 (20 mol %), Cs₂CO₃	DMSO	60	22	(95)	49
CuI	10	oxazolidin-2-one (20 mol %), NaOMe	DMSO	120	24	(89)	492

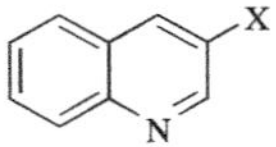

pyridine–pyrrolidinone product

X-substituted quinoline

CuI (2 mol %),
DMEDA (x mol %),
K₂CO₃, diglyme,
110°, 24 h

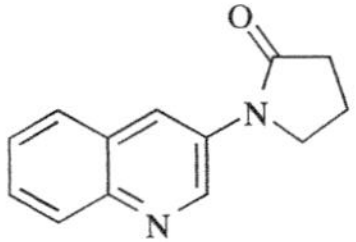

quinoline–pyrrolidinone product

X	x	Temp (°)	
Br	20	110	(98)	69
I	10	120	(99)	115

imidazopyridine, Ar = 4-FC₆H₄

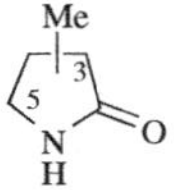

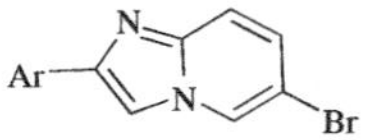

CuI (10 mol %),
DMEDA (10 mol %),
K₂CO₃, toluene, 110°, 20 h

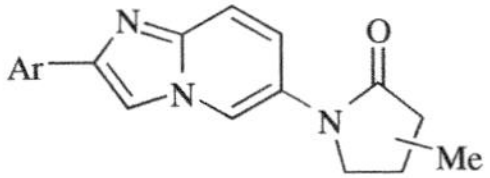

imidazopyridine–pyrrolidinone product (Ar)

Isomer		
3	(83)	500
5	(83)	

2-bromopyridine

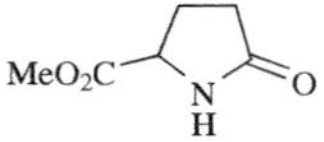

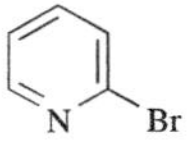

CuI (5 mol %), Cs₂CO₃,
dioxane, 60°, 36 h

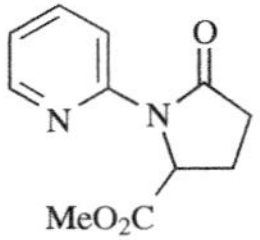

(73) 522

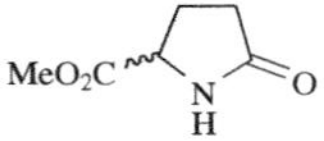

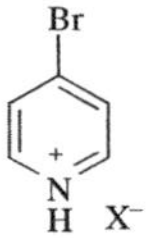

pyridine–pyrrolidinone (MeO₂C) product

4-bromopyridinium X⁻

CuI (50 mol %), Cs₂CO₃,
dioxane, 100°, 23 h

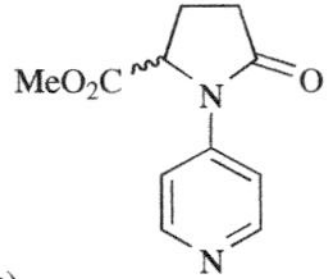

pyridine–pyrrolidinone (MeO₂C) product

X	Config.	Temp (°)	Time (h)		
Cl	(S)	reflux	24	(68)	530
Br	(R,S)	100	23	(68)	522

TABLE 18B. N-HETEROARYLATION OF LACTAMS, OXAZOLIDINONES, AND CYCLIC IMIDES (Continued)

Nitrogen Nucleophile	Heteroaryl Halide	Conditions	Product(s) and Yield(s) (%)	Refs.

Please refer to the charts preceding the tables for structures indicated by the **bold** numbers.

C5–6

Nitrogen Nucleophile: lactam $)_n$, NH

Heteroaryl Halide: ArX

Conditions: CuI (10 mol %), dioxane, 24 h

Product: $)_n$, N–Ar lactam

n	Ar	X	Additives	Time (h)		Refs.
1	3-furyl	Br	DMEDA (10 mol %), K$_2$CO$_3$	110	(19)	83
1	selenophen-2-yl	I	EDA (20 mol %), K$_3$PO$_4$	reflux	(23)	87
2	3-furyl	Br	DMEDA (10 mol %), K$_2$CO$_3$	110	(80)	83
2	selenophen-2-yl	I	EDA (20 mol %), K$_3$PO$_4$	reflux	(21)	87

C5

Nitrogen Nucleophile: δ-valerolactam

Heteroaryl Halide: imidazopyridine–Br, Ar = 4-FC$_6$H$_4$

Conditions: CuI (10 mol %), DMEDA (10 mol %), K$_2$CO$_3$, toluene, 110°, 20 h

Product: (80) — Refs. 500

C6

Nitrogen Nucleophile: pyrazolo-azepinedione (N–Me, N–R)

Heteroaryl Halide: ArBr

Conditions: CuSiO$_2$ (14 mol %), KOAc, reflux

Product:

Ar	R	
2-thienyl	Bn	(78)
3-thienyl	Bn	(80)
2-pyridyl	Me	(77)
2-pyridyl	Bn	(73)

Refs. 516

C6

Nitrogen Nucleophile: 4-isopropyloxazolidin-2-one

Heteroaryl Halide: 2-chloropyridine, BnO(...)$_6$

Conditions: CuI (15 mol %), DMEDA (30 mol %), K$_2$CO$_3$, 140°, 72 h

Product: (90) — Refs. 531

C10

Nitrogen Nucleophile: pyrrolidinone with (c-C$_5$H$_9$O, OMe)phenyl

Heteroaryl Halide: pyridine iodide, 2-, 3-, or 4-I

Conditions: CuI, CDA, K$_3$PO$_4$, DMF, 110°, 12 h

Product: (—) — Refs. 527

448

C$_{20}$

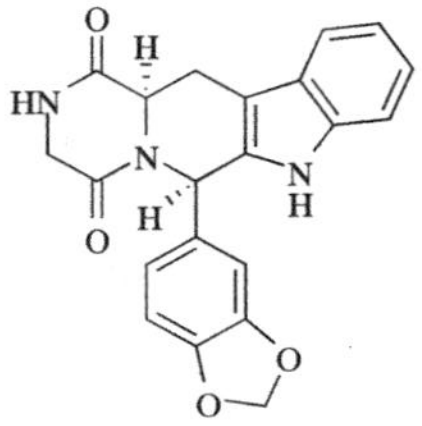

ArI

CuI (10 mol %),
CDA (10 mol %),
K$_3$PO$_4$, dioxane, 14°, 15 d

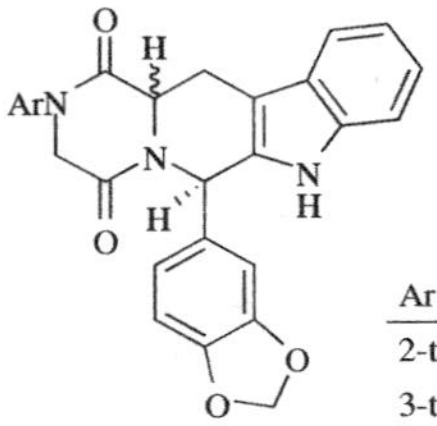

529

Ar		(R,R)/(R,S)
2-thienyl	(43)	58:42
3-thienyl	(77)	78:22
2-pyrazinyl	(35)	47:53
3-pyridyl	(66)	66:34

TABLE 19A. N-ARYLATION OF HETEROAROMATIC LACTAMS

*Please refer to the charts preceding the tables for structures indicated by the **bold** numbers.*

Nitrogen Nucleophile	Aryl Halide	Conditions	Product(s) and Yield(s) (%)	Refs.
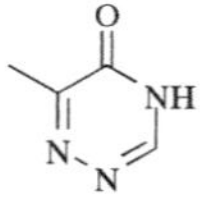 C4	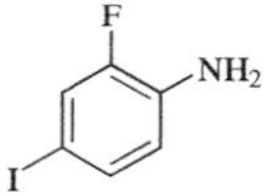	CuI (15 mol %), 8-HOquin (15 mol %), K₂CO₃, DMSO, 130°, 18 h	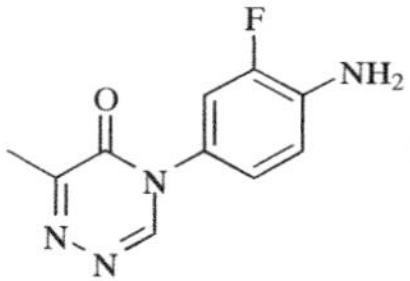 (22)	532
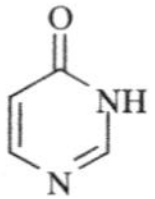	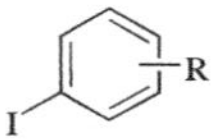	CuI (x mol %), K₂CO₃	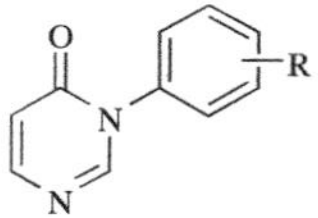	

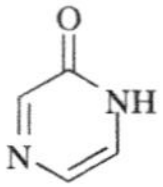

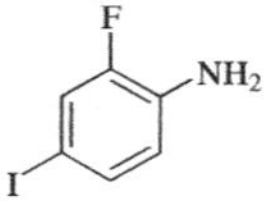

R	x	Additive	Solvent	Temp (°)	Time (h)		
H	1	none	DMF	150	6	(0)	519
3-F, 4-NH₂	15	8-HOquin	DMSO	130	18	(28)	532

C4-5		CuI (15 mol %), 8-HOquin (15 mol %), K₂CO₃, DMSO, 130°, 18 h	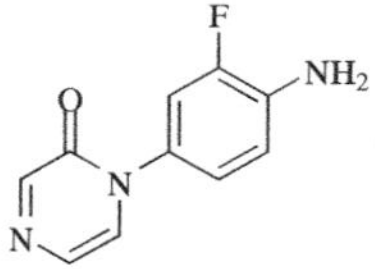 (43)	532
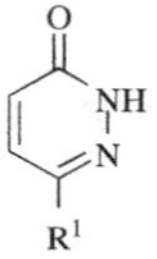	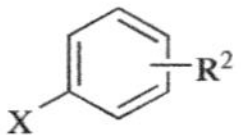	Catalyst (x mol %)	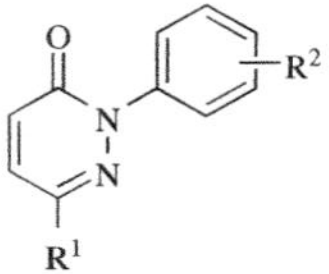	

R^1	R^2	X	Catalyst	x	Additive(s)	Solvent(s)	Temp (°)	Time (h)		Refs.
H	H	I	Cu₂O	5	L30 (20 mol %), Cs₂CO₃, 3 Å MS	MeCN	82	100	(89)	40
H	4-MeO	Br	13	5	K₂CO₃	DMF	140	18	(85)	533
H	3-NC–	I	13	5	K₂CO₃	DMF	140	18	(80)	533
H	4-NC–	I	13	5	K₂CO₃	DMF	140	18	(94)	533
H	3,5-Me₂	I	13	5	K₂CO₃	DMF	140	18	(80)	533
Me	3-F, 4-NH₂	I	CuI	15	8-HOquin (15 mol %), K₂CO₃	DMSO	130	18	(59)	532
Me	4-MeO	Br	13	5	K₂CO₃	DMF	140	18	(80)	533

533

534

303
516
49
519
535
535
138

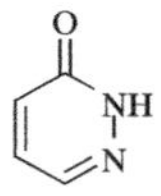

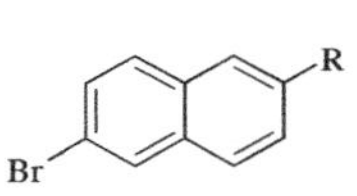

Cat. 13 (5 mol %),
K₂CO₃, DMF,
140°, 18 h

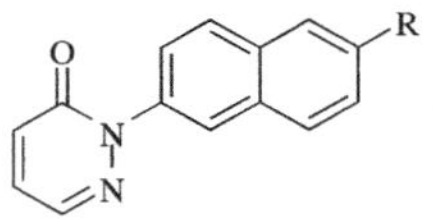

R	
Br	(85)
HOCH₂CH₂	(70)

(—)

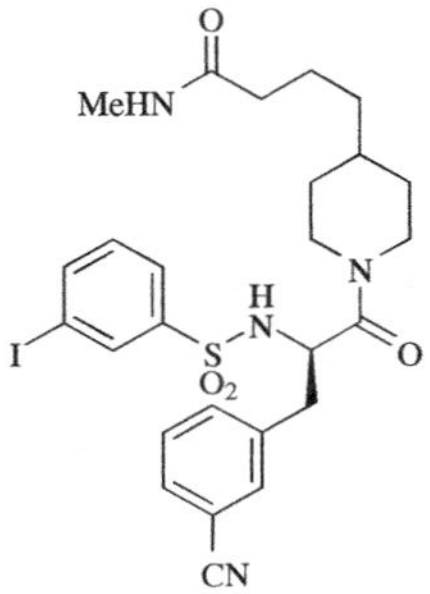

CuI (4 mol %),
DMCDA (10 mol %),
K₃PO₄, DMF, 110°, 1.5 h

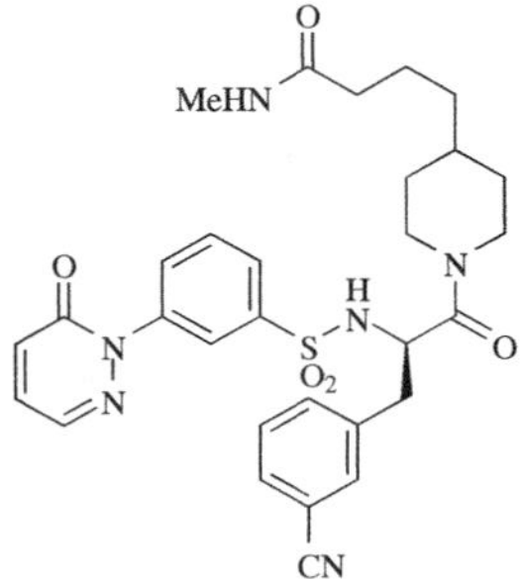

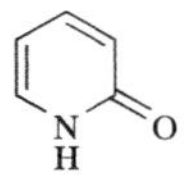

See table.

PhX

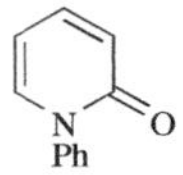

X	Catalyst	Additive(s)	Solvent(s)	Temp (°)	Time (h)
Br	Cu (3 eq)	none	—	180	12 (100)
Br	Cu/SiO₂ (14 mol %)	KOAc	—	reflux	— (57)
I	CuBr (10 mol %)	L9 (20 mol %), Cs₂CO₃	DMSO	60	22 (95)
I	CuI (1 mol %)	K₂CO₃	DMF	150	6 (72)
Br	CuI (20 mol%)	DMCDA (1 eq), K₂CO₃	toluene	reflux	24 (91)
I	CuI (20 mol %)	DMCDA (1 eq), K₂CO₃	toluene	reflux	24 (84)
I	CuI (5 mol %)	L43 (10 mol %), K₃PO₄	DMSO	110	5 (70)

C₄

C₅

TABLE 19A. N-ARYLATION OF HETEROAROMATIC LACTAMS (*Continued*)

Nitrogen Nucleophile	Aryl Halide	Conditions	Product(s) and Yield(s) (%)	Refs.

*Please refer to the charts preceding the tables for structures indicated by the **bold** numbers.*

C₅

Nitrogen Nucleophile: 2-pyridinone (N–H)

Aryl Halide: PhI

Conditions: Cu₂O (5 mol %), **L30** (20 mol %), Cs₂CO₃, 3 Å MS, MeCN, 82°, 24 h

Product(s) and Yield(s) (%): (**90**) + (**2**) ; 2-OPh pyridine

Refs.: 40

ArX

Catalyst (x mol %)

N-Ar pyridinone

Ar	X	Catalyst	x	Additives	Solvent(s)	Temp (°)	Time (h)		Refs.
4-ClC₆H₄	I	CuI	10	4,7-(MeO)₂phen (15 mol %), K₂CO₃	DMSO	110	30	(80)	131
4-ClC₆H₄	I	CuI	20	DMEDA (40 mol %), K₃PO₄	dioxane	110	24	(40)	53
3-H₂NC₆H₄	Br	CuI	5	4,7-(MeO)₂phen (7.5 mol %), K₂CO₃	DMSO	110	30	(78)	131
3-H₂NC₆H₄	I	Cu₂O	5	**L30** (20 mol %), Cs₂CO₃, 3 Å MS	MeCN	82	48	(82)	40
4-Ph₂NC₆H₄	Br	CuI	20	DMCDA (20 mol %), K₂CO₃	toluene	reflux	20	(59)	535
4-Ph₂NC₆H₄	I	CuI	20	DMCDA (20 mol %), K₂CO₃	toluene	reflux	20	(79)	535
2-MeOC₆H₄	Br	CuI	20	DMCDA (20 mol %), K₂CO₃	toluene	reflux	20	(2)	535
3-MeOC₆H₄	I	CuI	20	DMEDA (40 mol %), K₃PO₄	dioxane	110	24	(84)	53
4-MeOC₆H₄	Br	CuI	20	DMCDA (20 mol %), K₂CO₃	toluene	reflux	20	(89)	535
4-MeOC₆H₄	Br	CuI	20	DMEDA (40 mol %), K₃PO₄	dioxane	110	24	(82)	53
4-MeOC₆H₄	I	CuI	20	DMEDA (40 mol %), K₃PO₄	dioxane	110	24	(78)	53
3-Br, 4-MeOC₆H₄	Br	CuI	20	DMCDA (20 mol %), K₂CO₃	toluene	reflux	20	(37)	535
4-MeSC₆H₄	Br	CuI	20	DMCDA (20 mol %), K₂CO₃	toluene	reflux	20	(80)	535
2-MeC₆H₄	I	CuI	20	DMEDA (40 mol %), K₃PO₄	dioxane	110	24	(40)	53
4-MeC₆H₄	I	CuBr	10	**L9** (20 mol %), Cs₂CO₃	DMSO	60	24	(86)	49
4-MeC₆H₄	Br	CuI	20	DMEDA (40 mol %), K₃PO₄	dioxane	110	24	(88)	53
4-MeC₆H₄	I	CuI	20	DMEDA (40 mol %), K₃PO₄	dioxane	110	24	(81)	53
4-NCC₆H₄	Br	CuI	20	DMCDA (20 mol %), K₂CO₃	toluene	reflux	20	(70)	535
4-MeO₂CC₆H₄	I	CuI	15	8-HOquin (15 mol %), K₂CO₃	DMSO	130	18	(22)	532
4-EtO₂CC₆H₄	Br	CuI	20	DMEDA (40 mol %), K₃PO₄	dioxane	110	24	(91)	53
4-EtO₂CC₆H₄	I	CuI	20	DMEDA (40 mol %), K₃PO₄	dioxane	110	24	(82)	53
4-CH₂=CHC₆H₄	Br	CuI	20	DMCDA (20 mol %), K₂CO₃	toluene	reflux	20	(82)	535
4-PhC₆H₄	Br	CuI	20	DMCDA (20 mol %), K₂CO₃	toluene	reflux	20	(70)	535
1-Np	Br	CuI	20	DMCDA (20 mol %), K₂CO₃	toluene	reflux	20	(37)	535

452

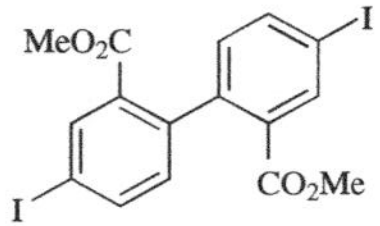

535

(81)

CO₂Me

MeO₂C

CuI (20 mol %),
DMCDA (20 mol %),
K₂CO₃, toluene,
reflux, 20 h

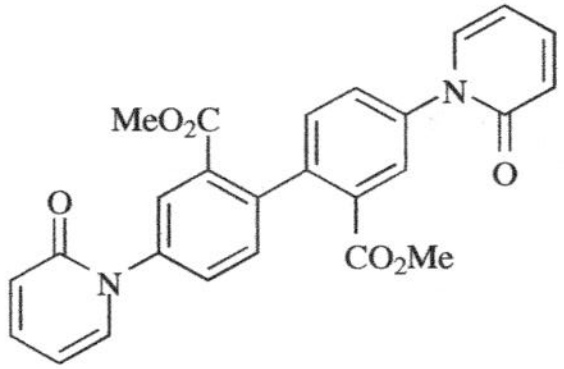

CO₂Me

MeO₂C

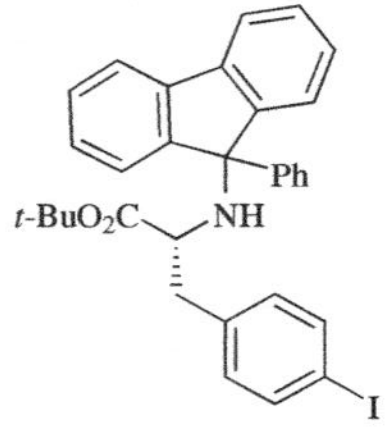

I

536

(66)

Ph
NH

t-BuO₂C

CuI, Cs₂CO₃, DMF, 150°

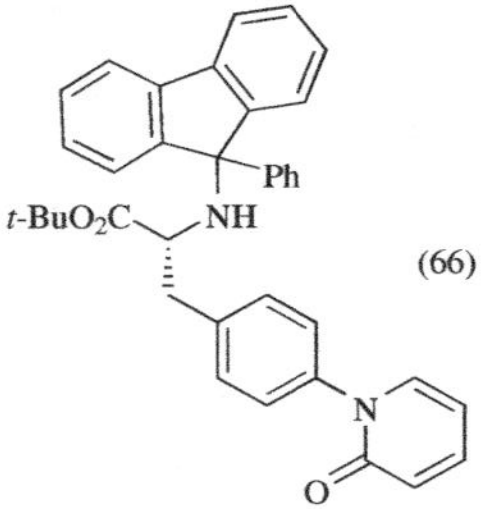

Ph
NH

t-BuO₂C

I

131

NMe₂

(71)

Cl

O

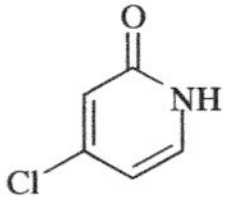

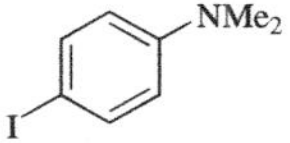

CuI (10 mol %),
4,7-(MeO)₂phen (15 mol %),
K₂CO₃, DMSO, 120°, 30 h

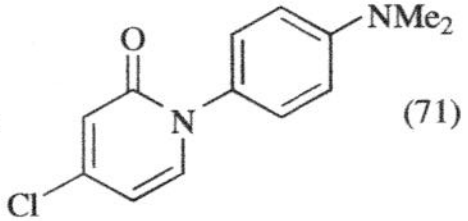

NMe₂

I

53

R

Cl

O

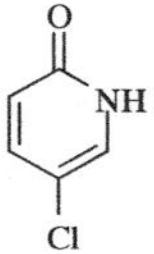

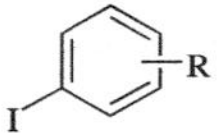

CuI (20 mol %),
DMEDA (40 mol %),
K₃PO₄, dioxane,
110°, 24 h

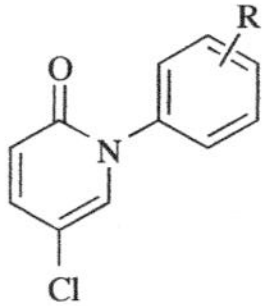

R

I

R	
4-Cl	(50)
3-MeO	(90)
4-MeO	(67)
4-Me	(68)
4-EtO₂C	(83)

NH

O

Cl

NH

O

Cl

TABLE 19A. N-ARYLATION OF HETEROAROMATIC LACTAMS (Continued)

Please refer to the charts preceding the tables for structures indicated by the **bold** numbers.

Nitrogen Nucleophile	Aryl Halide	Conditions	Product(s) and Yield(s) (%)	Refs.

C5

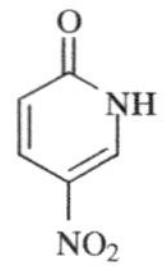

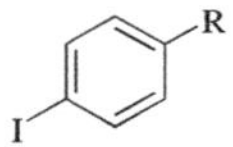

CuI (20 mol %), DMEDA (40 mol %), K₃PO₄, dioxane, 110°, 24 h

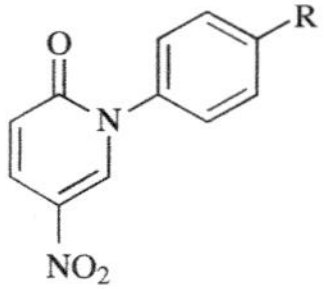

R	
MeO	(trace)
EtO₂C	(trace)

53

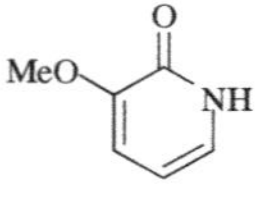

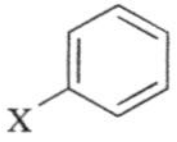

Cu(OAc)₂ (1 eq), DBU, DMSO, MW, 130°, 10 min

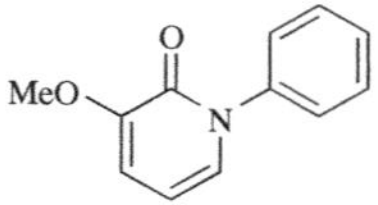

X	
Br	(81)
I	(84)

85

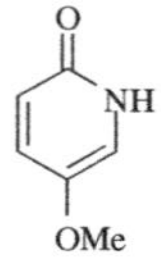

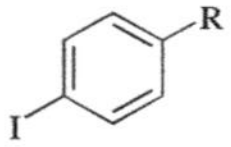

CuI (20 mol %), DMEDA (40 mol %), K₃PO₄, dioxane, 110°, 24 h

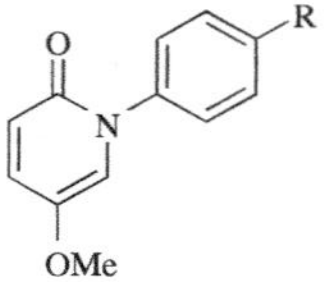

R	
4-Cl	(75)
3-MeO	(82)
4-MeO	(90)
4-Me	(87)
4-EtO₂C	(81)

53

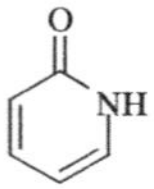

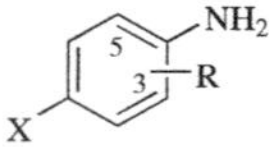

CuI (15 mol %), 8-HOquin (15 mol %), K₂CO₃, DMSO, 130°, 18 h

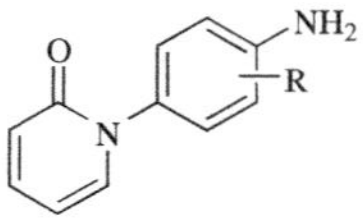

R	X	
3-F	I	(62)
3,5-F₂	Br	(41)
3-Cl	Br	(11)
3-Cl	I	(31)
3-i-PrO	Br	(56)

532

C$_{5-6}$

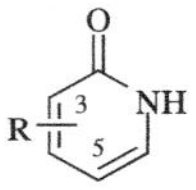

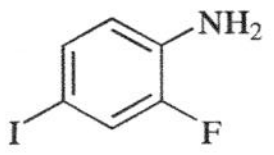

CuI (15 mol %),
8-HOquin (15 mol %),
K$_2$CO$_3$, DMSO. 130°, 18 h

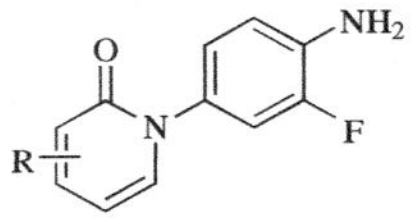

532

R		R	
3-F	(31)	3-PMBO	(55)
3,5-F$_2$	(58)	4-MeO	(74)
6-Cl	(0)	3-EtS	(51)
3-H$_2$N	(38)	3-Me	(78)
3-BocNH	(2)^a	4-Me	(81)
3-O$_2$N	(4)	5-Me	(81)
3-HO	(0)	6-Me	(6)
3-AcO	(0)	3-CF$_3$	(30)
3-BnO(CH$_2$)$_2$O	(49)	5-EtO$_2$C	(81)

C$_5$

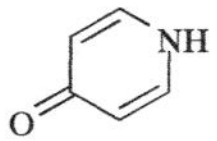

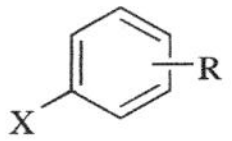

CuI (x mol %),
K$_2$CO$_3$, DMSO,
110°, 30 h

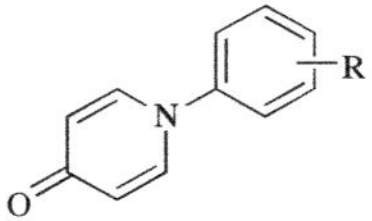

131

R	X	x	Additive	
3-MeO	I	5	4,7-(MeO)$_2$phen (7.5 mol %)	(95)
2-Me	I	5	TMHD (40 mol %)	(60)
3,5-Me$_2$	I	2	TMHD (8 mol %)	(90)
4-EtOC	Br	5	TMHD (20 mol %)	(92)

TABLE 19A. N-ARYLATION OF HETEROAROMATIC LACTAMS (*Continued*)

*Please refer to the charts preceding the tables for structures indicated by the **bold** numbers.*

Nitrogen Nucleophile	Aryl Halide	Conditions	Product(s) and Yield(s) (%)	Refs.
		See table.		

C$_6$

R^1	R^2	X	Catalyst	Additive(s)	Solvent	Temp (°)	Time		Refs.
H	H	Br	Cu(OAc)$_2$ (1 eq)	DBU	DMSO	MW, 130	10 min	(81)	85
H	H	I	Cu(OAc)$_2$ (1 eq)	DBU	DMSO	MW, 130	10 min	(84)	85
H	4-Cl	I	CuI (20 mol %)	DMEDA (40 mol %), K$_3$PO$_4$	dioxane	110	24 h	(85)	53
H	3-MeO	I	CuI (20 mol %)	DMEDA (40 mol %), K$_3$PO$_4$	dioxane	110	24 h	(95)	53
H	4-MeO	I	CuI (20 mol %)	DMEDA (40 mol %), K$_3$PO$_4$	dioxane	110	24 h	(97)	53
H	4-Me	I	CuI (20 mol %)	DMEDA (40 mol %), K$_3$PO$_4$	dioxane	110	24 h	(82)	53
H	3-H$_2$NCH$_2$	I	CuI (15 mol %)	8-HOquin (15 mol %), K$_2$CO$_3$	DMSO	130	18 h	(37)	532
H	4-NC–	I	CuI (15 mol %)	8-HOquin (15 mol %), K$_2$CO$_3$	DMSO	130	18 h	(58)	532
H	4-EtO$_2$C	I	CuI (20 mol %)	DMEDA (40 mol %), K$_3$PO$_4$	dioxane	110	24 h	(83)	53
NO$_2$	3,5-Me$_2$	I	CuI (10 mol %)	4,7-(MeO)$_2$phen (15 mol %), K$_2$CO$_3$	DMSO	150	30 h	(0)	131

CuI (1 mol %), K$_2$CO$_3$, DMF, 150°, 6 h

(87) 519

CuI (20 mol %), DMEDA (40 mol %), K$_3$PO$_4$, dioxane, 110°, 24 h

R	
4-Cl	(85)
3-MeO	(78)
4-MeO	(91)
4-Me	(94)
4-EtO$_2$C	(92)

53

CuI (20 mol %), DMEDA (40 mol %), K$_3$PO$_4$, dioxane, 110°, 24 h

R	
4-MeO	(trace)
4-EtO$_2$C	(trace)

53

456

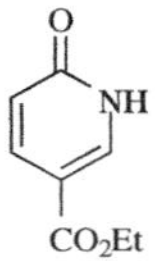

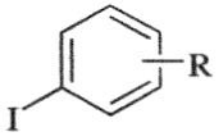

Substrate	Reagent	Conditions	Product	Refs.
C₇ NH-pyridinone, CO₂Et	aryl iodide, R	CuI (20 mol %), DMEDA (40 mol %), K₃PO₄, dioxane, 110°, 24 h	N-aryl pyridinone, CO₂Et — R (55) 4-Cl; (92) 3-MeO; (69) 4-MeO; (77) 4-Me; (95) 4-EtO₂C	53
C₇ furo[pyridin]-one	2-fluoro-4-iodoaniline	CuI (15 mol %), 8-HOquin (15 mol %), K₂CO₃, DMSO, 130°, 18 h	aryl product (92)	532
C₇₋₈ benzodiazinone, Y	iodobenzene	CuI (10 mol %), K₂CO₃, DMF, 150°, 6 h	N-phenyl product — Y (14) N; (78) CH	519
C₉ methylquinoxalinone	2-fluoro-4-iodoaniline	CuI (15 mol %), 8-HOquin (15 mol %), K₂CO₃, DMSO, 130°, 18 h	aryl product (40)	532

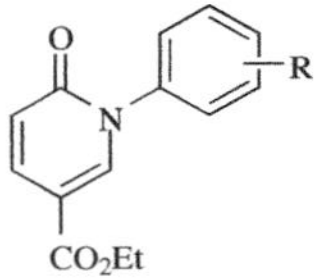

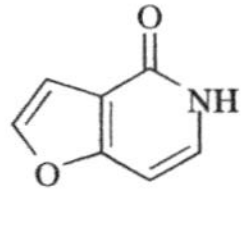

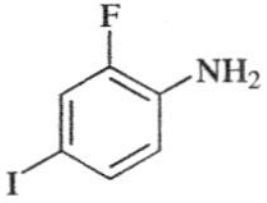

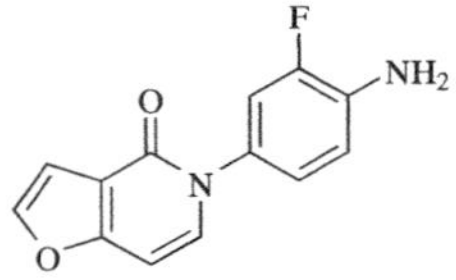

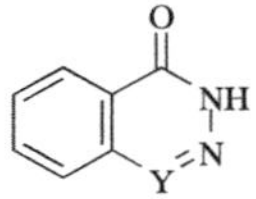

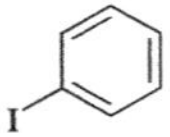

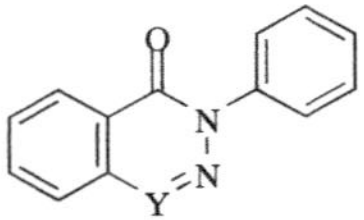

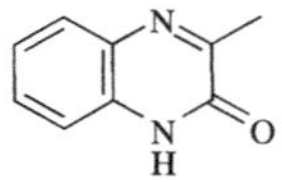

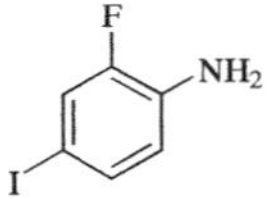

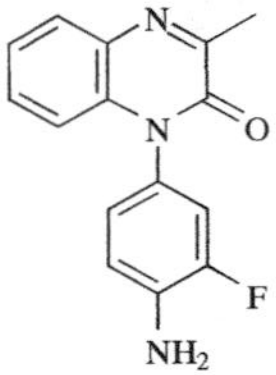

TABLE 19A. N-ARYLATION OF HETEROAROMATIC LACTAMS (Continued)

Nitrogen Nucleophile	Aryl Halide	Conditions	Product(s) and Yield(s) (%)	Refs.

*Please refer to the charts preceding the tables for structures indicated by the **bold** numbers.*

C₉

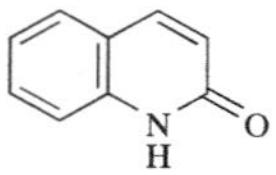

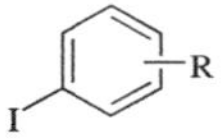

CuI (x mol %), K₂CO₃

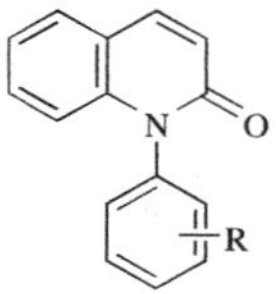

R	x	Additive	Solvent	Temp (°)	Time (h)	
H	10	none	DMF	150	6	(0)
3-F, 4-H₂N	15	8-HOquin (15 mol %)	DMSO	130	18	(54)

519
532

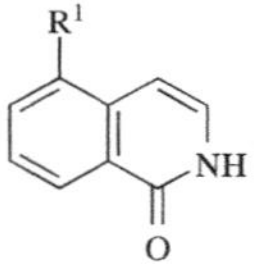

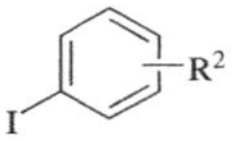

CuI (x mol %), K₂CO₃

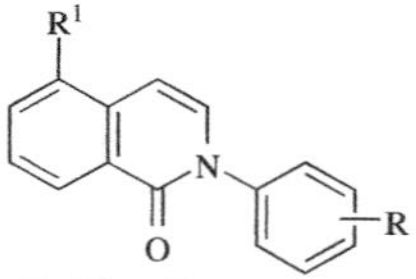

R¹	R²	x	Additive	Solvent	Temp (°)	Time (h)	
H	3-F, 4-H₂N	15	8-HOquin (15 mol %)	DMSO	130	18	(66)
HO	H	10	none	DMF	150	6	(32)

532
519

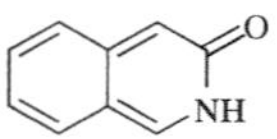

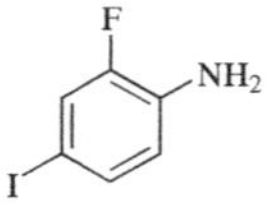

CuI (15 mol %),
8-HOquin (15 mol %),
K₂CO₃, DMSO, 130°. 18 h

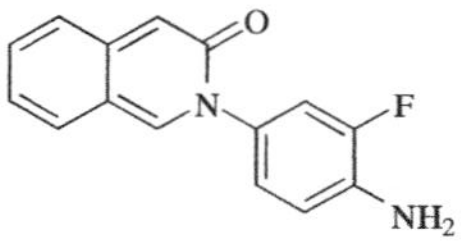

(0)

532

C₉₋₁₁

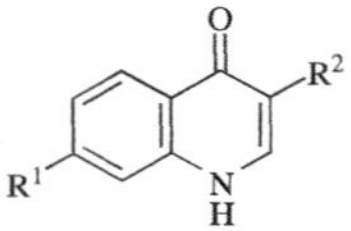

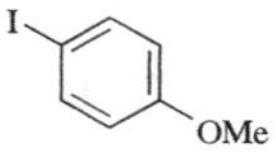

CuI (10 mol %),
K₃PO₄, DMSO,
TMHD (40 mol %), 30 h

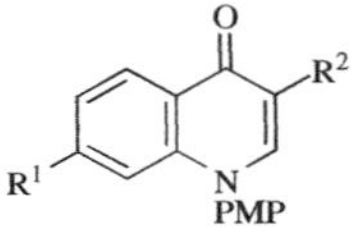

R¹	R²	Temp (°)	
Cl	H	120	(68)
CF₃	CO₂Et	140	(0)

131

PMP

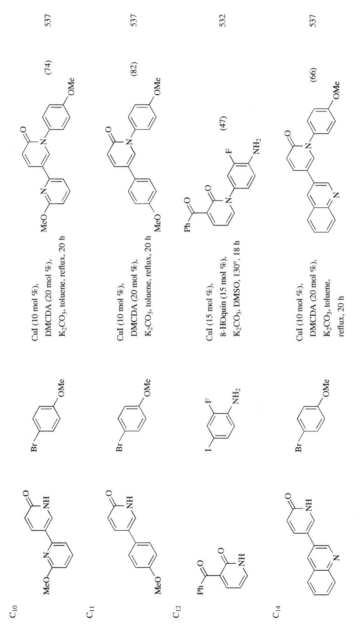

CuI (10 mol %),
DMCDA (20 mol %),
K₂CO₃, toluene, reflux, 20 h

(74)

537

CuI (10 mol %),
DMCDA (20 mol %),
K₂CO₃, toluene, reflux, 20 h

(82)

537

CuI (15 mol %),
8-HOquin (15 mol %),
K₂CO₃, DMSO, 130°, 18 h

(47)

532

CuI (10 mol %),
DMCDA (20 mol %),
K₂CO₃, toluene,
reflux, 20 h

(66)

537

C₁₀

C₁₁

C₁₂

C₁₄

TABLE 19A. *N*-ARYLATION OF HETEROAROMATIC LACTAMS (*Continued*)

Nitrogen Nucleophile	Aryl Halide	Conditions	Product(s) and Yield(s) (%)	Refs.

*Please refer to the charts preceding the tables for structures indicated by the **bold** numbers.*

C₁₄

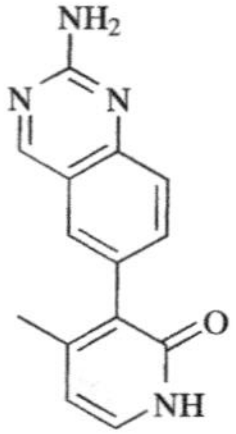

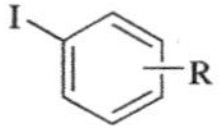

CuI (40 mol %),
DMEDA (1 eq),
K₃PO₄, NMP, 85°, 6 h

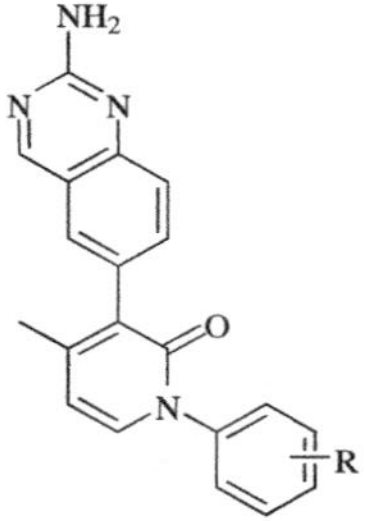

R	
3-F, 4-Cl	(—)
4-Cl	(—)
4-(4-morpholino)	(—)
3-MeO	(—)
4-MeO	(—)
3-F, 4-Me	(—)
3-Me, 4-F	(—)
3-CF₃	(91)
4-EtO₂C	(—)
4-Ac	(—)
4-*t*-Bu	(—)
4-(1-pyrazolyl)	(—)
4-(2-oxazolyl)	(—)

538

C₁₈

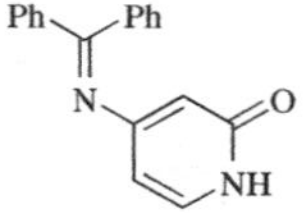

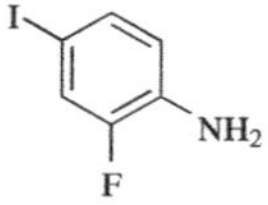

CuI (15 mol %),
8-HOquin (15 mol %),
K₂CO₃, DMSO, 130°, 18 h

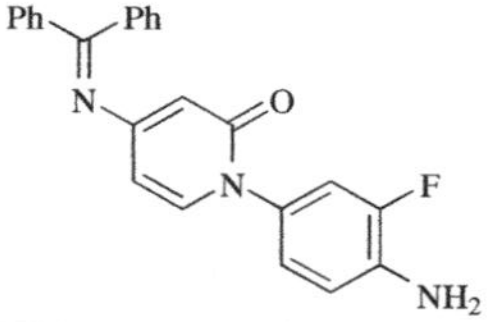

(81)

532

ᵃ The deprotected free amine was formed in 14% yield.

TABLE 19B. N-HETEROARYLATION OF HETEROAROMATIC LACTAMS

Nitrogen Nucleophile	Heteroaryl Halide	Conditions	Product(s) and Yield(s) (%)	Refs.

*Please refer to the charts preceding the tables for structures indicated by the **bold** numbers.*

C₄

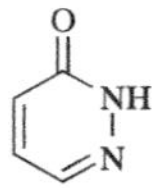

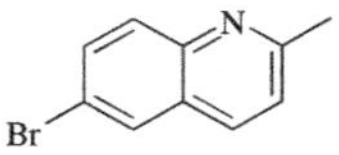

Nitrogen Nucleophile: (pyridazinone, O=, NH–N)

Heteroaryl Halide: (2-methyl-6-bromoquinoline)

Conditions: **13** (5 mol %), K₂CO₃, DMF, 140°, 18 h

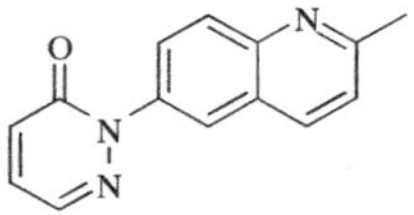

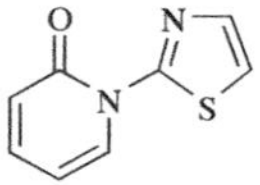

Product(s) and Yield(s): (90)

Refs.: 533

C₅

Nitrogen Nucleophile: (2-pyridone, O=, NH)

Heteroaryl Halide: (2-bromothiazole)

Conditions: CuI (x mol %)

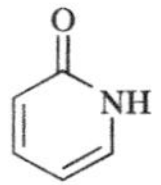

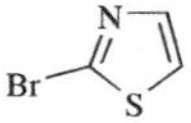

x	Additives	Solvent	Temp (°)	Time (h)		
5	4,7-(MeO)₂phen (7.5 mol %), K₂CO₃	DMSO	110	30	(85)	131
20	DMEDA (40 mol %), K₃PO₄	dioxane	110	24	(27)	53
20	DMCDA (20 mol %), K₂CO₃	toluene	reflux	16	(33)	535

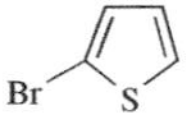

Heteroaryl Halide: (2-bromothiophene)

Conditions: CuI (20 mol %), DMEDA (40 mol %), K₃PO₄, dioxane, 110°, 24 h

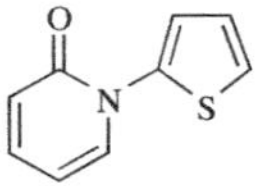

Product(s) and Yield(s): (81) Refs.: 53

Heteroaryl Halide: (3-bromothiophene)

Conditions: Catalyst (x mol %)

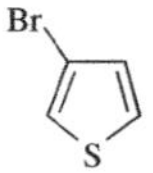

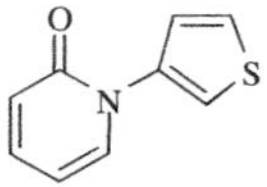

Catalyst	x	Additive(s)	Solvent	Temp (°)	Time (h)		
Cu/SiO₂	14	KOAc	—	reflux	—	(84)	516
CuI	20	DMEDA (40 mol %), K₃PO₄	dioxane	110	24	(80)	53
CuI	20	DMCDA (20 mol %), K₂CO₃	toluene	reflux	16	(74)	535

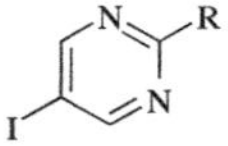

Heteroaryl Halide: (5-iodo-2-R-pyrimidine)

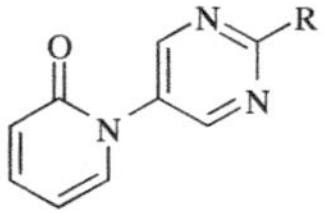

Conditions: CuI (x mol %), K₂CO₃, DMSO

R	x	Additive	Temp (°)	Time (h)		
H	5	4,7-(MeO)₂phen (7.5 mol %)	110	30	(47)	131
NH₂	15	8-HOquin (15 mol %)	130	18	(59)	532

461

TABLE 19B. N-HETEROARYLATION OF HETEROAROMATIC LACTAMS (Continued)

Nitrogen Nucleophile	Heteroaryl Halide	Conditions	Product(s) and Yield(s) (%)	Refs.

*Please refer to the charts preceding the tables for structures indicated by the **bold** numbers.*

C₅

(pyridinone nucleophile, NH lactam)

Heteroaryl Halide: X–pyridine with R at positions 3, 2

Conditions: Catalyst (x mol %), reflux

Product: pyridinone–pyridine with R

R	X	Catalyst	x	Additive(s)	Solvent	Time (h)	(yield)	Refs.
H	Br	Cu/SiO₂	14	KOAc	—	—	(76)	516
H	Cl	CuI	20	DMCDA (1 eq), K₂CO₃	toluene	24	(7)	535
H	Br	CuI	20	DMCDA (20 mol %), K₂CO₃	toluene	16	(68)	535
H	I	CuI	20	DMCDA (20 mol %), K₂CO₃	toluene	16	(44)	535
2-O₂N	Br	CuI	20	DMCDA (20 mol %), K₂CO₃	toluene	16	(62)	535
3-Me	Br	CuI	20	DMCDA (20 mol %), K₂CO₃	toluene	20	(36)	535
4-Me	Br	CuI	20	DMCDA (20 mol %), K₂CO₃	toluene	20	(57)	535

Conditions: CuI (x mol %)

Heteroaryl Halide: X–pyridine with R

Product: pyridinone with R

R	X	x	Additives	Solvent	Temp (°)	Time (h)	(yield)	Refs.
H	Br	20	DMEDA (40 mol %), K₃PO₄	dioxane	110	24	(8)	53
H	Br	20	DMCDA (20 mol %), K₂CO₃	toluene	reflux	16	(76)	535
H	Br	15	8-HOquin (15 mol %), K₂CO₃	DMSO	130	18	(46)	532
H₂N	I	15	8-HOquin (15 mol %), K₂CO₃	DMSO	130	18	(59)	532
t-BuO	Br	20	DMCDA (20 mol %), K₂CO₃	toluene	reflux	20	(63)	535
t-BuO	Br	20	DMCDA (20 mol %), 18-c-6 (20 mol %), K₂CO₃	toluene	reflux	24	(63)	539

Heteroaryl Halide: 2,6-dibromopyridine (Br–N–Br)

Conditions: CuI (20 mol %), DMCDA (20 mol %), K₂CO₃, toluene, reflux, 16 h

Product: bis(pyridinone)pyridine

| | | | | | | | (65) | 535 |

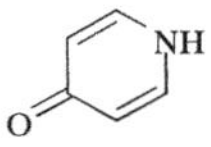

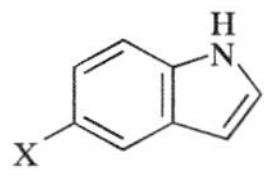

CuI (x mol%)

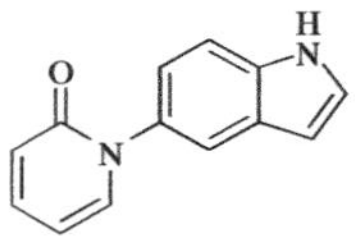

X	x	Additives	Solvent	Temp (°)	Time (h)		
Br	20	DMEDA (40 mol %), K$_3$PO$_4$	dioxane	110	24	(16)	53
I	10	DMCDA (20 mol %), K$_2$CO$_3$	DMSO	100	30	(75)	131

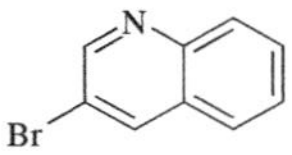

CuI (5 mol %),
4,7-(MeO)$_2$phen (7.5 mol %),
K$_2$CO$_3$, DMF, 110°, 30 h 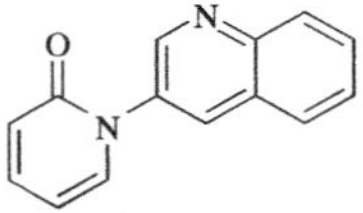
(82)   131

CuI (20 mol %),
DMEDA (40 mol %),
K$_3$PO$_4$, dioxane, 110°, 24 h 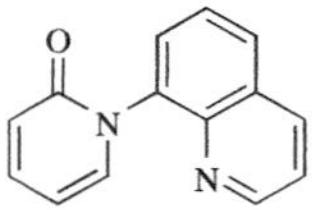
(10)   53

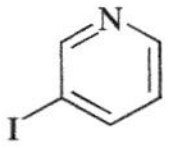

CuI (7 mol %),
4,7-(MeO)$_2$phen (7.5 mol %),
K$_2$CO$_3$, DMSO, 110°, 30 h 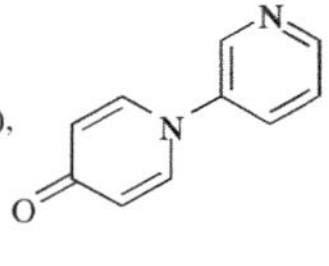
(89)   131

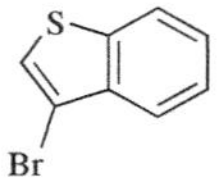

CuI (5 mol%),
TMHD (40 mol %),
K$_2$CO$_3$, DMSO, 110°, 30 h
(67)   131

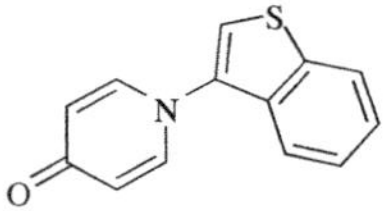

TABLE 19B. *N*-HETEROARYLATION OF HETEROAROMATIC LACTAMS (*Continued*)

*Please refer to the charts preceding the tables for structures indicated by the **bold** numbers.*

Nitrogen Nucleophile	Heteroaryl Halide	Conditions	Product(s) and Yield(s) (%)	Refs.
C₇				
		CuI (10 mol %), TMHD (40 mol %), K₃PO₄, DMSO, 120°, 30 h	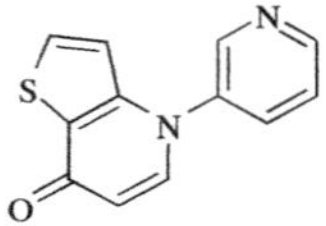 (65)	131
C₈				
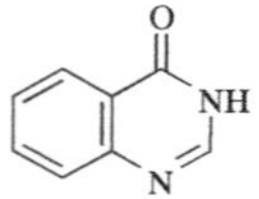	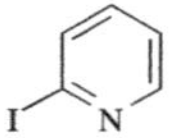	CuI (10 mol %), oxazolidinone (20 mol %), NaOMe, DMSO, 120°, 24 h	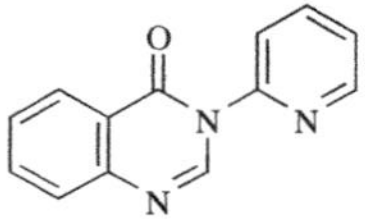 (93)	492
C₁₀				
		CuI (10 mol %), DMCDA (20 mol %), K₂CO₃, toluene, reflux, 20 h	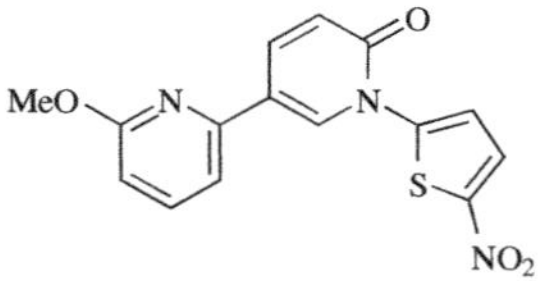 (60)	537
		CuI (10 mol %), DMCDA (20 mol %), K₂CO₃, toluene, reflux, 20 h		537
			X R
			I Br (51)
			Br NH₂ (40)
		CuI (20 mol %), DMCDA (20 mol %), K₂CO₃, toluene, reflux	(82)	535
		Additive Time (h)	
		none 20 (82)	
		18-c-6 24 (82)	

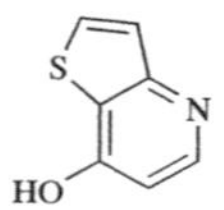

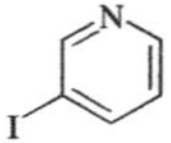

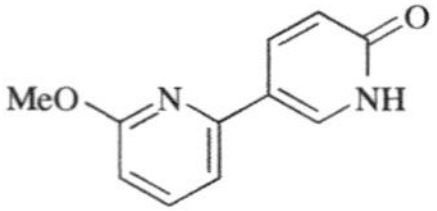

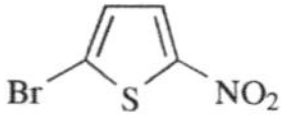

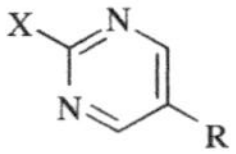

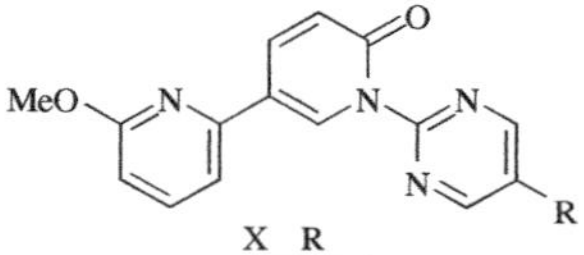

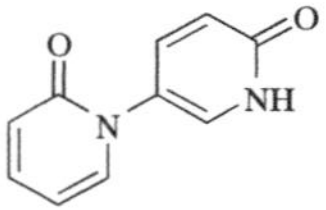

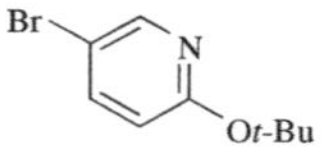

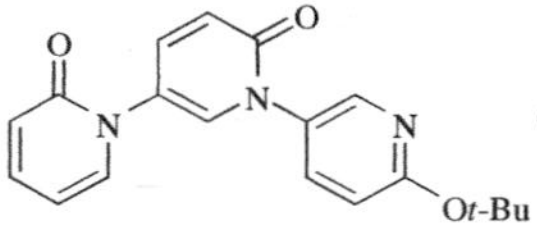

464

C₁₁

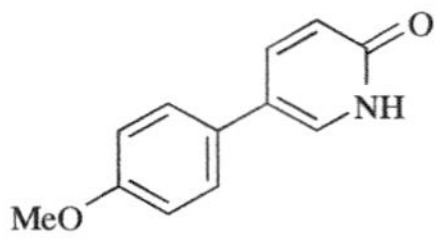

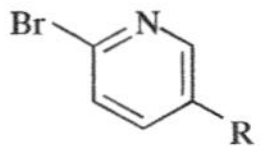

CuI (10 mol %),
DMCDA (20 mol %),
K₂CO₃, toluene, reflux, 20 h

R	
F	(77)
O₂N	(83)
CF₃	(80)

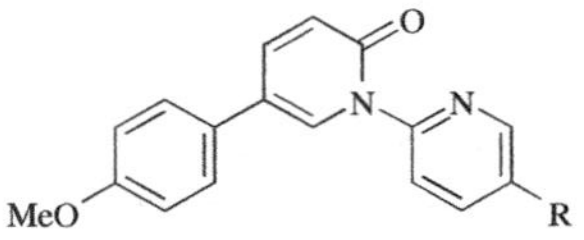

537

CuI (10 mol %),
DMCDA (20 mol %),
K₂CO₃, toluene, reflux, 20 h

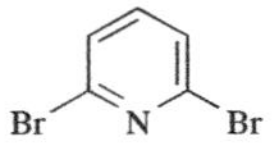

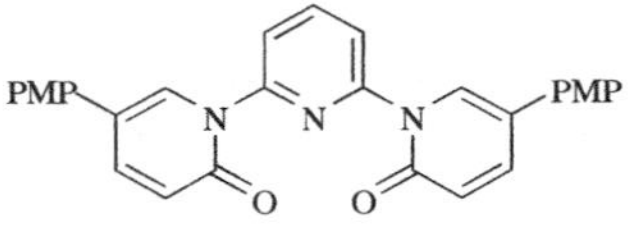

PMP (69)

537

C₁₂₋₁₄

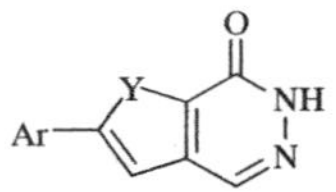

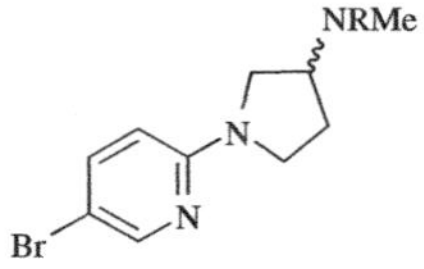

CuI (1 eq), DMCDA (1.4 eq),
dioxane, 110°, 9 h

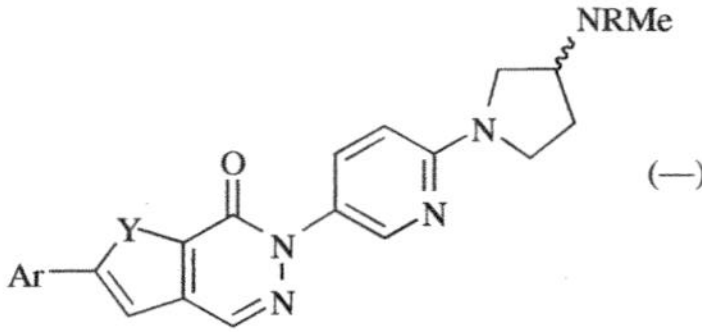

(—)

540

Ar	Y	R	Config.
4-FC₆H₄	S	H	(R)
4-FC₆H₄	S	H	(S)
4-ClC₆H₄	S	H	(S)
4-MeOC₆H₄	S	H	(S)
4-CF₃C₆H₄	NMe	Me	(R)
4-CF₃C₆H₄	O	Me	(R)
4-CF₃C₆H₄	S	H	(R)
4-CF₃C₆H₄	S	H	(S)
4-CF₃C₆H₄	S	Me	(R)
4-CF₃C₆H₄	Se	H	(R)
4-EtC₆H₄	S	H	(S)

465

TABLE 19B. N-HETEROARYLATION OF HETEROAROMATIC LACTAMS (Continued)

Nitrogen Nucleophile	Heteroaryl Halide	Conditions	Product(s) and Yield(s) (%)	Refs.

*Please refer to the charts preceding the tables for structures indicated by the **bold** numbers.*

C_{14}

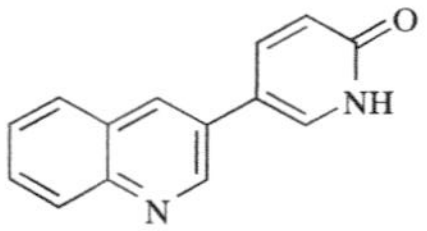

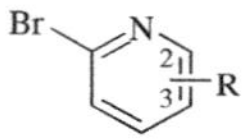

Conditions: CuI (10 mol %), DMCDA (20 mol %), K_2CO_3, toluene, reflux, 20 h

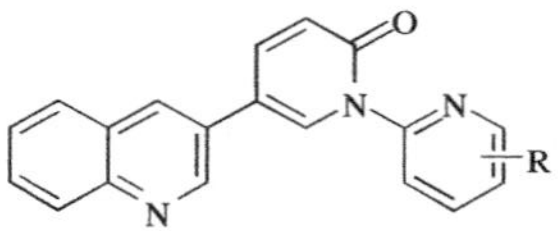

R	
2-MeO	(72)
3-CF_3	(72)

Refs. 537

C_{14-15}

Conditions: CuI (1 eq.), DMCDA (1.4 eq.), dioxane, 110°, 9 h

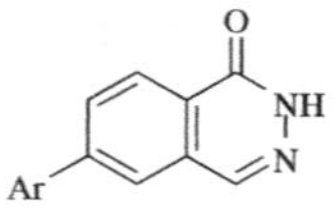

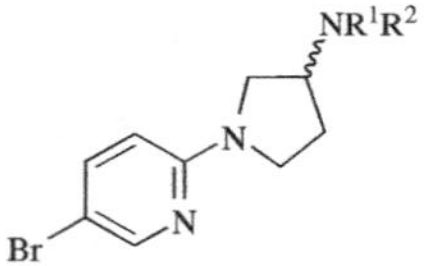

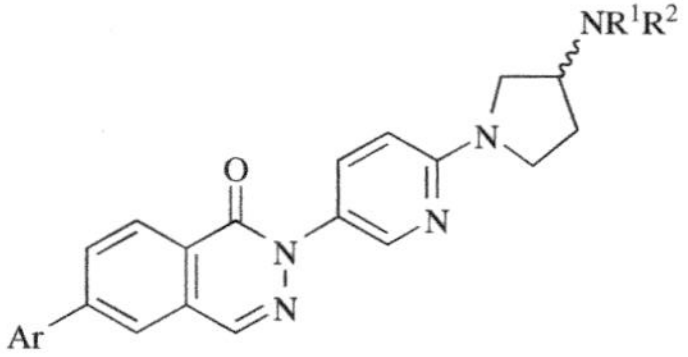

Refs. 540

Ar	R^1	R^2	Config.	
4-ClC_6H_4	Me	H	(R)	(—)
4-ClC_6H_4	Me	H	(S)	(—)
4-ClC_6H_4	Et	H	(R)	(—)
4-ClC_6H_4	Et	H	(S)	(—)
4-ClC_6H_4	Me	Me	(R)	(—)
4-ClC_6H_4	Me	Me	(S)	(39)
4-ClC_6H_4	—(CH_2)_4—		(R)	(—)
4-ClC_6H_4	—(CH_2)_4—		(S)	(—)
2-Me-3-MeOC_6H_3	Me	H	(R)	(—)
4-CF_3C_6H_4	Me	H	(R)	(—)
4-CF_3C_6H_4	Me	H	(S)	(—)

C$_{15-20}$

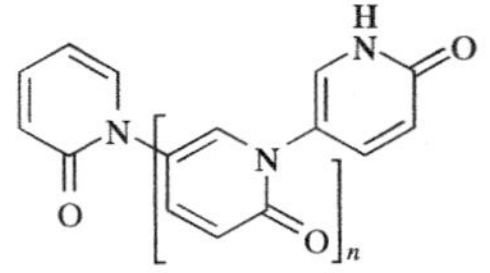

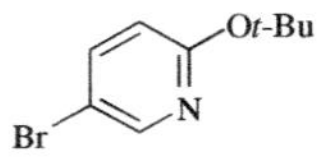

CuI (20 mol %),
DMCDA (20 mol %),
18-c-6, (20 mol %), K$_2$CO$_3$,
toluene, reflux, 24 h

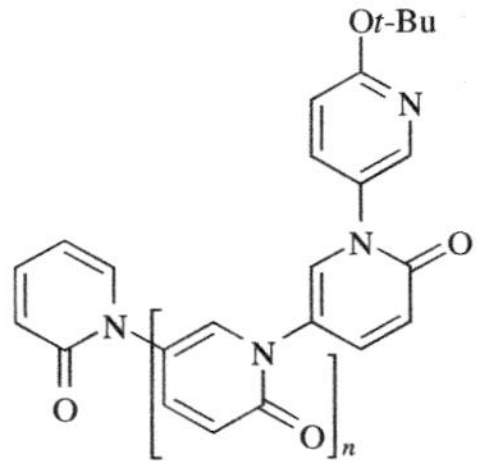

n	
1	(53)
2	(18)

539

TABLE 20. N-ARYLATION AND N-HETEROARYLATION OF HYDRAZINE DERIVATIVES

Nitrogen Nucleophile	(Hetero)aryl Halide	Conditions	Product(s) and Yield(s) (%)		Refs.
$H_2N-N(H)-Boc$	RX	CuI (x mol %), Cs_2CO_3, 80°	$H_2N-N(R)-Boc$		

R	X	x	Additive	Solvent	Time (h)		Refs
3-thienyl	Br	10	4-HOPro (20 mol %)	DMSO	24	(43)	133
3-pyridyl	Br	10	4-HOPro (20 mol %)	DMSO	15	(82)	133
3-pyridyl	I	5	2-picoline (10 mol %)	DMF	24	(52)	134
4-pyridyl	I	10	4-HOPro (20 mol %)	DMSO	24	(91)	133

ArX, CuI (x mol %), Cs_2CO_3 → $H_2N-N(Ar)-Boc$ (I) + $ArHN-N(H)-Boc$ (II)

Ar	X	x	Additive	Solvent	Temp (°)	Time (h)	I	II	Refs
Ph	I	1	phen (10 mol %)	DMF	80	21	(97)	(0)	135
3-FC6H4	I	5	none	DMSO	50	2.5	(92)	(0)	133
3-FC6H4	Br	10	4-HOPro (20 mol %)	DMSO	80	24	(91)	(0)	133
4-FC6H4	I	5	none	DMSO	50	4	(86)	(0)	133
4-FC6H4	Br	10	4-HOPro (20 mol %)	DMSO	80	15	(66)	(0)	133
4-FC6H4	I	5	2-picoline (10 mol %)	DMF	80	24	(65)	(0)	134
4-ClC6H4	I	5	2-picoline (10 mol %)	DMF	80	24	(65)	(0)	134
4-BrC6H4	I	5	none	DMSO	50	2.5	(75)	(0)	133
4-BrC6H4	I	5	phen (20 mol %)	DMF	80	21	(71)	(0)	135
3-H2NC6H4	Br	10	4-HOPro (20 mol %)	DMSO	80	16	(73)	(0)	133
4-H2NC6H4	I	5	none	DMSO	50	4	(76)	(0)	133
4-H2NC6H4	Br	10	4-HOPro (20 mol %)	DMSO	80	24	(64)	(0)	133
4-H2NC6H4	I	5	2-picoline (10 mol %)	DMF	80	24	(62)	(0)	134
4-H2NC6H4	I	1	phen (20 mol %)	DMF	80	21	(78)	(0)	135
3-O2NC6H4	I	5	none	DMSO	50	2.5	(75)	(0)	133
4-O2NC6H4	I	10	4-HOPro (20 mol %)	DMSO	80	4	(43)	(0)	133
4-HOC6H4	I	5	none	DMSO	50	2.5	(65)	(0)	133
4-HOC6H4	Br	10	4-HOPro (20 mol %)	DMSO	80	20	(70)	(0)	133
4-HOC6H4	I	1	phen (10 mol %)	DMF	80	21	(67)	(0)	135
3-MeOC6H4	I	5	none	DMSO	50	2.5	(92)	(0)	133
3-MeOC6H4	Br	10	4-HOPro (20 mol %)	DMSO	80	15	(78)	(0)	133
3-MeOC6H4	I	1	phen (10 mol %)	DMF	80	21	(80)	(0)	135

4-MeOC$_6$H$_4$	I	1	phen (10 mol %)	DMF	80	21	(85)	(0)	135
4-MeOC$_6$H$_4$	I	5	2-picoline (10 mol %)	DMF	80	24	(61)	(0)	134
2-MeC$_6$H$_4$	I	5	none	DMSO	50	2.5	(33)	(9)	133
2-MeC$_6$H$_4$	I	5	2-picoline (10 mol %)	DMSO	80	24	(28)	(0)	134
2-MeC$_6$H$_4$	Br	10	5-HOPro (20 mol %)	DMSO	80	15	(56)	(0)	133
3-MeC$_6$H$_4$	I	5	none	DMSO	50	2.5	(92)	(0)	133
3-MeC$_6$H$_4$	Br	10	5-HOPro (20 mol %)	DMSO	80	16	(90)	(0)	133
3-MeC$_6$H$_4$	I	5	2-picoline (10 mol %)	DMF	80	24	(70)	(0)	134
4-MeC$_6$H$_4$	Br	10	5-HOPro (20 mol %)	DMSO	80	15	(80)	(0)	133
4-MeC$_6$H$_4$	I	5	none	DMSO	50	2.5	(85)	(0)	133
4-MeC$_6$H$_4$	I	5	2-picoline (10 mol %)	DMF	80	24	(67)	(0)	134
3-CF$_3$C$_6$H$_4$	Br	10	5-HOPro (20 mol %)	DMSO	80	15	(79)	(0)	133
4-CF$_3$C$_6$H$_4$	I	5	none	DMSO	50	4	(86)	(0)	133
4-CF$_3$C$_6$H$_4$	I	5	2-picoline (10 mol %)	DMF	80	24	(67)	(0)	134
3-HOCH$_2$C$_6$H$_4$	Br	10	5-HOPro (20 mol %)	DMSO	80	24	(89)	(0)	133
3-HOCH$_2$C$_6$H$_4$	I	5	2-picoline (10 mol %)	DMF	80	24	(30)	(0)	134
3-HOCH$_2$C$_6$H$_4$	I	1	phen (10 mol %)	DMF	80	21	(89)	(0)	135
3-NCC$_6$H$_4$	I	5	2-picoline (10 mol %)	DMF	80	24	(52)	(0)	134
4-NCC$_6$H$_4$	I	1	phen (10 mol %)	DMF	80	4	(78)	(0)	135
3-MeO$_2$CC$_6$H$_4$	I	5	none	DMSO	50	4	(88)	(0)	133
3-EtO$_2$CC$_6$H$_4$	I	5	phen (10 mol %)	DMF	80	21	(78)	(0)	135
4-MeO$_2$CC$_6$H$_4$	I	5	none	DMSO	50	3	(90)	(0)	133
4-MeO$_2$CC$_6$H$_4$	Br	10	5-HOPro (20 mol %)	DMSO	80	24	(34)	(41)	133
4-EtO$_2$CC$_6$H$_4$	I	5	phen (10 mol %)	DMF	80	21	(88)	(0)	135
3,5-Me$_2$C$_6$H$_3$	I	5	2-picoline (10 mol %)	DMF	80	24	(73)	(0)	134
3,5-Me$_2$C$_6$H$_3$	I	1	phen (10 mol %)	DMF	80	21	(90)	(0)	135
4-AcC$_6$H$_4$	I	10	5-HOPro (20 mol %)	DMSO	80	24	(60)	(0)	133
4-AcC$_6$H$_4$	I	1	phen (10 mol %)	DMF	80	4	(43)	(0)	135
4-i-PrC$_6$H$_4$	I	1	phen (10 mol %)	DMF	80	21	(87)	(0)	135
4-PhC$_6$H$_4$	Br	10	5-HOPro (20 mol %)	DMSO	80	15	(80)	(0)	133

TABLE 20. N-ARYLATION AND N-HETEROARYLATION OF HYDRAZINE DERIVATIVES (*Continued*)

Nitrogen Nucleophile	(Hetero)aryl Halide	Conditions	Product(s) and Yield(s) (%)	Refs.
C6				
BocHN–N(Ph)(H)	aryl-X, R	CuI (10 mol %), phen (10 mol %), Cs$_2$CO$_3$, DMF, 80°	aryl–N(Boc)–N(H)Ph, R	541

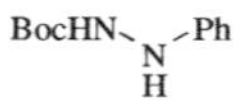

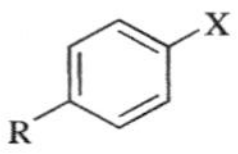

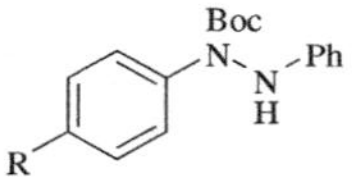

R	X	Time (h)	
H	I	22	(87)
O$_2$N	I	5	(51)
MeO	I	22	(87)
NC–	Br	47	(56)
EtO$_2$C	Br	48	(45)

| BocHN–N(Ph)(Boc) | I-aryl, R | CuI (1 eq), phen (10 mol %), Cs$_2$CO$_3$, DMF, 80° | aryl–N(Boc)–N(Ph)(Boc) | 541 |

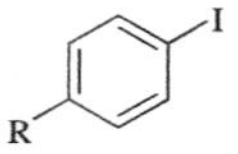

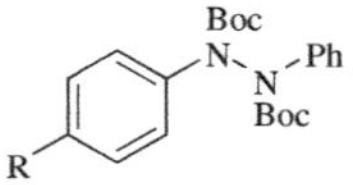

R	Time (h)	
H	3	(99)
I	18	(99)
O$_2$N	3	(99)
MeO	18	(86)
Me	3	(99)
HOCH$_2$	3	(95)
NC–	3	(99)
MeO$_2$C	3	(99)
Ph	34	(99)

| | I-aryl, R | CuI (1 eq), phen (10 mol %), Cs$_2$CO$_3$, DMF, 110° | aryl–N=N–Ph, R | 541 |

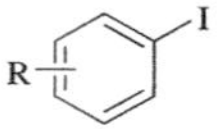

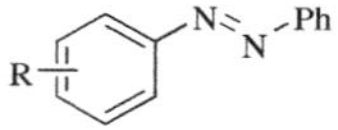

R	Time (d)	
H	6.5	(64)
4-I	3	(45)
4-O$_2$N	2	(50)
4-MeO	4	(43)
4-Me	6.5	(46)
4-HOCH$_2$	7	(trace)
3-NC–	3	(60)
4-MeO$_2$C	2	(69)
4-Ac	4	(50)
4-Ph	3	(67)

C₇

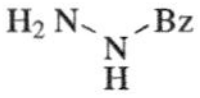

H₂N–N(H)–Bz

Ar–N(Ar)–N(H)–Bz (I) + Ar–N(H)–N(H)–Bz (II)

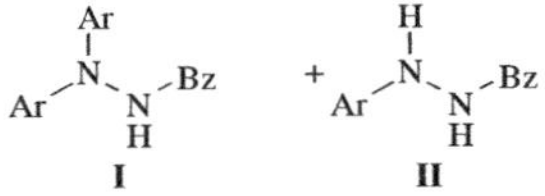

Ar	X	x	Additive(s)	Solvent	Temp (°)	Time (h)	I	II	Refs
Ph	I	1 eq	t-BuOK	HMPA	100	1	(30)	(0)	542
4-MeOC₆H₄	I	1 eq	t-BuOK	HMPA	100	1	(13)	(0)	542
2-MeC₆H₄	I	1 eq	t-BuOK	HMPA	100	1	(20)	(0)	542
2-MeC₆H₄	Br	10 mol %	PPAPM (20 mol %), K₂CO₃, LiCl	DMF	110	36	(0)	(74)	48
4-MeC₆H₄	I	1 eq	t-BuOK	HMPA	100	1	(32)	(0)	542
3,5-Me₂C₆H₃	I	1 mol %	CDA (10 mol %), K₂CO₃	dioxane	110	23	(0)	(62)	115
4-Ph	I	1 eq	t-BuOK	HMPA	100	1	(26)	(0)	542
1-Np	I	1 eq	t-BuOK	HMPA	100	1	(13)	(21)	542

C₈

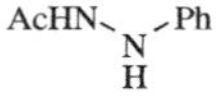

AcHN–N(H)–Ph

PhI

CuI (5 mol %), Gly (20 mol %), K₃PO₄, dioxane, 100°, 24 h

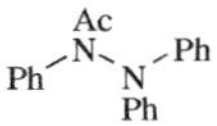

Ph–N(Ac)–N(Ph)–Ph   (91)   136

C₁₃

H₂N–N=C(Ph)Ph

(3,5-dimethylphenyl iodide)

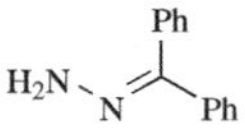

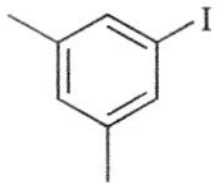

CuI (1 mol %), CDA (10 mol %), NaOt-Bu, dioxane, 110°, 23 h

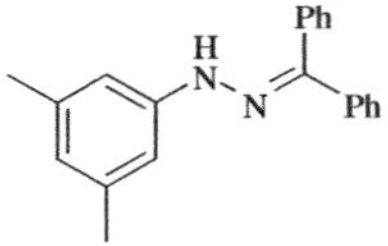

(83)   115

TABLE 21. N-ARYLATION OF HYDROXYLAMINE DERIVATIVES

Nitrogen Nucleophile	Aryl Halide	Conditions	Product(s) and Yield(s) (%)	Refs.
C_1 MeO–N(H)–Boc	R–C6H4–X	CuI (5 mol %), phen (50 mol %), Cs2CO3, DMF, 80°, 24 h	MeO–N(Boc)–C6H4–R R | X H | Br (0) H | I (89) H | OTf (0) 4-F | I (72) 3-Br | I (74) 4-O2N | I (77) R | X 4-MeO | I (69) 2-Me | I (0) 4-Me | I (72) 3-EtO2C | I (84) 4-Ac | I (73)	543
C_3 allyl-O–N(H)–Boc	I–C6H4–R	CuI (5 mol %), phen (50 mol %), Cs2CO3, DMF, 80°, 24 h	allyl-O–N(Boc)–C6H4–R R O2N (70) Me (57)	543
C_4 t-BuO–N(H)–Boc	I–C6H4–R	CuI (5 mol %), phen (50 mol %), Cs2CO3, DMF, 80°, 24 h	t-BuO–N(Boc)–C6H4–R R F (70) O2N (68) MeO (86) Me (74) NC⁻ (86)	543
C_5 THP-O–N(H)–Boc	I–C6H4–R	CuI (5 mol %), phen (50 mol %), Cs2CO3, DMF, 80°, 24 h	THP-O–N(Boc)–C6H4–R R Me (70) NC⁻ (76)	543

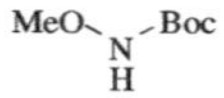

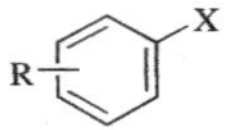

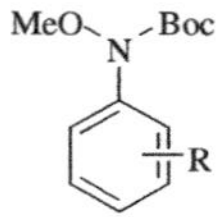

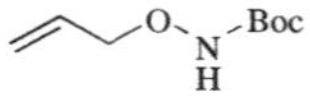

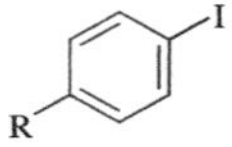

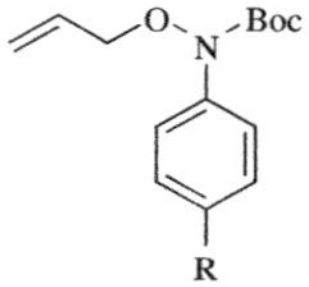

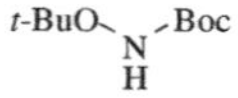

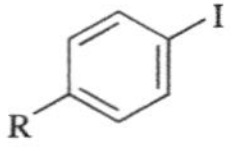

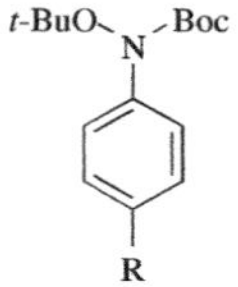

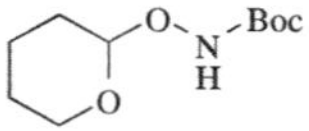

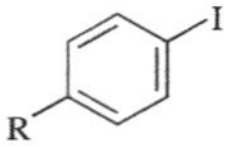

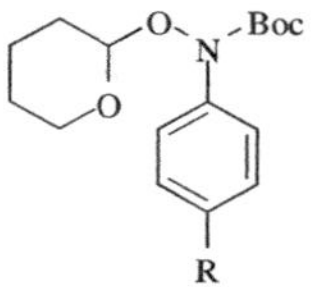

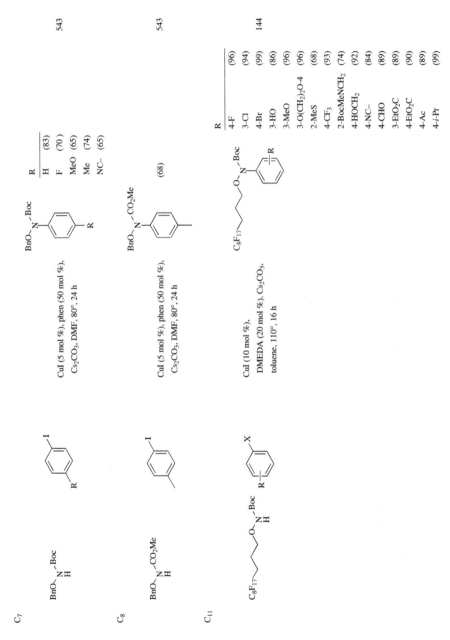

C$_7$

BnO–N(H)–Boc + 4-iodo(R)benzene → BnO–N(Boc)–C$_6$H$_4$–R

CuI (5 mol %), phen (50 mol %), Cs$_2$CO$_3$, DMF, 80°, 24 h

R	
H	(83)
F	(70)
MeO	(65)
Me	(74)
NC–	(65)

543

C$_8$

BnO–N(H)–CO$_2$Me + 4-iodotoluene → BnO–N(CO$_2$Me)–C$_6$H$_4$–Me   (68)

CuI (5 mol %), phen (50 mol %), Cs$_2$CO$_3$, DMF, 80°, 24 h

543

C$_{11}$

C$_8$F$_{17}$(CH$_2$)$_4$O–N(H)–Boc + R–C$_6$H$_4$–X → C$_8$F$_{17}$(CH$_2$)$_4$O–N(Boc)–C$_6$H$_4$–R

CuI (10 mol %), DMEDA (20 mol %), Cs$_2$CO$_3$, toluene, 110°, 16 h

R	
4-F	(96)
3-Cl	(94)
4-Br	(99)
3-HO	(86)
3-MeO	(96)
3-O(CH$_2$)$_2$O-4	(96)
2-MeS	(68)
4-CF$_3$	(93)
2-BocMeNCH$_2$	(74)
4-HOCH$_2$	(92)
4-NC–	(84)
4-CHO	(89)
3-EtO$_2$C	(89)
4-EtO$_2$C	(90)
4-Ac	(89)
4-i-Pr	(99)

144

TABLE 21. *N*-ARYLATION OF HYDROXYLAMINE DERIVATIVES (*Continued*)

Nitrogen Nucleophile	Aryl Halide	Conditions	Product(s) and Yield(s) (%)	Refs.
C₁₁				
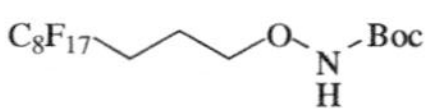 C₈F₁₇ ⟨chain⟩ O–N(H)–Boc	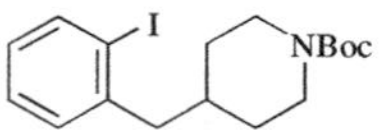 I, ⟨aryl⟩ NBoc	CuI (10 mol %), DMEDA (20 mol %), Cs₂CO₃, toluene, 110°, 16 h	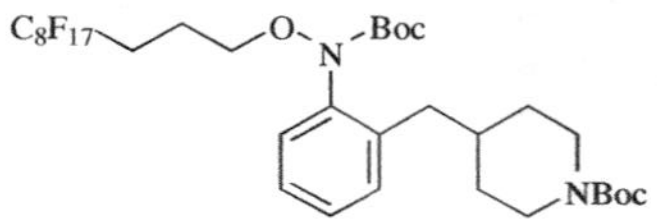 C₈F₁₇ ⟨chain⟩ O–N(Boc)–⟨aryl⟩ NBoc (96)	144

Please refer to the charts preceding the tables for structures indicated by the **bold** numbers.

Nitrogen Nucleophile	Aryl Halide	Conditions	Product(s) and Yield(s) (%)	Refs.

C_1

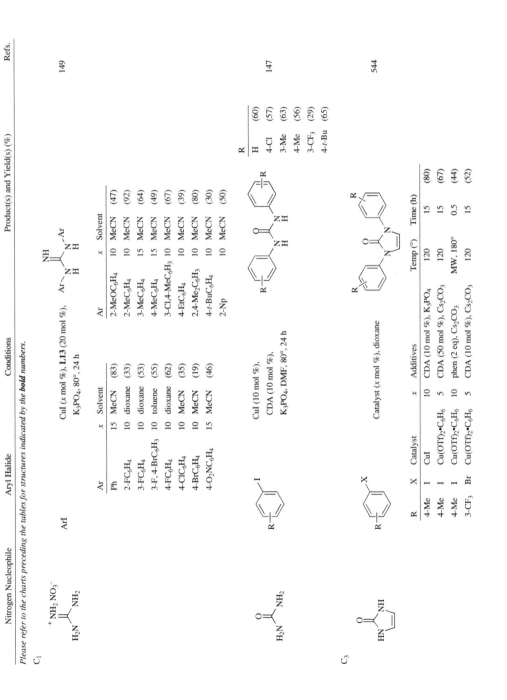

$\overset{+}{N}H_2\ NO_3^-$, H_2N , NH_2

ArI

CuI (x mol %), **L13** (20 mol %), K_3PO_4, 80°, 24 h

NH, $Ar\overset{}{N}\underset{H}{\overset{}{}}\overset{}{N}\underset{H}{}Ar$

Ar	x	Solvent		Ar	x	Solvent
Ph	15	MeCN (83)		2-MeOC$_6$H$_4$	10	MeCN (47)
2-FC$_6$H$_4$	10	dioxane (33)		2-MeC$_6$H$_4$	10	MeCN (92)
3-FC$_6$H$_4$	10	dioxane (53)		3-MeC$_6$H$_4$	15	MeCN (64)
3-F, 4-BrC$_6$H$_3$	10	toluene (55)		4-MeC$_6$H$_4$	15	MeCN (49)
4-FC$_6$H$_4$	10	dioxane (62)		3-Cl,4-MeC$_6$H$_3$	10	MeCN (67)
4-ClC$_6$H$_4$	10	MeCN (35)		4-EtC$_6$H$_4$	10	MeCN (39)
4-BrC$_6$H$_4$	10	MeCN (19)		2,4-Me$_2$-C$_6$H$_3$	10	MeCN (80)
4-O$_2$NC$_6$H$_4$	15	MeCN (46)		4-t-BuC$_6$H$_4$	10	MeCN (30)
				2-Np	10	MeCN (50)

149

CuI (10 mol %), CDA (10 mol %), K_3PO_4, DMF, 80°, 24 h

R—(ring)—I , H_2N—C(=O)—NH_2

R—(ring)—NH—C(=O)—NH—(ring)—R

R	
H	(60)
4-Cl	(57)
3-Me	(63)
4-Me	(56)
3-CF$_3$	(29)
4-t-Bu	(65)

147

C_3

R—(ring)—X , HN—(ring)—C(=O)—NH

Catalyst (x mol %), dioxane

R—(ring)—N(ring with C=O)N—(ring)—R

R	X	Catalyst	x	Additives	Temp (°)	Time (h)	
4-Me	I	CuI	10	CDA (10 mol %), K$_3$PO$_4$	120	15	(80)
4-Me	I	Cu(OTf)$_2$•C$_6$H$_6$	5	CDA (50 mol %), Cs$_2$CO$_3$	120	15	(67)
4-Me	I	Cu(OTf)$_2$•C$_6$H$_6$	10	phen (2 eq), Cs$_2$CO$_3$	MW, 180°	0.5	(44)
3-CF$_3$	Br	Cu(OTf)$_2$•C$_6$H$_6$	5	CDA (10 mol %), Cs$_2$CO$_3$	120	15	(52)

544

475

TABLE 22A. N-ARYLATION OF UREAS AND GUANIDINES (Continued)

| Nitrogen Nucleophile | Aryl Halide | | Conditions | | | | Product(s) and Yield(s) (%) | | Refs. |

Please refer to the charts preceding the tables for structures indicated by the **bold** numbers.

C₃ — nucleophile: imidazolidin-2-one (O=C with NH, HN ring)

Conditions: CuI (10 mol %), K₂CO₃

Product I: Ar–N(imidazolidinone) **I**; Product II: Ar–N / N–Ar imidazolidinone **II**

Ar	X	Additive	Solvent	Temp (°)	Time (h)	I	II	Refs.
Ph	I	DMEDA (30 mol %)	n-BuOH	100	5	(74)	(17)	148
Ph	Br	DMCDA (20 mol %)	toluene	reflux	24	(—)	(85)	545
2-FC₆H₄	I	DMEDA (30 mol %)	n-BuOH	100	5	(60)	(—)	148
3-FC₆H₄	I	DMEDA (30 mol %)	n-BuOH	100	5	(67)	(—)	148
4-FC₆H₄	I	DMEDA (30 mol %)	n-BuOH	100	5	(67)	(—)	148
2-ClC₆H₄	I	DMEDA (30 mol %)	n-BuOH	100	5	(49)	(—)	148
3-ClC₆H₄	I	DMEDA (30 mol %)	n-BuOH	100	5	(53)	(—)	148
4-ClC₆H₄	I	DMEDA (30 mol %)	n-BuOH	100	5	(65)	(—)	148
4-BrC₆H₄	I	DMEDA (30 mol %)	n-BuOH	100	5	(48)	(—)	148
4-H₂NC₆H₄	I	DMEDA (30 mol %)	n-BuOH	100	5	(54)	(—)	148
4-O₂NC₆H₄	Br	DMCDA (20 mol %)	toluene	reflux	24	(—)	(82)	545
4-HOC₆H₄	I	DMEDA (30 mol %)	n-BuOH	100	5	(61)	(—)	148
2-MeOC₆H₄	Br	DMCDA (20 mol %)	toluene	reflux	24	(—)	(82)	545
2-MeOC₆H₄	I	DMEDA (30 mol %)	n-BuOH	100	5	(70)	(—)	148
3-MeOC₆H₄	I	DMEDA (30 mol %)	n-BuOH	100	5	(71)	(—)	148
4-MeOC₆H₄	I	DMEDA (30 mol %)	n-BuOH	100	5	(54)	(—)	148
4-MeOC₆H₄	Br	DMCDA (20 mol %)	toluene	reflux	24	(—)	(88)	545
4-MeSC₆H₄	Br	DMCDA (20 mol %)	toluene	reflux	24	(—)	(76)	545
2-MeC₆H₄	I	DMEDA (30 mol %)	n-BuOH	100	5	(55)	(—)	148
3-MeC₆H₄	I	DMEDA (30 mol %)	n-BuOH	100	5	(65)	(—)	148
4-MeC₆H₄	I	DMEDA (30 mol %)	n-BuOH	100	5	(61)	(—)	148
4-NCC₆H₄	Br	DMCDA (20 mol %)	toluene	reflux	24	(—)	(84)	545
4-NCC₆H₄	I	DMEDA (30 mol %)	n-BuOH	100	5	(46)	(—)	148
4-EtO₂CC₆H₄	Br	DMCDA (20 mol %)	toluene	reflux	24	(—)	(73)	545
4-AcC₆H₄	Br	DMCDA (20 mol %)	toluene	reflux	24	(—)	(82)	545
4-AcC₆H₄	I	DMEDA (30 mol %)	n-BuOH	100	5	(45)	(—)	148
1-Np	Br	DMCDA (20 mol %)	toluene	reflux	24	(—)	(88)	545

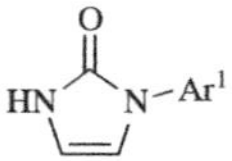

Ar²X

(CuOTf)₂•C₆H₆ (10 mol %), DMCDA (50 mol %), dba (10 mol %), Cs₂CO₃, dioxane, MW, 150°, 30 min

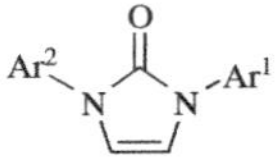

Ar¹	Ar²	X	
Ph	3-O₂NC₆H₄	I	(73)
Ph	4-O₂NC₆H₄	Br	(96)
Ph	4-O₂NC₆H₄	I	(100)
Ph	4-MeC₆H₄	I	(97)
Ph	4-CF₃C₆H₄	Br	(83)
4-MeOC₆H₄	4-O₂NC₆H₄	I	(74)
4-MeC₆H₄	4-MeC₆H₄	Br	(84)
4-MeC₆H₄	4-MeC₆H₄	I	(86)

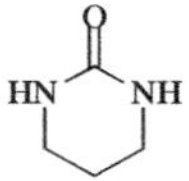

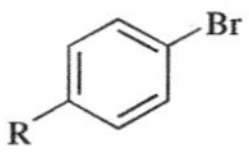

544

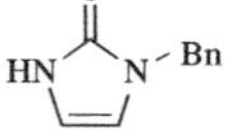

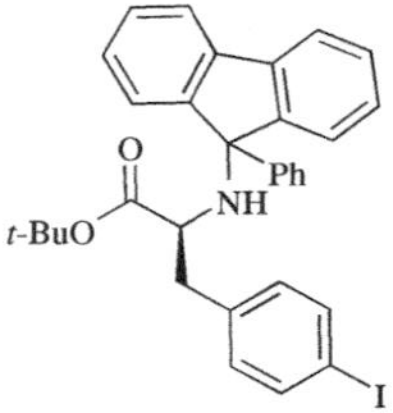

CuI (cat.), Cs₂CO₃, DMF, 150°

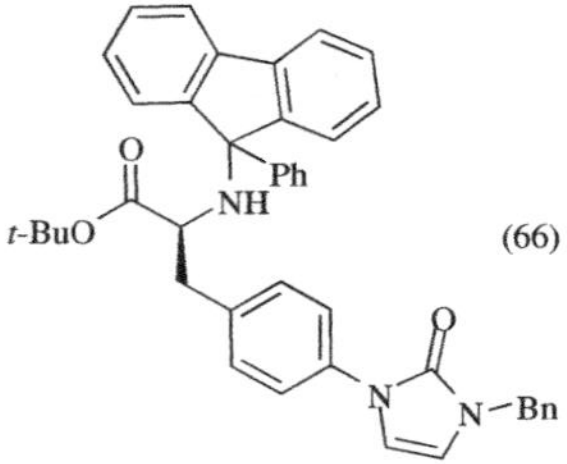

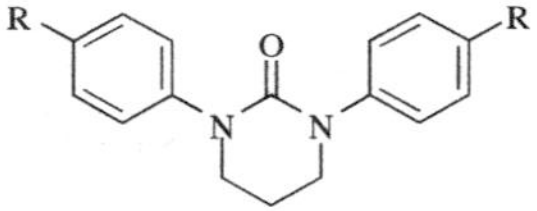

(66)   536

C₃₋₉

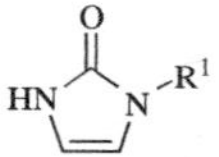

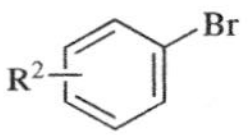

CuI (10 mol %), DMCDA (20 mol %), K₂CO₃, toluene, reflux, 24 h

R¹	R²	
t-Bu	SMe	(94)
4-MeSC₆H₄	EtO₂C	(76)

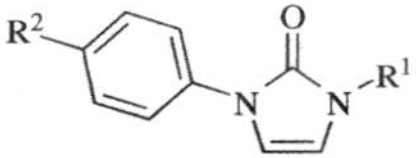

545

C₄

CuI (10 mol %), DMCDA (20 mol %), K₂CO₃, toluene, reflux, 24 h

R	
O₂N	(67)
MeO	(84)

545

TABLE 22A. N-ARYLATION OF UREAS AND GUANIDINES (Continued)

Nitrogen Nucleophile	Aryl Halide	Conditions	Product(s) and Yield(s) (%)	Refs.

*Please refer to the charts preceding the tables for structures indicated by the **bold** numbers.*

C₅

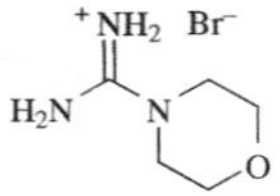

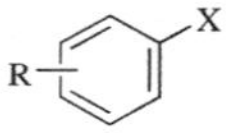

CuI (10 mol %), DMEDA (20 mol %), Cs₂CO₃, dioxane, 100°, 24 h

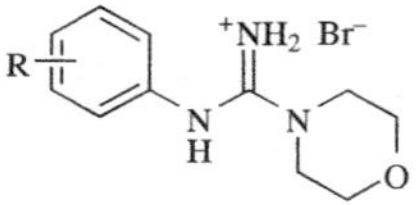

R	X	
3,4-Cl₂	I	(86)
4-O₂N	I	(93)
4-MeO	I	(75)
4-Me	Br	(35)
4-Me	I	(86)
4-NC–	I	(84)
4-Ac	I	(83)

Refs. 141

C₇

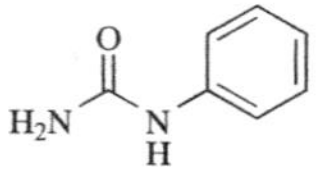

ArI

CuI (15 mol %), KF/Al₂O₃, BnNH(CH₂)₂NHBn (15 mol %), THF, 70°, 4 h

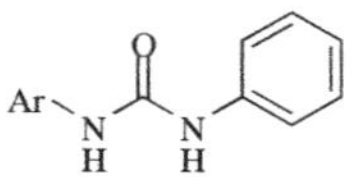

Ar	
Ph	(83)
4-BrC₆H₄	(88)
4-O₂NC₆H₄	(75)
2-MeOC₆H₄	(75)
4-MeOC₆H₄	(78)
2-MeC₆H₄	(75)
4-MeC₆H₄	(70)
3-CF₃C₆H₄	(73)
4-EtCO	(70)
1-Np	(67)

Refs. 146

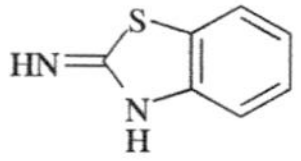

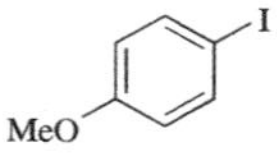

CuI (10 mol %), pyrrole-2-carboxylic acid (20 mol %), K₃PO₄, DMSO, 90°, 24 h

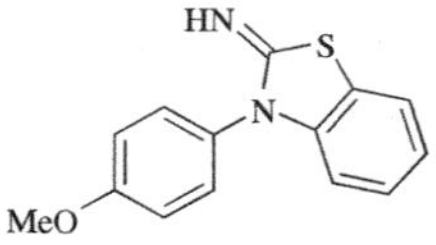

(82)

Refs. 50

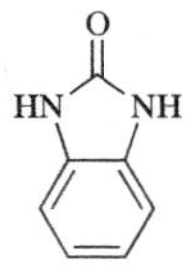

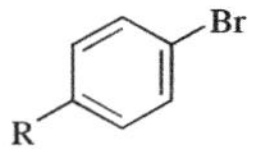

CuI (x mol %), K$_2$CO$_3$, 24 h

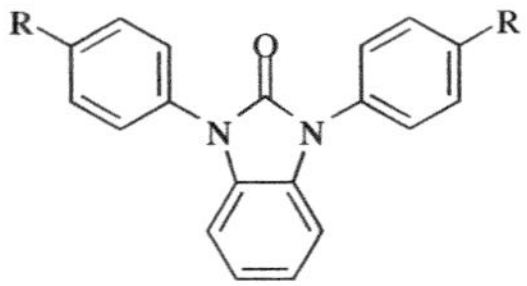

R	x	Additive	Solvent	Temp (°)		
Br	20	none	DMF	150	(28)	546
O$_2$N	10	DMCDA (20 mol %)	toluene	reflux	(82)	545
MeO	10	DMCDA (20 mol %)	toluene	reflux	(82)	545

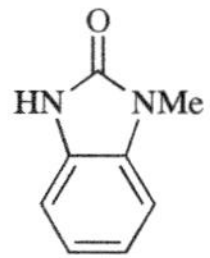

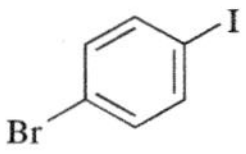

CuI (5 mol %),
phen (10 mol %), Cs$_2$CO$_3$,
DMF; 80°, 24 h

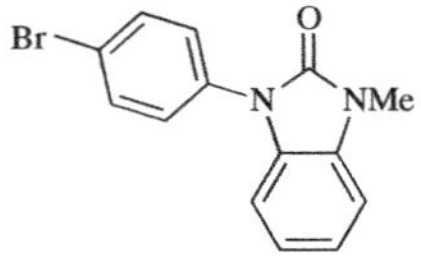

(81)

408

TABLE 22B. *N*-HETEROARYLATION OF UREAS AND GUANIDINES

*Please refer to the charts preceding the tables for structures indicated by the **bold** numbers.*

Nitrogen Nucleophile	Heteroaryl Halide	Conditions	Product(s) and Yield(s) (%)	Refs.
C₁				
		CuI (10 mol %), **L13** (20 mol %), K₃PO₄, dioxane, 80°, 24 h	(75)	149
C₃				
		(CuOTf)₂•C₆H₆ (10 mol %), phen (20 mol %), Cs₂CO₃, dioxane, MW, 150°, 30 min	(32)	544
		(CuOTf)₂•C₆H₆ (10 mol %), DMCDA (50 mol %), dba (10 mol %), Cs₂CO₃, dioxane, MW, 150°, 30 min	(96)	544
		CuI (10 mol %), DMEDA (30 mol %), K₂CO₃, *n*-BuOH, 100°, 5 h	Isomer 2 (45) 3 (46)	148
		CuI (10 mol %), DMCDA (20 mol %), K₂CO₃, toluene, reflux, 24 h	(82)	545
C₇				
		CuI (5 mol %), phen (10 mol %), Cs₂CO₃, DMF, 80°, 24 h	Y N (76) CH (94)	408
C₁₃				
		CuI (10 mol %), DMCDA (20 mol %), K₂CO₃, toluene, reflux, 24 h	(21)	545

480

TABLE 23A. N-ARYLATION OF SULFONAMIDES AND SULFONIMIDAMIDES

*Please refer to the charts preceding the tables for structures indicated by the **bold** numbers.*

Nitrogen Nucleophile	Aryl Halide	Conditions	Product(s) and Yield(s) (%)	Refs.

C₁

Nitrogen Nucleophile: $H_2N-S(O)_2-Me$

Aryl Halide: $R-C_6H_4-X$

Conditions: CuI (x mol %)

Product: $R-C_6H_4-N(H)-S(O)_2-Me$

R	X	x	Additive(s)	Solvent	Temp (°)	Time (h)	Yield (%)	Refs.
H	Br	10	K₂CO₃	NMP	MW, 195	2	(67)	547
H	I	5	NMG (20 mol %), K₃PO₄	DMF	100	24	(99)	150
H	Br	20	DMG (20 mol %), K₃PO₄	DMF	reflux	48	(99)	150
2-MeO	I	5	NMG (20 mol %), K₃PO₄	DMF	100	24	(60)	150
4-MeO	Br	20	DMG (20 mol %), K₃PO₄	DMF	reflux	48	(80)	150
2-Me	I	5	NMG (20 mol %), K₃PO₄	DMF	100	24	(80)	150
4-Me	Br	20	DMG (20 mol %), K₃PO₄	DMF	reflux	48	(89)	150
3-MeO₂C	I	5	NMG (20 mol %), K₃PO₄	DMF	100	24	(99)	151
4-MeO₂C	Br	20	L22 (20 mol %), K₂CO₃	DMF	110	40	(75)	151
4-EtO₂C	I	10	NMG (20 mol %), K₃PO₄	DMF	100	24	(95)	150
4-Ac	Br	20	DMG (20 mol %), K₃PO₄	DMF	reflux	48	(96)	150

C₂

Nitrogen Nucleophile: $MeHN-S(O)_2-Me$

Aryl Halide: 4-(MeO₂C)C₆H₄–Br

Conditions: CuI (20 mol %), **L22** (20 mol %), K₂CO₃, DMF, 110°, 40 h

Product: MeO₂C–C₆H₄–N(Me)–S(O)₂–Me (75)

Refs.: 151

C₃

Nitrogen Nucleophile: cyclic HN–S(O)₂ (sultam)

Aryl Halide: $R-C_6H_4-Br$

Conditions: CuI (20 mol %), **L22** (20 mol %), K₂CO₃, DMF

R	Temp (°)	Time (h)	Yield (%)
4-MeO	120	36	(40)
3-Me, 4-NH₂, 5-O₂N	110	40	(75)
4-MeO₂C	110	40	(63)

Refs.: 151

TABLE 23A. N-ARYLATION OF SULFONAMIDES AND SULFONIMIDAMIDES (*Continued*)

Nitrogen Nucleophile	Aryl Halide	Conditions	Product(s) and Yield(s) (%)	Refs.

*Please refer to the charts preceding the tables for structures indicated by the **bold** numbers.*

C4

Catalyst (x mol %)

R	X	Catalyst	x	Additives	Solvent	Temp (°)	Time (h)		Refs.
2-O₂N	I	Cu₂O	40	bpy (40 mol %), K₃PO₄	NMP	120	18	(44)	152
2-MeO	Br	CuI	20	**L22** (20 mol %), K₂CO₃	DMF	110	40	(10)	151
3-MeO	Br	CuI	20	**L22** (20 mol %), K₂CO₃	DMF	110	40	(43)	151
4-MeO	Br	CuI	20	**L22** (20 mol %), K₂CO₃	DMF	120	36	(43)	151
2-Me	Br	CuI	20	**L22** (20 mol %), K₂CO₃	DMF	110	40	(14)	151
2-Me, 5-O₂N	Br	Cu₂O	40	bpy (40 mol %), K₃PO₄	NMP	120	48	(50)	152
3-Me, 4-H₂N, 5-O₂N	Br	CuI	20	**L22** (20 mol %), K₂CO₃	DMF	110	40	(64)	151
3-NC–	Br	Cu₂O	40	bpy (40 mol %), K₃PO₄	NMP	120	24	(35)	152
4-NC–	Br	Cu₂O	40	bpy (40 mol %), K₃PO₄	NMP	120	15	(35)	152
2-MeO₂C	Br	CuI	20	**L22** (20 mol %), K₂CO₃	DMF	110	40	(0)	151
3-MeO₂C	Br	CuI	20	**L22** (20 mol %), K₂CO₃	DMF	110	40	(60)	151
4-EtO₂C	Br	CuI	20	**L22** (20 mol %), K₂CO₃	DMF	120	36	(85)	151
2,6-Me₂	Br	CuI	20	**L22** (20 mol %), K₂CO₃	DMF	110	40	(0)	151

C₆

Catalyst (x mol %)

R	X	Catalyst	x	Additive(s)	Solvent(s)	Temp (°)	Time (h)		Refs.
H	Br	CuI	10	K₂CO₃	NMP	MW, 195	3	(66)	547
H	I	CuI	10	K₂CO₃	NMP	MW, 195	3	(90)	547
H	I	CuI	10	NMG (20 mol %), K₃PO₄	DMF	100	24	(95)	150
H	Br	CuI	20	DMG (20 mol %), K₃PO₄	DMF	reflux	48	(95)	150
4-O₂N	I	CuO nanoparticles	5	KOH	t-BuOH/DMSO (3:1)	110	24	(54)	74
2-MeO	I	CuI	10	NMG (20 mol %), K₃PO₄	NMP	110	24	(55)	150
3-MeO	Br	CuI	10	K₂CO₃	NMP	MW, 195	3	(90)	547
3-MeO	I	CuI	10	K₂CO₃	NMP	MW, 195			547

4-MeO	Br	CuI	20	DMG (20 mol %), K$_3$PO$_4$	DMF	reflux	48	(75)	150
2-Me	I	CuI	5	NMG (20 mol %), K$_3$PO$_4$	DMF	100	24	(56)	150
4-Me	Br	CuI	20	DMG (20 mol %), K$_3$PO$_4$	DMF	reflux	48	(96)	150
4-NC–	Br	CuI	10	K$_2$CO$_3$	NMP	MW, 195	2	(88)	547
3-MeO$_2$C	I	CuI	5	NMG (20 mol %), K$_3$PO$_4$	DMF	100	24	(99)	150
4-EtO$_2$C	I	CuI	5	NMG (20 mol %), K$_3$PO$_4$	DMF	100	24	(99)	150
3,5-Me$_2$	Br	CuI	10	K$_2$CO$_3$	NMP	MW, 195	2	(81)	547
3,5-Me$_2$	I	CuI	10	K$_2$CO$_3$	NMP	MW, 195	3	(88)	547
4-Ac	Br	CuI	20	DMG (20 mol %), K$_3$PO$_4$	DMF	reflux	48	(91)	150
4-t-Bu	Br	CuI	10	K$_2$CO$_3$	NMP	MW, 195	2	(78)	547
4-t-Bu	I	CuI	10	K$_2$CO$_3$	NMP	MW, 195	2	(88)	547

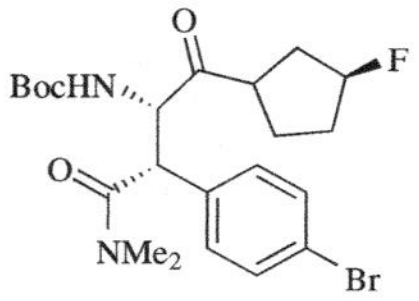

CuI (1 eq), DMEDA (3 eq), K$_2$CO$_3$, toluene, 110°, 48 h

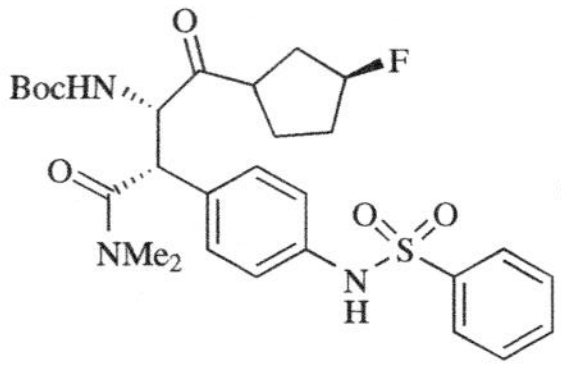

(80) 548

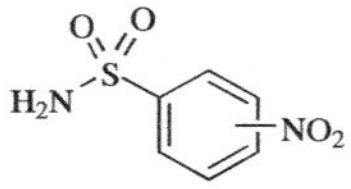

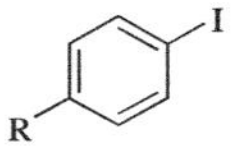

CuI (2 eq), Cs$_2$CO$_3$, DMSO, 90°, 18 h

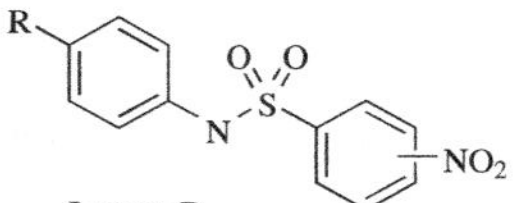

Isomer	R	
2	H	(91)
4	EtO$_2$C	(82)

274

Nitrogen Nucleophile	Aryl Halide				Conditions			Product(s) and Yield(s) (%)	Refs.

*Please refer to the charts preceding the tables for structures indicated by the **bold** numbers.*

C7

Nitrogen Nucleophile: H_2N–SO_2–(4-methylphenyl)

Aryl Halide: R–C_6H_4–X

R	X	Catalyst	x	Additive(s)	Solvent(s)	Temp (°)	Time (h)	Yield	Refs.
H	Br	CuI	10	K2CO3	NMP	MW, 195	3	(70)	547
H	Br	CuI	20	DMG (20 mol %), K3PO4	DMF	reflux	48	(99)	150
H	I	CuO nanoparticles	5	KOH	t-BuOH/DMSO (3:1)	110	24	(45)	74
H	I	Cu(OAc)2•H2O	15	sparteine (30 mol %), K2CO3	DMF	130	24	(80)	290
2-MeO	I	CuI	5	NMG (20 mol %), K3PO4	DMF	100	24	(76)	150
4-MeO	Br	CuI	20	DMG (20 mol %), K3PO4	DMF	reflux	48	(85)	150
2-Me	I	CuI	5	NMG (20 mol %), K3PO4	DMF	100	24	(50)	150
4-Me	Br	CuI	20	DMG (20 mol %), K3PO4	DMF	reflux	48	(91)	150
3-MeO2C	I	CuI	5	NMG (20 mol %), K3PO4	DMF	100	24	(99)	150
4-EtO2C	I	CuI	5	NMG (20 mol %), K3PO4	DMF	100	24	(99)	150
4-EtO2C	Br	CuI	20	L22 (20 mol %), K2CO3	DMF	110	40	(51)	151
4-Ac	Br	CuI	20	DMG (20 mol %), K3PO4	DMF	reflux	48	(95)	150

Product: (4-methylphenyl)SO_2–NH–(C_6H_4–R)

Nitrogen Nucleophile: MeHN–SO_2–Ph

Aryl Halide: R–C_6H_4–X

Catalyst: CuI (x mol %)

R	X	x	Additive(s)	Solvent	Temp (°)	Time (h)	Yield	Refs.
H	Br	10	K2CO3	NMP	MW, 195	4	(54)	547
H	I	5	NMG (20 mol %), K3PO4	DMF	100	24	(82)	150
H	Br	20	DMG (20 mol %), K3PO4	DMF	reflux	48	(82)	150
2-MeO	I	5	NMG (20 mol %), K3PO4	DMF	100	24	(trace)	150
4-MeO	Br	20	DMG (20 mol %), K3PO4	DMF	reflux	48	(68)	150
2-Me	I	5	NMG (20 mol %), K3PO4	DMF	100	24	(trace)	150
4-Me	Br	20	DMG (20 mol %), K3PO4	DMF	reflux	48	(99)	150
3-MeO2C	I	5	NMG (20 mol %), K3PO4	DMF	100	24	(99)	150
4-MeO2C	Br	20	L22 (20 mol %), K2CO3	DMF	110	40	(51)	151
4-EtO2C	I	5	NMG (20 mol %), K3PO4	DMF	100	24	(95)	150
4-Ac	Br	20	DMG (20 mol %), K3PO4	DMF	reflux	48	(90)	150

Product: (C_6H_4–R)–N(Me)–SO_2–Ph

CuI (20 mol %), DMCDA (40 mol %), K$_3$PO$_4$, dioxane, 110°, 22 h

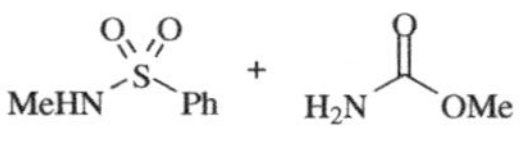

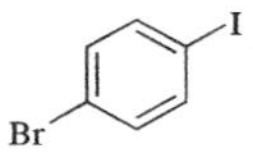

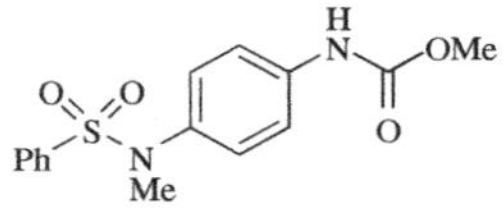

(83) 504

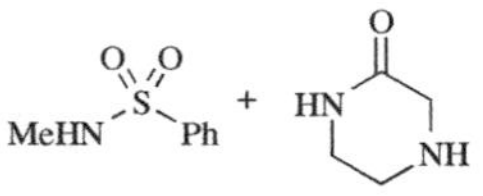

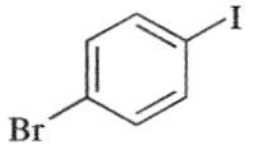

CuI (20 mol %), DMCDA (40 mol %), K$_3$PO$_4$, dioxane, 110°, 22 h

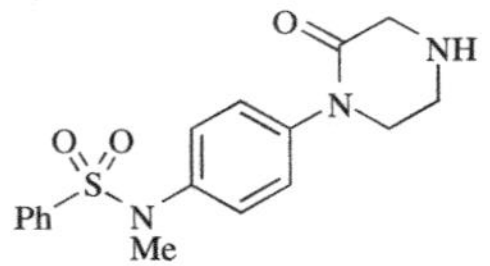

(81) 504

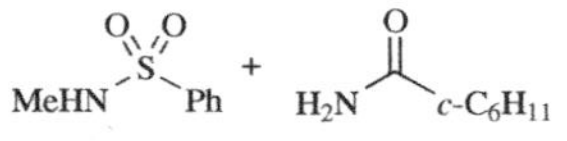

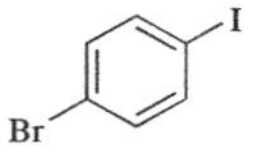

CuI (20 mol %), DMCDA (40 mol %), K$_3$PO$_4$, dioxane, 110°, 22 h

(70) 504

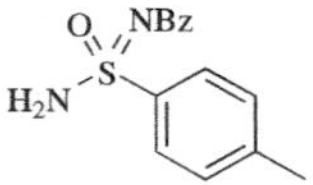

CuCl (1 eq), K$_2$CO$_3$, DMSO, 100°, 20 h

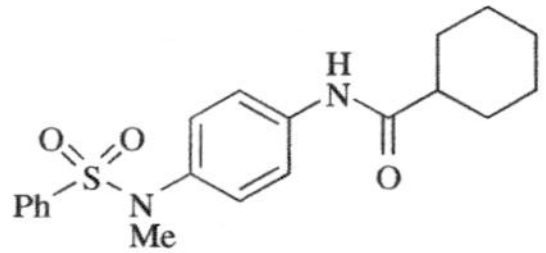

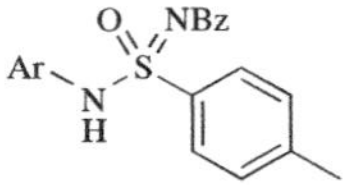

153

Ar	X	
Ph	I	(50)
4-FC$_6$H$_4$	I	(51)
2-ClC$_6$H$_4$	I	(55)
3-ClC$_6$H$_4$	I	(59)
4-ClC$_6$H$_4$	I	(75)
4-IC$_6$H$_4$	I	(65)
4-MeOC$_6$H$_4$	Br	(59)
4-MeOC$_6$H$_4$	I	(68)
3-MeC$_6$H$_4$	I	(62)
4-MeC$_6$H$_4$	I	(59)
2-Np	I	(66)

ArX

485

TABLE 23A. *N*-ARYLATION OF SULFONAMIDES AND SULFONIMIDAMIDES (*Continued*)

*Please refer to the charts preceding the tables for structures indicated by the **bold** numbers.*

Nitrogen Nucleophile	Aryl Halide	Conditions	Product(s) and Yield(s) (%)	Refs.
C_{13}				
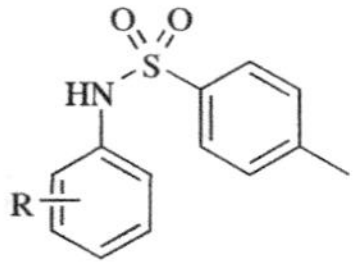	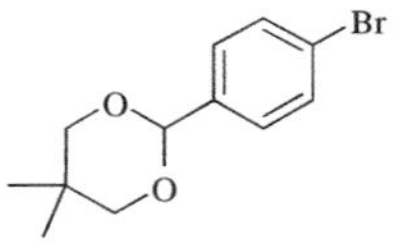	Cu (2.3 eq), K_2CO_3, reflux, 13 h	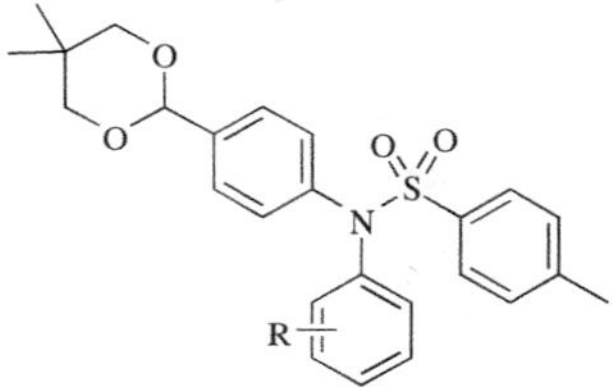	549
C_{13-15}				
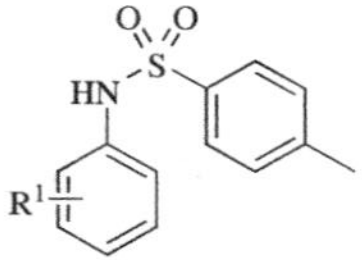	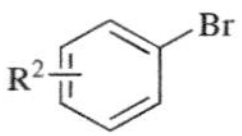	Cu (2.3 eq), K_2CO_3, reflux, 13 h	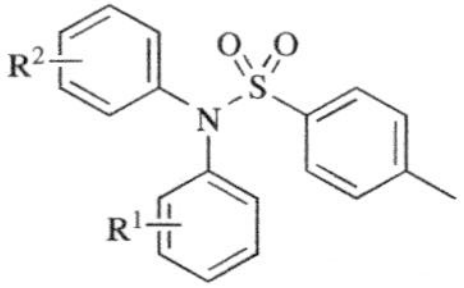	549

R	
2,3-$(MeO)_2$	(25)
3,4,5-$(MeO)_3$	(67)

R^1	R^2	
2-MeO	2,5-Me_2	(67)
4-MeO	3-Me, 5-MeO	(45)
3,5-$(MeO)_2$	2,5-Me_2	(21)
3,4-OCH_2O–	2,5-Me_2	(35)

R^1	R^2	
3,4,5-$(MeO)_3$	2,5-Me_2	(12)
2,5-Me_2	H	(17)
2,5-Me_2	3-MeO	(63)
2,5-Me_2	2,5-$(MeO)_2$	(52)

TABLE 23B. *N*-HETEROARYLATION OF SULFONAMIDES

*Please refer to the charts preceding the tables for structures indicated by the **bold** numbers.*

Nitrogen Nucleophile	Heteroaryl Halide	Conditions	Product(s) and Yield(s) (%)	Refs.

C₁₋₂

RHN–S(O)(O)–Me | 3-Br-pyridine | CuI (20 mol %), **L22** (20 mol %), K₂CO₃, DMF, 110°, 40 h | R-N(S(O)(O)Me)(pyridyl)

R
H (42)
Me (91) | 151 |

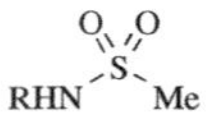

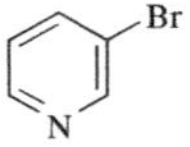

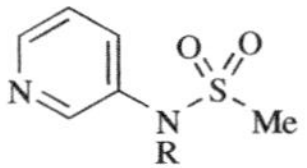

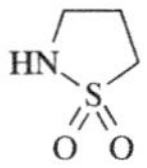

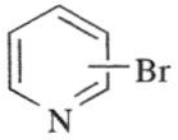

C₃

sultam (5-membered) | 2-Br-pyridine | CuI (20 mol %), **L22** (20 mol %), K₂CO₃, DMF, 110°, 40 h |
Isomer
2 (85)
3 (70)
4ᵃ (60) | 151 |

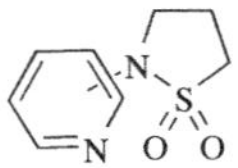

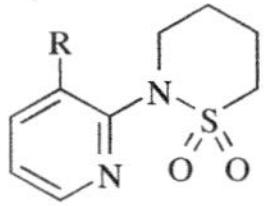

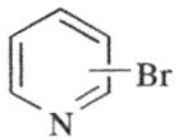

C₄

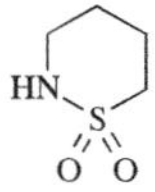

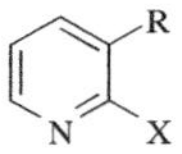

sultam (6-membered) | R-substituted 2-X-pyridine | Catalyst (*x* mol %) | | |

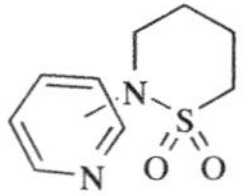

R	X	Catalyst	*x*	Additives	Solvent	Temp (°)	Time (h)		
H	Br	CuI	20	**L22** (20 mol %), K₂CO₃	DMF	110	40	(70)	151
H	Br	Cu₂O	40	bpy (40 mol %), K₃PO₄	NMP	120	6	(93)	152
NC–	Cl	Cu₂O	40	bpy (40 mol %), K₃PO₄	NMP	120	36	(35)	152

sultam (6-membered) | 3-Br-pyridine | CuI (20 mol %), **L22** (20 mol %), K₂CO₃, DMF, 110°, 40 h |
Isomer
3 (61)
4ᵃ (86) | 151 |

C₆₋₈

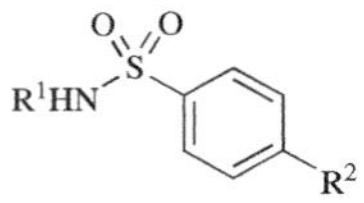

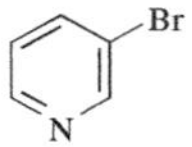

R¹HN–S(O)(O)–C₆H₄–R² | 3-Br-pyridine | CuI (20 mol %), **L22** (20 mol %), K₂CO₃, DMF, 120°, 36 h |
R¹ R²
H Cl (55)
H O₂N (33)
H MeO (54)
H Me (90)
Me Me (99) | 151 |

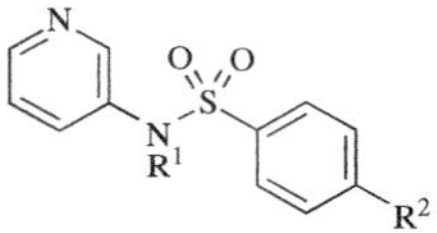

ᵃ The heteroaryl halide was the HCl salt.

TABLE 24A. *N*-ARYLATION OF SULFOXIMINES

Nitrogen Nucleophile	Aryl Halide			Conditions				Product(s) and Yield(s) (%)		Refs.
				Catalyst (x amount)						
R	X	Catalyst	x	Additive(s)	Solvent	Temp (°)	Time (h)			
H	Br	CuI	10 mol %	DMEDA (20 mol %), NaI, Cs$_2$CO$_3$	dioxane	110	22	(93)		159
H	I	CuI	10 mol %	DMEDA (20 mol %), Cs$_2$CO$_3$	toluene	110	22	(95)		159
H	Br	Cu$_2$O	10 mol %	Cs$_2$CO$_3$	DMF	110	18	(89)		221
H	I	Cu$_2$O	10 mol %	Cs$_2$CO$_3$	DMF	100	18	(95)		221
2-F	I	CuI	10 mol %	DMEDA (20 mol %), NaI, Cs$_2$CO$_3$	dioxane	110	22	(86)		159
2,4,6-Cl$_3$	I	CuI	1 eq	CsOAc	DMSO	90	—	(85)		158
2,4,6-Cl$_3$	I	CuI	10 mol %	DMEDA (20 mol %), Cs$_2$CO$_3$	toluene	110	22	(93)		159
2-Br	I	CuI	1 eq	Cs$_2$CO$_3$	DMSO	90	—	(68)		158
2-AcNH, 3-O$_2$N	I	CuI	1 eq	Cs$_2$CO$_3$	DMSO	90	—	(48)		158
2-AcNH, 5-O$_2$N	Br	CuI	1 eq	Cs$_2$CO$_3$	DMSO	90	—	(65)		158
2-O$_2$N	Br	CuI	1 eq	CsOAc	DMSO	90	—	(78)		158
2-O$_2$N	I	CuI	1 eq	CsOAc	DMSO	90	—	(83)		158
2-O$_2$N	I	CuI	10 mol %	DMEDA (20 mol %), Cs$_2$CO$_3$	toluene	110	22	(93)		159
2-O$_2$N, 4-F	I	CuI	1 eq	CsOAc	DMSO	90	—	(95)		158
2-O$_2$N, 4-MeO	I	CuI	1 eq	CsOAc	DMSO	90	—	(80)		158
2-O$_2$N, 4-CF$_3$	Br	CuI	1 eq	CsOAc	DMSO	90	—	(83)		158
3-O$_2$N	I	CuI	10 mol %	DMEDA (20 mol %), Cs$_2$CO$_3$	toluene	110	22	(98)		159
4-O$_2$N	I	CuI	1 eq	Cs$_2$CO$_3$	DMSO	90	—	(82)		158
4-O$_2$N	I	CuI	10 mol %	DMEDA (20 mol %), Cs$_2$CO$_3$	toluene	110	22	(92)		159
2-MeO	I	CuI	1 eq	Cs$_2$CO$_3$	DMSO	90	—	(72)		158
2-MeO	Br	CuI	10 mol %	DMEDA (20 mol %), NaI, Cs$_2$CO$_3$	dioxane	110	22	(97)		159
2-MeO	I	CuI	10 mol %	DMEDA (20 mol %), Cs$_2$CO$_3$	toluene	110	22	(99)		159
4-MeO	Br	CuI	10 mol %	DMEDA (20 mol %), NaI, Cs$_2$CO$_3$	dioxane	110	22	(84)		159

C$_7$

2-Me	I	CuI	1 eq	Cs$_2$CO$_3$	DMSO	90	—	(84)	158
2-Me	Br	CuI	10 mol %	DMEDA (20 mol %), NaI, Cs$_2$CO$_3$	dioxane	110	22	(93)	159
4-Me	I	CuI	1 eq	Cs$_2$CO$_3$	DMSO	90	—	(93)	158
4-Me	I	CuI	10 mol %	DMEDA (20 mol %), Cs$_2$CO$_3$	toluene	110	22	(95)	159
3-NC–	I	CuI	1 eq	CsOAc	DMSO	90	—	(94)	158
3-NC–	I	CuI	10 mol %	DMEDA (20 mol %), Cs$_2$CO$_3$	toluene	110	22	(99)	159
4-CHO	I	CuI	1 eq	Cs$_2$CO$_3$	DMSO	90	—	(85)	158
4-EtO$_2$C	I	CuI	1 eq	Cs$_2$CO$_3$	DMSO	90	—	(88)	158
4-EtO$_2$C	I	CuI	10 mol %	DMEDA (20 mol %), Cs$_2$CO$_3$	toluene	110	22	(92)	159
3,5-Me$_2$	Br	CuI	10 mol %	DMEDA (20 mol %), NaI, Cs$_2$CO$_3$	dioxane	110	22	(82)	159

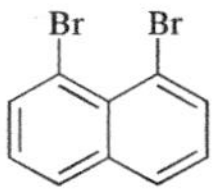

CuI (2 eq), CsOAc, DMSO, 90°   (56)   158

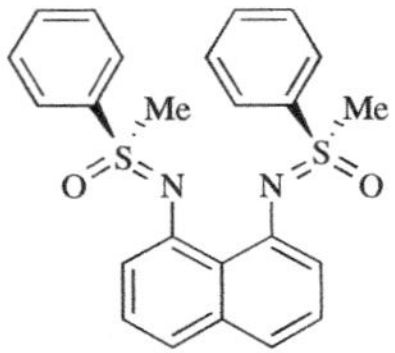

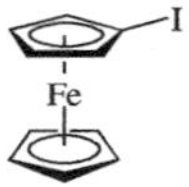

CuI (1 eq), K$_2$CO$_3$, DMSO, 90°, 18 h   (80)   372

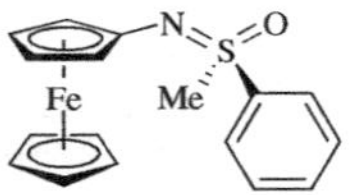

TABLE 24A. *N*-ARYLATION OF SULFOXIMINES (*Continued*)

Nitrogen Nucleophile	Aryl Halide	Conditions	Product(s) and Yield(s) (%)	Refs.

C₇

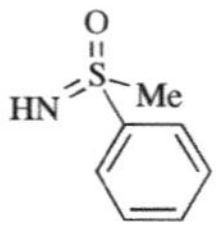

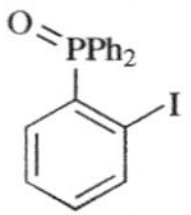

				Solvent	Temp (°)	Time (h)		
	x	Additive(s)						
	1 eq	CsOAc		DMSO	90	—	(71)	158
	10 mol %	DMEDA (20 mol %), Cs₂CO₃		toluene	110	22	(86)	159

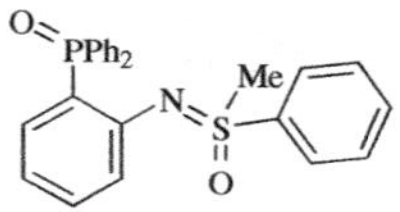

490

TABLE 24B. N-HETEROARYLATION OF SULFOXIMINES

Nitrogen Nucleophile	Heteroaryl Halide	Conditions	Product(s) and Yield(s) (%)	Refs.
C₇				
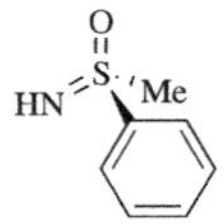	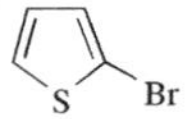	CuI (10 mol %), DMEDA (20 mol %), NaI, Cs₂CO₃, dioxane, 110°, 22 h	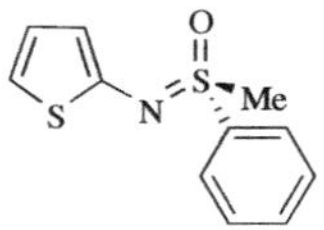 (89)	159
	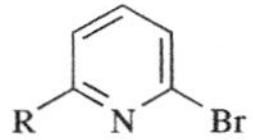	CuI (10 mol %), DMEDA (20 mol %), NaI, Cs₂CO₃, dioxane, 110°, 22 h	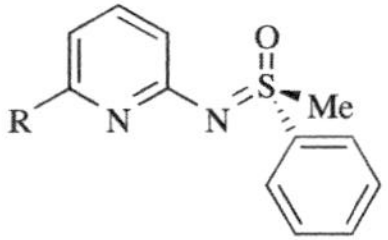 R ― H (96) Me (97)	159

TABLE 25. PREPARATION OF ARYL AND HETEROARYL AZIDES

C5-6

(Hetero)aryl Halide	Conditions	Product(s) and Yield(s) (%)		Refs.
pyridine-Br (R at 4,6)	NaN₃, CuBr (10 mol %), DMEDA (20 mol %), Cs₂CO₃, DMSO, 90°	R	Time (h)	247
		5-K⁺ F₃B⁻	1 (98)	
		3-Me, 5-K⁺ F₃B⁻	1 (95)	
		6-K⁺ F₃B⁻	3 (98)	

C6-8

NaN₃, catalyst (x mol %)   →   R-phenyl-N₃

R	X	Catalyst	x	Additives	Solvent(s)	Temp (°)	Time		Refs.
H	Br	CuI	10	DMCDA (15 mol %), Na ascorbate	EtOH/H₂O (7:3)	reflux	20 min	(96)	154
H	Br	CuI	10	L-Pro (20 mol %), NaOH	DMSO	60	10 h	(93)	155
H	I	CuI	10	L-Pro (20 mol %), NaOH	DMSO	60	5 h	(92)	155
H	I	CuSO₄•H₂O	20	L-Pro (20 mol %), Na ascorbate, Na₂CO₃	DMSO/H₂O (9:1)	70	24 h	(92)	75
2-K⁺ F₃B⁻	Br	CuBr	10	DMEDA (20 mol %), Cs₂CO₃	DMSO	90	24 h	(73)	247
3-K⁺ F₃B⁻	Br	CuBr	10	DMEDA (20 mol %), Cs₂CO₃	DMSO	90	9 h	(90)	247
3-K⁺ F₃B⁻	I	CuBr	10	DMEDA (20 mol %), Cs₂CO₃	DMSO	90	5 h	(93)	247
4-K⁺ F₃B⁻	Br	CuBr	10	DMEDA (20 mol %), Cs₂CO₃	DMSO	90	7 h	(95)	247
4-K⁺ F₃B⁻	I	CuBr	10	DMEDA (20 mol %), Cs₂CO₃	DMSO	90	1 h	(95)	247
2-F	I	CuI	10	L-Pro (20 mol %), NaOH	DMSO	60	24 h	(82)	155
3-F	I	CuI	10	L-Pro (20 mol %), NaOH	DMSO	60	10 h	(85)	155
4-Cl	Br	CuI	10	L-Pro (20 mol %), NaOH	DMSO	95	24 h	(91)	155
4-Cl	Br	CuI	10	L-Pro (20 mol %), NaOH	EtOH/H₂O	95	24 h	(90)	336
4-Cl	Br	CuI	10	DMCDA (15 mol %), Na ascorbate	EtOH/H₂O (7:3)	reflux	20 min	(84)	154
3-Br	I	CuI	10	DMCDA (15 mol %), Na ascorbate	EtOH/H₂O (7:3)	reflux	30 min	(77)	154
4-Br	I	CuI	10	L-Pro (20 mol %), NaOH	DMSO	60	10 h	(82)	155
3-H₂N	I	CuI	10	L-Pro (20 mol %), NaOH	DMSO	95	24 h	(77)	155
4-H₂N	I	CuI	10	L-Pro (20 mol %), NaOH	DMSO	60	10 h	(77)	155
3-HO	Br	CuI	10	DMCDA (15 mol %), Na ascorbate	EtOH/H₂O (7:3)	reflux	30 min	(99)	154
4-HO	Br	CuI	10	DMCDA (15 mol %), Na ascorbate	EtOH/H₂O (7:3)	reflux	40 min	(99)	154
4-HO	I	CuI	10	L-Pro (20 mol %), NaOH	DMSO	60	10 h	(87)	155
2-MeO	I	CuI	10	L-Pro (20 mol %), NaOH	DMSO	60	24 h	(67)	155
2,4-(MeO)₂	Br	CuI	10	L-Pro (20 mol %), NaOH	DMSO	95	30 h	(70)	155
3-MeO	Br	CuI	10	L-Pro (20 mol %), NaOH	DMSO	95	24 h	(92)	155

Substituent	Halide	Cu	10	Ligand, Base	Solvent	Temp	Time	Yield	Ref
3-MeO	Br	CuI	10	DMCDA (15 mol %), Na ascorbate	EtOH/H$_2$O (7:3)	reflux	40 min	(95)	154
3-K$^+$ F$_3$B$^-$, 5-MeO	Br	CuBr	10	DMEDA (20 mol %), Cs$_2$CO$_3$	DMSO	90	4 h	(92)	247
4-MeO	I	CuI	10	L-Pro (20 mol %), NaOH	DMSO	60	5 h	(92)	155
4-MeO	Br	CuI	10	L-Pro (20 mol %), NaOH	DMSO	95	10 h	(93)	155
4-MeO	Br	CuI	10	DMCDA (15 mol %), Na ascorbate	EtOH/H$_2$O (7:3)	reflux	20 min	(90)	154
2-Me	Br	CuI	10	L-Pro (20 mol %), NaOH	DMSO	95	24 h	(90)	155
3-Me, 4-H$_2$N	I	CuI	10	DMCDA (15 mol %), Na ascorbate	EtOH/H$_2$O (7:3)	reflux	50 min	(99)	154
3-K$^+$ F$_3$B$^-$, 5-Me	Br	CuBr	10	DMEDA (20 mol %), Cs$_2$CO$_3$	DMSO	90	8 h	(93)	247
4-Me	Br	CuI	10	L-Pro (20 mol %), NaOH	DMSO	95	10 h	(91)	155
4-Me	I	CuI	10	L-Pro (20 mol %), NaOH	DMSO	60	5 h	(92)	155
4-Me	Br	CuI	10	DMCDA (15 mol %), Na ascorbate	EtOH/H$_2$O (7:3)	reflux	10 min	(89)	154
4-Me	I	CuI	10	DMCDA (15 mol %), Na ascorbate	EtOH/H$_2$O (7:3)	reflux	15 min	(91)	154
3-H$_2$N, 4-Me	Br	CuI	10	DMCDA (15 mol %), Na ascorbate	EtOH/H$_2$O (7:3)	reflux	30 min	(90)	154
4-HOCH$_2$	I	CuI	10	L-Pro (20 mol %), NaOH	DMSO	60	24 h	(81)	155
4-HOCH$_2$	I	CuI	10	DMEDA, Na ascorbate	EtOH/H$_2$O (7:3)	MW, 100	40 min	(81)	550
2-HO$_2$C	I	CuI	10	L-Pro (20 mol %), NaOH	DMSO	60	7 h	(58)	155
2-HO$_2$C	I	CuI	10	DMCDA (15 mol %), Na ascorbate	EtOH/H$_2$O (7:3)	reflux	1 h	(84)	154
2-MeO$_2$C	I	CuI	10	DMCDA (15 mol %), Na ascorbate	EtOH/H$_2$O (7:3)	reflux	15 min	(91)	154
3-HO$_2$C	Br	CuI	10	L-Pro (20 mol %), NaOH	DMSO	95	30 h	(66)	155
3-HO$_2$C	I	CuI	10	DMCDA (15 mol %), Na ascorbate	EtOH/H$_2$O (7:3)	reflux	1 h	(99)	154
3,5-Me$_2$	I	CuI	10	L-Pro (20 mol %), NaOH	DMSO	60	5 h	(92)	155
2,5-Me, 4-K$^+$ F$_3$B$^-$	Br	CuBr	10	DMEDA (20 mol %), Cs$_2$CO$_3$	DMSO	90	12 h	(90)	247
3-Me, 4-HO$_2$C	Br	CuI	10	DMCDA (15 mol %), Na ascorbate	EtOH/H$_2$O (7:3)	reflux	1 h	(82)	154
3-Ac	Br	CuI	10	DMCDA (15 mol %), Na ascorbate	EtOH/H$_2$O (7:3)	reflux	30 min	(88)	154

TABLE 25. PREPARATION OF ARYL AND HETEROARYL AZIDES (*Continued*)

(Hetero)aryl Halide	Conditions	Product(s) and Yield(s) (%)		Refs.

C9

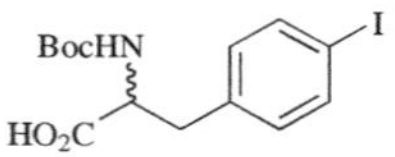

NaN₃, CuI (10 mol %), L-Pro (20 mol %), NaOH, DMSO, 60°, 5 h

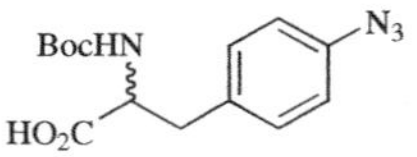

Config.	
(R)	(91)
(S)	(91)

155

C10

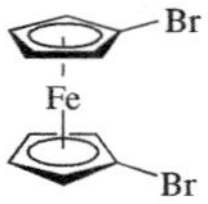

NaN₃, CuCl (1 eq), EtOH/H₂O (9:1), rt, 48 h

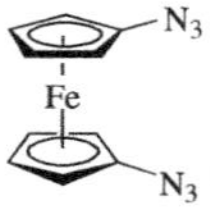

(59)

551

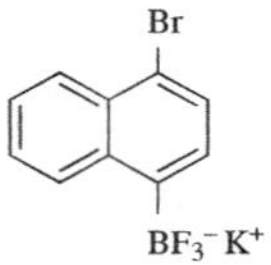

NaN₃, CuBr (10 mol %), DMEDA (20 mol %), Cs₂CO₃, DMSO, 90°, 4 h

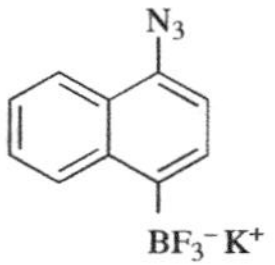

(92)

247

C12

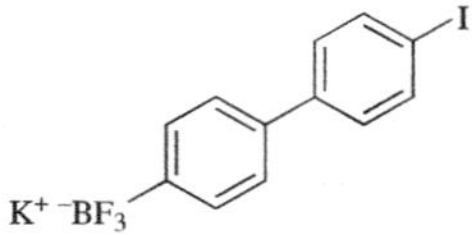

NaN₃, CuBr (10 mol %), DMEDA (20 mol %), Cs₂CO₃, DMSO, 90°, 5 min

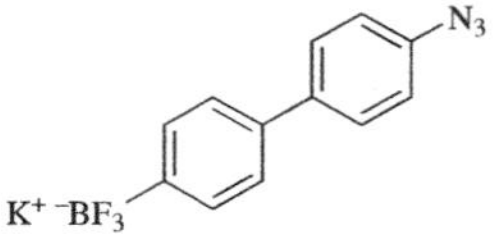

(96)

247

C25

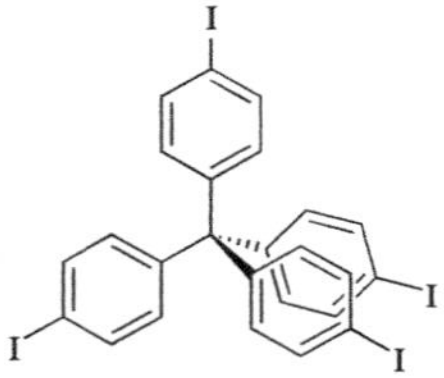

NaN₃, CuI (80 mol %), DMEDA (1.2 eq), Na ascorbate, DMSO/H₂O (5:1), 100°, 48 h

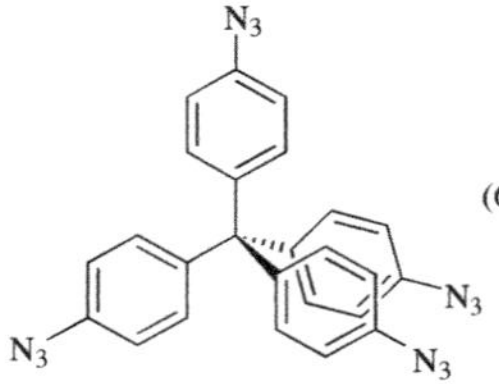

(65)

552

494

C$_{34}$

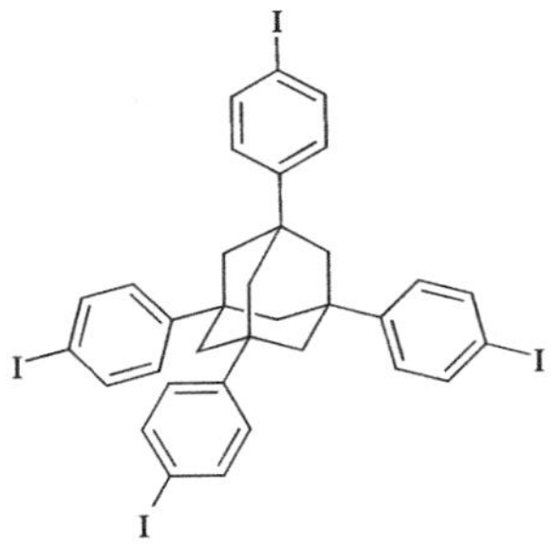

NaN$_3$, CuI (80 mol %), DMEDA (1.2 eq), Na ascorbate, DMSO/H$_2$O (5:1), 100°, 48 h

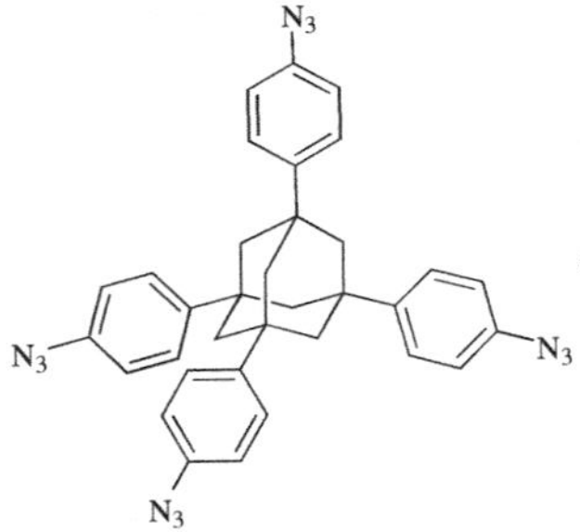

(34)

552

C$_{41}$

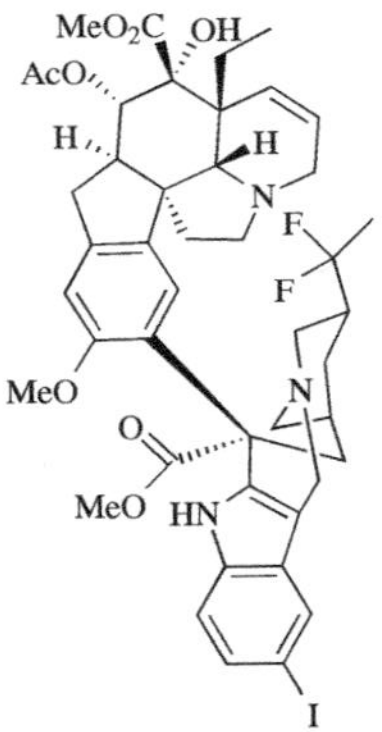

NaN$_3$, CuI, L-Pro, NaOH, DMSO

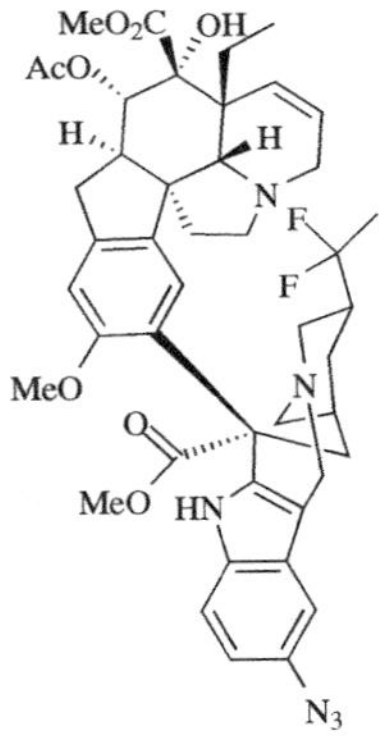

(68)

553

MeO$_2$C OH

AcO

H

MeO

MeO

HN

I

MeO$_2$C OH

AcO

H

MeO

MeO

HN

N$_3$

N$_3$

N$_3$

N$_3$

N$_3$

I

I

I

I

TABLE 26. INTRAMOLECULAR ARYLATIONS
A. SYNTHESIS OF FIVE-MEMBERED NITROGEN HETEROCYCLES

*Please refer to the charts preceding the tables for structures indicated by the **bold** numbers.*

Substrate	Conditions	Product(s) and Yield(s) (%)	Refs.

C$_{7-8}$

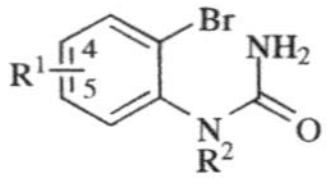

Conditions: CuI (8.5 mol %), TMEDA (3.5 eq), H$_2$O, 120°, 15 h. — Refs. 554

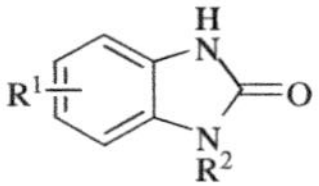

R^1	R^2	
H	Et	(71)
H	allyl	(71)
H	n-Bu	(60)
H	Ph	(69)
H	4-ClC$_6$H$_4$	(71)
H	Bn	(87)
H	4-FC$_6$H$_4$CH$_2$	(62)
H	3-MeOC$_6$H$_4$CH$_2$	(78)
H	4-MeOC$_6$H$_4$CH$_2$	(74)
H	2-NpCH$_2$	(72)
4-F	Ph	(60)
4-F	Bn	(74)
4,6-F$_2$	Bn	(51)
5-O$_2$N	Bn	(79)
4-MeO	Bn	(87)
4-Me	Bn	(76)
4-Me	4-FC$_6$H$_4$CH$_2$	(92)
4-CF$_3$	n-Bu	(80)
4-CF$_3$	Ph	(54)

C$_{8-15}$

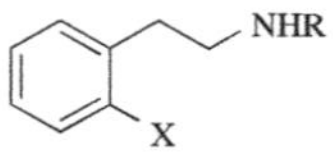

Catalyst (x amount)

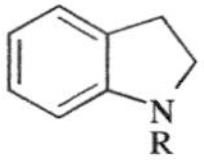

R	X	Catalyst	x	Additive(s)	Solvent(s)	Temp (°)	Time (h)		Refs.
H	Br	CuI	2 eq	CsOAc	DMSO	rt	1	(47)	197
H	I	CuI	2 eq	CsOAc	DMSO	rt	1	(44)	197
H	Br	CuI	5 mol %	DMEDA (10 mol %), Cs$_2$CO$_3$	DMSO	rt	1	(47)	274
H	Cl	CuI	10 mol %	L-proline (20 mol %), K$_2$CO$_3$	DMSO	70	45	(71)	47
H	I	CuI	5 mol %	DMEDA (10 mol %), Cs$_2$CO$_3$	DMSO	rt	1	(44)	274

496

R	X	Cu	amount	Ligand/Base	Solvent	Temp	Time	Yield	Ref
H	Br	CuI	2 mol %	L1 (2 mol %), KOt-Bu	—	100	12	(99)	288
H	Cl	CuI	5 mol %	L16 (20 mol %), Cs$_2$CO$_3$	DMF	rt	0.5	(90)	52
H	Br	CuI	5 mol %	L16 (20 mol %), Cs$_2$CO$_3$	DMF	rt	0.5	(90)	182
H	Br	CuOAc	5 mol %	L13 (20 mol %), K$_3$PO$_4$	DMF	35	12	(88)	120
H	Br	Cu(acac)$_2$	10 mol %	Fe$_2$O$_3$ (20 mol %), Cs$_2$CO$_3$	DMSO/H$_2$O	MW, 150	0.5	(52)	262
CHO	Cl	CuI	5 mol %	DMEDA (10 mol %), Cs$_2$CO$_3$	H$_2$O	100	—	(88)	69
CHO	Br	CuI	5 mol %	DMEDA (10 mol %), Cs$_2$CO$_3$	H$_2$O	rt	—	(100)	69
Ac	Br	CuI	2 eq	CsOAc	DMSO	90	24	(66)	197
Ac	I	CuI	2 eq	CsOAc	DMSO	90	24	(96)	197
Ac	Br	CuI	5 mol %	DMEDA (10 mol %), Cs$_2$CO$_3$	DMSO	90	24	(82)	274
Ac	I	CuI	5 mol %	DMEDA (10 mol %), Cs$_2$CO$_3$	DMSO	90	24	(96)	274
Boc	Br	CuI	5 mol %	DMEDA (10 mol %), Cs$_2$CO$_3$	DMSO	90	24	(82)	274
Boc	I	CuI	5 mol %	DMEDA (10 mol %), Cs$_2$CO$_3$	DMSO	90	24	(93)	274
CH$_2$=CHCH$_2$O$_2$C	Br	CuI	5 mol %	DMEDA (10 mol %), Cs$_2$CO$_3$	DMSO	90	24	(75)	274
CH$_2$=CHCH$_2$O$_2$C	I	CuI	5 mol %	DMEDA (10 mol %), Cs$_2$CO$_3$	DMSO	90	24	(81)	274
2-O$_2$NC$_6$H$_4$SO$_2$	Br	CuI	2 eq	CsOAc	DMSO	90	1	(92)	197
2-O$_2$NC$_6$H$_4$SO$_2$	I	CuI	2 eq	CsOAc	DMSO	90	24	(90)	197
2-O$_2$NC$_6$H$_4$SO$_2$	Br	CuI	5 mol %	DMEDA (10 mol %), Cs$_2$CO$_3$	DMSO	90	24	(92)	274
2-O$_2$NC$_6$H$_4$SO$_2$	I	CuI	5 mol %	DMEDA (10 mol %), Cs$_2$CO$_3$	DMSO	90	1	(87)	274
2-O$_2$NC$_6$H$_4$SO$_2$	Br	CuI	1 mol %	CsOAc	DMSO	90	24	(97)	274
Bn	Br	CuI	2 eq	CsOAc	DMSO	rt	4	(87)	197
Bn	Br	CuI	5 mol %	CsOAc	DMSO	90	24	(83)	274
Bn	Br	CuI	5 mol %	DMEDA (10 mol %), Cs$_2$CO$_3$	DMSO	rt	5	(87)	274
Cbz	Br	CuI	2 eq	CsOAc	DMSO	90	24	(64)	197
Cbz	I	CuI	2 eq	CsOAc	DMSO	90	24	(95)	197
Cbz	Br	CuI	5 mol %	DMEDA (10 mol %), Cs$_2$CO$_3$	DMSO	90	24	(77)	274
Cbz	I	CuI	5 mol %	DMEDA (10 mol %), Cs$_2$CO$_3$	DMSO	90	24	(95)	274

Substrate	Conditions	Product(s) and Yield(s) (%)	Refs.
Please refer to the charts preceding the tables for structures indicated by the bold numbers.			
C₈			
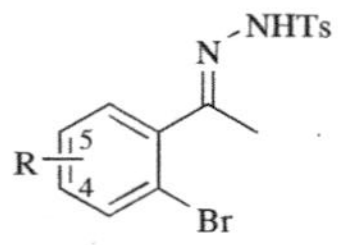 [structure: R (ring positions 5,4), Br, N–NHTs]	CuI (10 mol %), DMEDA (30 mol %), Na_2CO_3, $EtOH/H_2O$ (1:1), rt, 10 min	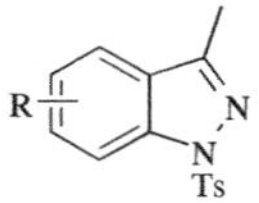 R H (98) 5-MeO (98) 4,5-$(MeO)_2$ (99) 4-$O(CH_2)_2O$-5 (98)	555
C₉			
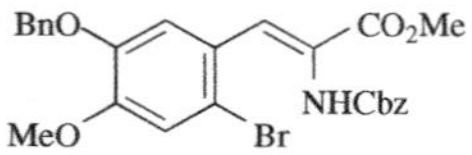 [structure: BnO, MeO ring, CO_2Me, NHCbz, Br]	CuI (5 mol %), Cs_2CO_3, DMSO, 90°, 24 h	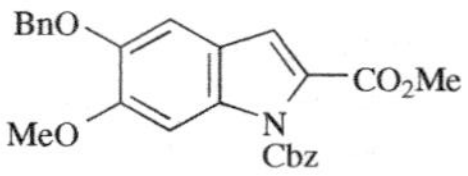 [indole: CO_2Me, N–Cbz, BnO, MeO] (80)	556
C₉₋₁₀			
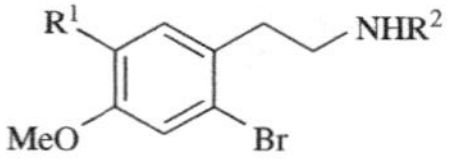 [structure: R^1, MeO ring, NHR^2, Br]	CuI (1 eq), NaH, DMF, 80°, 12 h	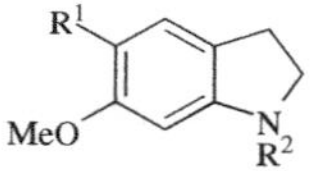 [indoline: R^1, MeO, N–R^2] R^1 R^2 H Ac (52) MeO CO_2Me (66) MeO Ac (74)	557
C₉			
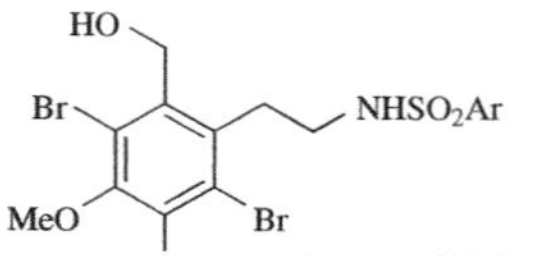 [structure: HO, Br, MeO, OMe ring, $NHSO_2Ar$, $Ar = 2\text{-}O_2NC_6H_4$]	CuI (10 mol %), CsOAc, DMSO, 80°, 24 h	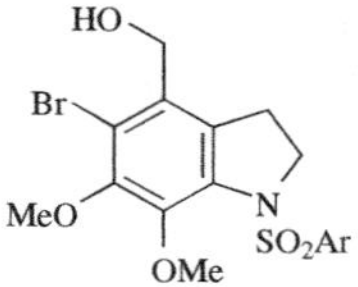 [indoline: HO, Br, MeO, OMe, N–SO_2Ar] (95)	556
C₉₋₂₀			
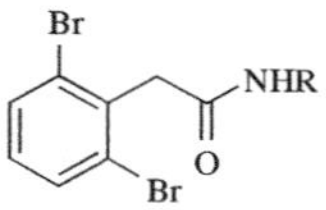 [structure: Br, NHR, O, Br]	CuCl (10 mol %), acac (25 mol %), K_2CO_3, NMP, 85°	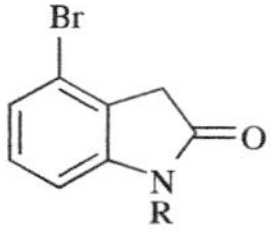 [oxindole: Br, N–R] R Time (h) Me 1 (84) t-Bu 22 (83) 4-pyridyl 1 (66) Ph 1 (83)	558

498

C₉

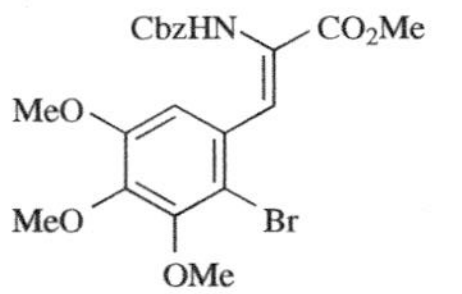

CbzHN, CO₂Me, Br, OMe, MeO, MeO

CuI (1.5 eq), CsOAc, DMSO, rt, 24 h

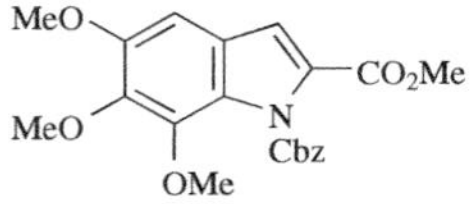

MeO, MeO, OMe, N–Cbz, CO₂Me (98)

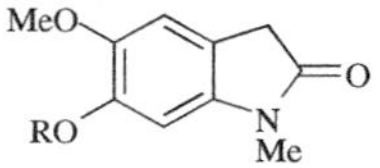

559

MeO, RO, Br, NHMe, O

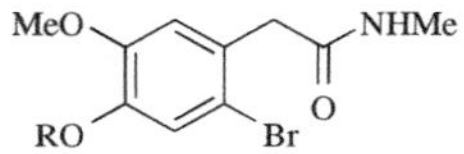

CuI (1 eq), NaH, DMF, 80°, 12 h

MeO, RO, N–Me, O

R	
Me	(63)
Bn	(73)

557

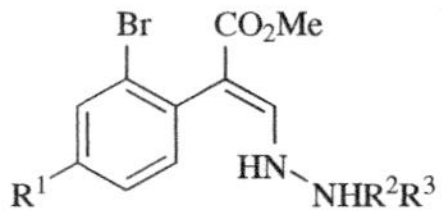

Br, CO₂Me, HN–NHR²R³, R¹

CuI (5 mol %), K₃PO₄

CO₂Me, NR²R³, R¹

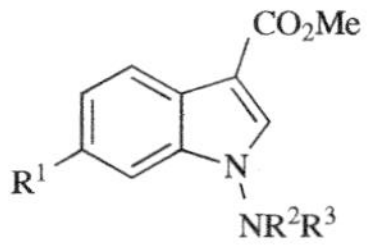

R¹	R²	R³	Solvent	Temp (°)	Time (h)		
H	Boc	H	DMF	85	2	(0)	560
H	Boc	Boc	DMF	85	2	(0)	560
H	Me	Me	DMF	85	2	(82)	560
H	c-C₃H₇	H	HOCH₂CH₂OH	75	4	(78)	561
H	t-Bu	H	DMF	75	20	(32)	561
H	c-C₆H₁₁	H	DMF	75	20	(57)	561
H	Ph	H	DMF	85	2	(0)	560
H	Ph	Boc	DMF	85	2	(78)	560
H	4-MeOC₆H₄	H	HOCH₂CH₂OH	75	4	(94)	561
H	2-MeC₆H₄	H	DMF	75	9	(65)	561
H	3-CF₃C₆H₄	H	HOCH₂CH₂OH	75	4	(74)	561
H	Bn	H	HOCH₂CH₂OH	75	4	(91)	561
H	3-MeC₆H₄CH₂	H	DMF	85	2	(68)	560
H	(R)-PhMeCH	H	DMF	75	9	(68)	561
H	PhCH₂CH₂	H	HOCH₂CH₂OH	75	6	(80)	561
H	2,4,6-Me₃C₆H₂	H	DMF	75	6	(77)	561
MeO	2-ClC₆H₄CH₂	Boc	DMF	85	2	(86)	560
MeO	3-MeC₆H₄CH₂	H	DMF	85	2	(65)	560

TABLE 26. INTRAMOLECULAR ARYLATIONS (*Continued*)

A. SYNTHESIS OF FIVE-MEMBERED NITROGEN HETEROCYCLES (*Continued*)

*Please refer to the charts preceding the tables for structures indicated by the **bold** numbers.*

Substrate	Conditions	Product(s) and Yield(s) (%)	Refs.

C$_{10-15}$

Substrate: R^1—NHR^2 (with X)
Conditions: CuI (20 mol %), DBU, DMSO, MW, 20 min
Product: R^1—benzoxazolone with R^2
Ref. 562

R^2	X	Additive	Temp (°)		R^1	R^2	X	Additive	Temp (°)	
allyl	I	none	140	(60)	H	3-NC–	I	none	120	(89)
2-thiazolyl	I	none	120	(72)	H	4-EtO$_2$C	I	none	120	(81)
i-Bu	Br	L-Pro (50 mol %)	140	(65)		4-ethenyl	Br	L-Pro (50 mol %)	120	(77)
i-Bu	I	none	140	(71)	H	4-ethenyl	I	none	120	(86)
4-pyridyl	I	none	120	(83)	H	3-ethynyl	Br	L-Pro (50 mol %)	120	(73)
4-BrC$_6$H$_4$	I	none	120	(92)	H	3-ethynyl	I	none	120	(85)
2-MeOC$_6$H$_4$	I	none	120	(93)	H	4-FC$_6$H$_4$CH$_2$	I	none	140	(85)
4-MeOC$_6$H$_4$	I	none	120	(95)	H	*n*-C$_8$H$_{17}$	I	none	140	(59)
4-MeC$_6$H$_4$	Br	L-Pro (50 mol %)	120	(79)	5-F	4-pyridyl	I	none	120	(81)
4-Me	I	none	120	(90)	4-Cl, 6-F	*i*-Bu	I	none	120	(66)
2-CF$_3$C$_6$H$_4$	I	none	120	(89)	4-Cl, 6-F	4-FC$_6$H$_4$	I	none	120	(77)
4-CF$_3$C$_6$H$_4$	Br	L-Pro (50 mol %)	120	(79)	5-MeO	Bn	I	none	120	(70)
4-CF$_3$C$_6$H$_4$	I	none	120	(93)	3-CF$_3$	4-FC$_6$H$_4$	I	none	120	(73)
					4-MeO$_2$C	4-FC$_6$H$_4$	I	none	120	(75)

C$_{10}$

Substrate: MOMO / MeO aryl with Cl and NH$_2$
Conditions: CuI (50 mol %), DMEDA (1 eq), K$_2$CO$_3$, DMF, reflux, 12 h
Product: MOMO / MeO oxindole (96)
Ref. 563

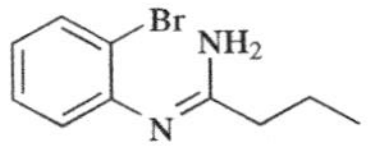

CuI (50 mol %), NaH, THF, reflux, 1 h (37) 139

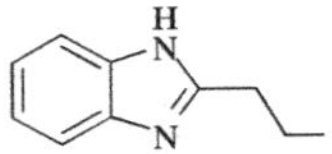

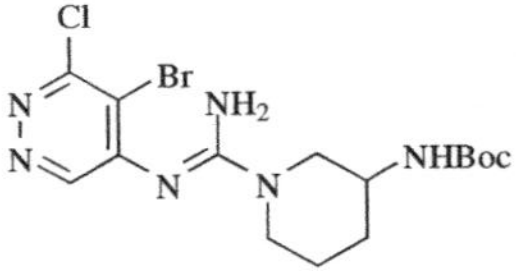

CuI (50 mol %), NaH, THF, reflux, 1 h (73) 139

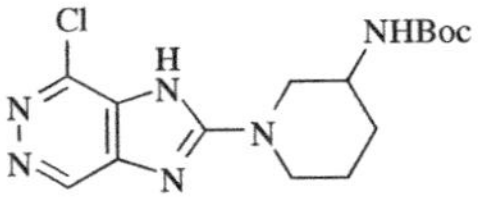

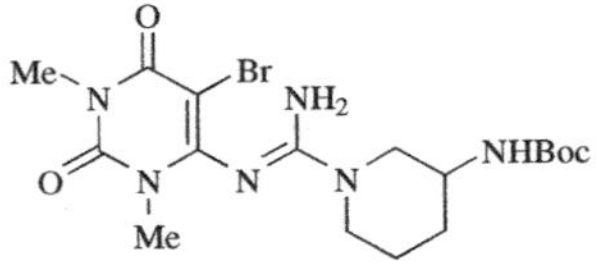

CuI (50 mol %), NaH, THF, reflux, 1 h (77) 139

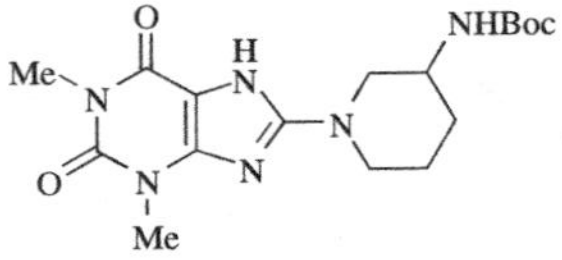

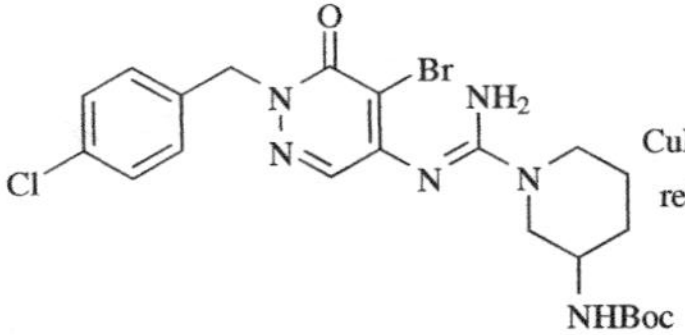

CuI (50 mol %), NaH, THF, reflux, 1 h (81) 139

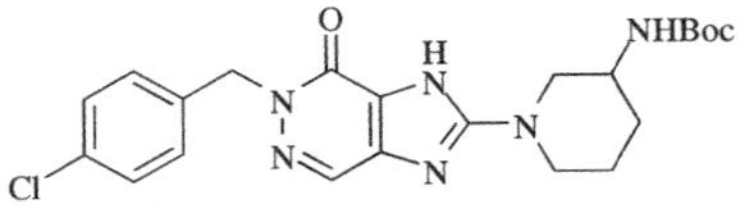

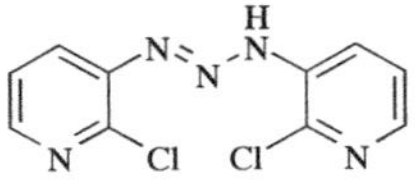

CuI (10 mol %), K$_2$CO$_3$, PEG, H$_2$O, 110°, 6.5 h (84) 564

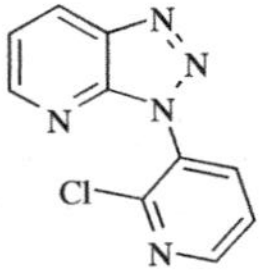

TABLE 26. INTRAMOLECULAR ARYLATIONS (*Continued*)

A. SYNTHESIS OF FIVE-MEMBERED NITROGEN HETEROCYCLES (*Continued*)

Substrate	Conditions	Product(s) and Yield(s) (%)	Refs.

*Please refer to the charts preceding the tables for structures indicated by the **bold** numbers.*

C_{10-12}

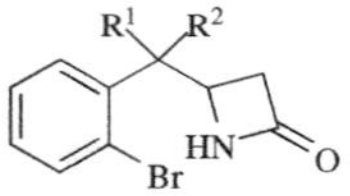

Catalyst (5 eq)

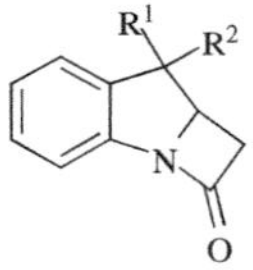

R^1	R^2	Catalyst	Solvent	Temp (°)	Time		
H	H	Cu	DMF	135	8 h	(8)	565
H	H	Cu (active)	—	MW (100 W)	6 min	(69)	280
EtO_2C	H	Cu (active)	—	MW (100 W)	8 min	(78)	280
BnO_2C	H	Cu	DMF	135	8 h	(19)	565
BnO_2C	BnO_2C	Cu	DMF	135	8 h	(23)	565
BnO_2C	BnO_2C	Cu (active)	—	MW (100 W)	6 min	(69)	280

C_{10-11}

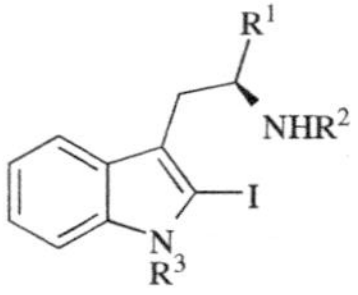

CuI (10 mol %),
DMEDA (20 mol %), K_3PO_4,
toluene, 110°, 16 h

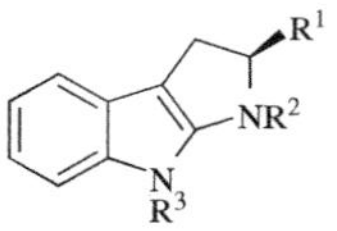

R^1	R^2	R^3	
H	Boc	Ac	(50)
$TBSOCH_2$	Boc	Ac	(51)
MeO_2C	Ac	Boc	(98)
MeO_2C	Boc	Ac	(87)
MeO_2C	Cbz	H	(85)
MeO_2C	Cbz	Ac	(93)
MeO_2C	Cbz	Boc	(73)
BnO_2C	Boc	Boc	(62)

566

C_{11}

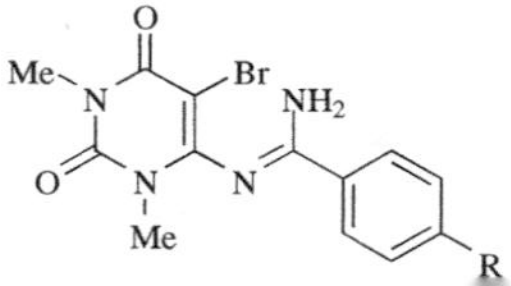

CuI (50 mol %), NaH, THF,
reflux, 1 h

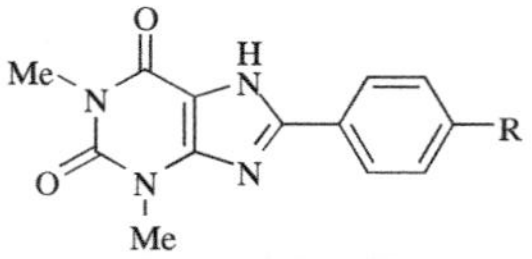

R	
H	(82)
Br	(53)

139

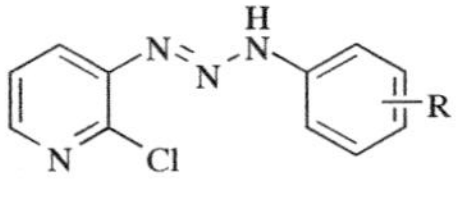

CuI (10 mol %), K_2CO_3, PEG, H_2O, 110°, 6 h

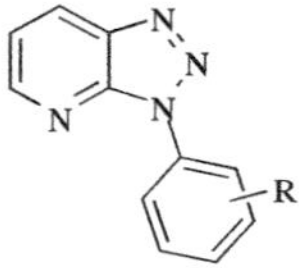

R	Time (h)	
4-Cl	6	(88)
2,4-Cl_2	6	(88)
4-O_2N	6.5	(75)
4-Me	6	(86)

564

C_{11-12}

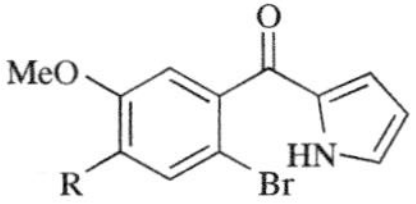

CuBr (1.3 eq), NaOH, H_2O, DMF, rt, 3 h

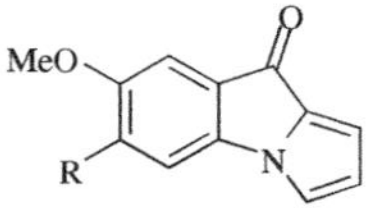

R	
MeO	(52)
Me	(53)

567

C_{11-20}

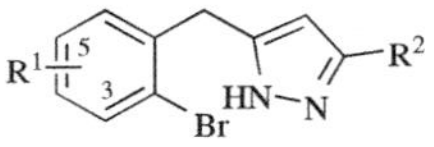

CuI (5 mol %), phen (10 mol %), K_2CO_3, dioxane, 110°, 20 h

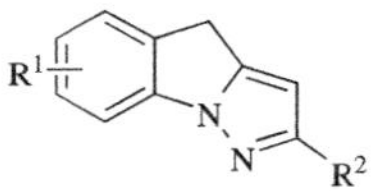

R^1	R^2	
H	Me	(89)
H	Ph	(98)
H	4-$O_2NC_6H_4$	(97)
H	4-$MeOC_6H_4$	(86)
5-F	Ph	(95)
5-Cl	Ph	(83)
5-MeO	Ph	(96)
3-$(CH{=}CH)_2$–4	Ph	(45)

568

TABLE 26. INTRAMOLECULAR ARYLATIONS (Continued)
A. SYNTHESIS OF FIVE-MEMBERED NITROGEN HETEROCYCLES (Continued)

Substrate	Conditions	Product(s) and Yield(s) (%)	Refs.

*Please refer to the charts preceding the tables for structures indicated by the **bold** numbers.*

C11

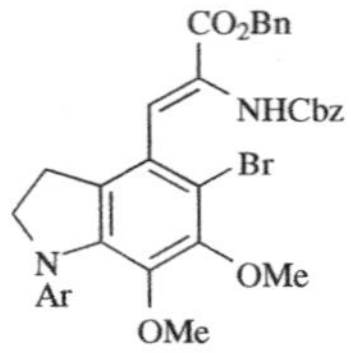

CuI (1 eq), CsOAc, DMSO, rt, 12 h

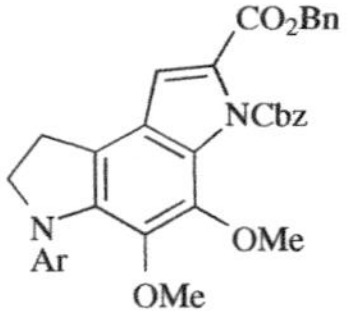

(77)

569, 556

Ar = 2-O$_2$NC$_6$H$_4$

C12

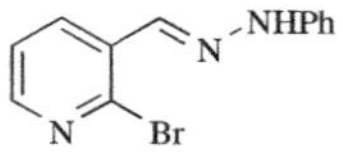

CuI (5 mol %), phen (10 mol %), KOH, dioxane, 100°, 5 h

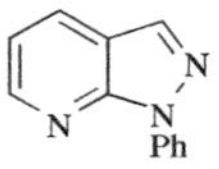

(71)

570

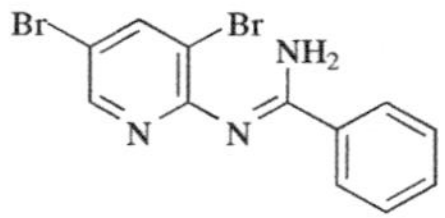

CuI (50 mol %), NaH, THF, reflux, 1 h

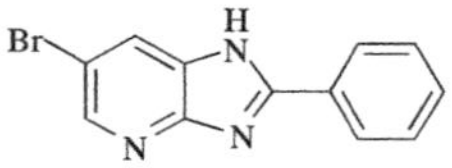

(73)

139

R^1	R^2	Temp (°)	Time
H	2-pyridyl	120	1 h (87)
H	2-BrC$_6$H$_4$	110	1.3 h (99)
H	2,4,6-Me$_3$C$_6$H$_2$	120	4.5 h (85)
H	2,6-(i-Pr)$_2$C$_6$H$_3$	120	5 h (96)
H	1-adamantyl	140	6 h (95)
H	9-anthracenyl	120	2 h (92)
4-Cl	2,4-Cl$_2$C$_6$H$_4$	120	190 h (53)
4,6-Cl$_2$	2,4,6-Cl$_3$C$_6$H$_2$	120	171 h (49)

C12-21

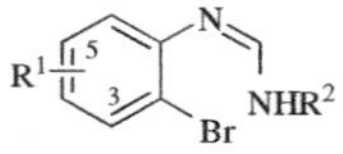

CuI (20 mol %), DBU, DMSO, 120°

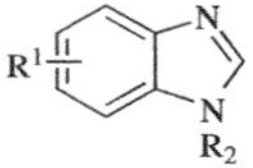

571

R^1	R^2	Temp (°)	Time
4,6-Br$_2$	2,4,6-Br$_3$C$_6$H$_2$	120	70 min (93)
4,6-Br$_2$	2,4,6-Me$_3$C$_6$H$_2$	120	75 min (98)
5-CF$_3$	2-Br-5-CF$_3$C$_6$H$_3$	120	75 min (92)
3,5-Me$_2$	2-Br-3,5-Me$_2$C$_6$H$_2$	120	23 h (93)
4,6-Me$_2$	2-t-BuC$_6$H$_4$	120	96.5 h (85)
4,6-Me$_2$	2-Br-4,6-Me$_2$C$_6$H$_2$	120	70 min (97)
4,6-Me$_2$	2,4,6-Me$_3$C$_6$H$_2$	120	90 min (99)
3,5-Me$_2$, 4-Br	2,4-Br$_2$-3,5-Me-C$_6$H	120	24 h (99)

C$_{12-17}$

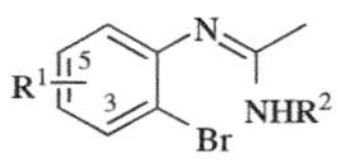

CuO nanoparticles (5 mol %),
KOH, DMSO, 110°

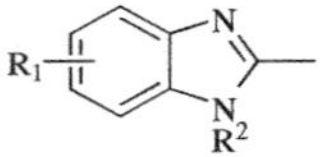

R^1	R^2	Time (h)	
H	n-Bu	5	(83)
H	Ph	4	(94)
H	Bn	6	(95)
H	2,4-(MeO)$_2$C$_6$H$_4$CH$_2$	6	(82)
4-Cl	Bn	4	(92)
4-Cl, 6-Br	Ph	4	(94)
4-MeO	Bn	3	(93)

R^1	R^2	Time (h)	
4-Me	c-C$_6$H$_{11}$	16	(91)
4-Me	Ph	5	(95)
4-Me	Bn	8	(93)
4,5-Me$_2$	Ph	24	(93)
4,5-Me$_2$	Bn	13	(92)
4,6-Me$_2$	Ph	4	(90)
4,6-Me$_2$	Bn	15	(88)

C$_{12-19}$

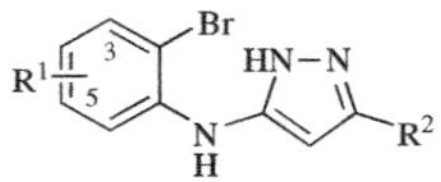

CuI (10 mol %),
L-Pro (20 mol %), NaH,
DMF, 90°

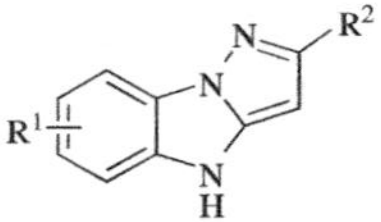

R^1	R^2	Time (h)	
H	2-furyl	1.5	(94)
H	2-thienyl	2.5	(80)
H	2-pyridyl	0.5	(78)
H	4-ClC$_6$H$_4$	1.5	(73)
H	4-MeOC$_6$H$_4$	1	(84)
4-F	4-FC$_6$H$_4$	1	(53)

R^1	R^2	Time (h)	
4-MeO	Ph	1.5	(0)
4,5-(MeO)$_2$	i-Pr	0.75	(81)
4-Me	Ph	1.25	(83)
4-i-Pr	4-pyridyl	2.5	(80)
4-i-Pr	4-CF$_3$C$_6$H$_4$	1	(84)

C$_{12}$

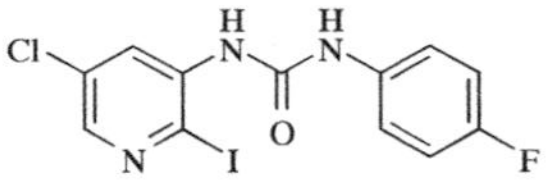

CuI (20 mol %), DBU, DMSO,
MW, 120°, 40 min

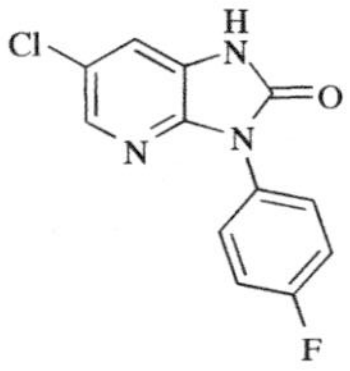

(54)

TABLE 26. INTRAMOLECULAR ARYLATIONS (*Continued*)
A. SYNTHESIS OF FIVE-MEMBERED NITROGEN HETEROCYCLES (*Continued*)

*Please refer to the charts preceding the tables for structures indicated by the **bold** numbers.*

Substrate	Conditions	Product(s) and Yield(s) (%)	Refs.
C$_{12-13}$	CuI (10 mol %), K$_2$CO$_3$, PEG, H$_2$O, 110°, 2.5 h	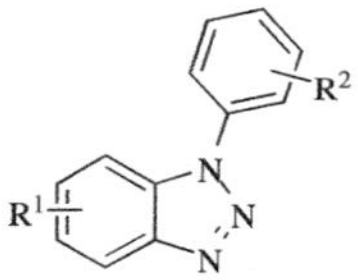	564
C$_{12}$	CuI (10 mol %), CDA (20 mol %), K$_3$PO$_4$, toluene, 110°, 16 h	(55)	566
	1. CuI (5 mol %), DMG (20 mol %), K$_2$CO$_3$, MeCN, reflux, 12 h 2. HCl	(82)	568

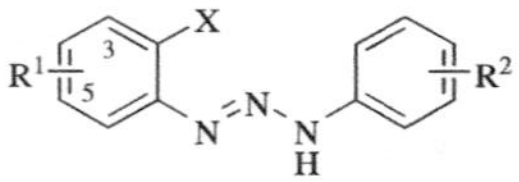

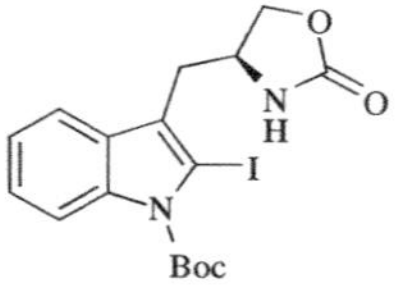

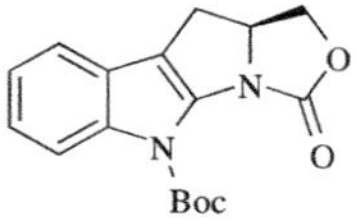

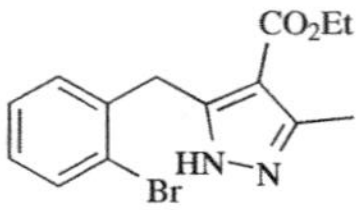

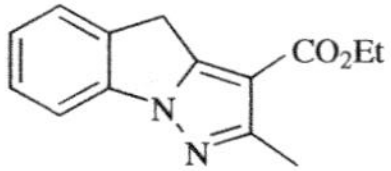

R^1	R^2	X	Time (h)
H	H	I	2.5 (92)
H	2,5-Cl$_2$	Br	3 (90)
H	2-Br	I	2.5 (90)
H	2-I	I	2.5 (92)
H	3-O$_2$N	I	3 (89)
H	4-Me	I	2.5 (92)
5-Cl	2,4-Cl$_2$	Br	3 (91)

R^1	R^2	X	Time (h)
4-Cl	4-Me	I	2.5 (92)
5-Cl	2-O$_2$N	Cl	6 (82)
5-Cl	3-O$_2$N	Cl	6 (84)
5-Cl	4-O$_2$N	Cl	6.5 (84)
4-Me	4-Cl	Br	2.5 (92)
4-Me	2,4-Cl$_2$	Br	3 (91)

559

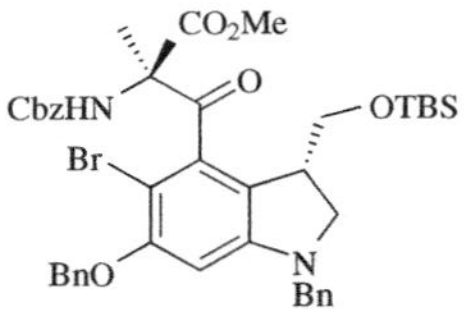

570

572

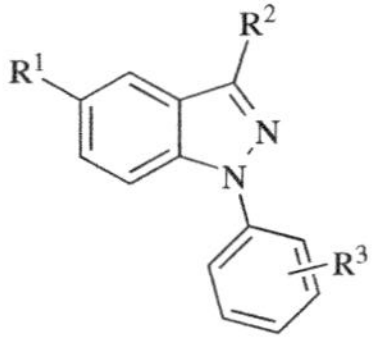

R^1	X	R^2	R^3	
H	Cl	H	H	(10)
H	Br	H	H	(82)
H	Br	H	4-F	(70)
H	Br	H	4-O$_2$N	(85)
H	Br	H	4-MeO	(80)
H	Br	H	4-Me	(90)
H	Br	Me	H	(71)
F	Br	H	H	(77)
MeO	Br	H	H	(63)

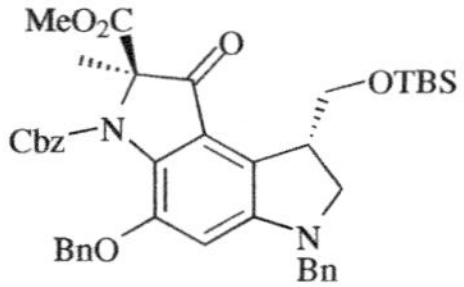

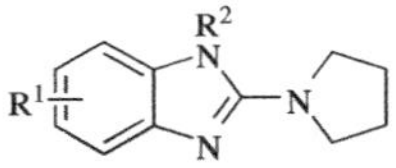

R^1	R^2	Time (h)	
H	Ph	5	(95)
H	2-MeOC$_6$H$_4$	5	(88)
H	4-MeOC$_6$H$_4$	5	(78)
H	4-MeC$_6$H$_4$	4	(93)
H	Bn	4	(93)
H	2,4-Me$_2$C$_6$H$_4$	5	(83)
H	2,6-Me$_2$C$_6$H$_4$	5	(80)
H	3,5-Me$_2$C$_6$H$_4$	4	(93)
4-Me	Bn	14	(89)
4,6-Me$_2$	Bn	12	(90)

(100)

CuI (2 eq), CsOAc,
DMSO, rt, 8 h

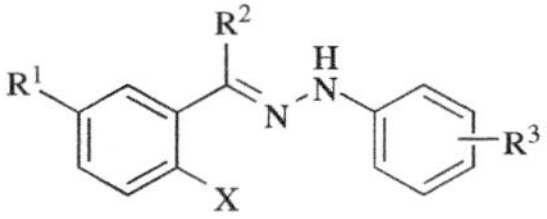

CuI (5 mol %), phen (10 mol %),
KOH, dioxane, 100°, 5 h

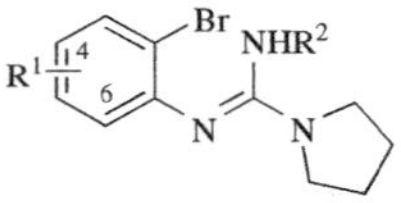

CuO nanoparticles (5 mol %),
KOH, DMSO, 110°

C$_{13}$

C$_{13-14}$

C$_{13-18}$

TABLE 26. INTRAMOLECULAR ARYLATIONS (*Continued*)

A. SYNTHESIS OF FIVE-MEMBERED NITROGEN HETEROCYCLES (*Continued*)

Substrate	Conditions	Product(s) and Yield(s) (%)	Refs.

*Please refer to the charts preceding the tables for structures indicated by the **bold** numbers.*

C$_{13-14}$

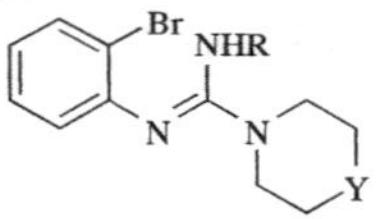

CuO nanoparticles (5 mol %), KOH, DMSO, 110°

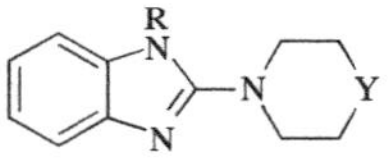

R	Y	Time (h)	
Ph	BocN	4	(90)
Bn	O	18	(88)

572

C$_{13-18}$

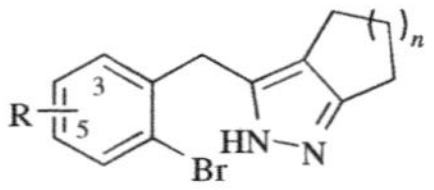

CuI (5 mol %), phen (10 mol %), K$_2$CO$_3$, dioxane, 110°, 20 h

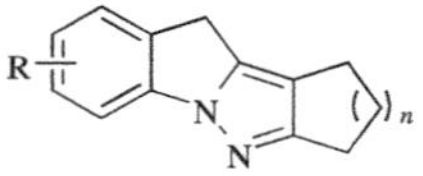

R	n	
H	1	(89)
H	2	(91)
H	3	(78)
5-F	2	(77)
5-Cl	2	(81)
3-Me	2	(88)
3-(CH=CH)$_2$-4	2	(44)

568

C$_{13-17}$

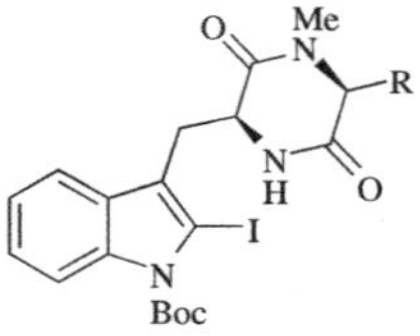

CuI (10 mol %), CDA (20 mol %), K$_3$PO$_4$, toluene, 110°, 16 h

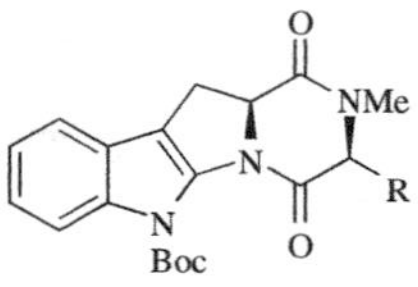

R	
H	(40)
i-Bu	(40)
(*sec*-butyl)	(65)

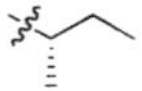

566

C$_{14}$

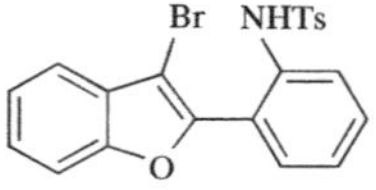

CuI (8.5 mol %), DMEDA (3.5 eq), H$_2$O, 120°

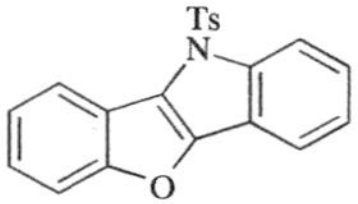

(85)

575

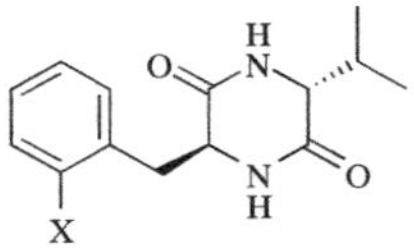

C$_{14-16}$

CuI (1 eq), CsOAc, DMSO, 90°, 12 h

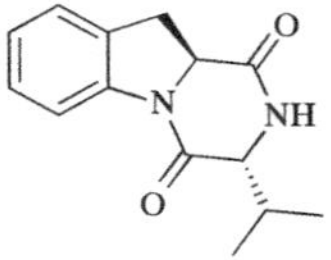

Config.	X		Time (h)
(3S,6R)	Br	(84)	
(3S,6R)	I	(99)	
(3R,6R)	Br	(13)	

C$_{14-16}$

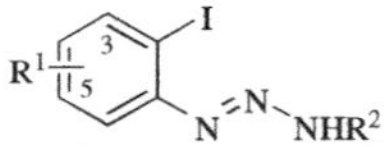

CuI (10 mol %), K$_2$CO$_3$, PEG, H$_2$O, 110°, 3 h

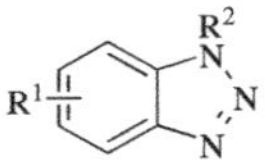

R^1	R^2	Time (h)	
H	3-quinolyl	3	(82)
H	6-quinolyl	3	(75)
5-Cl	3-quinolyl	2.5	(88)
4-Me	Bn	2.5	(92)
4-Me	3-quinolyl	3	(90)

C$_{14-15}$

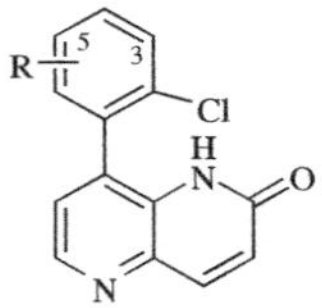

CuI (5 mol %), DMEDA (10 mol %), Cs$_2$CO$_3$, H$_2$O, dioxane, reflux, 1 h

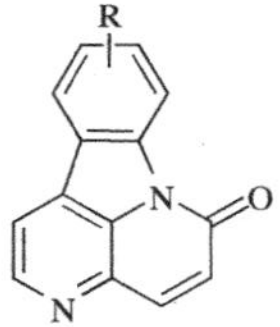

R	
H	(99)
4-Cl	(96)
5-Cl	(97)
6-Cl	(94)
3-MeO	(91)
5-MeO	(98)
4-CF$_3$	(99)

C$_{15-16}$

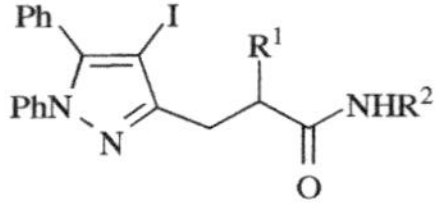

CuI (20 mol %), EDA (40 mol %), Cs$_2$CO$_3$, DMF, 100°

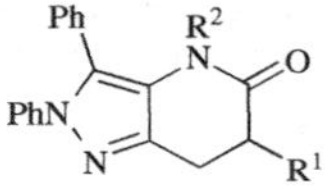

R^1	R^2	Time (h)	
H	i-Pr	24	(69)
H	c-Pr	24	(69)
Me	i-Pr	48	(49)
Me	c-Pr	48	(78)

TABLE 26. INTRAMOLECULAR ARYLATIONS (*Continued*)

A. SYNTHESIS OF FIVE-MEMBERED NITROGEN HETEROCYCLES (*Continued*)

Substrate	Conditions	Product(s) and Yield(s) (%)	Refs.

*Please refer to the charts preceding the tables for structures indicated by the **bold** numbers.*

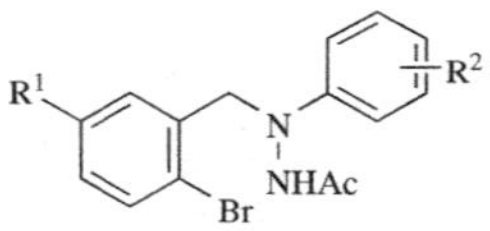

C15–16

CuI (5 mol %),
Cs$_2$CO$_3$, 105°, 20 h

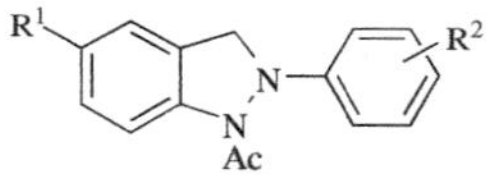

R^1	R^2	
H	H	(95)
H	4-F	(98)
H	4-Cl	(95)
H	4-MeO	(85)
H	4-Me	(92)
H	3-CF$_3$	(81)
H	3-NC–	(96)
F	H	(89)
MeO	H	(85)

578

C15

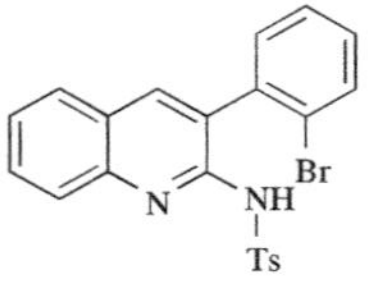

CuI (5 eq), NaH, diglyme,
rt, 30 min

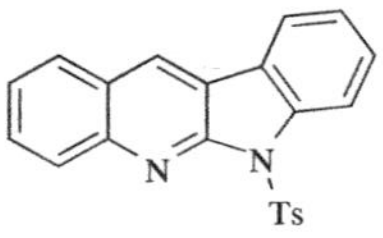

(86)

579

C16–17

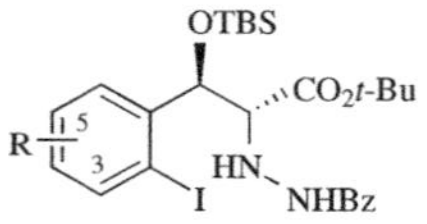

CuI (10 mol %),
DMEDA (10 mol %), Cs$_2$CO$_3$,
DMSO, rt, 10 min

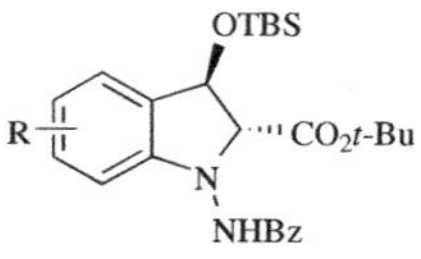

R	
6-F	(71)
4-Cl	(71)
4,5-(MeO)$_2$	(70)
3-Me	(29)

580

C16

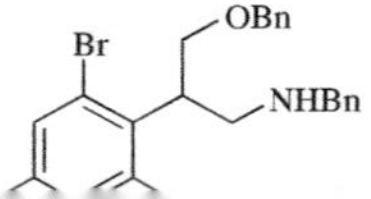

CuI (2 eq), CsOAc,
DMF, rt, 20 min

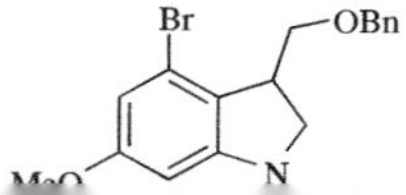

(60)

274

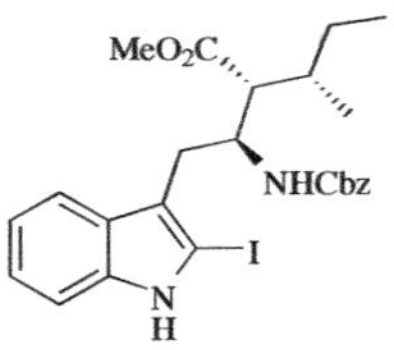

MeO₂C... NHCbz (iodoindole)	CuI (10 mol %), DMEDA (20 mol %), K₃PO₄, toluene, 110°, 16 h	product (CO₂Me, N–Cbz indole) (67)	566

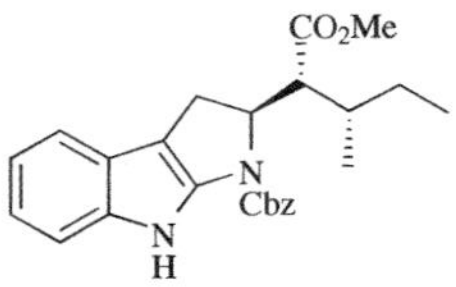

C₁₇₋₂₂

proline diketopiperazine (N-Boc iodoindole)	CuI (10 mol %), CDA (20 mol %), K₃PO₄, toluene, 110°, 16 h	product (81)	566

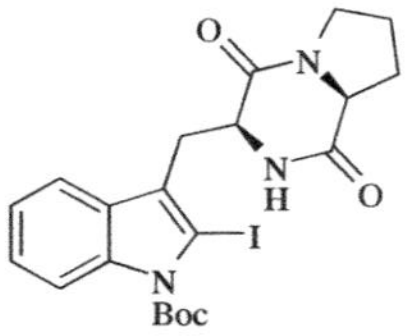

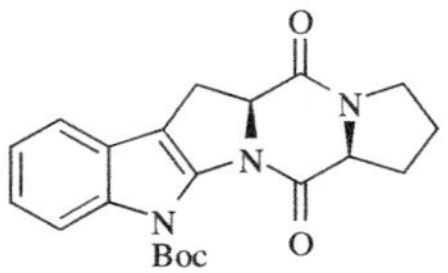

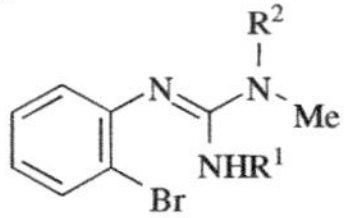

CuI (5 mol %), phen (10 mol %), Cs₂CO₃, DME, 80°, 16 h

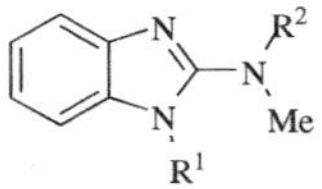

R^1	R^2	
Ph	Bn	(83)
Bn	NC(CH₂)₂	(87)
Bn	(S)-PhMeCH	(97)

581

C₁₇₋₂₀

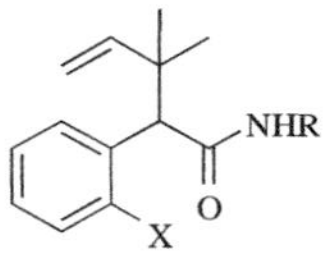

Cu(phen)(PPh₃)Br (10 mol %), K₃PO₄, toluene, 115°, 72 h

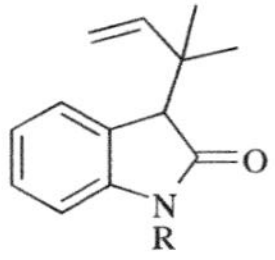

R	X	
n-Bu	Br	(53)
n-Bu	I	(75)
i-Pr	Br	(18)
i-Pr	I	(79)
n-C₅H₁₁	Br	(43)
n-C₅H₁₁	I	(85)
Ph	I	(78)
Bn	Br	(78)
Bn	I	(87)

582

511

TABLE 26. INTRAMOLECULAR ARYLATIONS (*Continued*)

A. SYNTHESIS OF FIVE-MEMBERED NITROGEN HETEROCYCLES (*Continued*)

Substrate	Conditions	Product(s) and Yield(s) (%)	Refs.
*Please refer to the charts preceding the tables for structures indicated by the **bold** numbers.*			
C_{18-19} 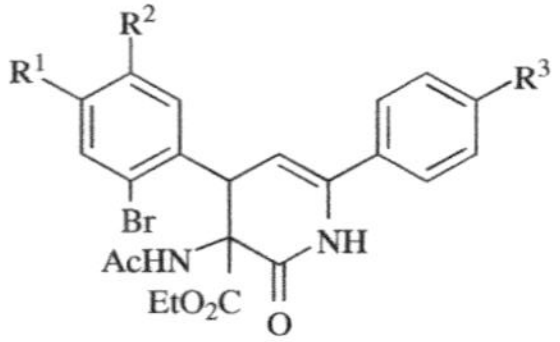	1. CuCl (2 eq), NaH, DME, 90°, 2 h 2. aq NH_4OH	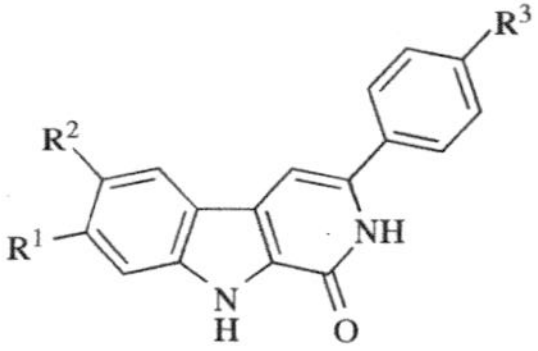 R^1 / R^2 / R^3 H / H / H (68) H / H / Cl (67) H / H / Me (70) MeO / MeO / H (65) MeO / MeO / Cl (77) MeO / MeO / Me (60) —OCH_2O— / H (60) —OCH_2O— / Cl (64) —OCH_2O— / Me (64)	583
C_{18} 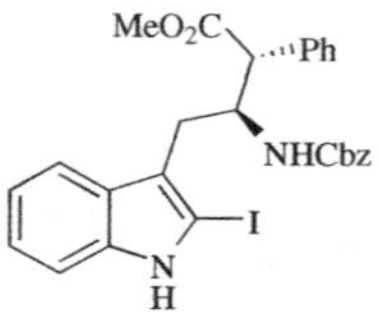	CuI (10 mol %), DMEDA (20 mol %), K_3PO_4, toluene, 110°, 16 h	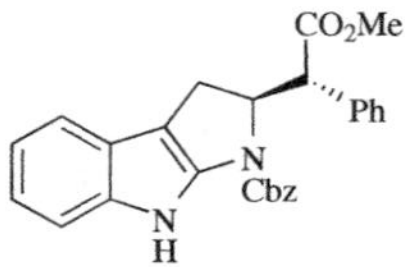 (45)	566
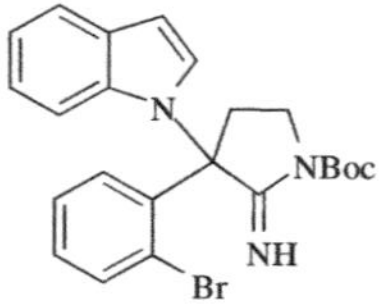	CuI (1 eq), K_3PO_4, DMSO, 80°, 1.5 h	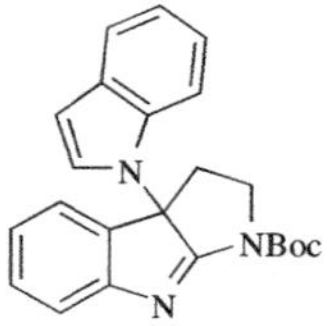 (91)	169
C_{18-19} 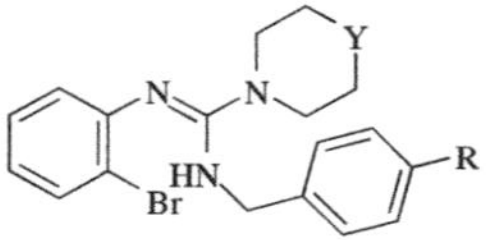	CuI (5 mol %), phen (10 mol %), Cs_2CO_3, DME, 80°, 16 h	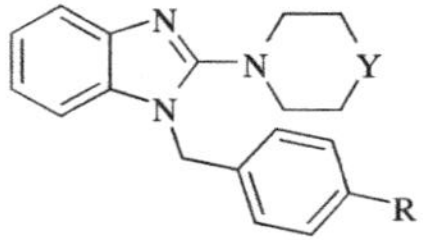 R / Y MeO / BocN (97) H / 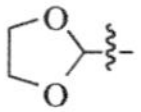 (96)	581

C$_{19-21}$

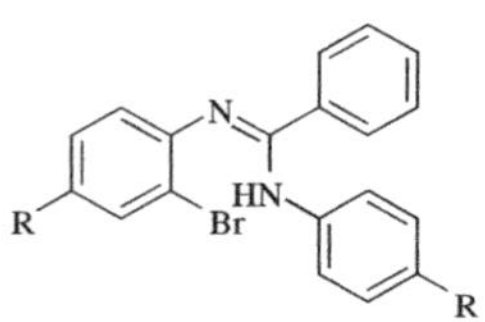

C$_{19}$

C$_{21-24}$

R	
H	(97)
Me	(96)

CuO nanoparticles (5 mol %), KOH, DMSO, 110°, 3 h   572

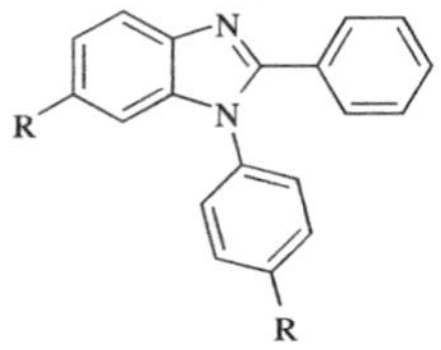

(86)

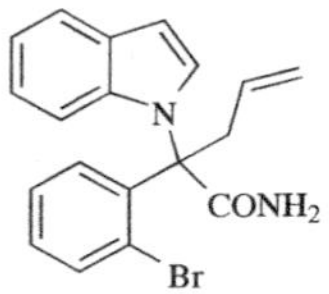

CuI (1 eq), Cs$_2$CO$_3$, DMSO, 90°   169

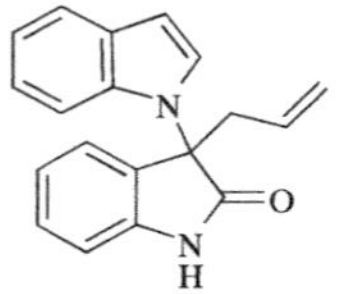

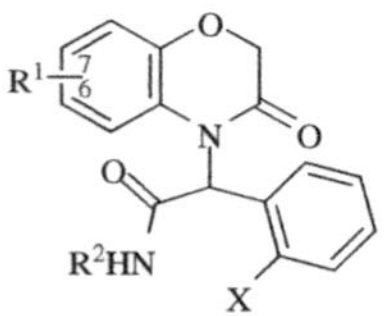

CuI (10 mol %), MeCN, K$_3$PO$_4$·3H$_2$O, MW

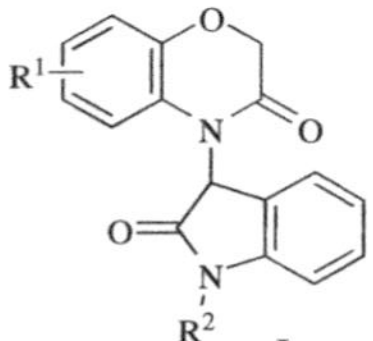

R^1	R^2	X	Additive	Solvent	Temp (°)	Time (min)	I	II	
6-Cl	Bn	Br	L-Pro (20 mol %)	MeCN	150	50	(59)	(0)	584
6-Cl	Bn	Br	none	DMF	150	35	(80)	(0)	585
6-Me	Bn	Br	L-Pro (20 mol %)	MeCN	150	50	(85)	(0)	584
6-Me	Bn	Br	none	DMF	150	35	(89)	(0)	585
7-Me	t-Bu	Br	L-Pro (20 mol %)	MeCN	180	60	(60)	(17)	584
7-Me	t-Bu	I	L-Pro (20 mol %)	MeCN	180	60	(25)	(15)[a]	584
7-Me	c-C$_6$H$_{11}$	Br	L-Pro (20 mol %)	MeCN	180	80	(40)	(11)	584
7-Me	Bn	Br	L-Pro (20 mol %)	MeCN	150	35	(90)	(0)	584
7-Me	Bn	Br	none	DMF	150	35	(93)	(0)	585

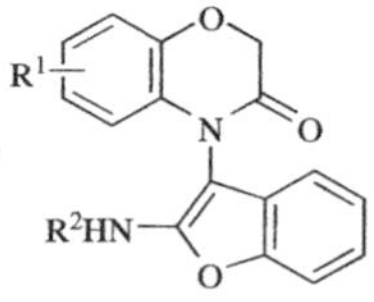

TABLE 26. INTRAMOLECULAR ARYLATIONS (Continued)
A. SYNTHESIS OF FIVE-MEMBERED NITROGEN HETEROCYCLES (Continued)

*Please refer to the charts preceding the tables for structures indicated by the **bold** numbers.*

Substrate	Conditions	Product(s) and Yield(s) (%)	Refs.
C₂₂	CuI (50 mol %), NaH, THF. reflux, 1 h	(84)	139
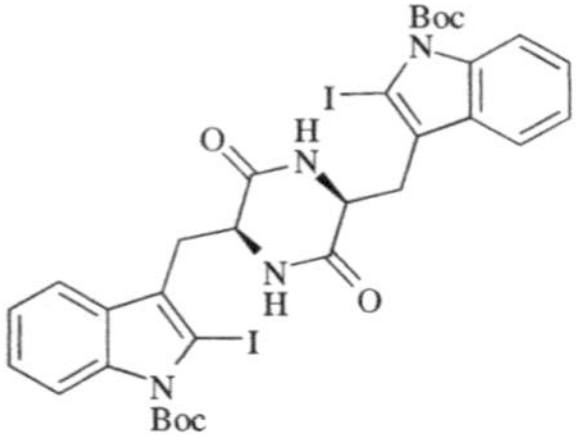	CuI (10 mol %), DMEDA (20 mol %), K$_3$PO$_4$, toluene, 110°, 16 h	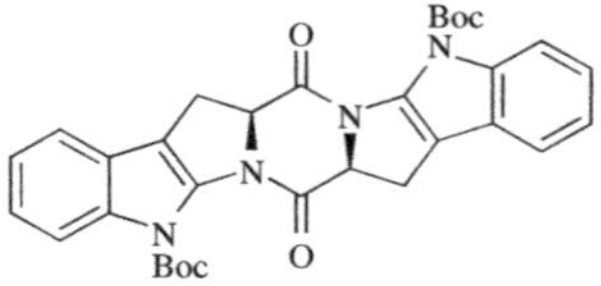 (72)	566
C_{22–24}	CuI (5 mol %), phen (10 mol %), Cs$_2$CO$_3$, DME, 80°, 16 h	(see table)	581

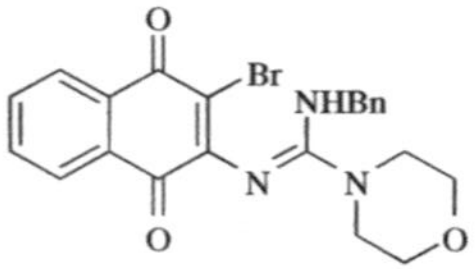

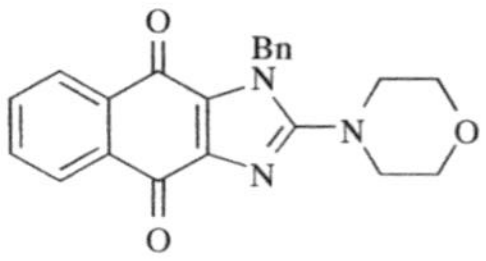

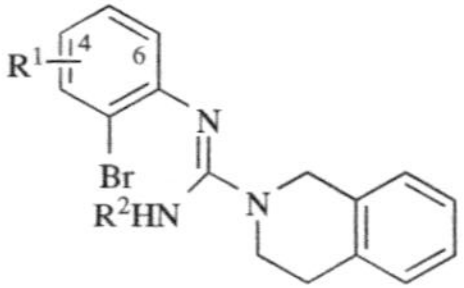

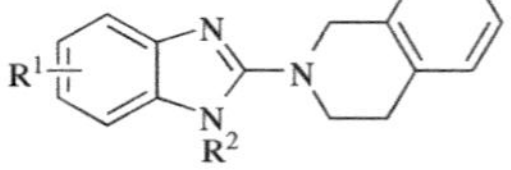

R^1	R^2	
H	Ph	(58)
H	Bn	(83)
H	4-MeOC$_6$H$_4$CH$_2$	(95)
4-Cl	Bn	(90)
4-Me	Bn	(90)
4-Me, 6-Br	Bn	(90)
5-CF$_3$	Bn	(90)

Substrate	Conditions	Product(s) and Yield(s) (%)	Refs.
C₂₄ Ar = 2-H$_2$N-4-RC$_6$H$_3$	Cu (50 mol %), CuI (10 mol %), K$_2$CO$_3$, (n-Bu)$_2$O, reflux, 12 h		586

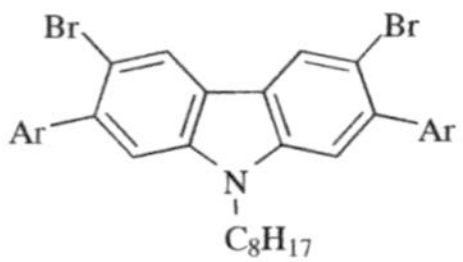

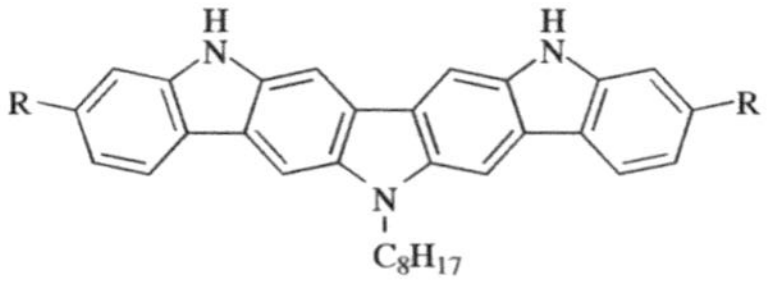

R	
H	(57)
Cl	(68)

587

588

(75)

(35)

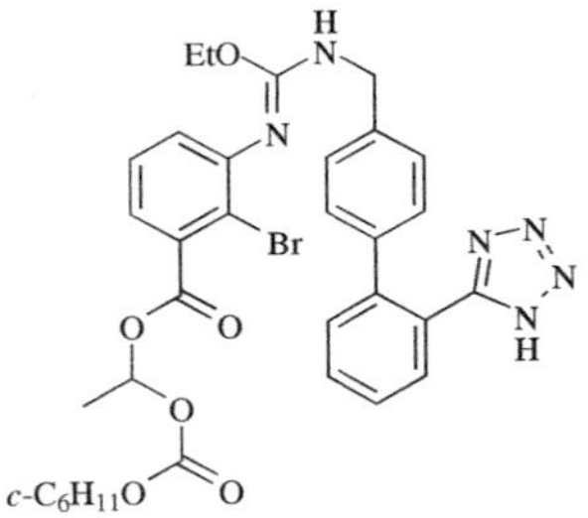

CuI (50 mol %), L-Pro (1 eq), K₂CO₃, DMSO, 70°, 3 h

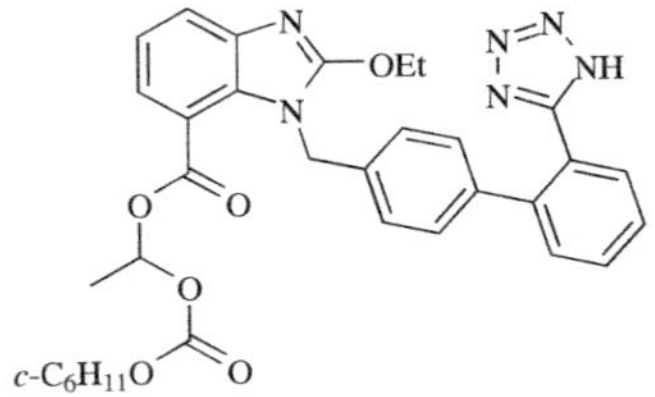

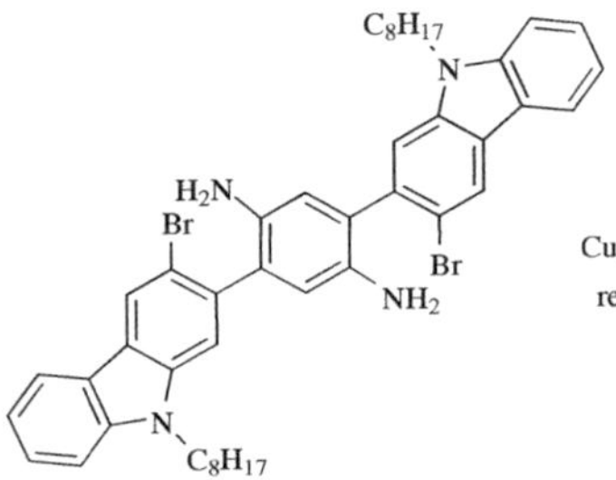

Cu, CuI, K₂CO₃, (n-Bu)₂O, reflux, 24 h

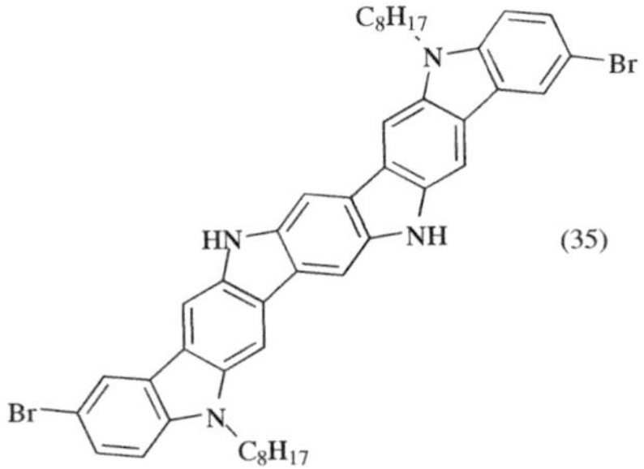

C₂₅

C₃₀

TABLE 26. INTRAMOLECULAR ARYLATIONS (*Continued*)
A. SYNTHESIS OF FIVE-MEMBERED NITROGEN HETEROCYCLES (*Continued*)

Substrate	Conditions	Product(s) and Yield(s) (%)	Refs.

*Please refer to the charts preceding the tables for structures indicated by the **bold** numbers.*

C30

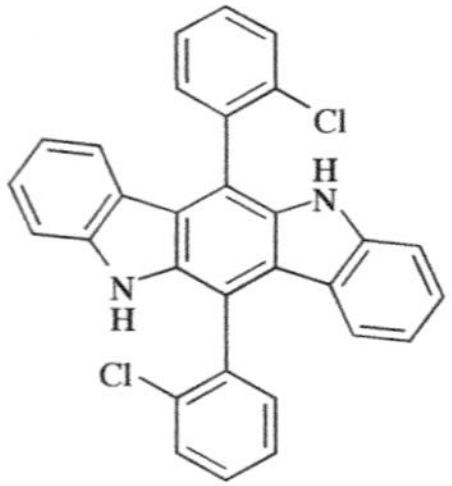

CuI (10 mol %), TBAH, DMF, 120°, 20 h

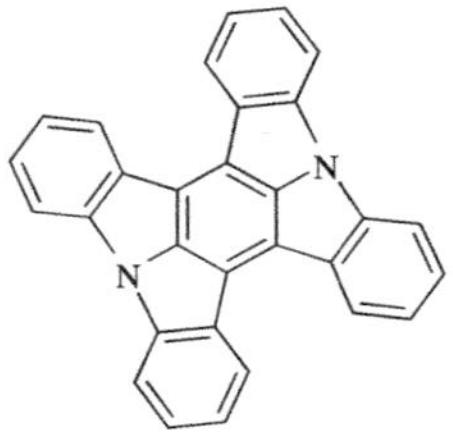

(74)

589

---

a The dehalogenated substrate was formed in 22% yield.

TABLE 26. INTRAMOLECULAR ARYLATIONS (Continued)
B. SYNTHESIS OF SIX-MEMBERED NITROGEN HETEROCYCLES

*Please refer to the charts preceding the tables for structures indicated by the **bold** numbers.*

Substrate	Conditions	Product(s) and Yield(s) (%)	Refs.
C$_{7-18}$	CuI (20 mol %), EDA (40 mol %), Cs$_2$CO$_3$, DMF, 100°, 18 h		577
C$_{8-14}$	CuI (10 mol %), phen (20 mol %), Cs$_2$CO$_3$, DMF, 80°, 16 h		590
C$_9$ Ar = 2-O$_2$NC$_6$H$_4$	CuI (50 mol %), DMSO		569 556

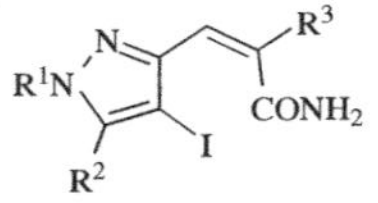

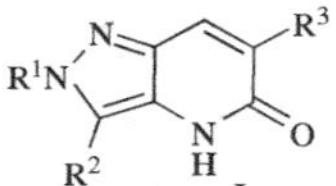

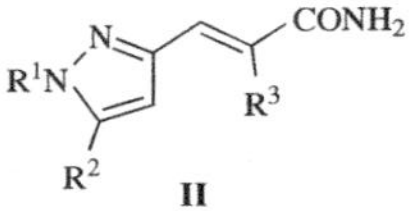

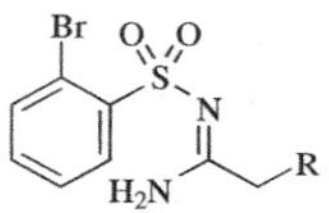

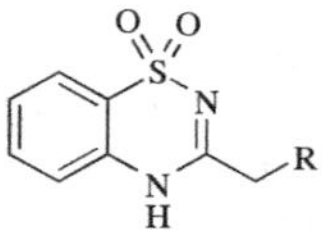

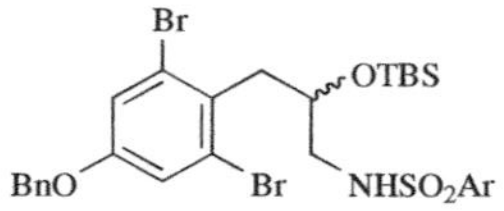

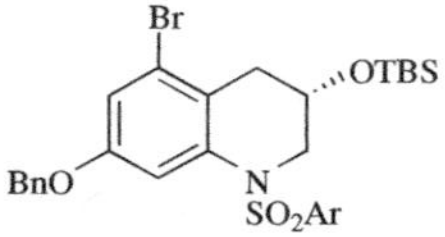

Product table for first entry:

R^1	R^2	R^3	I	II
Me	H	Me	(70)	(7)
Ph	Ph	Me	(86)	(8)
Ph	4-ClC$_6$H$_4$	Me	(87)	(6)
Ph	4-MeC$_6$H$_4$	Me	(90)	(5)
Ph	Ph	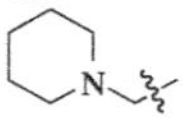	(88)	(7)

Product table for second entry:

R	
TMS	(72)
n-Bu	(86)
Ph	(75)

Product table for third entry:

Isomer	Additive	Temp (°)	Time (h)	
(R)	CsOAc	60	24	(83)
(S)	Cs$_2$CO$_3$	50	15	(89)

TABLE 26. INTRAMOLECULAR ARYLATIONS (Continued)
B. SYNTHESIS OF SIX-MEMBERED NITROGEN HETEROCYCLES (Continued)

*Please refer to the charts preceding the tables for structures indicated by the **bold** numbers.*

C9–16

Substrate				Conditions (Catalyst (x amount))					Product(s) and Yield(s) (%)	Refs.
R^1	R^2	X	Catalyst	x	Additive(s)	Solvent(s)	Temp (°)	Time (h)		
H	H	Br	CuI	10 mol %	CsOAc	DMSO	90	24	(54)	197
H	H	Br	CuI	2 eq	CsOAc	DMSO	rt	1	(54)	274
H	H	I	CuI	2 eq	CsOAc	DMSO	rt	24	(56)	197
H	H	I	CuI	2 eq	CsOAc	DMSO	rt	24	(56)	274
H	H	Br	CuOAc	5 mol %	**L13** (20 mol %), K_3PO_4	DMF	40	18	(80)	120
H	$SO_2C_4F_9$	Br	CuI	2 eq	CsOAc	DMSO	90	5	(98)	274
H	$SO_2C_4F_9$	Br	CuI	2 eq	CsOAc	DMSO	90	4.5	(98)	197
H	Bn	Br	CuI	2 eq	CsOAc	DMSO	90	9	(71)	197, 274
H	Bn	I	CuI	2 eq	CsOAc	DMSO	90	9	(54)	197, 274
H	$2\text{-}O_2NC_6H_4O_2S$	I	CuI	2 eq	CsOAc	DMSO	90	4	(99)	274
H	$2\text{-}O_2NC_6H_4O_2S$	I	CuI	2 eq	CsOAc	DMSO	90	4	(81)	197
I	$C_4F_9O_2S$	Br	CuI	2 eq	CsOAc	DMSO	90	24	(81)	274

C11–12

CuI (x amount), DMSO

R^1	R^2	X	x	Additive(s)	Temp (°)	Time		Refs.
H	H	Br	10 mol %	DMEDA (10 mol %), Cs_2CO_3	rt	10 min	(89)	580
H	H	I	10 mol %	DMEDA (10 mol %), Cs_2CO_3	rt	10 min	(89)	
H	TBS	Br	2 eq	NaOAc	50	19 h	(65)	
H	TBS	I	10 mol %	DMEDA (10 mol %), Cs_2CO_3	rt	10 min	(93)	
6-F	TBS	I	10 mol %	DMEDA (10 mol %), Cs_2CO_3	rt	10 min	(100)	
4-Cl	TBS	I	10 mol %	DMEDA (10 mol %), Cs_2CO_3	rt	10 min	(100)	
4,5-(MeO)$_2$	TBDPS	I	10 mol %	DMEDA (10 mol %), Cs_2CO_3	rt	10 min	(61)	
3-Me	TBS	I	10 mol %	DMEDA (10 mol %), Cs_2CO_3	rt	10 min	(58)	

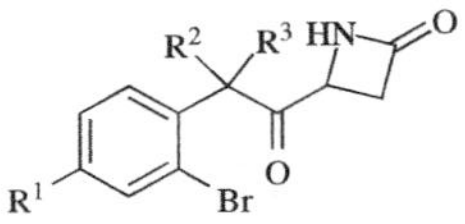

C$_{12}$

Catalyst (5 eq)

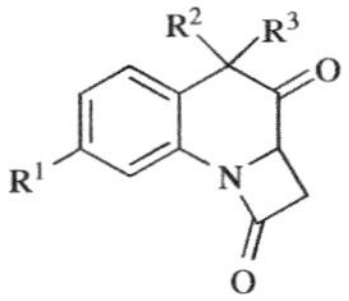

R^1	R^2	R^3	Catalyst	Solvent	Temp (°)	Time		
H	H	H	Cu	DMF	135	8 h	(13)	565
H	H	H	Cu (active)	—	MW (100 W)	10 min	(72)	280
H	BnO$_2$C	BnO$_2$C	Cu (active)	—	MW (100 W)	10 min	(68)	280
MeO$_2$C	H	H	Cu	DMF	135	8 h	(10)	565

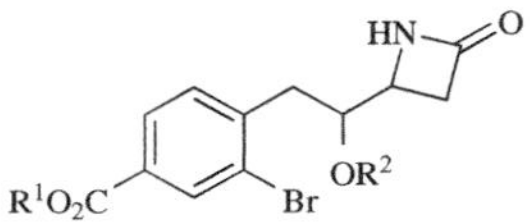

Catalyst (5 eq)

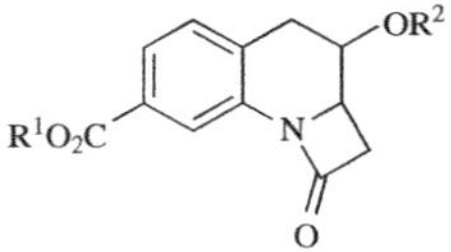

R^1	R^2	Catalyst	Solvent	Temp (°)	Time		
Me	Ac	Cu	DMF	135	8 h	(65)	565
Me	Ac	Cu (active)	—	MW (100 W)	10 min	(84)	280
t-Bu	H	Cu	DMF	135	8 h	(41)	565
t-Bu	Ac	Cu	DMF	135	8 h	(50)	565
Bn	H	Cu	DMF	135	8 h	(68)	565
Bn	H	Cu (active)	—	MW (100 W)	8 min	(83)	280

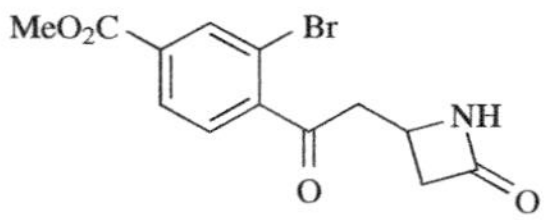

Active Cu (5 eq), MW (100 W), 8 min 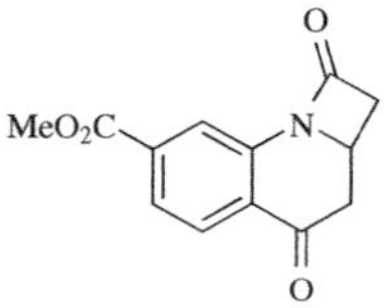   (75)   280

TABLE 26. INTRAMOLECULAR ARYLATIONS (Continued)
B. SYNTHESIS OF SIX-MEMBERED NITROGEN HETEROCYCLES (Continued)

*Please refer to the charts preceding the tables for structures indicated by the **bold** numbers.*

Substrate	Conditions	Product(s) and Yield(s) (%)	Refs.
C$_{12}$	CuI (5 mol %), DMEDA (10 mol %), K$_2$CO$_3$, toluene, reflux, 18 h	(70)	591
C$_{13-21}$	CuI (20 mol %), EDA (40 mol %), Cs$_2$CO$_3$, DMF, 100°, 24 h		577
C$_{13-14}$	Cu (5 eq), DMF, 135°, 8 h		565

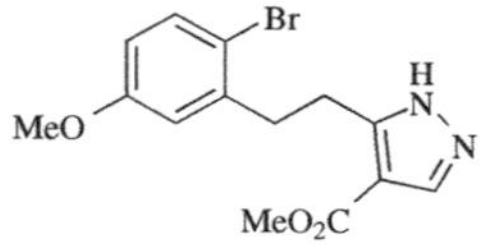

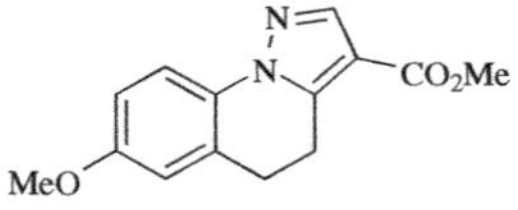

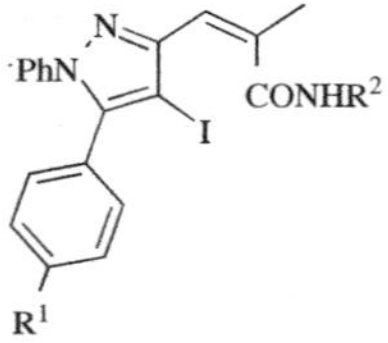

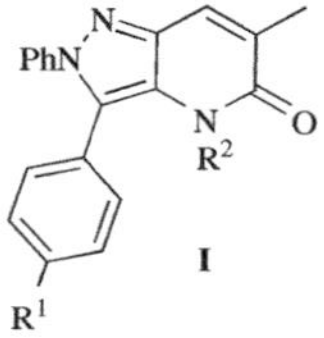

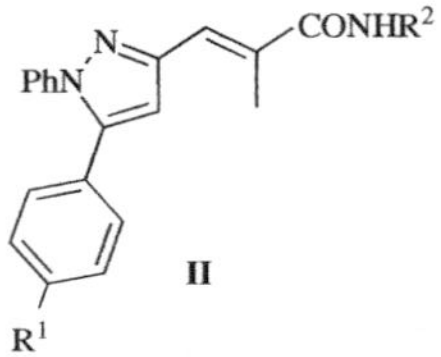

For C$_{13-21}$:

R^1	R^2	I	II
Ph	H	(39)	(21)
Ph	n-Pr	(54)	(14)
Ph	i-Pr	(21)	(43)
Ph	c-Pr	(57)	(21)
Ph	s-Bu	(10)	(47)
Ph	c-C$_6$H$_{11}$	(14)	(49)
Ph	Ph	(39)	(27)

R^1	R^2	I	II
Ph	3-ClC$_6$H$_4$	(28)	(20)
Ph	4-ClC$_6$H$_4$	(13)	(22)
Ph	4-MeOC$_6$H$_4$	(58)	(18)
Ph	4-MeC$_6$H$_5$	(62)	(22)
Ph	Bn	(47)	(22)
Ph	Ph(CH$_2$)$_2$	(38)	(31)
Cl	H	(32)	(18)

R^1	R^2	I	II
Cl	c-Pr	(50)	(28)
Cl	Ph	(37)	(19)
Cl	Bn	(46)	(33)
Me	H	(35)	(16)
Me	c-Pr	(54)	(21)
Me	Ph	(33)	(23)
Me	Bn	(57)	(24)

For C$_{13-14}$:

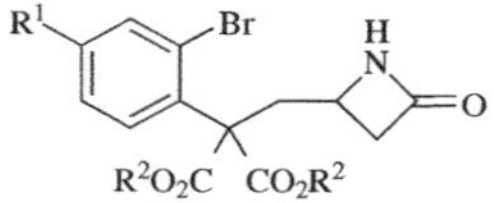

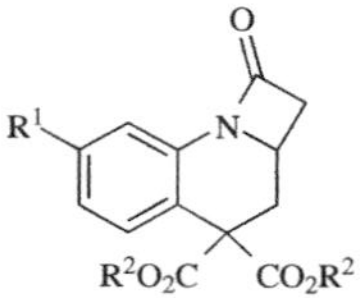

R^1	R^2	
H	Et	(60)
H	Bn	(5)
EtO$_2$C	Et	(45)

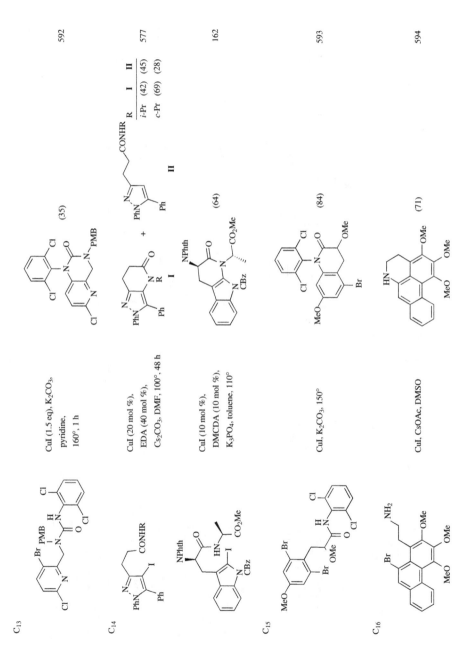

592

577

R	I	II
i-Pr	(42)	(45)
c-Pr	(69)	(28)

162

593

594

521

TABLE 26. INTRAMOLECULAR ARYLATIONS (*Continued*)

B. SYNTHESIS OF SIX-MEMBERED NITROGEN HETEROCYCLES (*Continued*)

Substrate	Conditions	Product(s) and Yield(s) (%)	Refs.

*Please refer to the charts preceding the tables for structures indicated by the **bold** numbers.*

C_{18-23}

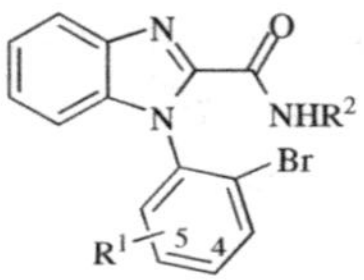

CuI (20 mol %),
DMG (20 mol %),
K_3PO_4, DMF, 150°, 12 h

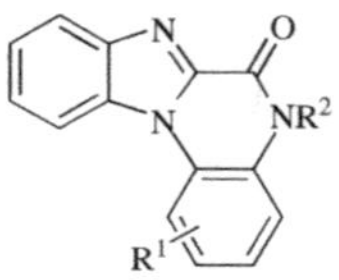

R^1	R^2	
H	$PhCH_2CH_2$	(56)
$4\text{-}CF_3O$	$n\text{-}Bu$	(trace)
$4\text{-}CF_3O$	Bn	(66)
$4\text{-}CF_3O$	$2,4\text{-}Cl_2C_6H_4CH_2$	(44)
$4\text{-}CF_3O$	$4\text{-}MeOC_6H_4CH_2$	(71)
$4\text{-}CF_3O$	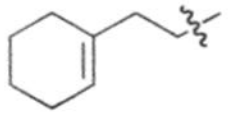	(42)
$4\text{-}CF_3O$	$PhCH_2CH_2$	(56)
$5\text{-}CF_3$	$PhCH_2CH_2$	(40)

595

C_{23-24}

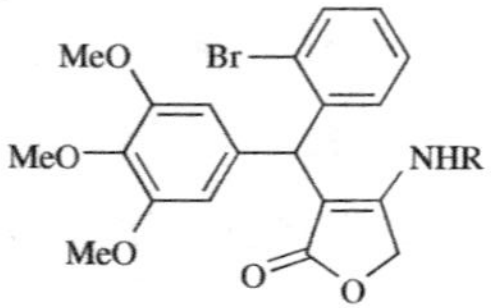

CuI (1.2 eq), Cs_2CO_3, DMF,
90°, 16 h

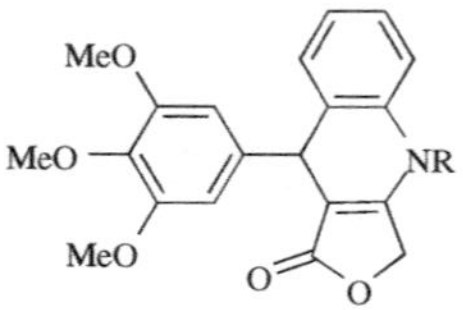

R	
Ph	(82)
Bn	(100)
$4\text{-}MeOC_6H_4CH_2$	(84)

596

Substrate	Conditions	Product(s) and Yield(s) (%)	Refs.
C_{9-10} 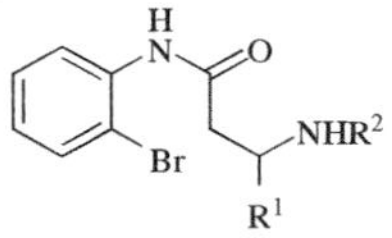	CuI (5 mol %), L-Pro (10 mol %), K_2CO_3, toluene, 110°, 60 h	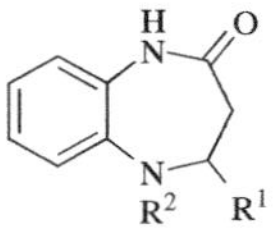 $\begin{array}{ll} R^1 & R^2 \\ \hline H & DIPP\ (63) \\ Me & Boc\ (60) \end{array}$	597
C_9 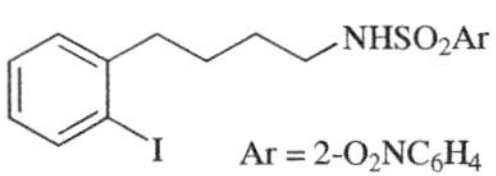 Ar = $2\text{-}O_2NC_6H_4$	CuI (1 eq), CsOAc, DMSO, 120°, 24 h	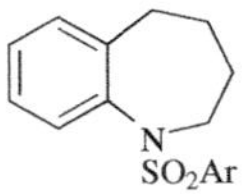 (74)	274, 197
C_{13} 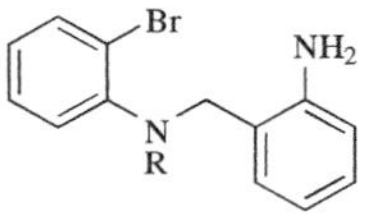	CuI (10 mol %), Cs_2CO_3, $HOCH_2CH_2OH$, 100°, 3 h	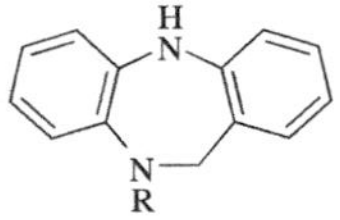 $\begin{array}{ll} R & \\ \hline H & (82) \\ Me & (94) \end{array}$	598
C_{13-17} 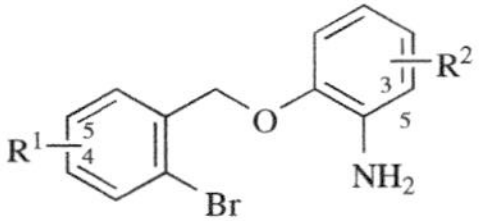	CuI (10 mol %), L-Pro (20 mol %), DABCO·6H₂O, DMSO, H₂O, 90°	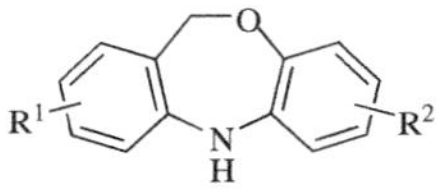	599

Conditions table (for C_{13-17}):

R^1	R^2	Time (h)	
H	4-Cl	72	(40)
H	3-Me	72	(93)
H	4-Me	72	(73)
H	6-Me	72	(66)
5-MeO	H	48	(35)
5-MeO	3-Me	24	(90)

Product table (for C_{13-17}):

R^1	R^2	Time (h)	
3-t-BuO$_2$C	H	24	(81)
3-t-BuO$_2$C	3-Me	24	(71)
3-t-BuO$_2$C	4-Me	24	(65)
5-t-BuO$_2$C	H	24	(80)
5-t-BuO$_2$C	3-Me	24	(92)
4-Ac	4-Me	48	(63)
3-(CH=CH)$_2$-4	H	48	(75)

TABLE 26. INTRAMOLECULAR ARYLATIONS (Continued)

C. SYNTHESIS OF SEVEN-MEMBERED NITROGEN HETEROCYCLES (Continued)

Substrate	Conditions	Product(s) and Yield(s) (%)			Refs.

C₁₄₋₁₉

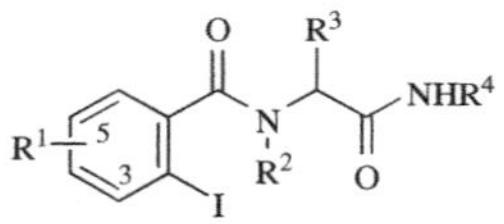

(structure)

CuI (10 mol %), thiophene-2-carboxylic acid (20 mol %), K₂CO₃, DMSO, 100°

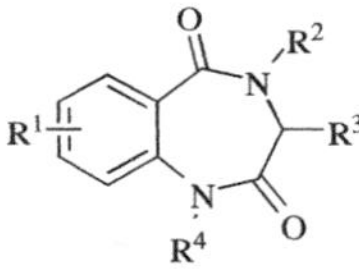

(product structure)

600

R¹	R²	R³	R⁴	Temp (°)	Time (h)
H	Me	H	2-IC₆H₄CH₂	100	15 (52)
H	n-Bu	i-Pr	MeO₂CCH₂	110	16 (88)
H	Bn	i-Pr	1-morpholinyl-(CH₂)₂	110	16 (99)
H	Bn	i-Pr	2,4-(MeO)₂C₆H₄CH₂	110	16 (95)
H	Bn	CbzHN(CH₂)₄	MeO₂CCH₂	110	16 (88)
H	Bn	AcO(CH₂)₄	MeO₂CCH₂	110	16 (74)
H	Bn	n-C₆H₁₃	MeO₂CCH₂	110	16 (86)
H	2,4-(MeO)₂C₆H₄CH₂	i-Pr	4-morpholinyl-(CH₂)₂	110	16 (97)
5-Me	Bn	i-Pr	4-morpholinyl-(CH₂)₂	110	16 (95)
4-(CH=CH₂)-5	Bn	i-Pr	MeO₂CCH₂	110	16 (50)

C₁₅

(structure)

Cu (80 mol %), K₂CO₃, DMSO, 160°, 24 h

(product structure)

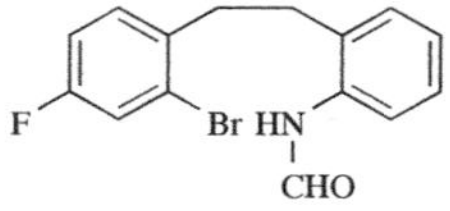

(61)

601

(structure)

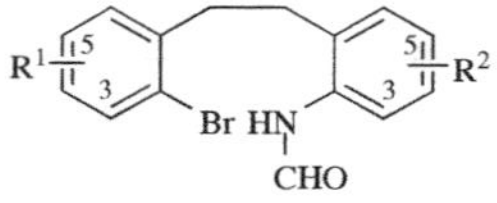

1. Cu (80 mol %), K₂CO₃, DMSO, 160°, 24 h
2. 6 M NaOH, EtOH, reflux, 1 h

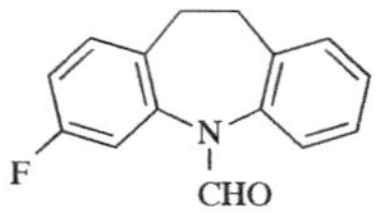

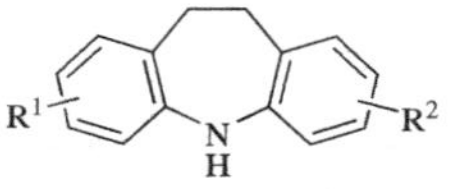

(product structure)

601

R¹	R²	
H	4-Cl	(53)
H	4-MeO	(46)
3-F	3-F	(61)
4-Cl	4-Cl	(75)

C$_{19}$

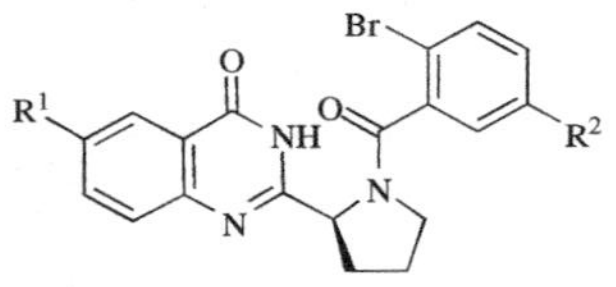

CuI (20 mol %),
L-Pro (30 mol %),
NaH, DMF, 120°, 10 h

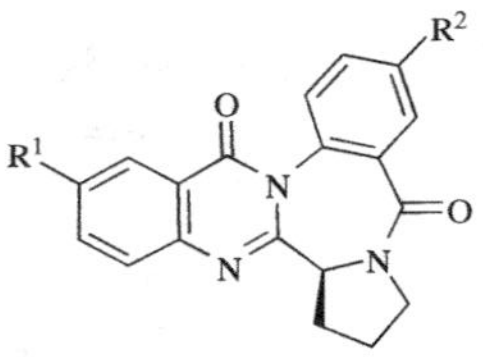

R^1	R^2	
H	H	(84)
H	MeO	(91)
MeO	H	(89)
MeO	MeO	(90)

602

TABLE 26. INTRAMOLECULAR ARYLATIONS (Continued)

D. SYNTHESIS OF EIGHT- AND HIGHER-MEMBERED NITROGEN HETEROCYCLES

Substrate	Conditions	Product(s) and Yield(s) (%)	Refs.
C_{10-18}	CuI (5 mol %), L-Pro (10 mol %), K$_2$CO$_3$, toluene, 110°, 72 h	 R, n DIPP, 1 (68) Boc, 1 (62) DIPP, 2 (64) DIPP, 3 (63) DIPP, 5 (60) DIPP, 9 (61)	597
C_{11} Ar = 2-O$_2$NC$_6$H$_4$	CuI (x eq), CsOAc, DMSO, 120°	 x, Time (h) 1, 24 (0) 2, 4 (0)	274 197
C_{12}	1. CuI (5 mol %), DMG (20 mol %), K$_2$CO$_3$, MeCN, reflux, 12 h 2. HCl	(62)	603
C_{14-15}	CuI (10 mol %), Cs$_2$CO$_3$, HOCH$_2$CH$_2$OH, 100°, 3 h	 R^1, R^2 H, H (64) H, Me (91) H, allyl (97) H, Bn (80) 5-Cl, H (53) 5-Cl, Me (86) 5-Br, Me (84) 5-HO, Me (62) 4,5-(MeO)$_2$, Me (77) 4-CF$_3$, Me (96)	598

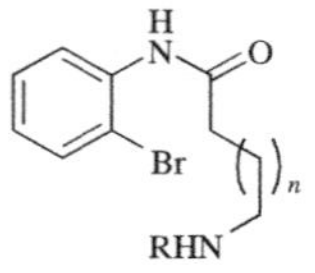

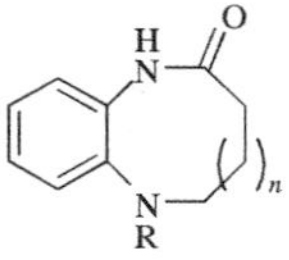

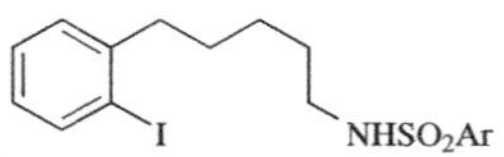

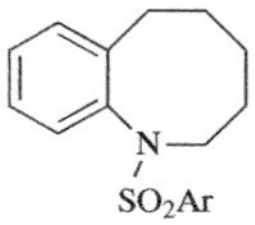

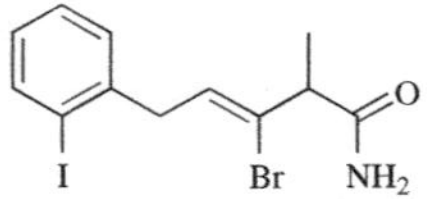

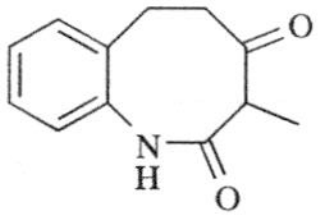

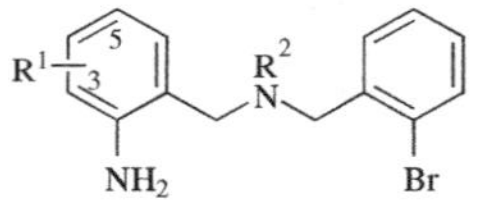

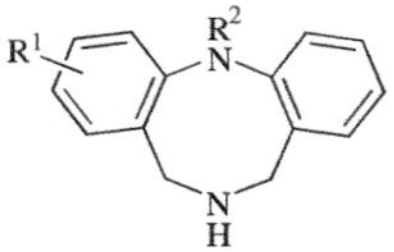

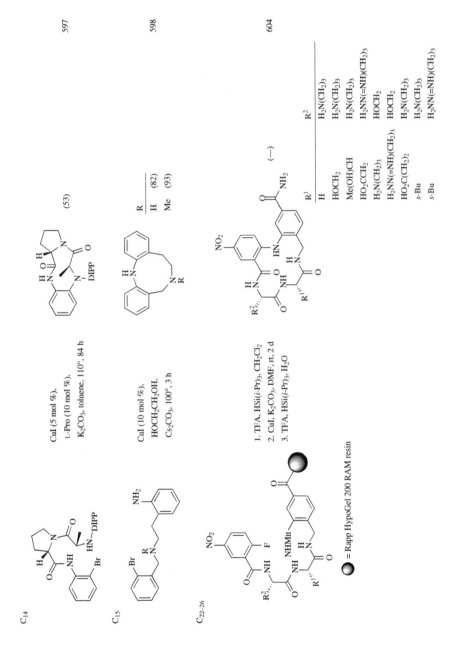

597

(53)

C₁₄

CuI (5 mol %),
L-Pro (10 mol %),
K₂CO₃, toluene, 110°, 84 h

598

R	
H	(82)
Me	(93)

C₁₅

CuI (10 mol %),
HOCH₂CH₂OH,
Cs₂CO₃, 100°, 3 h

604

C₂₂₋₂₆

1. TFA, HSi(i-Pr)₃, CH₂Cl₂
2. CuI, K₂CO₃, DMF, rt, 2 d
3. TFA, HSi(i-Pr)₃, H₂O

= Rapp HypoGel 200 RAM resin

R^1	R^2
H	H₂N(CH₂)₃
HOCH₂	H₂N(CH₂)₃
Me(OH)CH	H₂N(CH₂)₃
HO₂CCH₂	H₂NN(=NH)(CH₂)₃
H₂N(CH₂)₃	HOCH₂
H₂NN(=NH)(CH₂)₃	HOCH₂
HO₂C(CH₂)₂	H₂N(CH₂)₃
s-Bu	H₂N(CH₂)₃
s-Bu	H₂NN(=NH)(CH₂)₃

TABLE 26. INTRAMOLECULAR ARYLATIONS (*Continued*)

D. SYNTHESIS OF EIGHT- AND HIGHER-MEMBERED NITROGEN HETEROCYCLES (*Continued*)

Substrate	Conditions	Product(s) and Yield(s) (%)	Refs.
C_{25}			
	CuI (50 mol %), DMEDA (1 eq), K_2CO_3, toluene, 100°, 36 h	R / Bond **a**: H, single (82); MeO, single (>80); MeO, double, (*Z*) (82)	605
C_{26}			
	CuI (50 mol %), DMEDA (1 eq), K_2CO_3, toluene, 100°, 36 h	(81)	164

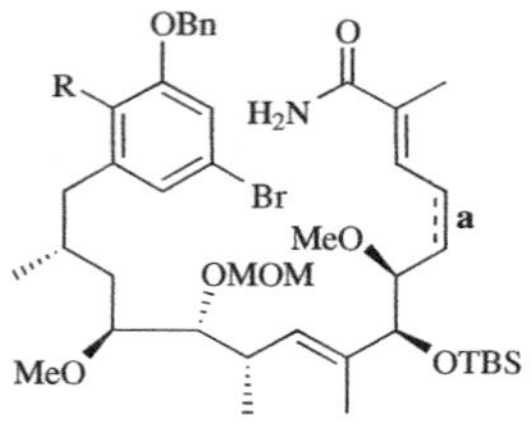

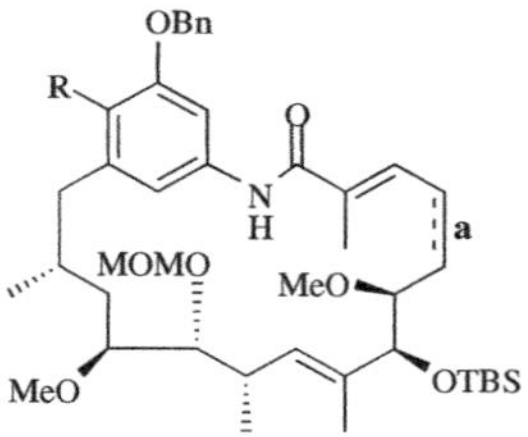

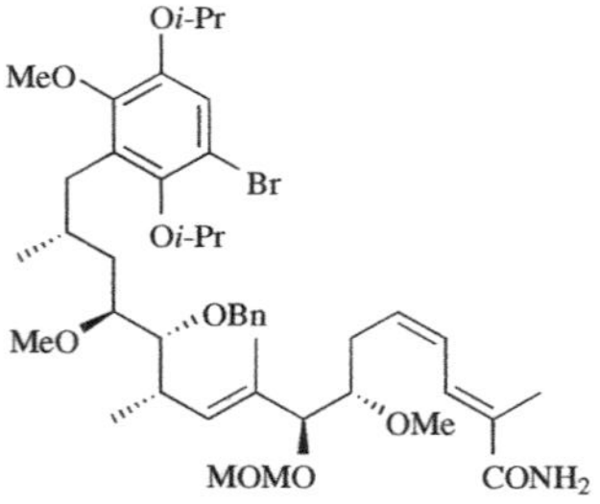

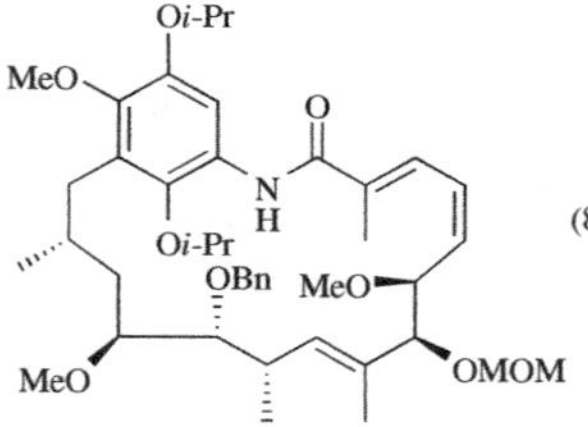

TABLE 27. N-ARYLATIONS IN MULTI-STEP REACTIONS
A. SYNTHESIS OF FIVE-MEMBERED RINGS

*Please refer to the charts preceding the tables for structures indicated by the **bold** numbers.*

Nitrogen Nucleophile	Electrophile	Conditions	Product(s) and Yield(s) (%)	Refs.
C$_0$				
NaN$_3$	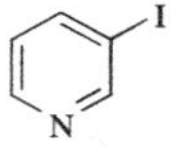 (3-iodopyridine)	 NEt$_2$, CuSO$_4$•5H$_2$O (5 mol %), Na ascorbate, L-Pro (20 mol %), Na$_2$CO$_3$, DMSO/H$_2$O (9:1), 60°, 18 h	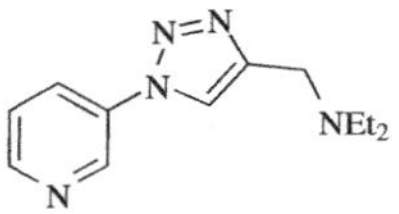 (triazolylpyridine, NEt$_2$) (94)	606
	ArI	R, CuSO$_4$•5H$_2$O (5 mol %), Na ascorbate, L-Pro (20 mol %), Na$_2$CO$_3$, DMSO/H$_2$O (9:1), 60°, 18 h	(R-triazole, Ar)	606

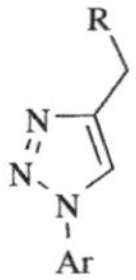

Ar	R	
Ph	Et$_2$N	(90)
Ph	4-ClC$_6$H$_4$O	(98)
Ph	4-MeOC$_6$H$_4$O	(76)
Ph	4-AcC$_6$H$_4$O	(87)
Ph	HO(CH$_2$)$_3$	(98)
Ph	HO$_2$C(NH$_2$)CH	(52)
Ph	1-cyclohexenyl	(74)
Ph	Ph	(84)
Ph	3-H$_2$NC$_6$H$_4$	(74)
4-BrC$_6$H$_4$	Et$_2$N	(78)

Nitrogen Nucleophile	Electrophile	Conditions	Product(s) and Yield(s) (%)	Refs.
	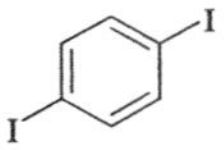 (1,4-diiodobenzene)	 NEt$_2$, CuSO$_4$•5H$_2$O (5 mol %), Na ascorbate, L-Pro (20 mol %), Na$_2$CO$_3$, DMSO/H$_2$O (9:1), 60°, 18 h	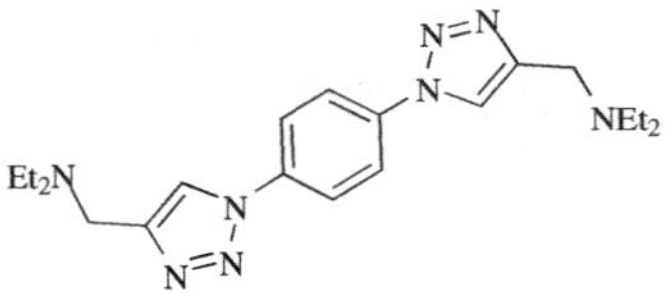 (bis-triazole, Et$_2$N) (66)	606
H$_2$NBoc	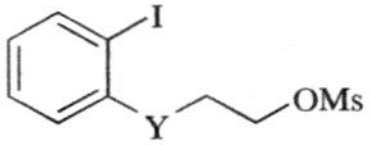 (2-iodophenyl Y CH$_2$CH$_2$OMs)	CuI (5 mol %), DMEDA (20 mol %), Cs$_2$CO$_3$, THF, 80°, 16 h	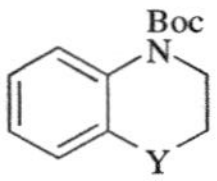 (Boc dihydro ring Y) Y O (76) CH$_2$ (74)	607

TABLE 27. N-ARYLATIONS IN MULTI-STEP REACTIONS (Continued)
A. SYNTHESIS OF FIVE-MEMBERED RINGS (Continued)

Please refer to the charts preceding the tables for structures indicated by the bold numbers.

Nitrogen Nucleophile	Electrophile	Conditions	Product(s) and Yield(s) (%)	Refs.

C_{0-6}

Nitrogen Nucleophile	Electrophile	Conditions	Product(s) and Yield(s) (%)		Refs.
H_2NR^1	$R^2 \overset{5}{\underset{4}{\rule{0pt}{1em}}}$ with Br, Br groups	CuOAc (10 mol %), K_3PO_4, toluene, DMEDA (20 mol%), reflux, 24 h	R^1 / R^2 structure (indole)		106

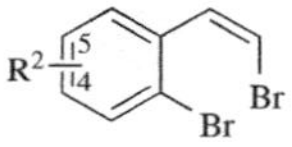

Product table for indole (R^1, R^2):

R^1	R^2	Yield
EtO_2C	H	(81)
BnO_2C	H	(84)
$Et(O)C$	H	(89)
$n\text{-}C_5H_{11}(O)C$	H	(87)
$n\text{-}C_6H_{13}(O)C$	H	(21)
Ph	H	(80)
Boc	5-F	(80)
Boc	5-Br	(53)
Boc	4,5-$(MeO)_2$	(82)
Boc	4-(OCH_2O)-5	(62)
Boc	4-Me	(80)
Boc	5-Ac	(70)

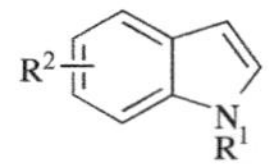

C_0

Nitrogen Nucleophile	Electrophile	Conditions	Product(s) and Yield(s) (%)	Refs.
H_2NBoc	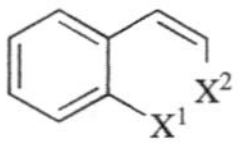 X^1, X^2 substituted arene	CuOAc (10 mol %), K_3PO_4, toluene, DMEDA (20 mol%), reflux, 24 h	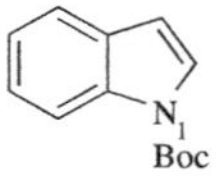 N-Boc indole; X^1/X^2: Cl/Br (56), Br/Cl (80)	106
	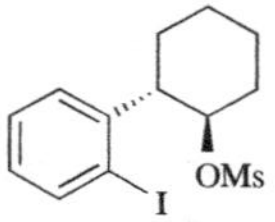 trans cyclohexane with OMs and aryl-I	CuI (5 mol %), DMEDA (20 mol %), Cs_2CO_3, THF, 80°, 16 h	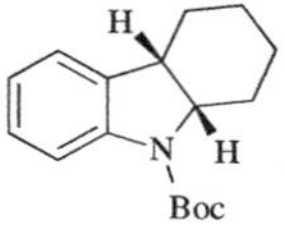 (94)	607

530

NH$_3$

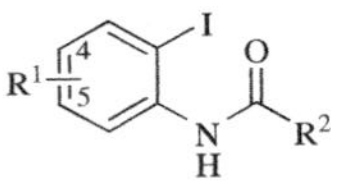

R^1	R^2	
H	BocHNCH$_2$	(82)
H	2-furyl	(77)
H	Ph	(83)
H	Bn	(79)
4-F	2-furyl	(81)
4-F	Ph	(74)
4-Cl	Ph	(68)

R^1	R^2	
4-Cl	4-ClC$_5$H$_4$	(82)
4-Cl	Bn	(72)
4-Br	*n*-Pr	(81)
4-I, 6-Cl	Ph	(61)
4-O$_2$N	MeO	(78)
4-O$_2$N	Ph	(66)
4-MeO	Ph	(75)

1. CuI (10 mol %), L-Pro (20 mol %), NaOH, DMSO, rt, 7 h
2. HOAc, 80°, 12 h

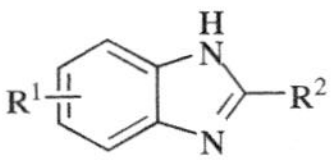

R^1	R^2	
5-MeO	Ph	(77)
4-Me	2-furyl	(65)
4-*n*-PrHNCO	MeO	(74)
4-MeO$_2$C	*n*-Pr	(86)
4-Ac	MeO	(72)
4-Ac	Ph	(79)
4-6-Me$_2$	MeO	(76)

178

H$_2$NBoc

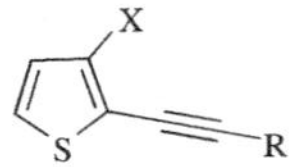

Cu (5 mol %), DMEDA (20 mol %), Cs$_2$CO$_3$, THF, 80°, 15 h

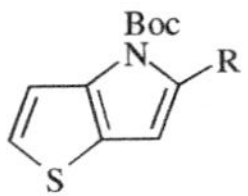

R	X	
n-Bu	Br	(83)
n-Bu	I	(83)
3-thienyl	Br	(84)

608

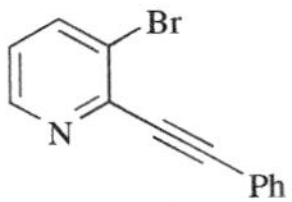

CuI (10 mol %), DMEDA (30 mol %), K$_2$CO$_3$, toluene, 105°, 24 h

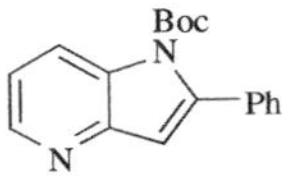

(63)

609

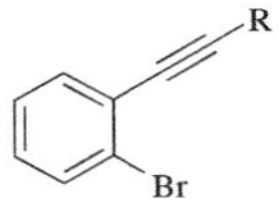

CuI (10 mol %), DMEDA (30 mol %), K$_2$CO$_3$, toluene, 105°, 24 h

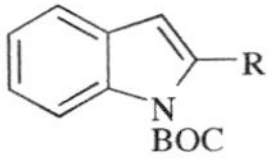

R	
2-thienyl	(61)
Ph	(73)
3-MeOC$_6$H$_4$	(65)
4-MeC$_6$H$_4$	(63)
3-NCC$_6$H$_4$	(63)
4-Ac	(52)

609

TABLE 27. *N*-ARYLATIONS IN MULTI-STEP REACTIONS (*Continued*)
A. SYNTHESIS OF FIVE-MEMBERED RINGS (*Continued*)

*Please refer to the charts preceding the tables for structures indicated by the **bold** numbers.*

C₀

Nitrogen Nucleophile	Electrophile	Conditions	Product(s) and Yield(s) (%)	Refs.
H_2NBoc	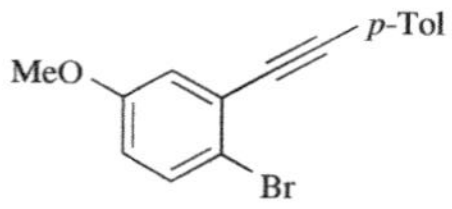	CuI (10 mol %), DMEDA (30 mol %), K_2CO_3, toluene, 105°, 24 h	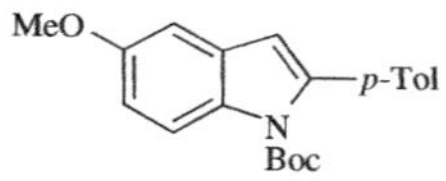 (58)	609
H_2NCOPh	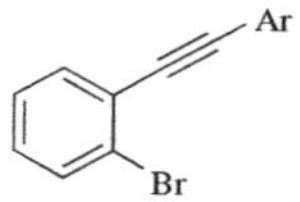	CuI (10 mol %), DMEDA (30 mol %), K_2CO_3, toluene, 105°, 24 h	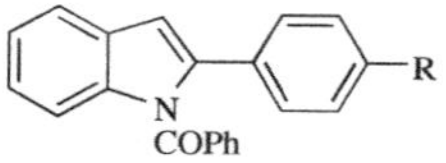 Ar Ph (62) PMP (71)	609
H_2NCO_2Bn	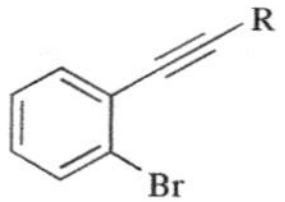	CuI (10 mol %), DMEDA (30 mol %), K_2CO_3, toluene, 105°, 24 h	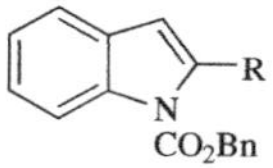 R *n*-Bu (52) Ph (57) 4-MeOC₆H₄ (56) 4-CF₃C₆H₄ (55)	609
H_2NBoc	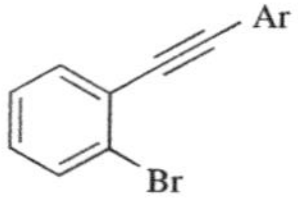	1. CuI (10 mol %), DMEDA (30 mol %), K_2CO_3, toluene, 105°, 24 h 2. TFA, MeCl₂, rt	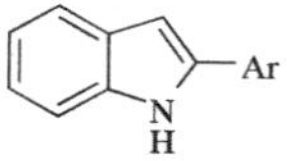 Ar Ph (69) 4-AcHNC₆H₄ (55) 2-MeOC₆H₄ (50) 3-MeOC₆H₄ (64) 4-MeOC₆H₄ (65) 4-MeC₆H₄ (63) 4-CF₃C₆H₄ (72) 3-NCC₆H₄ (63) 4-AcC₆H₄ (54) (64)	609

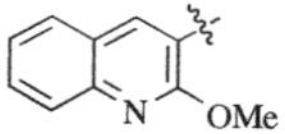

H₂NCOR¹

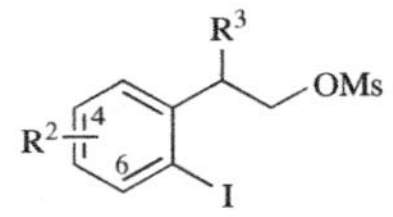

CuI (5 mol %),
DMEDA (20 mol %),
Cs₂CO₃, THF,
80°, 15 h

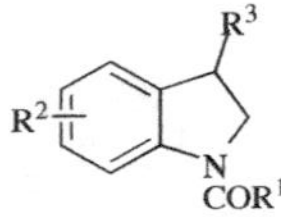

R¹	R²	R³	
MeO	H	H	(90)
t-BuO	H	Me	(95)
t-BuO	H	Ph(CH₂)₂	(92)
t-BuO	6-F	H	(95)
t-BuO	4,5-(MeO)₂	H	(75)
t-BuO	4-(OCH₂O)-5	H	(82)
BnO	H	H	(90)
Me	H	H	(87)

607

H₂NBoc

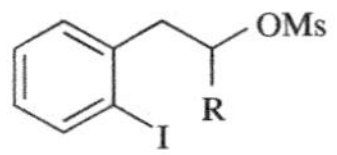

CuI (5 mol %),
DMEDA (20 mol %),
Cs₂CO₃, THF,
80°, 15 h

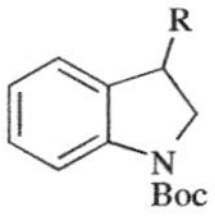

R	
H	(89)
Me	(96)
TBSOCH₂	(91)
BzOCH₂	(88)
Et (S)	(94)

607

C₀₋₇

H₂NCOR¹

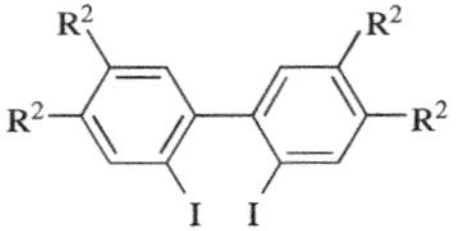

CuI (20 mol %),
DMCDA (20 mol %),
K₂CO₃, toluene, 110°, 48 h

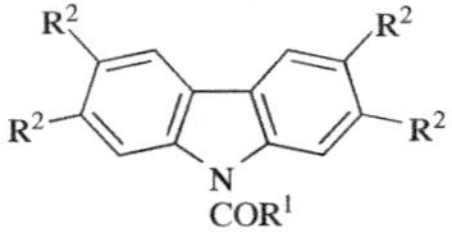

R¹	R²	
EtO	H	(58)
EtO	MeO	(47)
EtO	Me	(47)
n-Bu	H	(85)
n-Bu	MeO	(55)

R¹	R²	
Ph	Me	(41)
4-H₂NC₆H₄	H	(65)
4-MeOC₆H₄	H	(84)
Bn	H	(85)

610

TABLE 27. *N*-ARYLATIONS IN MULTI-STEP REACTIONS (*Continued*)

A. SYNTHESIS OF FIVE-MEMBERED RINGS (*Continued*)

Nitrogen Nucleophile	Electrophile	Conditions	Product(s) and Yield(s) (%)	Refs.

*Please refer to the charts preceding the tables for structures indicated by the **bold** numbers.*

C_0

H_2NBoc	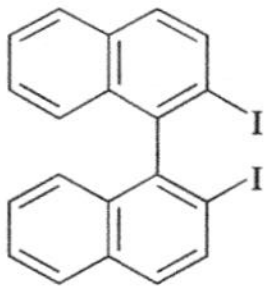	CuI (5 mol %), DMEDA (20 mol %), Cs$_2$CO$_3$, toluene, 110°, 24 h	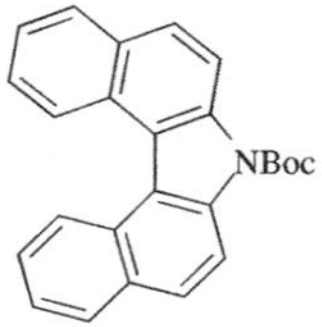 (75)	100
NaN$_3$	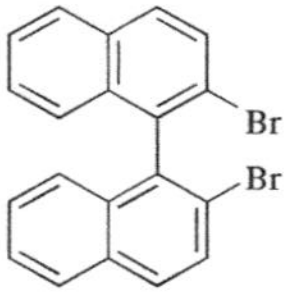	CuI (1 eq), DMEDA (1 eq), DMSO, 100°, 48 h	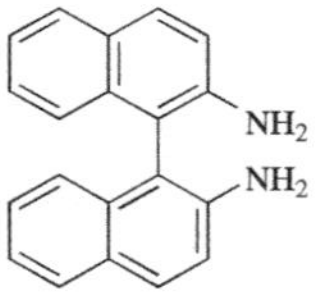 (48) 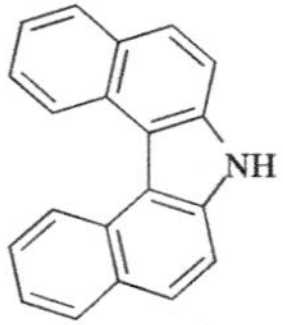 + (48)	156

C_1

H_2NNHMe	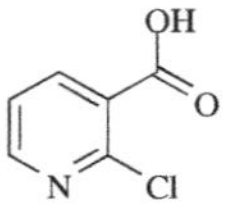	CuO (2 mol %), K$_2$CO$_3$, C$_6$H$_6$, 110°, 20 h	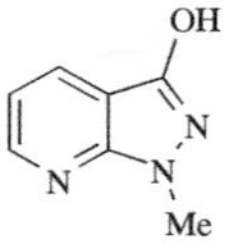 (54)	611
		CuO (2 mol %), K$_2$CO$_3$, C$_6$H$_6$, 110°, 20 h		611

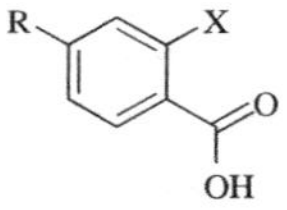

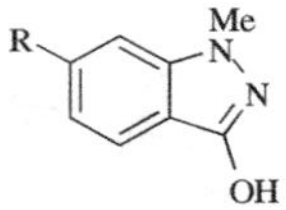

R	X	
H	Cl	(20)
H	Br	(22)
Cl	Cl	(16)
O$_2$N	Cl	(26)
MeO	Cl	(22)

C$_{1-6}$

H$_2$NNHR^1

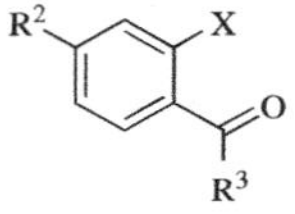

CuO (2 mol %), K$_2$CO$_3$, C$_6$H$_6$, 110°, 20 h

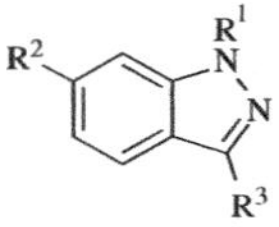

R^1	R^2	R^3	X	
Me	O$_2$N	Ph	Cl	(80)
Me	MeO	Me	F	(83)
HO(CH$_2$)$_2$	H	Me	F	(30)
t-Bu	H	Me	F	(32)
Ph	H	Me	F	(40)
Me	H	Me	F	(50)
Me	H	Me	Cl	(54)
Me	H	Et	F	(63)
Me	H	Ph	Cl	(43)
Me	Cl	Me	F	(30)

611

C$_1$

H$_2$NNHMe

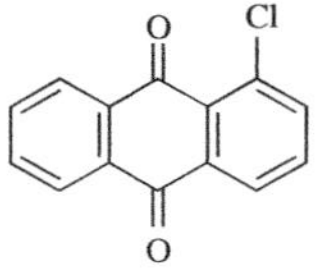

CuO (2 mol %), K$_2$CO$_3$, DMA/THF, 110°, 20 h

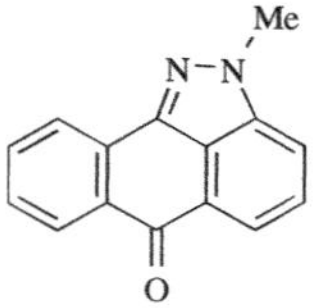

(43)

611

C$_2$

CNCH$_2$Ts

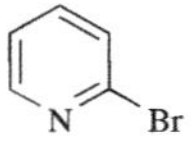

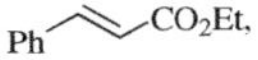

Ph⌒CO$_2$Et,
CuI (5 mol %),
DMEDA (10 mol %),
Cs$_2$CO$_3$, NaOt-Bu,
toluene, −30 to 110°, 5 h

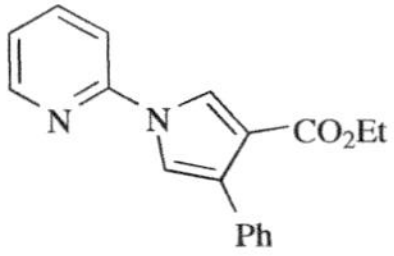

(62)

612

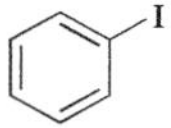

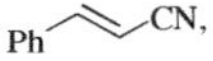

CNCH$_2$Ts

Ph⌒CN,
CuI (5 mol %),
DMEDA (10 mol %),
Cs$_2$CO$_3$, NaOt-Bu,
toluene, −30 to 110°, 8 h

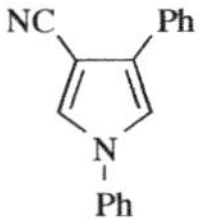

(70)

612

TABLE 27. N-ARYLATIONS IN MULTI-STEP REACTIONS (Continued)
A. SYNTHESIS OF FIVE-MEMBERED RINGS (Continued)

Nitrogen Nucleophile	Electrophile	Conditions	Product(s) and Yield(s) (%)	Refs.

*Please refer to the charts preceding the tables for structures indicated by the **bold** numbers.*

C2

CNCH₂Ts — Electrophile: PhI — Conditions: R^1⌒COR^2, CuI (5 mol %), DMEDA (10 mol %), Cs₂CO₃, NaOt-Bu, toluene, –30 to 110°

Product: pyrrole with R^1, COR^2, N–Ph

R^1	R^2	Time (h)		Refs.
Me	EtO	8	(45)	
EtO₂C	EtO	6	(52)	612
Ph	Me	10	(50)	
Ph	Ph	8	(68)	
2-BrC₆H₄	Ph	8	(50)	
4-O₂NC₆H₄	Ph	7	(62)	
4-MeC₆H₄	Ph	7	(62)	
Bn	EtO	8	(50)	

Electrophile: R–C₆H₄–X — Conditions: Ph⌒CO₂Et, CuI (5 mol %), DMEDA (10 mol %), Cs₂CO₃, NaOt-Bu, toluene, –30 to 110°

Product: pyrrole Ph, CO₂Et, N-aryl-R

R	X	Time (h)		Refs.
H	Br	24	(55)	
H	I	8	(50)	612
4-F	Br	24	(56)	
3-Cl	Br	5	(62)	
4-O₂N	I	5	(61)	
3-MeO	Br	5	(62)	
4-MeO	Br	24	(52)	
2-Me	Br	24	(0)	

C₂₋₇

Nitrogen Nucleophile: H₂N–C(=NH)–R — Electrophile: 1,2-diiodobenzene — Conditions: CuI (15 mol %), DMEDA (30 mol %), Cs₂CO₃, DMA

Products: **I** (2-R-benzimidazole) + **II** (amidine aryl)

R	Temp (°)	Time (h)	I	II	Refs.
NHMe	165	16	(36)	(0)	
NMe₂	130	16	(59)	(0)	
NHAc	165	16	(29)	(0)	
i-Pr	170	24	(66)	(22)	
t-Bu	170	24	(64)	(23)	
1-pyrrolidinyl	150	16	(47)	(0)	
2-furyl	170	24	(64)	(23)	141
4-pyridyl	170	24	(32)	(—)	
Ph	170	24	(64)	(23)	
3-O₂NC₆H₄	170	24	(45)	(31)	
3-(OCH₂O)-4-C₆H₃	170	24	(45)	(31)	
PhHN	130	16	(32)	(0)	

C₂₋₄

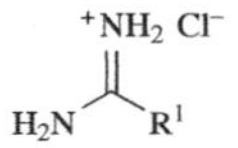

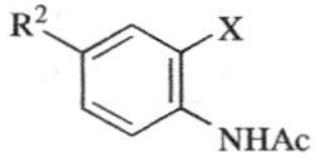

H_2N–C(R^1)=NH_2^+ Cl⁻

+

R²–C₆H₃(X)(NHAc)

CuBr (10 mol %), Cs₂CO₃,
DMSO, 90–120°, 72 h

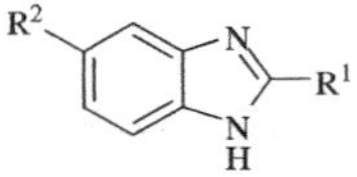

→

R^2-benzimidazole (R^1, NH)

R^1	R^2	X	
Me	H	Br	(82)
Me	H	I	(80)
Me	Br	Br	(89)
Me	O₂N	Br	(74)
Me	Me	Br	(75)
n-Pr	H	Br	(63)
n-Pr	H	I	(65)
n-Pr	Br	Br	(65)
n-Pr	O₂N	Br	(80)
n-Pr	Me	Br	(62)
c-Pr	H	Br	(65)
c-Pr	Br	Br	(62)
c-Pr	Me	Br	(70)

613

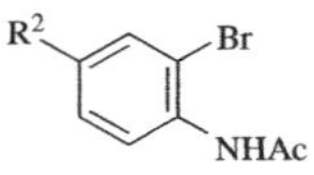

Br–C₆H₃(R²)(NHAc)

1. CuBr (10 mol %), Cs₂CO₃,
   DMSO, 70–120°, 72 h
2. ArI, **L46** (20 mol %)

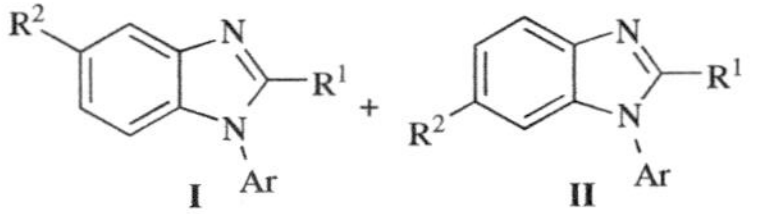

→ **I** + **II**

613

R^1	R^2	Ar	**I**	**II**
Me	H	Ph	(65)	(—)
Me	H	4-ClC₆H₄	(70)	(—)
Me	H	3-O₂NC₆H₄	(49)	(—)
Me	H	4-MeC₆H₄	(62)	(—)

R^1	R^2	Ar	**I**	**II**
Me	Br	Ph	(62)	(0)
Me	Me	Ph	(33)	(24)
Me	Me	4-ClC₆H₄	(33)	(27)
n-Pr	H	Ph	(65)	(—)
n-Pr	H	4-ClC₆H₄	(57)	(—)

C₂

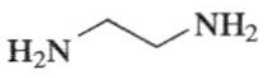

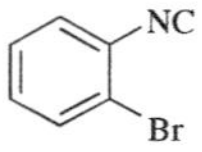

H_2N–CH₂CH₂–NH_2

+

NC–C₆H₄–Br

CuBr (5 mol %),
phen (10 mol %), Cs₂CO₃,
DMF, 20–90°, 16 h

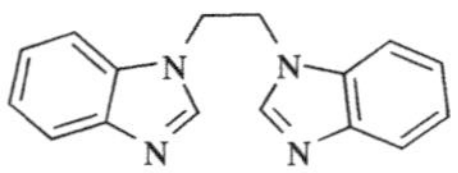

→

(42)

614

TABLE 27. N-ARYLATIONS IN MULTI-STEP REACTIONS (Continued)
A. SYNTHESIS OF FIVE-MEMBERED RINGS (Continued)

*Please refer to the charts preceding the tables for structures indicated by the **bold** numbers.*

Nitrogen Nucleophile	Electrophile	Conditions	Product(s) and Yield(s) (%)	Refs.
C_{2-6}				
H_2NR^1	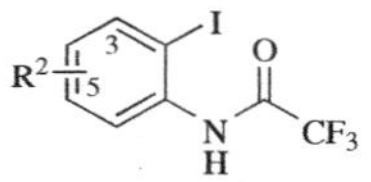	CuI (10 mol %), L-Pro (20 mol %), K_2CO_3, DMSO, rt	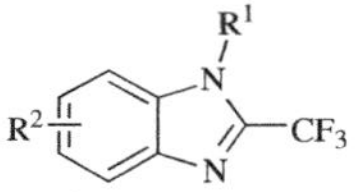	251

R^1	R^2	Time (h)	
$HOCH_2CH_2$	H	24	(92)
$HOCH_2CH_2$	6-MeO	15	(89)
$HOCH_2CH_2$	$6\text{-}MeO_2C$	24	(72)
$HOCH_2CH_2$	$4,6\text{-}Me_2$	17	(92)
$HOCH_2CH_2$	6-Ac	24	(90)
allyl	H	10	(92)
$n\text{-}C_6H_{13}$	H	24	(90)

Nitrogen Nucleophile	Electrophile	Conditions	Product(s) and Yield(s) (%)	Refs.
C_{2-9}				
H_2NR^1	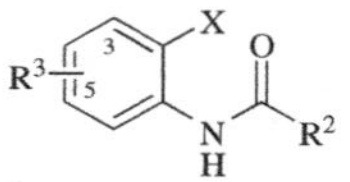	1. CuI (10 mol %), L-Pro (20 mol %), K_2CO_3, DMSO, temp 1, time 1 2. AcOH, temp 2, time 2	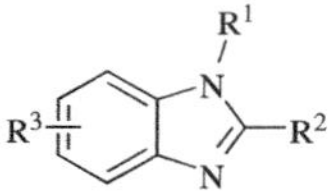	251

R^1	R^2	R^3	X	Temp 1 (°)	Time 1 (h)	Temp 2 (°)	Time 2 (h)	
$HOCH_2CH_2$	Me	H	I	40	2	40	4	(75)
$HOCH_2CH_2$	CF_3	H	Br	40	36	70	12	(90)
allyl	CF_3	H	Br	50	10	50	6	(80)
allyl	CF_3	H	I	40	2	40	4	(81)
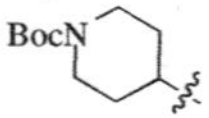 BocN	CF_3	4-Ac	Br	rt	24	50	1	(85)
$n\text{-}C_6H_{13}$	CF_3	H	Br	rt	10	50	2	(80)
$n\text{-}C_6H_{13}$	Bn	H	I	40	6	40	10	(90)
$c\text{-}C_6H_{11}$	CF_3	H	Br	rt	10	50	5	(75)
$c\text{-}C_6H_{11}$	CF_3	H	I	rt	24	50	5	(94)

R^1	R^2		X			Time 1	Time 2	
c-C$_6$H$_{11}$	2-furyl	H	I	40	60	10	12	(88)
c-C$_6$H$_{11}$	2-furyl	5-MeO$_2$C	I	40	60	8	12	(75)
c-C$_6$H$_{11}$	2-furyl	4-Ac	Br	50	80	12	12	(70)
c-C$_6$H$_{11}$	Ph	4-MeO	Br	50	90	10	10	(62)
c-C$_6$H$_{11}$	Bn	H	Br	45	50	10	6	(70)
(S)-BnCH(CO$_2$Me)	CF$_3$	H	I	40	70	36	12	(61)

Br–[R^2 ring, positions 4,6]–NHCO$_2$Me

1. CuI (10 mol %),
   4-HOPro (20 mol %),
   K$_2$CO$_3$, DMSO, 70°, time 1
2. 130°, time 2

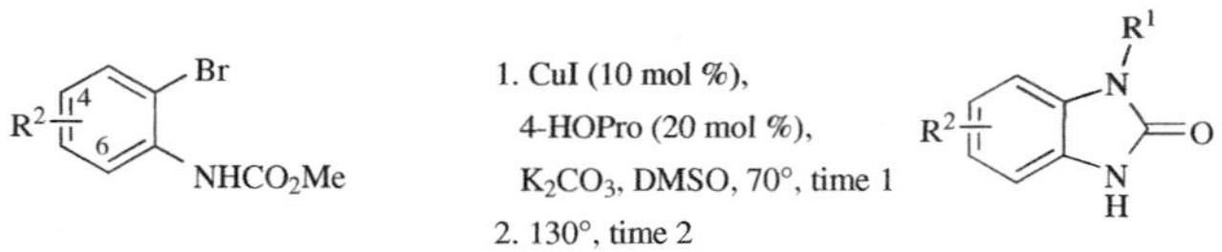

[product: benzimidazolone with R^1 on N, R^2 on ring, N–H]

R^1	R^2	Time 1 (h)	Time 2 (h)	R^1	R^2	Time 1 (h)	Time 2 (h)
HO$_2$CCH$_2$	H	5	4 (73)	n-C$_6$H$_{13}$	4-Ac	10	9 (70)
i-Pr	4-TBSOCH$_2$	11	14 (75)	c-C$_6$H$_{11}$	4-n-PrNHCO	21	12 (66)
allyl	4-MeO	11	8 (62)	Bn	H	4	6 (77)
2-furylmethyl	4-O$_2$N	4	10 (82)	Bn	4-MeO$_2$C	12	12 (74)
2-furylmethyl	4-n-PrHNCO	5	4 (82)	(S)-Bn(CONHMe)CH	4,6-Me$_2$	22	24 (76)

Nitrogen Nucleophile	Electrophile	Conditions	Product(s) and Yield(s) (%)	Refs.

*Please refer to the charts preceding the tables for structures indicated by the **bold** numbers.*

C₂₋₇

Nitrogen Nucleophile: H_2NR^1

Electrophile: (aryl with I, N=C(Cl)–CF_2R^2, R^3 at positions 14, 15)

Conditions: CuI (10 mol %), K_2CO_3, DMF

Product: (indazole with R^1, CF_2R^2, R^3); Refs. 616

R^1	R^2	R^3	Temp (°)		R^1	R^2	R^3	Temp (°)		R^1	R^2	R^3	Temp (°)
$HOCH_2CH_2$	F	H	40 (87)		n-Bu	F	4-O_2N	80 (73)		4-$O_2NC_6H_4$	F	H	80 (64)
$HOCH_2CH_2$	Br	H	60 (68)		n-Bu	F	5-O_2N	80 (76)		4-$O_2NC_6H_4$	Br	H	80 (58)
allyl	F	H	60 (70)		n-Bu	Br	4-Me	80 (68)		4-$MeOC_6H_4$	F	H	80 (91)
allyl	Br	H	60 (68)		c-C_6H_{11}	F	H	80 (69)		4-$MeOC_6H_4$	Br	H	60 (67)
n-Bu	F	H	rt (78)		c-C_6H_{11}	Br	H	60 (76)		4-MeC_6H_4	F	H	60 (93)
n-Bu	Br	H	60 (76)		4-ClC_6H_4	F	H	80 (62)		4-MeC_6H_4	Br	H	60 (74)
n-Bu	Br	4-Cl	80 (91)		4-ClC_6H_4	Br	H	80 (31)		Bn	Br	H	60 (82)

C₃₋₁₂

Nitrogen Nucleophile: H_2NR^1

Electrophile: (aryl with I, $NHCO_2Me$, R^2 at positions 14, 15)

Conditions: 1. CuI (20 mol %), 4-HOPro (40 mol %), K_3PO_4, DMSO, 50°, time 1 2. 130°, time 2

Product: (benzimidazolone with R^1, R^2, =O); Refs. 615

R^1	R^2	Time 1 (h)	Time 2 (h)		R^1	R^2	Time 1 (h)	Time 2 (h)
allyl	H	4	(82)		Bn	H	3	3 (81)
allyl	4-Br	7	(66)		Bn	5-MeO	7	3 (80)
2-furylmethyl	4-Ac	12	(82)		Ph (4-cyclohexyl)	H	7	3 (85)
BocN-(piperidin-4-yl)	H	7	(74)					

C₃₋₁₀

Nitrogen Nucleophile: H_2NR

Electrophile: (thiophene with NC, Br, S)

Conditions: CuBr (5 mol %), phen (10 mol %), Cs_2CO_3, DMF, 20–90°, 16 h

Product: (thienopyrimidine with N, N, R, S); Refs. 614

R	Yield
$BnO(CH_2)_3$	(44)
Bn	(49)

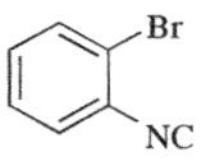

Br / NC

CuBr (5 mol %),
phen (10 mol %), Cs$_2$CO$_3$,
DMF: 20–90°, 16 h

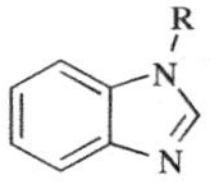

$R-N$ benzimidazole

R	
n-Pr	(65)
c-Pr	(40)
BnO(CH$_2$)$_3$	(66)
2-furylmethyl	(46)
c-C$_6$H$_{11}$	(46)
4-MeC$_6$H$_4$	(41)
Bn	(70)
2-MeOC$_6$H$_4$CH$_2$	(67)
3,5-(MeO)$_2$C$_6$H$_3$CH$_2$	(65)
4-CF$_3$C$_6$H$_4$CH$_2$	(55)
3-indolylethyl	(59)

C$_{3–8}$

HNR^{1}R^2

R^3 X / =N–N–Ar

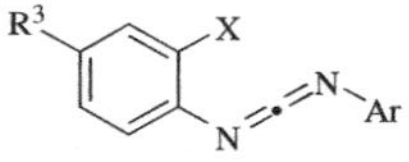

CuI (10 mol %),
additive (20 mol %),
Cs$_2$CO$_3$, dioxane, 80°

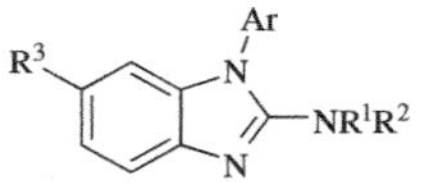

R^3 benzimidazole N-Ar, NR^{1}R^2

R^1	R^2	R^3	Ar	X	Additive	Temp (°)	Time (h)
Me	HOCH$_2$CH$_2$	Me	Ph	I	L-Pro	80	30 (78)
H	c-C$_6$H$_{11}$	H	4-MeOC$_6$H$_4$	I	phen	80	20 (86)
H	Bn	H	4-MeOC$_6$H$_4$	I	phen	80	20 (88)
H	(S)-PhMeCH	H	Ph	I	phen	80	20 (91)
Me	Bn	Me	Ph	Br	phen	80	20 (85)
n-Bu	n-Bu	Me	Ph	I	L-Pro	70	30 (83)

TABLE 27. N-ARYLATIONS IN MULTI-STEP REACTIONS (*Continued*)

A. SYNTHESIS OF FIVE-MEMBERED RINGS (*Continued*)

*Please refer to the charts preceding the tables for structures indicated by the **bold** numbers.*

Nitrogen Nucleophile	Electrophile	Conditions	Product(s) and Yield(s) (%)	Refs.

C₃₋₄

Nitrogen Nucleophile: imidazole (R¹, N–H)

Electrophile: R^2, X, N=•=N–R³

Conditions: CuI (10 mol %), phen (20 mol %), Cs₂CO₃, dioxane

Electrophile:

R¹	R²	R³	X	Temp (°)	Time (h)	
H	H	Ph	Br	80	20	(90)
H	H	4-MeOC₆H₄	I	100	24	(85)
H	H	4-MeC₆H₄	I	80	20	(86)
H	H	Bn	I	80	20	(92)
H	Cl	Ph	I	80	20	(82)
H	Me	Ph	Br	100	24	(90)

Product:

R¹	R²	R³	X	Temp (°)	Time (h)	
H	Me	Ph	I	80	24	(91)
H	Me	3-ClC₆H₄	Br	80	24	(87)
H	Me	4-MeC₆H₄	I	80	20	(86)
Me	H	4-MeC₆H₄	I	100	36	(73)
Me	Cl	Ph	Br	100	36	(75)

Refs. 183

C₄₋₈

Nitrogen Nucleophile: H₂NR

Electrophile: X, Cl, C(CF₃)=N

Conditions: CuI (10 mol %), TMEDA (20 mol %), Cs₂CO₃, toluene, 110°, 8 h

Product: N–R, CF₃

R	X		R	X	
n-Bu	I	(79)	Bn	Cl	(76)
4-ClC₆H₄	I	(92)	Bn	Br	(89)
4-O₂NC₆H₄	I	(62)	2-FC₆H₄CH₂	I	(89)
2-MeOC₆H₄	I	(70)	4-FC₆H₄CH₂	I	(95)
3-MeOC₆H₄	I	(73)	2-ClC₆H₄CH₂	I	(98)
4-MeOC₆H₄	I	(86)	4-ClC₆H₄CH₂	I	(88)
2-MeC₆H₄	I	(76)	4-MeOC₆H₄CH₂	I	(93)
4-MeC₆H₄	Cl	(70)	4-MeC₆H₄CH₂	I	(98)
4-MeC₆H₄	Br	(77)	PhMeCH	I	(83)

Refs. 617

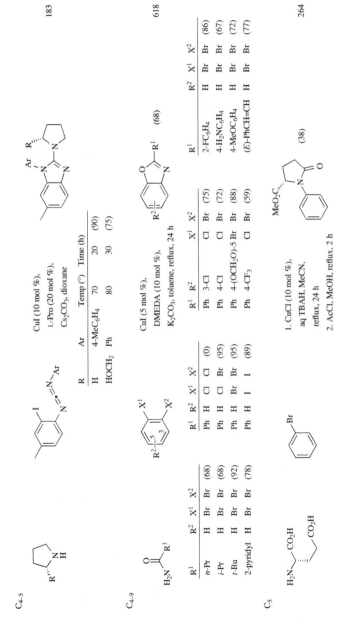

C4-5

CuI (10 mol %),
L-Pro (20 mol %),
Cs_2CO_3, dioxane

R	Ar	Temp (°)	Time (h)	
H	4-MeC$_6$H$_4$	70	20	(90)
HOCH$_2$	Ph	80	30	(75)

183

C4-9

CuI (5 mol %),
DMEDA (10 mol %),
K_2CO_3, toluene, reflux, 24 h

R^1	R^2	X^1	X^2	
Ph	H	Cl	Cl	(0)
Ph	H	Cl	Br	(95)
Ph	H	Br	Br	(95)
Ph	H	I	I	(89)

R^1	R^2	X^1	X^2	
Ph	3-Cl	Cl	Br	(75)
Ph	4-Cl	Cl	Br	(72)
Ph	4-(OCH$_2$O)-5-	Br	Br	(88)
Ph	4-CF$_3$	Cl	Br	(59)

(68)

R^1	R^2	X^1	X^2	
2-FC$_6$H$_4$	H	Br	Br	(86)
4-H$_2$NC$_6$H$_4$	H	Br	Br	(67)
4-MeOC$_6$H$_4$	H	Br	Br	(72)
(E)-PhCH=CH	H	Br	Br	(77)

R^1	R^2	X^1	X^2	
n-Pr	H	Br	Br	(68)
i-Pr	H	Br	Br	(68)
t-Bu	H	Br	Br	(92)
2-pyridyl	H	Br	Br	(78)

618

C5

1. CuCl (10 mol %),
aq TBAH, MeCN,
reflux, 24 h
2. AcCl, MeOH, reflux, 2 h

(38)

264

Nitrogen Nucleophile	Electrophile	Conditions	Product(s) and Yield(s) (%)	Refs.

*Please refer to the charts preceding the tables for structures indicated by the **bold** numbers.*

C$_{5-12}$

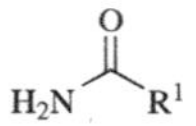

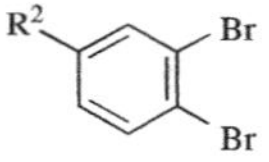

Nucleophile: H_2N–C(=O)–R^1

Electrophile: dibromobenzene bearing R^2, 2 Br

Conditions: CuI (10 mol %), DMEDA (20 mol %), K$_3$PO$_4$, toluene, 110°, 48 h

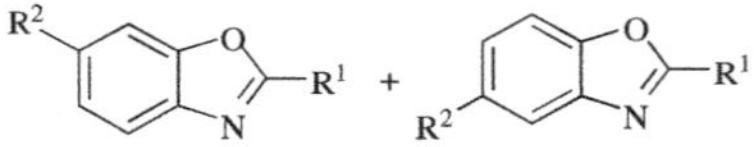

Product: benzoxazole **I** + **II**

R^1	R^2	I	II
2-thienyl	H	(58)	(—)
Ph	H	(74)	(—)
Ph	Me	(36)	(36)
2-MeC$_6$H$_4$	H	(58)	(—)
n-C$_{11}$H$_{23}$	H	(77)	(—)

Refs. 619

C$_{5-8}$

Nucleophile: pyridine (I, NHMe, SBn)

Conditions: CuI (10 mol %), DMCDA (40 mol %), Cs$_2$CO$_3$, dioxane, 90°

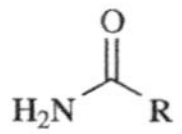

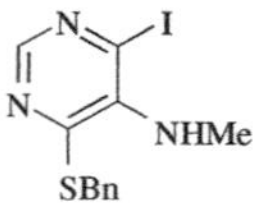

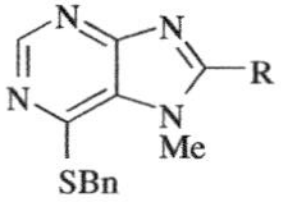

R	
n-Bu	(45)
2-pyridyl	(66)
3-pyridyl	(61)
Ph	(61)
2-FC$_6$H$_4$	(73)
3-FC$_6$H$_4$	(67)
4-FC$_6$H$_4$	(70)
3-ClC$_6$H$_4$	(68)
4-ClC$_6$H$_4$	(75)
3-O$_2$NC$_6$H$_4$	(68)
4-O$_2$NC$_6$H$_4$	(75)
3-MeOC$_6$H$_4$	(75)
4-MeOC$_6$H$_4$	(65)
2-MeC$_6$H$_4$	(80)

Refs. 620

C$_5$

Nucleophile: H_2N–pyridine

Electrophile: dibromopyridine

Conditions: Cu (7.5 mol %), phen (15 mol %), Cs$_2$CO$_3$, DME, 95°, 17 h

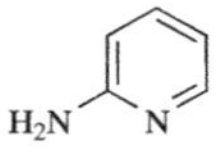

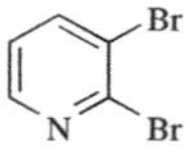

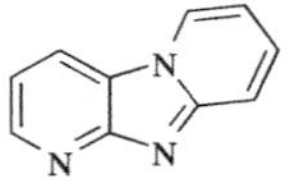

Product: (95)

Refs. 621

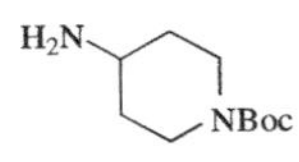

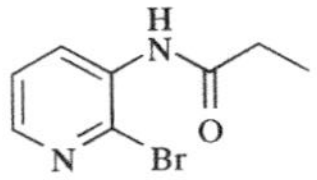

(76)

(16)

1. CuI (10 mol %),
   L-Pro (20 mol %),
   K$_2$CO$_3$, DMSO, 45°, 6 h
2. 150°, 12 h

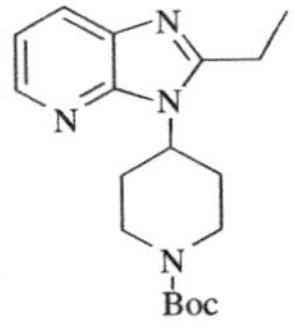

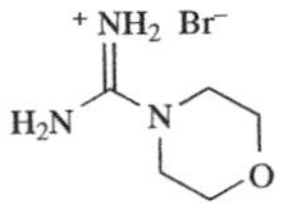

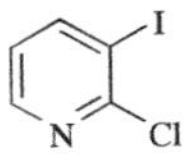

CuI (15 mol %),
DMEDA (30 mol %),
Cs$_2$CO$_3$, DMA, 150°, 16 h

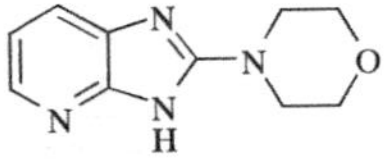

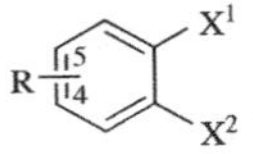

CuI (x mol %), Cs$_2$CO$_3$

R	X^1	X^2	x	Additive	Solvent	Temp (°)	Time	
H	Cl	Cl	15	DMEDA (30 mol %)	DMA	150	16 h	(0)
H	Br	Cl	15	DMEDA (30 mol %)	DMA	150	16 h	(51)
H	I	Cl	15	DMEDA (30 mol %)	DMA	150	16 h	(25)
H	I	Br	15	DMEDA (30 mol %)	DMA	150	16 h	(76)
H	Br	Br	15	DMEDA (30 mol %)	DMA	150	16 h	(40)
H	I	I	15	DMEDA (30 mol %)	DMA	150	16 h	(52)
4-Cl	I	Cl	15	DMEDA (30 mol %)	DMA	150	16 h	(26)
4-O$_2$N	I	Cl	20	DBU	DMSO	MW, 130	10 min	(15)
5-Me	I	Br	15	DMEDA (30 mol %)	DMA	165	16 h	(53)
5-CF$_3$	I	Cl	15	DMEDA (30 mol %)	DMA	165	16 h	(18)

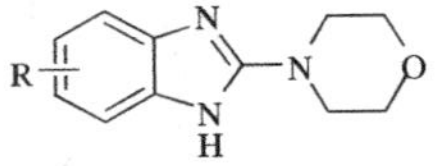

TABLE 27. N-ARYLATIONS IN MULTI-STEP REACTIONS (*Continued*)

A. SYNTHESIS OF FIVE-MEMBERED RINGS (*Continued*)

Nitrogen Nucleophile	Electrophile	Conditions	Product(s) and Yield(s) (%)	Refs.

*Please refer to the charts preceding the tables for structures indicated by the **bold** numbers.*

C_5

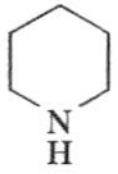

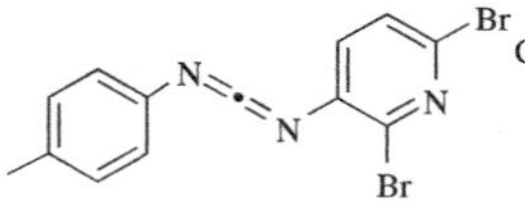

| | | CuI (10 mol %), DMG (20 mol %), K$_3$PO$_4$, DMSO, 90°, 24 h | (92) | 183 |
| | | CuI (10 mol %), additive (20 mol %), Cs$_2$CO$_3$, dioxane | 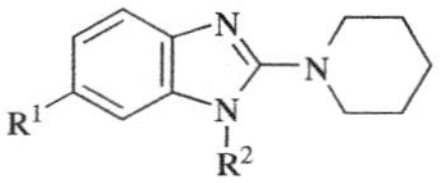 | 183 |

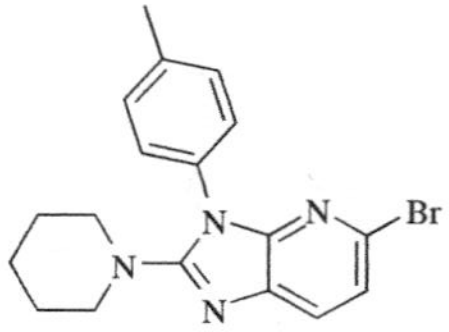

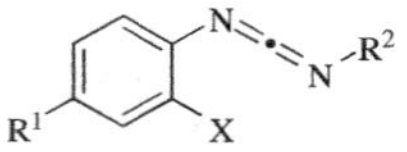

R^1	R^2	X	Additive	Temp (°)	Time (h)	
H	Bn	I	phen	80	20	(92)
Cl	Ph	I	L-Pro	70	20	(93)
O$_2$N	4-MeC$_6$H$_4$	I	L-Pro	70	30	(59)
Me	Ph	Br	phen	80	20	(91)
Me	Ph	I	L-Pro	70	20	(95)
Me	3-ClC$_6$H$_4$	I	L-Pro	70	20	(84)
Me	4-MeOC$_6$H$_4$	I	L-Pro	70	30	(96)
Me	4-MeC$_6$H$_4$	I	L-Pro	80	30	(68)
Me	3-CF$_3$C$_6$H$_4$	I	L-Pro	80	30	(70)

| | | CuI (15 mol %), DMEDA (30 mol %), Cs$_2$CO$_3$ | | 143 |

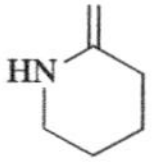

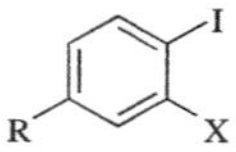

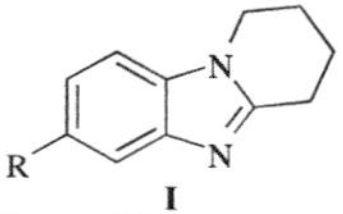

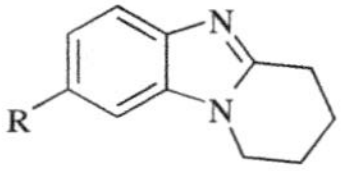

R	X	Solvent	Temp (°)	Time (h)	I	II
H	I	DMA	170	24	(52) (I + II)	

C₆₋₇

H_2NR^1

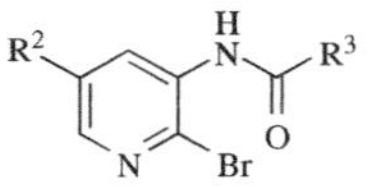

1. CuI (10 mol %),
   L-Pro (20 mol %), K₂CO₃,
   DMSO, temp, time
2. See table.

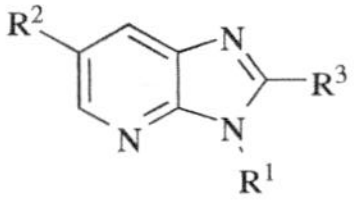

R¹	R²	R³	Temp (°)	Time (h)	Conditions 2
$n\text{-}C_6H_{13}$	H	CF₃	50	10	AcOH, 90°, 4 h (72)
Bn	H	Et	45	7	150°, 12 h (78)
Bn	Br	CF₃	40	7	AcOH, 90°, 4 h (78)

C₆₋₁₂

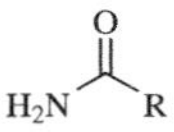

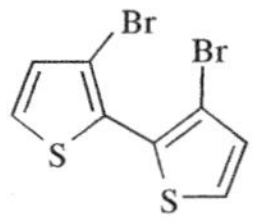

CuI (5 mol %),
DMEDA (20 mol %),
K₂CO₃, toluene,
reflux, 24 h

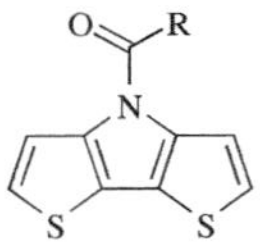

R	
$n\text{-}C_5H_{11}$	(24)
$c\text{-}C_6H_{11}$	(24)
Ph	(36)
$n\text{-}C_7H_{15}$	(38)
$n\text{-}C_{11}H_{23}$	(19)

C₆₋₉

$O=C(R)-NH_2$

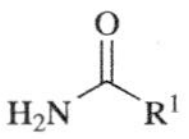

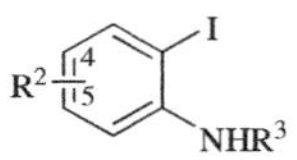

1. CuI (5 mol %),
   DMEDA (10 mol %),
   Cs₂CO₃, dioxane,
   110°, 17 h
2. See table.

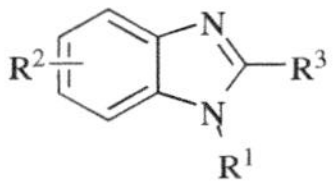

R¹	R²	R³	Conditions 2
$n\text{-}C_5H_{11}$	H	$c\text{-}C_6H_{11}CH_2$	AcOH, 75° (85)
$n\text{-}C_5H_{11}$	4-Me	Et	K₃PO₄, t-BuOH, 110° (68)
$n\text{-}C_5H_{11}$	5-Me	Et	K₃PO₄, t-BuOH, 110° (76)
$n\text{-}C_5H_{11}$	4-CF₃	Et	K₃PO₄, t-BuOH, 110° (75)
$n\text{-}C_5H_{11}$	5-CF₃	Et	K₃PO₄, t-BuOH, 110° (75)
$c\text{-}C_6H_{11}$	H	$c\text{-}C_6H_{11}CH_2$	AcOH, 75° (80)

R¹	R²	R³	Conditions 2
Ph	H	Me	K₃PO₄, t-BuOH, 110° (80)
Ph	H	$c\text{-}C_6H_{11}CH_2$	K₃PO₄, t-BuOH, 110° (68)
Ph	4-Me	Et	K₃PO₄, t-BuOH, 110° (85)
Ph	5-Me	Et	K₃PO₄, t-BuOH, 110° (79)
Ph	4-CF₃	Et	K₃PO₄, t-BuOH, 110° (82)
Ph	5-CF₃	Et	K₃PO₄, t-BuOH, 110° (82)
(E)-PhCH=CH	H	$c\text{-}C_6H_{11}CH_2$	AcOH, 75° (60)

*Please refer to the charts preceding the tables for structures indicated by the **bold** numbers.*

C₆

Nitrogen Nucleophile	Electrophile	Conditions	Product(s) and Yield(s) (%)	Refs.
NH_2Ar	CO₂H, Br	Cu₂O (30 mol %), K₂CO₃, 240°, 4 h	Ar: 3-ClC₆H₄ (75) 3-MeOC₆H₄ (82)	326
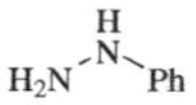 H₂N–N–Ph (H on N)	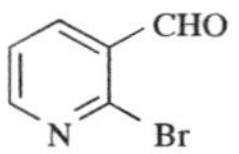 CHO, Br, N	1. NMP, MW, 160°, 10 min 2. CuI (5 mol %), DMCDA (10 mol %), K₂CO₃, MW, 160°, 10 min	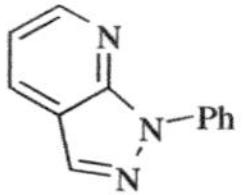 (95)	624
	X, CHO, R (positions 14, 15)	1. NMP, MW, 160°, 10 min 2. CuI (5 mol %), DMCDA (10 mol %), K₂CO₃, MW, 10 min	R, Ph	624

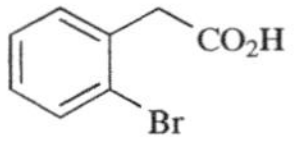

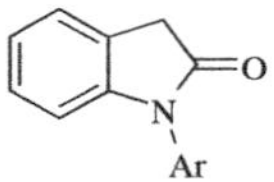

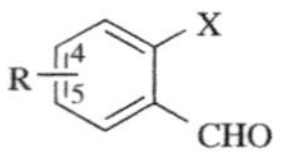

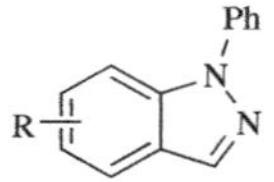

R	X	Temp (°)	
H	Cl	160	(87)
H	Br	160	(91)
H	I	120	(87)
5-F	Br	160	(84)
5-O₂N	Br	120	(95)
4,5-(MeO)₂	Br	160	(82)

Nitrogen Nucleophile	Electrophile	Conditions	Product(s) and Yield(s) (%)	Refs.
Cl, H₂N–	Br, CHO, F	1. NMP, MW, 160°, 10 min 2. CuI (5 mol %), DMCDA (10 mol %), K₂CO₃, MW, 160°, 10 min	Cl, F (82)	624
H₂N–N–Ph (H on N)	R¹, O, X, R² (positions 14, 15)	1. NMP, MW, 160°, 10 min 2. CuI (5 mol %), DMCDA (10 mol %), K₂CO₃, MW, 10 min	R¹, R², Ph	624

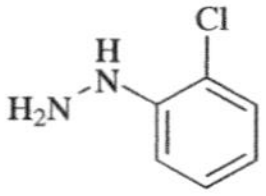

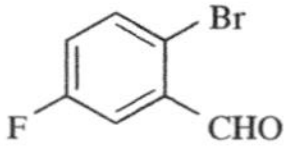

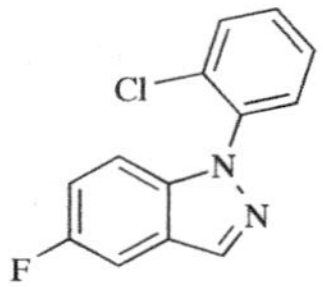

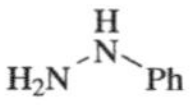

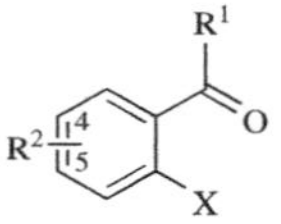

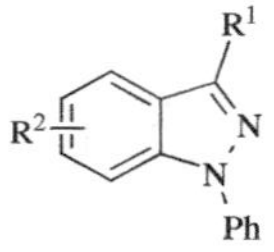

R¹	R²	X	Temp (°)	
Me	H	Br	160	(75)
Me	5-CF₃	Cl	160	(78)
Et	4-Cl	Cl	120	(69)

C_{6–10}

H₂NAr

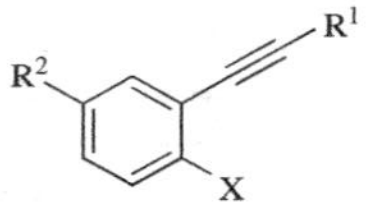

CuI (10 mol %), KO*t*-Bu,
toluene, 105°

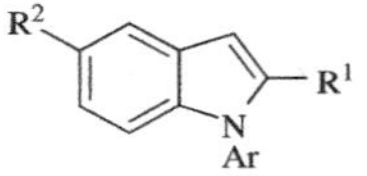

Ar	R¹	R²	X	Time (h)		
Ph	n-C₆H₁₃	H	Br	2	(84)	626, 609
4-ClC₆H₄	n-C₆H₁₃	H	Br	12	(70)	609
2-MeOC₆H₄	n-Bu	H	Br	12	(69)	626, 609
4-MeOC₆H₄	n-C₆H₁₃	H	Br	12	(60)	609
2-Cl, 5-MeOC₆H₃	n-C₆H₁₃	H	Br	12	(60)	609
4-MeC₆H₄	n-Bu	H	Br	2	(75)	626, 609
4-MeC₆H₄	n-C₆H₁₃	H	Br	12	(67)	609
4-MeC₆H₄	n-C₆H₁₃	CF₃	Cl	2	(53)	626, 609
4-MeC₆H₄	Ph	H	Cl	2	(21)	626
2-Cl, 4-MeC₆H₃	n-Bu	H	Br	12	(72)	609
1-Np	n-C₆H₁₃	H	Br	12	(77)	609

C₇

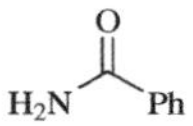

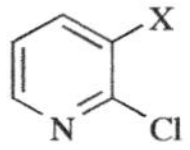

CuI (5 mol %),
DMEDA (10 mol %),
K₂CO₃, toluene,
reflux, 24 h

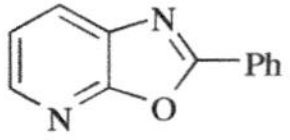

X	
Cl	(0)
Br	(77)

618

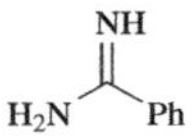

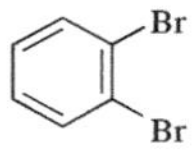

CuI (10 mol %),
DMEDA (20 mol %),
K₃PO₄, xylenes,
150°, 48 h

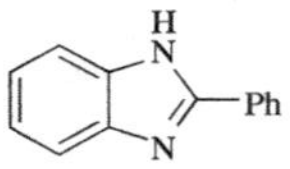

(27)

619

Nitrogen Nucleophile	Electrophile	Conditions	Product(s) and Yield(s) (%)	Refs.

*Please refer to the charts preceding the tables for structures indicated by the **bold** numbers.*

C7-15

Nitrogen Nucleophile: $R^1-N=C=N-R^2$

Electrophile: 2-iodophenol (R^3 substituted)

Conditions: CuI (x mol %), 24 h

Product: benzoxazol-2-imine

R^1	R^2	R^3	x	Additive(s)	Solvent	Temp (°)	
i-Pr	i-Pr	H	15	Cs_2CO_3	MeCN	100	(—)
i-Pr	Ph	H	15	Cs_2CO_3	MeCN	100	(83)
Ph	Ph	H	15	Cs_2CO_3	MeCN	100	(93)
Ph	Ph	F	15	Cs_2CO_3	MeCN	100	(85)
Ph	Ph	Cl	15	Cs_2CO_3	MeCN	100	(86)
Ph	Ph	Me	10	sparteine (20 mol %), K_3PO_4	NMP	130	(92)
Ph	Ph	t-Bu	15	Cs_2CO_3	MeCN	100	(85)
4-ClC$_6$H$_4$	4-ClC$_6$H$_4$	H	15	Cs_2CO_3	MeCN	100	(83)
4-ClC$_6$H$_4$	4-ClC$_6$H$_4$	Cl	10	sparteine (20 mol %), K_3PO_4	NMP	130	(74)
4-CF$_3$OC$_6$H$_4$	4-CF$_3$OC$_6$H$_4$	H	15	Cs_2CO_3	MeCN	100	(83)
2-MeC$_6$H$_4$	2-MeC$_6$H$_4$	H	10	sparteine (20 mol %), K_3PO_4	NMP	130	(75)
2-MeC$_6$H$_4$	2-MeC$_6$H$_4$	Cl	15	Cs_2CO_3	MeCN	100	(75)
4-MeC$_6$H$_4$	4-MeC$_6$H$_4$	H	15	Cs_2CO_3	MeCN	100	(89)
4-MeC$_6$H$_4$	4-MeC$_6$H$_4$	t-Bu	15	Cs_2CO_3	MeCN	100	(83)
3-CF$_3$C$_6$H$_4$	3-CF$_3$C$_6$H$_4$	H	15	Cs_2CO_3	MeCN	100	(78)

Refs.: 184

C7

Nitrogen Nucleophile: MeHN—Ph

Conditions: CuI (10 mol %), phen (20 mol %), Cs_2CO_3, dioxane, 85°, 18 h

Product: (72)

Refs.: 183

C8

Nitrogen Nucleophile: H$_2$N—CH$_2$CH$_2$—Ph

Conditions: CuI (20 mol %), 4-HOPro (40 mol %),

Product: (62)

Refs.: 615

C_8

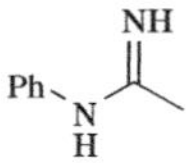

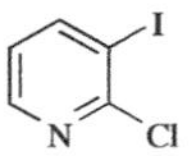

Ph–NH–C(CH$_3$)=NH + 2-chloro-3-iodopyridine

CuI (15 mol %),
DMEDA (30 mol %),
Cs$_2$CO$_3$, DMA, 150°, 16 h

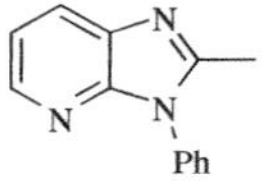

product (imidazo[4,5-b]pyridine, N-Ph, 2-Me)   (39)

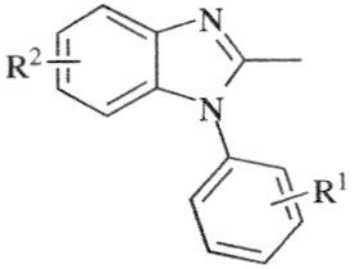

C_{8-9}

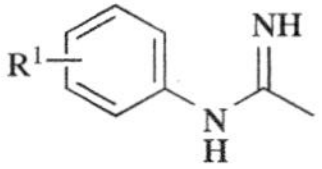

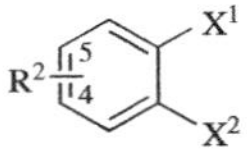

R^1-C$_6$H$_4$–NH–C(CH$_3$)=NH + dihalobenzene (X^1, X^2, R^2)

CuI (15 mol %),
DMEDA (30 mol %),
Cs$_2$CO$_3$

R^1	R^2	X^1	X^2	Solvent	Temp (°)	Time (h)	
H	5-CF$_3$	I	Cl	NMP	150	16	(24)
H	4-NC–	I	Br	NMP	130	16	(31)
H	3,5-Me$_2$	I	I	NMP	150	24	(51)
4-MeO	4-Me	I	Br	NMP	150	16	(48)
2-Me	4-Me	I	Br	NMP	150	16	(49)
4-NC–	4-Me	I	Br	DMA	100	16	(41)

R^1	R^2	X^1	X^2	Solvent	Temp (°)	Time (h)	
H	3-Cl	I	Cl	NMP	150	24	(24)
H	4-Cl	I	Cl	NMP	150	16	(41)
H	4-O$_2$N	I	Cl	NMP	150	16	(0)
H	3-Me	Br	Cl	NMP	150	24	(10)
H	4-Me	I	Br	NMP	150	24	(53)
H	5-Me	I	Br	NMP	150	24	(57)
H	6-Me	I	Cl	NMP	150	24	(38)

C_8

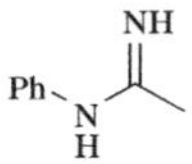

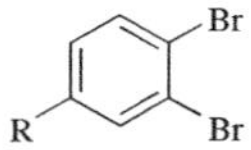

Ph–NH–C(CH$_3$)=NH + 2,4-dibromo(R)benzene

CuI (15 mol %),
DMEDA (30 mol %),
Cs$_2$CO$_3$, NMP,
150°, 24 h

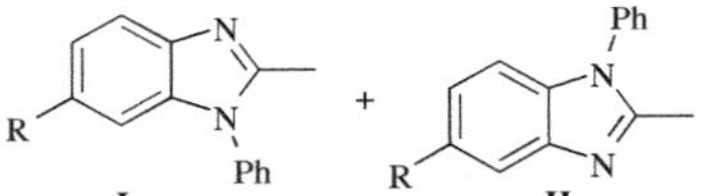

I + II

R	I	II
Me	(6)	(5)
t-Bu	(8)	(6)

TABLE 27. *N*-ARYLATIONS IN MULTI-STEP REACTIONS (*Continued*)

A. SYNTHESIS OF FIVE-MEMBERED RINGS (*Continued*)

Nitrogen Nucleophile	Electrophile	Conditions	Product(s) and Yield(s) (%)	Refs.

*Please refer to the charts preceding the tables for structures indicated by the **bold** numbers.*

C$_{8-14}$

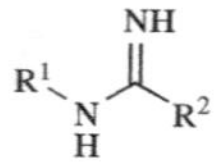

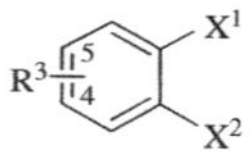

Nitrogen Nucleophile: R^1–N(H)–C(=NH)–R^2

Electrophile: R^3 benzene with X^1, X^2

Conditions: CuI (15 mol %), DMEDA (30 mol %), Cs$_2$CO$_3$, 150°, 16 h

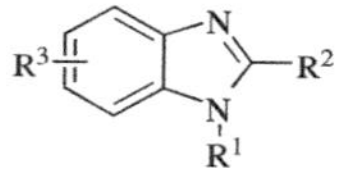

Product: benzimidazole (R^3, R^2, R^1)

R^1	R^2	R^3	X^1	X^2	Solvent	Temp (°)	Time (h)	Yield	Refs.
Me	Ph	4-Me	I	Br	NMP	150	16	(48)	143
i-Pr	Ph	4-Me	I	Br	NMP	150	16	(24)	143
Ph	4-pyridyl	4-Me	I	Br	DMA	150	16	(56)	143
Ph	Ph	H	I	I	DMA	170	24	(58)	141
Ph	Ph	3-Cl	I	Cl	NMP	150	16	(29)	143
Ph	Ph	4-Me	I	Br	NMP	150	16	(56)	143
Ph	Ph	5-Me	Br	Cl	NMP	150	16	(20)	143
Ph	4-MeOC$_6$H$_4$	4-Me	I	Br	NMP	150	16	(0)	143
Ph	4-O$_2$NC$_6$H$_4$	4-Me	I	Br	NMP	150	16	(11)	143
Bn	Ph	4-Me	I	Br	NMP	150	16	(16)	143

C$_8$

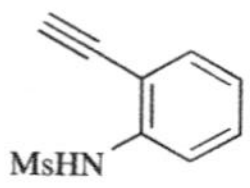

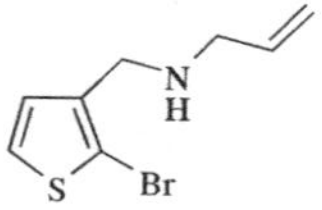

MsHN (ethynylbenzene)	thiophene–CH$_2$NH–allyl, Br	1. CuI (2.5 mol %), (H$_2$CO)$_n$, dioxane, MW, 170°, 40 min 2. NaOMe, MW, 170°, 20 min	(56)	627
	pyridine–CH$_2$NH–allyl, Br	1. CuI (2.5 mol %), (H$_2$CO)$_n$, dioxane, MW, 170°, 40 min 2. NaOMe, MW, 170°, 20 min	(71)	627

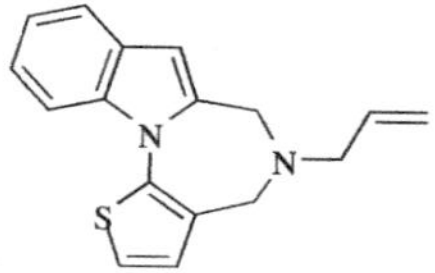

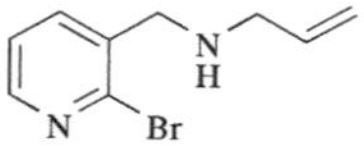

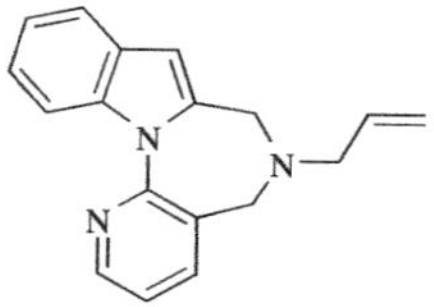

C$_{8-9}$

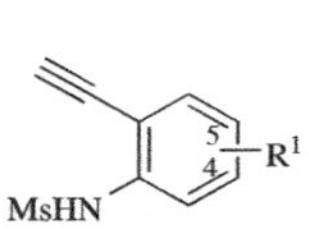

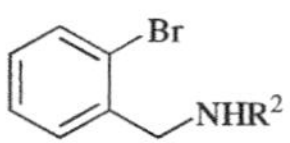

1. CuI (2.5 mol %),
   (H$_2$CO)$_n$, dioxane,
   MW, 170°, 40 min
2. NaOMe, MW, 170°, 20 min

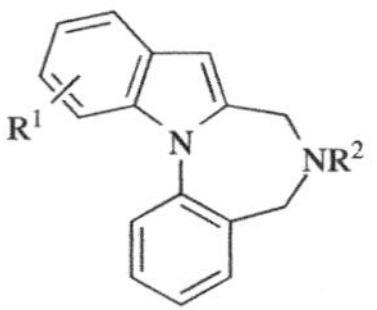

R^1	R^2	
H	Me	(51)
H	allyl	(81)
H	n-Bu	(88)
H	Bn	(83)
5-Me	allyl	(85)
4-CF$_3$	allyl	(53)
5-CF$_3$	allyl	(81)
5-MeCO$_2$	allyl	(23)

627

C$_8$

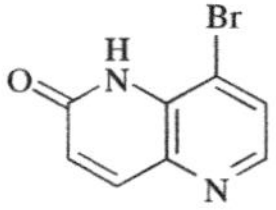

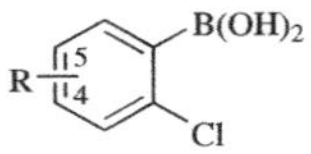

1. Pd(dppf)Cl$_2$•MeCl$_2$ (2 mol %),
   K$_2$CO$_3$, dioxane/H$_2$O (3:1),
   reflux, 30 min
2. CuI (5 mol %),
   DMEDA (x mol %), 15 min

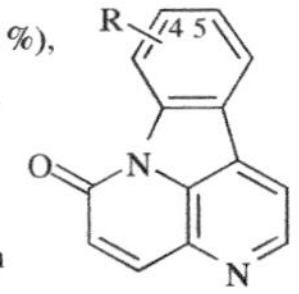

R	x	
H	10	(95)
4-F	10	(88)
3-Cl	10	(82)
4-Cl	10	(82)
5-Cl	10	(77)
6-Cl	10	(0)
4-MeO	20	(92)
6-MeO	20	(trace)
4-Me	20	(92)
4-CF$_3$	20	(78)
5-CF$_3$	20	(71)

185

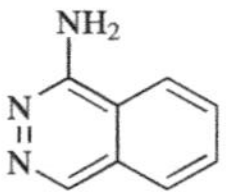

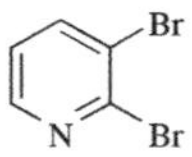

1. Pd$_2$(dba)$_3$ (4 mol %), Cs$_2$CO$_3$,
   DME, xantphos (9 mol %),
   reflux, 17 h
2. CuI (10 mol %),
   CDA (20 mol %), reflux, 8 h

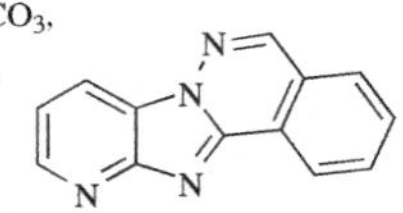

(65)

628

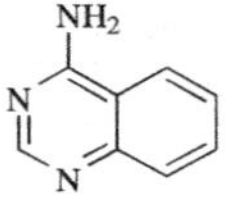

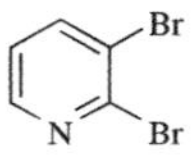

1. Pd$_2$(dba)$_3$ (6 mol %), Cs$_2$CO$_3$,
   DME, xantphos (13 mol %),
   reflux, 17 h
2. CuI (10 mol %),
   CDA (20 mol %), reflux, 8 h

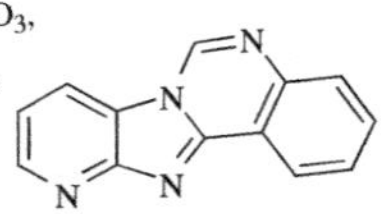

(93)

628

TABLE 27. *N*-ARYLATIONS IN MULTI-STEP REACTIONS (*Continued*)
A. SYNTHESIS OF FIVE-MEMBERED RINGS (*Continued*)

Nitrogen Nucleophile	Electrophile	Conditions	Product(s) and Yield(s) (%)	Refs.
Please refer to the charts preceding the tables for structures indicated by the ***bold*** *numbers.*				
C_8				
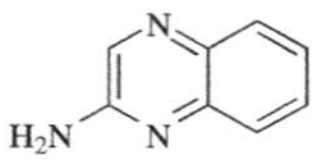	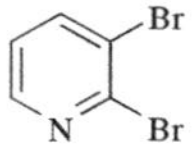	1. $Pd_2(dba)_3$ (4 mol %), Cs_2CO_3, DME, xantphos (9 mol %), reflux, 17 h 2. CuI (10 mol %), CDA (20 mol %), reflux, 8 h	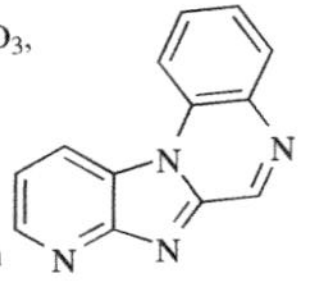 (99)	628
C_{10-16}				
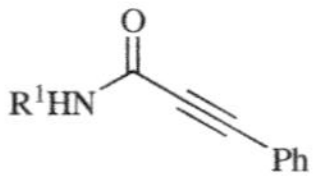	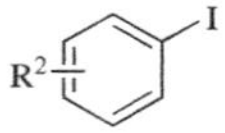	1. CuI (10 mol %), DMEDA (20 mol %), K_2CO_3, DMF/MeCN (1:1), 110°, 12 h 2. $Pd(OAc)_2$ (10 mol %), 100°, 12 h	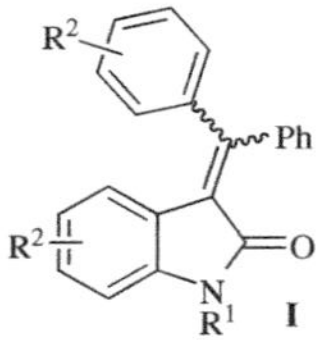 **I** + 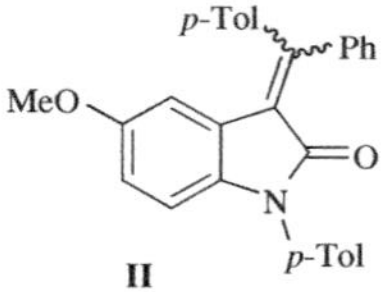 **II**	630

R^1	R^2	**I**	(*E*)/(*Z*)	**II**
Me	4-MeO	(33)	3:1	(—)
Ph	H	(45)	—	(—)
2-$MeOC_6H_4$	2-MeO	(32)	1:1	(—)
4-$MeOC_6H_4$	H	(27)	—	(—)
4-$MeOC_6H_4$	4-Me	(37)	2:1	(27)

R^1	R^2	**I**	(*E*)/(*Z*)	**II**
2-MeC_6H_4	2-Me	(21)	3:1	(—)
3-MeC_6H_4	3-Me	(55)	2:1	(—)
4-MeC_6H_4	4-Me	(53)	2:1	(—)
Bn	4-MeO	(36)	2:1	(—)

C_{11-17}

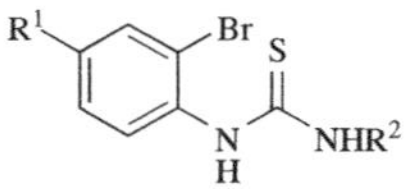

Intramolecular

1. Et₃N, R³X
2. CuI (5 mol %), phen (10 mol %), K₂CO₃, dioxane, 85°, time

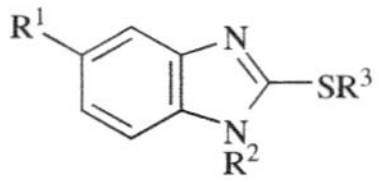

631

R¹	R²	R³X	Time (h)
H	4-morpholinyl	MeI	8 (86)
H	2-furylmethyl	MeI	12 (78)
H	2-pyridyl	MeI	8 (79)
H	Ph	CH₂=CHCH₂I	8 (88)
H	Ph	BnBr	12 (87)
H	Ph	4-MeOC₆H₄CH₂Br	20 (48)
H	4-FC₆H₄	MeI	4 (83)
H	4-MeOC₆H₄	MeI	4 (93)
H	3,4-(MeO)₂C₆H₃CH₂CH₂	MeI	20 (48)
H	3,5-(MeO)₂C₆H₃	MeI	4 (88)
H	4-MeC₆H₄	MeI	4 (91)

R¹	R²	R³X	Time (h)
H	Bn	MeI	8 (81)
H	2,3-Me₂C₆H₃	MeI	4 (74)
H	4-t-BuC₆H₄	MeI	4 (78)
Br	Ph	CH₂=CHCH₂I	12 (87)
MeO	Ph	MeI	4 (82)
MeO	Ph	4-O₂NC₆H₄CH₂Br	20 (84)
MeO	Ph	4-MeOC₆H₄CH₂Br	20 (48)
MeO	Bn	MeI	12 (74)
Me	Ph	EtI	4 (85)
Me	4-FC₆H₄	EtI	4 (81)
Me	4-MeC₆H₄	EtI	4 (78)

C_{13-14}

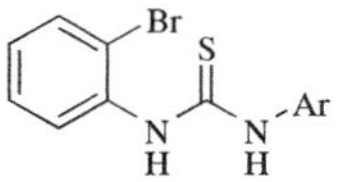

Intramolecular

1. Et₃N, 4-MeOC₆H₄CH₂Br
2. CuI (5 mol %), phen (10 mol %), K₂CO₃, dioxane, 85°
3. TFA, reflux

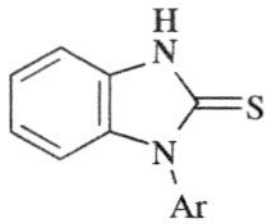

631

Ar	
Ph	(68)
4-FC₆H₄	(71)
4-MeC₆H₄	(72)

C_{13}

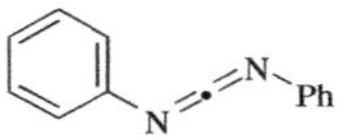

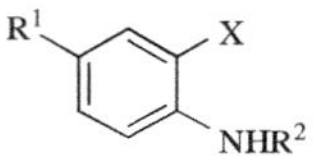

CuI (15 mol %), Cs₂CO₃, MeCN, 100°, 24 h

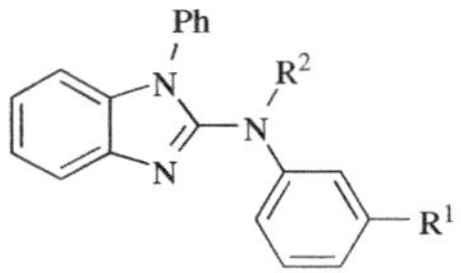

184

R¹	R²	X	
H	H	Br	(62)
H	H	I	(81)
H	Me	I	(63)
Cl	H	I	(72)
Me	H	Br	(64)
Me	H	I	(70)

555

Nitrogen Nucleophile	Electrophile	Conditions	Product(s) and Yield(s) (%)	Refs.

*Please refer to the charts preceding the tables for structures indicated by the **bold** numbers.*

C$_{13-14}$

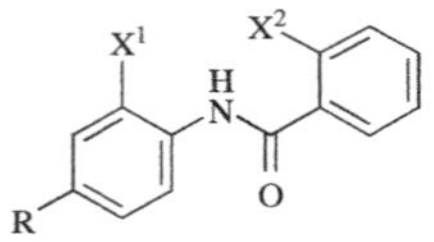		CuCl (10 mol %), Na$_2$CO$_3$, DMSO, 100°, 12 h	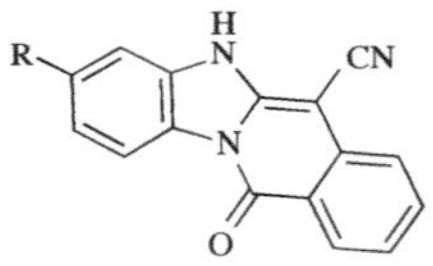	632

Electrophile table:

R^1	R^2	X^1	X^2	R^3	
H	H	Br	Br	Me	(87)
H	H	Cl	Br	Et	(75)
H	H	Br	Cl	Et	(82)
H	H	Br	Br	Et	(90)
H	H	Br	Br	n-Bu	(87)

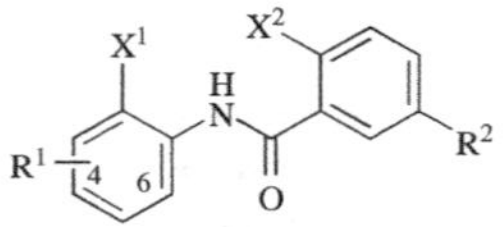

Product table:

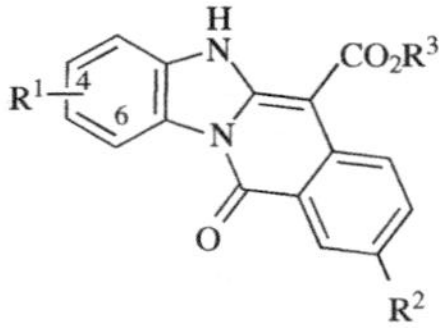

R^1	X^1	X^2	
H	Br	Cl	(71)
H	Cl	Br	(70)
H	Br	Br	(85)
Me	Br	Br	(80)

		CuCl (10 mol %), Na$_2$CO$_3$, DMSO, 100°, 12 h		632

Electrophile table:

R^1	R^2	X^1	X^2	R^3	
H	H	Cl	Br	Me	(65)
H	H	Cl	Br	Et	(78)
H	H	Cl	Br	n-Bu	(65)
H	4-Me	H	Br	Me	(89)
H	4-Me	H	Br	Et	(91)

Product table:

R^1	R^2	X^1	X^2	R^3	
4-Me	H	Br	Br	n-Bu	(63)
4-Me		Cl	Br	Me	(81)
5-MeO$_2$C	H	Br	Br	Me	(81)
5-MeO$_2$C	H	Br	Br	Et	(72)
5-MeO$_2$C	H	Br	Br	n-Bu	(72)

C$_{14-15}$

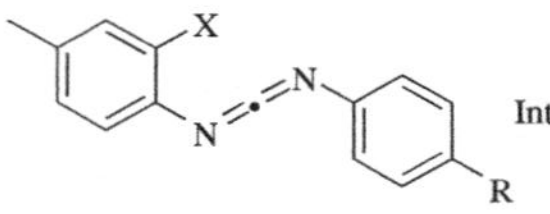

	Intramolecular	ArOH, CuI (10 mol %), DMG (20 mol %), K$_3$PO$_4$, DMSO, 90°, 24 h		183

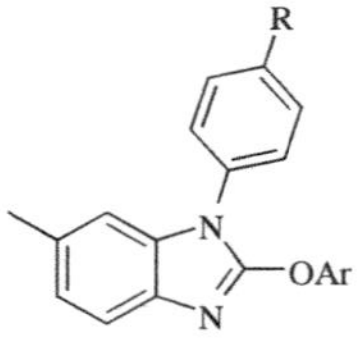

Product table:

R	X	Ar	
H	I	Ph	(76)
H	I	4-MeC$_6$H$_4$	(71)
H	Br	2-Np	(68)
Me	I	Ph	(72)

C₁₄

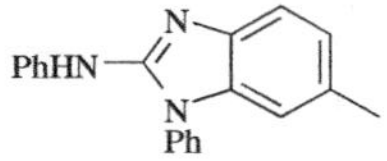

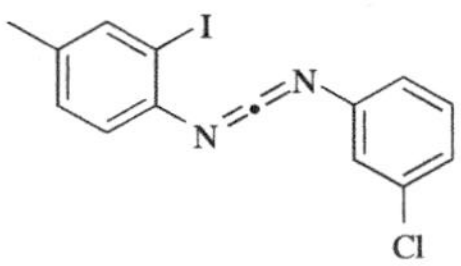

CuI (15 mol %), Cs$_2$CO$_3$, MeCN, 100°, 24 h

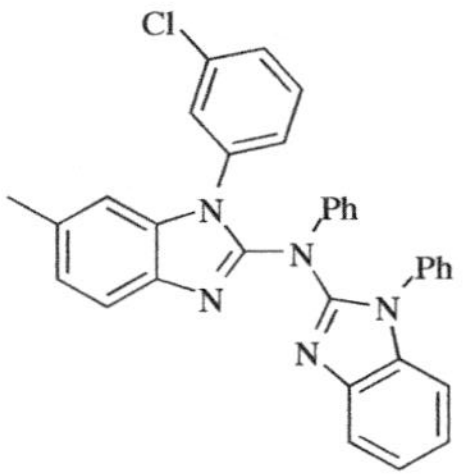

(62)

184

C₂₂

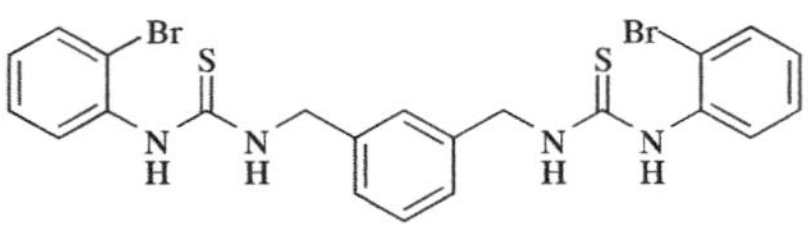

1. Et$_3$N, MeI
2. CuI (5 mol %), phen (10 mol %), K$_2$CO$_3$, dioxane, 85°, 20 h

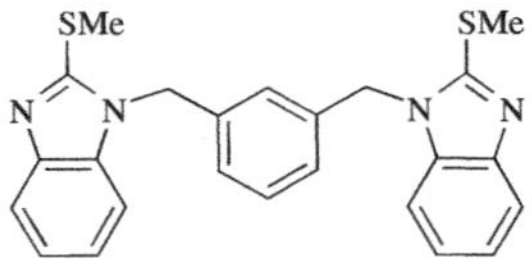

(54)

631

TABLE 27. N-ARYLATIONS IN MULTI-STEP REACTIONS (Continued)
B. SYNTHESIS OF SIX-MEMBERED RINGS

Nitrogen Nucleophile	Electrophile	Conditions	Product(s) and Yield(s) (%)	Refs.
NH$_4$Cl	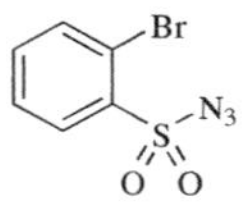	1. Ph—≡≡, CuI (10 mol %), Et$_3$N, MeCl$_2$, rt, 1 h 2. Phen (10 mol %), Cs$_2$CO$_3$, DMF, 80°, 16 h	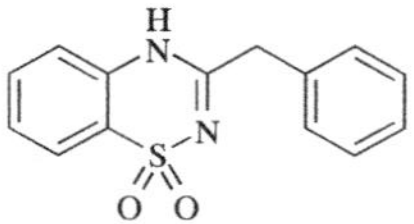 (77)	590
NaN$_3$	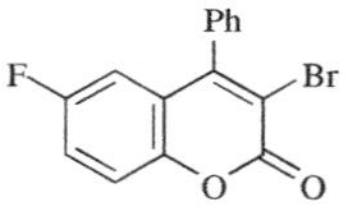	Cu (10 mol %), piperidine-2-carboxylic acid (30 mol %), ascorbic acid (20 mol %), EtOH, 100°, 3 h	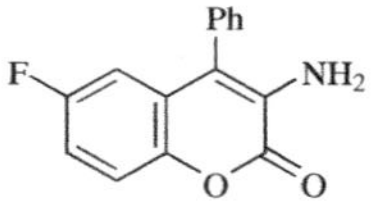 (43)	250
	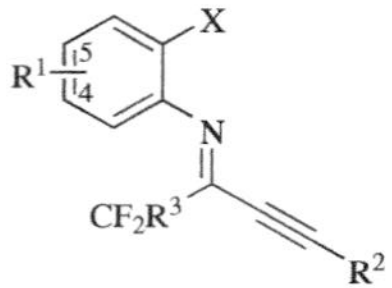	CuI (10 mol %), L-Pro (20 mol %), DMSO	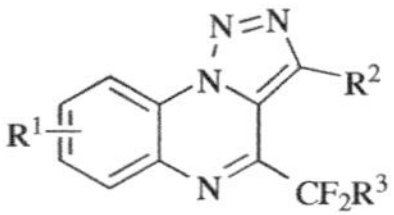	634

Electrophile sub-table (third row):

R^1	R^2	R^3	X	Temp (°)	Time (h)	
H	AcOCH$_2$	F	I	rt	10	(65)
H	MeO$_2$C	F	I	60	10	(56)
H	n-Bu	F	I	rt	0.5	(97)
H	n-Bu	Cl	I	60	2	(70)
H	t-Bu	F	I	rt	1	(91)
H	2-thienyl	F	I	60	10	(92)
H	Ph	F	Cl	130	10	(65)
H	Ph	F	Br	60	10	(95)

Product sub-table (third row):

R^1	R^2	R^3	X	Temp (°)	Time (h)	
H	Ph	Cl	I	60	10	(70)
H	4-ClC$_6$H$_4$	F	I	80	10	(87)
H	4-MeC$_6$H$_4$	F	I	rt	0.5	(94)
4-F	Ph	F	I	60	10	(80)
5-O$_2$N	Ph	F	I	80	10	(70)
5-MeO	Ph	F	I	60	4	(87)
4-Me	Ph	F	I	rt	1.5	(90)

C$_0$

C₁

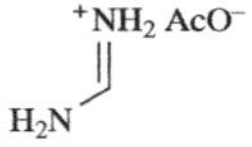

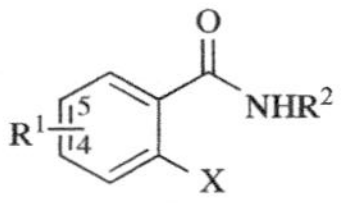

Reactant: H₂N–CH=CH–⁺NH₂ AcO⁻

and aroyl amide:

R¹	R²	X	
H	H	I	(82)
H	Me	Br	(3)
H	Me	I	(81)
H	Et	I	(63)
H	i-Pr	I	(55)
H	2-thiazolyl	I	(91)
H	Ph	I	(77)

CuI (10 mol %), K₂CO₃, DMF, 80°, 6 h

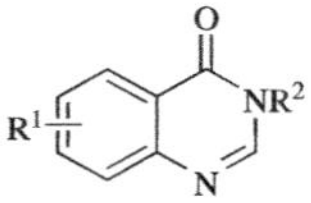

R²	X	
4-ClC₆H₄	I	(56)
4-BrC₆H₄	I	(71)
4-O₂NC₆H₄	I	(20)
2-MeOC₆H₄	I	(72)
3-MeOC₆H₄	I	(73)
4-MeOC₆H₄	I	(83)
2-MeC₆H₄	I	(67)

R¹	R²	X	
H	3-MeC₆H₄	I	(62)
H	4-MeC₆H₄	I	(69)
H	4-NCC₆H₄	I	(66)
H	4-AcC₆H₄	I	(75)
H	Bn	I	(84)
4-Cl	H	I	(66)
5-Me	H	I	(69)

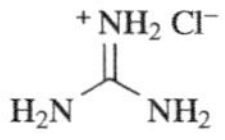

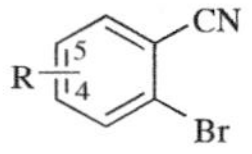

635

C₂₋₇

Reactant: H₂N–C(=⁺NH₂ Cl⁻)–NH₂ and bromobenzonitrile

CuI (10 mol %), DMEDA (20 mol %), K₂CO₃, DMF, 80°, 12 h

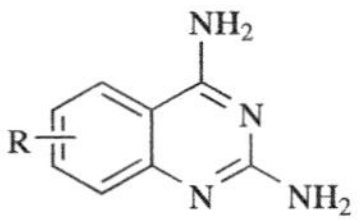

R	
H	(57)
4-O₂N	(85)
5-Me	(66)

636

Reactant: H₂N–C(=⁺NH₂ Cl⁻)–R¹ and bromobenzonitrile

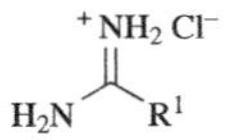

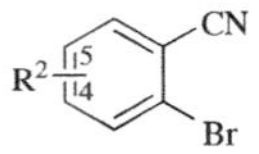

CuI (10 mol %), DMEDA (20 mol %), Cs₂CO₃, DMF, 80°, 12 h

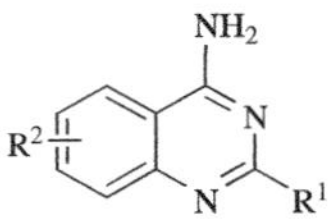

R¹	R²	
Me	H	(69)
Me	4-O₂N	(63)
Me	5-Me	(75)
n-Pr	H	(79)
n-Pr	4-O₂N	(76)
n-Pr	5-Me	(79)
c-Pr	H	(72)
c-Pr	5-Me	(55)
Ph	H	(85)
Ph	5-Me	(67)

636

TABLE 27. N-ARYLATIONS IN MULTI-STEP REACTIONS (*Continued*)
B. SYNTHESIS OF SIX-MEMBERED RINGS (*Continued*)

Nitrogen Nucleophile

C_{2-7}

H_2N / $^+NH_2$ Cl^- — R^1

Electrophile

$R^2 \overset{15}{\underset{14}{\text{-}}}$ —CHO / X

Conditions: CuI (10 mol %), Cs_2CO_3

Product(s) and Yield(s) (%):

$R^2 \overset{\parallel}{\underset{N}{\bigcirc}} \overset{N}{\underset{N}{\parallel}} R^1$

R^1	R^2	X	Additive	Solvent	Temp (°)	Time (h)	Product(s) and Yield(s) (%)	Refs.
Me	H	Br	L-Pro (20 mol %)	DMF	110	24	(75)	140
Me	4-(OCH$_2$O)-5	Br	L-Pro (20 mol %)	DMF	110	24	(92)	140
n-Pr	H	Br	L-Pro (20 mol %)	DMF	110	24	(82)	140
n-Pr	4-(OCH$_2$O)-5	Br	L-Pro (20 mol %)	DMF	110	20	(95)	140
c-Pr	H	I	none	MeOH	60	18	(87)	637
5-pyrimidinyl	H	I	none	MeOH	60	18	(53)	637
2-pyridyl	H	I	none	MeOH	60	18	(59)	637
4-pyridyl	H	I	none	MeOH	60	18	(80)	637
Ph	H	Br	L-Pro (20 mol %)	DMF	110	24	(55)	140
Ph	H	I	none	MeOH	60	18	(89)	637
Ph	4-F	I	none	MeOH	60	18	(62)	637
Ph	4,5-(MeO)$_2$	I	none	MeOH	60	18	(89)	637
Ph	4-(OCH$_2$O)-5	Br	L-Pro (20 mol %)	DMF	110	20	(89)	140
4-FC$_6$H$_4$	H	I	none	MeOH	60	18	(82)	637
4-ClC$_6$H$_4$	H	I	none	MeOH	60	18	(88)	637
3-O$_2$NC$_6$H$_4$	H	I	none	MeOH	60	18	(83)	637
2-MeOC$_6$H$_4$	H	I	none	MeOH	60	18	(86)	637
3-MeOC$_6$H$_4$	H	I	none	MeOH	60	18	(83)	637
4-MeOC$_6$H$_4$	H	I	none	MeOH	60	18	(61)	637
2-MeC$_6$H$_4$	H	I	none	MeOH	60	18	(86)	637
3-MeC$_6$H$_4$	H	I	none	MeOH	60	18	(94)	637
4-MeC$_6$H$_4$	H	I	none	MeOH	60	18	(89)	637
4-CF$_3$C$_6$H$_4$	H	I	none	MeOH	60	18	(70)	637

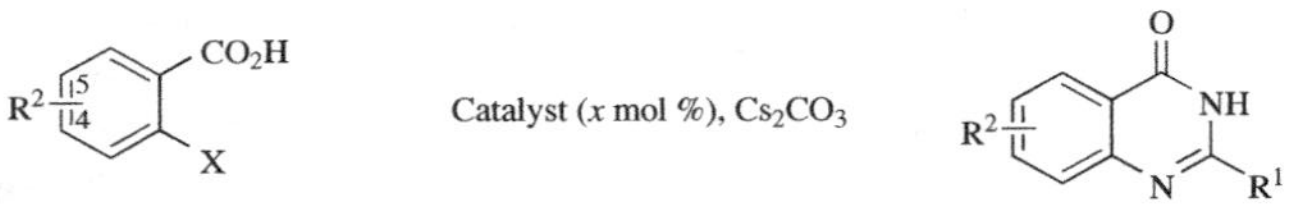

Catalyst (x mol %), Cs_2CO_3

R^1	R^2	X	Catalyst	x	Additive	Solvent	Temp (°)	Time (h)	
Me	H	Br	Cu(acac)$_2$	10	Fe$_2$O$_3$ (20 mol %)	DMSO	MW, 130	0.5	(48)	262
Me	H	Br	CuI	20	none	DMF	rt	12	(81)	142
Me	H	I	CuI	20	none	DMF	rt	12	(89)	142
Me	4-Cl	Br	CuI	20	none	DMF	rt	12	(83)	142
Me	5-O$_2$N	Br	CuI	20	none	DMF	rt	12	(69)	142
n-Pr	H	Cl	CuI	20	none	DMF	rt	12	(79)	142
n-Pr	H	Br	CuI	20	none	DMF	rt	12	(79)	142
n-Pr	H	I	CuI	20	none	DMF	rt	12	(90)	142
n-Pr	4-Cl	Br	CuI	20	none	DMF	rt	12	(87)	142
n-Pr	5-O$_2$N	Br	CuI	20	none	DMF	rt	12	(89)	142
n-Pr	4-(OCH$_2$O)-5	Br	CuI	20	none	DMF	rt	12	(40)	142
c-Pr	H	Br	CuI	20	none	DMF	rt	12	(84)	142
c-Pr	4-Cl	Br	CuI	20	none	DMF	rt	12	(75)	142
c-Pr	5-O$_2$N	Br	CuI	20	none	DMF	rt	12	(68)	142
1-pyrrolidinyl	H	I	CuI	20	none	DMF	rt	12	(72)	142
1-piperidinyl	H	I	CuI	20	none	DMF	rt	12	(70)	142
Ph	H	Cl	CuI	20	none	DMF	rt	12	(79)	142
Ph	H	Br	CuI	20	none	DMF	rt	12	(79)	142
Ph	H	I	CuI	20	none	DMF	rt	12	(97)	142
Ph	4-Cl	Br	CuI	20	none	DMF	rt	12	(66)	142
Ph	5-O$_2$N	Br	CuI	20	none	DMF	rt	12	(79)	142
Ph	4-(OCH$_2$O)-5	Br	CuI	20	none	DMF	rt	12	(79)	142

TABLE 27. N-ARYLATIONS IN MULTI-STEP REACTIONS (*Continued*)

B. SYNTHESIS OF SIX-MEMBERED RINGS (*Continued*)

Nitrogen Nucleophile	Electrophile	Conditions	Product(s) and Yield(s) (%)	Refs.

C$_{2-7}$

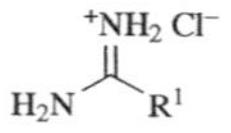

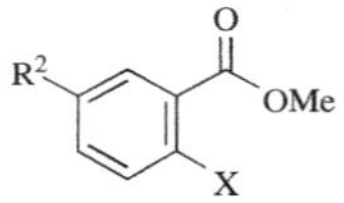

CuI (10 mol %), L-Pro (20 mol%), Cs$_2$CO$_3$, DMF, 110°

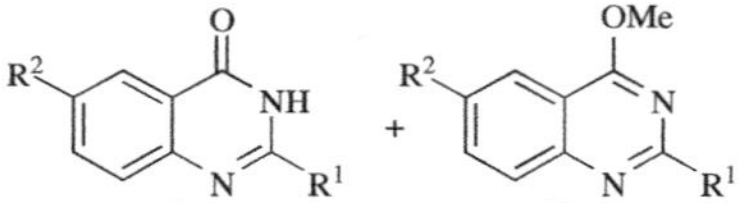

R^1	R^2	X	Time (h)	I	II
Me	H	Cl	20	(90)	(0)
Me	H	Br	16	(91)	(0)
Me	Cl	Br	16	(87)	(0)
n-Pr	H	Cl	20	(77)	(18)
n-Pr	H	Br	16	(74)	(22)
n-Pr	Cl	Br	16	(95)	(0)
Ph	Cl	Br	16	(86)	(0)

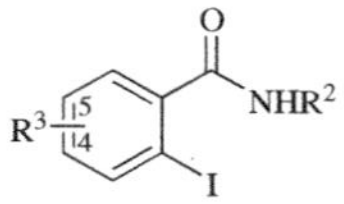

Products **I** and **II**; Refs. 140

CuI (10 mol %), K$_2$CO$_3$, DMF, 80°

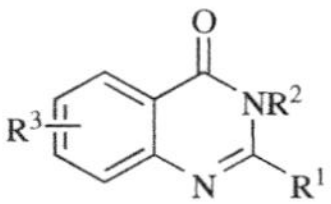

R^1	R^1	R^3	Time (h)	
Me	H	H	6	(78)
Me	H	5-Cl	6	(65)
Me	H	5-Me	6	(60)
Me	Me	H	6	(52)
Me	Ph	H	6	(69)
Ph	H	H	6	(81)
Ph	H	4-Cl	16	(86)
Ph	H	6-MeO	6	(65)
Ph	H	6-O$_2$N	16	(86)
Ph	H	5-Me	6	(66)

Refs. 635

C₂₋₄

R¹HN=C(NH₂⁺ Cl⁻) ... H₂N structure + R²-[ring]-SO₂NH₂ (X)

CuI (15 mol %),
DMEDA (30 mol %), Cs₂CO₃,
DMA, 170°

R¹	R²	X	Time (h)
Me	H	Br	24 (77)
Me	5-Br	Br	12 (70)
Me	5-Me	I	24 (81)
Me	4-CF₃	Br	10 (77)
Me	4,5-Me₂	Br	10 (70)

R¹	R²	X	Time (h)
n-Pr	H	Br	24 (80)
n-Pr	5-Br	Br	12 (68)
n-Pr	5-Me	I	10 (78)
n-Pr	4-CF₃	Br	10 (75)
n-Pr	4,5-Me₂	Br	10 (65)

R¹	R²	X	Time (h)	
c-Pr	H	Br	24 (75)	179
c-Pr	5-Br	Br	12 (65)	179
c-Pr	5-Me	I	10 (80)	179
c-Pr	4-CF₃	Br	10 (80)	180
c-Pr	4,5-Me₂	Br	10 (80)	180

C₂₋₉

R¹HN-C(=O)-CH(Cl)-R² + R³-[ring]-OH (X)

CuI (x mol %), Cs₂CO₃

R¹	R²	R³	X	x	Additive(s)	Solvent	Temp (°)	Time	
H	H	Cl	Br	10	phen (10 mol %)	dioxane	90	24 h	(55) 179
H	H	Me	I	10	phen (10 mol %)	dioxane	90	6 h	(71) 179
H	H	t-Bu	I	10	phen (10 mol %)	dioxane	90	24 h	(80) 179
HOCH₂CH₂	H	H	I	20	DBU	DMSO	MW, 130	10 min	(70) 180
n-Pr	H	H	I	20	DBU	DMSO	MW, 130	10 min	(82) 180
c-PrCH₂	H	Me	I	20	DBU	DMSO	MW, 130	10 min	(65) 180
n-Bu	H	H	Br	10	phen (10 mol %)	dioxane	reflux	24 h	(55) 179
n-Bu	H	H	I	10	phen (10 mol %)	dioxane	90	24 h	(84) 179
n-Bu	H	Cl	I	10	phen (10 mol %)	dioxane	reflux	24 h	(70) 179
n-Bu	H	Me	I	10	phen (10 mol %)	dioxane	90	24 h	(51) 179
n-Bu	H	t-Bu	I	10	phen (10 mol %)	dioxane	reflux	24 h	(98) 179
n-Bu	Et	H	I	10	phen (10 mol %)	dioxane	90	24 h	(73) 179
2-pyridyl	H	H	I	20	DBU	DMSO	MW, 130	10 min	(67) 180
2-pyridyl	H	H	I	10	phen (10 mol %)	dioxane	90	6 h	(45) 179

C$_{2-9}$

Nitrogen Nucleophile	Electrophile				Conditions CuI (x mol %), Cs$_2$CO$_3$					Product(s) and Yield(s) (%)		Refs.
R^1	R^2	R^3	X	x	Additive	Solvent	Temp (°)	Time				
Ph	H	H	I	10	phen (10 mol %)	dioxane	90	24 h	(97)			179
Ph	H	H	I	20	DBU	DMSO	MW, 130	10 min	(68)			180
Ph	H	Cl	I	10	phen (10 mol %)	dioxane	90	24 h	(83)			179
Ph	H	Me	I	10	phen (10 mol %)	dioxane	90	24 h	(98)			179
Ph	H	*t*-Bu	I	10	phen (10 mol %)	dioxane	90	24 h	(99)			179
Ph	Me	H	Br	10	phen (10 mol %)	dioxane	reflux	24 h	(97)			179
Ph	Me	F	I	10	phen (10 mol %)	dioxane	reflux	24 h	(64)			179
3-ClC$_6$H$_4$	H	H	I	10	phen (10 mol %)	dioxane	90	24 h	(71)			179
4-BrC$_6$H$_4$	H	H	I	20	DBU	DMSO	MW, 130	10 min	(75)			180
4-BrC$_6$H$_4$	H	*t*-Bu	I	20	DBU	DMSO	MW, 130	10 min	(82)			180
2-MeOC$_6$H$_4$	H	H	Br	20	DBU	DMSO	MW, 130	20 min	(86)			180
2-MeOC$_6$H$_4$	H	H	I	20	DBU	DMSO	MW, 130	10 min	(82)			180
2-MeOC$_6$H$_4$	H	*t*-Bu	I	20	DBU	DMSO	MW, 130	10 min	(75)			180
3-MeOC$_6$H$_4$	H	H	Br	20	DBU	DMSO	MW, 130	20 min	(51)			180
3-MeOC$_6$H$_4$	H	H	I	20	DBU	DMSO	MW, 130	10 min	(82)			180
3-MeOC$_6$H$_4$	H	*t*-Bu	I	20	DBU	DMSO	MW, 130	10 min	(82)			180
4-MeOC$_6$H$_4$	H	H	Br	10	phen (10 mol %)	dioxane	reflux	24 h	(86)			179
4-MeOC$_6$H$_4$	H	H	I	20	DBU	DMSO	MW, 130	10 min	(86)			180
4-MeOC$_6$H$_4$	Me	H	I	20	DBU	DMSO	MW, 130	10 min	(73)			180
4-MeOC$_6$H$_4$	Me	Me	Br	20	DBU	DMSO	MW, 130	10 min	(61)			180
2-MeC$_6$H$_4$	H	H	I	20	DBU	DMSO	MW, 130	10 min	(85)			180
2-MeC$_6$H$_4$	H	Me	Br	20	DBU	DMSO	MW, 130	10 min	(55)			180
2-MeC$_6$H$_4$	Me	H	I	20	DBU	DMSO	MW, 130	10 min	(70)			180
3-MeC$_6$H$_4$	H	H	I	20	DBU	DMSO	MW, 130	10 min	(84)			180
3-MeC$_6$H$_4$	H	Me	Br	20	DBU	DMSO	MW, 130	10 min	(56)			180

R¹			X		Additive	Solvent	Conditions	Time	Yield	Ref
4-MeC₆H₄	H	H	Br	10	phen (10 mol %)	dioxane	reflux	24 h	(80)	179
4-MeC₆H₄	H	H	Br	20	DBU	DMSO	MW, 130	20 min	(89)	180
4-MeC₆H₄	H	H	I	10	phen (10 mol %)	dioxane	90	24 h	(98)	179
4-MeC₆H₄	H	H	I	20	DBU	DMSO	MW, 130	10 min	(89)	180
4-MeC₆H₄	H	Me	Br	20	DBU	DMSO	MW, 130	10 min	(60)	180
4-MeC₆H₄	Me	H	I	20	DBU	DMSO	MW, 130	10 min	(82)	180
4-CF₃C₆H₄	H	H	I	20	DBU	DMSO	MW, 130	10 min	(72)	180
4-EtO₂CC₆H₄	H	H	I	20	DBU	DMSO	MW, 130	10 min	(75)	180
4-EtO₂CC₆H₄	H	t-Bu	I	20	DBU	DMSO	MW, 130	10 min	(82)	180
4-EtO₂CC₆H₄	Me	H	I	20	DBU	DMSO	MW, 130	10 min	(82)	180
4-EtO₂CC₆H₄	Me	Me	Br	20	DBU	DMSO	MW, 130	10 min	(59)	180
Bn	H	H	I	10	phen (10 mol %)	dioxane	90	24 h	(71)	179

CuI (x mol %),
DMCDA (y mol %)

R¹	R²	X	x	y	Additive	Solvent	Temp (°)	Time (h)		Ref
H	Me	Br	20	40	K₃PO₄	DMF	115	36	(52)	639
n-Pr	H	I	7	14	Cs₂CO₃	dioxane	100	9	(62)	
n-Pr	Cl	I	20	40	K₃PO₄	DMF	115	36	(60)	
n-Bu	Me	I	7	14	Cs₂CO₃	dioxane	100	9	(68)	
Ph	Cl	I	7	14	K₃PO₄	dioxane	100	13	(76)	
3-ClC₆H₄	Me	I	7	14	K₃PO₄	dioxane	100	13	(60)	
4-MeOC₆H₄	H	I	7	14	K₃PO₄	dioxane	100	13	(86)	
4-MeOC₆H₄	CF₃O	Br	20	40	K₃PO₄	dioxane	115	36	(55)	

TABLE 27. N-ARYLATIONS IN MULTI-STEP REACTIONS (Continued)
B. SYNTHESIS OF SIX-MEMBERED RINGS (Continued)

Nitrogen Nucleophile	Electrophile	Conditions	Product(s) and Yield(s) (%)	Refs.

Continued from previous page.

C$_{2-9}$

Nitrogen Nucleophile: R^{1}HN–C(=O)–CH$_2$Cl

Electrophile: R^2-C$_6$H$_3$ with X, NHMs

Conditions: CuI (x mol %), DMCDA (y mol %)

Product: quinoxalinone with R^1, R^2

R^1	R^2	X	x	y	Additive	Solvent	Temp (°)	Time (h)	Yield
4-MeOC$_6$H$_4$	Me	I	7	14	K$_3$PO$_4$	dioxane	100	13	(91)
4-MeC$_6$H$_4$	H	Br	20	40	K$_3$PO$_4$	dioxane	115	24	(58)
4-MeC$_6$H$_4$	H	I	7	14	K$_3$PO$_4$	dioxane	100	13	(91)
4-MeC$_6$H$_4$	Cl	I	7	14	K$_3$PO$_4$	dioxane	100	13	(87)
4-MeC$_6$H$_4$	CF$_3$O	Br	20	40	K$_3$PO$_4$	DMF	115	36	(61)
4-MeC$_6$H$_4$	Me	I	7	14	K$_3$PO$_4$	dioxane	100	13	(94)
Bn	H	I	7	14	Cs$_2$CO$_3$	dioxane	100	9	(80)
Bn	Cl	I	20	40	K$_3$PO$_4$	DMF	reflux	36	(70)
Bn	Me	I	7	14	Cs$_2$CO$_3$	dioxane	100	9	(88)

Refs. 639

C$_3$

Nitrogen Nucleophile: 2-aminothiazole (H$_2$N–)

Electrophile: 2-chlorobenzoic acid (CO$_2$H, Cl), R$^{14-3}$

Conditions: Cu (7.5 mol %), K$_2$CO$_3$, DMF, rt

R	Time (min)
H	15 (88)
3-F	25 (90)
6-F	25 (86)
3-Cl	25 (86)
6-Me	15 (85)
3,5-Me$_2$	15 (90)

Refs. 242

C$_{4-5}$

Nitrogen Nucleophile: thiazole with R^1, R^2 (H$_2$N–)

Electrophile: 2-chlorobenzoic acid (CO$_2$H, Cl), R^{3}$^{14-3}$

Conditions: Cu (7.5 mol %), K$_2$CO$_3$, DMF, rt

R^1	R^2	R^3	Time (min)
H	Me	6-F	25 (91)
Me	H	H	15 (88)

Refs. 242

C4–7

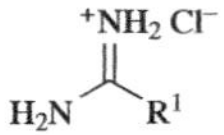

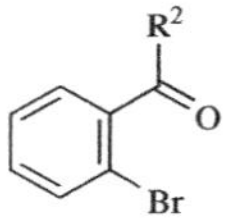

CuI (10 mol %),
L-Pro (20 mol %),
Cs$_2$CO$_3$, DMF, 110°

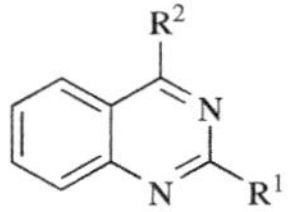

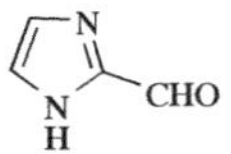

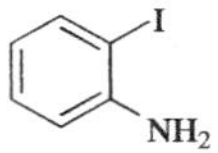

R^1	R^2	Time (h)	
n-Pr	Me	20	(86)
n-Pr	Ph	24	(81)
Ph	Me	20	(84)
Ph	Ph	24	(81)

140

C4

CuI (10 mol %),
sparteine (20 mol %), K$_3$PO$_4$,
NMP, 130°, 24 h

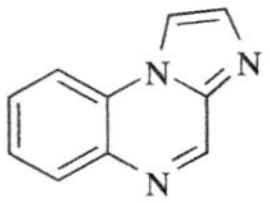

(51)

640

C5

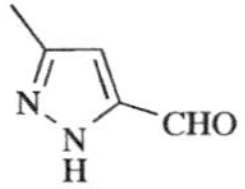

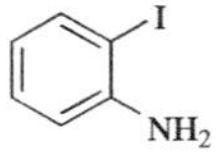

CuI (10 mol %),
sparteine (20 mol %), K$_3$PO$_4$,
NMP, 130°, 24 h

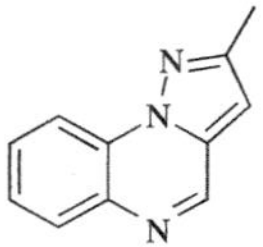

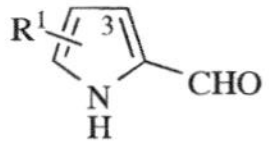

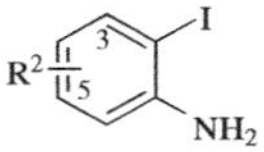

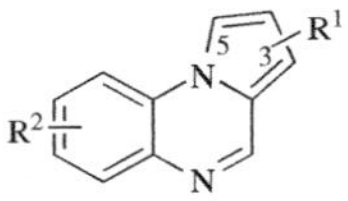

(83)

640

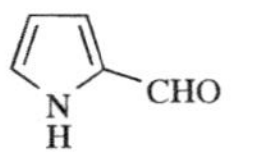

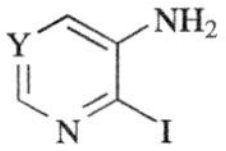

CuI (10 mol %),
sparteine (20 mol %), K$_3$PO$_4$,
NMP, 130°, 24 h

Y	
N	(63)
CH	(78)

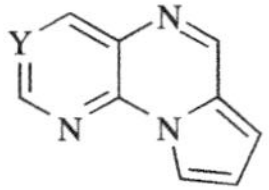

640

C5–11

CuI (10 mol %),
sparteine (20 mol %), K$_3$PO$_4$,
NMP, 130°, 24 h

R^1	R^2	Time (h)	
H	H	24	(83)
H	5-F	24	(83)
H	4-CF$_3$	24	(69)
H	4-NC–	24	(48)
H	4-MeO$_2$C	24	(61)
3,5-Me$_2$	H	36	(82)
5-Ph	H	36	(78)

640

TABLE 27. N-ARYLATIONS IN MULTI-STEP REACTIONS (Continued)
B. SYNTHESIS OF SIX-MEMBERED RINGS (Continued)

Nitrogen Nucleophile	Electrophile	Conditions	Product(s) and Yield(s) (%)	Refs.

C5-11

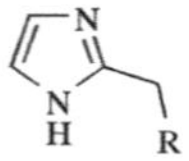

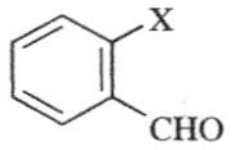

Nitrogen Nucleophile: imidazole with CH2R

Electrophile: benzaldehyde bearing X and CHO

Conditions: CuI (10 mol %), L-Pro (20 mol %), K2CO3, DMSO, 8 h

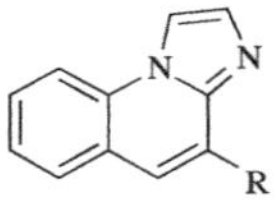

Product:

R	X	Temp (°)	
NC–	Cl	80	(35)
NC–	I	rt	(53)
Bz	Cl	110	(23)
Bz	Br	80	(62)
Bz	I	rt	(75)

Refs: 641

C5-7

Nitrogen Nucleophile: R^1 pyrrole CO_2Me

Electrophile: R^2-substituted arene (positions 4,5), I, NHC(O)CF3

R^1	R^2	Time 1 (h)	Time 2 (h)	
H	H	23	11	(93)
H	4-F	40	10	(75)
H	4-MeO	23	12	(97)
H	4-Me	23	11	(93)
H	4-Ac	36	15	(44)
Cl	H	23	11	(85)
Cl	4-MeO	20	15	(84)
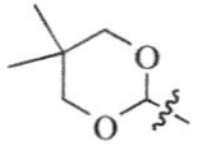 (dioxane acetal)	5-Ac	36	15	(72)

Conditions: 1. CuI (10 mol %), L-Pro (20 mol %), K2CO3, DMSO, 80°, time 1; 2. H2O, 60°, time 2

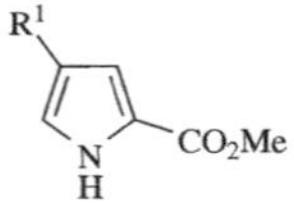

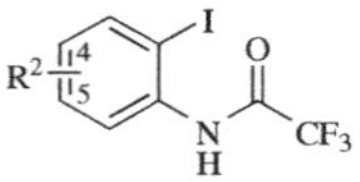

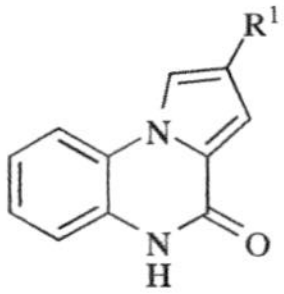

Product: pyrrole-fused quinolinone bearing R^1, R^2, NH, O

R^1	R^2	Time 1 (h)	Time 2 (h)	
MeO_2C	H	35	12	(86)
MeO_2C	5-MeO	29	13	(72)
Et	H	23	11	(83)
Et	4-MeO	20	12	(81)
Ac	H	25	11	(84)
Ac	4-Me	26	15	(88)
Ac	5-Et	38	12	(89)

Refs: 642

C$_5$

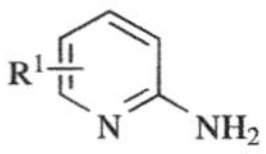

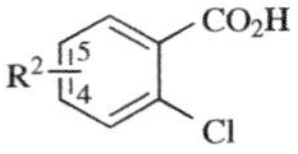

Cu (8 mol %), K$_2$CO$_3$,
DMF, reflux, 6 h

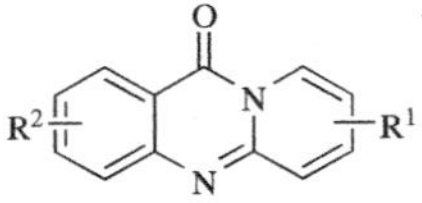

R^1	R^2	
H	H	(64)
H	4-Cl	(49)	643
H	3-O$_2$N	(52)
H	4-O$_2$N	(25)
H	5-O$_2$N	(72)
H	3.5-(O$_2$N)$_2$	(84)
H	5-MeO	(72)
3-O$_2$N	5-O$_2$N	(45)
5-O$_2$N	5-O$_2$N	(77)

C$_{5-8}$

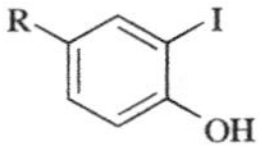

CuI (2.5 mol %),
EDA (5 mol %), K$_2$CO$_3$,
DMF, 110°

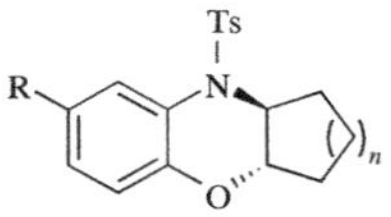

n	R	Time (h)	
1	H	20	(78)
1	Cl	12	(89)	182
2	H	15	(99)
2	Cl	14	(94)
2	t-Bu	16	(85)
3	H	13	(80)
3	Cl	15	(92)
3	t-Bu	16	(89)
3	Ph	16	(92)
4	H	26	(48)

Nitrogen Nucleophile	Electrophile	Conditions	Product(s) and Yield(s) (%)	Refs.
C$_{5-8}$				
	CuI (2.5 mol %), EDA (5 mol %), K$_2$CO$_3$, dioxane, 100°, 48 h			181

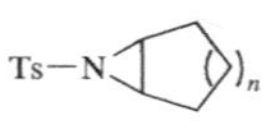

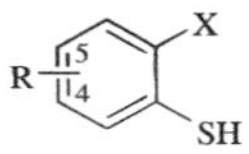

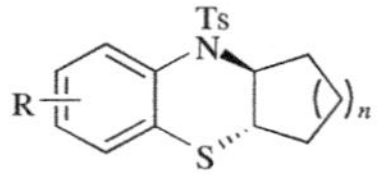

n	R	X	Time (h)		*n*	R	X	Time (h)
1	H	Cl	48	(60)	2	H	I	12	(97)
1	H	Br	28	(92)	2	4-Cl	Cl	48	(82)
1	4-Cl	Cl	38	(70)	2	5-Me	Br	24	(92)
1	5-Me	Br	21	(85)	3	H	Br	48	(74)
2	H	Cl	32	(74)	3	H	I	24	(74)
2	H	Br	17	(97)	4	H	Br	48	(94)
					4	5-Me	Br	28	(65)

C$_6$				
		CuI (10 mol %), L-Pro (20 mol %), Cs$_2$CO$_3$, DMSO, 50°, 8 h		641

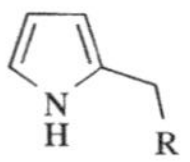

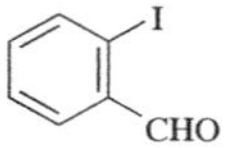

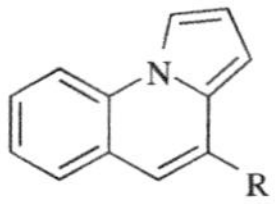

R	
NC–	(30)
EtO$_2$C	(25)

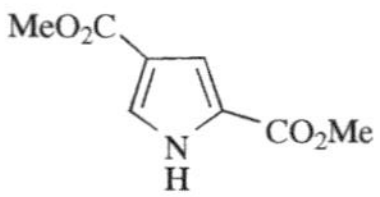	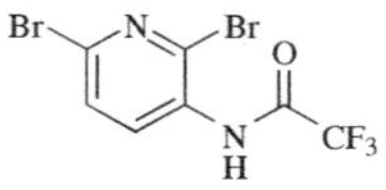	1. CuI (10 mol %), L-Pro (20 mol %), K$_2$CO$_3$, DMSO, 80°, 13 h 2. H$_2$O, 60°, 13 h	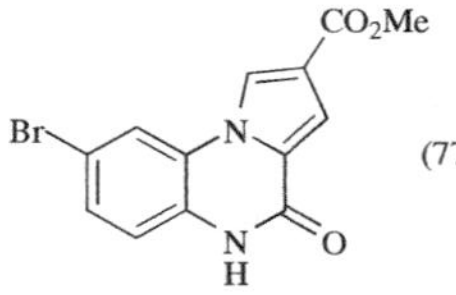 (77)	642

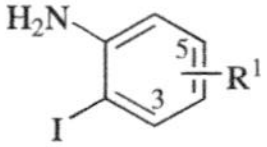	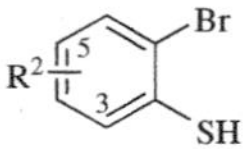	CuI (20 mol %), L-Pro (40 mol %), K$_2$CO$_3$, DMSO, 90°	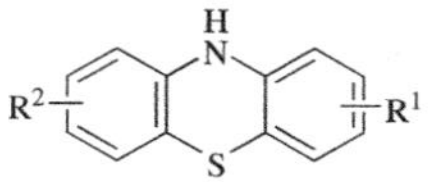	644

R^1	R^2	Time (h)	
H	3,5-Me$_2$	70	(46)
4-Cl	6-F	H	48	(80)

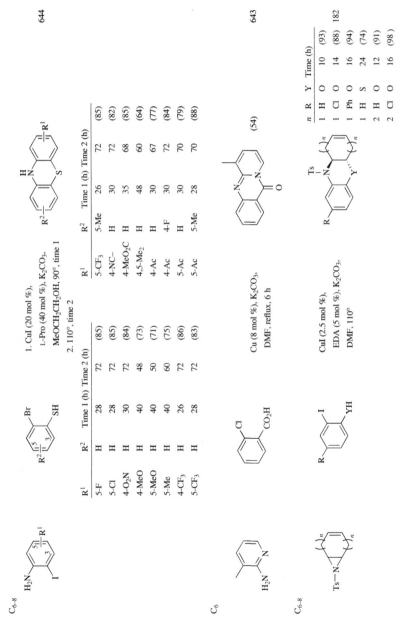

C6-8

1. CuI (20 mol %),
   L-Pro (40 mol %), K₂CO₃,
   MeOCH₂CH₂OH, 90°, time 1
2. 110°; time 2

644

R¹	R²	Time 1 (h)	Time 2 (h)		R¹	R²	Time 1 (h)	Time 2 (h)	
5-F	H	28	72	(85)	5-CF₃	5-Me	26	72	(85)
5-Cl	H	28	72	(85)	4-NC–	H	30	72	(82)
4-O₂N	H	30	72	(84)	4-MeO₂C	H	35	68	(85)
4-MeO	H	40	48	(73)	4,5-Me₂	H	48	60	(64)
5-MeO	H	40	50	(71)	4-Ac	H	30	67	(77)
5-Me	H	40	60	(75)	4-Ac	4-F	30	72	(84)
4-CF₃	H	26	72	(86)	5-Ac	H	30	70	(79)
5-CF₃	H	28	72	(83)	5-Ac	5-Me	28	70	(88)

C6

Cu (8 mol %), K₂CO₃,
DMF, reflux, 6 h

(54)

643

C6-8

CuI (2.5 mol %),
EDA (5 mol %), K₂CO₃,
DMF, 110°

182

n	R	Y	Time (h)	
1	H	O	10	(93)
1	Cl	O	14	(88)
1	Ph	O	16	(94)
1	H	S	24	(74)
2	H	O	12	(91)
2	Cl	O	16	(98)

571

TABLE 27. *N*-ARYLATIONS IN MULTI-STEP REACTIONS (*Continued*)

B. SYNTHESIS OF SIX-MEMBERED RINGS (*Continued*)

Nitrogen Nucleophile	Electrophile	Conditions	Product(s) and Yield(s) (%)	Refs.

C$_{7-8}$

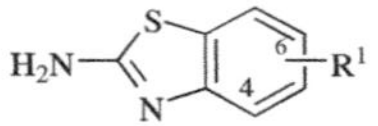

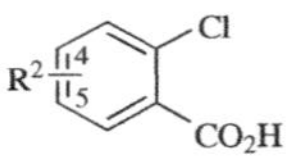

Cu (7.5 mol %), K$_2$CO$_3$, DMF, rt, 25 min

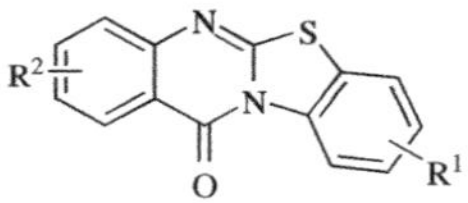

R^1	R^2	
H	H	(85)
H	4-F	(93)
H	4-Cl	(83)
H	5-Cl	(95)
H	4-O$_2$N	(86)
H	5-O$_2$N	(92)
6-MeO	H	(92)
6-EtO	H	(90)
4-Me	H	(94)
6-Me	H	(94)

242

C$_8$

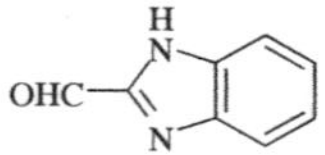

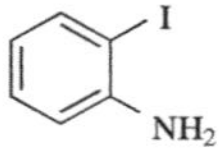

CuI (10 mol %), sparteine (20 mol %), K$_3$PO$_4$, NMP, 130°, 24 h

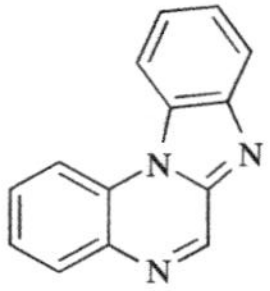

(50)

640

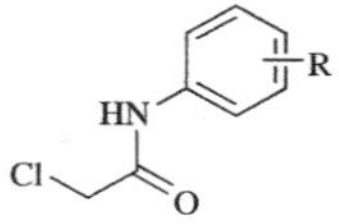

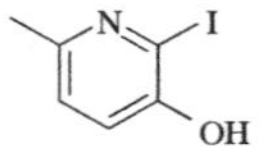

CuI (20 mol %), Cs$_2$CO$_3$, DBU, DMSO, MW, 130°, 10 min

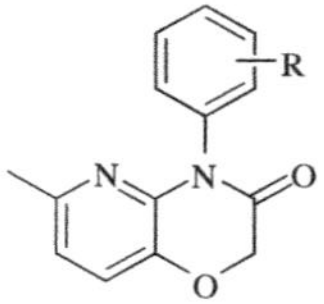

R	
4-MeO	(59)
2-Me	(78)
3-Me	(65)
4-Me	(71)

180

C$_{9-11}$

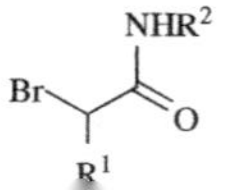

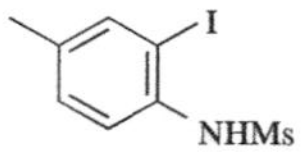

CuI (20 mol %), DMCDA (40 mol %), Cs$_2$CO$_3$, dioxane, reflux, 36 h

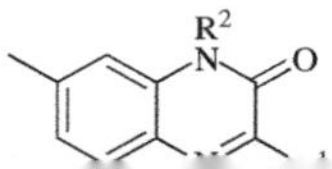

R^1	R^2	
Me	Ph	(51)
Et	Bn	(35)

639

	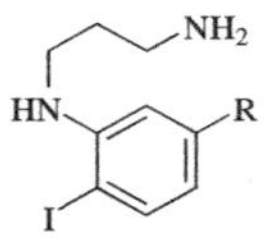	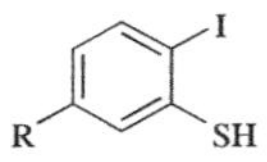	1. CuI (20 mol %), L-Pro (40 mol %), K_2CO_3, $MeOCH_2CH_2OH$, 90° 2. 110°	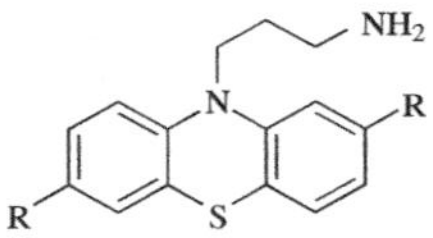 R: Cl (78); CF_3 (81); Ac (75)	644
C_{9-10}	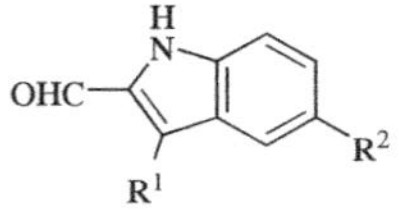	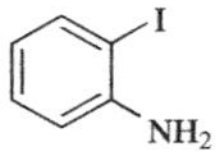	CuI (10 mol %), sparteine (20 mol %), K_3PO_4, NMP, 130°, 24 h	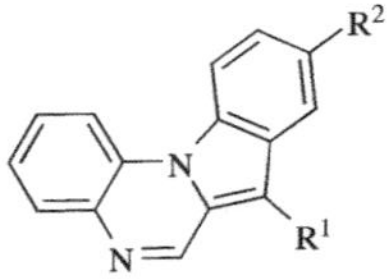 R^1 = H, R^2 = H (71); R^1 = H, R^2 = OMe (64); R^1 = Me, R^2 = F (78)	640
C_9	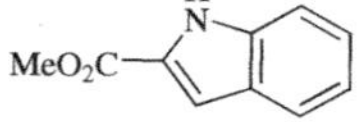	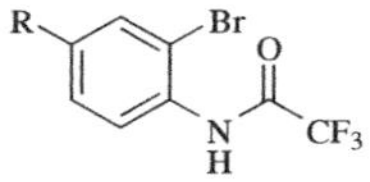	1. CuI (10 mol %), L-Pro (20 mol %), K_2CO_3, DMSO, 80°, 38 h 2. H_2O, 60°, 12 h	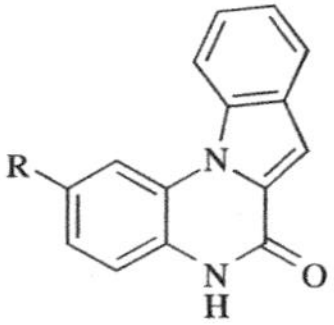 R: Me (74); Ac (63)	642
	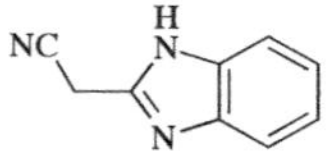	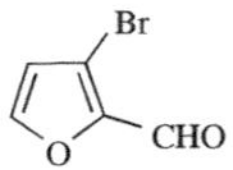	CuI (10 mol %), L-Pro (20 mol %), K_2CO_3, DMSO, 80°, 6 h	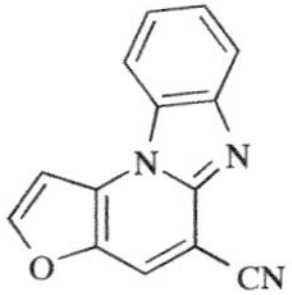 (62)	641
		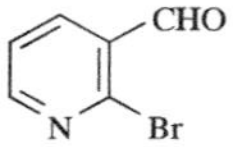	CuI (10 mol %), L-Pro (20 mol %), K_2CO_3, DMSO, 80°, 6 h	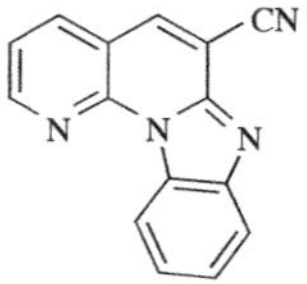 (93)	641

TABLE 27. *N*-ARYLATIONS IN MULTI-STEP REACTIONS (*Continued*)
B. SYNTHESIS OF SIX-MEMBERED RINGS (*Continued*)

Nitrogen Nucleophile	Electrophile	Conditions	Product(s) and Yield(s) (%)	Refs.

C9–10

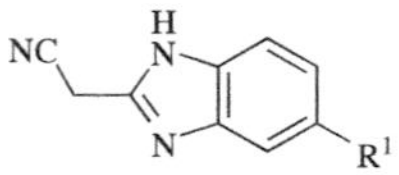

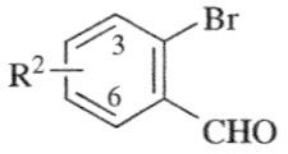

1. Piperidine, dioxane, rt
2. CuI (10 mol %), phen (20 mol %), K2CO3, 90°, 7 h

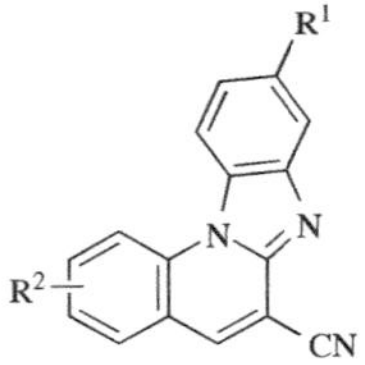

R^1	R^2	
H	H	(84)
H	5-Cl	(79)
H	5-HO	(43)
H	5-MeO	(75)
H	4-Me	(68)
Cl	H	(81)
Me	H	(88)
Me	5-MeO	(72)

645

C9

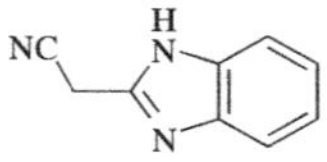

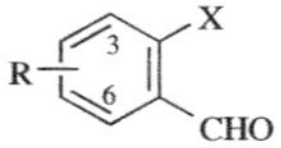

CuI (10 mol %), L-Pro (20 mol %), K2CO3, DMSO

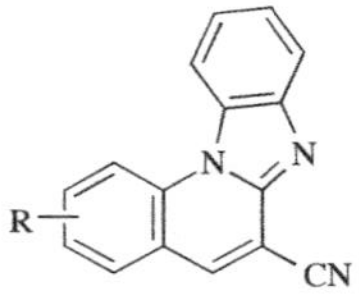

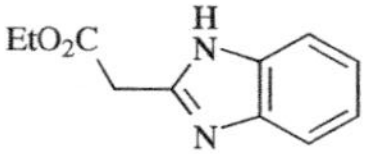

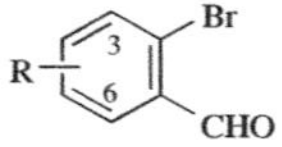

R	X	Temp (°)	Time (h)		R	X	Temp (°)	Time (h)	
H	Br	80	6	(87)	4,5-(MeO)2	Br	80	6	(43)
4-Cl	Cl	80	6	(50)	5-Me	Br	80	6	(45)
5-Cl	Br	80	6	(91)	5-CF3	Br	80	6	(81)
5-Br	I	rt	8	(82)	5-NC–	I	rt	8	(90)
5-O2N	I	rt	8	(85)	4-MeO2C	I	rt	8	(85)
5-HO	I	50	8	(89)	4,5-Me2	I	50	8	(85)
5-MeO	I	50	8	(89)	3-(CH=CH)2-4	Br	80	6	(80)
3,4-(MeO)2	Cl	110	8	(47)					

641

1. Piperidine, dioxane, rt
2. CuI (10 mol %), phen (20 mol %), K2CO3, 90°, 7 h

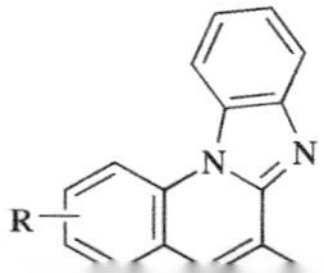

R	
H	(40)
4-Cl	(45)
5-Cl	(37)
5-MeO	(20)

645

C11		

Substrate: structure with X, N-NHAc, CO₂t-Bu, OTBS

Reagent: R— / X / CHO

CuI (10 mol %),
L-Pro (20 mol %), K$_2$CO$_3$,
DMSO, 8 h

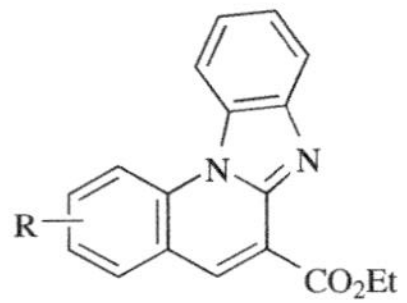

Product: benzimidazole, CO$_2$Et, R—

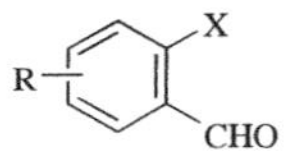

X	Temp (°)	
Cl	110	(30)
Br	80	(60)
I	rt	(72)

641

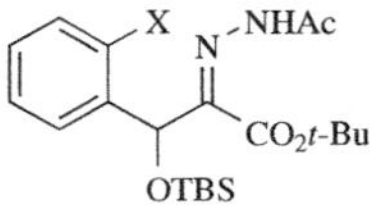

Intramolecular

CuI (2 eq), CsOAc, DMSO, rt

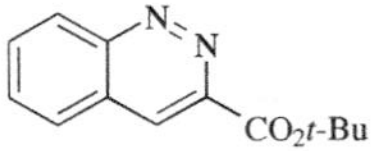

Product: CO$_2$t-Bu

X	Time	
Br	15 h	(30)
I	10 min	(100)

580

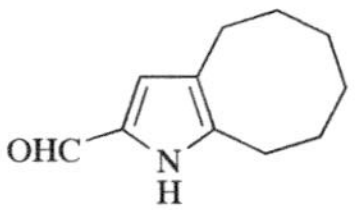

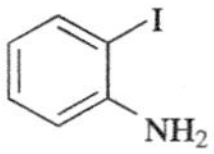

C13		

Substrate: cyclooctane-fused indole, OHC

Reagent: I / NH$_2$

CuI (10 mol %),
sparteine (20 mol %), K$_3$PO$_4$,
NMP, 130°, 236 h

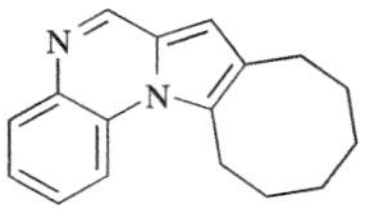

(74)

640

Substrate: OMe indole, OHC

Reagent: NH$_2$ / I

CuI (10 mol %),
sparteine (20 mol %), K$_3$PO$_4$,
NMP, 130°, 48 h

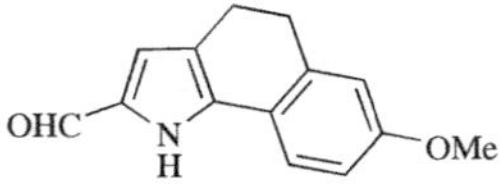

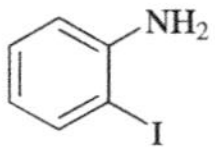

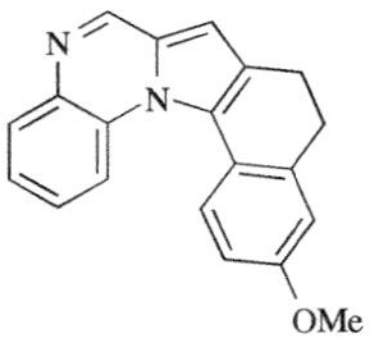

(63) OMe

640

C14		

Substrate: benzimidazole, H, N, S(=O), phenyl

Reagent: X / CHO

CuI (10 mol %),
L-Pro (20 mol %), K$_2$CO$_3$,
DMSO, 80°, 8 h

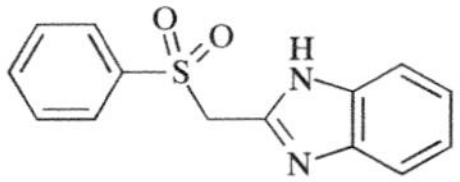

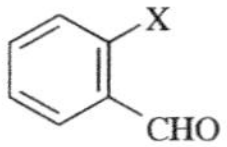

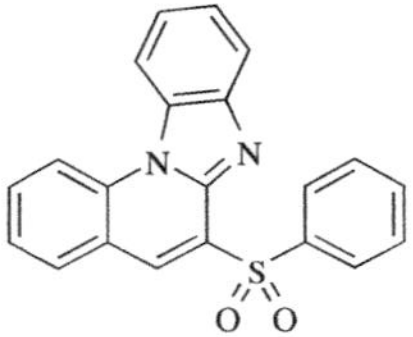

Product: phenylsulfonyl benzimidazole

X	
Br	(62)
I	(65)

641

Nitrogen Nucleophile	Electrophile	Conditions	Product(s) and Yield(s) (%)	Refs.
C$_{14}$				
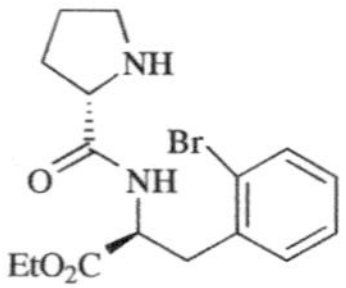	Intramolecular	CuI (1 eq), Cs$_2$CO$_3$, DMSO, 90°, 12 h	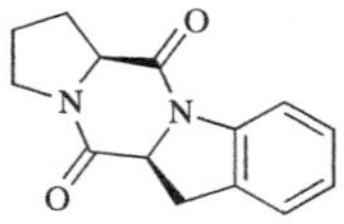 (48) 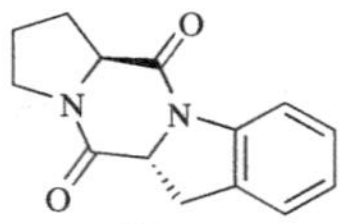 + (12)	576
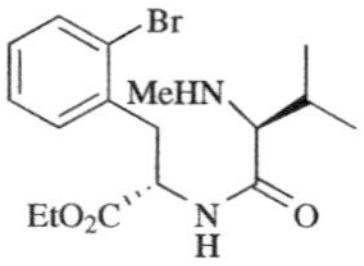	Intramolecular	CuI (1 eq), Cs$_2$CO$_3$, DMSO, 90°, 12 h	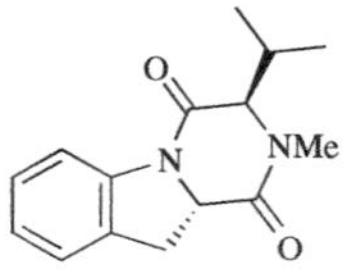 (59) 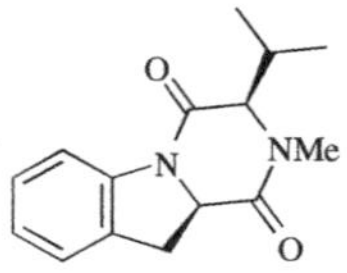 + (18)	576
C$_{16}$				
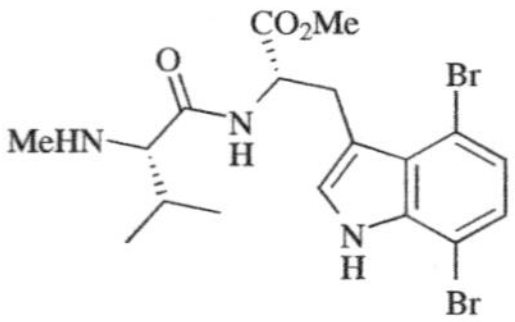	Intramolecular	CuI (1 eq), Cs$_2$CO$_3$, DMSO, 90°, 12 h	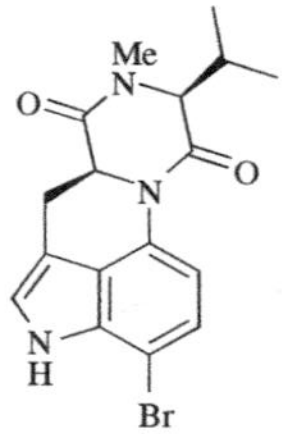 (63)	576

Nitrogen Nucleophile	Electrophile	Conditions	Product(s) and Yield(s) (%)	Refs.

C$_0$

H$_2$NBoc — electrophile: structure with OMs and I — Conditions: CuI (5 mol %), DMEDA (20 mol %), Cs$_2$CO$_3$, THF, 80°, 16 h — Product (57) — Refs. 607

C$_{2-8}$

H$_2$NR^1 — electrophile: proline-derived structure with R^3, CO$_2$i-Bu, I, R^2 — Conditions: 1. Cu$_2$O (20 mol %), L-Pro (20 mol %), K$_2$CO$_3$, DMSO, 80°, 24 h; 2. 1 N HCl, 110°, 5 h

R^1	R^2	R^3		R^1	R^2	R^3	
BocHNCH$_2$CH$_2$	H	H	(62)	Bn	5-Cl	H	(81)
BocHNCH$_2$CH$_2$	5-Cl	TBSO	(50)	Bn	4,5-(MeO)$_2$	H	(85)
t-BuO$_2$CCH$_2$	H	TBSO	(66)	Bn	5-Me	H	(80)
n-Pr	H	H	(76)	4-MeOC$_6$H$_4$CH$_2$	H	H	(60)
allyl	H	H	(82)	3,4-(OCH$_2$O)C$_6$H$_3$CH$_2$	H	H	(81)
c-Pr	H	H	(63)	3,4,5-(MeO)$_3$C$_6$H$_2$	H	H	(64)
TBSO(CH$_2$)$_3$	H	H	(46)	3,4,5-(MeO)$_3$C$_6$H$_2$	4,5-(MeO)$_2$	H	(75)
c-PrCH$_2$	H	H	(70)	4-CF$_3$C$_6$H$_4$CH$_2$	H	H	(75)
Bn	5-F	H	(56)				

Refs. 646

C$_{3-11}$

H$_2$N–CO$_2$H (R^1) — electrophile: structure with Br, NHR^3, R^2 — Conditions: 1. CuI (10 mol %), Cs$_2$CO$_3$, DMF, 90°, 24 h; 2. DPPA, 5°, 8 h

R^1	R^2	R^3		R^1	R^2	R^3	
Me	H	Bn	(50)	i-Pr	MeO	Bn	(35)
i-Pr	H	allyl	(52)	i-Pr	MeO$_2$C	Bn	(48)
i-Pr	H	n-Bu	(57)	Bn	H	Bn	(51)
i-Pr	H	Bn	(58)	4-HOC$_6$H$_4$CH$_2$	H	Bn	(52)
i-Pr	F	Bn	(48)	3-indolylmethyl	H	Bn	(40)
i-Pr	O$_2$N	Bn	(44)				

Refs. 176

TABLE 27. *N*-ARYLATIONS IN MULTI-STEP REACTIONS (*Continued*)
C. SYNTHESIS OF SEVEN- AND HIGHER-MEMBERED RINGS (*Continued*)

Nitrogen Nucleophile	Electrophile	Conditions	Product(s) and Yield(s) (%)	Refs.
C_3				
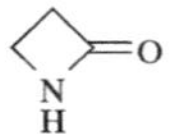	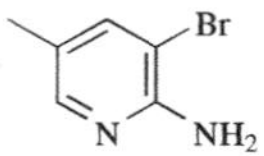	1. Cu (5 mol %), K_2CO_3, toluene, DMEDA (10 mol %), 100°, 24 h 2. Ti(O*i*-Pr)$_4$ (50 mol %), dioxane, 110°, 24 h	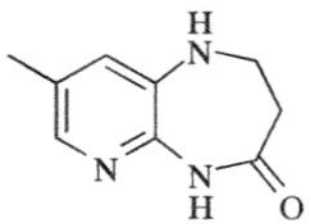 (59)	177
	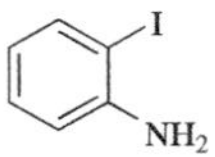	1. Cu (5 mol %), K_2CO_3, toluene, DMEDA (10 mol %), 100°, 24 h 2. Ti(O*i*-Pr)$_4$ (50 mol %), dioxane, 110°, 24 h	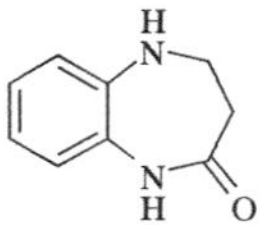 (92)	177
C_{3-9}				
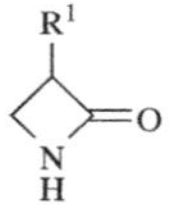	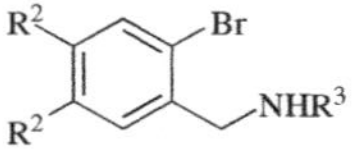	CuI (5 mol %), DMEDA (10 mol %), K_2CO_3, toluene, 110°, 24 h	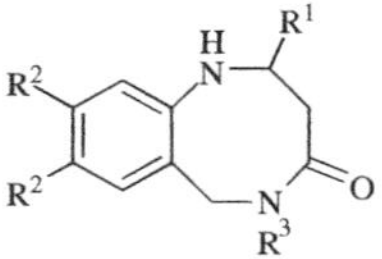 R^1 R^2 R^3 H H H (96) H H HO(CH$_2$)$_5$ (90) H MeO Bn (90) Ph H H (88)	177
C_3				
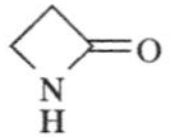	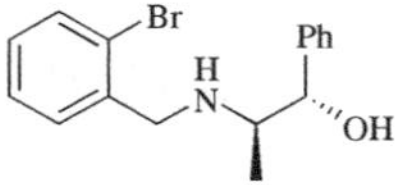	1. CuI (5 mol %), K_2CO_3, toluene, DMEDA (10 mol %), 110°, 24 h 2. AcOH, THF, 60°, 4 h	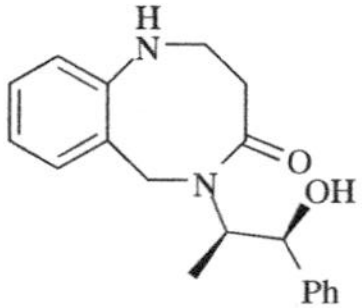 (90)	177
	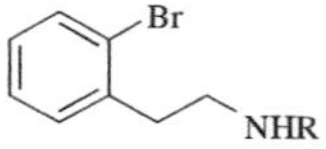	CuI (5 mol %), K_2CO_3, toluene, 110°, 24 h	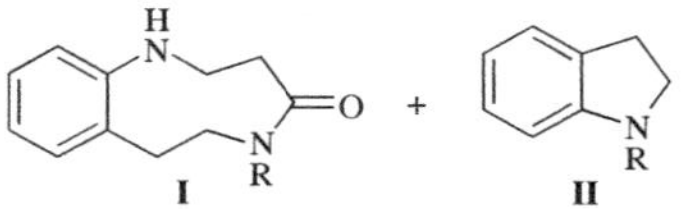 I + II R Additive I II H none (68) (18) Bn DMEDA (10 mol %) (78) (17)	177

646

C7 H2NBn

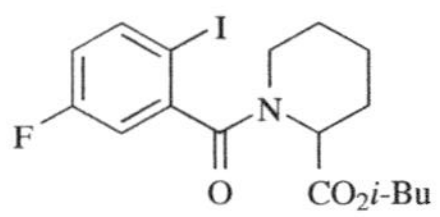

1. Cu$_2$O (20 mol %),
   L-Pro (20 mol %), K$_2$CO$_3$,
   DMSO, 80°, 24 h
2. 1 N HCl, 110°, 5 h

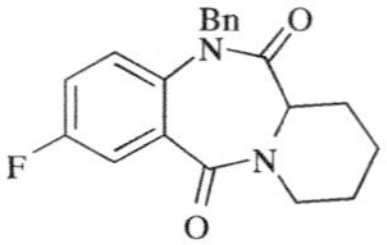

(65)

C7–14

346

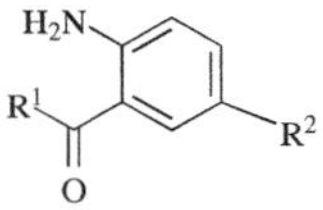

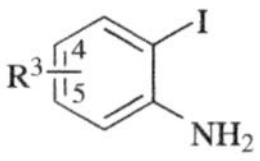

Cu$_2$O (10 mol %), K$_2$CO$_3$,
xylene, 145°

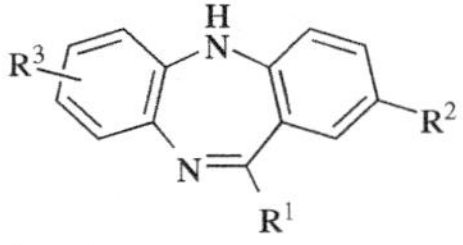

R^1	R^2	R^3	Time (h)	
H	H	4-Me	7	(60)
Me	H	4-Cl	4.5	(85)
Me	H	4-Br	6	(75)
Me	H	4-Me	8	(88)
Me	H	4,5-Me$_2$	8	(90)
Ph	H	4-Me	5	(80)

R^1	R^2	R^3	Time (h)	
Ph	Cl	4-Me	8	(80)
Ph	Cl	4,5-Me$_2$	2.5	(78)
Ph	Me	4-Cl	7	(80)
Ph	Me	4,5-Me$_2$	2.5	(76)
3-MeC$_6$H$_4$	Cl	H	20	(40)
3-MeC$_6$H$_4$	Cl	4-Br	13	(68)
3-MeC$_6$H$_4$	Cl	4-Me	6	(82)

C12

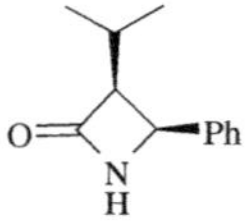

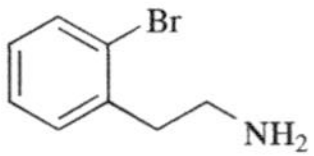

177

1. CuI (5 mol %), K$_2$CO$_3$,
   toluene, DMEDA
   (10 mol %), 110°, 24 h
2. AcOH, NEt$_3$, dioxane,
   110°, 24 h

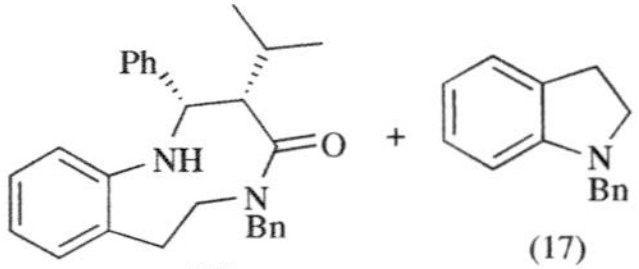

(17)

+

(78)

Nitrogen Nucleophile	Electrophile	Conditions	Product(s) and Yield(s) (%)	Refs.
C_{13}				
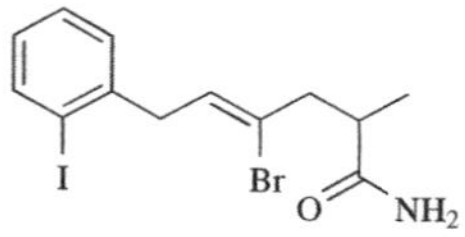	Intramolecular	1. CuI (10 mol %), DMG (20 mol %), K_2CO_3, MeCN, reflux, 12 h 2. 2 N HCl, THF	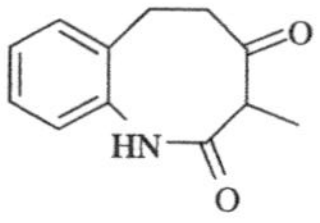 (90)	603
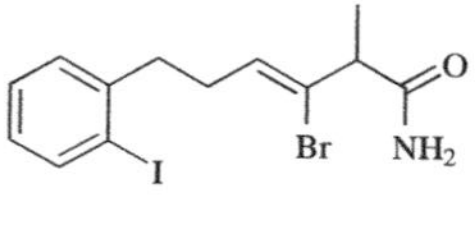	Intramolecular	1. CuI (10 mol %), DMG (20 mol %), K_2CO_3, MeCN, reflux, 12 h 2. 2 N HCl, THF	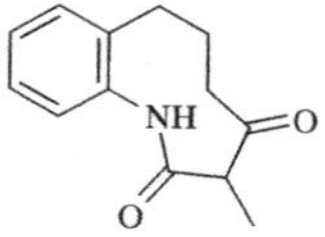 (99)	603
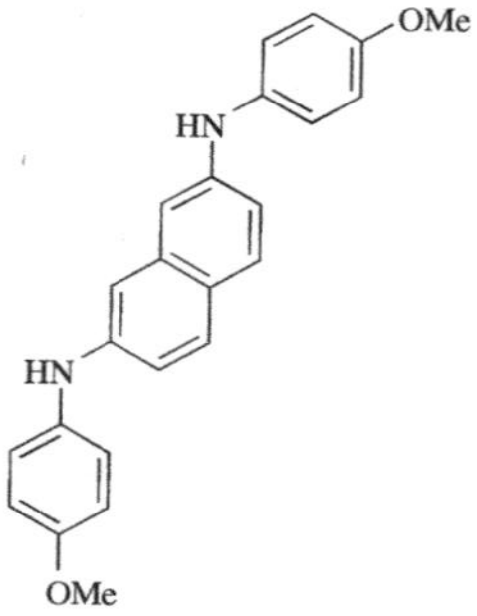	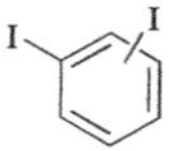	Cu (4 eq), K_2CO_3, Ph_2O, 190°, 48 h	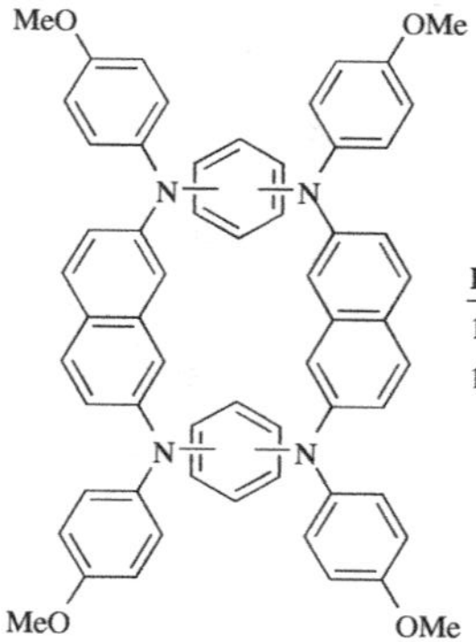 Isomer 1,3 (21) 1,4 (28)	633

TABLE 28. N-VINYLATION OF AMINES

Nitrogen Nucleophile	Electrophile	Conditions	Product(s) and Yield(s) (%)	Refs.

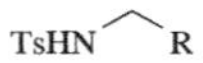

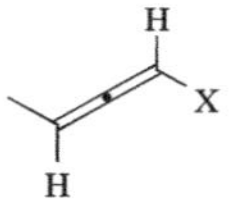

*Please refer to the charts preceding the tables for structures indicated by the **bold** numbers.*

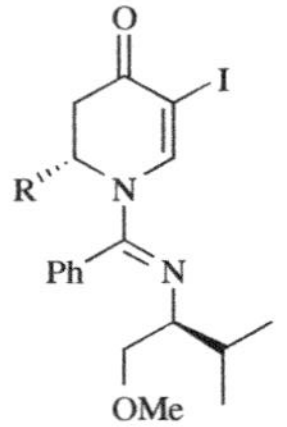

C_0

H₂NBoc

CuI (10 mol %), DMEDA (20 mol %), Cs₂CO₃, toluene, 90°

R	
Et	(93)
CH₂=CH(CH₂)₅	(93)

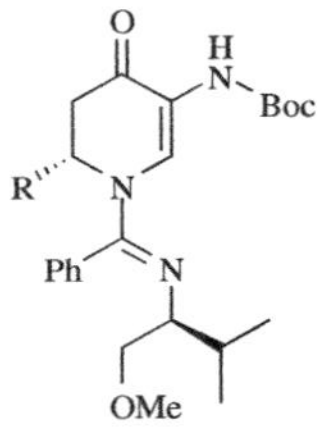

647

C_{2-10}

TsNHR¹

8 (15 mol %), DMEDA (13 mol %), Cs₂CO₃, 24 h

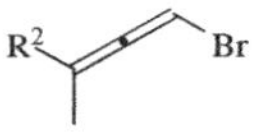

625

R¹	R²	Solvent	Temp (°)
t-BuO₂CCH₂	Me	dioxane	reflux (41)
CH₂=CHCH₂	Me	toluene	80 (89)
(E)-MeCH=CHCH₂	Me	toluene	80 (82)
Me₂C=CHCH₂	Me	toluene	80 (74)
Me₂C=CHCH₂	Et	toluene	80 (67)
(E)-n-PrCH₂CH=CH	Me	toluene	80 (72)

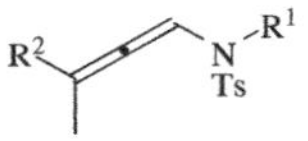

R	R²	Solvent	Temp (°)
Ph	Me	dioxane	reflux (56)
2-AcC₆H₄	Me	dioxane	reflux (63)
(E)-PhCH₂CH=CH	Me	toluene	40 (53)
(E)-4-MeC₆H₄CH₂CH=CH	Me	toluene	40 (20)
(E)-4-CF₃C₆H₄CH₂CH=CH	Me	toluene	40 (52)

C_{5-7}

TsHN⌒R

Catalyst (x mol %), DMEDA (y mol %), Cs₂CO₃, toluene

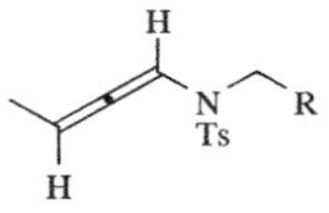

R	X	Catalyst	x	y	Temp (°)	Time (h)		Refs.
Me₂C=CH₂	Br	**8**	15	30	80	24	(35)	625
Ph	I	CuCN	10	20	50	—	(60)	91

TABLE 28. N-VINYLATION OF AMINES (*Continued*)

Nitrogen Nucleophile	Electrophile	Conditions	Product(s) and Yield(s) (%)	Refs.

*Please refer to the charts preceding the tables for structures indicated by the **bold** numbers.*

C_{6–7}

| 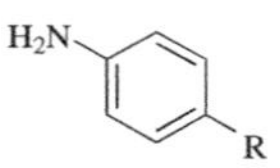 | 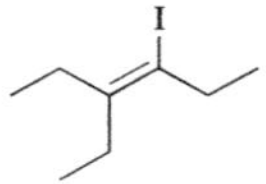 | CuI (10 mol %), DMEDA (20 mol %), NaO*t*-Bu, toluene, 140°, 24 h | 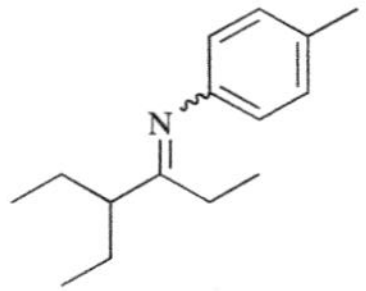 **R** _(E)/(Z)_
 H (64) 1:1
 Me (88) 1:1 | 88 |

C_{7–14}

HNR¹R²

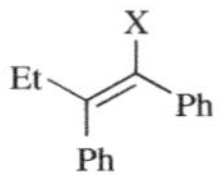

| | | CuI (10 mol %), DMEDA (20 mol %), NaO*t*-Bu, toluene, 140° | | 88 |

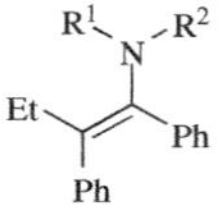

R¹	R²	X	Time (h)	
Me	Ph	I	36	(85)
Et	Ph	I	36	(85)
Ph	Ph	Br	16	(60)
Ph	Ph	I'	16	(89)
Ph	1-Np	I	16	(87)
4-ClC₆H₄	4-ClC₆H₄	I	48	(81)
4-MeC₆H₄	4-MeC₆H₄	I	48	(81)

C₁₂

HNPh₂

| | | CuI (10 mol %), DMEDA (20 mol %), NaO*t*-Bu, toluene, 140° | | 88 |

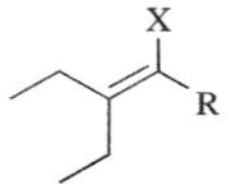

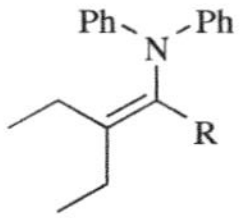

R	X	Time (h)	
Et	I	16	(63)
n-Pr	Br	36	(67)
n-Pr	I	36	(42)

C_{12–14}

HNAr₂

| | | CuI (10 mol %), DMEDA (20 mol %), NaO*t*-Bu, toluene, 140° | | 88 |

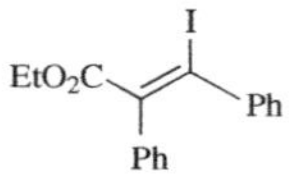

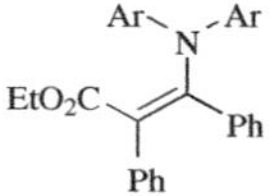

Ar	Time (h)	
Ph	16	(55)
4-MeC₆H₄	24	(67)

| | 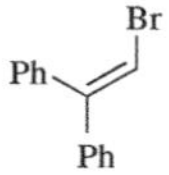 | CuI (10 mol %), DMEDA (20 mol %), NaO*t*-Bu, toluene, 140°, 24 h 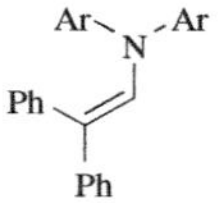 | **Ar**
 Ph (89)
 4-MeC₆H₄ (89) | 88 |

582

TABLE 29A. N-VINYLATION OF PYRROLES, PYRAZOLES, TRIAZOLES, AND TETRAZOLES

Nitrogen Nucleophile	Electrophile	Conditions	Product(s) and Yield(s) (%)	Refs.

*Please refer to the charts preceding the tables for structures indicated by the **bold** numbers.*

C₂

| | Ph vinyl Br | CuI (10 mol %), L30 (5 mol %), Cs₂CO₃, DMF, 110°, 24 h | (triazole) (38) + (triazole) (50) | 190 |
| | Ph vinyl Br | CuI (10 mol %), L30 (5 mol %), Cs₂CO₃, DMF, 110°, 24 h | (imidazole) (94) | 190 |

C₃

| | isobutenyl Br | CuI (10 mol %), ligand (20 mol %), Cs₂CO₃, MeCN, 80° | | |

Ligand	Time (h)	
L30	36	(70)
L47	24	(47)

Refs: 648, 190

C₃₋₄

3-R-pyrazole + Ph vinyl Br → Catalyst (10 mol %), Cs₂CO₃ → **I** + **II**

R	Catalyst	Additive	Solvent	Temp (°)	Time (h)	I	II	Refs.
H	CuI	none	DMF	120	36	(95)	(—)	103
H	CuI	L25 (10 mol %)	MeCN	80	15	(90)	(—)	395
H	CuI	L30 (5 mol %)	MeCN	50	30	(94)	(—)	190
H	CuI	L47 (10 mol %)	MeCN	80	2	(100)	(—)	413
H	CuI	L48 (0.21 mol %)	MeCN	80	2	(100)	(—)	413
H	9	none	MeCN	82	4	(100)	(—)	27
Me	CuI	L30 (5 mol %)	MeCN	50	30	(62)	(27)	190
CF₃	CuI	L30 (5 mol %)	MeCN	50	30	(98)	(0)	190

C₄

| | EtO₂C vinyl I, R | CuI (50 mol %), DMEDA (1 eq), K₃PO₄, toluene, 65° | (pyrrole product) | 649 |

R	
t-Bu	(77)
n-C₅H₁₁	(63)
Ph	(54)

Nitrogen Nucleophile	Electrophile	Conditions	Product(s) and Yield(s) (%)	Refs.

*Please refer to the charts preceding the tables for structures indicated by the **bold** numbers.*

C₄₋₅

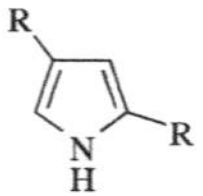

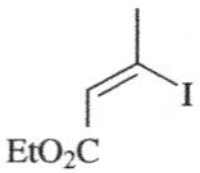

CuI (50 mol %),
DMEDA (1 eq), K₃PO₄,
toluene, 65°

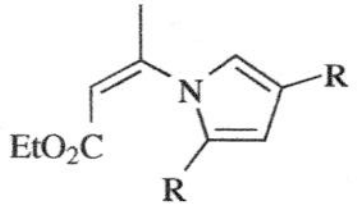

R	
H	(70)
Me	(40)

649

C₄

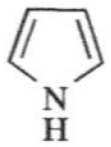

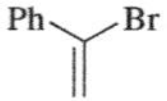

CuI (20 mol %),
EDA (40 mol %), K₃PO₄,
dioxane, 110°, 48 h

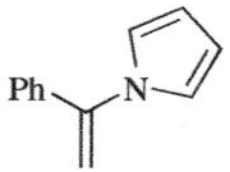

(21)

102

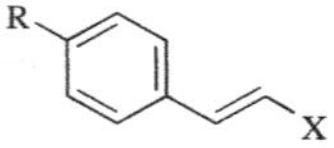

CuI (10 mol %), Cs₂CO₃

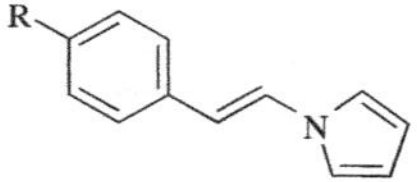

R	X	Additive	Solvent	Temp (°)	Time (h)	
H	Cl	none	DMF	120	50	(7)
H	Br	none	DMF	120	36	(75)
H	Br	**L30** (5 mol %)	MeCN	80	24	(89)
MeO	Cl	none	DMF	120	50	(13)
MeO	Br	none	DMF	120	36	(93)

103
103
190
103
103

C₇

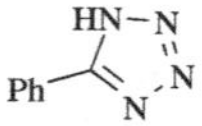

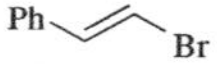

CuI (10 mol %),
L30 (5 mol %), Cs₂CO₃,
MeCN, 80°, 24 h

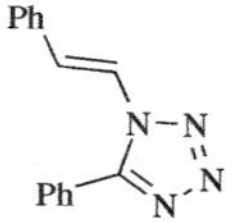

(0)

190

TABLE 29B. N-VINYLATION OF IMIDAZOLES

*Please refer to the charts preceding the tables for structures indicated by the **bold** numbers.*

Nitrogen Nucleophile	Electrophile	Conditions	Product(s) and Yield(s) (%)	Refs.
 C₃ imidazole (NH)	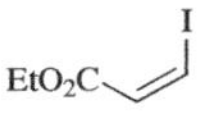 EtO₂C–CH=CH–I	CuO nanoparticles (1.5 eq), KOH, DMSO, 80°, 5 h	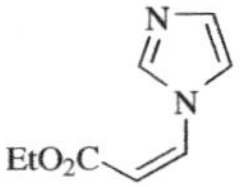 (0)	94
	isopropenyl bromide	CuI (10 mol %), EDA (20 mol %), K₃PO₄, dioxane, 110°, 48 h	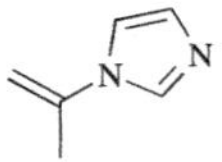 (58)	102
	2-(2-bromovinyl)furan	Cu₂O (5 mol %), **L9** (10 mol %), Cs₂CO₃, MeCN, 85°, 24 h	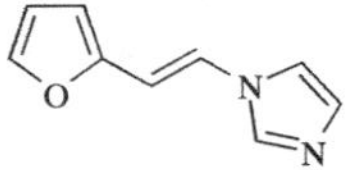 (70)	89
C₃₋₄ R-imidazole (NH)	α-bromostyrene	CuI (10 mol %), EDA (20 mol %), K₃PO₄, dioxane, 110°, 48 h	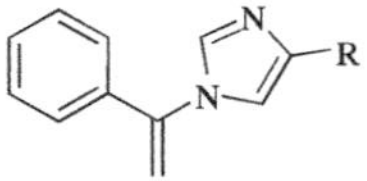 R: H (41); Me (25)	102
C₃ imidazole (NH)	(Z)-PhCH=CH–Br	CuI (10 mol %), Cs₂CO₃	see below	103, 650

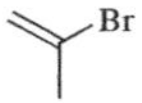

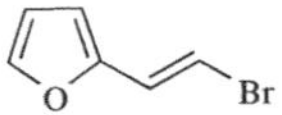

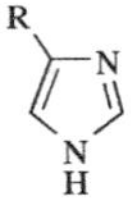

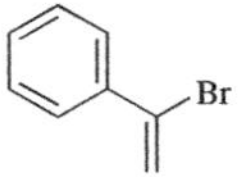

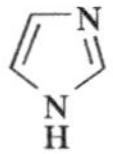

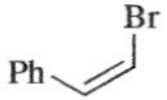

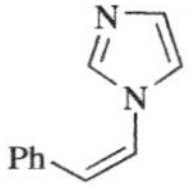

Additive	Solvent	Temp (°)	Time (h)	
none	DMF	120	50	(61)
L9 (30 mol %)	DMSO	80	24	(0)

585

TABLE 29B. N-VINYLATION OF IMIDAZOLES (Continued)

Please refer to the charts preceding the tables for structures indicated by the **bold** numbers.

C₃

Nitrogen Nucleophile	Electrophile		Conditions						Product(s) and Yield(s) (%)	Refs.
				Catalyst (x amount)						
	R	X	Catalyst	x	Additive(s)	Solvent	Temp (°)	Time (h)		
	H	Cl	CuI	10 mol %	Cs₂CO₃	DMF	120	50	(22)	103
	H	Br	CuI	10 mol %	L-Pro (20 mol%), K₂CO₃	[C₄mim][BF₄]	110	20	(87)	231
	H	Br	CuI	15 mol %	L9 (30 mol %), Cs₂CO₃	DMSO	60	24	(88)	650
	H	Br	CuI	10 mol %	L30 (5 mol %), Cs₂CO₃	MeCN	60	24	(100)	190
	H	Cl	Cu₂O	5 mol %	L9 (10 mol %), Cs₂CO₃	MeCN	85	24	(20)	89
	H	Cl	Cu₂O	5 mol %	L9 (10 mol %), Cs₂CO₃	MeCN	130	45	(53)	89
	H	Br	Cu₂O	5 mol %	L9 (10 mol %), Cs₂CO₃	MeCN	85	24	(98)	89
	H	Cl	CuO nanoparticles	2 eq	KOH	DMSO	80	11	(20)	94
	H	Br	CuO nanoparticles	2 eq	KOH	DMSO	80	11	(89)	94
	H	I	CuO nanoparticles	1.5 eq	KOH	DMSO	80	5	(98)	94
	3-F	I	CuO nanoparticles	1.5 eq	KOH	DMSO	80	5	(88)	94
	4-F	Br	CuI	15 mol %	L9 (30 mol %), Cs₂CO₃	DMSO	60	24	(90)	650
	4-F	Br	Cu₂O	5 mol %	L9 (10 mol %), Cs₂CO₃	MeCN	85	24	(99)	89
	4-Cl	Br	CuI	10 mol %	L-Pro (20 mol%), K₂CO₃	[C₄mim][BF₄]	110	20	(93)	231
	4-Cl	Br	CuO nanoparticles	2 eq	KOH	DMSO	80	11	(82)	94
	4-Cl	Br	CuI	15 mol %	L9 (30 mol %), Cs₂CO₃	DMSO	60	24	(87)	650
	4-Cl	Br	Cu₂O	5 mol %	L9 (10 mol %), Cs₂CO₃	MeCN	85	24	(99)	89
	4-Cl	I	CuO nanoparticles	1.5 eq	KOH	DMSO	80	5	(91)	94
	4-MeO	Cl	CuI	10 mol %	Cs₂CO₃	DMF	140	36	(66)	103
	4-MeO	Br	CuI	10 mol %	Cs₂CO₃	DMF	120	36	(57)	103
	4-MeO	Br	Cu₂O	5 mol %	L9 (10 mol %), Cs₂CO₃	MeCN	85	24	(92)	89
	4-MeO	I	CuO nanoparticles	1.5 eq	KOH	DMSO	80	5	(98)	94
	3,4-(MeO)₂	Br	Cu₂O	5 mol %	L9 (10 mol %), Cs₂CO₃	MeCN	85	24	(72)	89
	4-Me	Br	CuI	10 mol %	L-Pro (20 mol %), K₂CO₃	[C₄mim][BF₄]	110	20	(86)	231
	4-Me	Br	CuI	15 mol %	L9 (30 mol %), Cs₂CO₃	DMSO	60	24	(85)	650
	4-Me	Br	Cu₂O	5 mol %	L9 (10 mol %), Cs₂CO₃	MeCN	85	24	(91)	89
	4-Me	Br	CuO nanoparticles	1.5 eq	KOH	DMSO	80	5	(80)	94
	4-Me	I	CuO nanoparticles	1.5 eq	KOH	DMSO	90	5	(90)	94

C3–9

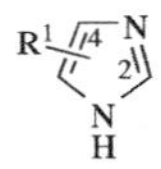

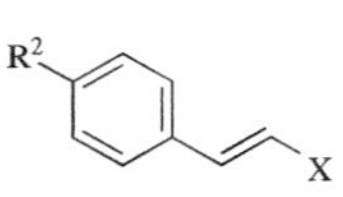

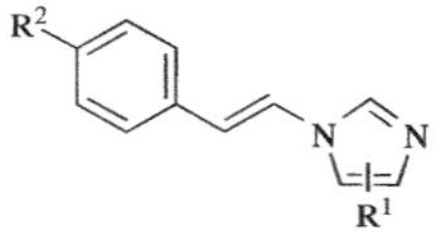

R¹	R²	X	Catalyst	x	Additive(s)	Solvent	Temp (°)	Time (h)			
4-O$_2$N	H	I	CuO nanoparticles	1.5 eq	KOH	DMSO	80	5	(10)		94
2-Me	H	Br	CuI	10 mol %	L-Pro (20 mol %), K$_2$CO$_3$	[C$_4$mim][BF$_4$]	110	30	(60)		231
2-Me	H	Br	Cu$_2$O	5 mol %	L9 (10 mol %), Cs$_2$CO$_3$	MeCN	85	24	(62)		89
4-NC—	H	I	CuO nanoparticles	1.5 eq	KOH	DMSO	80	5	(0)		94
4-Ph	H	Br	CuO nanoparticles	2 eq	KOH	DMSO	80	11	(79)		94
4-Ph	H	I	CuO nanoparticles	1.5 eq	KOH	DMSO	80	5	(87)		94
4-Ph	Cl	I	CuO nanoparticles	1.5 eq	KOH	DMSO	80	5	(89)		94
4-Ph	MeO	Br	CuO nanoparticles	2 eq	KOH	DMSO	80	11	(84)		94
4-Ph	MeO	I	CuO nanoparticles	1.5 eq	KOH	DMSO	80	5	(91)		94
4-Ph	Me	I	CuO nanoparticles	1.5 eq	KOH	DMSO	80	5	(84)		94

C$_3$

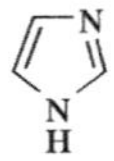

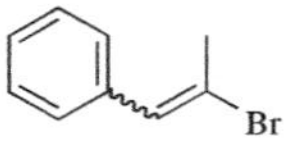

CuI (10 mol %)

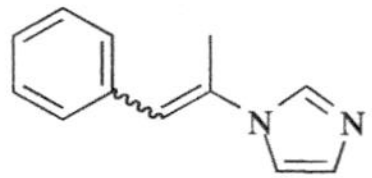

(E)/(Z)	Additive(s)	Solvent	Temp (°)	Time (h)		(E)/(Z)	
58:42	Cs$_2$CO$_3$	DMF	120	50	(70)	67:33	103
—	L-Pro (20 mol %), K$_2$CO$_3$	[C$_4$mim][BF$_4$]	110	20	(87)	—	231

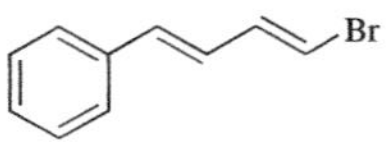

Cu$_2$O (5 mol %), L9 (10 mol %), Cs$_2$CO$_3$, MeCN, 85°, 24 h

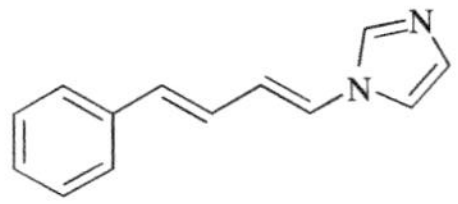

(60) 89

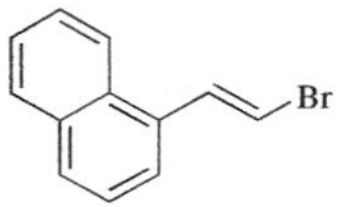

Cu$_2$O (5 mol %), L9 (10 mol %), Cs$_2$CO$_3$, MeCN, 85°, 24 h

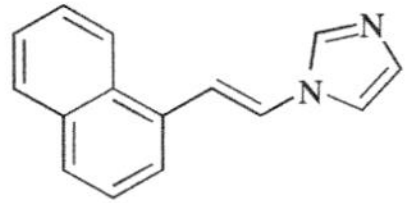

(98) 89

TABLE 29C. N-VINYLATION OF INDOLES

Nitrogen Nucleophile	Electrophile	Conditions	Product(s) and Yield(s) (%)	Refs.
C₈		CuI (10 mol %), EDA (20 mol %), K₃PO₄, dioxane, 110°, 24 h	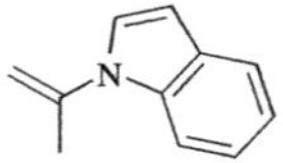 (85)	102
	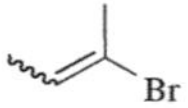	CuI (10 mol %), EDA (20 mol %), K₃PO₄, dioxane, 110°, 24 h	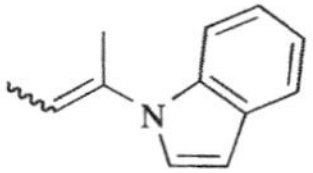 (63)	102
	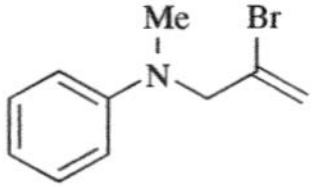	CuI (10 mol %), EDA (20 mol %), K₃PO₄, dioxane, 110°, 24 h	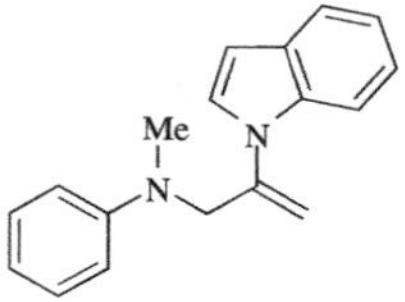 (88)	102
	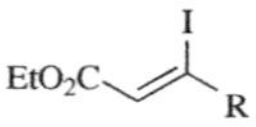	CuI (50 mol %), DMEDA (1 eq), K₃PO₄, toluene, 65°	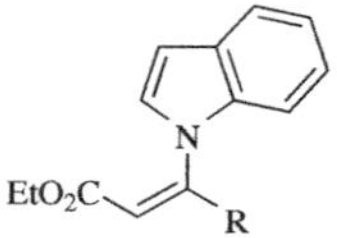 $\dfrac{R}{Me \quad (60)}$ $n\text{-C}_5H_{11} \quad (69)$	649
	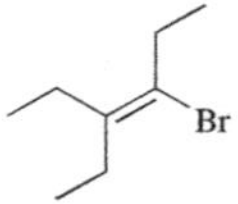	CuI (10 mol %), EDA (20 mol %), K₃PO₄, dioxane, 110°, 24 h	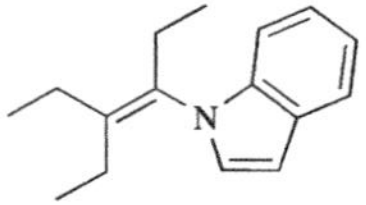 (78)	102
	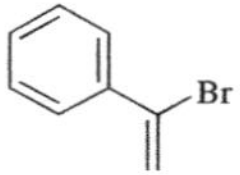	CuI (10 mol %), EDA (20 mol %), K₃PO₄, dioxane, 110°, 24 h	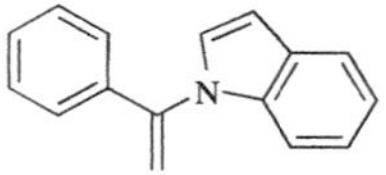 (85)	102
	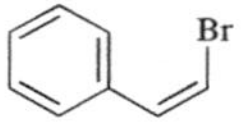	CuI (10 mol %), EDA (20 mol %), K₃PO₄, dioxane, 110°, 24 h	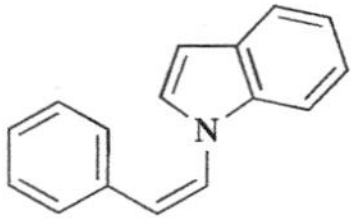 (76)	102

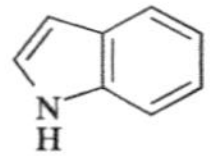

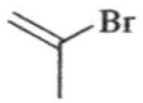

Please refer to the charts preceding the tables for structures indicated by the **bold** numbers.

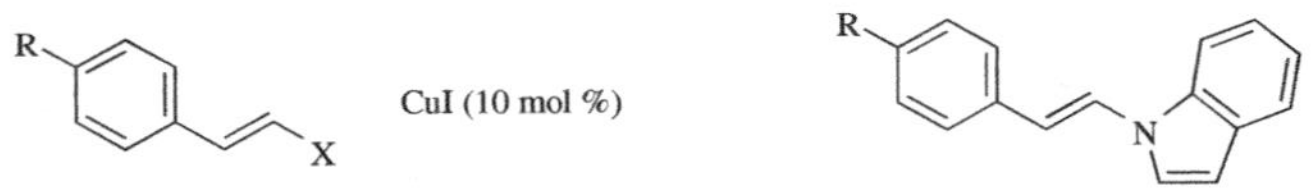

CuI (10 mol %)

R	X	Additive(s)	Solvent	Temp (°)	Time (h)		
H	Cl	Cs$_2$CO$_3$	DMF	120	50	(7)	103
H	Br	Cs$_2$CO$_3$	DMF	120	36	(80)	103
H	Br	EDA (20 mol %), K$_3$PO$_4$	dioxane	110	24	(80)	102
H	Br	L30 (5 mol %), Cs$_2$CO$_3$	MeCN	80	24	(83)	190
H	Br	L-Pro (20 mol %), K$_2$CO$_3$	[C$_4$mim][BF$_4$]	110	20	(91)	231
MeO	Cl	Cs$_2$CO$_3$	DMF	120	50	(35)	103

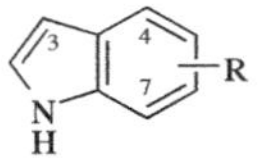

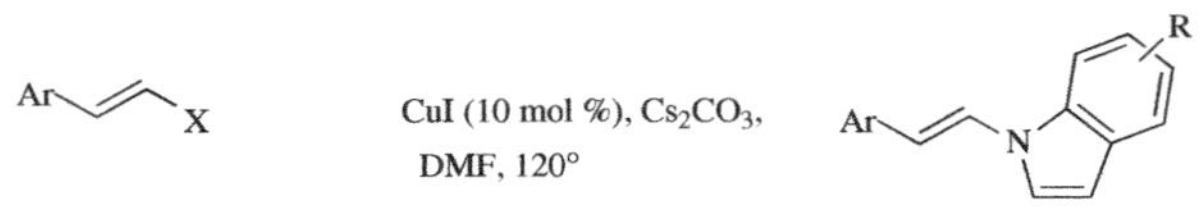

CuI (10 mol %), Cs$_2$CO$_3$,
DMF, 120°

Ar	R	X	Time (h)		
4-BnOC$_6$H$_4$	H	Br	36	(71)	103
4-BnOC$_6$H$_4$	MeO	Br	36	(91)	103
3-MeC$_6$H$_4$	H	Br	36	(70)	
3-MeC$_6$H$_4$	MeO	Br	36	(82)	
3-OHCC$_6$H$_4$	H	Cl	50	(17)	
3-OHCC$_6$H$_4$	H	Br	36	(63)	
3-OHCC$_6$H$_4$	MeO	Cl	50	(25)	
3-OHCC$_6$H$_4$	MeO	Br	36	(80)	
3-AcC$_6$H$_4$	H	Br	36	(48)	
3-AcC$_6$H$_4$	MeO	Br	36	(93)	

TABLE 29D. N-VINYLATION OF INDAZOLES, BENZIMIDAZOLES, BENZOTRIAZOLES, AND CARBAZOLES

Nitrogen Nucleophile	Electrophile	Conditions	Product(s) and Yield(s) (%)	Refs.

*Please refer to the charts preceding the tables for structures indicated by the **bold** numbers.*

C₆

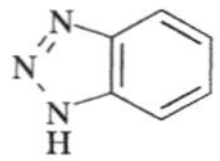

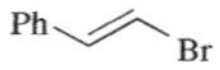

CuI (10 mol %),
L30 (5 mol %), Cs₂CO₃,
DMF, 110°, 24 h

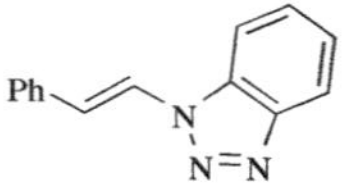

(87) + (7) 190

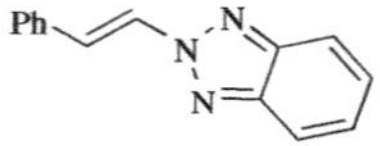

C₇

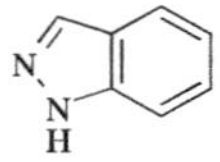

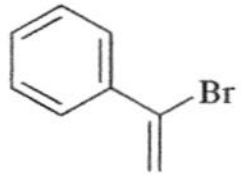

CuI (10 mol %),
EDA (20 mol %), K₃PO₄,
dioxane, 110°, 48 h

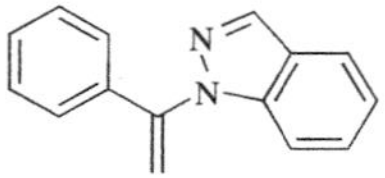

(57) 102

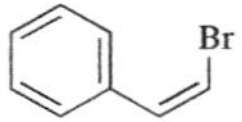

CuI (10 mol %),
EDA (20 mol %), K₃PO₄,
dioxane, 110°, 48 h

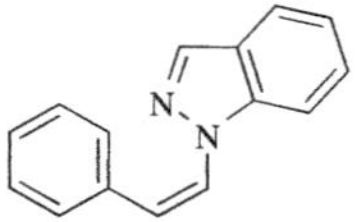

(72) 102

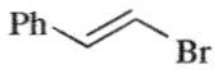

CuI (10 mol %),
L30 (5 mol %), Cs₂CO₃,
MeCN, 80°, 24 h

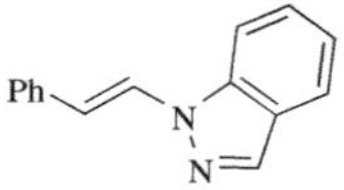

(80) + (9) 190

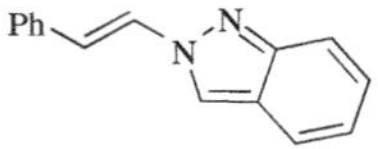

R	X	Catalyst	x	Additive(s)	Solvent	Temp (°)	Time (h)	
H	Br	CuI	10 mol %	Cs$_2$CO$_3$	DMF	120	36 (69)	103
H	Br	CuI	10 mol %	L-Pro (20 mol %), K$_2$CO$_3$	[C$_4$mim][BF$_4$]	110	20 (88)	231
H	Br	CuI	15 mol %	L9 (30 mol %), Cs$_2$CO$_3$	DMSO	80	24 (82)	650
H	Br	Cu$_2$O	5 mol %	L9 (10 mol %), Cs$_2$CO$_3$	MeCN	85	24 (82)	89
H	Br	CuO nanoparticles	2 eq	KOH	DMSO	80	11 (78)	94
H	I	CuO nanoparticles	1.5eq	KOH	DMSO	80	5 (90)	94
3-F	I	CuO nanoparticles	1.5 eq	KOH	DMSO	80	5 (85)	94
4-F	Br	CuI	15 mol %	L9 (30 mol %), Cs$_2$CO$_3$	DMSO	60	24 (88)	650
4-F	Br	Cu$_2$O	5 mol %	L9 (10 mol %), Cs$_2$CO$_3$	MeCN	85	24 (91)	89
4-Cl	Br	CuI	15 mol %	L9 (30 mol %), Cs$_2$CO$_3$	DMSO	60	24 (84)	650
4-Cl	Br	Cu$_2$O	5 mol %	L9 (10 mol %), Cs$_2$CO$_3$	MeCN	85	24 (78)	89
4-MeO	Br	CuI	10 mol %	Cs$_2$CO$_3$	DMF	120	36 (91)	103
4-MeO	Br	CuI	10 mol %	L-Pro (20 mol %), K$_2$CO$_3$	[C$_4$mim][BF$_4$]	110	20 (85)	231
4-MeO	Br	Cu$_2$O	5 mol %	L9 (10 mol %), Cs$_2$CO$_3$	MeCN	85	24 (54)	89
4-MeO	I	CuO nanoparticles	1.5 eq	KOH	DMSO	80	5 (88)	94
3,4-(MeO)$_2$	Br	Cu$_2$O	5 mol %	L9 (10 mol %), Cs$_2$CO$_3$	MeCN	85	24 (60)	89
4-Me	Br	CuI	10 mol %	L-Pro (20 mol %), K$_2$CO$_3$	[C$_4$mim][BF$_4$]	110	20 (86)	231
4-Me	Br	CuI	15 mol %	L9 (30 mol %), Cs$_2$CO$_3$	DMSO	80	24 (85)	650
4-Me	Br	Cu$_2$O	5 mol %	L9 (10 mol %), Cs$_2$CO$_3$	MeCN	85	24 (71)	89
4-Me	I	CuO nanoparticles	1.5 eq	KOH	DMSO	80	5 (88)	94

TABLE 29D. *N*-VINYLATION OF INDAZOLES, BENZIMIDAZOLES, BENZOTRIAZOLES, AND CARBAZOLES (*Continued*)

Nitrogen Nucleophile	Electrophile	Conditions	Product(s) and Yield(s) (%)	Refs.

*Please refer to the charts preceding the tables for structures indicated by the **bold** numbers.*

C_{12}

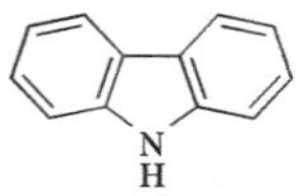	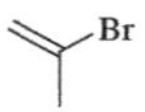	CuI (10 mol %), EDA (20 mol %), K_3PO_4, dioxane, 110°, 24 h	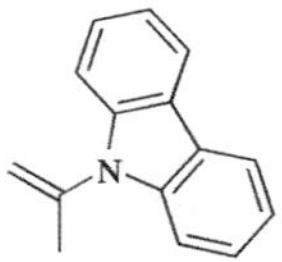 (78)	102
		**8** (7 mol %), DMCDA (15 mol %), K_3PO_4, toluene, 85°, 16 h	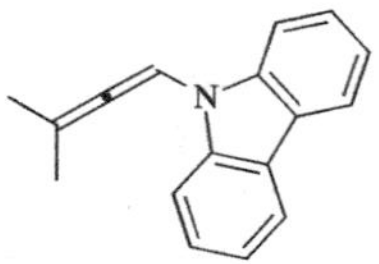 (94)	651
	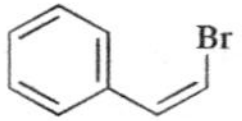	CuI (10 mol %), EDA (20 mol %), K_3PO_4, dioxane, 110°, 39 h	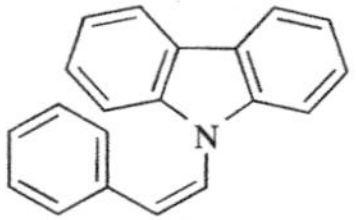 (90)	102
	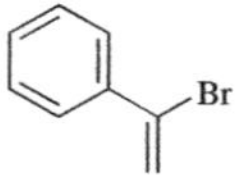	CuI (10 mol %), EDA (20 mol %), K_3PO_4, dioxane, 110°, 30 h	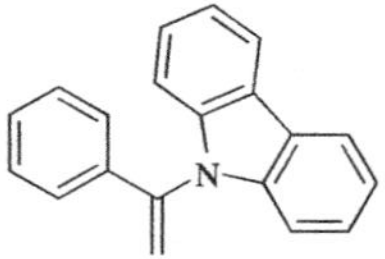 (78)	102

592

TABLE 30A. N-VINYLATION OF PRIMARY ACYCLIC AMIDES

Nitrogen Nucleophile	Electrophile	Conditions	Product(s) and Yield(s) (%)	Refs.

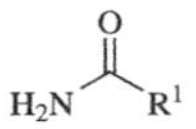

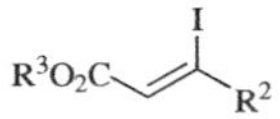

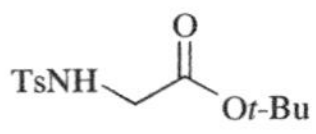

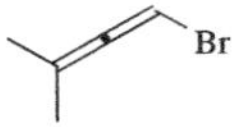

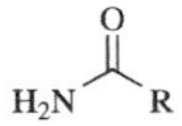

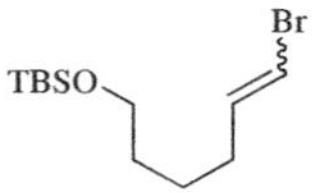

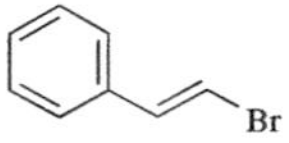

*Please refer to the charts preceding the tables for structures indicated by the **bold** numbers.*

C2-6

Conditions: CuI (x mol %), Cs₂CO₃

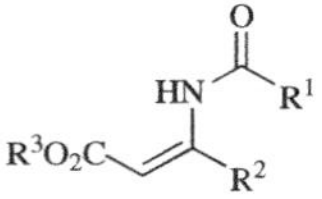

R¹	R²	R³	x	Additive	Solvent	Temp (°)	Time (h)		Refs.
Me	H	Et	10	DMG (20 mol %)	dioxane	60	12	(78)	90
Me	H	Et	5	DMEDA (10 mol %)	THF	70	5	(67)	101
Me	Me	Et	5	DMEDA (10 mol %)	THF	70	6	(62)	101
(E,E)-Me(CH=CH)₂	H	Et	10	DMG (20 mol %)	dioxane	60	12	(73)	90
(E,E)-Me(CH=CH)₂	H	allyl	10	DMG (20 mol %)	dioxane	60	12	(87)	90

C2

	8 (15 mol %). DMEDA (30 mol %). Cs₂CO₃, dioxane, reflux, 24 h	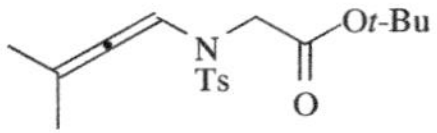 (41)	625

C2-8

1. CuI (10 mol %), DMEDA (20 mol%). Cs₂CO₃, toluene, 70°, 20 h
2. TBAF, THF, 24 h

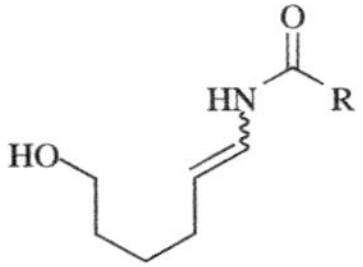

R	Config.	
Me	(E)	(85)
Me	(Z)	(86)
c-Pr	(E) + (Z)	(86)
Bn	(E) + (Z)	(86)

652

C2-7

CuI (1 eq), HMPA, KH, 130°

R	Time (h)	
Me	3	(45)
CF₃	2	(83)
Ph	1	(38)

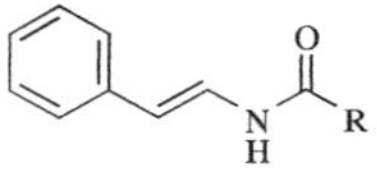

653

TABLE 30A. N-VINYLATION OF PRIMARY ACYCLIC AMIDES (Continued)

Nitrogen Nucleophile	Electrophile	Conditions	Product(s) and Yield(s) (%)	Refs.

*Please refer to the charts preceding the tables for structures indicated by the **bold** numbers.*

C$_2$

| 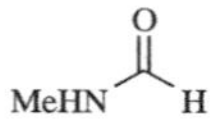 | 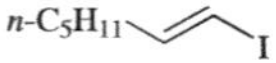 n-C$_5$H$_{11}$ ⟶ I | **8** (30 mol %), Cs$_2$CO$_3$, NMP, 90°, 12 h | 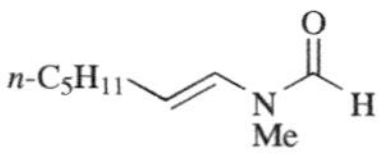 n-C$_5$H$_{11}$ ⟶ N(Me)H amide (75) | 165 |

MeHN–C(=O)–H

C$_{2–11}$

H$_2$N–C(=O)–CH(NH$_2$)R

Electrophile: furan–(CH$_2$)$_4$ ⟶ I

R	Temp (°)	
H	70	(32)
Me	70	(82)
MeSCH$_2$CH$_2$	70	(77)
i-Pr	70	(58)
H$_2$NCOCH$_2$CH$_2$	50	(31)

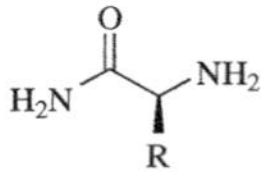

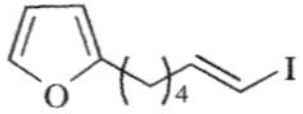

Conditions: CuI (5 mol %), DMEDA (10 mol %), Cs$_2$CO$_3$, THF, 16 h

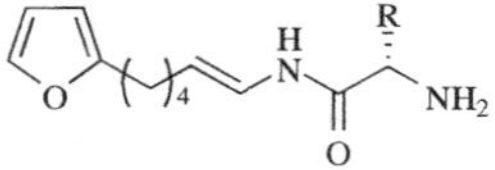

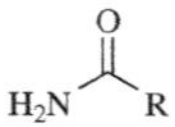

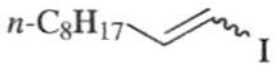

Product: furan–(CH$_2$)$_4$ ⟶ N(H)–C(=O)–CH(NH$_2$)R

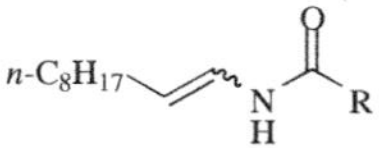

R	Temp (°)	
i-Bu	70	(87)
(S)-EtMeCH	70	(63)
Bn	70	(87)
4-HOC$_6$H$_4$CH$_2$	50	(30)
3-indolylCH$_2$	50	(50)

Refs: 244

C$_{2–9}$

H$_2$N–C(=O)–R

Electrophile: n-C$_8$H$_{17}$ ⟶ I

Conditions: CuI (5 mol %), DMEDA (10 mol %), Cs$_2$CO$_3$, THF,

Product: n-C$_8$H$_{17}$ ⟶ N(H)–C(=O)–R

R	Config.	Temp (°)	Time (h)	
Me	(E)	rt	15	(82)
Me	(Z)	50	12	(84)
(E)-PhCH=CH	(E)	70	14	(86)

Refs: 101

C_{2-11}

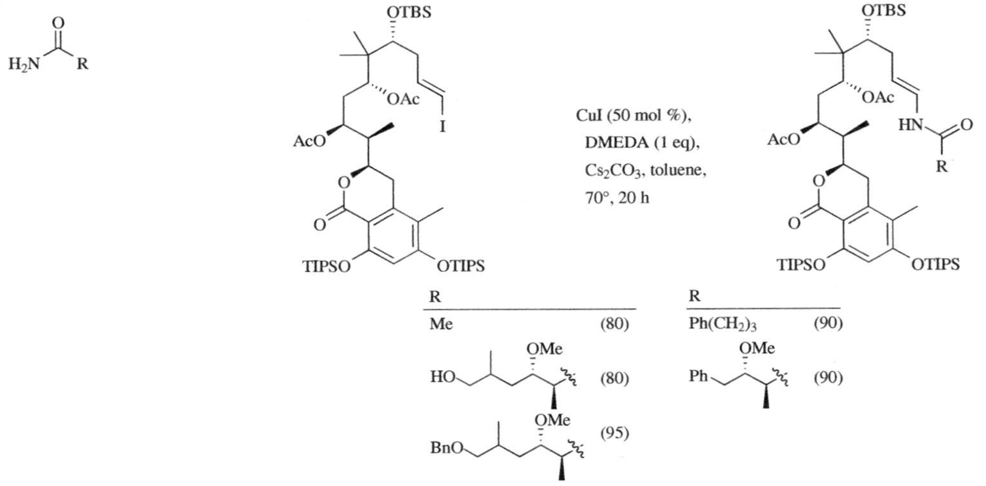

OTBS
OAc
AcO
I
O
TIPSO
OTIPS
CuI (50 mol %),
DMEDA (1 eq),
Cs_2CO_3, toluene,
70°, 20 h
HN
R
R
Me (80)
HO OMe (80)
BnO OMe (95)
R
$Ph(CH_2)_3$ (90)
Ph OMe (90)

654

TABLE 30A. *N*-VINYLATION OF PRIMARY ACYCLIC AMIDES (*Continued*)

Nitrogen Nucleophile	Electrophile	Conditions	Product(s) and Yield(s) (%)	Refs.

*Please refer to the charts preceding the tables for structures indicated by the **bold** numbers.*

C$_{2-7}$

Conditions: CuI (1.5 eq), DMEDA (*x* eq), K$_2$CO$_3$, DMF, rt, 1 h

			Configuration						I	II	
R^1	R^2	C–7	C–10	C–11	C–19	C–20	*x*	I	I	II	Refs.
Me	H$_2$NCO	(S)	(S)	(S)	(R)	(R)	3	(47)	(—)	98
i-Bu	H$_2$NCO	(S)	(S)	(S)	(R)	(R)	3	(41)	(—)	98
Me$_2$C=CH	H	(S)	(S)	(S)	(R)	(R)	3	(51)	(—)	655
Me$_2$C=CH	H$_2$NCO	(R)	(S)	(S)	(R)	(R)	3	(46)	(<5)	655
Me$_2$C=CH	H	(R)	(R)	(R)	(R)	(R)	2	(45)	(—)	655
Me$_2$C=CH	H$_2$NCO	(R)	(R)	(R)	(R)	(R)	3	(31)	(<5)	655
Me$_2$C=CH	H	(R)	(R)	(R)	(S)	(R)	3	(45)	(—)	655
Me$_2$C=CH	H$_2$NCO	(R)	(R)	(R)	(S)	(R)	3	(37)	(10)	655
Me$_2$C=CH	H	(R)	(R)	(R)	(S)	(S)	3	(54)	(—)	655
Me$_2$C=CH	H$_2$NCO	(R)	(R)	(R)	(S)	(S)	3	(55)	(<5)	655
Me$_2$C=CH	H	(R)	(R)	(R)	(R)	(S)	3	(42)	(—)	655
Me$_2$C=CH	H$_2$NCO	(R)	(R)	(R)	(R)	(S)	2	(44)	(10)	166, 655
2-methyl-4-thiazolyl	H$_2$NCO	(S)	(S)	(S)	(R)	(R)	3	(40)	(—)	98
2-pyridyl	H$_2$NCO	(S)	(S)	(S)	(R)	(R)	3	(31)	(—)	98
3-pyridyl	H$_2$NCO	(S)	(S)	(S)	(R)	(R)	3	(54)	(—)	98
Ph	H$_2$NCO	(S)	(S)	(S)	(R)	(R)	3	(44)	(—)	98

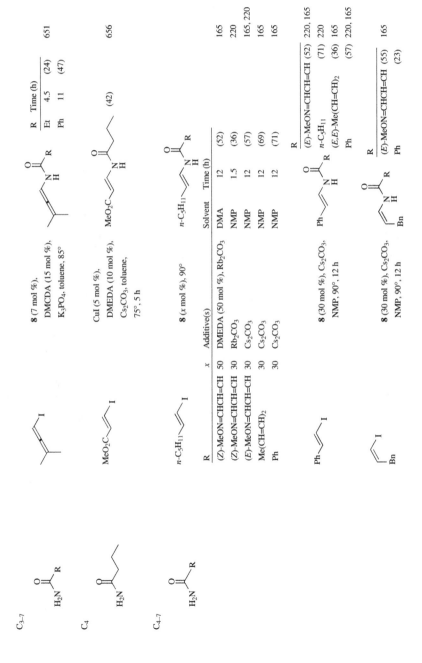

C_{3–7}

8 (7 mol %),
DMCDA (15 mol %),
K₃PO₄, toluene, 85°

R	Time (h)	
Et	4.5	(24)
Ph	11	(47)

651

C₄

CuI (5 mol %),
DMEDA (10 mol %),
Cs₂CO₃, toluene,
75°, 5 h

(42)

656

C₄–₇

8 (x mol %), 90°

R	x	Additive(s)	Solvent	Time (h)		
(Z)-MeON=CHCH=CH	50	DMEDA (50 mol %), Rb₂CO₃	DMA	12	(52)	165
(Z)-MeON=CHCH=CH	30	Rb₂CO₃	NMP	1.5	(36)	220
(E)-MeON=CHCH=CH	30	Cs₂CO₃	NMP	12	(57)	165, 220
Me(CH=CH)₂	30	Cs₂CO₃	NMP	12	(69)	165
Ph	30	Cs₂CO₃	NMP	12	(71)	165

8 (30 mol %), Cs₂CO₃,
NMP, 90°, 12 h

R		
(E)-MeON=CHCH=CH	(52)	220, 165
n-C₅H₁₁	(71)	220
(E,E)-Me(CH=CH)₂	(36)	165
Ph	(57)	220, 165

8 (30 mol %), Cs₂CO₃,
NMP, 90°, 12 h

R		
(E)-MeON=CHCH=CH	(55)	165
Ph	(23)	

TABLE 30A. N-VINYLATION OF PRIMARY ACYCLIC AMIDES (*Continued*)

Nitrogen Nucleophile	Electrophile	Conditions	Product(s) and Yield(s) (%)	Refs.

*Please refer to the charts preceding the tables for structures indicated by the **bold** numbers.*

C₄

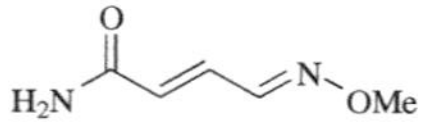

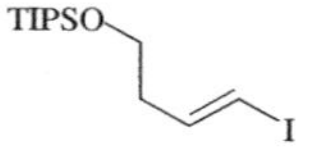

8 (30 mol %), Cs₂CO₃, NMP, 90°, 12 h — (50) — 165

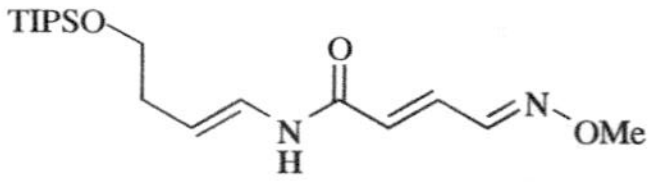

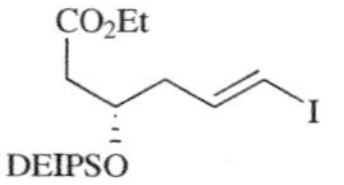

8 (50 mol %), phen, dba, Cs₂CO₃, DMA, 65°, 20 h — (52) — 657, 165

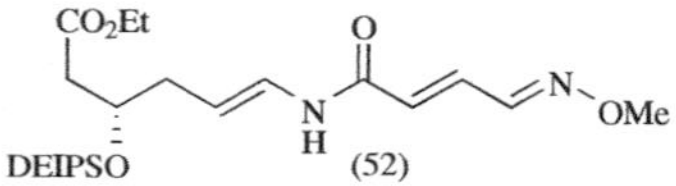

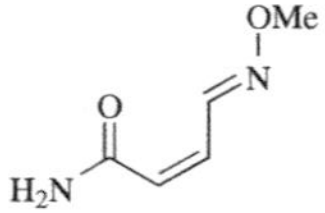

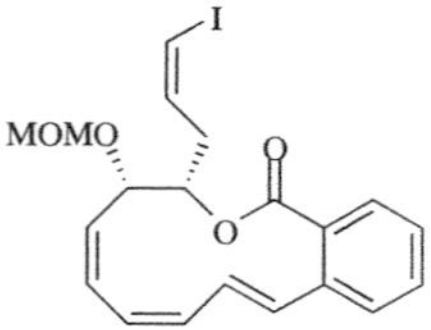

8 (57 mol %), DMEDA (1.1 eq), K₂CO₃, DMA, rt, 1 h — (44) — 658

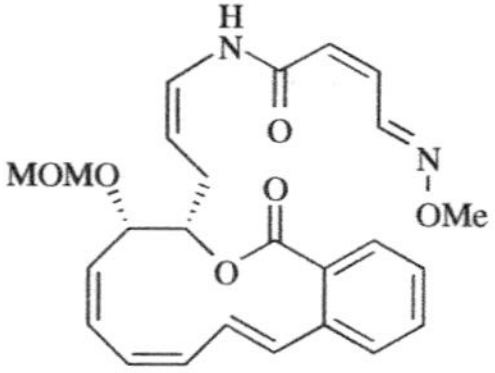

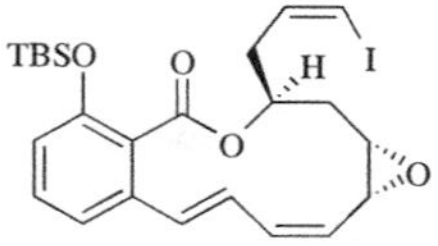

1. TBAF, THF, rt
2. **8**, DMEDA, K₂CO₃, DMA, 50°, 1 h — (45) — 659

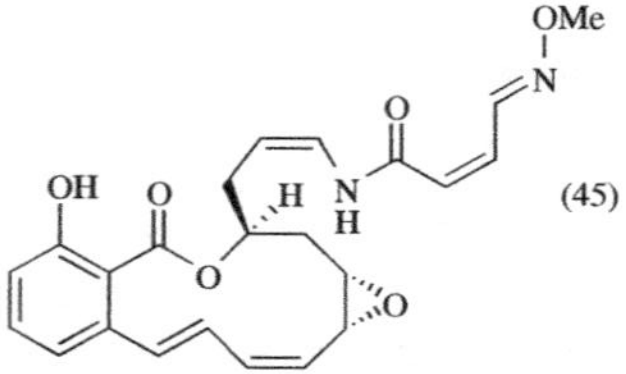

598

660

C_{4-7}

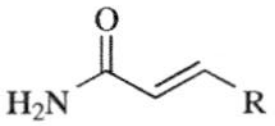

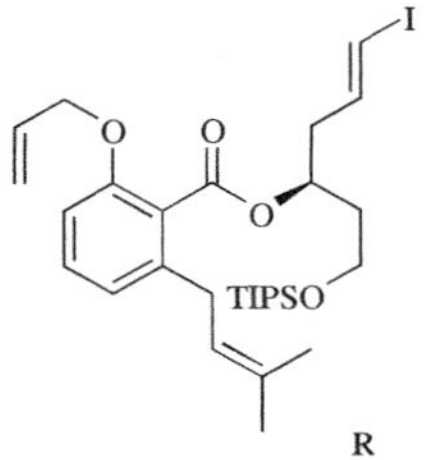

8 (45 mol %),
DMEDA (90 mol %),
Cs$_2$CO$_3$, THF,
DMA, 50°, 11 h

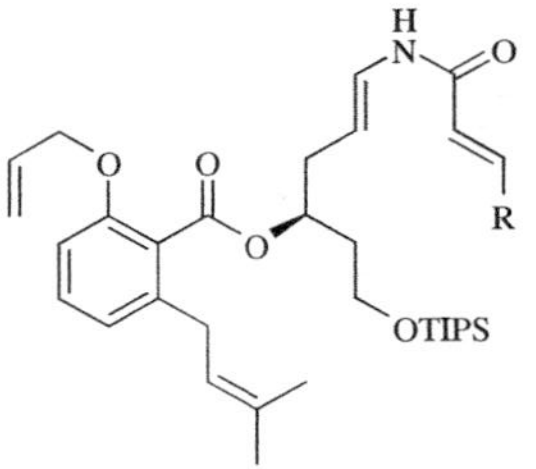

R	
HON=CH	(50)
(E)-CH$_2$=CHCH$_2$O$_2$CNHCH$_2$CH=CH	(51)

OTIPS

C_4

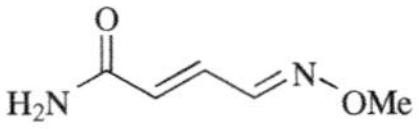

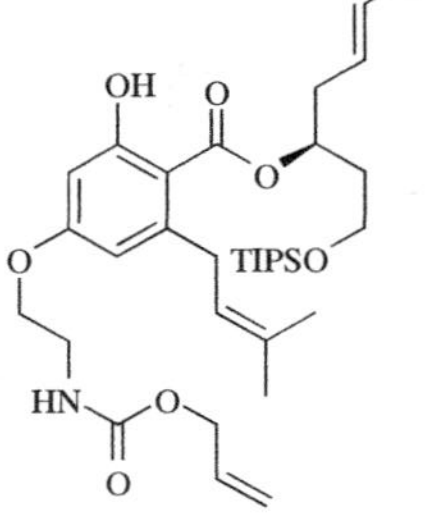

1. NaHMDS,
allyl bromide
2. **8** (45 mol %),
DMEDA (90 mol %),
Cs$_2$CO$_3$, THF, DMA,
50°, 11 h

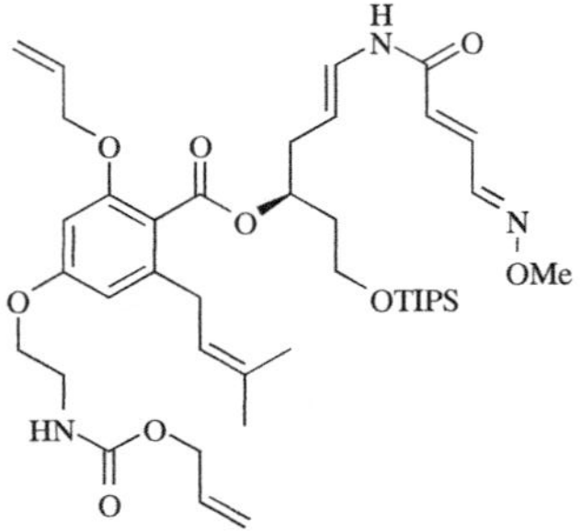

N–OMe   (50)   660

OTIPS

TABLE 30A. N-VINYLATION OF PRIMARY ACYCLIC AMIDES (*Continued*)

Nitrogen Nucleophile	Electrophile	Conditions	Product(s) and Yield(s) (%)	Refs.

*Please refer to the charts preceding the tables for structures indicated by the **bold** numbers.*

C_{5-6}

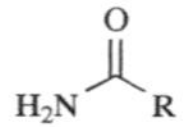

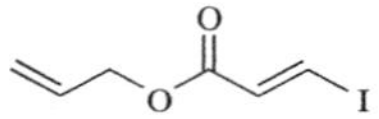

| | Cu(MeCN)₄PF₆ (10 mol %), phen (20 mol %), Rb₂CO₃, DMA, 12 h | 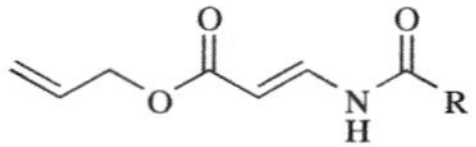 | 92 |

R	Temp (°)	
2-tetrahydrofuryl	rt	(61)
(*E*,*E*)-Me(CH=CH)₂	45	(58)
c-C₅H₁₁	45	(6)
2-pyridyl	45	(43)

R	Temp (°)	
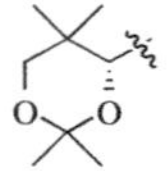	45	(90)
Ph	45	(70)

C₅

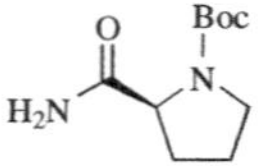

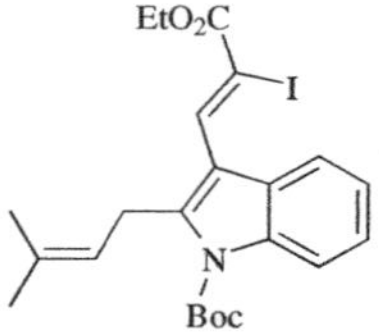

| | CuI (10 mol %), DMEDA (20 mol %), Cs₂CO₃, toluene, 100°, 13 h | | (57) | 661 |

| | CuI (5 mol %), DMEDA (10 mol %), Cs₂CO₃, toluene, 75°, 5 h | | (42) | 656 |

| | CuI (5 mol %), DMEDA (10 mol %), Cs₂CO₃, THF, 50°, 24 h | | (75) | 662 |

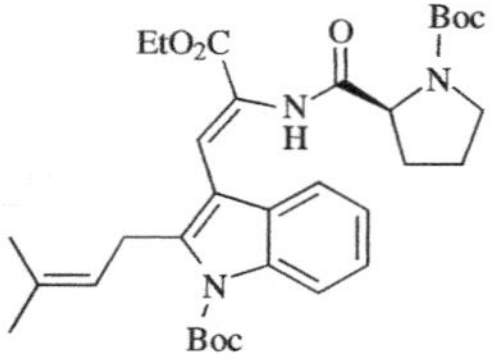

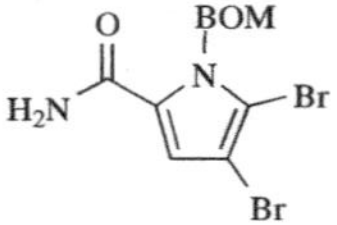

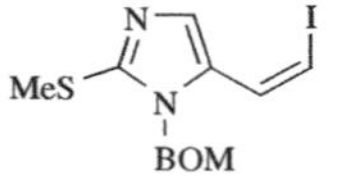

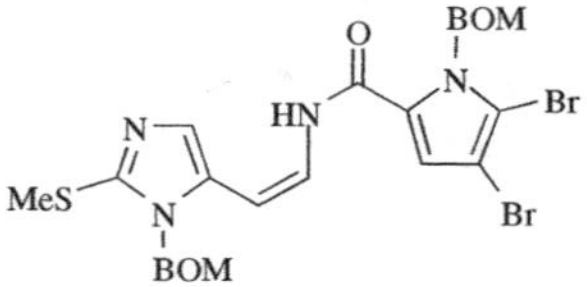

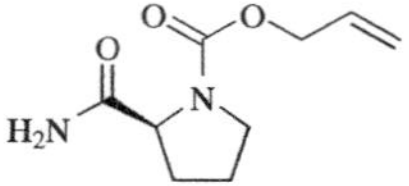

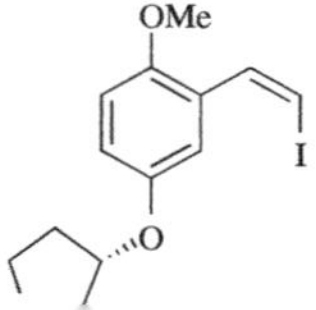

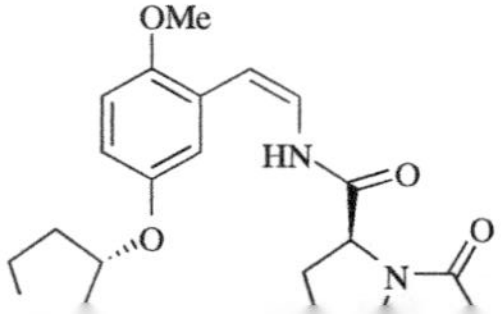

 	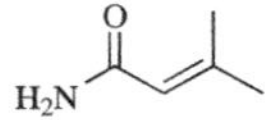 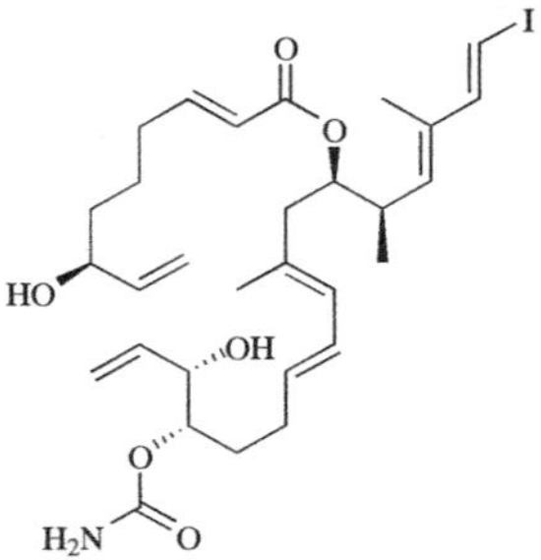 CuI (1.5 eq), DMEDA (3 eq), K$_2$CO$_3$, DMF, rt, 1 h	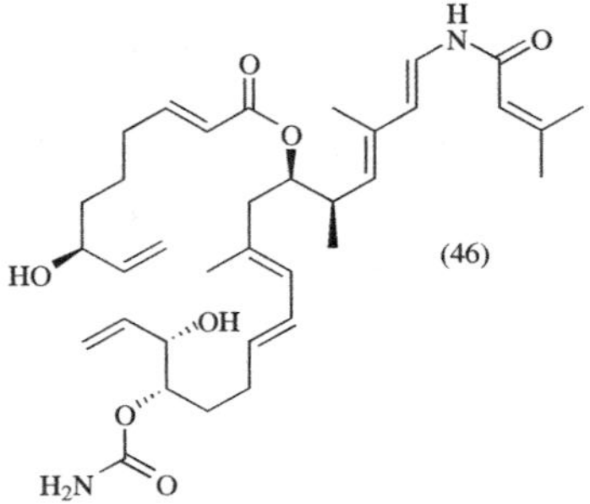 (46)	655
C_{6-7}	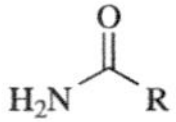 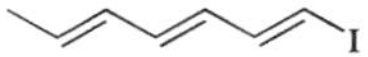 Cat. **8** (30 mol %), Cs$_2$CO$_3$, NMP, 90°, 12 h	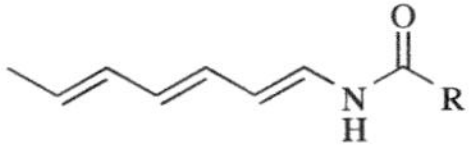 $\dfrac{R}{\begin{array}{l} n\text{-C}_5\text{H}_{11}\ (69)\\ \text{Ph}\ (58)\end{array}}$	220
C_6	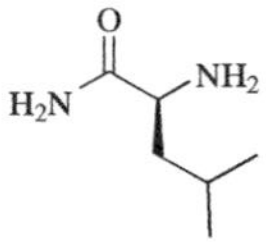 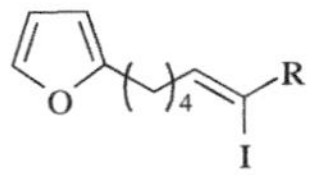 CuI (5 mol %), DMEDA (10 mol %), Cs$_2$CO$_3$, THF, 70°, 16 h	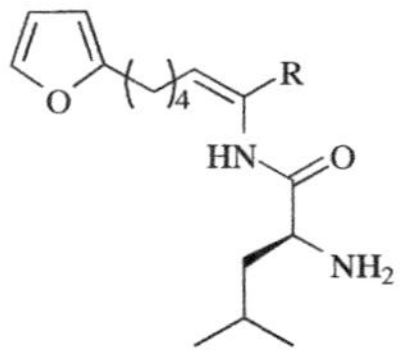 $\dfrac{R}{\begin{array}{l} \text{H}\ (50)\\ \text{Me}\ (41)\end{array}}$	244
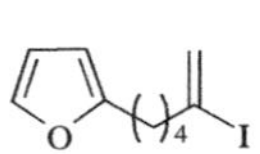	CuI (5 mol %), DMEDA (10 mol %), Cs$_2$CO$_3$, THF, 70°, 16 h	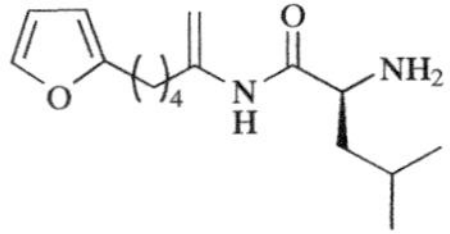 (20)	244

601

TABLE 30A. N-VINYLATION OF PRIMARY ACYCLIC AMIDES (Continued)

*Please refer to the charts preceding the tables for structures indicated by the **bold** numbers.*

Nitrogen Nucleophile	Electrophile	Conditions	Product(s) and Yield(s) (%)	Refs.

C₆

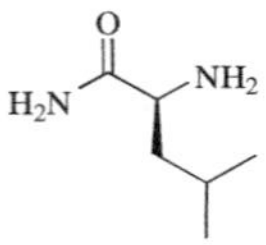

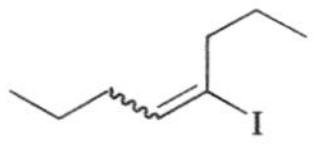

| | | CuI (5 mol %), DMEDA (10 mol %), Cs₂CO₃, THF, 70°, 16 h | 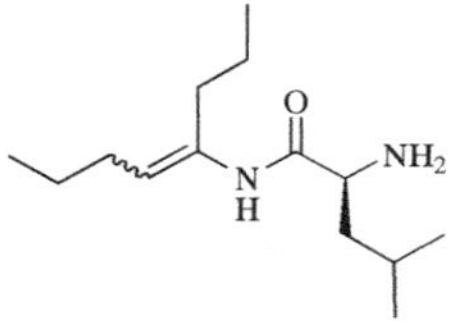 R Config.
 H (E) (35)
 H (Z) (34)
 Cbz (E) (54)
 Boc (E) (90) | 244 |

| | | CuI (10 mol %), DMG (20 mol %), Cs₂CO₃, dioxane, 60°, 12 h | R Y
 Me(CH=CH)₂ CH₂ (74)
 Ph CH₂ (81)
 Ph C=O (82) | 90 |

| | | CuI (10 mol %), DMG (20 mol %), Cs₂CO₃, dioxane, 80°, 24 h | 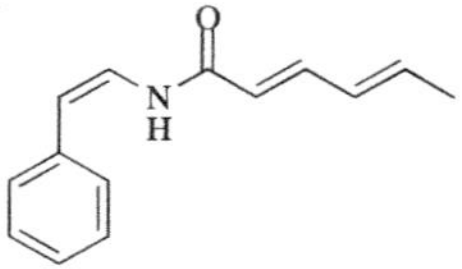 (65) | 90 |

| | | Cu(MeCN)₄PF₆ (10 mol %), phen (20 mol %), Rb₂CO₃, DMA, 60°, 12 h | 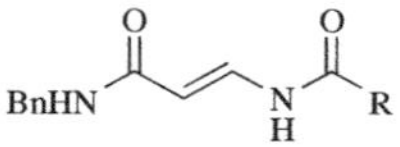 R
 (E,E)-Me(CH=CH)₂ (18) 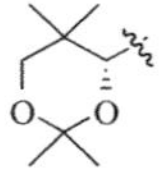
  (70)
 Ph (72) | 92 |

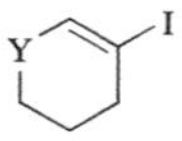

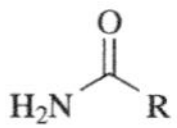

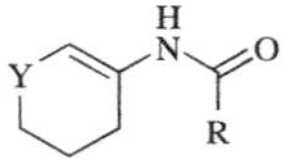

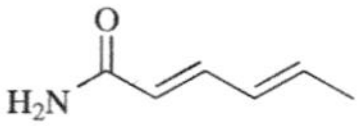

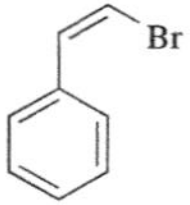

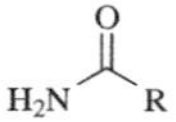

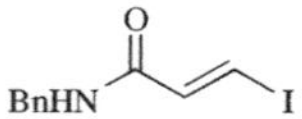

602

C_6

CuI, DMG, Cs$_2$CO$_3$, dioxane

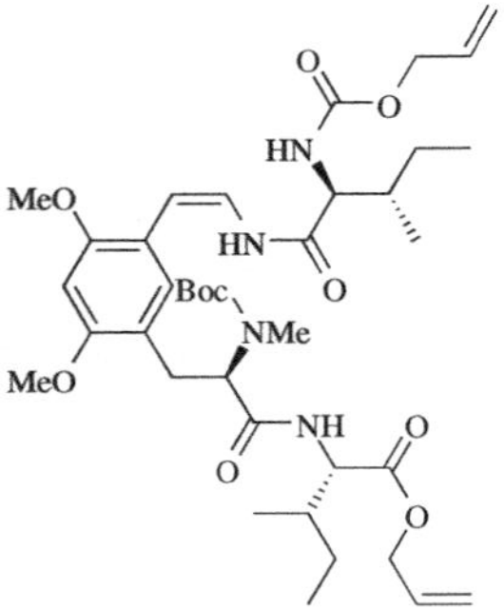

(60) 451

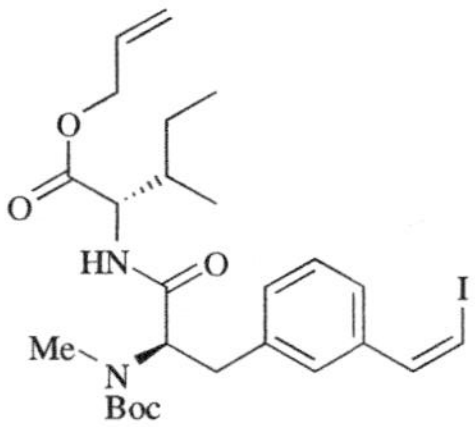

CuI, DMG, Cs$_2$CO$_3$, dioxane

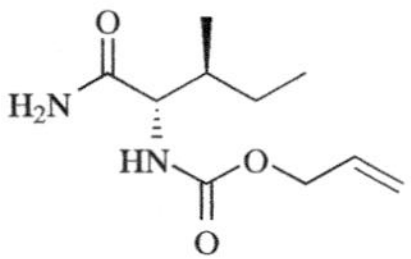

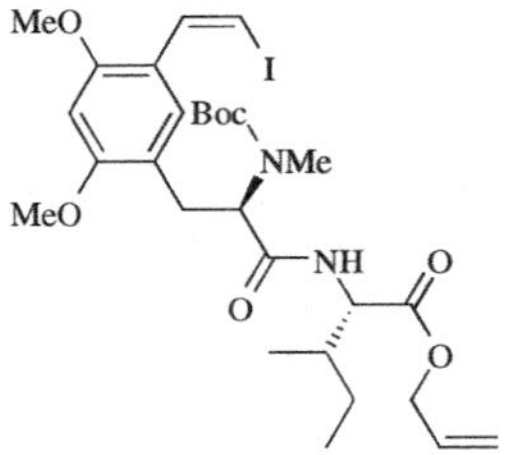

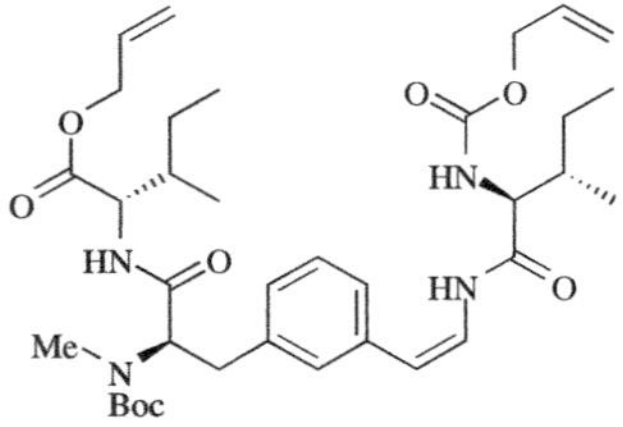

(61) 451

C_7

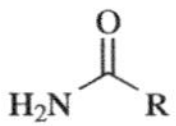

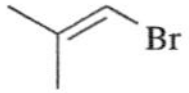

CuI (5 mol %), DMEDA (10 mol %), K$_2$CO$_3$, toluene, 110°, 24 h

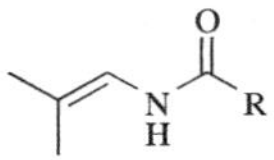

R	
c-C$_6$H$_{11}$	(84)
2-H$_2$NC$_6$H$_4$	(81)

101

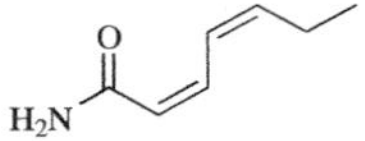

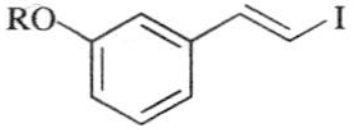

8 (50 mol %), Rb$_2$CO$_3$, DMA, 90°, 2 h

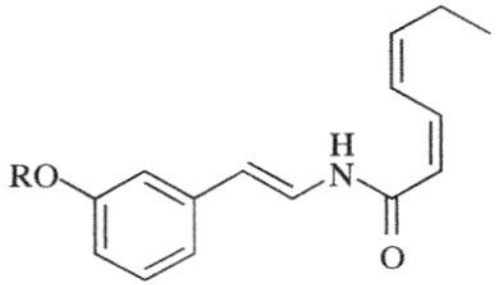

R	
H	(86)
Me	(89)

95

TABLE 30A. *N*-VINYLATION OF PRIMARY ACYCLIC AMIDES (*Continued*)

Nitrogen Nucleophile	Electrophile	Conditions	Product(s) and Yield(s) (%)	Refs.

*Please refer to the charts preceding the tables for structures indicated by the **bold** numbers.*

C₇

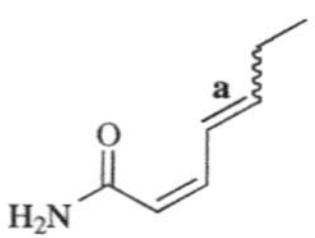

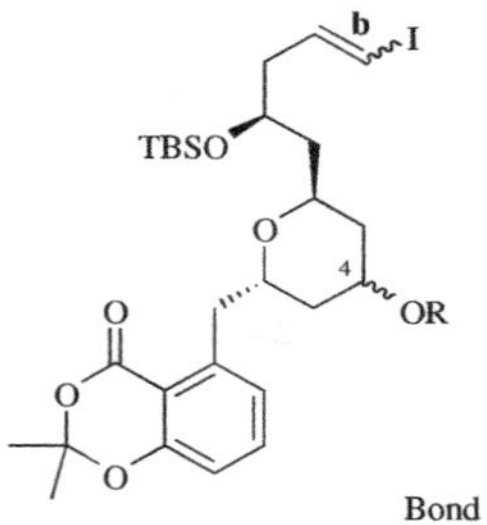

8 (50 mol %),
Rb₂CO₃, DMA,
90°, 12 h

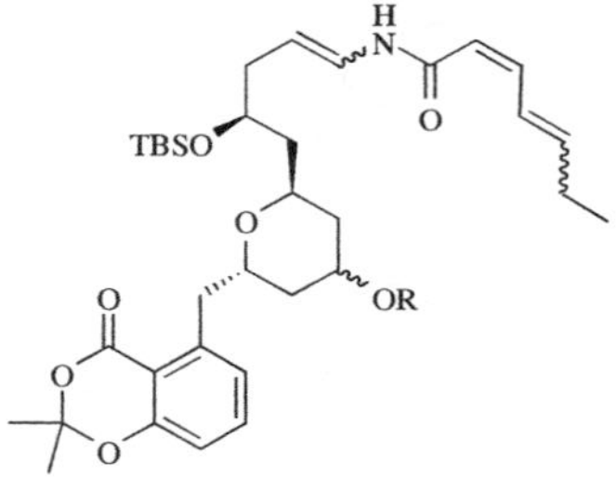

Bond a	Bond b	C-4	R	
(*E*)	(*E*)	(*R*)	Ac	(41)
(*Z*)	(*E*)	(*R*)	Ac	(90)
(*E*)	(*Z*)	(*S*)	TBS	(41)

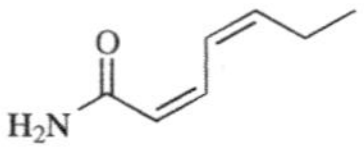

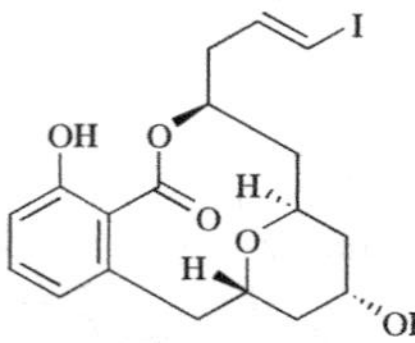

97

8 (68 mol %),
phen (1.4 eq), Rb₂CO₃,
DMA, 58°, 8 h

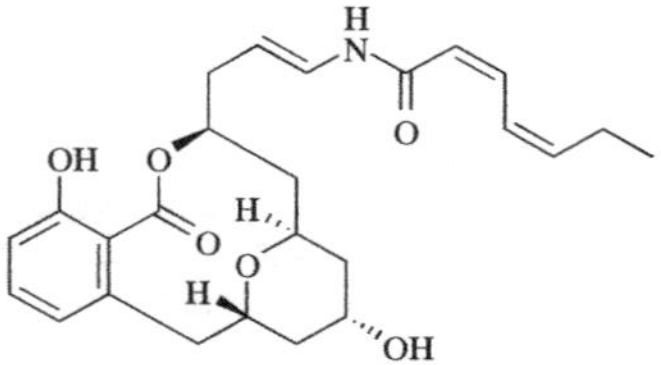

(40) 167

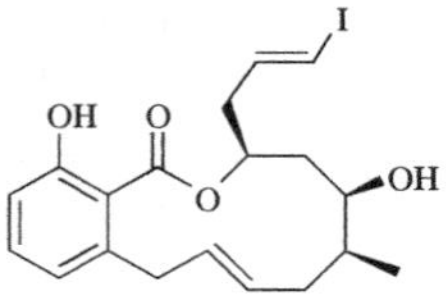

8 (50 mol %), Rb₂CO₃,
DMA, 90°, 2 h

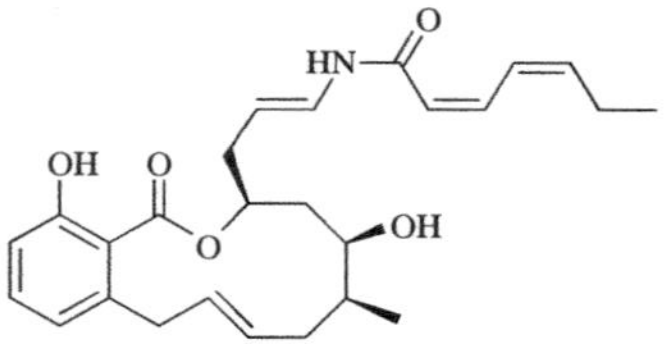

(57) 95

604

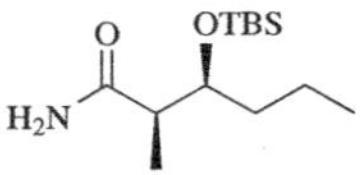

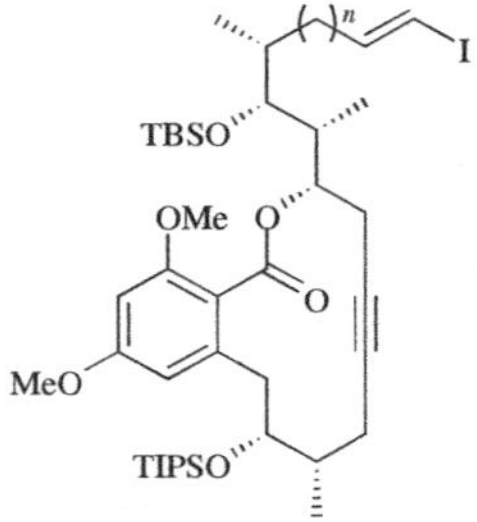

663

CuI (50 mol %),
DMEDA (1 eq),
Cs$_2$CO$_3$, THF, 60°

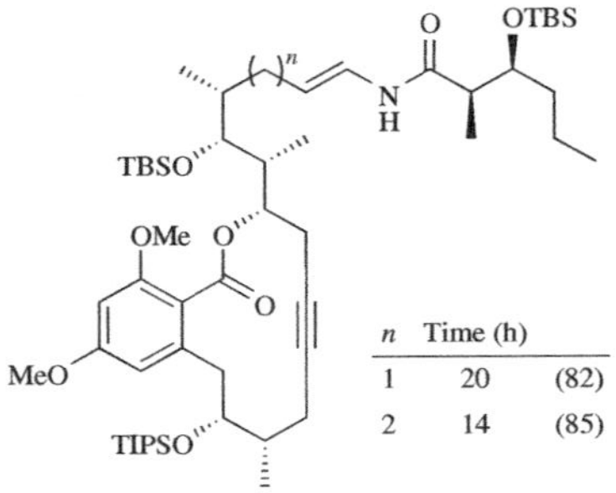

n	Time (h)	
1	20	(82)
2	14	(85)

664

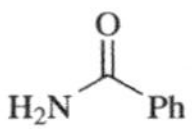

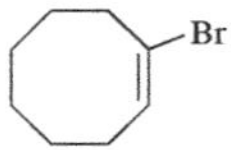

CuI (5 mol %),
DMEDA (20 mol %),
K$_2$CO$_3$, toluene, 100°

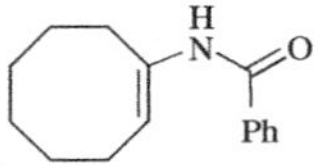

(94)

95

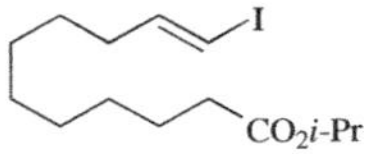

8 (30 mol %), Cs$_2$CO$_3$,
NMP, 90°, 12 h

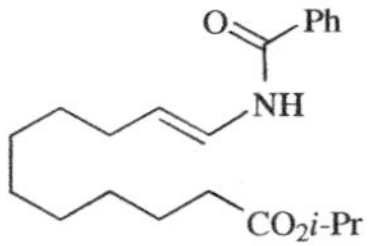

(70)

93

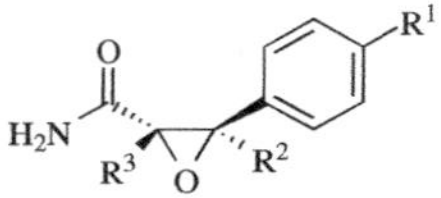

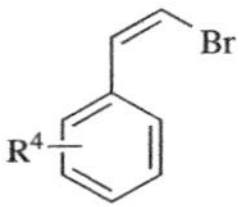

CuI (40 mol %),
DMG (80 mol %),
Cs$_2$CO$_3$, dioxane,
reflux, 3 h

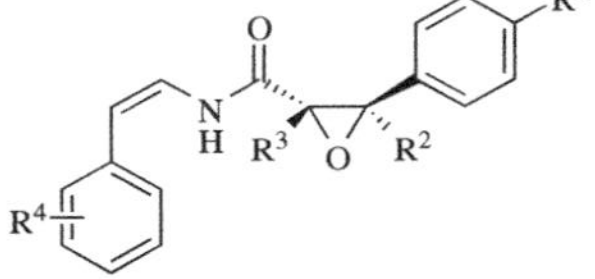

R^1	R^2	R^3	R^4		dr
H	H	H	H	(82)	93:7
H	H	H	4-Cl	(64)	84:16
H	H	H	2-Me	(49)	84:16
H	H	H	3-Me	(62)	92:8
H	H	H	4-Me	(54)	88:12

R^1	R^2	R^3	R^4		dr
H	H	H	H	(62)	78:22
H	Me	H	H	(79)	95:5
H	H	Me	H	(91)	83:17
Cl	H	H	H	(54)	93:7
Me	H	H	H	(74)	83:17

C$_{9-10}$

TABLE 30A. *N*-VINYLATION OF PRIMARY ACYCLIC AMIDES (*Continued*)

Nitrogen Nucleophile	Electrophile	Conditions	Product(s) and Yield(s) (%)	Refs.

*Please refer to the charts preceding the tables for structures indicated by the **bold** numbers.*

C₉

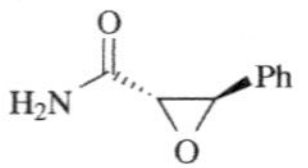

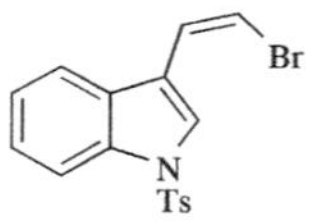

| | | CuI (40 mol %), DMG (80 mol %), Cs₂CO₃, dioxane, reflux, 3 h | 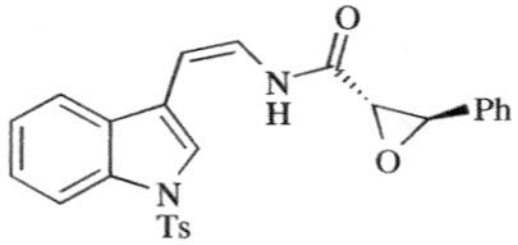 (63) | 93 |
| | | CuI (5 mol %), DMEDA (10 mol %), Cs₂CO₃, THF, 70°, 16 h | 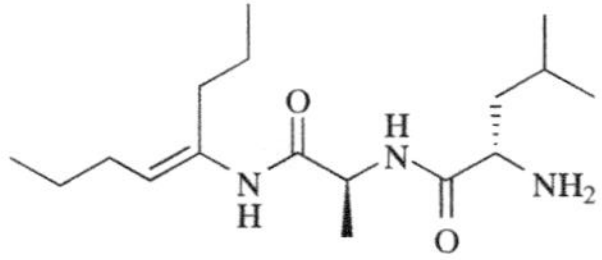 (54) | 244 |

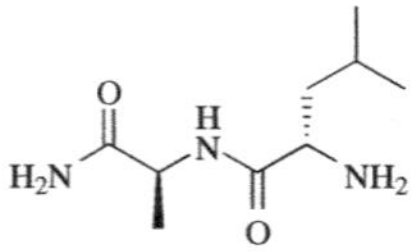

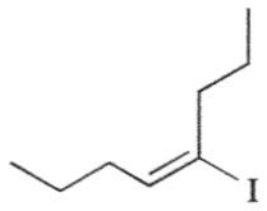

C₂₀

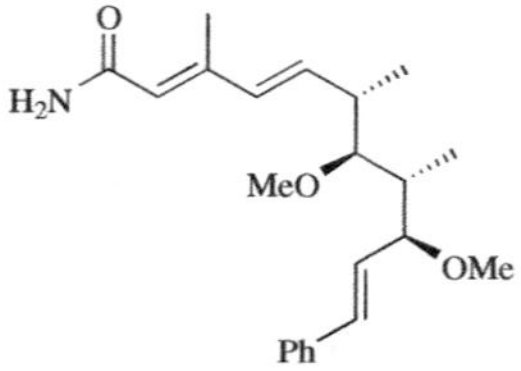

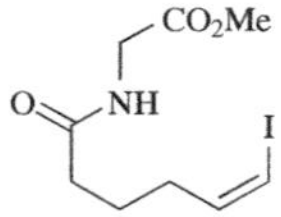

| | | CuI (5 mol %), DMEDA (10 mol %), Cs₂CO₃, THF, 70°, 15 h | 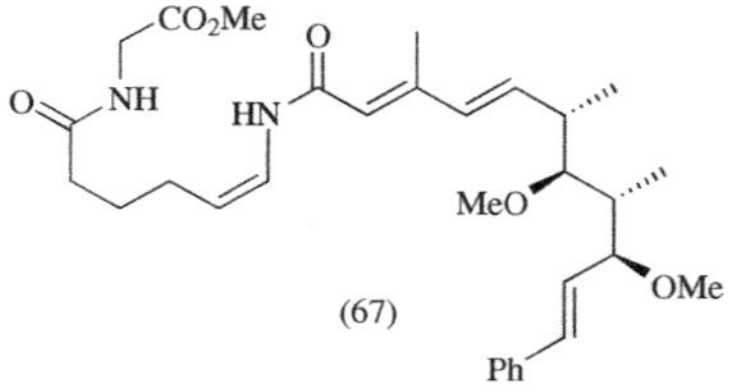 (67) | 665 |

606

TABLE 30B. *N*-VINYLATION OF SECONDARY ACYCLIC AMIDES

Nitrogen Nucleophile	Electrophile	Conditions	Product(s) and Yield(s) (%)	Refs.

*Please refer to the charts preceding the tables for structures indicated by the **bold** numbers.*

C$_2$

| 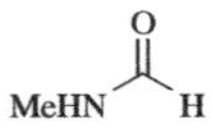 | 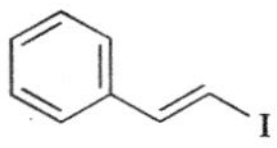 | **8** (30 mol %), Cs$_2$CO$_3$, NMP, 90°, 12 h | 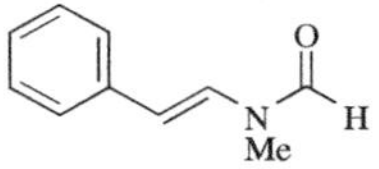 (70) | 165 |
| | 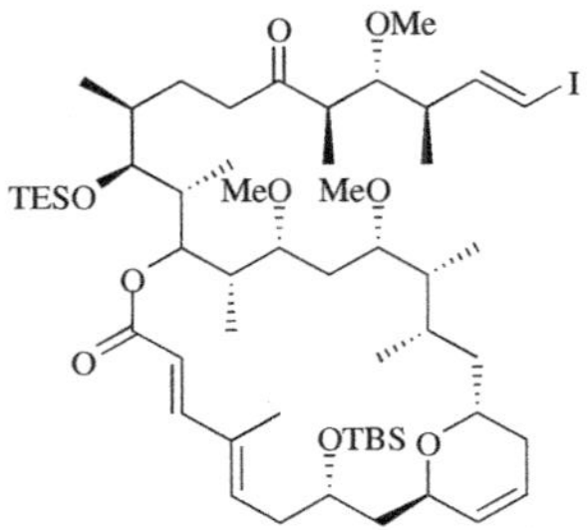 | CuI (50 mol %), DMEDA (1.1 eq). dioxane, 60°, 13 h | 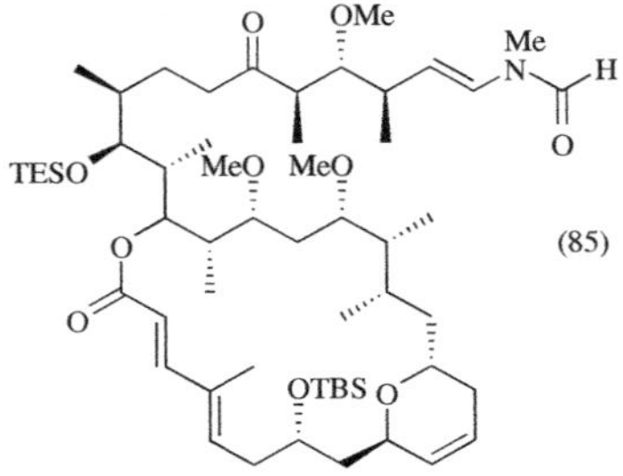 (85) | 666 |

C$_{7-8}$

| 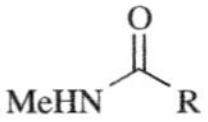 | 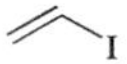 | **8** (30 mol %), Cs$_2$CO$_3$, NMP, 90°, 12 h | 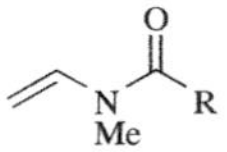 R   *n*-C$_5$H$_{11}$ (75) Ph (70) | 220 |

C$_8$

| 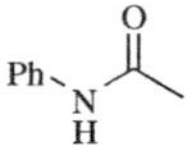 |  | **8** (10 mol %), DMEDA (20 mol %), Cs$_2$CO$_3$, toluene, 50° | 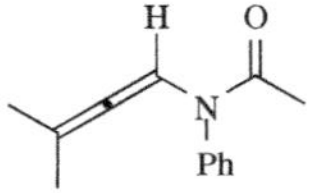 (52) | 91 |
| | 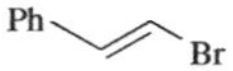 | CuI (1 eq). KH, HMPA, 130°, 3 h | 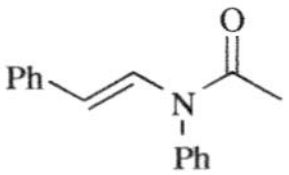 (38) | 653 |

TABLE 30B. *N*-VINYLATION OF SECONDARY ACYCLIC AMIDES (*Continued*)

Nitrogen Nucleophile	Electrophile	Conditions	Product(s) and Yield(s) (%)	Refs.

*Please refer to the charts preceding the tables for structures indicated by the **bold** numbers.*

C₉

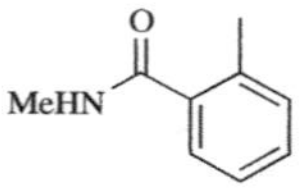

8 (7 mol %),
DMCDA (15 mol %),
K₃PO₄, toluene,
85°, 22 h

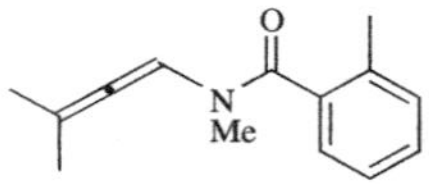

(20)

651

Nitrogen Nucleophile	Electrophile	Conditions	Product(s) and Yield(s) (%)	Refs.

*Please refer to the charts preceding the tables for structures indicated by the **bold** numbers.*

C₃

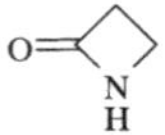

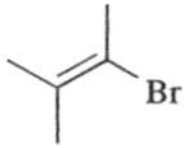

| | | CuI (5 mol %), DMEDA (10 mol %), K₂CO₃, toluene, 110°, 25 h | 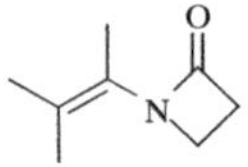 (88) | 101 |
| | | CuI (20 mol %), DMEDA (40 mol %), K₃PO₄, toluene, 65° | | 649 |

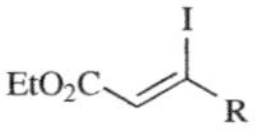

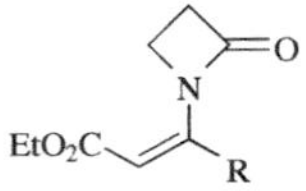

R	Time (h)	
Me	3	(66)
n-C₅H₁₁	24	(79)

| | | CuI (10 mol %), DMG (20 mol %), K₃PO₄, NH₄OAc, MeCN, reflux, 24 h | 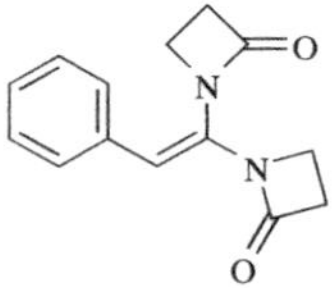 (84) | 574 |
| | | CuI (20 mol %), DMEDA (40 mol %), K₃PO₄, toluene, 65°, 23 h | 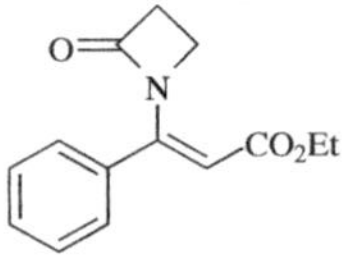 (65) | 649 |

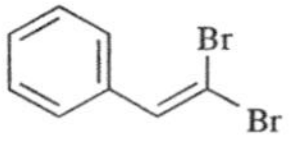

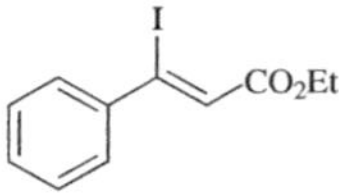

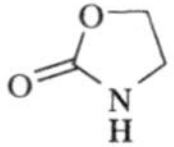

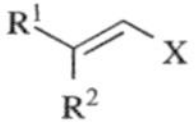

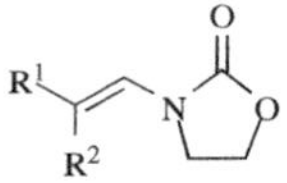

R¹	R²	X	x	Additives	Solvent	Temp (°)	Time (h)	
H	EtO₂C	I	10	DMG (20 mol %), Cs₂CO₃	dioxane	60	12	(82)
H	4-MeOC₆H₄	I	10	DMG (20 mol %), Cs₂CO₃	dioxane	60	12	(78)
H	n-C₁₁H₂₃	I	10	DMG (20 mol %), Cs₂CO₃	dioxane	60	12	(78)
Me	Me	Br	5	DMEDA (10 mol %), K₂CO₃	toluene	90	22	(94)

| 90 |
| 90 |
| 90 |
| 101 |

Nitrogen Nucleophile	Electrophile	Conditions	Product(s) and Yield(s) (%)	Refs.

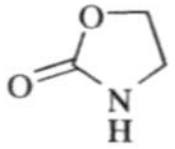

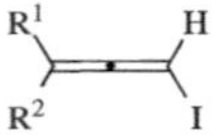

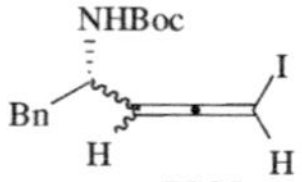

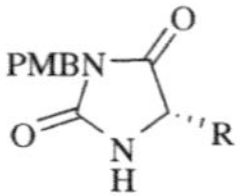

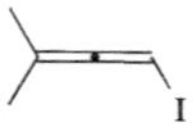

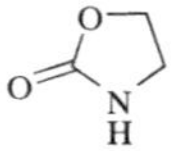

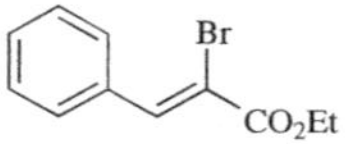

*Please refer to the charts preceding the tables for structures indicated by the **bold** numbers.*

C₃

Catalyst (x mol %), toluene

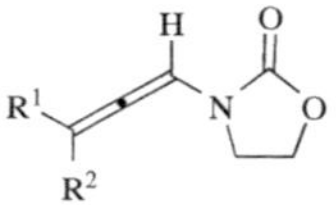

R¹	R²	Catalyst	x	Additives	Temp (°)	Time (h)		
H	Me	CuCN	10	DMEDA (20 mol %), Cs₂CO₃	50	—	(90)	91
H	n-Pr	**8**	7	DMCDA (15 mol %), K₃PO₄	85	5	(100)	651
Me	Me	**8**	7	DMCDA (15 mol %), K₃PO₄	85	7	(97)	651

Electrophile: NHBoc structure, er 75:25

CuCN (10 mol %), DMEDA (20 mol %), Cs₂CO₃, toluene, 50°

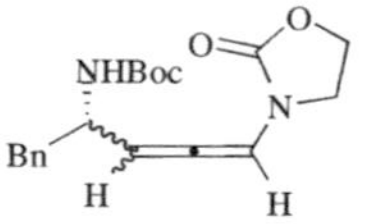

Config.	
(P)	(73)
(M)	(65)

91

C₃₋₇ (PMBN imide nucleophile)

8 (7 mol %), K₃PO₄, toluene, DMCDA (15 mol %), 85°

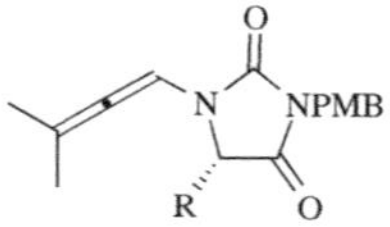

R	Time (h)	
H	6	(99)
i-Bu	7	(100)

651

C₃

CuI (10 mol %), DMG (20 mol %), Cs₂CO₃, dioxane, 80°, 24 h

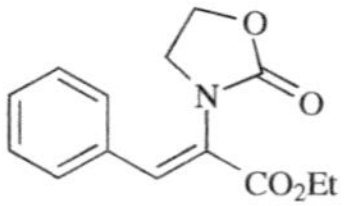

(63)

90

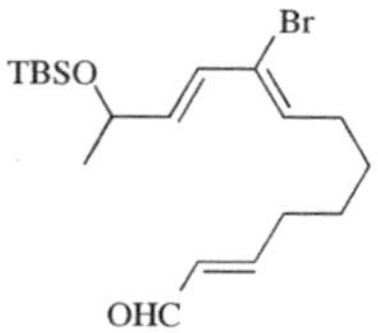

CuI (50 mol %), DMEDA (2.5 eq), K₂CO₃, toluene, reflux, 21 h

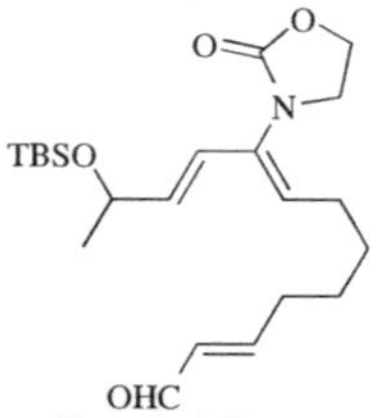

(65)

667

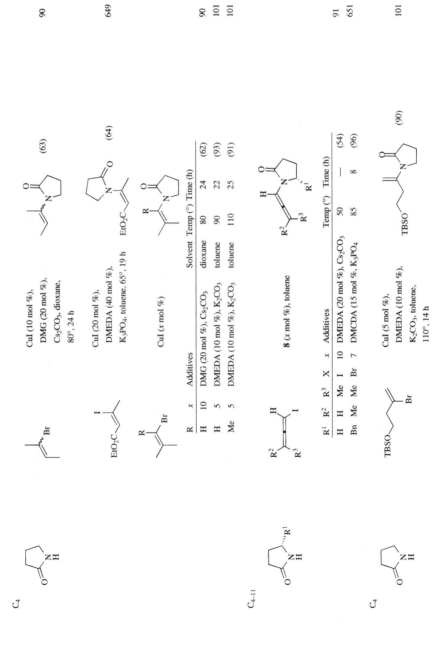

C₄

	CuI (10 mol %), DMG (20 mol %), Cs₂CO₃, dioxane, 80°, 24 h	(63)	90

| | CuI (20 mol %), DMEDA (40 mol %), K₃PO₄, toluene, 65°, 19 h | (64) | 649 |

CuI (x mol %)

R	x	Additives	Solvent	Temp (°)	Time (h)	
H	10	DMG (20 mol %), Cs₂CO₃	dioxane	80	24	(62)	90
H	5	DMEDA (10 mol %), K₂CO₃	toluene	90	22	(93)	101
Me	5	DMEDA (10 mol %), K₂CO₃	toluene	110	25	(91)	101

C₄₋₁₁

8 (x mol %), toluene

R¹	R²	R³	X	x	Additives	Temp (°)	Time (h)	
H	H	Me	I	10	DMEDA (20 mol %), Cs₂CO₃	50	—	(54)	91
Bn	Me	Me	Br	7	DMCDA (15 mol %), K₃PO₄	85	8	(96)	651

C₄

| | CuI (5 mol %), DMEDA (10 mol %), K₂CO₃, toluene, 110°, 14 h | (90) | 101 |

TABLE 30C. *N*-VINYLATION OF LACTAMS, OXAZOLIDINONES, AND CYCLIC IMIDES (*Continued*)

Nitrogen Nucleophile	Electrophile	Conditions	Product(s) and Yield(s) (%)	Refs.

*Please refer to the charts preceding the tables for structures indicated by the **bold** numbers.*

C$_4$

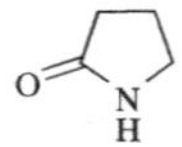

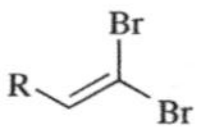

CuI (10 mol %),
DMG (20 mol %),
K$_3$PO$_4$, NH$_4$OAc.
MeCN, reflux, 24 h

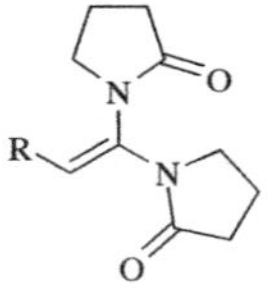

R	
TBDPSOCH$_2$	(60)
2-thienyl	(73)
Ph	(94)
3-O$_2$NC$_6$H$_4$	(60)
4-MeOC$_6$H$_4$	(56)
(*E*)-PhCH=CH	(85)
2-Np	(58)

668

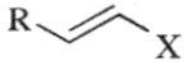

Catalyst (*x* amount)

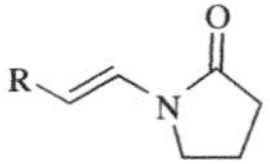

R	X	Catalyst	*x*	Additive	Solvent	Temp (°)	Time (h)	
n-C$_5$H$_{11}$	I	**8**	30 mol %	Cs$_2$CO$_3$	NMP	90	12	(59)
Ph	Br	CuI	1 eq	KH	HMPA	130	22	(43)

165
653

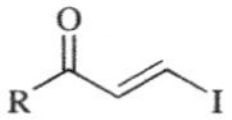

Catalyst (*x* amount), 12 h

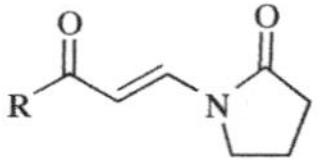

R	Catalyst	*x*	Additives	Solvent	Temp (°)	
CH$_2$=CHCH$_2$O$_2$C	CuI	1 eq	DMG (1 eq), Cs$_2$CO$_3$	dioxane	45	(70)
CH$_2$=CHCH$_2$O$_2$C	Cu(MeCN)$_4$PF$_6$	10 mol %	phen (20 mol %), Rb$_2$CO$_3$	DMA	rt	(92)
BnNHCO	Cu(MeCN)$_4$PF$_6$	10 mol %	phen (20 mol %), Rb$_2$CO$_3$	DMA	60	(72)

90
92
92

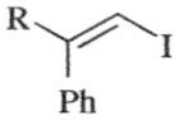

CuI (5 mol %),
DMEDA (10 mol %).
Cs$_2$CO$_3$, THF,
70°, 24 h

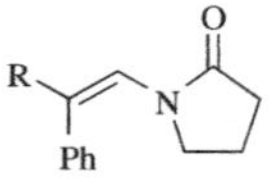

R	
Cl	(92)
Br	(21)

669

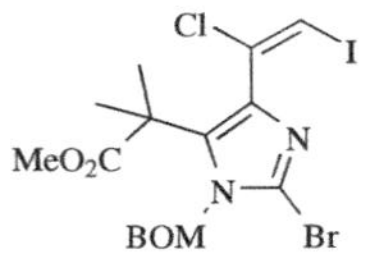

Substrate	Reagent	Conditions	Product (yield)	Ref.
(OTBS-dihydropyrrolone, NH)	(vinyl iodide, Cl, CO₂Me, BOM, Br imidazole)	CuI (10 mol %), DMEDA (20 mol %), Cs₂CO₃, THF, 50°, 45 h	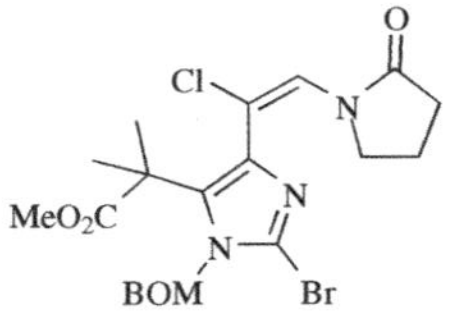 (43)	669
(succinimide K)	(dienyl iodide, isopropyl)	8 (10 mol %), DMCDA (20 mol %), K₃PO₄, dioxane, 90°, 24 h	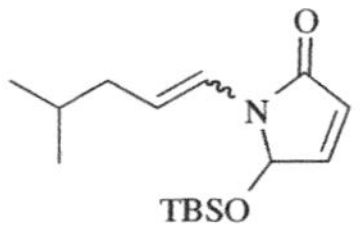 Config. (E) 72 (Z) 56	670
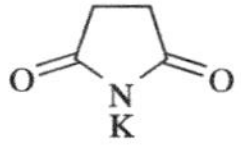 (succinimide NH)	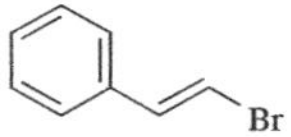 (β-bromostyrene)	CuI (1 eq), HMPA, 130°, 3 h	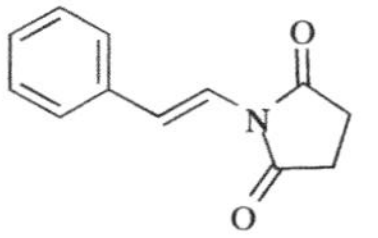 (72)	653
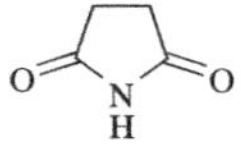 (succinimide NH)	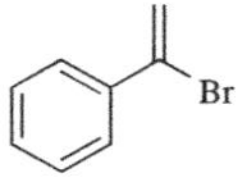 (α-bromostyrene)	CuI (1 eq), HMPA, KH, 130°, 23 h	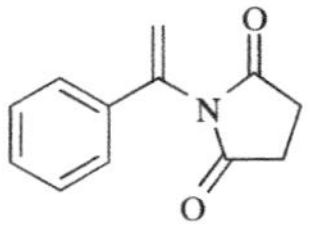 (28)	653
(β-lactam, HN, AcO, OTBS)	(allenyl iodide, R¹, R²)	8 (7 mol %), DMCDA (15 mol %), K₃PO₄, toluene, 85°, 5 h	R¹ R² H n-Pr (36) Me Me (47)	651

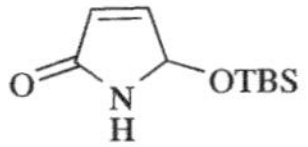

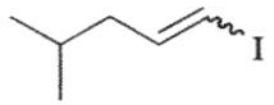

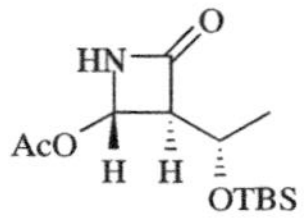

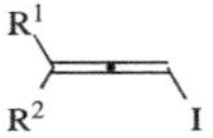

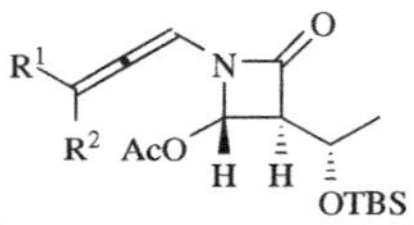

Nitrogen Nucleophile	Electrophile	Conditions	Product(s) and Yield(s) (%)	Refs.

*Please refer to the charts preceding the tables for structures indicated by the **bold** numbers.*

C$_{5-10}$

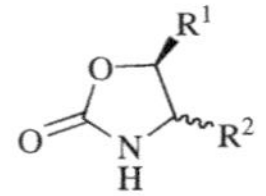

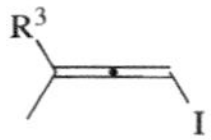

Catalyst (x mol %), toluene

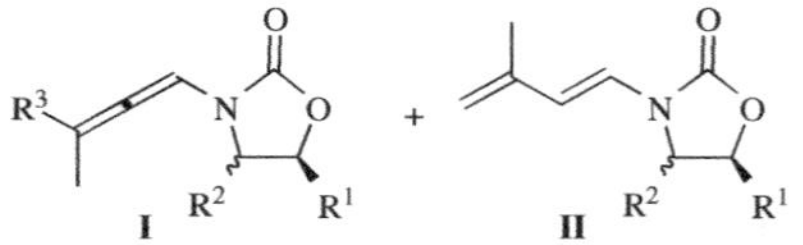

R¹	R²	Config.	R³	Catalyst	x	Additives	Temp (°)	Time (h)	I	II	Refs.
H	CH₂=CH	(R)	Me	8	7	DMCDA (15 mol %), K₃PO₄	85	7	(53)	(26)	651
H	i-Pr	(S)	H	CuCN	10	DMEDA (20 mol %), Cs₂CO₃	50	—	(60)	(0)	91
H	(2-furyl)-CH₂CH₂	(R)	H	CuCN	10	DMEDA (20 mol %), Cs₂CO₃	50	—	(55)	(0)	91
H	Ph	(R)	Me	CuCN	10	DMEDA (20 mol %), Cs₂CO₃	50	—	(51)	(0)	91
H	Bn	(R)	H	CuCN	10	DMEDA (20 mol %), Cs₂CO₃	rt	—	(84)	(0)	91
H	Bn	(R)	Me	CuCN	10	DMEDA (20 mol %), Cs₂CO₃	50	—	(30)	(0)	91
H	Bn	(S)	Me	8	7	DMCDA (15 mol %), K₃PO₄	85	6	(95)	(0)	651
Me	Ph	(R)	Me	8	7	DMCDA (15 mol %), K₃PO₄	85	6	(99)	(0)	651

C$_5$

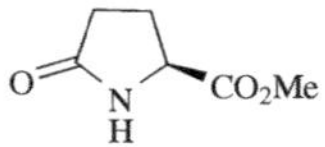

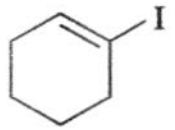

		CuI (10 mol %), Cs₂CO₃, dioxane, 60°, 12 h		90
		Additive Me₂NCH₂CH₂CO₂H (20 mol %) (78) DMG (20 mol %) (78)		

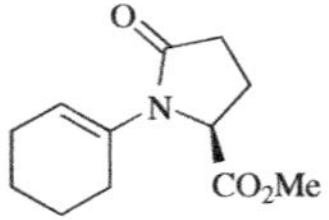

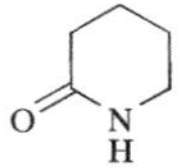

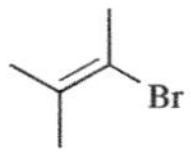

		CuI (5 mol %), DMEDA (10 mol %), K₂CO₃, toluene, 115°, 30 h	(76)	101

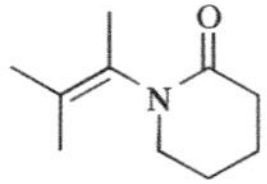

8 (x mol %), toluene

101

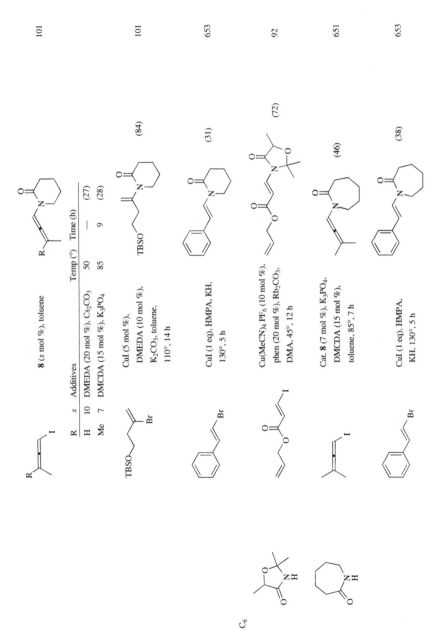

R	x	Additives	Temp (°)	Time (h)	
H	10	DMEDA (20 mol %), Cs_2CO_3	50	—	(27)
Me	7	DMCDA (15 mol %), K_3PO_4	85	9	(28)

CuI (5 mol %), DMEDA (10 mol %), K_2CO_3, toluene, 110°, 14 h

(84)

101

CuI (1 eq), HMPA, KH, 130°, 5 h

(31)

653

C_6

$Cu(MeCN)_4PF_6$ (10 mol %), phen (20 mol %), Rb_2CO_3, DMA, 45°, 12 h

(72)

92

Cat. **8** (7 mol %), K_3PO_4, DMCDA (15 mol %), toluene, 85°, 7 h

(46)

651

CuI (1 eq), HMPA, KH, 130°, 5 h

(38)

653

Nitrogen Nucleophile	Electrophile	Conditions	Product(s) and Yield(s) (%)	Refs.

*Please refer to the charts preceding the tables for structures indicated by the **bold** numbers.*

C$_7$

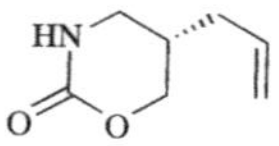

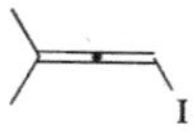

CuCN (10 mol %), DMEDA (20 mol %), Cs$_2$CO$_3$, toluene, 50°

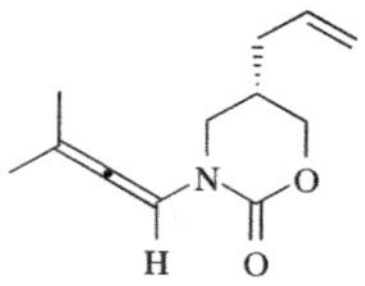

(35)

91

C$_{7-8}$

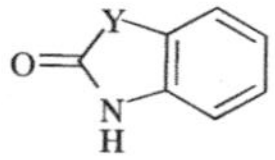

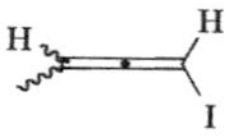

CuCN (10 mol %), DMEDA (20 mol %), Cs$_2$CO$_3$, toluene

Y	Config.	er	Temp (°)	
NMe$_2$	(P)	75:25	50	(63)
NMe$_2$	(M)	87.5:24.5	50	(79)
NMe$_2$	(P,M)	—	50	(65)
CH$_2$	(P)	75:25	rt	(75)
CH$_2$	(M)	87.5:24.5	rt	(71)
CH$_2$	(P,M)	—	50	(60)

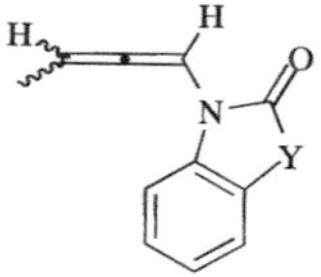

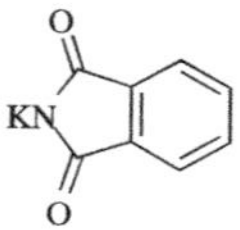

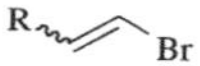

91

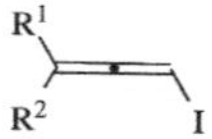

8 (7 mol %), K$_3$PO$_4$, DMCDA (15 mol %), toluene, 85°, 6 h

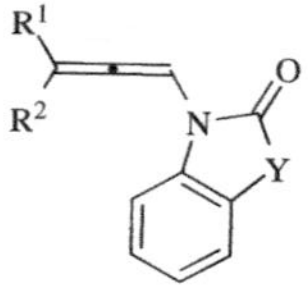

Y	R^1	R^2	
O	H	n-Pr	(100)
CH$_2$	Me	Me	(68)

651

C$_8$

CuI (1 eq), DMA, 165°, 24 h

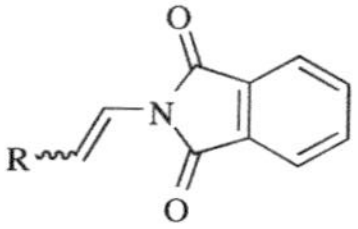

R		
H	(73)	
Me	(63)	

515

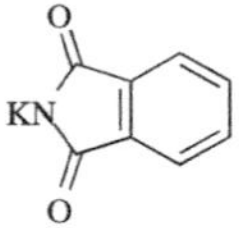

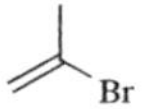

				515

CuI (1 eq), DMA, 165°, 24 h → (80)

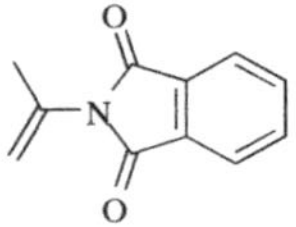

				653
				515
				515
				515

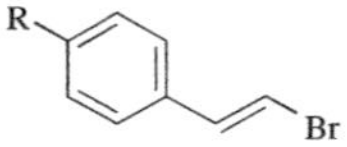

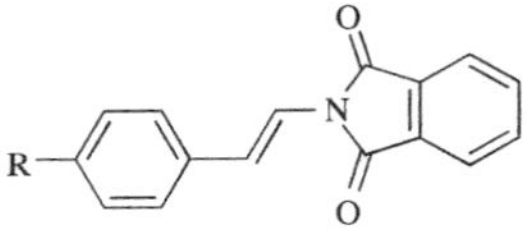

CuI (1 eq)

R	Solvent	Temp (°)	Time (h)	
H	TMU	130	7	(95)
H	DMA	165	24	(72)
O$_2$N	DMA	165	24	(82)
MeO	DMA	165	24	(93)

653

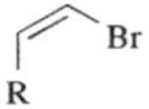

CuI (1 eq), HMPA, 130°

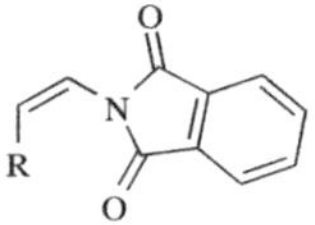

R	Time (h)	
n-C$_6$H$_{13}$	42	(69)
Ph	4	(77)
4-ClC$_6$H$_4$	24	(76)
4-MeOC$_6$H$_4$	21	(57)

651

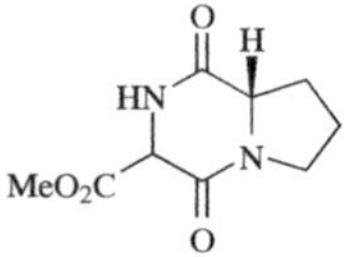

8 (7 mol %), DMCDA (15 mol %), K$_3$PO$_4$, toluene, 85°, 12 h → (63)

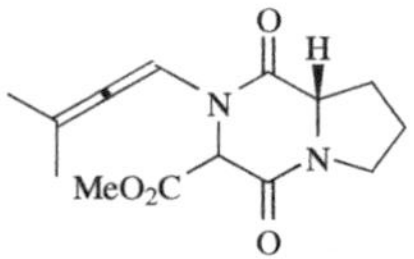

90

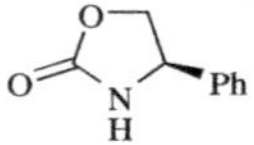

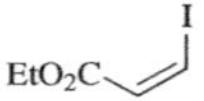

CuI (10 mol %), DMG (20 mol %), Cs$_2$CO$_3$, dioxane, 60°, 12 h → (76)

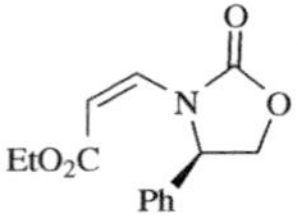

C$_9$

TABLE 30C. *N*-VINYLATION OF LACTAMS, OXAZOLIDINONES, AND CYCLIC IMIDES (*Continued*)

Nitrogen Nucleophile	Electrophile	Conditions	Product(s) and Yield(s) (%)	Refs.

*Please refer to the charts preceding the tables for structures indicated by the **bold** numbers.*

C_{9-10}

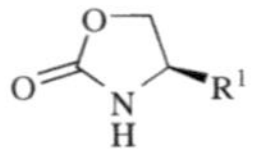

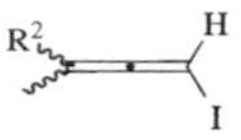

CuCN (10 mol %), DMEDA (20 mol %), Cs$_2$CO$_3$, toluene

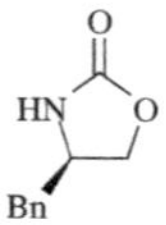

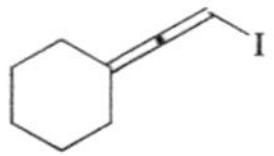

R^1	R^2	Config.	er	
Ph	Me	(*M*)	85:15	(68)
Ph	Me	(*P*)	75:25	(68)
Ph	(*S*)-*i*-PrCH(NHBoc)	(*M*)	—	(44)
Ph	(*S*)-*i*-PrCH(NHBoc)	(*P*)	—	(45)
Bn	Me	(*M*)	85:15	(62)

Refs. 91

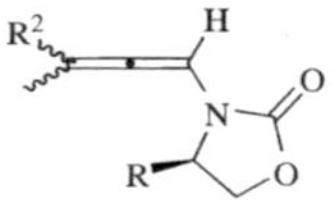

C_{10}

CuCN (10 mol %), DMEDA (20 mol %), Cs$_2$CO$_3$, toluene, 50°

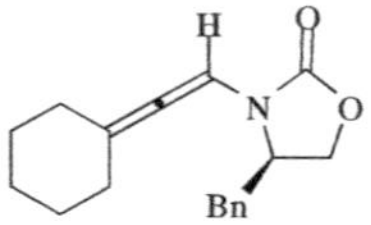

(41)

Refs. 91

C_{10-11}

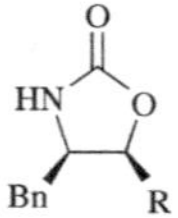

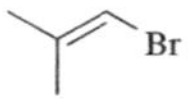

CuI (5 mol %), DMEDA (10 mol %), K$_2$CO$_3$, toluene, 110°, 16 h

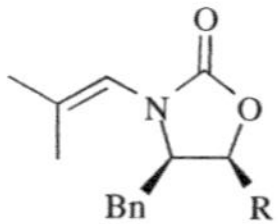

R	
H	(95)
Me	(94)

Refs. 101

C_{10}

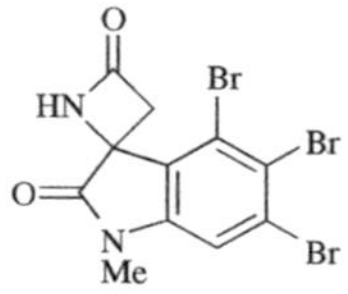

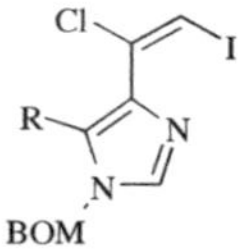

CuI (5 mol %), DMEDA (10 mol %), Cs$_2$CO$_3$, THF, 70°, 24 h

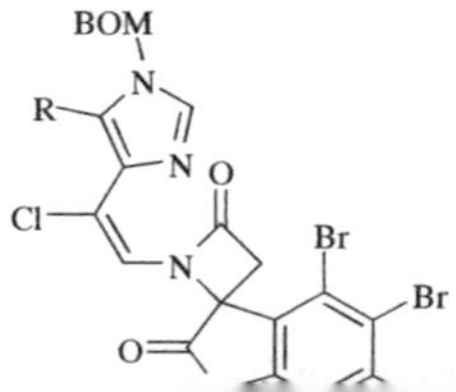

R	Temp (°)	
H	50	(91) 669
MeO$_2$C(Me$_2$)C	70	(45)

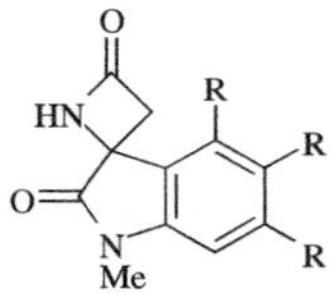	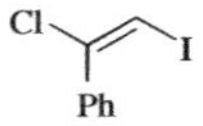	CuI (10 mol %), DMEDA (20 mol %), Cs_2CO_3, THF, 70°, 24 h	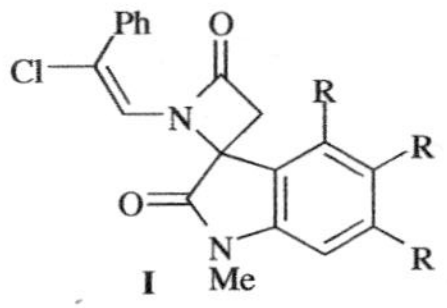 I + 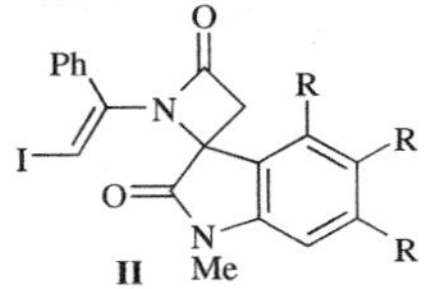 II R / I / II H / (75) / (16) Br / (43) / (0)	669

TABLE 31. *N*-VINYLATION OF HYDRAZINE DERIVATIVES

Nitrogen Nucleophile	Electrophile	Conditions	Product(s) and Yield(s) (%)	Refs.
C₀				
BocHN—NHBoc	(image)	CuI (5 mol %), phen (10 mol %), Cs₂CO₃, DMF, 80°, 13 h	(image) R: Me (75), n-C₈H₁₇ (75)	671
	(image)	CuI (5 mol %), phen (10 mol %), Cs₂CO₃, DMF, 80°, 13 h	(image) (70)	671
	(image)	CuI (5 mol %), phen (10 mol %), Cs₂CO₃, DMF, 80°, 13 h	(image) (87)	671
	(image)	CuI (5 mol %), DMEDA (20 mol %), K₂CO₃, toluene, 100°	(image) (85)	671
	(image)	CuI (5 mol %), phen (10 mol %), Cs₂CO₃, DMF, 80°, 13 h	(image) (85)	671
	(image)	CuI (5 mol %), phen (10 mol %), Cs₂CO₃, DMF, 80°, 13 h	(image) (94)	671

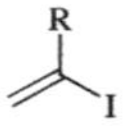

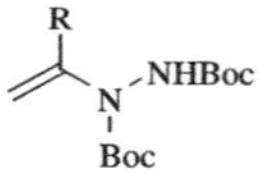

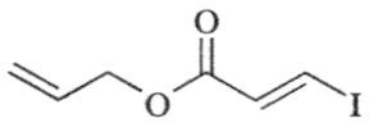

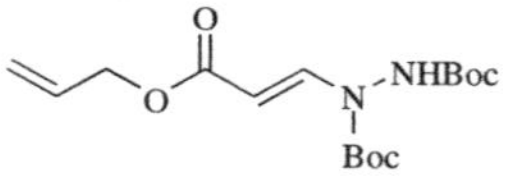

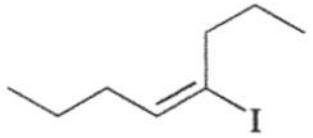

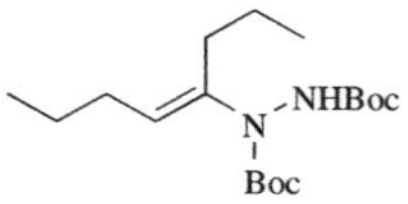

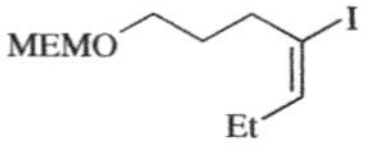

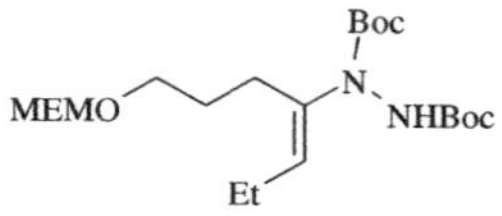

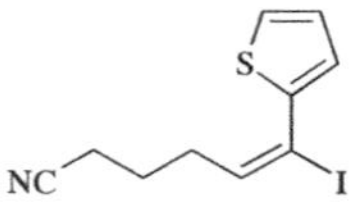

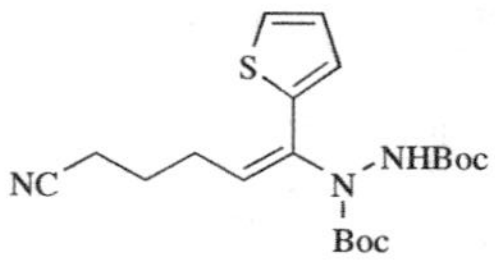

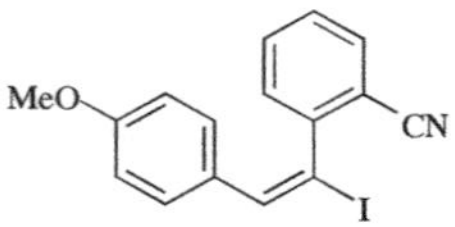

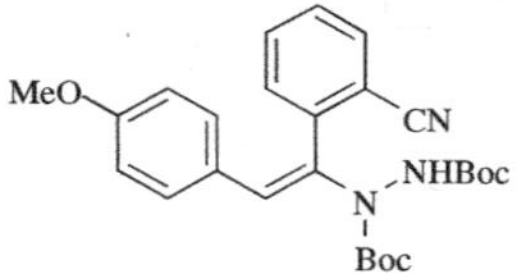

TABLE 32. N-VINYLATION OF SULFOXIMINES

Nitrogen Nucleophile	Electrophile	Conditions	Product(s) and Yield(s) (%)	Refs.
C₄		CuI (1 eq), DMEDA (2 eq), K₂CO₃, toluene, reflux, 24 h	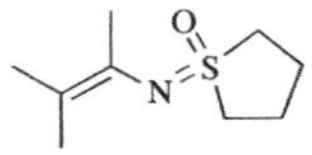 (91)	672
C₇		CuI (1 eq), DMEDA (2 eq), K₂CO₃, toluene, reflux, 24 h	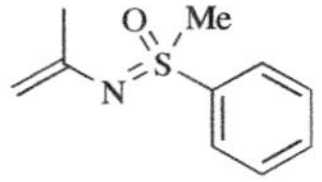 (97)	672
C₇₋₈ 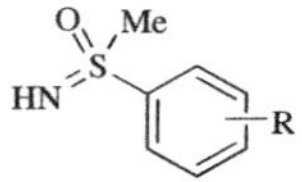	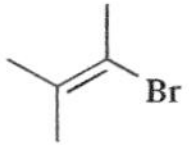	CuI (1 eq), DMEDA (2 eq), K₂CO₃, toluene, reflux, 24 h	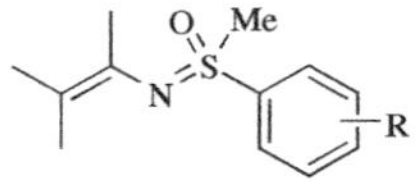 R / H (94) / 2-MeO (93) / 4-Me (92)	672
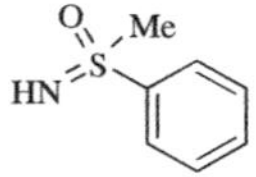	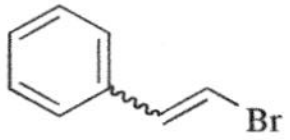	CuI (1 eq), DMEDA (2 eq), K₂CO₃, toluene, reflux, 24 h	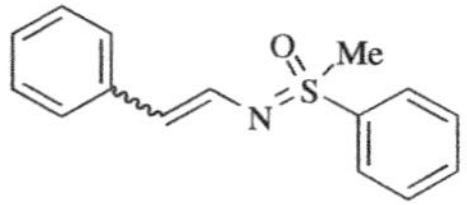 (90)	672
C₇	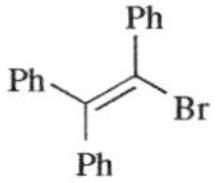	CuI (1 eq), DMEDA (2 eq), K₂CO₃, toluene, reflux, 24 h	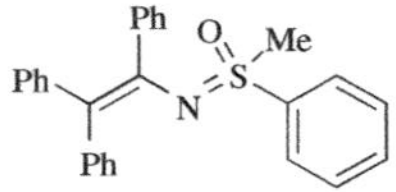 (78)	672

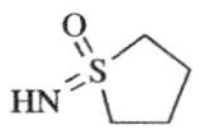

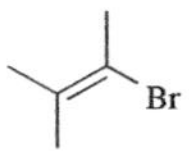

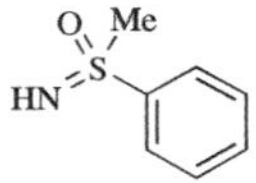

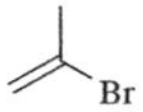

621

TABLE 33. PREPARATION OF VINYL AZIDES

Vinyl Halide	Conditions	Product(s) and Yield(s) (%)	Refs.
C₆ (structure)	NaN₃, CuI (10 mol %), ʟ-Pro (20 mol %), DMSO, 70°, 4 h	(structure) N₃ (78)	155
C₈₋₁₃ (structure)	NaN₃, CuI (10 mol %), ʟ-Pro (20 mol %), DMSO, 70°, 4 h	R (structure) N₃ R Ph (70) 4-ClC₆H₄ (76) 4-MeOC₆H₄ (64) n-C₁₁H₂₃ (82)	155

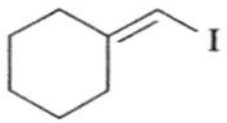

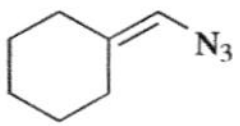

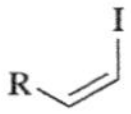

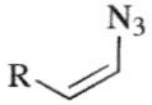

TABLE 34. INTRAMOLECULAR VINYLATIONS

A. SYNTHESIS OF FOUR-MEMBERED NITROGEN HETEROCYCLES

Substrate	Conditions	Product(s) and Yield(s) (%)	Refs.
C$_{4-10}$ (TsHN, R^1, X, R^3, R^2 substrate structure)	CuI (x mol %), DMEDA (y mol %), Cs$_2$CO$_3$	(R^2, R^3, R^1, N–Ts product structure)	104
C$_6$ (TsHN, Br, isopropylidene substrate)	CuI (10 mol %), DMEDA (40 mol %), Cs$_2$CO$_3$, THF, 68°, 6 h	(isopropylidene azetidine, N–Ts) (99)	104
C$_8$ (Br, NHTs, Br diene substrate)	CuI (10 mol %), DMEDA (20 mol %), Cs$_2$CO$_3$, dioxane, 68°, 3 h	(methylene azetidine N–Ts with butenyl-Br chain) (66) + (methylene pyrrolidine N–Ts with Br) (33)	105
C$_9$ (Br, TsHN, C$_5$H$_{11}$ substrate)	CuI (20 mol %), DMEDA (40 mol %), Cs$_2$CO$_3$, THF, 68°	(azetidine N–Ts, =CH–C$_5$H$_{11}$)	104
C$_{10-17}$ (Br, O, R^{1}HN, R^2 substrate)	CuI (5 mol %), DMG (10 mol %), K$_2$CO$_3$, THF, reflux	(β-lactam: R^2, O, N–R, methylene)	603

For C$_{4-10}$ substrate:

R^1	R^2	R^3	X	x	y	Solvent	Temp (°)	Time (h)	
H	H	H	I	10	20	THF	40	1	(94)
H	Me	H	Cl	20	40	dioxane	100	4	(99)
H	Me	Me	Cl	20	40	dioxane	100	3	(99)
Me	H	H	Br	10	20	THF	68	1	(99)

For C$_{4-10}$ product:

R^1	R^2	R^3	X	x	y	Solvent	Temp (°)	Time (h)	
n-Pr	H	H	Cl	20	40	dioxane	100	2	(99)
n-Pr	H	H	Br	10	20	THF	68	2	(99)
Ph	H	H	Cl	20	40	dioxane	100	2	(99)
Ph	H	H	Br	10	20	THF	68	2	(99)

For C$_9$ product:

Config.	Time (h)	
(E)	3	(86)
(Z)	12	(89)

For C$_{10-17}$ product:

R^1	R^2	Time (h)	
c-C$_6$H$_{11}$	H	24	(70)
Ph	Bn	17	(94)
Ph	H	18	(99)
4-MeOC$_6$H$_4$	H	16	(99)
4-MeC$_6$H$_4$	H	18	(98)
Bn	H	16	(92)

TABLE 34. INTRAMOLECULAR VINYLATIONS (*Continued*)
A. SYNTHESIS OF FOUR-MEMBERED NITROGEN HETEROCYCLES (*Continued*)

Substrate	Conditions	Product(s) and Yield(s) (%)	Refs.
C_{12–18}	CuI (5 mol %), DMG (10 mol %), K₂CO₃, THF, reflux	 R — Time (h) Ph — 18 (99) 4-MeO₂CC₆H₄ — 20 (89) n-C₁₂H₂₅ — 35 (95)	603
C₁₂	CuI (5 mol %), DMG (10 mol %), K₂CO₃, THF, reflux, 24 h	(99)	603
C_{14–16}	CuI (5 mol %), DMG (10 mol %), K₂CO₃, THF, reflux	 R — n — Time (h) Ph — 1 — 18 (99) Ph — 2 — 21 (92) Bn — 1 — 25 (70) Bn — 2 — 17 (98)	603
C_{14–15}	CuI (5 mol %), DMG (10 mol %), K₂CO₃, THF, reflux	 R — Time (h) Ph — 19 (99) Bn — 17 (98)	603
C₁₄	CuI (5 mol %), DMG (10 mol %), K₂CO₃, THF, reflux, 21 h	(99)	603

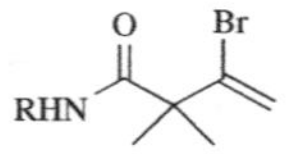

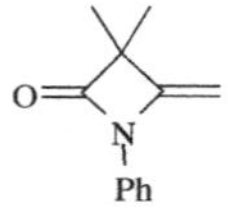

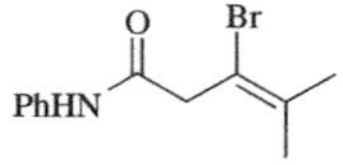

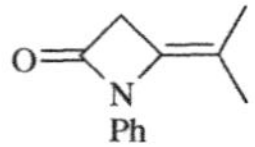

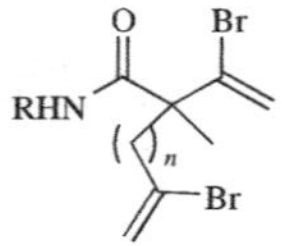

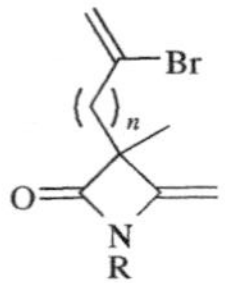

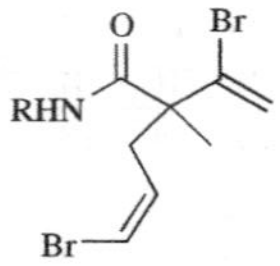

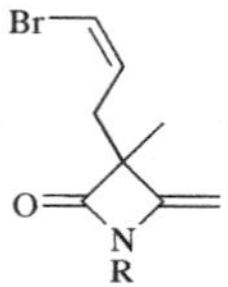

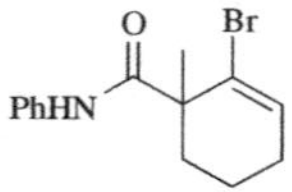

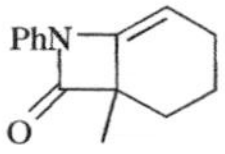

C17

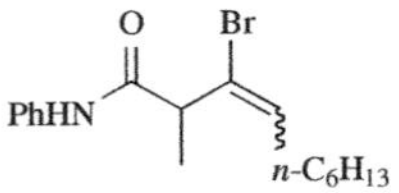

PhHN─C(O)─C(Br)=CH─*n*-C6H13 (with methyl substituent)

CuI (5 mol %), DMG (10 mol %),
K2CO3, THF, reflux

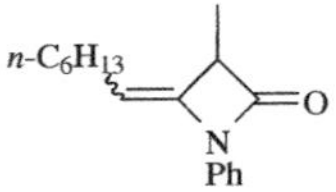

Config.	Time (h)	
(E)	28	(97)
(Z)	32	(99)

603

TABLE 34. INTRAMOLECULAR VINYLATIONS (*Continued*)

B. SYNTHESIS OF FIVE-MEMBERED NITROGEN HETEROCYCLES

Substrate	Conditions	Product(s) and Yield(s) (%)	Refs.

C$_{5-7}$

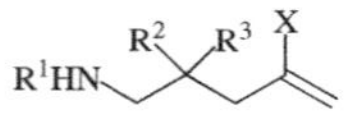

CuI (20 mol %), DMEDA (40 mol %),
Cs$_2$CO$_3$, dioxane, 100°

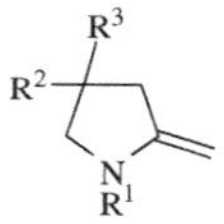

R^1	R^2	R^3	X	Temp (°)	Time (h)	
Ms	H	H	Cl	100	4	(99)
Ts	H	H	Cl	100	4	(99)
Ts	H	H	Br	68	2	(99)
Ts	H	H	I	20	24	(92)
Ts	Me	H	Cl	100	4	(99)
Ts	Me	Me	Cl	100	4	(99)

105

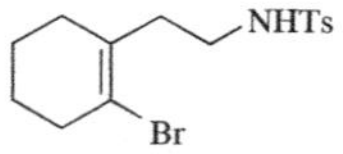

CuI (10 mol %), DMEDA (20 mol %),
Cs$_2$CO$_3$, dioxane, 68°, 2 h

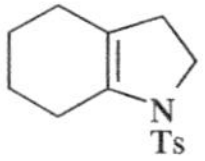

(99)

105

C$_8$

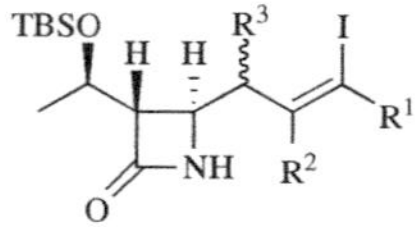

CuI (20 mol %), bpy (40 mol %),
K$_3$PO$_4$, toluene, 40°, 18 h

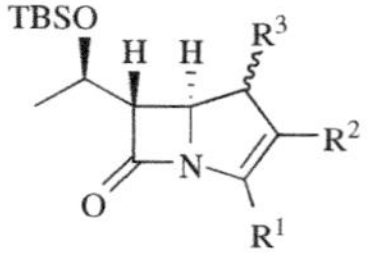

R^1	R^2	R^3	Time (h)	
H	I	H	18	(97)
Me	I	H	22	(96)
AcOCH$_2$	I	H	22	(91)
BnOCH$_2$	I	H	24	(95)
EtO$_2$C	Cl	H	12	(85)
EtO$_2$C	Br	H	13	(92)
EtO$_2$C	I	H	21	(94)
EtO$_2$C	I	Me (R)	24	(92)
EtO$_2$C	I	Me (S)	28	(85)
PNBO$_2$C	I	H	18	(94)

629

C$_{8-10}$

C$_{10}$

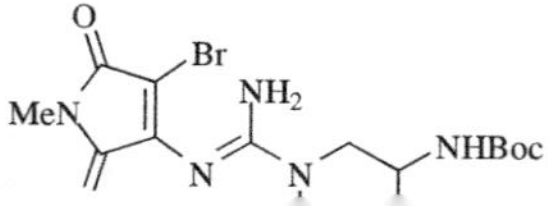

CuI (50 mol %), NaH, THF, reflux, 1 h

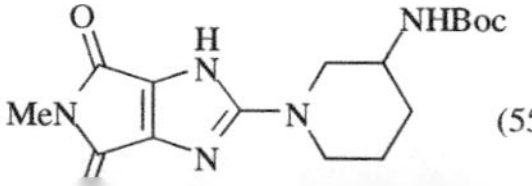

(55)

139

Substrate	Conditions	Product	Refs.
C₁₁ (PhHN-CO-CH₂CH₂-C(=CH₂)-X)	CuI (20 mol %), DMEDA (40 mol %), Cs₂CO₃, dioxane, 100°, 20 h	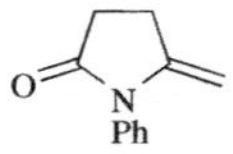 $\dfrac{\text{X}}{\text{Br (trace)}}$; I (95)	673
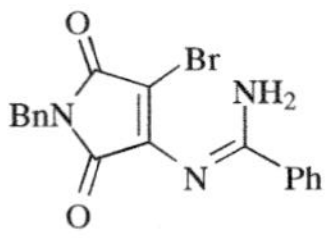 (BnN-maleimide with Br and C(=NH)Ph=NH₂)	CuI (50 mol %), NaH, THF, reflux, 1 h	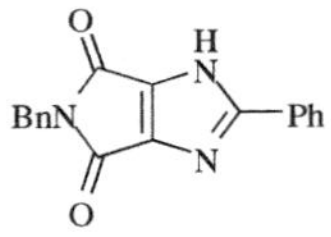 (94)	139
C₁₅ (2-bromovinyl aniline with N(Ph)Me carbamoyl)	CuI (10 mol %), DMEDA (20 mol %), K₂CO₃, toluene, 120°, 24 h	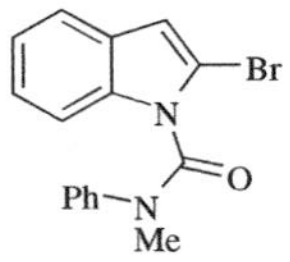 (92)	186

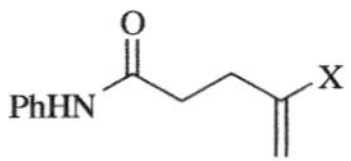

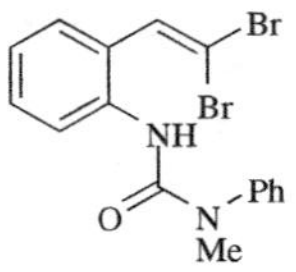

TABLE 34. INTRAMOLECULAR VINYLATIONS (*Continued*)
C. SYNTHESIS OF SIX-MEMBERED NITROGEN HETEROCYCLES

Substrate	Conditions	Product(s) and Yield(s) (%)	Refs.
C₅	CuI (10 mol %), DMEDA (20 mol %), Cs₂CO₃, dioxane, 68°, 8 h	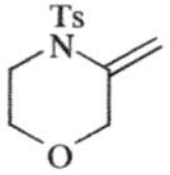 (99)	105
C₆	CuI (20 mol %), DMEDA (40 mol %), Cs₂CO₃, dioxane, 100°, 20 h	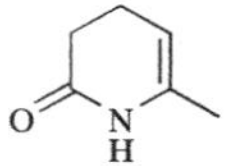 (86)	673
	CuI (20 mol %), DMEDA (40 mol %), Cs₂CO₃, dioxane, 100°, 144 h	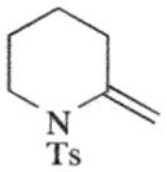 (19)	105
C₉	CuI (10 mol %), DMEDA (20 mol %), Cs₂CO₃, dioxane, 100°, 3 h	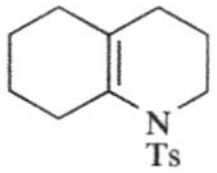 (91)	105
C₁₁	CuI (20 mol %), DMEDA (40 mol %), Cs₂CO₃, dioxane, 100°, 20 h	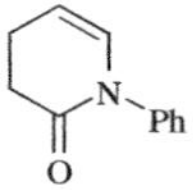 (91)	673
	CuI (10 mol %), DMEDA (20 mol %), Cs₂CO₃, dioxane, 68°, 3 h	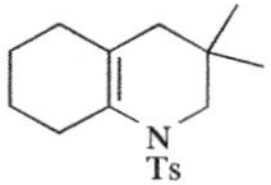 (99)	105
C₁₂	CuI (20 mol %), DMEDA (40 mol %), Cs₂CO₃, dioxane, 100°, 20 h	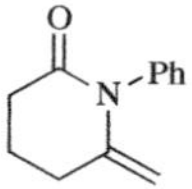 (86)	673

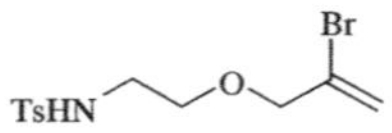

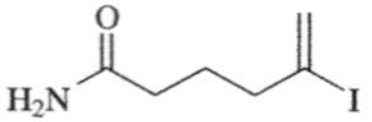

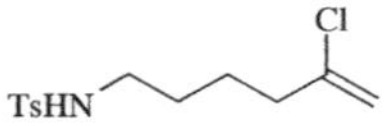

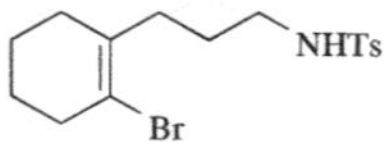

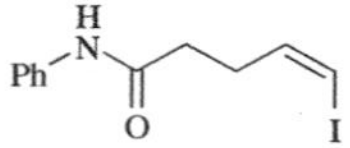

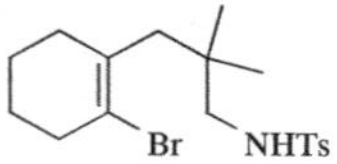

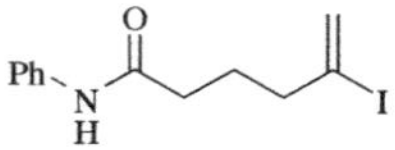

628

Substrate	Conditions	Product(s) and Yield(s) (%)	Refs.
C_{6-7}	CuI (20 mol %), DMEDA (40 mol %), Cs_2CO_3, dioxane, 100°, 20 h	R / H (44) / Me (91)	673
C_7	CuI (20 mol %), DMEDA (40 mol %), Cs_2CO_3, dioxane, 100°, 20 h	(46)	673
C_8	CuI (15 mol %), DMEDA (30 mol %), Cs_2CO_3, THF, 70°, 12 h	(82)	669
C_9	CuI (10 mol %), DMEDA (20 mol %), Cs_2CO_3, dioxane, 100°, 3 h	(90)	105
C_{10}	CuI (50 mol %), DMEDA (1 eq), Cs_2CO_3, dioxane, 68°, 12 h	(66)	105
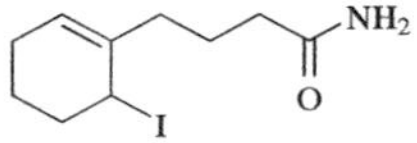	CuI (20 mol %), DMEDA (40 mol %), Cs_2CO_3, dioxane, 100°, 20 h	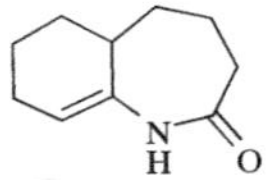 (45)	673

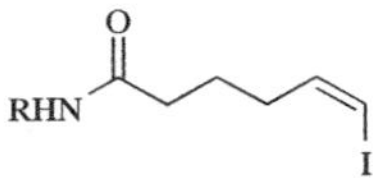

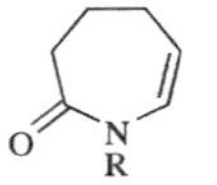

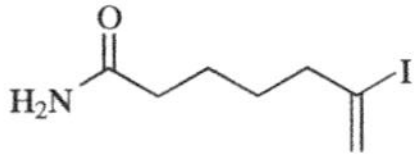

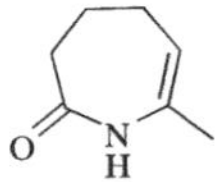

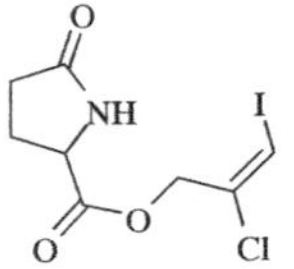

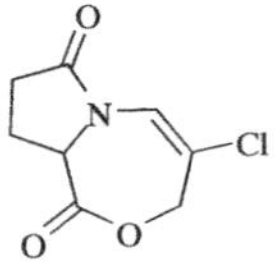

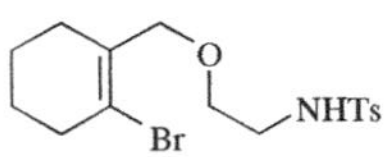

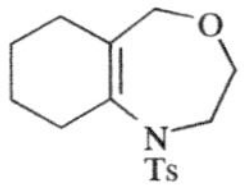

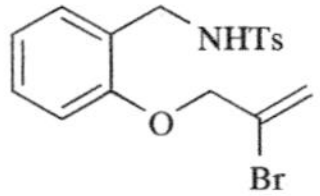

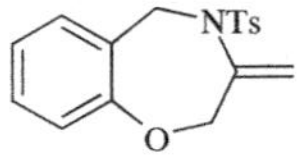

TABLE 34. INTRAMOLECULAR VINYLATIONS (*Continued*)

D. SYNTHESIS OF SEVEN-MEMBERED NITROGEN HETEROCYCLES (*Continued*)

Substrate	Conditions	Product(s) and Yield(s) (%)	Refs.
C$_{12}$	CuI (10 mol %), DMEDA (20 mol %), Cs$_2$CO$_3$, dioxane, 100°, 3 h	(96)	105
C$_{13}$	CuI (20 mol %), DMEDA (40 mol %), Cs$_2$CO$_3$, dioxane, 100°, 20 h	(91)	673
C$_{15}$	CuI (1 eq), DMEDA (2 eq), Cs$_2$CO$_3$, THF, rt, 24 h	R Temp (°) Cl 75 (80) MeS rt (92)	674 675, 674
C$_{16}$	CuI (20 mol %), DMEDA (40 mol %), Cs$_2$CO$_3$, dioxane, 100°, 20 h	(83)	673

TABLE 34. INTRAMOLECULAR VINYLATIONS (*Continued*)

E. SYNTHESIS OF EIGHT- AND HIGHER-MEMBERED NITROGEN HETEROCYCLES

Substrate	Conditions	Product(s) and Yield(s) (%)	Refs.
C$_{14}$	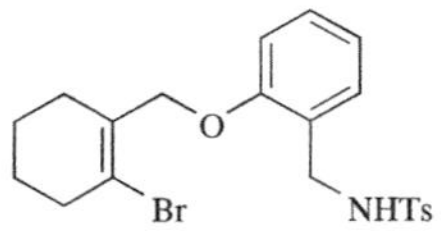 CuI (20 mol %), DMEDA (40 mol %), Cs$_2$CO$_3$, dioxane, 153°, 48 h	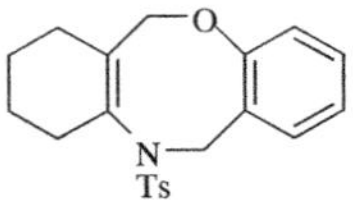 (62)	105
C$_{18}$	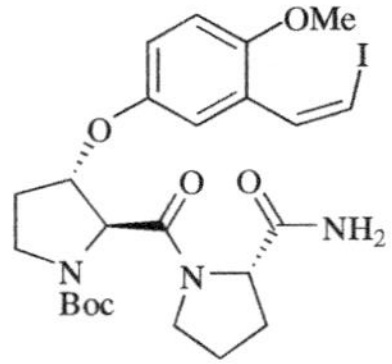 CuI (10 mol %), DMEDA (20 mol %), Cs$_2$CO$_3$, THF, 60°	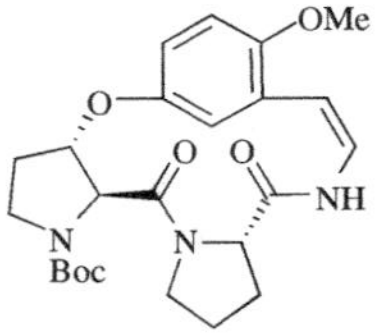 (82)	168
C$_{19}$	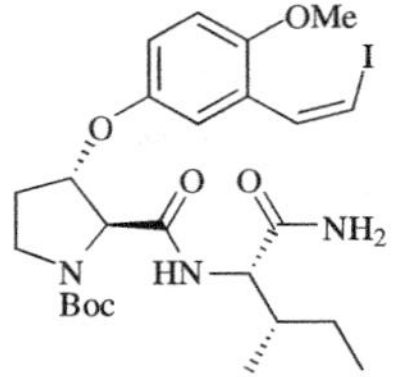 CuI (10 mol %), DMEDA (20 mol %), Cs$_2$CO$_3$, THF, 60°	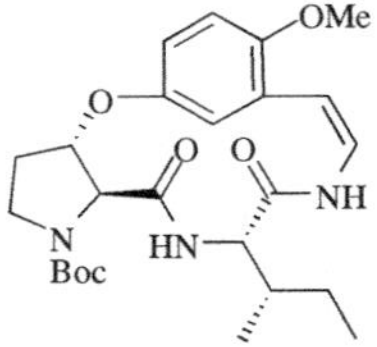 (70)	676, 168
C$_{23}$	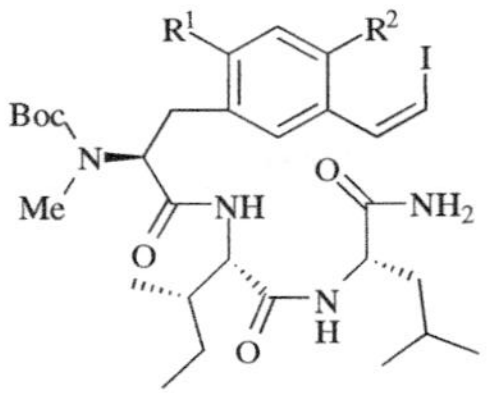 CuI (x mol %), DMEDA (y mol %), Cs$_2$CO$_3$, THF	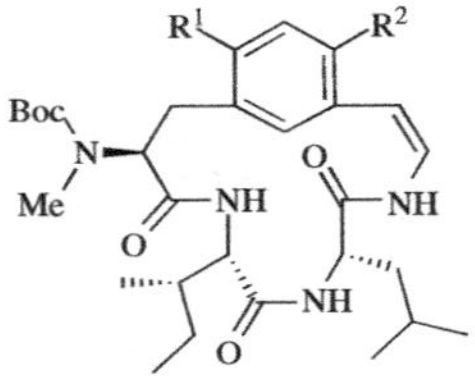	

R^1	R^2	x	y	Temp (°)	Time (h)		
HO	MeO	10	20	60	—	(84)	168
MeO	H	20	40	reflux	72	(83)	99, 677
MeO	MeO	10	20	60	—	(82)	168

631

TABLE 35. N-VINYLATIONS IN MULTI-STEP REACTIONS

Nitrogen Nucleophile	Electrophile	Conditions	Product(s) and Yield(s) (%)	Refs.

C_0

H$_2$NBoc

Electrophile:

R^2, R^3, R^1, X (alkyne/vinyl halide)

R^1	R^2	R^3	X	Time (h)	
TBS	4-MeC$_6$H$_4$	4-MeC$_6$H$_4$	I	14	(95)
TIPSOCH$_2$	H	n-C$_5$H$_{11}$	I	7	(83)
MeO$_2$C	H	n-C$_5$H$_{11}$	Br	14	(82)
MeO$_2$C	H	1-cyclohexenyl	Br	14	(81)

Conditions: CuI (5 mol %), DMEDA (20 mol %), Cs$_2$CO$_3$, THF, 80°

Product:

R^2, R^3, R^1, N, H (pyrrole)

R^1	R^2	R^3	X	Time (h)	
n-Pr	H	n-Pr	Br	14	(70)
n-Pr	H	n-Pr	I	8	(74)
n-Pr	n-Pr	H	I	8	(68)
n-Bu	Et	n-C$_8$H$_{17}$	I	14	(85)
1-cyclohexenyl	H	n-Pr	I	5	(84)

Refs: 608

Electrophile: structure with I, I, R, S (thiophene)

Conditions: CuI (5 mol %), DMEDA (20 mol %), K$_2$CO$_3$, toluene, 80°, 15 h

Product: Boc-N, R, S

R	
n-Bu	(83)
3-thienyl	(74)

Refs: 608

Electrophile: n-C$_5$H$_{11}$ alkyne, I, N

Conditions: CuI (5 mol %), DMEDA (20 mol %), K$_2$CO$_3$, toluene, 80°, 15 h

Product: n-C$_5$H$_{11}$, N-Boc (71)

Refs: 608

Electrophile: R, I, cyclopentene

Conditions: CuI (5 mol %), DMEDA (20 mol %), Cs$_2$CO$_3$, THF, 80°

Product: R, N-Boc

R	Time (h)	
Cl(CH$_2$)$_3$	4	(78)
Ph	6	(52)

Refs: 608

Electrophile: BocN, BocN, I, I

Conditions: CuI (5 mol %), DMEDA (20 mol %), Cs$_2$CO$_3$, THF, 80°, 16 h

Product: BocN, BocN, NBoc (78)

Refs: 100

Electrophile: S, thiophene, I

Conditions: CuI (5 mol %), DMEDA (20 mol %),

Product: S, N (78)

Refs: 100

C$_5$	(TsHN—prenyl)	**8** (15 mol %), DMEDA (30 mol %), Cs$_2$CO$_3$, toluene, 80°, 24 h	(65)	625

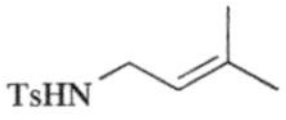

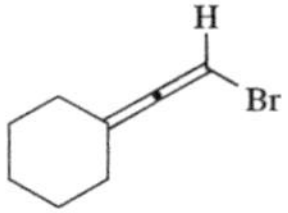

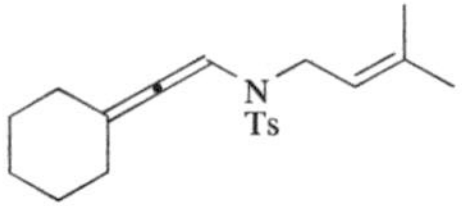

C$_6$	(NHTs, NHTs)	**8** (15 mol %), DMEDA (30 mol %), Cs$_2$CO$_3$, THF, reflux, 24 h	(63)	625

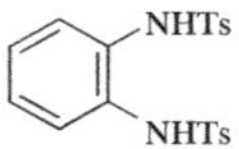

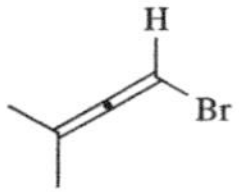

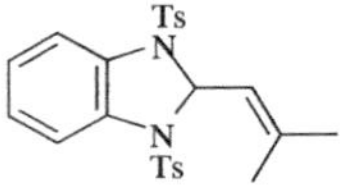

C$_{7-13}$

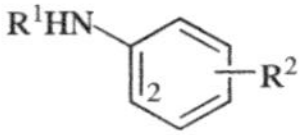

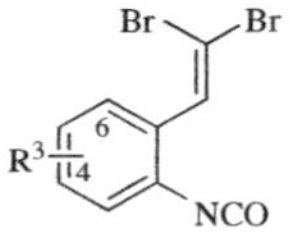

(R^{1}HN—aryl—R^2) + (Br$_2$C=CH—aryl—NCO)

1. CuI (10 mol %),
DMEDA (20 mol %),
K$_2$CO$_3$, toluene,
120°, time 1
2. Pd(dppf)Cl$_2$ (10 mol %),
KOAc, toluene, 120°, time 2

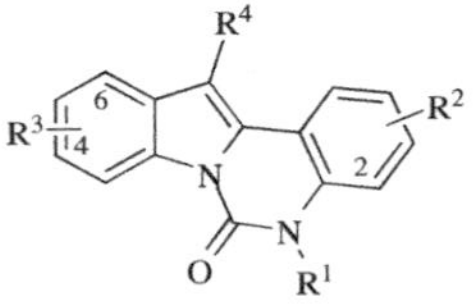

186

R^1	R^2	R^3	R^4	Time 1 (h)	Time 2 (h)	
Me	H	5-Cl	H	24	24	(63)
Me	H	4-Br	H	24	24	(57)
Me	H	5-Br	H	24	24	(78)
Me	H	4,5-(MeO)$_2$	H	18	24	(65)
Me	H	H	CF$_3$	24	24	(67)
Me	4-Cl	H	H	26	24	(77)
Me	4-Br	H	H	24	24	(71)
Me	4-MeO	H	H	18	24	(67)
Me	4-CF$_3$O	H	H	30	36	(69)
Me	2-Me	H	H	30	24	(62)
Me	3-Me	H	H	30	36	(68)
Me	4-Me	H	H	24	36	(80)
Me	3,5-Me$_2$	H	H	26	36	(64)
Me	2-(CH=CH)$_2$-3	H	H	16	36	(69)
Bn	H	H	H	24	24	(87)

TABLE 35. N-VINYLATIONS IN MULTI-STEP REACTIONS (Continued)

Nitrogen Nucleophile	Electrophile	Conditions	Product(s) and Yield(s) (%)	Refs.

C₅

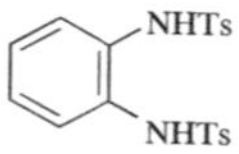

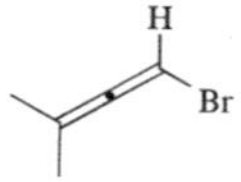

Row 1:

Nucleophile: 1,2-bis(NHTs)benzene

Electrophile: (H)(Br)C=C–CH=C(CH₃)₂ type

Conditions: **8** (15 mol %), DMEDA (30 mol %), Cs₂CO₃, THF, reflux, 24 h

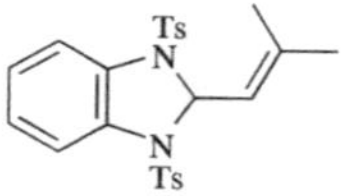

Product: (63)

Refs: 625

Row 2:

Conditions: CuI (5 mol %), DMEDA (20 mol %), Cs₂CO₃, THF, 80°, 6 h

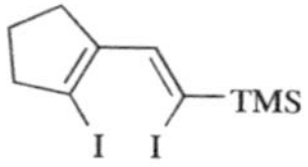

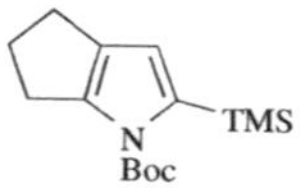

Product: TMS, Boc (94)

Refs: 100

Row 3:

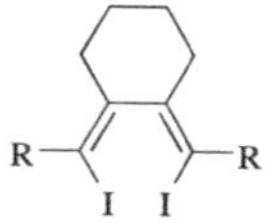

Conditions: CuI (5 mol %), DMEDA (20 mol %), Cs₂CO₃, THF, 80°

Product:

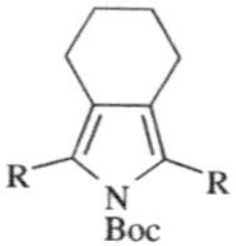

R	Time (h)	
H	7	(99)
TMS	10	(99)

Refs: 100

Row 4:

Nucleophile: NaN₃

Electrophile: n-C₅H₁₁–CH=CH–I

Conditions: CuSO₄•5H₂O (5 mol %), Na ascorbate, L-Pro (20 mol %), Na₂CO₃, DMSO/H₂O (9:1), 60°, 18 h

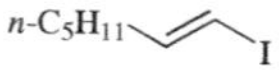

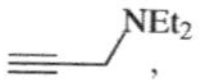

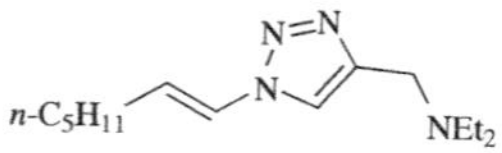

Product: n-C₅H₁₁ ... NEt₂ (73)

Refs: 606

Row 5:

Nucleophile: Boc–N(H)–N(H)–Boc

Electrophile:

R¹	R²	R³	Time (h)	
H	H	n-C₅H₁₁	9	(92)
H	H	Ph	9	(93)
TIPSOCH₂	H	n-C₅H₁₁	11	(78)
n-Pr	H	n-Pr	9	(83)

Conditions: 1. CuI (5 mol %), DMEDA (20 mol %), Cs₂CO₃, THF, 80°, time 2. TFA, MeCl₂, rt

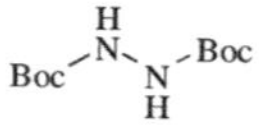

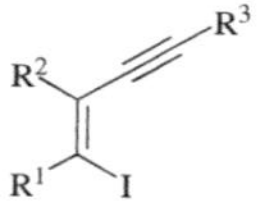

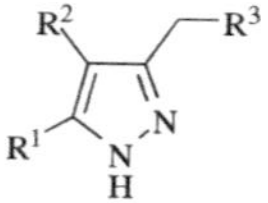

Product:

R¹	R²	R³	Time (h)	
n-Bu	H	EtO₂C	16	(81)
n-Bu	Et	n-C₈H₁₇	14	(66)
Ph	Et	n-C₅H₁₁	14	(72)
Bn	H	BnO(CH₂)₂	9	(72)

Refs: 608

C₀

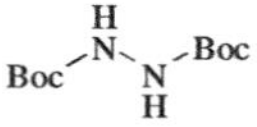

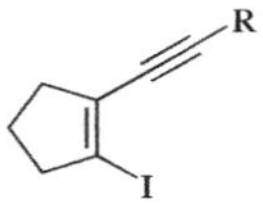

Boc–N(H)–N(H)–Boc

1. CuI (5 mol %),
   DMEDA (20 mol %),
   Cs₂CO₃, THF, 80°, time
2. TFA, MeCl₂, rt

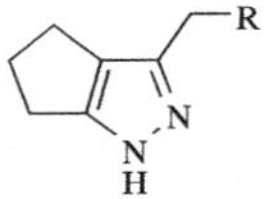

R	Time (h)	
Cl(CH₂)₃	13	(89)
Ph	6	(86)

608

H₂NBoc

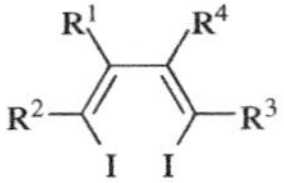

CuI (5 mol %),
DMEDA (20 mol %),
Cs₂CO₃, THF, 80°

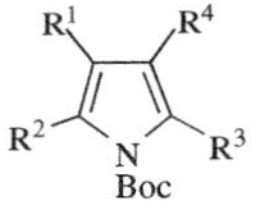

R¹	R²	R³	R⁴	Time (h)	
H	H	n-Pr	n-Pr	5	(73)
TMS	H	n-Pr	n-Pr	6	(89)
TMS	Me	Me	TMS	8	(97)
TMS	n-Bu	H	THPO(CH₂)₂	10	(86)
TMS	n-Bu	n-Pr	n-Pr	14	(78)

R¹	R²	R³	R⁴	Time (h)	
TMS	n-Bu	H	Cl(CH₂)₄	9	(92)
Me	Me	Me	Me	14	(97)
n-Bu	H	t-BuO₂C	TBS	8	(98)
3-thienyl	H	CH₂=C(Me)	TMS	10	(97)

100

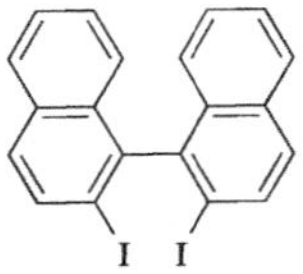

CuI (5 mol %),
DMEDA (20 mol %),
Cs₂CO₃, toluene, 110°, 24 h

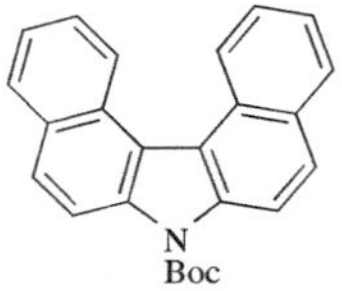

(75)

100

C₂₋₇

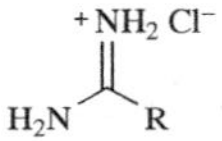

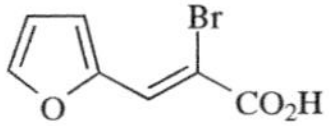

Cu₂O (20 mol %), Cs₂CO₃,
DMF, 100°, 12 h

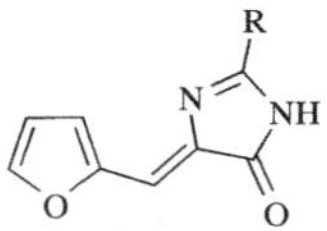

R	
Me	(48)
n-Pr	(45)
c-Pr	(94)
Ph	(67)

678

TABLE 35. *N*-VINYLATIONS IN MULTI-STEP REACTIONS (*Continued*)

Nitrogen Nucleophile	Electrophile	Conditions	Product(s) and Yield(s) (%)	Refs.
C$_{2-7}$				
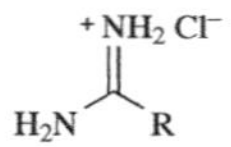	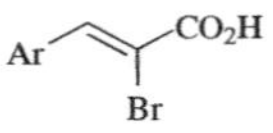	Cu$_2$O (20 mol %), Cs$_2$CO$_3$, DMF, 80°, 12 h	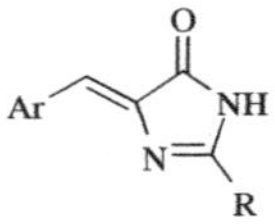	678

Ar	R	
4-MeC$_6$H$_4$	H	(53)
4-MeC$_6$H$_4$	MeO	(70)
4-MeC$_6$H$_4$	Me	(44)
4-*n*-PrC$_6$H$_4$	H	(66)
4-*n*-PrC$_6$H$_4$	Cl	(60)
4-*n*-PrC$_6$H$_4$	MeO	(75)
4-*n*-PrC$_6$H$_4$	Me	(52)
4-*c*-PrC$_6$H$_4$	H	(55)
4-*c*-PrC$_6$H$_4$	Cl	(62)
4-*c*-PrC$_6$H$_4$	MeO	(68)
4-*c*-PrC$_6$H$_4$	Me	(86)
4-PyC$_6$H$_4$	Cl	(62)
4-PyC$_6$H$_4$	MeO	(82)
4-PyC$_6$H$_4$	Me	(44)
4-PhC$_6$H$_4$	H	(75)
4-PhC$_6$H$_4$	MeO	(87)
4-PhC$_6$H$_4$	Me	(71)

Nitrogen Nucleophile	Electrophile	Conditions	Product(s) and Yield(s) (%)	Refs.
C$_{3-7}$				
H$_2$NR		CuI (10 mol %), DMG (20 mol %), K$_3$PO$_4$, NH$_4$OAc, MeCN, reflux	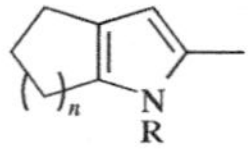	574

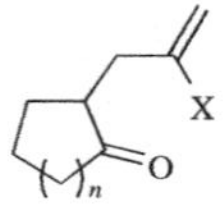

n	R	X	Time (h)	
1	Ph	Br	24	(95)
2	allyl	Br	15	(99)
2	*n*-C$_5$H$_{11}$	Br	15	(99)
2	*c*-C$_6$H$_{11}$	Br	22	(78)
2	4-MeOC$_6$H$_4$	Br	12	(92)
2	Bn	Cl	24	(86)
2	Bn	Br	10	(99)

C₄

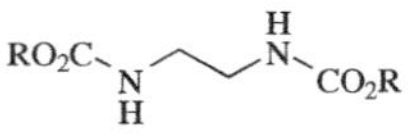

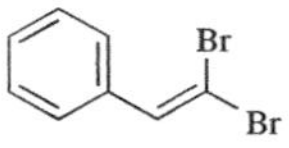

Substrate: ROC₂C–NH–NH–CO₂R

CuI (10 mol %), DMG (20 mol %), K₃PO₄, NH₄OAc, toluene, 80°, 24 h

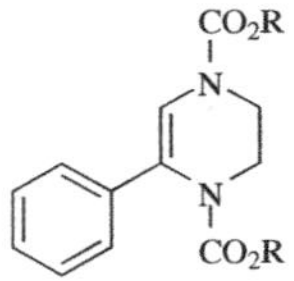

Product: dihydropyrazine bearing CO₂R groups and phenyl

R	
Me	(43)
allyl	(37)

668

C₄₋₈

Substrate: BocHN–N=C(R¹)(R²) + PhCH=CBr₂

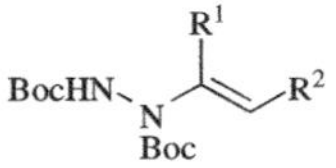

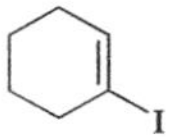

1. CuI (10 mol %), phen (20 mol %), Cs₂CO₃, DMF, 80°, 24 h
2. Xylene, 140°, 24 h

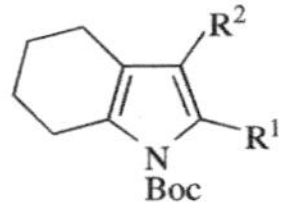

Product: tetrahydroindole with R¹, R², Boc

R¹	R₂
Me	MeO₂C (49)
MEMO(CH₂)₄	Et (61)

671

C₅₋₈

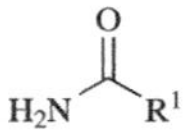

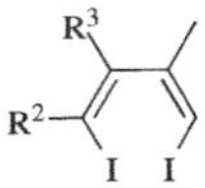

Substrate: H₂N–CO–R¹ + diiodoalkene

CuI (20 mol %), DMCDA (20 mol %), Cs₂CO₃, dioxane, 100°

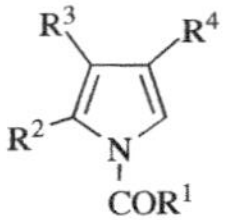

Product: pyrrole with R², R³, R⁴, COR¹

R¹	R²	R³	R⁴	Time (h)	
n-Bu	H	n-Bu	n-Bu	24	(95)
Ph	H	n-Bu	n-Bu	18	(54)
Ph	H	Ph	Ph	3	(54)
Ph	n-Pr	Ph	n-Pr	24	(54)
Ph	Ph	Ph	n-Bu	20	(32)

R¹	R²	R³	R⁴	Time (h)	
4-H₂NC₆H₄	H	n-Bu	n-Bu	18	(71)
4-H₂NC₆H₄	n-Pr	n-Pr	n-Bu	24	(70)
4-MeC₆H₄	H	n-Bu	n-Bu	24	(71)
Bn	H	n-Bu	n-Bu	24	(73)
Bn	H	Ph	Ph	4	(48)

679, 610

C₅₋₇

Substrate: H₂NR + vinyl bromide ketone

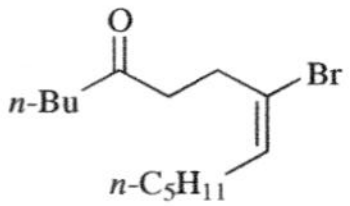

CuI (10 mol %), DMG (20 mol %), K₃PO₄, NH₄OAc, MeCN, reflux, 24 h

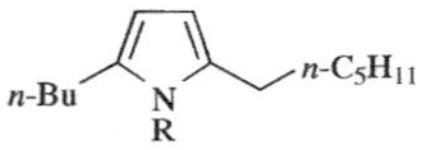

Product: pyrrole

R	
n-C₅H₁₁	(93)
Bn	(93)

574

TABLE 35. *N*-VINYLATIONS IN MULTI-STEP REACTIONS (*Continued*)

Nitrogen Nucleophile	Electrophile	Conditions	Product(s) and Yield(s) (%)	Refs.

C$_{5-8}$

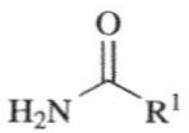

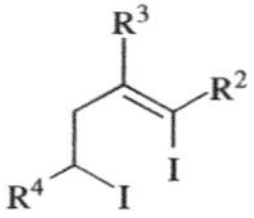

CuI (20 mol %), DMCDA (20 mol %), K$_2$CO$_3$, 48 h

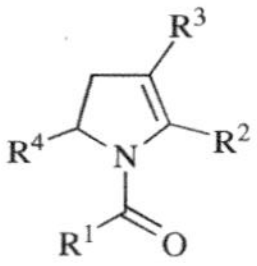

R^1	R^2	R^3	R^4	Solvent	Temp (°)		
n-Bu	Et	Et	H	toluene	reflux	(70)	680, 610
n-Bu	*n*-Pr	*n*-Pr	H	toluene	reflux	(69)	680
n-Bu	*n*-Pr	*n*-Pr	H	dioxane	110	(69)	610
n-Bu	Ph	Ph	TMS	dioxane	110	(41)	680, 610
Ph	*n*-Pr	*n*-Pr	H	toluene	reflux	(65)	680
Ph	*n*-Pr	*n*-Pr	H	dioxane	110	(46)	610
Ph	Ph	Ph	H	toluene	reflux	(65)	680
Ph	Ph	Ph	H	dioxane	reflux	(65)	610
4-MeC$_6$H$_4$	Et	Et	H	toluene	reflux	(64)	680, 610
4-MeC$_6$H$_4$	Ph	Ph	H	dioxane	reflux	(47)	610
4-MeC$_6$H$_4$	Ph	Ph	H	toluene	reflux	(47)	680

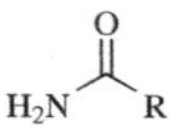

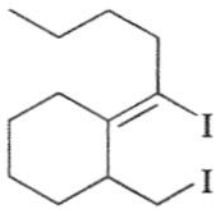

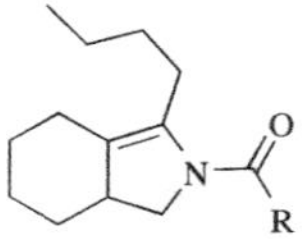

C$_{5-7}$

CuI (20 mol %), DMCDA (20 mol %), K$_2$CO$_3$, 48 h

R	Solvent	Temp (°)		
n-Bu	toluene	reflux	(57)	680
n-Bu	dioxane	110	(57)	610
PMP	toluene	reflux	(32)	680
PMP	dioxane	110	(47)	610

C$_{5-6}$

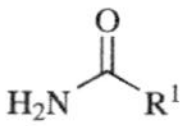

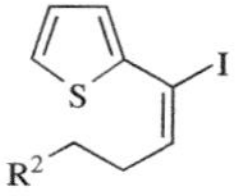

1. CuI (5 mol %), DMEDA (20 mol %), K$_2$CO$_3$, toluene, 80°, 8 h
2. I$_2$, DBU, rt to 80°, 10 h

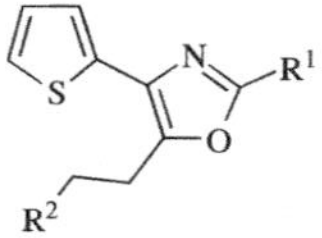

R^1	R^2		
2-furyl	NCCH$_2$	(63)	664
Ph	NCCH$_2$	(68)	
Ph	TIPSO	(72)	

Substrate	Reagent/Conditions	Product	Reference
C_{5-7} H_2NR^1	CuI (10 mol %), DMG (20 mol %), K_3PO_4, NH_4OAc, MeCN, reflux, 24 h	R¹: $n\text{-}C_5H_{11}$ (68 Bn), Bn (90 Me), Bn (85 Bn)	574
C_6 Intramolecular	CuI (1 eq), DMEDA (2 eq), Cs_2CO_3, dioxane, reflux, 4 h	(86)	673
	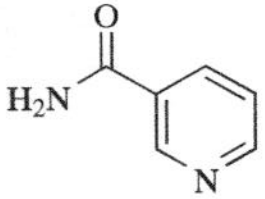 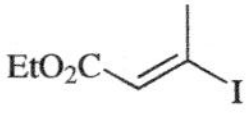 1. CuI (5 mol %), DMEDA (20 mol %), K_2CO_3, toluene, 80°, 8 h 2. I_2, DBU, rt to 80°, 10 h	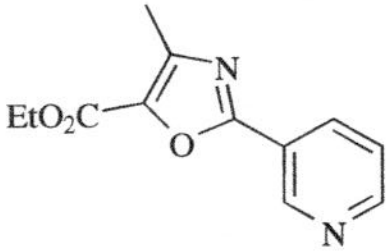 (70)	664
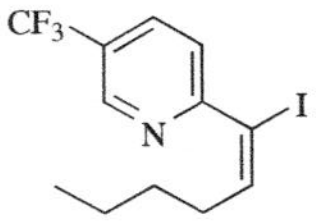	1. CuI (5 mol %), DMEDA (20 mol %), K_2CO_3, toluene, 80°, 8 h 2. I_2, DBU, rt to 80°, 10 h	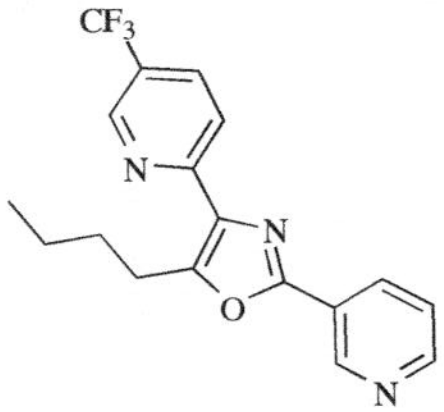 (74)	664
C_7	1. CuI (10 mol %), K_2CO_3, DMEDA (20 mol %), toluene, 120°, 24 h 2. Pd(dppf)Cl₂ (10 mol %), KOAc, toluene, 120°, 24 h	(78)	157

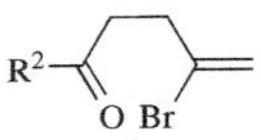

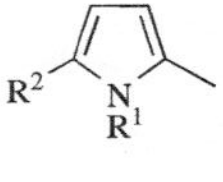

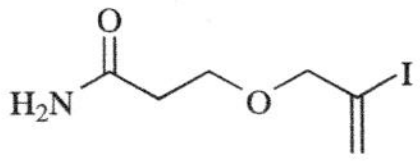

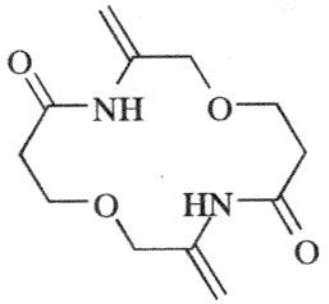

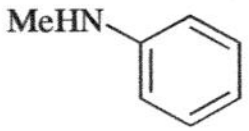

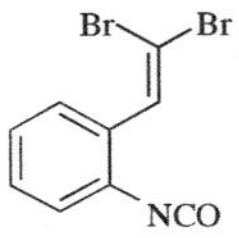

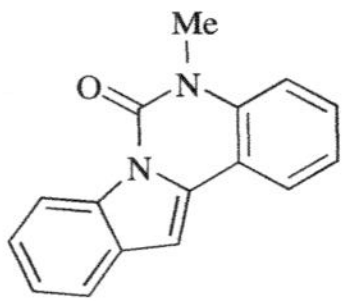

R¹ / R² table:

R¹	R²	
$n\text{-}C_5H_{11}$	Bn	(68)
Bn	Me	(90)
Bn	Bn	(85)

TABLE 35. *N*-VINYLATIONS IN MULTI-STEP REACTIONS (*Continued*)

Nitrogen Nucleophile	Electrophile	Conditions	Product(s) and Yield(s) (%)	Refs.
C$_{7-8}$ 2-chlorobenzamide	EtO$_2$C— alkenyl iodide	1. CuI (5 mol %), DMEDA (20 mol %), K$_2$CO$_3$, toluene, 80°, 8 h 2. I$_2$, DBU, rt to 80°, 10 h	oxazole (77)	664
4-R benzamide	OTIPS iodo/bromo	CuI (5 mol %), DMEDA (20 mol %), K$_2$CO$_3$, toluene, 80°	oxazole with OTIPS, R — R Time (h); H 8 (98); Cl 8 (99); CF$_3$ 10 (99)	664
C$_7$ H$_2$NBn	Br— aldehyde	CuI (10 mol %), DMG (20 mol %), K$_3$PO$_4$, NH$_4$OAc, MeCN, reflux, 24 h	pyrrole (83)	574
benzamide (H$_2$N—C(=O)Ph)	diiodo tetralin	CuI (20 mol %), DMCDA (20 mol %), Cs$_2$CO$_3$, dioxane, 100°, 20 h	isoindole N—C(=O)Ph (89) benzoxazole	679, 610
	iodo cyclooctene	1. CuI (5 mol %), DMEDA (20 mol %), K$_2$CO$_3$, toluene, 80°, 8 h 2. I$_2$, DBU, rt to 80°, 10 h	X Br (74); I (78)	664
2-chlorobenzamide	iodo benzothiophene	CuI (5 mol %), DMEDA (20 mol %), Cs$_2$CO$_3$, toluene, 110°, 6 h	(83)	100

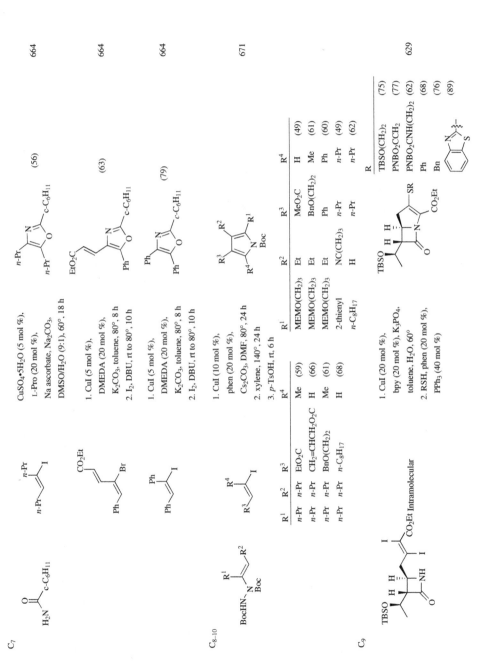

C_7

CuSO$_4$•5H$_2$O (5 mol %),
L-Pro (20 mol %),
Na ascorbate, Na$_2$CO$_3$,
DMSO/H$_2$O (9:1), 60°, 18 h

(56)

664

1. CuI (5 mol %),
DMEDA (20 mol %),
K$_2$CO$_3$, toluene, 80°, 8 h
2. I$_2$, DBU, rt to 80°, 10 h

(63)

664

1. CuI (5 mol %),
DMEDA (20 mol %),
K$_2$CO$_3$, toluene, 80°, 8 h
2. I$_2$, DBU, rt to 80°, 10 h

(79)

664

C_{8-10}

1. CuI (10 mol %),
phen (20 mol %),
Cs$_2$CO$_3$, DMF, 80°, 24 h
2. xylene, 140°, 24 h
3. p-TsOH, rt, 6 h

671

R^1	R^2	R^3	R^4		R^1	R^2	R^3	R^4	
n-Pr	EtO$_2$C	MEMO(CH$_2$)$_3$	Me	(59)	MEMO(CH$_2$)$_3$	Et	MeO$_2$C	H	(49)
n-Pr	CH$_2$=CHCH$_2$O$_2$C	MEMO(CH$_2$)$_3$	H	(66)	MEMO(CH$_2$)$_3$	Et	BnO(CH$_2$)$_2$	Me	(61)
n-Pr	BnO(CH$_2$)$_2$	MEMO(CH$_2$)$_3$	Me	(61)	MEMO(CH$_2$)$_3$	Et	Ph	Ph	(60)
n-Pr	n-C$_8$H$_{17}$	2-thienyl	H	(68)	NC(CH$_2$)$_3$	n-Pr	n-Pr	(49)
					n-C$_8$H$_{17}$	H	n-Pr	n-Pr	(62)

C_9

1. CuI (20 mol %),
bpy (20 mol %), K$_3$PO$_4$,
toluene, H$_2$O, 60°
2. RSH, phen (20 mol %),
PPh$_3$ (40 mol %)

629

R	
TBSO(CH$_2$)$_2$	(75)
PNBO$_2$CCH$_2$	(77)
PNBO$_2$CNH(CH$_2$)$_2$	(62)
Ph	(68)
Bn	(76)
	(89)

641

TABLE 35. N-VINYLATIONS IN MULTI-STEP REACTIONS (Continued)

Nitrogen Nucleophile	Electrophile	Conditions	Product(s) and Yield(s) (%)	Refs.

C_{10-16}

Nitrogen Nucleophile: (structure with R^1, R^2, Br, NH, NHBoc, R^3, O; positions 1,4,5)

Electrophile: Intramolecular

Conditions: CuI (2.5 mol %), CDA (5 mol %), K_2CO_3, toluene, 120°

Products **I** and **II**:

R^1	R^2	R^3	Time (h)	I	er	II
H	H	H	22	(74)	—	(0)
H	H	Me	13	(84)	93:7	(0)
H	H	BnO₂CCH₂CH₂	15	(71)	53:47	(0)
H	H	i-Pr	22	(63)	52:48[a]	(0)
H	H	BocHN(CH₂)₄	15	(69)	96.5:3.5	(0)
H	H	i-Bu	22	(68)	94.5:5.5	(0)
H	H	Bn	18	(73)	88:12	(0)
H	4-F	Me	13	(65)	94.5:5.5	(0)

R^1	R^2	R^3	Time (h)	I		er	II	
H	3-MeO	Me	49	(56)	64:36	(0)	681
H	4,5-(BnO)₂	Me	14	(69)	94.5:5.5	(0)	
H	6-Me	Me	13	(78)	86.5:13.5	(0)	
H	4-CF₃	Me	12	(67)	82:12	(0)	
H	5-MeO,6-BnO	Me	13	(67)	90.5:9.5	(0)	
H	4-FC₆H₄	H	14	(52)[b]	—	(0)	
H	4-FC₆H₄	H	13	(40)	—	(28)	

C_{14}

Nitrogen Nucleophile: H₂N–C(O)–CHPh₂

Electrophile: (vinyl halide) R^1, X, R^2

Conditions:
1. CuI (5 mol %), DMEDA (20 mol %), K_2CO_3, toluene, 80°, 8 h
2. I₂, DBU, rt to 80°, 10 h

Product: (oxazole with R^1, R^2, CHPh₂)

R^1	R^2	X		
Me	EtO₂C	I	(82)	664
NC(CH₂)₃	2-thienyl	I	(72)	
Ph	Ph	Br	(79)	
Ph	Ph	I	(79)	

C_{15}

Nitrogen Nucleophile: (structure Boc, BocHN, PMP, NC, CHPh₂)

Electrophile: n-C₅H₁₁ / I (vinyl iodide)

Conditions:
1. CuI (10 mol %), phen (20 mol %), Cs₂CO₃, DMF, 80°, 24 h
2. xylene, 140°, 24 h
3. p-TsOH, rt, 6 h

Product: (pyrrole with n-C₅H₁₁, PMP, N-Boc, NC) (54)   671

[a] The er of the substrate was 52:48.

642

REFERENCES

1 Ullmann, F. *Ber. Dtsch. Chem. Ges.* **1903**, *36*, 2382.
2 Goldberg, I. *Ber. Dtsch. Chem. Ges.* **1906**, *39*, 1691.
3 Lindley, J. *Tetrahedron* **1984**, *40*, 1433.
4 Hartwig, J. F. *Angew. Chem., Int. Ed.* **1998**, *37*, **2046**.
5 Wolfe, J. P.; Wagaw, S.; Marcoux, J.-F.; Buchwald, S. L. *Acc. Chem. Res.* **1998**, *31*, 805.
6 Hartwig, J. F. *Acc. Chem. Res.* **2008**, *41*, 1534.
7 Surry, D. S.; Buchwald, S. L. *Angew. Chem., Int. Ed.* **2008**, *47*, 6338.
8 Chan, D. M. T.; Monaco, K. L.; Wang, R.-P.; Winters, M. P. *Tetrahedron Lett.* **1998**, *39*, 2933.
9 Lam, P. Y. S.; Clark, C. G.; Saubern, S.; Adams, J.; Winters, M. P.; Chan, D. M. T.; Combs, A. *Tetrahedron Lett.* **1998**, *39*, 2941.
10 Collman, J. P.; Zhong, M. *Org. Lett.* **2000**, *2*, 1233.
11 Kiyomori, A.; Marcoux, J.-F.; Buchwald, S. L. *Tetrahedron Lett.* **1999**, *40*, 2657.
12 Ley, S. V.; Thomas, A. W. *Angew. Chem., Int. Ed.* **2003**, *42*, 5400.
13 Corbet, J.-P.; Mignani, G. *Chem. Rev.* **2006**, *106*, 2651.
14 Beletskaya, I. P.; Cheprakov, A. V. *Coord. Chem. Rev.* **2004**, *248*, 2337.
15 Dehli, J. R.; Legros, J.; Bolm, C. *Chem. Commun.* **2005**, 973.
16 Evano, G.; Blanchard, N.; Toumi, M. *Chem. Rev.* **2008**, *108*, 3054.
17 Ma, D.; Cai, Q. *Acc. Chem. Res.* **2008**, *41*, 1450.
18 Sorokin, V. I. *Mini-Rev. Org. Chem.* **2008**, *5*, 323.
19 Sperotto, E.; van Klink, G. P. M.; van Koten, G.; de Vries, J. G. *Dalton Trans.* **2010**, *39*, 10338.
20 Paine, A. J. *J. Am. Chem. Soc.* **1987**, *109*, 1496.
21 Aalten, H. L.; van Koten, G.; Grove, D. M.; Kullman, T.; Piekstra, O. G.; Hulshof, L. A.; Sheldon, R. A. *Tetrahedron* **1989**, *45*, 5565.
22 Resnik, R.; Cohen, T.; Fernando, Q. *J. Am. Chem. Soc.* **1961**, *83*, 3344.
23 Weingarten, H. *J. Org. Chem.* **1964**, *29*, 3624.
24 Kondratov, S. A.; Shein, S. M. *Zh. Org. Khim* **1979**, *15*, 2387.
25 Röttger, S.; Sjöberg, P. J. R.; Larhed, M. *J. Comb. Chem.* **2007**, *9*, 204.
26 Strieter, E. R.; Bhayana, B.; Buchwald, S. L. *J. Am. Chem. Soc.* **2009**, *131*, 78.
27 Kaddouri, H.; Vicente, V.; Ouali, A.; Quazzani, F.; Taillefer, M. *Angew. Chem., Int. Ed.* **2009**, *48*, 333.
28 Zhang, S.-L.; Liu, L.; Fu, Y.; Guo, Q.-X. *Organometallics* **2007**, *26*, 4546.
29 Bacon, R. G. R.; Karim, A. *J. Chem. Soc., Perkin Trans. 1* **1973**, 272.
30 Yamamoto, T.; Ehara, Y.; Kubota, M.; Yamamoto, A. *Bull. Chem. Soc. Jpn.* **1980**, *53*, 1299.
31 Delp, S. A.; Goj, L. A.; Pouy, M. J.; Munro-Leighton, C.; Lee, J. P.; Gunnoe, T. B.; Cundari, T. R.; Peterson, J. L. *Organometallics* **2011**, *30*, 55.
32 Tye, J. W.; Weng, Z.; Johns, A. M.; Incarvito, C. D.; Hartwig, J. F. *J. Am. Chem. Soc.* **2008**, *130*, 9971.
33 Strieter, E. R.; Blackmond, D. G.; Buchwald, S. L. *J. Am. Chem. Soc.* **2005**, *127*, 4120.
34 Giri, R.; Hartwig, J. F. *J. Am. Chem. Soc.* **2010**, *132*, 15860.
35 Arai, S.; Hida, M.; Yamagishi, T. *Bull. Chem. Soc. Jpn.* **1978**, *51*, 277.
36 Bowman, W. R.; Heaney, H.; Smith, P. H. G. *Tetrahedron Lett.* **1984**, *25*, 5821.
37 Cohen, T.; Wood, J.; Dietz, A. G., Jr. *Tetrahedron Lett.* **1974**, *15*, 3555.
38 Bethell, D.; Jenkins, I. L.; Quan, P. M. *J. Chem. Soc., Perkin Trans. 2* **1985**, 1789.
39 Bacon, R. G. R.; Hill, H. A. O. *J. Chem. Soc.* **1964**, 1097.
40 Cristau, H.-J.; Cellier, P. P.; Spindler, J.-F.; Taillefer, M. *Chem.—Eur. J.* **2004**, *10*, 5607.
41 Casitas, A.; King, A. E.; Parella, T.; Costas, M.; Stahl, S. S.; Ribas, X. *Chem. Sci.* **2010**, *1*, 326.
42 Huffman, L. M.; Stahl, S. S. *J. Am. Chem. Soc.* **2008**, *130*, 9196.
43 Zhang, S.; Ding, Y. *Organometallics* **2011**, *30*, 633.
44 Yu, H.-Z.; Jiang, Y.-Y.; Fu, Y.; Liu, L. *J. Am. Chem. Soc.* **2010**, *132*, 18078.
45 Franc, G.; Cacciuttolo, Q.; Lefévre, G.; Adamo, C.; Ciofini, I.; Jutand, A. *ChemCatChem* **2011**, *3*, 305.
46 Jones, G. O.; Liu, P.; Houk, K. N.; Buchwald, S. L. *J. Am. Chem. Soc.* **2010**, *132*, 6205.
47 Zhang, H.; Cai, Q.; Ma, D. *J. Org. Chem.* **2005**, *70*, 5164.
48 Rao, H.; Jin, Y.; Fu, H.; Jiang, Y.; Zhao, Y. *Chem.—Eur. J.* **2006**, *12*, 3636.
49 Lv, X.; Bao, W. *J. Org. Chem.* **2007**, *72*, 3863.
50 Altman, R. A.; Anderson, K. W.; Buchwald, S. L. *J. Org. Chem.* **2008**, *73*, 5167.
51 Zhu, X.; Su, L.; Huang, L.; Chen, G.; Wang, J.; Song, H.; Wan, Y. *Eur. J. Org. Chem.* **2009**, 635.
52 Shafir, A.; Buchwald, S. L. *J. Am. Chem. Soc.* **2006**, *128*, 8742.

[53] Li, C. S.; Dixon, D. D. *Tetrahedron Lett.* **2004**, *45*, 4257.
[54] Gauthier, S.; Fréchet, J. M. *J. Synthesis* **1987**, 383.
[55] Liu, Y.; Bai, Y.; Zhang, J.; Li, Y.; Jiao, J.; Qi, X. *Eur. J. Org. Chem.* **2007**, 6084.
[56] Patil, N. M.; Kelkar, A. A.; Nabi, Z.; Chaudhari, R. V. *Chem. Commun.* **2003**, 2460.
[57] Ran, C.; Dai, Q.; Harvey, R. G. *J. Org. Chem.* **2005**, *70*, 3724.
[58] Alcalde, E.; Dinarés, I.; Rodríguez, S.; Garcia de Miguel, C. *Eur. J. Org. Chem.* **2005**, 1637.
[59] Zhu, L.; Li, G.; Luo, L.; Guo, P.; Lan, J.; You, J. *J. Org. Chem.* **2009**, *74*, 2200.
[60] Kim, J.; Chang, S. *Chem. Commun.* **2008**, 3052.
[61] Yang, C.-T.; Fu, Y.; Huang, Y.-B.; Yi, J.; Guo, Q.-X.; Liu, L. *Angew. Chem., Int. Ed.* **2009**, *48*, 7398.
[62] Jiang, D.; Fu, H.; Jiang, Y.; Zhao, Y. *J. Org. Chem.* **2007**, *72*, 672.
[63] Phillips, D. P.; Zhu, X.-F.; Lau, T. L.; He, X.; Yang, K.; Liu, H. *Tetrahedron Lett.* **2009**, *50*, 7293.
[64] Kantam, M. L.; Yadav, J.; Laha, S.; Sreedhar, B.; Jha, S. *Adv. Synth. Catal.* **2007**, *349*, 1938.
[65] Maheswaran, H.; Krishna, G. G.; Srinivas, V.; Prasanth, K. L.; Rajasekhar, C. V. *Bull. Chem. Soc. Jpn.* **2008**, *81*, 515.
[66] Choudary, B. M.; Sridhar, C.; Kantam, M. L.; Venkanna, G. T.; Sreedhar, B. *J. Am. Chem. Soc.* **2005**, *127*, 9948.
[67] Guo, X.; Rao, H.; Fu, H.; Jiang, Y.; Zhao, Y. *Adv. Synth. Catal.* **2006**, *348*, 2197.
[68] Tang, B.-X.; Guo, S.-M.; Zhang, M.-B.; Li, J.-H. *Synthesis* **2008**, 1707.
[69] Klapars, A.; Huang, X.; Buchwald, S. L. *J. Am. Chem. Soc.* **2002**, *124*, 7421.
[70] Wolf, C.; Liu, S.; Mei, X.; August, A. T.; Casimir, M. D. *J. Org. Chem.* **2006**, *71*, 3270.
[71] Maradolla, M. B.; Amaravathi, M.; Kumar, V. N.; Mouli, G. V. P. C. *J. Mol. Catal. A: Chem.* **2007**, *266*, 47.
[72] Sarrafi, Y.; Mohadeszadeh, M.; Alimohammadi, K. *Chin. Chem. Lett.* **2009**, *20*, 784.
[73] Kantam, M. L.; Ramani, T.; Chakrapani, L. *Synth. Commun.* **2008**, *38*, 626.
[74] Jammi, S.; Sakthivel, S.; Rout, L.; Mukherjee, T.; Mandal, S.; Mitra, R.; Saha, P.; Punniyamurthy, T. *J. Org. Chem.* **2009**, *74*, 1971.
[75] Goriya, Y.; Ramana, C. V. *Tetrahedron* **2010**, *66*, 7642.
[76] Kang, S.-K.; Lee, S.-H.; Lee, D. *Synlett* **2000**, 1022.
[77] Verma, A. K.; Singh, J.; Larock, R. C. *Tetrahedron* **2009**, *65*, 8434.
[78] Zhu, Y.; Shi, Y.; Wei, Y. *Monatsh. Chem.* **2010**, *141*, 1009.
[79] Xi, Z.; Liu, F.; Zhou, Y.; Chen, W. *Tetrahedron* **2008**, *64*, 4254.
[80] Xie, Y.-X.; Pi, S.-F.; Wang, J.; Yin, D.-L.; Li, J.-H. *J. Org. Chem.* **2006**, *71*, 8324.
[81] Lee, H.-G.; Won, J.-E.; Kim, M.-J.; Park, S.-E.; Jung, K.-J.; Kim, B. R.; Lee, S.-G.; Yoon, Y.-J. *J. Org. Chem.* **2009**, *74*, 5675.
[82] Cristau, H.-J.; Cellier, P. P.; Spindler, J.-F.; Taillefer, M. *Eur. J. Org. Chem.* **2004**, 695.
[83] Crawford, K. R.; Padwa, A. *Tetrahedron Lett.* **2002**, *43*, 7365.
[84] Altman, R. A.; Koval, E. D.; Buchwald, S. L. *J. Org. Chem.* **2007**, *72*, 6190.
[85] Huang, H.; Yan, X.; Zhu, W.; Liu, H.; Jiang, H.; Chen, K. *J. Comb. Chem.* **2008**, *10*, 617.
[86] Zhu, L.; Cheng, L.; Zhang, Y.; Xie, R.; You, J. *J. Org. Chem.* **2007**, *72*, 2737.
[87] do Rêgo Barros, O. S.; Nogueira, C. W.; Stangherlin, E. C.; Menezes, P. H.; Zeni, G. *J. Org. Chem.* **2006**, *71*, 1552.
[88] Wang, Y.; Liao, Q.; Xi, C. *Org. Lett.* **2010**, *12*, 2951.
[89] Shen, G.; Lv, X.; Qiu, W.; Bao, W. *Tetrahedron Lett.* **2008**, *49*, 4556.
[90] Pan, X.; Cai, Q.; Ma, D. *Org. Lett.* **2004**, *6*, 1809.
[91] Shen, L.; Hsung, R. P.; Zhang, Y.; Antoline, J. E.; Zhang, X. *Org. Lett.* **2005**, *7*, 3081.
[92] Han, C.; Shen, R.; Su, S.; Porco, J. A., Jr. *Org. Lett.* **2004**, *6*, 27.
[93] Yang, L.; Deng, G.; Wang, D.-X.; Huang, Z.-T.; Zhu, J.-P.; Wang, M.-X. *Org. Lett.* **2007**, *9*, 1387.
[94] Reddy, V. P.; Kumar, A. V.; Rao, K. R. *Tetrahedron Lett.* **2010**, *51*, 3181.
[95] Fürstner, A.; Dierkes, T.; Thiel, O. R.; Blanda, G. *Chem.—Eur. J.* **2001**, *7*, 5286.
[96] Nicolaou, K. C.; Kim, D. W.; Baati, R. *Angew. Chem., Int. Ed.* **2002**, *41*, 3701.
[97] Nicolaou, K. C.; Kim, D. W.; Baati, R.; O'Brate, A.; Giannakakou, P. *Chem.—Eur. J.* **2003**, *9*, 6177.
[98] Nicolaou, K. C.; Leung, G. Y. C.; Dethe, D. H.; Guduru, R.; Sun, Y.-P.; Lim, C. S.; Chen, D. Y.-K. *J. Am. Chem. Soc.* **2008**, *130*, 10019.
[99] Toumi, M.; Couty, F.; Evano, G. *J. Org. Chem.* **2007**, *72*, 9003.
[100] Martín, R.; Larsen, C. H.; Cuenca, A.; Buchwald, S. L. *Org. Lett.* **2007**, *9*, 3379.
[101] Jiang, L.; Job, G. E.; Klapars, A.; Buchwald, S. L. *Org. Lett.* **2003**, *5*, 3667.
[102] Liao, Q.; Wang, Y.; Zhang, L.; Xi, C. *J. Org. Chem.* **2009**, *74*, 6371.
[103] Mao, J.; Hua, Q.; Guo, J.; Shi, D.; Ji, S. *Synlett* **2008**, 2011.
[104] Lu, H.; Li, C. *Org. Lett.* **2006**, *8*, 5365.
[105] Lu, H.; Yuan, X.; Zhu, S.; Sun, C.; Li, C. *J. Org. Chem.* **2008**, *73*, 8665.

106 Hodgkinson, R. C.; Schulz, J.; Willis, M. C. *Org. Biomol. Chem.* **2009**, *7*, 432.
107 Huang, X.; Anderson, K. W.; Zim, D.; Jiang, L.; Klapars, A.; Buchwald, S. L. *J. Am. Chem. Soc.* **2003**, *125*, 6653.
108 Rao, H.; Fu, H.; Jiang, Y.; Zhao, Y. *J. Org. Chem.* **2005**, *70*, 8107.
109 Mei, X.; August, A. T.; Wolf, C. *J. Org. Chem.* **2006**, *71*, 142.
110 Aubin, Y.; Fischmeister, C.; Thomas, C. M.; Renaud, J.-L. *Chem. Soc. Rev.* **2010**, *39*, 4130.
111 Cohen, T.; Tirpak, J. G. *Tetrahedron Lett.* **1975**, 143.
112 Ntaganda, R.; Dhudshia, B.; Macdonald, C. L. B.; Thadani, A. N. *Chem. Commun.* **2008**, 6200.
113 Xia, N.; Taillefer, M. *Angew. Chem., Int. Ed.* **2009**, *48*, 337.
114 Xu, H.; Wolf, C. *Chem. Commun.* **2009**, 3035.
115 Klapars, A.; Antilla, J. C.; Huang, X.; Buchwald, S. L. *J. Am. Chem. Soc.* **2001**, *123*, 7727.
116 Ma, D.; Zhang, Y.; Yao, J.; Wu, S.; Tao, F. *J. Am. Chem. Soc.* **1998**, *120*, 12459.
117 Likhar, P. R.; Arundhathi, R.; Kantam, M. L. *Tetrahedron Lett.* **2007**, *48*, 3911.
118 Zhang, Z.; Mao, J.; Zhu, D.; Wu, F.; Chen, H.; Wan, B. *Catal. Commun.* **2005**, *6*, 784.
119 Zhang, Z.; Mao, J.; Zhu, D.; Wu, F.; Chen, H.; Wan, B. *Tetrahedron* **2006**, *62*, 4435.
120 Kwong, F. Y.; Buchwald, S. L. *Org. Lett.* **2003**, *5*, 793.
121 Brackman, F.; Es-Sayed, M.; de Meijere, A. *Eur. J. Org. Chem.* **2005**, 2250.
122 Egger, M.; Li, X.; Müller, C.; Bernhardt, G.; Buschauer, A.; König, B. *Eur. J. Org. Chem.* **2007**, 2643.
123 Likhar, P. R.; Roy, S.; Roy, M.; Kantam, M. L.; De, R. L. *J. Mol. Catal. A: Chem.* **2007**, *271*, 57.
124 Zhu, D.; Wang, R.; Mao, J.; Xu, L.; Wu, F.; Wan, B. *J. Mol. Catal. A: Chem.* **2006**, *256*, 256.
125 Xu, L.; Zhu, D.; Wu, F.; Wang, R.; Wan, B. *Tetrahedron* **2005**, *61*, 6553.
126 Enguehard, C.; Allouchi, H.; Gueiffier, A.; Buchwald, S. L. *J. Org. Chem.* **2003**, *68*, 4367.
127 Bellina, F.; Calandri, C.; Cauteruccio, S.; Rossi, R. *Eur. J. Org. Chem.* **2007**, 2147.
128 Antilla, J. C.; Baskin, J. M.; Barder, T. E.; Buchwald, S. L. *J. Org. Chem.* **2004**, *69*, 5578.
129 Zhu, R.; Xing, L.; Wang, X.; Cheng, C.; Su, D.; Hu, Y. *Adv. Synth. Catal.* **2008**, *350*, 1253.
130 Taillefer, M.; Xia, N.; Oualli, A. *Angew. Chem., Int. Ed.* **2007**, *46*, 934.
131 Altman, R. A.; Buchwald, S. L. *Org. Lett.* **2007**, *9*, 643.
132 Tao, C.-Z.; Li, J.; Fu, Y.; Liu, L.; Guo, Q.-X. *Tetrahedron Lett.* **2008**, *49*, 70.
133 Jiang, L.; Lu, X.; Zhang, H.; Jiang, Y.; Ma, D. *J. Org. Chem.* **2009**, *74*, 4542.
134 Lam, M. S.; Lee, H. W.; Chan, A. S. C.; Kwong, F. Y. *Tetrahedron Lett.* **2008**, *49*, 6192.
135 Wolter, M.; Klapars, A.; Buchwald, S. L. *Org. Lett.* **2001**, *3*, 3803.
136 Deng, W.; Wang, Y.-F.; Zou, Y.; Liu, L.; Guo, Q.-X. *Tetrahedron Lett.* **2004**, *45*, 2311.
137 Hosseinzadeh, R.; Golchoubian, H.; Masoudi, M. *J. Chin. Chem. Soc.* **2008**, *55*, 649.
138 Mino, T.; Harada, Y.; Shindo, H.; Sakamoto, M.; Fujita, T. *Synlett* **2008**, 614.
139 Szczepankiewicz, B. G.; Rohde, J. J.; Kurukulasuriya, R. *Org. Lett.* **2005**, *7*, 1833.
140 Huang, C.; Fu, Y.; Fu, H.; Jiang, Y.; Zhao, Y. *Chem. Commun.* **2008**, 6333.
141 Deng, X.; McAllister, H.; Mani, N. S. *J. Org. Chem.* **2009**, *74*, 5742.
142 Liu, X.; Fu, H.; Jiang, Y.; Zhao, Y. *Angew. Chem., Int. Ed.* **2009**, *48*, 348.
143 Deng, X.; Mani, N. S. *Eur. J. Org. Chem.* **2010**, 680.
144 Nielsen, S. D.; Smith, G.; Begtrup, M.; Kristensen, J. L. *Eur. J. Org. Chem.* **2010**, 3704.
145 Ghosh, A.; Sieser, J. E.; Caron, S.; Couturier, M.; Dupont-Gaudet, K.; Girardin, M. *J. Org. Chem.* **2006**, *71*, 1258.
146 Hossienzadeh, R.; Sarrafi, Y.; Mohadjerani, M.; Mohammadpourmir, F. *Tetrahedron Lett.* **2008**, *49*, 840.
147 Nandakumar, M. V. *Tetrahedron Lett.* **2004**, *45*, 1989.
148 Stabile, P.; Lamonica, A.; Ribecai, A.; Castoldi, D.; Guercio, G.; Curcuruto, O. *Tetrahedron Lett.* **2010**, *51*, 3232.
149 Cortes-Salva, M.; Nguyen, B.-L.; Cuevas, J.; Pennypacker, K. R.; Antilla, J. C. *Org. Lett.* **2010**, *12*, 1316.
150 Deng, W.; Liu, L.; Zhang, C.; Liu, M.; Guo, Q.-X. *Tetrahedron Lett.* **2005**, *46*, 7295.
151 Han, X. *Tetrahedron Lett.* **2010**, *51*, 360.
152 Steinhuebel, D.; Palucki, M.; Askin, D.; Dolling, U. *Tetrahedron Lett.* **2004**, *45*, 3305.
153 Worch, C.; Bolm, C. *Synthesis* **2008**, 739.
154 Andersen, J.; Madsen, U.; Björkling, F.; Liang, X. *Synlett* **2005**, 2209.
155 Zhu, W.; Ma, D. *Chem. Commun.* **2004**, 888.
156 Markiewicz, J. T.; Wiest, O.; Helquist, P. *J. Org. Chem.* **2010**, *75*, 4887.
157 Monguchi, Y.; Maejima, T.; Mori, S.; Maegawa, T.; Sajiki, H. *Chem.—Eur. J.* **2010**, *16*, 7372.
158 Cho, G. Y.; Rémy, P.; Jansson, J.; Moessner, C.; Bolm, C. *Org. Lett.* **2004**, *6*, 3293.
159 Sedelmeier, J.; Bolm, C. *J. Org. Chem.* **2005**, *70*, 6904.

[160] Evano, G.; Schaus, J. V.; Panek, J. S. *Org. Lett.* **2004**, *6*, 525.
[161] Evano, G.; Toumi, M.; Coste, A. *Chem. Commun.* **2009**, 4166.
[162] Toumi, M.; Couty, F.; Marrot, J.; Evano, G. *Org. Lett.* **2008**, *10*, 5027.
[163] Wrona, I. E.; Gabarda, A. E.; Evano, G.; Panek, J. S. *J. Am. Chem. Soc.* **2005**, *127*, 15026.
[164] Qin, H.-L.; Panek, J. S. *Org. Lett.* **2008**, *10*, 2477.
[165] Shen, R.; Lin, C. T.; Bowman, E. J.; Bowman, B. J.; Porco, J. A., Jr. *J. Am. Chem. Soc.* **2003**, *125*, 7889.
[166] Nicolaou, K. C.; Guduru, R.; Sun, Y.-P.; Banerji, B.; Chen, D. Y.-K. *Angew. Chem., Int. Ed.* **2007**, *46*, 5896.
[167] Su, Q.; Panek, J. S. *J. Am. Chem. Soc.* **2004**, *126*, 2425.
[168] Toumi, M.; Rincheval, V.; Young, A.; Gergeres, D.; Turos, E.; Couty, F.; Mignotte, B.; Evano, G. *Eur. J. Org. Chem.* **2009**, 3368.
[169] Matsuda, Y.; Kitajima, M.; Takayama, H. *Org. Lett.* **2008**, *10*, 125.
[170] Liu, Y.-F.; Wang, C.-L.; Bai, Y.-J.; Han, N.; Jiao, J.-P.; Qi, X.-L. *Org. Proc. Res. Dev.* **2008**, *12*, 490.
[171] Ma, D.; Xia, C.; Jiang, J.; Zhang, J.; Tang, W. *J. Org. Chem.* **2003**, *68*, 442.
[172] Li, Z. H.; Wong, M. S. *Org. Lett.* **2006**, *8*, 1499.
[173] Song, D.; Wu, Q.; Hook, A.; Kozin, I.; Wang, S. *Organometallics* **2001**, *20*, 4683.
[174] Yang, W.-Y.; Chen, L.; Wang, S. *Inorg. Chem.* **2001**, *40*, 507.
[175] Selby, T. D.; Blackstock, S. C. *J. Am. Chem. Soc.* **1998**, *120*, 12155.
[176] Wang, H.; Jiang, Y.; Gao, K.; Ma, D. *Tetrahedron* **2009**, *65*, 8956.
[177] Klapars, A.; Parris, S.; Anderson, K. W.; Buchwald, S. L. *J. Am. Chem. Soc.* **2004**, *126*, 3529.
[178] Diao, X.; Wang, Y.; Jiang, Y.; Ma, D. *J. Org. Chem.* **2009**, *74*, 7974.
[179] Chen, D.; Shen, G.; Bao, W. *Org. Biomol. Chem.* **2009**, *7*, 4067.
[180] Feng, E.; Huang, H.; Zhou, Y.; Ye, D.; Jiang, H.; Liu, H. *J. Org. Chem.* **2009**, *74*, 2846.
[181] Prasad, D. J. C.; Sekar, G. *Org. Biomol. Chem.* **2009**, *7*, 5091.
[182] Rao, R. K.; Naidu, A. B.; Sekar, G. *Org. Lett.* **2009**, *11*, 1923.
[183] Lv, X.; Bao, W. *J. Org. Chem.* **2009**, *74*, 5618.
[184] Shen, G.; Bao, W. *Adv. Synth. Catal.* **2010**, *352*, 981.
[185] Gollner, A.; Koutentis, P. *Org. Lett.* **2010**, *12*, 1352.
[186] Wang, Z.-J.; Yang, J.-G.; Yang, F.; Bao, W. *Org. Lett.* **2010**, *12*, 3034.
[187] Arterburn, J. B.; Pannala, M.; Gonzalez, A. M. *Tetrahedron Lett.* **2001**, *42*, 1475.
[188] de Lange, B.; Lambers-Verstappen, M. H.; Schmieder-van de Vondervoort, L.; Sereinig, N.; de Rijk, R.; de Vries, A. H. M.; de Vries, J. G. *Synlett* **2006**, 3105.
[189] Okano, K.; Tokuyama, H.; Fukuyama, T. *Org. Lett.* **2003**, *5*, 4987.
[190] Taillefer, M.; Ouali, A.; Renard, B.; Spindler, J.-F. *Chem.—Eur. J.* **2006**, *12*, 5301.
[191] Antilla, J. C.; Klapars, A.; Buchwald, S. L. *J. Am. Chem. Soc.* **2002**, *124*, 11684.
[192] Ma, D.; Cai, Q.; Zhang, H. *Org. Lett.* **2003**, *5*, 2453.
[193] Lang, F.; Zewge, D.; Houpis, I. N.; Volante, R. P. *Tetrahedron Lett.* **2001**, *42*, 3251.
[194] Kwong, F. Y.; Klapars, A.; Buchwald, S. L. *Org. Lett.* **2002**, *4*, 581.
[195] Lu, Z.; Twieg, R. J.; Huang, S. D. *Tetrahedron Lett.* **2003**, *44*, 6289.
[196] Shafir, A.; Lichtor, P. A.; Buchwald, S. L. *J. Am. Chem. Soc.* **2007**, *129*, 3490.
[197] Yamada, K.; Kubo, T.; Tokuyama, H.; Fukuyama, T. *Synlett* **2002**, 231.
[198] Bouteiller, C.; Becerril-Ortega, J.; Marchand, P.; Nicole, O.; Barré, L.; Buisson, A.; Perrio, C. *Org. Biomol. Chem.* **2010**, *8*, 1111.
[199] Ross, S. D.; Finkelstein, M. *J. Am. Chem. Soc.* **1963**, *85*, 2603.
[200] Heaney, H. *Chem. Rev.* **1962**, *62*, 81.
[201] Roberts, J. D.; Vaughan, C. W.; Carlsmith, L. A.; Semenow, D. A. *J. Am. Chem. Soc.* **1956**, *78*, 611.
[202] Seeboth, H. *Angew. Chem., Int. Ed. Engl.* **1967**, *6*, 307.
[203] Kienle, M.; Dubbaka, S. R.; Brade, K.; Knochel, P. *Eur. J. Org. Chem.* **2007**, 4166.
[204] Commodity cash prices on April 25, 2011.
[205] Altman, R. A.; Hyde, A. M.; Huang, X.; Buchwald, S. L. *J. Am. Chem. Soc.* **2008**, *130*, 9613.
[206] Houpis, I. N.; Weerts, K.; Nettekoven, U.; Canters, M.; Tan, H.; Liu, R.; Wang, Y. *Adv. Synth. Catal.* **2011**, *353*, 538.
[207] Lam, P. Y. S.; Vincent, B.; Clark, C. G.; Deudon, S.; Jadhav, P. K. *Tetrahedron Lett.* **2001**, *42*, 3415.
[208] Grushin, V. V.; Alper, H. *Chem. Rev.* **1994**, *94*, 1047.
[209] Gao, C.-Y.; Cao, X.; Yang, L.-M. *Org. Biomol. Chem.* **2009**, *7*, 3922.
[210] Iglesias, M. J.; Prieto, A.; Nicasio, M. C. *Adv. Synth. Catal.* **2010**, *352*, 1949.
[211] Correa, A.; Bolm, C. *Angew. Chem., Int. Ed.* **2007**, *46*, 8862.
[212] Correa, A.; Carril, M.; Bolm, C. *Chem.—Eur. J.* **2008**, *14*, 10919.
[213] Guo, D.; Huang, H.; Xu, J.; Jiang, H.; Liu, H. *Org. Lett.* **2008**, *10*, 4513.

[214] Buchwald, S. L.; Bolm, C. *Angew. Chem., Int. Ed.* **2009**, *48*, 5586.
[215] Toma, G.; Fujita, K.-i.; Yamaguchi, R. *Eur. J. Org. Chem.* **2009**, 4586.
[216] Sapountzis, I.; Knochel, P. *J. Am. Chem. Soc.* **2002**, *124*, 9390.
[217] Kopp, F.; Sapountzis, I.; Knochel, P. *Synlett* **2003**, 885.
[218] Sapountzis, I.; Knochel, P. *Angew. Chem., Int. Ed.* **2004**, *43*, 897.
[219] Manifar, T.; Rohani, S.; Bender, T. P.; Goodbrand, H. B.; Gaynor, R.; Saban, M. *Ind. Eng. Chem. Res.* **2005**, *44*, 789.
[220] Shen, R.; Porco, J. A., Jr. *Org. Lett.* **2000**, *2*, 1333.
[221] Correa, A.; Bolm, C. *Adv. Synth. Catal.* **2007**, *349*, 2673.
[222] Larsson, P.-F.; Correa, A.; Carril, M.; Norrby, P.-O.; Bolm, C. *Angew. Chem., Int. Ed.* **2009**, *48*, 5691.
[223] Son, S. U.; Park, I. K.; Park, J.; Hyeon, T. *Chem. Commun.* **2004**, 778.
[224] Rout, L.; Jammi, S.; Punniyamurthy, T. *Org. Lett.* **2007**, *9*, 3397.
[225] Jammi, S.; Krishnamoorthy, S.; Saha, P.; Kundu, D. S.; Sakthivel, S.; Ali, M. A.; Paul, R.; Punniyamurthy, T. *Synlett* **2009**, 3323.
[226] Kim, A. Y.; Lee, H. J.; Park, J. C.; Kang, H.; Yang, H.; Song, H.; Park, K. H. *Molecules* **2009**, *14*, 5169.
[227] Kantam, M. L.; Venkanna, G. T.; Sridhar, C.; Kumar, K. B. S. *Tetrahedron Lett.* **2006**, *47*, 3897.
[228] Sreedhar, B.; Arundhathi, R.; Reddy, P. L.; Reddy, M. A.; Knatam, M. L. *Synthesis* **2009**, 2517.
[229] Surry, D. S.; Buchwald, S. L. *Chem. Sci.* **2010**, *1*, 13.
[230] Lv, X.; Wang, Z.; Bao, W. *Tetrahedron* **2006**, *62*, 4756.
[231] Wang, Z.; Bao, W.; Jiang, Y. *Chem. Commun.* **2005**, 2849.
[232] Yan, J.-C.; Zhou, L.; Wang, L. *Chin. J. Chem.* **2008**, *26*, 165.
[233] Wang, Y.; Wu, Z.; Wang, L.; Li, Z.; Zhou, X. *Chem.—Eur. J.* **2009**, *15*, 8971.
[234] Liang, L.; Li, Z.; Zhou, X. *Org. Lett.* **2009**, *11*, 3294.
[235] Li, X.; Yang, D.; Jiang, Y.; Fu, H. *Green Chem.* **2010**, *12*, 1097.
[236] Xie, J.; Zhu, X.; Huang, M.; Meng, F.; Chen, W.; Wan, Y. *Eur. J. Org. Chem.* **2010**, 3219.
[237] Guo, Z.; Guo, J.; Song, Y.; Wang, L.; Zou, G. *Appl. Organomet. Chem.* **2009**, *23*, 150.
[238] Wu, Z.; Jiang, Z.; Wu, D.; Xiang, H.; Zhou, X. *Eur. J. Org. Chem.* **2010**, 1854.
[239] Lange, J. H. M.; Hofmeyer, L. J. F.; Hout, F. A. S.; Osnabrug, S. J. M.; Verveer, P. C.; Kruse, C. G.; Feenstra, R. W. *Tetrahedron Lett.* **2002**, *43*, 1101.
[240] Martín, A.; Pellón, R. F.; Mesa, M.; Docampo, M. L.; Gómez, V. *J. Chem. Res.* **2005**, *9*, 561.
[241] Pellón, R. F.; Estévez-Braun, A.; Docampo, M. L.; Ravelo, A. G. *Synlett* **2005**, 1606.
[242] Pellón, R. F.; Docampo, M. L.; Fascio, M. L. *Synth. Commun.* **2007**, *37*, 1853.
[243] Faler, C. A.; Joullié, M. M. *Tetrahedron Lett.* **2006**, *47*, 7229.
[244] Cesati, R. R., III; Dwyer, G.; Jones, R. C.; Hayes, M. P.; Yalamanchili, P.; Casebier, D. S. *Org. Lett.* **2007**, *9*, 5617.
[245] Wang, D.; Cai, Q.; Ding, K. *Adv. Synth. Catal.* **2009**, *351*, 1722.
[246] Wu, X.-F.; Darcel, C. *Eur. J. Org. Chem.* **2009**, 4753.
[247] Cho, Y. A.; Kim, D.-S.; Ahn, H. R.; Canturk, B.; Molander, G. A.; Ham, J. *Org. Lett.* **2009**, *11*, 4330.
[248] Gao, X.; Fu, H.; Qiao, R.; Jiang, Y.; Zhao, Y. *J. Org. Chem.* **2008**, *73*, 6864.
[249] Zhao, H.; Fu, H.; Qiao, R. *J. Org. Chem.* **2010**, *75*, 3311.
[250] Messaoudi, S.; Brion, J.-D.; Alami, M. *Adv. Synth. Catal.* **2010**, *352*, 1677.
[251] Zou, B.; Yuan, Q.; Ma, D. *Angew. Chem., Int. Ed.* **2007**, *46*, 2598.
[252] Anderson, C. A.; Taylor, P. G.; Zeller, M. A.; Zimmerman, S. C. *J. Org. Chem.* **2010**, *75*, 4848.
[253] McBriar, M. D.; Guzik, H.; Shapiro, S.; Paruchova, J.; Xu, R.; Palani, A.; Clader, J. W.; Cox, K.; Greenlee, W. J.; Hawes, B. E.; Kowalski, T. J.; O'Neill, K.; Spar, B. D.; Weig, B.; Weston, D. J.; Farley, C.; Cook, J. *J. Med. Chem.* **2006**, *49*, 2294.
[254] Vedejs, E.; Trapencieris, P.; Suna, E. *J. Org. Chem.* **1999**, *64*, 6724.
[255] Elmkaddem, M. K.; Fischmeister, C.; Thomas, C. M.; Renaud, J.-L. *Chem. Commun.* **2010**, *46*, 925.
[256] Gaillard, S.; Elmkaddem, M. K.; Fischmeister, C.; Thomas, C. M.; Renaud, J.-L. *Tetrahedron Lett.* **2008**, *49*, 3471.
[257] Nagao, Y.; Hirota, K.; Tokumaru, M.; Kozawa, K. *Heterocycles* **2007**, *73*, 593.
[258] Zeng, L.; Fu, H.; Qiao, R.; Jiang, Y.; Zhao, Y. *Adv. Synth. Catal.* **2009**, *351*, 1671.
[259] Yang, M.; Liu, F. *J. Org. Chem.* **2007**, *72*, 8969.
[260] Lu, Z.; Twieg, R. *J. Tetrahedron Lett.* **2005**, *46*, 2997.
[261] Wang, D.; Ding, K. *Chem. Commun.* **2009**, 1891.
[262] Guo, D.; Huang, H.; Zhou, Y.; Xu, J.; Jaing, H.; Chen, K.; Liu, H. *Green Chem.* **2010**, *12*, 276.
[263] Arai, S.; Yamagishi, T.; Ototake, S.; Hida, M. *Bull. Chem. Soc. Jpn.* **1977**, *50*, 547.

264 Clement, J.-B.; Hayes, J. F.; Sheldrake, H. M.; Sheldrake, P. W.; Wells, A. S. *Synlett* **2001**, 1423.

265 Narendar, N.; Velmathi, S. *Tetrahedron Lett.* **2009**, *50*, 5159.

266 Xu, H.; Wolf, C. *Chem. Commun.* **2009**, 1715.

267 Zhu, X.; Ma, Y.; Su, L.; Song, H.; Chen, G.; Liang, D.; Wan, Y. *Synthesis* **2006**, 3955.

268 Rolfe, A.; Hanson, P. R. *Tetrahedron Lett.* **2009**, *50*, 6935.

269 Sperotto, E.; van Klink, G. P. M.; de Vries, J. G.; van Koten, G. *Tetrahedron* **2010**, *66*, 3478.

270 Ma, D.; Xia, C. *Org. Lett.* **2001**, *3*, 2583.

271 Jiang, Q.; Jiang, D.; Jiang, Y.; Fu, H.; Zhao, Y. *Synlett* **2007**, 1836.

272 Gan, J.; Ma, D. *Org. Lett.* **2009**, *11*, 2788.

273 Tao, C.-Z.; Liu, W.-W.; Sun, J.-Y.; Cao, Z.-L.; Li, H.; Zhang, Y.-F. *Synthesis* **2010**, 1280.

274 Kubo, T.; Katoh, C.; Yamada, K.; Okano, K.; Tokuyama, H.; Fukuyama, T. *Tetrahedron* **2008**, *64*, 11230.

275 Kurokawa, M.; Nakanishi, W.; Ishikawa, T. *Heterocycles* **2007**, *71*, 847.

276 Arundhathi, R.; Kumar, D. C.; Sreedhar, B. *Eur. J. Org. Chem.* **2010**, 3621.

277 Wang, Z.; Fu, H.; Jiang, Y.; Zhao, Y. *Synlett* **2008**, 2540.

278 Liu, S.; Pestano, J. P. C.; Wolf, C. *Synthesis* **2007**, 3519.

279 Job, G. E.; Buchwald, S. L. *Org. Lett.* **2002**, *4*, 3703.

280 Yadav, L. D. S.; Yadav, B. S.; Rai, V. K. *Synthesis* **2006**, 1868.

281 Feng, Y.-S.; Man, Q.-S.; Pan, P.; Pan, Z.-Q.; Xu, H.-J. *Tetrahedron Lett.* **2009**, *50*, 2585.

282 Xu, H.-J.; Zheng, F.-Y.; Liang, Y.-F.; Cai, Z.-Y.; Feng, Y.-S.; Che, D.-Q. *Tetrahedron Lett.* **2010**, *51*, 669.

283 Jadhav, V. H.; Dumbre, D. K.; Phapale, V. B.; Borate, H. B.; Wakharkar, R. D. *Catal. Commun.* **2006**, *8*, 65.

284 Wang, H.; Li, Y.; Sun, F.; Feng, Y.; Jin, K.; Wang, X. *J. Org. Chem.* **2008**, *73*, 8639.

285 Sreedhar, B.; Arundhathi, R.; Reddy, P. L.; Lakshmi Kantam, M. *J. Org. Chem.* **2009**, *74*, 7951.

286 Alakonda, L.; Periasamy, M. *J. Organomet. Chem.* **2009**, *694*, 3859.

287 Sperotto, E.; de Vries, J. G.; van Klink, G. P. M.; van Koten, G. *Tetrahedron Lett.* **2007**, *48*, 7366.

288 Gajare, A. S.; Toyota, K.; Yoshifuji, M.; Ozawa, F. *Chem. Commun.* **2004**, 1994.

289 Mao, J.; Guo, J.; Song, H.; Ji, S.-J. *Tetrahedron* **2008**, *64*, 1383.

290 Mao, J.; Hua, Q.; Guo, J.; Shi, D. *Catal. Commun.* **2008**, *10*, 341.

291 Jerphagnon, T.; van Link, G. P. M.; de Vries, J. G.; van Koten, G. *Org. Lett.* **2005**, *7*, 5241.

292 Yeh, V. S. C.; Wiedeman, P. E. *Tetrahedron Lett.* **2006**, *47*, 6011.

293 Lin, X.-F.; Li, Y.; Ma, D. *Chin. J. Chem.* **2004**, *22*, 932.

294 Reynolds, D. J.; Hermitage, S. A. *Tetrahedron* **2001**, *57*, 7765.

295 Lohmann, S.; Andrews, S. P.; Burke, B. J.; Smith, M. D.; Attfield, J. P.; Tanaka, H.; Kaneko, K.; Ley, S. V. *Synlett* **2005**, 1291.

296 Lu, Z.; Twieg, R. J. *Tetrahedron* **2005**, *61*, 903.

297 Kazock, J.-Y.; Théry, I.; Chezal, J.-M.; Chavignon, O.; Teulade, J.-C.; Gueiffier, A.; Enguehard-Gueiffier, C. *Bull. Chem. Soc. Jpn.* **2006**, *79*, 775.

298 George, T. G.; Endeshaw, M. M.; Morgan, R. E.; Mahasenan, K. V.; Delfín, D. A.; Mukherjee, M. S.; Yakovich, A. J.; Fotie, J.; Li, C.; Werbovetz, K. A. *Bioorg. Med. Chem.* **2007**, *15*, 6071.

299 Veverková, E.; Toma, Š. *Chem. Pap.* **2008**, *62*, 334.

300 Sun, H.; Cheung, W. S.; Fung, B. M. *Liq. Cryst.* **2000**, *27*, 1473.

301 Chandrasekhar, S.; Sultana, S. S.; Yaragorla, S. R.; Reddy, N. R. *Synthesis* **2006**, 839.

302 Kamenecka, T. M.; Lanza, T., Jr.; de Laszlo, S. E.; Li, B.; McCauley, E. D.; Van Riper, G.; Egger, L. A.; Kidambi, U.; Mumford, R. A.; Tong, S.; MacCoss, M.; Schmidt, J. A.; Hagmann, W. K. *Bioorg. Med. Chem.* **2002**, *12*, 2205.

303 Yamamoto, T.; Kurata, Y. *Can. J. Chem.* **1983**, *61*, 86.

304 Kelkar, A. A.; Patil, N. M.; Chaudhari, R. V. *Tetrahedron Lett.* **2002**, *43*, 7143.

305 Patil, N. M.; Kelkar, A. A.; Chaudhari, R. V. *J. Mol. Catal. A: Chem.* **2004**, *223*, 45.

306 Liu, Y.-H.; Chen, C.; Yang, L.-M. *Tetrahedron Lett.* **2006**, *47*, 9275.

307 Bender, T. P.; Graham, J. F.; Duff, J. M. *Chem. Mater.* **2001**, *13*, 4105.

308 Nandurkar, N. S.; Bhanushali, M. J.; Bhor, M. D.; Bhanage, B. M. *Tetrahedron Lett.* **2007**, *48*, 6573.

309 Sawant, S. K.; Gaikwad, G. A.; Sawant, V. A.; Yamgar, B. A.; Chavan, S. S. *Inorg. Chem. Commun.* **2009**, *12*, 632.

310 Haider, J.; Kunz, K.; Scholz, U. *Adv. Synth. Catal.* **2004**, *346*, 717.

311 Goodbrand, H. B.; Hu, N.-X. *J. Org. Chem.* **1999**, *64*, 670.

312 Scherrer, R. A.; Beatty, H. R. *J. Org. Chem.* **1980**, *45*, 2127.

313 Pellón, R. F.; Mamposo, T.; Carasco, R.; Rodés, L. *Synth. Commun.* **1996**, *26*, 3877.

314 Docampo Palacios, M. L.; Pellón Comdom, R. F. *Synth. Commun.* **2003**, *33*, 1771.

315 Carrasco, R.; Pellón, R. F.; Elguero, J.; Goya, P.; Páez, J. A. *Synth. Commun.* **1989**, *18*, 2077.
316 Allen, C. F. H.; McKee, G. H. W. *Org. Synth.* **1939**, *19*, 6.
317 Girisha, H. R.; Srinivasa, G. R.; Gowda, D. C. *J. Chem. Res.* **2006**, 342.
318 Maradolla, M. B.; Mandha, A.; Garimella, C. M. *J. Chem. Res.* **2007**, 587.
319 Field, J. E.; Venkataraman, D. *Chem. Mater.* **2002**, *14*, 962.
320 Rewcastle, G. W.; Denny, W. A.; Baguely, B. C. *J. Med. Chem.* **1987**, *30*, 843.
321 Ozaki, K.-i.; Yamada, Y.; Oine, T.; Ishizuka, T.; Iwasawa, Y. *J. Med. Chem.* **1985**, *28*, 568.
322 Kaltenbronn, J. S.; Scherer, R. A.; Short, F. W.; Jones, E. M.; Beatty, H. R.; Saka, M. M.; Winder, C. W.; Wax, J.; Williamson, W. R. N. *Arzneim.-Forsch./Drug Res.* **1983**, *33*, 621.
323 Baqi, Y.; Müller, C. E. *Org. Lett.* **2007**, *9*, 1271.
324 Glänzel, M.; Bültmann, R.; Starke, K.; Frahm, A. W. *Eur. J. Med. Chem.* **2003**, *38*, 303.
325 Pearson, J. C.; Burton, S. J.; Lowe, C. R. *Anal. Biochem.* **1986**, *158*, 382.
326 Przheval'skii, N. M.; Grandberg, I. I. *Chem. Heterocycl. Compd.* **1982**, *7*, 716.
327 Pellón, R. F.; Carrasco, R.; Márquez, T.; Mamposo, T. *Tetrahedron Lett.* **1997**, *38*, 5107.
328 Mei, T.-S.; Wang, D.-H.; Yu, J.-Q. *Org. Lett.* **2010**, *12*, 3140.
329 Wu, J.; Li, X.; Yang, M.; Gao, Y.; Lv, Q.; Chen, B. *Can. J. Chem.* **2008**, *86*, 871.
330 Selby, T. D.; Kim, K.-Y.; Blackstock, S. C. *Chem. Mater.* **2002**, *14*, 1685.
331 Bushby, R. J.; McGill, D. R.; Ng, K. M.; Taylor, N. *J. Mater. Chem.* **1997**, *7*, 2343.
332 Pellón, R. F.; Carasco, R.; Rodés, L. *Synth. Commun.* **1993**, *23*, 1447.
333 Hanoun, J. P.; Galy, J.-P.; Tenaglia, A. *Synth. Commun.* **1996**, *25*, 2443.
334 Reisch, J.; Probst, W. *Arch. Pharm.* **1989**, *322*, 31.
335 Zhao, Y.; Wang, Y.; Sun, H.; Li, L.; Zhang, H. *Chem. Commun.* **2007**, 3186.
336 Cai, Q.; Zhu, W.; Zhang, H.; Zhang, Y.; Ma, D. *Synthesis* **2005**, 496.
337 Bueno, M. A.; Silva, L. R. S. P.; Corrêa, A. G. *J. Braz. Chem. Soc.* **2008**, *19*, 1264.
338 Wong, K.-T.; Ku, S.-Y.; Yen, F.-W. *Tetrahedron Lett.* **2007**, *48*, 5051.
339 Gujadhur, R.; Venkataraman, D.; Kintigh, J. T. *Tetrahedron Lett.* **2001**, *42*, 4791.
340 Wolf, C.; Mei, X. *J. Am. Chem. Soc.* **2003**, *125*, 10651.
341 Field, J. E.; Hill, T. J.; Venkataraman, D. *J. Org. Chem.* **2003**, *68*, 6071.
342 Tanaka, H.; Tokito, S.; Taga, Y.; Okada, A. *Chem. Commun.* **1996**, 2175.
343 Wilkinson, J. H.; Finar, L. L. *J. Chem. Soc.* **1948**, 32.
344 Baudoin, O.; Teulade-Fichou, M.-P.; Vigneron, J.-P.; Lehn, J.-M. *J. Org. Chem.* **1997**, *62*, 5458.
345 Clemo, G. R.; Perkin, W. H., Jr.; Robinson, R. J. *Chem. Soc., Trans.* **1924**, *125*, 1751.
346 Yang, M.; Wu, L.; She, D.; Hui, H.; Zhao, Q.; Chen, M.; Huang, G.; Liang, Y. *Synlett* **2008**, 448.
347 Rewcastle, G. W.; Denny, W. A. *Synth. Commun.* **1987**, *17*, 309.
348 Hellwinkel, D.; Melan, M. *Chem. Ber.* **1971**, *104*, 1001.
349 Glaser, R.; Blount, J. F.; Mislow, K. *J. Am. Chem. Soc.* **1980**, *102*, 2777.
350 Dokorou, V.; Kovala-Demertzi, D.; Jasinski, J. P.; Galani, A.; Demertzis, M. A. *Helv. Chim. Acta.* **2004**, *87*, 1940.
351 Teulade-Fichou, M.-P.; Vigneron, J.-P.; Lehn, J.-M. *Supramol. Chem.* **1995**, *5*, 139.
352 Jacquelin, C.; Saettel, N.; Hounsou, C.; Teulade-Fichou, M.-P. *Tetrahedron Lett.* **2005**, *46*, 2589.
353 Cañas Rogdriguez, A.; Mateo, B. A. *Anal. Quim.* **1987**, *83*, 21.
354 Miller, R. D.; Lee, V. Y.; Twieg, R. J. *Chem. Commun.* **1995**, 245.
355 Hellwinkel, D.; Schmidt, W. *Chem. Ber.* **1980**, *113*, 358.
356 Mei, X.; Wolf, C. *Chem. Commun.* **2004**, 2078.
357 Mei, X.; Wolf, C. *J. Am. Chem. Soc.* **2004**, *126*, 14736.
358 Huang, C.-Y.; Su, Y. O. *Dalton Trans.* **2010**, *39*, 8306.
359 Belfield, K. D.; Schafer, K. J.; Mourad, W.; Reinhardt, B. A. *J. Org. Chem.* **2000**, *65*, 4475.
360 Docampo Palacios, M. L.; Pellón Comdom, R. F. *Synth. Commun.* **2003**, *33*, 1777.
361 Price, C.; Roberts, R. *J. Org. Chem.* **1946**, *11*, 463.
362 Robin, M.; Faure, R.; Périchaud, A.; Galy, J.-P. *Synth. Commun.* **2002**, *32*, 981.
363 Meesala, R.; Nagarajan, R. *Tetrahedron Lett.* **2010**, *51*, 422.
364 Meesala, R.; Nagarajan, R. *Tetrahedron* **2009**, *65*, 6050.
365 Martić, S.; Liu, X.; Wang, S.; Wu, G. *Chem.—Eur. J.* **2008**, *14*, 1196.
366 Lovering, F.; Kirincich, S.; Wang, W.; Combs, K.; Resnick, L.; Sabalski, J. E.; Butera, J.; Liu, J.; Parris, K.; Telliez, J. B. *Bioorg. Med. Chem.* **2009**, *17*, 3342.
367 Hager, F. D. *Org. Synth.* **1928**, *8*, 116.
368 Gujadhur, R. K.; Bates, C. G.; Venkataraman, D. *Org. Lett.* **2001**, *3*, 4315.
369 Koene, B. E.; Loy, D. E.; Thompson, M. E. *Chem. Mater.* **1998**, *10*, 2235.
370 Strohriegl, P.; Jesberger, G.; Heinze, J.; Moll, T. *Makromol. Chem.* **1992**, *193*, 909.
371 He, M.; Twieg, R. J.; Gubler, U.; Wright, D.; Moerner, W. E. *Chem. Mater.* **2003**, *15*, 1156.

372 Özçubukçu, E.; Schmitt, E.; Leifert, A.; Bolm, C. *Synthesis* **2007**, 389.
373 Grigalevičius, S.; Getautis, V.; Gražulevičius, J. V.; Gaidelis, V.; Jankauskas, V.; Montrimas, E. *Mater. Chem. Phys.* **2001**, *72*, 395.
374 Low, P. J.; Paterson, M. A. J.; Goeta, A. E.; Yufit, D. S.; Howard, J. A. K.; Cherryman, J. C.; Tackley, D. R.; Brown, B. *J. Mater. Chem.* **2004**, *14*, 2516.
375 Yano, M.; Inoue, K.; Motoyama, T.; Azuma, Y.; Tatsumi, M.; Yamauchi, O.; Oyama, M.; Sato, K.; Takui, T. *Polyhedron* **2005**, *24*, 2112.
376 Moriwaki, K.; Satoh, K.; Takada, M.; Ishino, Y.; Ohno, T. *Tetrahedron Lett.* **2005**, *46*, 7559.
377 Jia, W. L.; Feng, X. D.; Bai, D. R.; Lu, Z. H.; Wang, S.; Vamvounis, G. *Chem. Mater.* **2005**, *17*, 164.
378 Liebermann, H. *Liebigs Ann. Chem.* **1935**, *518*, 245.
379 Sasaki, S.; Murakami, F.; Yoshifuji, M. *Tetrahedron Lett.* **1997**, *38*, 7095.
380 Herberhold, M.; Ellinger, M.; Kremnitz, W. *J. Organomet. Chem.* **1983**, *241*, 227.
381 Pang, J.; Tao, Y.; Freiberg, S.; Yang, X.-P.; D'Iorio, M.; Wang, S. *J. Mater. Chem.* **2002**, *12*, 206.
382 Pang, J.; Marcotte, E. J.-P.; Seward, C.; Brown, R. S.; Wang, S. *Angew. Chem., Int. Ed.* **2001**, *40*, 4042.
383 Yang, W.; Schmider, H.; Wu, Q.; Zhang, Y.-s.; Wang, S. *Inorg. Chem.* **2000**, *39*, 2397.
384 Bedworth, P. V.; Cai, Y.; Jen, A.; Marder, S. R. *J. Org. Chem.* **1996**, *61*, 2242.
385 Zhang, Q.; Hu, Y.; Cheng, Y. X.; Su, G. P.; Ma, D. G.; Wang, L. X.; Jing, X. B.; Wang, F. S. *Synth. Met.* **2003**, *137*, 1111.
386 Zhu, L.; Guo, P.; Li, G.; Lan, J.; Xie, R.; You, J. *J. Org. Chem.* **2007**, *72*, 8535.
387 Tao, C. Z.; Li, J.; Cui, X.; Fu, Y.; Guo, Q. X. *Chin. Chem. Lett.* **2007**, *18*, 1199.
388 Deng, W.; Wang, Y. F.; Zhang, C.; Liu, L.; Guo, Q. X. *Chin. Chem. Lett.* **2006**, *17*, 313.
389 Sreedhar, B.; Shiva Kumar, K. B.; Srinivas, P.; Balasubrahmanyam, V.; Venkanna, G. T. *J. Mol. Catal. A: Chem.* **2007**, *265*, 183.
390 Hosseinzadeh, R.; Tajbakhsh, M.; Alikarami, M.; Mohadjerani, M. *J. Heterocycl. Chem.* **2008**, *45*, 1815.
391 Chang, J. W. W.; Xu, X.; Chan, P. W. H. *Tetrahedron Lett.* **2007**, *48*, 245.
392 Ma, H. C.; Wu, S.; Sun, Q.; Lei, Z. *Lett. Org. Chem.* **2010**, *7*, 212.
393 Suresh, P.; Pitchumani, K. *J. Org. Chem.* **2008**, *73*, 9121.
394 Cheng, C.; Sun, G.; Wan, J.; Sun, C. *Synlett* **2009**, 2663.
395 Oshovsky, G. V.; Ouali, A.; Xia, N.; Zablocka, M.; Boeré, R. T.; Duhayon, C.; Taillefer, M.; Majoral, J. P. *Organometallics* **2008**, *27*, 5733.
396 Ma, H.-C.; Jiang, X.-Z. *J. Org. Chem.* **2007**, *72*, 8943.
397 Wan, J.-P.; Chai, Y.-F.; Wu, J.-M.; Pan, Y.-J. *Synlett* **2008**, 3068.
398 Huang, Y.-Z.; Miao, H.; Zhang, Q.-H.; Chen, C.; Xu, J. *Catal. Lett.* **2008**, *122*, 344.
399 Huang, Y.-Z.; Gao, J.; Ma, H.; Miao, H.; Xu, J. *Tetrahedron Lett.* **2008**, *49*, 948.
400 Xue, F.; Cai, C.; Sun, H.; Shen, Q.; Rui, J. *Tetrahedron Lett.* **2008**, *49*, 4386.
401 Ma, D.; Cai, Q. *Synlett* **2004**, 128.
402 Khan, M. A.; Polya, J. B. *J. Chem. Soc., C* **1970**, 85.
403 Verma, A. K.; Kesharwani, T.; Singh, J.; Tandon, V.; Larock, R. C. *Angew. Chem., Int. Ed.* **2009**, *48*, 1138.
404 Kantam, M. L.; Rao, B. P. C.; Choudary, B. M.; Reddy, R. S. *Synlett* **2006**, 2195.
405 Kang, S.-K.; Kim, D.-H.; Pak, J.-N. *Synlett* **2002**, 427.
406 Wu, Y.-J.; He, H.; L'Heureux, A. *Tetrahedron Lett.* **2003**, *44*, 4217.
407 Purecha, V. H.; Nandurkar, N. S.; Bhanage, B. M.; Nagarkar, J. M. *Tetrahedron Lett.* **2008**, *49*, 1384.
408 Siddle, J. S.; Batsanov, A. S.; Bryce, M. R. *Eur. J. Org. Chem.* **2008**, 2746.
409 Enguehard, C.; Allouchi, H.; Gueiffier, A.; Buchwald, S. L. *J. Org. Chem.* **2003**, *68*, 5614.
410 Hosseinzadeh, R.; Tajbakhsh, M.; Alikarami, M. *Tetrahedron Lett.* **2006**, *47*, 5203.
411 Hosseinzadeh, R.; Tajbakhsh, M.; Alikarami, M. *Synlett* **2006**, 2124.
412 Chen, W.; Zhang, Y.; Zhu, L.; Lan, J.; Xie, R.; You, J. *J. Am. Chem. Soc.* **2007**, *129*, 13879.
413 Ouali, A.; Laurent, R.; Caminade, A.-M.; Majoral, J.-P.; Taillefer, M. *J. Am. Chem. Soc.* **2006**, *128*, 15990.
414 Maheswaran, H.; Krishna, G. G.; Prasanth, K. L.; Srinivas, V.; Chaitanya, G. K.; Bhanuprakash, K. *Tetrahedron* **2008**, *64*, 2471.
415 Li, F.; Hor, T. S. A. *Chem.—Eur. J.* **2009**, *15*, 10585.
416 Lexy, H.; Kauffmann, T. *Chem. Ber.* **1980**, *113*, 2755.
417 Chouhan, G.; Wang, D.; Alper, H. *Chem. Commun.* **2007**, 4809.
418 Chow, W. S.; Chan, T. H. *Tetrahedron Lett.* **2009**, *50*, 1286.
419 Arsenyan, P.; Paegle, E. *Tetrahedron Lett.* **2010**, *51*, 5052.
420 Tubaro, C.; Biffis, A.; Scattolin, E.; Basato, M. *Tetrahedron* **2008**, *64*, 4187.

421 Biffis, A.; Tubaro, C.; Scattolin, E.; Basato, M.; Papini, G.; Santini, C.; Alvarez, E.; Conejero, S. *Dalton Trans.* **2009**, 7223.

422 Hanamoto, T.; Iwamoto, Y.; Yamada, K.; Anno, R. *J. Fluorine Chem.* **2007**, *128*, 1126.

423 Tironi, C.; Fruttero, R.; Garrone, A. *Il Farmaco* **1990**, *45*, 473.

424 Young, R. J.; Borthwick, A. D.; Brown, D.; Burns-Kurtis, C. L.; Campbell, M.; Chan, C.; Charbaut, M.; Convery, M. A.; Diallo, H.; Hortense, E.; Irving, W. R.; Kelly, H. A.; King, N. P.; Kleanthous, S.; Mason, A. M.; Pateman, A. J.; Patikis, A. N.; Pinto, I. L.; Pollard, D. R.; Senger, S.; Shah, G. P.; Toomey, J. R.; Watson, N. S.; Weston, H. E.; Zhou, P. *Bioorg. Med. Chem. Lett.* **2008**, *18*, 28.

425 Guillou, S.; Bonhomme, F. J.; Chahine, D. B.; Nesme, O.; Janin, Y. L. *Tetrahedron* **2010**, *66*, 2654.

426 Ribecai, A.; Bacchi, S.; Delpogetto, M.; Guelfi, S.; Manzo, A. M.; Perboni, A.; Stabile, P.; Westerduin, P.; Hourdin, M.; Rossi, S.; Provera, S.; Turco, L. *Org. Proc. Res. Dev.* **2010**, *14*, 895.

427 Reddy, R. C.; Kumar, N. S.; Sreedhar, B.; Kantam, M. L. *J. Mol. Catal. A: Chem.* **2006**, *252*, 136.

428 Ali, M. A.; Saha, P.; Punniyamurthy, T. *Synthesis* **2010**, 908.

429 Cheng, D.; Gan, F.; Qian, W.; Bao, W. *Green Chem.* **2008**, *10*, 171.

430 Altman, R. A.; Buchwald, S. L. *Org. Lett.* **2006**, *8*, 2779.

431 Verma, A. K.; Singh, J.; Sankar, V. K.; Chaudhary, R.; Chandra, R. *Tetrahedron Lett.* **2007**, *48*, 4207.

432 Vargas, V. C.; Rubio, R. J.; Hollis, T. K.; Salcido, M. E. *Org. Lett.* **2003**, *5*, 4847.

433 Collman, J. P.; Wang, Z.; Zhong, M.; Zeng, L. *J. Chem. Soc., Perkin Trans. 1* **2000**, 1217.

434 Liu, L.; Frohn, M.; Xi, N.; Dominguez, C.; Hungate, R.; Reider, P. J. *J. Org. Chem.* **2005**, *70*, 10135.

435 Maryanoff, B. E.; McComsey, D. F.; Martin, G. E.; Shank, R. P. *Bioorg. Med. Chem. Lett.* **1998**, *8*, 983.

436 Cozzi, P.; Carganico, G.; Fusar, D.; Grossoni, M.; Meinichinchert, M.; Pinciroli, V.; Tonant, R.; Vaght, F.; Salvati, P. *J. Med. Chem.* **1993**, *36*, 2964.

437 Wu, Q.; Lavigne, J. A.; Tao, Y.; D'Iorio, M.; Wang, S. *Chem. Mater.* **2001**, *13*, 71.

438 Wan, Y.; Wallinder, C.; Plouffe, B.; Beaudry, H.; Mahalingam, A. K.; Wu, X.; Johansson, B.; Holm, M.; Botoros, M.; Karlén, A.; Pettersson, A.; Nyberg, F.; Fändriks, L.; Gallo-Payet, N.; Hallberg, A.; Alterman, M. *J. Med. Chem.* **2004**, *47*, 5995.

439 Quan, M. L.; Lam, P. Y. S.; Han, Q.; Pinto, D. J. P.; He, M. Y.; Li, R.; Ellis, C. D.; Clark, C. G.; Teleha, C. A.; Sun, J.-H.; Alexander, R. S.; Bai, S.; Luettgen, J. M.; Knabb, R. M.; Wong, P. C.; Wexler, R. R. *J. Med. Chem.* **2005**, *48*, 1729.

440 Yue, W.; Lewis, S. I.; Koen, Y. M.; Hanzlik, R. P. *Bioorg. Med. Chem. Lett.* **2004**, *14*, 1637.

441 Altman, R. A.; Buchwald, S. L. *Nature Protoc.* **2007**, *2*, 2474.

442 Sundberg, R. J.; Mente, D. C.; Yilmaz, I.; Gupta, G. *J. Heterocycl. Chem.* **1977**, *14*, 1279.

443 Liu, Y.; Yan, W.; Chen, Y.; Petersen, J. L.; Shi, X. *Org. Lett.* **2008**, *10*, 5389.

444 Tokmakov, G. P.; Grandberg, I. I. *Tetrahedron* **1995**, *51*, 2091.

445 Zhou, T.; Chen, Z.-C. *Synth. Commun.* **2002**, *32*, 903.

446 Rao, R. K.; Naidu, A. B.; Jaseer, E. A.; Sekar, G. *Tetrahedron* **2009**, *65*, 4619.

447 Periasamy, M.; Vairaprakash, P.; Dalai, M. *Organometallics* **2008**, *27*, 1963.

448 Khan, M. A.; Rocha, E. K. *Chem. Pharm. Bull.* **1977**, *25*, 3110.

449 Haneda, S.; Adachi, Y.; Hayashi, M. *Tetrahedron* **2009**, *65*, 10459.

450 Colacino, E.; Villebrun, L.; Martinez, J.; Lamaty, F. *Tetrahedron* **2010**, *66*, 3730.

451 Wang, J.; Schaeffler, L.; He, G.; Ma, D. *Tetrahedron Lett.* **2007**, *48*, 6717.

452 Kuil, M.; Bekedam, E. K.; Visser, G. M.; van den Hoogenband, A.; Terpstra, J. W.; Kamer, P. C. J.; van Leeuwen, P. W. N. M.; van Strijdonck, G. P. F. *Tetrahedron Lett.* **2005**, *46*, 2405.

453 Murakami, Y.; Watanabe, T.; Hagiwara, T.; Akiyama, Y.; Ishii, H. *Chem. Pharm. Bull.* **1995**, *43*, 1281.

454 Unanast, P. C.; Connor, D. R.; Stabler, S. R.; Weikert, R. J. *J. Heterocycl. Chem.* **1987**, *24*, 811.

455 Wu, T. Y. H.; Schultz, P. G. *Org. Lett.* **2002**, *4*, 4033.

456 Hendi, S. B.; Basanagoudar, L. D. *Ind. J. Chem.* **1981**, *20B*, 330.

457 Newhouse, T.; Lewis, C. A.; Eastman, K. J.; Baran, P. S. *J. Am. Chem. Soc.* **2010**, *132*, 7119.

458 Dessole, G.; Branca, D.; Ferrigno, F.; Kinzel, O.; Muraglia, E.; Palumbi, M. C.; Rowley, M.; Serafini, S.; Steinkühler, C.; Jones, P. *Bioorg. Med. Chem. Lett.* **2009**, *19*, 4191.

459 Salovich, J. M.; Lindsley, C. W.; Hopkins, C. R. *Tetrahedron Lett.* **2010**, *51*, 3796.

460 Zhou, T.; Chen, Z.-C. *Heteroatom. Chem.* **2002**, *13*, 617.

461 Wu, J.; Li, H.-Y.; Xu, Q.-L.; Zhu, Y.-C.; Tao, Y.-M.; Li, H.-R.; Zheng, Y.-X.; Zuo, J.-L.; You, X.-Z. *Inorg. Chim. Acta* **2010**, *363*, 2394.

462 Kimoto, A.; Cho, J.-S.; Higuchi, M.; Yamamoto, K. *Macromolecules* **2004**, *37*, 5531.

463 Xu, T.; Lu, R.; Liu, X.; Zheng, X.; Qiu, X.; Zhao, Y. *Org. Lett.* **2007**, *9*, 797.

464 Kato, Y.; Conn, M. M.; Rebek, J., Jr. *J. Am. Chem. Soc.* **1994**, *116*, 3279.

465 Albrecht, K.; Yamamoto, K. *J. Am. Chem. Soc.* **2009**, *131*, 2244.

466 Albrecht, K.; Kasai, Y.; Kimoto, A.; Yamamoto, K. *Macromolecules* **2008**, *41*, 3793.

467 Wu, Q.; Lavigne, J. A.; Tao, Y.; D'Iorio, M.; Wang, S. *Inorg. Chem.* **2000**, *39*, 5248.

468 Promarak, V.; Ichikawa, M.; Sudyoadsuk, T.; Saengsuwan, S.; Jungsuttiwong, S.; Keawin, T. *Synth. Met.* **2007**, *157*, 17.

469 Zhang, S.; Chen, R.; Yin, J.; Liu, F.; Jiang, H.; Shi, N.; An, Z.; Ma, C.; Liu, B.; Huang, W. *Org. Lett.* **2010**, *12*, 3438.

470 Zhao, Z.; Xu, B.; Yang, Z.; Wang, H.; Wang, X.; Lu, P.; Tian, W. *J. Phys. Chem. C* **2008**, *112*, 8511.

471 Hu, B.; Wang, Y.; Chen, X.; Zhao, Z.; Jiang, Z.; Lu, P.; Wang, Y. *Tetrahedron* **2010**, *66*, 7583.

472 Jian, H.; Tour, J. M. *J. Org. Chem.* **2003**, *68*, 5091.

473 Martínez-Palau, M.; Perea, E.; López-Calahorra, F.; Velasco, D. *Lett. Org. Chem.* **2004**, *1*, 231.

474 Hameurlaine, A.; Dehaen, W. *Tetrahedron Lett.* **2003**, *44*, 957.

475 Ding, J.; Gao, J.; Cheng, Y.; Xie, Z.; Wang, L.; Ma, D.; Jing, X.; Wang, F. *Adv. Funct. Mater.* **2006**, *16*, 575.

476 Jiao, G.-S.; Loudet, A.; Lee, H. B.; Kalinin, S.; Johansson, L. B.-Å.; Burgess, K. *Tetrahedron* **2003**, *59*, 3109.

477 Grigalevičius, S.; Gražulevičius, J. V.; Gaidelis, V.; Jankauskas, V. *Polymer* **2002**, *43*, 2603.

478 Xu, T.; Lu, R.; Jin, M.; Qiu, X.; Xue, P.; Bao, C.; Zhao, Y. *Tetrahedron Lett.* **2005**, *46*, 6883.

479 Xu, T. H.; Lu, R.; Qiu, X. P.; Liu, X. L.; Xue, P. C.; Tan, C. H.; Bao, C. Y.; Zhao, Y. Y. *Eur. J. Org. Chem.* **2006**, 4014.

480 Xing, Y.; Lin, H.; Wang, F.; Lu, P. *Sensors and Actuators* **2006**, *114B*, 28.

481 McClenaghan, N. D.; Passalacqua, R.; Loiseau, F.; Campagna, S.; Verheyde, B.; Hameurlaine, A.; Dehaen, W. *J. Am. Chem. Soc.* **2003**, *125*, 5356.

482 Lu, J.; Xia, P. F.; Lo, P. K.; Tao, Y.; Wong, M. S. *Chem. Mater.* **2006**, *18*, 6194.

483 Hong, C. S.; Seo, J. Y.; Yum, E. K. *Tetrahedron Lett.* **2007**, *48*, 4831.

484 Kang, Y.; Wang, S. *Tetrahedron Lett.* **2002**, *43*, 3711.

485 Wu, Q.; Hook, A.; Wang, S. *Angew. Chem., Int. Ed.* **2000**, *39*, 3933.

486 Jia, W.-L.; Wang, R.-Y.; Song, D.; Ball, S. J.; McLean, A. B.; Wang, S. *Chem.—Eur. J.* **2005**, *11*, 832.

487 Harbert, C. A.; Plattner, J. J.; Welch, W. M. *J. Med. Chem.* **1980**, *23*, 635.

488 Robarge, M. J.; Bom, D. C.; Tumey, L. N.; Varga, N.; Gleason, E.; Silver, D.; Song, J.; Murphy, S. M.; Ekema, G.; Doucett, C.; Hanniford, D.; Palmer, M.; Pawloski, G.; Danzing, J.; Loftus, M.; Hunady, K.; Sherf, B. A.; Mays, R. W.; Stricker-Krongrad, A.; Brunden, K. R.; Harrington, J. J.; Bennani, Y. L. *Bioorg. Med. Chem. Lett.* **2005**, *15*, 1749.

489 Nguyen, D. M.; Miles, D. H. *Synth. Commun.* **2009**, *39*, 2829.

490 Hosseinzadeh, R.; Tajbakhsh, M.; Mohadjerani, M.; Mehdinejad, H. *Synlett* **2004**, 1517.

491 Hosseinzadeh, R.; Tajbakhsh, M.; Mohadjerani, M.; Ghorbani, E. *Chin. J. Chem.* **2008**, *26*, 2120.

492 Ma, H. C.; Jiang, X. Z. *Synlett* **2008**, 1335.

493 Wang, C.; Liu, L.; Wang, W.; Ma, D.-S.; Zhang, H. *Molecules* **2010**, *15*, 1154.

494 Deng, W.; Zhang, C.; Liu, M.; Zou, Y.; Liu, L.; Guo, Q.-X. *Chin. J. Chem.* **2005**, *23*, 1241.

495 Chen, W.; Li, J.; Fang, D.; Feng, C.; Zhang, C. *Org. Lett.* **2008**, *10*, 4565.

496 Wiedemann, S. H.; Ellman, J. A.; Bergman, R. G. *J. Org. Chem.* **2006**, *71*, 1969.

497 Jones, C. P.; Anderson, K. W.; Buchwald, S. L. *J. Org. Chem.* **2007**, *72*, 7968.

498 Finerty, M. J.; Bingham, J. P.; Hartley, J. A.; Shipman, M. *Tetrahedron Lett.* **2009**, *50*, 3648.

499 Satyanarayana, K.; Srinivas, K.; Himabindu, V.; Reddy, G. M. *Org. Proc. Res. Dev.* **2007**, *11*, 842.

500 Enguehard-Gueiffier, C.; Thery, I.; Gueiffier, A.; Buchwald, S. L. *Tetrahedron* **2006**, *62*, 6042.

501 Padwa, A.; Crawford, K. R.; Rashatasakhon, P.; Rose, M. *J. Org. Chem.* **2003**, *68*, 2609.

502 Ramalingam, V.; Bhagirath, N.; Muthyala, R. S. *J. Org. Chem.* **2007**, *72*, 3976.

503 Toto, P.; Gesquière, J.-C.; Deprez, B.; Willand, N. *Tetrahedron Lett.* **2006**, *47*, 1181.

504 Toto, P.; Gesquière, J.-C.; Coussaert, N.; Deprez, B.; Willand, N. *Tetrahedron Lett.* **2006**, *47*, 4973.

505 Greiner, A. *Synthesis* **1989**, 312.

506 Dharmasena, P. M.; Oliveira-Campos, A. M.-F.; Raposo, M. M. M.; Shannon, P. V. R. *J. Chem. Res.* **1994**, 296.

507 Goldberg, I. *Ber. Dtsch. Chem. Ges.* **1907**, *40*, 4541.

508 Ishii, H.; Sugiura, T.; Akiyama, Y.; Ichikawa, X.; Watanabe, T.; Murakami, Y. *Chem. Pharm. Bull.* **1990**, *38*, 2118.

509 Nodiff, E. A.; Lipshutz, S.; Craig, P. N.; Gordon, M. *J. Org. Chem.* **1960**, *25*, 60.

510 Ito, A.; Saito, T.; Tanaka, K.; Yamabe, T. *Tetrahedron Lett.* **1995**, *36*, 8809.

511 Thu-Cuc, T.; Buu-Hoi, N. P.; Xuong, N. D. *J. Heterocycl. Chem.* **1964**, *1*, 28.

512 Freeman, H. S.; Butler, J. R.; Freedman, L. D. *J. Org. Chem.* **1978**, *43*, 4975.

513 Ishii, H.; Sugiura, T.; Kogusuri, K.; Watanabe, T.; Murakami, Y. *Chem. Pharm. Bull.* **1991**, *39*, 572.

514 Ferreira, I. C. F. R.; Queiroz, M.-J. R. P.; Kirsch, G. *Tetrahedron* **2002**, *58*, 7943.

515 Bacon, R. G. R.; Karim, A. *J. Chem. Soc., Perkin Trans. 1* **1973**, 278.

516 Renger, B. *Synthesis* **1985**, 856.

517 Mallesham, B.; Rajesh, B. M.; Rajamohan Reddy, P.; Srinivas, D.; Trehan, S. *Org. Lett.* **2003**, *5*, 963.

518 Phillips, D. P.; Hudson, A. R.; Nguyen, B.; Lau, T. L.; McNeill, M. H.; Dalgard, J. E.; Chen, J.-H.; Penuliar, R. J.; Miller, T. A.; Zhi, L. *Tetrahedron Lett.* **2006**, *47*, 7137.

519 Sugahara, M.; Ukita, T. *Chem. Pharm. Bull.* **1997**, *45*, 719.

520 Chen, Y.-J.; Chen, H.-H. *Org. Lett.* **2006**, *8*, 5609.

521 Haldón, E.; Álvarez, E.; Nicasio, M. C.; Pérez, P. *J. Organometallics* **2009**, *28*, 3815.

522 Ghinet, A.; Oudir, S.; Hénichart, J.-P.; Rigo, B.; Pommery, N.; Gautret, P. *Tetrahedron* **2010**, *66*, 215.

523 Sato, M.; Ebine, S.; Akabori, S. *Synthesis* **1981**, 472.

524 Andresen, B. M.; Caron, S.; Couturier, M.; DeVries, K. M.; Do, N. M.; Dupont-Gaudet, K.; Ghosh, A.; Girardin, M.; Hawkins, J. M.; Makowski, T. M.; Riou, M.; Sieser, J. E.; Tucker, J. L.; Vanderplas, B. C.; Watson, T. J. N. *Chimia* **2006**, *60*, 554.

525 Sarges, R.; Howard, H. R.; Koe, B. K.; Weissman, A. *J. Med. Chem.* **1989**, *32*, 437.

526 Beadle, C. D.; Boot, J.; Camp, N. P.; Dezutter, N.; Findley, J.; Hayhurst, L.; Masters, J. J.; Penariol, R.; Walter, M. W. *Bioorg. Med. Chem. Lett.* **2005**, *15*, 4432.

527 Wu, B.; Kuhen, K.; Ngyuen, T. N.; Ellis, D.; Anaclerio, B.; He, X.; Yang, K. Y.; Karanewski, D.; Yin, H.; Wolff, K.; Bieza, K.; Caldwell, J.; He, Y. *Bioorg. Med. Chem. Lett.* **2006**, *16*, 3430.

528 Wannberg, J.; Dallinger, D.; Kappe, C. O.; Larhed, M. *J. Comb. Chem.* **2005**, *7*, 574.

529 Beghyn, T.; Hounsou, C.; Deprez, B. P. *Bioorg. Med. Chem. Lett.* **2007**, *17*, 789.

530 Kawabata, T.; Muramatsu, W.; Nishio, T.; Shibata, T.; Schedel, H. *J. Am. Chem. Soc.* **2007**, *129*, 12890.

531 Scheiper, B.; Glorius, F.; Leitner, A.; Fürstner, A. *Proc. Natl. Acad. Sci. USA* **2004**, *101*, 11960.

532 Filipski, K. J.; Kohrt, J. T.; Casimiro-Garcia, A.; Van Huis, C. A.; Dudley, D. A.; Cody, W. L.; Bigge, C. F.; Desiraju, S.; Sun, S.; Maiti, S. N.; Jaber, M. R.; Edmunds, J. J. *Tetrahedron Lett.* **2006**, *47*, 7677.

533 Pu, Y.-M.; Ku, Y.-Y.; Grieme, T.; Henry, R.; Bhatia, A. V. *Tetrahedron Lett.* **2006**, *47*, 149.

534 Schweinitz, A.; Dönnecke, D.; Ludwig, A.; Steinmetzer, P.; Schulze, A.; Kotthaus, J.; Wein, S.; Clement, B.; Steinmetzer, T. *Bioorg. Med. Chem. Lett.* **2009**, *19*, 1960.

535 Wang, P.-S.; Liang, C.-K.; Leung, M.-k. *Tetrahedron* **2005**, *61*, 2931.

536 Yang, G. X.; Chang, L. L.; Truong, Q.; Doherty, G. A.; Magriotis, P. A.; de Laszlo, S. E.; Li, B.; Mac-Coss, M.; Kidambi, U.; Egger, L. A.; McCauley, E. D.; Van Riper, G.; Mumford, R. A.; Schmidt, J. A.; Hagmann, W. K. *Bioorg. Med. Chem. Lett.* **2002**, *12*, 1497.

537 Siddle, J. S.; Batsanov, A. S.; Caldwell, S. T.; Cooke, G.; Bryce, M. R. *Tetrahedron* **2010**, *66*, 6138.

538 Hu, E.; Tasker, A.; White, R. D.; Kunz, R. K.; Human, J.; Chen, N.; Bürli, R.; Hungate, R.; Novak, P.; Itano, A.; Zhang, X.; Yu, V.; Nguyen, Y.; Tudor, Y.; Plant, M.; Flynn, S.; Xu, Y.; Meagher, K. L.; Whittington, D. A.; Ng, G. Y. *J. Med. Chem.* **2008**, *51*, 3065.

539 Liang, C.-K.; Wang, P.-S.; Leung, M.-k. *Tetrahedron* **2009**, *65*, 1679.

540 Dyck, B.; Markison, S.; Zhao, L.; Tamiya, J.; Grey, J.; Rowbottom, M. W.; Zhang, M.; Vickers, T.; Sorensen, K.; Norton, C.; Wen, J.; Heise, C. E.; Saunders, J.; Conlon, P.; Madan, A.; Schwarz, D.; Goodfellow, V. S. *J. Med. Chem.* **2006**, *49*, 3753.

541 Kim, K.-Y.; Shing, J.-T.; Lee, K.-S.; Cho, C.-G. *Tetrahedron Lett.* **2004**, *45*, 117.

542 Suzuki, H.; Yamamoto, A. *J. Chem. Res.* **1992**, 280.

543 Jones, K. L.; Porzelle, A.; Hall, A.; Woodrow, M. D.; Tomkinson, N. C. O. *Org. Lett.* **2008**, *10*, 797.

544 Hafner, T.; Kunz, D. *Synthesis* **2007**, 1403.

545 Lee, C.-C.; Wang, P.-S.; Viswanath, M. B.; Leung, M.-k. *Synthesis* **2008**, 1359.

546 Ward, R. E.; Meyer, T. Y. *Macromolecules* **2003**, *36*, 4368.

547 He, H.; Wu, Y.-J. *Tetrahedron Lett.* **2003**, *44*, 3385.

548 Duffy, J. L.; Kirk, B. A.; Wang, L.; Eiermann, G. J.; He, H.; Leiting, B.; Lyons, K. A.; Patel, R. A.; Patel, S. B.; Petrov, A.; Scapin, G.; Wu, J. K.; Thornberry, N. A.; Weber, A. E. *Bioorg. Med. Chem. Lett.* **2007**, *17*, 2879.

549 Hall, A. J.; Marchant, J.; Oliveira-Campos, A. M.-F.; Queiroz, M.-J. R. P.; Shannon, P. V. R. *J. Chem. Soc., Perkin Trans. 1* **1992**, 3439.

550 Mukherjee, S.; Robinson, C. A.; How, A. G.; Mazor, T.; Wood, P. A.; Urgaonkar, S.; Hebert, A. M.; RayChaudhuri, D.; Shaw, J. T. *Bioorg. Med. Chem. Lett.* **2007**, *17*, 6651.

551 Shafir, A.; Power, M. P.; Whitener, G. D.; Arnold, J. *Organometallics* **2000**, *19*, 3978.

552 Schilling, C. I.; Bräse, S. *Org. Biomol. Chem.* **2007**, *5*, 3586.
553 Sheng, L. X.; Da, Y. X.; Long, Y.; Hong, L. Z.; Cho, T. P. *Bioorg. Med. Chem. Lett.* **2008**, *18*, 4602.
554 Barbero, N.; Carril, M.; SanMartin, R.; Dominguez, E. *Tetrahedron* **2008**, *64*, 7283.
555 Kylmälä, T.; Udd, S.; Tois, J.; Franzén, R. *Tetrahedron Lett.* **2010**, *51*, 3613.
556 Okano, K.; Tokuyama, H.; Fukuyama, T. *Chemistry—Asian J.* **2008**, *3*, 296.
557 Kametani, T.; Ohsawa, T.; Ihara, M. *Heterocycles* **1980**, *14*, 277.
558 van den Hoogenband, A.; Lange, J. H. M.; den Hartog, J. A. J.; Henzen, R.; Terpstra, J. W. *Tetrahedron Lett.* **2007**, *48*, 4461.
559 Yamada, K.; Kurokawa, T.; Tokuyama, H.; Fukuyama, T. *J. Am. Chem. Soc.* **2003**, *125*, 6630.
560 Melkonyan, F.; Topolyan, A.; Yurovskaya, M.; Karchava, A. *Eur. J. Org. Chem.* **2008**, 5952.
561 Melkonyan, F. S.; Karchava, A. V.; Yurovskaya, M. A. *J. Org. Chem.* **2008**, *73*, 4275.
562 Li, Z.; Sun, H.; Jiang, H.; Liu, H. *Org. Lett.* **2008**, *10*, 3263.
563 Asakawa, K.; Noguchi, N.; Takashima, S.; Nakada, M. *Tetrahedron: Asymm.* **2008**, *19*, 2304.
564 Mukhopadhyay, C.; Tapaswi, P. K.; Butcher, R. J. *Org. Biomol. Chem.* **2010**, *8*, 4720.
565 Joyeau, R.; Yadav, L. D. S.; Wakselman, M. *J. Chem. Soc., Perkin Trans. 1* **1987**, 1899.
566 Coste, A.; Toumi, M.; Wright, K.; Razafimahaléo, V.; Couty, F.; Marrot, J.; Evano, G. *Org. Lett.* **2008**, *10*, 3841.
567 Kametani, T.; Ohsawa, T.; Ihara, M.; Fukumoto, K. *J. Chem. Soc., Perkin Trans. 1* **1978**, 460.
568 Zhu, Y.-M.; Qin, L.-N.; Liu, R.; Ji, S.-J.; Katayama, H. *Tetrahedron Lett.* **2007**, *48*, 6262.
569 Okano, K.; Tokuyama, H.; Fukuyama, T. *J. Am. Chem. Soc.* **2006**, *128*, 7136.
570 Liu, R.; Zhu, Y.; Qin, L.; Ji, S. *Synth. Commun.* **2008**, *38*, 249.
571 Hirano, K.; Biju, A. T.; Glorius, F. *J. Org. Chem.* **2009**, *74*, 9570.
572 Saha, P.; Ramana, T.; Purkait, N.; Ali, M. A.; Paul, R.; Punniyamurthy, T. *J. Org. Chem.* **2009**, *74*, 8719.
573 Kumar, S.; Ila, H.; Junjappa, H. *J. Org. Chem.* **2009**, *74*, 7046.
574 Pan, Y.; Lu, H.; Fang, Y.; Fang, X.; Chen, L.; Qian, J.; Wang, J.; Li, C. *Synthesis* **2007**, 1242.
575 Carril, M.; SanMartin, R.; Domínguez, E.; Tellitu, I. *Green Chem.* **2007**, *9*, 219.
576 Lim, H. J.; Gallucci, J. C.; RajanBabu, T. F. *Org. Lett.* **2010**, *12*, 2162.
577 Nag, S.; Nayak, M.; Batra, S. *Adv. Synth. Catal.* **2009**, *351*, 2715.
578 Liu, R.; Zhu, Y.-m.; Qin, L.-n.; Ji, S.-j.; Katayama, H. *Heterocycles* **2007**, *71*, 1755.
579 Molina, P.; Fresneda, P. M.; Delgado, S. *Synthesis* **1999**, 326.
580 Hasegawa, K.; Kimura, N.; Arai, S.; Nishida, A. *J. Org. Chem.* **2008**, *73*, 6363.
581 Evindar, G.; Batey, R. A. *Org. Lett.* **2003**, *5*, 133.
582 Ignatenko, V. A.; Deligonul, N.; Viswanathan, R. *Org. Lett.* **2010**, *12*, 3954.
583 Wang, S.; Sun, J.; Yu, G.; Hu, X.; Liu, J. O.; Hu, Y. *Org. Biomol. Chem.* **2004**, *2*, 1573.
584 Feng, G.; Wu, J.; Dai, W.-M. *Tetrahedron Lett.* **2007**, *48*, 401.
585 Xing, X.; Wu, J.; Feng, G.; Dai, W.-M. *Tetrahedron* **2006**, *62*, 6774.
586 Wakim, S.; Bouchard, J.; Blouin, N.; Michaud, A.; Leclerc, M. *Org. Lett.* **2004**, *6*, 3413.
587 Wang, P.; Zheng, G.-j.; Wang, Y.-p.; Wang, X.-j.; Li, Y.; Xiang, W.-S. *Tetrahedron* **2010**, *66*, 5402.
588 Wakim, S.; Leclerc, M. *Synlett* **2005**, 1223.
589 Niebel, C.; Lokshin, V.; Ben-Asuly, A.; Marine, W.; Karapetyan, A.; Khodorkovsky, V. *New J. Chem.* **2010**, *34*, 1243.
590 Kim, J.; Lee, S. Y.; Lee, J.; Do, Y.; Chang, S. *J. Org. Chem.* **2008**, *73*, 9454.
591 Shen, H. C.; Ding, F.-X.; Deng, Q.; Wilsie, L. C.; Krsmanovic, M. L.; Taggart, A. K.; Carballo-Jane, E.; Ren, N.; Cai, T.-Q.; Wu, T.-J.; Wu, K. K.; Cheng, K.; Chen, Q.; Wolff, M. S.; Tong, X.; Holt, T. G.; Waters, M. G.; Hammond, M. L.; Tata, J. R.; Colletti, S. L. *J. Med. Chem.* **2009**, *52*, 2587.
592 Natarajan, S. R.; Wisnoski, D. D.; Singh, S. B.; Stelmach, J. E.; O'Neill, E. A.; Schwartz, C. D.; Thompson, C. M.; Fitzgerald, C. E.; O'Keefe, S. J.; Kumar, S.; Hop, C. E. C. A.; Zaller, D. M.; Schmatz, D. M.; Doherty, J. B. *Bioorg. Med. Chem. Lett.* **2003**, *13*, 273.
593 Chen, M.-H.; Fitzgerald, P.; Singh, S. B.; O'Neill, E. A.; Schwartz, C. D.; Thompson, C. M.; O'Keefe, S. J.; Zaller, D. M.; Doherty, J. B. *Bioorg. Med. Chem. Lett.* **2008**, *18*, 2222.
594 Fürstner, A.; Mamane, V. *Chem. Commun.* **2003**, 2112.
595 Meng, T.; Zhang, Y.; Li, M.; Wang, X.; Shen, J. *J. Comb. Chem.* **2010**, *12*, 222.
596 Madec, D.; Mingoia, F.; Prestat, G.; Poli, G. *Synlett* **2008**, 1475.
597 Yang, T.; Lin, C.; Fu, H.; Jiang, Y.; Zhao, Y. *Org. Lett.* **2005**, *7*, 4781.
598 Kenwright, J. L.; Galloway, W. R. J. D.; Blackwell, D. T.; Isidro-Llobet, A.; Hodgkinson, J.; Workmann, L.; Bowden, S. D.; Welch, M.; Spring, D. R. *Chem.—Eur. J.* **2011**, *17*, 2981.
599 Guo, L.; Li, B.; Huang, W.; Pei, G.; Ma, D. *Synlett* **2008**, 1833.
600 Cuny, G.; Bois-Choussy, M.; Zhu, J. *J. Am. Chem. Soc.* **2004**, *126*, 14475.

[601] Jørgensen, T. K.; Andersen, K. E.; Lau, J.; Madsen, P.; Huusefel, P. O. *J. Heterocycl. Chem.* **1999**, *36*, 57.

[602] Kshirsagar, U.; Argade, N. *Org. Lett.* **2010**, *12*, 3716.

[603] Zhao, Q.; Li, C. *Org. Lett.* **2008**, *10*, 4037.

[604] Pattarawarapan, M.; Zaccaro, M. C.; Saragovi, U. H.; Burgess, K. *J. Med. Chem.* **2002**, *45*, 4387.

[605] Wrona, I. E.; Gozman, A.; Taldone, T.; Chiosis, G.; Panek, J. S. *J. Org. Chem.* **2010**, *75*, 2820.

[606] Feldman, A. K.; Colasson, B.; Fokin, V. V. *Org. Lett.* **2004**, *6*, 3897.

[607] Minatti, A.; Buchwald, S. L. *Org. Lett.* **2008**, *10*, 2721.

[608] Martín, R.; Rodriguez Rivero, M.; Buchwald, S. L. *Angew. Chem., Int. Ed.* **2006**, *45*, 7079.

[609] Ackermann, L.; Barfüßer, S.; Potukuchi, H. K. *Adv. Synth. Catal.* **2009**, *351*, 1064.

[610] Li, E.; Xu, X.; Li, H.; Zhang, H.; Xu, X.; Yuan, X.; Li, Y. *Tetrahedron* **2009**, *65*, 8961.

[611] Viña, D.; del Olmo, E.; López-Pérez, J. L.; San Feliciano, A. *Org. Lett.* **2007**, *9*, 525.

[612] Zhu, R.; Xing, L.; Liu, Y.; Deng, F.; Wang, X.; Hu, Y. *J. Organomet. Chem.* **2008**, *693*, 3897.

[613] Yang, D.; Fu, H.; Hu, L.; Jiang, Y.; Zhao, Y. *J. Org. Chem.* **2008**, *73*, 7841.

[614] Lygin, A. V.; de Meijere, A. *Eur. J. Org. Chem.* **2009**, 5138.

[615] Zou, B.; Yuan, Q.; Ma, D. *Org. Lett.* **2007**, *9*, 4291.

[616] Zhu, J.; Xie, H.; Chen, Z.; Li, S.; Wu, Y. *Chem. Commun.* **2009**, 2338.

[617] Chen, M.-W.; Zhang, X.-G.; Zhong, P.; Hu, M.-L. *Synthesis* **2009**, 1431.

[618] Altenhoff, G.; Glorius, F. *Adv. Synth. Catal.* **2004**, *346*, 1661.

[619] Viirre, R. D.; Evindar, G.; Batey, R. A. *J. Org. Chem.* **2008**, *73*, 3452.

[620] Ibrahim, N.; Legraverend, M. *J. Org. Chem.* **2009**, *74*, 463.

[621] Loones, K. T. J.; Maes, B. U. W.; Meyers, C.; Deruytter, J. *J. Org. Chem.* **2006**, *71*, 260.

[622] Evenson, S. J.; Rasmussen, S. C. *Org. Lett.* **2010**, *12*, 4054.

[623] Zheng, N.; Buchwald, S. L. *Org. Lett.* **2007**, *9*, 4749.

[624] Pabba, C.; Wang, H.-J.; Mulligan, S. R.; Chen, Z.-J.; Stark, T. M.; Gregg, B. T. *Tetrahedron Lett.* **2005**, *46*, 7553.

[625] Persson, A. K. Å.; Johnston, E. V.; Bäckvall, J.-E. *Org. Lett.* **2009**, *11*, 3814.

[626] Ackermann, L. *Org. Lett.* **2005**, *7*, 439.

[627] Ohta, Y.; Chiba, H.; Oishi, S.; Fujii, N.; Ohno, H. *Org. Lett.* **2008**, *10*, 3535.

[628] Rauws, T. R. M.; Biancalani, C.; De Schutter, J. W.; Maes, B. U. W. *Tetrahedron* **2010**, *66*, 6958.

[629] Jiang, B.; Tian, H.; Huang, Z.-G.; Xu, M. *Org. Lett.* **2008**, *10*, 2737.

[630] Song, R.-J.; Liu, Y.; Li, R.-J.; Li, J.-H. *Tetrahedron Lett.* **2009**, *50*, 3912.

[631] Murru, S.; Patel, B. K.; Le Bras, J.; Muzart, J. *J. Org. Chem.* **2009**, *74*, 2217.

[632] Lu, J.; Gong, X.; Yang, H.; Fu, H. *Chem. Commun.* **2010**, *46*, 4172.

[633] Selby, T. D.; Blackstock, S. C. *Org. Lett.* **1999**, *1*, 2053.

[634] Chen, Z.; Zhu, J.; Xie, H.; Li, S.; Wu, Y.; Gong, Y. *Adv. Synth. Catal.* **2010**, *352*, 1296.

[635] Zhou, J.; Fu, L.; Lv, M.; Liu, J.; Pei, D.; Ke, D. *Synthesis* **2008**, 3974.

[636] Yang, X.; Liu, H.; Fu, H.; Qiao, R.; Jiang, Y.; Zhao, Y. *Synlett* **2010**, 101.

[637] Truong, V. L.; Morrow, M. *Tetrahedron Lett.* **2010**, *51*, 758.

[638] Yang, D.; Liu, H.; Yang, H.; Fu, H.; Hu, L.; Jiang, Y.; Zhao, Y. *Adv. Synth. Catal.* **2009**, *351*, 1999.

[639] Chen, D.; Bao, W. *Adv. Synth. Catal.* **2010**, *352*, 955.

[640] Reeves, J. T.; Fandrick, D. R.; Tan, Z.; Song, J. J.; Lee, H.; Yee, N. K.; Senanayake, C. H. *J. Org. Chem.* **2010**, *75*, 992.

[641] Cai, Q.; Li, Z.; Wei, J.; Fu, L.; Ha, C.; Pei, D.; Ding, K. *Org. Lett.* **2010**, *12*, 1500.

[642] Yuan, Q.; Ma, D. *J. Org. Chem.* **2008**, *73*, 5159.

[643] Pellón, R. F.; Carasco, R.; Rodés, L. *Synth. Commun.* **1996**, *26*, 3869.

[644] Ma, D.; Geng, Q.; Zhang, H.; Jiang, Y. *Angew. Chem., Int. Ed.* **2010**, *49*, 1291.

[645] Zhou, B. W.; Gao, J.-R.; Jiang, D.; Jia, J.-H.; Yang, Z.-P.; Jin, H.-W. *Synthesis* **2010**, 2794.

[646] Lu, X.; Shi, L.; Zhang, H.; Jiang, Y.; Ma, D. *Tetrahedron* **2010**, *66*, 5714.

[647] Focken, T.; Charette, A. B. *Org. Lett.* **2006**, *8*, 2985.

[648] Ouali, A.; Spindler, J.-F.; Jutand, A.; Taillefer, M. *Adv. Synth. Catal.* **2007**, *349*, 1906.

[649] Rainka, M. P.; Aye, Y.; Buchwald, S. L. *Proc. Natl. Acad. Sci. USA* **2004**, *101*, 5821.

[650] Bao, W.; Liu, Y.; Lv, X. *Synthesis* **2008**, 1911.

[651] Trost, B. M.; Stiles, D. T. *Org. Lett.* **2005**, *7*, 2117.

[652] Huang, X.; Shao, N.; Palani, A.; Aslanian, R. *Tetrahedron Lett.* **2007**, *48*, 1967.

[653] Ogawa, T.; Kiji, T.; Hayami, K.; Suzuki, H. *Chem. Lett.* **1991**, 1443.

[654] Huang, X.; Shao, N.; Huryk, R.; Palani, A.; Aslanian, R.; Seidel-Dugan, C. *Org. Lett.* **2009**, *11*, 867.

[655] Nicolaou, K. C.; Sun, Y.-P.; Guduru, R.; Banerji, B.; Chen, D. Y.-K. *J. Am. Chem. Soc.* **2008**, *130*, 3633.

656 Smith, A. B., III; Duffey, M. O.; Basu, K.; Walsh, S. P.; Suennemann, H. W.; Frohn, M. *J. Am. Chem. Soc.* **2008**, *130*, 422.
657 Shen, R.; Lin, C. T.; Porco, J. A., Jr. *J. Am. Chem. Soc.* **2002**, *124*, 5650.
658 Wang, X.; Porco, J. A., Jr. *J. Am. Chem. Soc.* **2003**, *125*, 6040.
659 Wang, X.; Bowman, E. J.; Bowman, B. J.; Porco, J. A., Jr. *Angew. Chem., Int. Ed.* **2004**, *43*, 3601.
660 Shen, R.; Inoue, T.; Forgac, M.; Porco, J. A., Jr. *J. Org. Chem.* **2005**, *70*, 3686.
661 Wagger, J.; Svete, J.; Stanovnik, B. *Synthesis* **2008**, 1436.
662 He, G.; Wang, J.; Ma, D. *Org. Lett.* **2007**, *9*, 1367.
663 Vintonyak, V. V.; Calà, M.; Lay, F.; Kunze, B.; Sasse, F.; Maier, M. E. *Chem.—Eur. J.* **2008**, *14*, 3709.
664 Martín, R.; Cuenca, A.; Buchwald, S. L. *Org. Lett.* **2007**, *9*, 5521.
665 Dias, L. C.; de Oliveira, L. G.; Vilcachagua, J. D.; Nigsch, F. *J. Org. Chem.* **2005**, *70*, 2225.
666 Nakamura, R.; Tanino, K.; Miyashita, M. *Org. Lett.* **2003**, *5*, 3583.
667 Movassaghi, M.; Hunt, D. K.; Tjandra, M. *J. Am. Chem. Soc.* **2006**, *128*, 8126.
668 Coste, A.; Couty, F.; Evano, G. *Org. Lett.* **2009**, *11*, 4454.
669 Sun, C.; Camp, J. E.; Weinreb, S. M. *Org. Lett.* **2006**, *8*, 1779.
670 Coleman, R. S.; Liu, P.-H. *Org. Lett.* **2004**, *6*, 577.
671 Rivero, M. R.; Buchwald, S. L. *Org. Lett.* **2007**, *9*, 973.
672 Dehli, J. R.; Bolm, C. *Adv. Synth. Catal.* **2005**, *347*, 239.
673 Hu, T.; Li, C. *Org. Lett.* **2005**, *7*, 2035.
674 Nodwell, M.; Pereira, A.; Riffell, J. L.; Zimmermann, C.; Patrick, B. O.; Roberge, M.; Andersen, R. J. *J. Org. Chem.* **2009**, *74*, 995.
675 Nodwell, M.; Riffell, J. L.; Roberge, M.; Andersen, R. *J. Org. Lett.* **2008**, *10*, 1051.
676 Toumi, M.; Couty, F.; Evano, G. *Angew. Chem., Int. Ed.* **2007**, *46*, 572.
677 Toumi, M.; Couty, F.; Evano, G. *Synlett* **2008**, 29.
678 Gong, X.; Yang, H.; Liu, H.; Jiang, Y.; Zhao, Y.; Fu, H. *Org. Lett.* **2010**, *12*, 3128.
679 Yuan, X.; Xu, X.; Zhou, X.; Yuan, J.; Mai, L.; Li, Y. *J. Org. Chem.* **2007**, *72*, 1510.
680 Zhou, X.; Zhang, H.; Yuan, J.; Mai, L.; Li, Y. *Tetrahedron Lett.* **2007**, *48*, 7236.
681 Yuen, J.; Fang, Y.-Q.; Lautens, M. *Org. Lett.* **2006**, *8*, 653.

Supplemental References for Table 1A

682 Chen, J.; Yuan, T.; Hao, W.; Cai, M. *Tetrahedron Lett.* **2011**, *52*, 3710.
683 Huang, M.; Wang, L.; Zhu, X.; Mao, Z.; Kuang, D.; Wan, Y. *Eur. J. Org. Chem.* **2012**, 4897.
684 Jha, A. K.; Jain, N. *Tetrahedron Lett.* **2013**, *54*, 4738.
685 Ji, P.; Atherton, J. H.; Page, M. I. *J. Org. Chem.* **2012**, *77*, 7471.
686 Keßler, M. T.; Robke, S.; Sahler, S.; Prechtl, M. H. G. *Catal. Sci. Technol.* **2014**, *4*, 102.
687 Komati, R.; Jursic, B. S. *Tetrahedron Lett.* **2014**, *55*, 1523.
688 Li, Y.; Zhu, X.; Meng, F.; Wan, Y. *Tetrahedron* **2011**, *67*, 5450.
689 Maejima, T.; Shimoda, Y.; Nozaki, K.; Mori, S.; Sawama, Y.; Monguchi, Y.; Sajiki, H. *Tetrahedron* **2012**, *68*, 1712.
690 Quan, Z.; Xia, H.; Zhang, Z.; Da, Y.; Wang, X. *Chin. J. Chem.* **2013**, *31*, 501.
691 Srivastava, A.; Jain, N. *Tetrahedron* **2013**, *69*, 5092.
692 Thakur, K. G.; Ganapath, D.; Sekar, G. *Chem. Commun.* **2011**, *47*, 5076.
693 Wang, Y.; Luo, J.; Hou, T.; Liu, Z. *Aust. J. Chem.* **2013**, *66*, 586.
694 Yang, B.; Liao, L.; Zeng, Y.; Zhu, X.; Wan, Y. *Catal. Commun.* **2014**, *45*, 100.
695 Zeng, X.; Huang, W.; Qiu, Y.; Jiang, S. *Org. Biomol. Chem.* **2011**, *9*, 8224.
696 Zhu, Y.; Wei, Y. *Can. J. Chem.* **2011**, *89*, 645.
697 Albadi, J.; Mansournezhad, A. *Chin. J. Chem.* **2014**, *32*, 396.
698 Albadi, J.; Shiran, J. A.; Mansournezhad, A. *Acta Chim. Slov.* **2014**, *61*, 900.
699 Berzina, B.; Sokolovs, I.; Suna, E. *ACS Catalysis* **2015**, *5*, 7008.
700 Fan, M.; Zhou, W.; Jiang, Y.; Ma, D. *Org. Lett.* **2015**, *17*, 5934.
701 Wang, Y.; Ling, J.; Zhang, Y.; Zhang, A.; Yao, Q. *Eur. J. Org. Chem.* **2015**, *2015*, 4153.

Supplemental References for Table 1B

682 Chen, J.; Yuan, T.; Hao, W.; Cai, M. *Tetrahedron Lett.* **2011**, *52*, 3710.
688 Li, Y.; Zhu, X.; Meng, F.; Wan, Y. *Tetrahedron* **2011**, *67*, 5450.

689 Maejima, T.; Shimoda, Y.; Nozaki, K.; Mori, S.; Sawama, Y.; Monguchi, Y.; Sajiki, H. *Tetrahedron* **2012**, *68*, 1712.
690 Quan, Z.; Xia, H.; Zhang, Z.; Da, Y.; Wang, X. *Chin. J. Chem.* **2013**, *31*, 501.
691 Srivastava, A.; Jain, N. *Tetrahedron* **2013**, *69*, 5092.
693 Wang, Y.; Luo, J.; Hou, T.; Liu, Z. *Aust. J. Chem.* **2013**, *66*, 586.
694 Yang, B.; Liao, L.; Zeng, Y.; Zhu, X.; Wan, Y. *Catal. Commun.* **2014**, *45*, 100.
695 Zeng, X.; Huang, W.; Qiu, Y.; Jiang, S. *Org. Biomol. Chem.* **2011**, *9*, 8224.
700 Fan, M.; Zhou, W.; Jiang, Y.; Ma, D. *Org. Lett.* **2015**, *17*, 5934.
702 Aillerie, A.; Pellegrini, S.; Bousquet, T.; Pélinksi, L. *New J. Chem.* **2014**, *38*, 1389.
703 Mastalir, M.; Rosenberg, E. E.; Kirchner, K. *Tetrahedron* **2015**, *71*, 8104.

Supplemental References for Table 2A

686 Keßler, M. T.; Robke, S.; Sahler, S.; Prechtl, M. H. G. *Catal. Sci. Technol.* **2014**, *4*, 102.
699 Berzina, B.; Sokolovs, I.; Suna, E. *ACS Catalysis* **2015**, *5*, 7008.
701 Wang, Y.; Ling, J.; Zhang, Y.; Zhang, A.; Yao, Q. *Eur. J. Org. Chem.* **2015**, *2015*, 4153.
704 Ahmadi, S. J.; Sadjadi, S.; Hosseinpour, M.; Abdollahi, M. *Monatsh. Chem.* **2011**, *142*, 801.
705 Anokhin, M. V.; Averin, A. D.; Beletskaya, I. P. *Eur. J. Org. Chem.* **2011**, *2011*, 6240.
706 Bach, A.; Stroemgaard, K. *Synthesis* **2011**, 807.
707 Bahlaouan, Z.; Thibonnet, J.; Duchene, A.; Parrain, J.-L.; Elhilali, M.; Abarbri, M. *Synlett* **2011**, 2509.
708 Bardajee, G. R. *Tetrahedron Lett.* **2013**, *54*, 4937.
709 Chen, B.; Li, F.; Huang, Z.; Xue, F.; Lu, T.; Yuan, Y.; Yuan, G. *ChemCatChem* **2012**, *4*, 1741.
710 Chen, D.; Yang, K.; Xiang, H.; Jiang, S. *Tetrahedron Lett.* **2012**, *53*, 7121.
711 Costa, M. V.; Viana, G. M.; de Souza, T. M.; Malta, L. F. B.; Aguiar, L. C. S. *Tetrahedron Lett.* **2013**, *54*, 2332.
712 Huang, L.; Yu, R.; Zhu, X.; Wan, Y. *Tetrahedron* **2013**, *69*, 8974.
713 Huang, M.; Lin, X.; Zhu, X.; Peng, W.; Xie, J.; Wan, Y. *Eur. J. Org. Chem.* **2011**, 4523.
714 Jiao, J.; Zhang, X.-R.; Chang, N.-H.; Wang, J.; Wei, J.-F.; Shi, X.-Y.; Chen, Z.-G. *J. Org. Chem.* **2011**, *76*, 1180.
715 Khatri, P. K.; Jain, S. L. *Tetrahedron Lett.* **2013**, *54*, 2740.
716 Meng, F.; Wang, C.; Xie, J.; Zhu, X.; Wan, Y. *Appl. Organomet. Chem.* **2011**, *25*, 341.
717 Nasir Baig, R. B.; Varma, R. S. *RSC Adv.* **2014**, *4*, 6568.
718 Tao, C.; Liu, F.; Liu, W.; Zhu, Y.; Li, Y.; Liu, X.; Zhao, J. *Tetrahedron Lett.* **2012**, *53*, 7093.
719 Xie, R.; Fu, H.; Ling, Y. *Chem. Commun.* **2011**, *47*, 8976.
720 Yang, H.; Xi, C.; Miao, Z.-W.; Chen, R.-Y. *Eur. J. Org. Chem.* **2011**, *2011*, 3353.
721 Yang, K.; Qiu, Y.; Li, Z.; Wang, Z.; Jiang, S. *J. Org. Chem.* **2011**, *76*, 3151.
722 Yang, Q.; Wang, Y.; Lin, D.; Zhang, M. *Tetrahedron Lett.* **2013**, *54*, 1994.
723 Yin, H.; Jin, M.; Chen, W.; Chen, C.; Zheng, L.; Wei, P.; Han, S. *Tetrahedron Lett.* **2012**, *53*, 1265.
724 Anokhin, M. V.; Averin, A. D.; Panchenko, S. P.; Maloshitskaya, O. A.; Beletskaya, I. P. *Mendeleev Commun.* **2015**, *25*, 245.
725 Arai, S.; Nakajima, M.; Nishida, A. *Angew. Chem., Int. Ed.* **2014**, *53*, 5569.
726 Baig, R. B. N.; Vaddula, B. R.; Nadagouda, M. N.; Varma, R. S. *Green Chem.* **2015**, *17*, 1243.
727 Bodhak, C.; Kundu, A.; Pramanik, A. *Tetrahedron Lett.* **2015**, *56*, 419.
728 Kurandina, D. V.; Eliseenkov, E. V.; Khaibulova, T. S.; Petrov, A. A.; Boyarskiy, V. P. *Tetrahedron* **2015**, *71*, 7931.
729 Martina, K.; Rinaldi, L.; Baricco, F.; Boffa, L.; Cravotto, G. *Synlett* **2015**, *26*, 2789.
730 Miao, D.; Shi, X.; He, G.; Tong, Y.; Jiang, Z.; Han, S. *Tetrahedron* **2015**, *71*, 431.
731 Modi, A.; Ali, W.; Mohanta, P. R.; Khatun, N.; Patel, B. K. *ACS Sustainable Chem. Eng.* **2015**, *3*, 2582.
732 Panchenko, S. P.; Averin, A. D.; Anokhin, M. V.; Maloshitskaya, O. A.; Beletskaya, I. P. *Beilstein J. Org. Chem.* **2015**, *11*, 2297.
733 Reddy, P. L.; Arundhathi, R.; Rawat, D. S. *RSC Adv.* **2015**, *5*, 92121.
734 Shah, D. R.; Lakum, H. P.; Chikhalia, K. H. *Heterocycl. Commun.* **2014**, *20*, 305.
735 Su, J.; Qiu, Y.; Jiang, S.; Zhang, D. *Chin. J. Chem.* **2014**, *32*, 685.
736 Sung, S.; Braddock, D. C.; Armstrong, A.; Brennan, C.; Sale, D.; White, A. J. P.; Davies, R. P. *Chem.—Eur. J.* **2015**, *21*, 7179.
737 Wang, D.; Kuang, D.; Zhang, F.; Yang, C.; Zhu, X. *Adv. Synth. Catal.* **2015**, *357*, 714.

738 Yan, N.-N.; Wu, F.-T.; Zhang, J.; Wei, Q.-B.; Liu, P.; Xie, J.-W.; Dai, B. *Asian J. Org. Chem.* **2014**, *3*, 1159.
739 Zhou, W.; Fan, M.; Yin, J.; Jiang, Y.; Ma, D. *J. Am. Chem. Soc.* **2015**, *137*, 11942.

Supplemental References for Table 2B

703 Mastalir, M.; Rosenberg, E. E.; Kirchner, K. *Tetrahedron* **2015**, *71*, 8104.
710 Chen, D.; Yang, K.; Xiang, H.; Jiang, S. *Tetrahedron Lett.* **2012**, *53*, 7121.
714 Jiao, J.; Zhang, X.-R.; Chang, N.-H.; Wang, J.; Wei, J.-F.; Shi, X.-Y.; Chen, Z.-G. *J. Org. Chem.* **2011**, *76*, 1180.
723 Yin, H.; Jin, M.; Chen, W.; Chen, C.; Zheng, L.; Wei, P.; Han, S. *Tetrahedron Lett.* **2012**, *53*, 1265.
724 Anokhin, M. V.; Averin, A. D.; Panchenko, S. P.; Maloshitskaya, O. A.; Beletskaya, I. P. *Mendeleev Commun.* **2015**, *25*, 245.
737 Wang, D.; Kuang, D.; Zhang, F.; Yang, C.; Zhu, X. *Adv. Synth. Catal.* **2015**, *357*, 714.
739 Zhou, W.; Fan, M.; Yin, J.; Jiang, Y.; Ma, D. *J. Am. Chem. Soc.* **2015**, *137*, 11942.
740 Fantasia, S.; Windisch, J.; Scalone, M. *Adv. Synth. Catal.* **2013**, *355*, 627.
741 Liu, Z.-J.; Vors, J.-P.; Gesing, E. R. F.; Bolm, C. *Adv. Synth. Catal.* **2010**, *352*, 3158.
742 Liu, Z.-J.; Vors, J.-P.; Gesing, E. R. F.; Bolm, C. *Green Chem.* **2011**, *13*, 42.
743 Toulot, S.; Heinrich, T.; Leroux, F. R. *Adv. Synth. Catal.* **2013**, *355*, 3263.
744 Verma, S. K.; Acharya, B. N.; Kaushik, M. P. *Org. Biomol. Chem.* **2011**, *9*, 1324.
745 Abel, A. S.; Averin, A. D.; Anokhin, M. V.; Maloshitskaya, O. A.; Butov, G. M.; Savelyev, E. N.; Orlinson, B. S.; Novakov, I. A.; Beletskaya, I. P. *Russ. J. Org. Chem.* **2015**, *51*, 301.
746 Anokhin, M. V.; Averin, A. D.; Panchenko, S. P.; Maloshitskaya, O. A.; Beletskaya, I. P. *Russ. J. Org. Chem.* **2014**, *50*, 923.
747 Anokhin, M. V.; Averin, A. D.; Panchenko, S. P.; Maloshitskaya, O. A.; Buryak, A. K.; Beletskaya, I. P. *Helv. Chim. Act.* **2015**, *98*, 47.
748 Knight, J. G.; Alnoman, R. B.; Waddell, P. G. *Org. Biomol. Chem.* **2015**, *13*, 3819.
749 Noji, T.; Okano, K.; Tokuyama, H. *Tetrahedron* **2015**, *71*, 3833.
750 Orrego-Hernández, J.; Cobo, J.; Portilla, J. *Eur. J. Org. Chem.* **2015**, *2015*, 5064.
751 Wang, L.; Liu, N.; Dai, B.; Hu, H. *Eur. J. Org. Chem.* **2014**, *2014*, 6493.

Supplemental References for Table 3A

686 Keßler, M. T.; Robke, S.; Sahler, S.; Prechtl, M. H. G. *Catal. Sci. Technol.* **2014**, *4*, 102.
687 Komati, R.; Jursic, B. S. *Tetrahedron Lett.* **2014**, *55*, 1523.
699 Berzina, B.; Sokolovs, I.; Suna, E. *ACS Catalysis* **2015**, *5*, 7008.
707 Bahlaouan, Z.; Thibonnet, J.; Duchene, A.; Parrain, J.-L.; Elhilali, M.; Abarbri, M. *Synlett* **2011**, 2509.
708 Bardajee, G. R. *Tetrahedron Lett.* **2013**, *54*, 4937.
714 Jiao, J.; Zhang, X.-R.; Chang, N.-H.; Wang, J.; Wei, J.-F.; Shi, X.-Y.; Chen, Z.-G. *J. Org. Chem.* **2011**, *76*, 1180.
717 Nasir Baig, R. B.; Varma, R. S. *RSC Adv.* **2014**, *4*, 6568.
723 Yin, H.; Jin, M.; Chen, W.; Chen, C.; Zheng, L.; Wei, P.; Han, S. *Tetrahedron Lett.* **2012**, *53*, 1265.
728 Kurandina, D. V.; Eliseenkov, E. V.; Khaibulova, T. S.; Petrov, A. A.; Boyarskiy, V. P. *Tetrahedron* **2015**, *71*, 7931.
729 Martina, K.; Rinaldi, L.; Baricco, F.; Boffa, L.; Cravotto, G. *Synlett* **2015**, *26*, 2789.
736 Sung, S.; Braddock, D. C.; Armstrong, A.; Brennan, C.; Sale, D.; White, A. J. P.; Davies, R. P. *Chem.—Eur. J.* **2015**, *21*, 7179.
737 Wang, D.; Kuang, D.; Zhang, F.; Yang, C.; Zhu, X. *Adv. Synth. Catal.* **2015**, *357*, 714.
752 Heo, Y.; Hyun, D.; Kumar, M. R.; Jung, H. M.; Lee, S. *Tetrahedron Lett.* **2012**, *53*, 6657.
753 Zhang, Y.; Yang, X.; Yao, Q.; Ma, D. *Org. Lett.* **2012**, *14*, 3056.
754 Bhosale, M. A.; Bhanage, B. M. *RSC Adv.* **2014**, *4*, 15122.

Supplemental References for Table 3B

703 Mastalir, M.; Rosenberg, E. E.; Kirchner, K. *Tetrahedron* **2015**, *71*, 8104.
737 Wang, D.; Kuang, D.; Zhang, F.; Yang, C.; Zhu, X. *Adv. Synth. Catal.* **2015**, *357*, 714.

[741] Liu, Z.-J.; Vors, J.-P.; Gesing, E. R. F.; Bolm, C. *Adv. Synth. Catal.* **2010**, *352*, 3158.
[744] Verma, S. K.; Acharya, B. N.; Kaushik, M. P. *Org. Biomol. Chem.* **2011**, *9*, 1324.
[747] Anokhin, M. V.; Averin, A. D.; Panchenko, S. P.; Maloshitskaya, O. A.; Buryak, A. K.; Beletskaya, I. P. *Helv. Chim. Act.* **2015**, *98*, 47.
[753] Zhang, Y.; Yang, X.; Yao, Q.; Ma, D. *Org. Lett.* **2012**, *14*, 3056.

Supplemental References for Table 4A

[686] Keßler, M. T.; Robke, S.; Sahler, S.; Prechtl, M. H. G. *Catal. Sci. Technol.* **2014**, *4*, 102.
[699] Berzina, B.; Sokolovs, I.; Suna, E. *ACS Catalysis* **2015**, *5*, 7008.
[701] Wang, Y.; Ling, J.; Zhang, Y.; Zhang, A.; Yao, Q. *Eur. J. Org. Chem.* **2015**, *2015*, 4153.
[704] Ahmadi, S. J.; Sadjadi, S.; Hosseinpour, M.; Abdollahi, M. *Monatsh. Chem.* **2011**, *142*, 801.
[708] Bardajee, G. R. *Tetrahedron Lett.* **2013**, *54*, 4937.
[714] Jiao, J.; Zhang, X.-R.; Chang, N.-H.; Wang, J.; Wei, J.-F.; Shi, X.-Y.; Chen, Z.-G. *J. Org. Chem.* **2011**, *76*, 1180.
[715] Khatri, P. K.; Jain, S. L. *Tetrahedron Lett.* **2013**, *54*, 2740.
[716] Meng, F.; Wang, C.; Xie, J.; Zhu, X.; Wan, Y. *Appl. Organomet. Chem.* **2011**, *25*, 341.
[717] Nasir Baig, R. B.; Varma, R. S. *RSC Adv.* **2014**, *4*, 6568.
[720] Yang, H.; Xi, C.; Miao, Z.-W.; Chen, R.-Y. *Eur. J. Org. Chem.* **2011**, *2011*, 3353.
[721] Yang, K.; Qiu, Y.; Li, Z.; Wang, Z.; Jiang, S. *J. Org. Chem.* **2011**, *76*, 3151.
[722] Yang, Q.; Wang, Y.; Lin, D.; Zhang, M. *Tetrahedron Lett.* **2013**, *54*, 1994.
[726] Baig, R. B. N.; Vaddula, B. R.; Nadagouda, M. N.; Varma, R. S. *Green Chem.* **2015**, *17*, 1243.
[728] Kurandina, D. V.; Eliseenkov, E. V.; Khaibulova, T. S.; Petrov, A. A.; Boyarskiy, V. P. *Tetrahedron* **2015**, *71*, 7931.
[733] Reddy, P. L.; Arundhathi, R.; Rawat, D. S. *RSC Adv.* **2015**, *5*, 92121.
[734] Shah, D. R.; Lakum, H. P.; Chikhalia, K. H. *Heterocycl. Commun.* **2014**, *20*, 305.
[735] Su, J.; Qiu, Y.; Jiang, S.; Zhang, D. *Chin. J. Chem.* **2014**, *32*, 685.
[736] Sung, S.; Braddock, D. C.; Armstrong, A.; Brennan, C.; Sale, D.; White, A. J. P.; Davies, R. P. *Chem.—Eur. J.* **2015**, *21*, 7179.
[739] Zhou, W.; Fan, M.; Yin, J.; Jiang, Y.; Ma, D. *J. Am. Chem. Soc.* **2015**, *137*, 11942.
[752] Heo, Y.; Hyun, D.; Kumar, M. R.; Jung, H. M.; Lee, S. *Tetrahedron Lett.* **2012**, *53*, 6657.
[753] Zhang, Y.; Yang, X.; Yao, Q.; Ma, D. *Org. Lett.* **2012**, *14*, 3056.
[754] Bhosale, M. A.; Bhanage, B. M. *RSC Adv.* **2014**, *4*, 15122.
[755] Kundu, D.; Bhadra, S.; Mukherjee, N.; Sreedhar, B.; Ranu, B. C. *Chem.—Eur. J.* **2013**, *19*, 15759.
[756] Yang, Q.; Ulysse, L. G.; McLaws, M. D.; Keefe, D. K.; Haney, B. P.; Zha, C.; Guzzo, P. R.; Liu, S. *Org. Proc. Res. Dev.* **2012**, *16*, 499.
[757] Yong, F.-F.; Teo, Y.-C.; Tan, K.-N. *Tetrahedron Lett.* **2013**, *54*, 5332.
[758] Rawat, V.; Press, K.; Goldberg, I.; Vigalok, A. *Org. Biomol. Chem.* **2015**, *13*, 11189.
[759] Truong, T.; Nguyen, C. V.; Truong, N. T.; Phan, N. T. S. *RSC Adv.* **2015**, *5*, 107547.

Supplemental References for Table 4B

[699] Berzina, B.; Sokolovs, I.; Suna, E. *ACS Catalysis* **2015**, *5*, 7008.
[701] Wang, Y.; Ling, J.; Zhang, Y.; Zhang, A.; Yao, Q. *Eur. J. Org. Chem.* **2015**, *2015*, 4153.
[703] Mastalir, M.; Rosenberg, E. E.; Kirchner, K. *Tetrahedron* **2015**, *71*, 8104.
[741] Liu, Z.-J.; Vors, J.-P.; Gesing, E. R. F.; Bolm, C. *Adv. Synth. Catal.* **2010**, *352*, 3158.
[743] Toulot, S.; Heinrich, T.; Leroux, F. R. *Adv. Synth. Catal.* **2013**, *355*, 3263.
[744] Verma, S. K.; Acharya, B. N.; Kaushik, M. P. *Org. Biomol. Chem.* **2011**, *9*, 1324.
[753] Zhang, Y.; Yang, X.; Yao, Q.; Ma, D. *Org. Lett.* **2012**, *14*, 3056.
[760] Shah, D. R.; Lakum, H. P.; Chikhalia, K. H. *Lett. Org. Chem.* **2015**, *12*, 237.

Supplemental References for Table 5A

[699] Berzina, B.; Sokolovs, I.; Suna, E. *ACS Catalysis* **2015**, *5*, 7008.
[712] Huang, L.; Yu, R.; Zhu, X.; Wan, Y. *Tetrahedron* **2013**, *69*, 8974.
[713] Huang, M.; Lin, X.; Zhu, X.; Peng, W.; Xie, J.; Wan, Y. *Eur. J. Org. Chem.* **2011**, 4523.

714   Jiao, J.; Zhang, X.-R.; Chang, N.-H.; Wang, J.; Wei, J.-F.; Shi, X.-Y.; Chen, Z.-G. *J. Org. Chem.* **2011**, *76*, 1180.
715   Khatri, P. K.; Jain, S. L. *Tetrahedron Lett.* **2013**, *54*, 2740.
716   Meng, F.; Wang, C.; Xie, J.; Zhu, X.; Wan, Y. *Appl. Organomet. Chem.* **2011**, *25*, 341.
717   Nasir Baig, R. B.; Varma, R. S. *RSC Adv.* **2014**, *4*, 6568.
719   Xie, R.; Fu, H.; Ling, Y. *Chem. Commun.* **2011**, *47*, 8976.
722   Yang, Q.; Wang, Y.; Lin, D.; Zhang, M. *Tetrahedron Lett.* **2013**, *54*, 1994.
733   Reddy, P. L.; Arundhathi, R.; Rawat, D. S. *RSC Adv.* **2015**, *5*, 92121.
734   Shah, D. R.; Lakum, H. P.; Chikhalia, K. H. *Heterocycl. Commun.* **2014**, *20*, 305.
738   Yan, N.-N.; Wu, F.-T.; Zhang, J.; Wei, Q.-B.; Liu, P.; Xie, J.-W.; Dai, B. *Asian J. Org. Chem.* **2014**, *3*, 1159.
744   Verma, S. K.; Acharya, B. N.; Kaushik, M. P. *Org. Biomol. Chem.* **2011**, *9*, 1324.
754   Bhosale, M. A.; Bhanage, B. M. *RSC Adv.* **2014**, *4*, 15122.
761   Chavan, S. S.; Sawant, S. K.; Sawant, V. A.; Lahiri, G. K. *Inorg. Chem. Comm.* **2011**, *14*, 1373.
762   Engel-Andreasen, J.; Shimpukade, B.; Ulven, T. *Green Chem.* **2013**, *15*, 336.
763   Islam, S. M.; Mondal, S.; Mondal, P.; Roy, A. S.; Tuhina, K.; Salam, N.; Mobarak, M. *J. Organomet. Chem.* **2012**, *696*, 4264.
764   Quan, Z.-J.; Xia, H.-D.; Zhang, Z.; Da, Y.-X.; Wang, X.-C. *Appl. Organomet. Chem.* **2014**, *28*, 81.
765   Wu, C.-C.; Lee, B.-H.; Liao, P.-K.; Fang, C.-S.; Liu, C. W. *J. Chin. Chem. Soc.* **2012**, *59*, 480.
766   Xu, Z.-L.; Li, H.-X.; Ren, Z.-G.; Du, W.-Y.; Xu, W.-C.; Lang, J.-P. *Tetrahedron* **2011**, *67*, 5282.
767   Zhang, P.; Yuan, J.; Li, H.; Liu, X.; Xu, X.; Antonietti, M.; Wang, Y. *RSC Adv.* **2013**, *3*, 1890.
768   Liu, Y.; Yang, L. *Chin. J. Chem.* **2015**, *33*, 473.
769   Liu, Y. S.; Gu, N. N.; Liu, Y.; Ma, X. W.; Liu, P.; Xie, J. W. *Asian J. Chem.* **2015**, *27*, 1075.
770   Nasrollahzadeh, M.; Zahraei, A.; Pourbasheer, E. *Monatsh. Chem.* **2015**, *146*, 1329.
771   Safaei-Ghomi, J.; Akbarzadeh, Z.; Khojastehbakht-Koopaei, B. *RSC Adv.* **2015**, *5*, 28879.
772   Shang, Z.; Yang, L.; Chang, G. *Macromol. Res.* **2015**, *23*, 937.
773   Sharma, R. K.; Gaur, R.; Yadav, M.; Rathi, A. K.; Pechousek, J.; Petr, M.; Zboril, R.; Gawande, M. B. *ChemCatChem* **2015**, *7*, 3495.

Supplemental References for Table 5B

699   Berzina, B.; Sokolovs, I.; Suna, E. *ACS Catalysis* **2015**, *5*, 7008.
703   Mastalir, M.; Rosenberg, E. E.; Kirchner, K. *Tetrahedron* **2015**, *71*, 8104.
741   Liu, Z.-J.; Vors, J.-P.; Gesing, E. R. F.; Bolm, C. *Adv. Synth. Catal.* **2010**, *352*, 3158.
742   Liu, Z.-J.; Vors, J.-P.; Gesing, E. R. F.; Bolm, C. *Green Chem.* **2011**, *13*, 42.
744   Verma, S. K.; Acharya, B. N.; Kaushik, M. P. *Org. Biomol. Chem.* **2011**, *9*, 1324.
750   Orrego-Hernández, J.; Cobo, J.; Portilla, J. *Eur. J. Org. Chem.* **2015**, *2015*, 5064.
751   Wang, L.; Liu, N.; Dai, B.; Hu, H. *Eur. J. Org. Chem.* **2014**, *2014*, 6493.
752   Heo, Y.; Hyun, D.; Kumar, M. R.; Jung, H. M.; Lee, S. *Tetrahedron Lett.* **2012**, *53*, 6657.
760   Shah, D. R.; Lakum, H. P.; Chikhalia, K. H. *Lett. Org. Chem.* **2015**, *12*, 237.
764   Quan, Z.-J.; Xia, H.-D.; Zhang, Z.; Da, Y.-X.; Wang, X.-C. *Appl. Organomet. Chem.* **2014**, *28*, 81.
771   Safaei-Ghomi, J.; Akbarzadeh, Z.; Khojastehbakht-Koopaei, B. *RSC Adv.* **2015**, *5*, 28879.
774   Chang, E.-C.; Chen, C.-Y.; Wang, L.-Y.; Huang, Y.-Y.; Yeh, M.-Y.; Wong, F. F. *Tetrahedron* **2013**, *69*, 570.
775   Zhang, M.; Xiong, B.; Wang, T.; Wang, X.; Yan, F.; Ding, Y. *Heterocycles* **2012**, *85*, 1393.
776   Wang, D.; Kuang, D.; Zhang, F.; Liu, Y.; Ning, S. *Tetrahedron Lett.* **2014**, *55*, 7121.
777   Zhang, Y.; Quan, Z.-J.; Gong, H.-P.; Da, Y.-X.; Zhang, Z.; Wang, X.-C. *Tetrahedron* **2015**, *71*, 2113.

Supplemental References for Table 5C

722   Yang, Q.; Wang, Y.; Lin, D.; Zhang, M. *Tetrahedron Lett.* **2013**, *54*, 1994.
748   Knight, J. G.; Alnoman, R. B.; Waddell, P. G. *Org. Biomol. Chem.* **2015**, *13*, 3819.
764   Quan, Z.-J.; Xia, H.-D.; Zhang, Z.; Da, Y.-X.; Wang, X.-C. *Appl. Organomet. Chem.* **2014**, *28*, 81.
766   Xu, Z.-L.; Li, H.-X.; Ren, Z.-G.; Du, W.-Y.; Xu, W.-C.; Lang, J.-P. *Tetrahedron* **2011**, *67*, 5282.
775   Zhang, M.; Xiong, B.; Wang, T.; Wang, X.; Yan, F.; Ding, Y. *Heterocycles* **2012**, *85*, 1393.
776   Wang, D.; Kuang, D.; Zhang, F.; Liu, Y.; Ning, S. *Tetrahedron Lett.* **2014**, *55*, 7121.
777   Zhang, Y.; Quan, Z.-J.; Gong, H.-P.; Da, Y.-X.; Zhang, Z.; Wang, X.-C. *Tetrahedron* **2015**, *71*, 2113.

[778] Sheremetev, A. B.; Palysaeva, N. V.; Struchkova, M. I.; Suponitsky, K. Y.; Antipin, M. Y. *Eur. J. Org. Chem.* **2012**, *2012*, 2266.

Supplemental References for Table 6A

[686] Keßler, M. T.; Robke, S.; Sahler, S.; Prechtl, M. H. G. *Catal. Sci. Technol.* **2014**, *4*, 102.
[753] Zhang, Y.; Yang, X.; Yao, Q.; Ma, D. *Org. Lett.* **2012**, *14*, 3056.
[754] Bhosale, M. A.; Bhanage, B. M. *RSC Adv.* **2014**, *4*, 15122.
[768] Liu, Y.; Yang, L. *Chin. J. Chem.* **2015**, *33*, 473.
[779] Kanazawa, Y.; Yokota, T.; Ogasa, H.; Watanabe, H.; Hanakawa, T.; Soga, S.; Kawatsura, M. *Tetrahedron* **2015**, *71*, 1395.
[780] Mitrofanov, A. Y.; Bessmertnykh-Lemeune, A. G.; Beletskaya, I. P. *Inorg. Chim. Acta* **2015**, *431*, 297.

Supplemental References for Table 6B

[686] Keßler, M. T.; Robke, S.; Sahler, S.; Prechtl, M. H. G. *Catal. Sci. Technol.* **2014**, *4*, 102.
[747] Anokhin, M. V.; Averin, A. D.; Panchenko, S. P.; Maloshitskaya, O. A.; Buryak, A. K.; Beletskaya, I. P. *Helv. Chim. Act.* **2015**, *98*, 47.
[753] Zhang, Y.; Yang, X.; Yao, Q.; Ma, D. *Org. Lett.* **2012**, *14*, 3056.

Supplemental References for Table 7A

[720] Yang, H.; Xi, C.; Miao, Z.-W.; Chen, R.-Y. *Eur. J. Org. Chem.* **2011**, *2011*, 3353.
[721] Yang, K.; Qiu, Y.; Li, Z.; Wang, Z.; Jiang, S. *J. Org. Chem.* **2011**, *76*, 3151.
[754] Bhosale, M. A.; Bhanage, B. M. *RSC Adv.* **2014**, *4*, 15122.
[759] Truong, T.; Nguyen, C. V.; Truong, N. T.; Phan, N. T. S. *RSC Adv.* **2015**, *5*, 107547.
[765] Wu, C.-C.; Lee, B.-H.; Liao, P.-K.; Fang, C.-S.; Liu, C. W. *J. Chin. Chem. Soc.* **2012**, *59*, 480.
[766] Xu, Z.-L.; Li, H.-X.; Ren, Z.-G.; Du, W.-Y.; Xu, W.-C.; Lang, J.-P. *Tetrahedron* **2011**, *67*, 5282.
[773] Sharma, R. K.; Gaur, R.; Yadav, M.; Rathi, A. K.; Pechousek, J.; Petr, M.; Zboril, R.; Gawande, M. B. *ChemCatChem* **2015**, *7*, 3495.
[781] Ganesh Babu, S.; Karvembu, R. *Ind. Eng. Chem. Res.* **2011**, *50*, 9594.
[782] Hosseini-Sarvari, M.; Moeini, F. *RSC Adv.* **2014**, *4*, 7321.
[783] Pai, G.; Chattopadhyay, A. P. *Tetrahedron Lett.* **2014**, *55*, 941.
[784] Teo, Y.-C.; Yong, F.-F.; Lim, G. S. *Tetrahedron Lett.* **2011**, *52*, 7171.
[785] Wang, Y.-L.; Luo, J.; Liu, Z.-L. *J. Chin. Chem. Soc* **2013**, *60*, 1007.
[786] Yadav, D. K. T.; Rajak, S. S.; Bhanage, B. M. *Tetrahedron Lett.* **2014**, *55*, 931.
[787] Yang, H.; Miao, Z.; Chen, R. *Lett. Org. Chem.* **2011**, *8*, 325.
[788] Yang, X.; Xing, H.; Zhang, Y.; Lai, Y.; Zhang, Y.; Jiang, Y.; Ma, D. *Chin. J. Chem.* **2012**, *30*, 875.
[789] Yong, F.-F.; Teo, Y.-C.; Tay, S.-H.; Tan, B. Y.-H.; Lim, K.-H. *Tetrahedron Lett.* **2011**, *52*, 1161.
[790] Larsson, P.-F.; Wallentin, C.-J.; Norrby, P.-O. *ChemCatChem* **2014**, *6*, 1277.
[791] Toummini, D.; Tlili, A.; Berges, J.; Ouazzani, F.; Taillefer, M. *Chem.—Eur. J.* **2014**, *20*, 14619.
[792] Wang, Y.; Zhang, Y.; Yang, B.; Zhang, A.; Yao, Q. *Org. Biomol. Chem.* **2015**, *13*, 4101.
[793] Wu, F.-T.; Yan, N.-N.; Liu, P.; Xie, J.-W.; Liu, Y.; Dai, B. *Tetrahedron Lett.* **2014**, *55*, 3249.

Supplemental References for Table 7B

[742] Liu, Z.-J.; Vors, J.-P.; Gesing, E. R. F.; Bolm, C. *Green Chem.* **2011**, *13*, 42.
[751] Wang, L.; Liu, N.; Dai, B.; Hu, H. *Eur. J. Org. Chem.* **2014**, *2014*, 6493.
[783] Pai, G.; Chattopadhyay, A. P. *Tetrahedron Lett.* **2014**, *55*, 941.
[792] Wang, Y.; Zhang, Y.; Yang, B.; Zhang, A.; Yao, Q. *Org. Biomol. Chem.* **2015**, *13*, 4101.
[794] Tber, Z.; Hiebel, M.-A.; Akssira, M.; Guillaumet, G.; Berteina-Raboin, S. *Synthesis* **2015**, *47*, 1780.

Supplemental References for Table 8A

709 Chen, B.; Li, F.; Huang, Z.; Xue, F.; Lu, T.; Yuan, Y.; Yuan, G. *ChemCatChem* **2012**, *4*, 1741.
713 Huang, M.; Lin, X.; Zhu, X.; Peng, W.; Xie, J.; Wan, Y. *Eur. J. Org. Chem.* **2011**, 4523.
721 Yang, K.; Qiu, Y.; Li, Z.; Wang, Z.; Jiang, S. *J. Org. Chem.* **2011**, *76*, 3151.
722 Yang, Q.; Wang, Y.; Lin, D.; Zhang, M. *Tetrahedron Lett.* **2013**, *54*, 1994.
735 Su, J.; Qiu, Y.; Jiang, S.; Zhang, D. *Chin. J. Chem.* **2014**, *32*, 685.
738 Yan, N.-N.; Wu, F.-T.; Zhang, J.; Wei, Q.-B.; Liu, P.; Xie, J.-W.; Dai, B. *Asian J. Org. Chem.* **2014**, *3*, 1159.
758 Rawat, V.; Press, K.; Goldberg, I.; Vigalok, A. *Org. Biomol. Chem.* **2015**, *13*, 11189.
759 Truong, T.; Nguyen, C. V.; Truong, N. T.; Phan, N. T. S. *RSC Adv.* **2015**, *5*, 107547.
765 Wu, C.-C.; Lee, B.-H.; Liao, P.-K.; Fang, C.-S.; Liu, C. W. *J. Chin. Chem. Soc.* **2012**, *59*, 480.
766 Xu, Z.-L.; Li, H.-X.; Ren, Z.-G.; Du, W.-Y.; Xu, W.-C.; Lang, J.-P. *Tetrahedron* **2011**, *67*, 5282.
770 Nasrollahzadeh, M.; Zahraei, A.; Pourbasheer, E. *Monatsh. Chem.* **2015**, *146*, 1329.
784 Teo, Y.-C.; Yong, F.-F.; Lim, G. S. *Tetrahedron Lett.* **2011**, *52*, 7171.
786 Yadav, D. K. T.; Rajak, S. S.; Bhanage, B. M. *Tetrahedron Lett.* **2014**, *55*, 931.
788 Yang, X.; Xing, H.; Zhang, Y.; Lai, Y.; Zhang, Y.; Jiang, Y.; Ma, D. *Chin. J. Chem.* **2012**, *30*, 875.
790 Larsson, P.-F.; Wallentin, C.-J.; Norrby, P.-O. *ChemCatChem* **2014**, *6*, 1277.
791 Toummini, D.; Tlili, A.; Berges, J.; Ouazzani, F.; Taillefer, M. *Chem.—Eur. J.* **2014**, *20*, 14619.
792 Wang, Y.; Zhang, Y.; Yang, B.; Zhang, A.; Yao, Q. *Org. Biomol. Chem.* **2015**, *13*, 4101.
795 Abele, E.; Abele, R. *Chem. Heterocycl. Compd. (N. Y., NY, U.S.)* **2013**, *49*, 1384.
796 Boswell, M. G.; Yeung, F. G.; Wolf, C. *Synlett* **2012**, *23*, 1240.
797 Huang, L.; Jin, C.; Su, W. *Chin. J. Chem.* **2012**, *30*, 2394.
798 Huang, Z.; Li, F.; Chen, B.; Xue, F.; Chen, G.; Yuan, G. *Appl. Catal., A–Gen.* **2011**, *403*, 104.
799 Larsson, P.-F.; Astvik, P.; Norrby, P.-O. *Beilstein J. Org. Chem.* **2012**, *8*, 1909.
800 Wang, D.; Zhang, F.; Kuang, D.; Yu, J.; Li, J. *Green Chem.* **2012**, *14*, 1268.
801 Zhang, Q.; Luo, J.; Wei, Y. *Synth. Commun.* **2012**, *42*, 114.
802 Zou, Y.; Lin, H.-S.; Maggard, P. A.; Deiters, A. *Eur. J. Org. Chem.* **2011**, *2011*, 4154.
803 Beyer, A.; Castanheiro, T.; Busca, P.; Prestat, G. *ChemCatChem* **2015**, *7*, 2433.
804 Heidarizadeh, F.; Majdi-nasab, A. *Tetrahedron Lett.* **2015**, *56*, 6360.
805 Liu, Y.-S.; Liu, Y.; Ma, X.-W.; Liu, P.; Xie, J.-W.; Dai, B. *Chin. Chem. Lett.* **2014**, *25*, 775.
806 Nasrollahzadeh, M.; Sajadi, S. M.; Maham, M. *RSC Adv.* **2015**, *5*, 40628.

Supplemental References for Table 8B

741 Liu, Z.-J.; Vors, J.-P.; Gesing, E. R. F.; Bolm, C. *Adv. Synth. Catal.* **2010**, *352*, 3158.
742 Liu, Z.-J.; Vors, J.-P.; Gesing, E. R. F.; Bolm, C. *Green Chem.* **2011**, *13*, 42.
743 Toulot, S.; Heinrich, T.; Leroux, F. R. *Adv. Synth. Catal.* **2013**, *355*, 3263.
751 Wang, L.; Liu, N.; Dai, B.; Hu, H. *Eur. J. Org. Chem.* **2014**, *2014*, 6493.
784 Teo, Y.-C.; Yong, F.-F.; Lim, G. S. *Tetrahedron Lett.* **2011**, *52*, 7171.
792 Wang, Y.; Zhang, Y.; Yang, B.; Zhang, A.; Yao, Q. *Org. Biomol. Chem.* **2015**, *13*, 4101.
794 Tber, Z.; Hiebel, M.-A.; Akssira, M.; Guillaumet, G.; Berteina-Raboin, S. *Synthesis* **2015**, *47*, 1780.
803 Beyer, A.; Castanheiro, T.; Busca, P.; Prestat, G. *ChemCatChem* **2015**, *7*, 2433.

Supplemental References for Table 9A

699 Berzina, B.; Sokolovs, I.; Suna, E. *ACS Catalysis* **2015**, *5*, 7008.
704 Ahmadi, S. J.; Sadjadi, S.; Hosseinpour, M.; Abdollahi, M. *Monatsh. Chem.* **2011**, *142*, 801.
709 Chen, B.; Li, F.; Huang, Z.; Xue, F.; Lu, T.; Yuan, Y.; Yuan, G. *ChemCatChem* **2012**, *4*, 1741.
712 Huang, L.; Yu, R.; Zhu, X.; Wan, Y. *Tetrahedron* **2013**, *69*, 8974.
713 Huang, M.; Lin, X.; Zhu, X.; Peng, W.; Xie, J.; Wan, Y. *Eur. J. Org. Chem.* **2011**, 4523.
720 Yang, H.; Xi, C.; Miao, Z.-W.; Chen, R.-Y. *Eur. J. Org. Chem.* **2011**, *2011*, 3353.
721 Yang, K.; Qiu, Y.; Li, Z.; Wang, Z.; Jiang, S. *J. Org. Chem.* **2011**, *76*, 3151.
722 Yang, Q.; Wang, Y.; Lin, D.; Zhang, M. *Tetrahedron Lett.* **2013**, *54*, 1994.
733 Reddy, P. L.; Arundhathi, R.; Rawat, D. S. *RSC Adv.* **2015**, *5*, 92121.
735 Su, J.; Qiu, Y.; Jiang, S.; Zhang, D. *Chin. J. Chem.* **2014**, *32*, 685.

[738] Yan, N.-N.; Wu, F.-T.; Zhang, J.; Wei, Q.-B.; Liu, P.; Xie, J.-W.; Dai, B. *Asian J. Org. Chem.* **2014**, *3*, 1159.

[754] Bhosale, M. A.; Bhanage, B. M. *RSC Adv.* **2014**, *4*, 15122.

[759] Truong, T.; Nguyen, C. V.; Truong, N. T.; Phan, N. T. S. *RSC Adv.* **2015**, *5*, 107547.

[762] Engel-Andreasen, J.; Shimpukade, B.; Ulven, T. *Green Chem.* **2013**, *15*, 336.

[766] Xu, Z.-L.; Li, H.-X.; Ren, Z.-G.; Du, W.-Y.; Xu, W.-C.; Lang, J.-P. *Tetrahedron* **2011**, *67*, 5282.

[767] Zhang, P.; Yuan, J.; Li, H.; Liu, X.; Xu, X.; Antonietti, M.; Wang, Y. *RSC Adv.* **2013**, *3*, 1890.

[770] Nasrollahzadeh, M.; Zahraei, A.; Pourbasheer, E. *Monatsh. Chem.* **2015**, *146*, 1329.

[773] Sharma, R. K.; Gaur, R.; Yadav, M.; Rathi, A. K.; Pechousek, J.; Petr, M.; Zboril, R.; Gawande, M. B. *ChemCatChem* **2015**, *7*, 3495.

[781] Ganesh Babu, S.; Karvembu, R. *Ind. Eng. Chem. Res.* **2011**, *50*, 9594.

[782] Hosseini-Sarvari, M.; Moeini, F. *RSC Adv.* **2014**, *4*, 7321.

[784] Teo, Y.-C.; Yong, F.-F.; Lim, G. S. *Tetrahedron Lett.* **2011**, *52*, 7171.

[785] Wang, Y.-L.; Luo, J.; Liu, Z.-L. *J. Chin. Chem. Soc.* **2013**, *60*, 1007.

[786] Yadav, D. K. T.; Rajak, S. S.; Bhanage, B. M. *Tetrahedron Lett.* **2014**, *55*, 931.

[787] Yang, H.; Miao, Z.; Chen, R. *Lett. Org. Chem.* **2011**, *8*, 325.

[788] Yang, X.; Xing, H.; Zhang, Y.; Lai, Y.; Zhang, Y.; Jiang, Y.; Ma, D. *Chin. J. Chem.* **2012**, *30*, 875.

[789] Yong, F.-F.; Teo, Y.-C.; Tay, S.-H.; Tan, B. Y.-H.; Lim, K.-H. *Tetrahedron Lett.* **2011**, *52*, 1161.

[791] Toummini, D.; Tlili, A.; Berges, J.; Ouazzani, F.; Taillefer, M. *Chem.—Eur. J.* **2014**, *20*, 14619.

[792] Wang, Y.; Zhang, Y.; Yang, B.; Zhang, A.; Yao, Q. *Org. Biomol. Chem.* **2015**, *13*, 4101.

[793] Wu, F.-T.; Yan, N.-N.; Liu, P.; Xie, J.-W.; Liu, Y.; Dai, B. *Tetrahedron Lett.* **2014**, *55*, 3249.

[795] Abele, E.; Abele, R. *Chem. Heterocycl. Compd. (N. Y., NY, U.S.)* **2013**, *49*, 1384.

[796] Boswell, M. G.; Yeung, F. G.; Wolf, C. *Synlett* **2012**, *23*, 1240.

[797] Huang, L.; Jin, C.; Su, W. *Chin. J. Chem.* **2012**, *30*, 2394.

[798] Huang, Z.; Li, F.; Chen, B.; Xue, F.; Chen, G.; Yuan, G. *Appl. Catal., A–Gen.* **2011**, *403*, 104.

[800] Wang, D.; Zhang, F.; Kuang, D.; Yu, J.; Li, J. *Green Chem.* **2012**, *14*, 1268.

[801] Zhang, Q.; Luo, J.; Wei, Y. *Synth. Commun.* **2012**, *42*, 114.

[802] Zou, Y.; Lin, H.-S.; Maggard, P. A.; Deiters, A. *Eur. J. Org. Chem.* **2011**, *2011*, 4154.

[804] Heidarizadeh, F.; Majdi-nasab, A. *Tetrahedron Lett.* **2015**, *56*, 6360.

[805] Liu, Y.-S.; Liu, Y.; Ma, X.-W.; Liu, P.; Xie, J.-W.; Dai, B. *Chin. Chem. Lett.* **2014**, *25*, 775.

[807] Cao, C.; Lu, J.; Cai, Z.; Pang, G.; Shi, Y. *Synth. Commun.* **2012**, *42*, 279.

[808] Davis, O. A.; Hughes, M.; Bull, J. A. *J. Org. Chem.* **2013**, *78*, 3470.

[809] Ghorbani-Vaghei, R.; Hemmati, S.; Veisi, H. *Tetrahedron Lett.* **2013**, *54*, 7095.

[810] Jiao, Y.; Yan, N.; Xie, J.; Ma, X.; Liu, P.; Dai, B. *Chin. J. Chem.* **2013**, *31*, 267.

[811] Kaswan, P.; Pericherla, K.; Kumar, A. *Synlett* **2013**, *24*, 2751.

[812] Liu, Y.; Zhang, Q.; Ma, X.; Liu, P.; Xie, J.; Dai, B.; Liu, Z. *Int. J. Org. Chem.* **2013**, *3*, 185.

[813] Lv, R.; Wang, Y.; Zhou, C.; Li, L.; Wang, R. *ChemCatChem* **2013**, *5*, 2978.

[814] Lv, T.; Wang, Z.; You, J.; Lan, J.; Gao, G. *J. Org. Chem.* **2013**, *78*, 5723.

[815] Mukhopadhyay, C.; Tapaswi, P. K. *Synth. Commun.* **2012**, *42*, 2217.

[816] Rad, M. N. S.; Behrouz, S.; Doroodmand, M. M.; Moghtaderi, N. *Synthesis* **2011**, 3915.

[817] Salam, N.; Kundu, S. K.; Roy, A. S.; Mondal, P.; Roy, S.; Bhaumik, A.; Islam, S. M. *Catal. Sci. Technol.* **2013**, *3*, 3303.

[818] Wu, F.-T.; Liu, P.; Ma, X.-W.; Xie, J.-W.; Dai, B. *Chin. Chem. Lett.* **2013**, *24*, 893.

[819] Wu, X.; Hu, W. *Chin. J. Chem.* **2011**, *29*, 2124.

[820] Ziegler, D. T.; Choi, J.; Muñoz-Molina, J. M.; Bissember, A. C.; Peters, J. C.; Fu, G. C. *J. Am. Chem. Soc.* **2013**, *135*, 13107.

[821] Huang, Q.; Zhou, L.; Jiang, X.; Zhou, Y.; Fan, H.; Lang, W. *ACS Appl. Mater. Interfaces* **2014**, *6*, 13502.

[822] Liu, Y.; Gu, N.; Liu, P.; Xie, J.; Dai, B.; Liu, Y. *Appl. Organomet. Chem.* **2015**, *29*, 468.

[823] Movahed, S. K.; Dabiri, M.; Bazgir, A. *Appl. Catal., A–Gen.* **2014**, *481*, 79.

[824] Nador, F.; Volpe, M. A.; Alonso, F.; Radivoy, G. *Tetrahedron* **2014**, *70*, 6082.

[825] Nguyen, T. T.; Phan, N. T. S. *Catal. Lett.* **2014**, *144*, 1877.

Supplemental References for Table 9B

[738] Yan, N.-N.; Wu, F.-T.; Zhang, J.; Wei, Q.-B.; Liu, P.; Xie, J.-W.; Dai, B. *Asian J. Org. Chem.* **2014**, *3*, 1159.

[742] Liu, Z.-J.; Vors, J.-P.; Gesing, E. R. F.; Bolm, C. *Green Chem.* **2011**, *13*, 42.

[751] Wang, L.; Liu, N.; Dai, B.; Hu, H. *Eur. J. Org. Chem.* **2014**, *2014*, 6493.

[787] Yang, H.; Miao, Z.; Chen, R. *Lett. Org. Chem.* **2011**, *8*, 325.

[792] Wang, Y.; Zhang, Y.; Yang, B.; Zhang, A.; Yao, Q. *Org. Biomol. Chem.* **2015**, *13*, 4101.

[793] Wu, F.-T.; Yan, N.-N.; Liu, P.; Xie, J.-W.; Liu, Y.; Dai, B. *Tetrahedron Lett.* **2014**, *55*, 3249.

[794] Tber, Z.; Hiebel, M.-A.; Akssira, M.; Guillaumet, G.; Berteina-Raboin, S. *Synthesis* **2015**, *47*, 1780.

[797] Huang, L.; Jin, C.; Su, W. *Chin. J. Chem.* **2012**, *30*, 2394.

[800] Wang, D.; Zhang, F.; Kuang, D.; Yu, J.; Li, J. *Green Chem.* **2012**, *14*, 1268.

[807] Cao, C.; Lu, Z.; Cai, Z.; Pang, G.; Shi, Y. *Synth. Commun.* **2012**, *42*, 279.

[808] Davis, O. A.; Hughes, M.; Bull, J. A. *J. Org. Chem.* **2013**, *78*, 3470.

[809] Ghorbani-Vaghei, R.; Hemmati, S.; Veisi, H. *Tetrahedron Lett.* **2013**, *54*, 7095.

[824] Nador, F.; Volpe, M. A.; Alonso, F.; Radivoy, G. *Tetrahedron* **2014**, *70*, 6082.

Supplemental References for Table 10A

[699] Berzina, B.; Sokolovs, I.; Suna, E. *ACS Catalysis* **2015**, *5*, 7008.

[766] Xu, Z.-L.; Li, H.-X.; Ren, Z.-G.; Du, W.-Y.; Xu, W.-C.; Lang, J.-P. *Tetrahedron* **2011**, *67*, 5282.

[770] Nasrollahzadeh, M.; Zahraei, A.; Pourbasheer, E. *Monatsh. Chem.* **2015**, *146*, 1329.

[782] Hosseini-Sarvari, M.; Moeini, F. *RSC Adv.* **2014**, *4*, 7321.

[788] Yang, X.; Xing, H.; Zhang, Y.; Lai, Y.; Zhang, Y.; Jiang, Y.; Ma, D. *Chin. J. Chem.* **2012**, *30*, 875.

[791] Toummini, D.; Tlili, A.; Berges, J.; Ouazzani, F.; Taillefer, M. *Chem.—Eur. J.* **2014**, *20*, 14619.

[792] Wang, Y.; Zhang, Y.; Yang, B.; Zhang, A.; Yao, Q. *Org. Biomol. Chem.* **2015**, *13*, 4101.

[797] Huang, L.; Jin, C.; Su, W. *Chin. J. Chem.* **2012**, *30*, 2394.

[800] Wang, D.; Zhang, F.; Kuang, D.; Yu, J.; Li, J. *Green Chem.* **2012**, *14*, 1268.

[806] Nasrollahzadeh, M.; Sajadi, S. M.; Maham, M. *RSC Adv.* **2015**, *5*, 40628.

[811] Kaswan, P.; Pericherla, K.; Kumar, A. *Synlett* **2013**, *24*, 2751.

[815] Mukhopadhyay, C.; Tapaswi, P. K. *Synth. Commun.* **2012**, *42*, 2217.

[826] Mangion, I. K.; Sherry, B. D.; Yin, J.; Fleitz, F. J. *Org. Lett.* **2012**, *14*, 3458.

[827] Kommu, N.; Ghule, V. D.; Kumar, A. S.; Sahoo, A. K. *Chem.—Asian J.* **2014**, *9*, 166.

Supplemental References for Table 10B

[794] Tber, Z.; Hiebel, M.-A.; Akssira, M.; Guillaumet, G.; Berteina-Raboin, S. *Synthesis* **2015**, *47*, 1780.

[827] Kommu, N.; Ghule, V. D.; Kumar, A. S.; Sahoo, A. K. *Chem.—Asian J.* **2014**, *9*, 166.

Supplemental References for Table 11A

[704] Ahmadi, S. J.; Sadjadi, S.; Hosseinpour, M.; Abdollahi, M. *Monatsh. Chem.* **2011**, *142*, 801.

[721] Yang, K.; Qiu, Y.; Li, Z.; Wang, Z.; Jiang, S. *J. Org. Chem.* **2011**, *76*, 3151.

[735] Su, J.; Qiu, Y.; Jiang, S.; Zhang, D. *Chin. J. Chem.* **2014**, *32*, 685.

[738] Yan, N.-N.; Wu, F.-T.; Zhang, J.; Wei, Q.-B.; Liu, P.; Xie, J.-W.; Dai, B. *Asian J. Org. Chem.* **2014**, *3*, 1159.

[754] Bhosale, M. A.; Bhanage, B. M. *RSC Adv.* **2014**, *4*, 15122.

[759] Truong, T.; Nguyen, C. V.; Truong, N. T.; Phan, N. T. S. *RSC Adv.* **2015**, *5*, 107547.

[765] Wu, C.-C.; Lee, B.-H.; Liao, P.-K.; Fang, C.-S.; Liu, C. W. *J. Chin. Chem. Soc. (Weinheim, Ger.)* **2012**, *59*, 480.

[766] Xu, Z.-L.; Li, H.-X.; Ren, Z.-G.; Du, W.-Y.; Xu, W.-C.; Lang, J.-P. *Tetrahedron* **2011**, *67*, 5282.

[769] Liu, Y. S.; Gu, N. N.; Liu, Y.; Ma, X. W.; Liu, P.; Xie, J. W. *Asian J. Chem.* **2015**, *27*, 1075.

[773] Sharma, R. K.; Gaur, R.; Yadav, M.; Rathi, A. K.; Pechousek, J.; Petr, M.; Zboril, R.; Gawande, M. B. *ChemCatChem* **2015**, *7*, 3495.

[781] Ganesh Babu, S.; Karvembu, R. *Ind. Eng. Chem. Res.* **2011**, *50*, 9594.

[782] Hosseini-Sarvari, M.; Moeini, F. *RSC Adv.* **2014**, *4*, 7321.

[783] Pai, G.; Chattopadhyay, A. P. *Tetrahedron Lett.* **2014**, *55*, 941.

[784] Teo, Y.-C.; Yong, F.-F.; Lim, G. S. *Tetrahedron Lett.* **2011**, *52*, 7171.

[785] Wang, Y.-L.; Luo, J.; Liu, Z.-L. *J. Chin. Chem. Soc.* **2013**, *60*, 1007.

[786] Yadav, D. K. T.; Rajak, S. S.; Bhanage, B. M. *Tetrahedron Lett.* **2014**, *55*, 931.

[787] Yang, H.; Miao, Z.; Chen, R. *Lett. Org. Chem.* **2011**, *8*, 325.

788 Yang, X.; Xing, H.; Zhang, Y.; Lai, Y.; Zhang, Y.; Jiang, Y.; Ma, D. *Chin. J. Chem.* **2012**, *30*, 875.
789 Yong, F.-F.; Teo, Y.-C.; Tay, S.-H.; Tan, B. Y.-H.; Lim, K.-H. *Tetrahedron Lett.* **2011**, *52*, 1161.
791 Toummini, D.; Tlili, A.; Berges, J.; Ouazzani, F.; Taillefer, M. *Chem.—Eur. J.* **2014**, *20*, 14619.
792 Wang, Y.; Zhang, Y.; Yang, B.; Zhang, A.; Yao, Q. *Org. Biomol. Chem.* **2015**, *13*, 4101.
795 Abele, E.; Abele, R. *Chem. Heterocycl. Compd. (N. Y., NY, U.S.)* **2013**, *49*, 1384.
797 Huang, L.; Jin, C.; Su, W. *Chin. J. Chem.* **2012**, *30*, 2394.
801 Zhang, Q.; Luo, J.; Wei, Y. *Synth. Commun.* **2012**, *42*, 114.
802 Zou, Y.; Lin, H.-S.; Maggard, P. A.; Deiters, A. *Eur. J. Org. Chem.* **2011**, *2011*, 4154.
804 Heidarizadeh, F.; Majdi-nasab, A. *Tetrahedron Lett.* **2015**, *56*, 6360.
806 Nasrollahzadeh, M.; Sajadi, S. M.; Maham, M. *RSC Adv.* **2015**, *5*, 40628.
815 Mukhopadhyay, C.; Tapaswi, P. K. *Synth. Commun.* **2012**, *42*, 2217.
820 Ziegler, D. T.; Choi, J.; Muñoz-Molina, J. M.; Bissember, A. C.; Peters, J. C.; Fu, G. C. *J. Am. Chem. Soc.* **2013**, *135*, 13107.
823 Movahed, S. K.; Dabiri, M.; Bazgir, A. *Appl. Catal., A–Gen.* **2014**, *481*, 79.
828 Lee, H.-Y.; Chang, J.-Y.; Chang, L.-Y.; Lai, W.-Y.; Lai, M.-J.; Shih, K.-H.; Kuo, C.-C.; Chang, C.-Y.; Liou, J.-P. *Org. Biomol. Chem.* **2011**, *9*, 3154.
829 Maligres, P. E.; Krska, S. W.; Dormer, P. G. *J. Org. Chem.* **2012**, *77*, 7646.
830 Reddy, K. H. V.; Satish, G.; Ramesh, K.; Karnakar, K.; Nageswar, Y. V. D. *Tetrahedron Lett.* **2012**, *53*, 3061.
831 Amadine, O.; Maati, H.; Abdelouhadi, K.; Fihri, A.; El Kazzouli, S.; Len, C.; El Bouari, A.; Solhy, A. *J. Mol. Catal. A–Chem.* **2014**, *395*, 409.
832 Kodicherla, B.; Perumgani C, P.; Mandapati, M. R. *Appl. Catal., A–Gen.* **2014**, *483*, 110.
833 Modha, S. G.; Greaney, M. F. *J. Am. Chem. Soc.* **2015**, *137*, 1416.

Supplemental References for Table 11B

742 Liu, Z.-J.; Vors, J.-P.; Gesing, E. R. F.; Bolm, C. *Green Chem.* **2011**, *13*, 42.
751 Wang, L.; Liu, N.; Dai, B.; Hu, H. *Eur. J. Org. Chem.* **2014**, *2014*, 6493.
783 Pai, G.; Chattopadhyay, A. P. *Tetrahedron Lett.* **2014**, *55*, 941.
784 Teo, Y.-C.; Yong, F.-F.; Lim, G. S. *Tetrahedron Lett.* **2011**, *52*, 7171.
788 Yang, X.; Xing, H.; Zhang, Y.; Lai, Y.; Zhang, Y.; Jiang, Y.; Ma, D. *Chin. J. Chem.* **2012**, *30*, 875.
792 Wang, Y.; Zhang, Y.; Yang, B.; Zhang, A.; Yao, Q. *Org. Biomol. Chem.* **2015**, *13*, 4101.
794 Tber, Z.; Hiebel, M.-A.; Akssira, M.; Guillaumet, G.; Berteina-Raboin, S. *Synthesis* **2015**, *47*, 1780.
831 Amadine, O.; Maati, H.; Abdelouhadi, K.; Fihri, A.; El Kazzouli, S.; Len, C.; El Bouari, A.; Solhy, A. *J. Mol. Catal. A–Chem.* **2014**, *395*, 409.
833 Modha, S. G.; Greaney, M. F. *J. Am. Chem. Soc.* **2015**, *137*, 1416.

Supplemental References for Table 12A

789 Yong, F.-F.; Teo, Y.-C.; Tay, S.-H.; Tan, B. Y.-H.; Lim, K.-H. *Tetrahedron Lett.* **2011**, *52*, 1161.
791 Toummini, D.; Tlili, A.; Berges, J.; Ouazzani, F.; Taillefer, M. *Chem.—Eur. J.* **2014**, *20*, 14619.
834 Chung, C. K.; Bulger, P. G.; Kosjek, B.; Belyk, K. M.; Rivera, N.; Scott, M. E.; Humphrey, G. R.; Limanto, J.; Bachert, D. C.; Emerson, K. M. *Org. Proc. Res. Dev.* **2014**, *18*, 215.

Supplemental References for Table 12B

751 Wang, L.; Liu, N.; Dai, B.; Hu, H. *Eur. J. Org. Chem.* **2014**, *2014*, 6493.
794 Tber, Z.; Hiebel, M.-A.; Akssira, M.; Guillaumet, G.; Berteina-Raboin, S. *Synthesis* **2015**, *47*, 1780.

Supplemental References for Table 13A

709 Chen, B.; Li, F.; Huang, Z.; Xue, F.; Lu, T.; Yuan, Y.; Yuan, G. *ChemCatChem* **2012**, *4*, 1741.
713 Huang, M.; Lin, X.; Zhu, X.; Peng, W.; Xie, J.; Wan, Y. *Eur. J. Org. Chem.* **2011**, 4523.
720 Yang, H.; Xi, C.; Miao, Z.-W.; Chen, R.-Y. *Eur. J. Org. Chem.* **2011**, *2011*, 3353.

721   Yang, K.; Qiu, Y.; Li, Z.; Wang, Z.; Jiang, S. *J. Org. Chem.* **2011**, *76*, 3151.
762   Engel-Andreasen, J.; Shimpukade, B.; Ulven, T. *Green Chem.* **2013**, *15*, 336.
766   Xu, Z.-L.; Li, H.-X.; Ren, Z.-G.; Du, W.-Y.; Xu, W.-C.; Lang, J.-P. *Tetrahedron* **2011**, *67*, 5282.
781   Ganesh Babu, S.; Karvembu, R. *Ind. Eng. Chem. Res.* **2011**, *50*, 9594.
785   Wang, Y.-L.; Luo, J.; Liu, Z.-L. *J. Chin. Chem. Soc.* **2013**, *60*, 1007.
787   Yang, H.; Miao, Z.; Chen, R. *Lett. Org. Chem.* **2011**, *8*, 325.
788   Yang, X.; Xing, H.; Zhang, Y.; Lai, Y.; Zhang, Y.; Jiang, Y.; Ma, D. *Chin. J. Chem.* **2012**, *30*, 875.
791   Toummini, D.; Tlili, A.; Berges, J.; Ouazzani, F.; Taillefer, M. *Chem.—Eur. J.* **2014**, *20*, 14619.
792   Wang, Y.; Zhang, Y.; Yang, B.; Zhang, A.; Yao, Q. *Org. Biomol. Chem.* **2015**, *13*, 4101.
795   Abele, E.; Abele, R. *Chem. Heterocycl. Compd. (N. Y., NY, U.S.)* **2013**, *49*, 1384.
796   Boswell, M. G.; Yeung, F. G.; Wolf, C. *Synlett* **2012**, *23*, 1240.
798   Huang, Z.; Li, F.; Chen, B.; Xue, F.; Chen, G.; Yuan, G. *Appl. Catal., A–Gen.* **2011**, *403*, 104.
800   Wang, D.; Zhang, F.; Kuang, D.; Yu, J.; Li, J. *Green Chem.* **2012**, *14*, 1268.
801   Zhang, Q.; Luo, J.; Wei, Y. *Synth. Commun.* **2012**, *42*, 114.
805   Liu, Y.-S.; Liu, Y.; Ma, X.-W.; Liu, P.; Xie, J.-W.; Dai, B. *Chin. Chem. Lett.* **2014**, *25*, 775.
811   Kaswan, P.; Pericherla, K.; Kumar, A. *Synlett* **2013**, *24*, 2751.
812   Liu, Y.; Zhang, Q.; Ma, X.; Liu, P.; Xie, J.; Dai, B.; Liu, Z. *Int. J. Org. Chem.* **2013**, *3*, 185.
813   Lv, R.; Wang, Y.; Zhou, C.; Li, L.; Wang, R. *ChemCatChem* **2013**, *5*, 2978.
815   Mukhopadhyay, C.; Tapaswi, P. K. *Synth. Commun.* **2012**, *42*, 2217.
816   Rad, M. N. S.; Behrouz, S.; Doroodmand, M. M.; Moghtaderi, N. *Synthesis* **2011**, 3915.
818   Wu, F.-T.; Liu, P.; Ma, X.-W.; Xie, J.-W.; Dai, B. *Chin. Chem. Lett.* **2013**, *24*, 893.
819   Wu, X.; Hu, W. *Chin. J. Chem.* **2011**, *29*, 2124.
820   Ziegler, D. T.; Choi, J.; Muñoz-Molina, J. M.; Bissember, A. C.; Peters, J. C.; Fu, G. C. *J. Am. Chem. Soc.* **2013**, *135*, 13107.
822   Liu, Y.; Gu, N.; Liu, P.; Xie, J.; Dai, B.; Liu, Y. *Appl. Organomet. Chem.* **2015**, *29*, 468.
823   Movahed, S. K.; Dabiri, M.; Bazgir, A. *Appl. Catal., A–Gen.* **2014**, *481*, 79.
825   Nguyen, T. T.; Phan, N. T. S. *Catal. Lett.* **2014**, *144*, 1877.
829   Maligres, P. E.; Krska, S. W.; Dormer, P. G. *J. Org. Chem.* **2012**, *77*, 7646.

Supplemental References for Table 13B

738   Yan, N.-N.; Wu, F.-T.; Zhang, J.; Wei, Q.-B.; Liu, P.; Xie, J.-W.; Dai, B. *Asian J. Org. Chem.* **2014**, *3*, 1159.
751   Wang, L.; Liu, N.; Dai, B.; Hu, H. *Eur. J. Org. Chem.* **2014**, *2014*, 6493.
781   Ganesh Babu, S.; Karvembu, R. *Ind. Eng. Chem. Res.* **2011**, *50*, 9594.
792   Wang, Y.; Zhang, Y.; Yang, B.; Zhang, A.; Yao, Q. *Org. Biomol. Chem.* **2015**, *13*, 4101.

Supplemental References for Table 14A

754   Bhosale, M. A.; Bhanage, B. M. *RSC Adv.* **2014**, *4*, 15122.
766   Xu, Z.-L.; Li, H.-X.; Ren, Z.-G.; Du, W.-Y.; Xu, W.-C.; Lang, J.-P. *Tetrahedron* **2011**, *67*, 5282.
781   Ganesh Babu, S.; Karvembu, R. *Ind. Eng. Chem. Res.* **2011**, *50*, 9594.
788   Yang, X.; Xing, H.; Zhang, Y.; Lai, Y.; Zhang, Y.; Jiang, Y.; Ma, D. *Chin. J. Chem.* **2012**, *30*, 875.
829   Maligres, P. E.; Krska, S. W.; Dormer, P. G. *J. Org. Chem.* **2012**, *77*, 7646.
835   Chen, F.; Liu, N.; Ji, E.; Dai, B. *RSC Adv.* **2015**, *5*, 51512.
836   Yoo, W.-J.; Tsukamoto, T.; Kobayashi, S. *Org. Lett.* **2015**, *17*, 3640.

Supplemental References for Table 15A

699   Berzina, B.; Sokolovs, I.; Suna, E. *ACS Catalysis* **2015**, *5*, 7008.
720   Yang, H.; Xi, C.; Miao, Z.-W.; Chen, R.-Y. *Eur. J. Org. Chem.* **2011**, *2011*, 3353.
762   Engel-Andreasen, J.; Shimpukade, B.; Ulven, T. *Green Chem.* **2013**, *15*, 336.
766   Xu, Z.-L.; Li, H.-X.; Ren, Z.-G.; Du, W.-Y.; Xu, W.-C.; Lang, J.-P. *Tetrahedron* **2011**, *67*, 5282.
783   Pai, G.; Chattopadhyay, A. P. *Tetrahedron Lett.* **2014**, *55*, 941.
784   Teo, Y.-C.; Yong, F.-F.; Lim, G. S. *Tetrahedron Lett.* **2011**, *52*, 7171.
787   Yang, H.; Miao, Z.; Chen, R. *Lett. Org. Chem.* **2011**, *8*, 325.

789 Yong, F.-F.; Teo, Y.-C.; Tay, S.-H.; Tan, B. Y.-H.; Lim, K.-H. *Tetrahedron Lett.* **2011**, *52*, 1161.
797 Huang, L.; Jin, C.; Su, W. *Chin. J. Chem.* **2012**, *30*, 2394.
806 Nasrollahzadeh, M.; Sajadi, S. M.; Maham, M. *RSC Adv.* **2015**, *5*, 40628.
811 Kaswan, P.; Pericherla, K.; Kumar, A. *Synlett* **2013**, *24*, 2751.
815 Mukhopadhyay, C.; Tapaswi, P. K. *Synth. Commun.* **2012**, *42*, 2217.
816 Rad, M. N. S.; Behrouz, S.; Doroodmand, M. M.; Moghtaderi, N. *Synthesis* **2011**, 3915.
823 Movahed, S. K.; Dabiri, M.; Bazgir, A. *Appl. Catal., A–Gen.* **2014**, *481*, 79.
837 Niu, H.-Y.; Xia, C.; Qu, G.-R.; Zhang, Q.; Jiang, Y.; Mao, R.-Z.; Li, D.-Y.; Guo, H.-M. *Org. Biomol. Chem.* **2011**, *9*, 5039.
838 Larsen, A. F.; Ulven, T. *Chem. Commun.* **2014**, *50*, 4997.

Supplemental References for Table 15B

751 Wang, L.; Liu, N.; Dai, B.; Hu, H. *Eur. J. Org. Chem.* **2014**, *2014*, 6493.
783 Pai, G.; Chattopadhyay, A. P. *Tetrahedron Lett.* **2014**, *55*, 941.
784 Teo, Y.-C.; Yong, F.-F.; Lim, G. S. *Tetrahedron Lett.* **2011**, *52*, 7171.
838 Larsen, A. F.; Ulven, T. *Chem. Commun.* **2014**, *50*, 4997.

Supplemental References for Table 16A

687 Komati, R.; Jursic, B. S. *Tetrahedron Lett.* **2014**, *55*, 1523.
709 Chen, B.; Li, F.; Huang, Z.; Xue, F.; Lu, T.; Yuan, Y.; Yuan, G. *ChemCatChem* **2012**, *4*, 1741.
764 Quan, Z.-J.; Xia, H.-D.; Zhang, Z.; Da, Y.-X.; Wang, X.-C. *Appl. Organomet. Chem.* **2014**, *28*, 81.
765 Wu, C.-C.; Lee, B.-H.; Liao, P.-K.; Fang, C.-S.; Liu, C. W. *J. Chin. Chem. Soc. (Weinheim, Ger.)* **2012**, *59*, 480.
829 Maligres, P. E.; Krska, S. W.; Dormer, P. G. *J. Org. Chem.* **2012**, *77*, 7646.
839 Cortes-Salva, M.; Garvin, C.; Antilla, J. C. *J. Org. Chem.* **2011**, *76*, 1456.
840 Dong, J.; Wang, Y.; Xiang, Q.; Lv, X.; Weng, W.; Zeng, Q. *Adv. Synth. Catal.* **2013**, *355*, 692.
841 Lawson, C. P. A. T.; Slawin, A. M. Z.; Westwood, N. J. *Chem. Commun.* **2011**, *47*, 1057.
842 Quan, Z.-J.; Xia, H.-D.; Zhang, Z.; Da, Y.-X.; Wang, X.-C. *Tetrahedron* **2013**, *69*, 8368.
843 Yong, F.-F.; Teo, Y.-C.; Chua, G.-L.; Lim, G. S.-Y.; Lin, Y.-Z. *Tetrahedron Lett.* **2011**, *52*, 1169.
844 Güell, I.; Ribas, X. *Eur. J. Org. Chem.* **2014**, *2014*, 3188.

Supplemental References for Table 16B

742 Liu, Z.-J.; Vors, J.-P.; Gesing, E. R. F.; Bolm, C. *Green Chem.* **2011**, *13*, 42.
748 Knight, J. G.; Alnoman, R. B.; Waddell, P. G. *Org. Biomol. Chem.* **2015**, *13*, 3819.
844 Güell, I.; Ribas, X. *Eur. J. Org. Chem.* **2014**, *2014*, 3188.

Supplemental References for Table 17A

765 Wu, C.-C.; Lee, B.-H.; Liao, P.-K.; Fang, C.-S.; Liu, C. W. *J. Chin. Chem. Soc. (Weinheim, Ger.)* **2012**, *59*, 480.
829 Maligres, P. E.; Krska, S. W.; Dormer, P. G. *J. Org. Chem.* **2012**, *77*, 7646.
842 Quan, Z.-J.; Xia, H.-D.; Zhang, Z.; Da, Y.-X.; Wang, X.-C. *Tetrahedron* **2013**, *69*, 8368.
845 Deprez-Poulain, R.; Cousaert, N.; Toto, P.; Willand, N.; Deprez, B. *Eur. J. Med. Chem.* **2011**, *46*, 3867.
846 Racine, E.; Monnier, F.; Vors, J.-P.; Taillefer, M. *Org. Lett.* **2011**, *13*, 2818.
847 Wang, M.; Yu, H.; You, X.; Wu, J.; Shang, Z. *Chin. J. Chem.* **2012**, *30*, 2356.
848 Wilson, R. J.; Rosenberg, A. J.; Kaminsky, L.; Clark, D. A. *J. Org. Chem.* **2014**, *79*, 2203.
849 Jiang, L. *Molecules* **2014**, *19*, 13448.
850 Kathiravan, S.; Ghosh, S.; Hogarth, G.; Nicholls, I. A. *Chem. Commun.* **2015**, *51*, 4834.
851 Sugimoto, K.; Tamura, K.; Toyooka, N.; Matsuya, Y. *Heterocycles* **2014**, *88*, 755.

Supplemental References for Table 17B

[845] Deprez-Poulain, R.; Cousaert, N.; Toto, P.; Willand, N.; Deprez, B. *Eur. J. Med. Chem.* **2011**, *46*, 3867.

[849] Jiang, L. *Molecules* **2014**, *19*, 13448.

[850] Kathiravan, S.; Ghosh, S.; Hogarth, G.; Nicholls, I. A. *Chem. Commun.* **2015**, *51*, 4834.

[852] Wang, M.-G.; Yu, H.; Wu, J.; Shang, Z.-C. *Synthesis* **2013**, *45*, 1955.

Supplemental References for Table 18A

[699] Berzina, B.; Sokolovs, I.; Suna, E. *ACS Catalysis* **2015**, *5*, 7008.

[755] Kundu, D.; Bhadra, S.; Mukherjee, N.; Sreedhar, B.; Ranu, B. C. *Chem.—Eur. J.* **2013**, *19*, 15759.

[791] Toummini, D.; Tlili, A.; Berges, J.; Ouazzani, F.; Taillefer, M. *Chem.—Eur. J.* **2014**, *20*, 14619.

[802] Zou, Y.; Lin, H.-S.; Maggard, P. A.; Deiters, A. *Eur. J. Org. Chem.* **2011**, *2011*, 4154.

[829] Maligres, P. E.; Krska, S. W.; Dormer, P. G. *J. Org. Chem.* **2012**, *77*, 7646.

[841] Lawson, C. P. A. T.; Slawin, A. M. Z.; Westwood, N. J. *Chem. Commun.* **2011**, *47*, 1057.

[842] Quan, Z.-J.; Xia, H.-D.; Zhang, Z.; Da, Y.-X.; Wang, X.-C. *Tetrahedron* **2013**, *69*, 8368.

[843] Yong, F.-F.; Teo, Y.-C.; Chua, G.-L.; Lim, G. S.-Y.; Lin, Y.-Z. *Tetrahedron Lett.* **2011**, *52*, 1169.

[853] Lange, J.; Bissember, A. C.; Banwell, M. G.; Cade, I. A. *Aust. J. Chem.* **2011**, *64*, 454.

[854] Li, J.; Zhang, Y.; Jiang, Y.; Ma, D. *Tetrahedron Lett.* **2012**, *53*, 3981.

[855] Mao, S.; Guo, F.; Li, J.; Geng, X.; Yu, J.; Han, J.; Wang, L. *Synlett* **2013**, *24*, 1959.

[856] Yang, W.; Coutinho, A. L.; Abdel-Hafez, A. A.; Jiang, C.; Xue, F. *Tetrahedron Lett.* **2015**, *56*, 5599.

Supplemental References for Table 18B

[742] Liu, Z.-J.; Vors, J.-P.; Gesing, E. R. F.; Bolm, C. *Green Chem.* **2011**, *13*, 42.

[743] Toulot, S.; Heinrich, T.; Leroux, F. R. *Adv. Synth. Catal.* **2013**, *355*, 3263.

[854] Li, J.; Zhang, Y.; Jiang, Y.; Ma, D. *Tetrahedron Lett.* **2012**, *53*, 3981.

[857] Kallemeyn, J. M.; Ku, Y.-Y.; Mulhern, M. M.; Bishop, R.; Pal, A.; Jacob, L. *Org. Proc. Res. Dev.* **2014**, *18*, 191.

[858] Wang, M.; Zhang, Z.; Xie, F.; Zhang, W. *Chem. Commun.* **2014**, *50*, 3163.

Supplemental References for Table 19A

[829] Maligres, P. E.; Krska, S. W.; Dormer, P. G. *J. Org. Chem.* **2012**, *77*, 7646.

[859] Walker, J. R.; Fairfull-Smith, K. E.; Anzai, K.; Lau, S.; White, P. J.; Scammells, P. J.; Bottle, S. E. *MedChemComm* **2011**, *2*, 436.

[860] Chen, J.; Zhang, Y.; Hao, W.; Zhang, R.; Yi, F. *Tetrahedron* **2013**, *69*, 613.

[861] Wu, W.; Li, X.-L.; Fan, X.-H.; Yang, L.-M. *Eur. J. Org. Chem.* **2013**, *2013*, 862.

[862] Xiong, X.; Jiang, Y.; Ma, D. *Org. Lett.* **2012**, *14*, 2552.

[863] Yavari, I.; Ghazanfarpour-Darjani, M.; Solgi, Y.; Ahmadian, S. *Synlett* **2011**, 1745.

Supplemental References for Table 21

[864] Kukosha, T.; Trufilkina, N.; Belyakov, S.; Katkevics, M. *Synthesis* **2012**, *44*, 2413.

Supplemental References for Table 22A

[865] Gavade, S. N.; Balaskar, R. S.; Mane, M. S.; Pabrekar, P. N.; Shingare, M. S.; Mane, D. V. *Chin. Chem. Lett.* **2011**, *22*, 675.

[866] Hammoud, H.; Schmitt, M.; Bihel, F.; Antheaume, C.; Bourguignon, J.-J. *J. Org. Chem.* **2012**, *77*, 417.

[867] Xing, H.; Zhang, Y.; Lai, Y.; Jiang, Y.; Ma, D. *J. Org. Chem.* **2012**, *77*, 5449.

868 Yavari, I.; Nematpour, M. *Mol. Diversity* **2015**, *19*, 703.
869 Zhang, C.; Huang, B.; Bao, A.-Q.; Li, X.; Guo, S.; Zhang, J.-Q.; Xu, J.-Z.; Zhang, R.; Cui, D.-M. *Org. Biomol. Chem.* **2015**, *13*, 11432.

Supplemental References for Table 22B

865 Gavade, S. N.; Balaskar, R. S.; Mane, M. S.; Pabrekar, P. N.; Shingare, M. S.; Mane, D. V. *Chin. Chem. Lett.* **2011**, *22*, 675.
866 Hammoud, H.; Schmitt, M.; Bihel, F.; Antheaume, C.; Bourguignon, J.-J. *J. Org. Chem.* **2012**, *77*, 417.
867 Xing, H.; Zhang, Y.; Lai, Y.; Jiang, Y.; Ma, D. *J. Org. Chem.* **2012**, *77*, 5449.

Supplemental References for Table 23A

870 Azzaro, S.; Desage-El Murr, M.; Fensterbank, L.; Lacôte, E.; Malacria, M. *Synlett* **2011**, 849.
871 Tan, B. Y.-H.; Teo, Y.-C.; Seow, A.-H. *Eur. J. Org. Chem.* **2014**, *2014*, 1541.
872 Teo, Y.-C.; Yong, F.-F. *Synlett* **2011**, 837.
873 Wang, X.; Guram, A.; Ronk, M.; Milne, J. E.; Tedrow, J. S.; Faul, M. M. *Tetrahedron Lett.* **2012**, *53*, 7.
874 Geng, X.; Mao, S.; Chen, L.; Yu, J.; Han, J.; Hua, J.; Wang, L. *Tetrahedron Lett.* **2014**, *55*, 3856.

Supplemental References for Table 23B

870 Azzaro, S.; Desage-El Murr, M.; Fensterbank, L.; Lacôte, E.; Malacria, M. *Synlett* **2011**, 849.
873 Wang, X.; Guram, A.; Ronk, M.; Milne, J. E.; Tedrow, J. S.; Faul, M. M. *Tetrahedron Lett.* **2012**, *53*, 7.
875 Lin, J.; Houpis, I. N.; Liu, R.; Wang, Y.; Zhang, J. *Org. Proc. Res. Dev.* **2014**, *18*, 205.

Supplemental References for Table 24A

876 Macé, Y.; Pégot, B.; Guillot, R.; Bournaud, C.; Toffano, M.; Vo-Thanh, G.; Magnier, E. *Tetrahedron* **2011**, *67*, 7575.
877 Vaddula, B.; Leazer, J.; Varma, R. S. *Adv. Synth. Catal.* **2012**, *354*, 986.

Supplemental References for Table 24B

742 Liu, Z.-J.; Vors, J.-P.; Gesing, E. R. F.; Bolm, C. *Green Chem.* **2011**, *13*, 42.
876 Macé, Y.; Pégot, B.; Guillot, R.; Bournaud, C.; Toffano, M.; Vo-Thanh, G.; Magnier, E. *Tetrahedron* **2011**, *67*, 7575.
877 Vaddula, B.; Leazer, J.; Varma, R. S. *Adv. Synth. Catal.* **2012**, *354*, 986.

Supplemental References for Table 25

878 Chen, Y.; Zhuo, Z.-J.; Cui, D.-M.; Zhang, C. *J. Organomet. Chem.* **2014**, *749*, 215.
879 Lanke, S. R.; Bhanage, B. M. *Synth. Commun.* **2014**, *44*, 399.
880 Hajipour, A. R.; Karimzadeh, M.; Ghorbani, S. *Synlett* **2014**, *25*, 2903.
881 Hajipour, A. R.; Mohammadsaleh, F. *Tetrahedron Lett.* **2014**, *55*, 6799.

Supplemental References for Table 26A

882 El Akkaoui, A.; Hiebel, M.-A.; Mouaddib, A.; Berteina-Raboin, S.; Guillaumet, G. *Tetrahedron* **2012**, *68*, 9131.

883 Gao, M.; Liu, X.; Wang, X.; Cai, Q.; Ding, K. *Chin. J. Chem.* **2011**, *29*, 1199.

884 Hu, Z.; Li, S.-D.; Hong, P.-Z.; Wu, Z. *ARKIVOC* **2011**, 147.

885 Jhan, Y.-H.; Kang, T.-W.; Hsieh, J.-C. *Tetrahedron Lett.* **2013**, *54*, 1155.

886 Khoumeri, O.; Giuglio-Tonolo, G.; Crozet, M. D.; Terme, T.; Vanelle, P. *Tetrahedron* **2011**, *67*, 6173.

887 Kukosha, T.; Trufilkina, N.; Katkevics, M. *Synlett* **2011**, 2525.

888 Lee, H. K.; Cho, C. S. *Synth. Commun.* **2013**, *43*, 915.

889 Liubchak, K.; Tolmachev, A.; Nazarenko, K. *J. Org. Chem.* **2012**, *77*, 3365.

890 Peng, J.; Ye, M.; Zong, C.; Hu, F.; Feng, L.; Wang, X.; Wang, Y.; Chen, C. *J. Org. Chem.* **2011**, *76*, 716.

891 Yang, W.; Long, Y.; Zhang, S.; Zeng, Y.; Cai, Q. *Org. Lett.* **2013**, *15*, 3598.

892 Zhou, F.; Guo, J.; Liu, J.; Ding, K.; Yu, S.; Cai, Q. *J. Am. Chem. Soc.* **2012**, *134*, 14326.

893 Asthana, M.; Kumar, R.; Gupta, T.; Singh, R. M. *Tetrahedron Lett.* **2015**, *56*, 907.

894 Hoshi, M.; Kaneko, O.; Nakajima, M.; Arai, S.; Nishida, A. *Org. Lett.* **2014**, *16*, 768.

895 Vijay Kumar, S.; Saraiah, B.; Parameshwarappa, G.; Ila, H.; Verma, G. K. *J. Org. Chem.* **2014**, *79*, 7961.

896 Zhou, S.; Chen, H.; Luo, Y.; Zhang, W.; Li, A. *Angew. Chem., Int. Ed.* **2015**, *54*, 6878.

Supplemental References for Table 26B

712 Huang, L.; Yu, R.; Zhu, X.; Wan, Y. *Tetrahedron* **2013**, *69*, 8974.

897 Jiang, J.-Q.; Xu, Z.-R.; Jia, Y.-X. *J. Chin. Pharm. Sci.* **2012**, *21*, 113.

898 Nayak, M.; Batra, S. *RSC Adv.* **2012**, *2*, 3367.

899 Zhang, X.; Guo, X.; Fang, L.; Song, Y.; Fan, X. *Eur. J. Org. Chem.* **2013**, *2013*, 8087.

900 Doušová, H.; Horák, R.; Ružičková, Z.; Šimunek, P. *Beilstein J. Org. Chem.* **2015**, *11*, 884.

901 He, N.; Huo, Y.; Liu, J.; Huang, Y.; Zhang, S.; Cai, Q. *Org. Lett.* **2015**, *17*, 374.

902 Long, Y.; Shi, J.; Liang, H.; Zeng, Y.; Cai, Q. *Synthesis* **2015**, *47*, 2844.

903 Zhou, F.; Cheng, G.-J.; Yang, W.; Long, Y.; Zhang, S.; Wu, Y.-D.; Zhang, X.; Cai, Q. *Angew. Chem., Int. Ed.* **2014**, *53*, 9555.

Supplemental References for Table 26C

902 Long, Y.; Shi, J.; Liang, H.; Zeng, Y.; Cai, Q. *Synthesis* **2015**, *47*, 2844.

903 Zhou, F.; Cheng, G.-J.; Yang, W.; Long, Y.; Zhang, S.; Wu, Y.-D.; Zhang, X.; Cai, Q. *Angew. Chem., Int. Ed.* **2014**, *53*, 9555.

904 Ghorai, M. K.; Sahoo, A. K.; Bhattacharyya, A. *J. Org. Chem.* **2014**, *79*, 6468.

Supplemental References for Table 27A

862 Xiong, X.; Jiang, Y.; Ma, D. *Org. Lett.* **2012**, *14*, 2552.

878 Chen, Y.; Zhuo, Z.-J.; Cui, D.-M.; Zhang, C. *J. Organomet. Chem.* **2014**, *749*, 215.

893 Asthana, M.; Kumar, R.; Gupta, T.; Singh, R. M. *Tetrahedron Lett.* **2015**, *56*, 907.

895 Vijay Kumar, S.; Saraiah, B.; Parameshwarappa, G.; Ila, H.; Verma, G. K. *J. Org. Chem.* **2014**, *79*, 7961.

905 Ali, M. A.; Suri, M.; Punniyamurthy, T. *Synthesis* **2013**, *45*, 501.

906 Bae, Y. K.; Cho, C. S. *Appl. Organomet. Chem.* **2013**, *27*, 224.

907 Besandre, R.; Jaimes, M.; May, J. A. *Org. Lett.* **2013**, *15*, 1666.

908 Chen, D.; Chen, Q.; Liu, M.; Dai, S.; Huang, L.; Yang, J.; Bao, W. *Tetrahedron* **2013**, *69*, 6461.

909 Chen, D.-S.; Dou, G.-L.; Li, Y.-L.; Liu, Y.; Wang, X.-S. *J. Org. Chem.* **2013**, *78*, 5700.

910 He, H.-F.; Dong, S.; Chen, Y.; Yang, Y.; Le, Y.; Bao, W. *Tetrahedron* **2012**, *68*, 3112.

911 Jin, H.; Zhou, B.; Wu, Z.; Shen, Y.; Wang, Y. *Tetrahedron* **2011**, *67*, 1178.

912 Kavala, V.; Janreddy, D.; Raihan, M. J.; Kuo, C.-W.; Ramesh, C.; Yao, C.-F. *Adv. Synth. Catal.* **2012**, *354*, 2229.
913 Kim, Y.; Kumar, M. R.; Park, N.; Heo, Y.; Lee, S. *J. Org. Chem.* **2011**, *76*, 9577.
914 Kolarovič, A.; Schnürch, M.; Mihovilovic, M. D. *J. Org. Chem.* **2011**, *76*, 2613.
915 Liu, Y.; Wan, J.-P. *Org. Biomol. Chem.* **2011**, *9*, 6873.
916 Melkonyan, F.; Topolyan, A.; Karchava, A.; Yurovskaya, M. *Tetrahedron* **2011**, *67*, 6826.
917 Melkonyan, F. S.; Kuznetsov, D. E.; Yurovskaya, M. A.; Karchava, A. V. *RSC Adv.* **2013**, *3*, 8388.
918 Okano, K.; Mitsuhashi, N.; Tokuyama, H. *Tetrahedron* **2013**, *69*, 10946.
919 Pericherla, K.; Jha, A.; Khungar, B.; Kumar, A. *Org. Lett.* **2013**, *15*, 4304.
920 Wang, F.; Cai, S.; Liao, Q.; Xi, C. *J. Org. Chem.* **2011**, *76*, 3174.
921 Wang, H.; Li, Y.; Jiang, L.; Zhang, R.; Jin, K.; Zhao, D.; Duan, C. *Org. Biomol. Chem.* **2011**, *9*, 4983.
922 Wang, Z.-J.; Yang, F.; Lv, X.; Bao, W. *J. Org. Chem.* **2011**, *76*, 967.
923 Wu, X.-F.; Neumann, H.; Neumann, S.; Beller, M. *Tetrahedron Lett.* **2013**, *54*, 3040.
924 Xia, Z.; Wang, K.; Zheng, J.; Ma, Z.; Jiang, Z.; Wang, X.; Lv, X. *Org. Biomol. Chem.* **2012**, *10*, 1602.
925 Zhu, Z.; Yuan, J.; Zhou, Y.; Qin, Y.; Xu, J.; Peng, Y. *Eur. J. Org. Chem.* **2014**, 511.
926 Gao, J.; Zhu, J.; Chen, L.; Shao, Y.; Zhu, J.; Huang, Y.; Wang, X.; Lv, X. *Tetrahedron Lett.* **2014**, *55*, 3367.
927 Gao, L.; Song, Y.; Zhang, X.; Guo, S.; Fan, X. *Tetrahedron Lett.* **2014**, *55*, 4997.
928 Kiruthika, S. E.; Perumal, P. T. *Org. Lett.* **2014**, *16*, 484.
929 Li, B.; Guo, S.; Zhang, J.; Zhang, X.; Fan, X. *J. Org. Chem.* **2015**, *80*, 5444.
930 Lou, Z.; Wu, X.; Yang, H.; Zhu, C.; Fu, H. *Adv. Synth. Catal.* **2015**, *357*, 3961.
931 Mahy, W.; Plucinski, P. K.; Frost, C. G. *Org. Lett.* **2014**, *16*, 5020.

Supplemental References for Table 27B

731 Modi, A.; Ali, W.; Mohanta, P. R.; Khatun, N.; Patel, B. K. *ACS Sustainable Chem. Eng.* **2015**, *3*, 2582.
899 Zhang, X.; Guo, X.; Fang, L.; Song, Y.; Fan, X. *Eur. J. Org. Chem.* **2013**, *2013*, 8087.
915 Liu, Y.; Wan, J.-P. *Org. Biomol. Chem.* **2011**, *9*, 6873.
924 Xia, Z.; Wang, K.; Zheng, J.; Ma, Z.; Jiang, Z.; Wang, X.; Lv, X. *Org. Biomol. Chem.* **2012**, *10*, 1602.
932 Chen, Y.-F.; Wu, Y.-S.; Jhan, Y.-H.; Hsieh, J.-C. *Org. Chem. Front.* **2014**, *1*, 253.
933 Dai, C.; Sun, X.; Tu, X.; Wu, L.; Zhan, D.; Zeng, Q. *Chem. Commun.* **2012**, *48*, 5367.
934 Ge, Z.-Y.; Xu, Q.-M.; Fei, X.-D.; Tang, T.; Zhu, Y.-M.; Ji, S.-J. *J. Org. Chem.* **2013**, *78*, 4524.
935 Guo, H.; Liu, J.; Wang, X.; Huang, G. *Synlett* **2012**, *23*, 903.
936 Hu, Q.; Xia, Z.; Fan, L.; Zheng, J.; Wang, X.; Lv, X. *ARKIVOC* **2012**, 129.
937 Liu, Q.; Zhao, Y.; Fu, H.; Cheng, C. *Synlett* **2013**, *24*, 2089.
938 Omar, M. A.; Conrad, J.; Beifuss, U. *Tetrahedron* **2014**, *70*, 3061.
939 Qian, W.; Wang, H.; Allen, J. *Angew. Chem. Int. Ed.* **2013**, *52*, 10992.
940 Reeves, J. T.; Tan, Z.; Lu, B. Z.; Senanayake, C. H. *J. Heterocycl. Chem.* **2013**, *50*, 680.
941 Sang, P.; Xie, Y.; Zou, J.; Zhang, Y. *Org. Lett.* **2012**, *14*, 3894.
942 Sreenivas, D. K.; Ramkumar, N.; Nagarajan, R. *Org. Biomol. Chem.* **2012**, *10*, 3417.
943 Sun, L.-L.; Liao, Z.-Y.; Tang, R.-Y.; Deng, C.-L.; Zhang, X.-G. *J. Org. Chem.* **2012**, *77*, 2850.
944 Wang, M.; Jin, Y.; Yang, H.; Fu, H.; Hu, L. *RSC Adv.* **2013**, *3*, 8211.
945 Xu, W.; Fu, H. *J. Org. Chem.* **2011**, *76*, 3846.
946 Xu, W.; Jin, Y.; Liu, H.; Jiang, Y.; Fu, H. *Org. Lett.* **2011**, *13*, 1274.
947 Yuan, H.; Li, K.; Chen, Y.; Wang, Y.; Cui, J.; Chen, B. *Synlett* **2013**, *24*, 2315.
948 Fan, X.; Li, B.; Guo, S.; Wang, Y.; Zhang, X. *Chem.—Asian J.* **2014**, *9*, 739.
949 Fan, X.-S.; Zhang, J.; Li, B.; Zhang, X.-Y. *Chem.—Asian J.* **2015**, *10*, 1281.
950 Ghorai, M. K.; Sayyad, M.; Nanaji, Y.; Jana, S. *Chem.—Asian J.* **2015**, *10*, 1480.
951 Guo, S.; Li, Y.; Tao, L.; Zhang, W.; Fan, X. *RSC Adv.* **2014**, *4*, 59289.
952 Guo, S.; Tao, L.; Zhang, W.; Zhang, X.; Fan, X. *J. Org. Chem.* **2015**, *80*, 10955.
953 Guo, S.; Wang, J.; Li, Y.; Fan, X. *Tetrahedron* **2014**, *70*, 2383.
954 Hu, B.-Q.; Wang, L.-X.; Xiang, J.-F.; Yang, L.; Tang, Y.-L. *Chin. Chem. Lett.* **2015**, *26*, 369.
955 Li, B.; Guo, C.; Fan, X.; Zhang, J.; Zhang, X. *Tetrahedron Lett.* **2014**, *55*, 5944.
956 Li, C.; Zhang, L.; Shu, S.; Liu, H. *Beilstein J. Org. Chem.* **2014**, *10*, 2441.

[957] Li, C.; Zhang, W.-T.; Wang, X.-S. *J. Org. Chem.* **2014**, *79*, 5847.
[958] Obulesu, O.; Babu Nanubolu, J.; Suresh, S. *Org. Biomol. Chem.* **2015**, *13*, 8232.
[959] Omar, M. A.; Conrad, J.; Beifuss, U. *Tetrahedron* **2014**, *70*, 3061.
[960] Songsichan, T.; Promsuk, J.; Rukachaisirikul, V.; Kaeobamrung, J. *Org. Biomol. Chem.* **2014**, *12*, 4571.
[961] Tian, H.; Qiao, H.; Zhu, C.; Fu, H. *RSC Adv.* **2014**, *4*, 2694.
[962] Yang, D.; An, B.; Wei, W.; Tian, L.; Huang, B.; Wang, H. *ACS Comb. Sci.* **2015**, *17*, 113.
[963] Zhang, W.-T.; Chen, D.-S.; Li, C.; Wang, X.-S. *Synthesis* **2015**, *47*, 562.
[964] Zhang, X.-Y.; Guo, X.-J.; Fan, X.-S. *Chem.—Asian J.* **2015**, *10*, 106.

Supplemental References for Table 27C

[915] Liu, Y.; Wan, J.-P. *Org. Biomol. Chem.* **2011**, *9*, 6873.
[965] Wang, H.-J.; Wang, Y.; Camara, F.; Paquette, W. D.; Csakai, A. J.; Mangette, J. E. *Tetrahedron Lett.* **2011**, *52*, 541.
[966] Chen, W.; Li, H.; Gu, X.; Zhu, Y. *Synlett* **2015**, *26*, 785.
[967] Gawande, S. D.; Kavala, V.; Zanwar, M. R.; Kuo, C.-W.; Huang, W.-C.; Kuo, T.-S.; Huang, H.-N.; He, C.-H.; Yao, C.-F. *Adv. Synth. Catal.* **2014**, *356*, 2599.
[968] Hu, X.; Dong, Y.; Liu, G. *Mol. Diversity* **2015**, *19*, 695.

Supplemental References for Table 29A

[969] Kabir, M. S.; Namjoshi, O. A.; Verma, R.; Lorenz, M.; Phani Babu Tiruveedhula, V. V. N.; Monte, A.; Bertz, S. H.; Schwabacher, A. W.; Cook, J. M. *J. Org. Chem.* **2012**, *77*, 300.
[970] Kotovshchikov, Y. N.; Latyshev, G. V.; Lukashev, N. V.; Beletskaya, I. P. *Eur. J. Org. Chem.* **2013**, *2013*, 7823.

Supplemental References for Table 29B

[738] Yan, N.-N.; Wu, F.-T.; Zhang, J.; Wei, Q.-B.; Liu, P.; Xie, J.-W.; Dai, B. *Asian J. Org. Chem.* **2014**, *3*, 1159.
[970] Kotovshchikov, Y. N.; Latyshev, G. V.; Lukashev, N. V.; Beletskaya, I. P. *Eur. J. Org. Chem.* **2013**, *2013*, 7823.

Supplemental References for Table 29C

[969] Kabir, M. S.; Namjoshi, O. A.; Verma, R.; Lorenz, M.; Phani Babu Tiruveedhula, V. V. N.; Monte, A.; Bertz, S. H.; Schwabacher, A. W.; Cook, J. M. *J. Org. Chem.* **2012**, *77*, 300.
[970] Kotovshchikov, Y. N.; Latyshev, G. V.; Lukashev, N. V.; Beletskaya, I. P. *Eur. J. Org. Chem.* **2013**, *2013*, 7823.

Supplemental References for Table 29D

[969] Kabir, M. S.; Namjoshi, O. A.; Verma, R.; Lorenz, M.; Phani Babu Tiruveedhula, V. V. N.; Monte, A.; Bertz, S. H.; Schwabacher, A. W.; Cook, J. M. *J. Org. Chem.* **2012**, *77*, 300.
[970] Kotovshchikov, Y. N.; Latyshev, G. V.; Lukashev, N. V.; Beletskaya, I. P. *Eur. J. Org. Chem.* **2013**, *2013*, 7823.

Supplemental References for Table 30A

[971] Pandey, A. K.; Sharma, R.; Shivahare, R.; Arora, A.; Rastogi, N.; Gupta, S.; Chauhan, P. M. S. *J. Org. Chem.* **2013**, *78*, 1534.

[972] Pereira, G.; Vilaca, H.; Ferreira, P. M. T. *Amino Acids* **2013**, *44*, 335.
[973] Zhou, C.; Ma, D. *Chem. Commun.* **2014**, *50*, 3085.
[974] Kuranaga, T.; Mutoh, H.; Sesoko, Y.; Goto, T.; Matsunaga, S.; Inoue, M. *J. Am. Chem. Soc.* **2015**, *137*, 9443.
[975] Yamashita, T.; Matoba, H.; Kuranaga, T.; Inoue, M. *Tetrahedron* **2014**, *70*, 7746.

Supplemental References for Table 30B

[973] Zhou, C.; Ma, D. *Chem. Commun.* **2014**, *50*, 3085.
[976] Jouvin, K.; Coste, A.; Bayle, A.; Legrand, F.; Karthikeyan, G.; Tadiparthi, K.; Evano, G. *Organometallics* **2012**, *31*, 7933.

Supplemental References for Table 30C

[755] Kundu, D.; Bhadra, S.; Mukherjee, N.; Sreedhar, B.; Ranu, B. C. *Chem.—Eur. J.* **2013**, *19*, 15759.
[829] Maligres, P. E.; Krska, S. W.; Dormer, P. G. *J. Org. Chem.* **2012**, *77*, 7646.
[970] Kotovshchikov, Y. N.; Latyshev, G. V.; Lukashev, N. V.; Beletskaya, I. P. *Eur. J. Org. Chem.* **2013**, *2013*, 7823.
[976] Jouvin, K.; Coste, A.; Bayle, A.; Legrand, F.; Karthikeyan, G.; Tadiparthi, K.; Evano, G. *Organometallics* **2012**, *31*, 7933.
[977] Huang, W.-S.; Xu, R.; Dodd, R.; Shakespeare, W. C. *Tetrahedron Lett.* **2014**, *55*, 441.

Supplemental References for Table 32

[976] Jouvin, K.; Coste, A.; Bayle, A.; Legrand, F.; Karthikeyan, G.; Tadiparthi, K.; Evano, G. *Organometallics* **2012**, *31*, 7933.

Supplemental References for Table 33

[978] Yamashita, T.; Kuranaga, T.; Inoue, M. *Org. Lett.* **2015**, *17*, 2170.

Supplemental References for Table 34B

[928] Kiruthika, S. E.; Perumal, P. T. *Org. Lett.* **2014**, *16*, 484.
[979] Xiao, X.; Chen, T.-Q.; Ren, J.; Chen, W.-D.; Zeng, B.-B. *Tetrahedron Lett.* **2014**, *55*, 2056.

Supplemental References for Table 35

[910] He, H.-F.; Dong, S.; Chen, Y.; Yang, Y.; Le, Y.; Bao, W. *Tetrahedron* **2012**, *68*, 3112.
[922] Wang, Z.-J.; Yang, F.; Lv, X.; Bao, W. *J. Org. Chem.* **2011**, *76*, 967.
[924] Xia, Z.; Wang, K.; Zheng, J.; Ma, Z.; Jiang, Z.; Wang, X.; Lv, X. *Org. Biomol. Chem.* **2012**, *10*, 1602.
[928] Kiruthika, S. E.; Perumal, P. T. *Org. Lett.* **2014**, *16*, 484.
[980] Li, Y.; Cheng, L.; Shao, Y.; Jiang, S.; Cai, J.; Qing, N. *Eur. J. Org. Chem.* **2015**, *2015*, 4325.

INDEX

Copper-Catalyzed Amination of Aryl and Alkenyl Electrophiles, by Keven H. Shaughnessy,
Engelbert Ciganek, and Rebecca B. DeVasher
© 2017 Organic Reactions, Inc. Published 2017 by John Wiley & Sons, Inc.